McGraw-Hill

Encyclopedia of
ENERGY

Encyclopedia of
ENERGY

DANIEL N. LAPEDES Editor in Chief

McGRAW-HILL BOOK COMPANY

NEW YORK	KUALA LUMPUR	PARIS
SAN FRANCISCO	LONDON	SAO PAULO
ST. LOUIS	MEXICO	SINGAPORE
AUCKLAND	MONTREAL	SYDNEY
DUSSELDORF	NEW DELHI	TOKYO
JOHANNESBURG	PANAMA	TORONTO

Library of Congress Cataloging in Publication Data

McGraw-Hill encyclopedia of energy.

    Includes index.
    1. Power   resources—Dictionaries. 2. Power   (Mechanics)—Dictionaries. 3. Power   resources—Addresses,   essays,   lectures. 4. Power (Mechanics)—Addresses,   essays,   lectures. I. Lapedes,   Daniel   N. II. Title: Encyclopedia of energy.
TJ163.2.M3   333.7   76-19026
ISBN 0-07-045261-X

Table of Contents

Editorial Staff

Daniel N. Lapedes, Editor in Chief

Sybil P. Parker, Senior editor

Jonathan Weil, Staff editor

Edward J. Fox, Art director

Joe Faulk, Editing manager

Ellen Okin, Editing supervisor

Patricia Albers, Editing assistant

Ann D. Bonardi, Art production supervisor

Richard A. Roth, Art editor

Cecelia M. Giunta, Art/traffic

Consulting Editors

Preface

Energy, the mainspring that keeps life's clock ticking on Earth, is present in many forms. It is in the heat and light radiated by the Sun: in the carbohydrates and wood in plants; in coal, oil, and natural gas, and in oil shale and tar sands; in geothermal wells; in the wind that sweeps over the land and sea; in the water coursing to the oceans; and in the atomic nucleus.

Some of these forms of energy are the fossil fuels — coal and oil — that are the end result of a process that has been taking place under the Earth's surface for several hundred million years. We have been using coal and oil for some time — slowly at first and then from the time of the Industrial Revolution at an increasing pace. Projections based on present consumption of oil indicate that 80% of that energy resource will be used up worldwide by the beginning of the 21st century. The projection made for coal foresees a supply for the United States until the middle of the 21st century.

The situation became critical for the industrialized nations when the OPEC (Organization of Petroleum Exporting Countries) embargo in 1973 effectively reduced the flow of crude oil to their refineries. This affected the supply and price of electric power, heating oil, and gasoline, as well as the output of industries that use crude oil or fractions of it as a raw material. The embargo accelerated a worldwide recession and changed the manner in which people and their governments regard energy.

It is generally accepted that we have been profligate in our use of energy resources — particularly those such as coal, oil, and natural gas that are finite and exhaustible — and have neglected development of other resources. Society through its governmental institutions is attempting to make decisions that will ensure an energy supply base for the future. The individuals involved in the decision-making process face a complex problem that has deep sociological, economic, and technological overtones. They must weigh curtailment of certain energy-intensive activities on one side of the scale against achievement of desirable conservation practices on the other — activities such as automobile use with a mandate for lower-horsepower cars, and modernization of factories to use less energy; they must weigh the effect on the environment of using high-sulfur coal and oil, and of strip mining to recover coal deposited near the surface; and they must explore the development of other forms of energy.

The *McGraw-Hill Encyclopedia of Energy* with its more than 300 articles written by specialists is designed to aid the student, librarian, scientist, engineer, teacher, and lay reader with information on any aspect of energy from the economic and political to the environmental and technological. The Encyclopedia is arranged in two parts. The first part, "Energy Perspectives," has six feature articles: Energy Consumption, Outlook for Fuel Reserves, Exploring Energy Choices, World Energy Economy, U.S. Policies and Politics, and Protecting the Environment. The second part, "Energy Technology," with its 300 alphabetically arranged articles, contains information on such subjects as coal mining, nuclear power, laser-induced fusion, wind power, solar power, and hydroelectric power. The articles, some drawn from the *McGraw-Hill Encyclopedia of Science and Technology* (3d ed., 1971) and its Yearbooks (1971–1976), and some written especially for this volume, were selected or suggested by the Board of Consultants. All articles are signed, and the authors and their affiliations are provided in the List of Contributors beginning on page 739. Almost every article opens with a definition of the subject and ends with a bibliography for further reading. The Appendix, beginning on page 747, aids the reader in converting U.S. Customary Units to metric and System International units. The cross-references to other articles and the analytic index beginning on page 759 interrelate the articles.

DANIEL N. LAPEDES
Editor in Chief

Energy Perspectives

People have adapted energy to a wide range of personal and industrial uses. The most significant personal uses are for cooking, comfort heating and cooling, illumination, transportation, hot water, refrigeration, and communication. These uses extend far beyond the bare essentials for life, and they provide increasingly for comfort and convenience. The most significant industrial uses are for heat and power.

Nonindustrialized societies still are heavily dependent on the traditional energy sources—local solar energy that is made available through the agencies of food, work-animal feed, nonmineral fuels (wood, dung, and agricultural wastes), wind power, and direct waterpower. Energy consumption per person is very small, only a few times the food energy required to sustain life.

In contrast, industrialized societies use large quantities of fossil fuel (coal, oil, and natural gas) and electricity, and consumption of energy per person is as much as a hundred times the energy contained in food. Figure 1 illustrates the tremendous per capita consumption of fossil fuels and hydropower in the industrialized nations compared with the rest of the world. These two forms of energy provide a twelvefold increase in energy for the industrialized regions, compared with a twofold increase for the nonindustrialized regions. When one speaks of energy in an industrialized society, one ordinarily refers only to energy for heat, light, power, and communication, leaving aside the energy content of food. In keeping with this custom, food energy will not be further considered in this article.

Fairly accurate records exist for the overall energy consumption of the United States, particularly in recent decades, since it is known how much coal, oil, natural gas, hydropower, nuclear power, and other forms of energy are consumed each year. But the records are incomplete with respect to energy consumption for most specific purposes or end uses. There are good records for some, for example, energy in the form of gasoline for automobiles. Suppliers know how much energy in the form of electricity is delivered to each home, but the proportions that are used for cooking, heating, light, refrigeration, television, and other purposes can only be estimated.

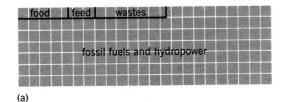

(a)

(b)

Fig. 1. Per capita consumption of energy in 1970 for (*a*) industrialized regions (30% of world population) and (*b*) nonindustrial regions (70% of world population). Each square represents 1,000,000 Btu per person per year. (*From J. C. Fisher, Energy Crises in Perspective, 1974*)

Energy Consumption

- **flow of energy**
- **trends and patterns**
- **quantification of energy**
- **work and heat**
- **electricity consumption**

Dr. John C. Fisher

Table 1. Consumption pattern of energy for significant end uses, as an approximate percentage of total consumption, in the United States during the mid-1970s*

| End use | Segment of the Economy | | | |
	Industrial	Residential and personal	Commercial and public	Total
Transportation	1	16	9	26
Comfort heat	2	11	7	20
Process steam	16	—	—	16
Direct heat	11	—	—	11
Electric drive	9	—	—	9
Lighting	1	1	3	5
Hot water	—	3	1	4
Air conditioning	—	1	2	3
Refrigeration	—	1	1	2
Cooking	—	1	—	1
Electrochemistry	1	—	—	1
Other (mostly electric)	—	1	1	2
Total	41	35	24	100

*Based on a study by the Stanford Research Institute updated by a task force of the National Academy of Engineering, and on data obtained by the U.S. Bureau of the Census.

Table 1 shows the approximate pattern of energy consumption in the United States during the mid-1970s. Energy can be transformed to electricity before it is used, as for lighting and for powering machine tools in industry. Wherever this is done, the table shows the energy content of the fuel required to make the electricity. There is no doubt that the major features of the nation's energy consumption pattern are correctly portrayed in the table, but individual percentage entries are probably not accurate to better than one percentage point. Wherever there is a dash in the table, the energy consumption for that segment of the economy is estimated to be less than ½% of the nation's total consumption.

THE FLOW OF ENERGY

A number of different sources have provided significant energy inputs to the United States at one time or another. In approximate order of their historical development, they are:

Solar energy: Conversion via fuel wood, work-animal feed, wind power, waterpower.

Fossil fuel: Combustion of coal, petroleum, natural gas.

Nuclear fuel: Fission of uranium.

Other sources of potential significance for large-scale energy production include:

Solar energy: Conversion via new technologies.

Fossil fuel: Combustion of hydrocarbons from oil shale, tar sand.

Nuclear fuel: Fission of thorium.

Sources of energy are judged to be potentially significant where the available quantities are large and where technological and economic considerations show that costs are competitive or close to competitive. Other potential sources such as tidal power, geothermal power, fusion power, and trash combustion are likely to be of less significance for large-scale energy production because of limited availability or because of economic or technological barriers, although they may have limited applications at special locations or in special situations.

Some energy sources are more abundant than others. Solar energy is dilute, but large in magnitude and unlimited in time. Fossil fuels are con-

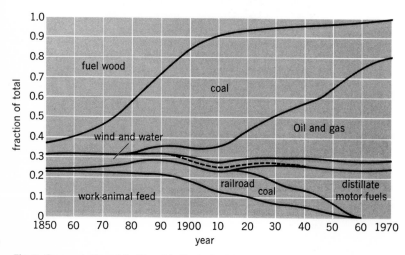

Fig. 2. Segmentation of fuel input to the United States, 1850 to 1970. (*From J. C. Fisher, Energy Crises in Perspective, 1974*)

Table 2. Sources of energy for the United States in 1970

Source	Percent
Fossil fuel	
Coal	20
Petroleum	41
Natural gas	33
Solar energy	
Hydroelectricity	4
Other	2
	100

Table 3. Sources of energy for the United States in 1850

Source	Percent
Solar energy	
Fuel wood	64
Work-animal feed	22
Wind and water	7
Fossil fuel	
Coal	7
	100

centrated and inexpensive to recover, but can become exhausted after several centuries. Nuclear fuels are practically inexhaustible, particularly if breeder reactors are able to utilize the common isotopes of uranium and thorium. Broadly speaking, for the industrialized societies of the world, the years of significance for fuel wood, work-animal feed, and wind power have passed; and the years of significance for nuclear fuels are just beginning. The energy sources of current significance are fossil fuels and waterpower (Table 2).

SHIFTS IN EMPHASIS

The 1970 emphasis on fossil fuels in the United States represents a strong shift from the 1850 emphasis on wood (for heat) and animal feed (for farm work and transportation) (Table 3). The years from 1850 to 1970 saw five major substitutions of the new energy forms for the old, as shown in Fig. 2. Fuel wood, used primarily for heating, was largely replaced by coal between 1850 and 1910. Since 1910, coal has been progressively replaced by fluid hydrocarbons (gas and oil). Work-animal feed, used primarily for motive power in transportation and on farms, was partially replaced by railroad coal in the late 1800s and early 1900s. Then, as the country adopted automobiles and tractors and as railroads converted to oil, both animal feed and railroad coal were largely replaced by gasoline and other distillate motor fuels in the years 1920–1950. Direct wind power and waterpower were replaced by hydroelectricity in the years 1890–1940.

A final substitution, not shown in Fig. 2, is a steady increase in the proportion of energy converted to electricity prior to its ultimate consump-

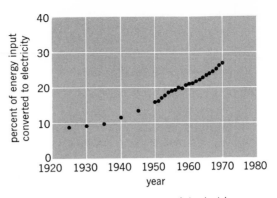

Fig. 3. Conversion of primary energy into electric energy in the United States, 1925 to 1970. (*From J. C. Fisher, Energy Crises in Perspective, 1974*)

tion. The percentage of primary energy input converted to electricity grew steadily from about 9% in 1925 to nearly 27% in 1970 (Fig. 3). It is important to keep in mind that most of the growth in electric energy consumption has resulted from this substitution of electric energy for other forms of energy. Waterpower used to be harnessed directly to factory machinery by waterwheels, pulleys, and belts, but now it is harnessed indirectly through electricity. Hydroelectric power has substituted for direct waterpower. Fuels used to be burned on the users' premises for illumination, stationary work, and heat, but increasingly fuels are burned off the users' premises in electric power plants.

Energy flows through the United States economy from the sources shown in Fig. 2 to the end

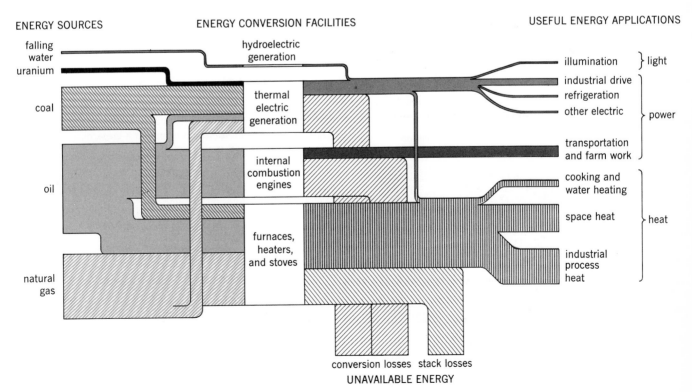

Fig. 4. Flow of energy through the United States economy in the mid-1970s, from major energy sources (left) through conversion facilities (middle) to useful applications (right), and unavailable energy at the bottom. Width of a channel is proportional to the amount of energy.

uses in Table 1. The relative proportions of the major energy flows are illustrated in Fig. 4. The nation's energy sources are converted for useful applications by means of various energy conversion facilities. These include furnaces, heaters, and stoves for generating heat, internal combustion engines for generating power, and steam engines and other heat engines for generating power in electric power plants. In the process of conversion there is a flow of unavailable energy in the form of low-temperature heat which is lost up stacks and chimneys and is also lost in the conversion of high-temperature heat to mechanical power. *See* ENERGY FLOW.

QUANTIFICATION OF ENERGY

When considering the quantitative aspects of energy consumption, miners deal in tons of coal, oil suppliers in barrels of oil, gas suppliers in cubic feet of gas, and electric utility people in kilowatt-hours of hydroelectricity. Some uniform standard of measurement is required for comparing the quantities of energy from these various sources. Sources that customarily are used for the production of heat can be quantified by the amounts of heat they are capable of generating. More specifically, the numerical energy values for fossil fuels, fuel wood, and animal feed are the amounts of heat they would generate during combustion. The values for nuclear fuels are the amounts of heat generated by nuclear fission in electric power plants.

Table 4. Approximate energy contents for selected energy sources

Source	Approximate energy content
Coal	12,500 Btu per pound
Crude oil	5,800,000 Btu per barrel
Natural gas	1,035 Btu per cubic foot
Hydroelectricity	3,412 Btu per kilowatt-hour
Fuel equivalent	10,500 Btu per kilowatt-hour

Table 5. Energy consumption in the United States in 1970

Source	Conventional quantity	Energy content, quad (=10^{15} Btu)
Fossil fuel		
Coal	525×10^6 tons	12.8
Petroleum	5.36×10^9 bbl	26.5
Natural gas	21.4×10^{12} ft³	21.3
Solar energy		
Hydroelectricity	253×10^9 kWhr	2.6
Other		1.3
		64.5

Table 6. Energy consumption in the world in 1970

Source	Energy content	
	Quad (=10^{15} Btu)	%
Fossil fuel		
Coal	65	30
Petroleum	77	36
Natural gas	38	18
Solar energy		
Hydroelectricity	13	6
Traditional		
Wood, waste, feed	22	10
	215	100

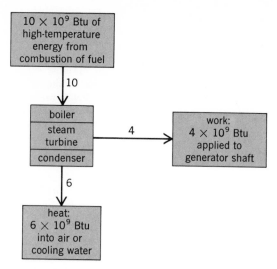

Fig. 5. Flow of energy through a modern boiler-turbine-condenser heat engine. (*From J. C. Fisher, Energy Crises in Perspective, 1974*)

Hydroelectricity presents a special problem. Its energy content can be measured either by the amount of heat it would generate in an electric heater or by the larger amount of heat that would be required to generate the same amount of electricity in a fuel-burning power plant. Except for Figs. 4 and 14, the second of these measures—the fuel equivalent of hydroelectricity—is used in this article because it more accurately reflects hydroelectricity's economic significance.

Heat and other forms of energy can be measured in terms of British thermal units (Btu). One Btu is the amount of energy it takes to warm up 1 lb of water (approximately 1 pint of water) 1°F. In the metric system, energy is measured by joules, with a joule equal to 1 watt-second of energy. Since there are 3412 Btu in a kilowatt-hour, 1 Btu represents 1055 joules. Approximate energy contents for different energy sources, measured in Btu, are shown in Table 4.

The United States consumes so much energy (Table 5) that the annual amount in Btu is a very large number. To bring such large numbers down to size, it is more convenient to measure energy in quads (1 quad = 1 quadrillion Btu = 10^{15} Btu). Overall in 1970, the energy input to the United States amounted to 64.5 quads. World energy consumption in the same year amounted to about 215 quads, comprising 193 quads of mineral fuels and hydroelectricity and an estimated 22 quads of traditional fuels (Table 6).

WORK AND HEAT

As mentioned, the major uses for energy are for the production of work or heat, and it is important to understand the relationships between them. The flow of energy is called work when it exerts a force. The flow of energy is called heat when it does not exert a force (for example, the flow of energy from a hot oven to a cold potato).

Because work and heat are alternate modes for the flow of energy, they can be quantified by the amount of energy that flows via each mode. Consider as an example the operation of a modern steam turbine used to drive an electric generator, shown in Fig. 5. The fuel burned in the boiler generates 10×10^9 Btu of energy each hour of operation. A portion of this energy, amounting to 4×10^9 Btu per hour, flows through the rotating turbine shaft in the form of work, where it is used to turn the shaft of an electric generator. The rest of the energy, amounting to 6×10^9 Btu per hour, is discharged into the air or into cooling water. Thus, only 40% of the energy in the fuel was actually used for the purpose for which it was intended. The engine therefore has an efficiency of 40%. *See* HEAT; WORK.

Reversible heat engines. Many engines, including jet engines, automobile engines, and steam turbines, receive energy at high temperature, transform some to work, and discharge the rest of it as heat at a lower temperature. Much study has been devoted to the potential efficiency of these engines, and the concept of a "reversible heat engine" has emerged as an idealization against which the lesser performance of real engines can be measured. Imagine a reservoir of energy at a high temperature T_1 and a second reservoir of energy at a low temperature T_2. Imagine that a reversible heat engine, with a rotatable shaft along which work can flow (Fig. 6), is in contact with both reservoirs and is able to exchange heat with both.

When utilized to generate work, a reversible heat engine draws heat from the high-temperature reservoir, delivers work along the rotating shaft, and discharges heat to the low-temperature reservoir. It is the idealization of a steam turbine power plant. Figure 7 shows what happens to a quantity Q_1 of heat that flows into the engine from the hotter reservoir: part of it, Q_2, goes to the colder reservoir and part of it, $Q_1 - Q_2$, comes out as work.

Now suppose that the shaft is twisted in the opposite direction by means of some outside agency, so that work flows into the engine instead of out. The engine then draws heat from the low-temperature reservoir and delivers heat to the high-temperature reservoir. It operates as a heat pump, the idealization of an air conditioner or refrigerator. Figure 8 shows the flows of work and heat associated with the delivery of a quantity Q_1 of energy to the hotter reservoir: part of it, Q_2, comes from the colder reservoir and part, $Q_1 - Q_2$, comes in through the shaft as work. This picture is the same as the previous one except that everything is flowing in the opposite direction. This is the meaning of reversibility.

Heat engine efficiency. The efficiency of a reversible heat engine depends only on the absolute temperatures T_1 and T_2, as shown in Eq. (1). No

$$\text{Theoretical efficiency} = \frac{T_1 - T_2}{T_1} \qquad (1)$$

practical heat engine can actually achieve this theoretical maximum efficiency. For a steam turbine where T_1 is the temperature of the steam in the boiler ($540°C = 1460°R$) and T_2 is the temperature in the condenser where cold steam is converted back to water ($50°C = 580°R$), the theoretical maximum efficiency is about 60%. Modern engineering practice has achieved a respectable 40%, but there is still room for improvement.

Heat pump performance. In addition to their role in generating mechanical power, heat engines

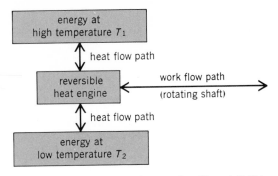

Fig. 6. Idealized reversible heat engine. (*From J. C. Fisher, Energy Crises in Perspective, 1974*)

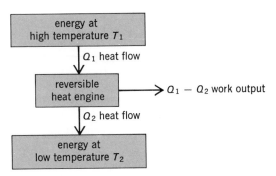

Fig. 7. Reversible heat engine used to perform work. (*From J. C. Fisher, Energy Crises in Perspective, 1974*)

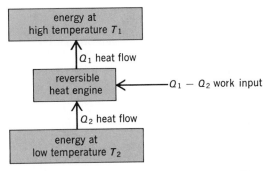

Fig. 8. Reversible heat engine used to pump heat. (*From J. C. Fisher, Energy Crises in Perspective, 1974*)

are used to pump energy from a cooler place to a warmer place. When the purpose is to cool an area, the heat engine is an air conditioner or a refrigerator. When the purpose is to warm an area, the heat engine is simply a heat pump.

The performance of a heat pump is evaluated by the coefficient of performance (COP), that is, the ratio of the energy delivered to the warmer reservoir (the desired result) to the work required to operate the pump (the necessary input). When a reversible heat engine is used as a heat pump, its COP is simply the reciprocal of its efficiency as an engine. Hence, for a reversible heat pump, COP is determined by Eq. (2). This relationship shows the

$$\text{COP} = \frac{T_1}{T_1 - T_2} \qquad (2)$$

theoretical maximum amount of energy that can be

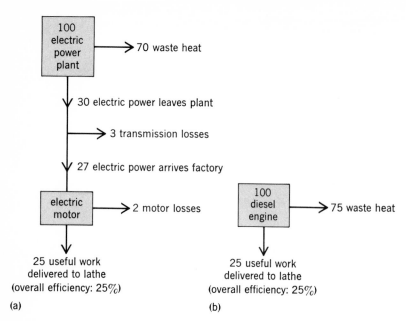

Fig. 9. Overall efficiencies of (a) electric power and (b) diesel engine power (illustrative). (From J. C. Fisher, Energy Crises in Perspective, 1974)

a commercial heat pump working between 7 and 21°C is able to deliver only about three times as much energy in the form of heat as the energy content of the electricity that drives it.

ROLE OF ELECTRICITY

Electricity is not a primary source of energy, but rather the most highly refined form of energy. There is no alternative to electricity for some purely electrical and electronic end uses of energy. But for most other end uses of energy, consumers have a choice of burning fuel on their own premises or of utilizing electricity generated in an electric power plant, and over the years consumers have opted increasingly for electricity.

Stationary work. The flow of electric energy is equivalent to the flow of work. Work can be transmitted from one place to another by a long rotating shaft, or by a belt stretched over pulleys, or by electricity along conducting wires. Once generated, electricity can be utilized with very little additional waste.

Consider two alternatives for delivering work to a lathe in a factory: (1) burning oil in a power plant to make electricity, then transmitting the electricity to a factory where it turns an electric motor that turns a lathe; and (2) burning oil in a diesel engine that turns the lathe directly. The two alternatives are compared in Fig. 9. The comparison uses an older power plant with only 30% efficiency, which is characteristic of the average around the country. It uses a small diesel engine with only 25% efficiency, also characteristic of the average around the country. As far as generating unavailable heat is concerned, there is an approximate standoff. The power plant is more efficient than the diesel engine, but there are additional losses in transmission and in the electric motor that tend to even things up. Hence electrification of industrial drive does not increase or decrease the amount of unavailable or waste heat associated with powering industry, but merely shifts its location from factories to power plants.

pumped into a reservoir at temperature T_1 per unit work input. See HEAT PUMP.

As an example, consider the possibility of heating a house by means of a heat pump. Suppose that the outside air temperature (T_2) is $-7°C$ (480°R) and the inside temperature (T_1) is 21°C (530°R). Then, using Eq. (2), the maximum theoretical ratio of heat (delivered into the house) to work (required for doing it) would be 10.6 The work put into running the pump would be multiplied more than tenfold in delivering heat to the house. For each 10.5 units of heat delivered indoors, 1.0 unit would come from the work expended in driving the heat pump and 9.5 units would come from the air outdoors. However, this is the theoretical limit. So far,

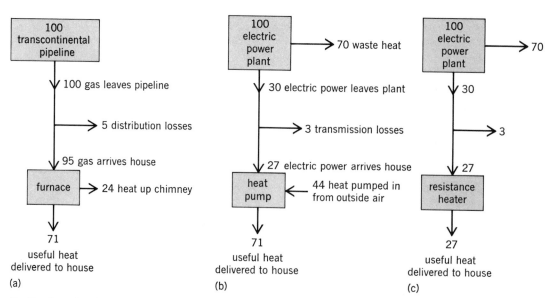

Fig. 10. Overall performance of (a) direct combustion of natural gas, (b) electric heat pump, and (c) electric resistance heating (illustrative). (From J. C. Fisher, Energy Crises in Perspective, 1974)

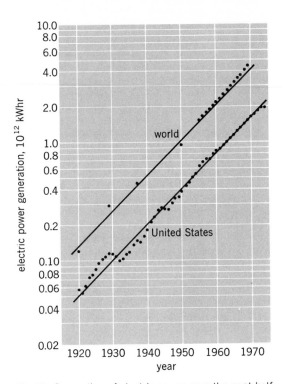

Fig. 11. Generation of electric power over the past half century, for selected years worldwide and annually for the United States. (*From J. C. Fisher, Energy Crises in Perspective, 1974*)

Table 7. Approximate electrification of the United States in 1968

End uses	Percent electrified	Percent of United States energy	Percent of United States electricity
Electrified			
Electric drive			
Lighting			
Air conditioning	nearly	22.0	84.4
Refrigeration	100%		
Electrochemistry			
Other			
Heating	8% average		
Clothes drying	70%	0.4	1.0
Cooking	40%	1.4	2.1
Hot water	38%	4.2	6.2
Direct heat	6%	11.0	2.6
Comfort heat	5%	20.8	3.3
Process steam	–	14.6	–
		52.4	15.2
Transportation			
All forms	½%	25.6	0.4

Overall cost. From the standpoint of overall cost, the comparison favors electricity. Fuel is cheaper for the power plant, and per unit of work delivered to a lathe, the cost of a power plant plus transmission system plus electric motor is less than the cost of a diesel engine. This is largely because the power plant is more fully utilized. Whereas a diesel engine might run 40 hr a week at an average 30% of its maximum rating for an overall utilization factor of about 7%, the power plant, by providing electricity to a number of users whose demands occur at different and partly overlapping times, might have an overall utilization factor of 65%, thereby achieving a much better utilization of the invested capital. Operating and maintenance costs are less for the power plant for much the same reason. Factory layout can be rearranged much more easily and cheaply when electricity provides the power. And, of increasing importance as the country turns its concern to pollution abatement, large power plants generally are able to burn fuel more thoroughly than many small engines operating independently, substantially lessening pollution.

Illumination. Illumination is another instance of substantial benefits and savings through electrification. Compared with the alternative of direct combustion, electricity gives more light at less cost with less bother, less pollution, and greater safety. Fuel is also conserved, for a gallon of oil burned in a power plant gives more light from an electric lamp than a gallon burned directly in an oil lamp.

Heating. Electric resistance heating is of particular value for producing very high temperatures, including the 5000°C plasma in a mercury-vapor lamp and even the 100,000,000°C plasmas being studied in thermonuclear fusion research. Resistance heating stays competitive down to lower temperatures such as 1540°C at which iron melts, even though such temperatures can be reached by combustion, because combustion heating tends to become less efficient as the temperature increases through loss of hot combustion products up the chimney.

At low-to-moderate temperatures, however, combustion heating tends to be more efficient because the products of combustion can give up a larger proportion of their energy as they cool down before going up the chimney. For end uses such as space heating, hot-water supply, cooking, and much industrial heating, direct combustion is inexpensive and efficient. Electricity has found only limited application to these end uses, and it is instructive to compare the two methods of electric heating—resistance heating and the electrically

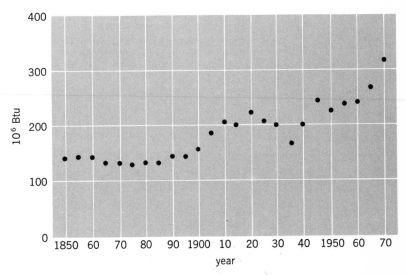

Fig. 12. Per capita consumption of energy in the United States, 1850 to 1970. (*From J. C. Fisher, Energy Crises in Perspective, 1974*)

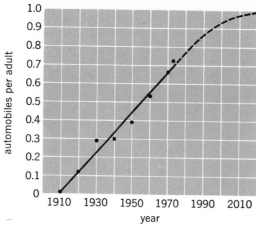

Fig. 13. Average number of automobiles per adult (age 18 and over) in the United States, 1910 to 1973 and projected to 2020.

driven heat pump—with direct combustion to see why electricity has made so little headway (Fig. 10).

From the standpoint of fuel conservation and overall cost, electric resistance heating is not attractive. More fuel is required, and the cost is greater. Yet where only small amounts of heat are desired and where convenience is an important consideration, resistance heating is frequently chosen. Heat pumps and direct combustion stand more or less at a draw when fuel conservation is considered, but heat pumps are more expensive than furnaces. In centrally air-conditioned buildings this does not matter as much, because a reversible heat pump can both cool and heat, but cost nevertheless remains a factor in favor of direct combustion.

Electrification patterns. In viewing the pattern of electrification of the United States, it is instructive to classify end uses according to the degree of electrification, as is done in Table 7. Three groupings emerge: completely electrified uses, heating uses, and transportation.

First consider the electrified end uses. These uses include those that are possible only through the agency of electricity, and those in which electricity has clear advantages over direct combustion.

Heating applications have not been electrified to so great an extent, and overall the penetration of electricity is small. When convenience is particularly important and energy consumption is modest, as in clothes drying, cooking, and hot-water supplying, electricity has won moderate acceptance. But when consumption is greater and cost looms larger relative to convenience, direct combustion of fuel on the premises is ordinarily the rule. For electric heating to make solid inroads against direct combustion, technological progress along two lines is essential: the efficiency of electric generation from fossil fuels must be increased, and the performance of heat pumps must be improved.

Transportation resists electrification because it is difficult to devise an arrangement of wires to supply electricity to a moving motor. If the vehicle runs on rails or some other well-defined track,

electric wires can be run parallel to the track and electrification is possible. But the trend in the United States has been in the opposite direction for half a century, with vehicles of all sorts—automobiles, trucks, aircraft, ships—increasingly free to go their own ways. Electric storage batteries carried on a vehicle offer one possibility for electrification of transportation, but aside from low-speed, short-haul uses such as forklift trucks and golf carts, this method has not made much headway. For highway transportation, the development of the internal combustion engine pulled ahead of battery development in the 1890s, when electric automobiles started losing ground.

Largely through the electrification of stationary engines and of illumination, the past century has seen a steady worldwide growth in the consumption of electricity. When measured in terms of kilowatt-hours of electric energy as is the practice of the electric utility industry, United States and world consumption have been growing at an average annual rate of about 7% for half a century (Fig. 11). This growth rate corresponds to a doubling of electric energy consumption every decade. However, since the proportion of energy turned to electricity in the United States is already approaching 30% and cannot increase beyond 100%—it may indeed level off well short of 100%—the growth rate can be expected to slow down in future decades. *See* ELECTRIC POWER GENERATION.

TRENDS IN ENERGY CONSUMPTION

A hundred years ago it took about the same amount of energy to heat a house as it takes today. It took about half as much energy to feed the family horse as it now takes to power the family car. It took about the same amount of energy to cook a meal. People use more energy today, partly because they drive more and partly because they work in offices and factories instead of in open fields, but they still only use about 2½ times as much per person.

Figure 12 shows how energy consumption per person grew between 1850 and 1970. It has been growing very rapidly in recent years. If this growth were to continue, the supplying of the required coal, oil, gas, and uranium would create a strain. The supply problem would not be so serious if energy consumption per person were to level off.

Personal automobile driving is likely to level off by the time every adult has a car to drive. Figure 13 shows how the average number of cars per adult has increased from practically nothing in 1910 to over 0.7 car per adult in the 1970s.

Job-related energy consumption has gone up as more factories and offices have been built. The fraction of the population employed in factories and offices amounted to only about 10% in 1850, but it rose to about 30% a hundred years later. Since 1960 it has risen to about 36% as more and more women have taken jobs outside the home. This trend has a natural limit at about 45% of the population, when everybody of working age will have a job in an office or factory. Growth of the nonfarm labor force will slow down to match overall population growth, and growth of job-related energy consumption will tend to do the same.

As affluence increases, partly through more jobs per family, more energy tends to be consumed in

ENERGY SOURCES   ENERGY CONVERSION FACILITIES   USEFUL ENERGY APPLICATIONS

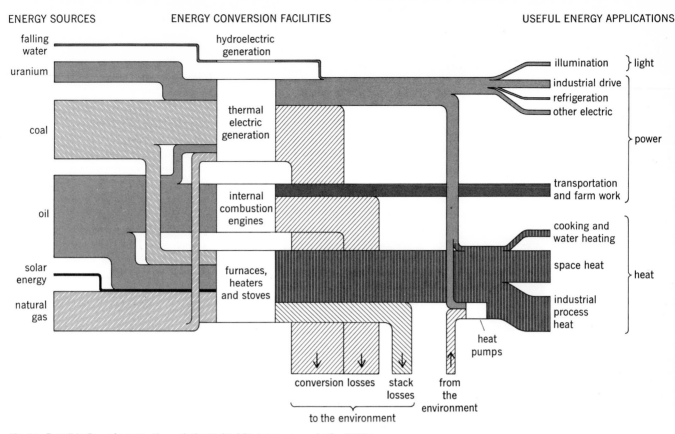

Fig. 14. Possible flow of energy through the United States economy in the 1990s.

the home, mostly for hot water and for comfort heating and cooling. When these basic energy needs are met, energy consumption in the home rises more slowly with increased affluence.

It is anticipated that overall per capita energy consumption may level off. The conservation movement is a welcome expression of people's desire to limit and control energy consumption; indeed, the vitality of the movement may be a symptom, as much as a cause, of the growing achievement of sufficient energy for personal use in society.

The pattern of energy consumption may change in still other ways over the next several decades. Progressive electrification of energy usage is likely to continue. This is the best way to make use of nuclear energy, and improved technology can be expected to increase the efficiencies of electric power generation and application so that electricity will be chosen more often over direct fuel combustion. There is a potential for limited use of solar power, primarily for supplying hot water and for comfort heating.

A possible future pattern for the flow of energy through the United States economy is shown in Fig. 14. Compared with the present as shown in Fig. 4, uranium and coal may provide more of the energy. The efficiency of conversion facilities may improve. Heat pumps, by drawing heat from the air, may augment the effectiveness of electrical heat. More efficient use of energy, as projected in Fig. 14, combined with a leveling off of per capita energy consumption and slower population growth, will tend to moderate the nation's overall energy consumption.

[JOHN C. FISHER]

Bibliography: J. C. Fisher, *Energy Crises in Perspective*, 1974; Stanford Research Institute, for the Office of Science and Technology, Executive Office of the President, Washington, D.C., *Patterns of Energy Consumption in the United States*, January 1972; Task Force on Energy of the National Academy of Engineering, *U.S. Energy Prospects: An Engineering Viewpoint*, 1974; U.S. Bureau of the Census, *Statistical Abstract of the United States: 1974*.

The significance of energy in human affairs can best be appreciated when it is realized that energy is involved in everything that happens on the Earth — everything that moves. The Earth is essentially a closed material system composed of the naturally occurring 92 chemical elements, all but a minute fraction of which are nonradioactive and hence obey the rules of conservation of matter and nontransmutability of the elements of classical chemistry. Into and out of the Earth's surface environment there occurs a continuous influx, degradation, and efflux of energy in consequence of which the mobile materials of the Earth's surface undergo either continuous or intermittent circulation. In addition, there are certain large chemical, thermal, and nuclear stores of energy within minable or drillable depths beneath the Earth's surface.

EARTH'S ENERGY SYSTEM

This total energy system of the Earth's surface is depicted graphically in Fig. 1. The horizontal bar near the bottom of the chart represents the surface of the Earth, below which are the energy stores of the fossil fuels and of geothermal, gravitational, and nuclear energy. The upper part of the chart is an energy flow diagram. The main energy influxes into the Earth's surface environment are three: the solar radiation intercepted by the Earth's diametral plane; tidal energy derived from the combined potential and kinetic energy of the Earth-Moon-Sun system; and terrestrial (especially geothermal) energy from inside the Earth. The magnitudes of these three inputs are: solar, $174,000 \times 10^{12}$ thermal watts; geothermal, 32×10^{12} thermal watts; and tidal, 3×10^{12} thermal watts. Thus, it is seen that the rate of energy influx from the Sun is roughly 5000 times the sum of the other two.

Of the solar power influx, about 30%, the albedo, is reflected and scattered into outer space as short-wavelength visible radiation. The remaining solar-energy flux of approximately $120,000 \times 10^{12}$ thermal watts, and the tidal and geothermal sources, are effective in terrestrial processes. With one small exception, the energy from all of these sources undergoes a series of transformations and degradations until it becomes heat at the lowest environmental temperature, after which it leaves the Earth as low-temperature thermal radiation.

The greater part of this energy flux serves to warm the atmosphere, the oceans, and the ground, and to produce atmospheric, oceanic, and hydrologic circulations. Of particular significance, however, is the 40×10^{12} W of solar power which is captured by the green leaves of plants and which by the process of photosynthesis drives the reaction whereby the inorganic compounds H_2O and CO_2 are synthesized into carbohydrates in which the solar energy becomes stored chemically. This then becomes the basic energy source for the physiological requirements of the entire plant and animal kingdoms, including the human species. *See* SOLAR ENERGY.

Nearly all the plant and animal material decays by oxidation and returns to its original constituents, H_2O and CO_2, at the same average rate as it is formed, and the stored energy is released as heat. The small exception pertains to the minute quantities of biologic materials which become deposited in peat bogs or other oxygen-deficient

Outlook for Fuel Reserves

- total energy system of the Earth
- depletion cycle for exhaustible resources
- worldwide reserves
- production capabilities

Dr. M. King Hubbert

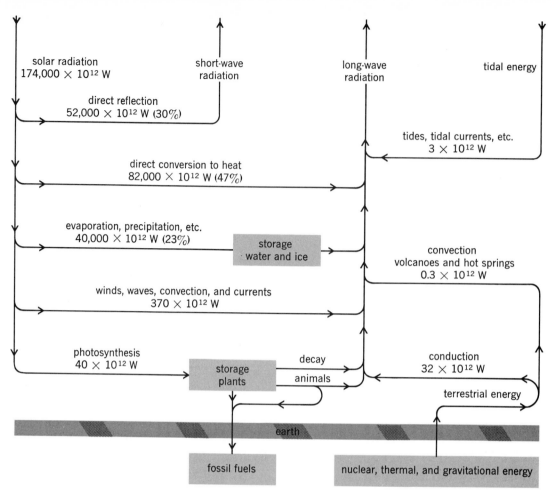

Fig. 1. Energy flow-sheet for the Earth. (*From M. K. Hubbert, U.S. energy resources: A review as of 1972, pt. 1, in A National Fuels and Energy Policy Study, U.S. 93d Congress, 2d Session, Senate Committee on Interior and Insular Affairs, ser. no. 93-40 (92-75), 1974)*)

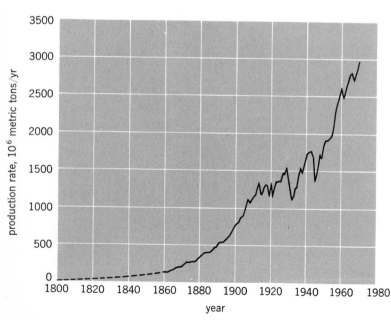

Fig. 2. World production of coal and lignite. Annual statistics are difficult to assemble for years prior to 1860 and have been estimated based on 2% average growth rate during the preceding 8 centuries. (*From Hubbert, op. cit., 1974*)

environments where complete oxidation is impossible and the energy of the material is preserved. This process has been occurring during the last several hundred million years of geologic time, and the accumulated organic debris, after burial under great thicknesses of sedimentary sands, muds, and limes, has been transformed into the Earth's present supply of fossil fuels. *See* FUEL, FOSSIL.

FOSSIL FUELS

The basic energy for the physiological requirements of the human species—its food supply—is obtained from the photosynthetic channel. However, during the last 2,000,000 years or so, the ancestors of the present human species have been progressively tampering with the Earth's energy system. Initially this consisted in the use of tools and weapons, and clothing and housing, whereby ever-larger fractions of the energy of the photosynthetic channel could be converted to human uses. Later, the ancient Egyptians, Greeks, and Romans began using the channel of wind power, and the Romans that of water power. This made possible a continuous increase in the human population, both in areal density and in geographical extent, but only a slight increase in the energy use per capita.

See the feature article ENERGY CONSUMPTION.

Exploitation of fossil fuels. A large increase in the energy per capita was not possible until exploitation of the large, concentrated quantities of energy stored in the fossil fuels was begun. The exploitation of coal as a continuing enterprise began in northeast England near Newcastle-upon-Tyne about 900 years ago; and the production of petroleum, the second major fossil fuel, was begun in Rumania in 1857 and in the United States in 1859.

World production. A graph of the rate of world production of coal is shown in Fig. 2. Scattered statistics exist to show that the cumulative production by 1860 was about 7×10^9 metric tons. Cumulative coal production by 1970 amounted to 139×10^9 metric tons. Of this, the amount of coal produced since 1940 exceeds somewhat all of the coal produced during the preceding 9 centuries.

During the period from 1860 to World War I, annual coal production increased steadily at an average growth rate of 4.2% per year, with a doubling period of 16.5 years. From the beginning of World War I to the end of World War II, the growth rate was only about 0.8% per year. Since World War II it has been at an intermediate rate of about 3% per year.

World production of crude oil from 1880 to 1970 is shown in Fig. 3. From 1890 to 1970 this increased at a uniform rate of growth of 7% per year, with a doubling period of 10 years. At such a growth rate, the cumulative production also doubles every 10 years, so that the cumulative production from 1960 to 1970 was approximately equal to all the oil produced before 1960.

In terms of their energy contents as measured by the heats of combustion, the contribution of crude oil as compared with that of coal was barely significant until about 1900. Subsequently, the energy contribution of crude oil increased more rapidly than that of coal, and became greater than that of coal by 1970. Were the additional energy contributions of natural gas and natural-gas liquids to be added to that of crude oil, the energy of petroleum fluids would represent about two-thirds and coal about one-third of the total rate of energy production from the fossil fuels.

United States production. Coal production and crude oil production in the United States are shown in Figs. 4 and 5. Coal mining in the United States began about 1820 and increased exponentially until about 1910 at a mean rate of 6.7% per year, with a doubling period of 10.4 years. Since World War I, United States coal production has fluctuated about a constant rate of 500×10^6 metric tons per year. *See* COAL MINING.

Figure 5 shows the growth in the annual production of crude oil in the United States since 1860. From 1880 to 1929 the production rate increased at a steady rate of 8.3% per year, with a doubling period of 8.4 years. After 1929 there was a drop in production during the Depression, and then a gradual slowing down in the growth rate until the peak in the production rate was reached in 1970. After that annual production has declined each succeeding year.

In the United States, as in the world, the rate of energy production from crude oil, natural gas, and natural-gas liquids has increased much faster than

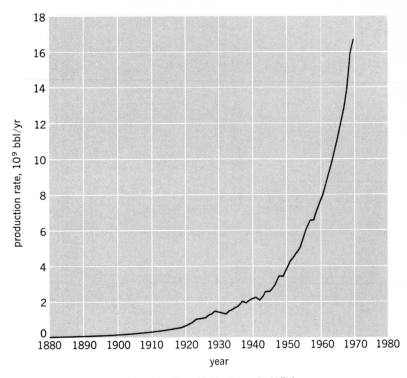

Fig. 3. World crude oil production. (*From Hubbert, op. cit., 1974*)

that of coal. By 1973, of the total energy produced in the United States from the fossil fuels and from nuclear and water power, only 17.9% was contributed by coal and 5% by nuclear and water power. The remainder, 77.1%, was contributed by oil and natural gas.

In the light of the rates of growth in world coal and oil production, the question unavoidably arises: About how long can the rates of production of

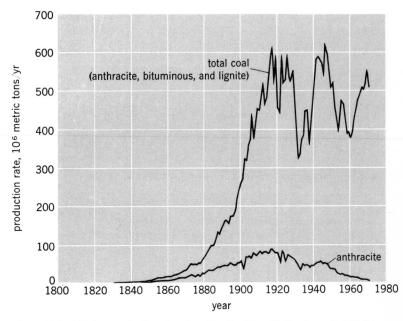

Fig. 4. United States production of coal and lignite. (*From Hubbert, op. cit., 1974*)

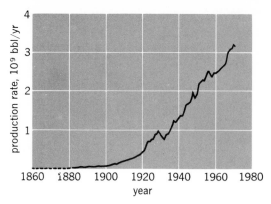

Fig. 5. United States crude oil production; figures are estimated between 1860 and 1880. (*From Hubbert, op. cit., 1974*)

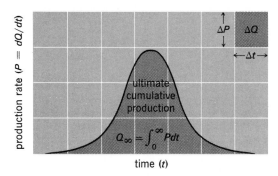

Fig. 6. Mathematical properties of arithmetical graph of production rate P versus time t for the complete production cycle of an exhaustible resource. (*After M. K. Hubbert, Nuclear Energy and the Fossil Fuels: Drilling and Production Practice, American Petroleum Institute, 1956; and Hubbert, op. cit., 1974*)

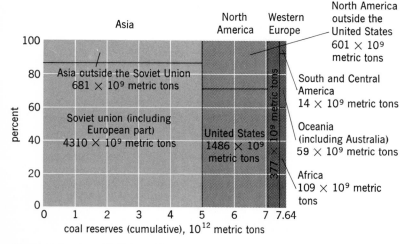

Fig. 7. Estimates of initial world resources of recoverable coal in beds 12 or more inches thick occurring at depths of 6000 ft or less. (*From P. Averitt, Coal Resources of the United States, Jan. 1, 1967, U.S. Geological Surv. Bull. 1275, 1969; and Hubbert, op. cit., 1974*)

these fuels continue to increase, and over what period of time can the fossil fuels serve as major sources of industrial energy?

Various methods of analysis have been de-

veloped which, when applied individually or conjointly, are capable of giving reasonably reliable answers to these questions. These methods are: (1) the complete cycle analysis; (2) analysis of cumulative statistical data of production, reserves, and recovery; and (3) discoveries per foot of exploratory drilling. These methods are not applicable to all types of fossil fuels.

COMPLETE CYCLE ANALYSIS

One of these methods is based upon the properties of the complete cycle of production of any exhaustible and nonrenewable resource. The fossil fuels are ideal examples of such a resource. The Earth's present deposits of both coal and oil are of finite magnitudes and have required hundreds of millions of years for their geological accumulation, whereas the time required for their depletion is measurable in centuries, or at most, millennia. Although the same natural processes by which the fossil fuels were initially accumulated are still operative, the rates are so slow that no significant additions to the world's coal and oil resources can occur within the next few thousand years. Hence, the exploitation of the fossil fuels amounts to the progressive depletion of an initial stockpile.

Because of this, for any given region or for the entire world, the curve of the rate of production of coal or oil as a function of time must, during the complete cycle of production and depletion, have the following properties: The curve must begin initially at zero, and then rise until it reaches one or more maxima. Then, as the resource approaches exhaustion, the curve must gradually decline back to zero. A simplified illustration of such a complete cycle is shown in Fig. 6.

A fundamental property of such a curve may be seen as follows. At some time t, let a small amount of time Δt (say one year) be taken on the time axis, and upon Δt as a base let a narrow vertical band be erected to the production-rate curve. The altitude of this band at time t will be

$$P = \Delta Q / \Delta t, \qquad (1)$$

where ΔQ is the quantity produced during Δt. Then the area of this band will be the product of its base by its altitude, or

$$\Delta A = P \Delta t. \qquad (2)$$

But from Eq. (1), $\Delta Q = P \Delta t$. Therefore, the area ΔA is a graphical measure of the quantity ΔQ produced during the time interval Δt. Hence, the area between the curve and the time axis to any given time is a measure of the cumulative production to that time. Likewise, the total area beneath the curve for the complete production cycle will be a measure of the total quantity Q_∞ produced during the entire cycle of production.

A graphical scale relating an area ΔA to the resource quantity ΔQ is given by the grid square in the upper-right-hand corner of Fig. 6. The quantity ΔQ represented by this area is

$$\Delta Q = \Delta t \times \Delta P. \qquad (3)$$

This signifies that if a constant production rate ΔP were to be sustained for a period Δt, the quantity produced would be ΔQ.

If for a complete cycle of production the area beneath the rate of production should be n grid squares, or rectangles, then the ultimate cumulative production would be

$$Q_\infty = n\Delta Q. \qquad (4)$$

Conversely, if from geological or other information the magnitude of the ultimate quantity Q_∞ to be produced in a given region can be estimated, the number of grid squares beneath the complete-cycle curve would be $n = Q_\infty/\Delta Q$, and the curve must be drawn subject to this constraint.

World coal estimates. This principle can be applied to the world production of coal. Because coal occurs in stratified seams which often crop out on the surface and may be continuous underground for tens of kilometers, reasonably good estimates of the quantities of coal in given regions can be made by surface geological mapping and a small number of deep drill holes. Such studies of coal resources have been made during the present century in all the countries of the world, and the results of these studies were compiled by P. Averitt in 1969.

Figure 7 is a graphical representation of Averitt's estimates of the recoverable coal (assuming a 50% extraction) initially present in major geographical regions of the world. The areas of the columns are proportional to the quantities of recoverable coal initially present. It will be noted that the total for the world is given as 7.64×10^{12} metric tons, and for the United States as 1.486×10^{12}, or 19% of the world total.

These estimates, however, may be unrealistically high in terms of coal mining because they include seams as thin as 12 in. (0.3 m) and occurring at depths to 4000 ft and in some cases to 6000 ft (1200 and 1800 m). In view of this fact, Averitt compiled a separate estimate in 1972 of the amount of coal in the United States occurring at depths of 1000 ft (305 m) or less and in seams of not less than 28 in. (0.71 m) thick for anthracite and bituminous coal, and not less than 5 ft (1.5 m) thick for subbituminous coal and lignite. The initial amount of recoverable coal in these categories was reduced to 390×10^9 metric tons as compared with the earlier figure of 1.486×10^{12} metric tons—a reduction of 74%. Assuming that the same reduction ratio would also be valid for the world, a reduced figure of 2.0×10^{12} metric tons is obtained for world coal of the specified minimum thickness occurring at depths of 1000 ft or less.

Using Averitt's high and low figures of 7.64×10^{12} and 2.0×10^{12} metric tons for Q_∞, two complete-cycle curves for world coal production can be drawn. These are given in Fig. 8. In this figure, for one grid square, $\Delta Q = 10^{10}$ metric ton/yr $\times 10^2$ yr $= 10^{12}$ metric tons. Therefore, for $Q_\infty = 7.6 \times 10^{12}$ metric tons, the area beneath the complete-cycle curve will be $(7.6 \times 10^{12})/10^{12} = 7.6$ squares. For the smaller value of 2.0×10^{12} metric tons for Q_∞, the number of squares will be but 2. The curves of Fig. 8 are constructed accordingly. Obviously, the shapes are not unique, but for a fixed number of grid squares, the larger the peak rate of production the shorter the time span for the complete cycle.

In the complete cycle of coal production, long periods of time—possibly a thousand years—will be required to produce the first and last 10 percentiles of Q_∞. A much briefer time, however, will be required to produce the middle 80%. According to Fig. 8, for any likely magnitude of the maximum production rate, the date of the peak of production will probably occur within the next 100 to 200 years, and the time span for the middle 80% will probably be not more than about 3 centuries.

Complete-cycle curves for United States coal production are not shown graphically, but based upon Averitt's high and low estimates for Q_∞, the time scales for the U.S. production are about the same as those for the world.

Oil and gas estimates. The problem of estimating the ultimate amounts of crude oil and natural gas to be produced in a given region is much more difficult than for estimates of coal, because accumulations of oil or gas occur in porous sedimentary rocks in limited regions of underground space with horizontal dimensions from 100 m to more than 100 km, and at depths ranging from about 100 m to 7.5 km. However, as exploration and drilling proceed in a given region, the eventual decline of the dis-

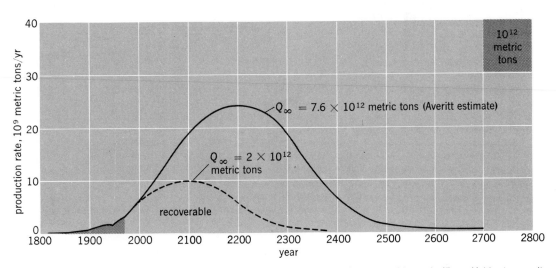

Fig. 8. Two complete cycles of world coal production based upon Averitt higher and lower estimates of initial resources of recoverable coal. (*From Hubbert, op. cit., 1974*)

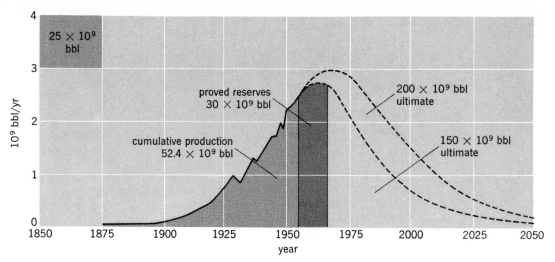

Fig. 9. Hubbert prediction of 1956 of future production of crude oil in the conterminous United States and adjacent continental shelves. (*After Hubbert, op. cit., 1956 and 1974*)

coveries per unit of exploratory effort affords a basis for estimates of the ultimate amounts of oil or gas that a given region is likely to produce.

In the United States, by 1956, the cumulative production since the initial oil discovery in 1859 amounted to 52.4×10^9 bbl (1 bbl = 0.159 m³) and the production rate was continuously increasing. Nevertheless, the cumulative experience in petroleum exploration led to a consensus among petroleum geologists and engineers that the value of Q_∞ for crude oil to be produced in the conterminous United States and adjacent continental shelves would probably be within the range of 150–200×10^9 bbl. Figure 9 shows two complete cycles for United States crude oil production based upon these two figures made in 1956 by M. K.

Hubbert. Since each grid square in this figure represents 25×10^9 bbl, for the lower figure of 150×10^9 bbl for Q_∞ there could be but six squares beneath the curve, two of which had already been used by cumulative production, leaving but four more for the future. To satisfy these conditions, the peak in the production rate would have to occur within about 10 years, or at about 1966. For the higher figure of 200×10^9 barrels, two more squares would be added, but the date of peak production would be retarded by only about 5 years, or to 1971. Hence, from prevailing estimates of 1956 for Q_∞, it was possible to predict that the peak in the rate of United States crude oil production would probably occur within the period 1966–1971.

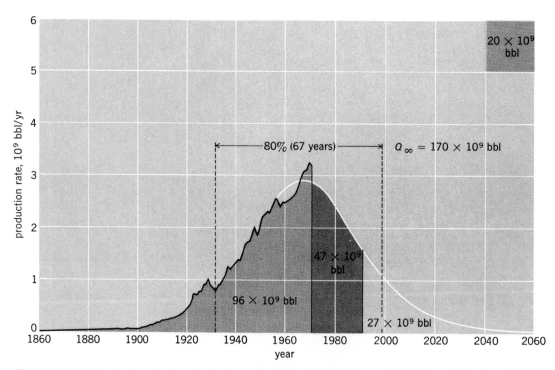

Fig. 10.   Complete cycle of crude oil production in conterminous United States as of 1971. (*From Hubbert, op. cit., 1974*)

Table 1. Estimates of ultimate amounts of energy contents of crude oil, natural-gas liquids, and natural gas to be produced in the United States and bordering continental shelves

	Conterminous United States	Alaska	Total United States
Crude oil, 10^9 bbl	170	43	213
Natural-gas liquids, 10^9 bbl	34	5	39
Total hydrocarbon liquids, 10^9 bbl	204	48	252
Natural gas, 10^{12} ft³	1050	134	1184
Energy contents			
Energy of liquids, 10^{18} thermal joules	1208	284	1492
Energy of natural gas, 10^{18} thermal joules	1143	146	1289
Total energy, 10^{18} thermal joules	2351	430	2781

SOURCE: M. K. Hubbert, in *A National Fuels and Energy Policy Study*, U.S. 93d Congress, 2d Session, Senate Committee on Interior and Insular Affairs, ser. no. 93-40 (92-75), 1974.

A difficulty in the application of the complete-cycle method to petroleum estimates is that it requires an independent estimate of Q_∞. Even so, for any reasonable estimates for Q_∞, the time scales obtained by this method are comparatively insensitive to error.

Crude oil production cycle. The complete cycle analysis has also been applied to the production of crude oil and ultimate amounts of crude oil, natural gas, and natural-gas liquids. The complete cycle of crude oil production for the conterminous United States, based upon 170×10^9 bbl for Q_∞, is shown in Fig. 10. Of particular significance is the time required to produce the middle 80% of Q_∞. This is estimated to be the 67-year period from about 1932 to 1999.

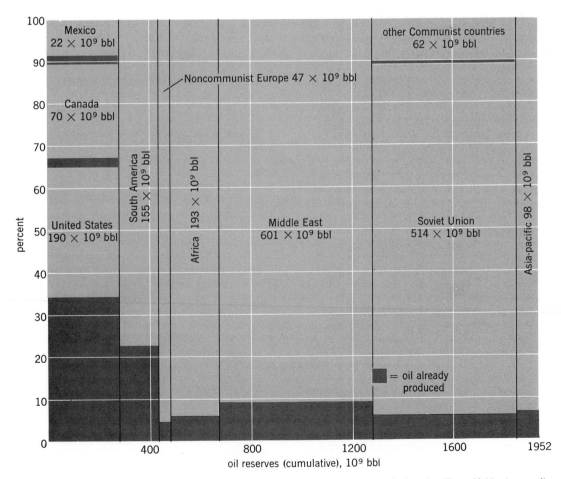

Fig. 11. Graphical representation of Jodry estimate of world ultimately recoverable crude oil. The shaded areas at the foot of each column or sector represent quantities consumed already. (*From Hubbert, op. cit., 1974*)

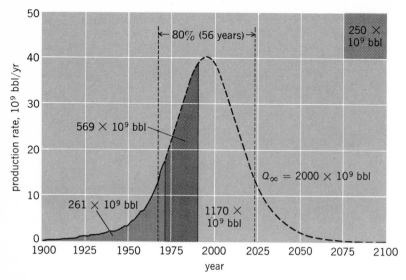

Fig. 12. Estimate as of 1972 of complete cycle of world crude oil production. (*From Hubbert, op. cit., 1974*)

Estimates of the ultimate cumulative production of natural gas in the United States have been made on the basis both of the ratio of gas discoveries to crude oil discoveries and the prior estimate of Q_∞ for oil, and of gas discoveries per foot of exploratory drilling. These methods give a range from about 1×10^{15} to 1.1×10^{15} ft³ (1 ft³ = 0.02832 m³) for Q_∞ for natural gas for the conterminous United States. The peak in natural-gas proved reserves was reached in 1967, and the peak in the production rate occurred in 1973.

Estimates of natural-gas liquids are obtained from the prior estimates for natural gas and the gas-to-liquids ratio. For the conterminous United States, the estimated value for Q_∞ for natural-gas liquids is about 34×10^9 bbl, of which 15.6×10^9 bbl had been produced by the end of 1974.

For Alaska, which is still in its early stages of petroleum exploration, only rough estimates can be given at present of the ultimate quantities of petroleum fluids that may be produced. Including both land and offshore areas, such rough estimates are the following: crude oil, 43×10^9 bbl; natural gas, 134×10^{12} ft; natural-gas liquids, 5×10^9 bbl.

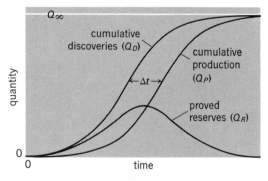

Fig. 13. Variation with time of proved reserves Q_R, cumulative production Q_p, and cumulative proved discoveries, Q_D, during a complete cycle of petroleum production. (*After M. K. Hubbert, Energy resources: A report to the Committee on Natural Resources, Nat. Acad. Sci.–Nat. Res. Counc. Publ. 1000-D, 1962; and Hubbert, op. cit., 1974*)

Table 1 gives a summary of the approximate magnitude of the ultimate quantities of crude oil, natural gas, and natural-gas liquids, and their energy contents, to be produced in the United States and the bordering continental shelves.

Ultimate world crude oil production. For the ultimate world production of crude oil, 15 estimates by geologists and international oil companies, published between 1959 and 1973, give a range from 1.2×10^{12} to 2.48×10^{12} bbl, and an average of 1.84×10^{12} bbl. Figure 11, based upon a study by R. L. Jodry, shows the estimated geographical distribution of the world's oil. The areas of the separate columns are proportional to the estimated ultimate oil production, totaling for the world 1.952×10^{12} bbl.

The North American column, with an estimated 282×10^9 bbl, is especially significant. This represents only 14.4% of the world total, of which about two-thirds is in the United States. Yet the United States, with only about 10% of the world's oil initially, has been until 1974 the world's largest producer as well as the world's largest consumer of oil. It is, accordingly, not surprising that the United States has already consumed half of its oil and is the farthest toward ultimate depletion of its oil of any of the major oil-producing countries.

World crude oil production cycle. Figure 12 shows the complete cycle of world oil production, based upon a round figure of 2×10^{12} bbl for Q_∞ and upon the assumption of an orderly future evolution of the petroleum industry. According to this figure, a peak production rate of 40×10^9 bbl/yr is due to occur about the year 1995, and the middle 80% of the world's oil will be consumed between about 1966 and 2022.

Should an orderly evolution not ensue, it is possible that the production rate might become stabilized at some nearly constant rate near that of 1975. In that case, the area in Fig. 12 above that constant rate would be shifted farther in time and distributed along the back slope, thus prolonging by a few years the time required for near-exhaustion.

CUMULATIVE STATISTICAL DATA

A different approach to petroleum estimation is based upon the use of cumulative statistical data on drilling, discovery, proved reserves, and production as a means of determining how far advanced the petroleum industry may be in its complete cycle. Figure 13 shows the evolution during the complete cycle of three statistical quantities, cumulative production Q_P, proved reserves Q_R, and cumulative proved discoveries Q_D. In the United States, statistical data on annual crude oil production are available from 1860 to the present. The sum of annual productions to any given year gives the cumulative production. Statistics on the proved reserves at the close of each year have been issued by a nationwide committee of petroleum engineers of the American Petroleum Institute since 1936, and approximate estimates are available annually since 1900. Proved reserves at any given time represent the amount of oil that almost certainly is present in fields already discovered, and producible by equipment already installed. It is, therefore, a working inventory; it is the difference between cumulative additions to reserves

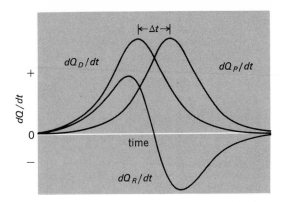

Fig. 14. Variation of rates of production, of proved discovery, and of rate of increase of proved reserves of crude oil or natural gas during a complete production cycle. (*After Hubbert, op. cit., 1962 and 1974*)

and withdrawals by means of production.

Cumulative proved discoveries is a derived quantity defined by

$$Q_D = Q_P + Q_R;$$ (5)

that is, all the oil that can be regarded as having been proved to be discovered by a given time is the oil produced to that time plus proved reserves.

The approximate nature of the variation with time of the three quantities, Q_D, Q_P, and Q_R, during a complete cycle is shown in Fig. 13. This is based upon the assumption that the complete cycle is one of a single maximum in the rate of production. Here, the Q_D and Q_P curves are logistic-type growth curves beginning at zero and ending asymptotically to the value of Q_∞. The Q_R curve

begins at zero, reaches a maximum at about midrange, and then declines to zero at the end of the cycle. Also, in the midrange there is a time delay of Δt years between the discovery curve and the production curve.

The rates of discovery, of production, and of the increase of proved reserves are equal to the slopes of the respective curves in Fig. 13. Mathematically, from Eq. (5), these are

$$dQ_D/dt = dQ_P/dt + dQ_R/dt.$$ (6)

Graphs of these curves are shown in Fig. 14. It is to be noted that the peak in the rate of production occurs approximately Δt years later than the peak in the rate of discovery. The curve of rate of increase of proved reserves, dQ_R/dt, has a positive loop while reserves are increasing, crosses the zero line when reserves are at their maximum, and has a negative loop while reserves are decreasing. At the time when reserves are at their maximum, the rate of increase of proved reserves is

$$dQ_R/dt = 0.$$ (7)

At that time, by Eq. (6), the rates of discovery and production are

$$dQ_D/dt = dQ_P/dt,$$ (8)

and these two curves cross one another, the rate of production still increasing but the rate of discovery already declining. The date of this event is about halfway between the discovery peak and the production peak.

In the earlier stages of the cycle, these two sets of curves are not very informative, but they become increasingly so from about the time of the peak in the rate of discovery onward.

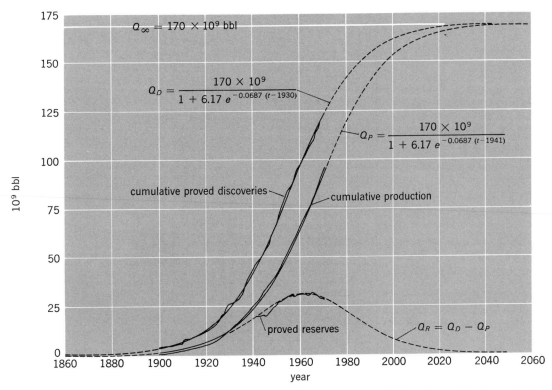

Fig. 15. Logistic equations and curves of cumulative production, cumulative discoveries, and proved reserves for crude oil from the conterminous United States, 1900–1971. (*From Hubbert, op. cit., 1974*)

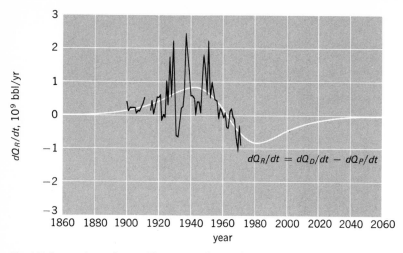

Fig. 16. Comparison of annual increases of proved reserves of conterminous United States, 1900–1971, with theoretical curve derived from logistic equations. (*From Hubbert, op. cit., 1974*)

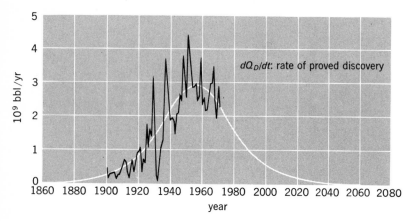

Fig. 17. Comparison of annual proved discoveries of crude oil in the conterminous United States, 1900–1971, with corresponding theoretical curve derived from logistic equation. (*From Hubbert, op. cit., 1974*)

These two sets of curves were constructed in 1962 by Hubbert, using the petroleum industry data to the end of 1961. Ten years later, the corresponding curves were constructed using cumulative data to the end of 1971 and reported by Hubbert in 1974. The results of these two separate sets of analyses are given in Table 2.

The actual data, as of 1972, for the Q_D, Q_P, and Q_R curves are shown in Fig. 15. The derivative, or rate, curves for the rate of increase of proved reserves and the rate of discovery, respectively, are given in Figs. 16 and 17.

DISCOVERIES PER FOOT OF EXPLORATORY DRILLING

A different kind of analysis has been used to estimate ultimate production of oil or gas. This con-

sists in determining the quantity of oil discovered per foot of exploratory drilling, dQ/dh, as a function of cumulative depth h of exploratory drilling. The area beneath this curve also is a measure of cumulative discoveries. Figure 18 shows the average numbers of barrels of crude oil discovered in the United States for each 10^8 ft of exploratory drilling from 1860 to 1972. The area of each column in the figure represents the quantity of oil discovered during each 10^8-ft interval of drilling.

As is seen from the figure, the discovery rate during the first four drilling units averaged about 225 bbl/ft. This was followed by a drastic decline to a final figure of only 30 bbl/ft by the last, or seventeenth, unit of drilling. The cumulative discoveries, defined as the sum of cumulative production plus proved reserves plus an estimated additional amount of oil in fields already discovered, by the 17×10^8 ft of drilling, amounted to 143×10^9 bbl. The rate of decline in discoveries per unit of drilling shown in Fig. 18 is roughly a negative exponential. The best fit for such a curve, as shown in the figure, is one that equalizes the excesses and deficiencies, and passes through the last point of 30.4 bbl/ft. Extrapolation of this decline curve for unlimited future drilling gives an additional 29×10^9 bbl as the estimated future discoveries. Adding this to the 143×10^9 bbl already discovered gives a sum of 172×10^9 bbl for Q_∞. This is practically identical with the figure of 170×10^9 bbl obtained by the previous, quite different method of analysis. *See* PETROLEUM.

OTHER FOSSIL FUELS

Besides coal and lignite and the petroleum fluids (crude oil, natural gas, and natural-gas liquids) the principal remaining fossil fuels are tar or heavy-oil sands, oil shales, and minor quantities of the solid hydrocarbon gilsonite.

Tar sands. The world's largest known deposits of tar or heavy-oil sands are the Athabasca sands of northern Alberta, Canada. These consist essentially of heavy crude oil, filling the pore spaces of coarse-grained quartz sands, which is too viscous to flow into wells. These tar sands occur at various depths ranging from surface outcrops along the Athabasca River in northeastern Alberta to depths up to 2000 ft (600 m) in deposits farther west. Estimates of recoverable quantities of these oils given by T. F. Scott in 1974 are the following: oil-in-place, 625×10^9 bbl; oil extractable, 148×10^9 bbl (by mining, 38×10^9, by at-site methods, 110×10^9). These raw oils must be coverted into synthetic crude oils by preliminary refining. According to Scott, only 70 bbl of synthetic crude oil are obtainable from 100 bbl of raw oil. Hence, the figure of 148×10^9 bbl of raw oil is equivalent to about 104×10^9 bbl of crude oil.

The only mining operation now going on is that

Table 2. Crude oil estimates as of 1962 and 1972 for conterminous United States, based upon analyses of Q_D, Q_P, and Q_R

Entity estimated	Date of estimation		Observed
	1962	1972	
Date of maximum discovery rate	1957	1957	1957
Time lag Δt between discovery and production	10.5 years	11.0 years	
Date of peak of proved reserves	1962	1962	1962
Date of maximum production rate	1968	1968	1970
Ultimate cumulative production Q_∞	170×10^9 bbl	170×10^9 bbl	

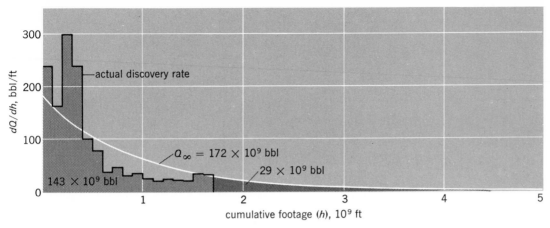

Fig. 18. Estimation of ultimate crude oil production of the conterminous United States by means of the curve of discoveries per foot versus cumulative footage of exploratory drilling. (*After Hubbert, op. cit., 1974*)

Table 3. Oil content of known shale oil resources of world land areas

	Recoverable under 1965 conditions, 10^9 bbl (10 to 100 gal/ton)	Marginal and submarginal, 10^9 bbl		
Continent		25 to 100 (gal/ton)	10 to 25 (gal/ton)	5 to 10 (gal/ton)
Africa	10	90	*	*
Asia	20	70	14	†
Australia and New Zealand	*	*	1	†
Europe	30	40	6	†
North America	80	520	1600	2200
South America	50	*	750	†
Total	190	720	2400‡	2200

SOURCE: D. C. Duncan and V. E. Swanson, *Organic-rich shale of the United States and World Land Areas*, U.S. Geol. Surv. Circ. 523, 1965; M. K. Hubbert, in *A National Fuels and Energy Policy Study*, U.S. 93d Congress, 2d Session, Senate Committee on Interior and Insular Affairs, ser. no. 93–40 (92–75), 1974.
*Small.   †No estimate.   ‡Rounded.

Table 4. Reserves of recoverable oil from the Green River Formation, in Colorado, Utah, and Wyoming

	Shale oil reserves (at 60% recovery), 10^9 bbl			
Location	Class 1	Class 2	Class 3	Total
Piceance Basin, Colorado	20	50	100	170
Uinta Basin, Colorado and Utah	—	7	9	16
Green River Basin, Wyoming	—	—	2	2
Total	20	57	111	188

SOURCE: National Petroleum Council, Oil Shale Task Group, 1972; M. K. Hubbert, in *A National Fuels and Energy Policy Study*, U.S. 93d Congress, 2d Session, Senate Committee on Interior and Insular Affairs, sec. no. 93–40 (92–75), 1974.

of the Great Canadian Oil Sands, Ltd., which began operation in 1966, with a capacity of 45,000 barrels per day (bpd).

According to reports in 1973 and 1974, a second major deposit of heavy oil underlies a region roughly 85 km wide by 600 km long extending east and west, north of and parallel to the Orinoco River in eastern Venezuela. Within this region there are four areas in which the principal quantities of this oil occur. In the westernmost area the thickness of the deposit is about 82 m; in the other three the thicknesses are about 100 m. These deposits are estimated to contain about 700×10^9 bbl of oil-in-place, of which about 10%, or 70×10^9 bbl, may be recoverable. *See* OIL SAND.

Oil shale. The "oil" in oil shales differs from ordinary petroleum oils in that it occurs in a solid form, kerogen, rather than as a liquid. When a few chips of an oil shale are heated in a test tube, a dense vapor is distilled off which condenses on the walls of the tube as an amber-colored liquid. This is raw shale oil. Like tar sand oil, this too must be refined into a synthetic crude oil before it can be sent to a conventional oil refinery.

Table 3 gives a summary of the world's known shale oil deposits as compiled by D. C. Duncan and V. E. Swanson in 1965. The oil contents of the deposits range from 5 to 100 gal of oil per ton of rock. The estimated total amount of oil within these grades is given as 5.3×10^{12} bbl. Of this, however, Duncan and Swanson consider only about 190×10^9 bbl as being recoverable under 1965 conditions.

The largest known oil shale deposits in the world

Table 5. Magnitudes and energy contents of the world's initial recoverable fossil fuels

Fuel and quantity	Energy content		Percent	
	10^{21} thermal joules	10^{15} thermal kilowatt-hours	Maximum	Minimum
Coal plus lignite				
Maximum 7.6×10^{12} metric tons	188.0	52.2	85.44	—
Minimum 2×10^{12} metric tons	49.6	13.8	—	60.78
Crude oil, 2×10^{12} bbl	12.2	3.4	5.56	14.95
Tar sand oil, 370×10^9 bbl	2.3	.6	.98	2.82
Shale oil, 300×10^9 bbl	1.8	.5	.82	2.21
Natural-gas liquids, 400×10^9 bbl	1.7	.5	.82	2.08
Natural gas, 12.8×10^{15} ft^3	14.0	3.9	6.38	17.16
Total: Maximum	220.0	61.1	100.00	—
Minimum	81.6	22.7	—	100.00

SOURCE: M. K. Hubbert, in *A National Fuels and Energy Policy Study*, U.S. 93d Congress, 2d Session, Senate Committee on Interior and Insular Affairs, sec. no. 93–40 (92–75), 1974.

are those of the Green River shales of Eocene age occurring in three localities: southwestern Wyoming, western Colorado, and northeastern Utah. The approximate quantities of oil in these basins in classes 1 to 3 of decreasing favorability, as given by the Oil Shale Task Group of the National Petroleum Council in 1972, are shown in Table 4.

Class 1 comprises beds at least 30 ft (9 m) thick having an average oil content of 35 gal/ton. Class 2 comprises beds at least 30 ft thick having an average oil content of at least 30 gal/ton. Class 3 comprises shales comparable to those of class 2, only less well defined. The Oil Shale Task Group considered only the shales of the class 1 group, occurring in western Colorado, and containing an estimated 20×10^9 bbl of oil as being suitable for exploitation at present.

In appraising this, it should be borne in mind that 20×10^9 bbl of oil is only about a 4-year's supply for the United States at present rates of consumption. In 1973 several Federal leases of these oil shales of 5000 acres (20.235 km²) each to different groups of oil companies were announced. These were to produce about 50,000 bpd each by the early 1980s.

By the end of 1975 none of these projects was in operation. In October 1974 one large consortium, the Colony Development Operation, discontinued plant development because of rapidly inflating costs. A second group, the Rio Blanco Oil Shale Project, by August 1975 was still in the planning stage.

To appraise the significance of shale oil as a means of meeting the domestic oil requirements of the United States, account needs to be taken of the fact that by the mid-1970s the rate of consumption of petroleum liquids in the United States was about 17×10^6 bpd. To produce even 1×10^6 bbl of shale oil per day would require 20 plants with capacities of 50,000 bpd each, and even this rate of production would be barely significant with respect to domestic requirements.

In the Rio Blanco Project, according to a report in 1975, it will require 10,000 metric tons of rock to produce 6000 bbl of oil. This is a ratio of 1.67 metric tons, or 0.73 m³, of rock to be mined per barrel of oil produced. At a production rate of 1×10^6 bpd, a volume of 730,000 m³ of rock would have to be mined each day. Upon retorting to extract the oil, this shale would expand to about 1×10^6 m³ of cinders produced per day, or about one-third of a cubic kilometer of cinders produced per year. For the estimated 20×10^9 bbl of shale oil obtainable from the Piceance Basin of western Colorado, the total volume of these wastes would amount to about 20 km³, one-quarter of which would comprise highly alkaline calcium and magnesium oxides. Through these wastes would occur a flow of

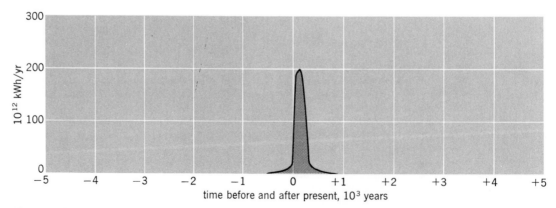

Fig. 19.   The epoch of fossil-fuel exploitation as it appears on a time scale of human history ranging from 5000 years ago to 5000 years into the future. (*From Hubbert, op. cit., 1974*)

groundwater, leaching the alkaline salts and discharging into the Colorado River drainage system.

It is questionable, therefore, whether the small amount of oil obtainable from the Green River oil shales can adequately compensate for the environmental damage that must inevitably accrue. *See* OIL SHALE.

SUMMARY OF WORLD FOSSIL FUELS

The approximate magnitudes of the world's recoverable fossil fuels and their energy contents are summarized in Table 5. It will be noted that coal is the largest energy source of any of the fossil fuels, representing about 85% of the total if Averitt's higher estimate is used, but only about 61% if the reduced estimate is used.

FOSSIL FUELS IN HUMAN HISTORY

The role of the fossil fuels in the longer span of human history can best be appreciated if one considers the period extending from 5000 years in the past to 5000 years in the future. On such a time scale the epoch of the fossil fuels is shown graphically in Fig. 19. This appears as a spike with a middle-80% width of about 3 centuries. It is thus seen that the epoch of the exploitation of the fossil fuels is but an ephemeral event in the totality of human history. It is a unique event, nonetheless, in geological history. Moreover, it is responsible for the world's present technological civilization and has exercised the most profound influence ever experienced by the human species during its entire biological existence.   [M. KING HUBBERT]

Bibliography: P. Averitt, *Coal Resources of the United States, Jan. 1, 1967*, U.S. Geol. Surv. Bull. 1275, 1969; Colony postpones plans for Colorado shale-oil project, *Oil Gas J.*, 72(41):52–53, 1974; D. C. Duncan and V. E. Swanson, *Organic-rich shale of the United States and World Land Areas*, U.S. Geol. Surv. Circ. 523, 1965; M. K. Hubbert, *Energy resources: A report to the Committee on Natural Resources*, Nat. Acad. Sci.–Nat. Res. Counc. Publ. 1000-D, 1962, and reprinted as U.S. Dep. Commer. Rep. PB-222401, 1973; M. K. Hubbert, *Nuclear Energy and the Fossil Fuels: Drilling and Production Practice*, American Petroluem Institute, 1956; M. K. Hubbert, U.S. energy resources: A review as of 1972, pt. 1, in *A National Fuels and Energy Policy Study*, U.S. 93d Congress, 2d Session, Senate Committee on Interior and Insular Affairs, ser. no. 93-40 (92-75), 1974; Large-scale action nearer for Orinoco, *Oil Gas J.*, 71(33): 44.–45, 1973; New data indicate Orinoco belt exceeding expectations, *Oil Gas J.*, 72(45):134, 1974; Rio Blanco oil-shale plan due by year-end, *Oil Gas J.*, 73(30):48, 1975; T. F. Scott, Athabasca oil sands to A.D. 2000, *Can. Min. Met. Bull.*, pp. 98–102, October 1974.

Energy became an everyday issue for most Americans because of the oil embargo by the Arab oil-producing countries in 1973, and the subsequent sudden rise in price of both gasoline and fuel oil. This action by the oil-producing countries is not an isolated one, unrelated to deeper causes. Even before the oil embargo, the United States government was sufficiently concerned about the future of national energy resources to commission a major study by the then chairman of the Atomic Energy Commission (AEC), Dixy Lee Ray, to explore future energy needs and the research and development that would be required to assure adequate future energy supplies to the United States. The very fact that unilateral action on the part of a small number of oil-producing countries could cause so much consternation in the Western world, and the inability of the industrialized nations to find substitutes for the oil imported from those countries, demonstrates that alterations of a very fundamental character have occurred.

There are many ways of characterizing the deep-seated nature of the energy issue for the United States and the rest of the industrialized world, but Fig. 1 is a sufficient introduction. It shows the consumption and domestic production of total energy in the United States from 1945 to 1973. According to this curve, the United States stopped being self-sufficient in energy production in the late 1950s. Until about 1970 the gap between total consumption and domestic production of energy was not large, but in 1970 production began to flatten out, although consumption continued to accelerate.

The United States imports more than one-third of its total annual oil consumption. There is no hope that the United States can regain its former state of self-sufficiency insofar as oil is concerned, but the issue is far deeper than that. As one examines the possibilities for substituting other energy

Exploring Energy Choices

- **resource availability versus consumption**
- **comparison of energy sources**
- **present and future**

Dr. Robb M. Thomson

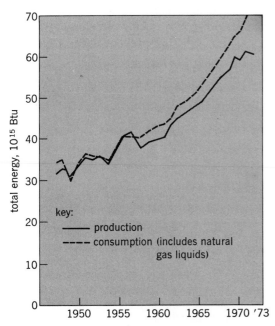

Fig. 1. United States energy production and consumption, 1947–1973; figures for 1973 are preliminary. (*From Energy Policy Project of the Ford Foundation, Preliminary Report, 1974*)

sources for oil, the great difficulties become evident. These difficulties stem partly from the finiteness of the world's energy resources, partly from the lack of technological and economic feasibility in the use of substitute resources, and partly from limitations due to the environmental and social consequences of resource use.

Another issue related to energy choices is energy conservation. This subject will not be discussed in this article, but an appropriate energy conservation program should considerably ameliorate energy shortages. Energy conservation in general will be much in the public eye in both the short- and the long-term future. *See* CONSERVATION OF ENERGY.

ENERGY PROBLEM IN PERSPECTIVE

Throughout human history, one form of fuel has supplanted another, as consumption has exceeded the supply of one and then another fuel source. Figure 2 illustrates this principle in the United States for the period from 1850 to 1975. The shifts from wood to coal to oil and gas, and, ultimately, to nuclear energy and solar energy, have all been prompted by increased demand for energy, the ultimate exhaustion of natural supplies (or the threat of exhaustion and the consequent rise in cost), and the development of new energy sources through the means of new technology. *See the feature article* ENERGY CONSUMPTION.

This suggests that there is an even larger context within which the energy crisis must be placed, namely, the overall pressure of population and economic growth on the ultimate capacity of the Earth and the human environment to sustain that growth. For example, the oil shortage has already had an impact on the production of food through the increased costs of nitrogenous fertilizer, most of which is produced from methane in the form of natural gas. Thus, increased production of food is actually being capitalized by limited and nonrenewable fossil fuel deposits.

This perspective would not be complete without some mention of the role of economics. Energy resources are compared to one another by means of cost. The economic price in the marketplace is the effective means by which scarcity of resources, difficulty of resource development, transportation, and even environmental impact are all joined together into the measure of the desirability of one energy source relative to another. But beyond that,

the availability of cheap energy is thought by many individuals to be an absolute requirement for economic growth. Over the long span of human development, energy costs in terms of the human labor necessary to produce a given unit of energy have decreased. For example, pumping a gallon of oil out of an Iranian oil well and transporting it by tanker to the east coast of the United States and thence through the refinery to its ultimate use costs less now per unit of energy in terms of real human labor costs than at any other time in the history of the world.

Some economists also point out the one-to-one relationship between economic development and energy use per capita. For example, the ranking of the nations of the world in a Gross National Product per capita basis corresponds very closely to the ranking of the same nations on the basis of energy consumption per capita. Likewise, on a somewhat smaller scale, energy growth can be correlated with economic growth in the United States on a year-by-year basis, as shown in Fig. 3.

Many people believe that the era of ever-lessening real energy costs is now coming to an end, for the various reasons which shall be explored in this article. There is thus a very real concern that these increased real costs will lead to decreased energy consumption and, ultimately, to a slowing or stopping of economic development itself. If advancing technology cannot provide cheaper energy, however, the impact on an industrialized country like the United States, where growth of services not based on raw industrial production is more important than it used to be, is likely to be far different from that on a developing country. Thus, for the industrialized part of the world, economic growth should not in the future be so strong a function of energy growth as it has been in the past. Nevertheless, for the developing countries, where industrial development forms the basis for new economic growth, a sustained increased cost of energy is expected to have important effects upon the prospects for future economic development.

Therefore the availability, cost, and ultimate limitations on the use of energy pose grave questions for the well-being of humankind as a whole, both in the short and in the long term. The energy crisis far transcends any local and temporary implications which derive from the oil embargo and the increases in the price of oil. The whole energy question poses a challenge of the most fundamental sort to world social structure, and to society's ability to make new technological and scientific responses. *See the feature article* WORLD ENERGY ECONOMY.

ENERGY RESOURCE AVAILABILITY

This section will review the general energy resource picture primarily as regards the United States, but also, when appropriate, in terms of the world situation. In order to do this, however, some basic points must be developed about a finite nonrenewable resource. The particular point of view described herein was developed by M. K. Hubbert, who noted that the cumulative production of a finite resource over time must begin at zero, grow fairly fast for a time, and, finally, approach a limiting value asymptotically with time. The derivative of the cumulative production is the yearly time

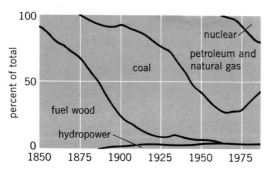

Fig. 2. Historical and projected shifts in United States energy consumption patterns. (*From D. C. Burnham, A Time To Choose, Energy Policy Project of the Ford Foundation, 1974*)

rate of production, and is bell-shaped. *See the feature article* OUTLOOK FOR FUEL RESERVES; *see also* ENERGY SOURCES.

These properties are characteristic of many growth phenomena, for example, biological growth. In the case of resource development, the limit to growth, of course, comes about because, as the resource is exhausted, it is necessary to explore for deeper wells, work smaller deposits, or work mines with lower-grade deposits. The cost of new production becomes higher, demand decreases, and production rate falls off.

One other general idea necessary for this discussion is the distinction between reserves and resources. The term "resource" is intended to imply the total recoverable quantity of an ore. A resource is thus an estimated quantity because there will, in all cases, be undiscovered resources as of any given date. The term "reserve," however, refers to that portion of the recoverable resource which has been discovered and measured ("proved"), though still in the ground.

Oil. When Hubbert's analysis is applied to the production of petroleum liquids in the conterminous United States, that is, not counting both Alaskan and outer continental shelf resources, it is shown that production peaked in 1970; later analysis has confirmed the turndown after that date. As of 1973, a total of 109.7×10^9 bbl (1 bbl = 0.16 m³) of oil was produced, and another 61×10^9 bbl was in reserve. From an analysis of the curves (and with considerable additional analysis of successful drilling for oil) Hubbert estimated in 1973 that another 33.4×10^9 bbl was yet to be discovered. This leads to a total cumulative oil resource, about half of which has now been produced. The future production predicted from the Hubbert curves suggests that well before the year 2000 the United States will have, for practical purposes, exhausted its petroleum resources. In reality, of course, the use rate can be modified, because the resource can be exploited more quickly than the simple model predicts by means of Federal policy actions. For example, Federal policy could conceivably reverse itself and act to conserve the petroleum resource of the United States for a longer period. *See the feature article* U.S. POLICIES AND POLITICS.

Various educated estimates and predictions have been made for the resources on the outer continental shelf and in Alaska, but these estimates differ, with some authorities projecting a much larger total resource than others. The latest responsible estimates for the total oil resource of the United States, including Alaska and the outer continental shelf, have, however, tended to converge in the vicinity of $200-400 \times 10^9$ bbl of recoverable oil.

The world picture is more uncertain than that of the United States. In 1972 R. L. Jodry of the Sun Oil Company evaluated the situation. When his estimates are fitted to Hubbert's analysis, the conclusion is that the world production rate will reach a peak in approximately 1995 and that 80% of the total recoverable resource will be exhausted by the year 2023.

The general conclusions are that American petroleum resources will have disappeared as a major factor in the nation's energy budget well before the year 2000, and that the worldwide petroleum

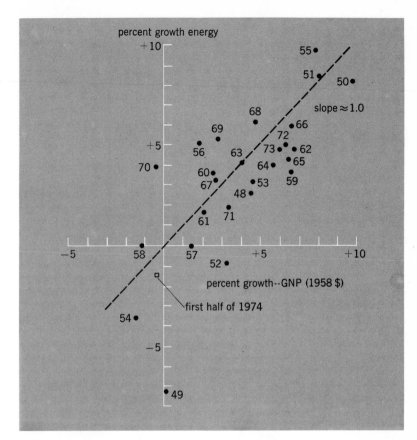

Fig. 3. The interdependence of energy and economic growth in the United States, 1950–1974. The percentage growth in Gross National Product closely parallels the percentage growth in energy. (*From D. C. Burnham, A Time To Choose, Energy Policy Project of the Ford Foundation, 1974*)

resource will disappear within the first quarter of the 21st century. These conclusions are, of course, based upon a continued high energy consumption rate and upon the admittedly inexact estimates of undiscovered oil. However, even if the estimates being made are incorrect by a factor of 2, or even more, and the problem of the ultimate finite character of the Earth's total resource still remains. *See* PETROLEUM.

Natural gas. If Hubbert's analysis is applied to natural-gas production in the conterminous United States, the result shows that 414×10^{12} ft³ (1 ft³ = 1.0283 m³) of gas has been produced, that another 428×10^{12} ft³ is in reserve, and that an estimated 200×10^{12} ft³ is still to be discovered; another 130×10^{12} ft³ is estimated to be in Alaska. Other estimates of the total natural gas are in the neighborhood of $1000-2000 \times 10^{12}$ ft³.

According to these various estimates, the supply of natural gas is somewhat greater than that of crude oil, but a disappearance of the resource of the United States can still be expected by the year 2000. *See* NATURAL GAS.

Coal. In contrast to oil and natural gas, coal is an ancient resource which is very widely distributed throughout the Earth in large amounts. Hubbert has estimated that worldwide coal production increased at approximately 2% per year from its initial mining, in 1060, to 1860. Since 1860 there have been several changes in this production curve, but for the past 30 years production has grown at ap-

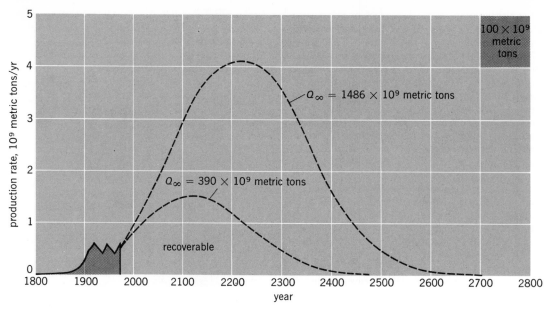

Fig. 4. Two complete United States coal production cycles based upon higher and lower estimates by P. Averitt of initial resources of recoverable coal. (*From M. K. Hubbert, U.S. Energy Resources, U.S. Senate, Serial 93-40 [92-75]*)

proximately 3% per year. The United States production is shown in Fig. 4.

The ultimately recoverable resources in coal are still enormous, both within the United States and worldwide. P. Averitt of the U. S. Geological Survey estimated that in the United States there are about 1500×10^9 metric tons, and that at present production rates this quantity would last several thousand years. Averitt also arrived at a somewhat more conservative figure of recoverable coal in the United States, in the neighborhood of 400×10^9 metric tons. A fit of Hubbert's curve to these estimates, as shown in Fig. 4, is, of course, rather arbitrary.

Future production of coal will depend entirely upon the extent to which the nation shifts back from petroleum to coal as a major source of energy. The uncertainties in such a shift are enormous; but if coal were to become the major source of energy for the United States by the year 2000, with the total energy consumption increasing at a rate of 2.5% per year, the production of coal would

have to increase by the year 2010 to the neighborhood of 3.5×10^9 tons of coal per year. This increase would represent a seven-fold increase over the present rate of production; under these circumstances, even the vast coal resources in the United States would last only until the mid-21st century (using Averitt's conservative figure). Well before that point would be reached, however, environmental and health effects would constitute a major issue.

Another important factor in the coal resource picture is the geographical distribution of coal in the United States. On a tonnage basis, about half of the country's coal is located east of the Mississippi, with the rest distributed to the west. The sulfur dioxide produced by burning high-sulfur coal is one of the major environmental problems with coal, necessitating specialized processing. One proposal is to process the coal into a synthetic fuel at the mine, but in the case of western coal, the large amounts of water required for the processing is difficult to obtain. *See* COAL; FUEL, SYNTHETIC.

United States uranium reserves

Concentration of U_3O_8 in ore, ppm	U_3O_8		Electricity producible, 10^3 MW-years	
	Cost, dollars per pound	Amount available,* 10^3 tons	Light-water reactor	Breeder reactor
1600	Up to 10‡	1,127	6,600	880,000
1000	Up to 15‡	1,630	9,500	1,270,000
200	Up to 30‡	2,400	14,000	1,860,000
60	Up to 50§	8,400	49,000	6,500,000
25	Up to 100§	17,400	102,000	13,500,000
3†	Several hundred	$10^6 - 10^7$		

SOURCE: From U.S. AEC Rep. WASH 1242, 1973; Rep. WASH 1243, 1973.
*The amount of U_3O_8 available comprises reasonably assured plus estimated additional reserves.
†Natural crustal abundance.
‡Includes copper leach residues and phosphates.
§Includes Chattanooga shale.

Uranium. The popular concept of nuclear power is that it is essentially an infinite resource. The actual picture is considerably more complicated, especially when the factor of time is considered. The estimated uranium resources of the United States are shown in the table, broken down according to the concentration of the particular ore. The concentration of uranium in its ore is important because of the cost of separating the metal from the lower-grade ores. In addition to the resources indicated in the table, the government has a stockpile of about 50,000 tons. *See* URANIUM MET-ALLURGY.

The problem with uranium arises because the additional costs involved in using lower grades of uranium ore become reflected in consequently higher costs of the electric power produced. If the nuclear power industry is limited to the higher grades of ore costing, say, less than $10 per pound, then by the year 2000 the supply will be exhausted if a major shift to nuclear power is made. This is true, however, only if the present form of nuclear power reactors, which operate on fuel enriched by the very rare ^{235}U isotope, continue to be used. ^{235}U exists at a level of about 1% in natural uranium. The breeder reactors being developed, however, are capable of using nearly all the uranium in generating nuclear power. This line of argument was used by the AEC in making its original proposals for the development of the breeder reactor, and for estimating an optimum "time slot" about the year 1990 for its deployment. An additional factor in this argument is that, with the breeder, the cost of refining the ore becomes negligible compared with other costs in providing electric power, and the lower grades of ore can be used with no cost penalty. *See* REACTOR, NUCLEAR; URANIUM.

The argument for introducing the breeder has become complicated. The cost of electric power has increased since the original estimates were made, and the lower-grade ores are now profitable; thus the time during which the breeder reactors and the light-water power reactors will be feasible has increased. D. J. Rose of MIT maintains that uranium costing as much as $100 per pound will be competitive with fossil fuels. But the United States has not been thoroughly prospected for the lower-grade ores that would produce uranium costing more than $10 per pound, and there are poor estimates as to the actual extent of the resource. Many geologists believe that the amount of lower-grade ore is much larger than the table indicates, thus in effect reducing the pressure for an early introduction of the breeder. Ultimately, of course, if the full potential of uranium is to be realized, some form of breeder reactor will be necessary.

Estimates can be made of the number of years the United States could rely on nuclear power alone at its present rate of energy use. Using present reactors with $100-per-pound uranium would mean about a 25-year supply. The breeder reactor, on the other hand, would provide nearly 3000 years of power. Thus, nuclear energy is an "infinite" resource only if the breeder reactor is brought into use.

Oil shale. One of the potential bright spots in discussions of the petroleum resources of the United States has always been the presence of huge deposits of oil shale in the so-called Green River Formation, spread over portions of Utah, Wyoming, and Colorado. Estimates by Project Independence, of the Federal Energy Administration, of the total resource indicate that there is in the neighborhood of 1.8×10^{12} bbl of oil. While this total resource is many times the total original petroleum deposits in the United States, oil shale has never been produced commercially because of a number of factors, the most important of which is cost.

The most discussed method for producing oil from oil shale is to mine the shale by traditional belowground methods and to separate the oil from the shale in aboveground processing plants. The waste (tailings) left over from the processing of the shale is a fine powder which in volume is one-third greater than the original shale itself. Large-scale mining of shale would then leave huge deposits of tailings aboveground, with no really adequate alternatives for avoiding significant adverse environmental impacts. Further, the processing of the shale and the compaction of the resulting tailings require enormous amounts of water in a region which has limited water resources. There have been suggestions for reducing the environmental impact by in-place separation of the oil from its shale underground, with consequently less impact on the aboveground environment. Major questions of technical and economic feasibility still remain, with the water requirements being significant in some methods.

Thus, with little possibility for near-term development, the Green River deposits still represent a possible future option. *See* OIL SHALE.

Solar energy. The energy incident upon the Earth from the sun provides a constantly renewable source of energy and, therefore, an ultimate source of energy against which all other forms must eventually be compared. In addition, this energy source is relatively benign from a health and environmental standpoint.

The flux of energy the Earth receives from the sun is large compared to that obtainable from any other energy source. In the United States roughly 150 W/m² are incident on the Earth on an average year-round basis. At 10% efficiency, the United States could supply all of its energy needs from the sunshine which falls on about 50,000 mi² (1 mi² = 2,590,000 m²) of land, which is about 3% of the total arable land of the country. Figure 5 shows how the average solar energy per day varies in the United States. The average incidence, of course, varies with cloud cover, with the angle the sun makes with the surface, and with other factors. *See* SOLAR ENERGY.

Other sources. Of the other possibilities which are often discussed as sources of energy, one is nuclear fusion, from which energy is obtained by the fusion of two light nuclei to form a heavier nucleus. Energy produced from a controlled fusion reaction is thought to have an advantage over ordinary nuclear power because it seems to generate smaller amounts of nuclear wastes. Also the hydrogen in the ocean might conceivably be used in nuclear fusion and, of course, the supply would then be inexhaustible.

Fusion has not yet been shown to be a feasible method of energy production; and in the best of

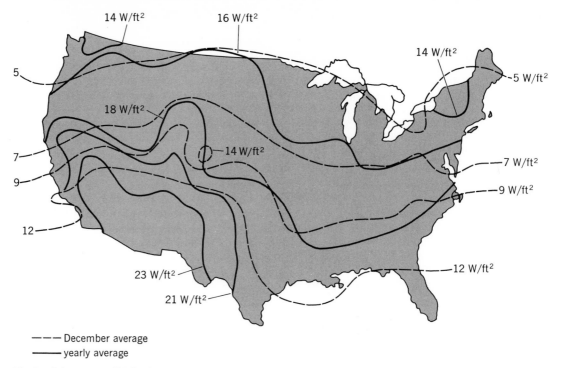

Fig. 5. Solar energy distribution over United States. (*From J. A. Merrigan, From Sunlight to Electricity, MIT Press, 1975*)

circumstances it is a very long-term alternative which could become important only well into the 21st century. *See* FUSION, NUCLEAR.

Energy from geothermal heat is another form which can be important locally. Since the interior of the Earth is very hot, there is, relatively speaking, an inexhaustible supply of energy in the heat of the Earth itself. Of course, the technical and economic problems of using this heat are enormous except in certain locations where steam and hot rock are close to the surface. Geothermal energy thus does not appear likely to be an important source of energy nationally. *See* GEOTHERMAL POWER.

Other schemes have also been discussed for producing additional energy, such as the use of tides and expansion of hydroelectric power generation, but their eventual impact on the total energy production is not expected to become significant. *See* HYDROELECTRIC GENERATOR; TIDAL POWER.

FACTORS AFFECTING ENERGY CHOICES

The basic availability of an energy resource is the most important underlying factor in making a national energy choice. Indeed, the impending depletion of oil and natural gas is the immediate cause for the present difficulties. Nevertheless, other factors can be equally limiting when practical alternatives are sought. Three often interrelated factors are also recognized: technological feasibility, economic feasibility, and environmental constraints. For instance, the time required to make basic changes in the energy system is a crucial factor. If technical feasibility is still unproved, then 30 years or more may be required to make a major impact on national energy-use patterns. Thus, midterm possibilities which can have an impact before about 1985–1990, when oil and gas

disappearance is the major concern, are distinguished from longer-term or ultimate choices which are available after that time.

Coal technology. In spite of the fact that coal production is still significant, simply reverting to an updated version of this energy commodity is not an easy task. The time required to increase coal production to the levels necessary is an important factor. It requires 5–9 years to bring a new underground coal mine into production. It requires about 1½ years to do the same for a strip mine, and orders for the capital equipment for this task faced a backlog of 5 years in mid-1975. In addition, manpower limitations, conversion of oil-burning equipment, and installation of "scrubbing" equipment to handle the high-sulfur eastern coals all cause delays.

Also, for many uses solid coal is not the most desirable form of fuel. Most plans for the shift from oil and gas to coal assume that a large portion of the coal will be converted to either a liquid or a gas form. The reasons for conversion are that the synthetic fuel can be made into a cleaner product, and that solid coal is not suitable for automotive use. Even for the generation of primary electricity, solid coal is more costly to transport great distances to the point of consumption than a gas or a liquid is. Given the existence of an external natural-gas pipeline in the United States, transporting the coal in the form of a gasified product would be far less costly, and more desirable from an environmental point of view, than burning the coal at the point of consumption.

During the 1930s the Germans developed processes which were used extensively at the time for the production of both liquid and gas as fuel from coal. The Lurgi process for producing high-Btu gas is currently a practical process, but it places rather severe requirements on the input coal, and there are other processes in the research stage which promise more efficient conversion from the coal to gas. Extensive use of the Lurgi process could be started now, and follow-up technology for the newer processes will probably be developed in the early 1980s. *See* COAL GASIFICATION.

In the matter of coal liquefaction, however, the situation is more problematic. Coal becomes a liquid by the addition of hydrogen to the coal under high pressure and temperature. In order to revive the German techniques, much new engineering will be required, and a feasible technology will probably not be available before the early 1980s. Thus, in both cases, large amounts of synthetic fuel will not be available until the late 1980s or 1990s, because of the dual constraints of technological feasibility and the capital costs for the building of new plants. *See* COAL LIQUEFACTION.

Other long-term possibilities exist for coal technology. For example, proposals have been made for converting coal to a slurry, transporting the slurry to the point of consumption, and burning the coal directly. One such direct burning process is the magnetohydrodynamic (MHD) electrical generator, in which finely powdered coal is burned at an extremely high temperature. The hot gases are then transported through a channel or duct and form a plasma. Under these conditions electricity is produced directly in the walls of the duct. Technological and economic feasibility of the MHD process, however, has not yet been demonstrated.

Nuclear reactor technology. Although reactors which burn ^{235}U are practical and are being established, the rate of construction is slower than anticipated. Original plans were to have on the order of 300 GW of nuclear electrical generating capacity installed by 1985 (about one-third of the total planned electrical capacity in 1985). However, a combination of circumstances has conspired to cut this estimate drastically, by 50% or more. The delay has been caused partly by the extremely long construction cycle (about 10 years), caused in turn by safety and environmental impact considerations. The problem is also magnified by inefficient licensing procedures, by difficulties in finding suitable sites for new power plants, and by lack of capital. *See* ELECTRICAL UTILITY INDUSTRY.

For the full potential of nuclear energy to be realized, the breeder reactor will have to be used. However, great controversy swirls not only about the breeder but about the whole nuclear option. In the first place, the United States breeder program has been strongly criticized because it has been too costly. Part of this cost arose because the development program was planned not only to assure an efficient breeder reactor design but also to test exhaustively various safety aspects of the design. In addition, technical difficulties have arisen as the program progressed. For example, the so-called breeding ratio, or the time required to convert the nonfissile ^{235}U into fissile plutonium, is still too long for an ultimate system.

Consequently, the American breeder is still some years from being technically ready; just how long depends on decisions still to be made concerning funding rates in the program. In the meantime, the French, English, and Russians have developed breeders which are now operating.

But these matters do not touch the main problems of the breeder program. It is widely criticized as being unsafe, for being vulnerable to theft of the fissile material, and for inadequate procedures for handling the high-level, long-term radioactive wastes. While the same criticisms have been leveled at the ^{235}U reactors, the main target is the breeder.

The safety question has received the greatest public attention. An exhaustive study, headed by N. C. Rasmussen of MIT, was carried out on the safety of ^{235}U reactors, with the general conclusion that the present reactors pose a smaller safety hazard to the public than nearly all other forms of human activity. This study has been criticized on various grounds, but the general conclusions appear to remain valid. The breeder reactor is another matter, however, because of the notorious pathogenicity of its plutonium fissile material. Whether the breeder reactor can be made adequately safe is a much more difficult matter than ensuring the safety of the ^{235}U reactors. This issue is, of course, at the heart of much of the design effort in the development of breeder reactors, and in addition forms the core of the debate about their widespread construction. In all these cases, incidentally, the safety problem does not arise from the potential of the reactor to become a nuclear bomb, but from the possible occurrence of incidents such as ordinary chemical explosions which might release portions of the radioactive core into the atmosphere.

The two other environmental questions surrounding the nuclear option have not received the attention given to safety. One concerns the ability to use the fissile material used in a breeder reactor for making bombs. This problem takes on many dimensions. For example, it is possible to develop a nuclear weapon from the plutonium materials produced in a commercial reactor. This was emphasized by the explosion of a nuclear device by India in 1975. Theft of fissile or highly radioactive reactor material is also a major hazard.

The second problem concerns the disposal of the long-term high-level radioactive wastes produced in any reactor. If nuclear reactors are used on a massive scale for hundreds of years, the human race will accumulate significant amounts of radioactive material which will remain radioactive for a time comparable to that which has elapsed since the appearance of the human race.

Solar technology. Heating by solar energy is achieved very simply by embedding pipes for carrying circulating water in panels set on the roof of a building (Fig. 6). As the water circulates through the solar panel, it absorbs the radiant energy and is heated to about 150–190°F (65–88°C). The warm water is then circulated through the building or stored in a large underground water tank for later use. In those parts of the country where there are many sunny days, and where traditional sources such as gas or oil are expensive or in short supply, solar heating is practical now. *See* SOLAR HEATING.

By far the most desirable space conditioning system using solar energy would, of course, provide cooling in the summer along with heating capability in the winter. Several technical possibilities exist for cooling by solar energy. One is based on

Fig. 6. National Bureau of Standards solar house. (*Institute for Applied Technology, NBS*)

the absorption-desorption system similar to that used in the old gas refrigerator. None of these systems is very efficient, however, and the cost is high because the temperature of the solar-heated water is relatively low. Thus, considerable refinement of present methods will be necessary before a practical system will be possible. After this, of course, extensive use of solar energy for heating and cooling of buildings will have to await the normal commercial build-up phase. *See* SOLAR COOLING.

The introduction of solar heating and cooling is analogous to the introduction of central air conditioning in buildings after its technical and economical feasibility was demonstrated in the 1930s. About 20 years was required to develop the industrial and marketing experience necessary for satisfactory mass use. A nontechnical issue illustrating the type of innovation which may be required follows from the fact that capital costs of installation of a solar heating and cooling system in a structure are expected to be much higher than for a conventional system. Thus, in order for the average person to be attracted to solar energy, the whole marketing structure may have to change. For example, the heating and cooling system, as installed in a house, may be owned and serviced by a utility rather than by the homeowner. The homeowners would then simply purchase their heating and cooling by the month.

Solar energy is viewed by many people, and especially by those critical of the nuclear option, as an answer to the energy crisis. However, prac-

tical solar energy for electricity and central station power generation remains in the distant future. The most serious proposals for generating electrical power from the sun involve (1) wind chargers, (2) solar furnaces set in high towers upon which a large number of ground-level mirrors focus the sun's rays, (3) photovoltaic systems for direct generation of electricity from light, (4) the use of thermal gradients between the surface and deep undersurface layers of water in the oceans, and (5) the direct burning of agricultural products and wastes. The technology closest to practical application at the moment is the wind generator. The idea is to construct large propellers which are turned by the prevailing winds in areas of the country in which steady winds of sufficient velocity occur. Megawatt-size and larger units are being planned and built for test and evaluation.

The major technical problem with solar radiation is that it is diffuse and of low intensity. Thus, solar energy systems are characterized by large installations and high capital costs. The object of presently accelerating work on all types of solar energy schemes is to develop designs which will generate power at a competitive cost level. For example, present photovoltaic systems are about a hundred times too costly to be practical. However, solar energy has been so little exploited technically that these costs are expected to yield to the intensive explorations now planned and under way. An especially important factor in systems such as solar furnaces and wind generators which receive widely varying energy inputs is the necessity to develop large energy storage facilities to accommodate the variations in input and thus match supply to demand. Adequate schemes for energy storage do not yet exist.

Because of these various technical limitations, production of solar electric power in significant quantities to make an impact on the national energy system is not a possibility before well into the 21st century, even if the research programs progress well.

CONCLUSIONS

And so, the question remains: what energy choices or combination of energy choices is best? In the overall complex picture, no one option is clearly and simply superior to all others. In the short term, especially, the problem is acute, but it must be solved. In sorting out the alternatives, certain conclusions stand out clearly, and the discussion ends with a simple listing of them.

1. As far as the United States is concerned, domestic oil and gas resources will begin to disappear as a major factor in the energy system sometime in the 1990s. Because of the long lead time involved, work on an alternative basis for the future energy system must be started now.

2. In the midterm, between approximately 1985 and 2010, the feasible alternatives are imported oil and gas, nuclear power, and coal. All entail serious problems. In the midterm, conventional ^{235}U reactors look promising, in terms of resources available, environmental impact, and technical feasibility. Both nuclear and coal options will be limited by capital. Imports up to some reasonable fraction (perhaps 25%) of the total energy base may also be admissible.

3. The long-term picture is relatively bright. Solar and nuclear breeders or nuclear fusion are the ultimate energy resources. In the long term, the question is whether technical progress can sufficiently lower solar energy costs, or allay the nuclear safety and environmental problems, and under what conditions will one or the other be preferable.

4. Given the complex of constraints predictable in short-, medium-, and long-term time frames, far greater attention will be given to energy conservation. This is a vast subject beyond the scope of this article, but its importance is easily perceived from the conclusions reached here. From the technical standpoint, a great potential exists for increasing the efficiency of energy use throughout the entire energy system, and the future will demand far more serious attention to this factor than has been given it in the past.   [ROBB M. THOMSON]

The United States is attempting to transform the set of energy policies it inherited from an earlier and very different period. This effort is being made under pressure, much of it self-generated, at a time when the legislative and executive branches of government are operating in an atmosphere of mutual mistrust and antagonism. More importantly, the attitudes and desires of the American people with respect to the available energy options are in many ways unknown. The government is struggling to create and impose a new energy policy without a clear mandate from the people concerning the scope, structure, and impact of the policy. In a nation which allows major changes to be managed by technocrats—or by others less expert, but more powerful—this would not be a severe problem. But in the United States, and particularly in Congress, it is a source of great consternation.

The United States is not alone in this effort. Nearly all other countries active in world commerce are in the process of working toward policies for accommodating major changes in the way their energy supply and consumption are structured. Politics is the mechanism which is driving this process of policy formation. Certainly politics dominates the energy situation in the United States, and there is still great uncertainty about the outcome. The situation is unprecedented, and few observers of the events of 1974–1975 are willing to commit themselves as to how it will all turn out.

The Ford Administration, the large energy companies, the environmental and consumer groups, and the various congressional factions all have perspectives on energy policy, and some have advanced programs of varying degrees of specificity. None has control over the key independent variables in the energy situation: the new supplies of primary energy which will be used at the margin, the energy consumption patterns in the United States, and the influence of energy use on the health and growth of the domestic economy.

SHORT-TERM VERSUS LONG-TERM OPTIONS

Discussions of energy supply and demand often involve both long- and short-term considerations. The imperatives of a democratically controlled policy process—as opposed to a technocratic one—force attention of key participants in the process toward the short run. The election cycle in the United States achieves a maximum every 2 years. Thus the hardest choices in energy politics in the United States inevitably focus on and emphasize options whose impacts occur within this period. Policies for the long term are somehow easier to deal with, seem far less binding, and do not attract the intense scrutiny given to the "gut issues" which influence elections.

The following will therefore not concentrate on what seems to many to be the most interesting features of energy policy. Discussions of energy research and development—the alternate energy technologies such as controlled thermonuclear fusion, solar electric power, giant windmills, electric automobiles, and oil produced from shale—make excellent reading, but are almost irrelevant to the central energy supply-and-demand questions with which Congress and the President must deal. It is

U.S. Policies and Politics

- **supply and conservation**
- **OAPEC embargo**
- **Emergency Petroleum Allocation Act**
- **93d and 94th Congresses**

Dr. Benjamin S. Cooper

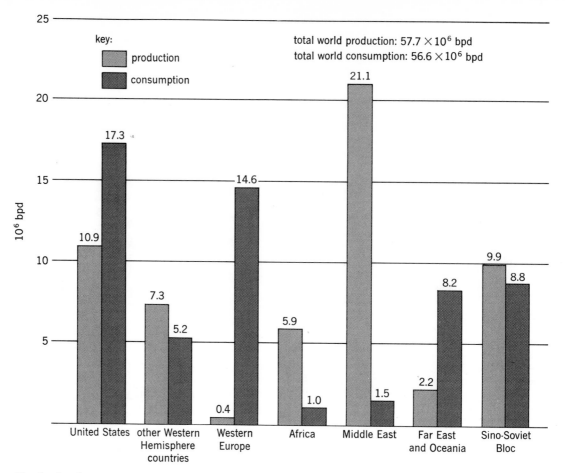

Fig. 1. Graph comparing petroleum production and consumption throughout the world for the year 1973.

(*Data from DeGolyer and MacNaughton, Inc., Twentieth Century Petroleum Statistics, 1974*)

essential that these technologies be the subject of research and investigation, but even the most intensive crash program could not milk from them the energy which is required within the time horizons of the current energy debate. Thus, long-term energy supply and conservation options receive serious attention only in the context of the short-term Federal budget support for the research designed to bring about these options. For better or worse, the basic Federal policy for energy research and development has been determined. It is, first, to have a large Federally funded program—involving billions of dollars on an annual basis; and, second, to move forward with research and development on all reasonable options at a rapid pace. No significant political faction opposes this policy.

At the heart of the real controversy in the present politics of energy are the conventional energy sources: oil, coal, natural gas, hydroelectric power, and light-water nuclear power reactors. It is on the issues surrounding these energy sources that the policy debate focuses. *See* ENERGY SOURCES.

CURRENT ENERGY SITUATION

Despite the uncertainty concerning the eventual shape of the policies emerging from the present period of change, some reasonably definite assertions can be made about the current energy situation which provide a useful starting point for this discussion.

1. The availability and price of all primary energy in the United States responds almost entirely to the politics of availability and price of oil—to the health and tractability of a diverse international cartel of oil-producing countries and an extraordinarily powerful collection of multinational oil companies.

2. There are no domestic policy options capable of enactment or implementation which will significantly free the United States over the next several years from the need to import large amounts of oil.

3. Despite this fact, the United States is now, and will remain for many years, a leading world producer of conventional energy: oil, natural gas, and coal. This asset provides the United States with the option to substantially limit the impact on its economy of high energy prices to a significant but still relatively small stream of imported energy. *See* COAL; NATURAL GAS; PETROLEUM.

4. In the energy debate, Congress and the President must reach an agreement on fundamental differences about the role of the Federal government in the economic life of the country. It is not at all clear that these differences will be resolved soon.

5. The expectations of a substantial portion of the public about the prospects of a technical fix to the energy problem far exceed the prospects that such a fix can be provided. The centerpiece of the United States energy research and development

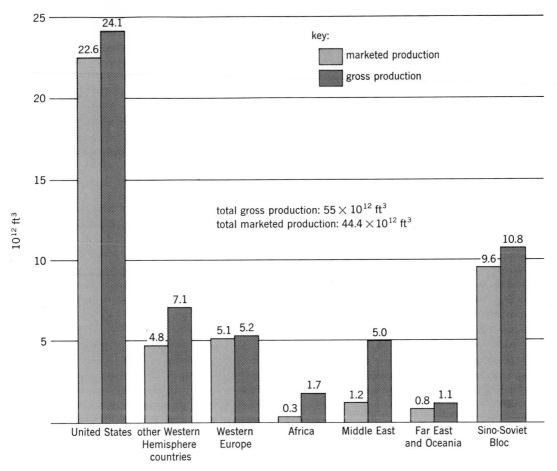

Fig. 2. World natural gas production, 1973. (*Data from U.S. Bureau of Mines*)

effort to date—electric power derived from nuclear fission—has attracted a dedicated, energetic, and now also formidable opposition.

6. The increases in energy prices experienced in 1974 fueled the inflation which reached as high as 14% in 1974 and exacerbated the worst recession since the 1930s. This economic deterioration has substantially threatened the disposable income of a middle class which has come to regard relatively abundant discretionary income as a right. For the poor and the elderly, the impact of these increases has been devastating. The prospect for the near-term future is for only moderate relief. On the other hand, the threat of new increases in energy prices comparable to those of 1973–1974 has receded. *See the feature article* WORLD ENERGY ECONOMY.

PREVIOUS ENERGY POLICIES

The overriding determinants of the role of energy in the United States economy have been the abundance and inexpensiveness of primary fuels and the fact that energy prices have, until recently, remained remarkably stable in current dollars—which is to say that the real value of energy has diminished with time. For 2 decades, the trend in the value of mineral fuels at the wellhead, mine mouth, and import dock as a percentage of Gross National Product (GNP) has been a steady decline, arrested only by the pivotal events of 1973–1974 (Table 1).

Table 1. Value of primary fuels as a percentage of Gross National Product*

Year	Percentage
1950	3.1
1960	2.7
1970	2.4
1971	2.5
1972	2.1
1973	2.5
1974	4.7

*From U.S. Bureau of Mines and U.S. Department of Commerce.

Table 2. Imports as a percentage of domestic consumption of fuels*

Year	Coal	Natural Gas	Petroleum	Total energy
1950	(5.9)†	(0.4)	9.1	1.3
1960	(10.2)	1.2	17.5	5.9
1970	(15.4)	3.5	23.1	8.4
1971	(11.4)	3.9	24.6	9.6
1972	(11.0)	4.2	28.2	11.7
1973	(10.9)	4.3	36.3	16.2
1974	(12.7)	4.0	37.5	16.3

*From U.S. Bureau of Mines.
†Parentheses indicate net exports.

The United States has had the option of maintaining cheap energy because it possesses an enormous productive capacity in oil, natural gas, and coal (Figs. 1–3). The nation is by far the world's leading producer of natural gas and is second only

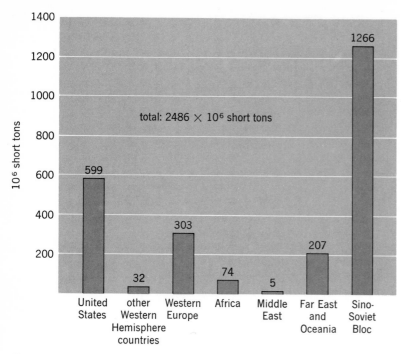

Fig. 3. World coal production, 1973. (*Data from U.S. Bureau of Mines*)

energy required to produce a fixed dollar of output.

As Fig. 5 shows, the variation in per capita energy consumption over time is also quite similar to the variation in per capita disposable personal income—again expressed in fixed dollars. (Personal income is the income remaining after capital consumption allowances, business tax liability, corporate profits, and transfer payments are deducted from GNP. Disposable personal income is obtained by subtracting taxes and social security contributions from personal income. Disposable personal income divides into personal consumption expenditures, interest payments, and savings.)

In no sense is the nature of the causal relationships, if any, behind these statistics clear. Nevertheless, the notion that growth in energy consumption is required to sustain economic growth and to maintain a high standard of living is a powerful one, and one which is widely accepted.

A rapid growth in energy consumption has in fact characterized the American economy. Gross energy consumption per capita in the United States increased from 40 bbl (1 bbl = 0.16 m³) of crude oil equivalent per person in 1950 to 44 bbl in 1960. During the next decade, per capita consumption accelerated. By 1973, domestic energy consumption amounted to the energy equivalent of nearly 64 bbl of crude oil for each person in the United States. The growth in per capita energy consumption—1.0% during the 1950s, 2.1% from 1960 to 1965, and 3.7% for the rest of that decade—coincided with a period of rapid economic growth. At the same time, the energy supply sector was experiencing low and stable energy prices, sharply declining exploratory activity within the United States, and the elimination of nearly all short-run excess producing capacity for the major fossil fuels—oil and natural gas.

After 1970, imported oil became the source of marginal supply for the added demand for energy. *See the feature article* ENERGY CONSUMPTION.

Oil-producing capacity. Excess producing capacity in the oil industry, which was 20% of total capacity in 1955 and 27% for most of the late 1950s and early 1960s, decreased rapidly after 1967 to under 5% in 1973. The Texas Railroad Commission, which had regulated Texas oil production for over 30 years, prorationing supply to match demand and maintain prices, increased allowable oil production from 28% of capacity in 1960 to 54% in 1969 and finally to 100% in 1973. In 1970, domestic production of crude oil peaked at 9.6 × 10⁶ barrels per day (bpd). Production declined at approximately 2% per year for the next 3 years.

The policies followed by the United States government and the oil industry during the 20-year period prior to the 1970s coincided with a transformation from a situation of continual potential oversupply domestically to one in which dependence on foreign sources solidified. The use of capital equipment and operating practices reflecting long-term experience with low energy prices became an integral feature of the American economy.

Prorationing and imports. The thrust of pre-1973 government policy is most clearly illustrated in the interaction between the practice of demand prorationing by oil-producing states such as Texas

to the Soviet Union in coal and oil production. As a consequence, serious United States dependence on foreign energy sources is a very recent phenomenon (Table 2).

Energy and economic growth. At the simplest level, the rationale for maintaining cheap and abundant supplies of energy has been the alleged strong coupling between energy consumption and economic growth. This rationale is often stated in terms of the existence of a steady trend over time in the ratio of energy consumption to GNP in fixed dollars. Figure 4 illustrates this trend for the period since 1950. Except for the late 1960s, when the efficiency with which energy was used to produce a dollar of GNP fell, the general trend has been for a steady and gradual decrease in the amount of

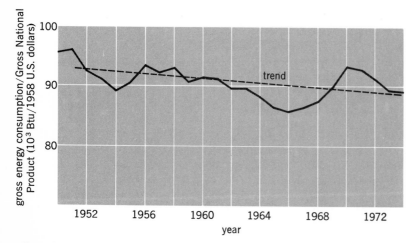

Fig. 4. Ratio of gross energy consumption to Gross National Product. (*Data from U.S. Bureau of Mines and U.S. Department of Commerce*)

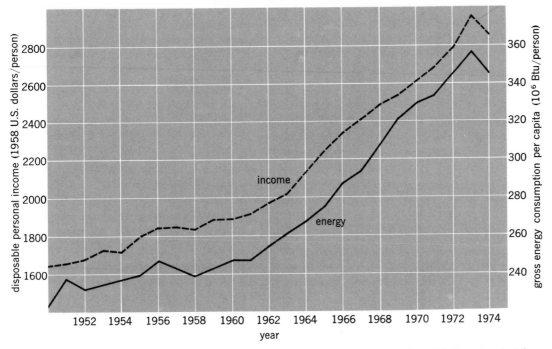

Fig. 5. Disposable personal income and gross energy consumption per capita in the United States for the years 1952–1974. (*Data from U.S. Department of Commerce and U.S. Bureau of Mines*)

and Louisiana and the oil import program administered by the U.S. Department of the Interior after 1959.

The problem historically in the domestic oil industry has been the tendency of excess supply to glut the market. Oversupply and depressed prices led to wasteful exploitation of resources and low profitability. In the aftermath of the Depression of the 1930s, producing-state governments, with encouragement from Washington, undertook to impose institutional restrictions on the extremely price-inelastic production of domestic crude oil. Because this demand prorationing kept prices stable, it did not discourage entry into the business, and domestic excess producing capacity remained high, as the number of producers sharing slowly rising production grew.

In the meantime, multinational oil companies, with government encouragement, were seeking handsome profits which could be made by exploiting the supergiant oil fields in the Mideast. American oil companies were drawn into the Mideast by incredibly low production costs, available consuming markets in Europe and later in Japan, favorable relationships with local governments, and generous United States tax incentives for foreign operations.

Eventually, domestic oil producers required import quotas to save the system of domestic demand prorationing from an overflow of world production into United States markets. These imports, produced at less than $0.10 per barrel in huge foreign fields, sold for under $2.00 per barrel at the import dock in the United States. Domestic producers were attempting to hold prices in the United States at levels approximately a dollar higher. In 1959 a mandatory import program was enacted restricting crude oil imports to 12.5% of

domestic production. Total imports, including imports of refined products, crept slowly upward from 22% of domestic production in 1959 to 30% in 1971. In the meantime, however, domestic demand growth was being filled by bringing on the unused domestic productive capacity which prorationing was designed to accommodate. This capacity was virtually exhausted in 12 years. The ability of the domestic producing industry to generate significant supplies of oil in the short run essentially vanished by 1971, while exploration and development of new domestic fields fell off dramatically. The attention of the industry was on the highly profitable overseas oil fields.

In April 1973, in recognition of the fact that domestic demand had caught up with and substantially surpassed domestic supply, the President removed the quota on imported oil. Within months, imports from the Eastern Hemisphere, chiefly from Arab countries, jumped by approximately a half million barrels per day.

THE ARAB EMBARGO

The embargo on petroleum shipments to the United States imposed on Oct. 17, 1973, by the Organization of Arab Petroleum Exporting Countries (OAPEC) catapulted energy policy to the center of the domestic political arena. The United States was denied direct imports from Arab nations, as well as crude oil sold to or exchanged with third parties by those oil companies transporting Arab crude oil. Saudi Arabia was principally responsible for all dramatic fluctuations in world crude oil production during 1973 and 1974, including the deep trough characterizing the embargo period (Fig. 6).

The multinational companies that were charged with administering the embargo generally attempt-

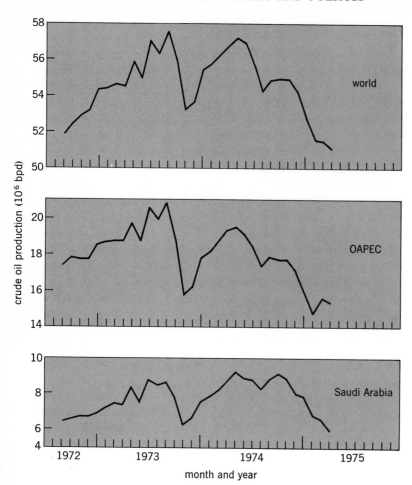

Fig. 6. Crude oil production for Saudi Arabia, OAPEC, and the world, 1973–1974. (*Data from Oil Gas J.*)

1974, United States imports dropped by 2.7×10^6 bpd, reducing total petroleum supply by 14% below expected consumption.

EMERGENCY PETROLEUM ALLOCATION ACT

The legal authority for management of the domestic shortages created by the loss in imports and the inability of domestic sources to increase production rested primarily in the Emergency Petroleum Allocation Act (Public Law 93-159), which was signed by the President on Nov. 27, 1973. The key goals of the act were to allocate shares of any shortages in a uniform manner while controlling prices, and to "preserve an economically sound and competitive petroleum industry." Since prices were to be controlled and not used as an allocation mechanism, a Federal allocation apparatus was required to perform the functions normally fulfilled by prices in the marketplace. The act further sought to prevent marketing practices which might involve controlling supplies to reduce competition, driving independent refiners or members of the independent marketing sector out of business.

Federal Energy Office. The Federal agency which received the task of implementing the petroleum policies embodied in the Allocation Act, as well as such other energy policy responsibilities as the President or Congress might assign it, was the Federal Energy Office (FEO). The policy which the FEO attempted to implement emphasized the maintenance of supply, to the greatest extent possible, to the industrial and commercial sectors of the economy. Refinery utilization was cut back to conserve crude oil, which was the imported commodity most affected by the embargo and worldwide production cut. Yields from refineries were altered to increase production of heating oil at the expense of motor gasoline.

Petroleum supply and demand for the first quarter of 1974 are described in Table 3. The bulk of the shortage was absorbed by consumers of motor gasoline. Although there were substantial reductions in the consumption of residual and distillate fuel oil below preembargo projections, these reductions resulted primarily from the totally unexpected warmth of the 1973–1974 winter.

ed, with considerable success, to allocate the shortage in world supply evenly, in percentage terms, among consuming nations. In the case of the United States, although overall supply was higher than might have been expected, the embargo was substantially effective. During January

Table 3. Petroleum supply and demand, first quarters of 1973 and 1974, in 10³ bpd

	Actual, 1974*	Preembargo projection for 1974†	Percentage of change	Actual, 1973*	Percentage of change, 1974–1973
Demand					
Motor gasoline	6,019	6,615	− 9.0	6,353	− 5.2
Aviation fuel	905	1,138	−20.5	1,063	−14.9
Distillates	3,592	4,629	−24.4	3,913	− 8.2
Residual fuel	2,848	3,629	−21.5	3,253	−12.5
Other	3,730	3,777	− 1.2	3,906	− 4.5
Total	17,094	19,788	−13.6	18,488	− 7.5
Supply					
Domestic	10,721	10,853	− 1.2	10,957	− 2.2
Imports, crude	2,368	3,713	−36.2	2,924	−19.0
Imports, refined	2,897	3,763	−23.0	3,325	−12.9
Stock withdrawal	614	939	−34.6	796	−22.9
Other	494	520	− 5.0	486	+ 1.6
Crude oil input to domestic refineries	11,323	12,831	−11.8	12,193	− 7.1

*From U.S. Bureau of Mines.
†From *Report of the Supply and Demand Committee of the Independent Petroleum Association of America, October 1973.*

Table 4. Year's highest posted price for selected crude oils, 1970–1974, current U.S. dollars*

Country	1970	1971	1972	1973	1974
Saudi Arabia	1.80	2.285	2.479	5.036	11.651
Iran	1.79	2.274	2.467	5.254	11.875
Kuwait	1.59	2.187	2.373	4.82	11.545
Libya	2.53	3.447	3.673	9.061	15.768
Nigeria	2.42	3.212	3.446	8.339	14.691
Indonesia	1.70	2.21	2.260	6.00	10.80
Venezuela	2.339	2.782	2.782	8.004	14.876
United States (east Texas, price-controlled)	3.60	3.60	3.60	5.20	5.20

*From *Oil Gas J.*, Nov. 11, 1974.

Oil prices. In late 1973 and early 1974, a widespread atmosphere of panic was created by the uncertainty about the severity and length of the embargo. As this fear that supply would be unavailable receded, a deep concern over the steep increases in oil prices emerged. In a series of coordinated actions, all members of the Organization of Petroleum Exporting Countries (OPEC) sharply increased both the price of oil purchased directly from the countries themselves and the revenue per barrel which the countries received from multinational companies operating within their borders. Table 4 illustrates these price trends for selected OPEC members, and Fig. 7 compares Mideast oil prices with world commodity prices.

The price control policies adopted in the United States exempted the price of approximately 40% of domestic oil production from regulation. With domestic producing capacity fully committed, the price of this oil rose to reflect the price of imported oil at dockside in the United States. The price of over half the approximately 12×10^6 bbl of crude oil per day refined in the United States rose from just under $4.00 per barrel to nearly $11.00 per barrel in the space of less than 4 months. These increases in prices of the basic resource commodity were quickly reflected in the prices of fuels and electricity purchased by consumers.

Coal, natural gas, and electricity. Price increases were not confined to oil. While the price of oil refined in the United States more than doubled, the average price of coal at the mine rose from $8.50 per short ton (1 short ton = 907 kg) in 1973 to over $15.00 per ton in 1974, an increase of 76%. The average price of natural gas at the wellhead increased by over 30% from $0.22 per thousand cubic feet (10^3 ft^3 = 28.3 m^3) in 1973 to $0.29 per thousand cubic feet in 1974. These trends continued during 1975.

Coal prices reflected the ability of producers to extract high economic rents from buyers whose only alternative was high-priced oil. The natural-gas market is more complicated and highly skewed. Well over half of the natural gas produced in the United States is sold in interstate commerce under prices administered by the Federal Power Commission, averaging around $0.25 per thousand cubic feet. On the other hand, prices in new contracts for natural gas in the interstate market are now allowed prices of $0.54 per thousand cubic feet, and in the unregulated intrastate market contracts were routinely settled in 1975 at prices well above $1.00 per thousand cubic feet.

Electric utilities fared particularly poorly in

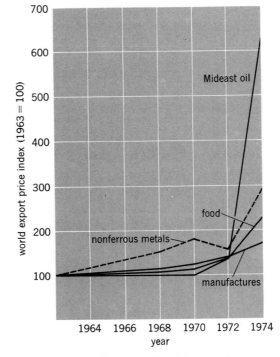

Fig. 7. Mideast oil prices and world commodity prices. (*Data from UN International Monetary Fund Conference Board Record*)

1974. Trends in the prices of the fossil fuels delivered to electric utilities in the United States are presented in Fig. 8. Total fossil fuel costs to United States electric utilities rose from $7,600,000,000 in 1973 to $14,100,000,000 in 1974. The increases in residential electricity bills as these costs were passed along sparked consumer resentment comparable to dismay over the prices of petroleum fuels. Growth in electricity consumption, which had been maintained at approximately 8% per year historically, was cut to zero in 1974.

Impact of the embargo. By the time the embargo was lifted in March 1974, 300,000 to 500,000 persons had been added to the unemployment roles, a $10- to $20,000,000,000 real dollar loss in GNP had been absorbed, and the United States was paying over $2,000,000,000 per month for imported energy—as opposed to approximately $500,000,000 per month in 1973. More importantly, the pervasive impact of high energy prices provided the major push which drove the inflation rate in the country to 14% and started the economy on a slide into the worst recession since the 1930s—adding nearly 4,000,000 additional unemployed within a year.

The embargo, the deepening economic recession, higher prices, a mild winter, and consumer conservation efforts all combined in 1974 to cause the first decline in gross energy consumption in the United States in over 20 years. In 1974, gross consumption of energy resources declined by 2.2%, bringing to an end a period of increases which had averaged 4.1% annually between 1960 and 1973. Table 5 presents figures compiled by the Bureau of Mines describing gross consumption of energy resources by major sources and consuming sectors.

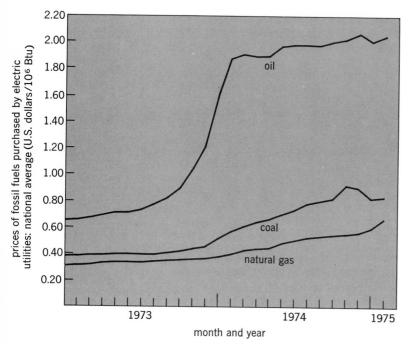

Fig. 8. Prices of fossil fuels purchased by electric utilities, 1973–1974. (*Data from Federal Power Commission*)

ENERGY AND THE 93D CONGRESS

For 3 years prior to the 1973 embargo, the executive branch, controlled by a Republican president, skirmished intermittently with a predominantly Democratic Congress over energy policy. The embargo eliminated any opportunity for a leisurely continuation of this skirmish.

Administration policy. In late 1973 and throughout 1974, the President's political resources in Congress were far from substantial, and one index of the weak leadership of the executive branch was the lack of real attention devoted to the energy situation. The initiatives proposed by the White House during this period scarcely went beyond a reiteration of fundamental philosophy: that energy problems will be most rapidly and efficiently solved by an energy industry which remains maximally free of government intervention. This philosophy implies a strategy for dealing with energy shortages which relies on prices to control and reduce demand, and profit accumulation to stimulate supply. The Allocation Act, which explicitly rejects this philosophy, was accepted only reluctantly by the Administration. In fall 1973, political considerations prevented serious advocacy of a public policy which would have permitted the oil companies to allocate supplies and determine prices during the embargo.

Table 5. United States gross consumption of energy resources by major sources and consuming sectors, in 10^{12} Btu[a]

Consuming sectors	Anthracite	Bituminous coal and lignite	Natural gas, dry[b]	Petroleum[c]	Hydropower[d]	Nuclear power[d]	Total gross energy inputs[e]	Utility electricity distributed[f]	Total net energy inputs[g]	Percentage change from 1973
Household and commercial										
1973	74	222	7,318	6,689	n.a.	n.a.	14,303	3,709	18,012	
1974 (preliminary)	62	229	7,116	6,390	n.a.	n.a.	13,797	3,687	17,484	−2.9
Industrial										
1973	33	4,344	10,970	6,059	34	n.a.	21,440	2,634	24,074	
1974 (preliminary)	32	4,176	11,129	5,826	34	n.a.	21,197	2,665	23,862	−0.9
Transportation[h]										
1973	—	3	743	18,164	n.a.	n.a.	18,910	15	18,925	
1974 (preliminary)	—	2	664	17,608	n.a.	n.a.	18,274	16	18,290	−3.4
Electricity generation, utilities[d]										
1973	37	8,581	3,681	3,656	2,964	888	19,807	6,358	n.a.	
1974 (preliminary)	38	8,630	3,328	3,448	3,018	1,173	19,635	6,368	n.a.	−0.9
Miscellaneous and unaccounted for										
1973	—	—	—	283	—	—	283	n.a.	283	
1974 (preliminary)	—	—	—	218	—	—	218	n.a.	218	
Total energy inputs										
1973	144	13,150	22,712	34,851	2,998	888	74,743	n.a.	61,294	−2.2
1974 (preliminary)	132	13,037	22,237	33,490	3,052	1,173	73,121	n.a.	59,854	

SOURCE: Division of Interfuels and Special Studies, Office of Assistant Director—Fuels, Bureau of Mines, U.S. Department of the Interior.

[a]Gross energy is the total of inputs into the economy of the primary fuels (petroleum, natural gas, and coal, including imports) of their derivatives plus the generation of hydro and nuclear power converted to equivalent energy inputs (see footnote d). 1 Btu (IST) = 1055.04 J.

[b]Excludes natural-gas liquids.

[c]Petroleum products including still gas, liquified refinery gas, and natural-gas liquids.

[d]Outputs of hydro power (adjusted for net imports or net exports) are converted to theoretical energy inputs calculated from national average heat rates for fossil-fueled steam electric plants provided by the Federal Power Commission using 10,389 Btu per net kilowatt-hour. Energy inputs for nuclear power are converted at an average heat rate of 10,660 Btu per kilowatt-hour based on information from the (former) Atomic Energy Commission. Excludes inputs for power generated by nonutility plants,

which are included within the other consuming sectors.

[e]Gross energy resource inputs with electricity generation shown as separate consuming sector.

[f]Utility electricity generated and imported, distributed to the other consuming sectors as energy resource inputs. Distribution to sectors is based on sales reported in the Edison Electric Institute's *Statistical Yearbook of the Electric Utility Industry for 1973*. Conversion of electricity to energy equivalent by sector was made at the value of contained energy corresponding to 100% thermal efficiency using a theoretical rate of 3412 Btu/kWhr.

[g]Energy inputs into the final consuming sectors—household and commercial, industrial and transportation—consisting of direct fuels and electricity distributed from the electricity generation sector. Conversion losses in the electric sector constitute the difference between net and gross energy.

[h]Includes bunkers and military transportation.

Table 6. Major energy legislation of the 93d Congress*

Public Law	Subject
P. L. 93-153	Rights of way through Federal lands (Trans-Alaska Pipeline), Nov. 6, 1973
P. L. 93-159	Emergency petroleum allocation, Nov. 27, 1973
P. L. 93-182	Emergency Daylight Saving Time energy conservation, Dec. 15, 1973
P. L. 93-239	Emergency highway energy conservation (national speed limit of 55 mph), Jan. 2, 1974
P. L. 93-245	Supplemental appropriations (provided funds to explore and develop Naval Petroleum Reserves), Jan. 3, 1974
P. L. 93-275	Federal Energy Administration, May 7, 1974
P. L. 93-319	Energy supply and environmental coordination (coal conversion, energy data and information, and automobile emission standards), June 22, 1974
P. L. 93-322	Appropriations for energy research and development, June 30, 1974
P. L. 93-409	Solar heating and cooling demonstration, Sept. 3, 1974
P. L. 93-410	Geothermal energy research, development, and demonstration, Sept. 3, 1974
P. L. 93-438	Energy reorganization (created the Energy Research and Development Administration and the Nuclear Regulatory Commission), Oct. 11, 1974
P. L. 93-473	Solar energy research, development, and demonstration, Oct. 26, 1974
P. L. 93-503	Urban mass transportation, Nov. 26, 1974
P. L. 93-511	Extension of the Emergency Petroleum Allocation Act until Aug. 31, 1975; Dec. 5, 1974
P. L. 93-577	Nonnuclear energy research and development policy, Dec. 31, 1974
P. L. 93-627	Deepwater ports policy, Jan. 3, 1975
Vetoed	
S. 2589	National Energy Emergency Act, Mar. 6, 1974
H. R. 15323	Extension of Price-Anderson Act, Oct. 12, 1974
H. R. 8193	Energy transportation security, Dec. 30, 1974
S. 425	Surface mining control and reclamation (pocket veto)

*From Congressional Research Service.

Congressional initiative. By contrast, in late 1973, key members of the 93d Congress had developed reasonably definite ideas about the management of energy shortages and about the need to create new institutions to consolidate the energy policy efforts of the executive branch. Legislation authorizing the construction of the Trans-Alaska Pipeline and the Allocation Act were both enacted in November 1973. An active debate on the issues surrounding the proposed creation of a Department of Energy and Natural Resources was postponed, but the knowledge and perspective developed during that debate smoothed the way for the statutory creation of two future components of such a department: the Federal Energy Administration and the Energy Research and Development Administration.

Major energy legislation of the 93d Congress is listed in Table 6.

The principal congressional energy initiative of the 1974 period was the Energy Emergency Act. This bill passed the Senate in mid-November 1973, and a companion bill was approved by the House less than a month later. The bill was a grab bag of measures designed to cushion the impact of the embargo on the domestic economy. Among other things, it authorized, subject to congressional review with right of disapproval, emergency mandatory energy conservation programs—for example, transportation controls, restrictions on the use of public buildings and on nonessential lighting, and gasoline rationing. Emergency measures were authorized to increase domestic energy supplies: stepped-up production in existing oil fields, adjustment of refinery yield patterns, export controls, allocation of materials needed for the production of fuels and energy. The bill attempted to clarify procedures of administration and judicial review during the emergency, provided special unemploy-

ment compensation to those put out of work by the shortages, and authorized grants to state government to assist in local conservation efforts.

The bill attempted to limit the "windfall" profits of domestic oil producers derived from the increases in domestic oil prices stimulated by OPEC pricing actions. In the form finally sent to the President, this provision was transformed into a ceiling on all domestic oil prices set substantially below the world market level. This, more than any other provision of a bill that the Administration found generally objectionable, prompted a veto. The veto was sustained in the Senate on Mar. 6, 1974.

Stalemate. The alliance which joined Republicans supporting the Administration to representatives of oil-producing and natural-gas-producing states stalemated the other factions of Congress during 1974. The significant energy legislation which was passed in 1974 placed the Federal Energy Administration on a firm statutory basis, abolished the Atomic Energy Commission and created the Nuclear Regulatory Commission (NRC) and the Energy Research and Development Administration (ERDA) in its place, and authorized a program for oil-to-coal conversion in major stationary oil-consuming installations (principally electric utilities). This legislation involved, if not active Administration support, at least Administration acquiescence. The principal piece of Administration-opposed energy legislation after the Energy Emergency Act—the Surface Mining Control and Reclamation Act—was pocket-vetoed at the close of the 93d Congress.

As the 93d Congress ended, energy-related legislation dealing with stand-by conservation authorities for a future embargo, land-use policy, outer continental shelf leasing, coal leasing, development of the Naval Petroleum Reserves, and mandatory energy conservation had each completed various stages of congressional action, but were bogged down in disagreements with the Administration and among key members of Congress.

The executive and legislative branches of the Federal government are very different in the resources which they can bring to bear on energy policy formulation, in their freedom to range over policy areas, and in the pressures under which they operate. "Normal" functioning of the legislative process for the development of detailed programs leans heavily on the analytical resources of the executive branch, represented by the enormous staff and base of experience and data residing in the executive departments and agencies. The congressional role is to modify and shape executive branch proposals to respond to popular and regional requirements. But the process can become nonfunctional when cooperation between the executive and legislative branches breaks down completely. Because of the cross-cancellation of powers, neither branch can overpower the other except in quite singular circumstances. Despite the great political pressures generated around energy policy issues in 1974, no concentration of power emerged that was sufficient to override the stalemate over energy policy.

ENERGY POLITICS IN THE 94th CONGRESS

The long-standing absence of an energy program developed and forcefully advanced by the executive branch was remedied by the President's announcement in the State of the Union message on Jan. 15, 1975, of the Energy Independence Act. This legislation, consisting of 13 separate titles (Table 7), embodied the approach to energy problems which the Administration had been advocating for well over a year: deregulation of the energy industry, postponement of environmental standards, accelerating incentives for energy production, and sharp energy price increases to curtail energy demand. In parallel with the legislation

Table 7. The Energy Independence Act of 1975

Title	Description
I	Naval Petroleum Reserves: exploration, development, and production
II	National strategic petroleum reserves
III	Deregulation of new natural gas
IV	Amendments to the Energy Supply and Environmental Coordination Act of 1974 (enhanced coal conversion authority)
V and VI	Amendments to the Clean Air Act of 1970 (relaxation or postponement of standards)
VII	Utilities (improve financial position of electric utilities)
VIII	Energy facilities planning and development (provide for enhanced Federal role in siting energy facilities)
IX	Energy development security (authorize price floor for domestically produced oil)
X	Building energy conservation standards (prescriptive and performance standards for new residential and commercial buildings designed to conserve energy)
XI	Winterization Assistance Act (grants to state governments to winterize homes of poor and elderly)
XII	National appliance and motor vehicle energy labeling
XIII	Stand-by authorities (for use during a future embargo)

submitted to Congress, the President announced a schedule of administrative actions, which could be taken without additional congressional authority, to substantially implement the energy pricing provisions of the Administration program. The President proposed, under existing law, to add a progressively increasing tariff to imported oil, with a partial rebate to reduce the impact on importers of refined petroleum products, and the removal of price controls from domestic oil.

Congressional reaction to this program was vehement. Representatives of states along the eastern seaboard, more dependent on oil because of the ready availability of imports in previous years, saw the program as discriminatory. Producing states opposed the excise tax on natural gas and the windfall profits tax on oil revenues which was proposed in the original Administration package. Environmentalists were incensed at proposals to defer implementation of clean air standards and at what was seen as an overemphasis on energy production as opposed to conservation. Very few elected representatives were willing to support steep price increases, especially for a commodity such as energy which has great potential to influence the cost of a very broad range of goods and services.

Economic impact. The economic impact of the President's January 1975 program would have been extremely painful. It was intended to be, in order to force self-denial over the short run. Few doubt that, imposed over several years, increases in energy prices of the magnitude contemplated in the President's program would be swamped by normal economic growth and would, to an extent which no one can predict with precision, induce slower growth in energy consumption.

However, the short-run elasticity—the ratio of percentage change in consumption to percentage change in price—for petroleum is estimated by the Federal Energy Administration at approximately -0.15. Thus a 50% price increase would be necessary to achieve a reduction of only 7.5% in consumption.

Because the equipment which consumes energy cannot be radically altered in the short run, truly substantial price increases may be required to achieve short-run savings. These increases must be imposed without a really clear picture of the size or secondary impact of the effects of sharply reducing consumption and transferring large amounts of money away from energy consumers. Because of the sheer volume of petroleum consumption, the amount of revenue generated by even a small petroleum price increase is quite substantial. For example, the United States consumes approximately 100×10^9 gal (1 gal = 3.785 liters) of motor gasoline annually. Thus an increase in gasoline prices of a penny a gallon, which would have a negligible effect on consumption, raises approximately $1,000,000,000 in annual revenues. Enactment of the President's program would have resulted in an enormous shift in funds from consumers to the energy industry and to government, increasing inflation while retarding economic recovery. Even if the President's target of daily savings equivalent to the energy contained in a million barrels of petroleum was realized within a year, and this entire savings is taken in oil imports

—which was by no means guaranteed under the program—the United States would still be importing a third of its oil and would remain vulnerable to Arab political blackmail.

It was never very likely that the President's State-of-the-Union program of January 15 would be enacted. Too many members of Congress felt—and their feelings were almost unanimously reinforced by extensive testimony at congressional hearings—that the proposed energy price increases would not accomplish conservation benefits which would be even remotely commensurate with their cost to the economy. Since much of the Administration's program depended on its energy tariff, tax, and price decontrol provisions to provide the driving force, the outlines of a compromise with Congress were very hard to discern.

Following the announcement of the President's initial proposal, the Administration gradually retreated from advocacy of sudden price increases. In midsummer of 1975 a gradual phase-in of higher energy prices was proposed to be accompanied by an escalating price ceiling for all domestic oil which had previously been free from price controls.

Congressional alternatives. Because of the very nature of the institution, it is unlikely that the diverse interests in Congress could compete with the executive branch in the creation from scratch of a comprehensive energy program. The congressional Democratic leadership did, however, attempt this, producing a Congressional Program of Economic Recovery and Energy Sufficiency in February 1975. The program advocated tax rebate and tax cut measures similar to, but larger in magnitude than, those proposed by the President in the 1975 State of the Union message, as well as accelerated Federal spending to stimulate economic recovery. Retention of energy price controls, mandatory fuel efficiency standards for new automobiles, a modest $0.05 per gallon tax on gasoline, and added funding for public transportation were features of the energy proposals of this congressional program, along with incentives for energy efficiency improvements in residences and commercial buildings, conservation investment incentives and energy savings targets for industry, and accelerated energy research and development. To stimulate energy supply, a National Energy Production Board, modeled on the War Production Board of World War II, was proposed. Accelerated conversion to coal, development of the Naval Petroleum Reserves, and leasing of the outer continental shelf were recommended, and a National Energy Trust Fund, supported by the proposed gasoline tax and by windfall profits taxes on oil revenues, was supposed to provide capital for the Federal government's activities in pursuit of energy self-sufficiency.

These proposals were forwarded to the congressional committees which exercise jurisdiction over the subject matter areas in question, and nearly all the concepts embodied in this program were cast in the form of legislation. Table 8 is a partial listing of major energy legislation before the 94th Congress.

Legislative action. Comparison of Table 7 and Table 8 for overlap provides an estimate of the energy proposals which stand reasonable chances of enactment. During 1975, the chances were considered good that laws would be written setting mandatory automobile efficiency standards, providing capital assistance for construction of plants for producing synthetic fuels, encouraging electric utility rate reform, inaugurating of a program of strategic reserves, and providing Federal incentives for improved residential and commercial energy efficiency. However, the chances were also rated good that surface mining legislation would be enacted in the 94th Congress. Despite modification of the legislation pocket-vetoed in the 93d Congress in response to Administration objections, the President again vetoed the surface mining bill passed by the 94th Congress. Responding to Administration claims that regulations on surface mining would substantially cut domestic coal production and trigger increases in residential electric utility bills, the House of Representatives sustained the veto by a narrow, four-vote margin.

A second veto battle in late summer 1975 settled the power balance for the fundamental issue dividing the Congress and the President. The President's veto of legislation extending the petroleum price control and allocation authority of the Allocation Act was sustained in the Senate by a margin of six votes. This watershed vote signaled the inability of the President's opposition in Congress to marshal two-thirds of the members to prevent imposition of a petroleum pricing policy which substantially incorporated the Administration's program of higher prices.

The Administration's victory on oil pricing coincided with the growth of substantial doubts as to the wisdom of the overall policy that the Administration had advocated since the embargo. By early fall, key Administration policymakers no longer regarded a policy of total deregulation of the oil industry and the concomitant shock of higher energy prices as politically acceptable. A compromise with the Democratic Congress was sought which would remove the contentious, confusing, and potentially dangerous energy issue from the center of the political arena.

The vehicle for the compromise was an omnibus energy bill, H. R. 7014, which had been moving ponderously through the House Committee on Interstate and Foreign Commerce. On October 9, with the acquiescence of congressional Republicans, a House-Senate conference opened to reconcile the provisions of this omnibus bill with no less than four separate Senate energy bills. Oil price controls were extended temporarily to permit the members of the Conference Committee to work out the compromise which all sides now desired.

The Conference Committee met in open session for over a month, and committee staff required another month of closed meetings with representatives of the Administration to reduce the policy decisions of the conference to legislative language. On Dec. 22, 1975, after a period of intense lobbying by the oil industry to obtain a veto, the President signed the Energy Policy and Conservation Act, which became Public Law 94-163.

The act extends regulation over the petroleum industry for a 3-year period and limits domestic oil prices to levels which are more in line with actual production costs. As part of the compromise, the President agreed to remove the highly controversial and possibly illegal tariff which had been

Table 8.   Major energy legislation in the 94th Congress*

Bill	Subject
H. R. 25	Surface mine control and reclamation; vetoed May 20, 1975
H. R. 49 (also H. R. 5919; S. 677 and S. 2173)	Authorize production in Naval Petroleum Reserves and establish strategic petroleum reserves; passed House July 8, 1974; passed Senate July 29, 1975; establishment of strategic petroleum reserves; incorporated in Public Law 94–163 Dec. 22, 1975.
H. R. 1767	Suspend President's authority to impose oil import fees; vetoed Mar. 4, 1975.
H. R. 2166	Tax reduction (increases investment tax credit for electric utilities, repeals oil depletion allowance and foreign tax credit); Public Law 94–12, Mar. 29, 1975
H. R. 6860	Energy Conservation and Conversion Act (establishes oil import quotas, a trust fund for new energy sources, and tax incentives for energy conservation); passed House June 19, 1975.
H. R. 7014	Energy Conservation and Oil Policy Act (sets oil price ceilings, authorizes gasoline rationing, restricts domestic gasoline consumption, fixes fuel economy standards for new automobiles, establishes strategic petroleum reserves, provides for Federal oil import purchasing agency, and authorizes mandatory energy conservation programs); passed House Sept. 23, 1975; incorporated in Public Law 94–163 Dec. 22, 1975
S. 349	Energy information and disclosure (energy efficiency labeling); passed Senate July 11, 1975; incorporated in Public Law 94–163 Dec 22, 1975
S. 391	Coal leasing; passed Senate July 31, 1975; passed House Jan. 21, 1976
S. 521	Outer continental shelf leasing; passed Senate July 30, 1975
S. 598 (and H. R. 3734)	Energy research and development authorization; passed House June 20, 1975; incorporated in Public Law 94–187, Dec. 31, 1975
S. 621 (and H. R. 4035)	Provides for congressional review of oil price decontrol and oil price floor decisions; passed Senate May 1, 1975; passed House June 5, 1975; vetoed July 21, 1975
S. 622	Stand-by energy authorities and mandatory energy conservation programs; passed Senate April 10, 1975; incorporated into Public Law 94–163 Dec. 22, 1975
S. 692	Natural-gas pricing reform and reregulation; set aside by the Senate Oct. 23, 1975
S. 740	National Energy Production Board
S. 1392	Energy conservation in buildings: research and demonstration
S. 1849	Extension of Emergency Petroleum Allocation Act of 1973 until Mar. 1, 1976; passed Senate July 15, 1975; passed House July 31, 1975; vetoed Sept. 9, 1975
S. 1864	Energy Information Act (data base)
S. 1883	Mandatory fuel economy standards for new automobiles; passed Senate July 15, 1975; incorporated in Public Law 94–163 Dec. 22, 1975
S. 1908	Industrial energy conservation; incorporated in Public Law 94–163 Dec. 22, 1975
S. 2310	Emergency Natural Gas Purchase Authority; passed Senate Oct. 22, 1975, with an amendment providing for long-term deregulation of the price of newly produced natural gas

*Partial list compiled January 1976.

placed by administrative action on all imported crude oil. This tariff had substantially raised the price of domestically consumed petroleum both because it made imported crude oil more expensive and because it artificially increased the price of the 40% of domestic production which was free from price controls. The pricing provisions of the Energy Policy and Conservation Act permit upward adjustments in oil prices only to account for inflation and only where higher prices are likely to enhance domestic supply.

The compromise permits the enactment of a number of useful and broadly accepted measures which had been stalled by the battle over oil industry regulation. Among other things, the act: (1) encourages the production of coal from new domestic mines; (2) extends authority to convert large stationary energy-consuming facilities from natural gas and oil to coal; (3) establishes mandatory mileage standards for new automobiles; (4) authorizes and directs the creation of a system of ready storage of petroleum to insulate the domestic economy from a future oil embargo; (5) authorizes the development of specific standby energy conservation measures for future supply emergencies to prevent the replication of the administrative confusion of the 1973–1974 embargo; and (6) provides for Federal assistance for the energy con-

servation efforts of state governments.

This legislation will not free the United States from the Arabs or substantially reduce the use of nonrenewable energy resources. It is designed to achieve the achievable in energy policy subject to the constraint that the economic impact of the policy be consistent with an early recovery from the depression of 1974–1975. Thus economic recovery is granted a higher priority than any of the policies designed to force substantial short-run changes in United States energy supply and consumption patterns.

CONCLUSIONS

Imported oil will provide the principal source of incremental energy supplies for the United States for the next several years, possibly for the next decade. Since 1974, oil prices have pulled up the price of nearly all energy produced domestically. This situation is in superficial accord with economic theory. When energy is priced at the level of marginal supply, incremental consumption must pay the full costs of producing the incremental supplies consumed. If prices are held artificially below the cost of marginal supply, consumption is artificially encouraged and shortages result. False signals, based on insufficiently high or overly stable prices, are sent to consumers, who buy more than they would at higher prices, while incorrect signals are sent to producers concerning the urgency of the need to develop added producing capacity.

In the case of oil, the reality is more complex than the theory. The world price of oil is set by a cartel of producing countries at a level which is at least an order of magnitude above the cost of producing incremental oil supplies in those countries. Domestically, the cost of incremental conventional oil and natural-gas supplies is also certainly far less than the cartel price, although domestic resources are probably significantly more expensive to develop now than they have ever been in the past, and they are certainly much harder to find. Because conventional noncartel resources have been exploited more intensively in the past, they are now also smaller in size, and therefore less likely to permit production levels which match historical rates of demand growth.

Alternatives to oil. The United States has nowhere else to turn for supplies in the short run. Abundant but high-priced alternative sources of hydrocarbons exist in undeveloped deep-mine coal reserves, in oil shale and, to lesser extent, in strippable coal reserves, but these resources cannot be made available as acceptable gaseous or liquid fuels in any significant amounts for at least 10 years. The lead times for any significant expansion of nuclear generating capacity, even if such an expansion could be agreed upon, are also of this magnitude. Moreover, these energy sources are very expensive, even compared to imported oil. Over the next several years, incremental energy supplies will be conventional oil, and if good fortune prevails, natural gas. The prices which are permitted for these energy sources will determine the prices of nearly all other forms of energy consumed. *See the feature article* EXPLORING ENERGY CHOICES.

What must be done. What is required is perhaps clearer than how it is to be accomplished:

1. More oil and natural gas must be produced, both domestically and in other countries which do not belong to the cartel.

2. The efficiency with which energy is consumed in the United States must be substantially increased. This effort will require a combination of approaches, including incentives, new and diligently enforced standards, and mandatory programs developed and agreed upon in the political arena. A beginning has been made with the automobile fuel economy standards contained in the Energy Policy and Conservation Act.

3. Insulation, in the form of standby energy reserves and standby programs to deal with future shortages and supply curtailments, must be developed and set in place. Authority to carry out these programs is in place.

4. Energy pricing must be rationalized in a way that, to the maximum extent possible, provides the proper signals to domestic producers, fosters economic recovery, and retains the benefits to the domestic economy of abundant domestic energy resources whose costs are well below arbitrary cartel-set world prices. The oil price debate appears to have been settled—temporarily. It is almost certain to return after the 1976 presidential election.

5. The acceptability of coal and nuclear energy must be upgraded, not by overpowering the legitimate objections to widespread use of these energy sources, but by intensive efforts to deal with the causes of these objections.

6. Alternative energy sources must be sought with ingenuity and urgency.

For too long, the debate over energy policy has been stalled by the unspoken assumption that a tough and comprehensive program must be rushed into place which accomplishes all of these goals at once and which, in some sense, involves sacrifice. Thus, rejection of the Administration's "energy program" has been held to require that an alternative just as radical and painful be imposed. This assumption is all the more unfortunate in that its superficiality lends itself to electoral politics, particularly presidential politics. Throughout 1975 counter-claims about "energy programs" impeded rather than fostered orderly policy development. The first priority of policy is restoration and maintenance of economic well-being. This will require energy, and it will require more energy than was being consumed in 1975. The short-run goal of economic recovery ought not to be sacrificed in an attempt to achieve aims which are really attainable over only much longer time periods.

The failure of energy politics to yield rapid progress toward solutions to the problem of energy dependence facing the United States has its roots in the nature of the political system. This failure reflects, to a significant extent, weaknesses which can arise in that system when it is faced with demands for rapid change. At the same time, this "failure" may not have served the country poorly. It is not clear that rapidity and progress are concepts which couple naturally in solutions to problems of energy availability. Few acceptable solutions with real shortrun impact are available, while numerous proposals which promise significant improvement over the intermediate and longer terms have been enacted and are now under consideration. This suggests the reasonableness of a less feverish approach to energy policy than has been evident in the past 2 years.

Much of the urgency of the energy debate has been based on one of two different concepts: world resource exhaustion and national self-sufficiency. The view which follows from the former emphasizes the need for a radical turning away from fossil fuels as a source of primary energy and the establishment of a priority effort to impose lifestyles based upon lower energy consumption and reliance on "renewable" energy sources. The second view emphasizes the need to place the highest priority of national policy on accelerated production of domestic energy resources and aims to remove any substantial reliance on foreign sources of supply as rapidly as possible.

Both of these visions of the world are less than complete in accounting for all available facts or in defining and organizing appropriate policy initiatives to deal with the energy problem. The adherents of one view generally regard adherents to the other as ideological foes. Yet both these groups have in common an apocalyptic view of the future.

What now needs to be debated is the validity and acceptability of this apocalyptic vision. Two years of center-stage energy policy debate have failed to clarify the issue, and because of this the energy policy the Federal government has chosen to impose pending the results of the 1976 presidential election is one of compromise. It rejects radical action in favor of a consensus policy which subordinates the imperatives of the energy issue to traditional economic concerns.

Does the nation really favor a rapid movement toward a new energy policy which is far more expensive to ordinary people than the one which is now evolving? Who is to bear the costs of the transition? What is the role of government and what is the role of the private sector in this new energy policy?

Americans have traditionally relied on the political process to organize debates such as these. But the performance of the past 2 years indicates that because basic decisions in energy policy have such a pervasive influence on the economy, the dialogue in the political arena polarizes at a very superficial level. This polarization has resulted in a rejection of extreme options, and has probably saved the economy from a good deal of short-run disruption.

But the energy crisis may not go away. If the way is to be prepared for a more fundamental transformation of the energy system, a substantial job of public education and debate must be undertaken. A consensus for a detailed program of radical action does not now exist, and in the postembargo period the political process has faithfully reflected this fact.

[BENJAMIN S. COOPER]

Few realize how much the production of fuels and energy can affect the quality of the environment. Beginning with the removal of fuels from the ground by mining or drilling and pumping, there are environmental effects associated with these activities and with the subsequent transportation, processing, conversion, transmission, and waste disposal activities. Uncontrolled production of fossil and nuclear fuels can rip up the land, release dangerous or objectionable pollutants into air and water, produce mountains of waste, and generally degrade the quality of the environment near the mines or wells. Waste heat from the generation of electricity or other industrial activities, which is usually dissipated into bodies of water, can have some impact upon water life, depending upon the amount of heat and the volume and dynamics of the receiving water body. Impoundments of water to generate electricity also can affect the environment, as can the use of geothermal steam. If solar energy becomes widely used, it too will affect the environment. Depending upon the size of energy activities, the environmental effects can range from negligible or unobjectionable to effects so intense that the environment becomes dangerous to the health of workers or the public.

The purpose of this article is to briefly identify the environmental effects of supplying fuels and energy, to indicate some measures for keeping these effects within socially acceptable limits, and to identify major legislative approaches to this end.

PRODUCTION OF FUELS

Aside from the small amount (4%) of energy supplied by hydroelectric plants, and some negligible inputs from animals, windmills, and solar units, the energy used in the United States comes from the burning of fossil fuels (coal, oil, and natural gas) or the fissioning of uranium. Of the energy used in the United States in 1974, 46.2% came from oil, 30.4% from natural gas, 17.8% from coal, and 1.9% from uranium. Over the next few decades, the energy is expected to come increasingly from coal and uranium, with the amount from oil and gas decreasing.

Coal. The mining, cleaning, and transportation of coal all have some environmental effects, the major effects being those from mining. Coal is mined in two different ways which have substantially different environmental effects: by underground mining and by surface mining. *See* COAL MINING.

Underground mining. Underground mining can cause unexpected cave-ins and subsidence of the surface above the mined area. In extreme cases this subsidence can cause buildings to collapse, disturb water supplies, prevent use of land for agriculture, and, if sudden, cause earth tremors. Approximately a third of the total area mined for coal already has experienced some subsidence. Experiments with backfilled mined areas have yet to produce an acceptable solution. However, the use of long-wall underground mining techniques promises controlled subsidence and stabilization of the surface.

Coal mining is also a source of water pollution. Water pumped from mines, or water that leaks from abandoned mines, may contain sulfuric acid and other water pollutants. The water pollutants

Protecting the Environment

- environmental impact and control
- fuel and energy production
- legislation
- air and water quality

Congressman Michael M. McCormack

from deep mines may also include trace quantities of elements and chemicals not found in the local environment. These pollutants can be toxic to plant and animal life in water and on land. The piles of waste that accumulate at deep mines also can be a source of water pollution from the leaching of pollutants into runoff water, or of air pollution if they catch fire and smolder.

Surface mining. Surface mining can virtually destroy the environment unless carefully carried out with positive reclamation measures. In area or strip mining, which is the method of surface mining used in flat areas, the overburden removed in one cut is usually dumped into the empty space left by a previous cut so that reclamation can be a continuing process. In addition to disruption of the surface, a major environmental impact from strip mining is the runoff of rainwater carrying silt which can pollute and clog streams and prematurely fill reservoirs. In contour mining, which is used in hilly regions such as Appalachia, the overburden is dumped downslope, destroying vegetation and property below, clogging streams, causing mud slides, and aggravating problems of runoff. Contour mining often is done in wilderness areas which are difficult to restore.

With the growing demand for coal, surface mining can be expected to affect large land areas. While reclamation is fairly straightforward in the eastern United States, where water is available to initiate quick growth of new plants, the scarcity of water in the arid western states is likely to make restoration much more difficult, expensive, and time-consuming.

Other effects. Coal mines can also be a local source of dust, particularly when compressed air is used to clean the coal or when the coal is heated to drive off moisture. Fine coal particles contained in wash water from coal preparation can adversely affect water life unless impounded or treated.

The concept of large coal refineries near mines promises economic benefits to mining areas. But such "complexes" are likely to be accompanied by objectionable emissions into local air and water unless they include pollution control systems.

Finally, the transportation of coal has some environmental impacts, but these are small in comparison with those from mining and processing.

Oil. Historically, most oil wells have been drilled on land, but an increasing number are now being drilled at sea. In the future, the outer continental shelf off both coasts is expected to supply an increasing share of the United States oil supply. Drilling for oil on land can have local environmental effects from the release of drilling fluid, or "mud," which may contain chemical agents that are water pollutants. Uncontrolled release of brine often found in oil deposits can cause water pollution on land, while blowouts and fires can cause local contamination. These environmental impacts are multiplied for wells drilled into the sea bed. There, any oil released from drilling accidents or from pipelines or tanks can contaminate distant shorelines. The environment at the interface between land and sea is particularly susceptible to adverse impacts from oil spills as are the fish, birds, and plant life of the coastal zone. *See* OIL AND GAS WELL DRILLING.

Oil refineries, unless properly designed and operated, can also be the source of objectionable air pollution from emission of hydrocarbon vapors, other chemicals, and combustion products from fuel burned to supply process heat. Oil refineries also can be sources of water pollution from oil leaks and chemical wastes. Some refining processes produce a heavy sludge which constitutes a solid-waste disposal problem.

The pumping out of oil fields along coastal areas can produce subsidence of the surface and increase the risk of flooding that land at high tides and during storms.

Past controversy over the anticipated environmental effects of building and operating the Trans-Alaska Pipeline highlights both the direct effects of such operations in remote regions and the indirect effects upon the environment caused by the sudden building of communities and increases in the local population.

Natural gas. Natural gas is the cleanest and most versatile of fossil fuels for stationary use. It is also the fuel in shortest domestic supply. The production of natural gas entails many of the environmental impacts associated with production of oil, for oil and gas are frequently found together and produced from the same wells. Transportation of natural gas has only minor environmental effects except for land taken for rights of way, and for local impacts from leaks which may injure local vegetation or cause explosions and fires. Pipeline explosions, however, can present a serious risk to public safety in populated areas. The growing transportation of liquefied natural gas also can pose potential risks to the public health and safety from accidental release and possible fire or explosion. *See* LIQUEFIED NATURAL GAS (LNG); NATURAL GAS.

Uranium. Uranium ores are mined by both underground and surface techniques. A unique environmental effect of mining and milling of uranium derives from the presence of radium in the ores and in the tailings from mills. This radium emits a radioactive gas, radon, which with its decay products can collect in underground mines and present an occupational health hazard. Overexposure is linked to lung cancer. The waste piles (tailings) from uranium mills, unless stabilized, can also become a source of contamination. Land areas may be contaminated by wind erosion of the tailings piles, and local streams may become polluted by runoff from the piles containing radioactive materials. Uranium mills also can produce liquid wastes which, unless processed or impounded, can introduce radioactive or toxic chemicals into streams.

The subsequent chemical and metallurgical processing of uranium compounds and the enrichment of uranium for nuclear fuel have environmental effects comparable to industrial operations in the chemical industries, for which control and collection technologies exist. These operations also generate some mildly radioactive wastes which are collected, packaged in steel drums, and sent to licensed disposal sites for burial. *See* URANIUM; URANIUM METALLURGY.

Other heat and fuel sources. Oil shales and tar sands are mentioned often as potential sources of liquid fuels. However, aside from still unfavorable economics, the production of liquid fuels from oil shales and tar sands would involve substantial environmental effects. The mining and processing of oil shale is complicated by the expansion of the volume of wastes during processing, so that the original mined areas cannot accommodate the return of all the solid wastes. Mining of oil shales and tar sands could also affect the environment through water runoff from open workings and waste piles. Also, unless controlled, oil shale refineries would be a source of dust and air pollutants. Restoration of mined areas in arid climates is likely to be difficult. Additionally, processing of oil shale requires large amounts of water, ranging from 1.4 to 4.5 barrels of water per barrel of oil produced. Finally, the establishment of large facilities to make oil from oil shale or tar sands would bring with them supporting communities and their impacts upon the environment. It should be noted that most of the environmental effects associated with the production of liquid fuels from oil shale are likely to be considerably reduced if in-place technologies can be commercially developed. *See* OIL SAND; OIL SHALE.

Geothermal energy, another potential energy source for some locations, will have some adverse environmental effects, including the possible release of foul-smelling gases, brines to contaminate water, waste heat, and subsidence of the surface. The reinjection of geothermal waters into the ground after use could cause contamination of underground waters and perhaps affect local geological stability. *See* GEOTHERMAL POWER.

PRODUCTION OF ENERGY

Heat energy from the fuel sources discussed above can be used directly as process heat by industry, to heat and cool buildings, or for conversion into electrical and mechanical energy. Common to all of these uses are the environmental effects from the burning of fossil fuels or the fissioning of uranium. *See* ENERGY FLOW.

Burning coal. Of all the fuels, coal is the dirtiest to burn, and the environmental effects of its uncontrolled burning has caused public concern, opposition, and regulation in both historical and recent times. The particulates create smoke and haze, soil buildings and materials, and increase cleaning costs. Fine particulates may be a health hazard and are suspected of accelerating the attack of corrosive gases upon buildings and structural materials. Emission of particulates can be controlled by collecting fly ash with electrostatic precipitators and filters. However, small particulates from some kinds of coal are difficult to trap.

Sulfur emission. Probably the most noticeable gas from the combustion of coal is sulfur dioxide, which can, in some cases, cause corrosion and damage plant life, and is also suspected of causing or aggravating respiratory illness in humans. The simplest way to keep sulfur dioxide emissions within acceptable limits is to burn coal of a low sulfur content. However, there is an insufficient supply of low-sulfur coal, so other control technologies must be used.

Constant emission limitation, or permanent control, involves imposing a fixed limit on the rate of sulfur dioxide emissions from a furnace or boiler. Constant emission limitation may be accomplished by burning low-sulfur coal to the extent available, or by installing stack-gas cleaning systems, or "scrubbers," to remove much of the sulfur dioxide before release to the atmosphere. Commercially, scrubber technology is relatively new, and there is still some controversy over how well it works.

An alternative to scrubbers is the supplementary control system (SCS), in which the emissions from a furnace are discharged through a stack high enough that the concentration at ground level usually is within acceptable limits. The electrical utility industry regards the supplementary control system as less expensive, less energy-consuming, and more reliable than scrubbers. The Environmental Protection Agency claims that sulfur dioxide in the air may be transformed into sulfates that may be injurious to the public health. Using SCS does not reduce the total atmospheric burden of sulfur, so EPA is strongly opposed to supplementary control systems, advocating permanent controls instead.

Another alternative is to trap sulfur during combustion. The fluidized-bed combustion process holds promise for doing so, but the engineering and economic practicability of large units has yet to be demonstrated.

Nitrogen oxides. Oxides of nitrogen are also emitted by furnaces and boilers. These gases, being both corrosive and one cause of photochemical smog, are suspected of having unfavorable health effects. Production of nitrogen oxides can be controlled by regulation of the combustion process and keeping the maximum temperature below 2800°F (1538°C), which is the threshold for formation of nitrogen oxide. Unfortunately, the drive to increase conversion efficiency of power plants requires going to higher temperatures. At present there are no well-established means of removing nitrogen oxide from combustion products.

Carbon dioxide. Another potentially troublesome product from burning any fossil fuel is carbon dioxide. Some scientists speculate that an increase in the carbon dioxide content of the atmosphere will increase its ability to absorb solar energy and lead to a gradual warming of the troposphere (the greenhouse effect). If the concentration of carbon dioxide were to become high enough, some scientists speculate, the temperature increase could melt the polar icecaps, with catastrophic flooding of coastal areas throughout the world. However, other scientists point to the increase of human-produced dust in the atmosphere which tends to reflect some of the solar energy, and thereby to decrease air temperatures. What the balance between these effects will be is not now known. There is no practical way to limit the emission of carbon dioxide from burning fuels.

Other emissions. Because many impurities are present in coal, a variety of trace metals and elements may occur in uncontrolled exhausts from furnaces and boilers. Those elements most likely to appear are mercury, lead, zinc, and beryllium. Incomplete combustion can also cause the presence of hydrocarbons in the exhaust from combustion. Some of these, such as benzo-alpha-pyrene, are known to cause cancer in experimental animals.

The ash from burning coal, including fly ash collected by emission control systems, forms a solid waste. Although some of this material is used to make cement or in road building or construction, most of it is piled on land or dumped at sea. Some very fine particles of ash are not removed by scrubbers, and there is concern that they can become trapped in the lung and impair health.

Burning liquid fuels. The burning of gasoline, jet fuels, fuel oils, and residual oils all produce some sulfur dioxide, nitrogen oxide, particulates, and carbon dioxide. In addition, burning of gasoline in internal combustion engines usually produces carbon monoxide. Such engines also discharge small amounts of incompletely burned hydrocarbons. For stationary use, emissions of sulfur dioxide can be controlled through use of low-sulfur fuel or control systems, while production of nitrogen oxides can be limited by control of combustion process and temperature. Systems also exist for control of carbon monoxide emissions.

Burning natural gas. The environmental effects from burning natural gas are mainly limited to production of carbon dioxide and nitrogen oxides, whose effects were mentioned above. Despite its desirable minimal environmental impacts, the growing shortage of natural gas is likely to make it increasingly reserved for special uses. Within a decade or so, supplies of synthetic natural gas made from coal are likely to appear. The environmental effects of burning this fuel should be indistinguishable from the burning of natural gas, although there will be some adverse effects resulting from the conversion of coal to synthetic gas. *See* COAL GASIFICATION.

Burning solid wastes. Urban and rural solid wastes may also be used as a fuel to produce process heat or steam. The environmental effects from burning such wastes approximate those from burning coal. In addition, the collection and holding of large volumes of solid wastes might create a public nuisance or public health hazard from decaying organic matter and from insects and rodents attracted by it.

Fissioning of uranium. In addition to heat, the fission process produces large amounts of intensely radioactive materials, which in routine operations are virtually all confined within the nuclear fuel. Small amounts do escape from the fuel and are collected, packaged, and sent to special burial grounds for disposal. Some small quantities of radioactive materials, mainly gases such as tritium, are released to the environment in concentrations that are low in comparison with emission limits set by the Nuclear Regulatory Commission (NRC). The environmental effects of small routine emissions from a nuclear power plant are believed to be negligible, although some observers have expressed concern that these small quantities might be concentrated in some animal or plant life form and thus become a health hazard to humans or to other creatures in the environment. *See* FISSION, NUCLEAR.

Accidental release. The radioactive fission products produced in nuclear power plants could have substantial environmental effects if a large amount were to be released in an accident or by some other mechanism. Various protective mechanisms are employed to provide multiple barriers against the release of hazardous quantities of radioactive materials. To date, there has been no such release in the United States or elsewhere in the world.

Reprocessing. After most of the usable energy has been extracted from the nuclear fuel, the spent fuel must be removed from the power plant. At that time it is highly radioactive because of the fission products it contains. In principle, the nuclear fuel is shipped to a reprocessing plant, where the remaining uranium and plutonium are recovered for reuse. Then the solidified wastes are permanently buried in a deep geologic stratum, probably salt, to remain isolated from the biosphere for many thousands of years.

In practice, however, only a small amount of nuclear fuel has so far been reprocessed. It is stored as a liquid in steel tanks. The remaining spent fuel is being kept in storage pools awaiting the time when commercial reprocessing will again be available. To date, there have been no releases to the environment of this high-level radioactive waste.

The fuel reprocessing plants at which high-level radioactive wastes are produced are themselves, in normal operation, likely to have environmental effects comparable to chemical plants of the same size. In addition, fuel reprocessing plants are the source of low-level radioactive wastes which are buried in shallow trenches at waste burial grounds. Reprocessing plants may emit small quantities of radioactive gases, within standards established by the Federal government.

Finally, fuel reprocessing plants could present some risk of accidental release of radioactive materials to the environment. As for nuclear power plants, extensive safety measures must be employed to prevent such occurrences.

On the whole, the risks of accidental release of radioactive materials from nuclear power plants and associated facilities are regarded as small and worth the benefits of the electricity supplied. This assessment, however, is disputed by some antinuclear advocates who claim that the potential effects of accidents are so severe that even a minute risk is unacceptable. *See* NUCLEAR POWER; REACTOR, NUCLEAR.

CONVERSION OF HEAT INTO MECHANICAL ENERGY

Heat can be converted into mechanical energy in internal combustion engines, gas turbines, and steam turbines. Heat from the fissioning of uranium can also be converted into mechanical energy through steam turbines. The environmental effects of these machines largely take the form of air pollution and noise, with some effects also caused by waste heat. Automobile engines are a major source of nitrogen oxide, carbon monoxide, and unburned hydrocarbons in the air. Their emissions are a major component of the photochemical smogs that plague some urban areas. Emission control systems and modification of engines to reduce emissions are two alternatives. A more radical alternative is to use electric vehicles which have virtually zero environmental effect in the vicinity of the vehicle, although supplying the electricity needed to power this mode of transportation would have some environmental impacts from the expanded use of power plants.

CONVERSION OF HEAT INTO ELECTRICITY

The principal impacts on the environment from the large-scale generation of electricity from heat energy result from the burning of fossil fuels or fission of nuclear fuels as described, and from the release of waste heat produced in these processes.

For a conventional steam electric power plant, about two-thirds of the heat produced by burning fuel or fissioning uranium is released to the environment, and only about one-third is converted to electricity. Engineers prefer to dissipate waste heat into a convenient body of water such as a river, lake, bay, or reservoir by pumping cooling water through the power plant. *See* ELECTRIC POWER GENERATION.

The effects of waste heat rejection systems upon bodies of water are twofold: the increase in temperature can affect plant and animal life in the waters, driving some species away and attracting others; and the enormous quantities of cooling water needed for a larger power plant require powerful pumps that can suck in marine life which may be killed or injured at the pump, or may succumb to or be injured by exposure to high temperatures in the plant's heat exchangers. Cooling systems also can be a source of toxic wastes from chemicals used to clean marine growth from the insides of pipes and other surfaces.

Alternative ways to dissipate waste heat into the environment are to discharge it directly into the air through evaporative or dry cooling towers, or into artificial ponds that are isolated from other bodies of water. However, these alternatives can produce undesirable effects of their own. Cooling towers are huge and unsightly. Evaporating waters of cooling towers or ponds can produce local mist or icing in cold weather. Those cooling towers that use salt water produce a salt-laden mist than can corrode buildings and structural materials and contaminate agricultural land downwind. Some cooling towers employ large, noisy fans.

Dry cooling towers, through still larger, avoid some of the problems. They have yet to be used on a large enough scale to observe their overall environmental effects. Some scientists theorize that the plumes of dry heated air above such towers could cause local atmospheric instability in some weather conditions and perhaps trigger tornadoes.

Another environmental impact of electric power plants is commitment of the land required. A large coal-fired power plant with controlled emission systems would require some 940 acres (3.8 km²) for the plant plus about another 17,000 acres (68.8 km²) for associated transmission lines, depending on the distance from its load center.

The diesel engines and gas turbines that are sometimes used to drive electric generators, because they are small in comparison with central power plants, produce comparatively small environmental effects.

OTHER ENERGY SOURCES

Looking ahead, other energy sources which may supply useful amounts of energy by the year 2000 include direct use of solar energy, and indirect use of it by tapping wind power, ocean heat and currents, and perhaps ocean wave energy. Some additional hydroelectric plants may be built as well as some tidal plants.

Hydroelectricity. Water power has long been valued as a clean and inexpensive source of electricity. However, hydroelectric plants have environmental effects which can be objectionable to some parts of society. Dams can interfere with the movement up- and downstream of water life-forms, increase the nitrogen content of the water with adverse effects upon fish, permanently flood productive, scenic, or historical lands, and alter water flow downstream. At deep reservoirs, water released from bottom layers may be very cold and devoid of oxygen. In addition, the enormous weight of dams and impounded waters could cause geological distortions in some locations. Of course, not all environmental effects of hydroelectric plants are considered detrimental. Reservoirs can provide recreational opportunities and expanded fishing opportunities. Flood control and irrigation water supplies from dams are usually also counted as positive benefits. *See* DAM; HYDROELECTRIC POWER.

Tidal power. The harnessing of tidal energy to generate electricity is a possibility for a few places such as Passamoquoddy Bay in Maine. However, the dams for such ventures and the system of locks would interfere with the normal flow of waters and would change the kinds and populations of water animals and plant life present. Scenic values could also be affected. The world's only operating large-scale tidal power plant is located on the estuary of the Rance River in France. While technically a success, it is not economical in comparison with conventional power sources because of its high capital costs. *See* TIDAL POWER.

Solar power. Solar energy, because it is so diffuse, requires large collecting surfaces to produce useful amounts of fuel or energy. A solar system involves large land areas. The land requirements for a solar electric plant of 1000 MW output range from 10 mi² (26 km²) for a thermal conversion system, to 30 mi² for photovoltaic conversion and up to 500 mi² for a fuel crop system. The land most likely to be used for central power stations is in the semidesert Southwest. Building the installations and access roads would disturb these lands. Once in place, the structures would shade the earth beneath them and probably change the kind of wildlife living there.

Large energy plantations for the growth of crops or trees would probably require land that otherwise could be used for agriculture or forestry and would entail potential effects from runoff from cultivated and fertilized fields. Burning of energy crops or wood would produce some air pollution. Large thermal steam systems that raise steam for turbogenerators also would produce waste heat, as do conventional power plants. This heat would have to be dissipated into the environment in one way or another. *See* SOLAR ENERGY.

Wind power. Many of the schemes proposed for large-scale use of wind power involve construction of chains of tall towers. Their principal known environmental effects would be esthetic. Such tall structures would also increase hazards to air navigation at low altitudes and to bird flights, and might affect radar transmission and other means of communication. Wind machines would have little effect upon air and water quality, but might require land for their structures and for rights-of-way of transmission lines. *See* WIND POWER.

LEGISLATION FOR ENVIRONMENTAL PROTECTION

In the United States much of the present control of objectionable effects of increasing supplies of fuels and energy is based upon legislation by the Federal and state governments. Today the entrepreneur who arranges for the financing of energy facilities and the engineers who design them face limitations set by society, limitations which often can be as demanding as those of nature, materials, or technology.

National environmental quality. A legislative landmark for preservation of the nation's environmental quality is the National Environmental Policy Act of 1969 (NEPA). The act is notable in three respects: it declared a governmental mandate and responsibility for environmental quality; it created the Council on Environmental Quality; and it required all Federal agencies to prepare an environmental impact statement on proposed actions which could significantly affect the environment. These concepts, particularly that of the environmental impact statement, have since been incorporated into similar legislation in many states.

NEPA requires Federal officials to examine many factors in the preparation of an environmental impact statement, among them: (1) the environmental impact of the proposed actions; (2) any adverse environment impacts that cannot be avoided, should the proposal be implemented; (3) alternatives to the proposed action; (4) the relationship of local short-term uses of the environment and the maintenance and enhancement of long-term productivity; (5) any irreversible and irretrievable commitments of resources which would be involved in the proposed action.

Preparation of environmental impact statements is now a recognized component of proceedings for many major construction projects at the Federal, state, and local government level, and has significantly contributed to protection of the environment. These benefits have, of course, been obtained in exchange for some costs—direct ones in the preparation of such statements, and indirect lost-opportunity costs during the additional time required to bring energy facilities into service.

Air quality. Federal legislation to protect air quality began in 1955 when Congress established the principle that state and local governments are responsible for air-pollution control, with the Federal government providing leadership, information, and support. The most recent revision of this legislation occurred in the Clean Air Act Amendments of 1970. Under this revision, while the state governments retain their fundamental responsibility for air-pollution control, the Federal government has a stronger voice. The Environmental Protection Agency now has the responsibility for establishing national standards. EPA in 1971 established standards for major air pollutants, including sulfur dioxide, particulates, carbon monoxide, hydrocarbons, nitrogen oxides, and photochemical oxidants. These standards form the basis for state control. In essence, the legislation requires polluters to install control technology or to take other steps which will permit air quality standards to be achieved.

The 1970 amendments also are the source of the controversial requirement for the substantial reduction of emissions of pollutants from motor vehicles.

Another legislative act affecting both the environment and the supply of fuels and energy is the Energy Supply and Environmental Coordination Act of 1974. This act directed the Federal Energy Administration to prohibit power plants or other major fuel-burning installations from burning oil if conversion to coal is practicable, if coal is available, and if reliability of power supply is not impaired. However, FEA cannot order a conversion until it has determined that the plant can comply with EPA air-pollution requirements and an environmental impact statement has been prepared. *See* AIR-POLLUTION CONTROL.

Water quality. The Water Quality Act of 1965 required the state governments to establish water-pollution control standards for all interstate waters, and placed most of the enforcement responsibility with the states. However, the state governments moved slowly. Congress, therefore, in 1972 revised the act with legislation which defined the respective responsibilities of the Federal and state governments and set 1985 as the date for attainment of a national goal of eliminating all discharges of pollutants into the waters. In this respect, waste heat is defined as a water pollutant. By 1977, all point sources of effluents, which include power plants, are to be limited to levels achievable through use of "best practicable" technology, and by 1985 by the "best available" technology.

Strip mining control. The Surface Mining Control and Reclamation Act, passed by Congress in 1975, would have established strict controls over strip mining practices and required reclamation of strip-mined land. However, the act, according to its critics, would also have cut coal production significantly, eliminated jobs, and increased consumer costs and oil imports. This legislation was vetoed by the President in June 1975.

Nuclear power control. The primary basis for regulation of the construction and operation of nuclear power plants is the Atomic Energy Act of 1954, as amended, and the Energy Reorganization Act of 1974. Together, these acts provide the framework of Federal authority and organization to regulate the safety and environmental impacts of nuclear power plants and associated facilities. The licensing process, which is administered by the Nuclear Regulatory Commission, includes the issuance of a construction permit and an operating license. One major part of the proceedings for both permits is the formulation, review, and consideration of environmental impact statements. Generally, the NRC analysis of a proposed nuclear power plant and the NRC regulations are intended to keep emissions of radioactive materials within acceptable limits and to control the manner in which waste heat is dissipated.

A major legislative proposal with environmental implications is one which would create nuclear parks, where many nuclear power plants would be built together with supporting industrial facilities, primarily fuel fabrication and reprocessing plants, and perhaps waste management and uranium enrichment plants. The nuclear park concept solves many problems relative to environmental protection.

Coastal zone management. Oil spills and discharge of waste heat are of particular concern in coastal zones. In addition to controls established by other legislation, the Coastal Zone Management Act requires that applicants for a Federal license or permit needed for an energy operation must certify that the proposed activity will be conducted in a manner consistent with the goals of the coastal zone management program. Another requirement forbids a Federal agency to issue a license or permit until the state involved has concurred with the applicant's certification.

CONCLUSION

The environmental effects inherent in the various forms of energy usage range from inconsequential to substantial. While it is obviously advantageous to control and reduce these effects to the maximum extent practicable, it must at the same time be recognized that environmental control measures are not without impacts of their own in terms of costs, jobs, resources, or other matters.

Control could conceivably be carried to the unacceptable extreme where society is denied the use of the fuel or energy source in question. On the other hand, uncontrolled production and use of fuels and energy without any concern for impacts on the environment is equally unacceptable. The problem, then, is to balance society's need for energy against the need for a livable environment, at the same time giving appropriate attention to important economic, technical, and social factors. *See the feature articles* EXPLORING ENERGY CHOICES; U.S. POLICIES AND POLITICS; *see also* ATMOSPHERIC POLLUTION; WATER POLLUTION.

[MICHAEL M. MC CORMACK]

Uncertainty is perhaps the chief characteristic of the world energy economy in the mid-1970s. The tripling of the world price for crude oil following the embargo of 1973–1974 has fundamentally altered the interrelationships among the various energy sources and the markets wherein they are traded.

There is little or no experience to serve as a guide in predicting either supply or demand for energy at current prices. This article can only establish working hypotheses and stress the severe limitations of the analysis and conclusions.

Data on reserves, production, and consumption all vary greatly in quality from country to country and resource to resource. Moreover, many of the data necessarily employed would not meet rigorous standards of uniformity of criteria employed in the data gathering and preparation. Nevertheless, even poor data can indicate trends and tentative conclusions.

Aside from the inability to predict the elasticities of demand for energy over time at current or higher prices, other unknowns present themselves. The rate at which political or environmental factors will hasten or retard energy development is not known. Perhaps more important yet, the development and impact of technology can only be alluded to; they cannot be predicted. In this regard, the retarding effect which cheap and abundant oil has had on technological innovation in alternative fuels and conservation techniques has resulted in a generally lower level of technology in nonpetroleum energy sources than in other areas of the economy. That gap can be expected to be narrowed under the pressures of high prices and the desire of many countries to develop secure, indigenous sources of energy. Thus, what may at least be hoped for is an unusually rapid evolution of energy and conservation technology in the intermediate run.

The success of the world energy market in reorienting itself toward nonpetroleum sources will be affected by many factors, some as yet unknown. But the transition can be materially facilitated by finding a means of stabilizing price so that it may guide research and investment decisions, and by the establishment of clear environmental criteria and expeditious means for resolving environmental disputes.

It is in this context that energy decisions must be made which will substantially affect the well-being of at least a generation.

FLOWS OF ENERGY

Two aspects are considered in the flows of energy. The first is the production and consumption of oil, the most important source of energy in world commerce. The second concerns itself with reserves of oil, gas, coal, and uranium.

Production and consumption. While oil is by far the most important energy source in world trade, in terms of consumption it constituted less than half of the primary global energy consumed in 1974 (see Fig. 1 and Table 1). But petroleum has by far provided the major part of the increase in global energy consumption since the mid-1960s. In 1962 oil accounted for a fraction over 35% of energy consumption. By 1974 it was almost 50%. This was the result of low price and convenience of use.

World Energy Economy

- **price of fuels**
- **worldwide available resources**
- **economics of exploiting energy sources**
- **constraints and conservation**

Mr. John K. Wilhelm

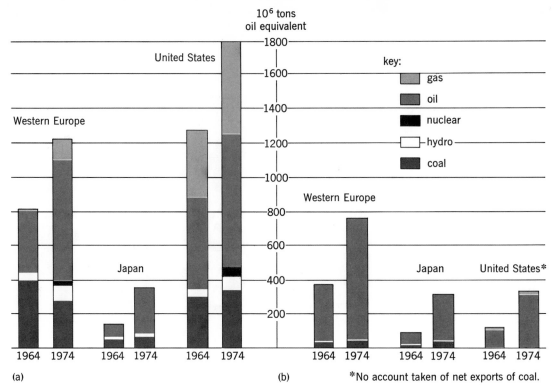

Fig. 1. Graphs of (a) total energy consumption and (b) net energy imports for Western Europe, Japan, and the United States in 1964 and 1974. (*From BP Statistical Review of the World Oil Industry 1974, British Petroleum Co. Ltd., Britannic House, London E. C. 249 Bu*)

Table 1. World primary energy consumption, in 10⁶ tons oil equivalent

Country/area	1974						1973					
	Oil	Natural gas	Solid fuels	Water power	Nuclear	TOTAL	Oil	Natural gas	Solid fuels	Water power	Nuclear	TOTAL
United States	785.4	560.4	331.9	76.9	28.8	1783.4	818.0	572.3	335.0	75.6	21.8	1822.7
Canada	88.1	64.1	15.0	59.4	4.9	231.5	83.7	63.8	16.4	55.5	4.4	223.8
Other Western Hemisphere	178.9	48.0	14.0	31.3	0.3	272.5	169.9	42.5	16.4	28.5	–	257.3
TOTAL WESTERN HEMISPHERE	1052.4	672.5	360.9	167.6	34.0	2287.4	1071.6	678.6	367.8	159.6	26.2	2303.8
Belgium and Luxembourg	27.5	10.0	11.6	0.1	–	49.2	31.5	8.2	11.0	0.1	–	50.8
Netherlands	35.4	32.9	3.3	–	0.8	72.4	41.3	32.2	3.1	–	0.3	76.9
France	120.1	17.0	27.0	12.6	3.0	179.7	127.3	15.7	26.3	10.6	3.0	182.9
West Germany	134.4	32.1	82.9	5.0	2.7	257.1	149.7	27.0	87.9	3.0	2.9	270.5
Italy	100.7	15.9	9.2	11.0	0.9	137.7	103.6	14.4	7.6	10.1	0.8	136.5
United Kingdom	105.8	30.8	68.9	1.3	7.3	214.1	113.4	26.1	78.5	1.2	5.9	225.1
Scandinavia	51.1	–	7.3	36.7	0.4	95.5	55.9	–	7.0	32.6	0.5	96.0
Spain	38.2	1.0	12.3	8.2	1.9	61.6	36.3	1.0	11.7	7.5	1.7	58.2
Other Western Europe	86.0	6.4	39.4	25.4	1.5	158.7	90.0	5.2	38.7	23.6	1.6	159.1
TOTAL WESTERN EUROPE	699.2	146.1	261.9	100.3	18.5	1226.0	749.0	129.8	271.8	88.7	16.7	1256.0
Japan	261.1	5.1	58.8	19.0	4.1	348.1	268.3	4.8	59.8	15.8	2.1	350.8
Australasia	33.9	4.1	26.1	2.4	–	66.5	32.8	3.9	29.4	2.2	–	68.3
Soviet Union	341.8	216.6	369.1	38.7	4.0	970.2	317.7	200.4	361.8	36.9	3.0	919.8
Eastern Europe	78.3	43.5	225.1	5.7	0.3	352.9	74.8	39.4	218.7	5.3	0.1	338.3
China*	48.8	4.4	331.5	9.5	–	394.2	41.7	4.0	316.3	9.0	–	371.0
Other Eastern Hemisphere	227.4	42.0	134.8	21.1	1.0	426.3	220.0	37.9	127.5	19.5	1.0	405.9
TOTAL EASTERN HEMISPHERE	1690.5	461.8	1407.3	196.7	27.9	3784.2	1704.3	420.2	1385.3	177.4	22.9	3710.1
WORLD	2742.9	1134.3	1768.2	364.3	61.9	6071.6	2775.9	1098.8	1753.1	337.0	49.1	6013.9

SOURCE: *BP Statistical Review of the World Oil Industry 1974, British Petroleum Co. Ltd., Britannic House, London E.C. 249 Bu.*
*Includes Albania, North Korea, and North Vietnam.

While oil will continue to be the predominant fuel in international trade for the foreseeable future, both price and security considerations have since 1973 presented powerful incentives for consumers to shift to other fuel sources in the intermediate to long run.

With the exception of the United States and the Soviet Union, the major industrial centers will con-

Table 2. Proved world reserves of oil and gas by region and selected countries, January 1974

Country	Oil, 10^3 bbl	Gas, 10^6 m³
Asia	35,635,040	3,801,700
Indonesia	10,500,000	424,929
Mainland China	20,000,000	566,572
Australia	2,300,000	1,067,989
Middle East	350,162,500	11,708,924
Abu Dhabi	21,500,000	354,108
Iran	60,000,000	7,648,725
Iraq	31,500,000	623,229
Kuwait	64,000,000	920,680
Neutral Zone	17,500,000	226,629
Saudi Arabia	132,000,000	1,441,926
Soviet Union and Eastern Europe	83,000,000	20,266,289
Soviet Union	80,000,000	20,000,000
Western Europe	15,990,500	5,490,000
Netherlands	251,000	2,606,232
Norway	4,000,000	651,558
United Kingdom	10,000,000	1,416,431
Africa	67,303,750	5,317,847
Algeria	7,640,000	3,001,275
Libya	25,500,000	764,873
Nigeria	20,000,000	1,133,144
Western Hemisphere	75,764,669	11,017,847
Canada	9,424,170	1,424,901
United States	34,700,249	7,005,949
Ecuador	5,675,000	141,643
Venezuela	14,000,000	1,189,802
TOTAL WORLD	627,856,459	57,602,606

SOURCE: *Oil Gas J.*, pp. 86-87, Dec. 31, 1973.

Table 3. Estimated world original coal resources by region, in 10^9 metric tons

	Coal resources determined by mapping and exploration	Probable additional resources (unmapped and unexplored)	Estimated total
North America	1,560.0	2,612.2	4,172.2
(United States)	(1,434.0)	(1,490.2)	(2,924.2)
Europe	562.4	190.4	752.8
Oceania	54.4	63.5	117.9
Soviet Union	5,895.5	2,721.0	8,616.5
Latin America	18.1	9.1	27.2
Asia	453.5	907.0	1,360.5
Africa	72.6	145.1	217.7
TOTAL	8,616.5	6,648.3	15,264.8

SOURCE: D. A. Brobst and W. P. Pratt (eds.), *United States Mineral Resources*, Geol. Surv. Prof. Pap. 20, pp. 137 and 140, 1973.

tinue to be highly dependent on imported fuels for the foreseeable future unless there is a dramatic change in the location of new reserves of oil, gas, coal and uranium (see Tables 2, 3, and 4). *See the feature article* ENERGY CONSUMPTION.

Reserves. The methodology used in estimating reserves of various energy resources makes a substantial difference in the resultant conclusions. Price is also an important determinant of what constitutes "recoverable" or actually available resources. Accordingly, all such data should be viewed as indicating rough orders of magnitude rather than precise calculations.

Table 2 indicates "proved" oil and gas reserves. But it is costly to prove oil reserves; therefore, they are "proved" to be commercially recoverable only when effective demand presents itself. These estimates are thus very conservative and valid only for the short and intermediate run. Estimates of "unproved" reserves vary widely, and projections of "unproved" reserves at three times those indicated in Table 2 are not uncommon.

Coal reserve data based upon mapping and exploration are somewhat more accurate (see Table 3). Probable coal reserve estimates are far more speculative. Because the recovery rate for coal is about 50%, usable coal availabilities are, on the basis of mapping and exploration, equivalent to approximately 4.3×10^{12} metric tons. These resources are immense when compared to a global consumption in 1973 of about 2.7×10^9 tons. The coal equivalent of all primary energy consumption in 1973 would be 9.5×10^9 metric tons. Coal can meet energy needs for well over a century, even if

high consumption growth rates are assumed and the estimates of probable reserves prove to be optimistic.

As with petroleum, the resource tends to be concentrated. The Soviet Union accounts for the vast majority of established reserves, with the United States in second place. In Asia, China accounts for all but a small portion of the coal reserves.

Uranium resources are by comparison more widely distributed than coal or petroleum as Table 4 indicates. (Reliable data for Communist countries are unavailable.) As the table also indicates, uranium availability is closely associated with price. Because the cost of uranium is a relatively insignificant portion of the cost of electricity produced by nuclear power, there is little impediment to use of the more expensive ores. They may therefore be viewed as part of the readily available resource base. *See the feature article* OUTLOOK FOR FUEL RESERVES; *see also* ENERGY FLOW.

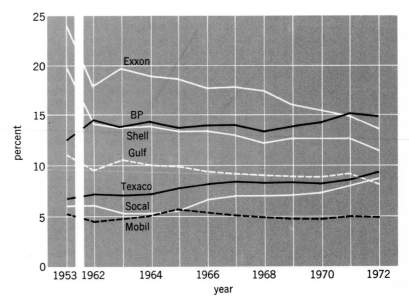

Fig. 2. Shares of crude oil production in the foreign oil industry by each of the seven largest oil companies, annually, 1962–1972. (*From N. Jacoby, Multinational Oil: A Study in Industrial Dynamics, Macmillan, 1974*)

Table 4. Uranium resources by major countries, 1973, in 10³ short tons

	Price range up to $10/lb U_3O_8		Price range $10-15/lb U_3O_8	
	Reasonably assured resources	Estimated additional resources	Reasonably assured resources	Estimated additional resources
Argentina	12.0	18.0	10.0	30.0
Australia	92.0	102.0	38.3	38.0
Brazil	–	3.3	0.9	–
Canada	241.0	247.0	158.0	284.0
Central African Republic	10.5	10.5	–	–
Denmark (Greenland)	7.0	13.0	–	–
Finland	–	–	1.7	–
France	47.5	31.5	26.0	32.5
Gabon	26.0	6.5	–	6.5
India	–	–	3.0	1.0
Italy	1.6	–	–	–
Japan	3.6	–	5.4	–
Mexico	1.3	–	1.2	–
Niger	52.0	26.0	13.0	13.0
Portugal (Europe)	9.3	7.7	1.3	13.0
(Angola)		–	–	17.0
South Africa	263.0	10.4	80.6	33.8
Spain	11.0	–	10.0	–
Sweden	–	–	351.0	52.0
Turkey	2.8	–	0.6	–
United States	337.0	700.0	183.0	300.0
Yugoslavia	7.8	13.0	–	–
Zaire	2.3	2.2	–	–
TOTAL	1127.7	1191.1	884.0	820.8

SOURCE: Organisation for Economic Co-operation and Development, *Uranium Resources Production and Demand*, a joint report by the OECD Nuclear Energy Agency and the International Atomic Energy Agency, p. 14, Paris, August 1973.

INDUSTRIES IN ENERGY

Some of the energy commodities just discussed are produced by companies that are national or international, while other energy sources, such as uranium, are produced by government agencies.

Oil and gas. The world petroleum industry has been dominated by a small number of fully integrated companies operating on a global basis. Standard Oil of California, Texaco, Mobil, Gulf, and Exxon are United States firms, British Petroleum is United Kingdom, and Royal Dutch Shell is British and Dutch. Figure 2 indicates the market positions of these firms. This group, sometimes referred to as the "Seven Sisters," has by some standards been joined by a French company, CFP (Compagnie Française des Petroles). The companies are distinguished from all others by their comprehensive efforts ranging from exploration to marketing of finished products. Their worldwide production refining and logistics systems are among the most sophisticated in existence, and even their detractors admit that they are enormously efficient.

In recent years a substantial number of national companies of other countries as well as independents (private nonintegrated) have entered the international oil trade. None has the resources of the "majors," but they are highly competitive and in some instances have strong political and financial support from their home governments.

Control of access to crude oil has been a major element in the success of the integrated companies, and the vast overseas concessions which they owned gave them a substantial advantage over new entrants. Extraordinary technical and capital requirements further limited possible new competitors.

With the virtual elimination of overseas concessions, future access to crude oil reserves will be determined by the governments of oil-exporting countries, and the crucial historical advantage of the major companies can no longer be assumed. Political considerations can be expected to play a more important role in the determination of who has access to oil.

While some have argued that the major oil companies were historically oligopolists and others have replied that they were competitive, neither description fully fits reality.

The major companies have historically been fierce competitors for concessions and markets. On the other hand, the baseline product which constituted the bulk of their trade, fuel oil, was until recently sold at that price where it successfully competed with coal, that is, at a much higher price than a "competitive" price which theoretically would have been the cost of production of the marginal barrel of fuel oil. The price charged emanated from the nature of the resource, and collusion was not necessary to establish it. The nonfuel oil products were, by and large, sold quite competitively.

Thus, the market for petroleum was at one and the same time characterized by economic rents (indicating a higher than purely competitive price) and intense competition.

In the 1950s the advent of the smaller national and independent nonintegrated companies accelerated the development of producer surpluses and began to exert strong downward pressures on prices. The realization by the producing countries that the "majors" could not control prices as so often alleged, resulted in the establishment of the Organization of Petroleum Exporting Countries (OPEC) to fulfill that function.

Coal. The coal industry, in large part because coal is usually more expensive to handle and transport than petroleum, tends to be more national in character. Also, specific coal reserves tend to be more closely tied to their consumers, such as the steel industry. The post-1973 petroleum price explosion has given powerful incentives to the coal industry. Currently, environmental considerations are the main restraint to extensive expansion. The high capital investment required for consumers to switch from petroleum to coal is another. Nevertheless, at present coal is the world's most abundant and readily available energy resource, given current prices, technology, and environmental factors.

Uranium. Because of the obvious security problems associated with nuclear power, governments have dominated this industry and its technology, and will most likely continue to do so despite efforts in the United States to further involve the private sector. Should the nuclear-related industry overcome its major environmental impediments, it may be expected to undergo substantial growth. Like coal, nuclear energy with current technology is very attractive at present petroleum prices.

PRICING OF ENERGY

Not only did the OPEC petroleum price increases following October 1973 fundamentally alter the relationships of various energy sources in terms of their profitability and attractiveness, but they also changed the very form of competition.

Market structure. In 1970 OPEC began to exert control over the price of oil, and in 1973 it dramatically reoriented almost all energy price relationships. Whereas oil previously exerted a downward pressure on price because it is the low-cost fuel, the new OPEC price now has petroleum propelling all energy prices upward. Moreover, whereas petroleum once eroded the market shares of other energy sources, the opposite is now true, and alternative fuels are beginning to compete for petroleum's market share.

This central role of petroleum stems from the fact that the international market for petroleum has historically, with the exception of the 1973–1974 embargo and a short period in the 1930s, been constrained by demand. That is, certain owners of oil have always had a price below which they preferred to withhold supply rather than lower price. This is because the demand for petroleum has been only slightly responsive to changes in price. A lowering of price did not sufficiently increase industry sales to augment or even maintain aggregate income. Petroleum has thus been the marginal energy source whose supply can be, and is, expanded or contracted to meet market demand at a given price.

Pricing. While it is true that the price of petroleum has not been genuinely competitive, it is also true that there is wide agreement that a competitive price would be so low as to force recovery techniques which are destructive of many oil-bearing formations and wasteful of the resource.

The issue, therefore, is not whether the price of petroleum should be administered, but rather who should administer it and how. Because it determines the price of all other energy resources, the pricing of crude oil constitutes perhaps the most important single economic decision that is made in the world.

Economic rents. An administered petroleum price decision results in a price higher than a rigorously defined "competitive" price, and economic rents (excess profits) are created. Who should get these rents and to what use they should be put constitute one of the more lively debates. In the past the great oil companies took a good portion of these rents and used them to develop the exceedingly complicated and costly global infrastructure of the petroleum industry. Today the oil-exporting countries are taking most if not all of the petroleum associated rents. It is thus very unclear how the future expansion of the industry will be financed.

The appropriation of economic rents from the companies to the exporting governments also removes a very strong incentive which once induced the oil companies to undertake substantial risks in exploring for oil. Unless another way is found to induce new exploration, the rate of discovery of petroleum can be expected to fall off.

CONSTRAINTS AND CONSERVATION

One would think that a society that was able to tap the power of the atom and was able to put a human being into outer space would be able to marshal the scientific and technological resources necessary to solve the energy problem. That viewpoint assumes that the circumstances that produced a Manhattan Project and a NASA are present today. Unfortunately this is not the case. The energy problem has political, ecological, and economic aspects that did not exist either for the Manhattan Project or for the space effort.

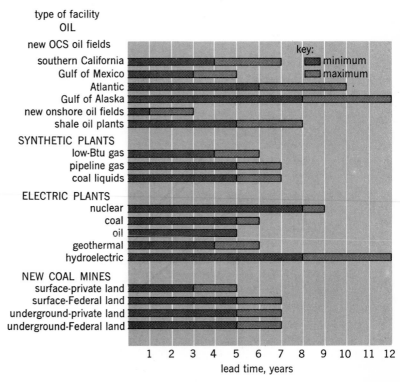

Fig. 3. United States energy facility production lead times. (*From U. S. Department of the Interior, U. S. Energy Prospects*)

Constraints. Uncertainty is a significant impediment to energy resource development which enervates the development process. Major political decisions on the ecology, on the terms of access to resources once a resource is discovered and developed, and on who gets what have bogged down the decision-making process at enormous cost to the ultimate energy consumer. The exceptionally long lead times associated with energy production further exacerbate this problem (see Fig. 3). *See the feature article* U.S. POLICIES AND POLITICS.

Ecology. Massive energy projects such as dams, pipelines, and open-pit coal mining have profound ecological implications which must be accommodated.

Essentially, the political processes of the world are beginning to recognize that there are real trade-offs between the ecology and energy availabilities and that there are real costs associated with the decisions taken.

Another major problem is that people are usually willing to strip-mine somebody else's coal, drill off others' shores, and build refineries on other people's beaches. Just how to encourage localities to accept environmental and esthetic costs which the larger community finds acceptable is one of the great political challenges for democracies. Beyond this, a more basic value judgment embodying profound political and economic consequences remains to be made in determining whether there should be more economic growth and, if so, at what rate. *See the feature article* PROTECTING THE ENVIRONMENT.

Access. The terms of access to energy resources are also in the process of change and thus a cause for further cost, delay, and uncertainty. Petroleum exploration, with the demise of the concession system, is particularly constrained. How one encourages capital to undertake very great risks in the absence of antipitation of great rewards remains to be determined. In energy forms such as hydroelectric, the river often traverses several countries and presents enormously complicated problems in the allocation of project costs and benefits. In other instances, the remoteness of the resource — whether it be inland or underwater — presents great technical and financial impediments.

Equipment production. The availability of both highly specialized capital equipment and skills which entail long lead times to develop are constraints on the rate of energy exploitation. The production of coal drag lines, necessary for open-pit mining, and of rigs for offshore oil drilling is difficult to accelerate. Both are examples of real

short-to-intermediate-run constraints on development.

Capital. Capital availability is a problem, both in terms of source and magnitude. Oil companies once generated the massive capital for needed exploration and development from profits. Now a new source must be found to generate the estimated 1.5×10^{12} dollars of new investment in energy between 1975 and 1985. Experts are skeptical about the ability of capital markets to generate the required huge amounts of venture capital in the face of uncertain and declining profits. Even if this proves possible, the drain on capital available for other purposes is cause for genuine concern. Likewise, it is not clear that democratic governments will find themselves able to divert huge sums from short-term, politically sensitive ends to long-term energy investment whose political benefits are far in the future and difficult to capture in the present.

Conservation. Using less energy is a legitimate alternative to producing more. However, in the short run it has proved to be difficult when efforts have been made to conserve energy through administrative methods rather than market prices. This is in large part because the infrastructure of the developed world was designed to use given quantities of energy at a certain price. Using less can have expensive side effects. For example, slowing a truck, from 70 mph to 55 mph may save fuel, but it also implies less round trips per annum — a reduction in the output of both the capital (truck) and labor (driver) associated with the fuel.

Other conservation measures such as adjusting temperatures or otherwise exerting greater vigilance to reduce energy consumption have been less controversial because they are voluntary. Their impact is real, but difficult to measure. Sluggish response to conservation efforts in the short run can also be accounted for by the need for capital expenditures such as insulation, or the desire to avoid the expense of premature disposal of existing equipment such as heavy automobiles.

In the longer run, however, conservation is more promising. Building and vehicle design can materially affect energy consumption. Mass transit, city planning, and other basic reorientation of energy uses are likewise promising.

PRESENT AND FUTURE ENERGY RESOURCES

Because costs and the direction and rate of technological changes are difficult to predict, one can only guess what the future composition of the world's energy supply will be. Table 5 and Fig. 4 offer an informed view and indicate that throughout the foreseeable future fossil fuels will continue

Table 5. World energy consumption by source, 1960–1990, in 10¹⁵ BTU*

Energy source	1960	1965	1970	1972	1980	1985	1990
Coal	61.5	62.6	66.8	66.3	79.2	85.7	92.0
Petroleum	45.3	64.4	96.9	107.5	132.3	147.2	165.0
Natural gas	18.0	26.6	40.6	45.8	56.8	67.3	77.1
Hydropower and geothermal	6.9	9.6	11.8	12.9	15.4	17.1	18.8
Nuclear		.2	.8	1.4	12.6	33.6	63.6
TOTAL	131.7	163.4	216.9	233.9	296.3	350.9	416.5

SOURCE: Historical Data, United Nations, 1974.

*10^{15} Btu = 500,000 bbl petroleum per day for a year, or 40×10^6 tons of bituminous coal, or 10^{12} ft³ of natural gas, or 100×10^9 kWhr (based on a 10,000 Btu/kWh heat rate).

to supply the vast bulk of the world's energy needs. Toward the end of the century, nuclear power begins to accelerate its accretion of market share. The energy resource base is ample; the challenge is to exploit it rationally while developing renewable sources of energy (solid waste, wood, geothermal, hydroelectricity, nonconventional nuclear, solar, wind and tides, and hydrogen) to exploit as the others are depleted (oil and gas, coal, conventional nuclear, oil shale, and tar sands).

Oil and gas. At current high prices for oil it is possible to substantially intensify exploration in promising but remote and expensive formations such as deeper wells on land and those found offshore and in the Arctic. New technology applied to old wells should also increase recoveries. *See* NATURAL GAS; PETROLEUM.

Coal. Coal is abundant, especially in the Soviet Union and the United States. Aside from current conventional uses, it has potential for conversion into various synthetic fuels, "syncrude," in both liquid and gaseous form. Syncrude, however, must overcome cost problems and is not expected to be in significant production until 1990. *See* COAL; COAL LIQUEFACTION.

Conventional nuclear. Nuclear power is commercially feasible at current prices. Environmental concerns, the need to generate very large sums of capital, and security problems have all inhibited the expansion of nuclear power. Only in the 1970s did nuclear power in the United States pass firewood as a percentage of total energy consumed. Nevertheless, the desire of countries to diversify their fuel sources should spur nuclear power development, and nuclear power is expected to experience substantial growth throughout the remainder of the century. *See* NUCLEAR POWER.

Oil shale. Currently known oil shale deposits are largely in the United States where the majority of the experience in exploitation has been concentrated. The Soviet Union and Germany also have shale deposits. At current energy prices shale offers only a marginal contribution. Moreover, there are severe environmental constraints on shale development. *See* OIL SHALE.

Tar sands. The largest and most promising tar sands are in Venezuela. Estimated at 700×10^9 bbl of oil equivalent, about 10%, or 70×10^9 bbl are recoverable at current price and technology. The only significant tar sand development to date is in Canada, where the Canadian government has invested $1,000,000,000 in a 125,000-barrel-per-day project. Lesser tar sand deposits are found in the United States and elsewhere.

Solid waste. Current high prices and the disappearance of convenient land fill areas have rekindled interest in solid waste (garbage) as a fuel for electric power and thermal heating and cooling. Current experiments are promising. The high cost of capital conversion and some environmental problems are impediments to rapid expansion of solid-waste fuel utilization. The need to establish an entirely new nexus of municipal and other agencies having little prior coordination with one another also inhibits growth.

Wood. Wood is a major fuel for cooking in rural parts of the developing world. In the industrial world it offers promise as a boiler fuel and for conversion to other fuels such as methyl alcohol, which is a partial substitute for gasoline. Thus, despite its relatively low Btu-to-bulk ratio, difficulty in harvesting, and high alternative value, wood could significantly increase its contribution to the world's energy needs.

Geothermal. This source has worldwide potential and is being exploited in many countries, including Italy, New Zealand, Iceland, and Mexico.

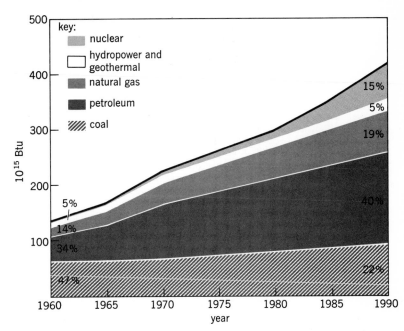

Fig. 4. World energy consumption by source, 1960–1990. (*From Historical Data, United Nations, 1974*)

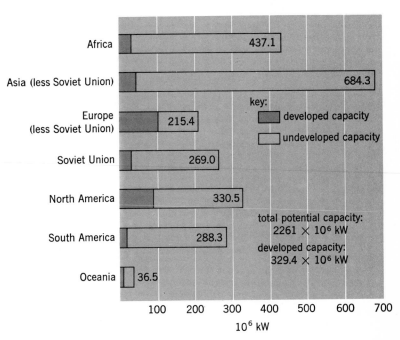

Fig. 5. World hydroelectric power, developed and undeveloped, 1973. (*From Federal Energy Administration, 1974*)

Despite high cost and certain environmental problems, research and development continue. Ocean "hot spots" or geopressured brines which occur in the oceans and various seas are currently being examined as potential energy sources for power generation. *See* GEOTHERMAL POWER.

Hydroelectricity. Only about 15% of the world's potential hydro capacity is currently being used. Hydroelectricity is environmentally clean, and its cost can be distributed among flood control, irrigation, and recreation, as well as energy. Offsetting this are high initial capital costs, very long lead times (see Fig. 5), disputes over water rights, and changes caused in the environment as a result of damming rivers. For these reasons, and because the best sites for hydro installations have already been exploited, hydro power will remain a small part of the total world energy supply for many years. *See* HYDROELECTRIC POWER.

Nonconventional nuclear. This includes the breeder reactor, which produces its own fuel; and fusion, which has no source limitations. Both are still experimental and may have very serious technical problems. Should these problems, however, be overcome, the long-run cost of such energy would be low. There is debate in the scientific community as to whether these virtually unlimited energy sources will be available by the year 2000. *See* REACTOR, NUCLEAR.

Solar. As with advanced nuclear sources, the sun offers a virtually limitless source of energy if it can be harnessed. Solar radiation has limited commercial application at present, but is currently used for heating and cooling of space and water as well as experimental production of electricity. Unless there is an unforeseen technological breakthrough, the contribution of solar power for the remainder of the century will be limited. *See* SOLAR ENERGY.

Wind and tides. Wind-driven generators are at work in some parts of the world but only on a small scale. As with solar energy, storage of electricity for use when the energy source is inoperative presents a very great problem. Tidal energy is currently in the experimental stage, but it is not expected to contribute greatly in the foreseeable future. *See* TIDAL POWER; WIND POWER.

Hydrogen. An excellent fuel, available in practically infinite quantity, hydrogen is easily transported by pipeline, is noncorrosive, is environmentally clean, and has a high Btu-to-bulk ratio. However, its productuion by hydrolysis of water requires a great deal of energy and is not commercially feasible at present. *See the feature article* EXPLORING ENERGY CHOICES.   [JOHN K. WILHELM]

Energy Technology

Absorption of electromagnetic radiation

The process whereby the intensity of a beam of electromagnetic radiation is attenuated in passing through a material medium by conversion of the energy of the radiation to an equivalent amount of energy which appears within the medium; the radiant energy is converted into heat or some other form of molecular energy. A perfectly transparent medium permits the passage of a beam of radiation without any change in intensity other than that caused by the spread or convergence of the beam, and the total radiant energy emergent from such a medium equals that which entered it, whereas the emergent energy from an absorbing medium is less than that which enters, and, in the case of highly opaque media, is reduced practically to zero.

No known medium is opaque to all wavelengths of the electromagnetic spectrum, which extends from radio waves, whose wavelengths are measured in kilometers, through the infrared, visible, and ultraviolet spectral regions, to x- and γ- rays, of wavelengths down to 10^{-11} cm. Similarly, no material medium is transparent to the whole electromagnetic spectrum. A medium which absorbs a relatively wide range of wavelengths is said to exhibit general absorption, while a medium which absorbs only restricted wavelength regions of no great range exhibits selective absorption for those particular spectral regions. For example, the substance pitch shows general absorption for the visible region of the spectrum, but is relatively transparent to infrared radiation of long wavelength. Ordinary window glass is transparent to visible light, but shows general absorption for ultraviolet radiation of wavelengths below about 3100 A, while colored glasses show selective absorption for specific regions of the visible spectrum. The color of objects which are not self-luminous and which are seen by light reflected or transmitted by the object is usually the result of selective absorption of portions of the visible spectrum. Many colorless substances, such as benzene and similar hydrocarbons, selectively absorb within the ultraviolet region of the spectrum, as well as in the infrared. *See* ELECTROMAGNETIC RADIATION.

Laws of absorption. The capacity of a medium to absorb radiation depends on a number of factors, mainly the electronic and nuclear constitution of the atoms and molecules of the medium, the wavelength of the radiation, the thickness of the absorbing layer, and the variables which determine the state of the medium, of which the most important are the temperature and the concentration of the absorbing agent. In special cases, absorption may be influenced by electric or magnetic fields. The state of polarization of the radiation influences the absorption of media containing certain oriented structures, such as crystals of other than cubic symmetry.

Lambert's law. Lambert's law, also called Bouguer's law or the Lambert-Bouguer law, expresses the effect of the thickness of the absorbing medium on the absorption. If a homogeneous medium is thought of as being constituted of layers of uniform thickness set normally to the beam, each layer absorbs the same fraction of the radiation incident on it. If I is the intensity to which a monochromatic parallel beam is attenuated after trav-

A - Z

Absorption of electromagnetic radiation

ersing a thickness d of the medium, and I_0 is the intensity of the beam at the surface of incidence (corrected for loss by reflection from this surface), the variation of intensity throughout the medium is expressed by Eq. (1), in which α is a constant for

$$I = I_0 e^{-\alpha d} \tag{1}$$

the medium called the absorption coefficient. This exponential relation can be expressed in an equivalent logarithmic form as in Eq. (2), where

$$\log_{10}(I_0/I) = (\alpha/2.303)\, d = kd \tag{2}$$

$k = \alpha/2.303$ is called the extinction coefficient for radiation of the wavelength considered. The quantity $\log_{10}(I_0/I)$ is often called the optical density, or the absorbance of the medium.

Equation (2) shows that as monochromatic radiation penetrates the medium, the logarithm of the intensity decreases in direct proportion to the thickness of the layer traversed. If experimental values for the intensity of the light emerging from layers of the medium of different thicknesses are available (corrected for reflection losses at all reflecting surfaces), the value of the extinction coefficient can be readily computed from the slope of the straight line representing the logarithms of the emergent intensities as functions of the thickness of the layer.

Equations (1) and (2) show that the absorption and extinction coefficients have the dimensions of reciprocal length. The extinction coefficient is equal to the reciprocal of the thickness of the absorbing layer required to reduce the intensity to one-tenth of its incident value. Similarly, the absorption coefficient is the reciprocal of the thickness required to reduce the intensity to $1/e$ of the incident value, where e is the base of the natural logarithms, 2.718.

Beer's law. This law refers to the effect of the concentration of the absorbing medium, that is, the mass of absorbing material per unit of volume, on the absorption. This relation is of prime importance in describing the absorption of solutions of an absorbing solute, since the solute's concentration may be varied over wide limits, or the absorption of gases, the concentration of which depends on the pressure. According to Beer's law, each individual molecule of the absorbing material absorbs the same fraction of the radiation incident upon it, no matter whether the molecules are closely packed in a concentrated solution or highly dispersed in a dilute solution. The relation between the intensity of a parallel monochromatic beam which emerges from a plane parallel layer of absorbing solution of constant thickness and the concentration of the solution is an exponential one, of the same form as the relation between intensity and thickness expressed by Lambert's law. The effects of thickness d and concentration c on absorption of monochromatic radiation can therefore be combined in a single mathematical expression, given in Eq. (3), in which k' is a constant for a

$$I = I_0 e^{-k'cd} \tag{3}$$

given absorbing substance (at constant wavelength and temperature), independent of the actual concentration of solute in the solution. In logarithms, the relation becomes Eq. (4). The values

of the constants k' and ϵ in Eqs. (3) and (4) depend on the units of concentration. If the concentration

$$\log_{10}(I_0/I) = (k'/2.303)\, cd = \epsilon cd \tag{4}$$

of the solute is expressed in moles per liter, the constant ϵ is called the molar extinction coefficient. Some authors employ the symbol a_M, called the molar absorbance index, instead of ϵ.

If Beer's law is adhered to, the molar extinction coefficient does not depend on the concentration of the absorbing solute, but usually changes with the wavelength of the radiation, with the temperature of the solution, and with the solvent.

The dimensions of the molar extinction coefficient are reciprocal concentration multiplied by reciprocal length, the usual units being liters/(mole) (cm). If Beer's law is true for a particular solution, the plot of $\log(I_0/I)$ against the concentrations for solutions of different concentrations, measured in cells of constant thickness, will yield a straight line, the slope of which is equal to the molar extinction coefficient.

While no true exceptions to Lambert's law are known, exceptions to Beer's law are not uncommon. Such exceptions arise whenever the molecular state of the absorbing solute depends on the concentration. For example, in solutions of weak electrolytes, whose ions and undissociated molecules absorb radiation differently, the changing ratio between ions and undissociated molecules brought about by changes in the total concentration prevents solutions of the electrolyte from obeying Beer's law. Aqueous solutions of dyes frequently deviate from the law because of dimerization and more complicated aggregate formation as the concentration of dye is increased.

Absorption measurement. The measurement of the absorption of homogeneous media is usually accomplished by absolute or comparative measurements of the intensities of the incident and transmitted beams, with corrections for any loss of radiant energy caused by processes other than absorption. The most important of these losses is by reflection at the various surfaces of the absorbing layer and of vessels which may contain the medium, if the medium is liquid or gaseous. Such losses are usually automatically compensated for by the method of measurement employed. Losses by reflection not thus compensated for may be computed from Fresnel's laws of reflection.

Scattering. Absorption of electromagnetic radiation should be distinguished from the phenomenon of scattering, which occurs during the passage of radiation through inhomogeneous media. Radiant energy which traverses media constituted of small regions of refractive index different from that of the rest of the medium is diverted laterally from the direction of the incident beam. The diverted radiation gives rise to the hazy or opalescent appearance characteristic of such media, exemplified by smoke, mist, and opal. If the centers of inhomogeneity are sufficiently dilute, the intensity of a parallel beam is diminished in its passage through the medium because of the sidewise scattering, according to a law of the same form as the Lambert-Bouguer law for absorption, given in Eq. (5), where I is the intensity of the primary beam of initial intensity I_0, after it has traversed a distance d through the scattering medium.

$$I = I_0 e^{-\tau d} \tag{5}$$

The coefficient τ, called the turbidity of the medium, plays the same part in weakening the primary beam by scattering as does the absorption coefficient in true absorption. However, in true scattering, no loss of total radiant energy takes place, energy lost in the direction of the primary beam appearing in the radiation scattered in other directions. In some inhomogeneous media, both absorption and scattering occur together.

Physical nature. Absorption of radiation by matter always involves the loss of energy by the radiation and a corresponding gain in energy by the atoms or molecules of the medium.

The energy of an assembly of gaseous atoms consists partly of kinetic energy of the translational motion which determines the temperature of the gas (thermal energy), and partly of internal energy, associated with the binding of the extranuclear electrons to the nucleus, and with the binding of the particles within the nucleus itself. Molecules, composed of more than one atom, have, in addition, energy associated with periodic rotations of the molecule as a whole and with oscillations of the atoms within the molecule with respect to one another.

The energy absorbed from radiation appears as increased internal energy, or in increased vibrational and rotational energy of the atoms and molecules of the absorbing medium. As a general rule, translational energy is not directly increased by absorption of radiation, although it may be indirectly increased by degradation of electronic energy or by conversion of rotational or vibrational energy to that of translation by intermolecular collisions.

Quantum theory. In order to construct an adequate theoretical description of the energy relations between matter and radiation, it has been necessary to amplify the wave theory of radiation by the quantum theory, according to which the energy in radiation occurs in natural units called quanta. The value of the energy in these units, expressed in ergs or calories, for example, is the same for all radiation of the same wavelength, but differs for radiation of different wavelengths. The energy E in a quantum of radiation of frequency ν (where the frequency is equal to the velocity of the radiation in a given medium divided by its wavelength in the same medium) is directly proportional to the frequency, or inversely proportional to the wavelength, according to the relation given in Eq. (6), where h is a universal constant known as

$$E = h\nu \tag{6}$$

Planck's constant. The value of h is 6.62×10^{-27} erg-sec, and if ν is expressed in $\sec^{-1}$, E is given in ergs per quantum.

The most energetic type of change that can occur in an atom involves the nucleus, and increase of nuclear energy by absorption therefore requires quanta of very high energy, that is, of high frequency or low wavelength. Such rays are the γ-rays, whose wavelength varies downward from 10^{-9} cm. Next in energy are the electrons nearest to the nucleus and therefore the most tightly bound. These electrons can be excited to states of higher energy by absorption of x-rays, whose range in wavelength is from about 10^{-7} to 10^{-9} cm. Less energy is required to excite the more loosely bound valence electrons. Such excitation can be accomplished by the absorption of quanta of visible radiation (wavelength 7×10^{-5} cm for red light to 4×10^{-5} cm for blue) or of ultraviolet radiation, of wavelength down to about 10^{-5} cm. Absorption of ultraviolet radiation of shorter wavelengths, down to those on the border of the x-ray region, excites electrons bound to the nucleus with intermediate strength.

The absorption of relatively low-energy quanta of wavelength from about 10^{-3} to 10^{-4} cm suffices to excite vibrating atoms in molecules to higher vibrational states, while changes in rotational energy, which are of still smaller magnitude, may be excited by absorption of radiation of still longer wavelength, from the microwave radio region of about 1 cm to far-infrared radiation, some hundredths of a centimeter in wavelength.

Gases. The absorption of gases composed of atoms is usually very selective. For example, monatomic sodium vapor absorbs very strongly over two narrow wavelength regions in the yellow part of the visible spectrum (the so-called D lines), and no further absorption by monatomic sodium vapor occurs until similar narrow lines appear in the near-ultraviolet. The valence electron of the sodium atom can exist only in one of a series of energy states separated by relatively large energy intervals between the permitted values, and the sharp-line absorption spectrum results from transitions of the valence electron from the lowest energy which it may possess in the atom to various excited levels. Line absorption spectra are characteristic of monatomic gases in general.

The visible and ultraviolet absorption of vapors composed of diatomic or polyatomic molecules is much more complicated than that of atoms. As for atoms, the absorbed energy is utilized mainly in raising one of the more loosely bound electrons to a state of higher energy, but the electronic excitation of a molecule is almost always accompanied by simultaneous excitation of many modes of vibration of the atoms within the molecule and of rotation of the molecule as a whole. As a result, the absorption, which for an atom is concentrated in a very sharp absorption line, becomes spread over a considerable spectral region, often in the form of bands. Each band corresponds to excitation of a specific mode of vibration accompanying the electronic change, and each band may be composed of a number of very fine lines close together in wavelength, each of which corresponds to a specific rotational change of the molecule accompanying the electronic and vibrational changes. Band spectra are as characteristic of the absorption of molecules in the gaseous state, and frequently in the liquid state, as line spectra are of gaseous atoms.

Liquids. Liquids usually absorb radiation in the same general spectral region as the corresponding vapors. For example, liquid water, like water vapor, absorbs infrared radiation strongly (vibrational transitions), is largely transparent to visible and near-ultraviolet radiation, and begins to absorb strongly in the far-ultraviolet. A universal difference between liquids and gases is the disturbance in the energy states of the molecules in a liquid caused by the great number of intermolecular collisions; this has the effect of broadening the very fine lines observed in the absorption spectra of vapors, so that sharp-line structure disappears in the absorption bands of liquids.

Solids. Substances which can exist in solid, liquid, and vapor states without undergoing a temperature rise to very high values usually absorb in the same general spectral regions for all three states of aggregation, with differences in detail because of the intermolecular forces present in the liquid and solid. Crystalline solids, such as rock salt or silver chloride, absorb infrared radiation of long wavelength, which excites vibrations of the electrically charged ions of which these salts are composed; such solids are transparent to infrared radiations of shorter wavelengths. In colorless solids, the valence electrons are too tightly bound to the nuclei to be excited by visible radiation, but all solids absorb in the near- or far-ultraviolet region.

The use of solids as components of optical instruments is restricted by the spectral regions to which they are transparent. Crown glass, while showing excellent transparency for visible light and for ultraviolet radiation immediately adjoining the visible region, becomes opaque to radiation of wavelength about 3000 A and shorter, and is also opaque to infrared radiation longer than about 20,000 A in wavelength. Quartz is transparent down to wavelengths about 1800 A in the ultraviolet, and to about 40,000 A in the infrared. The most generally useful material for prisms and windows for the near-infrared region is rock salt, which is highly transparent out to about 150,000 A (15 μ).

Fluorescence. The energy acquired by matter by absorption of visible or ultraviolet radiation, although primarily used to excite electrons to higher energy states, usually ultimately appears as increased kinetic energy of the molecules, that is, as heat. It may, however, under special circumstances, be reemitted as electromagnetic radiation. Fluorescence is the reemission, as radiant energy, of absorbed radiant energy, normally at wavelengths the same as or longer than those absorbed. The reemission, as ordinarily observed, ceases immediately when the exciting radiation is shut off. Refined measurements show that the fluorescent reemission persists, in different cases, for periods of the order of 10^{-9} to 10^{-6} sec. The simplest case of fluorescence is the resonance fluorescence of monatomic gases at low pressure, such as sodium or mercury vapors, in which the reemitted radiation is of the same wavelength as that absorbed. In this case, fluorescence is the converse of absorption: Absorption involves the excitation of an electron from its lowest energy state to a higher energy state by radiation, while fluorescence is produced by the return of the excited electron to the lower state, with the emission of the energy difference between the two states as radiation. The fluorescent radiation of molecular gases and of nearly all liquids, solids, and solutions contains a large component of wavelengths longer than those of the absorbed radiation, a relationship known as Stokes' law of fluorescence. In these cases, not all of the absorbed energy is reradiated, a portion remaining as heat in the absorbing material. The fluorescence of iodine vapor is easily seen on projecting an intense beam of visible light through an evacuated bulb containing a few crystals of iodine, but the most familiar examples are provided by certain organic compounds in solution, such as quinine sulfate, which absorbs ultraviolet radiation and reemits blue, or fluorescein, which absorbs blue-green light and fluoresces with an intense, bright-green color.

Phosphorescence. The radiant reemission of absorbed radiant energy at wavelengths longer than those absorbed, for a readily observable interval after withdrawal of the exciting radiation, is called phosphorescence. The interval of persistence, determined by means of a phosphoroscope, usually varies from about 0.001 sec to several seconds, but some phosphors may be induced to phosphorescence by heat days or months after the exciting absorption. An important and useful class of phosphors is the impurity phosphors, solids such as the sulfides of zinc or calcium which are activated to the phosphorescent state by incorporating minute amounts of foreign material (called activators), such as salts of manganese or silver. So-called fluorescent lamps contain a coating of impurity phosphor on their inner wall which, after absorbing ultraviolet radiation produced by passage of an electrical discharge through mercury vapor in the lamp, reemits visible light. The receiving screen of a television tube contains a similar coating, excited not by radiant energy but by the impact of a stream of electrons on the surface.

Luminescence. Phosphorescence and fluorescence are special cases of luminescence, which is defined as light emission that cannot be attributed merely to the temperature of the emitting body. Luminescence may be excited by heat (thermoluminescence), by electricity (electroluminescence), by chemical reaction (chemiluminescence), or by friction (triboluminescence), as well as by radiation.

Absorption and emission coefficients. The absorption and emission processes of atoms were examined from the quantum point of view by Albert Einstein in 1916, with some important results that have been realized practically in the invention of the maser and the laser. Consider an assembly of atoms undergoing absorption transitions of frequency ν sec^{-1} from the ground state 1 to an excited state 2 and emission transitions in the reverse direction, the atoms and radiation being at equilibrium at temperature T. The equilibrium between the excited and unexcited atoms is determined by the Boltzmann relation $N_2/N_1 = \exp(-h\nu/kT)$, where N_1 and N_2 are the equilibrium numbers of atoms in states 1 and 2, respectively, and the radiational equilibrium is determined by equality in the rate of absorption and emission of quanta. The number of quanta absorbed per second is $B_{12}N_1\rho(\nu)$, where $\rho(\nu)$ is the density of radiation of frequency ν (proportional to the intensity), and B_{12} is a proportionality constant called the Einstein coefficient for absorption. Atoms in state 2 will emit radiation spontaneously (fluorescence), after a certain mean life, at a rate of $A_{21}N_2$ per second, where A_{21} is the Einstein coefficient for spontaneous emission from state 2 to state 1. To achieve consistency between the density of radiation of frequency ν at equilibrium calculated from these considerations and the value calculated from Planck's radiation law, which is experimentally true, it is necessary to introduce, in addition to the spontaneous emission, an emission of intensity proportional to the radiation density of frequency ν in which the atoms are immersed. The radiational equilibrium is then determined by Eq. (7), where B_{21} is the Einstein

$$B_{12}N_1\rho\,(\nu) = A_{21}N_2 + B_{21}N_2\rho\,(\nu) \qquad (7)$$

coefficient of stimulated emission. The Einstein radiation coefficients are found to be related by Eqs. (8a) and (8b).

$$B_{12} = B_{21} \qquad (8a)$$
$$A_{21} = (8\pi h\nu^3/c^3) \cdot B_{21} \qquad (8b)$$

In the past when one considered radiation intensities available from terrestrial sources, stimulated emission was very feeble compared with the spontaneous process. Stimulated emission is, however, the fundamental emission process in the laser, a device in which a high concentration of excited molecules is produced by intense illumination from a "pumping" source, in an optical system in which excitation and emission are augmented by back-and-forth reflection until stimulated emission swamps the spontaneous process.

There are also important relations between the absorption characteristics of atoms and their mean lifetime τ in the excited state. Since A_{21} is the number of times per second that a given atom will emit a quantum spontaneously, the mean lifetime before emission in the excited state is $\tau = 1/A_{21}$. It can also be shown that A_{21} and τ are related, as shown in Eq. (9), to the f number or oscillator strength for the transition that occurs in the dispersion equations shown as Eqs. (13) to (17). The

$$A_{21} = 1/\tau = \frac{(8\pi^2\nu^2 e^2)}{mc^3} \cdot f$$
$$= 7.42 \times 10^{-22}\, f\nu^2 \qquad (\nu \text{ in sec}^{-1}) \qquad (9)$$

value of f can be calculated from the absorption integrated over the band according to Eq. (18).

Dispersion. A transparent material does not abstract energy from radiation which it transmits, but it always decreases the velocity of propagation of such radiation. In a vacuum, the velocity of radiation is the same for all wavelengths, but in a material medium, the velocity of propagation varies considerably with wavelength. The refractive index μ of a medium is the ratio of the velocity of light in vacuum to that in the medium, and the effect of the medium on the velocity of radiation which it transmits is expressed by the variation of refractive index with the wavelength λ of the radiation, $d\mu/d\lambda$. This variation is called the dispersion of the medium. For radiation of wavelengths far removed from those of absorption bands of the medium, the refractive index increases regularly with decreasing wavelength or increasing frequency; the dispersion is then said to be normal.

In regions of normal dispersion, the variation of refractive index with wavelength can be expressed with considerable accuracy by Eq. (10), known as

$$\mu = A + \frac{B}{\lambda^2} + \frac{C}{\lambda^4} \qquad (10)$$

Cauchy's equation, in which A, B, and C are con-

stants with positive values. As an approximation, C may be neglected in comparison with A and B, and the dispersion, $d\mu/d\lambda$, is then given by Eq. 11.

$$\frac{d\mu}{d\lambda} = \frac{-2B}{\lambda^3} \qquad (11)$$

Thus, in regions of normal dispersion, the dispersion is approximately inversely proportional to the cube of the wavelength.

Dispersion by a prism. The refraction, or bending, of a ray of light which enters a material medium obliquely from vacuum or air (the refractive index of which for visible light is nearly unity) is the result of the diminished rate of advance of the wavefronts in the medium. Since, if the dispersion is normal, the refractive index of the medium is greater for violet than for red light, the wavefront of the violet light is retarded more than that of the red light. Hence, white light entering obliquely into the medium is converted within the medium to a continuously colored band, of which the red is least deviated from the direction of the incident beam, the violet most, with orange, yellow, green, and blue occupying intermediate positions. On emergence of the beam into air again, the colors remain separated. The action of the prism in resolving white light into its constituent colors is called color dispersion.

The angular dispersion of a prism is the ratio, $d\theta/d\lambda$, of the difference in angular deviation $d\theta$ of two rays of slightly different wavelength which pass through the prism to the difference in wavelength $d\lambda$ when the prism is set for minimum deviation.

The angular dispersion of the prism given in Eq. (12) is the product of two factors, the variation, $d\theta/d\mu$, of the deviation θ with refractive index μ, and the variation of refractive index with wavelength, the dispersion of the material of which the

$$\frac{d\theta}{d\lambda} = \frac{d\theta}{d\mu} \cdot \frac{d\mu}{d\lambda} \qquad (12)$$

prism is made. The latter depends solely on this material, while $d\theta/d\mu$ depends on the angle of incidence and the refracting angle of the prism. The greater the dispersion of the material of the prism, the greater is the angular separation between rays of two given wavelengths as they leave the prism. For example, the dispersion of quartz for visible light is lower than that of glass; hence the length of the spectrum from red to violet formed by a quartz prism is less than that formed by a glass prism of equal size and shape. Also, since the dispersion of colorless materials such as glass or quartz is greater for blue and violet light than for red, the red end of the spectrum formed by prisms is much more contracted than the blue.

The colors of the rainbow result from dispersion of sunlight which enters raindrops and is refracted and dispersed in passing through them to the rear surface, at which the dispersed rays are reflected and reenter the air on the side of the drop on which the light was incident.

Anomalous dispersion. The regular increase of refractive index with decreasing wavelength expressed by Cauchy's equation breaks down as the wavelengths approach those of strong absorption bands. As the absorption band is approached from the long-wavelength side, the refractive index be-

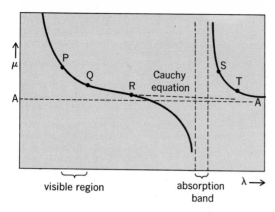

Fig. 1. Curve showing anomalous dispersion of quartz. A is limiting value of μ as λ approaches infinity. *(From F. A. Jenkins and H. E. White, Fundamentals of Optics, 3d ed., McGraw-Hill, 1957)*

comes very large, then decreases within the band to assume abnormally small values on the short-wavelength side, values below those for radiation on the long-wavelength side. A hollow prism containing an alcoholic solution of the dye fuchsin, which absorbs green light strongly, forms a spectrum in which the violet rays are less deviated than the red, on account of the abnormally low refractive index of the medium for violet light. The dispersion of media for radiation of wavelengths near those of strong absorption bands is said to be anomalous, in the sense that the refractive index decreases with decreasing wavelength instead of showing the normal increase. The theory of dispersion, for which reference must be made to treatises such as those cited in the bibliography, shows, however, that both the normal and anomalous variation of refractive index with wavelength can be satisfactorily described as aspects of a unified phenomenon, so that there is nothing fundamentally anomalous about dispersion in the vicinity of an absorption band.

Normal and anomalous dispersion of quartz are illustrated in Fig. 1. Throughout the near-infrared, visible, and near-ultraviolet spectral regions (between P and R on the curve), the dispersion is normal and adheres closely to Cauchy's equation, but it becomes anomalous to the right of R. From S to T, Cauchy's equation is again valid.

Relation to absorption. Figure 1 shows there is an intimate connection between dispersion and absorption; the refractive index rises to high values as the absorption band is approached from the long-wavelength side and falls to low values on the short-wavelength side of the band. In fact, the theory of dispersion shows that the complete dispersion curve as a function of wavelength is governed by the absorption bands of the medium. In classical electromagnetic theory, electric charges are regarded as oscillating, each with its appropriate natural frequency ν_0, about positions of equilibrium within atoms or molecules. Placed in a radiation field of frequency ν per second, the oscillator in the atom is set into forced vibration, with the same frequency as that of the radiation. When ν is much lower or higher than ν_0, the amplitude of the forced vibration is small, but the amplitude becomes large when the frequency of the

radiation equals the natural frequency of the oscillator. In much the same way, a tuning fork is set into vibration by sound waves corresponding to the same note emitted by another fork vibrating at the same frequency. To account for the absorption of energy by the medium from the radiation, it is necessary to postulate that in the motion of the atomic oscillator some frictional force, proportional to the velocity of the oscillator, must be overcome. For small amplitudes of forced oscillation, when the frequency of the radiation is very different from the natural period of the oscillator, the frictional force and the absorption of energy are negligible. Near resonance between the radiation and the oscillator, the amplitude becomes large, with a correspondingly large absorption of energy to overcome the frictional resistance. Radiation of frequencies near the natural frequency therefore corresponds to an absorption band.

To show that the velocity of the radiation within the medium is changed, it is necessary to consider the phase of the forced vibration, which the theory shows to depend on the frequency of the radiation. The oscillator itself becomes a source of secondary radiation waves within the medium which combine to form sets of waves moving parallel to the original waves. Interference between the secondary and primary waves takes place, and because the phase of the secondary waves, which is the same as that of the atomic oscillators, is not the same as that of the primary waves, the wave motion resulting from the interference between the two sets of waves is different in phase from that of the primary waves incident on the medium. But the velocity of propagation of the waves is the rate of advance of equal phase; hence the phase change effected by the medium, which is different for each frequency of radiation, is equivalent to a change in the velocity of the radiation within the medium. When the frequency of the radiation slightly exceeds the natural frequency of the oscillator, the radiation and the oscillator become 180° out of phase, which corresponds to an increase in the velocity of the radiation and accounts for the observed fall in refractive index on the short-wavelength side of the absorption band.

The theory leads to Eqs. (13) through (17) for the refractive index of a material medium as a function of the frequency of the radiation. In the equations the frequency is expressed as angular frequency, $\omega = 2\pi\nu \; \sec^{-1} = 2\pi c/\lambda$, where c is the velocity of light. When the angular frequency ω of the radiation is not very near the characteristic frequency of the electronic oscillator, the refractive index of a homogeneous medium containing N molecules per cubic centimeter is given by Eq. (13a), where e and

$$\mu^2 = 1 + \frac{4\pi Ne^2}{m} \cdot \frac{f}{\omega_0{}^2 - \omega^2} \qquad (13a)$$

$$\mu^2 = 1 + 4\pi Ne^2 \sum_i \frac{f_i/m_i}{\omega_i{}^2 - \omega^2} \qquad (13b)$$

m are the charge and mass of the electron, and f is the number of oscillators per molecule of characteristic frequency ω_0. The f value is sometimes called the oscillator strength. If the molecule contains oscillators of different frequencies and mass (for example, electronic oscillators of frequency corresponding to ultraviolet radiation and ionic oscillators corresponding to infrared radiation), the frequency term becomes a summation, as in Eq. (13b), where ω_i is the characteristic frequency of the ith type of oscillator, and f_i and m_i are the corresponding f value and mass. In terms of wavelengths, this relation can be written as Eq. (14),

$$\mu^2 = 1 + \sum_i \frac{A_i \lambda^2}{\lambda^2 - \lambda_i^2} \qquad (14)$$

where A_i is a constant for the medium, λ is the wavelength of the radiation, and $\lambda_i = c/\nu_i$ is the wavelength corresponding to the characteristic frequency ν_i per second (Sellmeier's equation).

If the medium is a gas, for which the refractive index is only slightly greater then unity, the dispersion formula can be written as Eq. (15).

$$\mu = 1 + 2\pi N e^2 \sum_i \frac{f_i/m_i}{\omega_i^2 - \omega^2} \qquad (15)$$

So long as the absorption remains negligible, these equations correctly describe the increase in refractive index as the frequency of the radiation begins to approach the absorption band determined by ω_i or λ_i. They fail when absorption becomes appreciable, since they predict infinitely large values of the refractive index when ω equals ω_i, whereas the refractive index remains finite throughout an absorption band.

The absorption of radiant energy of frequency very close to the characteristic frequency of the medium is formally regarded as the overcoming of a frictional force when the molecular oscillators are set into vibration, related by a proportionality constant g to the velocity of the oscillating particle; g is a damping coefficient for the oscillation. If the refractive index is determined by a single electronic oscillator, the dispersion equation for a gas at radiational frequencies within the absorption band becomes Eq. (16). At the same time an ab-

$$\mu = 1 + \frac{2\pi N e^2}{m} \frac{f(\omega_0^2 - \omega^2)}{(\omega_0^2 - \omega^2)^2 + \omega^2 g^2} \qquad (16)$$

sorption constant κ enters the equations, related to the absorption coefficient α of Eq. (1) by the expression $\kappa = \alpha c/2\omega\mu$. Equation (17) shows the

$$\kappa = \frac{2\pi N e^2}{m} \frac{f\omega g}{(\omega_0^2 - \omega^2)^2 + \omega^2 g^2} \qquad (17)$$

relationship. For a monatomic vapor at low pressure, Nf is about 10^{17} per cubic centimeter, ω_0 is about 3×10^{15} per second, and g is about 10^{11} per second. These data show that, when the frequency of the radiation is not very near ω_0, ωg is very small in comparison with the denominator and the absorption is practically zero. As ω approaches ω_0, κ increases rapidly to a maximum at a radiational frequency very near ω_0 and then falls at frequencies greater than ω_0. When the absorption is relatively weak, the absorption maximum is directly proportional to the oscillator strength f. In terms of the molar extinction coefficient ϵ of Eq. (4), it can be shown that this direct relation holds, as seen in Eq. 18. The integration in Eq. (18) is carried out

$$f = 4.319 \times 10^{-9} \int \epsilon d\bar{\nu} \qquad (18)$$

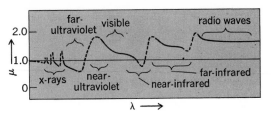

Fig. 2. Complete dispersion curve through the electromagnetic spectrum for a substance. *(From F. A. Jenkins and H. E. White, Fundamentals of Optics, 3d ed., McGraw-Hill, 1957)*

over the whole absorption spectrum. The integral can be evaluated from the area under the curve of ϵ plotted as a function of wave number $\bar{\nu}$ cm^{-1} $= \nu(\sec^{-1})/c = 1/\lambda$.

The width of the absorption band for an atom is determined by the value of the damping coefficient g; the greater the damping, the greater is the spectral region over which absorption extends.

The general behavior of the refractive index through the absorption band is illustrated by the dotted portions of Fig. 2. The presence of the damping term $\omega^2 g^2$ in the denominator of Eq. (17) prevents the refractive index from becoming infinite when $\omega = \omega_0$. Its value increases to a maximum for a radiation frequency less than ω_0, then falls with increasing frequency in the center of the band (anomalous dispersion) and increases from a relatively low value on the high-frequency side of the band.

Figure 2 shows schematically how the dispersion curve is determined by the absorption bands throughout the whole electromagnetic spectrum. The dotted portions of the curve correspond to absorption bands, each associated with a distinct type of electrical oscillator. The oscillators excited by x-rays are tightly bound inner electrons; those excited by ultraviolet radiation are more loosely bound outer electrons which control the dispersion in the near-ultraviolet and visible regions, whereas those excited by the longer wavelengths are atoms or groups of atoms.

It will be observed in Fig. 2 that in regions of anomalous dispersion the refractive index of a substance may assume a value less than unity; the velocity of light in the medium is then greater than in vacuum. The velocity involved here is that with which the phase of the electromagnetic wave of a single frequency ω advances, for example, the velocity with which the crest of the wave advances through the medium. The theory of wave motion, however, shows that a signal propagated by electromagnetic radiation is carried by a group of waves of slightly different frequency, moving with a group velocity which, in a material medium, is always less than the velocity of light in vacuum. The existence of a refractive index less than unity in a material medium is therefore not in contradiction with the theory of relativity.

In quantum theory, absorption is associated not with the steady oscillation of a charge in an orbit but with transitions from one quantized state to another. The treatment of dispersion according to quantum theory is essentially similar to that outlined, with the difference that the natural frequencies ν_0 are now identified with the frequencies of radiation which the atom can absorb in undergoing quantum transitions. These transition frequencies are regarded as virtual classical oscillators, which react to radiation precisely as do the oscillators of classical electromagnetic theory.

Selective reflection. Nonmetallic substances which show very strong selective absorption also strongly reflect radiation of wavelengths near the absorption bands, although the maximum of reflection is not, in general, at the same wavelength as the maximum absorption. The infrared rays selectively reflected by ionic crystals are frequently referred to as *reststrahlen*, or residual rays.

[WILLIAM WEST]

Bibliography: M. Born and E. Wolf, *Principles of Optics*, 3d ed., 1965; R. W. Ditchburn, *Light*, 2d ed., 1963; F. A. Jenkins and H. E. White, *Principles of Optics*, 3d ed., 1957; B. Rossi, *Optics*, 1957; A. Sommerfeld, *Lectures on Theoretical Physics*, vol. 4, 1954; J. A. Stratton, *Electromagnetic Theory*, 1941; J. Strong, *Concepts of Classical Optics*, 1958; R. W. Wood, *Physical Optics*, 3d ed., 1934.

Air conditioning

The maintenance of certain aspects of the environment within a defined space to facilitate the intended function of that space. Environmental conditions generally encompassed by the term air conditioning include air temperature and motion, radiant heat energy level, moisture level, and concentration of various pollutants, including dust, germs, and gases. Because these environmental factors are associated with air itself, and because air temperature and motion are the factors most readily sensed, simultaneous control of all these factors is called air conditioning, although space conditioning is more descriptive of the activity.

Comfort air conditioning refers to control of spaces inhabited by people to promote their comfort, health, or productivity. Spaces in which air is conditioned for comfort include residences, offices, institutions, sports arenas, hotels, and factory work areas. Process air conditioning systems are designed to facilitate the functioning of a production, manufacturing, or operational activity. For example, heat-producing electronic equipment in an airplane cockpit must be kept cool to function properly, while the occupants of the cockpit are maintained at comfortable conditions. The environment around a multicolor printing press must have constant relative humidity to avoid paper expansion or shrinkage for accurate registration, while press heat and ink mists must be conducted away for the health of pressmen. Maintenance of conditions within surgical suites of hospitals and in "clean" or "white" rooms of manufacturing plants, where an almost germ- or dust-free atmosphere must be maintained, has become a specialized subdivision of process air conditioning.

Calculation of loads. Engineering of an air-conditioning system starts with selection of design conditions; air temperature and relative humidity are principal factors. Next, loads on the system are calculated. Finally, equipment is selected and sized to perform the indicated functions and to carry the estimated loads.

Design conditions are selected on the bases discussed above. Each space is analyzed separately. A cooling load will exist when the sum of heat released within the space and transmitted to the space is greater than the loss of heat from the space. A heating load occurs when the heat generated within the space is less than the loss of heat from it. Similar considerations apply to moisture.

Heat generated within the space consists of body heat, approximately 250 Btu/hr/person, heat from all electrical appliances and lights, 3.41 Btu/hr/watt, and heat from other sources such as gas cooking stoves and industrial ovens. Heat is transmitted through all parts of the space envelope, which includes walls, floor, ceiling, and windows. Whether heat enters or leaves the space depends upon whether the outside surfaces are warmer or cooler than the inside surfaces. The rate at which heat is conducted through the space envelope is a function of the temperature difference across the envelope and the thermal conductance of the envelope. Conductances, which depend on materials of construction and their thicknesses along the path of heat transmission, are a large factor in walls and ceilings exposed to the outdoors in cold winters and hot summers. In these cases insulation is added to decrease the overall conductance of the envelope.

Solar heat loads are an especially important part of load calculation because they represent a large percentage of heat gain through walls and roofs, but are very difficult to estimate because solar irradiation is constantly changing. Intensity of radiation varies with the seasons (it rises to 457 Btu/hr/ft² in midwinter and drops to 428 in midsummer). Intensity of solar irradiation also varies with surface orientation. For example, the half-day total for a horizontal surface at 40 degrees north latitude on January 21 is 353 Btu/hr/ft² and on June 21 it is 1121 Btu, whereas for a south wall on the same dates comparable data are 815 and 311 Btu, a sharp decrease in summer. Intensity also varies with time of day and cloud cover and other atmospheric phenomena.

The way in which solar radiation affects the space load depends also upon whether the rays are transmitted instantly through glass or impinge on opaque walls. If through glass, the effect begins immediately but does not reach maximum intensity until the interior irradiated surfaces have warmed sufficiently to reradiate into the space, warming the air. In the case of irradiated walls and roofs, the effect is as if the outside air temperature were higher than it is. This apparent temperature is called the sol-air temperature, of which tables are available.

In calculating all these heating effects, the object is proper sizing and intelligent selection of equipment; hence, a design value is sought which will accommodate maximums. However, when dealing with climatic data, which are statistical, historical summaries, record maximums are rarely used. For instance, if in a particular locality the recorded maximum outside temperature was 100°, but 95°F was exceeded only four times in the past 20 years, 95°F may be chosen as the design summer outdoor temperature for calculation of heat transfer through walls. In practice, engineers use tables of design winter and summer outdoor temperatures which list winter temperatures exceeded more than 99% and 97.5% of the time during the coldest winter months, and summer temperatures not exceeded 1%, 2.5%, and 5% of the warmest months. The designer will select that value which represents the conservatism required for the particular type of occupancy. If the space contains vital functions where impairment by virtue of occasional departures from design space conditions cannot be tolerated, the more severe design outdoor conditions will be selected.

In the case of solar load through glass, but even more so in the case of heat transfer through walls and roof, because outside climate conditions are so variable, there may be a considerable thermal lag. It may take hours before the effect of extreme high or low temperatures on the outside of a thick masonry wall is felt on the interior surfaces and space. In some cases the effect is never felt on the inside, but in all cases the lag exists, exerting a leveling effect on the peaks and valleys of heating and cooling demand; hence, it tends to reduce maximums and can be taken advantage of in reducing design loads.

Humidity as a load on an air conditioning system is treated by the engineer in terms of its latent heat, that is, the heat required to condense or evaporate the moisture, approximately 1000 Btu/lb of moisture. People at rest or at light work generate about 200 Btu/hr. Steaming from kitchen activities and moisture generated as a product of combustion of gas flames, or from all drying processes, must be calculated. As with heat, moisture travels through the space envelope, and its rate of transfer is calculated as a function of the difference in vapor pressure across the space envelope and the permeability of the envelope construction. To decrease permeability where vapor pressure differential is large, vapor barriers (relatively impermeable membranes) are incorporated in the envelope construction.

Another load-reducing factor to be calculated is the diversity among the various spaces within a building or building complex served by a single system. Spaces with east-facing walls experience maximum solar loads when west-facing walls have no solar load. In cold weather, rooms facing south may experience a net heat gain due to a preponderant solar load while north-facing rooms require heat. An interior space, separated from adjoining spaces by partitions, floor, and ceiling across which there is no temperature gradient, experiences only a net heat gain, typically from people and lights. Given a system that can transfer this heat to other spaces requiring heat, the net heating load may be zero, even on cold winter days.

Air conditioning systems. A complete air conditioning system is capable of adding and removing heat and moisture and of filtering dust and odorants from the space or spaces it serves. Systems that heat, humidify, and filter only, for control of comfort in winter, are called winter air conditioning systems; those that cool, dehumidify, and filter only are called summer air conditioning systems, provided they are fitted with proper controls to maintain design levels of temperature, relative humidity, and air purity.

Design conditions may be maintained by multiple independent subsystems tied together by a single control system. Such arrangements, called

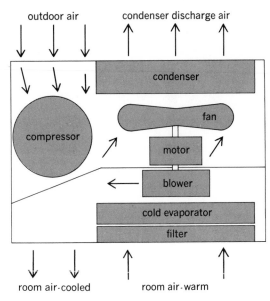

Fig. 1. Schematic of room air conditioner. (*American Society of Heating, Refrigerating and Air-Conditioning Engineers, Inc., Guide and Data Book, 1967*)

split systems, might consist, for example, of hot-water baseboard heating convectors around a perimeter wall to offset window and wall heat losses when required, plus a central cold-air distribution system to pick up heat and moisture gains as required and to provide filtration for dust and odor.

Air conditioning systems are either unitary or built-up. The window or through-the-wall air conditioner (Fig. 1) is an example of a unitary summer air conditioning system; the entire system is housed in a single package which contains heat removal, dehumidification, and filtration capabilities. When an electric heater is built into it with suitable controls, it functions as a year-round air conditioning system. Unitary air conditioners are manufactured in capacities as high as 100 tons (1 ton of air conditioning equals 12,000 Btu/hr) and are designed to be mounted conveniently on roofs, on the ground, or other convenient location, where they can be connected by ductwork to the conditioned space.

Built-up or field-erected systems are composed of factory-built subassemblies interconnected by means such as piping, wiring, and ducting during final assembly on the building site. Their capacities range up to thousands of tons of refrigeration and millions of Btu per hr of heating. Most large buildings are so conditioned.

Another important and somewhat parallel distinction can be made between incremental and central systems. An incremental system serves a single space; each space to be conditioned has its own, self-contained heating-cooling-dehumidifying-filtering unit. Central systems serve many or all of the conditioned spaces in a building. They range from small, unitary packaged systems to serve single-family residences to large, built-up or field-erected systems serving large buildings.

When many buildings, each with its own air conditioning system which is complete except for a refrigeration and a heating source, are tied to a central plant that distributes chilled water and hot

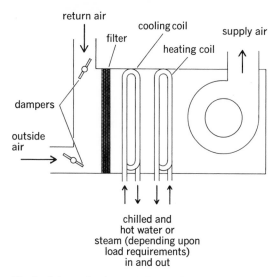

Fig. 2. Schematic of central air-handling unit.

water or steam, the interconnection is referred to as a district heating and cooling system. This system is especially useful for campuses, medical complexes, and office complexes under a single management.

Conditioning of spaces. Air temperature in a space can be controlled by radiant panels in floor, walls, or ceiling to emit or absorb energy, depending on panel temperature. Such is the radiant panel system. However, to control humidity and air purity, and in most systems for controlling air temperature, a portion of the air in the space is withdrawn, processed, and returned to the space to mix with the remaining air. In the language of the engineer, a portion of the room air is returned (to an air-handling unit) and, after being conditioned, is supplied to the space. A portion of the return air is spilled (exhausted to the outdoors) while an equal quantity (of outdoor air) is brought into the system and mixed with the remaining return air before entering the air handler.

Typically, the air-handling unit contains a filter, a cooling coil, a heating coil, and a fan in a suitable casing (Fig. 2). The filter removes dust from both return and outside air. The cooling coil, either containing recirculating chilled water or boiling refrigerant, lowers air temperature sufficiently to dehumidify it to the required degree. The heating coil, in winter, serves a straightforward heating function, but when the cooling coil is functioning, it serves to raise the temperature of the dehumidified air (to reheat it) to the exact temperature required to perform its cooling function. The air handler may perform its function, in microcosm, in room units in each space, as part of a self-contained, unitary air conditioner, or it may be a huge unit handling return air from an entire building. *See* AIR COOLING.

There are three principal types of central air-conditioning systems: all-air, all-water, and air-water. In the all-air system all return air is processed in a central air-handling apparatus. In one type of all-air system, called dual-duct, warm air and chilled air are supplied to a blending or mixing unit in each space. In a single-duct all-air

system air is supplied at a temperature for the space requiring the coldest air, then reheated by steam or electric or hot-water coils in each space.

In the all-water system the principal thermal load is carried by chilled and hot water generated in a central facility and piped to coils in each space; room air then passes over the coils. A small, central air system supplements the all-water system to provide dehumidification and air filtration. The radiant panel system, previously described, may also be in the form of an all-water system.

In an air-water system both treated air and hot or chilled water are supplied to units in each space. In winter hot water is supplied, accompanied by cooled, dehumidified air. In summer chilled water is supplied with warmer (but dehumidified) air. One supply reheats the other.

All-air systems preceded the others. Primary motivation for all-water and air-water systems is their capacity for carrying large quantities of heat energy in small pipes, rather than in larger air ducts. To accomplish the same purpose, big-building all-air systems are designed for high velocities and pressures, requiring much smaller ducts. *See* SOLAR COOLING.   [RICHARD L. KORAL]

Air cooling

Lowering of air temperature for comfort, process control, or food preservation. Air and water vapor occur together in the atmosphere. The mixture is commonly cooled by direct convective heat transfer of its internal energy (sensible heat) to a surface or medium at lower temperature. In the most compact arrangement, transfer is through a finned (extended surface) coil, metallic and thin, inside of which is circulating either chilled water, antifreeze solution, brine, or boiling refrigerant. The fluid acts as the heat receiver. Heat transfer can also be directly to a wetted surface, such as water droplets in an air washer or a wet pad in an evaporative cooler. *See* AIR CONDITIONING; HEAT TRANSFER.

Evaporative cooling. For evaporative cooling, nonsaturated air is mixed with water. Some of the sensible heat transfers from the air to the evaporating water. The heat then returns to the airstream as latent heat of water vapor. The exchange is thermally isolated (adiabatic) and continues until the air is saturated and air and water temperatures are equal. With suitable apparatus, air temperature approaches within a few degrees

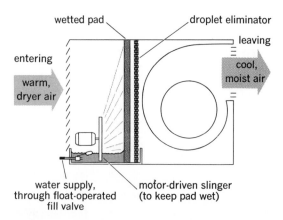

Fig. 1. Schematic view of simple evaporative air cooler.

of the theoretical limit, the wet-bulb temperature. Evaporative cooling is frequently carried out by blowing relatively dry air through a wet mat (Fig. 1). The technique is employed for air cooling of machines where higher humidities can be tolerated; for cooling of industrial areas where high humidities are required, as in textile mills; and for comfort cooling in hot, dry climates, where partial saturation results in cool air at relatively low humidity.

Air washer. In the evaporative cooler the air is constantly changed and the water is recirculated, except for that portion which has evaporated and which must be made up. Water temperature remains at the adiabatic saturation (wet-bulb) temperature. If water temperature is controlled, as by refrigeration, the leaving air temperature can be controlled within wide limits. Entering warm, moist air can be cooled below its dew point so that, although it leaves close to saturation, it leaves with less moisture per unit volume of air than when it entered. An apparatus to accomplish this is called an air washer (Fig. 2). It is used in many industrial and comfort air conditioning systems, and performs the added functions of cleansing the airstream of dust and of gases that dissolve in water, and in winter, through the addition of heat to the water, of warming and humidifying the air.

Air-cooling coils. The most important form of air cooling is by finned coils, inside of which circulates a cold fluid or cold, boiling refrigerant (Fig. 3). The latter is called a direct-expansion (DX) coil. In most applications the finned surfaces become wet as condensation occurs simultaneously with sensible cooling. Usually, the required amount of dehumidification determines the temperature at which the surface is maintained and, where this results in air that is colder than required, the air is reheated to the proper temperature. Droplets of condensate are entrained in the airstream, removed by a suitable filter (eliminator), collected in a drain pan, and wasted.

In the majority of cases, where chilled water or boiling halocarbon refrigerants are used, aluminum fins on copper coils are employed. Chief advantages of finned coils for air cooling are (1) complete separation of cooling fluid from airstream, (2) high velocity of airstream limited only by the need to separate condensate that is entrained in the airstream, (3) adaptability of coil configuration to requirements of different apparatus, and (4) compact heat-exchange surface.

Defrosting. Wherever air cooling and dehumidification occur simultaneously through finned coils, the coil surface must be maintained above 32°F to prevent accumulation of ice on the coil. For this reason, about 35°F is the lowest-temperature air that can be provided by coils (or air washers) without ice accumulation. In cold rooms, where air temperature is maintained below 32°F, provision is made to deice the cooling coils. Ice buildup is sensed automatically; the flow of cold refrigerant to. the coil is stopped and replaced, briefly, by a hot fluid which melts the accumulated frost. In direct-expansion coils, defrosting is easily accomplished by bypassing hot refrigerant gas from the compressor directly to the coil until defrosting is complete.

Cooling coil sizing. Transfer of heat from warm

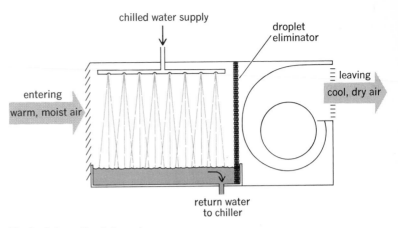

Fig. 2. Schematic of air washer.

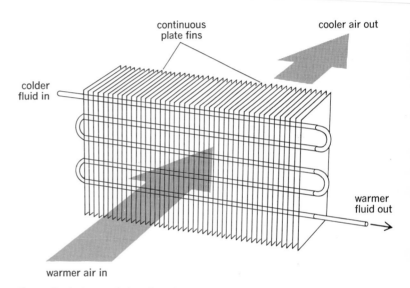

Fig. 3. Typical extended-surface air-cooling coil.

air to cold fluid through coils encounters three resistances: air film, metal tube wall, and inside fluid film. Overall conductance of the coil, U, is shown in the equation below, where K_o is film conductance

$$\frac{1}{U} = \frac{1}{K_o} + r_m + \frac{R}{K_i}$$

of the outside (air-side) surface in Btu per (hr)(sq ft) (F); r_m is metal resistance in (hr)(sq ft)(F) per Btu, where area is that of the outside surface; K_i is film conductance of the inside surface (water, steam, brine, or refrigerant side) in Btu per (hr)(sq ft)(F); U is overall conductance of transfer surface in Btu per (hr)(sq ft)(F), where area again refers to the outside surface; and R is the ratio of outside surface to inside surface.

Values of K_o are a function of air velocity and typically range from about 4 Btu per (hr)(sq ft)(F) at 100 feet per minute (fpm) to 12 at 600 fpm. If condensation takes place, latent heat released by the condensate is in addition to the sensible heat transfer. Then total (sensible plus latent) K_o increases by the ratio of total to sensible heat to be transferred, provided the coil is well drained.

Values of r_m range from 0.005 to 0.030 (hr)(sq ft) (F) per Btu, depending somewhat on type of metal but primarily on metal thickness.

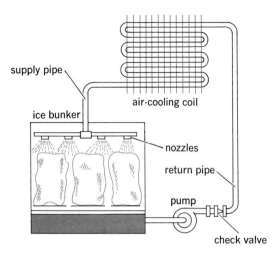

Fig. 4. Air cooling by circulating ice-cooled water.

Typical values for K_i range from 250 to 500 Btu per (hr)(sq ft)(F) for boiling refrigerant. In 40°F chilled water, values range from about 230 Btu per (hr)(sq ft)(F) when water velocity is 1 foot per second (fps) to 1250 when water velocity is 8 fps.

Use of well water. Well water is available for air cooling in much of the world. Temperature of water from wells 30 to 60 ft deep is approximately the average year-round air temperature in the locality of the well, although in some regions overuse of available supplies for cooling purposes and recharge of ground aquifers with higher-temperature water has raised well water temperature several degrees above the local normal. When well water is not cold enough to dehumidify air to the required extent, an economical procedure is to use it for sensible cooling only, and to pass the cool, moist air through an auxiliary process to dehumidify it. Usually, well water below 50°F will dehumidify air sufficiently for comfort cooling. Well water at these temperatures is generally available in the northern third of the United States, except the Pacific Coast areas.

Ice as heat sink. For installations that operate only occasionally, such as some churches and meeting halls, water recirculated and cooled over ice offers an economical means for space cooling (Fig. 4). Cold water is pumped from an ice bunker through an extended-surface coil. In the coil the water absorbs heat from the air, which is blown across the coil. The warmed water then returns to

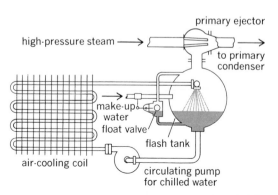

Fig. 5. Air cooling by circulating water that is cooled, in turn, by evaporation in flash tank.

the bunker, where its temperature is again reduced by the latent heat of fusion (144 Btu/lb) to 32°F. Although initial cost of such an installation is low, operating costs are usually high.

Refrigeration heat sink. Where electric power is readily available, the cooling function of the ice, as described above, is performed by a mechanical refrigerator. If the building complex includes a steam plant, a steam-jet vacuum pump can be used to cause the water to evaporate, thereby lowering its temperature by the latent heat of evaporation (about 1060 Btu/lb, depending on temperature and pressure). High-pressure steam, in passing through a primary ejector, aspirates water vapor from the evaporator, thereby maintaining the required low pressure that causes the water to evaporate and thus to cool itself (Fig. 5).

Where electric power is costly compared to low-temperature heat, such as by gas, absorption refrigeration may be used. Two fluids are used: an absorbent and a refrigerant. The absorbent is chosen for its affinity for the refrigerant when in vapor form, for example, water is used as the absorber with ammonia as the refrigerant. Concentrated ammonia water is pumped to a high pressure and then heated to release the ammonia. The high-pressure ammonia then passes through a condenser, an expansion valve, and an evaporator, as in a mechanical system, and is reabsorbed by the water. The cycle cools air circulated over the evaporator.

[RICHARD L. KORAL]

Bibliography: American Society of Heating and Air Conditioning Engineers, *Guide and Data Book*, annual; S. Elonka and Q. W. Minich, *Standard Refrigeration and Air Conditioning Questions and Answers*, 1961; F. W. Hutchinson, *Design of Air Conditioning Systems*, 1958; W. R. Woolrich and W. R. Woolrich, Jr., *Air Conditioning*, 1957.

Air-pollution control

Air pollution, according to the definition developed by the Engineers Joint Council, means the presence in the outdoor atmosphere of one or more contaminants, such as dust, fumes, gas, mist, odor, smoke, or vapor, in quantities, of characteristics, and of duration such as to be injurious to human, plant, or animal life or to property, or to interfere unreasonably with the comfortable enjoyment of life and property. The sources of airborne wastes are many. They may be roughly divided into natural, industrial, transportation, agricultural activity, commercial and domestic heat and power, municipal activities, and fallout. *See* ATMOSPHERIC POLLUTION.

Sources of pollution. Natural sources include the pollen from weeds, water droplet or spray evaporation residues, wind storm dusts, meteoritic dusts, and surface detritus. Industrial sources include ventilation products from local exhaust systems, process waste discharges, and heat, power, and waste disposal by combustion processes. Transportation sources include motor vehicles, rail-mounted vehicles, airplanes, and vessels. Agricultural activity sources include insecticidal and pesticidal dusting and spraying, and burning of vegetation. Commercial heat and domestic heat and power sources include gas-, oil-, and coal-fired furnaces used to produce heat or power for individ-

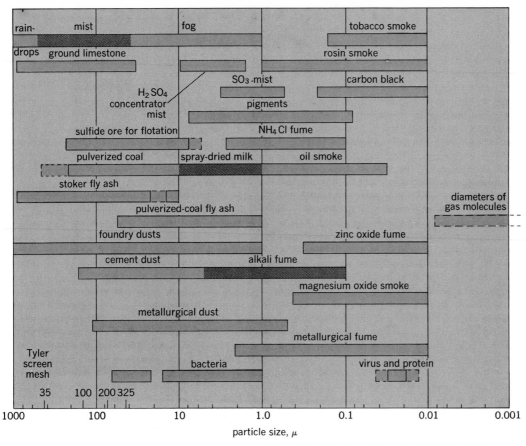

Fig. 1. Particle size ranges for aerosols, dusts, and fumes. (*From W. L. Faith, Air Pollution Control, Wiley, 1959*)

ual dwellings, multiple dwellings, commercial establishments, utilities, and industry. Municipal activity sources include refuse disposal, liquid waste disposal, road and street plant operations, and fuel-fired combustion operations. Fallout is a term applied to radioactive pollutants in mass atmosphere resulting from thermonuclear explosion.

The sources are so varied that pollution of the atmosphere is a matter of degree. Pollution from natural sources is in effect a base line of pollution. The major problems of pollution are associated with community activity as opposed to rural activity, because community air is generally more grossly polluted and may contain harmful and dangerous substances affecting property, plant life, and, on occasion, health. Environment is made less desirable by the polluting influence, and there is ample reason to conserve the air resource in many ways parallel to the need for conservation of the water resource. In actuality, the engineer is concerned with engineering management of the air resource, a broader concept than the control of air pollution.

Control. Air-pollution control suggests in its simplest form a background of knowledge concerning ideal atmospheres and criteria of clean air, the existence of specific standards setting limits on the allowable degree of pollution, means of precise measurement of pollutants, and practical means of treating polluting sources to maintain the desired degree of air cleanliness. There are many areas in the above listing that are under research at the

present time. University research foundations, Federal, state, and municipal air-pollution control agencies, and all of the professional engineering societies are actively engaged in the development of criteria, standards, design factors, and equipment for the control of air pollution.

Reduced visibility has been a focal point of air pollution for over 700 years. The burning of soft coal in England combined with the fog of the atmosphere forms a particularly opaque mixture which may at times reduce visibility to zero. The word smog has been coined to describe this mixture.

Microscopic water droplets condense about nucleating substances in the air to form aerosols. An aerosol is a liquid or solid submicron particle dispersed in a gaseous medium. An atmosphere having an aerosol concentration of about 1 mg/m³ has been estimated to limit visibility to 1600 ft. The mass would contain perhaps 16,000 particles/ml. Restriction in visibility is the result of light scattering by these particles. Chemical condensation of reaction products in the air may also nucleate and grow to size that will bring about light scattering. Sulfur dioxide is also a nucleating substance as it oxidizes and hydrolyzes to form sulfuric acid mist.

Elimination of sources of pollution has been one of the favored means of controlling pollution. There are many means of accomplishing the reduction of pollution, but complete elimination is not always practicable. Sulfur dioxide release can be reduced by choosing low sulfur-bearing fuel. An industrial process with a gaseous effluent can be changed to eliminate the gaseous waste. Gases and particulates can be removed from a gas stream by air-cleaning equipment.

Air-cleaning devices. Air-cleaning devices to remove particulates are selected to remove particles and aerosols on the basis of their size (Fig. 1). Screens will remove coarse solids. Settling chambers are containers which by expanded cross section reduce velocity below 10 feet per second (fps) and thereby allow particles to settle. Particles down to 10 μ in size may be recovered with such chambers. Cyclone separators operate by injecting a gas stream tangentially at the top of a cylindrical

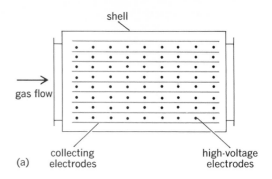

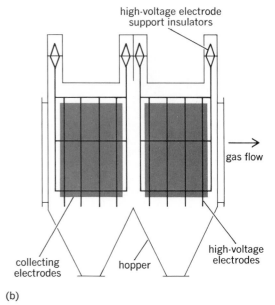

(b)

Fig. 4. Diagram of horizontal-flow electrostatic precipitator. (*a*) Plan. (*b*) Elevation. (*From W. L. Faith, Air Pollution Control, Wiley, 1959*)

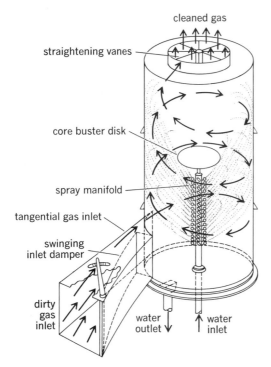

Fig. 2. Typical cyclonic spray scrubber. (*From W. L. Faith, Air Pollution Control, Wiley, 1959*)

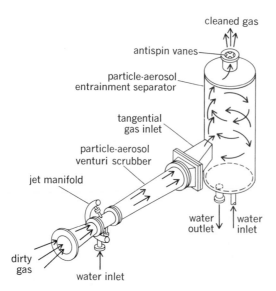

Fig. 3. Typical venturi scrubber. (*From W. L. Faith, Air Pollution Control, Wiley, 1959*)

chamber. A high-velocity spiral motion is created. Particles are centrifuged out of the gas stream, hit the side wall, and fall to a conical bottom out of the airflow, which turns up through the core or vortex beginning at the bottom and flows to the top through a pipe inserted into the core and extending into the body of the cyclone. Particles from 10 to 200 μ are removed with 50–90% efficiency. Filters are made of cloth, fiber, or glass. Air velocities are low and efficiency is about 50% for dry fiber filters. Efficiency is increased by using a low volatile oil viscous coating. Cloth filters are usually tubular and a number of bags are enclosed in a large chamber. Particles are trapped as air passes through the cloth from inside to outside. Dust is knocked down by shaking and falls to a hopper. Bag filters remove 99% of particles above 10-μ size. Wet collectors, or scrubbers, operate by passing and contacting the gas with a liquid. Water is sprayed, atomized, or distributed over a geometric shape. Deflectors may be added to provide an impinging surface. Scrubbers are efficient on 1- to 5-μ size particles (Figs. 2 and 3). Electrostatic precipitators operate by charging or ionizing particles as the gas flow passes through the unit (Fig. 4). Opposite-pole high-voltage plates, or electrodes, are provided to trap particles. Precipitators operate at 80–99% efficiency of ionizable aerosols down to 0.1-μ size.

Scrubbers may also remove water-soluble gases. Chemicals may be added to the liquid to provide improved absorption. Filters packed with activated charcoal are used to adsorb gases.

Packed towers, plate towers, and spray towers are also used to absorb gaseous pollutants from a gas stream. These devices provide for mixing a gas stream under treatment with water or a chemical solution, so that gases are taken into solution and possibly converted chemically as well.

Atmospheric dilution. This provides another means of reducing air pollution. Meteorology of a region, local topography, and building configuration are critical factors in determining suitability of atmosphere as a dispersal, diffusion, and dilution medium. Basic meteorological conditions of atmosphere that must be considered include wind speed and direction, gustiness of wind, and vertical temperature distribution. Humidity is also important under certain circumstances of emission.

In general, diffusion theories predict that the ground concentration of a gas or a fine particle effluent with very low subsidence velocity is inversely proportional to the mean wind speed. Vertical temperature distribution is an important factor, determining the distance from stack of known height at which maximum ground concentration occurs. Temperature of the stack gas has the effect of increasing stack height, as does stack gas velocity. Gas does not normally come to the ground under inversion conditions, but may accumulate aloft under calm or near calm conditions and be brought down to the surface as the Sun heats the ground in the early morning. Effect of building configuration is shown in Fig. 5. The turbulence introduced by buildings and topography is so com-

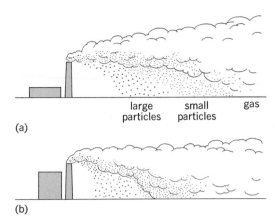

Fig. 5. Effect of building configuration on dispersal of gas plume. (*a*) Favorable configuration. (*b*) Unfavorable configuration. (*Research Division, New York University School of Engineering and Science*)

plex that it is difficult to make theoretical calculations of effect. Model studies in wind tunnels have been used successfully to make predictions based on measurements of gas concentration and visible pattern of smoke (Fig. 6).

Nonventilating conditions may be present over an area for several days as a result of certain meteorological phenomena. During such periods the pollution emitted from various sources, such as fuel-fired combustion and automobile exhaust, continues to increase in concentration until ventilation sufficient to dilute the accumulated gases and particulates takes place. Figure 7 illustrates the record of sulfur dioxide–concentration meas-

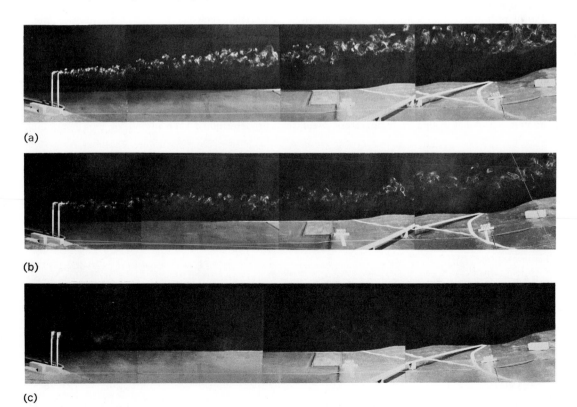

Fig. 6. Wind tunnel demonstration of dispersal patterns at specified wind speeds. (*a*) 20 mph. (*b*) 25 mph. (*c*) 30 mph.

urement in the atmosphere over New York City during one such period of poor ventilation lasting for several days.

Incineration. The need for municipalities to find a means of disposing of refuse when land values are high and little land is available for sanitary landfill has resulted in increased use of incineration for refuse disposal. Incineration introduces problems of air pollution that are quite different from those of fuel-fired combustion. The material is not homogeneous, and has a wide variation in fuel value ranging from 600 to 6500 Btu/lb of refuse as fired. Volatiles are driven off by destructive distillation and ignite from heat of the combustion chamber. Gases pass through a series of oxidation changes in which time-temperature relationship is important. The gases must be heated above 1200°F to destroy odors. End products of refuse combustion pass out of the stack at 800°F or less after passing through expansion chambers, fly ash collectors, wet scrubbers, and in some instances electrostatic precipitators. The end products include carbon dioxide; carbon monoxide; water; oxides of nitrogen; aldehydes; unoxidized or unburned hydrocarbons; particulate matter comprising unburned carbon, mineral oxides, and unburned refuse; and unused or excess air. Particulates are reduced in quantity. Normally only micron-size and submicron particles should escape with the flue gases. Care in operation is required to hold down particulate loading. Dust emissions in stacks may be in the range of 2–3 lb/ton/hr of refuse charged at a well-operated unit equipped with air-cleaning devices.

Incinerator design. There are several types of incinerator design promoted by manufacturers of incinerator equipment. Kiln shape may be round, rectangular, or rotary. The hearth may be horizontal fixed with grates, traveling with grates, multiple, step movement, or barrel-type rotary (Fig. 8). Drying hearths are provided on some types. Feed into the incinerator may be continuous, stoker, gravity, or batch.

It is necessary to know or estimate water content, percentage combustible material and inert material, Btu content, and weight of refuse to complete a rational design of incinerators. Heat balance can be calculated from several estimates based on averages. Available heat from the refuse must be balanced against the heat losses due to radiation, as well as from moisture, excess air, flue gas, and ash. Each type design has recommended sizings suggested by the manufacturer. There is fair agreement on the need for over 100% excess air. An allowance of 20,000 Btu/ft³ has been suggested for approximating chamber volume, and an allowance of 300,000 Btu/ft² for grate area. Incinerator loading rates of 40–70 lb/(ft² grate area) (hr) have been used. Small incinerators for apartment houses and institutions are loaded at much lower rates. The Incinerator Institute of America in its standards has suggested loading rates for household or domestic-type refuse from 20 lb/(ft²) (hr) in 100 lb/hr burning units up to 30 lb/(ft²) (hr) in 1000 lb/hr units.

The Building Research Advisory Board (National Academy of Sciences, National Research Council) suggests that apartment house single-chamber incinerators should be sized on the basis of 0.375 ft³ capacity per person, 0.075 ft² grate area per person, and heat release rate of not more than 18,000 Btu/ft³ of capacity, where the burning period is 10 hr or less.

Air-monitoring instruments. Air-sampling methods may be classified generally as those for sampling particulates or gases or both concurrently. The samples may be analyzed for specific pollutants or for general pollution levels. Sampling devices have been constructed with many variants. Generally, however, they follow reasonably well-defined principles which include gravity and suction-type collection, with passage through thermal and electrostatic precipitators; impingers and impactors; cyclones; absorption and adsorption trains; scrubbing apparatus; filters of various materials, such as paper, glass, plastic, membrane, and wool; glass plates; and impregnated papers. Combination instruments that measure wind direction and velocity and direct air samples into multiple sample units, each of which represents a wind sector, are used for general sampling and locating of emission sources. Samples may be taken as single samples, or as a composite over a predetermined time period, or as a continuing

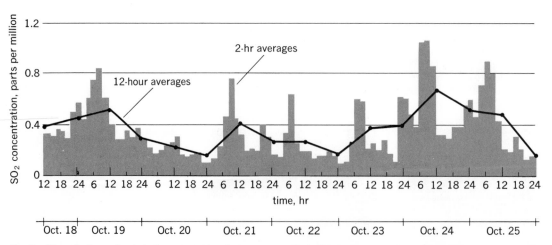

Fig. 7. Air-pollution episode in New York City, Oct. 18–25, 1963. The SO₂ values were at Christodora Station, 189 ft above ground. (*Research Division, New York University School of Engineering and Science*)

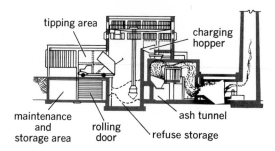

tipping area

charging hopper

maintenance and storage area

rolling door

ash tunnel

refuse storage

Fig. 8. Diagram of incinerator with rectangular grate. (*American Society of Civil Engineers*)

monitoring operation. Some instruments are designed to extract a sample from the air, analyze it automatically, and record the result on a chart. Others take a sample which must be examined in a laboratory.

Many instruments have been developed during the 1950s and 1960s that are mechanized, automatic, and recording, so that they can be used with a minimum of attendance and manipulation. Such instruments require careful initial calibration with standard test substances that are to be measured, and a continuing field check with recalibration at frequent intervals to maintain accuracy.

Particulate samples may be analyzed for weight, size range, effect on visibility, chemical character, shape, and other specific information required. Gas samples may be analyzed to determine the presence of a specific gas or of a group of gases of the same chemical family; that is, nitrogen oxide may be determined or the concentration of all oxides of nitrogen may be established by analysis, or total hydrocarbons may be measured, or by more specific analysis the fractions of several specific hydrocarbons making up the total may be found.

Several types of units with air pumps drawing air through paper tapes mounted on a spool have been developed. The tape is moved automatically so that successive samples on fresh paper are taken at timed intervals.

High-volume samplers are used at many sampling network stations in the United States. The electron microscope has been employed for the examination of aerosols and fine particles. Spectrographic instruments are used for analyzing hydrocarbons and oxides of nitrogen and carbon. Automatically operated units take a sample and then pump chemicals into it at the appropriate time to produce a succession of chemical reactions; they are used to obtain a continuous record of concentration fluctuation of gaseous pollutants, such as sulfur dioxide and others where a wet-chemical analytical method is appropriate. Other instruments use the principle of conductance for measurement of a gas dissolved in a liquid medium. The sample is passed into the liquid medium and a change in electrical energy is measured and recorded. Such instruments measure the effect of any substance that is ionizable in the medium. Orsat analyses are made on flue gases. Photoelectric cells are used to control alarm systems connected to stacks.

Analytical instruments that use several principles of measurement have made it possible to take more data at less cost and manpower. Systems are being developed whereby data from a number of monitoring stations can be transmitted to a central point, transferred to computer program operations, and become statistical information concerning air-pollution concentration. Other systems under development provide for measurement of certain index pollutants while in motion by utilizing automatic instruments mounted in vehicles.

Methods of sampling and methods of analysis have yet to achieve widespread agreement or standardization. The American Society for Testing and Materials (ASTM) has published some 17 standard methods of tests applicable to atmospheric analysis. ASTM has also published definitions of terms relating to air sampling and analysis. Numerous industrial associations and professional organizations are in the process of bringing together in published form the multitude of sampling and testing methods in use.

Air-quality control. This is predicated on standards or guides that take form in official regulations and laws according to three control approaches: restriction of all sources of pollution so that pollution levels in community air are not in excess of certain levels chosen as a safe standard of air quality; limitation on the amount of a specific pollutant that may be present in the exhaust gas from a duct or a stack; or limitation on the amount of impurity in raw materials whose residues reach the community air. These approaches are frequently combined in an attempt to achieve maximum control. The guides or standards may consist of ambient air-quality standards, emission standards, or material-quality standards.

Official control agencies at local, state, and Federal levels are now in the process of establishing ambient air-quality levels for such pollutants as sulfur dioxide and particulates. Many municipal control agencies have specified the limits of pollutants such as sulfur dioxide, particulates, and solvents that may be emitted from a single source. Many control agencies at municipal and state levels have adopted standards limiting the amount of sulfur in fuel. The U.S. Department of Health, Education, and Welfare publishes from time to time a digest of state air-pollution control laws. All major cities of the United States have adopted laws and regulations based on one or more of the quality control approaches mentioned.

Standards of ambient air quality may vary. In the state of New York, for example, there is a recognized difference in the kind and quantity of pollutants that may be emitted in rural areas, as opposed to highly urbanized areas. Within any region, land use may vary as to its industrial, commercial, residential, or rural components. Subregions based on the predominant land use and on air-quality objectives to be obtained may be established, each having its own air-quality guides. *See the feature article* PROTECTING THE ENVIRONMENT.

[WILLIAM T. INGRAM]

Bibliography: *Air Conservation*, Amer. Ass. Advan. Sci. Publ. no. 80, 1965; Air Quality Committee of the Manufacturing Chemists Association, *Source Materials for Air Pollution Control Laws*, 1968; American Industrial Hygiene Association, Air Pollution Committee, *Air Pollution Manual*, pt. 1: Evaluation, 1960, pt. 2: Control equipment, 1968; American Society for Testing and Materials Standards, *ASTM Standards on Methods of Atmospheric Sampling and Analysis*, pt. 23, 1967; *Apartment House Incinerators*, Nat. Acad. Sci.–Nat. Res. Counc. Publ. no. 1280, 1965; W. T Ingram et al., Adaption of Technicon Auto-Analyzer for continuous measurement while in motion, *Technicon Symposia, 1967*, vol. 1: *Automation in Analytical Chemistry*, 1968; W. T. Ingram and L. C. McCabe, *The Effects of Air Pollution on Airport Visibility*, Amer. Soc. Chem. Eng., J. Sanit. Eng. Div., Pap. no. 1543, 1958; W. T. Ingram, C. Simon, and J. McCarroll, *Air Research Monitoring Station System*, J. Sanit. Eng. Div., Proc. Amer. Soc. Chem. Eng. 93, no. SA2, 1967; Interbranch Chemical Advisory Committee, U.S. Department of Health, Education, and Welfare, *Selected Methods for the Measurement of Air Pollutants*, Environmental Health Series, 1965; A. J. Johnson and G. H. Auth, *Fuels and Combustion Handbook*, 1951; W. C. McCrone, R. G. Draftz, and J. G. Gustav, *The Particle Atlas*, 1967; Metropolitan Engineers Council on Air Resources, *Incineration of Solid Wastes*, Proc. MECAR Symp., Mar. 21, 1967; A. C. Stern (ed.), *Air Pollution*, vols. 1 and 2, 2d ed., 1968; C. D. Yaffe et al. (eds.), *Air Sampling Instruments for Evaluation of Atmospheric Contaminants*, 2d ed., 1962.

Aircraft engine

A component of an aircraft that develops either shaft horsepower or thrust and incorporates design features most advantageous for aircraft propulsion. An engine developing shaft horsepower requires an additional means to convert this power to useful thrust for aircraft, such as a propeller, a fan, or a helicopter rotor. It is common practice in this case to designate the unit developing shaft horsepower as the aircraft engine, and the combination of engine and propeller, for example, as an aircraft power plant. In case thrust is developed directly as in a turbojet engine, the terms engine and power plant are used interchangeably.

The characteristics primarily emphasized in an aircraft engine are high takeoff thrust and low specific weight, low specific fuel consumption, and low drag of the installed power plant at the aircraft speeds and altitudes desired. Reliability and durability are essential, as is emphasis on high output and light weight, so that a premium is placed on quality materials and fuels, as well as on design and manufacturing skills and practices. *See* AIRCRAFT FUEL; PROPULSION.

Air-breathing types of aircraft engines use oxygen from the atmosphere to combine chemically with fuel carried in the vehicle, providing the energy for propulsion, in contrast to rocket types in which both the fuel and oxidizer are carried in the aircraft. Air-breathing engines suffer decreased power or thrust output with altitude increase, due to decreasing air density, unless supercharged. *See* AIRCRAFT ENGINE, RECIPROCATING; INTERNAL COMBUSTION ENGINE.

[RONALD HAZEN]

Aircraft engine, reciprocating

A fuel-burning piston internal combustion engine specially designed and built for light weight in proportion to developed shaft horsepower. The reciprocating engine is currently the principal aircraft engine used for such low-flying, low-speed aircraft as small private and crop-spraying airplanes and helicopters. The reciprocating engine drives a propeller (or rotor) that, in turn, accelerates the surrounding air rearward (or downward), thereby imparting forward momentum (or lift) to the airplane (or helicopter).

Predominantly, aircraft reciprocating engines operate on a four-stroke Otto cycle. These spark-ignition engines burn hydrocarbon fuels and develop shaft power through connecting rods and a crankshaft. Major parts are the crankcase, crankshaft, connecting rods, pistons, cylinders with intake and exhaust valves, camshafts, and such operating auxiliaries as ignition, carburetor, and fuel and oil pumps.

Two-stroke spark ignition and two- or four-stroke compression ignition (diesel) reciprocating engines have been developed and successfully flown, but they have not been widely applied nor have they significantly improved aircraft performance.

The reciprocating engine powered all aircraft for the first 40 years of heavier-than-air flight, including all military, commercial, and private types. The advent of the turbojet engine near the end of World War II started a rapid conversion to turbine power. This conversion has been primarily directed toward military needs and is limited chiefly by availability of government financing and the time required to develop suitable types and sizes of turbine engines and applicable aircraft. The cold war and Korean War accelerated the changeover appreciably. As would be expected under these conditions, engines for combat-type aircraft received primary emphasis, so that funds for further development of reciprocating engines of new types or major model changes were minimized after the end of World War II. Some commercial aircraft were supplied with turbojet and turboprop power in the mid-1950s; however, the major swing to turbines occurred with advanced equipment in late 1958. With availability of small jet and turboprop engines, private aircraft are beginning to employ them, particularly in planes for business purposes for which speed is needed to compete with commercial transportation.

The reciprocating engine has been built for aircraft with numerous arrangements and numbers of cylinders, with a variety of fuels and antiknock ratings, with various cooling systems, with various means and amounts of supercharging, with several types of fuel systems, and with and without reduction gears. The military premium on increased power and performance, the high cost of development, and a relatively rapid rate of obsolescence have combined to discourage purely commercial sponsorship of new types or models of aircraft engines except for low-power applications.

Variety of engines. Through World War I, aircraft had relatively low operating speeds with resultant low velocity head available for cooling. A limited knowledge of air cooling of cylinders also tended to limit use of air-cooled fixed cylinders. Major combat aircraft were therefore mostly powered by water-cooled 6-cylinder in-line, 8-cylinder V, and 12-cylinder V types. There was, however, considerable use by the French of air-cooled rotary radial engines for fighter or pursuit aircraft and some use by the British of air-cooled V engines.

World War I demonstrated the great importance of aircraft as a military weapon. This resulted in the financing and exploration of a great variety of engine types and forms, particularly those promising higher power and improved aircraft performance, during this war and the following decade. Water-cooled engines built and tested included 12-, 18-, and 24-cylinder W, 12- and 16-cylinder parallel-vertical, 16-cylinder X, V, and fan engines. Fixed-radial engines, both air and water cooled, with 3–20 cylinders and one or two rows, were explored. Air-cooled 9-, 11-, and 18-cylinder rotary and 2-cylinder opposed engines were also built and run. As early as 1925, some 31 cylinder combinations in 36 types and nearly 300 models in horsepowers from 30 to 1000 had been or were being developed.

Before 1930 a few outstanding types of engines had emerged from this continued exploration of types, from new invention and design, and from better materials and processes both in the engines and the installations. These were the single-row 9-cylinder and two-row 14-cylinder air-cooled fixed-radial types in medium and high horsepowers, the liquid-cooled 12-cylinder V for high-speed aircraft, and the small 4 and 6 in-line and opposed air-cooled engines in low horsepower. Reciprocating engine development was concentrated during the first half of the 1930s on refinement and performance improvement of the above types. During the last half of the 1930s larger and much higher powered two-row engines of both 14 and 18 cylinders of the radial air-cooled type were developed. Design also emphasized higher outputs and use of higher-temperature cooling on liquid-cooled 12-cylinder V engines. A 24-cylinder double-V of 3420-in.^{3} displacement was the largest liquid-cooled engine developed prior to World War II. Flown in experimental bomber and fighter aircraft, it was scheduled for large production in a fighter airplane in 1943 but was discontinued before production quantities were made, in favor of using the facilities for turbojet engine development and production in late 1944.

The last type of reciprocating engine of medium to high horsepower initiated was a four-row 28-cylinder air-cooled fixed-radial type of 4360-in.^{3} displacement, started in 1940. This engine reached substantial postwar military production and was the only such new engine type to do so.

Fixed-radial air-cooled engines. The 9-cylinder single-row fixed air-cooled engine combined the inherently good features of a short and therefore light crankcase and crankshaft with the maximum number of cylinders arranged peripherally. It also provided uniform airflow at every cylinder and space at the cylinder heads for widely canted valves and the increased head thickness and deep finning needed with aluminum heads. However, considerable research, novel engineering design and development, as well as process improvement and installation-cooling knowledge were needed to utilize fully these basic advantages.

Some of these features, which have been applied to all aircraft engines, included forged and cast aluminum parts, cooled valves, supercharger, and low-drag cowl.

Forged aluminum pistons reduced the reciprocating mass. A one-piece master rod reduced rotating mass; it also decreased distortion of, and improved loading uniformity on, a one-piece master rod bearing. This required a two-piece crankshaft. Forged aluminum crankcases with separate oil sump permitted more compact and lighter structure.

Cast aluminum cylinder heads were screwed and shrunk to steel barrels with machined integral fins. The two-valve head with highly canted valves had steel or bronze valve-seat inserts shrunk in. This, in combination with a major improvement in casting technique, permitted more uniform valve, spark plug, and head cooling. Use of integral rocker boxes decreased weight, improved cooling, mechanical strength, and reliability, and simplified forced lubrication without leakage.

Internally cooled exhaust valves permitted higher outputs from a given fuel and improved reliability and durability.

A supercharger impeller, either crankshaft or (preferably) gear driven, improved mixture uniformity and distribution to the various cylinders.

Use of Townend ring or NACA cowl reduced drag in the aircraft.

The above features in combination with many detail refinements in design, such as improved cam design, temperature-compensated valve gears, and avoidance of torsional resonance in the normal operating range, resulted in the late 1920s in specific engine weights for single-row 9-cylinder radials of less than 1.5 lb/hp with a new order of reliability and durability. Reduction gearing when added increased specific weights but improved aircraft performance.

Two-row 14-cylinder air-cooled radial engines with features similar to those above were 0.1–0.2 lb/hp heavier than the comparable single-row engine of equal output. Their smaller diameter for the same displacement or power rating as the single-row type and the smaller cylinder size due to the larger number of cylinders tended to compensate for the higher specific weight and initial cost except in the smaller engine sizes. However, some structural problems and particularly difficulties with torsional vibration slowed two-row engine development in the larger cylinder sizes for several years, the 1830-in.^{3} size being the largest cylinder size undertaken in the United States until 1935.

Liquid-cooled 12-cylinder V engine. The development of liquid-cooled engines was sponsored throughout the 1930s by the military because of marked advantages for single-seat fighter or interceptor applications. Schneider trophy contests demonstrated the high specific power outputs obtainable for short periods with liquid cooling and the use of high engine rotational speed, the low drag associated with the low frontal area (form drag), and the low cooling drag obtained with high-temperature coolants and efficient radiators in strategic locations with the 12-cylinder V engine (Fig. 1). The result was that this type of engine predominated in fighters of the British, German, and U.S. air forces until late in World War II, when some large radial air-cooled fighters were introduced. However, the U.S. Navy and the Japanese, both relying primarily on carrier-based

Fig. 1. Cross section of 1710-in.³ liquid-cooled 12-cylinder engine used in fighters during 1937–1947. (*Detroit Diesel Allison Division, General Motors Corp.*)

fighters, used radial air-cooled engines throughout, the lighter weight and short length being considered of greater carrier utility than high speed.

Lightweight simple reduction gears of the offset spur type centered the propeller shaft in the V engine for low drag and good pilot visibility, and were necessary for the high engine speed inherently available with the adequate bearings and natural balance of the 12-cylinder V engine. High-temperature liquid cooling using either ethylene glycol or ethylene glycol—water mixtures at temperatures of 250–275°F reduced radiator requirements and drag and provided antifreeze protection. Because of the need for high burst performance, as well as high normal performance in fighter aircraft, many

Fig. 2. Cutaway of 18-cylinder twin-row radial air-cooled engine of 2804-in.³ total piston displacement. (*Pratt and Whitney Aircraft*)

of the detailed features in connection with radial air-cooled engines were first developed or applied on the liquid-cooled V engines. In general, a radial air-cooled engine of 1.5–2.0 times the engine displacement is required to give overall fighter aircraft performance comparable with a given liquid-cooled engine if equal skill is applied in the installation of each.

With the end of World War II, a number of additional liquid-cooled engines were manufactured for military use until suitable jet aircraft could be developed. A relatively small number of foreign liquid-cooled V engines were also converted for commercial use in modified aircraft. However, no liquid-cooled engines were in production and available after the war at comparable cost in the sizes required to compete with the large air-cooled engines, so the liquid-cooled type had been dropped in favor of developing turbojet or turboprop engines with apparent advantages over either the air- or liquid-cooled types of reciprocating engines.

Two-row radial air-cooled engines. In the period 1935–1937 manufacturers in the United States introduced large two-row radial air-cooled engines in sizes of 2600-, 2800-, and 3350-in.³ displacement. The first of these had 14 cylinders and the latter two had 18 cylinders. The 2600-in.³ engine was used in sizable quantities during World War II, but was not used in new military or commercial postwar applications. The two large 18-cylinder engines were used extensively during World War II and powered almost all advanced postwar reciprocating-engine commercial and military transports in the United States (Fig. 2). The principal exception was the large 28-cylinder radial air-cooled engine of 4360 in.³ initiated in 1940. Production of this engine was started at the end of World War II, and it was used after the war in both military bomber and cargo aircraft.

The 2800- and 3350-in.³ 18-cylinder engines represent the ultimate for radial air-cooled types both in variety of features or models and in detail design improvement for high specific outputs. These engines, initiated at power outputs of less than 0.5 hp/in.³ displacement on 87 octane fuel, approached 1.0 hp/in.³ in the late models on 115–145 octane fuel without use of exhaust turbines. In the case of the 3350-in.³ engine with three blowdown turbines feeding power back to the engine, specific outputs of over 1.1 hp/in.³ were attained for takeoff along with specific fuel consumptions under 0.4 lb/hp at cruise, while providing a major reduction in exhaust noise. The approximate doubling of specific output was partly a result of major improvement in the octane or antiknock rating of the fuel available and partly a result of major improvement in detail design of the engine, installation, and propeller.

High-output engine features. Some of the features provided in various models of piston engines during World War II for optimum characteristics in various aircraft were (1) various reduction gear ratios as engine speeds were increased and propeller activity factor or disk loadings were changed, and (2) various centrifugal compressors and drives for optimum supercharging, such as single stage, two speed; two stage, two speed; two stage, two speed with intercooler; and turbosupercharging with or without intercooling.

Following are some of the major design, process, and material details and component improvements that have been fed into these engines to permit higher speeds and higher cylinder temperatures and pressures, and to reduce local hot spots for higher overall engine performance on a given fuel: (1) higher-strength and temperature-resistant materials resulting from alloy improvement; (2) use of induction hardening, carburizing, or nitriding steels for higher strength and long wear characteristics, particularly where rubbing occurs, such as on cylinder barrel walls, piston rings, crankshafts, piston pins, link pins, and gears; (3) use of shot peening or surface rolling on highly stressed parts to eliminate residual stresses and add light compressive stresses for maximum uniformity, particularly at stress concentrations such as valve springs, connecting rods, rocker arms, and welded areas; (4) surface coating for improved functioning, which included such items as steel-backed, silver-plated master rod bearings with lead-indium coating; flame plating of high-temperature alloys on valve seat faces and insert seats; and silver, copper, or other coating, plating, or surface treatment to eliminate fretting, fretting erosion, or other action leading to fatigue cracks; (5) closer spacing and thinner fins on aluminum cylinder heads and steel barrels, with improved baffling for more uniform and better cylinder cooling; (6) tapered piston rings for higher-temperature piston operation without ring sticking; (7) availability of smaller-diameter spark plugs with ceramic insulation in place of mica; and (8) use of pressure type or floatless carburetors or fuel-injection systems for fuel metering to improve distribution, minimize icing troubles, reduce hazards of backfires (eliminated in the case of fuel injection), and avoid engine cutout, such as occurred with float-type carburetors, in negative-g maneuvering.

The result of these features and of other detail refinements in combination with an increase in engine speed and improved supercharger efficiency was a gain of 0.25–0.3 hp/in.³ exclusive of the gain from improved fuel. Similar improvement by detail refinement and reduction of drag in the engine installation in the nacelle itself further contributed to the satisfactory history of the reciprocating engine in both military and commercial aircraft.

All of these high-output features with the exception of fuel injection and blowdown exhaust turbines were in production on both air- and liquid-cooled engines during World War II in the United States, and many German engines used fuel injection.

Opposed air-cooled engines. The 4-, 6-, and 8-cylinder opposed air-cooled engines in both horizontal and vertical arrangements took over the aircraft power field up to about 400 hp shortly after World War II, largely eliminating in-line and radial engines in this rating both on commercial and military applications. Developed for commercial single- and twin-engine personal and business aircraft, these engines are also used for military training, reconnaissance, and drone and helicopter applications requiring high reliability and low cost (Fig. 3). They have achieved an excellent balance between automotive cost-saving features

Fig. 3. A 6-cylinder opposed engine of 471-in.³ displacement rated for 260 hp at 2625 rpm at sea level; dry weight is 426 lb. (*Continental Motors Corp.*)

and lightweight aircraft construction. An illustration of this is the availability of manifold fuel-injection systems providing good distribution and major reduction in icing and backfiring while avoiding the large expense of cylinder-head injection and while eliminating float-type carburetors.

As a result of improved design and rotor-casting techniques, small turbochargers have become available at reasonable cost and have been applied to several models of opposed piston engines to give much improved altitude performance. Annual production of opposed engines has increased in a major way because of military requirements for the Vietnam War and a large increase in sales of personal aircraft in the mid-1960s.

However, seeking to take advantage of the very low specific weights, simplified maintenance, and the capability of using lower-grade fuels, the military has sponsored and applied a variety of low-powered fan jets and turboprop-and-shaft power turbines which are directly competitive with available piston engines for helicopters, training craft, and personal aircraft. As production of the turbines increases, costs come down and the future of small reciprocating engines becomes less secure. See AIRCRAFT FUEL; INTERNAL COMBUSTION ENGINE; OTTO CYCLE.

[RONALD HAZEN]

Bibliography: K. S. Brown et al., *U.S. Army and Air Force Fighters, 1916–1961,* 1962; S. D. Heron, *History of the Aircraft Piston Engine,* 1961; K. Munson, *Military Aircraft,* 1966; Piston engines, *American Aviation Annual World Aviation Encyclopedia,* latest edition; J. W. R. Taylor (ed.), *Jane's All The World's Aircraft,* revised periodically; P. H. Wilkenson, *Aircraft Engines of the World,* 1966–1967, also 1964–1965, 1944–1962, and 1941.

Aircraft fuel

The source of energy required for the propulsion of airborne vehicles. This energy is released when the fuel reacts with the oxygen in air and so differs from rocket propellants, in which both fuel and oxidant are carried. Aircraft, more than other forms of transport, require that fuel weight be kept to a minimum.

Range equation. The Breguet range equation shows that the distance an aircraft can fly is directly proportional to the heat of combustion of the fuel. The relationships are shown in the equation below, where R is the range; k a constant; M the flight Mach number; SFC the power-plant specific fuel consumption; L/D the lift-drag ratio of the aircraft; W_i the initial aircraft weight; and W_f the final aircraft weight.

$$R = k \frac{M}{SFC} \frac{L}{D} \log \frac{W_i}{W_f}$$

Assuming a constant efficiency in the conversion of heat to thrust, the specific fuel consumption is inversely proportional to the heat of combustion of the fuel and, since the other terms in the range equation are not related to the fuel, the range becomes directly proportional to the heat of combustion. Therefore fuels must have the highest heats of combustion consistent with other requirements. These highest heats are found in the lower-molecular-weight elements: hydrogen, lithium, beryllium, boron, carbon, magnesium, and aluminum and in compounds of these elements. In practice, compounds of carbon and hydrogen (hydrocarbons) are used almost exclusively because of their excellent properties, low cost, and ready availability from petroleum. Fuels containing boron and magnesium have been considered for military aircraft, and liquid hydrogen is a candidate for possible future hypersonic flight.

Piston engine fuels. While light aircraft still use reciprocating engines, these have been almost completely replaced by the turbojet, turbofan, or turboshaft engines in commercial and military aircraft. Therefore piston engine fuels will be treated briefly. These are essentially high-grade gasolines.

While both reciprocating and turbine-powered aircraft have similar needs as to the physical properties of fuels, the two types of engines differ widely in their combustion requirements. The reciprocating engine is the more demanding. Just as in automobiles, but even more important in aircraft, the fuel should have a sufficiently high octane number so that the engine will not knock. Automobile engines often knock during a brief period of acceleration with no harm done. Aircraft engines use full power for much longer periods during takeoff and climb, and the sustained knocking may destroy the engine. Hydrocarbon fuels vary widely in their tendency to knock so that knock motor methods and ratings have been established to rate this property. *See* OCTANE NUMBER.

The fuel-air ratio fed to an aircraft engine can be varied from lean (for maximum economy) to rich (for maximum power). Engines are more prone to knock at the leaner conditions and, therefore, two numbers are used to characterize the various grades of aviation gasoline to represent lean and rich capabilities. For example, a 115/145 grade fuel expresses the performance numbers when tested lean and rich, respectively.

Aviation gasoline is usually a blend of selected virgin (uncracked) naphtha, alkylate, and cata-

lytically cracked gasoline. Alkylate is a synthetic product largely composed of highly branched paraffins, while catalytic gasoline is rich in aromatics. Up to 4.6 ml/gal of tetraethyllead may be added.

Aviation gasolines also must be sufficiently volatile to evaporate quickly and blend with air in the engine manifolds and be distributed evenly among all cylinders. They must not be too volatile or the fuel will boil in the tanks and lines. Gasolines boiling over a range of about 110–325°F and having a vapor pressure of $5\frac{1}{2}$–7 psi[2] meet these requirements. Aviation gasolines also must have low freezing points (−76°F is usually specified) and can neither contain a significant amount of nonvolatile residue (gum) nor form such residues in storage.

Jet fuels. These fuels are used in turbojet, turbofan, and turboshaft engines, and may be used someday in ramjets. While piston engine fuels all have the same volatility but differ in combustion characteristics, the jet fuels differ primarily in volatility; the differences in combustion qualities are minor. The volatility characteristics of typical samples of the several grades of jet fuel are shown in the table with aviation gasoline for comparison. Where the Reid vapor pressure (RVP, approximately the vapor pressure of the fuel at 100°F) is too low to be measured accurately, the flash point is given. This is the temperature to which a fuel must be heated to generate sufficient vapor to form a flammable mixture.

The fuels JP-1 and JP-3 are no longer used, and JP-2 never achieved specification status. JP-1 was the kerosine fuel first used by the military and was substantially the same as jet A, the commercial jet fuel now most widely used. JP-3 was a fuel having the maximum availability, but it was too volatile for use at very high altitudes: The fuel boiled out of the tanks with large losses. JP-4 and jet B are the military and commercial fuels, respectively, with substantially identical specifications. JP-4 is used in Air Force subsonic aircraft. Jet B has seen little use in the United States but is used by some overseas airlines. Jet B (and JP-4) is sufficiently volatile so that explosive mixtures are present at most ground storage conditions and at many flight conditions. Therefore it is generally agreed that there are greater flammable hazards with jet B than with jet A. If it were not for this factor, jet B would be the most-used commercial fuel since it is cheaper than jet A. JP-5 is the Navy service fuel and is the least volatile (highest flash

Volatility characteristics of jet fuels

Jet fuel grade	Distillation range, °F	RVP, psia	Flash, °F
JP-1	325–450		120
JP-3	100–500	6	
JP-4	150–500	$2\frac{1}{2}$	
JP-5	350–500		150
JP-6	300–500		100
Jet A	325–500		120
Jet B	150–500	$2\frac{1}{2}$	
Aviation gasoline	110–325	6	

point) of all the turbine fuels. JP-6 is the Air Force supersonic jet fuel.

Unlike reciprocating engines, jet engines can burn a wide variety of fuels efficiently. The major combustion problem is to keep smoke and combustor coking at low levels. Coke is a carbonaceous deposit that adheres to the internal parts of the combustor and can reduce engine life. Smoke can be an atmospheric pollution nuisance. Neither coke nor smoke represent an appreciable loss in combustion efficiency; under the worst conditions less than 0.25% of the fuel's heating value is lost. Increasing the aromatic content and increasing the concentrations of high-boiling components increases the smoking and coking tendencies of fuels. Therefore the aromatic content of jet fuels is restricted to less than 25% in the military and less than 20% in the commercial fuel specifications, and the concentrations of high-boiling material are controlled by specifying a maximum for either the 90% distilled temperature or the final boiling temperature. The final boiling temperature must be below 600°F for all jet fuels and is usually about 500°F.

Temperature stability. An increasingly important requirement is to provide a fuel that is stable to relatively high temperatures. In subsonic jets the fuel is used to cool the engine lubricant, and about 100 Btu is added to each pound of fuel under the severest conditions. This raises the fuel temperature by about 200°F. In supersonic jets the fuel is used as a heat sink not only for the engine lubricant but also for cabin air conditioning and cooling hydraulic systems. For very-high-speed flight the fuel may be used to cool additional engine components such as the turbine and exhaust nozzle. Fuel may be used to cool critical airframe areas such as the leading edges of wings at these high speeds. The fuel also is subjected to aerodynamic heating in the tanks. Therefore, depending on flight speeds, fuels can be heated to 300–500°F before they are burned. When fuels are so heated, small amounts of solids may be formed. These solids can foul heat exchangers, clog filters and fuel injectors, and cause controls to stick. These thermal instability problems are due to the presence of very small amounts of nonhydrocarbon impurities in the fuel.

The temperatures at which solids are first formed and the amounts of solids depend on fuel quality. In a specification test for this aspect of fuel quality, fuel is preheated and then passed through a heated filter for 5 hr. No significant amounts of solid may deposit on the preheater surface, and the pressure drop across the filter must stay within limits. The test temperatures are 300°F for the preheater and 400°F for the filter for all jet fuels except JP-6; these temperatures are 425 and 525°F for JP-6, showing the increased thermal stability requirement for the Air Force supersonic fuel.

Jet fuels also must have low freezing points (ranging from −40°F for jet A to −76°F for JP-4), be stable in storage, and meet other less important specification requirements. The most important criterion of fuel quality, heat of combustion, must exceed 18,300 Btu/lb for JP-5 and 18,400 Btu/lb for all the others. However, hydrocarbon blends suitable for jet fuels are nearly all the same in this regard, so that all grades of jet fuel have heats of combustion between 18,400 and 18,800 Btu/lb.

Special jet fuels. Sustained flight at speeds greater than about Mach 4 will require engine cycles other than the turbojet. The ramjet and the supersonic combustion ramjet (scramjet) are being considered along with hybrid types that combine rotating machinery and ramjet modes of operation. However, no matter what engine cycle is used, flight speeds will be limited by the amount of heat that can be added to a fuel prior to its combustion. Since JP or kerosine-type fuels are limited in this regard, two cryogenic, normally gaseous fuels (liquid hydrogen and liquid methane) are candidates for these advanced applications.

Liquid hydrogen, even though it boils at −423°F, is likely to be used in the scramjet at Mach numbers above 6. It not only can absorb 3–5 times more heat than the JP types, but its heat of combustion (51,570 Btu/lb) is 2.8 times greater than that of the JPs. It also is more readily mixed with air and burned in the very severe scramjet combustion environment. The other cryogenic fuel is liquid methane, or liquefied natural gas; most natural gases are greater than 90% methane. This fuel boils at −259°F, has a heat of combustion (21,500 Btu/lb) slightly greater than JP, and has a heat sink capacity intermediate between liquid hydrogen and JP. Liquefied natural gas is now being shipped by boat from Africa to Great Britain, and its cost has been lowered so that it is cheaper per Btu than JP-type fuels. For this reason, it is being considered for commercial air transport.

There also has been a past military interest in two types of nonhydrocarbon fuels. Next to hydrogen the boron hydrides have the highest heats of combustion. There was a major United States effort between 1952 and 1959 toward producing an alkyl decaborane fuel with handling qualities similar to those of the JPs but with heats of combustion of 26,000–27,000 Btu/lb. Experimental quantities of such fuels were produced and these, along with diborane and pentaborane, were found to be excellent fuels for ramjets and afterburners. However, the deposition of boric oxide caused serious problems in their use in turbojets. From 1947 to 1957, the use of concentrated slurries of boron and magnesium in liquid hydrocarbons was studied as fuels for ramjets and afterburners. Magnesium was used because it gives greater thrust (it has a higher flame temperature) than hydrocarbons; boron should give greater thrust and range. Slurries containing more than 50% metal were prepared. The magnesium slurries burned well, and small ramjets were flown successfully with this fuel. The boron slurries did not burn efficiently and formed objectionable deposits in the combustion chambers. *See* PROPULSION.

[ROBERT R. HIBBARD]

Bibliography: W. Francis, *Fuels and Fuel Technology*, 1965; N. A. Ragozin, *Jet Propulsion Fuels*, 1962.

Alkylation, petroleum

As applied to the petroleum industry, the reaction of isoparaffins with olefins to produce higher isoparaffinic compounds. In commercial operations C_3, C_4, and C_5 olefins alkylated with isobutane yield alkylates which distill in the gasoline

Table 1. Some commercial alkylation processes and their products

Olefin feed:	Ethylene	Propylene			Butenes		Pentenes	
Catalyst:	AlCl₃	AlCl₃	H₂SO₄	HF	H₂SO₄	HF	H₂SO₄	HF0
Temperature, °F:	100–150	70–150	30–80	15–100	30–70	15–120	30–70	40–10
Principal products				% volume of total alkylate				
Isopentane					10–1	10–1	10–50	10–?
2,3-Dimethylbutane	40–85							
2,4-Dimethylpentane		20–?	20–5	10–3				
2,3-Dimethylpentane		26–?	30–70	25–80				
Dimethylhexanes					15–3	40–3		
2,2,4-Trimethylpentane		4–?	5–15	20–5	25–45	9–45	10–20	24–?
2,3,4-Trimethylpentane					10–18	14–35	5–8	5–?
2,3,3-Trimethylpentane					12–21	0–15	5–8	4–?
2,2,3-Trimethylpentane					3–7	0–1	1–3	1–?
2,2,5-Trimethylhexane							25–5	29–?
2,3,5-Trimethylhexane							10–3	6–?

Table 2. Effects of alkylation conditions

Properties of product	Results under	
	Good conditions	Poor conditions
Quality specifications, C₆ alkylate (end point 338°F):		
ASTM motor octane number (clear)	96	90
ASTM research octane number (clear)	98	92
Rich aviation index number, 4.6 ml tetraethyllead	171	142
Lean aviation octane number, 4.6 ml tetraethyllead	110	105
Total yield of alkylate, vol. % on olefin	179	157
Selectivity, vol. % on C₅ + alkylate:		
iC_5H_{12}	2	11
$iC_6 + iC_7$	3	12
Trimethylpentanes	84	30
Dimethylhexanes	7	11
C_9+	4	36

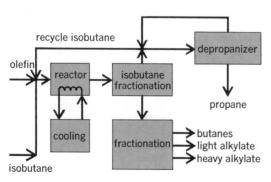

Alkylation, a petroleum fuels process.

range and which are principal components of high-octane fuel which is used for internal combustion engines.

The opposite reaction, dealkylation, is also important in the petroleum industry. Aromatic compounds such as naphthalene are now produced from refinery reformate stocks containing alkyl-aryl compounds by hydrodealkylation. *See* CRACKING.

Alkylation processes. Although thermal alkylation is possible, most commercial alkylation processes are catalytic because they give higher yields of the desired products. Some of the commercial catalytic processes for alkylating olefins with isobutane are summarized in Table 1. In most instances, with these processes highly branched hydrocarbons can be selectively produced from any olefin under the proper conditions of time, temperature, pressure, and olefin-isoparaffin ratio. The most selective alkylation processes operate at relatively low temperatures and at pressures sufficiently high to maintain liquid-phase operation. Thermal alkylation requires higher temperatures and rather high pressures. Alkylation of butenes with isobutane and a sulfuric acid catalyst surpasses other alkylation processes in total capacity.

Alkylation plants are simple. As the illustration shows, they include a reactor, an isoparaffin recycle system, and a product fractionation system. Cooling is normally employed because the alkylation reactions are highly exothermic.

Alkylation conditions. The alkylation reaction is complex, and operating conditions have a marked influence on the products that are formed. The reaction is usually explained by the formation of carbonium ions as reaction intermediates. Catalysts employed in alkylation have considerable polymerization activity for olefins. Isomerization, cracking of the feed or products, and self-alkylation can also occur.

Olefin polymerization results in the production of high-molecular-weight products that lead to loss of catalyst activity and high catalyst consumption. Polymerization is controlled by maintaining a high ratio of isobutane to olefin through recycle of isobutane and good dispersion of the olefin in the hydrocarbon and by control of temperature and residence time. The effects of alkylation conditions on product quality, yield, and selectivity are shown in Table 2.

Catalysts are subject to deactivation or destruction by several extraneous components inadvertently present in the feed. Components such as dienes, acetylenes, hydrogen sulfide, mercaptans, and excess water will adversely affect most alkyla-

tion catalysts. *See* BUTANE; PETROLEUM PRO-
CESSING.   [HAROLD C. RIES]

Bibliography: B. T. Brooks et al. (eds.), *The
Chemistry of Petroleum Hydrocarbons*, vols. 1–3,
1954–1955; A. E. Dunstan (ed.), *The Science of
Petroleum*, vol. 5, pt. 2, 1953: G. Egloff and G. Hul-
la, *Alkylation of Alkanes*, vols. 1–3, 1948; R. E.
Payne, Alkylation: What you should know about
this process, *Petrol. Refiner*, 37(9):316–329, 1958;
H. Steiner (ed.), *Introduction to Petroleum Chemistry*, 1961.

Alternating current

Electric current that reverses direction periodically, usually many times per second. Electrical energy is ordinarily generated by a public or a private
utility organization and provided to a customer,
whether industrial or domestic, as alternating current.

One complete period, with current flow first
in one direction and then in the other, is called a
cycle, and 60 cycles per second (60 hertz, or Hz) is
the customary frequency of alternation in the United States and in all of North America. In Europe
and in many other parts of the world, 50 Hz is the
standard frequency. On aircraft a higher frequency, often 400 Hz, is used to make possible lighter
electrical machines.

When the term alternating current is used as an
adjective, it is commonly abbreviated to ac, as in
ac motor. Similarly, direct current as an adjective
is abbreviated dc.

Advantages. The voltage of an alternating current can be changed by a transformer. This simple,
inexpensive, static device permits generation of
electric power at moderate voltage, efficient transmission for many miles at high voltage, and distribution and consumption at a conveniently low voltage. With direct (unidirectional) current it is not
possible to use a transformer to change voltage.
On a few power lines, electric energy is transmitted for great distances as direct current,
but the electric energy is generated as alternating
current, transformed to a high voltage, then
rectified to direct current and transmitted, then
changed back to alternating current by an inverter,
to be transformed down to a lower voltage for distribution and use.

In addition to permitting efficient transmission
of energy, alternating current provides advantages
in the design of generators and motors, and for
some purposes gives better operating characteristics. Certain devices involving chokes and transformers could be operated only with difficulty, if at
all, on direct current. Also, the operation of large
switches (called circuit breakers) is facilitated

because the instantaneous value of alternating
current automatically becomes zero twice in each
cycle and an opening circuit breaker need not in-
terrupt the current but only prevent current from
starting again after its instant of zero value.

Sinusoidal form. Alternating current is shown
diagrammatically in Fig. 1. Time is measured horizontally (beginning at any arbitrary moment) and
the current at each instant is measured vertically.
In this diagram it is assumed that the current is
alternating sinusoidally; that is, the current i is
described by Eq. (1), where I_m is the maximum in-

$$i = I_m \sin 2\pi ft \qquad (1)$$

stantaneous current, f is the frequency in cycles
per second (hertz), and t is the time in seconds.

A sinusoidal form of current, or voltage, is usually approximated on practical power systems
because the sinusoidal form results in less expensive construction and greater efficiency of operation of electric generators, transformers, motors,
and other machines.

Measurement. Quantities commonly measured
by ac meters and instruments are energy, power,
voltage, and current. Other quantities less commonly measured are reactive volt-amperes, power
factor, frequency, and demand (of energy during a
given interval such as 15 min).

Energy is measured on a watt-hour meter. There
is usually such a meter where an electric line enters a customer's premises. The meter may be
single-phase (usual in residences) or three-phase
(customary in industrial installations), and it displays on a register of dials the energy that has
passed, to date, to the system beyond the meter.
The customer frequently pays for energy consumed according to the reading of such a meter.
See WATT-HOUR METER.

Power is measured on a wattmeter. Since power
is the rate of consumption of energy, the reading of
the wattmeter is proportional to the rate of increase of the reading of a watt-hour meter. The
same relation is expressed by saying that the reading of the watt-hour meter, which measures energy, is the integral (through time) of the reading
of the wattmeter, which measured power. A wattmeter usually measures power in a single-phase
circuit, although three-phase wattmeters are sometimes used. *See* WATTMETER.

Current is measured by an ammeter. Current is
one component of power, the others being voltage
and power factor, as in Eq. (5). With unidirectional
(direct) current, the amount of current is the rate
of flow of electricity; it is proportional to the number of electrons passing a specified cross section of
a wire per second. This is likewise the definition of
current at each instant of an alternating-current
cycle, as current varies from a maximum in one
direction to zero and then to a maximum in the
other direction (Fig. 1.) An oscilloscope will indicate instantaneous current, but instantaneous current is not often useful. A dc (d'Arsonval-type) ammeter will measure average current, but this is
useless in an ac circuit, for the average of sinusoidal current is zero. A useful measure of alternating current is found in the ability of the current to
do work, and the amount of current is correspondingly defined as the square root of the average of
the square of instantaneous current, the average

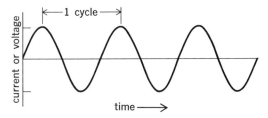

Fig. 1. Diagram of sinusoidal alternating current.

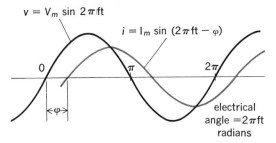

Fig. 2. The phase angle φ.

being taken over an integer number of cycles. This value is known as the root-mean-square (rms) or effective current. It is measured in amperes. It is a useful measure for current of any frequency. The rms value of direct current is identical with its dc value. The rms value of sinusoidally alternating current is $I_m/\sqrt{2}$, where I_m is the maximum instantaneous current. See Fig. 1 and Eq. (1).

Voltage is measured by a voltmeter. Voltage is the electrical pressure. It is measured between one point and another in an electric circuit, often between the two wires of the circuit. As with current, instantaneous voltage in an ac circuit reverses each half cycle and the average of sinusoidal voltage is zero. Therefore the root-mean-square (rms) or effective value of voltage is used in ac systems. The rms value of sinusoidally alternating voltage is $V_m/\sqrt{2}$, where V_m is the maximum instantaneous voltage. This rms voltage, together with rms current and the circuit power factor, is used to compute electrical power, as in Eqs. (4) and (5).

The ordinary voltmeter is connected by wires to the two points between which voltage is to be measured, and voltage is proportional to the current that results through a very high electrical resistance within the voltmeter itself. The voltmeter, actuated by this current, is calibrated in volts.

Phase difference. Phase difference is a measure of the fraction of a cycle by which one sinusoidally alternating quantity leads or lags another. Figure 2 shows a voltage v which is described in Eq. (2) and a current i which is described in Eq. (3).

$$v = V_m \sin 2\pi ft \tag{2}$$

$$i = I_m \sin (2\pi ft - \varphi) \tag{3}$$

The angle φ is called the phase difference between the voltage and the current; this current is said to lag (behind this voltage) by the angle φ. It would be equally correct to say that the voltage leads the current by the phase angle φ. Phase difference can be expressed as a fraction of a cycle or in degrees of angle, or as in Eq. (3), in radians of angle, with corresponding minor changes in the equations.

If there is no phase difference, and $\varphi = 0$, voltage and current are in phase. If the phase difference is a quarter cycle, and $\varphi = \pm 90$ degrees, the quantities are in quadrature.

Power factor. Power factor is defined in terms of the phase angle. If the rms value of sinusoidal current from a power source to a load is I and the rms value of sinusoidal voltage between the two wires connecting the power source to the load is V, the average power P passing from the source to the load is shown as Eq. (4). The cosine of the phase

angle, $\cos \varphi$, is called the power factor. Thus the rms voltage, the rms current, and the power factor are the components of power.

$$P = VI \cos \varphi \tag{4}$$

The foregoing definition of power factor has meaning only if voltage and current are sinusoidal. Whether they are sinusoidal or not, average power, rms voltage, and rms current can be measured, and a value for power factor is implicit in Eq. (5).

$$P = VI \, (\text{power factor}) \tag{5}$$

This gives a definition of power factor when V and I are not sinusoidal, but such a value for power factor has limited use.

If voltage and current are in phase (and of the same waveform), power factor equals 1. If voltage and current are out of phase, power factor is less than 1. If voltage and current are sinusoidal and in quadrature, power factor equals zero.

The phase angle and power factor of voltage and current in a circuit that supplies a load are determined by the load. Thus a load of pure resistance, as an electric heater, has unity power factor. An inductive load, such as an induction motor, has a power factor less than 1 and the current lags behind the applied voltage. A capacitive load, such as a bank of capacitors, also has a power factor less than 1, but the current leads the voltage, and the phase angle φ is a negative angle.

If a load that draws lagging current (such as an induction motor) and a load that draws leading current (such as a bank of capacitors) are both connected to a source of electric power, the power factor of the two loads together can be higher than that of either one alone, and the current to the combined loads may have a smaller phase angle from the applied voltage than would currents to either of the two loads individually. Although power to the combined loads is equal to the arithmetic sum of power to the two individual loads, the total current will be less than the arithmetic sum of the two individual currents (and may, indeed, actually be less than either of the two individual currents alone). It is often practical to reduce the total incoming current by installing a bank of capacitors near an inductive load, and thus to reduce power lost in the incoming distribution lines and transformers, thereby improving efficiency.

Three-phase system. Three-phase systems are commonly used for generation, transmission, and distribution of electric power. A customer may be supplied with three-phase power, particularly if he uses a large amount of power or if he wishes to use three-phase loads. Small domestic customers are usually supplied with single-phase power.

A three-phase system is essentially the same as three ordinary single-phase systems (as in Fig. 2, for instance) with the three voltages of the three single-phase systems out of phase with each other by one-third of a cycle (120 degrees) as shown in Fig. 3. The three voltages may be written as Eqs. (6), (7), and (8), where $V_{an(\max)}$ is the maximum

$$v_{an} = V_{an(\max)} \sin 2\pi ft \tag{6}$$

$$v_{bn} = V_{bn(\max)} \sin 2\pi (ft - 1/3) \tag{7}$$

$$v_{cn} = V_{cn(\max)} \sin 2\pi (ft - 2/3) \tag{8}$$

value of voltage in phase an, and so on. The

three-phase system is balanced if relation (9) holds,

$$V_{an(\max)} = V_{bn(\max)} = V_{cn(\max)} \qquad (9)$$

and if the three phase angles are equal, 1/3 cycle each as shown.

If a three-phase system were actually three separate single-phase systems, there would be two wires between the generator and the load of each system, requiring a total of six wires. In fact, however, a single wire can be common to all three systems, so that it is only necessary to have three wires for a three-phase system (a, b, and c of Fig. 4) plus a fourth wire n to serve as a common return or neutral conductor. On some systems the earth is used as the common or neutral conductor.

Each phase of a three-phase system carries current and conveys power and energy. If the three loads on the three phases of the three-phase system are equal and the voltages are balanced, then the currents are balanced also. Figure 2 can then apply to any one of the three phases. It will be recognized that the three currents in a balanced system are equal in rms (or maximum) value and that they are separated one from the other by phase angles of 1/3 cycle and 2/3 cycle. Thus the currents (in a balanced system) are themselves symmetrical, and Fig. 3 could be applied to line currents i_a, i_b, and i_c as well as to the three voltages indicated in the figure. Note, however, that the three currents will not necessarily be in phase with their respective voltages; the corresponding voltages and currents will be in phase with each other only if the load is pure resistance and the phase angle between voltage and current is zero; otherwise some such relation as that of Fig. 2 will apply to each phase.

It is significant that, if the three currents of a three-phase system are balanced, the sum of the three currents is zero at every instant. Thus if the three curves of Fig. 3 are taken to be the currents of a balanced system, it may be seen that the sum of the three curves at every instant is zero. This means that if the three currents are accurately balanced, current in the common conductor (n of Fig. 4) is always zero, and that conductor could theoretically be omitted entirely. In practice, the three currents are not usually exactly balanced, and either of two situations obtains. Either the common neutral wire n is used, in which case it carries little current (and may be of high resistance compared to the other three line wires), or else the common neutral wire n is not used, only three line wires being installed, and the three phase currents are thereby forced to add to zero even though this requirement results in some inbalance of phase voltages at the load.

It is also significant that the total instantaneous power from generator to load is constant (does not vary with time) in a balanced, sinusoidal, three-phase system. Power in a single-phase system that has current in phase with voltage is maximum when voltage and current are maximum and it is instantaneously zero when voltage and current are zero; if the current of the single-phase system is not in phase with the voltage, the power will reverse its direction of flow during part of each half cycle. But in a balanced three-phase system, regardless of phase angle, the flow of power is unvarying from instant to instant. This results in

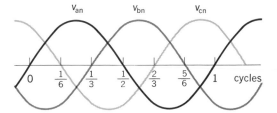

Fig. 3. Voltages of a balanced three-phase system.

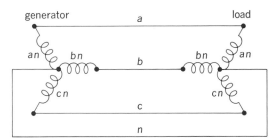

Fig. 4. Connections of a simple three-phase system.

smoother operation and less vibration of motors and other ac devices.

Three-phase systems are almost universally used for large amounts of power. In addition to providing smooth flow of power, three-phase motors and generators are more economical than single-phase machines. Polyphase systems with two, four, or other numbers of phases are possible, but they are little used except when a large number of phases, such as 12, is desired for economical operation of a rectifier.

Ideal circuit. An ideal power circuit should provide the customer with electric energy always available at unchanging voltage of constant waveform and frequency, the amount of current being determined by the customer's load. High efficiency is greatly desired. *See* CIRCUIT (ELECTRICITY); CURRENT (ELECTRICITY). [H. H. SKILLING]

Alternating-current generator

A machine which converts mechanical power into alternating-current electric power. The most common type, sometimes called an alternator, is the synchronous generator, so named because its operating speed is proportional to system frequency. Another type is the induction generator, the speed of which varies somewhat with load for constant output frequency.

Synchronous generators. These generators usually have the field winding mounted on the rotor and a stationary armature winding mounted on the stator. In small ratings where high reliability is imperative, the magnetic field may be created by permanent magnets. Small alternators are also being built with stationary field windings and rotating armatures with their leads brought out through collector rings or fed directly to rotating rectifiers, which supply direct current to the rotating field of a much larger alternator, as in the brushless excitation system. Still another type, the inductor alternator described below, has both its field and armature windings in the stator. Although synchronous generators may be single-phase, they are usually two- or three-phase; most are three-phase.

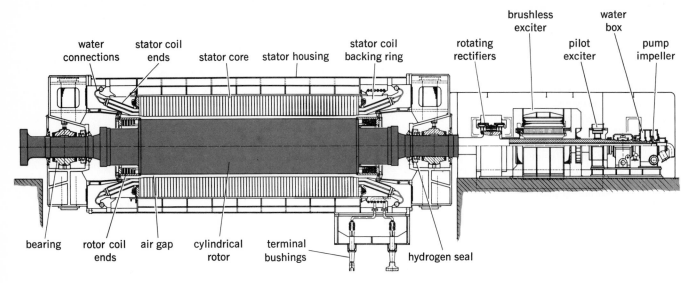

water connections   stator coil ends   stator core   stator housing   stator coil backing ring   rotating rectifiers   brushless exciter   pilot exciter   water box   pump impeller

bearing   rotor coil ends   air gap   cylindrical rotor   terminal bushings   hydrogen seal

Fig. 1. An 1800-MVA four-pole (operating at 1800 rpm) synchronous generator; it has a water-cooled stator and rotor windings, and brushless excitation equipment. (*Allis-Chalmers Power Systems*)

Single-phase generators are rare because they are larger than polyphase machines of the same kilovolt-ampere ratings, because they have a pulsating torque and are noisy, and because single-phase power is not well suited to self-starting ac motors other than those of fractional horsepower sizes. For theory of synchronous machines *see* SYNCHRONOUS MOTOR.

In the usual type with rotating field windings, a pair of field poles must pass a given point on the armature in 1 cycle. Hence the number of poles required are determined from the frequency *f* in hertz (Hz) and speed by Eq. (1). The speed se-

$$\text{Number of poles} = 120f/\text{rpm} \qquad (1)$$

lected is that best suited to the prime mover. Steam turbines operate most economically at high speed, hence two- or four-pole generators are used for this service, running at 3600 or 1800 rpm, respectively, for an output frequency of 60 Hz. For hydraulic turbine or engine drive, slow-speed machines having many poles are customary.

High-speed synchronous generators. Generators of this type have cylindrical rotors of solid alloy-steel forgings, with radial slots machined along their length to receive the field windings, as shown in Fig. 1. The field coils are of bare, hard-drawn, strip copper installed turn by turn, within an insulating channel in each slot. Mica plate, epoxy-bonded woven-glass laminate, or similar insulation is commonly used between turns.

The slot portion of the windings is supported against centrifugal force by nonmagnetic wedges, while the coil ends are retained by nonmagnetic metal rings lined with insulation. Since most high-speed generators over 15,000 kVA are now either hydrogen-cooled or water-cooled, the insulated field leads are brought out to the collector rings through an axial hole in the shaft, with the aid of gastight radial studs. Fans or blowers are mounted on the rotor in most cases, employing hydrogen as a cooling medium.

The stator of a large, high-speed, synchronous generator uses a steel yoke, which also serves as an enclosure for the ventilating medium. A cylindrical core of laminated electrical sheet steel is stacked on dovetail bars within the yoke and tightly clamped to minimize magnetic noise. The insulated armature coils of large machines are usually of one turn, lap wound in rectangular slots around the inner periphery of the core. Coil ends are securely lashed to supporting brackets, and both ends of each phase are brought out through terminal bushings. Hydrogen seals are mounted in the end covers, and the entire stator is designed to resist explosion pressures.

Cooling methods. Air-cooled and hydrogen-cooled generators under 100 MVA are most commonly indirectly cooled, that is, the cooling medium contacts metal surfaces and exterior surfaces of the coil insulation. Heat generated in the windings must flow through the major insulation, which is a poor heat conductor. In 1948 S. Beckwith designed a 60-MW 3600-rpm rotor in which hydrogen was forced by a powerful blower to flow at high velocity through ducts within the conductors. A tremendous increase in output was made possible by this highly effective cooling method, which is now called direct cooling. When applied also to the stator windings and augmented with higher hydrogen pressures, much greater gains in generator output were achieved.

In parallel with this development, direct water cooling of stator windings was also adopted in the 1950s, and later applied to rotors as well. Figure 1 illustrates an 1800-MVA four-pole generator with water-cooled stator and rotor windings, together with its brushless excitation equipment.

Slow-speed synchronous generators. The field poles of these generators are usually the salient-pole type. When driven by reciprocating engines, they sometimes require flywheels to minimize the pulsations transmitted to the power system. In hydroelectric generators the rotors are sometimes required to withstand high overspeed because of the time delay in closing the gates.

Generated voltage. The voltage generated per phase in a synchronous generator can be derived from Faraday's law. If it is assumed the flux distribution over each pole is sinusoidal, a sinusoidal voltage is induced in each coil side. However, the coil voltages must be added vectorially because of

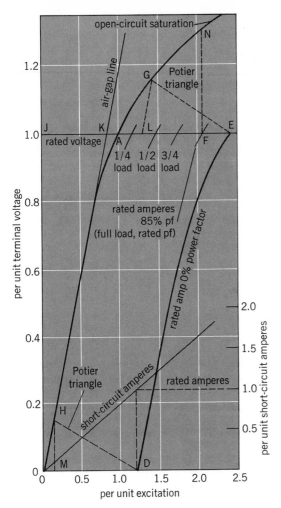

Fig. 2. Characteristic curves of a synchronous generator.

are adjusted at each successive test voltage point to circulate rated amperes between them, overexcited on the unit being tested and underexcited on the other machine. The curves show how field current varies with load and voltage, and indicate some of the constants. The Potier reactance drop, used in determining saturation, is the height MH of the Potier triangle OHD.

The triangle is found from the 0% power factor saturation curve DE as follows: (a) Lay off length OD to left of point E to find L. (b) Draw line LG parallel to the airgap line to find G, its intersection with the open circuit saturation curve. (c) Complete the triangle by drawing line GE. (d) Make OH equal to LG, thus establishing the Potier reactance drop MH.

An approximate value of the Potier reactance may be calculated from Eq. (3), where $x_l =$ leakage reactance and $x'_d =$ transient reactance.

$$x_p = x_l + 0.63(x'_d - x_l) \qquad (3)$$

Unsaturated per unit synchronous reactance is OD/JK, and the short circuit ratio, whose reciprocal is often used in steady-state stability studies, is JA/OD. The voltage regulation is given by FN/OJ for 85% power-factor lagging load.

Inductor alternator. A synchronous generator in which the field winding is fixed in magnetic position relative to the armature conductors is known as an inductor alternator. There are two types: the homopolar, in which the dc field coil is concentric with the shaft, and the heteropolar, in which the dc windings are distributed. In both types the ac windings are distributed, and generate their induced voltage from the pulsation in the flux caused by the change in position of the salient poles on the rotor. Inductor alternators are used for high-frequency power and, in conjunction with static rectifiers, as a maintenance-free power source for ac excitation systems.

Induction generators. These nonsynchronous ac generators are driven above synchronous speed by external sources of mechanical power. The construction of these machines is identical to that of induction motors. They can operate either as motors or generators, depending on whether the speed is below or above synchronous speed. In some frequency-changer sets, induction generators may operate as a motor part of the time. This is accomplished by coupling them to a two-speed synchronous machine capable of operating at system frequency as a motor at the higher speed and as a generator at the lower speed. Induction generators are not common in large sizes because of their poor power factor. They can deliver only leading currents. Moreover, they require a power supply from which to obtain their magnetizing current. This normally requires that they be operated in parallel with a synchronous source. Capacitors may be used to minimize the current taken from the source. Under such conditions the frequency of the induction generator is determined by the frequency of the synchronous source and the output of the generator is determined by the mechanical input to its shaft. Induction generators do not require any dc excitation and their control can be very simple. For this reason they are well suited to small, unattended hydroelectric units, where they are easily operated by remote control. *See* INDUCTION MOTOR.   [LEON T. ROSENBERG]

their time-phase displacement. The effective root-mean-square voltage per phase can be shown to be that in Eq. (2), where Φ_m is peak flux per

$$E = 2.22(a/c)\Phi_m f k_d k_p \qquad (2)$$

pole in webers, a is total conductors per phase, f is frequency in Hz, c is number of parallel circuits, k_d is distribution factor, and k_p is pitch factor.

If the flux distribution is not a perfect sine wave, any irregularites appear as odd harmonics in the generated voltage. Although seldom harmful to the generator, harmonics sometimes interfere with telephone communication. Perhaps the most troublesome harmonics result from pulsation in airgap reluctance as the poles move across the slots and teeth. These are known as slot harmonics. They occur in pairs at frequencies equal to (the slots per pair of poles $\pm$ 1) times the fundamental. Slot harmonics can be minimized by careful design of pole contour, fractional-slot windings, and skewed stator slots. Interference effects are often reduced with external resonant shunts or wave traps.

Characteristics. The characteristic curves of a synchronous generator are shown in Fig. 2. The open-circuit and short-circuit curves are readily found from tests at no-load. The saturation curve at rated amperes and 0% power factor is obtained by electrical connection to another synchronous machine. The field currents of both machines

(a)

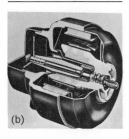

(b)

Fig. 1. Typical ac motors.
(a) One-half-hp split-
phase induction motor,
about 6 in. in diameter.
(b) One-hundredth-hp
motor, about 1 in. in
diameter.

Bibliography: D. G. Fink and J. M. Carroll (eds.),
Standard Handbook for Electrical Engineers,
1975; A. E. Fitzgerald et al., *Electrical Machinery*,
1971; D. G. Gehmlich and S. B. Hammond, *Electromechanical Systems*, 1967; E. Levi and M. Panzer, *Electromechanical Power Conversion*, 1966; J.
Rosenblatt and M. H. Friedman, *Direct and Alternating Current Machinery*, 1963; G. J. Thaler and
M. L. Wilcox, *Electric Machines: Dynamics and
Steady State*, 1966.

Alternating-current motor

An electric rotating machine which converts
alternating-current (ac) electric energy to mechanical energy; one of two general classifications of
electric motor. Because ac power is widely available, ac motors are commonly used. They are
made in sizes from a few watts to thousands of
horsepower (hp) (Fig. 1). *See* DIRECT-CURRENT
MOTOR.

CLASSIFICATIONS OF AC MOTORS

Each type of ac motor has special properties.
These motors are generally classified by application, construction, principle of operation, or operating characteristics.

Induction motor. The most common type of ac
motor is the induction motor. Current is induced
in a rotor as its conductors cut lines of magnetic
flux created by currents in a stator. Three-phase
induction motors are simple, reliable motors with
a fairly constant speed over the rated load range.
They are self-starting and are widely used in
industry. Single-phase induction motors require
special means for starting, but are widely used in
fractional and small integral horsepower sizes,
especially in homes. *See* INDUCTION MOTOR.

Synchronous motor. Where constant speed is
essential, a synchronous motor is used. It runs
at a fixed speed in synchronism with the frequency
of the power supply. Large synchronous motors
used in industry employ dc fields on their rotors
and three-phase armature (or stator) windings.
Efficiency and power factor of these motors are
high. The reluctance motor and the Permasyn
motor, either single- or three-phase, come in fractional and lower integral hp sizes. The reluctance
motor has low efficiency and power factor, but is
simple and inexpensive. The Permasyn has permanent magnets embedded in the squirrel-cage rotor
to provide the equivalent of a dc field. The hysteresis motor is used only in small sizes where its
quiet operation is especially desired, such as in
phonographs. *See* SYNCHRONOUS MOTOR.

AC series motor. For operation from either ac or
dc power, the series motor has its field winding

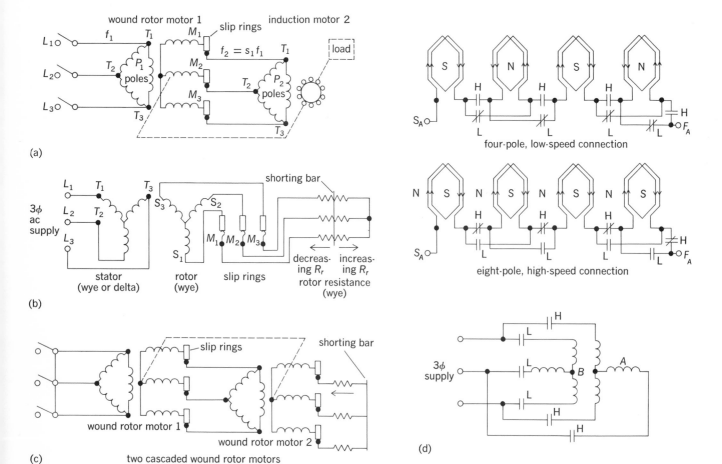

(a)

(b)

(c)   two cascaded wound rotor motors

(d)

Fig. 2. Schematics of methods of speed control for polyphase ac motors. (a) Frequency control using cascade-
wound rotor and squirrel-cage motors. (b) Slip control
using wound-rotor induction motor external rotor resistance. (c) Foreign voltage slip control using secondary
concatenation. (d) Phase A of a three-phase winding
showing consequent pole connections for multispeed
induction motors.

Standard ranges of hp, speeds, and slips at 60 Hz*

No. of phases	Type of motor	Power output, hp	Synchronous speed, rpm	Slip, %
Single	Capacitor	1–10	1800–1200	0–5
	Split-phase	0–0.5	1800–1200	0–5
	Shaded-pole	0–0.25	1800–900	11–14
	Repulsion-start	0–25	1800–1200	0–5
Three	Induction	0–100,000	3600–450	0–5
Three	Synchronous	20–30,000	3600–80	0

*National Electrical Manufacturers Association.

connected in series with the armature winding through a commutator and brush arrangement, as in a dc series motor. Field and armature iron are laminated. This universal motor has high starting torque. Speed can be controlled by adjusting the applied voltage. They are used in sizes up to 2000 hp in electric railways, and in small sizes for domestic appliances.

Repulsion motor. This is also a commutator motor, but the brushes and commutator are short-circuited and not connected in series with the stator. Armature current is set up by induction from the stator rather than by conduction, as in the series motor. The rotor of a repulsion motor differs from the squirrel-cage rotor of the single-phase induction motor; it is similar to a dc armature with commutator and brushes. Torque is developed by action of induced armature current on stator flux. The repulsion motor has high starting torque. Its speed can be controlled by changing the applied voltage or by shifting the brushes or both. Its no-load speed is above synchronism, but is lower than the no-load speeds of ac series and universal motors. In normal operation the brushes are located 15–25 electrical degrees off the stator position. Rotation is reversed by shifting the brushes to the opposite side of the stator axis. Although repulsion motors have been used on single-phase electric railways, by far the widest application in the past has been as a starting arrangement for single-phase induction motors. When used for this purpose, a centrifugal device short-circuits the commutator bars at about 70% of full speed. Combinations of series and repulsion connections have been used for controlling speed, commutation, and characteristics; occasionally a squirrel-cage winding is added on the armature. For starting duty, the repulsion motor draws low starting current, but despite this advantage the repulsion motor has been superseded by the simpler, cheaper capacitor motor. *See* REPULSION MOTOR.

The table gives a comparison of available sizes and characteristics of some common ac motors.

For general principles *see* ELECTRIC ROTATING MACHINERY; MOTOR, ELECTRIC.

[ALBERT F. PUCHSTEIN]

SPEED CONTROL OF AC MOTORS

The polyphase synchronous motor is a constant-speed (synchronous), variable-torque, doubly excited machine. The stator armature is excited with polyphase ac of a given frequency, and the rotor field is excited with dc. The rotor speed of the synchronous motor is a direct function of the number of stator and rotor field poles and the frequency of the ac applied to the stator. Since the number of

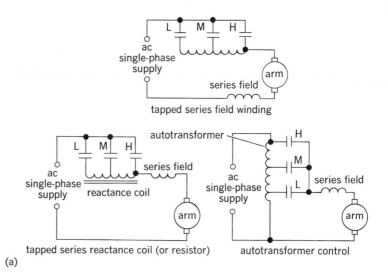

(a)

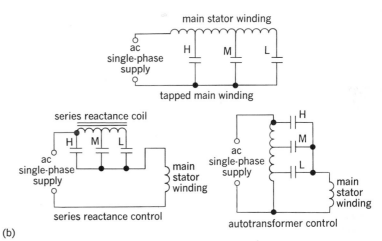

(b)

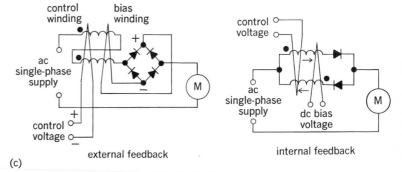

(c)

Fig. 3. Schematics for various methods of speed control for single-phase ac motors operating from fixed supplies. (*a*) Controls for ac series motors or universal motors. (*b*) Slip control for shaded-pole, reluctance, and split-phase motors. (*c*) Magnetic amplifier for voltage control of single-phase motors.

rotor poles of the polyphase synchronous motor is not easily modified, the change of frequency method is the only way to control synchronous speed of the motor.

The polyphase induction motor is also a doubly excited machine, whose stator armature is excited with polyphase ac of line frequency and whose rotor is excited by induced ac of variable frequency, depending on rotor slip. The speed of the polyphase induction motor is an asynchronous speed

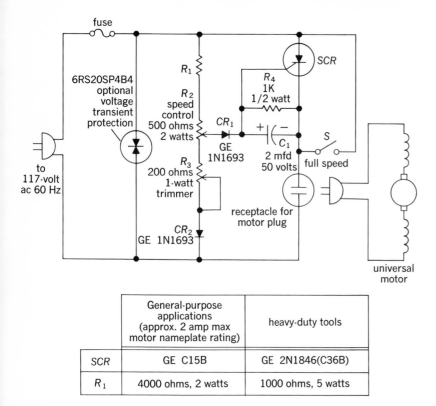

Fig. 4. Plug-in speed control with feedback using a silicon-controlled rectifier (SCR) for voltage control of single-phase motors.

	General-purpose applications (approx. 2 amp max motor nameplate rating)	heavy-duty tools
SCR	GE C15B	GE 2N1846(C36B)
R_1	4000 ohms, 2 watts	1000 ohms, 5 watts

varied by one or more of the following methods: (1) by changing the applied frequency to the stator (same method as for a synchronous motor, already discussed); (2) by controlling the rotor slip by means of rheostatic rotor resistance control (used for wound-rotor induction motors); (3) by changing the number of poles of both stator and rotor; and (4) by "foreign voltage" control, obtained by conductively or inductively introducing applied voltages of the proper frequency to the rotor. Figure 2 illustrates with schematics these methods of speed control.

It should be noted that a number of electromechanical or purely mechanical methods also serve to provide speed control of ac motors. One is the Rossman drive, in which an induction motor stator is mounted on trunnion bearings and driven with an auxiliary motor, providing the desired change in slip between stator and rotor. Polyphase induction and synchronous motors having essentially constant speed characteristics at rated voltage are also assembled in packaged drive units employing gears, cylindrical and conical pulleys, and even hydraulic pumps to produce a variable speed output.

In addition to reversal of rotation, some of these units employ magnetic slip clutches and solenoids to control the various mechanical and hydraulic arrangements through which a relatively smooth control of speed is achieved. A discussion of such electromechanical and mechanical speed-

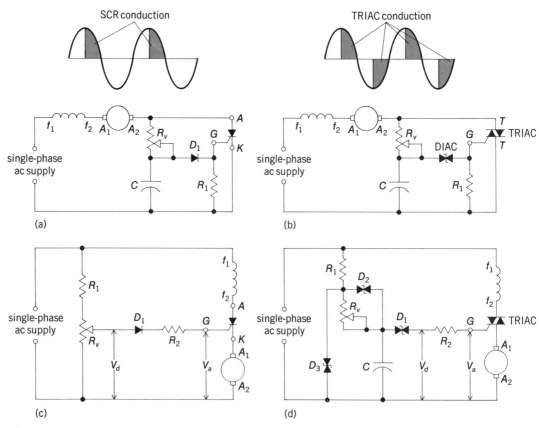

Fig. 5. Single-phase, universal or dc motor control using half- and full-wave (ac) controls, with and without voltage feedback. (a) Half-wave circuit. (b) Full-wave circuit. (c) Half-wave circuit with voltage feedback. (d) Full-wave circuit with voltage feedback. (From I. L. Kosow, Control of Electric Machines, 1973)

control methods is beyond the scope of this article.

The principal method of speed control used for fractional hp single-phase induction-type, shaded-pole, reluctance, series, and universal motors is the method of primary line-voltage control. It involves a reduction in line voltage applied to the stator winding (of the induction type) or to the armature of series and universal motors. In the former this produces a reduction of torque and increase in rotor slip. In the latter it is simply a means of controlling speed by armature voltage control or field flux control or both. The reduction in line voltage is usually accomplished by any one of five methods shown in Figs. 3 and 4; these are autotransformer control, series reactance control, tapped main-winding control, saturable reactor (or magnetic amplifier) control, and silicon-controlled rectifier feedback control.

Electronic speed control techniques. The development of the thyristor, or silicon-controlled rectifier (SCR), has created unlimited possibilities for control of virtually all types of motors (single-phase, dc, and polyphase). SCRs in sizes up to 1600 amperes (rms) with voltage ratings up to 1600 volts are available.

SCR control of series motors. Fractional horsepower universal, ac and dc series motors may be speed-controlled from a single-phase 110- or 220-volt supply using the half-wave circuit involving a diode D_1 and SCR (Fig. 5a). The trigger point of the SCR is adjusted via potentiometer R_v, which phase-shifts the gate turn-on voltage of the SCR whenever its anode A is positive with respect to cathode K. During the negative half-cycle of input voltage, SCR conduction is off and the gating pulse is blocked by diode D_1. When the positive half-cycle is initiated once again, capacitor C charges to provide the required gating pulse at the time preset by the time constant R_vC.

Replacing diode D_1 by a DIAC and the SCR by a TRIAC converts the circuit from half-wave to full-wave operation and control (Fig. 5b). Compared with Fig. 5a, this circuit provides almost twice the torque and improved speed regulation. Neither circuit, however, is capable of automatic maintenance of desired speed because of the inherently poor speed regulation of series-type ac, universal, or dc motors, with application or variation of motor loading.

Automatic speed regulation is achieved by using the half-wave feedback circuit shown in Fig. 5c. This circuit requires, however, that both field leads (f_1, f_2) and both armature leads (A_1, A_2) are separately brought out for connection as shown in Fig. 5c. The desired speed is set by potentiometer R_v, in terms of reference voltage V_d across diode D_1. The actual motor speed is sensed in terms of the armature voltage V_a. Diode D_1 conducts only when the reference voltage V_d (desired speed) exceeds the voltage V_a to restore the motor speed to its

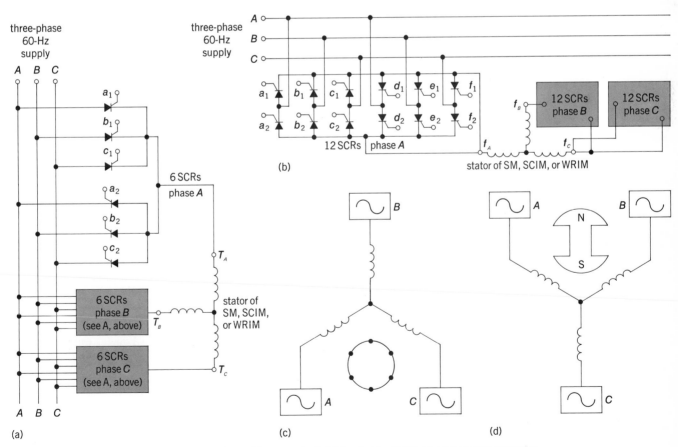

Fig. 6. Half-wave and full-wave cycloconverters with graphical symbols for simplified representation. (a) Half-wave cycloconverter (exclusive of triggering of SCR gates). (b) Full-wave cycloconverter (exclusive of SCR gating). (c) Symbol for SCIM driven by a cycloconverter. (d) Symbol for SM driven by a cycloconverter. (*From I. L. Kosow, Control of Electric Machines, 1973*)

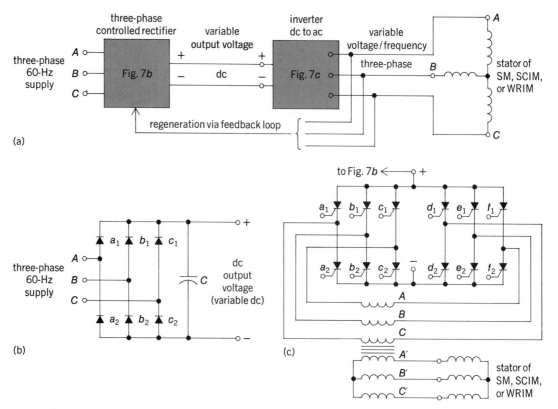

(a)

(b)

(c)

stator of
SM, SCIM,
or WRIM

Fig. 7.   Rectifier-inverter circuitry for variable voltage/
frequency control of three-phase motors. (a) Block diagram of rectifier-inverter. (b) Three-phase full-wave controlled rectifier. (c) Basic inverter dc to three-phase ac
for phase-shift voltage control. (*From I. L. Kosow, Control
of Electric Machines, 1973*)

desired setting. Thus, when the motor load is
increased and the speed drops, the SCR is gated
earlier in the cycle because diode D_1 conducts
whenever V_d exceeds V_a.

As with the previous half-wave circuit, the feedback circuit may be converted to full-wave operation by replacing diode D_1 with a DIAC and the
SCR with a TRIAC (Fig. 5d). This last circuit also
is shown with additional minor modifications to
improve regulation of speed at lower speeds, namely addition of DIACs D_2, D_3 and capacitor C. The
advent of high-power TRIACs and SCRs has
extended use of the circuit shown in Fig. 5d to
larger, integral horsepower ac series motors as
well.

When dc motors are operated by using electronic techniques shown in Fig. 5, it is customary to
derate motors proportionally because of heating
effect created by ac components in both half- and
full-wave waveforms. Note that there is no reversal
of direction in either the ac or dc series motor
under full-wave operation because both the armature and field currents have been reversed.

SCR control of polyphase ac motors. There are
fundamentally only three types of polyphase ac
motors, namely synchronous motors (SMs),
squirrel-cage induction motors (SCIMs), and
wound-rotor induction motors (WRIMs). All three
employ identical stator constructions. As a result,
larger polyphase motors (up to 10,000 hp) are
presently being controlled by SCR packages which
employ some so-called universal method of speed
control. In these methods, both the frequency and
stator voltages are varied (in the same proportion)
to maintain constant polyphase stator flux densities

and saturation, thus eliminating the possibility
of overheating.

Two major classes of solid-state adjustable
voltage/frequency drives have emerged, namely
the cycloconverter (an ac/ac package) and the
rectifier-inverter (an ac-dc-ac package). Both
packages convert three-phase fixed-frequency ac
to three-phase variable voltage/variable frequency.

The half-wave cycloconverter shown in Fig. 6a
is capable of supplying from zero to 20 Hz to an
ac polyphase motor. It uses 6 SCRs per phase but
is incapable of either phase-sequence reversal or
frequencies above 20 Hz. The full-wave converter,
shown in Fig. 6b, uses twice the number of SCRs
(12 per phase) and possesses advantages of wider
frequency variation (from +30 Hz down to 0 and
up to −30 Hz), potentialities for dynamic braking,
and capability of power regeneration.

The symbol for a polyphase SCIM driven by a
cycloconverter is shown in Fig. 6c, and an SM in
Fig. 6d. Note that the symbol implies frequency/
voltage control of identical stators, and the nature
of the motor is determined solely by its rotor.

The second class of solid-state drive package is
the rectifier-inverter, which converts fixed-frequency polyphase ac to dc (variable voltage). The
dc is then inverted by a three-phase inverter (using
a minimum of 12 SCRs, commercially) to produce
variable-voltage, variable-frequency three-phase
ac for application to the motor stator. The block
diagram of a solid-state rectifier-inverter package
is shown in Fig. 7a. The three-phase 60-Hz supply
is first rectified to produce variable dc (Fig. 7b) by
appropriate phase shift of SCR gates. The variable
dc is then applied to the dc bus of Fig. 7c. Inversion

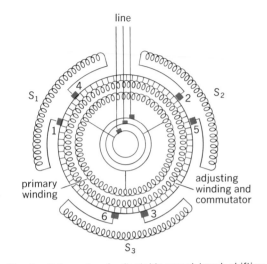

Fig. 8. Schematic of adjustable-speed brush-shifting motor. (*From A. E. Fitzgerald and C. Kingsley, Jr., Electric Machinery, McGraw-Hill, 1952*)

is accomplished by appropriate phase shift of gates of the 12 SCRs to produce phase and line voltages displaced, respectively, by 120°. This three-phase output voltage is applied to a transformer whose secondary is applied to the stator of the motor (Fig. 7c).

The rectifier-inverter is capable of power regeneration as noted by the feedback loop shown in Fig. 7a. There is no clear-cut choice between cycloconverters and rectifier-inverter packages, as of this writing.

Brush-shifting motors. By use of a regulating winding, a commutator, and a brush-shifting device, speed of a polyphase induction motor can be controlled similarly to that of a dc shunt motor. Such motors are used for knitting and spinning machines, paper mills, and other industrial services that require controlled variable-speed drive. The primary winding on the rotor is supplied from the line through slip rings. The stator windings are the secondary windings (S_1, S_2, S_3), and the third winding, also in the rotor, is an adjusting winding provided with a commutator (Fig. 8). Voltages collected from the commutator are fed into the secondary circuit. Brushes 1, 2, and 3 are mounted 120 electrical degrees apart on a movable yoke. Brushes 4, 5, and 6 are similarly mounted on a separate movable yoke. Each set of brushes can be moved as a group. Thus both the spacing between sets of brushes and the angular position of the brushes are adjustable. Brush spacing determines the magnitude of the voltage applied to the secondary. When brush sets are so adjusted that pairs of brushes are in contact with the same commutator segment, the secondary is short-circuited and no voltage is supplied. Under these conditions the motor behaves as an ordinary induction motor. The speed can be reduced by separating the brushes so that secondary current produces a negative torque. The machine can be operated above synchronism by interchanging the position of the brushes, so the voltage collected is in a direction to produce a positive torque. The motor can be reversed by reversing two of the leads supplying the primary.

[IRVING L. KOSOW]

Bibliography: D. G. Fink and J. M. Carroll (eds.), *Standard Handbook for Electrical Engineers*, 10th ed., 1968; A. E. Fitzgerald and C. Kingsley, *Electric Machinery*, 1961; I. L. Kosow, *Electric Machinery and Transformers*, 1972; I. L. Kosow, *Control of Electric Machines*, 1973.

Atmospheric pollution

Alteration of the atmosphere by the introduction of natural and artificial particulate contaminants. Most artificial impurities are injected into the atmosphere at or near the Earth's surface. The atmosphere cleanses itself of these quickly, for the most part. This occurs because in the troposphere, that part of the atmosphere nearest to the Earth, temperature decreases rapidly with increasing altitude (Fig. 1), resulting in rapid vertical mixing: the rainfall sometimes associated with these conditions also assists in removing the impurities. Exceptions, such as the occasional temperature inversion layer over the Los Angeles Basin, may have notably unpleasant results.

In the stratosphere, that is, above the altitude of the temperature minimum (tropopause), either temperature is constant or it increases with altitude, a condition that characterizes the entire stratosphere as a permanent inversion layer. As a result, vertical mixing in the stratosphere (and hence, self-cleansing) occurs much more slowly than that in the troposphere. Contaminants intro-

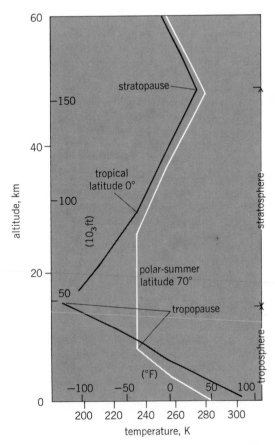

Fig. 1. The atmosphere's temperature-altitude profile. (*Adapted from R. E. Newell, Radioactive Contamination of the Upper Atmosphere, in Progress in Nuclear Energy, ser. 12: Health Physics, vol. 2, p. 538, Pergamon Press, 1969*)

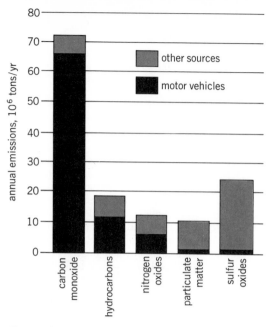

Fig. 2. Air pollutions' contribution to five atmospheric contaminants in the United States.

duced at a particular altitude remain near that altitude for periods as long as several years. Herein lies the source of concern: the turbulent troposphere cleanses itself quickly, but the relatively stagnant stratosphere does not. Nonetheless, the pollutants injected into the troposphere and stratosphere impact on humans and their habitable environment.

All airborne particulate matter, liquid and solid, and contaminant gases exist in the atmosphere in variable amounts. Typical natural contaminants are salt particles from the oceans or dust and gases from active volcanoes; typical artificial contaminants are waste smokes and gases formed by industrial, municipal, household, and automotive processes, and aircraft and rocket combustion processes. Another postulated important source of artificial contaminants is certain fluorocarbon compounds (gases) used widely as refrigerants, as propellants for aerosol products, and for other applications. Pollens, spores, rusts, and smuts are natural aerosols augmented artificially by humans' land-use practices.

Sources and types. Sources may be characterized in a number of ways. A frequent classification is in terms of stationary and moving sources. Examples of stationary sources are power plants, incinerators, industrial operations, and space heating. Examples of moving sources are motor vehicles, ships, aircraft, and rockets. Another system of classification describes the sources as point (a single stack), line (a line of stacks), or area (a city).

Different types of pollution are conveniently specified in various ways: gaseous, such as carbon monoxide, or particulate, such as smoke, pesticides, and aerosol sprays; inorganic, such as hydrogen fluoride, or organic, such as mercaptans; oxidizing substances, such as ozone, or reducing substances, such as oxides of sulfur and oxides of nitrogen; radioactive substances, such as iodine-131, or inert substances, such as pollen or fly

ash; or thermal pollution, such as the heat produced by nuclear power plants.

Air contaminants are produced in many ways and come from many sources. It is difficult to identify all the various producers. For example, it is estimated that in the United States 60% of the air pollution comes from motor vehicles and 14% from plants generating electricity. Industry produces about 17% and space heating and incineration the remaining 9%. Other sources, such as pesticides and earth-moving and agricultural practices, lead to vastly increased atmospheric burdens of fine soil particles, and of pollens, pores, rusts, and smuts; the latter are referred to as aeroallergens because many of them induce allergic responses in sensitive persons.

The annual emission over the United States of many contaminants is very great (Fig. 2). As mentioned, motor vehicles contribute about 60% of total pollution: nearly all the carbon monoxide, two-thirds of the hydrocarbons, one-half of the nitrogen oxides, and much smaller fractions in other categories.

Pollution in the stratosphere. Sources of contaminants in the stratosphere are effluents from high-altitude aircraft such as the supersonic transport (SST), powerful nuclear explosions, and volcanic eruptions. There are also natural and artificial sources of gases which diffuse from the troposphere into the stratosphere.

Table 1 lists the natural burden of gases and particles injected into the stratosphere by high-flying aircraft, assuming the consumption of 2×10^{11} kg of fuel during a period of one year. It is to be noted that the percentage increase over the natural burden is substantial for NO, NO_2, HNO_3, and SO_2. Since the concentrations are substantial, they may adversely affect humans' living environment.

Other manufactured pollutants which diffuse from the troposphere into the stratosphere are the halogenated hydrocarbons, specifically dichloromethane (CF_2Cl_2) and trichlorofluoromethane ($CFCl_3$) gases which are used as propellants in many of the so-called aerosol spray cans for deodorants, pesticides, and such, and as refrigerants. These gases have an average residence time (residence time is the time required for a substance to reduce its concentration by $1/e$, approximately $1/3$) of 1000 years or more. $CFCl_3$ is one of a family of halogenated hydrocarbons (also known as fluorocarbons) which are widely used. The world production of these halogenated hydrocarbons was 1.7×10^9 lb (770×10^6 kg) in 1973, which represented an 11% growth over the 1972 production. Almost all are ultimately released to the atmosphere.

The nitrous oxide (N_2O) and the halogenated hydrocarbons reach the upper regions of the stratosphere, where they are photodissociated by the Sun's radiation to produce nitric oxide (NO) and chlorine (Cl). The NO and Cl react with the ozone, as in Eqs. (1)–(6). The ozone is destroyed by NO and Cl respectively, whereas NO and Cl are conserved. Stratospheric ozone is valuable as a filter for solar ultraviolet (uv) radiation. A decrease of its concentration results in an increase in the amount of uv impinging on the surface of the Earth. As an illustration based on theoretical

Table 1. The natural stratospheric background of several atmospheric gases from 13 to 24 km compared to engine emissions

Gas	Mass mixing ratio	Natural burden, kg		Increase in mass due to aircraft emission, %‡
		IDA*	Penndorf†	
CO_2	480 E-6	500 E-12§	480 E-12	0.1
H_2	2.7 E-6	2 E-12	2.7 E-12	9
CH_4	0.55 E-6	1 E-12	0.55 E-12	0.02
CO	0.05 – 0.1 E-6	30 E-9	50 – 100 E-9	0.6 – 1.2
NO	0.5 E-9	1 E-9	0.52 E-9	100
NO_2	1.6 E-9	3 E-9	1.8 E-9	100
HNO_3	4 E-9	<10 E-9	3.6 E-9	85
SO_2	1 E-9	4 E-9 (?)	1.4 E-9	10 – 40
Aerosol ($\alpha > 0.1\ \mu$m)	2 E-9	0.3 E-9	2 E-9	10

*Estimation by R. Oliver, Institute for Defense Analyses, 1974.

†Estimation by R. Penndorf, *CIAP Atmospheric Monitoring and Experiments, The Program and Results*, DOT-TST-75-106, pp. 4 – 7, 1975.

‡One-year fuel consumption by stratospheric aircraft of 2×10^{11} kg.

§Read 500 E-12 as 5×10^{12}.

considerations, an injection 2×10^9 kg/yr of $NO_x (NO_x = NO + NO_2)$ at 17 km (see Fig. 1) will result in a reduction of the total amount of ozone by about 3%. This represents an increase of vertically incident uv on the Earth of 6%. An increase of uv could adversely affect humans and plants. *See* OZONE, ATMOSPHERIC.

$$N_2O + O \rightarrow 2NO \qquad (1)$$

$$NO + O_3 \rightarrow NO_2 + O_2 \qquad (2)$$

$$NO_2 + O \rightarrow NO + O_2 \qquad (3)$$

$$CF_2Cl_3 + h\nu \text{ (solar energy)} \rightarrow CFCl_2 + Cl \qquad (4)$$

$$Cl + O_3 \rightarrow ClO + O_2 \qquad (5)$$

$$ClO + O \rightarrow Cl + O_2 \qquad (6)$$

SO_2-aerosol-climate relations. Aircraft engine effluents contain SO_2, as shown in Table 1, and could add a considerable amount of aerosol particles at the end of this century for the predicted aircraft fleet sizes. Table 1 indicates aircraft effluents could increase the natural background by 10 – 40% — seemingly large, yet small compared to volcanic injections, for which estimates range up to 10,000%.

Why are particles so important? They scatter and absorb (in specific wavelength regions) solar radiation, and thereby influence the radiative budget of the Earth-atmospheric system, and finally perhaps the climate on the ground. The particles formed from aircraft effluents may increase the optical thickness of the layer, the upwelling and downwelling infrared radiation, and the albedo of the Earth. The average global albedo of the Earth-atmospheric system has been measured as 28% with probably some short-term variation of unknown but small magnitude. It has been calculated that for an additional mass of 0.1 μg/m³ particles over a 10-km layer from 15 to 25 km (equivalent to 5.1×10^8 kg for the whole Earth, or about 20% of the "natural" background concentration), the albedo increases by about 0.05% (from 28 to 28.05%) at low latitudes all year and at high latitudes in summer, but by about 0.1 – 0.15% from September to February in high latitudes. If the added mass is larger than 0.1 μg/m³, the albedo increases proportionally to the

cited numbers. For the optical thickness of the stratosphere, a value of 0.02 is generally assumed. While the present subsonic flights increase this value by a very small amount (10^{-4}), a large fleet of high-flying aircraft (Table 1) could increase it by about 10%.

The chemistry of the natural stratospheric aerosols is dominated by sulfate, presumably of volcanic origin. These naturally occurring aerosols are concentrated in thin layers at altitudes between 15 and 25 km. R. Cadle has described the chemical composition of stratospheric particles at an altitude of 20 km during the period 1969 – 1973 as consisting of 48%, by mass, of sulfate; 24% of stony elements (such as silicon, aluminum, calcium, and magnesium); and 20% of chlorine; other constituents make up the remainder. The amount of sulfate introduced into the stratosphere as a result of a large fleet of SSTs operating at about 18 – 20 km may, in a worst-case estimate, equal the total amount occurring naturally. The influence of dynamic motions of the stratosphere on the distribution of aerosols is indicated by Fig. 2, which illustrates high correlation of the aerosol and ozone-rich layers in the 15-km region. Moreover, the water vapor mixing ratio also increases in layers at about 15 – 20 km. Since there is no known chemical link between the production of aerosols and that of water vapor and ozone, the observation illustrated by Fig. 3 may be a dynamic rather than a chemical effect, the implication being that the dynamic effects are far more important in determining the relative profiles of these constituents than any chemical effect at this altitude.

The perturbation of the lower stratosphere by the engine effluents of a large fleet of vehicles may strongly increase its optical thickness to visual-band solar radiation, which has a natural value of about 0.02. An increase in optical thickness results in a reduction of solar radiation, the principal source of atmospheric heating, by about the same fraction. This effect can be likened to a reduction in solar constant by about the same amount at the subsolar point, and about double that value when the solar zenith angle is 60° or greater, as it may be at high latitudes. The sensitivity of the troposphere to changes in the solar constant has been studied

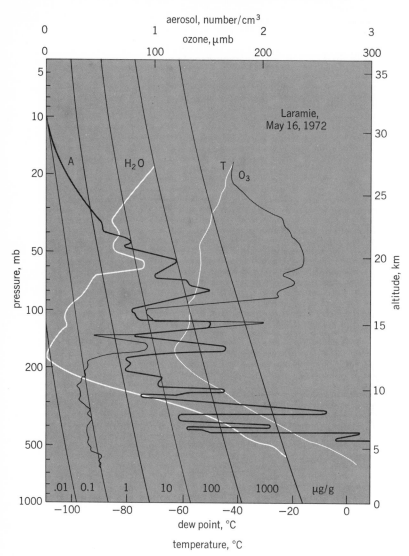

Fig. 3. A simultaneous measurement of the vertical distribution of ozone (O_3), water vapor (H_2O), and temperature (T). The aerosol (A) sounding was made about 5 hr before the ozone–water vapor soundings. The smooth curves are lines of constant water vapor mixing ratio. (*From T. J. Pepin, J. M. Rosen, and D. H. Hoffman, The University of Wyoming Global Monitoring Program, in Proceedings of the AIAA/AMS International Conference on the Environmental Impact of Aerospace Operations in the High Atmosphere, AIAA Pap. no. 73-521, 1973*)

by M. MacCracken. In the modeling for which the computed precipitation is illustrated by Fig. 4, a 10% reduction of solar constant leads to a reduction of average temperatures, from 3°C at the Equator to 10°C average from latitude 40° to the pole. The winds are substantially weakened, and total precipitation reduced, mainly in the summer. Snowfall increases, and the winter snow line moves lower in latitude by about 5°.

An increase in the solar constant of 13% results in an annual mean temperature increase of 2–5°C at all latitudes. Although the overall precipitation increases, as illustrated in Fig. 4, the relative humidity decreases slightly, and thus cloudiness decreases. The total snowfall decreases, and the snow line moves higher in latitude by 10°. Extensive melting of the polar ice caps also begins. The effects of doubling of the stratospheric optical thickness, for example, are an order of

magnitude smaller than the changes of precipitation and snowfall depicted in Fig. 4.

Sinks. A sink is defined as a process by which gases or particles are removed from a given volume of atmosphere. It could be chemical, homogeneous or heterogenous (gas-solid reactions), or dynamical, such as dispersion (transport), diffusion, gravity, or precipitation.

In the stratosphere, three important contaminants are NO, Cl, and SO_2. The sink for NO_2 is a chemical reaction by which NO_2 is combined by a complex method with (OH)—a derivative of water (H_2O) present in the stratosphere in parts per million (ppm)—to form nitric acid (HNO_3).

The SO_2 reacts chemically with oxygen (O), water, and its derivatives (OH, HO_2) to form H_2SO_4. Then, water molecules are absorbed by H_2SO_4 to form $nH_2O\cdot(H_2SO_4)$ cluster or aerosol (where n, number, is equal to about 2). Aggregates of these polymolecules, growing larger with each successive collision, eventually become aerosols; when the diameter is greater than 0.1 μm, they act as scatterers of the visual band of sunlight.

The chlorine (Cl) chemically reacts with oxygen and water to form Cl, ClO, and HCl. In the stratosphere, these reactions involving small concentrations have not been measured adequately.

In the troposphere, the predominant sinks are dispersion, transport, and precipitation. The chemical reactions in the troposphere are less active in general than the chemical reactions in the stratosphere.

Dispersion. Dispersion of pollution is dependent on atmospheric conditions. Winds transport and diffuse contaminants; rain may wash them to the surface; and under cloudless skies, solar radiation may induce important photochemical reactions.

Wind direction, speed, and turbulence influence atmospheric pollution. Wind direction determines the area into which the pollution is carried. Dilution of contaminants from a source is directly proportional, other factors being constant, to wind speed, which also determines the intensity of mechanical turbulence produced as the wind flows over and around surface objects, such as trees and buildings.

Eddy diffusion by wind turbulence is the primary mixing agency in the troposphere; molecular diffusion is negligible in comparison. In addition to mechanical turbulence, there is thermal turbulence which occurs in an unstable layer of air. Thermal turbulence and associated intense mixing develop in an unsaturated layer in which the temperature decreases with height at a rate greater than 1°C/100 m, the dry adiabatic rate of cooling. When the temperature decreases at a lower rate, the air is stable, and turbulence and mixing, now primarily mechanical, are less intense. If the temperature increases with height—its normal behavior in the stratosphere, creating a condition known as an inversion—the air is stable, and horizontal turbulence and mixing are still appreciable, but vertical turbulence and mixing are almost completely suppressed.

Inversion. Precipitation, fog, and solar radiation exert secondary meteorological influences. Falling raindrops may collect particles with radii greater than 1 μm or may entrain gases and smaller particles and carry them to the ground. Gas reactions with aerosols also occur; neutralizing cations in fog droplets or traces of ammonia (NH_3) in the air act as catalysts to accelerate reaction rates leading to rapid oxidation of sulfur dioxide (SO_2) in fog droplets. For highly polluted city air, it is estimated that in the presence of NH_3, the oxidation of the SO_2 to ammonium sulfate, $(NH_4)_2SO_4$, is completed in 1 hr for fog droplets 10 μm in radius. Photochemical oxidation of hydrocarbons in sunlight is frequent. Most hydrocarbons do not have appropriate absorption bands for a direct photochemical reaction; nitrogen dioxide (NO_2), when present, acts as an oxidation catalyst by absorbing solar radiation strongly and subsequently transferring the light energy to the hydrocarbon and thereby oxidizing it.

Natural ventilation in the atmosphere is best when the winds are strong and turbulent so that mixing is good, and when the volume in which mixing occurs is large so that dilution of pollution is rapid. As cities have grown in size, air pollution has become more widespread. It has become necessary to think of whole urban complexes as large area sources of pollution. The rate of natural ventilation of an urban area is dependent on two quantities: the wind speed and the mixing volume over the city. Active mixing upward is often limited by a stable layer, perhaps even a very stable inversion layer, aloft. The upward extent of this region of active mixing, known as the mixing height, determines the magnitude of the mixing volume of the city.

The number of air changes per unit time in this mixing volume specifies the rate of natural ventilation of the urban area. The problems of air pollution become highly complex, however, because the mixing height is rarely constant for long. Some of the factors causing it to vary are described below.

At night when the sky is clear and the wind light, Earth's surface loses heat by long-wave radiation to space. As a result, the ground cools and a surface radiation inversion is formed. The inversion inhibits mixing, so that pollution accumulates. Solar heating of the ground causes a reversal of the lapse rate, which may exceed the dry adiabatic rate of cooling and enhances active mixing in the unstable layer.

The mixing may bring pollution from aloft, causing a temporary peak in the surface concentrations, a process known as an inversion breakup fumigation. By midafternoon, the height of mixing is a maximum for the day, and surface concentrations tend to be low as the natural ventilation improves. In the evening, the lapse rate becomes stable, and accumulation of contaminants may begin again.

Subsidence inversion. The accumulation of pollution for longer periods of time is especially likely to occur if a persistent inversion aloft exists. Such an inversion aloft is the subsidence inversion formed by the sinking and vertical convergence of air in an anticyclone, illustrated in Fig. 5. A layer of air at high levels descends, diverging horizontally and hence converging in the vertical, and warms at the dry adiabatic rate of heating of 1°C/100 m. Figure 5a shows how a low-level inversion may result from this process, while Fig. 5b depicts how the mixing height H is limited in vertical extent by the subsidence inversion aloft, so that pollution accumulates within and just above the city. It is the presence of such a subsidence inversion aloft associated with the Pacific subtropical anticyclone which is the primary cause of Los Angeles and other California smogs; these are made even worse by local mountain and valley sides which prevent horizontal dispersion.

Fog. The worst pollution occurs when, in addition to subsidence inversions accompanying slowly moving or stationary anticyclones, fog also develops. All the major air pollution disasters, such as those listed in a later section, took place when fog persisted during protracted stagnant anticyclonic conditions. The reasons for the adverse influence of fog are shown in Fig. 6. When there is no fog (Fig. 6a), solar radiation heats the ground, which in turn causes a lapse rate equal to, or greater than, the dry adiababic rate of cooling, with good mixing and hence a substantial mixing height H. On the other hand, with a fog layer (Fig. 6b), up to 70% of the solar radiation incident at the top of the fog is reflected to space, with relatively little left to heat the fog and ground below. With the cloudless skies characteristic of anticyclonic weather, there is a continuous loss of heat to outer space from the upper surface of the fog bank, which acts radiatively as an elevated ground surface. More heat is lost to space than is gained from the Sun, and an inversion develops above the fog and persists night and day until the anticyclone dissipates or moves away. If the air is polluted, the fog particles may become acids and salts in solution; the saturation vapor pressure over such particles may decrease to 90 or 95% of the pure water value, so the smog becomes even more persistent than if it remained as a pure water fog. Disastrous concentrations of contaminants may accumulate during prolonged foggy conditions of this kind.

Warm fronts. Another significant inversion aloft is associated with a slowly moving warm frontal surface. Consider two cities, one lying to the southwest and the other lying to the northeast of a warm

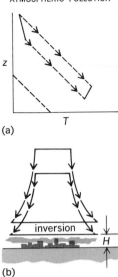

(a)

(b)

Fig. 5. Subsidence inversion. (a) Solid lines show temperature T and height z before and after dry adiabatic descent of air; dashed lines represent dry adiabatic rate of heating. (b) Inversion limits mixing height H over city.

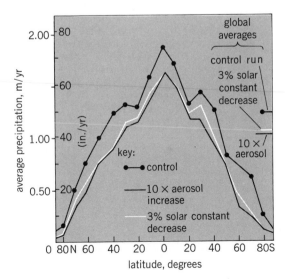

Fig. 4. Latitudinal distribution of total precipitation. *(From M. C. MacCracken, in Report of Findings: Effects of Stratospheric Pollution by Aircraft, DOT-TST-75-50, 1974)*

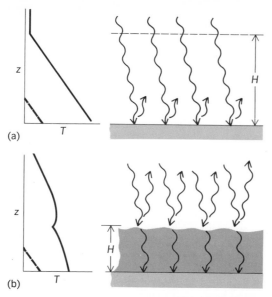

Fig. 6. The influence of fog in reducing mixing height H. (a) Without fog. (b) With fog. The dashed lines represent the dry adiabatic rate of cooling. Arrows represent solar and reflected radiation.

front extending from the southeast to the northwest (and moving in a northeast direction), as illustrated in Fig. 7. City B lies in the cool air, with the warm frontal above it. In the cool air ahead of the warm front, the pollution from City B is trapped below the warm frontal inversion and may travel for many miles with large surface concentrations. On the other hand, the prevailing southwest winds in the warm sector will carry pollution from City A up and above the warm frontal inversion, which effectively prevents its diffusion downward to the surface. This situation brings out an important point: an inversion layer may be advantageous, not disadvantageous, if it inhibits diffusion down to the ground.

Stack dispersion. Dispersion from an elevated point source, such as a stack, is conveniently expressed by Eq. (7), where χ is ground level concentration of contaminant in mass per unit volume;

$$\chi = \frac{Q}{\pi \sigma_y \sigma_z \bar{u}} \exp\left[-\frac{1}{2}\left(\frac{y^2}{\sigma_y^2} + \frac{h^2}{\sigma_z^2}\right)\right] \quad (7)$$

Q is the source strength in mass per unit time; $\bar{u}$ is mean wind speed; y is the horizontal direction perpendicular to the mean wind $\bar{u}$; σ_y and σ_z are diffusion coefficients expressed in length units in the y and z directions, respectively, the z direction being vertical; and h is the height of the source above ground.

This diffusion equation should be used only under the simplest conditions, for example, in flat uniform terrain and well away from hills, slopes, valleys, and shorelines. Table 2 lists various meteorological categories, and Table 3 gives values of the diffusion coefficients appropriate for each category. It should be noted that the values to be used depend on distance from the source at which concentrations are to be calculated. A variety of other forms is available for more complex conditions of terrain and meteorology.

Natural cleansing processes. Pollution is removed from the atmosphere in such ways as washout, rain-out, gravitational settling, and turbulent impaction. Washout is the process by which contaminants are washed out of the atmosphere by raindrops as they fall through the contaminants; in rain-out the contaminants unite with cloud droplets, which may later grow into precipitation.

Gravitational settling is significant mainly for large particles, those having a diameter greater than 20 μm. Agglomeration of finer particles may result in larger ones which settle out by gravitation. Fine particles may also impact on surfaces by centrifugal action in very small turbulent eddies. Gases may be converted to particulates, as by photochemical action of sunlight in Los Angeles, Denver, and Mexico City. These particulates may then be removed by settling or impaction.

The rate of natural cleansing may be slower than the rate of injection of pollutants into the atmosphere, in which case pollution may increase on a global scale. There is evidence that the concentration of atmospheric carbon dioxide has been increasing slowly since the beginning of the century because of combustion of fossil fuels. The tropospheric burden of very small particles and of Freon gas may also be increasing.

Effects of stratospheric pollution. Pollution of the stratosphere with nitrogen oxide causes reduction of stratospheric ozone. Ozone reduction in the stratosphere has been linked to biological effects such as skin cancer in two steps: (1) reduced ozone in the stratosphere causes an increase in uv radiation reaching the Earth's surface, and (2) increased uv radiation enhances the nor-

Table 2. Meteorological categories*

Surface wind speed, m/s	Daytime insolation			Thin overcast or $\geq$ 4/8 cloudiness†	$\leq$ 3/8 cloudiness
	Strong	Moderate	Slight		
<2	A	A-B	B		
2	A-B	B	C	E	F
4	B	B-C	C	D	E
6	C	C-D	D	D	D
<6	C	D	D	D	D

*A = Extremely unstable conditions.   D = Neutral conditions (applicable to heavy overcast, day or night).
 B = Moderately unstable conditions.   E = Slightly stable conditions.
 C = Slightly unstable conditions.   F = Moderately stable conditions.
 †The degree of cloudiness is defined as that fraction of the sky above the local apparent horizon which is covered by clouds.

Table 3. Values of diffusion coefficients

Distance from source, m	Diffusion coefficient, m², in various meteorological categories					
	A	B	C	D	E	F
$10^2, \sigma_y$	22	16	12	8	6	4
10^3	210	150	105	75	52	36
10^4	1700	1300	900	600	420	360
10^5	11,000	8500	6300	4100	2800	2000
$10^2, \sigma_z$	14	11	7.6	4.8	3.6	2.2
10^3	500	120	70	32	24	14
10^4	–	–	420	140	90	46
10^5	–	–	2100	440	170	92

mal biological effects of natural uv radiation.

Biological damage. In step 1, the relation between the reduction of stratospheric ozone and the increase in solar uv flux effective in causing sunburn and, presumably, also skin cancer is readily calculable. The factors of interest here are the uv wave band of 290–320 nanometers (nm), the middle latitudes (where the summer sun is nearly overhead and the worst cases of skin cancer occur), and small decreases in ozone. For these factors of interest, the percent increase in solar uv flux is about twice the percent decrease in ozone.

Step 2, from uv radiation to the enhancement of skin cancer incidence, involves the assumption, supported by some scientific evidence but not proven by experiments on humans, that skin cancer in humans is induced by exposure to uv radiation of the same wavelength (290–320 nm) that causes erythema (sunburn), and that the relative effectiveness of the various wavelengths for carcinogenesis is the same as that for sunburn. Estimates of biological damage to humans from exposure to uv radiation are based on inferences from the statistics of epidemiological surveys of humans and of laboratory experiments with animals. Nonmelanomic skin cancer, which is almost never fatal if given proper care, occurs primarily on sun-exposed areas of the skin, especially the face and hands. It is relatively common—about 250 cases per 100,000 persons occur in fair-skinned Caucasians in the United States. The incidence of nonmelanomic skin cancer is correlated with latitude—and, therefore, with sunlight, including uv flux, since average sunlight varies with latitude.

Climatic changes. Pollution of the stratosphere may involve also a climate chain of cause-and-effect relations by which aircraft engine effluents, notably sulfur dioxide (SO_2), and to a lesser degree water vapor (H_2O) and nitrogen oxides (NO_x), affect climatic variables such as temperature, wind, and rainfall.

If enough particles larger than 0.1 μm in diameter were added to the stratosphere, they could alter the radiative heat transfer of the Earth-Sun system, and thereby influence climate. Particles of this size are produced by several constituents of engine emissions, in particular those of SO_2. When considering the large numbers of aircraft postulated for future operations in the stratosphere, the amount of particles developed from the SO_2 engine emissions are potentially serious, unless the fuels used have a sulfur content smaller than that of today's fuels.

The sequence in the climate cause-and-effect chain is proposed as follows. Stratospheric SO_2, after first being oxidized, interacts with the abundant water vapor exhaust from jet engines to produce solid sulfuric acid particles that build up to sizes greater than 1 μm. These particles disperse within the stratosphere, principally within the hemisphere in which they are injected, where they remain for periods as long as 3 years, depending on their altitude.

The effects of SO_2 emissions are summarized in Table 4, where the stratosphere's opacity to sunlight is represented by optical thickness. The natural optical thickness of the stratosphere is about 0.02, which means that sunlight and heat are reduced by about 2% while passing through the stratosphere.

The present subsonic fleet of about 1700 aircraft operates at times in the low stratosphere, burning a total of about 2×10^{10} kg (2×10^7 tons)

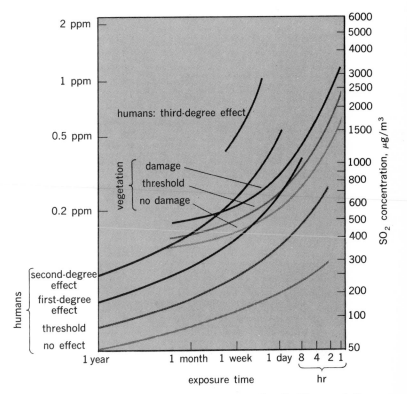

Fig. 7. Effects of sulfur dioxide on humans and vegetation. (*L. J. Brasser et al.*)

Table 4. Estimated increase in stratospheric optical thickness per 100 aircraft

Subsonic aircraft type*	Fuel burned, kg/yr†	Altitude, km (10³ ft)	Maximum SO_2 EI‡ without controls, g/kg fuel	Percent change in stratospheric optical thickness in Northern Hemisphere	
				Without controls	With future EI controls achieving only 1/20 of emission of present-day aircraft
707/DC-8	1×10^9	11 (36)	1	0.023	0.0012
DC-10/L-1011	1.5×10^9	11 (36)	1	0.032	0.0016
747	2×10^9	11 (36)	1	0.044	0.0022
747-SP	2×10^9	13.5 (44)	1	0.10	0.0050

*The present subsonic fleet consists of 1217 707/DC-8s, 232 DC-10/L-1011s, and 232 747s flying at a mean altitude of 11 km (36,000 ft), and is estimated to cause an increase in stratospheric optical thickness of 0.5%.

†Subsonics are assumed to operate at high altitude, 5.4 hr per day, 365 days per year.

‡EI is emission index, which is defined as grams of pollutants per kilogram of fuel used.

of fuel per year and producing an estimated increase in the stratosphere's optical thickness of about 0.0001 or 0.5% of the natural value.

As a matter of history, the variability of temperature due to natural causes is a substantial fraction of 1°C, even over a few decades. The year-to-year variation is several tenths of a degree. In the 1890–1940 period, a warming of ½°C occurred. During the 1940–1960 period, a general cooling of ⅕°C took place.

Effects of tropospheric pollution. A number of the many effects of tropospheric pollution are described briefly below.

Humans. The effects of many pollutants on human health under most ordinary circumstances of living, rural or urban, are difficult to specify with confidence. In the United States, a number of animal experiments under controlled conditions and epidemiological studies have been made, but the results are difficult to interpret in terms of human health. A study on the lower East Side of New York City indicated that, in children less than 8 years old, the occurrence of respiratory symptoms was associated with the levels of particulate matter and of carbon monoxide in the atmosphere. With heavy smokers, however, eye irritation and headache were directly related to increasing concentrations of carbon monoxide.

Air pollution is suspected as a causative agent in the occurrence of chronic bronchitis, emphysema, and lung cancer, but the evidence is not clear cut. On the other hand, from mid-August to late September the potent aeroallergen, ragweed pollen, is a substantial cause of allergic rhinitis and bronchial asthma for over 10,000,000 persons living east of the Rockies. In Los Angeles County the effects of smog on health are becoming better understood. For example, people living in less smoggy areas of the county survive heart attacks more readily than others: In 1958 in high-pollution areas the mortality rate per 100 hospital admissions for heart attacks averaged 27.3 in comparison with 19.1 for low-smog areas. Other studies show a small but significant relation between motor vehicle accidents and oxidant levels. Medical authorities in Los Angeles are becoming concerned about the long-term influences of various pollutants, including photochemical smog, despite the lack of comprehensive knowledge of the nature of such effects.

In Great Britain there is a similar lack of precise knowledge. There are indications that emphysema, bronchitis, and other respiratory diseases are not caused primarily by increased atmospheric pollution, but because more people are living longer. Despite a substantial reduction in the concentrations of atmospheric particulates since the Clean Air Act was passed in 1956, the respiratory disease rate continues to rise. These facts do not prove that air pollution is not a factor, but that its influence may be synergistic and therefore difficult to identify precisely. For example, it is known that the combined effect of sulfur oxides and particulates is substantially greater than the sum of the two separate effects, and many other such synergistic effects doubtless occur.

In the Netherlands special efforts have been made to relate SO_2, both concentration values and exposure times, to effects on humans and on vegetation. The results of these studies, based on investigations in England, the United States, West Germany, Italy, and the Netherlands, are illustrated in Fig. 7. Influences on humans are shown in the lower family of curves: A first-degree effect is a small increase in functional disturbances, symptoms, illnesses, diseases, and deaths; a second-degree effect is a more prevalent or more pronounced effect of the same kind; and a third-degree effect is a substantial increase in the number of deaths. It should be emphasized that the exposures were, in general, to SO_2 in dusty and sooty atmospheres.

Under extreme circumstances when stagnant atmospheric conditions with persistent low wind and fog exist, major disasters involving many deaths occurred, as in and around London in 1873, 1880, 1891, 1948, 1952, 1956, and 1962. Similar disasters occurred in the Meuse Valley of Belgium in 1930 and at Donora, Pa., in 1948.

Atmospheric pollution has a substantial influence on the social aspects of human life and activity. For example, the distribution of urban populations is being increasingly affected by such pollution, and recreational patterns are similarly influenced. The atmospheric burden of pollution is thus becoming more and more important as a determinant in social decision making.

Animals. Studies of the response of laboratory animals to specified concentrations of pollutants have been conducted for many years, but the interpretation of the results in terms of corresponding human response is most difficult. Assessment of the effects of certain contaminants on livestock is relatively straightforward, however. Thus contamination of forage by airborne fluorides and arsenicals from certain industrial operations has led to the loss of large numbers of cattle in the areas adjacent to such chemical industries.

Plants. Damage to vegetation by air pollution is of many kinds. Sulfur dioxide may damage such field crops as alfalfa, and trees such as pines, especially during the growing season; some general relations are presented in Fig. 7. Both hydrogen fluoride and nitrogen dioxide occurring in high concentrations have been demonstrated to be harmful to citrus trees and ornamental plants which are of economic importance in central Florida. Ozone and ethylene are examples of other contaminants which result in damage to certain kinds of vegetation.

Materials. Corrosion of materials by atmospheric pollution is a major problem. Damage occurs to ferrous metals; to nonferrous metals, such as aluminum, copper, silver, nickel, and zinc; to building materials; and to paint, leather, paper, textiles, dyes, rubber, and ceramics.

Weather. Tropospheric pollution may affect weather in a number of ways. Heavy precipitation at Laporte, Ind., is attributed to a substantial source of air pollution there, and similar but less pronounced effects have been observed elsewhere. Industrial smoke reduces visibility and also ultraviolet radiation from the Sun, and polluted fogs are more dense and more persistent than natural fogs occurring under similar conditions. Possible major effects of air pollution on Earth's climate have been mentioned earlier.

Costs. It is extremely difficult to estimate accurately the economic costs of air pollution. Dollar values are not readily established for losses due to illness caused by exposure to air pollution. At the present time a reasonable estimate appears to be that air pollution costs the United States about $10,000,000,000 a year. For Great Britain the corresponding figure—probably a conservative one—is £400,000,000 a year.

Controls. Four main methods of air-pollution control are indicated below.

Prevention. This method was originally applied mainly to reduce pollution from combustion processes. Improved equipment design and smokeless fuels have reduced pollution both from industrial and motor vehicle sources.

Collection. Collection of contaminants at the source has been one of the important methods of control. Many types of collectors have been employed successfully, such as settling chambers, cyclone units employing centrifugal action, bag filters, liquid scrubbers, gas-solid adsorbers, ultrasonic agglomerators, and electrostatic precipitators. The optimum choice for a given industrial process depends on many factors. A major problem is disposal of the collected materials. Sometimes these materials can be utilized in a by-product manufacture on a profitable or a break-even basis.

Containment. This method is useful for pollutants whose noxious characteristics may decrease with time, such as radioactive contaminants from nuclear power plants. For contaminants with a short half-life, containment may allow the radioactivity to decay to a level which permits their release to the atmosphere. Containment, with destruction or conversion of the offending substances, often malodorous or toxic, is used in certain chemical, oil refining, and metallurgical processes and in liquid scrubbing.

Dispersion. Atmospheric dispersion as a control method has a number of advantages, especially for industrial processes which can be varied to take advantage of the periods when dispersion conditions are so good that contaminants may be distributed very widely in such small concentrations that they inconvenience no one. Some coal-burning electrical power stations are building high stacks, up to 1000 ft (300 m), to lift the SO_2-bearing stack gases well above the ground. Some plants store low-sulfur anthracite coal for use when atmospheric dispersion is poor. *See* AIR-POLLUTION CONTROL.

Laws. Many laws designed to limit air pollution have been enacted. Major steps forward were taken by the Netherlands in 1952, by Great Britian in 1956, by Germany in 1959 and 1962, by France in 1961, by Norway in 1962, by the United States in 1963 and 1967, and by Belgium in 1964.

Efforts to control air pollution by legal means commenced many years ago in Great Britain. In 1906 the Alkali Act consolidated and extended previous similar acts, the first of which was passed in 1863. This calls for the annual registration of scheduled industrial processes, and requires that the escape of contaminants to the atmosphere from scheduled processes must be prevented by the "best practicable means." The Alkali Act functions by interpretation and not by statutory requirement, the Alkali Inspector being the sole judge of the "best practicable means." The Clean Air Act of 1956 provided more effective ways of limiting air pollution by domestic smoke, industrial particulates, gases and fumes from the processes registrable under the Alkali Act, and smoke from diesel engines. This legislative program has had considerable success in alleviating air-pollution problems in Great Britain.

In the United States, air-pollution control had been considered to be a matter of local concern only. By 1963 only one-third of the states had air-pollution control programs, most of which were relatively ineffective. Only in California were local programs, at the city and county level, supported adequately. The Clean Air Act of 1963 brought the Federal government into a regulatory position of increased scope by granting the Secretary of Health, Education, and Welfare specific abatement powers under certain circumstances. It also established a Federal program of financial assistance to local control agencies and recommended more vigorous action to combat pollution by motor vehicle exhausts and by smoke from incinerators.

The Air Quality Act of 1967 brought the Federal government into a more substantial regulatory role. One of its important effects has been to change the emphasis in legislation from standards based on emissions from sources, such as stacks, to standards based on concentrations of contaminants in the ambient air which result from such emissions. The Air Quality Act of 1967 consists of three main portions, listed below.

Title I: Air-Pollution Prevention and Control.

The first section of the Air Quality Act amends the Clean Air Act to encourage cooperative activities by states and local governments for the prevention and control of air pollution and the enactment of uniform state and local laws; to establish new and more effective programs of research, investigation, training, and related activities; to give special emphasis to research related to fuel and vehicles; to make grants to agencies to support their programs; to provide strong financial support for interstate air-quality agencies and commissions; to define atmospheric areas and to assist in establishing air quality control regions, criteria, and control techniques; to provide for abatement of pollution of the air in any state or states which endangers the health or welfare of any persons and to establish the necessary procedures; to establish the President's Air Quality Advisory Board and Advisory Committees; and to provide for control of pollution from Federal facilities.

Title II: National Emission Standards Act. This section is concerned mainly with pollution from motor vehicles which accounts for some 60% of the total for the United States. The act covers such matters related to motor vehicle emissions as the following: establishment of effective emission standards and of procedures to ensure compliance by means of prohibitions, injunction procedures, penalties, and programs of certification of new motor vehicles or motor vehicle engines and registration of fuel additives. The act also calls for a comprehensive report on the need for, and the effect of, national emission standards for stationary sources.

Title III. The final section is general and covers matters such as comprehensive economic cost studies, definitions, reports, and appropriations.

There is no doubt that this far-reaching legislative program, stimulating new approaches at the local, state, and Federal levels, will play a major role in controlling air pollution within the United States. The other industrial nations of the world are preparing to meet their growing air-pollution problems by initiatives appropriate to their own particular circumstances.

[A. J. GROBECKER; S. C. CORONITI; E. WENDELL HEWSON]

Bibliography: L. J. Brasser et al., *Sulphur Dioxide: To What Level Is It Acceptable?*, Research Institute of Public Health Engineering, Delft, Netherlands, Rep. no. G300, 1967; J. H. Chang and H. Johnston, *Proceedings of the 3d CIAP Conference*, DOT-TSC-OST-74-15, pp. 323–329, 1974; R. E. Dickinson, in *Proceedings of the AIAA/AMS International Conference on the Environmental Impact of Aerospace Operations in the High Atmosphere*, AIAA Pap. no. 73-527, 1973; Federal Task Force on Inadvertent Modification of the Stratosphere, *IMOS Report*, prepared for the Federal Council for Science and Technology, 1975; J. Friend, R. Liefer, and M. Tichon, *Atmos. Sci.*, 30:465–479, 1973; D. Garvin and R. F. Hampson, *Proceedings of the AIAA/AMS International Conference on the Environmental Impact of Aerospace Operations in the High Atmosphere*, AIAA Pap. no. 73-500, 1973; A. J. Grobecker, S. C. Coroniti, and R. H. Cannon, *Report of Findings: The Effects of Stratospheric Pollution by Aircraft*, U.S. Department of Transportation, DOT-TST-75-50, 1974;

P. A. Leighton, Geographical aspect of air pollution, *Geogr. Rev.*, 56:151, 1966; P. A. Leighton, *Photochemistry of Air Pollution*, 1961; M. C. MacCracken, *Tests of Ice Age Theories Using a Zonal Atmospheric Model*, UCRL-72803, Lawrence Livermore Laboratory, 1970; A. R. Meetham, *Atmospheric Pollution*, 1964; M. J. Molina and F. S. Rowland, *Geophys. Rev.*, pp. 810–812, 1974; National Academy of Sciences, *Environmental Impact of Stratospheric Flight*, pp. 128–129, 1975; R. Scorer, *Air Pollution*, 1968; A. R. Smith, *Air Pollution*, 1966; A. C. Stern (ed.), *Air Pollution*, 3 vols., 1968; R. S. Stolarski and R. J. Cicerone; *Can. J. Chem.*, 52:1610–1615, 1974; S. C. Wofsy and M. B. McElroy, *Can. J. Chem.*, 52:1582–1591; 1974; H. Wolozin (ed.), *The Economics of Air Pollution*, 1966.

Atomic energy

The energy released in the rearrangement of the particles making up the nucleus of an atom, popularly referred to as atomic energy, but preferably called nuclear energy. Atomic properly refers to phenomena involving the orbital electrons of the atom, but not involving any transformation of the nucleus. *See* NUCLEAR POWER.   [JAMES A. LANE]

Automotive engine

An integral major component of and source of power for automotive vehicles. Several types of engines are available for passenger and commercial vehicles, but most vehicles are driven by reciprocating internal combustion gasoline engines operating upon the four-cycle principle worked out by Nikolaus Otto in 1876. Many large commercial vehicles use diesel engines, with the fuel ignited by the heat caused by high air compression, for more economical operation and longer life, rather than gasoline engines with ignition by spark plugs. Diesel engines are also available in some imported passenger cars. *See* DIESEL ENGINE; OTTO CYCLE.

Domestic passenger car engines evolved from an assortment of electrics and steamers, designed with two or four cycles. Present four-, six-, and eight-cylinder engines are four-cycle types of in-line, V-block, slant, or horizontally opposed "pancake" configurations. In 1900, electrics and steamers far outnumbered gasoline engines, but within 2 decades they faded from the scene. Formerly popular in-line eights and V-12 and V-16 powerplants were discontinued when higher compression ratios were developed. Redesign of six- and eight-cylinder engines provided equivalent horsepower, with incidental reduction in weight.

As vehicles grew in size and type, and greater road speed was demanded, engines tended to increase in size and weight and had to improve in performance. In an attempt to reduce engine weight, aluminum cylinder blocks were tried, but found faulty. Present cylinder blocks and heads are gray iron castings, as are numerous other engine parts. Weight reduction has been achieved by decreasing cylinder wall sections and substituting stampings for castings, and plastics for metal parts.

Higher engine compression ratios produced greater combustion temperatures, and ignition knocks which could be eliminated only with higher-octane fuel, made possible by introduction of gasoline additives. Such developments created a new family of high-performance engines. This developmental trend was modified by the Clean Air Act of 1970, which mandated specific emission controls and use of nonleaded fuel, and encouraged manufacturers to get more miles per gallon.

Engine compartments have limited space available and this has been encroached upon by increasing "hang-on" hardware and by front suspension units and similar space-robbing components. The current tendency toward compact and subcompact cars accentuates this problem, which will undoubtedly require revolutionary changes. Rotary and other types of engines have been available for more than 2 decades, with approximately 40% fewer parts and 50% less bulk.

The immediate future promises a new engine vocabulary and a plethora of sophisticated devices: catalytic converters, lean-burn engines, electronic spark advance (ESA), computer controlled timing (CCT), stratified charge (SC), electronic fuel injection (EFI), electronic fuel metering (EFM), and a myriad of complicated systems still in development stages. *See* INTERNAL COMBUSTION ENGINE.

[PAUL A. OTT, JR.]

Battery, electric

A device which transforms chemical energy into electric energy. The term is usually applied to a group of two or more electric cells connected together electrically. In common usage the term battery is also applied to a single cell, such as a flashlight battery.

Types. There are in general two types of batteries, primary batteries and secondary storage or accumulator batteries. Primary types, although sometimes consisting of the same plate-active materials as secondary types, are constructed so that only one continuous or intermittent discharge can be obtained. Secondary types are constructed so that they may be recharged, following a partial or complete discharge, by the flow of direct current through them in a direction opposite to the current flow or discharge. By recharging after discharge, a higher state of oxidation is created at the positive plate or electrode and a lower state at the negative plate, returning the plates to approximately their original charged condition.

Primary and secondary cells may be constructed from several materials. For the more important of these types *see* PRIMARY BATTERY; STORAGE BATTERY.

Applications. Primary cells or batteries are used as a source of dc power where the following requirements are important.

1. Electrical charging equipment or power is not readily available.

2. Convenience is of major importance, as in the case of the hand or pocket flashlight.

3. Stand-by power is desirable without cell deterioration during periods of nonuse for days or years. Reserve-electrolyte designs may be necessary, as in torpedo, guided-missile, and some emergency light and power batteries. *See* RESERVE BATTERY.

4. The cost of a discharge is not of primary importance.

Secondary cells or batteries are used as a source of dc power where the following requirements are important.

1. The battery is the primary source of power and numerous discharge-recharge cycles are required, as in industrial hand or rider trucks, electric street trucks, mine or switching locomotives, and submarines.

2. The battery is used to supply large, short-time (or relatively small, longer-time), repetitive power requirements, as in automotive and airplane batteries.

3. Stand-by power is required and the battery is continuously connected to a voltage-controlled dc circuit. With proper voltage the battery is said to "float" (drawing from the dc circuit only sufficient current to compensate automatically for the battery's own internal self-discharge). Telephone exchange, central-station circuit breaker, and emergency light and power batteries are in this category.

4. Long periods of low-current-rate discharge followed subsequently by recharge are required, as in buoy service.

5. The very large capacitance is beneficial to the circuit, as in telephone exchanges.

Size. Both primary and secondary cells are manufactured in many sizes and designs, from the small electric wristwatch battery and the small penlight battery to the large submarine battery, where a single cell has weighed 1 ton. In all applications the cell must be constructed for its particular service, so that the best performance may be obtained consistent with cost, weight, space, and operational requirements. Automotive and aircraft batteries generally use thin positive and negative plates with thin separation to conserve space and weight and to provide high rates of current discharge at low temperatures. Stand-by batteries use thick plates and thick separators to provide long life. Notable size and weight reductions have been made through use of new plastic materials, active materials, and methods of construction.

Ratings. Since the power that can be obtained from a cell varies with its temperature and the rate of current discharge, the power-output rating is very important. Common secondary-battery practice is to rate cells in terms of ampere-hours (discharge rate in amperes times hours of discharge) and to specify the hourly rate of discharge. A popular automotive battery capable of giving 2.5 amp for 20 hr is rated at 50 amp-hr at the 20-hr rate. This same battery may provide an engine-cranking current of 150 amp for only 8 min at 80°F or for 4 min at 0°F, giving a service of only 20 and 10 amp-hr, respectively. By multiplying ampere-hours by average voltage during discharge, watt-hour rating is obtained. Ratings must be made to a specified final voltage, which is either at the point of rapid voltage drop or at minimum usable voltage. The rating of primary batteries is generally stated as the number of hours of discharge which can be obtained when discharging through a specified fixed resistance to a specified final voltage.

Life. Life of cells varies from the single discharge obtainable from primary types to 10,000 or more discharge-charge cycles obtainable from some secondary cells operating at very high rates for very short times. Automotive batteries may generally be expected to give approximately 300 cycles, or to last 2 years. Industrial-truck sizes may be expected to give 1500 to 3000 cycles in 5–10 years. Stand-by sizes may be expected to float across the dc bus 8–30 years. Generally the most costly, largest, heaviest cells are the longest-lived.

To obtain life from batteries, certain precautions are necessary. The stated shelf life and temperature of wet primary cells must not be exceeded. For dry reserve-electrolyte primary cells and secondary cells of the dry construction with charged plates, the cell or battery container must be protected against moisture, and storage must be within prescribed temperature limits. Wet, charged secondary batteries require periodic charging and water addition, depending upon the kind of construction.

Reliability. Batteries are probably the most reliable source of power known. In fact, most critical electric circuits are protected in some manner by battery power. There are no moving parts and, with good quality control in component materials and construction, one can be assured of power, particularly since adequate checks to indicate the condition of the cells usually exist. To ensure reliability, manufacturer's stipulations on storage and maintenance must be followed.

For other sources of electric energy known as batteries or cells see FUEL CELL; NUCLEAR BATTERY; SOLAR BATTERY. [HAROLD C. RIGGS]

Bergius process

Treatment of carbonaceous matter such as coal or cellulosic materials with hydrogen at elevated pressures and temperatures, in the presence of catalysts, to yield an oil similar to crude petroleum. F. Bergius, a German chemist, discovered the process in 1913; its subsequent industrial development in Germany resulted in 12 coal hydrogenation plants with a total capacity, in 1944, of 200,000,000 bbl/year of liquid fuel.

The process is conducted in two stages. First, the petroleumlike product is obtained from the hydrogenation of coal. Second, the product is further catalytically hydrorefined to produce chiefly a gasoline containing 35–50% of aromatic hydrocarbons, the remainder being a mixture of paraffins and naphthenes. In both stages the operating temperature is in the range 380–500°C (716–932°F), and the pressure range is 200–700 atm (3000–10,000 psi). The gasoline after addition of about 3 ml of tetraethyllead per gallon is suitable for aviation motors. Most of the oxygen, nitrogen, and sulfur in the coal appears as water, ammonia, and hydrogen sulfide in the gaseous products. If desired, tar acids (phenolic compounds) and tar bases (organic nitrogen compounds), equivalent in amount to 5–10% of the moisture-free and ash-free coal, can be extracted from the liquid product of the first stage.

Renewed interest in coal hydrogenation in the United States since 1960 resulted in pilot plant studies of an ebullating bed process and an extraction-hydrogenation process. The process consists of extraction of part of the coal with a solvent, low-temperature carbonization of the undissolved portion, and hydrogenation of the extract and liquid products of carbonization. The residue from these operations is recovered as a by-product char. See FISCHER-TROPSCH PROCESS.

[JOSEPH H. FIELD]

Bibliography: *Coal Age*, 72(6):26–27, 1967; *High Pressure Hydrogenation at Ludwigshagen-Heidelberg*, FIAT Rep. nos. 1317–1324, 1951.

Binding energy, nuclear

The amount by which the mass of an atom is less than the sum of the masses of its constituent protons, neutrons, and electrons expressed in units of energy. This energy difference accounts for the stability of the atom. In principle, the binding energy is the amount of energy which was released when the several atomic constituents came together to form the atom. Most of the binding energy is associated with the nuclear constituents (protons and neutrons), or nucleons, and it is customary to regard this quantity as a measure of the stability of the nucleus alone.

A widely used term, the binding energy (BE) per nucleon, is defined by the equation below, where

$$\text{BE/nucleon} = \frac{[ZH + (A - Z)n - {}_zM^A]c^2}{A}$$

${}_zM^A$ represents the mass of an atom of mass number A and atomic number Z, H and n are the masses of the hydrogen atom and neutron, respectively, and c is the velocity of light. The binding energies of the orbital electrons, here practically neglected, are not only small, but increase with Z in a gradual manner; thus the BE/nucleon gives an accurate picture of the variations and trends in nuclear stability. The figure shows the BE/nucleon (in million electron volts) plotted against mass number for $A > 40$.

The BE/nucleon curve at certain values of A suddenly changes slope in such a direction as to indicate that the nuclear stability has abruptly deteriorated. These turning points coincide with particularly stable configurations, or nuclear shells, to which additional nucleons are rather loosely bound. Thus there is a sudden turning of the curve over $A = 52$ (28 neutrons); the maximum occurs in the nickel region (28 protons, $\sim A = 60$); the stability rapidly deteriorates beyond $A = 90$ (50 neutrons); there is a slightly greater than normal stability in the tin region (50 protons, $\sim A = 118$); the stability deteriorates beyond $A = 140$ (82 neutrons) and beyond $A = 208$ (82 protons plus 126 neutrons).

The BE/nucleon is remarkably uniform, lying for most atoms in the range 5–9 Mev. This near constancy is evidence that nucleons interact only with near neighbors; that is, nuclear forces are saturated.

The binding energy, when expressed in mass units, is known as the mass defect, a term sometimes incorrectly applied to quantity $M - A$, where M is the mass of the atom. [H. E. DUCKWORTH]

The term binding energy is sometimes also used to describe the energy which must be supplied to a nucleus in order to remove a specified particle to infinity, for example, a neutron, proton, or α-particle. A more appropriate term for this energy is the separation energy. This quantity varies greatly from nucleus to nucleus and from particle to particle. For example, the binding energies for a neutron, a proton, and a deuteron in O^{16} are 15.67,

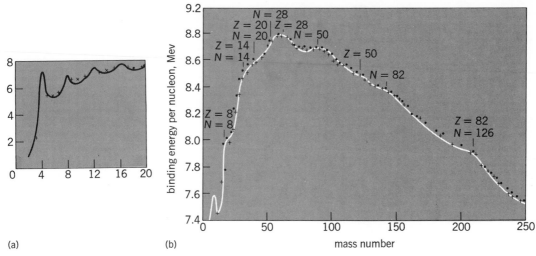

Graph of binding energy per nucleon. $N =$ number of neutrons. (a) Low mass numbers, from 2 to 20. (b) Mass numbers from 12 to 250. (*From A. H. Wapstra, Isotopic measure, part 1, where A is less than 34, Physica, 21:367–384, 1955*)

12.13, and 20.74 Mev, respectively, while the corresponding energies in O^{17} are 4.14, 13.78, and 14.04 Mev, respectively. The usual order of neutron or proton separation energy is $7-9$ Mev for most of the periodic table.   [D. H. WILKINSON]

Blackbody

An ideal energy radiator, which at any specified temperature emits in each part of the electromagnetic spectrum the maximum energy obtainable per unit time from any radiator due to its temperature alone. A blackbody also absorbs all the energy which falls upon it. The radiation properties of real radiators are limited by two extreme cases—a radiator which reflects all incident radiation, and a radiator which absorbs all incident radiation. Neither case is completely realized in nature. Carbon and soot are examples of radiators which, for practical purposes, absorb all radiation. Both appear black to the eye at room temperature, hence the name blackbody. Often a blackbody is also referred to as a total absorber. Such a total absorber constitutes a standard for the radiation of nonblackbodies, since Kirchhoff's law demands that the blackbody radiate the maximum amount of energy possible in any wavelength interval. For an extended discussion of blackbody radiation and Kirchhoff's law *see* HEAT RADIATION. *See also* GRAYBODY.

[HEINZ G. SELL; PETER J. WALSH]

Boring and drilling (mineral)

The drilling of holes for exploration, extractive development, and extraction of mineral deposits. Exploration holes are drilled to locate mineral deposits, define their extent, determine their ore dressing characteristics, determine mining or quarrying conditions to be met in their development or extraction, and otherwise to establish the commercial value of the deposit. Holes for extractive development are drilled for blasting, mine dewatering, access for utility lines, grouting for water control, pilot holes for shaft sinking, and as small shafts for backfilling, ventilation, portals, and escapeways.

Elements of drilling. The two essentials are a process to detach particles of soil or rock from the floor of the hole, and a way to remove the loosened particles from the hole. The principles that can be applied to these processes are limited by the confines of the drill hole, but they can be combined into an innumerable variety of drilling methods, applicable to mineral work.

Drills and equipment. Energy and action required for the two processes of drilling are furnished in various ways. Drills are designed to create motion, rotary, linear, or combined rotary and linear motion. Auxiliary drill-rig equipment (pumps, compressors, generators) may introduce additional energy into the hole in the form of hydraulic, pneumatic, or electric power. Another source of energy may be created at the bit by conversion of gases into heat.

The energy of motion created by the drill may be transmitted to the drill bit by a shaft (drill rods, drill pipe, drill stem, drill steel) or by a flexible cable (wire-line, rope). Drill rods, drill pipe, and drill steel are tubular so that they may be used for the injection of fluids, compressed air, or gases.

Types of drilling. Five general categories may be classified: rotary drilling, penetration by the abrasive action of a drill bit in rotary motion; core drilling, rotary drilling an annular groove leaving a central core; percussion drilling, penetration by the chipping or crushing action produced by a drill bit in linear motion; rotary-percussion drilling, combination of rotary and linear motions to produce penetration; fusion piercing, penetration by flaking or melting caused by applying heat.

Principal drilling methods. Many drills may be adapted to a variety of drilling methods, particularly those drills designed to produce penetration by rotary motion and which also incorporate dual linear motion mechanisms, one for control and one for hoisting. Some drills are specifically designed for a combination of drilling methods (Fig. 1).

Drilling methods have terminology generally indicative of the outstanding feature of the method, but the same term may be used for a method that is also used for a type of drilling, as in

Fig. 1. Combined auger-churn-core drill. (*Penndrill*)

Standard diamond-core bit sizes in inches*

Hole diameter	Design and core diameter		
	Lifter in bit	Lifter in case	Thin wall
$1\frac{5}{32}$			$\frac{3}{4}$
$1\frac{15}{32}$	$\frac{27}{32}$	$\frac{27}{32}$	$\frac{29}{32}$
$1\frac{7}{8}$	$1\frac{3}{16}$	$1\frac{3}{16}$	$1\frac{9}{32}$
$2\frac{11}{32}$	$1\frac{21}{32}$	$1\frac{21}{32}$	$1\frac{3}{4}$
$2\frac{31}{32}$	$2\frac{5}{32}$	$2\frac{5}{32}$	$2\frac{5}{16}$
$3\frac{37}{64}$	3	3	$3\frac{3}{16}$
$3\frac{7}{8}$		$2\frac{11}{16}$	
$5\frac{1}{2}$		$3\frac{15}{16}$	
$7\frac{3}{4}$		$5\frac{15}{16}$	

*Data from Diamond Core Drill Manufacturers Association.

the use of the term rotary drilling.

Rotary drilling. This is vertical drilling through a rotary table, and the typical drill is that used for drilling exploratory and producing wells for the petroleum minerals (oil and gas). The rotary table turns a square kelly, with mud swivel at the top and connection to drill pipe below. Penetration is by rotation of drill bits of two types: roller bits, which have rolling cutters with projecting hard teeth; and drag bits, with fixed chisel-type hard cutting edges. Cuttings removal is by circulation of drilling mud, which is also used to protect the wall of the hole and eliminate the need for casing. Controlled vertical motion of the tools is provided by wire-line drum-type hoist with multisheave blocks in the derrick for multiplying hoisting capacity, thus controlling weight on the bit by the partial support of the weight of drill pipe and heavy drill collars.

Smaller rotary drills, with a thrust mechanism added to react against the weight of the drill and increase bit pressure, are used for seismograph prospecting, core drilling, water wells, and blast holes. Circulation of a large volume of air, rather than drilling mud, is commonly used for cuttings removal in shallow rotary drill holes. Water is almost always used as the circulating medium with core drills, to dissipate heat as well as to remove cuttings.

Auger drilling. Rotary-type auger methods penetrate by the cutting or gouging action of chisel-type cutting edges forced into the formation by rotation of the bit. Cuttings are removed by mechanical action. Auger rods are made with a continuous helical projecting surface so that they act as screw conveyors to remove the cuttings. Auger drills are of various sizes, and usually incorporate a hydraulic cylinder to increase bit pressure and to withdraw the tools. The method is used for drilling blast holes and for reconnaissance prospecting. A special type is used as a mining tool.

Diamond core drilling. Another rotary type, this drilling utilizes the extreme hardness of the diamond to penetrate rock by abrasive action. Core drills rotate an annular bit to cut a narrow kerf around a central core and thus obtain unaltered samples of the formation drilled. Diamond drills are the principal tools for exploration of mineral deposits.

The diamond drill furnishes rotary motion for penetration, and generally also has dual linear motion mechanisms, a hydraulic cylinder or screw mechanism to control the feed of the bit against the rock, and a wire-line hoist or pneumatic cylinder for tool withdrawal. The drill head usually swivels for drilling at any angle. Tubular drill rods transmit rotary motion to the core barrel, which has a core-retaining device just above the diamond bit. The ring-shaped diamond bit is set with industrial-grade diamonds, those with imperfections that prohibit their use as gems. Water or mud circulation cools the bit and removes the cuttings. Double-tube core barrels may be used to keep water circulation from washing the core (see table).

A few variations are developed. Surface-set bits contain a single layer of diamonds. Impregnated bits have diamond fragments distributed through the crown so that fresh cutting points are exposed as the bit wears. Noncoring diamond bits may be used where a core is not desired, as in diamond blast-hole drilling.

Shot (abrasive) drilling. This method of rotary core drilling uses hard, chilled steel shot or other loose abrasive as a cutting medium. Penetration is accomplished by the grinding action of the abrasive when dragged over the rock by a rotating core barrel with a blank steel bit, much as any grinding process utilizing a loose abrasive.

The method is employed to obtain cores in large-diameter holes where abrasive bit cost is sufficiently lower than diamond bit cost to offset the increased operating costs of slower progress, and to drill shafts and other large-diameter holes where it is more economical or desirable to drill a groove and hoist out the core than to expend drilling energy over the entire cross-sectional area (Fig. 2).

Cuttings are removed from under the bit by water circulation but without sufficient velocity to carry them to the surface. Shot core barrels are therefore built with a chip box (calyx) on top so that the cuttings are retained when they settle out under the velocity drop at the top of the core barrel.

Air percussion drilling. Penetration of this drilling is by crushing action under pneumatically powered impact. The drill bit is chisel type, commonly with four cutting edges in the form of a cross. The drill produces linear impact motion by means of a reciprocating pneumatically powered piston to strike a rapid series of blows against a tubular drill steel, which transmits the energy to the bit. The drills also provide pneumatically powered rotary motion constantly to change the position of the bit against the floor of the hole. Cuttings are usually blown out of the hole by compressed air injected through the hollow drill steel and bit, although water may be substituted with drills designed for hazardous dust conditions. Dust collectors may be used to reduce hazard and also to collect samples.

Air percussion drills are primarily used for blast-hole drilling, in various sizes and mountings, for both surface and underground work. The method is limited in hole size and depth by two considerations—constant loss of bit gage and absorption of dynamic energy by the drill steel. Tungsten carbide cutting inserts help retain gage, and special down-the-hole drills are of aid in the latter problem. Detachable drill bits are commonly used in pneumatic drilling operations.

Cable tool (churn) drilling. A different percussion-type method, this penetrates by crushing impact of a falling heavy chisel bit. The drill creates an oscillating vertical linear motion, which is transmitted to the drilling tools by wire-line cable, so that they are alternately lifted and dropped. The cuttings are suspended in water by the churning action of the tools, and then periodically removed by bailing. Churn drills are sometimes used for prospecting where analysis of the cuttings gives sufficient information. In mineral exploitation work they are used for drilling blast holes in quarrying, for drilling water wells, and for drilling access holes for utility lines into mines.

Fusion piercing methods. For these, an oxygen-acetylene blow pipe applies intense heat to the floor of the hole and penetrates by flaking or melting. Drills must produce both linear and rotary motion to handle the blow pipe and break up slag formed in the process. The cuttings and slag are blown from the hole. Fusion piercing methods have been used successfully for drilling blast holes in hard rocks, such as taconite and granite.

Other drilling methods. Innumerable combinations of the basic drilling methods described above provide a wide range of variations in drilling methods.

Air or natural gas is employed as the circulating medium in rotary-type drilling. Faster rates of penetration have been obtained with these methods, but groundwater conditions may prohibit their use by creating too much fluid head for the available air pressure to work against. Techniques are under development to introduce foaming agents into the fluid column to reduce its weight.

Auger mining utilizes the conveyor action of continuous-flight augers for actual mining operations in large-diameter horizontal holes, such as for mining coal in the high wall beyond the economical limits of strip mining. The science of electronics has been applied to guide such an auger and keep it within the bed. *See* MINING EXCAVATION.

Auger stem drilling uses a short helix, run on a solid or telescoping stem, as a bit for gouging out and collecting cuttings. With a derrick higher than the hole is deep, or high enough to withdraw the entire string of tools in telescoped position, the auger can be withdrawn quickly and spun rapidly to throw off the cuttings by centrifugal force. The method is suitable for holes 12–72 in. or larger diameter. By substituting a bit incorporating a steel cylinder to confine the cuttings, the terminology is changed to bucket drilling.

Reverse-circulation drilling usually refers to a variation of the rotary drilling method in which the cuttings are pumped up and out of the drill pipe. This is advantageous in certain large-diameter holes. The term is also applied to diamond core drilling when the water is injected through a stuffing box into the annular space around the drill rods and thus forced up special large drill rods. Here the water forces the core up through the drill rods as a piston through a cylinder.

Rotary blast-hole drilling is a term commonly applied to two types of drilling. In quarrying and open-pit mining it implies rotary drilling with roller-type bits, using compressed air for cuttings removal, either conventional rotary table drive or hydraulic motor to produce rotation, and with hydraulic or wire-line mechanisms to add part of the weight of the drill to the weight of the tools and thereby increase bit pressure. In underground mining, and sometimes above ground, it implies the drilling of small-diameter blast holes with a diamond drill, using either coring or noncoring diamond bits.

Rotary-percussion drilling increases the penetration speed of a roller bit by adding pneumatic impact in linear motion to the normal abrading action of the bit in rotary motion.

Shaft drilling is shaft sinking by drilling a hole the size of the shaft, as contrasted with conven-

Fig. 2. Shot drill core. (*Pennsylvania Drilling Co.*)

tional shaft-sinking methods of drilling small holes and blasting. The perfect arch action of the smooth-cut circular wall of a drilled shaft makes shaft lining unnecessary in hard massive rocks, and much safer and more economical in others, because a blasted wall is shattered and requires more concrete for lining due to overbreak. Both shot core drilling and rotary drilling methods are used in shaft drilling, with variations such as the use of carbide or rolling cutters on the core barrel, rotary reaming a pilot hole in successively larger stages, use of reverse circulation, and down-the-hole-type rotary drills. A core barrel has been developed for shaft drilling that brings the core to the surface in the core barrel (Fig. 2).

Down-the-hole drilling is air percussion type (for large-diameter blast holes) with the reciprocating pneumatic piston placed in the drill tools close to the bit for minimizing energy losses.

Sonic drilling methods are percussion or rotary-percussion type, utilizing for impact the energy of a drill stem vibrating at sonic frequency. One method of producing sonic vibration is by use of eccentric weights driven by mud turbine.

Turbodrilling is rotary drilling with rotary motion created in the hole close to the bit by a turbine driven by the circulating mud.

Wash boring or jet drilling utilizes a chopping bit with a water jet run on a string of hollow drill rods to chop through soils and wash the cuttings to the surface. It is percussion-type drilling for which churn drills are ideally suited, but is often used on other drills because of the need for rotary coring methods in another part of the same hole. When a

combination churn and diamond drill is not available, the method may be used with a diamond drill with a cathead by snubbing a manila rope to create the churning action. By introducing a check valve at the bit, the churning action may also be used to pump the cuttings up the drill rods, using natural groundwater or water added to the hole. This prospecting method is known as hollow-rod drilling.

Wire-line coring is a method of removing core by pulling the inner tube of the core barrel, with the bit, core barrel, and drill rods remaining in the hole. The inner tube is dropped or pumped down through the drill rods, and recovered by running a retriever on a wire line, so the periodic removal of core can be done in deep drill holes in less time than required for removal and replacement of drill pipe or drill rods. In rotary drilling, the inner tube was designed to pass through the drill pipe and seat above a ring-type roller cutter core bit. In adapting the method to diamond core drilling, special drill rods were developed with maximum possible inside diameter so that the inner tube and the core could be kept as large as possible. Most of the deep diamond drill holes are now being drilled with wire-line diamond core drilling equipment. *See* CORE DRILLING; OIL AND GAS WELL DRILLING.

[FRANK C. STURGES]

Brayton cycle

A thermodynamic cycle (also variously called the Joule or complete expansion diesel cycle) consisting of two constant-pressure (isobaric) processes interspersed with two reversible adiabatic (isentropic) processes (Fig. 1). The ideal cycle performance, predicated on the use of perfect gases, is given by relationships (1) and (2). Thermal efficiency η_T, the work done per unit of heat added, is given by Eq. (3). In these relation-

$$V_3/V_2 = V_4/V_1 = T_3/T_2 = T_4/T_1 \qquad (1)$$

$$\frac{T_2}{T_1} = \frac{T_3}{T_4} = \left(\frac{V_1}{V_2}\right)^{k-1} = \left(\frac{V_4}{V_3}\right)^{k-1} = \left(\frac{p_2}{p_1}\right)^{\frac{k-1}{k}} \qquad (2)$$

$$\eta_T = [1 - (T_1/T_2)] = \left[1 - \left(\frac{1}{r^{k-1}}\right)\right] \qquad (3)$$

ships V is the volume in cubic feet, p is the pressure in pounds per square foot, T is the absolute temperature in degrees Rankine, k is the c_p/c_v, or ratio of specific heats at constant pressure and constant volume, and r is the compression ratio, V_1/V_2.

The thermal efficiency for a given gas, air, is solely a function of the ratio of compression (Fig. 2). This is also the case with the Otto cycle. For the diesel cycle with incomplete expansion, the thermal efficiency is lower, as shown for comparison in Fig. 2. The overriding importance of high compression ratio for intrinsic high efficiency is clearly demonstrated by these data.

A reciprocating engine, operating on the cycle of Fig. 1, was patented in 1872 by G. B. Brayton and was the first successful gas engine built in the United States. The Brayton cycle, with its high inherent thermal efficiency, requires the maximum volume of gas flow for a given power output. The Otto and diesel cycles require much lower gas flow rates, but have the disadvantage of higher peak pressures and temperatures. These conflicting elements led to many designs, all attempting to achieve practical compromises. With a piston and cylinder mechanism the Brayton cycle, calling for the maximum displacement per horsepower, led to proposals such as compound engines and variable-stroke mechanisms. They suffered overall disadvantages because of the low mean effective pressures. The positive displacement engine consequently preempted the field for the Otto and diesel cycles.

With the subsequent development of fluid acceleration devices for the compression and expansion of gases, the Brayton cycle found mechanisms which could economically handle the large volumes of working fluid. This is perfected today in the gas turbine power plant. The mechanism (Fig. 3) basically is a steady-flow device with a centrifugal or axial compressor, a combustion chamber where heat is added, and an expander-turbine element. Each of the phases of the cycle is accomplished with steady flow in its own mechanism rather than intermittently, as with the piston and cylinder mechanism of the usual Otto and diesel cycle engines. Practical gas-turbine engines have various recognized advantages and disadvantages which are evaluated by comparison with alternative engines available in the competitive market place. *See* GAS TURBINE; INTERNAL COMBUSTION ENGINE.

The net power output P_{net}, or salable power, of the gas-turbine plant (Fig. 3) can be expressed as shown by Eq. (4), where W_e is the ideal power out-

$$P_{net} = W_e \times \text{eff}_e - \frac{W_c}{\text{eff}_c} \qquad (4)$$

put of the expander (area b34a, Fig. 1), W_c the ideal power input to the compressor (area a12b, Fig. 1), eff_e the efficiency of expander, and eff_c the efficiency of compressor. This net power output for the ideal case, where both efficiencies are 1.0, is represented by net area (shaded) of the p-V cycle diagram of Fig. 1. The larger the volume increase from point 2 to point 3, the greater will be the net power output for a given size compressor. This volume increase is accomplished by utilizing the maximum possible temperature at point 3 of the cycle.

The difference in the two terms on the right-hand side of Eq. (4) is thus basically increased by the use of maximum temperatures at the inlet to the expander. These high temperatures introduce metallurgical and heat-transfer problems which must be properly solved.

The efficiency terms of Eq. (4) are of vital practical significance. If the efficiencies of the real compressor and of the real expander are low, it is entirely possible to vitiate the difference in the ideal powers W_e and W_c, so that there will be no useful output of the plant. In present practice this means that for adaptations of the Brayton cycle to acceptable and reliable gas-turbine plants, the engineering design must provide for high temperatures at the expander inlet and utilize high built-in efficiencies of the compressor and expander elements. No amount of cycle alteration, regeneration, or reheat can offset this intrinsic requirement for mechanisms which will safely operate at temperatures of 1500°±F. *See* CARNOT CYCLE;

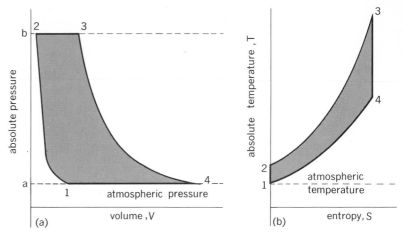

Fig. 1. Brayton cycle, air-card standard. Phases: 1–2, compression; 2–3, heat addition; 3–4, expansion; and 4–1, heat abstraction.

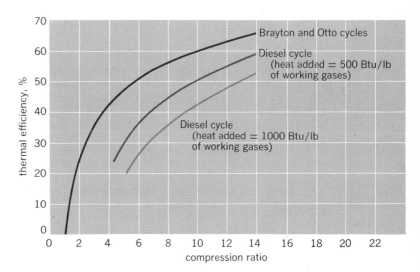

Fig. 2. Thermal efficiency of Brayton, Otto, diesel ideal cycles; air-card standard.

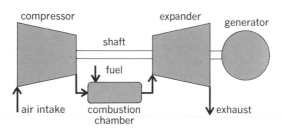

Fig. 3. Simple, open-cycle gas-turbine plant.

DIESEL CYCLE; OTTO CYCLE; THERMODYNAMIC CYCLE. [THEODORE BAUMEISTER]

Bibliography: T. Baumeister (ed.), *Standard Handbook for Mechanical Engineers*, 7th ed., 1967; J. B. Jones and G. A. Hawkins, *Engineering Thermodynamics*, 1960; J. H. Keenan, *Thermodynamics*, 1941; E. T. Vincent, *Theory and Design of Gas Turbines and Jet Engines*, 1950; M. J. Zucrow, *Principles of Jet Propulsion and Gas Turbines*, 1948.

British thermal unit (Btu)

A unit quantity of energy. Usually the expression is used in connection with energy as heat, but it may be used when referring to energy as work or energy in any other form. Before 1929 the 60° Btu was defined as that quantity of energy as heat required to increase the temperature of 1 lb of water from 59.5°F to 60.5°F, the water being held under a constant pressure of 1 atm. Because of the difficulty of experimentally determining the exact value of the Btu, and especially its relation to the foot-pound, the Btu is now defined in terms of electrical units. At the International Steam Table Conference in London in 1929, 1 Btu was defined as 251.996 IT (International Steam Table) cal, or 778.26 ft-lb. By definition 1 IT cal = 1/860 watt-hour; hence a Btu is equivalent to approximately 1/3 watt-hour. *See* ELECTRICAL MEASUREMENTS; HEAT.

[HAROLD C. WEBER]

Butane

An alkane having the formula C_4H_{10}, of which there are two isomers: *n*-butane, $CH_3CH_2CH_2CH_3$ (boiling point −0.5°C) and isobutane $(CH_3)_2CHCH_3$ (boiling point −11.7°C). Both butanes occur in petroleum and natural gas and are produced by the cracking of petroleum. The proportion of isobutane formed in catalytic cracking is greater than in thermal cracking. *See* CRACKING.

The butanes are components of the domestic fuel, liquefied petroleum gas (LPG). They are also important volatile constituents of gasoline.

Either of the butane isomers can be catalytically converted to an equilibrium mixture of the two. They can be dehydrogenated to the corresponding *n*-butylene (three isomers: 1-butene, *cis*-2-butene, and *trans*-2-butene) and isobutylene. Both butanes can be thermally alkylated, and isobutane can be catalytically alkylated with olefins to produce branched-chain alkanes.

[LOUIS SCHMERLING]

Cable-tool drill

In drilling oil and gas wells a cutting tool, or bit, is suspended on a wire line and pounds through rock formations as the line is alternately raised and lowered, generally 20–40 times per minute. A diagram of the cable-tool drill is given in the illustration. Water is poured into the bottom of the hole to keep the pulverized cuttings fluid. When the slurry gets so thick that drilling is impeded, the string of tools is pulled from the hole and a bucketlike bailer is used to remove the slurry. This percussion process of drilling and bailing is continued until the desired depth is reached.

The drilling string is about 40 ft long and consists of several parts. A socket connects the string to the wire line and is attached to a sinker, or short cylinder of steel, which is used for weight. This is followed by metal links, called jars, which permit movement in the string. These, in turn, connect with a long cylindrical steel drill stem, also added for weight. The drill bit, which cuts the hole, screws to the bottom of the drill stem. The bit can be of several types but generally is a heavy steel bar 7–8 ft long, dressed to a blunt end on the cutting edge, much like a chisel.

Power is transmitted from an engine to a band

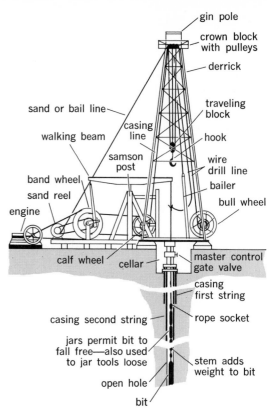

Cable-tool or percussion drilling rig.

wheel, which operates a connecting rod, or pitman, that is anchored to one end of a walking beam. The walking beam is supported at its center on a samson post, so that it resembles a seesaw in action. The drilling line is connected to the other end of the walking beam, directly over the well bore, and is raised and lowered alternately with each stroke of the walking beam. The drilling line passes up through the derrick to a crown block and down to a bull wheel, which is used to apply the power in hoisting or lowering the drill string.

A calf wheel lowers or lifts the heavy strings of casing, or pipe, used in lining the hole. The cable on the calf wheel passes up through the derrick and down, connecting to a hoisting or traveling block, a hook, and special clamps, called elevators. The elevators provide the means for suspending the casing from the hoisting block. The sand wheel driven from the band wheel bails out the cuttings by means of the sand line and bailer. *See* BORING AND DRILLING (MINERAL); OIL AND GAS WELL DRILLING.   [ADE L. PONIKVAR]

Carnot cycle

A hypothetical thermodynamic cycle originated by Sadi Carnot and used as a standard of comparison for actual cycles. The Carnot cycle shows that, even under ideal conditions, a heat engine cannot convert all the heat energy supplied to it into mechanical energy; some of the heat energy must be rejected.

In a Carnot cycle, an engine accepts heat energy from a high-temperature source, or hot body, converts part of the received energy into mechanical (or electrical) work, and rejects the remainder to a low-temperature sink, or cold body. The greater the temperature difference between the source and sink, the greater the efficiency of the heat engine.

The Carnot cycle (Fig. 1) consists first of an isentropic compression, then an isothermal heat addition, followed by an isothermal expansion, and concludes with an isentropic heat rejection process. In short, the processes are compression, addition of heat, expansion, and rejection of heat, all in a qualified and definite manner.

Processes. The air-standard engine, in which air alone constitutes the working medium, illustrates the Carnot cycle. A cylinder of air has perfectly insulated walls and a frictionless piston. The top of the cylinder, called the cylinder head, can either be covered with a thermal insulator, or, if the insulation is removed, can serve as a heat transfer surface for heating or cooling the cylinder contents.

Initially, the piston is somewhere between the top and the bottom of the engine's stroke, and the air is at some corresponding intermediate pressure but at low temperature. Insulation covers the cylinder head. By employing mechanical work from the surroundings, the system undergoes a reversible adiabatic, or an isentropic, compression. With no heat transfer, this compression process raises both the pressure and the temperature of the air, and is shown as the path a-b on Fig. 1.

After the isentropic compression carries the piston to the top of its stroke, the piston is ready to reverse its direction and start down. The second process is one of constant-temperature heat addition. The insulation is removed from the cylinder head, and a heat source, or hot body, applied that is so large that any heat flow from it will not affect its temperature. The hot body is at a temperature just barely higher than that of the gas it is to heat. The temperature gradient is so small it is considered reversible; that is, if the temperature changed slightly the heat might flow in the other direction, from the gas into the hot body. In the heat addition process, enough heat flows from the hot body into the gas to maintain the temperature of the gas while it slowly expands and does useful work on the surroundings. All the heat is added to the working substance at this constant top temperature of the cycle. This second process is shown as b-c on Fig. 1.

Part way down the cylinder, the piston is stopped; the hot body is removed from the cylinder head, and an insulating cover is put in its place. Then the third process of the cycle begins; it is a frictionless expansion, devoid of heat transfer, and carries the piston to the bottom of its travel. This isentropic expansion reduces both the pressure and the temperature to the bottom values of the cycle. For comparable piston motion, this isentropic expansion drops the pressure to a greater extent than the isothermal process would do. The path c-d on Fig. 1 represents this third process, and is steeper on the P-v plane than process b-c.

The last process is the return of the piston to the same position in the cylinder as at the start. This last process is an isothermal compression and simultaneous rejection of heat to a cold body which has replaced the insulation on the cylinder head. Again, the cold body is so large that heat

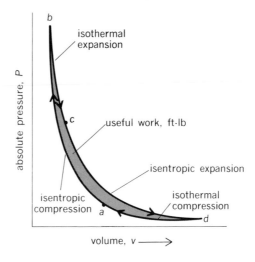

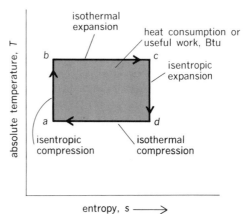

Fig. 1. Carnot cycle for air.

flow to it does not change its temperature, and its temperature is only infinitesimally lower than that of the gas in the system. Thus, the heat rejected during the cycle flows from the system at a constant low-temperature level. The path d-a on Fig. 1 shows this last process.

The net effect of the cycle is that heat is added at a constant high temperature, somewhat less heat is rejected at a constant low temperature, and the algebraic sum of these heat quantities is equal to the work done by the cycle.

Figure 1 shows a Carnot cycle when air is used as the working substance. The P-v diagram for this cycle changes somewhat when a vapor or a liquid is used, or when a phase change occurs during the cycle, but the T-s diagram always remains a rectangle regardless of phase changes or of working substances employed.

It is significant that this cycle is always a rectangle on the T-s plane, independent of substances used, for Carnot was thus able to show that neither pressure, volume, nor any other factor except temperature could affect the thermal efficiency of his cycle. Raising the hot-body temperature raises the upper boundary of the rectangular figure, increases the area, and thereby increases the work done and the efficiency, because this area represents the net work output of the cycle. Similarly, lowering the cold-body temperature increases the area, the work done, and the efficiency. In prac-

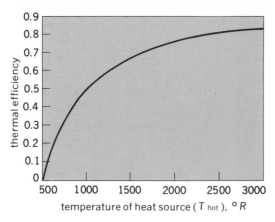

Fig. 2. Thermal efficiency of the Carnot cycle with heat-rejection temperature T_{cold} equal to 500°R.

and re-reflected with no loss of radiant energy and no change of wavelength.

The electromagnetic theory of radiation asserts that the radiant energy applies pressure P to the cylinder walls. This radiation pressure is equal to $u/3$, where u represents the radiant energy density, or the amount of radiant energy per unit volume.

The piston moves so that the cylinder boundaries expand, and additional radiant energy is supplied to the system so that the temperature remains constant. Then, cutting off any further supply of radiant energy, the system undergoes a further infinitesimal expansion with its associated pressure drop and temperature drop. The third process is an isothermal compression at the low-temperature level, accompanied by some rejection of radiant energy. One last process closes the cycle with a reversible compression that raises the temperature to the original level. The assumption is made throughout this analysis that energy density is a function of temperature alone. Thus, because this last compression process increases the energy density, it raises the temperature.

A record of all energy quantities in this radiant-energy Carnot cycle indicates that energy density u is proportional to the fourth power of the absolute temperature. Consequently the total rate of emission of radiant energy from the surface of a blackbody is also proportional to the fourth power of its absolute temperature, thereby using a Carnot cycle to provide a relationship by theoretical analysis, the same relationship which had previously been determined experimentally and labeled as Stefan's law.

Conversion of heat to electricity. Several physical phenomena convert heat energy directly into electrical energy. The extent of this direct conversion of heat to electricity is limited by the temperature levels between which the process operates. The ideal efficiency of such direct-conversion thermoelectric cycles equals the efficiency of a Carnot cycle that operates between the same temperature limits of heat source and heat sink. However, the conversion efficiency obtained in practice is only a small fraction of the ideal efficiency at the present stage of development.

The most widely known physical arrangement for direct thermoelectric generation is the thermocouple, which produces an electromotive force, or voltage, when one junction of two dissimilar conductors is heated while the other junction is kept cool. Thermocouples made of metals are inefficient converters of heat energy to electric energy, because metals that have good electrical conductivity unfortunately have equally good thermal conductivity, which permits heat loss by conduction from the hot to the cold junction. In contrast, thermocouples made of semiconductors offer the prospect of operation at high temperatures and with high temperature gradients, because semiconductors are good electrical conductors but poor heat conductors. Semiconductor thermocouples may have as large a junction potential as the metal couples do.

Thermionic emission is another phenomenon that permits the partial conversion of heat energy directly into electrical energy. Externally applied heat imparts kinetic energy to electrons, liberating them from a surface. The density of the emission

tice, nature establishes the temperature of the coldest body available, such as the temperature of ambient air or river water, and the bottom line of the rectangle cannot circumvent this natural limit.

The thermal efficiency of the Carnot cycle is solely a function of the temperature at which heat is added (phase b-c) and the temperature at which heat is rejected (phase d-a) (Fig. 1). The rectangular area of the T-s diagram represents the work done in the cycle so that thermal efficiency, which is the ratio of work done to the heat added, equals $(T_{hot} - T_{cold})/T_{hot}$. For the case of atmospheric temperature for the heat sink ($T_{cold} = 500°R$), the thermal efficiency, as a function of the temperature of the heat source, T_{hot}, is shown in Fig. 2.

Carnot cycle with steam. If steam is used in a Carnot cycle, it can be handled by the following flow arrangement. Let saturated dry steam at 500°F flow to the throttle of a perfect turbine where it expands isentropically down to a pressure corresponding to a saturation temperature of the cold body. The exhausted steam from the turbine, which is no longer dry, but contains several percent moisture, is led to a heat exchanger called a condenser. In this device there is a constant-pressure, constant-temperature, heat-rejection process during which more of the steam with a particular predetermined amount of condensed liquid is then handled by an ideal compression device. The isentropically compressed mixture may emerge from the compressor as completely saturated liquid at the saturation pressure corresponding to the hot-body temperature. The cycle is closed by the hot body's evaporating the liquid to dry saturated vapor ready to flow to the turbine.

The cycle is depicted in Fig. 3 by c-d as the isentropic expansion in the turbine; d-a is the constant-temperature condensation and heat rejection to the cold body; a-b is the isentropic compression; and b-c is the constant-temperature boiling by heat transferred from the hot body. *See* STEAM ENGINE.

Carnot cycle with radiant energy. Because a Carnot cycle can be carried out with any arbitrary system, it has been analyzed when the working substance was considered to be a batch of radiant energy in an evacuated cylinder. If the system boundaries are perfectly reflecting thermal insulators, the enclosed radiant energy will be reflected

CARNOT CYCLE

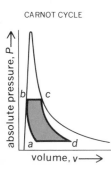

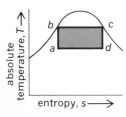

Fig. 3. Carnot cycle with steam.

current is a function of the absolute temperature and work function of the emitter material.

For many years thermionic emission received little attention as a source of power because of its very low conversion efficiency. However, interest has been stimulated by development of a contact potential thermionic emission cell. In this device, current flows between the surfaces of two materials which have different work functions. These materials are held at different temperatures, and the gap between the electrode surfaces is filled with gas at low pressure.

Such direct-conversion techniques show promise of becoming small-scale, if unconventional power sources. Although these techniques ideally can convert heat to electrical energy with the efficiency of a Carnot cycle operating between the same temperature levels, laboratory devices do not surpass about 8% efficiency, which is just a fraction of the ideal performance.

Reversed Carnot cycle. A Carnot cycle consists entirely of reversible processes; thus it can theoretically operate to withdraw heat from a cold body and to discharge that heat to a hot body. To do so, the cycle requires work input from its surroundings. The heat equivalent of this work input is also discharged to the hot body.

Just as the Carnot cycle provides the highest efficiency for a power cycle operating between two fixed temperatures, so does the reversed Carnot cycle provide the best coefficient of performance for a device pumping heat from a low temperature to a higher one. This coefficient of performance is defined as the ratio of the quantity of heat pumped to the amount of work required, or it equals $T_{hot}/(T_{hot} - T_{cold})$ for a warming machine, and $T_{cold}/(T_{hot} - T_{cold})$ for a cooling machine, where all temperatures are in degrees absolute. This is one of the few engineering indices with numerical values greater than unity. See HEAT PUMP; REFRIGERATION CYCLE.

Practical limitations. Good as the ideal Carnot cycle may be, there are serious difficulties that emerge when one wishes to make an actual Carnot engine. The method of heat transfer through the cylinder head either limits the operation of the engine to very low speeds, or requires an engine with a huge bore and small stroke. Moreover, the material of the cylinder head is subjected to the full top temperature of the cycle, imposing a metallurgical upper limit on the cycle's temperature.

A practical solution to the heat transfer difficulties which beset the Carnot cycle is to burn a fuel in the air inside the engine cylinder. The result is an internal combustion engine that consumes and replaces its working substance while undergoing a periodic sequence of processes.

The working substance in such an internal combustion engine can attain very high temperatures, far above the melting point of the metal of the cylinder walls, because succeeding lower-temperature processes will keep the metal parts adequately cool. Thus, as the contents change temperature rapidly between wide extremes, the metal walls hover near a median temperature. The fuel-air mixture can be ignited by a spark, or by the rise in temperature produced by the compression stroke. See DIESEL CYCLE; DIESEL ENGINE; INTERNAL COMBUSTION ENGINE; OTTO CYCLE.

Even so, the necessarily high peak pressures and temperatures limit the practical thermal efficiency that an actual engine can achieve. The same theoretical efficiency can be obtained from a cycle consisting of two isobaric processes interspersed by two isentropic processes. The isobaric process requires that the cycle handle large volumes of low-pressure gas, which can best be done in a rotating turbine. See BRAYTON CYCLE; GAS TURBINE.

Although the Carnot cycle is independent of the working substance, and hence is applicable to a vapor cycle, the difficulty of efficiently compressing a vapor-liquid mixture renders the cycle impractical. In a steam power plant the sequence of states assumed by the working substance is (1) condensate and feedwater are compressed and pumped into the boiler, (2) heat is added to the water first at constantly increasing temperature and then at constant pressure and temperature, (3) the steam expands in the engine, and (4) the cycle is completed by a constant-temperature heat-rejection process which condenses the exhausted steam.

By comparison to the Carnot cycle, in which heat is added only at the highest temperature, this steam cycle with heat added over a range of temperatures is necessarily less efficient than is theoretically possible. An analysis of engine operations in terms of thermodynamic cycles indicates what efficiencies can be expected and how the operations should be modified to increase engine performance, such as high compression ratios for reciprocating internal combustion engines and high steam temperatures for steam engines. See POWER PLANT; THERMODYNAMIC CYCLE; THERMODYNAMIC PRINCIPLES.   [THEODORE BAUMEISTER]

Bibliography: T. Baumeister, *Standard Handbook for Mechanical Engineers*, 1967; J. B. Jones and G. A. Hawkins, *Engineering Thermodynamics*, 1960; H. C. Weber and H. P. Meissner, *Thermodynamics for Chemical Engineers*, 2d ed., 1957; M. W. Zemansky, *Heat and Thermodynamics*, 4th ed., 1957.

Central heating

The use of a single heating plant to serve a group of buildings, facilities, or even a complete community through a system of distribution pipework that feeds each structure or facility. Central heating plants are basically of two types: steam or hot-water. The latter type uses high-temperature hot water under pressure and is becoming the more usual because of its considerable advantages. Steam systems are only used today where there is a specific requirement for high-pressure steam.

Benefits. Advantages of a central heating plant over individual ones for each building or facility in a group include reduced labor cost, lower fuel cost, and simpler maintenance. Even though a central plant may require a 24-hr shift of operators, the total number of employees can be substantially less than that required to operate and maintain a number of individual plants.

Firing efficiencies of 85–93%, dependent upon such factors as fuel, boiler, and plant design, are usual with large central heating plants. Corresponding efficiencies for small individual heating boiler plants average 60–70%. Fuel in bulk quantity has a lower unit cost, and single handling for one large plant as distinct from multiple handling for many small plants saves appreciably in labor and transportation. Maintenance costs on a single central plant are considerably lower than for the aggregate of small plants of equal total capacity.

The disadvantages of a central heating plant concern mainly the maintenance of the distribution system where steam is used. Corrosion of the condensate water return lines shortens their life, and the steam drainage traps need particular attention. These disadvantages do not occur with hot-water installations.

Where air conditioning is required, the central plant may be used to provide the energy source for the summer cooling refrigeration plant. Both hot-water or steam central plants are suitable for this purpose.

Design. Winter heat-load requirements are calculated by the addition of the following for each individual building or facility: (1) winter heat losses, (2) domestic water-heating requirements, and (3) industrial or other special heat requirements. To the sum of these for all buildings or facilities must be added the system distribution heat losses.

Where air conditioning is involved, the summer load on the central plant is calculated by the addition of the following for each individual building or facility: summer air conditioning requirements, and domestic water-heating requirements. To the sum of these for all buildings or facilities must be added the system summer distribution heat losses.

To the individual winter and summer totals a diversity factor of 70–80% is applied because not all heat loads peak simultaneously. Winter heat loss due to weather conditions must be taken at its maximum. Water-heating and industrial requirements vary throughout the daily cycle. The summer air conditioning load must also be taken at its maximum. System distribution losses must be calculated both for the winter and the summer loads. The individual characteristics of the system must be considered in the diversity factor used.

Divergencies between winter and summer loads must be calculated. Distribution losses are less for summer than for winter because the heat load carried through the distribution system is smaller. Individual boiler sizing should allow the best arrangement to meet load variations as between individual 24-hr peaks and summer and winter loads. Standby capacity is essential. Fuel selection, usually oil, coal, or gas, depends on local conditions and costs, taking into account labor and firing efficiencies to be expected with each fuel.

Distribution pipework sizing follows normal practice using suitable pressure drops with allowance for load variations and diversity factors as indicated. Such factors apply particularly to main distribution headers as distinct from branches. Numerous commercial forms of preinsulated, waterproofed, and structurally self-sufficient conduits are available for underground distribution mains.

Economics. The economics for each system must be individually computed.

The following factors directly affect each plant's economics: (1) system type such as low-pressure steam, high-pressure steam, hot water, high-temperature hot water, and so on; (2) fuel used, that is, coal, gas, or oil; (3) labor costs and conditions; (4) type and occupancy of facilities served; and (5) terrain.

Boiler plant. Both high-pressure (125 psi saturated) and low-pressure (15 psi saturated) steam plants are used, although the former is the more common. Both types follow conventional design. Boilers may be either water tube or fire tube, depending on the rate of load cycling. Feedwater and firing auxiliaries are of conventional type. Chemical treatment of feedwater is usual.

Either conventional hot-water heating plants or high-temperature hot-water plants, depending on size, are used. Design and components for high-temperature hot-water plants are more complex. However, standard manufactured equipment is available for both conventional and high-temperature plants. Hot-water circulation, due to losses in the distribution pipework, should be at maximum temperature-pressure limitation. Circulation through the distribution system is by centrifugal pump with standby equipment being furnished. High-temperature hot-water installations may operate in 400–500°F range.

Fuel handling, firing, and control arrangements also follow conventional design. Capacity of fuel storage should be on a minimal 3–4-week basis. Coal, oil, and gas are equally suitable as fuels, the individual choice being dependent upon local conditions and economics.

Physical size of plant depends on type and fuel used. A rule of thumb may be taken as 1.8 ft² of floor area per boiler horsepower (34.5 lb steam per hour from 212°F ≡ 33,500 Btu per hour). This does not include fuel storage space.

Distribution systems. Both overhead and underground pipework are used for distribution, although the latter is more usual except for industrial plants. Overhead mains must be strongly supported, insulated, and weatherproofed. Underground mains must be insulated and carefully waterproofed, particularly in damp areas. They must be structurally adequate. Steam distribution requires proper drainage of mains and often pumped return of condensate when gravity flow is not practical. In municipal distribution systems in larger cities where steam is used, it is frequently considered uneconomical to return the condensate to the central boiler plant, as pumping costs become excessive over large distances. For a plant serving a small community or medium-sized facility, however, this limitation does not apply.

Hot-water distribution systems have the advantage that they are not affected by grade variations, that is, they can be run both uphill and downhill. The circulating pump pressures must be calculated accordingly, but this presents no problem.

When one of the commercial forms of prefabricated conduit is used, proper cover must be arranged in depth of excavation to give adequate protection against surface loads. Granulated plastic minerals which harden when heated and have a high insulating value are also used around buried distribution mains for insulation and waterproofing purposes. Generally, this usage effects considerable economies in first cost. Where high-temperature hot-water is used, it may be necessary to provide heat exchangers at each building or facility to furnish secondary heat at more moderate temperatures for uses such as heating systems and hot-water supplies. Where low-pressure steam is required, such exchangers may furnish this on the secondary side, provided the primary hot water is at sufficiently high temperature.

Heat sales. Various methods of heat sale are in use where central plants service public communities or facilities. With steam distribution steam meters to each individual building or facility served are usual, and a utility type of sliding scale rate per pound of steam sold is charged. The use of condensate-measuring meters is in general not recommended because of their inaccuracy.

For hot-water distribution a combination meter measuring both water flow and temperature differential between supply and return mains may be used. This measures directly Btu per hour furnished. In certain cases where a constant temperature differential between supply and return water mains is maintained, metering may be by flow only, although this is not very accurate. *See* AIR CONDITIONING; HEATING, COMFORT; STEAM HEATING; WARM-AIR HEATING SYSTEM.

[JOHN K. M. PRYKE]

Bibliography: ASHRAE, *Guide and Data Book*: *Applications*, 1968, *Systems and Equipment*, 1967; ASHRAE, *Handbook of Fundamentals*, 1967; T. Baumeister (ed.), *Marks' Standard Handbook for Mechanical Engineers*, 7th ed., 1967; W. H. Carrier et al., *Modern Air Conditioning, Heating and Ventilating*, 3d ed., 1959; P. L. Geiringer, *High Temperature Water Heating*, 1963; National District Heating Association, *District Heating Handbook*, 3d ed., 1951.

Chain reaction, nuclear

A succession of generation after generation of acts of nuclear division such that the neutrons set free in the nuclear disruptions of the nth generation split the fissile nuclei (U^{233}, U^{235}, Pu^{239}) of the $(n+1)$st generation. The first few neutrons ordinarily come from the natural or spontaneous fission of admixed U^{238} or Pu^{240} or from other nuclear reactions. *See* FISSION, NUCLEAR.

The ratio of the average number of acts of fission in the $(n+1)$st and the nth generation is known as the multiplication factor k. Depending on whether k is greater than 1, equal to 1, or less than 1, the chain reaction is said to be divergent, self-sustaining, or convergent, and requires at least a mass of fissile material greater than, equal to, or less than the critical mass. *See* CRITICAL MASS.

The circumstance of several neutrons being given off in fission permits, in principle, a mulitplication factor as high as $k=2$ or more in a sufficiently large mass of metallic U^{235} or Pu^{239}. An atomic bomb and a runaway power reactor are distinguished not only by very different values of $k-1$ (several tenths versus 0.01) but also by the time between one generation and another (10^{-8} sec versus 10^{-3} sec in order of magnitude).

[JOHN A. WHEELER]

Bibliography: G. Friedlander, J. W. Kennedy, and J. M. Miller, *Nuclear and Radiochemistry*, 2d ed., 1964; A. M. Weinberg and E. P. Wigner, *The Physical Theory of Neutron Chain Reactors*, 1958.

Charge, electric

A basic property of elementary particles of matter. One does not define charge but takes it as a basic experimental quantity and defines other quantities in terms of it. The early Greek philosophers were aware that rubbing amber with fur produced properties in each that were not possessed before the rubbing. For example, the amber attracted the fur after rubbing, but not before. These new properties were later said to be due to "charge." The amber was assigned a negative charge and the fur was assigned a positive charge.

According to modern atomic theory, the nucleus of an atom has a positive charge because of its protons, and in the normal atom there are enough extranuclear electrons to balance the nuclear charge so that the normal atom as a whole is neutral. Generally, when the word charge is used in electricity, it means the unbalanced charge (excess or deficiency of electrons), so that physically there are enough "nonnormal" atoms to account for the positive charge on a "positively charged body" or enough unneutralized electrons to account for the negative charge on a "negatively charged body."

The rubbing process mentioned "rubs" electrons off the fur onto the amber, thus giving the amber a surplus of electrons, and it leaves the fur with a deficiency of electrons.

In line with the previously mentioned usage, the total charge q on a body is the total unbalanced charge possessed by the body. For example, if a sphere has a negative charge of 1×10^{-10} coulomb, it has 6.24×10^8 electrons more than are needed to neutralize its atoms. The coulomb is used as the unit of charge in the meter-kilogram-second (mks) system of units. *See* ELECTRICAL UNITS AND STANDARDS.

The surface charge density σ on a body is the charge per unit surface area of the charged body. Generally, the charge on the surface is not uniformly distributed, so a small area ΔA which has a magnitude of charge Δq on it must be considered. Then σ at a point on the surface is defined by the equation below.

$$\sigma = \lim_{\Delta A \to 0} \frac{\Delta q}{\Delta A}$$

The subject of electrostatics concerns itself with properties of charges at rest, while circuit analysis, electromagnetism, and most of electronics concern themselves with the properties of charges in motion.

[RALPH P. WINCH]

Chemical energy

In most chemical reactions, heat is either taken in or given out. By the law of conservation of energy, the increase or decrease in heat energy must be accompanied by a corresponding decrease or increase in some other form of energy. This other form is the chemical energy of the compounds involved in the reaction. The rearrangement of the atoms in the reacting compounds to produce new compounds causes a change in chemical energy. This change in chemical energy is equal numerically and of opposite sign to the heat change accompanying the reaction.

Most of the world's available power comes from

the combustion of coal or of petroleum hydrocarbons. The chemical energy released as heat when a specified weight, often 1 g, of a fuel is burned is called the calorific value of the fuel.

Depending on whether the pressure or volume of the system is kept constant, differing quantities of heat are liberated in a chemical reaction. The heat of reaction at constant pressure q_p is equal to minus the change in the chemical energy at constant pressure ΔH, called the change in the enthalpy; the heat of reaction at constant volume q_v is equal to minus the change in the chemical energy at constant volume ΔE, called the change in the internal energy. See ENTHALPY; INTERNAL ENERGY.

It is not possible to measure an absolute value for the chemical energy of a compound; only changes in chemical energy can be measured. It is therefore necessary to make some arbitrary assumption as a starting point. One such assumption would be to take the chemical energies of the free atoms as zero and measure the chemical energies of all elements and compounds relative to this standard. If this were done, all chemical energies would then be negative quantities. Heat is given out when all elements and compounds are formed from their atoms. Since chemical energy can be regarded as a form of potential energy, it is interesting that the formation of chemical bonds is always accompanied by a decrease of potential energy. The reason, in qualitative terms at least, lies in the quantum theory. When chemical bonds are formed between atoms, for example, in the formation of a chlorine molecule from two chlorine atoms, electrons are shared between the two atoms, and this electron-sharing produces a lowering of the potential energies of the shared electrons. This lowering of potential energy, in turn, causes a lowering of the potential energy of the molecule relative to the free atoms.

Although it would be more fundamental to take the chemical energies of the separated atoms as zero, in practice it is more convenient to take a more arbitrary starting point and to assume that the chemical energies of the elements are zero. For precision, the standard states of the elements at 25°C and 1 atm pressure are chosen. Thus, in the case of carbon the chemical energy of graphite, not diamond, is said to be zero at 25°C and 1 atm. Diamond then has a definite enthalpy value. See ENERGY SOURCES.

[THOMAS C. WADDINGTON]

Chemical fuel

The principal fuels used in internal combustion engines (automobiles, diesel, and turbojet) and in the furnaces of stationary power plants are organic fossil fuels. These fuels, and others derived from them by various refining and separation processes, are found in the earth in the solid (coal), liquid (petroleum), and gas (natural gas) phases.

Special fuels to improve the performance of combustion engines are obtained by synthetic chemical procedures. These special fuels serve to increase the fuel specific impulse of the engine (specific impulse is the force produced by the engine multiplied by the time over which it is produced, divided by the mass of the fuel) or to increase the heat of combustion available to the engine per unit mass or per unit volume of the fuel. A special fuel which possesses a very high heat of combustion per unit mass is liquid hydrogen. It has been used along with liquid oxygen in rocket engines. Because of its low liquid density, liquid hydrogen is not too useful in systems requiring high heats of combustion per unit volume of fuel ("volume-limited" systems). In combination with liquid fluorine, liquid hydrogen produces extremely large specific impulses, and rocket engines using this combination are under development. See AIRCRAFT FUEL.

A special fuel which produces high flame temperatures of the order of 5000°C is gaseous cyanogen, C_2N_2. This is used with gaseous oxygen as the oxidizer. The liquid fuel hydrazine, N_2H_4, and other hydrazine-based fuels, with the liquid oxidizer nitrogen tetroxide, N_2O_4, are used in many space-oriented rocket engines. The boron hydrides, such as diborane, B_2H_6, and pentaborane, B_5H_9, are high-energy fuels which are being used increasingly in advanced rocket engines.

For air-breathing propulsion engines (turbojets and ramjets), hydrocarbon fuels are most often used. For some applications, metal alkyl fuels which are pyrophoric (that is, ignite spontaneously in the presence of air), and even liquid hydrogen, are being used.

A partial list of additional currently used liquid fuels and their associated oxidizers is shown in the table.

Fuels which liberate heat in the absence of an oxidizer while decomposing either spontaneously or because of the presence of a catalyst are called monopropellants and have been used in rocket engines. Examples of these monopropellants are hydrogen peroxide, H_2O_2, and nitro-methane, CH_3NO_2.

Liquid fuels and oxidizers are used in most large-thrust (large propulsive force) rocket engines. When thrust is not a consideration, solid-propellant fuels and oxidizers are frequently employed because of the lack of moving parts such as valves and pumps, and the consequent simplicity of this type of rocket engine. Solid fuels fall into two broad classes, double-base and composites. Double-base fuels are compounded of nitroglycerin (glycerol trinitrate) and nitrocellulose, with no separate oxidizer required. The nitroglycerin plasticizes and swells the nitrocellulose, leading to a propellant of relatively high strength and low elongation. The double-base propellant is generally formed in a mold into the desired shape (called a grain) required for the rocket case. Composite propellants are made of a fuel and an oxidizer. The latter could be an inorganic perchlorate such as ammonium perchlorate, NH_4ClO_4, or potassium

Liquid fuels and their associated oxidizers

Fuel	Oxidizer
Ammonia	Liquid oxygen
95% Ethyl alcohol	Liquid oxygen
Methyl alcohol	87% Hydrogen peroxide
Aniline	Red fuming nitric acid
Furfural alcohol	Red fuming nitric acid

perchlorate, $KClO_4$, or a nitrate such as ammonium nitrate, NH_4NO_3, potassium nitrate, KNO_3, or sodium nitrate, $NaNO_3$. Fuels for composite propellants are generally the asphalt-oil-type, thermosetting plastics (phenol formaldehyde and phenolfurfural resins have been used) or several types of synthetic rubber and gumlike substances. Recently, metal particles such as boron, aluminum, and beryllium have been added to solid propellants to increase their heats of combustion and to eliminate certain types of combustion instability. *See* FUEL, LIQUID.

[WALLACE CHINITZ]

Bibliography: S. Penner, *Chemistry Problems in Jet Propulsion*, 1957; H. W. Ritchey and J. M. McDermott, Solid propellant rocket technology, F. I. Ordway, III (ed.), *Advances in Space Science and Technology*, 1963; G. P. Sutton, *Rocket Propulsion Elements*, 3d ed., 1963.

Circuit (electricity)

A general term referring to a system or part of a system of conducting parts and their interconnections through which an electric current is intended to flow. A circuit is made up of active and passive elements or parts and their interconnecting conducting paths. The active elements are the sources of electric energy for the circuit; they may be batteries, direct-current generators, or alternating-current generators. The passive elements are resistors, inductors, and capacitors. The electric circuit is described by a circuit diagram or map showing the active and passive elements and their connecting conducting paths.

Devices with an individual physical identity such as amplifiers, transistors, loudspeakers, and generators, are often represented by equivalent circuits for purposes of analysis. These equivalent circuits are made up of the basic passive and active elements listed above.

Electric circuits are used to transmit power as in high-voltage power lines and transformers or in low-voltage distribution circuits in factories and homes; to convert energy from or to its electrical form as in motors, generators, microphones, loudspeakers, and lamps; to communicate information as in telephone, telegraph, radio, and television systems; to process and store data and make logical decisions as in computers; and to form systems for automatic control of equipment.

Electric circuit theory. This includes the study of all aspects of electric circuits, including analysis, design, and application. In electric circuit theory the fundamental quantities are the potential differences (voltages) in volts between various points, the electric currents in amperes flowing in the several paths, and the parameters in ohms or mhos which describe the passive elements. Other important circuit quantities such as power, energy, and time constants may be calculated from the fundamental variables.

Electric circuit theory is an extensive subject and is often divided into special topics. Division into topics may be made on the basis of how the voltages and currents in the circuit vary with time; examples are direct-current, alternating-current, nonsinusoidal, digital, and transient circuit theory. Another method of classifying circuits is by the arrangement or configuration of the electric current paths; examples are series circuits, parallel circuits, series-parallel circuits, networks, coupled circuits, open circuits, and short circuits. Circuit theory can also be divided into special topics according to the physical devices forming the circuit, or the application and use of the circuit. Examples are power, communication, electronic, solid-state, integrated, computer, and control circuits.

Direct-current circuits. In dc circuits the voltages and currents are constant in magnitude and do not vary with time (Fig. 1). Sources of direct current are batteries, dc generators, and rectifiers. Resistors are the principal passive element.

Magnetic circuits. Magnetic circuits are similar to electric circuits in their analysis and are often included in the general topic of circuit theory. Magnetic circuits are used in electromagnets, relays, magnetic brakes and clutches, computer memory devices, and many other devices. For a detailed treatment *see* MAGNETIC CIRCUITS.

Alternating-current circuits. In ac circuits the voltage and current periodically reverse direction with time. The time for one complete variation is known as the period. The number of periods in 1 sec is the frequency in cycles per second. A cycle per second has recently been named a hertz (in honor of Heinrich Rudolf Hertz's work on electromagnetic waves).

Most often the term ac circuit refers to sinusoidal variations. For example, the alternating current in Fig. 2 may be expressed by $i = I_m \sin \omega t$. Sinusoidal sources are ac generators and various types of electronic and solid-state oscillators; passive circuit elements include inductors and capacitors as well as resistors. The analysis of ac circuits requires a study of the phase relations between voltages and currents as well as their magnitudes. Complex numbers are often used for this purpose.

Nonsinusoidal waveforms. These voltage and current variations vary with time but not sinusoidally (Fig. 3). Such nonsinusoidal variations are usually caused by nonlinear devices, such as saturated magnetic circuits, electron tubes, and transistors. Circuits with nonsinusoidal waveforms are analyzed by breaking the waveform into a series of sinusoidal waves of different frequencies known as a Fourier series. Each frequency component is analyzed by ac circuit techniques. Results are com-

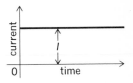

Fig. 1. Direct current.

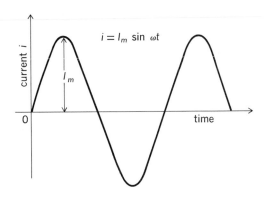

$$i = I_m \sin \omega t$$

Fig. 2. Alternating current.

Fig. 3. Nonsinusoidal voltage wave.

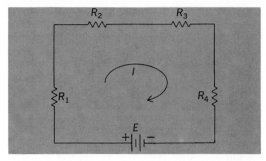

Fig. 5. Series circuit.

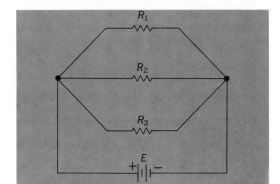

Fig. 6. Parallel circuit.

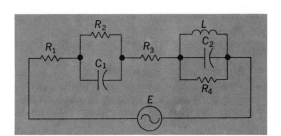

Fig. 7. Series-parallel circuit.

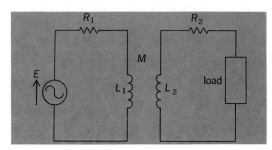

Fig. 8. A three-mesh electric network.

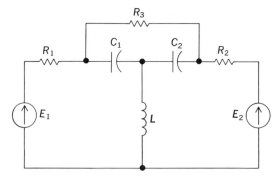

Fig. 9. Inductively coupled circuit.

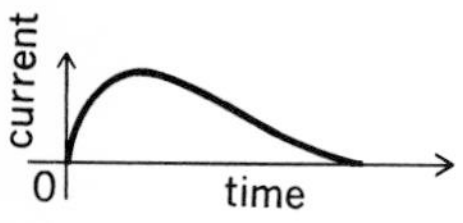

Fig. 4. Transient electric current.

bined by the principle of superposition to give the total response.

Electric transients. Transient voltage and current variations last for a short length of time and do not repeat continuously (Fig. 4). Transients occur when a change is made in the circuit, such as opening or closing a switch, or when a change is made in one of the sources or elements.

Series circuits. In a series circuit all the components or elements are connected end to end and carry the same current, as shown in Fig. 5.

Parallel circuits. Parallel circuits are connected so that each component of the circuit has the same potential difference (voltage) across its terminals, as shown in Fig. 6.

Series-parallel circuits. In a series-parallel circuit some of the components or elements are connected in parallel, and one or more of these parallel combinations are in series with other components of the circuit, as shown in Fig. 7.

Electric network. This is another term for electric circuit, but it is often reserved for the electric circuit that is more complicated than a simple series or parallel combination. A three-mesh electric network is shown in Fig. 8.

Coupled circuits. A circuit is said to be coupled if two or more parts are related to each other through some common element. The coupling may be by means of a conducting path of resistors or capacitors or by a common magnetic linkage (inductive coupling), as shown in Fig. 9.

Open circuit. An open circuit is a condition in an electric circuit in which there is no path for current flow between two points that are normally connected.

Short circuit. This term applies to the existence of a zero-impedance path between two points of an electric circuit.

Integrated circuit. The integrated circuit is a recent development in which the entire circuit is contained in a single piece of semiconductor material. Sometimes the term is also applied to circuits made up of deposited thin films on an insulating substrate.

[CLARENCE F. GOODHEART]

Bibliography: E. Brenner and M. Javid, *Analysis of Electric Circuits*, 1959; P. Chirlian, *Analysis and Design of Electronic Circuits*, 1965; A. E. Fitzgerald and D. E. Higginbotham, *Electrical and Electronic Engineering Fundamentals*, 1964; E. A. Guillemin, *Introductory Circuit Theory*, 1953; W. Hayt, Jr., and J. E. Kemmerly, *Engineering Circuit Analysis*, 1962; W. W. Lewis and C. F. Goodheart, *Basic Electric Circuit Theory*, 1957; R. E. Scott, *Linear Circuits*, 1960; R. Smith, *Circuits, Devices, and Systems*, 1966.

Coal

The general name for the natural, rocklike, brown to black derivative of forest-type plant material, usually accumulated in peat beds. By burial and subsequent geological processes, coal is progressively compressed and indurated, finally altering into graphite or graphitelike material. In American terminology, the rank varieties of coal comprising this carbonification series consist of lignitic coal, represented by brown coal and lignite; bituminous coal, also including subbituminous coals; and anthracite coals, consisting of semianthracite, anthracite, and metaanthracite. *See* LIGNITE; PEAT.

Formation. Coal may originate from isolated fragments of vegetation, but most coal represents the carbonification of woody plants accumulated in peat beds. These are mainly of two kinds: autochthonous deposits representing accumulations at the place of plant growth, such as those found in the Great Dismal Swamp of Virginia, and allochthonous deposits accumulated elsewhere than at the place of growth by the drifting action of stream, lake, or sea currents, such as the Red River "rafts." Generally autochthonous coal deposits overlie seat rock, or underclay containing traces of plant roots called Stigmaria in the case of coals of Paleozoic age.

Biochemical activity modifies the character of the unsubmerged, lightly submerged, or lightly buried peat. This process consists in part of general oxidation, but mainly of attack by aerobic bacteria and fungi that can live only where oxygen is available and of anaerobic bacteria where water or thin sediments cover the peat. Fires set by lightning or other causes may consume part of the peat from time to time, leaving in places a residue of charcoal which may eventually be incorporated into the coal bed in the form of fusain, known to miners as mineral charcoal, mother-of-coal, and mothercoal.

The forest fire origin of fusain is disputed by many botanists who believe the presence of certain combustible components in fusain, such as resins, indicates that chemical causes operating under special conditions bring about the formation of fusain. No completely satisfactory explanation of the origin of fusain has been stated. It is found in all ranks of coal with relatively little difference in composition. There is also transitional material, between normally coalified wood or bark and fusain, called semifusain. Because of its porosity, fusain is commonly mineralized into a hard and heavy substance; unmineralized fusain is soft and light. Fusain occurs in all sizes from particles of microscopic dimension to aggregates forming fairly continuous thin sheets or lenses several feet across and several inches thick.

The material composing the peat which is finally transformed into coal varies with differences in source material, in conditions of accumulation and diagenesis, and in the length of time involved prior to burial. Figure 1 shows the relative resistance of the principal peat-forming plant substances to microbic decomposition. Based upon the peat-forming components, which are more or less segregated into bands, several common types or varieties of coal can be recognized.

Unbanded coal is represented by cannel, or sapropelic, coal and includes common cannel and algal cannel, variously designated boghead cannel and torbanite. The microscope is used to determine the presence of algae in the latter type. Cannel coals are also distinguished by their greasy luster and blocky, conchoidal fracture.

Banded coal may be either bright or dull. Band-

ed bituminous coal is produced by thin lenses of highly lustrous coalified wood or bark called vitrain, if of megascopically distinguishable thickness (1/2 mm or 1/50 in.). Intervening between such black, lustrous bands are layers of more or less striated bright or dull coal (clarain or durain, respectively). The clarain contains a predominance of fine vitrainlike laminae or lenses (microvitrain); microvitrain is of minor importance in durain, and this material has a dull luster. Dullness may also result from a predominance of mineral matter or from a relatively high content of bituminous matter, such as spores, cuticles, resins, and waxes. Likewise the presence of opaque matter or fine fusain contributes to dullness in banded coal.

Carbonification. This term refers to the process of coal metamorphism brought about by increasing weight of overriding sediments, by tectonic movements, by an increase in temperature resulting from depth of burial, or from close approach to, or contact with, igneous intrusions or extrusions. Increase in pressure affects principally the physical properties of the coal, that is, hardness, strength, optical anisotropy, and porosity. Increase in temperature acts chiefly to modify the chemical composition by increasing the carbon content and decreasing the content of oxygen and hydrogen, decreasing the volatile matter and increasing the amount of fixed carbon, and increasing the calorific value to a maximum with about 20% volatile matter. Rapid metamorphism or carbonification of coal, effected by close approach to or contact with igneous intrustions or extrusions, may result in the formation of natural coke. Natural coke is found in some coal fields but is relatively rare.

Classification. The classification and description of coal depends largely upon information supplied by chemical analysis and the results of a number of empirical tests. Chemical analyses are of two kinds: the elementary or ultimate analysis and the proximate or commercial type of analysis. The ultimate analysis is usually limited to the per-

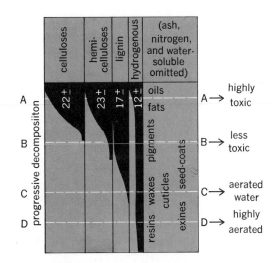

Fig. 1. Approximate proportions of principal peat-forming components in the dry-ingredient plant debris plotted according to relative resistance to microbic decomposition and order of disappearance. (*After D. White*)

cent of hydrogen, carbon, oxygen, nitrogen, and sulfur, exclusive of the mineral matter or ash. Phosphorus may be determined for those coals which are used for metallurgical purposes. The proximate analysis, which is empirical in character, is the prevailing form of analysis in North America. In this analysis, systematically standardized procedures are used to provide values for volatile matter, fixed carbon, moisture, and ash, all of which add up to 100%. Calorific and sulfur values are also determined, the latter sometimes in terms of forms of sulfur such as pyritic, organic, and sulfate sulfur.

It is common practice to use qualifying terms, such as "moisture-free" (mf), "moisture-and-ash-free" (maf), "pure coal," and "as received," in presenting the findings of proximate and ultimate analyses. Usually values are determined initially on a "moisture-free" or "dry" basis, the other forms of analysis being calculated with the use of determined moisture and ash values. The maf value is much used on the mistaken assumption that it represents the heat value or composition of the pure coal material, as though the formation of ash from the original mineral matter were without calorific or other effects. To make a rapid and convenient allowance for the errors inherent in the maf values, various procedures have been devised to arrive at a mineral-matter-free (mmf) basis of comparison and classification. The best known of these devices is that proposed by S. W. Parr whereby so-called unit coal, or the correction to be applied to obtain approximately correct values for mmf coal, is obtained:

$$\text{Unit coal} = 1.00 - (1.08 \text{ ash \%} + 0.55 \text{ sulfur \%})$$

Using this formula,

Moist mmf Btu

$$= \frac{\text{As rec'd Btu} - 5000 \text{ sulfur \%}}{1.00 - (1.08 \text{ ash \%} + 0.55 \text{ sulfur \%})}$$

and

Dry mmf fixed carbon

$$= \frac{\text{Fixed carbon} - 0.15 \text{ sulfur \%}}{1.00 - (\text{moisture \%} + 1.08 \text{ ash \%} + 0.55 \text{ sulfur \%})}$$

The calculations are all based upon the "as received" values, and both moist and dry mmf values are used in the standard classification of coal by rank (Table 1).

Mineral matter. The determination of mineral matter composition on an elementary basis requires a much more elaborate procedure than is necessary for the coal material itself, since the mineral substances found in coal show important local and regional variations. The following detrital minerals, other than clay minerals and quartz, are reported from an Illinois coal bed: feldspar, garnet, common hornblende, apatite, zircon, muscovite, epidote, biotite, augite, kyanite, rutile, staurolite, topaz, tourmaline, and chloritic material. Secondary minerals consist of kaolinite, calcite, pyrite, siderite, and ankerite. In specially mineralized areas, the range of secondary minerals in coal is almost unlimited.

The character of the nonorganic matter in the coal has a considerable effect upon the fusion temperature of the ash, the determination of which is one of the common subsidiary tests in coal analysis procedure. In general, ash that fuses at high temperatures is preferred to low-fusion ash, the temperature being below 2000°F for low-fusion ash and above 2400°F for high-fusion ash; the range is about 1600–2800°F.

Rank. The term rank refers to the stage of carbonification reached in the course of metamorphism. (In international usage the term type is sometimes given the meaning commonly given to rank in North America and in much of the world. No equivalent of the American term type is given.) Classification of coal by rank is fundamental in coal description and generally is based upon the chemical composition as stated in the previous section on carbonification, and upon proximate values based upon mmf coal obtained from standard face samples. In the case of low-rank bituminous and subbituminous coals, rank also is based upon agglomerating characteristics and, in the case of brown coal and lignite, upon calorific value, separation being at 6300 Btu. In Europe, coking properties are a more common basis of classification than in North America. The American Society of Testing Materials' standards of rank classification of coals of North America are shown in Table 1. The higher-rank coals are classified with respect to fixed carbon on the dry basis (mmf) and the lower-rank coals according to Btu on the moist basis (mmf).

The difference between common banded and the sapropelic, or canneloid, types of coal of the same rank, which occur in the same general region, is mainly in the higher hydrogen and slightly higher calorific value of the latter type.

Calorific values. The calorific values of coal are expressed in British thermal units (Btu) per pound in English-speaking countries, and in calories per gram where the decimal system prevails. Heat values, determined by a calorimeter, vary for coal material from about 6300 Btu for California lignite to about 16,300 Btu for some maf cannel coal. Proximate and ultimate analyses of representative United States coals by rank are given in Table 2, the analyses being on the "as received" basis. *See* BRITISH THERMAL UNIT (BTU).

Physical characteristics. The physical characteristics of coal concern the structural aspects of the coal bed and texture as determined by the megascopic and microscopic physical constitution of the coal itself. Structurally the coal bed (also called seam and, less appropriately, vein) is a geological stratum characterized by the same irregularities in thickness, uniformity, and continuity as other strata of sedimentary origin. Thickness varies greatly. German brown coals occasionally approach 300 ft. A drill hole in the Lake De Smet area in Wyoming penetrated 223 ft of lignite or subbituminous coal essentially in one bed.

Coals beds may consist of essentially uniform continuous strata or, like other sedimentary deposits, may be made up of distinctly different bands or benches of varying thickness. The benches may be separated by thin layers of clay, shale, fusain, pyrite, or other mineral matter, commonly called partings by the miner. Clay or shale bands may be called blue bands, as is the persistent clay parting in the Herrin (No. 6) coal bed of Illinois. Like other

Table 1. Classification of coals by rank[a]

Class	Group	Fixed carbon limits, % (dry, mineral-matter-free basis)		Volatile matter limits, % (dry, mineral-matter-free basis)		Calorific value limits, Btu/lb (moist,[b] mineral-matter-free basis)		Agglomerating character
		Equal or greater than	Less than	Greater than	Equal or less than	Equal or greater than	Less than	
I. Anthracitic	Metaanthracite	98	—	—	2	—	—	Nonagglomerating
	Anthracite	92	98	2	8	—	—	
	Semianthracite[c]	86	92	8	14	—	—	
II. Bituminous	Low-volatile bituminous coal	78	86	14	22	—	—	
	Medium-volatile bituminous coal	69	78	22	31	—	—	
	High-volatile A bituminous coal	—	69	31	—	14,000[d]	—	Commonly agglomerating[e]
	High-volatile B bituminous coal	—	—	—	—	13,000[d]	14,000	
	High-volatile C bituminous coal	—	—	—	—	11,500	13,000	
						10,500	11,500	Agglomerating
III. Subbituminous	Subbituminous A coal	—	—	—	—	10,500	11,500	
	Subbituminous B coal	—	—	—	—	9,500	10,500	Nonagglomerating
	Subbituminous C coal	—	—	—	—	8,300	9,500	
IV. Lignitic	Lignite A	—	—	—	—	6,300	8,300	
	Lignite B	—	—	—	—	—	6,300	

[a]This classification does not include a few coals, principally nonbanded varieties, which have unusual physical and chemical properties and which come within the limits of fixed carbon or calorific value of the high-volatile bituminous and subbituminous ranks. All these coals either contain less than 48% dry, mineral-matter-free fixed carbon or have more than 15,500 moist, mineral-matter-free Btu/lb.
[b]Moist refers to coal containing its natural inherent moisture but not including visible water on the surface of the coal.
[c]If agglomerating, classify in low-volatile group of the bituminous class.
[d]Coals having 69% or more fixed carbon on the dry, mineral-matter-free basis shall be classified according to fixed carbon, regardless of calorific value.
[e]There may be nonagglomerating varieties in these groups of the bituminous class, and there are notable exceptions in high volatile C bituminous group.
SOURCE: From *Book of ASTM Standards*, 1967.

Table 2. Proximate and ultimate analyses of samples of each rank of common banded coal in the United States*

Rank	State	Proximate analysis, %				Ultimate analysis, %					Heating value, Btu/lb
		Moisture	Volatile matter	Fixed carbon	Ash	S	H	C	N	O	
Anthracite	Pa.	4.4	4.8	81.8	9.0	0.6	3.4	79.8	1.0	6.2	13,130
Semianthracite	Ark.	2.8	11.9	75.2	10.1	2.2	3.7	78.3	1.7	4.0	13,360
Bituminous coal											
Low-volatile	Md.	2.3	19.6	65.8	12.3	3.1	4.5	74.5	1.4	4.2	13,220
Medium-volatile	Ala.	3.1	23.4	63.6	9.9	0.8	4.9	76.7	1.5	6.2	13,530
High-volatile A	Ky.	3.2	36.8	56.4	3.6	0.6	5.6	79.4	1.6	9.2	14,090
High-volatile B	Ohio	5.9	43.8	46.5	3.8	3.0	5.7	72.2	1.3	14.0	13,150
High-volatile C	Ill.	14.8	33.3	39.9	12.0	2.5	5.8	58.8	1.0	19.9	10,550
Subbituminous coal											
Rank A	Wash.	13.9	34.2	41.0	10.9	0.6	6.2	57.5	1.4	23.4	10,330
Rank B	Wyo.	22.2	32.2	40.3	4.3	0.5	6.9	53.9	1.0	33.4	9,610
Rank C	Colo.	25.8	31.1	38.4	4.7	0.3	6.3	50.0	0.6	38.1	8,580
Lignite	N. Dak.	36.8	27.8	30.2	5.2	0.4	6.9	41.2	0.7	45.6	6,960

Technology of Lignitic Coals, U.S. Bur. Mines Inform. Circ. no. 769, 1954. Sources of information omitted.

sedimentary strata, coal beds may be structurally disturbed by folding and faulting so that the originally approximate horizontality of position is lost to the extent that beds may become vertically oriented or even overturned, as in the anthracite fields of the eastern United States and in other places in the world where similar high-rank coals are found.

Texture. The texture of the coal itself is determined by the character, grain, and distribution of its megascopic and microscopic components. In general banded coals composed or relatively coarse highly lustrous vitrain lenses (1/4 in. or more in thickness) are considered coarsely textured. As the thickness of the vitrain bands progressively lessens, the texture becomes finebanded and then microbanded, with bright laminae composed of microvitrain. Coals with less than 5% vitrain or microvitrain are regarded as nonbanded or canneloid.

In general, the same textural units are observed in bituminous and anthracite coals. The names employed for the lithotypes, that is, vitrain, clarain, and durain, are somewhat less suitable for lignitic coals, because in these ranks of coal the bands of material that appear as vitrain in bituminous coals and anthracite commonly have the unmistakable appearance of tree trunks and pieces of wood or bark.

Microscopic texture is determined by the physical composition of the lithotypes as noted above, or of the anthraxylon and attritus, in terms of the microscopic constituents. The microscopy and petrology of coal are concerned very largely with these constituents. This textural aspect of coal is of imporatnce in accurate coal description and classification, and for an understanding of the behavior of coal in its preparation and utilization.

Botanical and petrologic entities. The fundamental physical constituents of coal have been investigated from two points of view. In North

America the conventional point of view established by Reinhardt Thiessen and long followed by the U.S. Bureau of Mines is microscopic, regarding coal as an aggregate of original botanical entities identifiable only by miscroscopic means. This approach has commonly been referred to as coal microscopy, although the use of the phrase coal petrography has become more frequent in recent years. The other point of view, that assumed by Marie C. Stopes of England and accepted in much of the world, is based upon the concept of the four megascopic ingredients now called lithotypes: vitrain, clarain, durain, and fusain. Since these concern coal materials as rock substances, the concept provides a basis for the petrologic study of coal. See the section below on petrology and petrography.

Microscopy. The microscopic study of coal, as developed in North America, is primarily concerned with the botanical entities or phyterals of coal, with fusain regarded as a coal substance of unique character and with appropriate consideration being given to mineral matter. By 1930 Thiessen had established three categories of microscopic components, anthraxylon, attritus, and fusain, and had recognized and described most of the botanical constituents of coal. This established a system of nomenclature, description, and classification which has been followed essentially in publications of the U.S. Bureau of Mines concerned with coal microscopy since that date. Anthraxylon consists of coal occurring in bands in which wood or bark structure is microscopically evident. All vitrain of the European classification, if more than 14 microns (μ) thick, is regarded as anthraxylon; below this threshold it is classified as attritus. The attritus consists of finely textured coalified plant entities or phyterals not classified as anthraxylon (less than 14 μ) or fusain (less than 40 μ). Attritus, therefore, may contain very fine shreds of anthraxylonlike material, fine particles of fusain, disintegrated or macerated humic material, or "humic degradation matter" (HDM) in addition to the following constituents: resins, waxes, cuticles, spore and pollen exines, algae, opaque matter, fungal bodies such as sclerotia, and fine mineral matter of various kinds with clay minerals usually predominating. The subdivision of attritus into translucent and opaque attritus on the basis of its content of opaque matter is the only subdivision made of the attritus with respect to variations in its heterogeneous constitution. In North America banded bituminous coals with 30% or more of opaque matter are classified as splint coals; those with 20–30% opaque matter are classified as semisplint coals. Generally, no equivalent classification is recognized in other parts of the world.

Analysis and classification of coals on the microscopic basis have usually been made in terms of the major components anthraxylon, attritus, and fusain, with consideration given to the quantity of opaque attritus and mineral matter. These distinctions in regard to opaque attritus have been applied because splint coals are generally not amenable to hydrogenation and commonly not to carbonization. Coal microscopy technique depends almost entirely on the use of thin, translucent sections of coal, a technique which is not adapted for use with high rank low-volatile and anthracitic coals. Maceration has also been used as a means of breaking down the coal and isolating the more resistant constituents, thus incidentally providing the fossil spores that are the basis of the science of palynology.

Petrology and petrography. The use of the word petrology as applied to coal assumes that coal can correctly be regarded as a rock substance; its description is therefore consistently regarded as petrographic, that is, as a field of petrography.

The initial contribution of Marie C. Stopes (1919) to the field of coal microscopy included the adoption of a petrographic concept of coal as a rock substance composed of banded "ingredients" now called lithotypes — vitrain, clarain, durain, and fusain and also mineral matter. She also introduced, in 1935, the "maceral" concept (Table 3), macerals being the individual components of the lithotypes, comparable to minerals of nonorganic rocks.

A primary dissatisfaction with the treatment of macerals as the equivalent of minerals arose from the realization that macerals possess no fixed chemical composition, such as that possessed by minerals. Each maceral becomes progressively modified chemically and physically as the rank of the coal advances. Hence it has become the practice in coal petrology to indicate the rank position of individual macerals solely by reliance upon measurement of some physical attribute, power of reflectance now being the most favored, because of the relative simplicity of its application.

Coal rank and coke. The usefulness of a bituminous coal for the production of metallurgical coke is primarily determined by its rank, that is, by its position in the lignite to anthracite series. The possibility of predicting the production of satisfactory metallurgical coke from coal on the basis of the rank of the coal determined by reflectance of certain macerals, particularly vitrinite, was introduced by I. Ammosov and associates in 1957. The usefulness of this method with respect to the coals of the United States was thoroughly investigated by N. Schapiro, R. J. Gray, and G. R. Eusner and further tested by J. A. Harrison, H. W. Jackman, J. A. Simon, and others. These investigations of many coals involved determination of reflectance, particularly of vitrinite; chemical analyses; and determination of stability or strength of laboratory cokes produced from the coal.

In the early years of activity in this field of investigation, emphasis was mainly on the determination of the reflectance of vitrinite because of its common occurrence and conspicuous reflectance. However, as investigations progressed, it was found that the presence of other macerals, and even of minerals, also affected coke strength even though some did not actually produce coke.

Reflectance procedure. Investigations concerned with the reflectance of coal between 1958 and 1968 consisted largely in the accumulation of petrographic and experimental data providing the basis for graphs used in predicting the suitability of a particular coal or blends of coal for the production of metallurgical coke suitable for use in the steel industry. Three such graphs (Figs. 2, 3, and 4) have come into frequent use in industrial and research laboratories concerned with the production of metallurgical coke.

For the preparation of these three graphs it was necessary to accumulate a large volume of laboratory data on reflectance, petrographic compositions, chemical analyses, and the stability of the coke made from many bituminous coals. Petrographic data were acquired by the point-count method (F. Chayes, 1956), using polished surfaces or broken coals (−8 mesh) mounted in a suitable medium, forming cylindrical blocks about 1 in. in diameter and 1 in. in height, size not being a critical consideration.

Of the three graphs that have come into use for predicting a coke stability, Fig. 2 is of fundamental importance. The experimentally determined position of the optimum coke for each reflectance class of vitrinite (classes 3 to 21; see Table 3) and the best ratio of reactives in inerts (R/I) is determined by the position of the heavy curved line. For all practical purposes, R/I varies between 0 and 25. When a coal or blend of coals contains reactives of several reflectance classes, as is usually the case, the inert index is derived from the equation

$$N = \frac{Q}{P_1/M_1 + P_2/M_2 + \cdots + P_{21}/M_{21}} = \text{inert index}$$

where $Q =$ total percent by volume of inerts in coal blend from analysis; P_1, P_2, etc. = percent of reactives (exinite, resinite, and 1/3 of total semifusinite) from analysis; and M_1, M_2, etc. = ratio of reactives to inerts to produce optimum coke. Inerts consist of fusinite, 2/3 of total semifusinite, micrinite, and ash. For the application of inert index see the explanation of Fig. 4.

Figure 3 consists of a set of curves, based upon experimental data, designed to show how the volume percent of the inerts (not the inert index) affects the strength of the coke (sometimes called the coking coefficient) made from coals of various reflectance classes. The position, spacing, and curvature of the curves for vitrinite reflectance classes are based upon experimental determinations with reference to an arbitrary scale of strength index (0−10) and to the volume of inerts (0−50% in steps of 5%). The strength index (coking coefficient) of a coal or blend of coals composed of reactives of various reflectance classes can be determined from the following equation in conjunction with Fig. 3:

$K_T =$ strength index of a coal or blends

$$= \frac{(K_1 \times P_1) + (K_2 \times P_2) + \cdots + K_{21} \times P_{21}}{P_T}$$

Here $P_T =$ total percentage of reactives; K_1, K_2, etc. = strength index of reactives in the reflectance classes present in the coal sample, obtained from the family of curves in Fig. 3; and P_1, P_2, etc. = percentage of reactives in reflectance classes present in the coal sample. The values K_1, K_2, etc. are obtained from Fig. 3 on the basis of the volume of inerts reported in the analytical data. A vertical line projected from the abscissa at this position will successively intersect the curves representing progressively higher reflectance classes present in the sample. By projecting lines horizontally from the points of intersection to the ordinate, what is designated as the strength index of the successive reflectance classes can be read, assuming the same percentage of inerts. The analytical data sup-

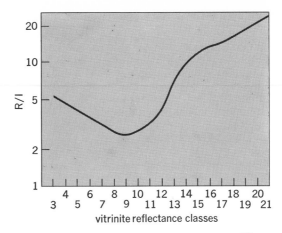

Fig. 2. Optimum ratio of the reactives to inerts (R/I) for each vitrinite reflectance class. (*Modification by J. A. Harrison, H. W. Jackman, and J. A. Simon, Predicting Coke Stability from Petrographic Analysis of Illinois Coals, Ill. State Geol. Surv. Circ. no. 366, 1964, of a figure by N. Schapiro, R. S. Gray, and G. R. Eusner, 1961*)

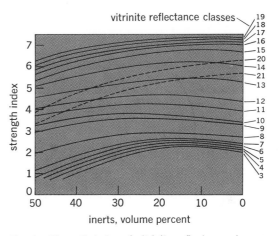

Fig. 3. Strength index of vitrinite reflectance classes depending on the amounts of inerts present. (*Modification by J. A. Harrison, H. W. Jackman, and J. A. Simon, Predicting Coke Stability from Petrographic Analysis of Illinois Coals, Ill. State Geol. Surv. Circ. no. 366, 1964, of curves by N. Schapiro, R. S. Gray, and G. R. Eusner, 1961*)

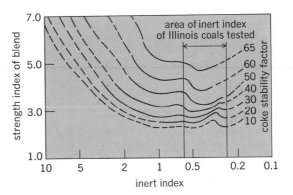

Fig. 4. Curves showing the relation between the strength index, inert index, and stability factor. (*Modification by J. A. Harrison, H. W. Jackman, and J. A. Simon, Predicting Coke Stability from Petrographic Analysis of Illinois Coals, Ill. State Geol. Surv. Circ. no. 366, 1964, of a figure by N. Schapiro, R. S. Gray, and G. R. Eusner, 1961*)

Table 3. Maceral reflectance classes and reactivity during carbonization*†

Reactives			Inerts		
Group macerals	Macerals	Reflectance class	Group macerals	Macerals	Reflectance class
Vitrinite		V0 to V21	Inert vitrinite		V22 to V80
	Collinite	C0 to C21		Inert resinite	R22 to R80
	Telinite	T0 to T21	Inertinite		I18 to I80
				Fusinite	F40 to F80
				Micrinite	M18 to M80
Exinite		E0 to E15		Semifusinite‡	SF22 to SF80
	Sporinite	St0 to St15		Sclerotinite	Sc22 to Sc80
	Cutinite	Ct0 to Ct15			
			Group minerals	Minerals	
	Alginite	At0 to At15	Sulfides	Pyrite, etc.	
	Resinite	R0 to R15	Carbonates	Calcite, etc.	
Fusible inertinite	Semifusinite‡	SF0 to SF21	Silicates	Illite, etc.	
	Micrinite	M0 to M18			

*From *Ill. State Geol. Surv. Circ.*, no. 366, 1964.

†Nomenclature as defined in Glossary of International Committee for Coal Petrology and based primarily on Stopes-Heerlen system of classification. Range of reflectance values of macerals based on values of N. Schapiro and R. J. Gray, Petrographic classification applicable to coals of all ranks, *Proc. Ill. Min. Inst.*, pp. 83–97, 1960.

‡Estimated values; reactive group is about one-third and inert group about two-thirds of semifusinite total. From I. I. Ammosov et al., Calculation of coking charges on the basis of petrographic characteristics of coke, *Koks i Khimiya*, no. 12, pp. 9–12, 1957.

ply values for P_1, P_2, etc. The products indicated by the calculation K_1, P_1, etc. are the results obtained by multiplying the strength index of a particular reflectance class by the percentage of such class as provided by the petrographic analysis. Adding the resulting values of the various reflectance classes provides a figure representing the total strength index (K_T) of the reactives. This value divided by the total amount of reactives (P_T) provides the calculated strength index, which is not the equivalent of stability (Fig. 4) based upon tumbler tests.

Many tests by N. Schapiro and associates and by J. A. Harrison and associates and by others provide the basis for graphs similar to Fig. 4. The purpose of such graphs is to predict the stability factor of coke made from various coals and from blends of coals. The curves are based on the results obtained in laboratory practice in terms of stability factors varying between 10 and 65%, strength indexes between 2 and 10, and inert indexes between 0.2 and 10.0. In order to apply this chart, the coal or coals must be subjected to the various petrographic tests and analyses already cited whereby the strength and inert indexes are obtained. By the use of these data and Fig. 4, a close approach to the actual stability of the coke can be forecast by using relatively small samples of coal in the laboratory.

Rank. There has been wide acceptance of the Stopes-Heerlen nomenclature of coal petrography. But the realization, at least by American coal geologists, that this nomenclature makes no provision for even the major subdivisions of coals by rank —anthracite, bituminous coal, and lignite—has led to at least one important series of proposals

for modification of the terminology by W. Spackman. The Spackman system accepts the general validity of the three maceral groups of the Stopes-Heerlen system—vitrinite, exinite, and inertinite —but prefers to designate these as suites and to use the term liptinite rather than exinite, as the former term includes both resinite and exinite. The suite subdivision, at least with respect to the vitrinite suite, on the basis of rank is subdivided into anthrinoid, vitrinoid, and xylinoid groups for anthracite, bituminous coal, and lignite ranks. Within the three maceral groups are various maceral types identified not by name but by physical properties, and particularly by reflectivity (Table 1).

Genetic varieties. Soviet petrographers and coal geologists have proposed the recognition petrographically of varieties of coal macerals produced not by diastrophism, that is, by changes in rank after the coal bed is buried, but resulting from "genetic changes" in peat accumulations before burial, and the subsequent geological effects of such burial. Although the Soviet coal geologists accept the maceral concept, they believe important sources of variability of coal reside in the dissimilar conditions of decay that may affect the accumulated peat prior to its permanent burial. The terminology proposed by the Soviet coal petrographers to indicate the varieties to be expected in coal of a relatively high woody origin as a result of "genetic processes" seems to be restricted to two general categories of alteration, which grade into one another. These are fusinization and gelification. Those varieties of coal upon which fusinization of woody material had worked the most distinctive effect had been subjected as peat to strong atmos-

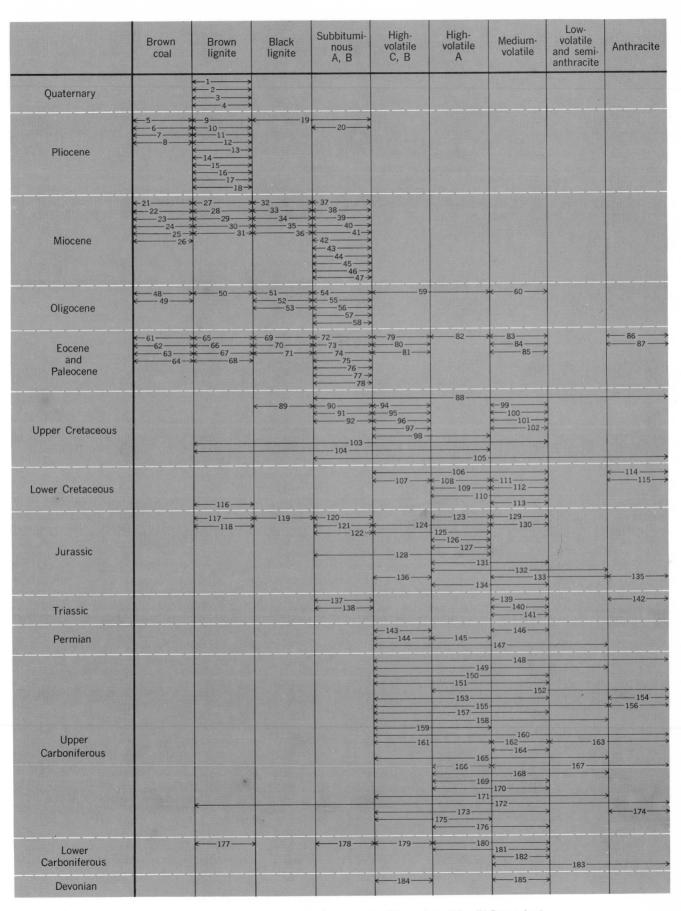

Fig. 5. Age and rank of coal in the better-known coal fields of the world. The index numbers are identified in the last paragraph of the article. (*After W. Petrascheck and W. E. Petrascheck, Lagerstättenlehre, Springer, 1950*)

pheric oxidation resulting in the formation of much fusain, whereas peat existing under conditions of prevailing but not too deep submergence would often be subjected to the process of gelification of the woody substances before burial. Types of coal are therefore recognized emphasizing either fusinization or gelification. These two types of coal, both more or less of the vitrinite class, are thought to provide six lithologic varieties depending mainly on the predominance of fusinite, on the one hand, and collinite, on the other hand, namely, fusinito-telinite, fusinito-precollinite, fusinito-post-telinite, fusinito-collinite, and gelito-telinite, gelito-precollinite, gelito-post-telinite, and gelito-collinite. Other varieties of coal derived from weathered peat in which the amount of woody matter was relatively small have names such as cutinite, resinite, or sporinite, depending upon the abundance (over 50%) of cuticular or resinous bodies or fossil spores.

Evaluation and uses. Although the utilization of coal varies widely with rank, three general fields may be distinguished. They are combustion (domestic, industrial, railroads, and public utilities), gasification, and carbonization (high-temperature coke for metallurgical uses and low-temperature coke for production of smokeless fuel). Anthracite is used mainly for domestic and related types of heating; some fine sizes are blended with bituminous coal to make metallurgical coke. Besides the main use of lignitic coal for combustion either in the raw state or in dehydrated and briqueted form, it is a source of industrial carbon, industrial gases, and montan wax.

The evaluation of coal for particular uses is usually based upon the information provided by the ultimate and proximate analyses, the values for sulfur, and the heat value. The proximate analysis is preferred in the United States. These criteria provide the means of classifying the coal by rank and, with the cost per ton known, the value of the coal in terms of 1,000,000 Btu. Ash fusion values are considered incidentally, and if the coal is to be used for making metallurgical coke, the information provided by certain coking tests is usually required. These tests consist of plasticity tests, swelling and swelling pressure tests, agglutinating tests, and agglomerating tests.

Occurrence. Coal is found on every continent, in the islands of Oceania, and in the West Indies. The amount of coal in the various countries varies greatly. The six countries with the greatest reserve (in millions of tons) are as follows: United States, China, U.S.S.R., Germany, Union of South Africa, and United Kingdom. The six countries with the greatest reserve (in millions of tons) of lignitic coal are United States, U.S.S.R., Alaska, Germany, Australia, and Canada.

The distribution of coal by rank, geologic age, and district is indicated in Fig. 5. The index numbers, which refer to coal districts, are grouped by continent and country in the following list:

Africa: Belgian Congo, 144; Natal, 147. *Asia:* China, 163, 132, 167, 135; India, 19; Siberia (U.S.S.R.), 118, 172, 119, 122, 128, 131, 133, 183, 171; Turkestan, 125. *Australia:* 8, 134; New Zealand, 31, 36, 89, 98. *Oceania:* Borneo, 81; Sumatra, 78. *Europe:* Austria, 1, 3, 14, 29, 33, 38, 72, 123, 139; Bavaria, 2, 24; Bulgaria, 7, 20, 100; Czechoslo-

vakia, 16, 28, 32, 37, 55, 91, 161, 164; England, 48, 153, 155, 158, 157, 180, 182; France, 53, 165, 166, 168; East Germany, 5, 21, 22, 23, 90, 121, 137, 106, 143; West Germany, 24, 49, 61, 63, 62, 64, 2, 27, 57, 59, 127, 150; Greece, 75, 86; Hungary, 30, 35, 42, 44, 74, 95, 130; Italy, 18, 45, 83; Poland, 6, 138, 149, 151; Ruhr district, 148; Romania, 17, 129, 162; Russia (U.S.S.R.), 117, 177, 124, 126, 179, 184, 185; Scotland, 120, 181; Spain, 160; Spitzbergen, 76, 178, 79, 82, 84, 107, 108; Switzerland, 4, 87, 154; Turkey, 58; Yugoslavia, 10, 11, 12, 13, 15, 50, 34, 51, 52, 39, 40, 41, 47, 54, 56, 60, 85, 99, 142, 156. *North America:* Canada, 68, 116, 104, 110, 176, 112, 113, 115; Mexico, 141; United States, 43, 77, 92, 65, 66, 67, 69, 70, 71, 80, 96, 97, 103, 101, 102, 88, 105, 140, 173, 174, 175. *South America:* Brazil, 145; Peru, 46, 111, 109, 114.

[GILBERT H. CADY]

Bibliography: American Society for Testing and Materials, *1967 Book of ASTM Standards*, pt. 19, 1967; Felix Chayes, *Petrographic Model Analysis: An Elemental Statistical Appraisal*, 1956; A. C. Fieldner and W. A. Selvig, *Methods of Analyzing Coal and Coke*, U.S. Bur. Mines Bull. no. 492, 1951; J. A. Harrison, H. W. Jackman, and J. A. Simon, *Predicting Coke Stability from Petrographic Analysis of Illinois Coals*, Ill. State Geol. Surv. Circ. no. 366, 1964; International Committee for Coal Petrology, *International Handbook of Coal Petrography*, 2d ed., 1963; D. W. van Krevelen and J. Schuyer, *Coal Science*, 1957; B. C. Parks and H. J. O'Donnell, *Petrography of American Coals*, U.S. Bur. Mines Bull. no. 550, 1956; N. Schapiro and R. J. Gray, *Petrographic Classification Applicable to Coals of All Ranks*, Ill. Mining Inst. 1960 Proc., 1960; N. Schapiro, R. J. Gray, and G. R. Eusner, Recent developments in coal petrography, in *Blast Furnace, Coke Oven and Raw Materials Conference*, Amer. Inst. Mining Engn. Proc., 20:89–112, 1961; W. H. Young and R. L. Anderson, *Thickness of Bituminous Coal and Lignite Seams Mined in 1960*, U.S. Bur. Mines Inform. Circ. no. 8118, 1962.

Coal gasification

The conversion of coal, coke, or char to gaseous products be reaction with air, oxygen, steam, carbon dioxide, or a mixture of these. Products consist of carbon dioxide, carbon monoxide, hydrogen, methane, and some other chemicals in a ratio dependent upon the particular reactants employed and the temperatures and pressures within the reactors, as well as upon the type of treatment which the gases from the gasifier undergo subsequent to their leaving the gasifier. Strictly speaking, reaction of coal, coke, or char with air or oxygen to produce heat plus carbon dioxide might be called gasification. However, that process is more properly classified as combustion, and thus is not included in this coal gasification summary. *See* COMBUSTION.

Industrial uses. Interest in coal gasification is widespread. The natural-gas industry views it as a means for the production of substitute natural gas (SNG) having a heating value of about 30 kilojoules/m³ (1000 Btu/ft³) and combustion properties sufficiently similar to those of natural gas that it may be used interchangeably with natural gas; the electric industry is investigating production of a clean, low-Btu (4.5 kilojoules/m³, or 150 Btu/ft³) gas from coal with the object of burning it in a combined power generation system using a gas turbine and a steam turbine to drive separate electric generators, thereby providing overall station efficiencies approaching 45%; part of the chemical industry is seeking an additional source of hydrogen and carbon monoxide to offset the diminishing supplies of these synthetic chemical building blocks, which were previously obtained primarily from steam reforming of natural gas, petroleum, natural-gas liquids, or petroleum derivatives; and members of certain large industrial complexes, occasionally in conjunction with natural gas distribution companies, are studying the feasibility of using low-Btu gas for many applications, thereby freeing critical volumes of natural gas, and for using the reducing properties of hydrogen and carbon monoxide in some metallurgical processes. *See* HEATING VALUE; NATURAL GAS; SUBSTITUTE NATURAL GAS (SNG).

Gasification processes. The basic technology for many of the coal gasification processes under active consideration today is quite old and well known. Much of it is derived directly from the large body of manufactured gas technology. In 1965 Bituminous Coal Research, Inc., completed a study for the Office of Coal Research (OCR) of the U.S. Department of the Interior and published a report containing the results of a literature search on 65 different coal gasification processes. In spite of the large number of processes described and identified at that time, there continues to be a variety of new processes under development. In nearly all of the processes the chemistry of the high-temperature gasification is the same (Fig. 1). The basic reactions are:

Coal reactions

$$Coal \xrightarrow{Heat} gases\ (CO, CO_2, CH_4, H_2) \\ + liquids + char \quad (1)$$

$$Coal + H_2 \xrightarrow{Catalyst} liquids + (char) \quad (2)$$

$$Coal + H_2\ (from\ a\ hydrogen\ donor) \rightarrow \\ liquids + (char) \quad (3)$$

$$Coal + H_2 \xrightarrow[destruction]{Noncatalytic} CH_4 + char \quad (4)$$

Char reactions

$$C\ (char) + 2H_2 \rightarrow CH_4 \quad exothermic \quad (5)$$
$$C\ (char) + H_2O \rightarrow CO + H_2 \quad endothermic \quad (6)$$
$$C\ (char) + CO_2 \rightarrow 2CO \quad endothermic \quad (7)$$
$$C\ (char) + O_2 \rightarrow CO_2 \quad exothermic \quad (8)$$

Gaseous reactions

$$CO + H_2O \xrightarrow{Catalyst} H_2 + CO_2 \quad exothermic \quad (9)$$
$$CO + 3H_2 \xrightarrow{Ni} CH_4 + H_2O \quad exothermic \quad (10)$$
$$CO_2 + 4H_2 \xrightarrow{Ni} CH_4 + 2H_2O \quad exothermic \quad (11)$$
$$xCO + yH_2 \xrightarrow{Fe} hydrocarbon\ gases\ and/or \\ liquids + zCO_2 \quad exothermic \quad (12)$$

Thermodynamics. From a thermodynamic standpoint in coal gasification, at least one simplifying assumption is customarily made; namely, coal can be considered as carbon. This assumption is made because coal is a chemically ill-defined material that does not fit into the regime of rigorous thermodynamic deduction. Errors associated with this

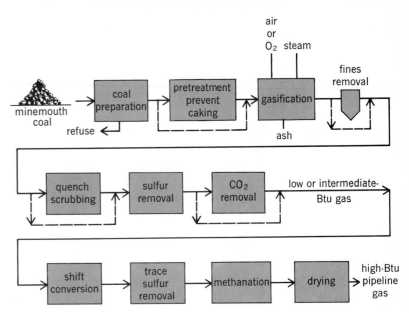

Fig. 1. Schematic representation of the processing steps in coal gasification.

assumption are not likely to be very large. Thus the individual high-Btu gas-producing step, reaction (5) is considered to be of great interest. Not only is this reaction exothermic, but it also has a large negative standard free energy change, indicating that it is a spontaneous reaction. Unfortunately, the rate of this reaction at ordinary temperatures and in the absence of a catalyst is nearly zero. In order to force the reaction to proceed at a fast rate, the temperature must be raised considerably. Some processes are designed to operate at as high as 1040°C. Another important consideration in regard to reaction (5) is that very large quantities of energy must be used to obtain the hydrogen for this reaction. The most widely available source of hydrogen is water, large quantities of which are necessary for a gasification plant. In fact, a plant producing 7,000,000 m³ (250,000,000 ft³) of pipeline-quality gas per day requires about 3,000,000 gal (11,300 m³) of water per day to supply hydrogen.

Reaction (13) is highly endothermic and also requires large quantities of energy. As a result,

$$H_2O \rightarrow H_2 + \frac{1}{2}O_2 \quad (13)$$

nearly all of the gasification processes under investigation today do not decompose water directly according to reaction (13) and then follow with a reaction of hydrogen with coal or char. Rather, the water is decomposed according to reaction (6). Products of this reaction are then treated as shown to reactions (9), (10), and (11). It is currently necessary to use a multistep process as outlined in Fig. 1, although some recent research indicates that coal can be reacted directly with water in a plasma arc to yield hydrocarbons including methane.

Gasification step. The various gasification processes which have already been developed or which are in the development stage vary with respect to the gasification step. In fact, to a large degree, the mechanical and engineering variations of this step, particularly the features designed for supplying the heat for the endothermic reaction C + H₂O and for handling solids in the gasifier, characterize the processes. The heat can be furnished by any one of several methods: by partial combustion of coal or char with air or oxygen, by an inert (pelletized ash, ceramic spheres, and so on) heat carrier, by heat released from the reaction of a metal oxide with CO_2, by high-temperature waste heat from a nuclear reactor, or by heat produced when an electric current passes through the coal or char bed between electrodes immersed in the bed. Processes have been investigated for gasification at atmospheric and elevated pressures by using fixed-bed, fluidized-bed, or entrained suspended, dilute-phase operations in the gasifier.

Research and development. The cost of gas made by coal gasification is sufficiently high to justify continuing research and development to try to decrease capital and operating costs of gasification plants. In the United States, calculated costs for a plant capable of producing 7,000,000 m³ (250,000,000 ft³) per day of high-Btu gas are as high as $900,000,000. Most development work on processes, or on parts of processes, is directed toward lowering the plant costs or increasing the efficiency of certain process steps. One of the major programs for developing improved coal gasification is funded jointly by the American Gas Association (AGA) and the OCR.

In addition to the AGA-OCR program, other research and development is underway. Slagging gasifier studies are being carried out on a Lurgi unit at Westfield, Scotland, and on a COGAS unit being developed by the British Coal Utilization Research Association (BCURA) at Leatherhead, England. This area is being emphasized because it is believed that slagging gasifiers will produce gas at a faster rate than nonslagging gasifiers can.

A discussion of some coal gasification processes and their pertinent features follows.

Battelle-Union Carbide process. In this agglomerating-bed process, coal is introduced near the bottom of the gasification reactor while ash agglomerates are preheated in the fluidized-bed combustor and introduced near the middle of the gasifier. The heavier ash particles fall through the gasifier, providing the heat required for the gasification reaction. After giving up part of its sensible heat, this pelletized ash is returned to the burner to be recycled. The raw gas exits from the top of the reactor.

Bi-gas process. The principal feature of the bi-gas process is the direct production of methane from coal in greater quantities than can be achieved in conventional steam-oxygen gasifiers. Coal and steam are introduced directly into a column of hydrogen-rich gas produced in the lower section of the gasifier by the slagging gasification of recycled char with oxygen and steam at temperatures of 2500°F (1370°C) or higher. In the upper section of the gasifier (stage 2), the volatile portion (about one-third) of the fresh coal is converted directly to methane at temperatures of 1700–1800°F (927–982°C) and at pressures of 1500 psig (1 psi = 6895 N/m²). Some of the unconverted coal falls directly to the slag bath, but most of it is carried out of the reactor in the product gas stream and fed to a cyclone where it is separated for recycling to the lower section. The raw gases leaving the cyclone are subsequently purified and subjected to methanation for the ultimate production of high-Btu pipeline gas.

COGAS process. In the COGAS process, low-pressure coal pyrolysis is combined with char gasification to produce liquid fuel and synthetic pipeline gas. Synthesis gas is produced from the char, with air used instead of oxygen to provide the heat for the steam-carbon reaction. The operation may be carried out at the relatively low pressure of 50 psig.

Commercially available process technology is reportedly used for hydrogen generation, gas purification and dehydration, sulfur and oil recovery, and oil hydrotreating.

Consolidated synthetic gas (CSG) process. A basic feature of this process, also known as the CO_2 acceptor process, is the use of lime or calcined dolomite to react exothermally with the CO_2 liberated in gasification to supply a portion of the heat needed for the carbon-steam reaction. The removal of CO_2 by this method enhances the water-gas shift reaction and methane formation, both of which are exothermic. The combination of these effects allows the gasification system to operate without an oxygen supply. The spent "acceptor" (lime or dolomite) is withdrawn from the gasification vessel and calcined separately, with residual char and air supplying the necessary heat. The process also embodies the feature of contacting the incoming raw coal with the hydrogen-rich synthesis gas to form a portion of the methane directly.

The operating conditions of the CO_2 acceptor process are limited to about 1600°F (870°C) and 300 psia because of melt-formation problems in the dolomite system. Since the rate of the uncatalyzed steam-carbon reaction for gasification of bituminous coals is prohibitively slow—1600°F or lower—the process is not competitive for the gasification of these types of coals.

Hydrane process. In the Hydrane process, raw coal is introduced into the dilute-phase reactor, where it is reacted with hot intermediate gas to prevent the coal from agglomerating. The feed gas to the dilute phase comes from the fluid-bed reactor, which further hydrogasifies the nonagglomerating char from the dilute phase. Nearly pure (95.6%) hydrogen is fed to this fluid-bed phase. The intermediate gases contain about equal amounts of hydrogen and methane, but after these gases react with the raw coal in the dilute phase, the off gas is enriched to a methane concentration of more than 60% before purification.

The Hydrane process differs from the Synthane process, discussed below, in that the bulk of the methane is produced by the direct reaction of hydrogen with coal char rather than by the production of an intermediate synthesis gas that is converted to methane. Reaction of the coal with hydrogen results in about 95% of the total methane being produced in the gasifier.

Hydrogen for the process can be produced in a number of ways. The preferred approach is to react hot char from the fluid bed with steam and oxygen in a pressurized fluid-bed gasifier. Synthesis gas from this gasifier is purified and shifted by conventional means to produce essentially pure hydrogen.

HYGAS. Prepared coal is slurried with a light oil (produced as a by-product of the process), and the slurry is pumped to 1000–1500 psi and injected into a fluidized bed at the top of the vertical multistage unit, where the oil is driven off and recovered for reuse.

The oil-free coal then passes downward through two hydrogasification stages which provide countercurrent treatment (coal passes down, gases produced pass upward and are drawn off).

In the first hydrogasification stage, dried coal is flash-heated to the reaction temperature (1200–1300°F; 649–704°C) by dilute-phase contact with hot reaction gas and recycled hot char. Volatile matter in the coal and active carbon are converted to methane in a few seconds. Active carbon, which is present during the first moments of gasification, gasifies at a rate in excess of 10 times that of the carbon in the less reactive char.

The solids then pass down to the second Hygas stage and enter a dense-phase, fluidized-bed reactor at temperatures of 1700–1800°F (927–982°C), where formation of methane from partially depleted coal char continues simultaneously with a steam-carbon reaction that produces hydrogen and carbon monoxide.

Hot gases produced in the lower hydrogasification stage rise, passing through the first-stage hydrogasification reactor and into the fluidized drying bed, where much of their heat is used to dry the feed coal. After leaving the hydrogasifier, the raw gas is shifted to the proper CO/H_2 ratio in preparation for methanation; the oils, carbon dioxide, unreacted steam, sulfur compounds, and other impurities are removed; and the purified gas is catalytically methanated. Sulfur is recovered in elemental form.

The partially depleted coal char is used to produce a hydrogen-rich stream required in the process by one of three different methods: electrothermal gasification, steam-oxygen gasification, or steam-iron-hydrogen system.

Lurgi. The Lurgi gasifier employs a bed of crushed coal traveling downward through the gasifier, and operates at pressures up to 450 psi. Steam and oxygen are admitted through a revolving steam-cooled grate which also removes the ash produced at the bottom of the gasifier. The gases are made to pass upward through the coal bed, carbonizing and drying the coal. Steam is used to prevent the ash from clinkering and the grate from overheating, and a hydrogen-rich gas is produced. Because of the pressure, some of the coal is hydrogenated into methane (in addition to that distilled from the coal), releasing heat which in turn is given up to the coal, minimizing the oxygen requirements.

Most gas produced by coal gasification comes from Lurgi reactors of the general type shown in Fig. 2. The internal design and operating conditions of these reactors may be altered for coals of different caking characteristics.

Synthane process. This process consists of fluid-bed gasification of the coal, followed by the gas-treatment steps of gas purification and methanation. Reported advantages of the system are that caking coals can be used directly.

For gasification of caking coals, the unit integrates three processing steps: pretreatment in free fall for the destruction of the caking quality of the coal, carbonization in a dense-fluid bed, and gasification of the residue in a dilute-fluid bed. Ash is removed from the bottom of the gasifier.

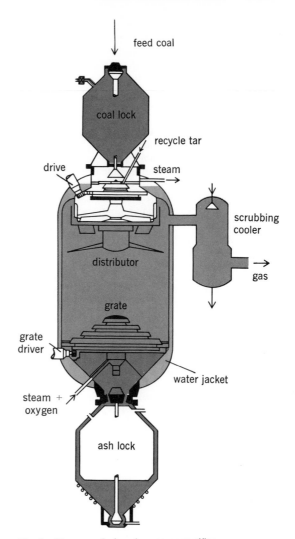

Fig. 2. Diagram of a Lurgi pressure gasifier.

The laboratory pilot plant is designed to operate at 600 psi and 1700°F in the gasification section and at 750°F (399°C) in the pretreatment section. As the coal falls through the tube leading to the gasifier, it is pretreated by oxygen (approximately 0.5 standard cubic foot/lb coal) plus a diluent, either steam or carbon dioxide. The pretreated coal then enters an expanded section of the gasifier where it is carbonized. The fluidization gas in the gasifier consists of oxygen (4 SCF/lb coal) plus steam (20–30 SCF/lb coal). For the methanation step, the U.S. Bureau of Mines has developed a tube-wall reactor that uses a flame-sprayed Raney nickel catalyst.

Underground gasification. At the request of the Bureau of Mines, a detailed survey of underground gasification technology was prepared in 1971. This survey of worldwide activities revealed no large-scale active program of either basic or applied research. However, there has been a renewal of interest in the United States on the part of industrial concerns, and there are indications of one or two exploratory experimental programs on a modest scale.

The Bureau of Mines has begun work in this field again, with particular emphasis on the utilization of new technology developed since the mid-

1950s. This new technology includes an understanding of the nature and direction of subsurface fracture systems and the means to calculate underground fluid movement. Directional drilling techniques have been advanced; and chemical explosive fracturing, a new method of preparing underground formations, has been introduced by the Bureau of Mines. Additionally, field experiments have established the technical possibility of in-place recovery of crude oil and shale oil, and modern methods of surface-processing of coal to high-Btu gases have been demonstrated. *See* OIL SHALE; PETROLEUM RESERVOIR ENGINEERING.

An interdisciplinary project team has been created to activate the bureau's program; a field site near Hanna, WY, has been selected; preliminary drilling has been completed; and exploratory gasification tests were expected to be carried out in 1976. At the same time, an active program on fracture characterization and gas flow in coal beds is under way at the Morgantown Energy Research Center. *See* COAL.

[ROBERT C. WEAST]

Bibliography: American Gas Association, *Proceedings of the 6th Synthetic Pipeline Gas Symposium*, 1974; Atomic Energy Commission, *Coal Processing: Gasification, Liquefaction, Desulfurization, Bibliography 1930–1974*, Rep. TID-3349, National Technical Information Service, 1974; Bituminous Coal Research, Inc., *Gas Generator Research and Development*, BCL Rep. L-156; Project Independence, *Synthetic Fuels from Coal*, U.S. Government Printing Office, Stock no. 4118-00010, 1974.

Coal liquefaction

The conversion of coal to synthetic crude oil (syncrude), a product suitable for use as a refinery feedstock and for petrochemical production. The technologies employed comprise one indirect approach, two direct pyrolytic methods, and four direct solvent extraction (dissolution) processes (see illustration). In recent years, coal liquefaction technology has shifted to the production of low-sulfur, low-ash fuel oil for power generation. This nonpolluting fuel could be used to replace natural gas and low-sulfur fuel oil, now used to generate one-half of United States electricity, if ample supplies were available. *See the feature article* OUTLOOK FOR FUEL RESERVES.

The single factor that distinguishes the indirect and direct technologies is conversion efficiency. The Fischer-Tropsch (indirect) process is about 40% efficient, whereas the direct liquefaction processes are 60–70% efficient. Based on a fixed output (in Btu) from a syncrude plant, the Fischer-Tropsch process would require about 50% more coal input. No commercial direct liquefaction plants are in operation.

Production constraints. In addition to the necessity to stay on stream a high percentage of the time, the plant product (syncrude) must have a consistent quality regardless of the quality of the coal. Because the input coal feedstock is a natural mineral, the syncrude plant must have sufficient flexibility to accommodate natural fluctuations in coal properties. These properties are minimized but certainly not eliminated by locating the plant at the mouth of a coal mine. Performance require-

ments limit the liquid product to 0.8 wt % sulfur and 0.1 wt % ash contents when produced from high-sulfur-content coals. If this quality level is attained, most of the nation's coal reserves can be converted to environmentally acceptable fuel oil and commercially acceptable synthetic crude oil feedstocks.

Technical approaches most likely to attain economical practicality in the 1975–1985 time period are indirect liquefacton, whereby anthracite, all ranks of bituminous coal, and lignite can be used as feedstock; and direct liquefaction, in which anthracite cannot be used as feedstock and in which the use of low-rank, high-oxygen-content bituminous coals and lignite is feasible but not economical.

Without governmental incentives, syncrude production from coal is expected to be zero through 1985 except for a total production of 1000 bbl (1 bbl = 0.159 m^3) per day from pilot plants, and then only if technical success is achieved in the design and pilot plant stages of research-and-development in fossil-energy programs. Although the Fischer-Tropsch process is an off-the-shelf technology, its syncrudes and upgraded liquid products are not price competitive. Syncrudes will be produced principally from advanced direct liquefaction processes now in development. Regardless of the type of technology selected for commercial plants, the major production constraints are the shortage of venture capital and facilities to produce alloy pressure-reactor vessels. Other constraints include air quality standards, water availability and quality, environmental control technology, community and socioeconomic impacts, and institutional barriers.

Fischer-Tropsch process. The Fischer-Tropsch process supplied the German armed forces during the last years of World War II with much of their requirements of gasoline, diesel fuel oil, and lubricants, while making available feedstocks for an emerging petrochemical demand. A modified Fischer-Tropsch process has been in commercial production since 1957 at Sasolburg in the Republic of South Africa. Domestic noncaking coal is crushed and screened. Minus-^{3}/$_8$-in. (1 in. = 25.4 mm) coal is used for electricity generation. The synthesis plant consumes 6600 tons (1 short ton = 0.9 metric ton) per day of dry ^{3}/$_8$ × 1^1/$_2$ in. coal.

In the South African plant, gases produced in Lurgi gasifiers supplied with coal and oxygen are purified before undergoing a shift conversion to attain an H$_2$/CO volume ratio of 1.8. This synthesis gas (syngas) together with recycled gases is passed through a fixed catalyst bed held at 430–490°F (221–254°C) and 360 psig (1 psi = 6895 N/m^2). The products are straight-chain high-boiling hydrocarbons, some medium-boiling oils, diesel oil, liquefied petroleum gas (LPG), and oxygenated compounds.

Fresh syngas, fixed-catalyst-bed tail gas, and tail gas from a fluid-bed synthesis operation are reformed with steam to increase the H$_2$/CO volume ratio. The reformed gas enters the fluid bed, in which catalyst is circulated along with syngas at 600–625°F (316–329°C) and 330 psig. The volumetric ratio of recycled gas to fresh feed gas is 2. Gas and catalyst leaving the reactor are separated by cyclones, and the catalyst is recycled. The products made by the fluid-bed-catalyst operation are mainly low-boiling C_1 to C_4 hydrocarbons and gasoline with minor amounts of medium- and high-boiling materials. Substantial amounts of oxygenated products and aromatics are also made. *See* FISCHER-TROPSCH PROCESS.

Direct pyrolytic methods. Two direct liquefaction processes heat coal out of contact with air (pyrolysis) to evolve the volatile matter content. The resultant gases and vapors are cooled to produce liquid hydrocarbons that are treated with hydrogen for desulfurization and quality improvement. Pyrolytic processes produce significant quantities of by-product gas and char that must be disposed of economically.

COED process. In the Char Oil Energy Development (COED) process minus-$\frac{1}{8}$-in. dry coal is heated to successively higher temperatures in a series of fluid-bed reactors at 6–10 psig. The initial temperature is 600°F (316°C); the second and third reactors are operated at 850°F (454°C) and 1000°F (538°C), respectively, by using gases leaving the fourth stage that is maintained at 1600°F (871°C) by combustion of part of the char with oxygen. Oils separated from water-quenched gases are filtered to remove char carryover. Gases separated from the oil recovery section are purified to remove NH_3, CO_2, and H_2S, and are then reformed with steam to produce hydrogen. The 4° API-gravity filtered syncrude is hydrotreated at about 750°F (400°C) and 1500–3000 psig to remove sulfur, nitrogen, and oxygen. About 1.04 bbl (43.7 gal) of syncrude was produced from 1 short ton of Illinois No. 6 seam coal during the operation of a 36-ton-per-day pilot plant by the Food Machinery Corporation near Princeton, NJ. Some research was directed to desulfurize the 1177 lb (1 lb = 0.454 kg) of residual char produced from 1 short ton of the same coal.

TOSCOAL process. The basic feature of the second major pyrolytic process (TOSCOAL) is the use of hot ceramic balls to supply heat for carbonization. Crushed dried coal is preheated by hot flue gas before the coal enters a rotating drum where it is heated to 800–1000°F (427–538°C) by contact with hot ceramic balls that are recycled to a heater. Char produced in the rotating drum equals about one-half the weight of the raw coal feed and contains about 80% of the raw coal's heating value. Obviously, economical use must be found for the char. The vapors leaving the drum are cooled, and the condensed oils are fractionated into gas oil, naphtha, and residuum. Gas fuels the ball heater. One short ton of Wyodak coal (8139 Btu/lb and 0.30 wt %, sulfur; 1 Btu = 1056 J) at 970°F (520°C) carbonization temperature yielded 21.8 gal (1 gal = 0.0038 m³) of 6–13° API-gravity oil having a high heating value (HHV) of 16,000 Btu/lb, and 10.5 gal of a similar quality oil at 800°F (427°C) carbonization temperature. The Oil Shale Corporation (TOSCO) has carbonized subbituminous coal in its 25-ton-per-day pilot plant at Golden, CO.

Direct solvent-extraction processes. Four solvent-extraction (dissolution) direct liquefaction processes are described.

Synthol process. The U.S. Bureau of Mines' Synthol process converts coal into fuel oil in a pilot plant operation. Dried crushed coal slurried in a recycled portion of its own product oil is fed to a fixed-bed catalytic reactor with turbulently flow-

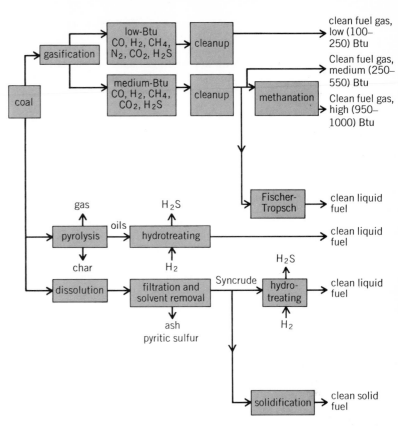

Clean fuels from coal. (*From W. W. Bodle and K. C. Vyas, Clean Fuels from Coal: Introduction to Modern Processes, paper presented at Clean Fuels from Coal Symposium, Institute of Gas Technology, Chicago, Sept. 10–14, 1973*)

ing hydrogen to desulfurize the coal. A commercially available catalyst, cobalt molybdate on silica-activated alumina, is maintained at 850°F (454°C) and 2000–4000 psig. The slurry residence time is less than 14 min. The violently moving slurry prevents plugging of the catalyst as the coal passes through a sticky plastic phase prior to becoming a liquid. After separation from the gas phase, the liquid product is centrifuged to remove ash and organic coal residues. After removal of H_2S and NH_3, the gas is recycled with fresh hydrogen to the catalytic reactor. A West Kentucky coal containing 4.6 wt % sulfur yielded 3 bbl (126 gal) of oil containing 0.19 wt % sulfur, 0.19 wt % nitrogen, 17,700 Btu/lb, 1.0 wt % ash, and having a viscosity ranging from 21 to 30 Saybolt Seconds Furol. The yield was based on 1 short ton of mineral ash-free (m.a.f.) coal.

H-coal process. The H-Coal process of Hydrocarbon Research, Inc. (HRI), produces syncrude using dried, minus-40 mesh coal slurried in a coal-derived oil. The slurry, mixed with hydrogen, is preheated and fed to the bottom of an ebullient bed reactor held at 850°F (454°C) and 3000 psig. The constant activity of the cobalt-molybdenum catalyst used is maintained by the continuous addition and withdrawal of the 0.063-in.-diameter pelletized catalyst that largely remains in the reactor bed, while the smaller particles of ash and unreacted coal, slurry oil, and gases are carried from the reactor into an atmospheric-pressure flash drum. The residence time of the coal in the reactor is 10–30 min.

The reactor products are separated into vapor

and liquid phases in a baffled vessel. After removal of NH_3, H_2S, and hydrocarbons, the gas phase is scrubbed using an amine solution before it is recycled with new hydrogen and coal oil slurry to another reactor whose liquid product is flashed at high and at low pressure to separate light hydrocarbons before the liquid is freed of solid residues by hydroclones and filters. The filtrate is stripped and fractionated into naphtha, middle oil, and fuel oil. The H_2S-containing liquid is treated in a Claus unit to recover elemental sulfur. The conversion efficiency of the H-Coal process reportedly is 90–94%. One ton of m.a.f. coal yields 3 bbl (126 gal) of C_4 to 975°F (524°C) distillate. The H-Coal process is the only advanced technology that offers the production of both refinery feedstock and an environmentally acceptable fuel oil.

Coalcon process. On Jan. 17, 1975, Coalcon, a joint venture of Union Carbide Corporation and Chemical Construction Corporation, received a $237,000,000 contract from the Department of the Interior's Office of Coal Research to design, construct, and operate a 2600-ton-per-day demonstration plant using a hydrocarbonization process to produce 3900 bbl (163,800 gal) of liquid product and 22×10^6 ft^3 (1 ft^3 = 0.028 m^3) of pipeline-quality gas per day.

Coalcon's process pyrolyzes high-sulfur-content coal to a primary product of clean boiler fuel. The pyrolysis in the presence of 1000-psig hydrogen approximately doubles the oil yield above that normally obtained from similar low-temperature carbonization at atmospheric pressure without addition of hydrogen. Raw coal is pulverized before drying by a hot flue gas containing 3 vol % oxygen obtained from a cyclone furnace burning process-generated by-product char. The warm, dried coal is pressurized to 200 psig, preheated to 662°F (350°C), and oxidized with 3%-by-weight oxygen.

The preoxidized coal is removed from the carrier gas by cyclones and is then added to hydrogen and recycled gas at 1000 psig in a lock hopper subsystem. The pressurized coal and gases move to the hydrocarbonization reactors in dense-phase fluid flow and are injected at 400 ft/sec (1 ft = 0.3 m) vertically into the reactors. The coal together with partly reacted char is detained in the reactors at 1040°F (560°C) for 25 min while being fluidized by the hydrogen feed gas.

Char comprising the nonreacted solids is removed continuously through standpipes whose contents are kept fluidized by blow-back steam generated by water injection to reduce the char temperature to 600°F (316°C). This design leads to less than a 1 wt % carryover of char and eliminates the necessity for filtration of the liquid product.

Gas and vapors from the reactors are partially quenched with a stream of recycled heavy oil in a system in which water vapor goes overhead along with the gas, gasoline-range hydrocarbons, some low-boiling phenols, and nitrogen bases. By further cooling and decantation, water and a light-oil product are separated. The light oil is mixed with heavy oil to make a total liquid primary-fuel product. The light oil can also be kept separate for its chemical values or for conversion to premium gasoline.

The gaseous product is stripped of CO_2, NH_3, and H_2S and other sulfur compounds prior to production of a syngas suitable for methanation into pipeline-quality gas. The clean gas also can be burned as a medium-Btu fuel.

The major products of the Coalcon process are clean boiler fuel and pipeline-quality gas. One ton of m.a.f. coal is expected to produce 1.6 bbl (67.2 gal) of hydrocarbon liquids and 9300 standard cubic feet (scf) of high-Btu gas, all of which is equivalent to 3.2 crude barrels equivalent (CBE).

SRC process. In the Solvent Refined Coal (SRC) process developed by the Pittsburg & Midway Coal Company, raw coal is dried, pulverized to 80 wt % minus-200 mesh and mixed with coal-derived solvents having a boiling range of 550–800°F (288–427°C). Slurry, which has a solvent/coal weight ratio of 1.5–4.0, is moved by reciprocating pumps to a preheater, at the entrance of which 1000–20,000 scf of hydrogen are added per ton of feed slurry.

The hydrogen-slurry mixture passes through the preheater and reactor vessel at 850°F (454°C) and 1000 psig. During its residence time of 0.2–1.8 hr, the coal is depolymerized and organic matter is dissolved in the solvent.

Gas from the reactor is separated from the slurry of undissolved solids and coal solution in a flash vessel operating at 625°F (329°C) and 995 psig. The gas is scrubbed for acid gas removal, light hydrocarbons are removed, and the resulting excess hydrogen together with fresh hydrogen is recycled to the slurry preheater.

The slurry from the flash vessel goes to a rotary filter for removal of undissolved solids. The dried solids containing 35–55 wt % undissolved carbon and 5–8 wt % sulfur are used for the production of steam in a combustor whose stack gas is reacted with H_2S in a Claus unit to produce elemental sulfur.

The solids-free coal solution from the rotary filter is subjected to vacuum flash distillation for the recovery of solvent used in the preparation of slurry feed. In addition to the recycle solvent recovered, the overhead feed solution also produces a light hydrocarbon by-product stream. The bottom fraction is a hot liquid called Solvent Refined Coal (SRC) which has a solidification temperature between 300 and 400°F (149 and 204°C), depending on its undistilled solvent content. SRC can be transported as a hot liquid, or it may be allowed to solidify into a relatively clean fuel containing 16,000 Btu/lb, 0.1 wt % ash and 0.8 wt % sulfur, when made from a West Kentucky coal containing 7.1 wt % ash and 3.4 wt % sulfur. The SRC product can be catalytically hydrotreated to produce a more valuable syncrude and other products. By this additional hydrotreatment, known as the Hybrid process, the yield of SRC will remain at 3 CBE per short ton m.a.f. coal, but the melting point of the product will be reduced to room temperature from 300°F (149°C) because of the deeper hydrogenation made possible by the use of an unspecified catalyst. *See* COAL; COAL GASIFICATION; FUEL, SYNTHETIC.

[FRANKLIN D. COOPER]

Bibliography: E. E. Donath, Chemicals from coal hydrogenation, *Gasification and Liquefaction of Coal*, AIMME, 1953; Energy and Research Development Administration, *Economic Evaluation of Coal Based Synthetic Crude*, 1975; B. K. Schmid, Status of the SRC project, *Chem. Eng. Progr.*,

71(4):75–78, 1975; L. J. Scotti et al., The project COED pilot plant, *Chem. Eng. Progr.*, 71(4):61–62, 1975; H. H. Storch, *Chemistry of Coal Utilization*, 1945; *Synthetic Fuels Commercialization Program*, vol. 3: *Technology and Recommended Incentives*, Synfuels Interagency Task Force to the President's Energy Resource Council, 1975.

Coal mining

The technical and mechanical job or removing coal from the earth and preparing it for market. Coal, the most abundant and traditionally the most economical source of power in the world, is found in varying amounts throughout the globe. It lies in veins of various thickness and richness beneath the crust of the Earth. The product of fossilized plant material mixed with various mineral matter, coal rests in giant subterranean sandwiches, shallow or deep, flat or pitched. At the present time, there are coal mines in the United States operating at the 1500-ft level in Virginia and shafts under development at the 1800-ft level in Alabama. Obtaining coal in sufficient quantities at a competitive expense and making it the proper grade for the market demand is, in its simplest terms, the science of coal mining. Mammoth mining machines require technically qualified manpower. The pick-and-shovel miner has all but disappeared from the American scene, along with the stereotyped characterization of him as an oppressed laborer living in a company town. Today's miner is a skilled, well-paid technician who handles complex, costly, and highly efficient machines.

Prospecting and planning. Of major importance in coal property development is the accumulation of seam information from borehole drilling. Chemical analysis of the cores will provide details concerning moisture, volatile material, fixed carbon, ash, sulfur, Btu per pound, and fusion temperature of the ash. Washability data will give the percentage of float (coal) and sink (foreign material) for each size coal at each specific gravity used (generally from 1.35 to 1.60), together with the amounts of ash and sulfur and the Btu value on a dry basis for each increment of specific gravity. *See* BORING AND DRILLING (MINERAL).

After preliminary analyses and correlations, the engineering details are planned. Production requirements and the extent of coal reserves generally determine the method of mining and the type and capacity of equipment to be used. Amortization of the investment and an adequate return on the investment are also influential factors. The cost of opening, developing, and operating the mine is estimated. A reasonable evaluation can then be based on the projected costs and markets available.

Development is commonly planned for the life of the property. This involves projecting and estimating the working sequence of the various parts of the area.

The type of mining equipment to be used; transportation to be required; the water drainage; ventilation and roof control (underground); overburden analysis (surface) are the other items that must be projected and estimated. The interrelations of these factors with costs and quality of the coal are a necessary part of the planning.

Coal is extracted from the earth by three basic

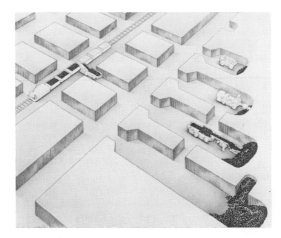

Fig. 1. Conventional coal mining system, for underground mining, shuttle car and train haulage. (*Joy Manufacturing Co.*)

Fig. 2. Coal mining system with continuous miner using shuttle car and belt haulage. (*Joy Manufacturing Co.*)

methods: (1) underground, (2) strip, and (3) auger. Approximately two-thirds of America's coal comes from underground mining. The remaining one-third is mined from the surface, either by strip or auger mining.

Underground mining. The systems of underground coal mining generally in use are room-and-pillar, longwall, and, in a few places in Europe, hydraulic. In the room-and-pillar system, tunnels are carved into the seam, leaving pillars of coal for support. In some mines these pillars are removed

Fig. 3. Continuous miner. (*Lee-Norse Co.*)

Fig. 4. Longwall mining equipment. Plow works back and forth across working face. (*Mining Progress, Inc.*)

in subsequent mining, allowing the overlying strata to collapse; in others the pillars are not recovered. In longwall mining, widely spaced tunnels are driven, leaving large blocks of coal. Later, these blocks are completely extracted, allowing the roof material to collapse behind the coal face as it is removed. In hydraulic mining, as now practiced, a stream of water is directed against the coal face with sufficient pressure to dislodge the coal. The water also acts as a transporting medium.

The room-and-pillar system of mining has been the most widely practiced method in the United States, whereas the longwall system has been used for many years in Europe. However, the demand for metallurgical coal has necessitated mining deeper into the earth. With the advent of mechanical plows and shearers, it has become economically feasible to use the longwall system in the United States.

Underground mining equipment. The type of equipment used at the mining face is governed by a complex of factors. Outstanding are the relative difficulties of supporting the immediate roof, the height of the seam, grades of coal, maintenance required on machinery, and productivity expected of manpower using different types of machines. Today, mechanical equipment falls into two classes, the so-called conventional and the continuous miner machine (Figs. 1 and 2).

In the conventional method several machines are used in a cycle of operation: undercutting or top-cutting the seam with cutting machines; blasting down the face; loading coal with a mechanical loader; and transporting it from the face by shuttle car or conveyor. The continuous miner (Fig. 3) has been used in many seams. It bores or rips to dislodge the coal from the face without blasting, then loads the coal and puts it into the transportation system.

The longwall mining machine (Fig. 4) employs a plow or shearer which is pulled back and forth across a working face several hundred feet long. The loosened coal is dropped into a conveyor. Self-advancing hydraulic jacks support the roof and follow the machine as it slices into the coal on a wide front. With longwall mining in overburdens of 1000 ft or more, it is possible to mine extremely gassy

seams. Because of the large tonnage being extracted from a single face, large volumes of air are employed to dilute the liberated gases. The massive jacks used in longwalling control the weight of the roof and also create the falls necessary to remove the pressure at the coal face, thereby providing protection from roof falls.

The power source for operating mining machinery can be liquid fuels, seldom used in coal mining, or electricity, either alternating or direct current. Power distribution for a mining property has to be planned for a whole mine in the same manner as the actual mining operation.

Prospecting information and actual experience determine the amount and type of roof control necessary for the protection of workmen. The old type of roof support is so-called timber; this may be simply wooden posts, or may consist of beams of steel or wood supported by posts or fastened to the walls. Currently, the use of roof bolts has replaced a great amount of timber support. Holes are drilled into the roof, and steel rods, held by expansion devices, are screwed in until the rod head is tight against the surface. The size, spacing, and length of the bolts are governed by the roof conditions to be controlled. Roof bolting combines many roof materials into one large beam instead of letting the various strata act separately. For permanent support at some mines the roof is gunited, particularly when the roof materials are subject to deterioration by weathering if exposed to the air currents.

Mined coal is transported from the working face to the main transportation system by shuttle cars (underground trucks) (Fig. 5) or by conveyor. Shuttle cars are used predominantly, although many properties with steep grades or thin coal find it more economical to employ conveyors. In hydraulic mining in Europe, the water used to break down the coal also conveys the coal away from the face.

The main transportation to the surface or to shafts or slopes is generally provided by mine cars which are pulled along tracks by locomotives. Although not as common, there are many instances where coal is transported on conveyor belts. Mainline belt transportation eliminates a great deal of grading required for track haulage but requires essentially the same amount of maintenance and roof protection.

The largest cars that seam height and width of working areas will permit are used in track systems. Modern mine cars hold 6–20 tons of material. Most have solid bodies that are emptied by tipping or rotary dumping. Motive power is generally provided by electric locomotives of 20-tons weight.

When the coal does not have a surface outcrop, it must be elevated by a slope or shaft. Almost all slopes are on inclinations that permit the use of belts to transport the coal, which is dumped at the bottom of the slope, to the surface. These belts are installed with sufficient capacity to carry the maximum production of the mine on a daily or shift basis. When the coal is brought to a shaft, it is dumped into skip buckets and hoisted to a dump at the surface. Modern installation in these shafts are completely automatic as to loading and dumping.

An underground coal mine consists of a great number of spaces and openings in which people

Fig. 5. Shuttle car hauling from working face area in a coal mine to the main transportation system. (*National Mine Service Co.*)

must work safely. In many cases, however, explosive gas is emitted. Hence, artificial ventilation over the entire mine is required to maintain a normal atmosphere and to dilute and carry away such gases. Numerous shafts and fans provide the necessary volume of air.

To prevent water from entering and to eliminate it from the mine are the aims of mine drainage. Removing excess water from mine properties may be difficult and expensive; in extreme cases over 30 tons of water have to be removed for each ton of coal mined. *See* MINING, UNDERGROUND; MINING MACHINERY.

Strip mining. Where the coal seam lies close to the surface, it is more economical to remove the overburden of earth and rock that covers the coal seam. For this job, power shovels, draglines, or wheel excavators are used. As a farmer plows his field in furrows, a shovel excavates a "furrow" and casts the overburden parallel to the cut. Draglines sit on the bank above the coal seam and remove the overburden from the seam. Bucket-wheel machines excavate the material from above the coal seam. It then flows in a continuous stream via a transfer to the conveyor system, which in turn transports it to the discharge point. When the seam is exposed, the coal is loaded by smaller power shovels into trucks and hauled to preparation plants or loading bins. *See* MINING, STRIP.

Strip mining is an efficient way to mine coal. Seventy-foot-thick seams of coal in Wyoming are now being strip-mined; seams such as these will furnish the United States with a substitute for natural gas. Gigantic machines have been developed to efficiently mine coal by the strip-mining method. The biggest shovel in 1968 weighed 27,000,000 lb, towering 220 ft above the coal seams in which it worked, moving 180 yd^3 of earth every 50 sec. A dragline is being erected which will take as much as 220 yd^3 of earth per bite (Fig. 6); it will employ a 310-ft boom. Off-the-highway haulage trucks in the 100-ton category are common; units capable of carrying 240 tons are available. Auxiliary equipment used to complement the excavating machines has correspondingly increased in capacity.

Augering. For coal seams which continue under rising land too thick for economical strip mining

and where underground mining cannot burrow further to the surface because of the shallow and more treacherous roof conditions, a relatively new procedure of mining, the auger method, was developed in 1951. The auger miner (Fig. 7) twists huge drills like carpenter's bits into a hillside coal seam, drawing out the coal to a conveyor which loads it into trucks. Section by section, the drills bore into the hillside. Augering in general will recover 40–60% of the coal seam. The development of the dual-headed auger and the multiheaded auger, with progressive increase in size until the giants of today have been manufactured with head diameters up to 96 in., has made it possible to penetrate to a depth of 300 ft in a level seam. Auger recovery gives additional tonnage at minimal cost for equipment and labor and permits recovery of coal that might not be recovered by other methods. Auger production rose from 205,000 tons in 1951 to 2,900,000 tons in 1967.

Cleaning, grading, and shipping. After the coal is mined, it may be loaded directly into transportation facilities for the market if the foreign material in it is not excessive. However, coal from most seams requires preparation to provide a desired and uniform quality.

A number of washing devices are used for cleaning coal, all of which operate on the basic principle of the difference of specific gravity between coal and foreign material. The coal is floated in a vessel and the foreign material, being heavier, drops to the bottom. Washed coal in sizes less than 3/8 in. carries sufficient water from the washing circuit so that it must be dewatered to be acceptable on the market. Where a market can accept a moisture content of 7–8%, mechanical devices can meet the requirements. If a lower moisture is required, the product is thermally dried of surface moisture to the required extent after being mechanically dewatered. Once the coal is ready for market, it is loaded for shipment to the customer by railroad, water transportation, or in some cases belt conveyor systems.

The 108-mi coal slurry pipe line between Cadiz and Cleveland, both in Ohio, pioneered a new concept in coal transportation. A 273-mi coal slurry pipe line is also in operation between Arizona and Nevada. This innovation in coal transportation has

Fig. 6. Sketch of 220-yd^3 dragline. (*Bucyrus-Erie Co.*)

Fig. 7. Single-head auger. (*Salem Tool Co.*)

proved to be a more economical method than building railroads for coal shipment.

One innovation in the industry has been the adoption of the unit train concept for coal shipments. A unit train, in its true sense, is a complete train of cars (usually privately owned) with assigned locomotives. It operates only on a regularly scheduled cycle movement between a single origin and a single destination each trip.

Future trends. In view of improvements in the past, the following future developments can be expected: (1) Conversion of coal into oil. Present coal reserves in the United States are known to be 3.2×10^{12} tons. With 1 ton of coal converting into approximately 4.5 bbl of oil, present reserves will yield more than 14×10^{12} bbl, which is more oil than presently known to exist in the world. (2) Conversion of coal into gas to be placed in the underground storage fields and delivered into the major pipelines which intersect the coal fields. (3) Further research into the possibilities of using coal tars and shales as a source for synthetic oils. *See* COAL GASIFICATION.

[JAMES D. REILLY]

Bibliography: I. A. Given, *Mechanical Loading of Coal Underground*, 1943; E. S. Moore, *Coal*, 2d ed., 1940; National Coal Association, *Yearbook*, 1968; P. Pfleider, *Surface Mining*, 1968.

Coking in petroleum refining

A process for thermally converting the heavy residual bottoms of crude oil entirely to lower-boiling petroleum products and by-product petroleum coke. The heavy residual bottoms cannot be catalytically cracked, principally because the metals present are potent catalyst poisons. Traditionally, the residual bottoms have been blended with lighter stocks and marketed as low-quality heavy fuel oils. However, in the United States the demands for gasoline, heating oil, and other products have increased substantially over the years. To meet this shift in demand, refiners have resorted to deeper vacuum flashing, vis-breaking, solvent deasphalting, and coking. The major products from coking are fuel gas, gasoline, gas oil, and petroleum coke. Generally speaking, the gasoline is of poor quality and must be further treated before use

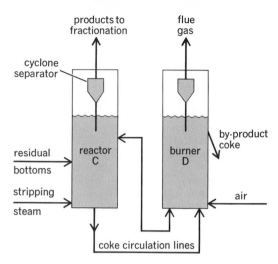

Fig. 2. Fluid coking, in simplified flow plan.

in modern premium fuels. The gas oil is commonly used as catalytic-cracking feed stock. *See* FUEL GAS; GASOLINE.

The two coking processes in commercial use are delayed coking and fluid coking. In delayed coking (Fig. 1), residual bottoms are heated to temperatures of about 900°F in a conventional furnace. The hot oil is charged directly to the bottom of a large coke drum A, where the coking reaction takes place. Coke is deposited in the drum while the lighter products are removed overhead to fractionation. Approximately every 24 hr, hot oil is diverted to the second coke drum B, and coke is removed from drum A by hydraulic drilling. In fluid coking (Fig. 2), residual bottoms are charged directly to the reactor C containing a bed of fluidized coke particles at about 950°F. The hot coke particles supply the heat of reaction and also a surface for coke deposition. Converted light products are removed overhead through cyclone separators to fractionation. The coke particles are continuously circulated between the reactor C and the burner D. In the burner D, air is used to burn a portion of the coke produced, thereby maintaining the system in heat balance. Normally, more coke is produced in the reactor than is required for heat balance. The excess coke is continuously withdrawn from the burner as by-product petroleum coke.

Petroleum coke is used principally as fuel or, after calcining, for carbon electrodes. The crude from which the coke is produced governs its chemical composition (high-sulfur crudes yield high-sulfur cokes). However, all petroleum cokes are characterized by low total ash contents. *See* CRACKING; PETROLEUM PROCESSING.

[J. F. MOSER, JR.]

Combustion

The burning of any substance, whether it be gaseous, liquid, or solid. In combustion, a fuel is oxidized, evolving heat and often light. The oxidizer need not be oxygen per se. The oxygen may be a part of a chemical compound, such as nitric acid (HNO_3) or ammonium perchlorate (NH_4ClO_4), and become available to burn the fuel during a com-

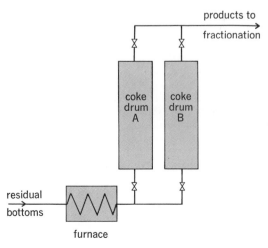

Fig. 1. Delayed coking, in simplified flow plan.

plex series of chemical steps. The oxidizer may even be a non-oxygen-containing material. Fluorine is such a substance. It combines with the fuel hydrogen, liberating light and heat. In the strictest sense, a single chemical substance can undergo combustion by decomposition, with emission of heat and light. Acetylene, ozone, and hydrogen peroxide are examples. The products of their decomposition are carbon and hydrogen for acetylene, oxygen for ozone, and water and oxygen for hydrogen peroxide.

Solids and liquids. The combustion of solids such as coal and wood occurs in stages. First, volatile matter is driven out of the solid by thermal decomposition of the fuel and burns in the air. At usual combustion temperatures, the burning of the hot, solid residue is controlled by the rate at which oxygen of the air diffuses to its surface. If the residue is cooled by radiation of heat, combustion ceases.

The first product of combustion at the surface of char, or coke, is carbon monoxide. This gas burns to carbon dioxide in the air surrounding the solid, unless it is chilled by some surface. Carbon monoxide is a poison and it is particularly dangerous because it is odorless. Its release from poorly designed, or malfunctioning, open heaters constitutes a serious hazard to human health.

Liquid fuels do not burn as liquids but as vapors above the liquid surface. The heat evolved evaporates more liquid, and the vapor combines with the oxygen of the air.

Spontaneous combustion. This occurs when certain materials are stored in bulk. The oxidizing action of microorganisms often produces the initial heat.

As the temperature increases, the air trapped in the material takes over the oxidation process, liberating more heat. Because the heat cannot be dissipated to the surroundings, the temperature of the material rises still more and the rate of oxidation increases. Eventually the material reaches an ignition point and bursts into flame. Coal is subject to spontaneous combustion and is generally stored in shallow piles to dissipate the heat of oxidation.

Gases. At ordinary temperatures, molecular collisions do not usually cause combustion. At elevated temperatures the collisions of the thermally agitated molecules are more frequent. More important as a cause of chemical reaction is the greater energy involved in the collisions. Moreover, it has been reasonably well established that there is very little combustion attributable to direct reaction between the molecules. Instead, a high-energy collision dissociates a molecule into atoms, or free radicals. These molecular fragments react with greater ease, and the combustion process proceeds generally by a chain reaction involving these fragments. An illustration will make this clear. The combustion of hydrogen and oxygen to form water does not occur in a single step, Eq. (1). In this seemingly simple case, some fourteen reactions have been identified. A hydrogen atom is first formed by collision; it then reacts with oxygen molecules, Eq. (2), forming an OH radical. The latter

$$2H_2 + O_2 = 2H_2O \tag{1}$$

ecules, Eq. (2), forming an OH radical. The latter

$$H + O_2 = OH + O \tag{2}$$

in turn reacts with a hydrogen molecule, Eq. (3),

$$OH + H_2 = H_2O + H \tag{3}$$

forming water and regenerating the H atom which repeats the process. This sequence of reactions constitutes a chain reaction. Sometimes the O atom reacts with a hydrogen molecule to form an OH radical and another H atom, Eq. (4). Thus

$$O + H_2 = OH + H \tag{4}$$

a single H atom can form a new H atom in addition to regenerating itself. This process constitutes a branched-chain reaction. Atoms and radicals recombine with each other to form a neutral molecule, either in the gas space or at a surface after being adsorbed. Thus, chain reactions may be suppressed by proximity of surfaces; and the number and length of the chains may be controlled by regulating the temperature, the composition of the mixtures, and other conditions.

Under certain conditions, where the rate of chain branching equals or exceeds the rate at which chains are terminated, the combustion process speeds up to explosive proportions; because of the rapidity of molecular events, a large number of chains are formed in a short time so that essentially all of the gas undergoes reaction at the same time; that is, an explosion results. The branched-chain type of explosion is similar in principle to atomic explosions of the fission type, where more than one neutron is generated by the reaction between a neutron and a uranium nucleus. Another cause of explosion in gaseous combustion arises when the rate at which heat is liberated in the reaction is greater than the rate at which the heat dissipates to the surroundings. The temperature increases, accelerates the reaction rate, liberates more heat, and so on, until the entire gas mixture reacts in a very short time. This type of explosion is known as a thermal explosion. There are cases intermediate between branched-chain and thermal explosions which depend upon the type and proportion of gases mixed, the temperature, and the density.

In slow combustion, intermediate products can be isolated. Aldehydes, acids, and peroxides are formed in the slow combustion of hydrocarbons, and hydrogen peroxide in the slow combustion of hydrogen and oxygen. At the relatively low temperature of combustion of paraffin hydrocarbons (propane, butane, ethers) a bluish glow is seen. This light from activated formaldehyde formed in the process is called a cool flame.

In the gaseous combustion and explosive reactions described above, the processes proceed simultaneously throughout the vessel. The gas mixture in a vessel may also be consumed by a combustion wave which, when initiated locally by a spark or a small flame, travels as a narrow intense reaction zone through the explosive mixture. The gasoline engine operates on this principle. Such combustion waves travel with moderate velocity, ranging from 1 ft/sec in hydrocarbons and air to 20–30 ft/sec in hydrogen and air. The introduction of turbulence or agitation accelerates the combustion wave. The accelerating wave sends out compression or shock waves which are reflected back and forth in the vessel. Under certain conditions these waves coalesce and change from a slow combustion wave to a high-velocity detonation wave. In hydrogen and oxygen mixtures, the speed is almost 2 mi/sec. The pressure created by detonation can be very high and dangerous.

Combustion mixtures can be made to react at lower temperatures by employing a catalyst. The molecules are adsorbed on the catalyst, where they may be dissociated into atoms or radicals, and thus brought to reaction condition. An example is the catalytic combination of hydrogen and oxygen at ordinary temperatures on the surface of platinum. The platinum glows as a result of the heat liberated in the surface combustion. *See* EXPLOSION AND EXPLOSIVE; FLAME.

[BERNARD LEWIS]

Bibliography: R. M. Fristom and A. A. Wistenberg, *Flame Structure*, 1965; B. Lewis and G. von Elbe, *Combustion, Flames and Explosions of Gases*, 1961. B. Lewis, R. N. Pease, and H. S. Taylor (eds.), *Combustion Processes*, vol. 2, 1956; Gabriel J. Minkoff and C. F. H. Tipper, *Chemistry of Combustion Reaction*, 1962; B. P. Mullins, *Spontaneous Ignition of Liquid Fuels*, AGARDograph no. 4, 1955; F. J. Weinberg, *Optics of Flames*, 1963; Forman A. Williams, *Combustion Theory*, 1965.

Combustion of light metals

The combustion of light metals in air or oxygen is a spectacular heat-release process, with a very hot and intensely luminous flame and copious "smoke" of finely divided metal oxide. These characteristics of burning metals make them useful in a broad range of applications: as additives to rocket fuels, as the light source in photographic flashbulbs, as constituents of commercial explosives, as well as in fireworks and military flares. Aluminum and magnesium are most commonly used in these ways, but the combustion of other metals such as boron, beryllium, titanium, and zirconium is also of interest. The light metals are very reactive and ignite easily when finely divided, shredded into thin foil, or milled into a powder. Large pieces of metal are very difficult to ignite and very rarely burn. However, accidental fires involving chips and shavings such as those produced in the machining of magnesium have occurred. Such fires must be put out with specially formulated extinguishing powders rather than water or even wet sand, because the burning metals are so reactive they can burn in steam. Even though burning metals release much energy, they cannot be exploited as an energy source because they are not found in pure form in nature and must be extracted by processes which absorb even more energy from other sources.

Fuel characteristics. Light metals generally have very large heats of combustion. Figure 1 shows a comparison of the light metals as fuels with a number of other substances on the basis of the energy released per unit weight of the fuel plus the necessary oxygen, when the substances burn to produce condensed oxides. This basis of comparison is particularly relevant to rockets, which must carry both fuel and oxygen on board. The figure shows clearly that lithium (Li), beryllium (Be), boron (B), magnesium (Mg), silicon (Si), calcium (Ca), and titanium (Ti) are all more energetic than gasoline, and zirconium (Zr) is only slightly less energetic. *See* METAL-BASE FUEL.

Flame temperature. The temperature of a flame depends on the heat of combustion, which can be thought of as the energy available to heat up the products of combustion from ambient temperature to flame temperature. Since light metals have very large heats of combustion, their flame temperatures are very high. For example, aluminum burning in oxygen at atmospheric pressure can produce a flame temperature as high as 3800 K (over 3500°C or 6300°F). Magnesium can burn at over 3300 K and zirconium at well over 4000 K. For comparison, the highest flame temperature for acetylene-oxygen in 3430 K and for propane in air, only 2265 K.

Light-metal oxides are refractory substances with very high melting points. Instead of boiling, they generally decompose at extremely high temperatures (depending on the pressure) and absorb much energy in the process. This property of the oxides generally limits the flame temperature in the combustion of metals to the decomposition temperature of the oxide formed. The heat of combustion of the metal is sufficient to heat all of the oxide to the decomposition temperature and then decompose only some portion of it. The remainder remains in a condensed state. For example, when aluminum burns in oxygen the products of combustion which exist in the flame are droplets of molten alumina (Al_2O_3) and various gaseous species such as Al, O, Al_2O, and AlO. Of course, as they cool down to ambient temperature, these fragments recombine, and outside the flame all of the oxide appears as a white smoke, actually an aerosol of alumina.

Light source. The particles of condensed metal oxide which exist in a flame act in the same way as the mantle in a gas lantern or the particles of burning soot in the luminous flame of a coal oil lamp. They all emit radiation which is much more like that from a solid surface than from a thin layer of hot gas. However, because of their much higher

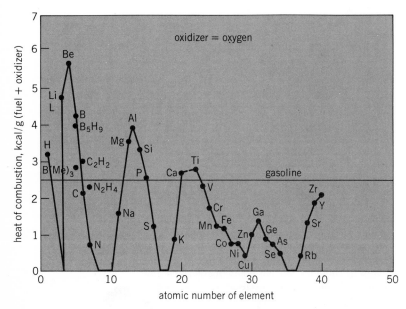

Fig. 1. Comparison of the light metals and other fuels on the basis of the heat of combustion per unit weight of the fuel and the required oxygen.

temperatures, the metal-oxygen flames are much more intense sources of light than even the most luminous hydrocarbon-air flames. The intensity of thermal radiation increases in proportion to the fourth power of the source temperature. The fraction of this radiation which is visible light also increases with source temperature. For example, a source at 2200 K can emit as much as 130 watts per square centimeter of surface, of which about 1% is visible light. A source at 3800 K can emit up to 1100 W/cm² of which 14% is visible. Therefore, as a source of light the aluminum-oxygen flame can be more than 100 times brighter than a hydrocarbon-air flame of the same size. The much hotter zirconium-oxygen flames are even brighter.

Ignition. When a metal is heated in oxygen, a layer of oxide grows on its surface. An oxide such as magnesium oxide (MgO), which has a smaller volume than the original metal, is porous and presents no barrier to the continuing diffusion of oxygen or metal vapor. By contrast, aluminum oxide (Al_2O_3) has a greater volume than the metal which was oxidized to form it. It forms a protective layer on the remaining metal and inhibits further oxidation. As a result of these differences in the properties of the oxide layer, magnesium and aluminum ignite in very different ways.

Ignition occurs when the rate of heat release in the oxidizing reaction first exceeds the rate of heat loss from the reaction zone. Magnesium continues to oxidize by the same mechanism from room temperature right up to ignition. The ignition temperature depends on the size of the metal particles, their concentration, and the amount of oxygen in the oxidizing atmosphere. For example, magnesium particles 35 μm in diameter ignite at 960 K in air at very low concentrations and at 920 K when their concentration is 160 mg/liter. Particles of 7-μm size at the same concentration ignite at 800 K. For comparison, the melting point of Mg is 923 K.

In the case of aluminum, ignition occurs when the protective layer of alumina melts at 2320 K, regardless of particle size and concentration. In this process, hot molten aluminum (melting point 932 K) is suddenly exposed to oxygen and ignites instantly.

Vapor phase. Metals generally burn on the surface of the original particle. However, the more volatile metals can also burn in the vapor phase. Magnesium is a very volatile metal. Its boiling point at 1 atm (101,325 N/m²) is 1381 K, far less than the decomposition temperature of magnesium oxide. Aluminum is less volatile, but its boiling point of 2740 K is still much less than the decomposition temperature of its oxide. As a result, it is possible for both aluminum and magnesium to burn by a vapor-phase diffusion mechanism, in which heat transfer from the flame evaporates the metal and the metal vapor diffuses to react with oxygen in a flame which surrounds the metal particle. Such a flame is shown in Fig. 2 for the case of magnesium burning in air at the low pressure of 0.08 atm. The resemblance between this flame and a candle flame is not a coincidence. Except for the details of chemistry, the processes are similar.

Energy relationship. Estimates show that on the average 54 kcal (1 kcal = 4184 joules) of energy are required to produce 1 g of pure aluminum from the ore in its natural state. Figure 1

Fig. 2. Magnesium ribbon burning in the vapor phase in air at 0.08 atm. The ribbon broke on igniting, and the two halves, now hanging down, are burning in separate but similar flames.

shows that the heat release on combustion of Al to $Al_2O_3(c)$ is 4 kcal per gram of Al + O_2, or about 15 kcal per gram of Al. As a result, a net debit of at least 40 kcal must be drawn from other sources of energy for each gram of aluminum burned. Thus, aluminum may be burned because its flame has characteristics which are useful in particular applications, but it is not a source of energy. The same conclusion holds for the other metals.

[T. A. BRZUSTOWSKI]

Bibliography: M. F. Elliott-Jones, Aluminum, in J. G. Myers et al. (eds.), *Energy Consumption in Manufacturing*, chap. 31, 1974; *12th International Symposium on Combustion*, Combustion Institute, Pittsburgh, pp. 39–81, 1969, *13th Symposium*, pp. 833–868, 1971, *14th Symposium*, pp. 1389–1412, 1973, *15th Symposium*, pp. 479–514, 1975; H. G. Wolfhard and I. Glassman (eds.), *Heterogeneous Combustion*, pp. 3–279, 1964.

Combustion turbine

A machine for generating mechanical power in rotary motion from the energy of high-pressure, high-temperature combustion gases. The combustion turbine is a versatile prime mover. In addition to the tens of thousands of units that serve daily in propjet and jet engines for military and commercial aircraft, and in the propulsion systems of many ships, boats, trains, buses, and trucks, combustion turbines also are a basic building block of industry. Thousands of combustion turbines, ranging in size from 25 to over 100,000 hp (1 hp = 746 W), drive pumps, compressors, and electric generators in base-load and peaking electrical service.

Turbine cycles and performance. The simplest combustion turbine arrangement consists of a compressor, combustion chamber, and turbine

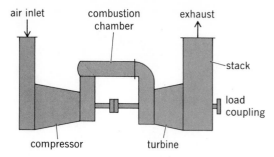

Fig. 1. Diagram of the simplest type of combustion turbine.

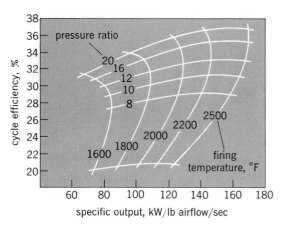

Fig. 2. Thermal efficiency of a typical aircraft-derivative combustion turbine.

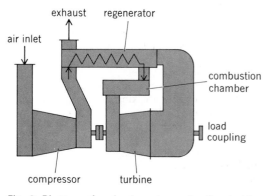

Fig. 3. Diagram of a simple-cycle combustion turbine modified by the addition of a regenerator.

(Fig. 1). The combustion turbine is designed to convert the heat energy of fuel into mechanical energy. Air enters the system and is compressed before entering the combustor, where it is heated at constant pressure. The compressor and combustion chamber produce a high-energy working fluid that can be expanded in the turbine to develop mechanical energy, just as steam is expanded in the more familiar steam turbine. *See* STEAM TURBINE.

Volume of the working fluid is smallest at the compressor outlet, although its temperature is higher than at the compressor inlet. A high excess–air ratio in the combustor keeps gas temper-

atures at levels which are compatible with turbine materials. At the turbine outlet, exhaust gas is at atmospheric pressure and gas volume is at maximum. The main elements of the combustion turbine must be proportioned to handle the flow of working fluid with minimum pressure loss through the system.

Work output. For a simple-cycle plant, the fuel represents net energy input. Work output (net useful output of the cycle) plus exhaust energy equals the fuel energy input. Compressor-shaft input work simply circulates within the cycle. In a sense, this is the catalyst that makes the combustion turbine cycle workable. The turbine must always be capable of generating the mechanical energy needed to drive the compressor and pressurize the air. Whatever it generates above this is available to drive an external load. Of the total shaft work developed by the turbine, roughly two-thirds goes to drive the compressor and one-third to drive the load. This may seem like a high proportion of mechanical work to make the cycle operate, but the same relation holds for any heat engine using hot gases directly to generate useful mechanical work output.

Efficiency. Thermal efficiency (the ratio of work output to heat input) of the cycle depends on its pressure ratio—that is, the ratio of the absolute compressor discharge pressure to the absolute compressor inlet pressure. Figure 2 shows how thermal efficiency rises with increasing pressure ratio and firing temperature. For a typical turbine with a pressure ratio of 8 : 1 and a firing temperature of 1600°F (870°C) the efficiency is approximately 27.2%. By increasing the firing temperature of this machine to 2500 °F (1370°C), a small efficiency gain is achieved (about 2%). But when the pressure ratio is doubled to 16 at the 2500°F temperature (conditions representative of today's most advanced machines), efficiency jumps to about 35%.

Regenerators. Regenerative cycles make use of the exhaust heat discharged from a simple-cycle combustion turbine by means of a heat exchanger called a regenerator (Fig. 3). The regenerator conserves energy by heating high-pressure compressor discharge air with low-temperature turbine exhaust gases. Savings can be considerable. A regenerator designed with an 80% air-side effectiveness, for example, reduces fuel consumption approximately 30% over a comparable simple-cycle unit. This corresponds to an improvement of about 25% in cycle efficiency for some large combustion turbines.

Combined cycles. Combined cycles are also becoming popular for increasing the efficiency of gas turbines (Fig. 4). They are employed where steam is required in addition to the mechanical energy of the combustion turbine. The exhaust gas from a combustion turbine flows to a fired or unfired heat exchanger called a waste-heat boiler. Heat is captured by finned tubes which transfer the heat from the high-temperature exhaust gas to water flowing through the tubes, thus producing steam. The steam is used to propel a conventional steam turbine to produce electricity.

Working parts. The three major components common to all gas turbines are compressors, combustors, and turbines.

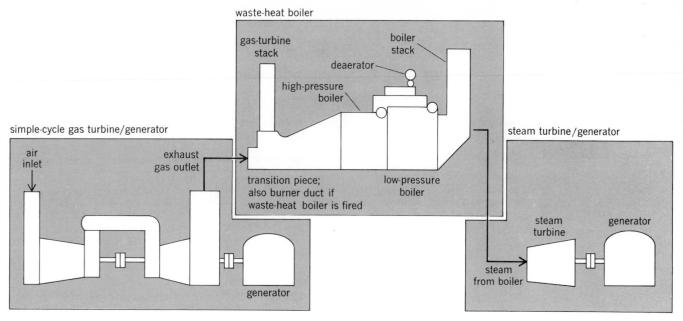

Fig. 4. Operation of a combined-cycle gas turbine.

Compressors. Efficient compression of a large-volume flow of air is the key to a successful gas turbine cycle. This has been achieved in two types of compressors, the axial-flow and the centrifugal. The axial-flow type is most common in industrial applications. In it, blades on the compressor rotor have airfoil shapes similar to those on aircraft propellers. They "bite" into the airflow, speed it up, and push it into the succeeding stationary-blade passages. These are shaped to form diffusers that slow up the incoming air to make it compress and pressurize itself by "catching up with the air ahead of it" in smooth fashion. In passing through both moving- and stationary-blade passages, the direction of airflow changes slightly. As a whole, the air follows a helical path through the compressor passages.

Blades of the axial compressor are twisted to accommodate the vortex-flow pattern of the air (higher circumferential velocity nearer the inner edge of the annular passage) and the different speeds of blade sections. Blade heights shrink as air flows through the compressor because of diminishing specific volume of the air at higher pressures. Blade shapes must be precise and clean to achieve high-compression efficiencies; surface fouling retards performance substantially.

Centrifugal compressors take in air at the center of the bladed impeller. The centrifugal force of the blades moves the air out radially at high speed into the stationary diffuser, where the slowing action pressurizes the air. Rotor blades and diffuser passages have varying shapes depending on the pressure characteristic desired with changing rotor speed.

Both types of compressors have pumping limits. When airflow drops below this limit at any rotor speed, the flow becomes unstable because of secondary circulation created in the blade passages. This causes the flow as a whole to pulse and may damage the blading. Axials are more liable to surge at low flows, so they are often bled at an intermediate stage during starting to induce a higher airflow and prevent damage. The air bled from an intermediate stage is wasted to the atmosphere.

Combustors. Combustors burn fuel in the pressurized air after it leaves the compressor. The lack of suitable high-temperature materials limits the exit temperatures to between 1900 and 2200°F (1040 and 1205°C) for base loading of the turbine. Since fuels burn with a flame at about 3000°F (1650°C), careful design is needed to ensure complete combustion. Only about 30% of the air takes part in oxidizing the fuel, with the remainder acting as a film coolant to preserve the combustor basket and as a diluent to cool the gas going to the turbine nozzles and buckets.

Combustors must be built to work with a low pressure drop of the working air, from compressor outlet to turbine inlet. To achieve this, adequate passage flow areas and a minimum number of changes in flow direction are required. These needs are met by arrangements varying from large single combustors to paralleled multiple combustors.

Ensuring continuous flame action depends largely on fuel nozzle design and proper circulation of combustion air to and through the burning zone. For turbine protection, the combustion gas must be mixed thoroughly with the remaining excess air to avoid large temperature differentials across the flow passages and consequent destructive thermal stresses.

Combustors have igniters to start combustion turbines from a cold condition. Once stable ignition is achieved, the igniters are removed from service.

Fuels for combustors range from natural gas to distillate and residual oils. Coal is still the subject of some experimental investigation since it presents serious problems with respect to disposal of the ash and the effect of ash on blading. Gasification or liquefaction of coal, however, probably will be the means by which these difficulties are over-

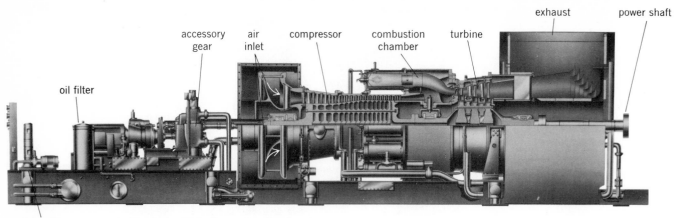

accessory
gear

air
inlet

compressor

combustion
chamber

turbine

exhaust

power shaft

oil filter

lube oil
pump

starting device (induction motor, diesel engine,
steam turbine or expansion turbine)

Fig. 5. Components of a simple-cycle single-shaft gas turbine, which is designed primarily for electric generation and produces more than 50,000 kW. (*General Electric Co.*)

come. *See* COAL GASIFICATION; COAL LIQUEFACTION.

Fuel must meet certain basic quality standards to ensure troublefree operation. Problems that can be caused by impurities in fuels include high-temperature corrosion, ash deposition on critical turbine components such as blades, and fuel system fouling. High-temperature corrosion is caused by some trace metallic elements found in fuel. They form compounds during combustion that melt on hot-gas-path components and dissolve the protective oxide coatings, leaving blade surfaces in a condition conducive to corrosion. Sodium, potassium, vanadium, and lead, in conjunction with sulfur, are the biggest offenders. Impurities may also form ash deposits which gradually decrease the cross-sectional area of the gas path. The rate of ash deposition depends on ash chemistry and other operating conditions, such as the quantity of additives used, particularly the vanadium-corrosion inhibitors. Fouling occurs in fuel systems when inorganic particulates, such as solid oxides, silicates, sulfides, and related compounds, are not adequately removed prior to reaching the turbine. These materials clog controls, fuel pumps, flow diverters, and nozzles.

Turbine. The turbine portion of the combustion turbine package, like a steam turbine, uses impulse and reaction principles, but because it works with lower pressure drops than the steam turbine, it has fewer stages, and less change in blade height from inlet to exhaust.

For best performance, the combustion turbine must work with high inlet-gas temperatures, and this poses severe materials design problems. By 1975, advanced alloys and coatings and claddings were commonly used to protect blades for turbine inlet temperatures below 2200°F (1040°C). It is expected that improvements in metallurgy will permit inlet-gas temperatures to increase to 2400°F (1315°C) by 1977–1978. Beyond that, ceramic materials or sophisticated cooling techniques probably will be required.

Advanced cooling techniques include film, water, liquid-metal, and transpiration cooling. Transpiration cooling is thought by some to show a great deal of promise for units at temperatures of 2500°F (1370°C) and above. In this technique, the cooling air effuses through pores in the airfoil into the boundary layer between the blade and the hot

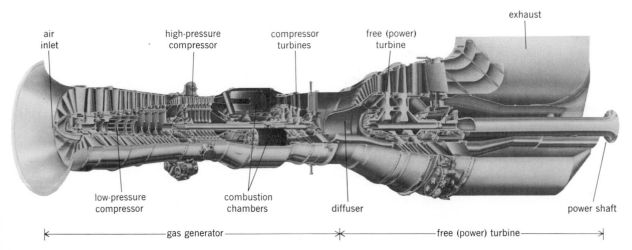

air
inlet

high-pressure
compressor

compressor
turbines

free (power)
turbine

exhaust

low-pressure
compressor

combustion
chambers

diffuser

power shaft

|← gas generator →|← free (power) turbine →|

Fig. 6. Components of a two-shaft aircraft-derivative gas turbine, which is used primarily for generating electricity in central stations. (*Turbo Power & Marine Systems Inc.*)

gases. Having cooled the airfoil structure, this air forms a relatively cool film that insulates the blade airfoil surface from the hot gas stream. Clean combustion gases and cooling air are essential to prevent air passages in the airfoil surface from becoming plugged.

Industrial designs. Combustion turbines used by industry are of two basic designs: the industrial and the aircraft-derivative. The so-called industrial combustion turbine was developed from the bed-plate up as a heavy-duty machine for industrial applications (Fig. 5). The aircraft-derivative combustion turbine is an adaptation of the aircraft jet engine to economical stationary service (Fig. 6). One distinguishing characteristic of aircraft-derivative engines is that they are of two-shaft design. That is, the external load—pump, compressor, or electric generator—is driven by turbine stages on a shaft separate from that carrying the machine's compressor and its turbine drive. This design offers advantages when the external load has varying power requirements. Industrial combustion turbines for electric generation service often are of single-shaft design—the compressor and the turbine driving the compressor, as well as the external load, are on one shaft; two-shaft units often drive pumps and compressors.

Environmental considerations. An outstanding feature of the combustion turbine is that it has little adverse impact on the environment. For example, the use of clean fuels precludes the formation of sulfur oxides and particulates. Oxides of nitrogen are formed, but these can be brought within the strictest limits imposed by regulatory bodies, with tight control of flame temperature in the combustor. Water cooling is one method used when clean air codes are extremely stringent, but redesigned combustors probably will eliminate the need for it in the future. Water pollution is not a problem since the combustion turbine requires no cooling water for operation. Noise is of some concern, but problems can be solved easily with soundproofed buildings, inlet silencers, and exhaust silencers. *See* ELECTRIC POWER GENERATION; TURBINE.

[ROBERT G. SCHWIEGER]

Bibliography: American Gas Association, *Gas Turbine Manual*, 1965; G. M. Dusinberre and J. C. Lester, *Gas Turbine Power*, 1958; J. W. Sawyer, *Gas Turbine Engineering Handbook*, 1966; B. G. A. Skrotzki, Gas turbines, *Power*, Special Report, December 1963; H. A. Sorensen, *Gas Turbines*, 1961; *Today's Technology*; *Gas Turbines*, a compilation of articles from *Power*, 1975.

Conduction (electricity)

Electrical conduction may be defined as the passage of electric charge. This can occur by a variety of processes.

In metals the electric current is carried by free electrons. These are not bound to any particular atom and can wander throughout the metal. In general, the conductivity of metals is higher than that of other materials. At very low temperatures certain metals become superconductors, possessing infinite conductivity. The free electrons are able to move through the crystal lattice without any resistance whatsoever.

In semiconductors (germanium, silicon, and so on) there are a limited number of free electrons and also "holes," which act as positive charges, available to carry current. The conductivity of semiconductors is much smaller than that of metals and, in contrast to most metals, increases with rising temperature.

Aqueous solutions of ionic crystals readily conduct electricity by means of the positive and negative ions present, for example, Na^+ and Cl^- in an ordinary solution of sodium chloride. Solid ionic crystals are themselves fair conductors. These crystals have sufficient lattice vacancies so that a few of the ions are able to migrate through the crystal under the influence of an applied electric field.

A strong electric field ionizes gas molecules, and thereby permits a flow of current through the gas in which the ions are the charge carriers. If sufficient ions are formed, there may be a spark.

Electric current can flow across a vacuum, for example, in a vacuum tube. The charge carriers are electrons emitted by the filament. The effective conductivity is low because of low available current densities at the normal temperatures of electron-emitting filaments.

[JOHN W. STEWART]

Conductor, electric

Metal wires, cables, rods, tubes, and bus-bars used for the purpose of carrying electric current. Although any metal assembly or structure can conduct electricity, the term conductor usually refers to the component parts of the current-carrying circuit or system.

Types of conductor. The most common forms of conductors are wires, cables, and bus-bars.

Wires. Wires employed as electrical conductors are slender rods or filaments of metal, usually soft and flexible. They may be bare or covered by some form of flexible insulating material. They are usually circular in cross section; for special purposes they may be drawn in square, rectangular, ribbon, or other shapes. Conductors may be solid or stranded, that is, built up by a helical lay or assembly of smaller solid conductors (Fig. 1).

Cables. Insulated stranded conductors in the larger sizes are called cables. Small, flexible, insulated cables are called cords. Assemblies of two or more insulated wires or cables within a common jacket or sheath are called multiconductor cables.

Bus-bars. Bus-bars are rigid, solid conductors and are made in various shapes, including rectangular, rods, tubes, and hollow squares. Bus-bars may be applied as single conductors, one bus-bar per phase, or as multiple conductors, two or more bus-bars per phase. The individual conductors of a multiple-conductor installation are identical.

Sizes. Most round conductors less than $1/2$ in. (1 in. = 2.54 cm) in diameter are sized according to the American wire gage (AWG)—also known as the Brown & Sharpe gage. AWG sizes are based on a simple mathematical law in which intermediate wire sizes between no. 36 (0.0050-in. diameter) and no. 0000 (0.4600-in. diameter) are formed in geometrical progression. There are 38 sizes between these two diameters. An increase of three gage sizes (for example, from no. 10 to no. 7) doubles the cross-sectional area, and an increase

CONDUCTOR, ELECTRIC

19-strand

7-strand

37-strand

Fig. 1. End views of stranded round conductor.

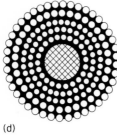

steel strands   aluminum strands
(a)

aluminum alloy   EC grade aluminum
(b)

(c)

(d)

Fig. 2. Aluminum conductors. (a) ACSR. (b) ACAR. (c) Compact concentric stranded conductor. (d) Expanded-core concentric stranded conductor.

of six gage sizes doubles the diameter of the wire.

Sizes of conductors greater than no. 0000 are usually measured in terms of cross-sectional area. Circular mil (cmil) is usually used to define cross-sectional area and is a unit of area equal to the area of a circle 1 mil (0.001 in.) in diameter.

Wire lengths are usually expressed in units of feet or miles in the United States. Bus-bar sizes are usually defined by their physical dimensions — height and width in inches or fractions of an inch, and length in feet.

Materials. Most wires, cables, and bus-bars are made from either copper or aluminum. Copper, of all the metals except silver, offers the least resistance to the flow of electric current. Both copper and aluminum may be bent and formed readily and have good flexibility in small sizes and in stranded constructions. Typical conductors are shown in Fig. 1.

Aluminum, because of its higher resistance, has less current-carrying capacity than copper for a given cross-sectional area. However, its low cost and light weight (only 30% that of the same volume of copper) permit wide use of aluminum for bus-bars, transmission lines, and large insulated-cable installations.

Metallic sodium conductors were used in 1965 on a trial basis for underground distribution insulated for both primary and secondary voltages. Sodium cable offered light weight and low cost for equivalent current-carrying rating compared with other conductor metals. Because of marketing problems and a few safety problems — the metal is reactive with water — the use of this cable was abandoned temporarily.

For overhead transmission lines where superior strength is required, special conductor constructions are used. Typical of these are aluminum conductors, steel reinforced (ACSR), a composite construction of electrical-grade aluminum strands surrounding a stranded steel core. Other constructions include stranded, high-strength aluminum alloy and a composite construction of aluminum strands around a stranded high-strength aluminum alloy core (ACAR).

For extra-high-voltage (EHV) transmission lines, conductor size is often established by corona performance rather than current-carrying capacity. Thus special "expanded" constructions are used to provide a large circumference without excessive weight. Typical constructions use helical lays of widely spaced aluminum strands around a stranded steel core. The space between the expanding strands is filled with paper twine, and outer layers of conventional aluminum strands are applied. In another construction the outer conductor stranding is applied directly over lays of widely separated helical expanding strands, without filler, leaving substantial voids between the stranded steel core and the closely spaced outer conductor layers. Diameters of 1.6 to 2.5 in. are typical. For lower reactance, conductors are "bundled," spaced 6–18 in. apart, and paralleled in groups of two, four, or more per phase. Figure 2 shows views of typical aluminum-conductor–steel-reinforced and expanded constructions.

Bare conductors. Bare wires and cables are used almost exclusively in outdoor power transmission and distribution lines. Conductors are

supported on or from insulators, usually porcelain, of various designs and constructions, depending upon the voltage of the line and the mechanical considerations involved. Voltages as high as 765 kV are in use, and research has been undertaken into the use of ehv transmission lines, with voltages as high as 1500 kV.

Bare bus-bars are used extensively in outdoor substation construction, in switchboards, and for feeders and connections to electrolytic and electroplating processes. Where dangerously high voltages are carried, the use of bare bus-bars is usually restricted to areas which are accessible only to authorized personnel. Bare bus-bars are supported on insulators of a design which is suitable for the voltage.

Insulated conductors. Insulated electric conductors are provided with a continuous covering of flexible insulating material. A great variety of insulating materials and constructions has been developed to serve particular needs and applications. The selection of an appropriate insulation depends upon the voltage of the circuit, the operating temperature, the handling and abrasion likely to be encountered in installation and operation, environmental considerations such as exposure to moisture, oils, or chemicals, and applicable codes and standards.

Magnet wires, used in the windings of motors, solenoids, transformers, and other electromagnetic devices, have relatively thin insulations, usually of enamel or cotton or both. Magnet wire is manufactured for use at temperatures ranging from 105 to 200°C.

Conductors in buildings. Building wires and cables are used in electrical systems in buildings to transmit electric power from the point of electric service (where the system is connected to the utility lines) to the various outlets, fixtures, and utilization devices. Building wires are designed for 600-volt operation but are commonly used at utilization voltages substantially below that value, typically 120, 240, or 480 volts. Insulations commonly used include thermoplastic, natural rubber, synthetic rubber, and rubberlike compounds. Rubber insulations are usually covered with an additional jacket, such as fibrous braid or polyvinyl chloride, to resist abrasion. Building wires are grouped by type in several application classifications in the National Electrical Code.

Classification is by a letter which usually designates the kind of insulation and, often, its application characteristics. For example, Type R indicates rubber or rubberlike insulation. TW indicates a thermoplastic, moisture-resistant insulation suitable for use in dry or wet locations; THW indicates a thermoplastic insulation with moisture and heat resistance.

Other insulations in commercial use include silicone, fluorinated ethylene, propylene, varnished cambric, asbestos, polyethylene, and combinations of these.

Building wires and cables are also available in duplex and multiple-conductor assemblies; the individual insulated conductors are covered by a common jacket. For installation in wet locations, wires and cables are often provided with a lead sheath (Fig. 3).

For residential wiring, the common construc-

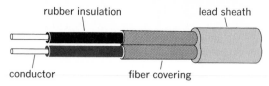

Fig. 3. Rubber-insulated, fiber-covered, lead-sheathed cable, useful for installation in wet places.

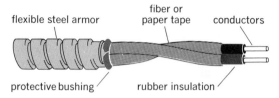

Fig. 4. Two-conductor armored cable.

tions used are nonmetallic-sheath cables, twin- and multiconductor assemblies in a tough abrasion-resistant jacket; and armored cable with twin-and multiconductor assemblies encased in a helical, flexible steel armor as in Fig. 4.

Power cables. Power cables are a class of electric conductors used by utility systems for the distribution of electricity. They are usually installed in underground ducts and conduits. Power cables are also used in the electric power systems of industrial plants and large buildings.

Power-cable insulations in common use include rubber, paper, varnished cambric, asbestos, and thermoplastic. Cables insulated with rubber (Fig. 5), polyethylene, and varnished cambric (Fig. 6) are used up through 69 kV, and impregnated paper to 138 kV. The type and thickness of the insulation for various voltages and applications are specified by the Insulated Power Cable Engineer Association (IPCEA).

Spaced aerial cable. Spaced aerial cable systems are employed for pole-line distribution at, typically, 5 to 15 kV, three-phase. Insulated conductors are suspended from a messenger, which may also serve as a neutral conductor, with ceramic or plastic insulating spacers, usually of diamond configuration. For 15 kV, a typical system may have conductors with 10/64-in. polyethylene insulation, 9-in. spacing between conductors, and 20 ft (6 m) between spacers.

High-voltage cable. High-voltage cable constructions and standards for installation and application are described by IPCEA. Insulations include (1) paper, solid type; (2) paper, low-pressure, gas-filled; (3) paper, low-pressure, oil-filled; (4) pipe cable, fully impregnated, oil-pressure; (5) pipe cable, gas-filled, gas-pressure; (6) rubber or plastic

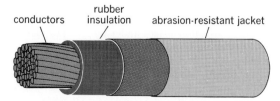

Fig. 5. Rubber-insulated power cable.

with neoprene or plastic jacket; (7) varnished cloth; (8) AVA and AVL (asbestos-varnished cloth).

Underground transmission cables are in service at voltages through 345 kV, and trial installations have been tested at 500 kV. Research has also been undertaken to develop cryoresistive and superconducting cables for transmitting power at high density. Preliminary designs have been prepared for a three-phase, 345-kV, 3660-MVA cable.

Enclosed bus-bar assemblies, or busways, are extensively used for service conductors and feeders in the electrical distribution systems of industrial plants and commercial buildings. They consist of prefabricated assemblies in standard lengths of bus-bars which are rigidly supported by solid insulation and enclosed in a sheet-metal housing.

Busways are made in two general types, feeder and plug-in. Feeder busways have no provision for taps or connections between the ends of the assembly. Low-reactance feeder busways are so constructed that conductors of different phases are in close proximity to minimize inductive reactance. Plug-in busways have provisions at intervals along the length of the assembly for the insertion of bus plugs.

Voltage drop in conductors. In electric circuits the resistance and (in ac circuits) the reactance of the circuit conductors result in a reduction in the voltage available at the load (except for capacitive loads). Since the line and load resistances are in series, the source voltage is divided proportionally. The difference between the source voltage and the voltage at the load is called voltage drop.

Electrical utilization devices of all kinds are designed to operate at a particular voltage or within a narrow range of voltages around a design center. The performance and efficiency of these devices are adversely affected if they are operated at a significantly lower voltage. Incandescent-lamp light output is lowered; fluorescent-lamp light output is lowered and starting becomes slow and erratic; the starting and pull-out torque of motors is seriously reduced.

Voltage drop in electric circuits caused by line resistance also represents a loss in power which appears as heat in the conductors. In excessive cases, the heat may rapidly age or destroy the insulation. Power loss also appears as a component of total energy use and cost. *See* COPPER LOSS.

Thus, conductors of electric power systems must be large enough to keep voltage drop at an acceptable value, or power-factor corrective devices—such as capacitors or synchronous condensers—must be installed. A typical maximum for a building wiring system for light and power is 3% voltage drop from the utility connection to any outlet under full-load condition.

[H. WAYNE BEATY]

Bibliography: Aluminum Association, *Aluminum Electrical Conductor Handbook*, September 1971; American National Standards Institute, *National Electrical Code*, ANSI C1, 1971; American National Standards Institute, *National Electrical Safety Code*, ANSI C2, 1973; Electric Power Research Institute, *Research Progress Report TD-3*, September 1975; D. Fink and J. Carroll (eds.), *Standard Handbook for Electrical Engineers*, 10th ed., 1968.

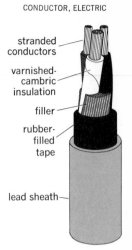

Fig. 6. Varnished cambric-insulated, lead-sheathed power cable, for voltages up to 28 kV.

Conservation of energy

The principle of conservation of energy states that energy cannot be created or destroyed, although it can be changed from one form to another. Thus in any isolated or closed system, the sum of all forms of energy remains constant. The energy of the system may be interconverted among many different forms—mechanical, electrical, magnetic, thermal, chemical, nuclear, and so on—and as time progresses, it tends to become less and less available; but within the limits of small experimental uncertainty, no change in total amount of energy has been observed in any situation in which it has been possible to ensure that energy has not entered or left the system in the form of work or heat. For a system that is both gaining and losing energy in the form of work and heat, as is true of any machine in operation, the energy principle asserts that the net gain of energy is equal to the total change of the system's internal energy. *See* THERMODYNAMIC PRINCIPLES.

Application to life processes. The energy principle as applied to life processes has also been studied. For instance, the quantity of heat obtained by burning food equivalent to the daily food intake of an animal is found to be equal to the daily amount of energy released by the animal in the forms of heat, work done, and energy in the waste products. (It is assumed that the animal is not gaining or losing weight.) Studies with similar results have also been made of photosynthesis, the process upon which the existence of practically all plant and animal life ultimately depends.

Conservation of mechanical energy. There are many other ways in which the principle of conservation of energy may be stated, depending on the intended application. Examples are the various methods of stating the first law of thermodynamics, the work-kinetic energy theorem, and the assertion that perpetual motion of the first kind is impossible. Of particular interest is the special form of the principle known as the principle of conservation of mechanical energy (kinetic E_k plus potential E_p) of any system of bodies connected together in any way is conserved, provided that the system is free of all frictional forces, including internal friction that could arise during collisions of the bodies of the system. Although frictional or other nonconservative forces are always present in any actual situation, their effects in many cases are so small that the principle of conservation of mechanical energy is a very useful approximation. Thus for a missile or satellite traveling high in space, the dissipative effects arising from such sources as the residual air and meteoric dust are so exceedingly small that the loss of mechanical energy $E_k + E_p$ of the body as it proceeds along its trajectory may, for many purposes, be disregarded. *See* ENERGY.

Mechanical equivalent of heat. The mechanical energy principle is very old, being directly derivable as a theorem from Newton's law of motion. Also very old are the notions that the disappearance of mechanical energy in actual situations is always accompanied by the production of heat and that heat itself is to be ascribed to the random motions of the particles of which matter is composed. But a really clear conception of heat as a form of energy came only near the middle of the 19th century, when J. P. Joule and others demonstrated the equivalence of heat and work by showing experimentally that for every definite amount of work done against friction there always appears a definite quantity of heat. The experiments usually were so arranged that the heat generated was absorbed by a given quantity of water, and it was observed that a given expenditure of mechanical energy always produced the same rise of temperature in the water. The resulting numerical relation between quantities of mechanical energy and heat is called the Joule equivalent, or mechanical equivalent of heat. The present accepted value is one 15° calorie = 4.1855 ± 0.0004 joules.

Conservation of mass-energy. In view of the principle of equivalence of mass and energy in the restricted theory of relativity, the classical principle of conservation of energy must be regarded as a special case of the principle of conservation of mass-energy. However, this more general principle need be invoked only when dealing with certain nuclear phenomena or when speeds comparable with the speed of light (3×10^{10} cm/sec) are involved.

If the mass-energy relation, $E = mc^2$, where c is the speed of light, is considered as providing an equivalence between energy E and mass m in the same sense as the Joule equivalent provides an equivalence between mechanical energy and heat, there results the relation, 1 kg = 9×10^{16} joules.

Laws of motion. The law of conservation of energy has been established by thousands of meticulous measurements of gains and losses of all known forms of energy. It is now known that the total energy of a properly isolated system remains constant. Some parts or particles of the system may gain energy but others must lose just as much. The actual behavior of all the particles, and thus of the whole system, obeys certain laws of motion. These laws of motion must therefore be such that the energy of the total system is not changed by collisions or other interactions of its parts. It is a remarkable fact that one can test for this property of the laws of motion by a simple mathematical manipulation that is the same for all known laws: classical, relativistic, and quantum mechanical.

The mathematical test is as follows. Replace the variable t, which stands for time, by $t + a$, where a is a constant. If the equations of motion are not changed by such a substitution, it can be proved that the energy of any system governed by these equations is conserved. For example, if the only expression containing time is $t_2 - t_1$, changing t_2 to $t_2 + a$ and t to $t_1 + a$ leaves the expression unchanged. Such expressions are said to be invariant under time displacement. When daylight-saving time goes into effect, every t is changed to $t + 1$ hr. It is unnecessary to make this substitution in any known laws of nature, for they are all invariant under time displacement.

Without such invariance laws of nature would change with the passage of time, and repeating an experiment would have no clear-cut meaning. In fact, science, as it is known today, would not exist.

[DUANE E. ROLLER/LEO NEDELSKY]

Bibliography: K. R. Atkins, *Physics*, 1965;

R. Benumof, *Concepts in Physics*, 1965; G. P. Harnwell and G. J. F. Legge, *Physics: Matter, Energy and the Universe*, 1967; E. M. Rogers, *Physics for the Inquiring Mind*, 1960; E. P. Wigner, Symmetry and conservation laws, *Phys. Today*, 17(3):34–40, March, 1964.

Conservation of resources

Conservation is concerned with the utilization of resources—the rate, purpose, and efficiency of use. This article emphasizes integrated conservation trends and policies.

Nature of resources. Universal natural resources are the land and soil, water, forests, grassland and other vegetation, fish and wildlife, rocks and minerals, and solar and other forms of energy. Some natural resources, such as metallic ores, coal, petroleum, and stone, are called fund or stock resources. They usually are referred to as nonrenewable natural resources because extraction from the stock depletes the usable quantity remaining and, even if some is being formed, the rate of formation is too slow for practical meaning. Other natural resources, such as living organisms and their products and solar and atomic radiation, are called flow resources. They usually are referred to as renewable natural resources because they involve organic growth and reproduction or because they are relatively quickly recycled or renewed in nature, as in the case of water in the hydrologic cycle and certain atmospheric phenomena. Some natural resources are difficult to fit into such a simple system. Soil, for example, is commonly thought of as renewable, as erosion and nutrient depletion can in some cases be rather quickly corrected, but if the upper layers of the soil are removed or bedrock is exposed, renewal may take thousands of years. Water also is commonly renewable, but rapid extraction of water by wells from deep aquifers may be equivalent to mining minerals.

Nature of conservation. Conservation has received many definitions because it has many aspects, concerns issues arising between individuals and groups, and involves private and public enterprise. Conservation receives impetus from the social conscience aware of an obligation to future generations and is viewed differently according to one's social and economic philosophy. To some extent, the meaning of conservation changes with the time and place. It is understood differently when approached from the natural sciences and technologies than when it is approached from the social sciences. Conservation for the petroleum engineer is largely the avoidance of waste from incomplete extraction; for the forester it may be sustained yield of products; and for the economist it is a change in the intertemporal distribution of use toward the future. In all cases, conservation deals with the judicious development and manner of use of natural resources of all kinds.

No definition of conservation exists that is satisfactory to all elements of the public and applicable to all resources. In its absence, an operational or functional definition can be arrived at by considering a series of conservation measures.

1. Preservation is the protection of nature from commercial exploitation to prolong its use for recreation, watershed protection, and scientific study. It is familiar in the establishment and protection of parks and reserves of many kinds.

2. Restoration, another widely familiar conservation measure, is essentially the correction of past willful and inadvertent abuses that have impaired the productivity of the resources base. This measure is familiar in modern soil and water conservation practices applied to agricultural land.

3. Beneficiation is the upgrading of the usefulness or quality of something, for instance, the utilization of ores that were formerly of uneconomic grade. Modern technology has provided many examples of this type of conservation.

4. Maximization includes all measures to avoid waste and increase the quantity and quality of production from resources.

5. Reutilization, in industry commonly called recycling, is the reuse of waste materials, as in the use of scrap iron in steel manufacture or of industrial water after it has been cleaned and cooled.

6. Substitution, an important conservation measure, has two aspects: the use of a common resource instead of a rare one when it serves the same end and the use of renewable rather than nonrenewable resources when conditions permit.

7. Allocation concerns the strategy of use—the best use of a resource. For many resources and products from them, the market price, as determined by supply and demand, establishes to what use a resource is put, but under certain circumstances the general welfare may dictate usage and resources may be controlled by government through the use of quotas, rationing, or outright ownership.

8. Integration in resources management is a conservation measure because it maximizes over a period of time the sum of goods and services that can be had from a resource or a resource complex, such as a river valley; this is preferable to maximizing certain benefits from a single resource at the expense of other benefits or other resources. This is one of the meanings of multiple use, and integration is a central objective of planning.

A generalized definition that fits many but not all meanings of conservation is "the maximization over time of the net social benefits in goods and services from resources." Although it is technologically based, conservation cannot escape socially determined values. There is an ethic involved in all aspects of conservation. Certain values are accepted in conservation, but they are the creation of society, not of conservation.

Conservation trends. There has been an important trend in conservation from an almost exclusive interest in production from individual natural resources to a balanced interest in that need and in the human resources and social goals for which resources are managed. The conservation movement originated with the realization that the economic doctrine of laissez-faire and quick profits—whether from forest, farm, or oil field—was resulting in tremendous waste that was socially harmful, even if it seemed to be good business. The beginning of the conservation movement stemmed from revulsion against destructive and wasteful lumbering. In time, the movement spread to farm and grazing lands, water, wildlife, and oil and gas. It was gradually learned that conservation management was good business in the long run.

The second trend in conservation was toward integrated management of resources. Students and administrators of resources — in colleges, business, and government — were discovering that the way one resource was handled affected the usability of others. Forestry broadened its interest to include forest influences on the watershed and the relations of the forest to wildlife and human uses for recreation. For many industries working directly with natural resources or their products, as in paper and chemical manufacturing and in coal mining, it was discovered that waste products produced costly and dangerous pollution of streams and air. Engineering on great river systems moved on from problems of flood hazard abatement and hydropower development to the design of structures with regard to fisheries and recreational values, and it was slowly realized that the way the land of a valley was managed affected erosion, siltation, water retention, and flooding.

Conservation in its third phase extends the ecological or integrated approach to resources management to include a more complete acceptance of the force of societal factors (such as economics, government, and social conditions) in determining resource management. There also is a closer examination of social costs and benefits and of human goals for which resources are employed.

Conservation policies. Individuals, corporations, and governmental entities at all levels have policies pertaining to resources. There could be a single national policy concerning a natural resource, such as water, only if the central government had complete authority over all governmental agencies and private enterprises. In the United States, however, many Federal and state departments, bureaus, agencies, and commissions have some authority over natural resources. There usually are several policies concerning the use and management of soil, water, forests, rangelands, fish, wildlife, minerals, and space, and they are not necessarily uniform as to objectives or program. One exception is the Atomic Energy Commission, which centralized authority in the creation and execution of policy regarding radioactive resources. This authority is granted by Congress and can be modified by congressional action.

Because each resource is capable of being utilized for a variety of goods and services, and because individuals, enterprises, and regions tend to value certain uses more than others, conflicts arise in the allocation of a resource among competing uses when the supply is inadequate for all desirable uses. The demand for water, for example, may be for rural domestic needs, urban and industrial needs, irrigation, power development, and recreation in its many aspects.

Because situations such as this exist or are potential with respect to every resource, two outstanding needs arise in the conservation of natural resources. Detailed information is needed concerning the location, quantity, and quality of each resource and of the interrelations among the resources. As a result, each agency of government needs to strengthen its own fact-finding, analysis, and programming machinery. The second need is for coordinating machinery that will improve the efficiency of allocation of resources so as to maximize the net private and public benefits from them. This is the goal of conservation policy, and it must, in a democratic society, be approached as far as possible through the voluntary cooperation of government and private enterprise. However, as pressures on society increase, whether because of actual depletion of resources or because of an increase in critical demand, as in war, authority to allocate resources tends to be delegated by the citizenry to the central government.

[STANLEY A. CAIN]

Bibliography: S. W. Allen and J. W. Leonard, *Conserving Natural Resources: Principles and Practice in a Democracy*, 2d ed., 1966; G. Borgstrom, *The Hungry Planet: The Modern World at the Edge of Famine*, 1965; C. H. Callison (ed.), *America's Natural Rsources*, 1957; S. V. Ciriacy-Wantrup, *Resource Conservation: Economics and Policies*, 1952; D. C. Coyle, *Conservation: An American Story of Conflict and Accomplishment*, 1957; S. T. Dana, *Forest and Range Policy*, 1956; R. F. Dasman, *Environmental Conservation*, 1959; E. S. Helfman, *Rivers and Watersheds in America's Future*, 1965; G.-H. Smith, *Conservation of Natural Resources*, 3d ed., 1965; E. W. Zimmermann, *World Resources and Industries*, rev. ed., 1951.

Cooling tower

A towerlike device in which atmospheric air (the heat receiver) circulates in direct or indirect contact with warmer water (the heat source), and the water is thereby cooled. A cooling tower may serve as the heat sink in a conventional thermodynamic process, such as refrigeration or steam power generation, or it may be used in any process in which water is used as the vehicle for heat removal, and when it is convenient or desirable to make final heat rejection to atmospheric air. Water, acting as the heat-transfer fluid, gives up heat to atmospheric air and, thus cooled, is recirculated through the system, affording economical operation of the process.

Basic types. Two basic types of cooling towers are commonly used. One transfers the heat from warmer water to cooler air mainly by an evaporation heat-transfer process and is known as the evaporative or wet cooling tower. The other transfers the heat from warmer water to cooler air by a sensible heat-transfer process and is known as the nonevaporative or dry cooling tower. These two basic types are sometimes combined, with the two cooling processes used in parallel, and are known as wet-dry cooling towers.

Cooling process. With the evaporative process, the warmer water is brought into direct contact with the cooler air. When the air enters the cooling tower, its moisture content is generally less than saturation; it emerges at a higher temperature and with a moisture content at or approaching saturation. Evaporative cooling takes place even when the incoming air is saturated, because as the air temperature is increased in the process of absorbing sensible heat from the water, there is also an increase in its capacity for holding water, and evaporation continues. The evaporative process accounts for about 65–75% of the total heat transferred; the remainder is transferred by the sensible heat-transfer process.

The wet-bulb temperature of the incoming air is

the theoretical limit of cooling. Cooling the water to within 5–20°F (1°F = 0.56°C) above wet-bulb temperature represents good practice. The amount of water evaporated is relatively small. Approximately 1000 Btu (1 Btu = 1055 J) is required to vaporize 1 lb (0.45 kg) of water at cooling tower operating temperatures. This represents a loss in water of approximately 0.75% of the water circulated for each 10°F cooling, taking into account the normal proportions of cooling by the combined evaporative and sensible heat-transfer processes. Drift losses may be as low as 0.01–0.05% of the water flow to the tower (recent performance of 0.001% has been achieved) and must be added to the loss of water by evaporation and losses from blowdown to account for the water lost from the system. Blowdown quantity is a function of makeup water quality, but it may be determined by regulations concerning its disposal. Its quality is usually expressed in terms of the allowable concentration of dissolved solids in the circulating cooling water and may vary from about two to six concentrations with respect to the dissolved solids content of the cooling water makeup.

With the nonevaporative process, the warmer water is separated from the cooler air by means of thin metal walls, usually tubes of circular cross section, but sometimes of elliptical cross section. Because of the low heat-transfer rates from a surface to air at atmospheric pressure, the air side of the tube is made with an extended surface in the form of fins of various geometries. The heat-transfer surface is usually arranged with two or more passes on the water side and a single pass, cross flow, on the air side. Sensible heat transfer through the tube walls and from the extended surface is responsible for all of the heat given up by the water and absorbed by the cooling air. The water temperature is reduced, and the air temperature increased. The nonevaporative cooling tower may also be used as an air-cooled vapor condenser and is commonly employed as such for condensing steam. The steam is condensed within the tubes at a substantially constant temperature, giving up its latent heat of vaporization to the cooling air, which in turn is increasing in temperature. The theoretical limit of cooling is the temperature of the incoming air. Good practice is to design nonevaporative cooling towers to cool the warm circulating water to within 25 to 35°F of the entering air temperature or to condense steam at a similar temperature difference with respect to the incoming air. Makeup to the system is to compensate for leakage only, and there is no blowdown requirement or drift loss.

With the combined evaporative-nonevaporative process, the heat-absorbing capacity of the system is divided between the two types of cooling towers, which are selected in some predetermined proportion and usually arranged so that adjustments can be made to suit operating conditions within definite limits. The two systems, evaporative and nonevaporative, are combined in a unit with the water flow arranged in a series relationship passing through the dry tower component first and the wet tower second. The airflow through the towers is in a parallel-flow relationship, with the discharge air from the two sections mixing before being expelled from the system. Since the evaporative process is employed as one portion of the cooling system,

Fig. 1. Counterflow natural-draft cooling tower at Trojan Power Plant in Spokane, WA. (*Research–Cottrel*)

Fig. 2. Cross-flow mechanical-draft cooling tower. (*Marley Co.*)

drift, makeup, and blowdown are characteristics of the combined evaporative-nonevaporative cooling tower system, generally to a lesser degree than in the conventional evaporative cooling towers.

Of the three general types of cooling towers, the evaporative tower as a heat sink has the greatest thermal efficiency but consumes the most water and has the largest visible vapor plume. When mechanical-draft cooling tower modules are arranged in a row, ground fogging can occur. This can be eliminated by using natural-draft towers, and can be significantly reduced with modularized mechanical-draft towers when they are arranged in circular fashion.

The nonevaporative cooling tower is the least efficient type, but it can operate with practically no consumption of water and can be located almost anywhere. It has no vapor plume.

The combined evaporative-nonevaporative cooling tower has a thermal efficiency somewhere between that of the evaporative and nonevaporative cooling towers. Most are of the mechanical-draft type, and the vapor plume is mitigated by mixing the dry warm air leaving the nonevaporative section of the tower with the warm saturated air leaving the evaporative section of the tower. This retards the cooling of the plume to atmospheric temperature; visible vapor is reduced and may be entirely eliminated. This tower has the advantage of flexibility in operation; it can accommodate variations in available makeup water or be adjusted to atmospheric conditions so that vapor plume formation and ground fogging can be reduced.

Evaporative cooling towers. Evaporative cooling towers are classified according to the means employed for producing air circulation through them: atmospheric, natural draft, and mechanical draft.

Atmospheric cooling. Some towers depend upon natural wind currents blowing through them in a substantially horizontal direction for their air supply. Louvers on all sides prevent water from being blown out of these atmospheric cooling towers, and allow air to enter and leave independently of wind direction. Generally, these towers are located broadside to prevailing winds for maximum sustained airflow.

Thermal performance varies greatly because it is a function of wind direction and velocity as well as wet- and dry-bulb temperatures. The normal loading of atmospheric towers is 1–2 gal/min of cooling water per square foot of cross section. They require considerable unobstructed surrounding ground space in addition to their cross-sectional area to operate properly. Because they need more area per unit of cooling than other types of towers, they are usually limited to small sizes.

Natural draft. Other cooling towers depend for their air supply upon the natural convection of air flowing upward and in contact with the water to be cooled. Essentially, natural-draft cooling towers are chimneylike structures with a heat-transfer section installed in their lower portion, directly above an annular air inlet in a counterflow relationship with the cooling air (Fig. 1), or with the heat-transfer section circumscribing the base of the tower in a cross-flow relationship with the cooling airflow (Fig. 2). Sensible heat absorbed by the air in passing over the water to be cooled increases the air temperature and its vapor content and thereby reduces its density so that the air is forced upward and out of the tower by the surrounding heavier atmosphere. The flow of air through the tower varies according to the difference in specific weights of the ambient air and the air leaving the

Fig. 3. Mechanical-draft cooling towers. (a) Conventional rectangular cross-flow evaporative induced-draft type; (b) circular cross-flow evaporative induced-draft type. (*Marley Co.*)

heat-transfer surfaces. Since the difference in specific weights generally increases in cold weather, the airflow through the cooling tower also increases, and the relative performance improves in reference to equivalent constant-airflow towers.

Normal loading of a natural-draft tower is 2–4 gal/(min)(ft²) of ground-level cross section. The natural-draft cooling tower does not require as much unobstructed surrounding space as the atmospheric cooling tower does, and is generally suited for both medium and large installations. The natural-draft cooling tower was first commonly used in Europe. Subsequently, a number of large installations were built in the United States, with single units 385 ft (1 ft = 0.3 m) in diameter by 492 ft high capable of absorbing the heat rejected from an 1100-MWe light-water-reactor steam electric power plant.

Mechanical draft. In cooling towers that depend upon fans for their air supply, the fans may be arranged to produce either a forced or an induced draft. Induced-draft designs are more commonly used than forced-draft designs because of lower initial cost, improved air-water contact, and less air recirculation (Fig. 3). With controlled airflow, the capacity of the mechanical-draft tower can be adjusted for economic operation in relation to heat load and in consideration of ambient conditions.

Normal loading of a mechanical-draft cooling tower is 2–6 gal/(min)(ft²) of cross section. The mechanical-draft tower requires less unobstructed surrounding space to obtain adequate air supply than the atmospheric cooling tower needs; however, it requires more surrounding space than natural-draft towers do. This type of tower is suitable for both large and small installations.

Nonevaporative cooling towers. Nonevaporative cooling towers are classified as air-cooled condensers and as air-cooled heat exchangers, and are further classified by the means used for producing air circulation through them. *See* HEAT EXCHANGER.

Air-cooled condensers and heat exchangers. Two basic types of nonevaporative cooling towers are in general use for power plant or process cooling. One type uses an air-cooled steam surface condenser as the means for transferring the heat rejected from the cycle to atmospheric cooling air. The other uses an air-cooled heat exchanger for this purpose. Heat is transferred from the air-cooled condenser, or from the air-cooled heat exchanger to the cooling air, by convection as sensible heat.

Nonevaporative cooling towers have been used for cooling small steam electric power plants since the 1930s. They have been used for process cooling since 1940; a complete refinery was cooled by the process in 1958. Until recently, the nonevaporative cooling tower was used almost exclusively with large steam electric power plants in Europe and South Africa. Interest in this type of tower is increasing in the United States, and a 330-MWe plant using nonevaporative cooling towers was nearing completion in Wyoming, as of 1976.

The primary advantage of nonevaporative cooling towers is that of flexibility of plant siting. There is seldom a direct economic advantage associated with the use of nonevaporative cooling tower systems in the normal context of power plant economics. They are the least efficient of the cooling systems used as heat sinks.

Cooling airflow. Each of the two basic nonevaporative cooling tower systems may be further classified with respect to type of cooling airflow. Both types of towers, the direct-condensing type and the heat-exchanger type, can be built as natural-draft or as mechanical-draft tower systems.

Design. The heat-transfer sections are constructed as tube bundles, with finned tubes arranged in banks two to five rows deep. The tubes are in a parallel relationship with each other and are spaced at a pitch slightly greater than their outside fin diameter, either in an in-line or a staggered pattern. For each section, two headers are used, with the tube ends secured in each. The headers may be of pipe or of a box-shaped cross section and are usually made of steel. The bundles are secured in an open metal frame.

The assembled tube bundles may be arranged in a V shape, with either horizontal or inclined tubes,

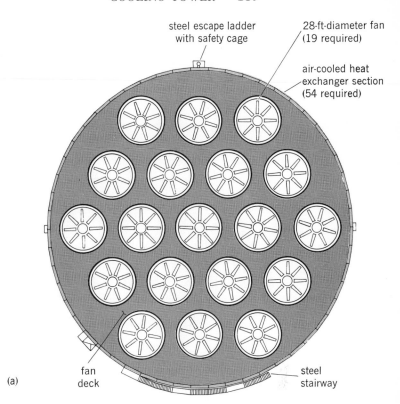

steel escape ladder with safety cage

28-ft-diameter fan (19 required)

air-cooled heat exchanger section (54 required)

(a)   fan deck   steel stairway

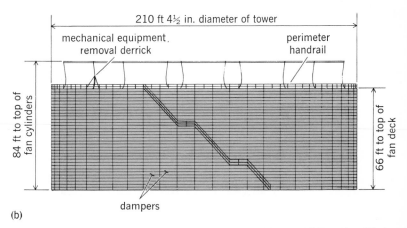

210 ft 4½ in. diameter of tower

mechanical equipment removal derrick   perimeter handrail

84 ft to top of fan cylinders

66 ft to top of fan deck

dampers

(b)

Fig. 4.   Multifan circular nonevaporative cooling tower. (*a*) Plan. (*b*) Elevation. (*Marley Co.*)

Fig. 5.   Wet-dry cooling tower at Atlantic Richfield Company. (*Marley Co.*)

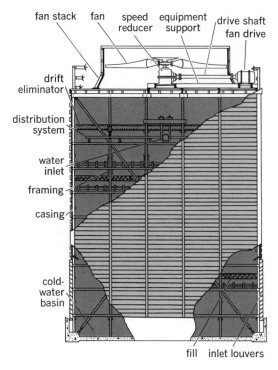

fan stack fan speed reducer equipment support drive shaft fan drive

drift eliminator

distribution system

water inlet

framing

casing

cold-water basin

fill inlet louvers

Fig. 6. Induced-draft counterflow cooling tower, showing the component parts.

or in an A shape, with the same tube arrangement. A similar arrangement may be used with vertical tube bundles. Generally, inclined tubes are shorter than horizontal ones. The inclined-tube arrangement is best suited to condensing vapor, the horizontal-tube arrangement best for heat-exchanger design. The A-shaped bundles are usually used with forced-draft airflow, the V-shaped bundles with induced-draft airflow. With natural-draft nonevaporative cooling towers there is no distinction made as to A- or V-shaped tube bundle arrangement; the bundles are arranged in a deck above the open circumference at the bottom of the tower, in a manner similar in principle to that used in the counterflow natural-draft evaporative cooling tower (Fig. 1). A natural-draft cooling tower with a vertical arrangement of tube bundles around the open circumference at the bottom has been built but has not been generally used.

Modularization of the bundle sections has become general practice with larger units; a recent installation of a circular module–type unit is shown in Fig. 4.

Combined evaporative-nonevaporative towers. The combination evaporative-nonevaporative cooling tower is arranged so that the water to be cooled first passes through a nonevaporative cooling section which is much the same as the tube bundle sections used with nonevaporative cooling towers

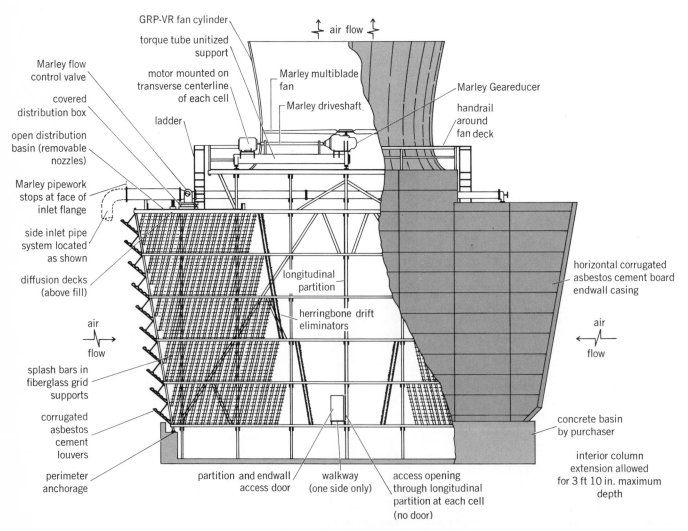

GRP-VR fan cylinder

torque tube unitized support

Marley flow control valve

motor mounted on transverse centerline of each cell

covered distribution box

open distribution basin (removable nozzles)

ladder

Marley pipework stops at face of inlet flange

side inlet pipe system located as shown

diffusion decks (above fill)

air flow

splash bars in fiberglass grid supports

corrugated asbestos cement louvers

perimeter anchorage

air flow

Marley multiblade fan

Marley driveshaft

Marley Geareducer

handrail around fan deck

longitudinal partition

herringbone drift eliminators

horizontal corrugated asbestos cement board endwall casing

air flow

concrete basin by purchaser

interior column extension allowed for 3 ft 10 in. maximum depth

partition and endwall access door

walkway (one side only)

access opening through longitudinal partition at each cell (no door)

Fig. 7. Transverse cross section of a cross-flow evaporative cooling tower.

Fig. 8. Nonevaporative cooling tower, Utrillas, Spain. (*GEA*)

of the heat-exchanger type. The hot water first passes through these heat exchangers, which are mounted directly above the evaporative cooling tower sections; the water leaving the nonevaporative section flows by force of gravity over the evaporative section.

The cooling air is divided into two parallel flow streams, one passing through the nonevaporative section, the other through the evaporative section to a common plenum chamber upstream of the induced-draft fans. There the two airflow streams are combined and discharged upward to the atmosphere by the fans. A typical cooling tower of this type is shown in Fig. 5.

In most applications of wet-dry cooling towers attempts are made to balance vapor plume suppression and the esthetics of a low silhouette in comparison with that of the natural-draft evaporative cooling tower. In some instances, these cooling towers are used in order to take advantage of the lower heat-sink temperature attainable with evaporative cooling when an adequate water supply is available, and to allow the plant to continue to operate, at reduced efficiency, when water for cooling is in short supply.

Tower components. The more important components of evaporative cooling towers are the supporting structure, casing, cold-water basin, distribution system, drift eliminators, filling and louvers, discharge stack, and fans (mechanical-draft towers only). The counterflow type is shown in Fig. 6; the cross-flow type is shown in Fig. 7.

Treated wood, especially Douglas fir, is the common material for atmospheric and small mechanical-draft cooling towers. It is used for structural framing, casing, louvers, and drift elimina-tors. Wood is commonly used as filler material for small towers. Plastics and asbestos cement are replacing wood to some degree in small towers and almost completely in large towers where fireproof materials are generally required. Framing and casings of reinforced concrete, and louvers of metal, usually aluminum, are generally used for large power plant installations. Fasteners for securing small parts are usually made of bronze, copper-nickel, stainless steel, and galvanized steel. Distribution system may be in the form of piping, in galvanized steel, or fiber glass–reinforced plastics; and they may be equipped with spray nozzles of noncorrosive material, in the form of troughs and weirs, or made of wood, plastic, or reinforced concrete. Structural framing may also be made of galvanized or plastic-coated structural steel shapes. Natural-draft cooling towers, especially in large sizes, are made of reinforced concrete. The cold-water basins for ground-mounted towers are usually made of concrete; wood or steel is usually used for roof-mounted towers. Fan blades are made of corrosion-resistant material such as monel, stainless steel, or aluminum; but most commonly fiber glass–reinforced plastic is used for fan blades.

The heat-transfer tubes used with nonevaporative cooling towers are of an extended-surface type usually with circumferential fins on the air side (outside). The tubes are usually circular in cross section, although elliptical tubes are sometimes used. Commonly used tube materials are galvanized carbon steel, ferritic stainless steel, and various copper alloys. They are usually made with wrapped aluminum fins, but steel fins are commonly used with carbon steel tubes and galvanized. Most designs employ a ratio of outside to inside surface of 20:25. Outside-diameter sizes range from ³/₄ to 1 ¹/₂ in. (1 in. = 2.54 cm).

Tube bundles are made with tube banks two to five rows deep. The tubes are in parallel relationship with each other and secured in headers, either of steel pipe or of weld-fabricated steel box headers. The tube bundle assemblies are mounted in steel frames which are supported by structural steel framework. The bundle assemblies may be arranged in a V pattern, requiring fans of the induced-draft type, or in an A pattern, requiring fans of the forced-draft type, Figure 8. Fans and louvers are similar to those described for evaporative cooling towers.

Nonevaporative cooling towers may also be used with natural airflow. In this case, the tube bundles are usually mounted on a deck within the tower and just above the top of the circumferential supporting structure for the tower. In this application, the tower has no cold-water basin, but otherwise it is identical with the natural-draft tower used for evaporative cooling with respect to materials of construction and design.

Performance. The performance of an evaporative cooling tower may be described by the generally accepted equation of F. Merkel, as shown below, where a = water-air contact area, ft²/ft³; h =

$$\frac{KaV}{L} = \int_{T_2}^{T_1} \frac{dT}{h'' - h}$$

enthalpy of entering air, Btu/lb; h'' = enthalpy of leaving air, Btu/lb; K = diffusion coefficient, lb/(ft²)(hr); L = water flow rate, lb/(hr)(ft²); T = water temperature, °F; T_1 = inlet water temperature, °F; T_2 = outlet water temperature, °F; and V = effective volume of tower, ft³/ft² of ground area. The Merkel equation is usually integrated graphically or by Simpson's rule.

The performance of a nonevaporative cooling tower may be described by the generally accepted equation of Fourier for steady-state unidirectional heat transfer, using the classical summation of resistances formula with correction of the logarithmic temperature difference for cross-counterflow design in order to calculate the overall heat transfer. It is usual practice to reference the overall heat-transfer coefficient to the outside (finned) tube surface.

Evaluation of cooling tower performance is based on cooling of a specified quantity of water through a given range and to a specified temperature approach to the wet-bulb or dry-bulb temperature for which the tower is designed. Because exact design conditions are rarely experienced in operation, estimated performance curves are frequently prepared for a specific installation, and provide a means for comparing the measured performance with design conditions.

[JOSEPH F. SEBALD]

Core drilling

The boring of a hole in the earth by drilling a circular groove and leaving a central core. It is usually done to obtain the core as a sample for visual examination and to test its physical and chemical properties. But core drilling is also done for making the hole itself whenever it is more economical to drill the annular groove and remove the core

Diamond drill core and diamond bit.

than to drill the entire cross-sectional area of the hole.

The greatest use for core drilling is in prospecting for mineral deposits. It is usually done with a diamond core drill, so called because it utilizes the extreme hardness of the diamond to cut the annular groove in rock. The cores carved out by the diamond bit hundreds of feet in the earth are brought to the surface as true and unaltered specimens of the rocks and their mineral contents (see illustration). Diamond core drilling is also widely used in civil engineering work to determine the strength of the rocks that must support the tremendous weights of buildings, bridges, and dams, and to determine how expensive it will be to excavate foundations and highway cuts.

In shaft sinking by the core-drill method, and in core drilling other large-diameter holes, chilled steel shot is used as a cutting medium instead of diamonds. Other cutting mediums sometimes used are hard alloy-steel teeth, sintered carbides, and corundum grit. *See* BORING AND DRILLING (MINERAL).

[FRANK C. STURGES]

Cracking

A process used in the petroleum industry to reduce the molecular weight of hydrocarbons by breaking molecular bonds. Cracking is carried out by thermal, catalytic, or hydrocracking methods. Increasing demand for gasoline and other middle distillates relative to demand for heavier fractions makes cracking processes important in balancing the supply of petroleum products.

Thermal cracking depends on a free-radical mechanism to cause scission of hydrocarbon carbon-carbon bonds and a reduction in molecular size, with the formation of olefins, paraffins, and some aromatics. Side reactions such as radical saturation and polymerization are controlled by regulating reaction conditions. In catalytic cracking, carbonium ions are formed on a catalyst surface, where bond scissions, isomerizations, hydrogen exchange, and so on, yield lower olefins, isoparaffins, isoolefins, and aromatics. Hydrocracking, a relative newcomer to the industry, is based on catalytic formation of hydrogen radicals to break carbon-carbon bonds and saturate olefinic bonds. Hydrocracking converts intermediate- and high-boiling distillates to middle distillates, high in paraffins and low in cyclics and olefins. Hydrocracking also causes hydrodealkylation of alkyl-aryl components in heavy reformate to produce benzene and naphthalene. *See* HYDROCRACKING.

Thermal cracking. This is a process in which carbon-to-carbon bonds are severed by the action of heat alone. It consists essentially in the heating of any fraction of petroleum to a temperature at which substantial thermal decomposition takes place through a thermal free-radical mechanism followed by cooling, condensation, and physical separation of the reaction products.

There are a number of refinery processes based primarily upon the thermal cracking reaction. They differ primarily in the intensity of the thermal conditions and the feedstock handled.

Visbreaking is a mild thermal cracking operation (850–950°F; 454–510°C) where only 20–25% of the residuum feed is converted to mid-distillate and lighter material. It is practiced to reduce the volume of heavy residuum which must be blended with low-grade fuel oils.

Thermal gas-oil or naphtha cracking is a more severe thermal operation (950–1100°F; 510–593°C) where 45% or more of the feed is converted to lower molecular weight. Attempts to crack residua under these conditions would coke the furnace tubes.

Steam cracking is an extremely severe thermal cracking operation (1100 to 1400°F; 593–760°C) in which steam is used as a diluent to achieve a very low hydrocarbon partial pressure. Primary products desired are olefins such as ethylene and butadiene.

Fluid coking is a thermal operation where the residuum is converted fully to gas-oil products boiling lower than 950°F (510°C) and coke. The thermal conversion is carried out on the surface of a fluidized bed of coke particles.

Delayed coking is a thermal cracking operation wherein a residuum is heated and sent to a coke drum, where the liquid has an infinite residence time to convert to lower-molecular-weight hydrocarbons which distill out of the coke drum, and to coke which remains in the drum and must be periodically removed.

In fluid coking and delayed coking, there is total conversion of the very heavy high-boiling end of the residuum feed.

Although there are many variations of visbreaking and thermal cracking, most commonly a feedstock that boils at higher temperatures than gasoline is pumped at inlet pressures of 75–1000 psig (1 psi = 6895 Pa) through steel tubes so placed in a furnace as to allow gradual heating of the coil to temperatures in the range 850–1100°F (454–593°C). The flow rate is controlled to provide sufficient time for the required cracking to lighter

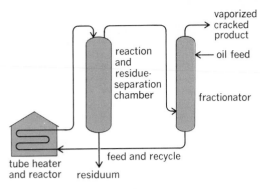

Fig. 1. Thermal cracking unit.

products; the time may be extended by subsequently passing the hot products through a reaction chamber that is maintained at a high temperature. To achieve optimum process efficiency, part of the overhead product ordinarily is returned to the cracking unit for further cracking (Fig. 1).

Crude oils differ in their compositions, both in molecular weight and molecular type of hydrocarbon. Since refiners must make products in harmony with market demand, they often need to alter the molecular structure of the hydrocarbons. The cracking of heavy distillates and residual oils increases the yields of gasoline and the light intermediate distillates used as diesel fuels and domestic heating oils, as well as providing low-molecular-weight olefins needed for the manufacture of chemicals and polymers.

Beginning in 1912, thermal cracking proved for many years to be eminently suitable for this purpose. During the period 1920–1940, more efficient automobile engines of higher compression ratios were developed. These engines required higher-octane-number gasolines, and thermal cracking operations in the United States were expanded to meet this need. Advantageously, thermal cracking reactions produce olefins and aromatics, leading to gasolines generally of higher octane number than those obtained by simple distillation of the same crude oils. The general nature of the hydrocarbon products and the basic mechanism of thermal cracking is well described by the free-radical theory of the pyrolysis of hydrocarbons.

In the early 1930s, the petrochemical industry began its growth. Olefinic gases from thermal cracking operations, especially propylene and butylenes, were used as the chief raw materials for the production of aliphatic chemicals. Simultaneously, practical catalytic processes were invented for polymerizing propylene and butylenes to gasoline components, and for dimerizing and hydrogenating isobutylene to isooctane (2,2,4-trimethylpentane), the prototype 100-octane fuel. Just prior to World War II, the alkylation of light olefins with isoparaffins to produce unusually high-octane gasoline components was discovered and extensively applied for military aviation use. These developments resulted in intense engineering efforts to bring thermal cracking operations to maximum efficiency, as exemplified by a number of commercial processes made available to the industry.

Since World War II, thermal cracking has been

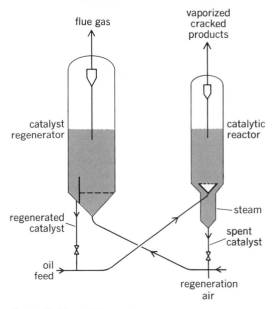

Fig. 2. Fluid catalytic cracking unit.

largely supplanted by catalytic cracking, both for the manufacture of high-octane gasoline and as a source of light olefins. It is, however, still used widely for the mild cracking of heavy residues to reduce their viscosities and for final cracking of gas oils derived from catalytic cracking.

Since carbon-to-hydrogen bonds are also severed in the course of thermal cracking, two hydrogen atoms can be removed from adjacent carbon atoms in a saturated hydrocarbon, producing molecular hydrogen and an olefin. This reaction prevails in ethane and propane cracking to yield ethylene and propylene, respectively. Methane cracking is a unique case wherein molecular hydrogen is obtained as a primary product and carbon as a coproduct. These processes generally operate at low pressures and high temperatures and in some cases utilize regenerative heating chambers lined with firebrick, or equipment through which preheated refractory pebbles continuously flow. Such conditions also favor the production of aromatics and diolefins from normally liquid feedstocks and are applied commercially to a limited extent despite relatively low yields of the desired products.

Catalytic cracking. This is the major process used throughout most of the world oil industry for the production of high-octane quality gasoline by the conversion of intermediate- and high-boiling petroleum distillates to lower-molecular-weight products. Oil heated to within the lower range of thermal cracking temperatures (850–1025°F; 454–551°C) reacts in the presence of an acidic inorganic catalyst under low pressures (10–35 psig). Gasoline of much higher octane number is obtained than from thermal cracking, a principal reason for the widespread adoption of catalytic cracking. All nonvolatile carbonaceous materials are deposited on the catalyst as coke and are burned off during catalyst regeneration.

In contrast to thermal cracking, residual oils are not generally processed, because excessive amounts of coke are deposited on the catalyst, and inorganic components of these oils contaminate the catalyst. Feedstocks usually are restricted to distillates boiling above gasoline.

Catalytic cracking, as conceived by E. J. Houdry in France, reached commercial status in 1936 after extensive engineering development by American oil companies. In its first form, the process used a series of fixed beds of catalyst in large steel cases. Each of these alternated between oil cracking and catalyst-coke burning at intervals of about 10 min and had suitable provision for heat and temperature control.

Successful operation led to major engineering improvements, and the goals of much improved efficiency, enlarged capacity, and ease of operation were achieved by two different systems. One employs a moving bed of small pellets or beads of catalyst traveling continuously through the oil-cracking vessel and subsequently through a regeneration kiln. The beads are lifted mechanically or by air to the top of the structure to flow down through the vessels again. This process has two commercially engineered embodiments, the Thermofor (or TCC) and the Houdriflow processes, which are similar in general process arrangements.

These moving-bed processes were limited in size and are now technically obsolete. They are being replaced by another type of unit, a fluid solids unit, as dictated by economic considerations, and no moving-bed catalytic cracking units are being constructed. In the fluid solids unit, a finely divided powdered catalyst is transported between oil-cracking and air-regeneration vessels in a fluidized state by gaseous streams in a continuous cycle. This system employs the principle of balanced hydrostatic heads of fluidized catalyst between the two vessels. Catalyst is moved by injecting heated oil vapors into the transport line from the regenerator to the reactor, and by injecting air into the transport line the reactor stripper to the regenerator. Large amounts of catalyst can be moved rapidly; cracking units of total oil intake as great as 180,000 bbl/day (28,800 m³/day) are in operation (Fig. 2).

In both the moving-bed and fluidized systems, the circulating catalyst provides the cracking heat. Coke deposited on the catalyst during cracking is burned at controlled air rates during regeneration; heat of combustion is converted largely to sensible heat of the catalyst, which supplies the endothermic heat of cracking in the reaction vessel.

Gasoline of 90–95 research octane number without tetraethyllead is rather uniformly produced by catalytic cracking of fractions from a wide variety of crude oils, compared with 65–80 research octane number via thermal cracking, the latter figures varying with crude oil source.

Although the primary objective of catalytic cracking is the production of maximum yields of gasoline concordant with efficient operation of the process, large amounts of normally gaseous hydrocarbons are produced at the same time. The gaseous hydrocarbons include propylene and butylenes, which are in great demand for chemical manufacture. Isobutane and isopentane are also produced in large quantities and are valuable for the alkylation of olefins, as well as for directly blending into gasoline as high-octane components.

The other chief product is the material boiling above gasoline, designated as catalytically cracked gas oil. It contains hydrocarbons relatively resistant to further cracking, particularly polycyclic aromatics. The lighter portion may be used di-

Representative yield structures for three different processing objectives in catalytic cracking

Process objective	Light gases	Gasoline	Light cycle oil
Feed	Light gas oil	Gas oil	Gas oil
Reactor temperature, °F	990	950–990	890–900
Light gases, wt %	4.5	2.8	1.6
Propane/propylene, vol %	15	10.0	7.5
Butane/butylene, vol %	22	16.4	11.2
Gasoline, vol %	46	69.5	32.6
Catalytic diesel oil, vol %	18	10	43.6
Bottom, vol %	5	5	5

rectly or blended with straight-run and thermally cracked distillates of the same boiling range for use as diesel and heating oils. Part of the heavier portion is recycled with fresh feedstock to obtain additional conversion to lighter products. The remainder is withdrawn for blending with residual oils so as to reduce the viscosity of heavy fuel, otherwise it is subjected to a final step of thermal cracking.

Thus, the catalytic cracking process is used in refineries to shift the production of products to match swings in market demand. It can process a wide variety of feeds to different product compositions. For example, light gases, gasoline, or diesel oil can be emphasized by varying process conditions, feedstocks, and boiling range of products as shown in the table.

To account for the difference between the product compositions obtained by catalytic and thermal cracking, the mechanism of cracking over acidic catalysts has been investigated intensively. In thermal cracking, free radicals are reaction intermediates, and the products are determined by their specific decomposition patterns. In contrast, catalytic cracking takes place through ionic intermediates, designated as carbonium ions (positively charged free radicals) generated at the catalyst surface. Although there is a certain parallelism between the modes of cracking of free radicals and carbonium ions, the latter undergo rapid intramolecular rearrangement reactions prior to cracking. This leads to more highly branched hydrocarbon structures than those from thermal cracking, and to important differences in the molecular weight distribution of the cracked products. Furthermore, the cracked products undergo much more extensive secondary reactions in the presence of the catalyst.

The catalytic cracking mechanism also favors the production of aromatics in the gasoline boiling range; these reach quite high concentrations in the higher-boiling portion. This characteristic, together with the copious production of branched aliphatic hydrocarbons especially in the lower-boiling portion, is largely responsible for the high octane rating of catalytically cracked gasoline.

Cracking catalysts must have two essential properties: (1) a chemical composition capable of maintaining a high degree of acidity, preferably as readily available hydrogen ions (protons); and (2) a physical structure of high porosity (high surface area). Mechanical durability is also necessary for industrial use.

Cracking catalysts are essentially silica-alumina compositions. A dramatic improvement in catalytic unit performance occurred with the switch from acid-treated clays (montmorillonite or kaolinite) to synthetic silica-aluminas. After 1960, a new group of aluminosilicates, molecular sieve zeolites, have been introduced into the catalyst formulation. These crystalline materials (Fig. 3) have cracking activity 50 to 100 times the previous amorphous catalyst. They permit cracking to greater conversion levels, producing more gasoline, less coke, and less gas.

As the catalyst particles pass through the reactor regenerator system every 3 to 15 min they are gradually deactivated, through loss of surface area by the effect of heat and steam and contamination through the effects of heat and steam, and the particles are contaminated by the trace metallic components on the feedstocks, mainly nickel, vanadium, and copper. They also undergo mechanical attrition, and fines are lost in the reactor and regenerator gases. To compensate, fresh catalyst is added.

The new zeolite catalysts have resulted in considerable change in the process itself. The catalysts are more resistant to thermal degradation, and regenerator temperatures can be safely raised to the 1350°F (732°C) level. The carbon on regenerated catalyst is reduced to the 0.05 wt % level resulting in improved gasoline yields. In addition, all the carbon monoxide produced at lower generation temperatures (10% concentration in the regenerator flue gas) can be combusted in the regenerator, making for a more efficient recovery of combustion heat and reduced atmospheric pollution. The effluent CO concentration can be reduced to less than 0.05 vol %.

The high-activity zeolite catalyst has permitted units to be designed with all riser cracking (Fig. 4), wherein all the cracking reaction takes place in a

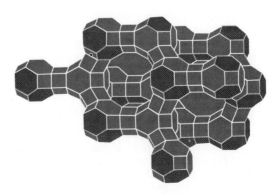

Fig. 3. Model of zeolite type Y.

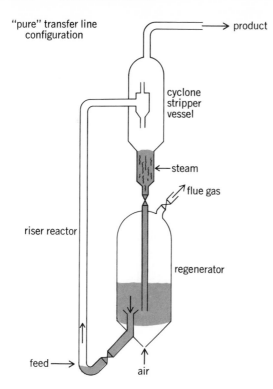

"pure" transfer line configuration

product

cyclone stripper vessel

steam

flue gas

riser reactor

regenerator

feed

air

Fig. 4. Riser catalytic cracker.

relatively dilute (less than 2 lb/ft³) catalyst suspension in a 2- or 5-sec residence time. No dense-bed (10–15 lb/ft³) cracking exists.

Many old units are being converted to riser crackers, and virtually all new units feature riser cracking. Some riser crackers are also provided with small dense beds to achieve an optimum yield pattern.

With added emphasis on protection of the environment, complex facilities are provided to remove pollutants from the effluent regeneration gases. Third-stage cyclone collectors, electrostatic precipitators, and scrubbers are used commercially to meet regulations.

To conserve energy in the process, flue gas expanders are used in the flue gas system to provide more than enough energy to compress the regeneration air.

With the coming emphasis on conservation of petroleum resources, catalytic cracking is assuming a more important role. The heavier products from a refinery will probably be replaced by coal-, shale-, or tar-sand-derived products. Thus, there will be a need for greater conversion of petroleum to gasoline. Emphasis will be on catalytic cracking since it is the cheapest major conversion process available. *See* ALKYLATION, PETROLEUM; PETROLEUM PROCESSING; REFORMING IN PETROLEUM REFINING.

[BERNARD S. GREENSFELDER/MOTT SOUDERS/ EDWARD LUCKENBACH]

Bibliography: F. H. Blanding, Reaction rates in catalytic cracking of petroleum, *I and EC*, 45: 1185–1197, June 1953; G. D. Hobson and W. Pohl *Modern Petroleum Technology*, pp. 278–344, 1973; K. A. Kobe and J. J. McKetta, Jr., *Advances in Petroleum Chemistry and Refining*, vols. 5 and 6, 1962; W. L. Nelson, *Petroleum Refinery Engineering*, pp. 759–818, 1958; M. Sittig, Catalytic cracking techniques in review, *Petrol. Refinery*, 37:263–316, 1952.

Critical mass

That amount of fissile material (U^{233}, U^{235}, or Pu^{239}) which permits a self-sustaining chain reaction. The critical mass ranges from about 950 g of U^{235} or a smaller amount of Pu^{239} for dissolved compounds through 16 kg for a solid metallic sphere of U^{235}, and up to hundreds of tons for some power reactors. It is increased by the presence of such neutron absorptive materials as admixed U^{238}, aluminum pipes for flow of coolant, and boron or cadmium control rods. It is reduced by a moderator, such as graphite or heavy water, which slows down the neutrons, inhibits their escape, and indirectly increases their chance to produce fission. *See* CHAIN REACTION, NUCLEAR; FISSION, NUCLEAR.   [JOHN A. WHEELER]

Current (electricity)

The net transfer of electric charge per unit time. It is usually measured in amperes. The passage of electric current involves a transfer of energy. Except in the case of superconductivity, a current always heats the medium through which it passes.

On the other hand, a stream of electrons or ions in a vacuum, which also may be regarded as an electric current, produces no local heating. Measurable currents range in magnitude from the nearly instantaneous 10^5 or so amperes in lightning strokes to values of the order of 10^{-16} amp, which occur in research applications.

All matter may be classified as conducting, semiconducting, or insulating, depending upon the ease with which electric current is transmitted through it. Most metals, electrolytic solutions, and highly ionized gases are conductors. Transition elements, such as silicon and germanium, are semiconductors, while most other substances are insulators.

Electric current may be direct or alternating. Direct current (dc) is necessarily unidirectional but may be either steady or varying in magnitude. By convention it is assumed to flow in the direction of motion of positive charges, opposite to the actual flow of electrons. Alternating current (ac) periodically reverses in direction.

Conduction current. This is defined as the transfer of charge by the actual motion of charged particles in a medium. In metals the current is carried by free electrons which migrate through the spaces between the atoms under the influence of an applied electric field. Although the propagation of energy is a very rapid process, the drift rate of the individual electrons in metals is only of the order of a few centimeters per second. In a superconducting metal or alloy the free electrons continue to flow in the absence of an electric field after once having been started. In electrolytic solutions and ionized gases the current is carried by both positive and negative ions. In semiconductors the carriers are the limited number of electrons which are free to move, and the "holes" which act as positive charges.

Displacement current. When alternating current traverses a condenser, there is no physical flow of charge through the dielectric (insulating material), but the effect on the rest of the circuit is as if there were a continuous flow. Energy can pass through the condenser by means of the so-called displacement current. James Clerk Maxwell introduced the concept of displacement current in order to make complete his theory of electromagnetic waves. *See* ALTERNATING CURRENT; CONDUCTION (ELECTRICITY); DIRECT CURRENT.

[JOHN W. STEWART]

Dam

A structure that bars or detains the flow of water in an open channel or watercourse. Dams are constructed for several principal purposes. Diversion dams divert water from a stream; navigation dams raise the level of a stream to increase the depth for navigation purposes (Fig. 1); power dams raise the level of a stream to create or concentrate hydrostatic head for power purposes; and storage dams store water for municipal and industrial use, irrigation, flood control, river regulation, recreation, or power production. A dam serving two or more purposes is called a multiple-purpose dam. Dams are commonly classified by the material from which they are constructed, such as masonry, concrete, earth, rock, timber, and steel. Most dams now are built either of concrete or of earth and rock.

Concrete dams. Concrete dams may be typed as gravity, arch, or buttress type. Gravity dams depend on weight for stability against overturning and for resistance to sliding on their foundations (Figs. 2 and 3). An arch dam may have a near-vertical face or, more usually, one that curves concave downstream (Figs. 4 and 5). The dam acts as an arch to transmit most of the horizontal thrust from the water pressure against the upstream face of the dam to the abutments of the dam. The buttress type of concrete dam includes the slab-and-buttress, or Ambursen, type; round- or diamond-head buttress type; multiple-arch type; and multiple-dome type. Buttress dams depend on the weight of the structure and of the water on the dam to resist overturning and sliding.

Forces acting on concrete dams. Principal forces acting on a concrete dam are (1) vertical forces from weight of the structure and vertical component of water pressure against the upstream and downstream faces of the dam, (2) uplift pressures under the base of the structure, (3) horizontal forces from the horizontal component of the water pressure against the upstream and downstream faces of the dam, (4) forces from earthquake accelerations in regions subject to earthquakes, (5) temperature stresses, (6) pressures from silt deposits and earth fills against the structure, and (7) ice pressures.

The uplift pressure under the base of a dam varies with the effectiveness of the foundation drainage system and with the perviousness of the foundation.

Earthquake loads are usually selected after consideration of the accelerations which may be expected at the site as indicated by the geology, proximity to major faults, and the earthquake history of the region. Conventionally, earthquake forces have been treated as static forces represent-

Fig. 1.   John Day Lock and Dam, looking upstream across the Columbia River at the Washington shore. In the foreground the navigation lock may be seen, beyond it the spillway dam, and then the powerhouse. The John Day multiple-purpose project boasts the highest single-lift navigation lock in the United States. (*U.S. Army Corps of Engineers*)

Fig. 2. Green Peter Dam, a concrete gravity type on the Middle Santiam River, Willamette River Basin, OR. Gate-controlled overflow-type spillway is constructed through crest of dam; powerhouse is at downstream toe of dam. (*U.S. Army Corps of Engineers*)

ing the effects of the acceleration of the dam itself and the hydrodynamic force produced against the dam by water in the reservoir. Such horizontal forces often are assumed to equal 0.05–0.10 the force of gravity, with a somewhat smaller vertical force. Dynamic analysis procedures have been developed which determine the structure's response to combined effects of the contemplated ground motion and the structure's dynamic properties.

Stresses resulting from temperature changes must be considered in analyzing arch dams. These stresses are usually disregarded in the design of concrete gravity dams, but must be controlled to acceptable limits by concreting and curing methods, discussed below.

Pressure from silt deposited in the reservoir against the dam is considered only after sedimentation studies indicate that it may be a significant factor. Backfill pressures are important where a concrete gravity dam ties into an embankment.

Ice pressure, applied at the maximum elevation at which the ice will occur in project operations, is considered when conditions indicate that it would be significant. The pressure, commonly assumed to be 10,000–20,000 lb per linear foot, results from the thermal expansion of the ice sheet and varies with the rate and magnitude of temperature rise and thickness of the ice.

Stability and allowable stresses. Stability of a concrete gravity dam is evaluated by analyzing the available resistance to overturning and sliding. To satisfy the former, the resultant of forces is required to fall within the middle third of the base

under normal load conditions. Sliding stability is assured by requiring available shear and friction resistance to be greater by a designated safety factor than the forces tending to produce sliding. The strengths used in computing resistance to sliding are based on investigation and tests of the foundation. Bearing strength of the foundation for a gravity dam is a controlling factor only for weak foundations or for high dams. Because an arch dam depends on the competency of the abutments, the rock bearing strength must be sufficient to provide an adequate safety factor for the compressive stresses, and the resistance to sliding along any weak surface must be great enough to provide an adequate safety factor.

Concrete stresses control the design of arch dams, but ordinarily not gravity dams. Stresses adopted for concrete arch dams are conservative. A safety factor of 4 on concrete compressive strength is commonly used for normal load conditions.

Concrete temperature control. Volume changes accompanying temperature changes in a concrete dam tend to cause the development of tensile stresses. A major factor in development of temperature changes within a concrete mass is the heat developed by chemical changes in the concrete after placement. Uncontrolled temperature changes can cause cracking which may endanger the stability of a dam, cause leakage, and reduce durability. Temperatures are controlled by using cementing materials having low heat of hydration, and by artificial cooling by precooling the concrete mix or circulating cold water through pipes embedded in the concrete or both.

Concrete dams are constructed in blocks, with the joints between the blocks serving as contraction joints (Fig. 6). In arch dams the contraction joints are filled with cement grout after maximum shrinkage has occurred to assure continuous bearing surfaces normal to the compressional forces set up in the arch when the water load is applied to the dam.

Quality control. During construction, continuing testing and inspection are performed to ensure that the concrete will be of required quality. Tests are also made on materials used in manufacture of the concrete, and concrete batching, mixing, transporting, placing, curing, and protection are continuously inspected.

Earth dams. Earth dams have been used for water storage since early civilizations. Improvements in earth-materials techniques, particularly the development of modern earth-handling equipment, have brought about a wider use of this type of dam, and today as in primitive times the earth embankment is the most common dam (Figs. 7 and 8). Earth dams may be built of rock, gravel, sand, silt, or clay in various combinations.

Most earth dams are constructed with an inner impervious core, with upstream and downstream zones of more pervious materials, sometimes including rock zones. Earth dams limit the flow of water through the dam by use of fine-grained soils. Where possible, these soils are formed into a relatively impervious core. When there is a sand or gravel foundation, the core may be connected to bedrock by a cutoff trench backfilled with compacted soil. If such cutoffs are not economically feasible because of the great depth of pervious

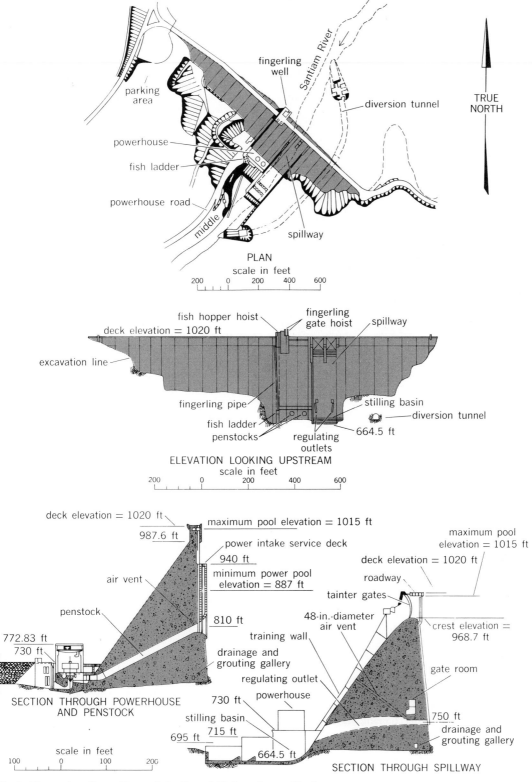

Fig. 3. Plan and sections of Green Peter Dam. *(U.S. Army Corps of Engineers)*

foundation soils, then the central impervious core is connected to a long horizontal upstream impervious blanket that increases the length of the seepage path. The impervious core is often encased in pervious zones of sand, gravel, or rock fill for stability. When there is a large difference in the particle sizes of the core and pervious zones, transition zones are required to prevent the core material from being transported into the pervious zones by seeping water. In some cases where pervious soils are scarce, the entire dam may be a homogeneous fill of relatively impervious soil. Downstream pervious drainage blankets are provided to collect seepage passing through, under,

Fig. 4.   East Canyon Dam, a thin-arch concrete structure on the East Canyon River, UT. Note uncontrolled overflow-type spillway through crest of dam at right center of photograph. (*U.S. Bureau of Reclamation*)

and around the abutments of the dam.

Materials can be obtained from required excavations for the dam and appurtenances or from borrow areas. Rock fill is generally used when large quantities of rock are available from required excavation or when soil borrow is scarce.

Earth-fill embankment is placed in layers and compacted by sheepsfoot rollers or heavy pneumatic-tire rollers. Moisture content of silt and clay soils is carefully controlled to facilitate optimum compaction. Sand and gravel fills are compacted in slightly thicker layers by pneumatic-tire rollers, vibrating steel drum rollers, or placement equipment. The placement moisture content of pervious fills is less critical than for silts and clays. Rock fill usually is placed in layers 1–3 ft deep and is compacted by placement equipment and vibrating steel drum rollers.

Spillways. A spillway releases water in excess of storage capacity so that the dam and its foundation are protected against erosion and possible failure. All dams must have a spillway, except small ones where the runoff can be safely stored in the reservoir without danger of overtopping the dam. Ample spillway capacity is of particular importance for large earth dams, which would be destroyed or severely damaged by being overtopped. Failure of a large dam could result in severe hazards to life and property downstream.

Types. Spillways are of two general types: the overflow type, constructed as an integral part of the dam; or the channel type, located as an independent structure discharging through an open chute or tunnel. Either type may be equipped with gates to control the discharge. Various control structures have been used for channel spillways, including the simple overflow weir, side-channel overflow weir, and drop or morning-glory inlet where the water flows over a circular weir crest and drops directly into a tunnel.

Unless the discharge end of a spillway is remote from the toe of the dam or erosion-resistant bedrock exists at shallow depths, some form of energy dissipator must be provided to protect the toe of the dam and the foundation from spillway discharges. For an overflow spillway the energy dissipator may be a stilling basin, a sloping apron downstream from the dam, or a submerged bucket. When a channel spillway terminates near the dam, it usually has a stilling basin. A flip bucket is used for both overflow and channel spillways when the flow can be deflected far enough downstream, usually onto rock, to prevent erosion at the toe of the dam or end of the spillway.

Gates. Several types of gates may be used to regulate and control the discharge of spillways (Fig. 9). Tainter gates are comparatively low in cost and require only a small amount of power for operation, being hydraulically balanced and of low friction. Drum gates, which are operated by reservoir pressure, are costly but afford a wide, unobstructed opening for passage of drift and ice over the gates. Vertical-lift gates of the fixed-wheel or roller type are sometimes used for spillway regulation, but are more difficult to operate than the others. Floating ring gates control the discharge of morning-glory spillways. Like the drum gate, this type offers a minimum of interference to the passage of ice or drift over the gate and requires no external power for operation.

Reservoir outlet works. These are used to regulate the release of water from the reservoir; they consist essentially of an intake and an outlet connected by a water passage, and are usually provided with gates. Outlet works usually have trashracks at the intake end to prevent clogging by debris. Bulkheads or stop logs are commonly provided to close the intakes so that the passages may be unwatered for inspection and maintenance. A stilling basin or other type of energy dissipator is usually provided at the outlet end.

Locations. Outlets may be sluices through concrete dams with control valves located in chambers in the dam or on the downstream end of the sluices, tunnels through the abutments of the dam, or cut-and-cover conduits extending along the foundation through an earth-fill dam. In the last case, the control valves are usually located within the dam or at the upstream end of the conduit, and special precautions must be taken to prevent leakage of water along the outside of the conduit.

Outlet control gates. Various gates and valves are used for regulating the release of water from reservoirs, including high-pressure slide gates, tractor gates (roller or wheel), and radial or tainter gates (Fig. 10); also needle valves of various kinds, butterfly valves, fixed cone dispersion valves, and cylinder or sleeve valves. They must be capable of operating, without excessive vibration and cavitation, at any opening and at any head up to the maximum to which they may be subjected. They also must be capable of opening and closing under the maximum operating head. Emergency gates generally are used upstream of the operating gates, where stored water is valuable, so that closure can be made if the service gate should fail to function.

The slide gate, which consists of a movable leaf that slides on a stationary seat, is the most commonly used control gate. The high-pressure slide

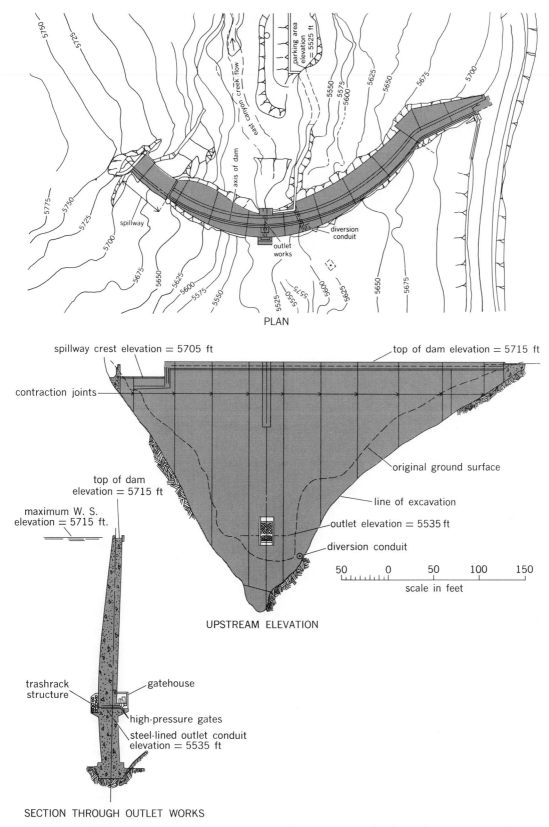

PLAN

spillway crest elevation = 5705 ft

top of dam elevation = 5715 ft

contraction joints

original ground surface

line of excavation

top of dam
elevation = 5715 ft

maximum W. S.
elevation = 5715 ft.

outlet elevation = 5535 ft

diversion conduit

50   0   50   100   150

scale in feet

UPSTREAM ELEVATION

trashrack
structure

gatehouse

high-pressure gates

steel-lined outlet conduit
elevation = 5535 ft

SECTION THROUGH OUTLET WORKS

Fig. 5. Plan and sections of East Canyon Dam. (*U.S. Bureau of Reclamation*)

gate is of rugged design, having corrosion-resisting metal seats on both the movable rectangular leaf and the fixed frame. This gate has been used for regulating discharges under heads of over 600 ft.

Provision of low-level outlet. The usual storage reservoir has low-level outlets near the elevation of the stream bed to enable release of all the stored water. Some power and multiple-purpose dams have relatively high-level dead storage pools and do not require low-level outlets for ordinary opera-

Fig. 6.   Block method of construction on a typical concrete gravity dam. *(U. S. Army Corps of Engineers)*

Fig. 7.   Aerial view of North Fork Dam, a combination earth and rock embankment on the North Fork of Pound River, VA. Channel-type spillway (left center) has simple overflow weir. *(U.S. Army Corps of Engineers)*

tion. In such a dam, provision of a capability for emptying the reservoir in case of an emergency must be weighed against the additional cost.

Penstocks. A penstock is a pipe that conveys water from a forebay, reservoir, or other source to a turbine in a hydroelectric plant. It is usually made of steel, but reinforced concrete and wood-stave pipe have also been used. Pressure rise and speed regulation must be considered in the design of a penstock.

Pressure rise, or water hammer, is the pressure change that occurs when the rate of flow in a pipe or conduit is changed rapidly. The intensity of this pressure change is proportional to the rate at which the velocity of the flow is accelerated or decelerated. Accurate determination of the pressure changes that occur in a penstock involves consideration of all operating conditions. For example, one important consideration is the pressure rise that occurs in a penstock when the turbine

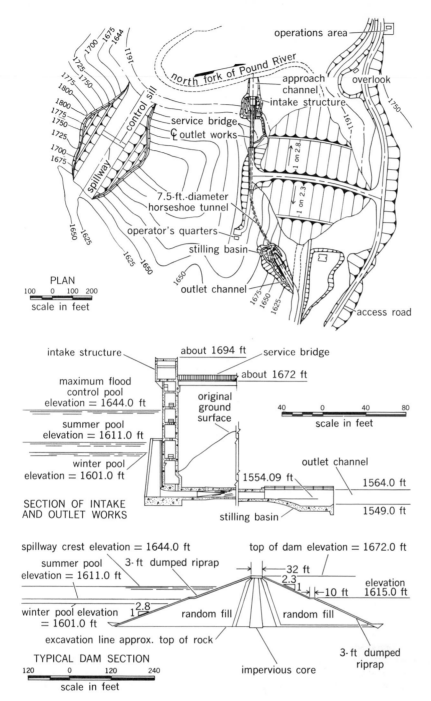

Fig. 8. Plan and sections of North Fork of Pound Dam. *(U.S. Army Corps of Engineers)*

wicket gates are closed subsequent to the loss of load.

Selection of dam site. This depends upon such factors as hydrologic, topographic, and geologic conditions; storage capacity of reservoir; accessibility; cost of lands and necessary relocations of prior occupants or uses; and proximity of sources of suitable construction materials. For a storage dam the objective is to select the site where the desired amount of storage can be most economically developed. Power dams must be located to develop the desired head and storage. For a diversion dam the site must be considered in conjunction with the location and elevation of the outlet

canal or conduit. Site selection for navigation dams involves special factors such as desired navigable depth and channel width, slope of river channel, natural river flow, amount of bank protection, amount of channel dredging, approach and exit conditions for tows, and locations of other dams in the system.

Unless topographic and geologic conditions for a proposed storage, power, or diversion dam site are satisfactory, hydrological features may need to be subordinated. Important topographic characteristics include width of the floodplain, shape and height of valley walls, existence of nearby saddles for spillways, and adequacy of reservoir rim to re-

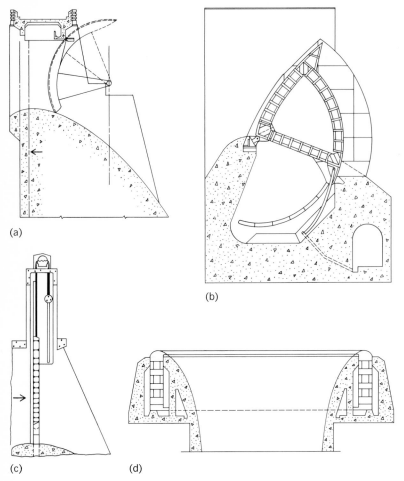

(a)

(b)

(c)   (d)

Fig. 9.   Spillway gates. (a) Tainter gate. (b) Drum gate. (c) Vertical lift gate. (d) Ring gate. (*U.S. Army Corps of Engineers and U.S. Bureau of Reclamation*)

tain impounded water. Controlling geologic conditions include the depth, classification, and engineering properties of soils and bedrock at the dam site, and the occurrence of sinks, faults, and major landslides at the site or in the reservoir area. The elevation of the groundwater table is also significant because it will influence the construction operations and suitability of borrow materials. The beneficial effect of reservoir water on groundwater recharge may become an important consideration, as well as the adverse effects on existing or potential mineral resources and developments that would be destroyed or require relocation at the site or within the reservoir.

Selection of type of dam. This is made on the basis of the estimated costs of various types. The most important factors are topography, foundation conditions, and the accessibility of construction materials. In general, a hard-rock foundation is suitable for any type of dam, provided the rock has no unfavorable jointing, there is no danger of movement in existing faults, and foundation underseepage can be controlled at reasonable cost. Rock foundations of high quality are essential for arch dams because the abutments receive the full thrust of the water pressure against the face of the dam. Rock foundations are necessary for all medium and high concrete dams. An earth dam may be built on almost any kind of foundation if properly

designed and constructed.

The chance of an embankment dam being most economical is improved if large spillway and outlet capacities are required and topography and foundation are favorable. In a wide valley a combination of an earth embankment dam and a concrete dam section containing the spillway and outlets often is economical. Availability of suitable construction materials frequently determines the most economical type of dam. A concrete dam requires adequate quantities of suitable concrete aggregate and reasonable availability of cement, while an earth dam requires sufficient quantities of both pervious and impervious earth materials. If quantities of earth materials are limited and enough rock is available, a rock-fill dam with an impervious earth core may be the most economical.

Determination of dam height. The dam must be high enough to (1) store water to the normal full-pool elevation required to meet intended functions of the project, (2) provide for the temporary storage needed to route the spillway design flood through the dam, and (3) provide sufficient freeboard height above the maximum surcharge elevation to assure an acceptable degree of safety against possible overtopping from waves and runup.

Physical characteristics of the dam and reservoir site or existing developments within the reservoir area may impose upper limits in selecting the normal full-pool level. In other circumstances economic considerations govern.

With the normal full-pool elevation established, flood flows of unusual magnitude may be passed by providing spillways and outlets large enough to discharge the probable maximum flood or other spillway design flood without raising the reservoir above the normal full elevation or, if it is more economical, by raising the height of the dam and obtaining additional lands to permit the reservoir to temporarily attain surcharge elevations above the normal pool level during extreme floods. Use of temporary surcharge storage capacity also serves to reduce the peak rates of spillway discharge.

Freeboard height is the distance between the maximum reservoir level and the top of the dam. Usually 3 ft or more of freeboard is provided to avoid overtopping the dam by wind-generated waves. Additional freeboard may be provided for possible effects of surges induced by earthquakes, landslides, or other unpredictable events.

Diversion of stream. During construction the dam site must be unwatered so that the foundation may be prepared properly and materials in the structure may be moved easily into position. The stream may be diverted around the site through tunnels, passed through or around the construction area by flumes, passed through openings in the dam, or passed over low sections of a partially completed concrete dam. Diversion may be conducted in one or more stages, with a different method used for each stage. Initial diversion is conducted during a period of low flow to avoid the necessity for passing large flows.

Foundation treatment. The foundation of a dam must support the structure under all operating conditions. For concrete dams, following removal of unsatisfactory materials to a sound foundation surface, imperfections such as adversely oriented rock joints, open bedding planes, localized soft seams, and faults lying on or beneath the foundation surface receive special treatment. Necessary foundation treatment prior to dam construction may include "dental excavation" of surface weaknesses, or shafting and mining to remove deeper localized weaknesses, followed by backfilling with concrete or grout. Such work is sometimes supplemented by pattern grouting of foundation zones after construction of the dam. Foundation features such as rock joints, bedding planes, or faults that do not require preconstruction treatment are made relatively water-tight by curtain grouting from a line of deep grout holes located near the upstream heel and extending the full length of the dam. Although a grout curtain controls seepage at depth, the effectiveness of the grouting or its permanency cannot be relied upon alone to reduce hydrostatic pressures acting on the base of the dam. As a result, drain holes are drilled into the foundation just downstream of the grout curtain to intercept seepage passing through it and to reduce hydrostatic pressure. Occasionally chemical solutions such as acrylamide, sodium silicate, chromelignin, and polyester and epoxy resins are used for consolidating soils or rocks with fine openings.

The foundation of an earth dam must safely support the weight of the dam, limit seepage of stored water, and prevent transportation of dam or foundation material away or into open joints or seams in the rock by seepage. Earth-dam foundation treatment may include removal of excessively weak surface soils to prevent both potential sliding and excessive settlement of the dam, excavation of a cutoff trench to rock, and grouting of joints and seams in the bottom and downstream side of the cutoff trench. The cutoff trenches and grouting extend up the abutments, which are first stripped of weak surface materials.

When weak soils in the foundation of an earth dam cannot be removed economically, the slopes of the embankment must be flattened to reduce shear stress in the foundation to a value less than the soil strength. Relief wells are installed in pervious foundations to control seepage uplift pressures and to reduce the danger of piping when the depth of the pervious material is such as to preclude an economical cutoff.

Instrumentation. Instruments are installed at dams to observe structural behavior and physical conditions during construction and after filling, to check safety, and to provide information for design improvement.

In concrete dams instruments are used to measure stresses either directly or to measure strains from which stresses may be computed. Plumb lines are used to measure bending, and clinometers to measure tilting. Contraction joint openings are measured by joint meters spanning between two adjacent blocks of a dam. Temperatures are measured either by embedded electrical resistance thermometers or by adapting strain, stress, and joint measuring instruments. Water pressure on the base of a concrete dam at the contact with the foundation rock is measured by uplift pressure cells. Interior pressures in a concrete dam are measured by embedded pressure cells. Measurements are also made to determine horizontal and

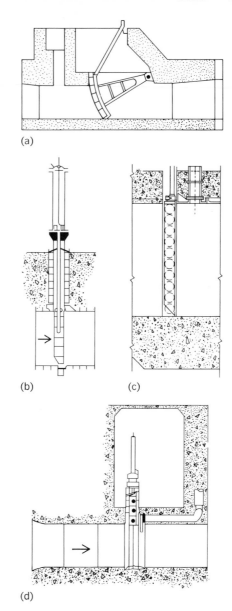

Fig. 10. Outlet gates. (*a*) Tainter gate. (*b*) High-pressure slide gate. (*c*) Tractor gate. (*d*) Jet flow gate. (*U.S. Army Corps of Engineers and U.S. Bureau of Reclamation*)

vertical movements; strong-motion accelerometers are being installed on and near dams in earthquake regions to record seismic data.

Instruments installed in earth-dam embankments and foundations are piezometers to determine pore water pressure in the soil or bedrock during construction and seepage after reservoir impoundments; settlement gages to determine settlements of the foundation of the dam under dead load; vertical and horizontal markers to determine movements, especially during construction; and inclinometers to determine horizontal movements along a vertical line.

Inspection of dams. Because failure of a dam may result in loss of life or property damage in the downstream area, it is essential that dams be inspected systematically both during construction and after completion. The design of dams should be reviewed to assure competency of the structure and its site, and inspections should be made during construction to ensure that the requirements of the design and specifications are incorporated in the structure.

After completion and filling, inspections may vary from cursory surveillance during day-to-day operation of the project to regularly scheduled comprehensive inspections. The objective of such inspections is to detect symptoms of possible distress in the dam at the earliest time. These symptoms include significant sloughs or slides in embankments; evidence of piping or boils near embankments; abnormal changes in flow from drains; unusual increases in seepage quantities; unexpected changes in pore water pressures or uplift pressures; unusual movement or cracking of embankments or abutments; significant cracking of concrete structures; appearance of sinkholes or localized subsidence near foundations; excessive deflection, displacement, erosion, or vibration of concrete structures; erratic movement or excessive deflection or vibration of outlet or spillway gates or valves; or any other unusual conditions in the structure or surrounding terrain.

Detection of any such symptoms of distress should be followed by an investigation of the causes, probable effects, and remedial measures required. Inspection of a dam and reservoir is particularly important following significant seismic events in the locality. Systematic monitoring of the instrumentation installed in dams is essential to the inspection program.

[JACK R. THOMPSON]

Bibliography: American Concrete Institute, *Symposium on Mass Concrete*, Spec. Publ. SP-6, 1963; W. P. Creager, J. D. Justin, and J. Hinds, *Engineering for Dams*, 1945; C. V. Davis, *Handbook of Applied Hydraulics*, 1952; J. L. Sherard et al., *Earth and Earth-Rock Dams*, 1963; G. B. Sowers and G. F. Sowers, *Introductory Soil Mechanics and Foundations*, 1961; U.S. Bureau of Reclamation, *Design of Small Dams*, 2d ed., 1973; U.S. Bureau of Reclamation, *Trial Load Method of Analyzing Arch Dams*, Boulder Canyon Proj. Final Rep., pt. 5, Bull. no. 1, 1938; H. M. Westergaard, Water pressures on dams during earthquakes, *Trans. ASCE*, 98:418–472, 1933.

Damping

A term broadly used to denote either the dissipation of energy in, and the consequent decay of, oscillations of all types or the extent of the dissipation and decay. The energy losses arise from frictional (or analogous) forces which are unavoidable in any system or from the radiation of energy to space or to other systems. For sufficiently small oscillations, the analogous forces are proportional to the velocity of the vibrating member and oppositely directed thereto; the ratio of force to velocity is $-R$, the mechanical resistance.

Damped oscillations. An undamped system of mass m and stiffness s oscillates at an angular frequency $\omega_0 = (s/m)^{1/2}$. The effect of a mechanical resistance R is twofold: It produces a change in the frequency of oscillation, and it causes the oscillations to decay with time. If u is one of the oscillating quantities (displacement, velocity, acceleration) of amplitude A, then Eq. (1) holds in the damped case, whereas in the undamped case

Eq. (2) holds. The reciprocal time $1/\alpha$ in Eq. (1) may be called the damping constant.

$$u = Ae^{-\alpha t} \cos \omega_d t \qquad (1)$$

$$u = A \cos \omega_0 t \qquad (2)$$

In Eqs. (1) and (2), the origin for the time t is chosen so that $t = 0$ when $u = A$. The damped angular frequency ω_d in Eq. (1) is always less than ω_0; its value will be given later. According to Eq. (1), the amplitude of the oscillation decays exponentially; the time, given in Eq. (3), is that required for

$$1/\alpha = 2m/R \qquad (3)$$

the amplitude to decrease to the fraction $1/e$ of its initial value.

A common measure of the damping is the logarithmic decrement δ, defined as the natural logarithm of the ratio of two successive maxima of the decaying sinusoid. If T is the period of the oscillation, then Eq. (4) holds, so that Eq. (1) becomes Eq. (5). Thus $1/\delta$ is the number of cycles required for

$$\delta = \alpha T \qquad (4)$$

$$u = Ae^{-\delta t/T} \cos \omega_d t \qquad (5)$$

the amplitude to decrease by the factor $1/e$ in the same way that $1/\alpha$ is the time required.

The Q of a system is a measure of damping usually defined from energy considerations. In the present case, the stored energy is partly kinetic and partly potential; when the displacement is a maximum, the velocity is zero and the stored energy is wholly potential, while at zero displacement, the energy is wholly kinetic. The Q is π times the ratio of peak energy stored to energy dissipated per cycle. In the present example, this reduces to Eq. (6). The damped frequency ω_d of

$$Q = \omega_0 m/R = \pi/\delta \qquad (6)$$

Eq. (1) is related to the undamped frequency ω_0 of Eq. (2) by Eq. (7), so that for high-Q (lightly

$$(\omega_d/\omega_0)^2 = 1 - (1/2Q^2) \qquad (7)$$

damped) systems, it is only slightly less than ω_0. *See* ENERGY.

Overdamping; critical damping. If α in Eq. (1) exceeds ω_0, then the system is not oscillatory and is said to be overdamped. If the mass is displaced, it returns to its equilibrium position without overshoot, and the return is slower as the ratio α/ω_0 increases. If $\alpha = \omega_0$ (that is, $Q = 1/2$), the oscillator is critically damped. In this case, the motion is again nonoscillatory, but the return to equilibrium is faster than for any overdamped case.

Distributed systems. An undamped, one-dimensional wave of frequency $\omega/2\pi$ propagated in the positive direction of x is represented by Eq. (8),

$$u = A \cos \omega(t - x/c) \qquad (8)$$

c being the velocity of the wave. If the vibration is maintained at $x = 0$ at the value $u = A \cos \omega t$, then the damping manifests itself as an exponential decrease of amplitude with distance x. Equation (8) is replaced by Eq. (9). The attenuation α' may

$$u = Ae^{-\alpha' x} \cos [\omega(t - x/c)] \qquad (9)$$

depend on frequency. If the medium is terminated, the wave will be reflected from the ends, and a system of standing waves will be set up. Examples are a rod carrying sound waves, a piece of electrical transmission line or waveguide, and a vibrating violin string. Such a system has a number of natural frequencies $\omega_n/2\pi$, at each of which it behaves like the lumped system of the previous sections. The decay of a vibration is characterized by Q in Eq. (10), where $\lambda_n = 2\pi c/\omega_n$ is the wavelength.

$$Q = \omega_n/2\alpha'c = \pi/\alpha'\lambda_n \qquad (10)$$

Hysteresis damping. At a given instant, the elongation (strain) of a metal bar which is under periodic, alternating stress is not determined exactly by the instantaneous value of the stress existing at that time. For example, the elongation is less at a given stress value when the stress is increasing than when it is decreasing. This phenomenon, which is known as mechanical hysteresis, causes an undesirable energy loss. A vibration problem of serious nature exists in the blades of jet engines and other steam and gas turbines. The blade material itself exhibits a mechanical hysteresis damping which holds the vibrations in check. When the stress is small, the hysteresis damping is very small in all metals, but it rises suddenly when the stress reaches a certain value. Unfortunately, in most metals the stress at which hysteresis damping becomes large and that at which the metal fails because of fatigue are very close together. However, a much higher hysteresis damping at safe stresses than that of ordinary steel is exhibited by certain alloys.

Oscillating electrical circuits. A simple series electrical circuit consisting of an inductance L, resistance R, and capacitance C is exactly analogous to the mechanical system described by Eqs. (1)–(6). The inductance, resistance, and elastance $(1/C)$ correspond to the mass, mechanical resistance, and stiffness, respectively. A distributed electrical circuit, such as a section of transmission line or wave guide, is analogous to a vibrating rod or disk.

In the ordinary electrical oscillator, the frequency is controlled by an electrical resonator (tank) lumped at the lower, and distributed at the higher, frequencies. Good frequency stability is associated with a high-Q tank. For frequencies of tens of megahertz (MHz) and below, mechanical resonators can be constructed which have a much higher Q than the equivalent electrical tanks. Thus, very stable electrical oscillators have mechanical resonators as their frequency-determining elements. Such an electromechanical system can operate only if there is some coupling between the electrical and mechanical aspects of the system.

The coupling can be arranged in various ways. In some materials, such as quartz, the constitutive relations involve the mechanical and electrical variables jointly; thus, for example, an electric field may produce a strain in the absence of any stress. Thus, the coupling is inherent in the quartz itself. Quartz crystals are used for frequency control of oscillators in the range from kilohertz to perhaps 100 MHz; the Q of a high-frequency crystal may be several million. Some low-frequency oscillators are controlled by tuning forks having a Q of several hundred thousand, the action being similar to that of the electric bell or buzzer.

The electrostatic motor-generator effect provides the coupling in such mixed systems as the condenser microphone and electrostatic loudspeaker, and the electromagnetic motor-generator effect plays the same role in the dynamic microphone, dynamic loudspeaker, and in various electrical instruments.

Reading of meters. Galvanometers and other electrical indicating instruments are examples of damped electromechanical systems. The free period depends on the moment of inertia of the rotating system (for example, of the coil in a galvanometer) and on the stiffness of the suspension or spring. If an electrical input is suddenly applied, the indicator will, if the system is highly underdamped, overshoot its equilibrium value and then execute a damped sinusoidal oscillation about it; if the system is highly overdamped, the indicator will approach its final reading sluggishly. If the reading time is taken as the time required to reach the equilibrium value $\pm 1\%$, then the minimum reading time is obtained if the logarithmic decrement is 83% of the critical value (relative damping = 0.83) and is equal to 67% of the free period.

In portable and switchboard instruments, the damping is either viscous or magnetic or both. Viscous damping is achieved with vanes attached to the movement which move in a narrow, air-filled space; magnetic damping is an eddy-current effect. The eddy currents are generated in the coil frame or in a metal plate moving between magnetic poles; this latter arrangement is used in magnetically damped analytical balances.

In the d'Arsonval galvanometer, the damping is largely due to the generator action of the moving coil, and it can be adjusted by varying the external circuit resistance. The same is true, to a lesser extent, of sensitive microammeters.

[MARTIN GREENSPAN]

Bibliography: F. K. Harris, *Electrical Measurements*, 1952; L. E. Kinsler and A. R. Frey, *Fundamentals of Acoustics*, 1950; C. Kittel, W. D. Knight, and M. A. Ruderman, *Mechanics*, 1965; N. W. McLachlan, *Theory of Vibrations*, 1951.

Decontamination of radioactive materials

The removal of radioactive contamination which is deposited on surfaces or may have spread throughout a work area. Personnel decontamination is also included. The presence of radioactive contamination is a potential health hazard, and in addition, it may interfere with the normal functioning of plant processes, particularly in those plants using radiation detection instruments for control purposes. Thus, the detection and removal of radioactive contaminants from unwanted locations to locations where they do not create a health hazard or interfere with production are the basic purposes of decontamination.

There are four ways in which radioactive contaminants adhere to surfaces, and these limit the decontamination procedures which are applicable. The contaminant may be (1) held more or less loosely by such physical forces as electrostatic or surface tension, (2) absorbed in porous materials, (3) adsorbed on or by the surface in the form of ions, atoms, or molecules, or (4) bonded to surfaces by oil, grease, tars, or paint.

Methods. Decontamination methods follow two broad avenues of attack, mechanical and chemical. Commonly used mechanical methods are vacuum cleaning, sand blasting, blasting with other abrasives, flame cleaning, scraping, ultrasonic radiation, and surface removal (for example, removal of concrete floors with an air hammer). The principal chemical methods of decontamination are water washing, steam cleaning, and scrubbing with detergents, acids, caustics, and solvents.

Another important method of handling contamination is to store the contaminated object, or temporarily abandon the contaminated space. This can be done when the use of the material or space is not necessary for a period of time and the half-life of the contaminant is relatively short. For example, tools contaminated with short-lived fission products may be stored, or a building contaminated with such material may be sealed off and barred from use, until the natural radioactive decay has reduced the contamination to an acceptable level.

Other methods involve covering the contamination by some means, such as painting, and disposing of part or all of the contaminated equipment or facility. Considerations which determine the methods used for decontamination or removal of contamination include (1) the hazards involved in the decontamination procedure, (2) the cost of removal of the contamination, and (3) the permanency of removal of the contamination (for example, painting over a surface contaminated with a long-lived radioactive material only postpones ultimate disposal considerations).

Personnel. Personnel decontamination methods differ from those used for materials primarily because of the possibilities of injury to the person being decontaminated. Procedures used for normal personal cleanliness usually will remove radioactive contaminants from the skin, and the method used will depend upon its form and associated dirt (grease, oil, soil, and so on). Soap and water (sequestrants and detergents) normally remove more than 99% of the contaminants. If it is necessary to remove the remainder, chemical methods which remove the outer layers of skin upon which the contamination has been deposited can be used. These chemicals—citric acid, potassium permanganate, and sodium bisulfite are examples—should be used with caution and preferably under medical supervision, because of the increased risk of injury to the skin surface. The use of coarse cleansing powders should be avoided for skin decontamination, because they may lead to scratches and abraded skin which can permit the radioactive material to enter the body. Similarly, the use of organic solvents should be avoided for skin decontamination because of the probability of penetration through the pores of the skin. It is very difficult to remove radioactive material once it is fixed inside the body, and the ensuing hazard depends very little on the method of entry into the body, that is, through wounds, through pores of the skin, by injection, or by inhalation. When certain of the more dangerous radioactive materials, such as radium or plutonium, have been taken into the body, various chemical treatments have been attempted to increase the body elimination, but the results of these treatments are not very encouraging. In the case of plutonium and certain other heavy metals, the most effective treatment for removal from the body is the administration of chelating agents, such as calcium ethylenediaminetetraacetate (CaEDTA) or a sodium citrate solution of zirconyl chloride. In any case, the safest and most reliable procedure for preventing internal exposure from radioactive material is the application of health physics procedures to prevent entry of radioactive material into the body.

Air and water. Air contaminants frequently are eliminated by dispersion into the atmosphere. Certain meteorological conditions, such as prevailing wind velocities, wind direction, and inversion layers, seriously limit the total amount of radioactive material that may be released safely to the environment. Consequently, decontamination of the airstream by filters, cyclone separators, scrubbing with caustic solutions, cryogenic removal, and entrapment on charcoal beds is often resorted to. The choice of method used for decontamination of air is guided by such things as the volume of airflow, the cost of heating and air conditioning, the hazards associated with the airborne radioactive material, and the isolation of the operation from other populated areas.

Water decontamination processes can use one or both of the two opposing philosophies of maximum dilution or maximum concentration (and subsequent removal) of the contaminant. Water concentration methods involve the use of water purification processes, that is, ion exchange, chemical precipitation, flocculation, filtration, and biological retention.

Certain phases of radioactive decontamination procedures are potentially hazardous to personnel. Health physics decontamination practices include the use of protective clothing, respiratory devices, localized shielding, isolation or restriction of an area, provisions for the proper disposal of the attendant wastes, and application of the recommended rules and procedures for limiting the internal and external doses of ionizing radiation. *See* RADIOACTIVE WASTE MANAGEMENT; RADIOACTIVITY.

[KARL Z. MORGAN]

Bibliography: *Control and Removal of Radioactive Contamination in Laboratories*, Natl. Bur. Std. Handb. no. 48, 1951; International Brotherhood of Electrical Workers Staff, *Radiation Hazards and Control*, 1965; S. Kinsman, *Radiological Health Handbook*, U.S. Department of Health, Education and Welfare, PB–121784, 1957.

Dehumidifier

Equipment designed to reduce the amount of water vapor in the atmosphere.

The atmosphere is a mechanical mixture of dry air and water vapor, the amount of water vapor being limited by air temperature. Water vapor is measured in either grains per pound of dry air or pounds per pound of dry air (7000 gr = 1 lb).

There are three methods by which water vapor may be removed: (1) the use of sorbent materials, (2) cooling to the required dew point, and (3) compression with aftercooling.

Sorbent type. Sorbents are materials which are hygroscopic to water vapor; they are available in both solid and liquid forms. Solid sorbents include silica gels, activated alumina, and aluminum bauxite. Liquid sorbents include halogen salts such as lithium chloride, lithium bromide, and calcium

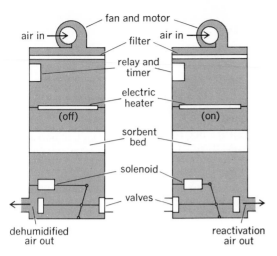

Fig. 1. Single-bed solid-sorbent dehumidifier. Dehumidifying cycle on left and reactivation cycle on right.

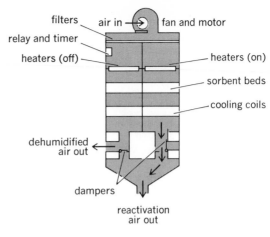

Fig. 2. Dual-bed solid-sorbent dehumidifier. Air is being dehumidified through the left bed at the same time that the right bed is being reactivated.

chloride, and organic liquids such as ethylene, diethylene, and triethylene glycols and glycol derivatives.

Solid sorbents may be used in static or dynamic dehumidifiers. Bags of solid sorbent materials within packages of machine tools, electronic equipment, and other valuable materials subject to moisture damage constitute static dehumidifiers. An indicator chemical may be included to show by a change in color when the sorbent is saturated. The sorbent then requires reactivation by heating at 300–350°F for 1–2 hr before reuse.

A dynamic dehumidifier for solid sorbent consists of a main circulating fan, one or more beds of sorbent material, reactivation air fan, heater, mechanism to change from dehumidifying to reactivation, and aftercooler.

A single-bed dehumidifier (Fig. 1) operates on an intermittent cycle of dehumidifying for 2–3 hr and then swtiches to the reactivation cycle for 15–45 min. No dehumidification is obtained during the reactivation cycle. A single-bed unit is used for small areas where moisture is a problem. The moist reactivation air is discharged to the outside.

The dual-bed machine is larger in capacity than the single-bed unit and has the advantage of providing a continuous supply of dehumidified air (Fig. 2). While one bed is dehumidifying, the other bed is reactivating. After a predetermined time interval, the air cycle is switched to pass the air through the reactivated bed for dehumidification and to reactivate the saturated bed.

The dew point of the effluent air of a fixed-bed machine is lowest at the start of a cycle immediately after the reactivated bed has been placed in service. The dew point gradually rises as the bed absorbs the water vapor and eventually would be the same as the entering dew point when the vapor pressure of the sorbent reached the vapor pressure of the air and could no longer absorb moisture from the air. The cutoff point at which the absorbing bed is changed over to reactivation is fixed by the maximum allowable effluent dew point.

A multibed unit with short operating cycles will reduce the range of effluent dew point to within a few degrees. A unit with rotating cylindrical bed

maintains a reasonably constant effluent dew point.

The liquid-sorbent dehumidifier consists of a main circulating fan, sorbent-air contactor, sorbent pump, and reactivator including contactor, fan, heater, and cooler (Fig. 3). This unit will control the effluent dew point at a constant level because dehumidification and reactivation are continuous operations with a small part of the sorbent constantly bled off from the main circulat-

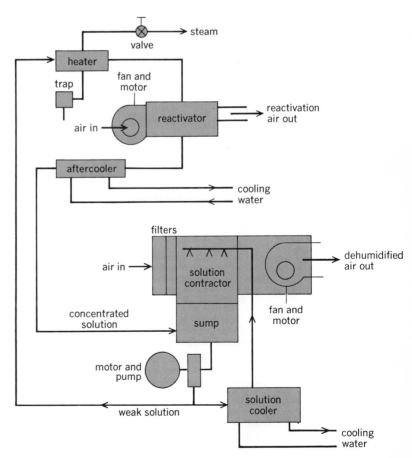

Fig. 3. Liquid-sorbent dehumidifier with continuous dehumidifying and reactivation.

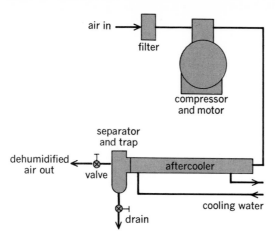

Fig. 4. Dehumidifying by compression and aftercooling.

ing system and reactivated to the concentration required for the desired effluent dew point.

Cooling type. A system employing the use of cooling for dehumidifying consists of a circulating fan and cooling coil. The cooling coil may use cold water obtained from wells or a refrigeration plant, or may be a direct-expansion refrigeration coil. In place of a coil, a spray washer may be used in which the air passes through two or more banks of sprays of cold water or brine, depending upon the dew-point temperature required.

When coils are used, the leaving dew point is seldom below 35°F because of possible buildup of ice on the coil. When it is necessary to use coils for temperatures below 35°F, as in cold-storage rooms, either two coils are used so one can be defrosted while the other is in operation, or only one coil is used and dehumidifying is stopped during the defrost period.

A brine-spray dehumidifier or brine-sprayed coil can produce dew-point temperatures below 35°F without frosting if properly operated and maintained.

Compression type. Dehumidifying by compression and aftercooling is used when the reduction of water vapor in a compressed-air system is required. This is particularly important, for example, if the air is used for automatic control instruments or cleaning of delicate machined parts.

If air is compressed and the heat of compression removed to bring the temperature of the air back to the temperature entering the compressor, condensation will take place and the remaining water vapor content will be directly proportional to the absolute pressure ratio of the compressed air (Fig. 4).

For example, if saturated air at 70°F (111 gr/lb of dry air) is compressed from atmospheric pressure (14.7 pounds per square inch absolute, psia) to 88 psia (6:1 compression ratio) and cooled to 70°F, the remaining water vapor in the compressed air will be 111/6 = 18.5 gr/lb of dry air. If the air is expanded back to atmospheric pressure and 70°F, the dew point will be 24°F.

The power required for compression systems is so high compared to power requirements for dehumidifying by either the sorbent or refrigeration method that the compression system is not an eco-

nomical one if dehumidifying is the only end result required.

[JOHN EVERETTS, JR.]

Bibliography: American Society of Heating, Refrigerating, and Air-Conditioning Engineers, *ASHRAE Guide and Data Book*, 1967; ASHRAE, *Heating, Ventilating, Air-Conditioning Guide*, annual; ASHRAE, Symposium on dehumidification: Journal section, *Heat. Piping Air Cond.*, 29(4):152–162, 1957; Carrier Air Conditioning Co., *Handbook of Air Conditioning System Design*, 1965; V. R. Dietz, *Bibliography of Solid Absorbents, 1943 to 1953*, Nat. Bur. Stand. Circ. no. 566, 1956; J. Everetts, Jr., Dehumidification methods and applications, *Heat. Piping Air Cond.*, 18(12): 121–124, 1964.

Destructive distillation

The primary chemical processing of materials such as wood, coal, oil shale, and some residual oils from refining of petroleum. It consists in heating material in an inert atmosphere at a temperature high enough for chemical decomposition. The principal products are (1) gases containing carbon monoxide, hydrogen, hydrogen sulfide, and ammonia, (2) oils, and (3) water solutions of organic acids, alcohols, and ammonium salts.

Crude shale oil obtained by destructive distillation of carboniferous shales, is being produced on a commercial scale in Scotland, Latvia, and Sweden and on a pilot plant scale in the United States. Crude shale oil may be subjected to a destructive, or coking, distillation to reduce its viscosity and increase its hydrogen content. Subsequent catalytic hydrogenation (cobalt molybdatealumina catalyst, about 400°C, 100–1500 psi pressure) lowers the nitrogen and sulfur contents so that the oil can then be refined by normal petroleum refinery operations. Residual oils from petroleum refinery operations are subjected to coking-distillation to reduce the carbon content. The coke is used for the manufacture of electrode carbon and the oil is returned to the feed for normal petroleum refining. *See* COKING IN PETROLEUM REFINING; OIL SHALE.

[H. H. STORCH/H. W. WAINWRIGHT]

Dewaxing of petroleum

The process of separating hydrocarbons which solidify readily (waxes) from petroleum fractions. Removal of wax is usually necessary to produce lubricating oil which will remain fluid down to the lowest temperature of use. It is therefore an important step in the manufacture of lubricating oils. The wax removed may be purified further to produce commercial paraffin or microcrystalline waxes.

Most commercial dewaxing processes utilize solvent dilution, chilling to crystallize the wax, and filtration. The MEK process (methyl ethyl ketone-toluene solvent) is most widely used. Wax crystals are formed by chilling through the walls of scraped surface chillers, and wax is separated from the resultant wax-oil-solvent slurry by using fully enclosed rotary vacuum filters. In a relatively new process modification, most of the chilling is accomplished by multistage injection of very cold solvent into the waxy oil with vigorous agitation, resulting in more uniform and compact wax crystals which filter faster.

In the propane process, part of the propane diluent is allowed to evaporate by reducing pressure, so as to chill the slurry to the desired filtration temperature, and rotary pressure filters are employed.

Other solvents in commercial use for dewaxing include MEK-MIBK (methyl isobutyl ketone), acetone-benzene, dichloroethane-methylene dichloride, and propylene-acetone.

Older dewaxing processes, still in limited use, are centrifugal dewaxing, applicable only to heavy residual stocks, utilizing naphtha dilution, indirect chilling, and centrifugal separation; and cold pressing, applicable only to low-viscosity light lube fractions, in which the crystallized wax is separated from the chilled, undiluted oil in plate-and-frame-type pressure filters.

Complex dewaxing requires no refrigeration, but depends upon the formation of a solid urea-*n*-paraffin complex which is separated by filtration and then decomposed. This process is used, to a limited extent, to make low-viscosity lubricants which must remain fluid at very low temperatures (refrigeration, transformer, and hydraulic oils). Similar use is anticipated for the catalytic dewaxing process, which is based on selective hydrocracking of the normal paraffins; it uses a molecular sieve-based catalyst in which the active hydrocracking sites are accessible only to the paraffin molecules. *See* PETROLEUM; PETROLEUM PROCESSING.

[STEPHEN F. PERRY]

Bibliography: R. N. Bennett, G. J. Elkes, and G. J. Wanless, New process produces low pour oils, *Oil Gas J.*, Jan. 6, 1975; J. F. Eagen et. al., Successful development of two new lubricating oil dewaxing processes, Paper at World Petroleum Congress, Japan, 1975; G. D. Hobson and W. Pohl, *Modern Petroleum Technology*, 1973; Hydrocarbon processing, *1974 Refining Process Handbook*, September 1974.

Diesel cycle

An internal combustion engine cycle in which the heat of compression ignites the fuel. Compression-ignition engines, or diesel engines, are thermodynamically similar to spark-ignition engines. The sequence of processes for both types is intake, compression, addition of heat, expansion, and exhaust. Ignition and power control in the compression-ignition engine are, however, very different from those in the spark-ignition engine.

Usually, a full unthrottled charge of air is drawn in during the intake stroke of a diesel engine. A compression ratio between 12 and 20 is used, in contrast to a ratio of 4 to 10 for the Otto spark-ignition engine. This high compression ratio of the diesel raises the temperature of the air during the compression stroke. Just before top center on the compression stroke, fuel is sprayed into the combustion chamber. The high temperature of the air ignites the fuel, which burns almost as soon as it is introduced, adding heat. The combustion products expand to produce power, and exhaust to complete the cyle.

Performance of a diesel engine is anticipated by analyzing the action of an air-standard diesel cycle. An insulated cylinder equipped with a frictionless piston contains a unit air mass. The metal cyl-

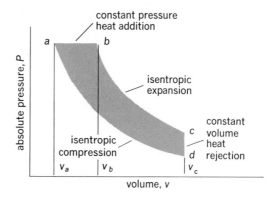

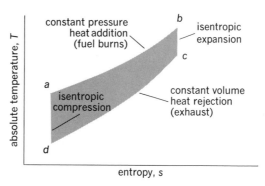

Fig. 1. Ideal diesel cycle, with pressure-volume and temperature-entropy bases.

inder head is alternately insulated and then uncovered for heat transfer.

Air is compressed until the piston reaches the top of the stroke. Then the air receives heat through the cylinder head and expands at constant pressure along path *a-b* as shown in Fig. 1, moving the piston part way down through the cylinder. Then the cylinder head is insulated, and the air completes its expansion along path *b-c* at constant entropy. The cylinder head is uncovered, and with the piston at the bottom of its stroke, a constant-volume heat rejection takes place on path *c-d*. The insulation is replaced, and the cycle is completed with an isentropic compression on path *d-a*.

An increase in compression ratio $r = v_d/v_a$ increases efficiency η, the increase becoming less at higher compression ratios. Another characteristic of the diesel cycle is the ratio of volumes at the end and at the start of the constant-pressure heat-addition process. This cutoff ratio $r_c = v_b/v_a$ measures the interval during which fuel is injected. For an engine to develop greater power output, the cutoff ratio is increased and heat continues to be added further into the expansion stroke. The air-standard cycle shows that, with less travel remaining during which to expend the additional heat energy as mechanical energy, the efficiency of the engine is reduced. Conversely, efficiency increases as the cutoff ratio decreases, so that a diesel engine is most efficient at light loads. Specifically, the equation below may be written, where $k = c_p/c_v$ the ratio

$$\eta = 1 - \frac{1}{r^{k-1}}\left[\frac{r_c^{\,k}-1}{k(r_c-1)}\right]$$

of specific heat of the working substance at constant pressure to its specific heat at constant vol-

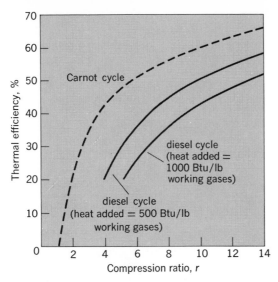

Fig. 2.   Thermal efficiency of an ideal diesel cycle.

ume (Fig. 2). In the limiting case when cutoff ratio r_c approaches unity, diesel cycle efficiency approaches Otto cycle efficiency for cycles of the same compression ratio.

In an acutal engine with a given compression ratio, the Otto engine has the higher efficiency. However, fuel requirements limit the Otto engine to a compression ratio of about 10, whereas a diesel engine can operate at a compression ratio of about 15 and consequently at a higher efficiency.

In addition, heat can be added earlier in the cycle by injecting fuel during the latter part of the compression process d-a. This mode of operation is the dual-combustion or semidiesel cycle. With most of the heat added near peak compression, semi-diesel efficiency approaches Otto cycle efficiency at a given compression ratio. See OTTO CYCLE.   [THEODORE BAUMEISTER]

Bibliography: T. Baumeister (ed.), *Standard Handbook for Mechanical Engineers*, 7th ed., 1967; A. W. Judge, *High-speed Diesel Engines*, 5th ed., 1957; L. Lichty, *Combustion Engine Processes*, 7th ed., 1967.

Diesel engine

An internal combustion engine operating on a thermodynamic cycle in which the ratio of compression $(R_v = 15\pm)$ of the air charge is sufficiently high to ignite the fuel subsequently injected into the combustion chamber. The engine differs essentially from the more prevalent mixture engine in which an explosive mixture of air and gas or air and the vapor of a volatile liquid fuel is made externally to the engine cylinder, compressed to a point some 200°F below the ignition temperature, and ignited at will as by an electric spark. The diesel engine utilizes a wider variety of fuels with a higher thermal efficiency and consequent economic advantage under many service applications. The true diesel engine, as projected by R. Diesel and as represented in most low-speed engines, such as about 300 rpm, uses a fuel-injection system where the injection rate is delayed and controlled to maintain constant pressure during combustion. Adaptation of the injection principle to

higher engine speeds, such as 1000–2000 rpm, has necessitated departure from the constant pressure specification because the time available for fuel injection is so short (milliseconds). Combustion proceeds with little regard to the constant-pressure specification. High peak pressures may be developed. Yet nonvolatile (distillate) fuels are burned to advantage in these engines which cannot be rigorously identified as true diesels but which properly should be called commercial diesels. In ordinary parlance all such engines are classified as diesels. See DIESEL CYCLE; OTTO CYCLE.

Identifying alternative features of diesel engine types include: (1) two-cycle or four-cycle operation; (2) horizontal or vertical piston movement; (3) single or multiple cylinder; (4) large (5000 hp) or small (50 hp); (5) cylinders in line, opposed, V, or radial; (6) single acting or double acting; (7) high (1000–2000 rpm), low (100–300 rpm), or medium speed; (8) constant speed or variable speed; (9) reversible or nonreversible; (10) air injection or solid injection; (11) supercharged or unsupercharged; and (12) single or multiple fuel. Section drawings of two representative engines are given in Figs. 1 and 2, and selected performance data are given in Table 1.

Maximum diesel engine sizes (5000 kw) are less than steam turbines (1,000,000 kw) and hydraulic turbines (300,000 kw). They give high instrinsic and actual thermal efficiency (20–40%); a sample comparative heat balance is shown in Table 2; variation in performance with load is shown in Fig. 3. Control of engine output is by regulation of the fuel supplied but without variation of the air supply (100±% excess air at full load). Supercharging (10–15 psi) increases cylinder weight charge and consequently power output for a given cylinder size and engine speed. With two-cycle

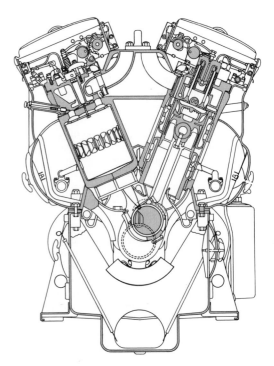

Fig. 1.   Section through a locomotive diesel engine. (*General Motors, Electromotive Division*)

Table 1. Performance of selected diesel engine plants

Type of plant	Shaft horse-power (shp)	Ratio of compression, R_v	Brake mean effective pressure, psi	Piston speed, ft/min	Weight, lb/in.³ displacement	Weight, lb/shp	Overall thermal efficiency, %
Air injection engine	300–5000	12–15	50–75	600–1000	3–8	25–200	30–35
Solid injection, compression ignition							
Automotive	20–300	12–15	75–100	800–1800	2.5–4	7–25	25–30
Railroad	200–2500	12–15	60–90	800–1800	2.5–4	10–40	30–35
Stationary							
Unsupercharged	50–2500	12–15	70–80	600–1500	2.5–5	10–100	30–35
Supercharged	60–4000	10–13	110–125	600–1500	2.5–5	7.5–75	32–40
Dual fuel, stationary							
Unsupercharged	50–2500	12–15	80–90	600–1500	2.5–5	10–100	30–35
Supercharged	60–4000	10–13	120–135	600–1500	2.5–5	7.5–75	32–40

constructions, scavenging air (approximately 5 psi) is delivered by crankcase compression, front end compression, or separate rotary, reciprocating, or centrifugal blowers. The engine cylinder may be without valves and with complete control of admission of scavenging air and release of spent gases in a two-port construction, the piston covering and uncovering the ports; or the cylinder may have a single port (for admission or release) uncovered by the main piston at the outer end of its stroke and conventional cam-operated valve in the cylinder head. The objective is to replace spent gases with fresh air by guided flow and high turbulence. The four-cycle engine, with its complement of admission and exhaust valves on each cylinder, is most effective in scavenging. But the sacrifice of one power stroke out of every two is a frequent deterrent to its selection. Valves are exclusively of the poppet type with the burden of tightness and cooling dominant in the exhaust valve designs. Cylinder heads become complicated structures because of valve porting, jacketing, and spray-valve locations and the accommodation of these to effective combustion, heat transfer, and internal bursting pressures.

Distillate fuel (40° API, 19,000–19,500 Btu/lb, 135,000–140,000 Btu/gal) prevails with locomotive, truck, bus, and automotive applications. Lower-speed engines (stationary and motorship service) burn heavier fuels (for example, 20° API, 18,500–19,000 Btu/lb, 145,000–150,000 Btu/gal). Alternative fuels are burned in dual-fuel and gas diesel engines for stationary service. The main fuel is typically natural gas (90–95%) with oil (5–10%) used to control burning and to stabilize ignition. In the more prevalent liquid-fuel-injection system, the technical problems are numerous and embrace such elements as pumps, spray nozzles, and com-

bustion chambers for the delivery, atomization, and burning of the fuel in the hot compressed air. There must be accurate timing (measured in milliseconds) for the entire process to give clean, complete combustion without undue excess air. Combustion characteristics of fuels are defined by rigorous specifications and include such factors as viscosity, flash point, pour point, ash, sulfur, basic sediment, water, Conradsen carbon number, cetane number, and diesel index.

Small size (<200 hp) engines are conveniently started by an electric motor and storage battery. Larger engines use compressed air (about 200 psi) introduced through valves in the cylinder

Table 2. Approximate allocation of losses in internal combustion engine plants

Type of loss	Mixture engines, %	Injection (diesel) engines, %
Output	20	33
Exhaust losses	40	33
Cooling system losses	40	33
Other	<1	1
Total (input)	100	100

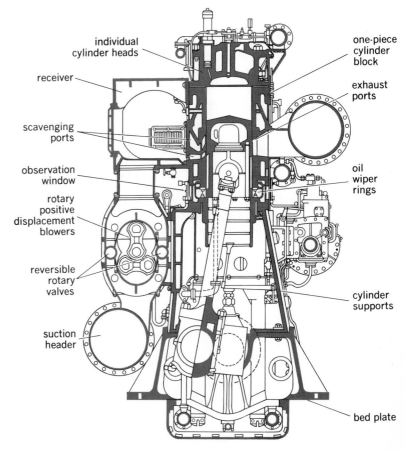

individual cylinder heads

one-piece cylinder block

receiver

exhaust ports

scavenging ports

observation window

oil wiper rings

rotary positive displacement blowers

reversible rotary valves

cylinder supports

suction header

bed plate

Fig. 2. Section through a Busch-Sulzer two-cycle diesel engine.

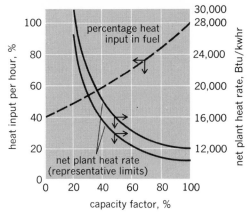

Fig. 3. Heat rate and heat input curves for selected diesel-engine generating plants.

head. Starting, with engine-driven generator sets, may be accomplished by motoring the generator.

Cooling systems use water at 120–180°F with radiators, cooling towers, and cooling ponds employed for conservation and reclamation. Lubrication costs can become prohibitive with inadequate engine maintenance. Foundations must be designed to handle stress loadings and to reduce vibration. Exhaust systems should be equipped with wavetrap silencers or mufflers. Filters on air and fuel supply are good insurance for engine reliability. *See* INTERNAL COMBUSTION ENGINE.

[THEODORE BAUMEISTER]

Bibliography: ASME, *Annual Oil Engine Power Cost Report*, 1968; T. Baumeister (ed.), *Standard Handbook for Mechanical Engineers*, 7th ed., 1967; Diesel Engine Manufacturers Association, *Standard Practices for Stationary Diesel and Gas Engines*, 1958; A. W. Judge, *High Speed Diesel Engines*, 5th ed., 1957; L. C. Lichty, *Combustion Engine Processes*, 7th ed., 1967.

Diesel fuel

A broad class of petroleum products which includes distillate or residual materials (or blends of these two) from the refining of crude oil. Diesel fuel generally has a distillation range between 190° and 380°C and a specific gravity 15°C/15°C range between 0.760 and 0.935. In addition to these two primary criteria, other properties which are used to define diesel fuel are: viscosity (1.4 to 26.5 mm²/s), sulfur content (usually < 1.0% by weight), cetane number (30 to 60), ash content (usually <0.10% by weight), water and sediment content (usually <0.5% by volume), flash point (usually >55°C), and cloud point (that temperature at which wax crystals begin to precipitate as the oil is cooled under prescribed conditions) or pour point (the lowest temperature at which the oil will flow when cooled under prescribed conditions). The properties of diesel fuel greatly overlap those of kerosine, jet fuels, and burner fuel oils; thus all these products are generally referred to as intermediate distillates. *See* FUEL OIL; JET FUEL; KEROSINE; PETROLEUM PROCESSING; PETROLEUM PRODUCTS.

The table contains the limiting requirements of the three grades of diesel fuel which are most commonly used in the United States. Although diesel engines usually operate on a wide range of fuels, they are designed to operate most cleanly and efficiently on one of the fuels described in this table; using a fuel which departs from that recommended by the engine manufacturer will have a negative effect on engine life, performance, exhaust emissions, noise, and so on. For example, increasing sulfur content causes an increase in corrosive wear of piston rings, cylinder walls, and bearings. Decreasing cetane number causes hard starting, rough operation, increased exhaust emissions, and increased noise. Increasing ash or wax content, or water and sediment, causes fuel system wear and fuel-handling problems. Some heavier (higher-boiling-range) fuels, such as Navy distillate fuel, MIL-F-24397 (ships), and marine diesel fuel oil, MIL-F-16884), are used in low- and medium-speed diesel engines, but they are generally not recommended for high-speed engines.

The diesel engine, with compression ratios ranging from 12:1 up to 22.1, ignites its fuel by injecting it into the charge air, which is heated to 500–550°C by the heat of compression; thus diesel engines are referred to as compression ignition engines. This method of ignition requires that the fuel ignite spontaneously and quickly (within 1 to 2 ms in a high-speed engine). The time lag between the initiation of injection and the initiation of combustion is called ignition delay. Two major factors characterize ignition delay: a mechanical factor which is influenced by such things as compression ratio, motion of the charge air during injection, and ability of the injector to atomize the fuel; and a chemical factor which is influenced by such things as the fuel's autoignition temperature, specific heat, density, thermal conductivity, surface tension, and coefficient of friction. *See* DIESEL ENGINE.

The fuel's effect on the chemical portion of ignition delay is expressed by a quantity called the cetane number. Cetane (hexadecane), which has a high-ignition quality (short chemical ignition delay), has arbitrarily been assigned a cetane number of 100, whereas heptamethylnonane has been assigned a cetane number of 0. The cetane number of a diesel fuel is determined by comparing it to a blend of cetane and heptamethylnonane which has the same ignition quality. This comparison is made with the use of an ASTM-CFR engine. The cetane number is the percentage by volume of cetane in the blend which has an ignition quality equal to the test fuel.

The cetane number of the paraffinic hydrocarbon compounds is generally high, whereas the cetane number of aromatic and naphthenic hydrocarbon compounds is low. Diesel fuel normally contains 60–80% by volume of paraffinic compounds and 20–40% by volume of aromatic and naphthenic compounds. When the paraffinic content is toward the low end of this range, the natural cetane number of the fuel is low. A low cetane number can be counteracted to some degree by the addition of a cetane improver additive, such as amyl nitrate or hexyl nitrate, which increases the cetane number of the fuel, in much the way antiknock additives increase the octane number of gasoline.

It has been found that cetane number can be approximated from the physical properties of the

Detailed requirements for diesel fuel oils

Grade of diesel fuel oil	Flash point, °F (°C) Min	Cloud point, °F (°C) Max	Water and sediment, vol % Max	Carbon residue on, 10% residuum, % Max	Ash, weight, % Max	Distillation temperatures, °F(°C) 90% point		Viscosity at 100°F (37.8°C) Kinematic, cSt (or SUS)		Sulfur,[d] wt % Max	Copper strip corrosion Max	Cetane number[e] Min
						Min	Max	Min	Max			
No. 1-D, a volatile distillate fuel oil for engines in service requiring frequent speed and load changes.	100 or legal (37.8)	[b]	0.05	0.15	0.01	—	550 (287.8)	1.4	2.5 (34.4)	0.50 or legal	No. 3	40[f]
No. 2-D, a distillate fuel oil of lower volatility for engines in industrial and heavy mobile service	125 or legal (51.7)	[b]	0.05	0.35	0.01	540[c] (282.2)	640 (338)	2.0[c] (32.6)	4.3 (40.1)	0.50 or legal	No. 3	40[f]
No. 4-D, a fuel oil for low- and medium-speed engines	130 or legal (54.4)	[b]	0.50	—	0.10	—	—	5.8 (45)	26.4 (125)	2.0	—	30[f]

SOURCE: American Society for Testing and Materials. *Standard Specifications for Diesel Fuel Oils*, ASTM D-975.

[a]To meet special operating conditions, modifications of individual limiting requirements may be agreed upon between purchaser, seller, and manufacturer.

[b]It is unrealistic to specify low-temperature properties that will ensure satisfactory operation on a broad basis. Satisfactory operation should be achieved in most cases if the cloud point (or wax appearance point) is specified at 10°F above the tenth percentile minimum ambient temperature for the area in which the fuel will be used. Some equipment designs use flow improver additives, or fuel properties; or operations may allow higher or require lower cloud point fuels. Appropriate low-temperature operability properties should be agreed on between the fuel supplier and purchaser for the intended use and expected ambient temperatures.

[c]When cloud point less than 10°F (−12.2°C) is specified, the minimum viscosity shall be 1.8 cSt. and the 90% point shall be waived.

[d]In countries outside the United States, other sulfur limits may apply.

[e]Where cetane number by Method D 613, is not available, ASTM Method D 976, Calculated Cetane Index of Distillate Fuels, may be used as an approximation. Where there is disagreement, Method D 613 shall be the referee method.

[f]Low-atmospheric temperatures as well as engine operation at high altitudes may require use of fuels with higher cetane ratings.

fuel. The best correlation is with the specific gravity and mid-boiling point.

[ROBERT TEASLEY, JR.]

Bibliography: American Society for Testing and Materials, *Annual Book of ASTM Standards*, pt. 23, 1975; American Society for Testing and Materials, *Diesel Fuel Oils*, STP 413, 1967; British Petroleum Co., Ltd., *Medium and High Speed Diesel Egines*, 1970; V. Gutherie, *Petroleum Products Handbook*, 1960; M. Popovich and C. Hering, *Fuels and Lubricants*, 1959.

Direct current

Electric current which flows in one direction only, as opposed to alternating current. The current may be of constant magnitude, as when produced by a battery, or it may vary with time, as rectified alternating current, or the output from a single-pole dc generator. The fluctuation of generated direct cur-

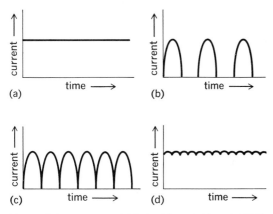

Types of direct current. (*a*) Output from battery. (*b*) Half-wave rectification. (*c*) Full-wave rectification. (*d*) Output from multiphase generator with ripple.

rent is called ripple (see illustration). In most applications the complete absence of ripple is not essential.

In parts of Europe direct current is still extensively used commercially, whereas in the United States it has largely been replaced by alternating current. Direct current cannot readily be changed from one voltage to another. As a result, it is not economically feasible to transmit direct current voltage for long distances over cross-country power lines. *See* ALTERNATING CURRENT; CURRENT (ELECTRICITY).

[JOHN W. STEWART]

Direct-current generator

A rotating electric machine which delivers a unidirectional voltage and current. An armature winding mounted on the rotor supplies the electric power output. One or more field windings mounted on the stator establish the magnetic flux in the air gap. A voltage is induced in the armature coils as a result of the relative motion between the coils and the air gap flux. Faraday's law states that the voltage induced is determined by the time rate of change of flux linkages with the winding. Since these induced voltages are alternating, a means of rectification is necessary to deliver direct current at the generator terminals. Rectification is accomplished by a commutator mounted on the rotor shaft. *See* ELECTRIC ROTATING MACHINERY; GENERATOR, ELECTRIC.

Carbon brushes, insulated from the machine frame and secured in brush holders, transfer the armature current from the rotating commutator to the external circuit. Brushes are held against the commutator under a pressure of $2-2\frac{1}{2}$ psi. Armature current passes from the brush to brush holder through a flexible copper lead. In multipolar machines all positive brush studs are connected

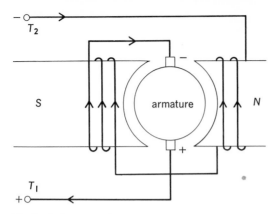

Fig. 1. Series generator.

together, as are all negative studs, to form the positive and negative generator terminals. In most dc generators the number of brush studs is the same as the number of main poles. In modern machines brushes are located in the neutral position where the voltage induced in a short-circuited coil by the main pole flux is zero. The brushes continuously pick up a fixed, instantaneous value of the voltage generated in the armature winding.

The generated voltage is dependent upon speed n in revolutions per minute, number of poles p, flux per pole Φ in webers, number of armature conductors z, and the number of armature paths a. The equation for the average voltage generated is given below.

$$E_g = \frac{np\Phi z}{60a} \quad \text{volts}$$

The field windings of dc generators require a direct current to produce a magnetomotive force (mmf) and establish a magnetic flux path across the air gap and through the armature. Generators are classified as series, shunt, compound, or separately excited, according to the manner of supplying the field excitation current. In the separately excited generator, the field winding is connected to an independent external source. Using the armature as a source of supply for the field current, dc generators are also capable of self-excitation. Re-

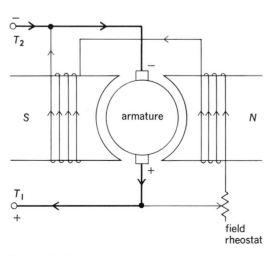

Fig. 2. Shunt generator.

sidual magnetism in the field poles is necessary for self-excitation. Series, shunt, and compound-wound generators are self-excited, and each produces different voltage characteristics.

When operated under load, the terminal voltage changes with change of load because of armature resistance drop, change in field current, and armature reaction. Interpoles and compensating, or pole-face, windings are employed in modern generators in order to improve commutation and to compensate for armature reaction.

Series generator. The armature winding and field winding of this generator are connected in series, as shown in Fig. 1. Terminals T_1 and T_2 are connected to the external load. The field mmf aids the residual magnetism in the poles, permitting the generator to build up voltage. The field winding is wound on the pole core with a comparatively few turns of wire of large cross section capable of carrying rated load current. The magnetic flux and consequently the generated emf and terminal voltage increase with increasing load current. Figure 4 shows the external characteristic or variation of terminal voltage with load current at constant speed. Series generators are suitable for special purposes only, such as a booster in a constant voltage system, and are therefore seldom employed.

Shunt generator. The field winding of a shunt generator is connected in parallel with the armature winding, as shown in Fig. 2. The armature supplies both load current I_t and field current I_f. The field current is 1–5% of the rated armature current I_a, the higher value applying to small machines. The field winding resistance is fairly high since the field consists of many turns of small-cross-section wire. For voltage buildup the total field-circuit resistance must be below a critical value; above this value the generator voltage cannot build up. The no-load voltage to which the generator builds up is varied by means of a rheostat in the field circuit. The external voltage characteristic (Fig. 4) shows a reduction of voltage with increases in load current, but voltage regulation is fairly good in large generators. The output voltage may be kept constant for varying load current conditions by manual or automatic control of the rheostat in the field circuit. A shunt generator will not maintain a large current in a short circuit in the external circuit, since the field current at short circuit is zero.

The shunt generator is suitable for fairly constant voltage applications, such as an exciter for ac generator fields, battery charging, and for electrolytic work requiring low-voltage and high-current capacity. Prior to the use of the alternating-current generator and solid-stage rectifying devices in automobiles, a shunt generator, in conjunction with automatic regulating devices, was used to charge the battery and supply power to the electrical system. Shunt-wound generators are well adapted to stable operation in parallel.

Compound generator. This generator has both a series field winding and a shunt field winding. Both windings are on the main poles with the series winding on the outside. The shunt winding furnishes the major part of the mmf. The series winding produces a variable mmf, dependent upon the load current, and offers a means of compensating for voltage drop. Figure 3 shows a cumulative-com-

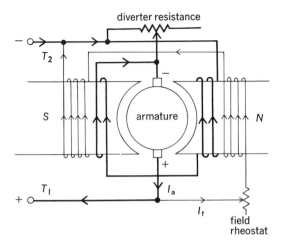

Fig. 3. Cumulative compound generator. The long-shunt connection is seen in this example.

pound connection with series and shunt fields aiding. A diverter resistance across the series field is used to adjust the series field mmf and vary the degree of compounding. By proper adjustment a nearly flat output voltage characteristic is possible. Cumulative-compound generators are overcompounded, flat-compounded, or undercompounded, as shown by the external characteristics in Fig. 4. The shunt winding is connected across the armature (short-shunt connection) or across the output terminals (long-shunt connection). Figure 3 shows the long-shunt connection.

Voltage is controllable over a limited range by a rheostat in the shunt field circuit. Compound generators are used for applications requiring constant voltage, such as lighting and motor loads.

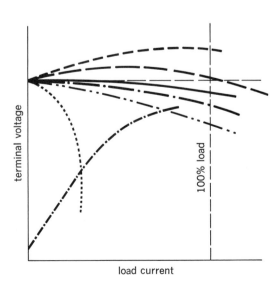

Fig. 4. External characteristics of dc generators.

Generators used for this service are rated at 125 or 250 volts and are flat or overcompounded to give a regulation of about 2%. An important application is in steel mills which have a large dc motor load. Cumulative-compound generators are capable of stable operation in parallel if the series fields are connected in parallel by an equalizer bus.

In the differentially compounded generator the series field is connected to oppose the shunt field mmf. Increasing load current causes a large voltage drop due to the demagnetizing effect of the series field. The differentially compounded generator has only a few applications, such as arc-welding generators and special generators for electrically operated shovels.

Separately excited generator. The field winding of this type of generator is connected to an independent dc source. The field winding is similar to that in the shunt generator. Separately excited generators are among the most common of dc generators, for they permit stable operation over a very wide range of output voltages. The slightly drooping voltage characteristic (Fig. 4) may be corrected by rheostatic control in the field circuit. Applications are found in special regulating sets, such as the Ward Leonard system, and in laboratory and commercial test sets.

Special types. Besides the common dc generators discussed in this article, there are a number of special types, including the homopolar, third-brush, diverter-pole, and Rosenberg generators. *See* DIRECT-CURRENT MOTOR.

Commutator ripple. The voltage at the brushes of dc generators is not absolutely constant. A slight high-frequency variation exists, which is superimposed upon the average voltage output. This is called commutator ripple and is caused by the cyclic change in the number of commutator bars contacting the brushes as the machine rotates. The ripple decreases as the number of commutator bars is increased and is usually ignored. In servomechanisms employing a dc tachometer for velocity feedback, the ripple frequency is kept as high as possible.

[ROBERT T. WEIL, JR.]

Bibliography: D. Fink and J. Carroll (eds.), *Standard Handbook for Electrical Engineers*, 10th ed., 1968; A. E. Fitzgerald and C. Kingsley, *Electric Machinery*, 1952; A. S. Langsdorf, *Principles of Direct-current Machines*, 5th ed., 1940; M. Liwschitz-Garik and R. T. Weil, Jr., *D-C and A-C Machines*, 1952; M. Liwschitz-Garik and C. C. Whipple, *Direct-current Machines*, 2d ed., 1956.

Direct-current motor

An electric rotating machine energized by direct current and used to convert electric energy to mechanical energy. It is characterized by its relative ease of speed control and, in the case of the series-connected motor, by an ability to produce large torque under load without taking excessive current. Output of this motor is given in horsepower, the unit of mechanical power. Normal full-load values of voltage, current, and speed are generally given.

Direct-current motors are manufactured in several horsepower-rating classifications: (1) subfractional, approximately 1–35 millihorsepower (mhp); (2) fractional, 1/40–1 horsepower (hp); and (3) in-

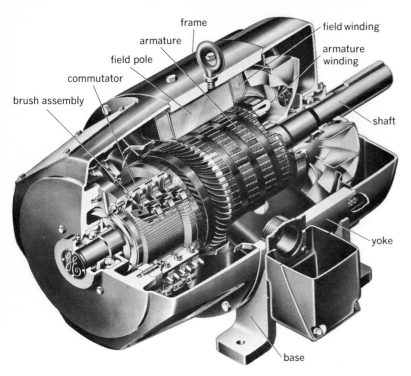

Fig. 1. Cutaway view of typical dc motor. (*General Electric*)

commutator and brush assemblies (Fig. 1). The frame consists of a steel yoke of open cylindrical shape mounted on a base. Salient field poles of sheet-steel laminations are fastened to the inside of the yoke. Field windings placed on the field poles are interconnected to form the complete field winding circuit. The armature consists of a cylindrical core of sheet-steel disks punched with peripheral slots, air ducts, and shaft hole. These punchings are aligned on a steel shaft on which is also mounted the commutator. The commutator, made of hard-drawn copper segments, is insulated from the shaft. Segments are insulated from each other by mica. Stationary carbon brushes in brush holders make contact with commutator segments. Copper conductors placed in the insulated armature slots are interconnected to form a reentrant lap or wave style of winding.

PRINCIPLES

Rotation of a dc motor is produced by an electromagnetic force exerted upon current-carrying conductors in a magnetic field. For basic principles of motor action *see* MOTOR, ELECTRIC.

In Fig. 2, forces act on conductors on the left path of the armature to produce clockwise rotation. Those conductors on the right path, whose current direction is reversed, also will have forces to produce clockwise rotation. The action of the commutator allows the current direction to be reversed as a conductor passes a brush.

The net force from all conductors acting over an average radial length to the shaft center produces a torque T given by Eq. (1), where K_t is a conver-

$$T = K_t \Phi I_a \qquad (1)$$

sion and machine constant, Φ is net flux per pole, and I_a is the total armature current.

The voltage E, induced as a counter electromotive force (emf) by generator action in the parallel paths of the armature, plus the voltage drop $I_a R_a$ through the armature due to armature current I_a and armature resistance R_a, must be overcome by the total impressed voltage V from the line. Voltage relations can be expressed by Eq. (2).

$$V = E + I_a R_a \qquad (2)$$

The counter emf and motor speed n are related by Eq. (3), where K is a conversion and machine constant.

$$n = \frac{E}{K\Phi} \qquad (3)$$

Mechanical power output can be expressed by Eq. (4), where n is the motor speed in rpm and T is the torque developed in pound-feet.

$$\mathrm{HP} = \frac{2\pi nT}{33,000} \quad \text{horsepower} \qquad (4)$$

By use of these four equations, the steady-state operation of the dc motor may be determined.

TYPES

Direct-current motors may be categorized as shunt, series, compound, or separately excited.

Shunt motor. The field circuit and the armature circuit of a dc shunt motor are connected in parallel (Fig. 3a). The field windings consist of many turns of fine wire. The entire field resistance, including a series-connected field rheostat, is rela-

tegral, 1/2 to several hundred horsepower.

The standard line voltages applied to dc motors are 6, 12, 27, 32, 115, 230, and 550 volts. Occasionally they reach higher values.

Normal full-load speeds are 850, 1140, 1725, and 3450 rpm. Variable-speed motors may have limiting rpm values stated.

Protection of the motor is afforded by several types of enclosures, such as splash-proof, dripproof, dust-explosion-proof, dust-ignition-proof, and immersion-proof enclosures. Some motors are totally enclosed.

The principal parts of a dc motor are the frame, the armature, the field poles and windings, and the

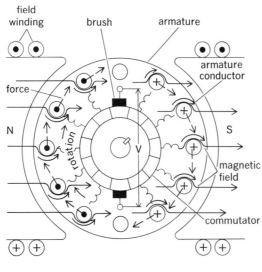

Fig. 2. Rotation in a dc motor.

tively large. The field current and pole flux are essentially constant and independent of the armature requirements. The torque is therefore essentially proportional to the armature current.

In operation an increased motor torque will be produced by a nearly equal increase in armature current, Eq. (1), since K_t and Φ are constant. Increased I_a produces an increase in the small voltage $I_a R_a$, Eq. (2). Since V is constant, E must decrease by the same small amount resulting in a small decrease in speed n, Eq. (3). The speed-load curve is practically flat, resulting in the term "constant speed" for the shunt motor. Typical characteristics are shown in Fig. 3b.

Typical applications are for load conditions of fairly constant speed, such as machine tools, blowers, centrifugal pumps, fans, conveyors, wood- and metalworking machines, steel, paper, and cement mills, and coal or coke plant drives.

Series motor. The field circuit and the armature circuit of a dc series motor are connected in series (Fig. 4a). The field winding has relatively few turns per pole. The wire must be large enough to carry the armature current. The flux Φ of a series motor is nearly proportional to the armature current I_a which produces it. Therefore, the torque, Eq. (1), of a series motor is proportional to the square of the armature current, neglecting the effects of core saturation and armature reaction. An increase in torque may be produced by a relatively small increase in armature current.

In operation the increased armature current, which produces increased torque, also produces increased flux. Therefore, speed must decrease to produce the required counter emf to satisfy Eqs. (1) and (3). This produces a variable speed characteristic. At light loads the flux is weak because of the small value of armature current, and the speed may be excessive. For this reason series motors are generally connected permanently to their loads through gearing.

The characteristics of the series motor are shown in Fig. 4b. Typical applications of this motor are to loads requiring high starting torques and variable speeds, for example, cranes, hoists, gates, bridges, car dumpers, traction drives, and automobile starters.

Compound motor. A compound motor has two separate field windings. One, generally the predominant field, is connected in parallel with the armature circuit; the other is connected in series with the armature circuit (Fig. 5).

The field windings may be connected in long or short shunt without radically changing the operation of the motor. They may also be cumulative or differential in compounding action. With both field windings, this motor combines the effects of the shunt and series types to an extent dependent upon the degree of compounding. In Fig. 6 its typical speed characteristics are compared with those of the shunt and series types. Applications of this motor are to loads requiring high starting torques and somewhat variable speeds, such as pulsating loads, shears, bending rolls, plunger pumps, conveyors, elevators, and crushers. *See* DIRECT-CURRENT GENERATOR.

Separately excited motor. The field winding of this motor is energized from a source different from that of the armature winding. The field winding may be of either the shunt or series type, and

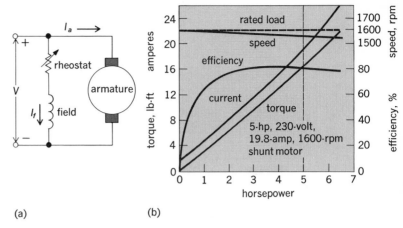

Fig. 3. Shunt motor. (a) Connections. (b) Typical operating characteristics.

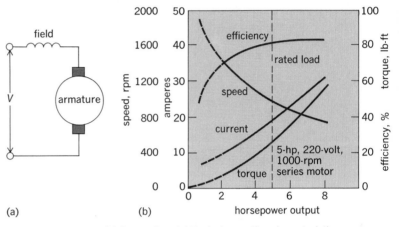

Fig. 4. Series motor. (a) Connection. (b) Typical operating characteristics.

adjustment of the applied voltage sources produces a wide range of speed and torque characteristics. Small dc motors may have permanent-magnet fields with armature excitation only. Such motors are used with fans, blowers, rapid-transfer switches, electromechanical activators, and programming devices.

STARTING AND SPEED CONTROL

Except in small dc motors, it is necessary to limit armature current when the motor is started. Therefore a load resistance must be in the armature circuit until the motor reaches full speed.

Starting. Direct-current motors are usually started with a rheostat in series with the armature circuit. This motor-starting resistor is of the proper rating in watts and ohms to withstand starting currents (Fig. 7).

When a dc motor is started, the field winding is fully excited. Since there is no rotation of the armature, no counter emf is generated. Therefore, the armature current would be dangerously high unless an additional starting resistance were placed in the armature circuit, Eq. (2). This rheostat is manually or automatically cut out of the circuit as the motor approaches full speed. Small motors which have low armature inertia reach full speed rapidly and they do not require starting resistors. Separately excited motors may be started by control of the voltages which are applied to the armature.

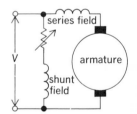

DIRECT-CURRENT MOTOR

Fig. 5. Connection of a compound motor.

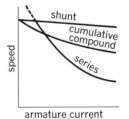

DIRECT-CURRENT MOTOR

Fig. 6. Comparative speed-current curves of dc motors.

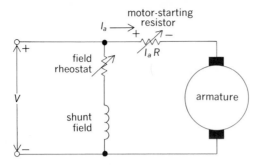

Fig. 7. Connection for starting a dc shunt motor.

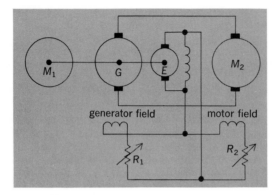

Fig. 8. Ward Leonard speed control system.

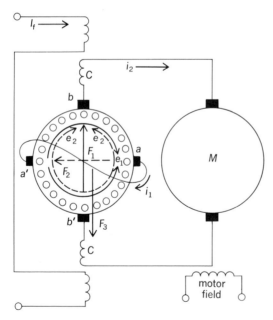

Fig. 9. Operation of the Amplidyne.

Speed control. Speed of a dc motor may be controlled by changing the flux or counter emf of the motor, Eq. (3). Adjustment of the armature voltage V will affect the counter emf E, Eq. (2), by approximately the same amount. The speed n is affected by the change in counter emf according to Eq. (3). Insertion of a resistor in the armature circuit would also affect the speed but is seldom used because of the large power losses in the resistor. Speed control by adjustment of the applied armature voltage is used extensively where separate, adjustable voltage sources are available.

A change in flux Φ will also affect speed n, Eq. (3). Flux may be changed by a variable resistor in series with the shunt field of a shunt or compound motor. This field rheostat should have a total resistance comparable to that of the shunt field and be of sufficient capacity to withstand the relatively small shunt-field current.

Ward Leonard. In this system the armature voltage of a separately excited dc motor is controlled by a motor-generator set. A typical circuit (Fig. 8) shows a prime mover M_1, often a three-phase induction motor, mechanically coupled to a dc generator G and to an exciter generator E. The latter provides field excitation for the dc machines. Control of the generator field rheostat R_1 affects the output voltage of the generator G. This voltage may be smoothly varied from a low value to a value above normal. When this voltage is applied to the armature of the motor M_2, the speed of this motor will be variable over a wide range. Additional speed control of motor M_2 may be gained by adjustment of rheostat R_2.

The disadvantages of this system are the added equipment and maintenance costs it entails. However, the wide range and fineness of control in a low-current circuit make it applicable to high-speed passenger elevators, large hoists, power shovels, steel-mill rolls, drives in paper or textile mills, and the propulsion of small ships.

Amplidyne. The dynamoelectric amplifier (Amplidyne) is a rotating, two-stage, power amplifier in which a small change in field power in a dc generator results in a large change in output armature power. A large motor connected to the output of the generator may be controlled in speed by adjustment of the relatively small field power of the Amplidyne.

In Fig. 9 the control field current I_f produces an mmf F_1. The resultant flux and the short circuit of brushes aa' cause the induced voltage e_1 to force a large current i_1 through armature circuit. Because of the magnetic core design, current i_1 produces mmf F_2 and its resultant flux which induces voltage e_2 between brushes bb'. Motor M connected across brushes bb' will draw a current i_2 which produces an mmf F_3 tending to weaken the original mmf F_1. However, compensating windings C energized by i_2 will produce an mmf to oppose F_3 and restore the value of F_1.

This dynamic amplifier may produce amplifications of 10,000 to 1 or higher. It is applied to a variety of servomotors to control starting, acceleration, and deceleration. Other typical applications include voltage regulation of large ac generators, dc voltage control in cold-strip mills, speed control of paper mills, positioning control of gun turrets, machine-tool drives, and power-factor control of synchronous generators.

Regulex. The regulating exciter (Regulex) is a dc generator acting as a power amplifier. By proper design of the machine magnetic core, an extensive linear portion of the voltage buildup curve is obtained (Fig. 10). A small change in mmf F will produce a large change in induced voltage E resulting in a degree of amplification. Critical-value adjustment of the field rheostat R (Fig. 11) will cause the generator to operate on this linear portion. A reference field F_2 and an opposing field F_3 combine with field F_1 to establish a point of operation, such as point a in Fig. 10.

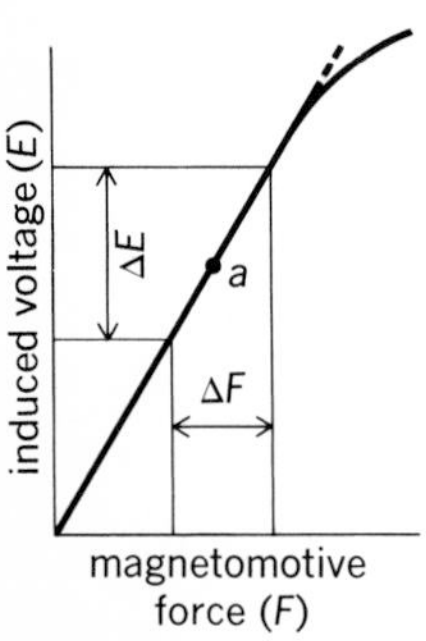

induced voltage (E)

ΔE

a

ΔF

magnetomotive force (F)

Fig. 10. Regulex magnetization curve.

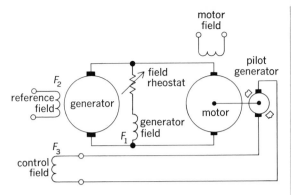

Fig. 11. Typical Regulex circuit.

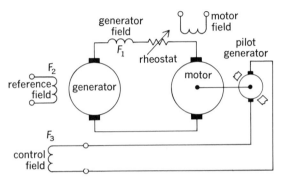

Fig. 12. Typical Rototrol circuit.

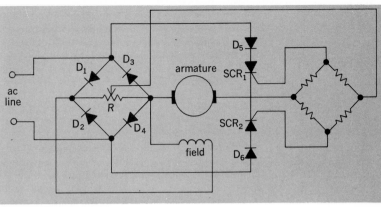

Fig. 14. SCR motor speed control.

Departure from this balance because of a variation in the control field F_3 will produce the large change in voltage E. The output of this device may be used to drive a dc motor M, which has its speed translated into voltage by means of a small pilot generator coupled to the motor shaft. By proper feedback of this voltage to control field F_3, the motor speed may be maintained at a constant value.

Rototrol. The rotating control (Rototrol) is a dc generator acting as a power amplifier. It is similar to the Regulex, but the self-excited field is a series type in contrast to the shunt-type field of the Regulex (Fig. 12).

Solid-state control for motor speed. Solid-state devices such as diodes and thyristors, including silicon-controlled rectifiers (SCRs), may be used in a number of circuit applications to control the speed of dc motors.

Pulse control. One such circuit supplies a number of unidirectional voltage pulses to the motor whose speed is adjusted by variation of pulse frequency or pulse width. In the circuit of Fig. 13, the thyristor acts as an ON-OFF switch. The relative on-to-off time determines the width of the voltage pulses. The start of thyristor conduction is deter-

mined by a gating voltage, but a negative voltage must be applied across the thyristor by the commutating circuit in order to stop its conduction. Controlling the width of the voltage pulses or the number of pulses per second (several hundred) determines the average value of the voltage applied to the motor and hence its speed. During the thyristor OFF period, diode D allows the energy stored in the motor field coils to be discharged through the motor armature. Current is continuous, increasing during the ON period and decreasing during the OFF period when the current is the result of the energy coming from the coils, but the average voltage is independent of the current.

Pulse control is applied to several types of electrically operated vehicles and affords an efficient and smooth control of speed.

Full-wave rectifier control. Thyristors (SCRs) are used in full-wave rectifier circuitry to control the motor speed, as indicated in Fig. 14.

Diodes D_1, D_2, D_3, and D_4 form a bridge rectifier to supply a full-wave rectified dc voltage to the field of the motor. A second bridge rectifier circuit also using diodes D_3 and D_4 as well as SCR_1 and SCR_2 provides adjustable current to the armature.

Control of the firing of the SCRs is accomplished by amplitude changes in the SCR gate voltages due to adjustment of potentiometer R. If the gate voltages exceed the back emf generated in the armature of the motor for a given load condition, the SCRs will fire. The average value of armature current and hence the speed is thus under control. Damage to the SCRs due to large values of reverse currents is prevented by diodes D_5 and D_6, which block the flow of such reverse currents.

[L. F. CLEVELAND]

Bibliography: A. E. Fitzgerald, D. E. Higginbotham, and A. Grabel, *Basic Electrical Engineering*, 4th ed., 1975; A. E. Fitzgerald, C. Kingsley, and A. Kusko, *Electric Machinery*, 3d ed., 1971; J. Rosenblatt and M. H. Friedman, *Direct and Alternating Current Machinery*, 1963; A. Kusko, *Solid-State DC Motor Drives*, 1969.

Distillate fuel

A broad term for any one of the wide variety of fuels obtained from fractions boiling above gasoline in the distillation of petroleum. The most important distillate fuels are kerosine, furnace oils, and diesel fuels. Formerly, heavy naphtha-kerosine distillates of low octane number (known as

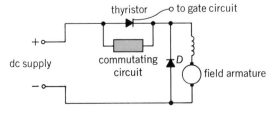

Fig. 13. Elementary Thymatrol circuit.

engine distillates or tractor fuels) were used as fuels for low-compression engines of farm tractors, farm lighting units, and small boats. Such power units have been largely replaced by high-compression gasoline engines or small diesels, requiring different fuels. *See* DIESEL FUEL; KEROSINE; PETROLEUM PRODUCTS.

[MOTT SOUDERS]

Dry cell

A primary cell in which the electrolyte is absorbed in a porous medium, or is otherwise restrained from flowing. Common practice limits the term dry cell to the Leclanche cell, which is the major commercial type. Other dry cells, not discussed in this article, include the mercury cell, the alkaline-zinc-manganese dioxide cell, and the air-depolarized cell. These cells all use aqueous electrolytes immobilized in absorbent materials or gels. By order of the Federal Trade Commission, "leakproof" must not be printed on any dry cell. *See* PRIMARY BATTERY.

Construction. The Leclanche cell is made in a variety of sizes in either round or flat shapes. American Standard Specifications for Dry Cells and Batteries, C.18.1–1965, lists the smallest cylindrical cell as the 0 cell, 0.02 in.^{3}, 0.032 oz, to the largest, the no. 6 cell, 29.45 in.^{3}, 35 oz. Flat-type cells, which are available as multiple-cell batteries only, range in size from the F 12 cell, 0.31 in.^{2} in cross section and 0.28 in. thick, to the F 100, which is 4.24 in.^{2} by 0.41 in. The energy-volume ratio of flat-cell batteries is about twice that of batteries made with round cells because of, first, the absence of an expansion chamber and carbon rod and, second, because of their rectangular form, which eliminates waste space in assembled batteries and voids between cells.

The negative electrode is usually solid zinc, either in cup form (cylindrical cell) or in sheet form (flat cell). The carbon positive electrode is embedded in a black mixture of manganese dioxide and carbon black. The carbon is either a rod located in the center of a cylindrical cell or a coating on the back of the flat zinc electrode in the flat cell. The separator between the black mix and the zinc electrode consists of a paper barrier coated with cereal or methyl cellulose. In the older construction the separator was composed of a gelatinous paste which also held the electrolyte. Not only does the paper barrier give about 30% more volume for the depolarizer mass, but it also decreases the internal resistance of the cell and reduces the number of manufacturing operations.

The electrolyte is a solution of ammonium chloride and zinc chloride in water. The cell enclosure consists of a top seal in the cylindrical cell or thin plastic wrappings for the flat cell.

Zinc electrode. Zinc cans (cup form of electrode) are made by drawing or impact extrusion. A typical alloy for extruded cans contains 1.0% lead, 0.05% cadmium, remainder zinc. The zinc used is of 99.99% purity. The alloying elements improve the mechanical properties and decrease wasteful corrosion. The cell capacity is less than that theoretically possible from the weight of zinc. This is because zinc is used as the container, the limiting factor being the manganese dioxide. For example, the no. 6 can weighs 110 g, giving a theoretical capacity of 90 amp-hr. The actual capacity is in the range of 40–50 amp-hr depending on the depolarizer composition and the rate of discharge.

Black mix. The black mix is composed of manganese dioxide mixed with carbon black. The manganese dioxide is usually obtained from natural ore (mainly from Gabon, Greece, and Mexico), but may be a synthetic product prepared by chemical precipitation or by electrolytic methods. Mixtures of the natural and synthetic oxides are also used. The carbon black is usually acetylene black made by the thermal decomposition of acetylene. Graphite is used to a lesser extent. Manganese dioxide has a theoretical capacity of about 0.3 amp-hr/g. The practical capacity is somewhat less. The carbon black is used in varying proportions, depending on design factors. It serves the double purpose of increasing the conductivity of the manganese dioxide and absorbing the electrolyte. For cells that require a high capacity at low current drains (for example, transistor use), the ratio of manganese dioxide to carbon may be 10:1 and, at the other extreme, cells requiring a very high flash current (for example, photoflash), the ratio may be as low as 1:1. In addition to these components, the black mix also contains electrolyte amounting to about 25% of the total weight.

Carbon rod. The carbon rod used in a cylindrical cell serves as the conductor of electricity for the positive electrode; it also serves as a vent to allow gas to escape. Carbon rods are usually made from petroleum coke which is calcined, ground, and mixed with pitch. The "green" rods are baked to form a hard carbon, having low electrical resistance. They may be partially waterproofed by impregnation with oil or paraffin wax to prevent capillary creepage of electrolyte out of the cell.

Flat cells are usually made with duplex electrodes. The zinc is coated on one side with a carbonaceous coating, which serves to conduct electricity between the zinc and the black mix of the adjacent cell.

Separator. The modern method of separating the two electrodes is to use kraft paper which has been coated on the side adjacent to the zinc with a film of cereal or methyl cellulose containing mercurous (or mercuric) chloride. The latter corresponds to a mercury concentration of 1 mg/in.^{2} of paper area.

In the older method a paste was made from a mixture of electrolyte with corn starch and wheat flour. The paste was added to the cell in liquid form and gelatinized by heating the cell in a water bath. By increasing the zinc chloride concentration in the electrolyte, the gelatinization of the paste could take place at ambient temperatures. Mercuric chloride was added to the paste, but it was not possible to control the intensity or uniformity of amalgamation to the same degree as with the paperlined cells. In a typical assembly line the cereal-coated paper is formed into a cylinder on a mandrel and inserted into the zinc can with a bottom washer. The calculated amount of wet depolarizer mix is injected into the lined can, followed by insertion of the carbon rod. Simultaneously with insertion of the carbon rod, the black mix is compressed. This gives a solid mass and forces sufficient electrolyte out of the mix to completely wet the cereal-coated paper barrier. A cardboard washer is placed on the black mix and an air space (expansion chamber) is enclosed over this washer by a further washer over which is poured a layer of bitumen. This seal is necessary to minimize moisture loss. Flat cells are made by placing treated paper containing the paste between the black-mix cake and the zinc of each cell.

Electrolyte. The electrolyte is made by dissolving ammonium chloride and zinc chloride in water. For paste cells a small amount of mercuric chloride is usually added. This component, however, converts to zinc chloride as soon as the zinc and electrolyte come into contact. Mercury then plates out on the zinc. The composition of the electrolyte depends on the cell designs.

During discharge the composition of the electrolyte changes. In one test in which a D-size cell was discharged through a 4-ohm resistance, the pH of the paste layer next to the zinc changed from 5.7 to 3.8 (more acid) while the pH of the innermost portion of the mix went from 5.8 to 10.1 (more alkaline).

Ordinary dry-cell electrolyte has a resistivity of 2.42 ohm-cm at +20°C. For low-temperature operation special electrolytes have been developed. An electrolyte of 12% zinc chloride, 15% lithium chloride, 8% ammonium chloride, and 65% water is fluid at −40°C. Other electrolytes for low-temperature operation use a mixture of calcium chloride, zinc chloride, and ammonium chloride solutions.

Cell enclosure. Whereas the cylindrical cell was originally wrapped in a paper jacket, modern methods are more sophisticated in order to resist leakage which could damage valuable equipment, for example, cameras, tape recorders, and record players. One method uses an absorbent board wrap and an outer jacket of sheet steel. The cells are finally sealed in these containers with tinplate top and bottom closures. The other method (see illustration) uses a jacket consisting of a laminate of absorbent paper, polyethylene, kraft paper covered with a cellulose acetate–coated label. As with the steel-jacketed cells, tinplate covers are placed top and bottom to make contact with the carbon rod and zinc can, respectively.

Flat cells use thin plastic wrappings around the edges of each cell. This confines the electrolyte to individual cells and avoids internal discharge in a stack. The wrappings are sufficiently gas-permeable to prevent the building up of pressure in the cell. After the requisite number of cells are stacked, the stack is bound together by tapes, and dipped in molten wax for further moistureproofing.

Cell chemistry. At the anode (zinc) the zinc oxidizes to zinc ion and simultaneously liberates electrons to the external circuit, at a rate proportional to the current. For each ampere which flows, 1.2 g of zinc per hour is converted to zinc ion.

At the cathode (manganese dioxide), the electrons from the external circuit reduce the manganese dioxide to three different substances, depending on circumstances which have not yet been thoroughly explained. Studies have shown, however, that the total ampere-hour output of the cell can be accounted for by analyzing the cathode mix for the following substances: soluble manganese (Mn^{++}), each gram of which accounts for nearly 1 amp-hr of discharge; insoluble manganite

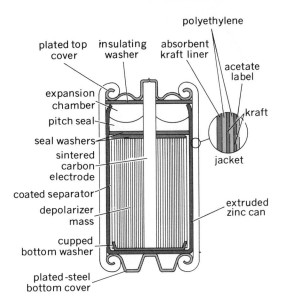

polyethylene
plated top cover
insulating washer
absorbent kraft liner
acetate label
expansion chamber
kraft
pitch seal
seal washers
sintered carbon electrode
jacket
coated separator
depolarizer mass
extruded zinc can
cupped bottom washer
plated-steel bottom cover

Modern Leclanche dry cell. (*Bright Star Industries Inc.*)

($MnOOH$), each gram of which accounts for about 0.3 amp-hr of discharge; insoluble hetaerolite ($ZnO \cdot Mn_2O_3$), each gram of which accounts for about 0.22 amp-hr of discharge.

The electrochemical reduction of the manganese dioxide (MnO_2) has been reported to occur as the reaction to form soluble manganese, shown as Eq. (1). This occurs only when the cell delivers current.

$$MnO_2 + 4H^+ + 2e^- = Mn^{++} + 2H_2O \qquad (1)$$

Two secondary reactions, Eqs. (2) and (3), can then occur.

$$MnO_2 + Mn^{++} + 2OH^- = 2MnOOH \qquad (2)$$

$$MnO_2 + Mn^{++} + 4OH^- + Zn^{++}$$
$$= ZnO \cdot Mn_2O_3 + 2H_2O \qquad (3)$$

Reaction (3) can occur only if zinc is in solution in the cathode mix.

Operating characteristics. The service capacity of dry cells is not a fixed number of ampere-hours, but varies with current drain, operating schedule, cutoff voltage, operating temperature, and storage conditions prior to use. Most cells are tailor-made for their rated end use. For example, D-size cells can be rated as general-purpose, industrial flashlight, transistor, electronic flash, and photoflash. No. 6 cells are specially formulated for bell ringing, telephone, protective alarm, ignition, and general purpose. Minimum outputs of D cells are specified by the General Service Administration are as follows:

(1) General-purpose flashlight — $2\frac{1}{4}$ ohms for 5-min periods at 24-hr intervals until the closed-circuit voltage drops to 0.65 volt; 400 min. (2) Light industrial flashlight — 4 ohms for 4-min periods beginning at hourly intervals for 8 consecutive hours each day until the closed circuit voltage drops to 0.9 volts; 950 min. (3) Heavy industrial flashlight — 4 ohms for 4-min periods beginning at 15 min intervals for 8 consecutive hours each day until the closed-circuit voltage drops to 0.9 volt; 800 min. (4) Photoflash bulb test — 0.15 ohm for 1

sec each minute for 1 hr at 24-hr intervals for 5 consecutive days each week until the closed-circuit voltage falls below 0.5 volts; 800 sec. (5) Electronic photoflash — 1.0 ohm for 15 sec each minute for 1 hr at 24-hr intervals for 5 consecutive days each week until the closed-circuit voltage falls below 0.75 volt; 275 15-sec discharges. (6) Transistor test — $83\frac{1}{3}$ ohms during a continuous period of 4 hr daily until the closed-circuit voltage falls below 0.9 volt; 200 hr.

Temperature effect. The higher the temperature during discharge, the greater is the energy output. Conversely, the lower the temperature, the lower is the output. At −10°F the battery is virtually inoperative. However, shelf life is influenced in the reverse direction by environmental temperatures.

Better low-temperature output can be obtained with special electrolytes and cell structures giving a high ratio of electrode area to mix thickness, and special types of manganese dioxide.

Shelf life. This is the period of time that a battery can be stored before it drops to 90% of its capacity when tested fresh at 70°F and 50% relative humidity.

Deterioration in a dry cell occurs in a number of ways: (1) Zinc can oxidize by reaction with the electrolyte; this reaction produces hydrogen. (2) Manganese dioxide can be reduced by carbon and by the organic materials used in the cells; this can produce carbon dioxide. (3) Water can be evaporated from the electrolyte; this increases cell resistance and alters the composition of the electrolyte unfavorably.

In general, shelf life decreases as the cell size becomes smaller: with well-constructed cells a shelf life of 3 years with a no. 6 telephone cell and 10 months with a penlight cell. Other sizes can be prorated. Flat cells have a shorter shelf life than cylindrical cells: 6–9 months for all sizes. American Standard Specifications quote 30 hr as the minimum initial output for a 9-volt transistor radio battery (six flat cells), and 28 hr after 6 months' delay. High temperatures reduce shelf life, and at 90°F the shelf life of a battery is about 1/3 that of one stored at 70°F. Low-temperature storage increases the shelf life of batteries considerably. Batteries stored at 40°F (sealed polyethylene bags should be used to prevent condensation with subsequent corrosion of the terminals and metal jacket or degradation of the paper jacket) have their shelf life increased two or three times. Tests have been conducted by military agencies in many countries showing that, when batteries are frozen, they suffer no deterioration for about 10 years. They must, however, be allowed to reach room temperature before use.

Recharging. It is possible to recharge dry-cell batteries for five or six cycles provided the following precautions are taken: The battery should be subject only to shallow discharges between cycles, used immediately after charging, charged over a 10- to 15-hr period at constant current, and not overcharged.

Modifications. In addition to those systems discussed in other sections, the following combinations have been developed: (1) The magnesium + magnesium perchlorate or bromide + manganese dioxide modification is more expensive than the Leclanche type. It exhibits the delayed voltage at the commencement of discharge characteristic of magnesium cells and has excellent storage properties. It has a higher initial voltage than the standard dry cell. (2) Another example of cell modification is magnesium + magnesium perchlorate or bromide + metadinitrobenzene. The organic depolarizer used can be produced at a cost that gives the same number of watt-hours per dollar as electrolytic manganese dioxide. This factor becomes important as supplies of battery-grade natural ore diminish. (3) The cell modification of aluminum + aluminum and chromic chlorides and ammonium chromate + manganese dioxide is attractive on account of the lower density and electrochemical equivalent of aluminum compared with zinc.

[JACK DAVIS]

Bibliography: *American Standard Specifications for Dry Cells*, C.18.1, 1965; Nicholas Branz, *Modern Primarbatterien*, 1951; D. H. Collins (ed.), *Power Sources*, biennial; *Eveready Battery Applications and Engineering Data*, 1968; Arthur Fleischer (ed.), *Proceedings of the Power Sources Conference*, annual; G. W. Vinal, *Primary Batteries*, 1950.

Dynamic braking

A technique of electric braking in which the retarding force is supplied by the same machine that originally was the driving motor. Dynamic braking is effective only in high inertial systems wherein the kinetic energy of the motor rotor with its connected load is converted into electrical energy and dissipated mainly as I^2R losses or returned to the source.

The commonest type of dynamic braking will be explained for a direct-current (dc) motor. To accomplish braking action, the supply voltage is removed from the armature of the motor but not from the field. The armature is then connected across a resistor. The electromotive force generated by the machine, now acting as a generator driven by the kinetic energy of the rotating system, forces current in the reverse direction through the armature. Thus a torque is produced to oppose rotation, and the load decelerates as its kinetic energy is dissipated, mostly in the external resistor but to some extent in core and copper losses of the machine.

Electric braking can also be accomplished by causing the kinetic energy of the rotating system to be converted in the armature to electrical energy and then returned to the supply lines. This mode of operation, called regenerative braking, occurs when the counter electromotive force exceeds the supply voltage.

Interchanging two of the lines supplying a three-phase alternating-current (ac) induction motor also produces braking. In this case, called plugging, the direction of the electromagnetic torque on the rotor is reversed to cause deceleration. Both the kinetic energy of the system and the energy drawn from the supply lines are expended in copper and core losses in the machine.

Sometimes the term dynamic braking is applied only when kinetic energy is dissipated in an external resistor, but in its more general interpretation the term includes regenerative braking and plugging.

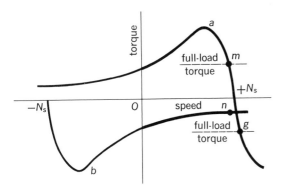

When the induction motor converts electric energy into mechanical torque for positive rotation, it operates along curve *a* in the first quadrant. When the induction motor converts mechanical torque into electrical energy, it operates along curves *a* or *b* in the fourth quadrant; curve *b* for plugging and curve *a* for regenerative braking.

The induction motor, when employed for dynamic braking in either the regenerative or plugging mode, represents an interesting departure from the normal operation of an induction motor. Curve *a* of the figure is a typical torque-speed characteristic of an induction motor, with N_s the synchronous speed. Curve *b* is the torque-speed characteristic when the rotation of the stator flux is opposite to the direction of rotation of the motor.

Interchanging two of the ac lines results in plugging due to reversal of the direction of rotation of the magnetic field. If the motor develops full-load torque at *m*, operation will be at point *n* on curve *b* immediately after plugging. The negative torque will decelerate the load, and the kinetic energy plus the energy drawn from the line will be dissipated in the rotor resistance. Unless the supply line is disconnected when the speed reaches zero, operation will be in the third quadrant along curve *b*, and the machine will become a motor operating in the reverse direction.

For regenerative braking to occur, the power transmitted across the air gap of the machine is also negative, and it results when the rotor speed is greater than synchronous speed. Point *g* on curve *a* corresponds to the condition of regenerative braking with rated torque. The occurrence of negative torque and power across the air gap means that the machine is an induction generator. The kinetic energy is converted into electrical energy and, less internal losses, is returned to the supply. When operating as a generator, an induction machine operates with a leading power factor; when operating as a motor, it operates with a lagging power factor. This behavior is a consequence of the fact that, for all conditions of operation, the power lines must supply the magnetizing current. *See* ALTERNATING-CURRENT GENERATOR: DIRECT-CURRENT MOTOR: INDUCTION MOTOR: SYNCHRONOUS MOTOR.

[ARTHUR R. ECKELS]

Dynamo

An electric machine for the conversion of electrical energy into mechanical energy or, conversely, mechanical energy into electrical energy. It is called a generator if it converts mechanical into electrical energy, and it is called a motor if it converts electrical into mechanical energy. *See* ELECTRIC ROTATING MACHINERY; GENERATOR, ELECTRIC: MOTOR, ELECTRIC.   [ARTHUR R. ECKELS]

Dynamometer

A special type of electric rotating machine used to measure the output torque or driving torque of rotating machinery. Most dynamometers consist of a direct-current (dc) machine with the stator cradle-mounted in antifriction bearings. The rotor is connected to the rotor of the machine under test. The field current is introduced through flexible leads. The stator is constrained from rotating by a radial arm of known length to which is attached a scale for measuring the force required to prevent rotation. The torque of the connected machine is found from the product of the lever arm length and the scale reading, after correcting the scale reading by the amount of the zero torque reading. By using a tachometer to measure the rotor speed, the power may be found from the equation hp $= 2\pi NT/33,000$, where N is the shaft speed in rpm and T is the torque in foot-pounds. *See* TORQUE.

If the machine under test is a motor, the dynamometer will act as a generator. The dynamometer output is absorbed by a loading resistance or by feeding it into a dc line. The amount of the output is easily adjusted by changing the loading resistance or by changing the field excitation. If the machine under test is a generator or mechanical load, the dynamometer will act as a motor. The speed is adjusted by changing the armature voltage or the field excitation. The dynamometer method is direct-reading and is more accurate than measuring the electrical output and correcting for the losses. Except for the inaccuracy caused by friction in the stator mounting bearings and by windage loss not reflected in the stator torque, all the shaft torque is accounted for in the scale reading. *See* DIRECT-CURRENT GENERATOR; DIRECT-CURRENT MOTOR; ELECTRIC ROTATING MACHINERY; GENERATOR, ELECTRIC; MOTOR, ELECTRIC.

When the machine under test is a motor, a mechanical device known as a prony brake may be employed to convert the output energy to heat through friction. A drum, which may be water-cooled, is driven by the machine under test. A brake arm, which is constrained from rotating by a scale at the outer end, is tightened around the drum to increase the friction and to produce the desired torque. The torque is the product of the scale reading and the length of the brake arm. This is an inexpensive and accurate method of measuring the torque of small motors. With high-horsepower motors, however, it becomes difficult to dissipate the large amounts of energy. Large-capacity units, which utilize a liquid brake in place of the friction drum, have been constructed. These are smaller and less expensive than electric units of like capacity but lack the flexibility and ease of recovering the energy.   [ARTHUR R. ECKELS]

Economizer, boiler

A component of a steam-generating unit that absorbs heat from the products of combustion after they have passed through the steam-generating and superheating sections. The name, accepted through common usage, is indicative of savings in the fuel required to generate steam.

An economizer is a forced-flow, once-through, convection heat-transfer device to which feedwater is supplied at a pressure above that in the steam-generating section and at a rate corresponding to the steam output of the unit. The economizer is in effect a feedwater heater, receiving water from the boiler feed pump and delivering it at a higher temperature to the steam generator or boiler. Economizers are used instead of additional steam-generating surface because the feedwater, and consequently the heat-receiving surface, is at a temperature below that corresponding to the saturated steam temperature; therefore, the economizer further lowers the flue gas temperature for additional heat recovery. *See* THERMODYNAMIC CYCLE.

Generally, steel tubes, or steel tubes fitted with externally extended surface, are used for the heat-absorbing section of the economizer; usually, the economizer is coordinated with the steam-generating section and placed within the setting of the unit.

The size of an economizer is governed by economic considerations involving the cost of fuel, the comparative cost and thermal performance of alternate steam-generating or air-heater surface, the feedwater temperature, and the desired exit gas temperature. In many situations it is economical to utilize an economizer and an air heater together.

[G. W. KESSLER]

Edison battery

A storage battery composed of cells having nickel and iron in an alkaline solution, also called a nickel-iron alkaline battery. The active material on the negative plates is iron and that on the positive plates is nickel oxide. The electrolyte is potassium hydroxide. During discharge the nickel oxide is reduced to a nickelous hydroxide and the iron is oxidized to ferrous hydrate. During charge the reverse process takes place. The Edison battery is lighter in weight than the more common lead storage battery and has longer life because there is no chemical deterioration during charge and discharge. Edison storage batteries are used chiefly in electric industrial trucks and tractors, mine locomotives, railroad signal and lighting service, telephone service, isolated airway beacons, and emergency power for lighting, police, and fire alarm systems. *See* STORAGE BATTERY.

[JOHN MARKUS]

Efficiency

The ratio, expressed as a percentage, of the power output to the power input. When only mechanical efficiency is concerned, the difference, or loss, between the input and output power is due to friction and is dissipated in the form of heat.

[RICHARD M. PHELAN]

Electric distribution systems

That part of an electric power system that supplies electric energy to the individual user or consumer. The distribution system includes the primary circuits and the distribution substations that supply them; the distribution transformers; the secondary circuits, including the services to the consumer; and appropriate protective and control devices. The four general classes of individual users are residential, industrial, commercial, and rural.

Systems. The three-phase, alternating-current (ac) system (Fig. 1) is practically universal, although a small amount of two-phase and direct-current systems from early days are still in operation. Three-phase transmission and substransmission lines require three wires, termed phase conductors. Most of the three-phase distribution systems consist of three phase conductors and a common or neutral conductor, making a total of four wires. Single-phase branches (consisting of two wires) supplied from the three-phase mains are used for single-phase utilization in residences, small stores, and farms. Loads are connected in parallel to common supply circuits. *See* ELECTRIC POWER SYSTEMS.

Substation. The distribution substation is an assemblage of equipment for the purpose of switching, changing, and regulating the voltage from subtransmission to primary distribution. More important substations are designed so that the failure of a piece of equipment in the substation or one of the subtransmission lines to the substation will not cause an interruption of power to the load.

Primary voltages. The primary system leaving the substation is most frequently in the 11,000–15,000 volt range. A particular voltage used is 12,-470-volt line-to-line and 7200-volt line-to-neutral (conventionally written 12,470Y/7200 volts). Some utilities use a lower voltage, such as 4160Y/2400 volts. The use of voltages above the 15-kv class is increasing. Several percent of primary distribution circuits are in the 25- and 35-kv classes; all are four-wire systems. Single-phase loads are connected line-to-neutral on the four-wire systems.

Secondary voltages. Secondary voltages are derived from distribution transformers connected to the primary system and they usually correspond to utilization voltages. Residential and most rural loads are supplied by 120/240-volt single-phase three-wire systems (Fig. 2). Commercial and small industrial needs are supplied by either 208Y/120-volt or 480Y/277-volt three-phase four-wire systems.

The secondary voltage is usually used to supply multiple street lights, in addition to service to consumers. Photoelectric controls are employed to turn the street lamps on and off.

Good voltage. Good voltage means that the average voltage level is correct, that variations do not exceed prescribed limits, and that sudden momentary changes in level do not cause objectionable light flicker. Utilization voltage varies with changing load on the system, but a voltage variation of less than 5% at the consumer's meter is common. To achieve this result, distribution systems are designed for a plus and minus voltage spread from the nominal voltages shown above. This is accomplished by proper wire size for the circuits, application of capacitors, both permanently connected and switched, and use of voltage regulators. Voltage regulators may be used at the substation or at a point along the circuit.

Good continuity. Service continuity is the providing of uninterrupted electric power to the con-

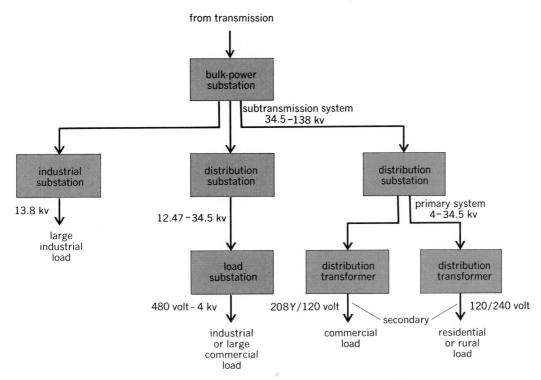

Fig. 1. Typical three-phase power system from bulk-power source to consumer's switch.

sumer; therefore, good continuity is doing this a high percentage of the time. This is accomplished for large industrial and commercial loads by use of some form of duplicate power supply. Downtown commercial areas are supplied from three-phase 208Y/120-volt grid networks (Fig. 3). These networks are fed from a number of primary feeders, stepping down through network transformers and automatic-reclosing secondary circuit breakers (protectors) to the secondary grid formed by cables under the streets. The system is arranged so that the failure of a primary feeder will not cause a loss of load on the secondary. Commercial buildings and shopping centers are often served by spot networks. All of the transformers and protectors are at the same location. Residential and rural loads are usually supplied by a radial system. Good continuity for them is obtained by sectionalizing the system with fuses, circuit breakers, and manual switches to reduce the extent of an outage due to a failure.

Elements of distribution systems. In distribution systems there are a number of elemental parts

or subsystems, which are discussed below.

Primary feeders. Power is carried by primary feeders from distribution substations to the load areas where the consumers are located.

Distribution transformers. The distribution transformer located near the consumers changes the voltage from the primary distribution voltage to the secondary distribution voltage.

Secondary mains. This element of electric distri-

Fig. 2. Single-phase three-wire secondary circuit.

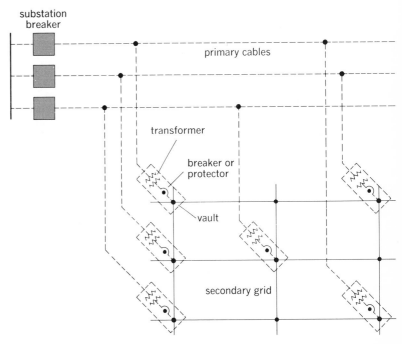

Fig. 3. Typical three-phase 208Y/120-volt secondary grid network system.

bution is a low-voltage system which connects the secondary winding of distribution transformers to the consumers' services.

Overhead construction. The majority of distribution systems already built in residential, industrial, and rural areas has been overhead construction. This construction utilizes poles treated with pentachlorophenol or creosote, distribution transformers mounted near the top of the pole, and bare primary and secondary conductors strung from pole to pole. Aluminum conductors have been used in place of copper because of their lower cost.

For aesthetic reasons, underground distribution systems are being installed in most new residential developments. *See* TRANSMISSION LINES.

Underground construction. Commercial areas and certain main thoroughfares usually have underground construction. Construction for these areas is usually a system of concrete or fiber ducts and vaults. Impregnated paper protected by lead sheaths or synthetic compounds is generally used as conductor insulation.

In residential areas the cable is usually insulated with polyethylene or chemically crosslinked polyethylene, which is a thermosetting material. Primary cables supplying distribution transformers in the development area are likely to be single-phase, consisting of an insulated phase conductor with a concentric bare neutral conductor taken from a 12,470-volt three-phase primary feeder. The cables are often directly buried in the earth. The distribution transformers used are either the submersible type installed in a hole with a liner or pad-mounted and located on the surface of the ground. *See* ELECTRIC POWER SUBSTATION.

[HAROLD E. CAMPBELL]

Bibliography: Edison Electric Institute, *Underground Systems Reference Book*, EEI Publ. no. 55-15, 1967; B. G. A. Skortizki (ed.), *Electric Transmission and Distribution*, 1954; U.S. Standards Institute, *Definition of Electrical Terms: Transmission and Distribution (Group 35)*, USASI C42.35, 1957; U.S. Standards Institute, *Preferred Voltage Ratings for A-C Systems and Equipment: Guide for (EEI R-6-1949) (IEC38)*, USASI C84.1, 1954; E. Vennard, *The Electric Power Business*, 1962.

Electric energy measurement

The measurement of the integral, with respect to time, of the power in an electric circuit. The absolute unit of measurement of electric energy is the joule. The joule, however, is too small (1 watt-second) for use in commercial practice, and the more common unit is the watt-hour (3.6×10^3 joules).

The most common measurement application is in the utility field, where it is estimated that over 1,693,659,000,000 kWhr were measured and sold during 1974 to industry and residential consumers in the United States.

Electric energy is one of the most accurately measured commodities sold to the general public. Many methods of measurement, with different degrees of accuracy, are possible. The choice depends on the requirements and complexities of the measurement problems. Basically, measurements of electric energy may be classified into two categories, direct-current power and alternating-current power. The fundamental concepts of measurement are, however, the same for both.

Methods of measurement. There are two types of methods of measuring electric energy: electric instruments and timing means, and electricity meters.

Electric instruments and timing means. These make use of conventional procedures for measuring electric power and time. The required accuracy of measurement dictates the type and quality of the measuring devices used (for example, portable instruments, laboratory instruments, potentiometers, stopwatches, chronographs, and electronic timers). Typical methods are listed below. *See* ELECTRIC POWER MEASUREMENT.

1. Measurement of energy on a direct-current circuit by reading the line voltage and load current at regular intervals over a measured period of time. The frequency of reading selected (such as one per second, one per 10 seconds) depends upon the steadiness of the load being measured, the time duration of the test, and the accuracy of measurement desired.

In electric energy measurements, the losses in the instruments must be considered. Unless negligible from a practical standpoint, they should be deducted from the total energy measured. If the voltmeter losses are included in the total energy measured, then watt-hours $= (VI - V^2R)t/3600$, where V is the average line voltage (volts), I is the average line current (amperes), R is the voltmeter resistance (ohms), and t is the time (seconds).

2. Measurement of energy on a direct-current circuit by controlling the voltage and current at constant predetermined values for a predetermined time interval. This method is common for controlling the energy used for a scientific experiment or for determining the accuracy of a watt-hour meter. For best accuracy, potentiometers and electronic timers are desirable.

3. Measurement of energy on an alternating-current circuit by reading the watts input to the load at regular intervals over a measured period of time. This method is similar to the first, except that the power input is measured by a wattmeter.

4. Measurement of energy on an alternating-current circuit by controlling the voltage, current, and watts input to the load at constant predetermined values. This method is similar to the second, except that the power input is measured by a wattmeter. A common application of this method is to determine the standard of measurement of electric energy, the watt-hour.

5. Measurement of energy by recording the watts input to the load on a linear chart progressed uniformly with time. This method makes use of a conventional power record produced by a recording wattmeter. The area under the load record over a period of time is the energy measurement.

Electricity meters. These are the most common devices for measuring the vast quantities of electric energy used by industry and the general public. The same fundamentals of measurement apply as for electric power measurement, but in addition the electricity meter provides the time-integrating means necessary for electric energy measurement.

A single meter is sometimes used to measure the energy consumed in two or more circuits. However, multistator meters are generally required for this purpose. Totalization is also accomplished with fair accuracy, if the power is from the same voltage source, by paralleling secondaries of instrument current transformers of the same ratio at the meter. Errors can result through unbalanced loading or use of transformers with dissimilar characteristics.

Watt-hour meters are generally connected to measure the losses of their respective current circuits. These losses are extremely small compared to the total energy being measured and are present only under load conditions.

Other errors result from the ratio and phase angle errors in instrument transformers. With modern transformers these errors can generally be neglected for commercial metering. If considered of sufficient importance, they can usually be compensated for in adjusting the calibration of the watt-hour meter. For particularly accurate measurements of energy over short periods of time, portable standard watt-hour meters may be used. *See* WATT-HOUR METER.

Watt-hour meters used for the billing of residential, commercial, and industrial loads are highly developed devices. Over the last decade many significant improvements have been made, including improvements in bearings, insulating materials, mechanical construction, and new sealing techniques which exclude dust and other foreign material. As a result of the higher degree of accuracy and dependability achieved by modern meters, the utility industries have adopted statistical sampling methods for testing of in-service accuracy as sanctioned by ANSI C12-1965.

Automatic remote reading. The energy crisis, high cost of capital, and increased concern for the environment have created widespread and growing national interest in the pricing structure for electric energy, historically based on total energy consumption. Proposals to conserve electric energy and improve utility system load factors by new price structures and by control of residential loads are being advanced by utility companies, utility commissions, and Federal agencies.

These new dimensions have spurred active development of automatic meter-reading systems with the functional capability of providing meter data for proposed new rate structures, for example, time-of-day pricing, and for initiating control of residential loads, such as electric hot-water heaters.

Automatic meter-reading systems under development generally consist of a utility-operated, minicomputer-controlled reading center, which initiates and transmits commands over a communication system to a terminal at each residential meter. The terminal carries out the command, transmitting the appropriate meter reading back to the reading center, or activating the control of a residential load.

Several communication media are being proposed and tested by system developers, including radio, CATV, use of the existing subscriber phone lines, and communication over the electric distribution system itself.

The rapid advances of technology, coupled with the increasing needs for improved management of energy usage, indicate that automatic meter reading and control systems may start replacing conventional manual meter reading within the next few years.

Quantities other than watt-hours. Included in the field of electric energy measurement are demand, var hours, and volt-ampere hours.

Demand. The American National Standards Institute defines the demand for an installation or system as "the load which is drawn from the source of supply at the receiving terminals, averaged over a suitable and specified interval of time. Demand is expressed in kilowatts, kilovolt-amperes, amperes, kilovars and other suitable units" (ANSI C12-1965).

This measurement provides the user with information as to the loading pattern or the maximum loading of equipments rather than the average loading recorded by the watt-hour meter. It is used by the utilities as a rate structure tool.

Var hour. ANSI defines the var hour (reactive volt-ampere hour) as the "unit for expressing the integral of reactive power in vars over an interval of time expressed in hours" (ANSI C12-1965).

This measurement is generally made by using reactors or phase-shifting transformers to supply to conventional meters a voltage equal to, but in quadrature with, the line voltage.

Volt-ampere hour. This is the unit for expressing the integral of apparent power in volt-amperes over an interval of time expressed in hours. Measurement of this unit is more complicated than for active or reactive energy and requires greater compromises in power-factor range, accuracy, or both. Typical methods include: (1) Conventional watt-hour meters with reactors or phase-shifting transformers tapped to provide an in-phase line voltage and current relationship applied to the meter at the mean of the expected range of power-factor variation. (2) A combination of a watt-hour and a var-hour meter mechanically acting on a rotatable sphere to add vectorially watt-hours and var-hours to obtain volt-ampere hours, volt-ampere demand, or both.

Measurement of volt-ampere hours is sometimes preferred over var-hours because it is a more direct measurement and possibly gives a more accurate picture of the average system power factor. This would not necessarily be true, however, where simultaneous active and reactive demand are measured and recorded. *See* ELECTRICAL MEASUREMENTS.   [J. ANDERSON]

Bibliography: T. S. Banghart and R. E. Riebs, Practical aspects of large-scale automatic meter reading using existing telephone lines, *Proceedings of the American Power Conference*, vol. 36, pp. 945–951, 1974; S. G. Hardy, *New Developments in Automatic Meter Reading*, paper presented to the EEI/AEIC Meter and Service Committee, Atlanta, Apr. 9, 1973; A new way to control a utility's load factor, *Bus. Week*, Aug. 24, 1974; A. W. Palmer and W. R. Germer, *A Novel Encoder/Register for a Self-Checking Automatic Meter Reading System*, IEEE Conf. Pap. C74-042-8, 1974; P. B. Robinson, Progress in automatic meter reading, *Proceedings of the American Power Conference*, vol. 36, pp. 959–964, 1974; Utilities Act of 1975, House of Representatives Bill, HR2650, Feb. 4, 1975.

Electric power generation

The production of bulk electric power for industrial, residential, and rural use. Although limited amounts of electricity can be generated by many

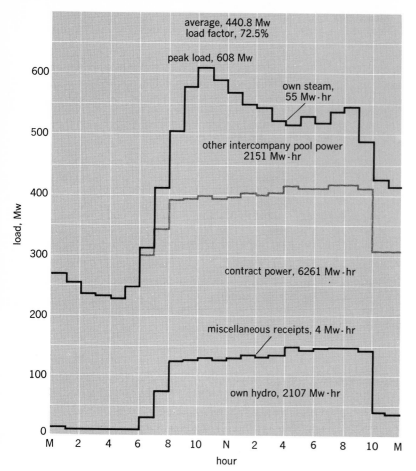

Fig. 1. Load graph indicates net system load of a metropolitan utility for typical 24-hr period (midnight to midnight), totaling 10,578 Mw-hr. Such graphs are made to forecast probable variations in power required.

means, including chemical reaction (as in batteries) and engine-driven generators (as in automobiles and airplanes), electric power generation generally implies large-scale production of electric power in stationary plants designed for that purpose. The generating units in these plants convert energy from falling water, coal, natural gas, oil, and nuclear fuels to electric energy. Most electric generators are driven either by hydraulic turbines, for conversion of falling water energy; or by steam or gas turbines, for conversion of fuel energy. Limited use is being made of geothermal energy, and developmental work is progressing in the use of solar energy in its various forms. Electric power generating plants are normally interconnected by a transmission and distribution system to serve the electric loads in a given area or region. *See* GENERATOR, ELECTRIC.

An electric load is the power requirement of any device or equipment that converts electric energy into light, heat, or mechanical energy, or otherwise consumes electric energy as in aluminum reduction, or the power requirements of electronic and control devices. The total load on any power system is seldom constant; rather, it varies widely with hourly, weekly, monthly, or annual changes in the requirements of the area served. The minimum

system load for a given period is termed the base load or the unity load-factor component. Maximum loads, resulting usually from temporary conditions, are called peak loads. Electric energy cannot be stored in large quantities; therefore the operation of the generating plants must be closely coordinated with fluctuations in the load.

Actual variations in the load with time are recorded, and from these data load graphs are made to forecast the probable variations of load in the future. A study of hourly load graphs (Fig. 1) indicates the generation that may be required at a given hour of the day, week, or month. A study of annual load graphs indicates the rate at which new generating stations must be built. Load graphs are an inseparable part of utility operation and are the basis for decisions that profoundly affect the financial requirements and overall development of a utility.

Generating plants. Often termed generating stations, these plants contain apparatus that converts some form of energy to electric energy in bulk. Three significant types of generating plants are hydroelectric, fossil-fuel-electric, and nuclear-electric.

Hydroelectric plant. This type of generating plant utilizes the potential energy released by the weight of water falling through a vertical distance called head. Ignoring losses, the power, in horsepower (hp) and kilowatts (kw), obtainable from falling water is shown in the equations below (metric quantities in brackets).

$$hp = \frac{\left(\begin{array}{c}\text{quantity of water} \\ \text{in ft}^3\text{/sec [m}^3\text{/sec]}\end{array}\right)\left(\begin{array}{c}\text{vertical head} \\ \text{in ft [m]}\end{array}\right)}{8.8\ [0.077]}$$

A plant consists basically of a dam to store the water in a forebay and create part or all of the head, a penstock to deliver the falling water to the turbine, a hydraulic turbine to convert the hydraulic energy released to mechanical energy, an alternating-current generator (alternator) to convert the mechanical energy to electric energy, and all accessory equipment necessary to control the power flow, voltage, and frequency, and to afford the protection required (Fig. 2). *See* DAM.

Pumped storage hydroelectric plants are being used increasingly. Under suitable geographical and geological conditions, electric energy can, in effect, be stored by pumping water from a low to a higher elevation and subsequently releasing this water to the lower elevation through hydraulic turbines. These turbines and their associated generators are reversible. The generators, operating in reverse direction as motors, drive their turbines as pumps to elevate the water. When this water is released through the turbines, electric power is produced by the generators. A relatively high overall cycle efficiency can be attained. *See* PUMPED STORAGE.

Since system peak loads are usually of relatively short duration (Fig. 1), the high output available for a short time from pumped storage can be used to supply this peak. During off-peak hours, that is, 10 P.M. to 6 A.M., the surplus generating capacity of the most economical system energy resources can be used to return the water by pumping to the elevated storage space for use on the next peak. This

type of operation assists in maintaining a high capacity factor on prime generation with resulting
best economy. *See* WATERPOWER.

Pumped storage plants can be brought up to
load much faster than large steam plants and,
hence, contribute to system reliability by providing
an immediately available reserve against the unscheduled loss of other generation.

Fossil-fuel-electric plant. This type utilizes the
energy of combustion from coal, oil, or natural gas.
A typical large plant consists of fuel processing
and handling facilities, a combustion furnace and
boiler to produce and superheat the steam, a
steam turbine, an alternator, and the accessory
equipment required for plant protection and for
control of voltage, frequency, and power flow. A
steam plant can frequently be built near a convenient load center, provided an adequate supply
of cooling water and fuel is available, and is usually
readily adaptable to either base loading or peak
loading. Environmental constraints require careful
control of stack emissions with respect to sulfur
oxides and particulates. Cooling towers or ponds
are often required for waste heat dissipation. Gas
turbine plants do not require condenser cooling
water (unless combined with a steam cycle), have
a relatively low unit capital cost and relatively high
unit fuel cost, and are widely used for peaking
service.

Nuclear electric plant. In this type of plant one
or more of the nuclear fuels are utilized in a suitable type of nuclear reactor, which takes the place
of the combustion furnace in the typical steam
electric plant. The heat exchangers and boilers (if
not combined in the reactor), the turbines, and
alternating-current generators, complete with controls, accessories, and auxiliaries, make up the
atomic electric plant. Large-scale fission reaction
plants have been developed to the point where
they are economically competitive in much of the
United States, and many millions of kilowatts of
capacity are under construction and more on order. The current and projected future growth of
the nuclear power industry in the United States is
shown graphically in Fig. 3. Although in 1975 only
approximately 8.5% of the total generating capacity was nuclear, by 1985 it was predicted to be
over 29%. Conventional hydro generation had
reached near saturation by 1975 and growth in
fossil-fuel generation showed a declining trend.
However, nuclear generation exhibits a doubling
time of less than 4 years during a considerable
period after 1975.

Three types of nuclear plants are in general use
and a fourth is being developed actively. The pressurized and boiling light-water types, shown schematically in Figs. 4 and 5, have been highly developed in the United States. The gas-cooled type,
one version of which is shown schematically in Fig.
6, has been highly developed and extensively used
in Great Britain. The liquid-metal (sodium) cooled
type is under intensive development and will be
featured prominently in the fast or breeder reactor
field. It is shown schematically in Fig. 7. As
in the high-temperature gas-cooled reactor system,
steam temperatures and pressures in the liquid-
metal systems can be essentially the same as those
recorded in the plants powered by fossil fuel and

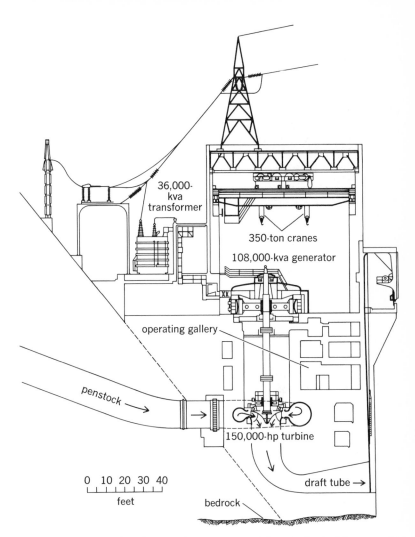

Fig. 2. Typical layout and apparatus arrangement in a hydroelectric generating
plant. Cross section is made through the powerhouse and dam at a main generator
position in the original Grand Coulee plant.

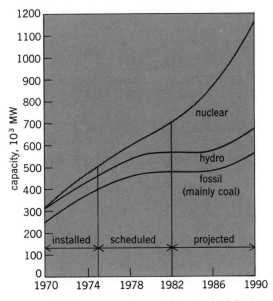

Fig. 3. Electric generating capacity in United States,
showing projected growth of nuclear power. (*USAEC*)

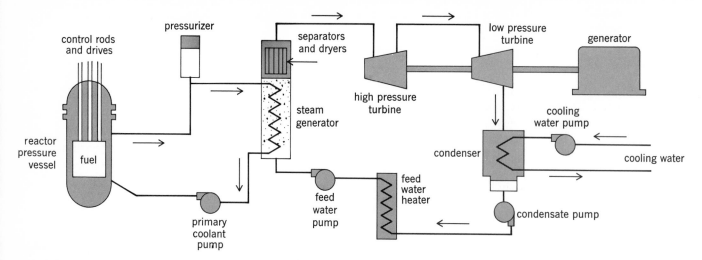

fuel: slightly enriched uranium oxide with zirconium alloy
moderator: water   coolant: water   pressure of primary system: 2250 psi
reactor outlet temperature: 605°F (318°C)

Fig. 4.   Schematic of a pressurized-water reactor plant. (*From The Nuclear Industry, USAEC, WASH 1174-73, 1973*)

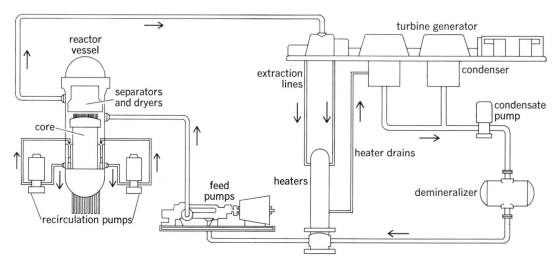

Fig. 5.   Single-cycle boiling-water reactor system flow diagram. (*General Electric Co.*)

efficiencies. *See* REACTOR, NUCLEAR.

Several other types of nuclear fission reactor systems are receiving attention and some may be expected to become commercially competitive. Fusion reaction plants are in the early research and development stage. Direct conversion from nuclear reaction energy to electric energy on a commercial scale for power utility service is a future possibility but is not economically feasible at present.

Power-plant circuits. Both main and accessory circuits in power plants can be classified as follows:

1. Main power circuits to carry the power from the generators to the step-up transformers and on to the station high-voltage terminals.

2. Auxiliary power circuits to provide power to the motors used to drive the necessary auxiliaries.

3. Control circuits for the circuit breakers and other equipment operated from the control room of the plant.

4. Lighting circuits for the illumination of the plant and to provide power for portable equipment required in the upkeep and maintenance of the plant. Sometimes special circuits are installed to supply the portable power equipment.

5. Excitation circuits, which are so installed that they will receive good physical and electrical protection because reliable excitation is necessary for the operation of the plant.

6. Instrument and relay circuits to provide values of voltage, current, kilowatts, reactive kilovolt-amperes, temperatures, and pressures, and to serve the protective relays.

7. Communication circuits for both plant and system communications. Telephone, radio, transmission-line carrier, and microwave radio may be involved.

It is important that reliable power service be provided for the plant itself, and for this reason station service is usually supplied from two or more sources. To ensure adequate reliability, aux-

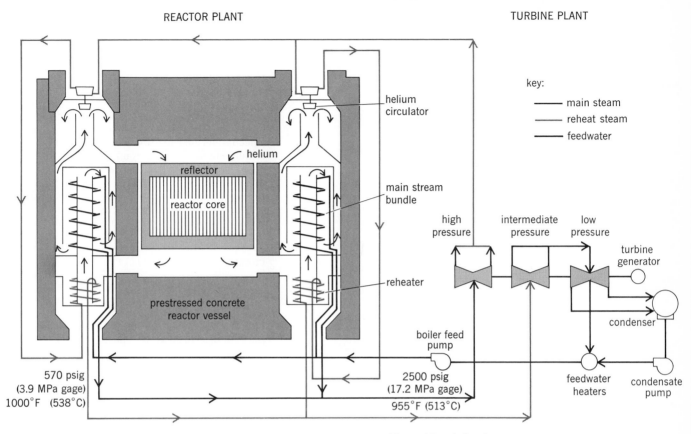

REACTOR PLANT

TURBINE PLANT

key:
— main steam
— reheat steam
— feedwater

helium
circulator

helium

reflector

reactor core

main stream
bundle

high
pressure

intermediate
pressure

low
pressure

turbine
generator

reheater

prestressed concrete
reactor vessel

condenser

boiler feed
pump

570 psig
(3.9 MPa gage)
1000°F (538°C)

2500 psig
(17.2 MPa gage)
955°F (513°C)

feedwater
heaters

condensate
pump

Fig. 6. Schematic diagram for a high-temperature gas-cooled reactor power plant. (*General Atomic Corp.*)

iliary power supplies are frequently provided for start-up, shut-down, and communication services.

Generator protection. Necessary devices are installed to prevent or minimize other damage in cases of equipment failure. Differential-current and ground relays detect failure of insulation, which may be due to deterioration or accidental overvoltage. Overcurrent relays detect overload currents that may lead to excessive heating; overvoltage relays prevent insulation damage. Loss-of-excitation relays may be used to warn operators of low excitation or to prevent pulling out of synchronism. Bearing and winding overheating may be detected by relays actuated by resistance devices or thermocouples. Overspeed and lubrication failure may also be detected.

Not all of these devices are used on small units or in every plant. The generator is immediately deenergized for electrical failure and shut down for any over-limit condition, all usually automatically.

Voltage regulation. This term is defined as change in voltage for specific change in load (usually from full load to no load) expressed as percentage of normal rated voltage. The voltage of an electric generator varies with the load; consequently, some form of regulating equipment is required to maintain a reasonably constant, and predetermined potential at the distribution stations or load centers. Since the inherent regulation of most alternating-current generators is rather poor (that is, high percentagewise), it is necessary to provide automatic voltage control. The rotating or magnetic amplifiers and voltage-sensitive cir-

cuits of the automatic regulators, together with the exciters, are all specially designed to respond quickly to changes in the alternaor voltage and to make the necessary changes in the main exciter output, thus providing the required adjustments in voltage. A properly designed automatic regulator acts rapidly, so that it is possible to maintain desired voltage with a rapidly fluctuating load without causing more than a momentary change in voltage even when heavy loads are thrown on or off.

Electronic voltage control has been adapted to some generator and synchronous condenser installations. Its main advantages are its speed of operation and its sensitivity to small voltage variations. As the reliability and ruggedness of electronic components are improved, this form of voltage regulator will become more common.

Generation control. Computer-assisted (or online controlled) load and frequency control and economic dispatch systems of generation supervision are being widely adopted, particularly for the larger new plants. Strong system interconnections greatly improve bulk power supply reliability but require special automatic controls to ensure adequate generation and transmission stability. Among the refinements found necessary in large, long-distance interconnections are special feedback controls applied to generator high-speed excitation and voltage regulator systems.

Synchronization of generators. Synchronization of a generator to a power system is the act of matching, over an appreciable period of time, the instantaneous voltage of an alternating-current generator (incoming source) to the instantaneous

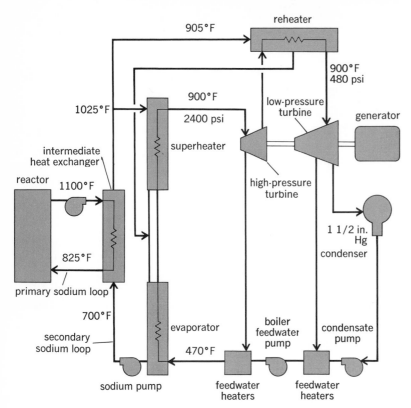

Fig. 7. Diagram for liquid-metal-cooled fast breeder reactor power plant. (*North American Rockwell Corp.*)

voltage of a power system of one or more other generators (running source), then connecting them together. In order to accomplish this ideally the following conditions must be met:

1. The effective voltage of the incoming generator must be substantially the same as that of the system.

2. In relation to each other the generator voltage and the system voltage should be essentially 180° out of phase; however, in relation to the bus to which they are connected, their voltages should be in phase.

3. The frequency of the incoming machine must be near that of the running system.

4. The voltage wave shapes should be similar.

5. The phase sequence of the incoming polyphase machine must be the same as that of the system.

Synchronizing of ac generators can be done manually or automatically. In manual synchronizing an operator controls the incoming generator while observing synchronizing lamps or meters and a synchroscope, or both. Voltage (potential) transformers may be used to provide voltages at lamp and instrument ratings. Lamps properly connected between the two sources are continuously dark when voltage, phase, and frequency are properly matched. Wave shape and phase sequence are determined by machine design and rotation or terminal sequence. Large units generally are provided with voltmeters and frequency meters for matching these quantities, and a synchroscope connected to both sources to indicate phase relationship. Lamps may also be included. The standard synchroscope needle revolves counterclock-

wise when the incoming machine is slow and clockwise when fast. The needle points straight up when the two sources are in phase. The operator closes the connecting switch or circuit breaker as the synchroscope needle slowly approaches the in-phase position.

Automatic synchronizing provides for automatically closing the breaker to connect the incoming machine to the system, after the operator has properly adjusted voltage (field current), frequency (speed), and phasing (by lamps or synchroscope). A fully automatic synchronizer will initiate speed changes as required and may also balance voltages as required, then close the breaker at the proper time, all without attention of the operator. Automatic synchronizers can be used in unattended stations or in automatic control systems where units may be started, synchronized, and loaded on a single operator command. *See* ALTERNATING-CURRENT GENERATOR; ELECTRIC POWER SYSTEMS.

[EUGENE C. STARR]

Bibliography: H. C. Barnes et al., Alternator-rectifier exciter for Cardinal Plant 724-MVA generation, *IEEE Trans. Power App. Syst.*, PAS-87 (4): 1189, 1968; J. G. Brown, *Hydro-Electric Engineering Practice*, 1958; P. H. Cootner, *Water Demand for Steam Electric Generation*, 1966; K. Fenton, *Thermal Efficiency and Power Production*, 1966; S. Glasstone and A. Sesonske, *Nuclear Reactor Engineering*, 1967; G. E. Kholodovskii, *Principles of Power Generation*, 1965; A. H. Lovell, *Generating Stations: Economic Elements of Electrical Design*, 4th ed., 1951; F. T. Morse, *Power Plant Engineering: The Theory and Practice of Stationary Electric Generating Plants*, 3d ed., 1953; F. R. Schleif et al., Excitation control to improve powerline stability, *IEEE Trans. Power App. Sys.*, PAS-87(6):1426–1432, 1968; B. G. A. Skrotzki, *Electric Generation: Hydro, Diesel, and Gas-Turbine Stations*, 1956; P. Sporn, *Energy: Its Production, Conversion and Use in the Service of Man*, 1963; *Standard Handbook for Electrical Engineers*, 10th ed., sect. 6, 8, 9, and 10, 1968; C. D. Swift, *Steam Power Plants: Starting, Testing and Operation*, 1959; H. G. Tak, *Economic Choice Between Hydroelectric and Thermal Power Developments*, 1966.

Electric power measurement

The measurement of the time rate at which work is done in an electric circuit. The work done in moving an electric charge is proportional to the charge and the voltage drop through which it moves. Charge per unit time defines electric current; electric power p is therefore defined as the product of the current i in a circuit and the voltage e across its terminals at a given instant. Expressed symbolically, $p = ei$.

A second important definition of power follows directly from Ohm's law. $p = i^2R$, where R is the resistance of the circuit.

The practical unit of electric power is the watt. The watt represents a rate of expending energy, and thus it is related to all other units of power; for example, in mechanics 1 watt $= 10^7$ ergs/sec and 746 watts $= 1$ horsepower. Commonly used small units are the milliwatt (0.001 watt) and the microwatt (0.000001 watt). Large units are the kilowatt (1000 watts) and the megawatt (1,000,000 watts).

Power measurements must cover the frequency spectrum from direct current through the conventional power frequencies, the audio and the lower radio frequencies, to the highest frequencies (up to 25,000 gigahertz). In general, different techniques are required in each frequency range, and this article is divided into sections dealing with these frequency ranges. *See* ELECTRICAL MEASUREMENTS.

DC AND AC POWER FREQUENCIES

In the measurement of power in a dc circuit not subject to rapid fluctuations, there is usually no difficulty in making simultaneous observations of the true values of voltage and current using common types of dc voltmeters and ammeters. The product of these observations then gives a sufficiently accurate measure of power in the given circuit, except that, if great accuracy is required, allowance must be made for the power used by the instruments themselves.

If in a circuit the voltage e, or current i, or both are subject to rapid variations, instantaneous values of power are difficult to measure and are usually of no interest. The important value is the average value, which is expressed mathematically in Eq. (1), where T is the period or time interval

$$P = \frac{1}{T}\int_0^T ei\,dt \qquad (1)$$

and t is time. This relation holds true for any waveform of current and voltage. In circuits with rapidly varying direct currents, pulsating rectified current, or, in general, alternating currents, the continuous averaging over short periods of time and the automatic multiplication of current and voltage values is accomplished by the wattmeter. *See* WATTMETER.

In ac circuits with steady effective values of voltage and current, the voltmeter-ammeter method may be used as in the dc case, except that, of course, ac meters are used, and a phase meter is also required to measure phase angle unless current and voltage are in phase. Because ac ammeters and voltmeters actually measure root-mean-square, or effective, values, these lead directly to values of average power.

Sinusoidal ac waves. Figure 1 illustrates the case of a sinusoidal voltage and current in a circuit containing only a resistive load. Here the current wave is entirely symmetrical with the voltage wave, and the power curve formed from the product of the voltage and current at each instant appears as a double-frequency wave on the positive side of the zero axis.

In Fig. 2, an inductance is assumed in the measured circuit. The current wave lags behind the voltage wave in what is called the negative out-of-phase, or quadrature, condition. A pure capacitance, however, would produce a positive out-of-phase, or quadrature, condition in which the current wave leads the voltage wave. In general, circuits contain elements of resistance, inductance, and capacitance in varying amounts and they must, therefore, assume some intermediate condition of phase angle between voltage and current. In this case the power wave, as shown in Fig. 2, dips below the zero line and becomes negative, indicating that during that part of the cycle power feeds back into the circuit. Measurements

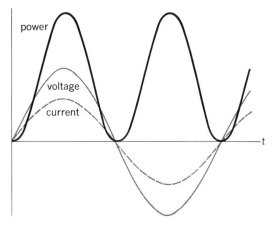

Fig. 1. Curves of instantaneous current, voltage, and power in an ac circuit; current and voltage in phase.

based on readings of conventional ac ammeters and voltmeters do not account for these negative excursions, and therefore, in general, the product of steady effective voltage and current readings in an ac circuit differs from, and is greater than, the reading of a wattmeter in such a circuit.

These relationships in the general ac circuit may be expressed as Eqs. (2) and (3), where I_m and E_m

$$i = I_m \int_0^T \sin 2\pi ft\,dt \qquad (2)$$

$$e = E_m \int_0^T \sin\,(2\pi ft \pm \phi)\,dt \qquad (3)$$

are maximum values of current and voltage, f is frequency in hertz, and ϕ is the phase angle by which the current leads (+) or lags behind (−) the voltage in the circuit.

But by definition Eq. (4) holds. Substituting and carrying out the indicated operations, Eq. (5) is obtained where E and I are effective values of voltage and current.

$$P = \frac{1}{T}\int_0^T ei\,dt \qquad (4)$$

$$P = EI\cos\phi \qquad (5)$$

The expression for P in Eq. (5) is the real or active power in the circuit and is distinguished from the simple product EI, which is called the appar-

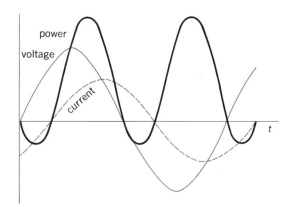

Fig. 2. Curves of instantaneous current, voltage, and power in an ac circuit; current and voltage out of phase.

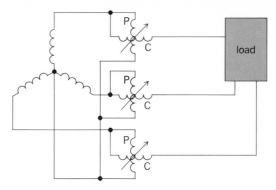

Fig. 3. Three wattmeters in three-phase, three-wire circuit. C and P refer to current and potential coils.

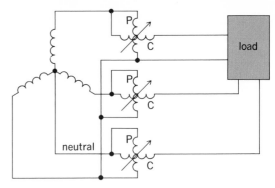

Fig. 5. Wattmeters in three-phase, four-wire circuit. C and P refer to current and potential coils.

ent or virtual power, by the factor cos ϕ, which is called the power factor. It is obvious from the previous formula that Eq. (6) holds.

$$\cos \phi = P/EI \qquad (6a)$$

or

$$\text{Power factor} = \frac{\text{real power}}{\text{apparent power}} \qquad (6b)$$

Negative, or reactive, power due to inductance and capacitance in a circuit, is given by the relation $EI \sin \phi$.

The units for these quantities are, for real power, watts; for apparent power, volt-amperes; and, for reactive power, reactive volt-amperes or vars.

Polyphase power measurement. Summation of power in the separate phases of a polyphase circuit is accomplished by combinations of single-phase wattmeters, or wattmeter elements, disposed according to the general rule, called Blondel's theorem, as follows: If energy is supplied to any system of conductors through N wires, the total power in the system is given by the algebraic sum of N wattmeters, so arranged that each of the N wires contains one current coil, the corresponding potential coil being connected between that wire and some point on the system which is common to all the potential circuits. If this common point is on one of the N wires and coincides with the point of attachment of the potential lead to that wire, the measurement may be effected by the use of $N-1$ wattmeters.

Considering measurement in three-phase circuits as an example of polyphase practice in wide application in the power industry, the two most common systems are the three-phase, three-wire system in which the source may be Y-connected or delta-connected to three load wires; or the three-

phase, four-wire system which also has three load wires and, in addition, a fourth wire, or neutral, which may or may not carry current to the load. If the neutral does not carry current, the circuit may be treated as a three-wire system.

Before applying the rule to commonly used circuits, an exception may be noted in the case of a balanced circuit in which the effective values of the currents and voltages and the phase relationships between them remain constant. In other words, the loads on the separate phases are equal. In this special case, power may be measured by a single wattmeter connected in one phase and the reading multiplied by three.

To measure power in either three-wire or four-wire systems, a wattmeter may be connected in each of the power-receiving circuits, as in Fig. 3. The sum of the three readings gives the total power.

Alternatively, in a three-wire system total power may be measured by two wattmeters, each having its current coil connected in one of the line conductors and its potential circuit connected between the line conductor in which its current coil is connected and the third line conductor (Fig. 4). The algebraic sum of the readings of the two wattmeters indicates the total power in the three power-receiving circuits.

In a four-wire system, three wattmeters may also be effectively used by connecting the current coils in each of two of the line conductors and in the neutral conductor, as in Fig. 5. The potential coils are connected between each of the line conductors and the neutral conductor in which the respective current coils are connected and the third line conductor.

In the last three cases the methods are correct for any value of balanced or unbalanced load and for any value of power factor.

A variety of other circuit connections is available for polyphase power measurement for various special conditions of use.

AUDIO AND RADIO FREQUENCIES

At frequencies above those used in the ordinary power-distribution systems, dynamometer-type wattmeters become inaccurate. For measurements of transmitted power at audio and the lower radio frequencies, no generally satisfactory substitute has been developed and measurements are therefore confined to determinations of power dissipat-

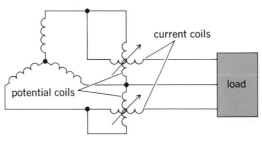

Fig. 4. Wattmeters in three-phase, three-wire circuit.

ed in a load, or available from a source, and are deduced from measurements of impedance and current or voltage.

Power-output meters. Power-output meters combine resistive loads, which can be adjusted to various known values, and voltmeters calibrated to indicate the power dissipated therein. They are used at audio frequencies to determine the maximum power output that can be obtained from a source, or to measure output power in studies of harmonic distortion, intermodulation, overload, frequency characteristic, and so forth.

Figure 6 shows an illustrative schematic. A voltmeter V is fed from an attenuator having a constant input resistance R and a voltage ratio k. The voltage V at the meter corresponds to a voltage kV at the attenuator input, and the meter scale is calibrated in power input to the attenuator, $P = (kV)^2/R$. By adjusting the attenuator one can obtain convenient scale multiplying factors k^2. The tapped transformer T provides different turns ratios n_1/n_2 to adjust the effective input resistance of the instrument over a range of values.

By adjusting this ratio the user can obtain a maximum meter reading when the corresponding input impedance is approximately equal to the output impedance of the source. If the output impedance is purely resistive, this condition occurs when the impedances are exactly equal, or matched, and the power indicated is the maximum power output that can be obtained from the source.

Standard-signal generators. At radio frequencies the principle of impedance matching is frequently used to determine the power input to a receiver, with a standard-signal generator serving as source. The circuit of Fig. 7 shows the arrangement.

A standard-signal generator produces a known voltage E "behind" an output resistance R. Impedances at radio frequencies frequently have significant reactive components; thus the impedance-matching network serves the dual function of tuning out the reactive component X_x of the receiver input impedance Z_x and multiplying the resistive component R_x by a factor that depends upon the values of the component inductance and capacitances. When the matching network is so adjusted that the receiver output is a maximum, the reactive component is nullified and the input resistance of the matching network equals the output resistance R of the standard-signal generator. The power input is therefore $P = E^2/4R$.

Calorimetric method. The power dissipated in a resistive load can be determined from the heat developed therein, and several ingenious schemes for measuring this heat have been derived.

A widely used technique depends upon dissipating the power in a resistive element of relatively high temperature coefficient of resistivity. The power is deduced from the change in resistance by one of several methods. Measurements that are of this general nature are known as bolometric methods.

Another method depends upon using as the load an incandescent lamp that will emit visible light at power inputs of the order of those to be measured. The temperature of the filament can be determined by pyrometric techniques, or the light can be measured with a photocell. The indicating sys-

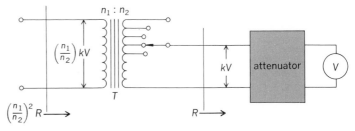

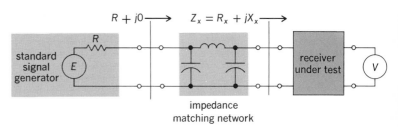

Fig. 6. Schematic representation of audio-frequency power-output meter.

Fig. 7. Schematic showing standard-signal generator and impedance matching set up to produce known power input to receiver.

tem can then be calibrated in terms of power input at direct current or low-frequency alternating current, where accurate moving-coil wattmeters can be used. Such calibrations will continue to hold good as the frequency of the source is raised until dielectric losses in the glass and eddy-current losses in the metal parts other than the filament become significant compared with the heating of the filament proper, or until the wavelength becomes so short that the current flow in the filament becomes nonuniform along its length. Reasonable accuracy can be attained at frequencies as high as a few hundred megahertz.

A variant of this method uses a diode in which a pure-metal filament is heated by the power to be measured. When not space-charge-limited, the dc plate current is a sensitive indicator of filament temperature. The temperature rise in a heater

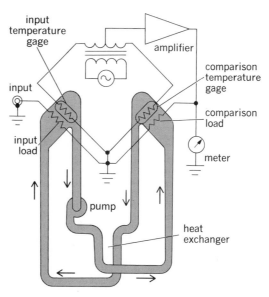

Fig. 8. Commercial self-balancing calorimeter.

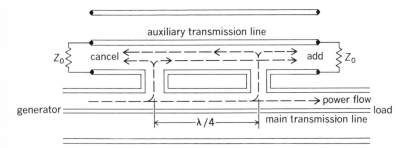

Fig. 9. Cross-sectional view of simplified two-hole directional coupler.

element can also be measured by thermocouples.

The classical calorimetric method depends upon measuring the temperature rise in a liquid coolant circulated through an enclosure containing the load and so insulated that heat losses other than to the coolant are negligible.

Measurements of rate of temperature rise can be used for absolute determination of the power input in terms of the specific heat and volume of the liquid, or the liquid can be used as a transfer mechanism to compare the radio-frequency power input with a known dc or low-frequency power developed elsewhere in the system.

Figure 8 shows a commercial instrument, which uses a self-balancing bridge to provide a comparison temperature rise that can be accurately calibrated in terms of a low-frequency source.

MICROWAVE FREQUENCIES

Measurement of dissipated power at microwave frequencies can be carried out by methods similar to those used at lower frequencies. Because voltage and impedance become increasingly difficult to define precisely as the frequency is increased, however, calorimetric methods generally, and bolometric methods specifically, are almost universally used.

In contrast to the lower radio frequencies, convenient methods of measuring transmitted power are available at microwave frequencies. These use the properties of waves propagated along transmission lines. A commonly used device for this purpose is the directional coupler, a simple example of which is shown in Fig. 9. This comprises an auxiliary line, coupled at two points spaced one-fourth wavelength apart to the main line down which power flows. For the sake of simplicity no coupling means is shown between the main and auxiliary lines other than holes in the outer conductors; in practice, either capacitive probes or small coupling loops may be used to transfer energy from one line to the other. For the flow of power illustrated, a wave entering the right-hand hole and propagating to the left in the auxiliary line arrives at the left-hand hole 180° out of phase with the wave entering the left-hand hole. If there are no losses in the system, and the couplings through the two holes are equal, the two signals cancel, and no signal reaches the left-hand termination. On the other hand, a wave entering the left-hand hole and propagating to the right in the auxiliary line arrives at the right-hand hole in phase with the wave entering the right-hand hole, and the two signals add and propagate to the right in the auxiliary line.

Terminating the auxiliary line in a resistance equal to the characteristic impedance Z_0 prevents any of this power from reflecting in the auxiliary line and reaching the left-hand termination. Power traveling to the right in the main line therefore produces a signal voltage only at the right-hand auxiliary-line termination.

Conversely, a signal that is propagating to the left in the main line reaches only the left-hand termination. If the two terminations are used as bolometers, for instance, they can then measure the individual powers traveling each way in the main line and the net power reaching the load, which is equal to the difference between them. *See* TRANSMISSION LINES.

This measurement can also be made with a standing-wave detector. The voltage maximum is equal to the sum of the voltages of the incident and reflected waves, $V_I + V_R$, whereas the voltage minimum is equal to the difference, $V_I - V_R$. The power transmitted is the difference in power given by expression (7), which is equal to $V_{max}V_{min}/Z_0$. The

$$\frac{(V_I)^2}{Z_0} - \frac{(V_R)^2}{Z_0} \qquad (7)$$

relation between the probe voltage and the line voltage can be established by measuring power in matched loads. In contrast to methods using directional couplers, this has the advantage of requiring substantially no power; however, it requires manipulation to obtain a measurement.

[DONALD B. SINCLAIR]

Bibliography: H. Buckingham and E. M. Price, *Principles of Electrical Measurements*, 1957; E. L. Ginzton, *Microwave Measurements*, 1957; E. W. Golding, *Electrical Measurements and Measuring Instruments*, 1955; F. K. Harris, *Electrical Measurements*, 1952; F. A. Laws, *Electrical Measurements*, 2d ed., 1938; C. G. Montgomery, *Technique of Microwave Measurements*, vol. 2, 1948; F. E. Terman and J. M. Pettit, *Electronic Measurements*, 2d ed., 1952; M. Wind (ed.), *Handbook of Electronic Measurements*, 2 vols., 1958; M. Wind and A. Rapaport (eds.), *Handbook of Microwave Measurements*, 2 vols., 2d ed., 1958.

Electric power substation

An assembly of equipment in an electric power system through which electric energy is passed for transmission, transformation, distribution, or switching. *See* ELECTRIC POWER SYSTEMS.

Specifically, substations are used for one or more of the following purposes: (1) transformation from one voltage level to another; (2) isolating faulted circuits; (3) controlling transmission voltage and power factor; (4) switching to change power flows to or from generating plants or interconnected power companies; (5) converting alternating current to direct current or vice versa; (6) converting from one frequency to another; (7) regulating voltage on feeder circuits; (8) automatic transfer to alternate sources of power; and (9) measuring and controlling electric power.

Classification. Substations are classified by the duty they perform. Transmission, bulk-power, or switching substations (Fig. 1) are those associated with the higher voltage or transmission portion of the power system. Distribution substations are associated with the distribution system and the primary feeders for supply to residential, commercial, and industrial loads. Customer substations are on the premises of utility customers, such as shopping centers, large office or commercial buildings, and industrial plants. Load-center substations are located within large buildings or industrial plants as part of the distribution system within a large plant. One large industrial system has an installed transformer capacity of 2000 Mw (megawatts) and over 80 load-center substations as well.

Indoor substation equipment today is generally metal-clad (Fig. 2). The high-voltage buses, instrument transformers, and power circuit breakers are enclosed within a standard heavy-duty metal (steel or aluminum) housing. The meters, relays, and control switches are mounted on a hinged door on the front of the equipment. The power circuit breakers, up to 500 or 750 Mva (megavolt-amperes) interrupting rating, are mounted on wheeled carriages that permit them to be withdrawn from their metal housings for maintenance. They are mechanically and electrically interlocked to ensure that they are open before being moved. In the higher voltage indoor substations, such as 115,000 or 138,000 volts, open construction is used. This is protected by a screen or wall-type enclosure.

Outdoor substation equipment is generally of the open type in the higher voltages, and either open or metal-clad in the lower or distribution voltages. In the outdoor metal-clad substations, the control panels containing meters and relays are located on the equipment, which in turn is enclosed in a weatherproof metal housing. In the open-type substations (Fig. 3), the modern tendency is toward a low-profile design with a walled enclosure for appearance as well as safety. The walls may be of brick, stone, or precast concrete (Fig. 4). The transformers, power circuit breakers, and other high-voltage equipment are located outdoors, but the metering and control panels are located within a control room or house built within the walls. In substations in outlying areas or in indus-

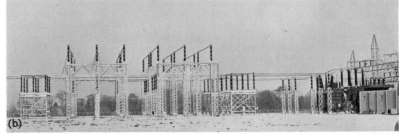

Fig. 1.   Typical transmission substations. (*a*) A 138-kv substation showing latticed structure with strain buses. (*b*) A 345,000-volt substation showing low-profile aluminum-pipe buses.

trial neighborhoods, the substation yard is enclosed within a woven wire or ornamental fence for safety and for protection against unauthorized entry.

In the smaller outdoor distribution substation area, the unit substation is used (Fig. 5). This is a lineup of metal-clad switch gear, with one or more transformers connected to the switch gear by means of a metallic throat connection. In populat-

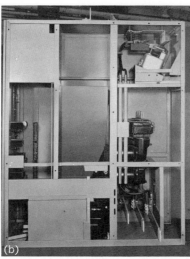

Fig. 2.   Indoor metal-clad switch gear. (*a*) Front panel on which is mounted the metering, relaying, and control equipment. (*b*) Interior view of 4.16-kv switch gear. (*c*) A 4.16-kv air circuit breaker.

Fig. 3. Low-profile outdoor substation. (a) From the left, lightning arresters, cable potheads, cable disconnect switches, and 120- to 24-kv transformer. (b) A 24-kv structure showing 24-kv bus instrument transformers, outgoing feeder power circuit breakers, transformer power circuit breaker, and transformer. (c) Control and relay panels for outdoor equipment.

ed areas the residential-type distribution substation may be enclosed within a residential-type building or the shell of a building which conforms to the homes in the neighborhood and has compatible landscaping. In the downtown areas of larger cities, even transmission substations are totally enclosed within attractive buildings conforming to the neighborhood. They may also be installed in the basement of a building or under an adjacent parking lot. In some cases, a combination of outdoor metal-enclosed switch gear and a building is used.

The converter substation is a new development

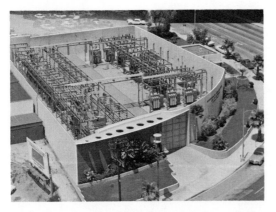

Fig. 4. Typical enclosed outdoor substation with pre-cast-concrete walls and low-profile structure.

in substation design associated with direct-current transmission of extra-high voltage (EHVDC). A new substation near the Columbia River in Oregon is designed to transmit 2880 Mw of power more than 800 mi south to Los Angeles and to Mead, Nev., over two 750,000-volt lines. The converter substation is designed to operate at a maximum continuous voltage of 800,000 volts between line conductors, or 400,000 volts to ground from the positive or negative conductor. The transmission is a two-wire system.

The input to the high-voltage dc substation is 230,000 volts ac. Power is rectified in one line terminal by six electronic bridges, three per pole, rated at 133,000 volts and 1800 amp and each made up of seven electronic valves.

Converter transformers, lightning arresters, switches, filter capacitors, and measuring devices are installed outdoors in the switchyard. The control switchboard, electronic valves, and maintenance rooms are located in the substation building.

This substation, the first of its kind in the United States, and the largest in the world, occupies a site of about 50 acres.

Mobile substations. For emergency replacement or maintenance of substation transformers. mobile substations are used by many utilities. They may range in size from 100 to 25,000 kva (kilovolt-amperes) and from 2400 up to 220,000 volts. Most units are designed for travel on public roads, but larger substations are built for travel on railroad tracks.

Mobile substations of 15,000 kva and smaller are

Fig. 5. Residential type of outdoor unit substation which has a double-ended underground supply con- nected to transformers at either end of the metal-clad switch gear.

generally complete with primary disconnect switches, fuses, or circuit interrupters such as vacuum switches on the primary side. The second-ary is protected with a power circuit breaker and line-disconnect switch. Lightning arresters are mounted on both primary and secondary of the transformer. The complete metering, relaying, and control panel is mounted on the unit. The whole package is mounted on a rubber-tired trailer de-signed to be moved by a gas-engined tractor.

The mobility of these substations is such that they can be moved and installed anywhere in an urban utility substation in 3–4 hr. Heavy-duty, insulated, shielded cables provide the secondary connections to the permanent substation, gener-ally laid on the ground. The primary connections are made with bare wire.

For the higher voltages and higher capacities, a mobile transformer is built (Fig. 6). Because of size and weight restrictions on public highways, this unit generally is limited to a transformer, with lightning arresters on primary and secondary sides, and cooling equipment.

Operation. Substations may be attended or unat-tended. Their operation may be automatic, semi-automatic, or remotely controlled from a manned substation or a dispatch center. Automatically controlled substations have switching operations initiated by relays or contact-making meters so that generators, synchronous condensers, capaci-tors, transformers, converters, or feeders may be placed in or removed from service as system con-ditions require.

Substation equipment. Basic substation equip-ment consists of one or more three-phase trans-formers (or banks of single-phase transformers). In the higher voltage substations autotransformers are used. Modern substations are provided with a disconnecting switch on the incoming circuit, fol-lowed by a power circuit breaker or other switching device to remove the transformer from the trans-mission system. The transformer circuit is gener-ally provided with a power circuit breaker and dis-connecting switch on the secondary side.

Distribution substations generally are designed with more than one transformer, and in most cases with two or more distribution feeders per trans-former. Feeder switching is accomplished with oil, air, or vacuum circuit breakers where fault current is high, or with oil or vacuum circuit reclosers where fault current is lower. Voltage is controlled by means of tap-changing underload equipment on the transformers, by bus or feeder voltage regula-tors, or by capacitor switching, which also controls the power factor. Apparatus and auxiliaries for handling electric power are classified as switch-gear. This general term covers switching, inter-rupting, control metering, relaying. and regulating devices and assemblies of this equipment. See WATTMETER.

Control switchboards. Control switchboards consist of one or more panels on which are mount-ed electrical devices to control and monitor the operation of remotely located electrical apparatus, such as power circuit breakers, transformers, gen-erators, or associated substation equipment. They differ from power switchboards in that main-cir-cuit switching or interrupting devices are not in-cluded.

System control switchboards are attended by system operators who observe the performance of the system, as indicated by meters and indicating lights, operators initiate changes by means of the controls or automatic equipment provided. A number of automatic-system control centers are in operation and more are in the planning stage.

Devices mounted on control switchboards· in-clude control switches, protective relays, control relays, and indicating and recording meters. These devices are grouped physically to facilitate opera-tion and maintenance. Wiring between devices on the switchboard and between panels and the appa-ratus being controlled is designed so that the phys-ical separation normally is limited to a few thou-sand feet.

Many substations and their control switch-boards are unattended because of the develop-ment of automatic control and remote controls, which may include supervisory control and micro-wave relaying.

Fig. 6. Mobile transformer.

Automatic control. Electric controls are designed to provide for opening or closing power circuit breakers or switches in an automatic sequence and under predetermined conditions. Reliable service is maintained and adequate protection is provided against usual operating emergencies.

Automatic or semiautomatic operation without the normal presence of an operator was first used in 1914, on ac-to-dc conversion substations supplying electric railways. In the 1920s the development of automatic reclosing relays, protective relays, and automatic voltage regulators enabled ac substations to be made almost completely automatic. Automatic control is applied to large distribution and transmission substations, up to and including

500,000-volt facilities. It is also used on individual circuits and equipment in many attended substations. With the addition of automatic relaying and control equipment, many of the older attended substations are being automated.

Supervisory control. Supervisory control is used in those substations of such importance that some human control or judgment is deemed necessary. This is quite common in the higher voltage or transmission substations and downtown distribution substations. This control may be designed to operate over leased telephone lines, private utility communication lines, power line carrier equipment, or microwave radio channels. It usually includes opening and closing operation of main power circuit breakers, voltage control, indication of the position of the power circuit breakers, and alarm and metering equipment. Supervisory control also includes a means of checking or testing the channels.

Protective relay systems. Protective relays are used to isolate any element of a power system that is faulted or starts to operate in a manner that may adversely affect the operation of the system. Protective relays do this by opening power circuit breakers or other switching devices that are capable of disconnecting the faulted element.

Power circuit breakers or other power switching equipment, such as circuit switchers or vacuum switches, are generally used to disconnect major electrical equipment or lines from the power system. Smaller units of equipment, generally under 5000 kva, are protected with power fuses capable of interrupting fault currents anticipated.

The principal function of protective relaying is to minimize the effect of fault currents on the units of equipment as well as the overall system. Other uses of protective relaying are to detect overloads and thus protect against thermal damage to insula-

Fig. 7. Lightning arresters mounted on both primary and secondary sides of 345- to 138-kv transformer.

tion, to detect abnormal voltage or single phasing, and to indicate the location and type of fault.

Generally, protective relays are operated from current and potential transformers or substitute devices to provide low-voltage secondary wiring for safety. Through direct or relative changes in current voltage, frequency, speed, pressure, temperature, and so on, protective relays can identify the type and location of a fault.

Lightning protection. Substations normally are protected against lightning where the expense of protection is less than the cost of repairing or replacing damaged equipment. Indirectly this protection assures more reliable service to the customer.

Protection against direct lightning strokes is generally provided by overhead ground wires, lightning masts, or combinations of both. Overhead power circuits leaving a substation, especially in the heavier duty and higher voltage areas, have overhead ground wires to protect against direct strokes to the line. The remainder of the circuit is usually protected by line-type lightning arresters or flashover protectors.

An additional benefit of lightning arresters is the protection against switching surge or remote lightning surge voltages. Since the effectiveness of this protection increases with proximity to the equipment, lightning arresters are mounted directly on large transformers (Fig. 7), or as close as possible to the equipment being protected.

Substation grounding systems. Grounding systems are installed (1) to provide safety to personnel by connecting to ground noncurrent-carrying parts such as transformer tanks, equipment enclosures, substation structures, fence, and auxiliary low-voltage circuits; (2) to stabilize circuit potentials with respect to ground and to provide relaying currents to remove ground faults from the power system; and (3) to provide effective lightning and switching surge protection.

A grounding system generally consists of a combination of driven ground rods and buried copper wire connecting the elements into a complete system; to this, ground connections are made for the substation neutral (if any), overhead ground wires, equipment, structure, fence, and auxiliary control wiring. To minimize operator touch potentials, a ground mat or grid is located below any manually operated disconnect switch at the position where the substation operator would stand while operating the switch. Since this is an important consideration in substation design, a means of testing the system is usually installed and the ground grid is tested every few years.

[R. A. HUNT]

Bibliography: A. M. Baker and G. E. Hertig, GPU-Penelec's 460kv substations, *IEEE Trans.,* no. 61–782, 1961; The compact substation is on the way, *Elec. World,* Dec. 18, 1967; A guide for minimum electrical clearances for standard basic insulation levels, *IEEE Tech. Pap.,* no. 54–80, 1954; G. E. Hertig, High and extra high voltage substation design and economic comparisons, *IEEE Trans.,* no. 62–246, 1962; G. E. Hertig and W. B. Kelley, Switching surge test results EHV substation bus configurations, *IEEE Trans.,* no. 31-TP66-101, 1966; IEEE application guide on methods of substation grounding, *IEEE Trans.,*

73(III-A):271–275, 1954; *IEEE Bibliography on Substation Appearance,* 1966; Minimum electrical clearances for substations based on switching surge requirements, *IEEE Trans.,* no. 31-TP65-39, 2d interim rep., 1965; Structure design gains public acceptance, *Transmission Distribution,* May, 1968; J. A. Turgeon, Insulation system: Key to 700kv switch, *Elec. World,* June 28, 1965; Westinghouse Electric Corp., *Electric Utility Engineering Reference Book Distribution Systems,* 1959.

Electric power systems

A complex assemblage of equipment and circuits for generating, transmitting, transforming, and distributing electrical energy. In the United States, electrical energy is generated to serve more than 81,000,000 customers. The investment represented by these facilities has grown rapidly over the years until in 1975 it was close to $215,000,000,000, with about 75% spent on the power systems of investor-owned utilities and the remainder spread among systems built by governmental agencies, municipal electric departments, and rural electric cooperatives. Principal elements of a typical power system are shown in Fig. 1.

Generation. Electricity in the large quantities required to supply electric power systems is produced in generating stations, commonly called power plants. Such generating stations, however, should be considered as conversion facilities in which the heat energy of fuel (coal, oil, gas, or uranium) or the hydraulic energy of falling water is converted to electricity. *See* ELECTRIC POWER GENERATION; POWER PLANT.

Steam stations. About 81% of the electric power used in the United States is obtained from generators driven by steam turbines. The largest such unit in service in 1975 was rated 1,300,000 kW, equivalent to about 1,730,000 hp. But many units of 500,000 to nearly 1,200,000 kW are in service or under construction. *See* STEAM TURBINE.

Coal was the fuel for nearly 55% of the steam turbine generation in 1974, and its share should increase somewhat because of the growing shortage of natural gas and the sharp rise in the cost of fuel oil stemming from the Arab oil embargo during the winter of 1973–1974. Natural gas, used extensively in the southern part of the United States, fueled about 20% of the steam turbine generation, and heavy fuel oil, 18%, largely in power plants able to take delivery from oceangoing tankers or river barges. The remaining 7% was generated from the radioactive energy of slightly enriched uranium, which, for many power systems, produces electricity at a lower total cost than either coal or fuel oil at current delivered prices. As a consequence, some 52% of the generating capability additions planned as of Jan. 1, 1975, according to an *Electrical World* survey, will be nuclear and largely in the 900,000 to 1,300,000-kW-per-unit range. As these nuclear units go into commercial operation, the contribution of uranium to the United States electrical energy supply will rise, probably to more than 25% of the total fuel generated output by the mid-1980s.

Nuclear steam systems used by United States utilities are mostly of the water-cooled-and-moderated type, in which the heat of a controlled nuclear reaction is used to convert water into steam to

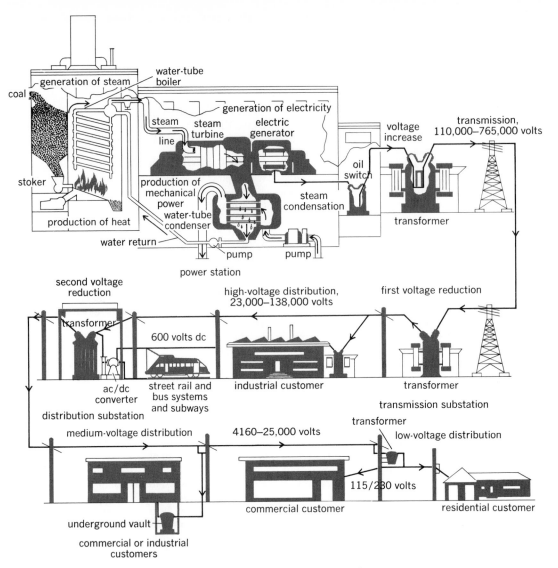

Fig. 1. Major steps in the generation, transmission, and distribution of electricity.

drive a conventional turbine generator; such units are presently limited to about 1,300,000 kW by the thermal limit placed on nuclear reactors by the former Atomic Energy Commission (some of whose functions are now the responsibility of the Nuclear Regulatory Commission). A small but increasing number of installations, however, may be moderated by graphite and cooled by helium gas, which, in turn, will heat water to supply steam for a conventional turbine generator; the ceiling for such units will probably be about 1,500,000 kW because of their somewhat higher conversion efficiency. *See* REACTOR, NUCLEAR.

Hydroelectric plants. Waterpower during 1974 supplied about 17.5% of the electric power consumed in the United States. But this share can only decline in the years ahead because very few sites remain undeveloped where sufficient water drops far enough in a reasonable distance to drive reasonably sized hydraulic turbines. Consequently, the generating capability of hydro plants, 11.5% of the utility industry's total as of Dec. 31, 1974, is slated to fall off to 8% by the mid-1980s because of

its very small share of the planned additions. Much of this additional hydro capability will be used at existing plants to increase their effectiveness in supplying peak power demands, and as a quickly available source of emergency power. *See* HYDRAULIC TURBINE; HYDROELECTRIC POWER.

Some hydro plants, totaling 8,819,000 kW as of Dec. 31, 1974, actually draw power from other generating facilities during light system-load periods to pump water from a river or lake into an artificial reservoir at a higher elevation from which it can be drawn through a hydraulic station when the power system needs additional generation. These pumped-storage installations consume about 50% more energy than they return to the power system and, accordingly, cannot be considered energy sources. Their use is justified, however, by their ability to convert surplus power that is available during low-demand periods into prime power to serve system needs during peak-demand intervals—a need that otherwise would require building more generating stations for operation during the relatively few hours of high system demand. *See*

PUMPED STORAGE; WATERPOWER.

Combustion turbine plants. Gas-turbine-driven generators, now commonly called combustion turbines because of the growing use of light oil as fuel, have gained wide acceptance as an economical source of additional power for heavy-load periods. In addition, they offer the fastest erection time and the lowest investment cost per kilowatt of installed capability. Offsetting these advantages, however, is their relatively less efficient consumption of more costly fuel. Typical unit ratings in the United States have climbed rapidly in recent years until some units operating in 1975 approached 100,000 kW. Many installations, however, involve a group of smaller units totaling, in one case, 256,000 kW. Combustion turbine units, even in the larger ratings, offer extremely flexible operation and can be started and run up to full load in as little as 10 min. Thus they are extremely useful as emergency power sources, as well as for operating during the few hours of daily load peaks. Combustion turbines totaled 8.2% of the total installed capability of United States utility systems at the close of 1974 and supplied less than 3% of the total energy generated. *See* COMBUSTION TURBINE.

In the years ahead, however, combustion turbines are slated for an additional role. Several installations in the 1970s have used their exhaust gases to heat boilers that generate steam to drive steam turbine generators. Such combined-cycle units offer fuel economy comparable to that of modern steam plants and at considerably less cost per kilowatt. In addition, because only part of the plant uses steam, the requirement for cooling water is considerably reduced. A number of additional combined-cycle installations have been planned, but wide acceptance is inhibited by the doubtful availability of light fuel oil for them. This barrier should be resolved, in time, by the successful development of systems for fueling them with gas derived from coal. *See* COAL GASIFICATION.

Internal combustion plants. Internal combustion engines of the diesel type drive generators in many small power plants. In addition, they offer the ability to start quickly for operation during peak loads or emergencies. However, their small size, commonly about 2000 kW per unit although a few approach 10,000 kW, has limited their use. Such installations account for about 1% of the total power-system generating capability in the United States, and make an even smaller contribution to total electric energy consumed. *See* INTERNAL COMBUSTION ENGINE.

Three-phase output. Because of their simplicity and efficient use of conductors, three-phase 60-Hz alternating-current systems are used almost exclusively in the United States. Consequently, power-system generators are wound for three-phase output at a voltage usually limited by design features to a range from about 11,000 V for small units to 30,000 V for large ones. The output of modern generating stations is usually stepped up by transformers to the voltage level of transmission circuits used to deliver power to distant load areas.

Transmission. The transmission system carries electric power efficiently and in large amounts from generating stations to consumption areas. Such transmission is also used to interconnect adjacent power systems for mutual assistance in

Table 1. Power capability of typical three-phase open-wire transmission lines

Line-to-line voltage	Capability, kVA
115,000 ac	60,000
138,000 ac	90,000
230,000 ac	250,000
345,000 ac	600,000
500,000 ac	1,200,000
765,000 ac	2,500,000
800,000 dc*	1,500,000

*Bipolar line with grounded neutral.

case of emergency and to gain for the interconnected power systems the economies possible in regional operation. Interconnections have expanded to the point where most of the generation east of the Rocky Mountains, except for a large part of Texas, regularly operates in parallel, and over 90% of all generation in the United States, exclusive of Alaska and Hawaii, and in Canada can be linked.

Transmission circuits are designed to operate up to 765,000 V, depending on the amount of power to be carried and the distance to be traveled. The permissible power loading of a circuit depends on many factors, such as the thermal limit of the conductors and their clearances to ground, the voltage drop between the sending and receiving end and the degree to which system service reliability depends on it, and how much the circuit is

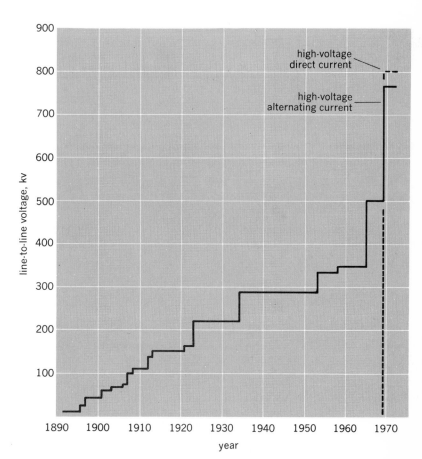

Fig. 2. Growth of ac transmission voltages from 1890.

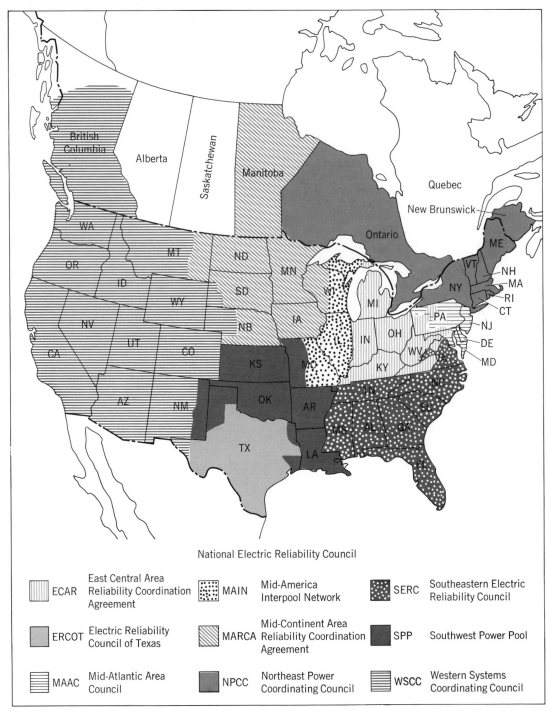

Fig. 3. Areas served by the nine regional reliability councils, coordinated by the National Electric Reliability Council, that guide the coordination and operation of generation and transmission facilities.

needed to hold various generating stations in synchronism.

A widely accepted approximation to the voltage appropriate for a transmission circuit is that the permissible load-carrying ability varies as the square of the voltage. Typical ratings are listed in Table 1.

Transmission as a distinct function began about 1886 with a 17-mi (1mi = 1.6 km) 2000-V line in Italy. Transmission began at about the same time in the United States, and by 1891 a 10,000-V line was operating (Fig. 2). In 1896 an 11,000-V three-

phase line brought electrical energy generated at Niagara Falls to Buffalo, 20 mi away. Subsequent lines were built at successively higher levels until 1936, when the Los Angeles Department of Water and Power energized two lines at 287,000 V to transmit 240,000 kW the 266 mi from Hoover Dam on the Colorado River to Los Angeles. A third line was completed in 1940.

For nearly 2 decades these three 287,000-V lines were the only extra-high-voltage (EHV) lines in North America, if not in the entire world. But in 1946 the American Electric Power (AEP) System inaugurated, with participating manufacturers, a test program up to 500,000 V. From this came the basic design for a 345,000-V system, the first link of which went into commercial operation in 1953 as part of a system overlay that finally extended from Roanoke, VA, to the outskirts of Chicago. By the late 1960s the 345,000-V level had been adopted by many utilities interconnected with the AEP System, as well as others in Illinois, Wisconsin, Minnesota, Kansas, Oklahoma, Texas, New Mexico, Arizona, and across New York State into New England.

The development of 500,000-V circuits began in 1964, even as the 345,000-V level was gaining wide acceptance. One reason for this was that many utilities that had already adopted 230,000 V could gain only about 140% in capability by switching to 345,000 V, but the jump to 500,000 V gave them nearly 400% more capability per circuit. The first line energized at this new level was by Virginia Electric & Power Company to carry the output of a new mine mouth station in West Virginia to its service area. A second line completed the same year provided transmission for a 1,500,000-kW seasonal interchange between the Tennessee Valley Authority and a group of utilities in the Arkansas-Louisiana area. Lines at this voltage level now extend from New Jersey to Texas and from New Mexico via California to British Columbia.

The next and latest step-up occurred in 1969 when the AEP System, after another cooperative test program, completed the first line of an extensive 765,000-V system to overlay its earlier 345,000-V system. The first installation in this voltage class, however, was by the Quebec Hydro-Electric Commission to carry the output at 735,000 V (the expected international standard) from a vast hydro project to its load center at Montreal, some 375 mi away.

Transmission engineers are anticipating even higher voltages, 1,100,000 or perhaps 2,000,000 V, but they are fully aware that this objective may prove too costly in space requirements and funds to gain wide acceptance. Experience already gained at 500,000 V and 765,000 V verifies that the prime requirement no longer is insulating the lines to withstand lightning discharges, but to tolerate continuous operation at rated voltage and occasional voltage surges caused by the operation of circuit breakers.

Experience has shown that, within about 10 years after the introduction of a new voltage level for overhead lines, it becomes necessary to begin connecting underground cable. This has already occurred for 345,000 V; the first overhead line was completed in 1953, and by 1967 about 100 mi of pipe-type cable had been installed to take power received at this voltage level into metropolitan areas. The first 500,000-V cable is slated for service in 1976 to take power generated at the enormous Grand Coulee hydro plant to a major switchyard several thousand feet away. And 765,000-V cable, now deemed possible after extensive testing above the 500,000-V level at the Waltz Mill Cable Test Center operated by Westinghouse Electric Corporation for the Electric Power Research Institute (EPRI), may be needed before 1980.

In anticipation of the need for transmission circuits of higher load capability, an extensive research program is in progress, spread among several large and elaborately equipped research centers. Among these are: Project UHV for ultra-high-voltage overhead lines operated by the General Electric Company near Pittsfield, MA, for EPRI; the Frank B. Black Research Center built and operated by the Ohio Brass Company near Mansfield; and the above-mentioned Waltz Mill Cable Test Center. All three include equipment for testing full-scale line or cable components at well over 1,000,000 V. In addition, many utilities and specialty manufacturers have test facilities related to their fields of operation.

A relatively new approach to high-voltage long-distance transmission is high-voltage direct current (HVDC), which offers the advantages of less costly lines, lower transmission losses, and insensitivity to many system problems that restrict alternating-current systems. Its greatest disadvantage is the need for costly equipment for converting the sending-end power to direct current, and for converting the receiving-end direct-current power to alternating current for distribution to consumers. Starting in the late 1950s with a 65-mi 100,000-V system in Sweden, HVDC has been applied successfully in a series of special cases around the world, each one for a higher voltage and greater power capability. The first, and only, such installation in the United States, built by a group of utilities and the Federal government, is operating at 800,000 V line-to-line to interchange 1,440,000 kW between the Pacific Northwest and California. System engineers look upon this system as an attractive alternative to EHV alternating current for some applications.

In addition to these high-capability circuits, every large utility has many miles of lower-voltage transmission, usually operating at 110,000 to 345,000 V, to carry bulk power to numerous cities, towns, and large industrial plants. These circuits often include extensive lengths of underground cable where they pass through densely populated areas. Their design, construction, and operation are based upon research done some years ago, augmented by extensive experience. *See* TRANSMISSION LINES.

Interconnections. As systems grow as the number and size of generating units increase, and as transmission networks expand, higher levels of bulk-power-system reliability are attained through properly coordinated interconnections among separate systems. This practice began more than 50 years ago with such voluntary pools as the Connecticut Valley Power Exchange and the Pennsylvania–New Jersey–Maryland Interconnection. Most of the electric utilities in the contiguous United States and a large part of Canada now operate as members of power pools, and these pools (except one in Texas) in turn are interconnected into one gigantic power grid known as the North American Power Systems Interconnection. The operation of this interconnection, in turn, is coordinated by the North American Power Systems Interconnection Committee (NAPSIC). Each individual utility in such pools operates independently, but has contractual arrangements with other members in respect to generation additions and scheduling of operation. Their participation in a power pool af-

Fig. 4. Matrix mimic board and CRT displays. (*Courtesy of Ferranti-Packard Inc. and Cleveland Electric Illuminating Co.*)

fords a higher level of service reliability and important economic advantages.

Regional and national coordination. The blackout of Nov. 9, 1965, stemmed from the unexpected trip-out of a key transmission circuit carrying emergency power into Canada and cascaded throughout the northeastern states to cut off electric service to some 30,000,000 people. It spurred the utilities into a chain reaction affecting the planning, construction, operation, and control procedures for their interconnected systems. They soon organized regional coordination councils, eventually nine in number, to cover the entire contiguous United States and four Canadian provinces (Fig. 3). Their stated objective was "to further augment reliability of the parties' planning and operation of their generation and transmission facilities."

Then, in 1968, the National Electric Reliability Council (NERC) was established to serve as an effective body for the collection, unification, and dissemination of various reliability criteria for use by individual utilities in meeting their planning and operating responsibilities. NAPSIC was reorganized shortly afterward to function as an advisory group to NERC.

Increased interconnection capability among power systems reduces the required generation reserve of each of the individual systems. In most utilities the loss-of-load probability (LOLP) is used to measure the reliability of electric service, and it is based on the application of probability theory to unit-outage statistics and load forecasts. A common LOLP criterion is 1 day in 10 years when load may exceed generating capability. The LOLP decreases (that is, reliability increases) with increased interconnection between two areas until a saturation level is reached which depends upon the amount of reserve, unit sizes, and annual load shape in each area. Any increase in interconnection capability beyond that saturation level will not cause a corresponding improvement in the level of system reliability.

Traditionally, systems were planned to withstand all reasonably probable contingencies, and operators seldom had to worry about the possible effect of unscheduled outages. Operators' normal security functions were to maintain adequate generation on-line and to ensure that such system variables as line flows and station voltages remained within the limits specified by planners. However, stronger interconnections, larger generating units, and rapid system growth spread the transient effects of sudden disturbances and increased the responsibilities of operators for system security.

System security is concerned with service continuity at standard frequency and voltage levels. The system is said to be insecure if a contingency would result in overloading some system components, in abnormal voltage levels at some stations, in change of system frequency, or in system instability, even if there is adequate capability as indicated by some reliability index.

Computer control centers. Numerous utilities have installed computers in their control centers to improve system operation and system security.

Economic considerations, such as fuel and start-up costs, which are a function of the operational phase of these centers, become secondary under emergency operating conditions. The security-related functions of these control centers have the primary objective of reducing the probability that system-wide emergencies ever occur. In all cases, although the functions carried by the computer are constantly being expanded, the decision-making role still lies with the human system operator.

Functions of the computer at the control center include collecting data, making computations, monitoring system conditions, alarming for abnormalities, logging data, controlling frequency, dispatching generation to match load requirements, planning and scheduling operation, analyzing security, recording prefault and postfault data, and communicating with other centers and their computers.

Computer functions are still limited mainly to assisting system operators during the normal operating state, and hence are of the preventive type. Continuing research and emerging concepts of security control will produce more effective methods for such preventive functions, develop and introduce means for detecting trouble during emergency operation, and develop methods by which computers can assist in (and perhaps automate) emergency and restorative control actions to return the power system to its normal operating state.

One good example of a control system was built in the early 1970s for the Cleveland Electric Illuminating Company. It was planned to improve system security through the use of several innovations in security analysis, trouble analysis, automatic circuit restoration, generalized load flow, interconnection modeling, precise economic dispatch, man-machine interfaces and displays, and operator training.

The Cleveland installation uses a large-matrix dynamic mimic board as an integral part of the to-

tal control complex (Fig. 4). Mimic boards have been used for many years, but with far less automation, to help system operators analyze system conditions. The Cleveland board is part of a fully integrated system that combines cathode-ray-tube (CRT) displays and the dynamic mimic board with a central computer to provide a flexible dynamic capability. The CRT screens provide detailed information not readily provided by the mimic board.

The CRTs in such installations are controlled by circuit elements that accept inputs from a central computer and convert them to visual images, either as coded symbols or line diagrams, upon command given by the operator, usually with a keyboard terminal. An extensive system of this general type has been installed by the Bonneville Power Administration; it uses 30 seven-color CRT consoles and displays to achieve control of generation and transmission in its vast hydro system.

Substations. Power delivered by transmission circuits must be stepped down in facilities called substations to voltages more suitable for use in industrial and residential areas. On transmission systems, these facilities are often called bulk-power substations; at or near factories or mines, they are termed industrial substations; and where they supply residential and commercial areas, distribution substations.

Basic equipment in a substation includes circuit breakers, switches, transformers, lightning arresters and other protective devices, instrumentation, control devices, and other apparatus related to specific functions in the power system. See ELECTRIC POWER SUBSTATION.

Distribution. That part of the electric power system that takes power from a bulk-power substation to customers' switches, commonly about 35% of the total plant investment, is called distribution. This category includes distribution substations, subtransmission circuits that feed them, primary circuits that extend from distribution substations to every street and alley, distribution transform-

Table 2. Electric utility industry statistics, 1968–1974*

Category	1968	1969	1970	1971	1972	1973	1974	1975
Generating capacity installed at year's end, 10^3 kW†	290,366	312,612	340,353	367,396	399,606	438,492	474,574	504,393
Electric energy output, 10^6 kWhr	1,330,400	1,449,600	1,540,300	1,617,500	1,754,900	1,878,500	1,884,000	1,919,912
Energy sales, 10^6 kWhr								
Total	1,202,321	1,307,178	1,391,359	1,466,440	1,577,714	1,703,203	1,700,769	1,735,906
Residential	367,692	407,922	447,795	479,080	511,423	554,171	554,960	591,108
Small light and power	265,151	286,686	312,750	333,752	361,859	396,903	392,716	421,088
Large light and power	518,834	557,220	572,522	592,699	639,467	687,235	689,435	656,440
Other	50,644	55,350	58,336	60,909	64,965	64,894	63,558	67,270
Customers at year's end × 10^3	69,716	70,929	72,485	74,265	76,150	78,461	80,102	81,892
Revenue, 10^6	18,580	20,130	22,066	24,725	27,921	31,663	39,127	47,874
Average residential use, kWhr/yr	6,057	6,571	7,066	7,380	7,691	8,079	7,907	8,246
Average residential rate, cents/kWhr	2.12	2.09	2.10	2.19	2.29	2.38	2.83	3.25
Coal (and equivalent) burned, lb/kWhr‡	0.870	0.880	0.909	0.918	0.911	0.918	0.946	0.953
Capital expenditures, 10^6	9,311	10,744	12,776	15,130	16,651	18,723	20,556	20,155

*From *Elec. World*, p. 35, Sept. 15, 1975; p. 39, Mar. 15, 1976; and Edison Electric Institute.
†Fuel and hydro.   ‡1 lb = 454 g.

ers, secondary lines, house service drops or loops, metering equipment, street and highway lighting, and a wide variety of associated devices.

Primary distribution circuits usually operate at 4160 to 34,500 V line-to-line, and may be overhead open wire on poles, overhead or aerial cable, or underground cable. These circuits supply large commercial, institutional, and some industrial customers directly. Smaller customers are supplied through numerous distribution transformers. There is a growing movement to place most of these facilities underground for new construction, although it would take many years to replace all of the existing overhead plant.

At conveniently located distribution transformers in residential and commercial areas, the voltage is stepped down again to 120 and 240 V for secondary lines, from which service drops extend to every customer's building to supply lights and appliances. Such low voltages, which include 125/216 in some commercial areas and sometimes include 480 in industrial areas, are known as utilization voltages. *See* ELECTRIC DISTRIBUTION SYSTEMS.

Electric utility industry. In the United States, which has the third highest per-capita use of electricity in the world and more electric power capability than any other nation, the electric utility systems are deemed by some criteria the largest industry (Table 2). The total plant investment, as of Dec. 31, 1975, was about $215,000,000,000. The electric utility industry spends about $20,000,000,000 a year for plant additions to supply the growing load, and collects nearly $48,000,000,000 per year from more than 81,000,000 customers. This industry comprised about 3115 public and investor-owned systems producing electricity in about 4000 generating plants with a combined operating capability of 504,393,000 kW at the close of 1975.

The 3115 systems included 301 investor-owned companies operating 78.8% of the generation and serving 78.3% of the ultimate customers. The remaining 21.7% were served by 1769 municipal systems, about 928 rural electric cooperatives, 58 public power districts, 7 irrigation districts, 40 United States government systems, 9 state-owned authorities, 1 county authority, and 2 mutual systems.

The industry's annual output reached 1,919,912,000,000 kWhr in 1975, and its sales to ultimate customers totaled 1,735,906,000,000 kWhr at an average of 2.76 cents/kWhr. Of this, about 38% was consumed by industrial and other large power customers, the remainder going mostly to residential and commercial customers. Residential usage has climbed steadily for many years, reaching a record 8079 kWhr/average customer in 1973. But energy conservation and rate increases to cover sharply higher fuel costs cut the 1974 residential average to 7907 kWhr at 2.83 cents/k Whr. *See* ELECTRICAL UTILITY INDUSTRY.

[LEONARD M. OLMSTED]

Bibliography: T. E. Di Liacco *IEEE Proc.*, 62(7): 884–891, 1974; T. E. Di Liacco, B. F. Wirtz, and D. A. Wheeler, *Proceedings of the 7th Power Industry Computer Applications Conference*, Boston, May 24–26, 1971; *EHV Transmission Line Reference Book*, Edison Electric Institute, 1968; A. H. El-Abiad, *Proceedings of the 3d Annual Southeastern Symposium on System Theory*, April 5–6, 1971; The electric century 1874–1974, *Elec. World*, 181(11):43–431, June 1, 1974; Electric Research Council, *Electric Transmission Structures*, Edison Electric Institute, 1968; *Electrical World Directory of Electrical Utilities 1974–1975*, 1975; O. I. Elgard, *Electric Energy Systems Theory: An Introduction*, 1971; C. F. Ham and K. B. Rennie, Control center ups system performance, *Elec. World*, 180(4):64–66, Aug. 15, 1973; Huge pumped-hydro nears completion, *Elec. World*, 180(5):60–63, Sept. 1, 1973; E. B. Kurtz, *The Lineman's and Cableman's Handbook*, 1969; 1976 Annual statistical report, *Elec. World*, 185(6):39–70, Mar. 15, 1976; L. M. Olmsted, 19th Steam station cost survey, *Elec. World*, 184(10)43–58, Nov. 15, 1975; L. M. Olmsted, 13th Steam station design survey, *Elec. World*, 182(10):41–64, Nov. 15, 1974; L. M. Olmsted, Transmission goals: Maximum rating with minimum environmental impact, *Elec. World*, 177(11), June 1, 1972; L. M. Olmsted and A. J. Stegeman, EHV: Today's designs, tomorrow's plans, *Elec. World*, 164(20):95–118, 1965; Power shortage seen for next summer, *Elec. World*, 179(1):42–43, Jan. 1, 1973; 750 kV dc intertie rescheduled for April 1969, *Elec. World*, 164(8): 98–101, 1965; S. Sherr, *Automatisme*, 19(5):227–283, 1974; Tie gives Con Edison 400 MW more, *Elec. World*, 179(4):58–59, Feb. 15, 1973; 26th Annual electrical industry forecast, *Elec. World*, 184(6):35–50, Sept. 15, 1975.

Electric rotating machinery

Any form of apparatus, having a rotating member, which generates, converts, transforms, or modifies electric power. The most common forms are motors, generators, synchronous condensers, synchronous converters, rotating amplifiers, phase modifiers, and combinations of these in one machine. The capacity, or rating, is usually indicated on a nameplate and denotes the maximum continuous duty which can be sustained without overheating or other injury. Motors are rated in horsepower. They are built in sizes from a small fraction of a horsepower to more than 230,000 hp. Generators are rated in kilowatts or kilovolt-amperes (kva). The maximum output of alternating-current generators being built exceeds 1,250,000 kva. Other types of rotating machines fall within these limits.

Construction. Most rotating machines consist of a stationary member, called the stator, and a rotating member, called the rotor. The rotor may be supported in bearings at both ends, or it may be supported at one or both ends by the shaft of another machine. *See* GENERATOR, ELECTRIC; MOTOR, ELECTRIC.

The illustration shows a typical rotating machine having a bracket bearing at one end and an arrangement for coupling to a turbine shaft at the other. Although small machines sometimes employ antifriction bearings, larger units are built with sleeve bearings generally lined with babbitt. Vertical shaft machines use thrust bearings to support the rotating member. Lubrication in slow- or medium-speed units is often supplied from an oil reservoir contained within the bearing housing. Where bearing losses are high, water-cooling coils may be immersed in the oil to prevent overheating. High-speed machines are often lubricated from a

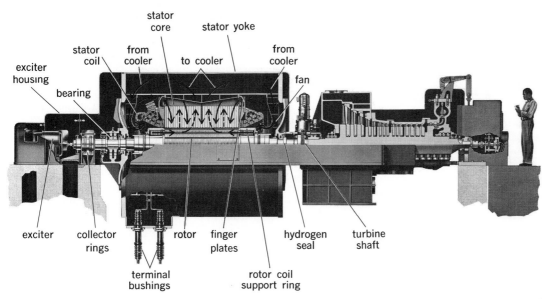

stator core · stator yoke · from cooler · to cooler · from cooler · stator coil · fan · exciter housing · bearing · exciter · collector rings · rotor · finger plates · hydrogen seal · turbine shaft · terminal bushings · rotor coil support ring

Cross section of a typical electric rotating machine. (*Allis-Chalmers*)

pressurized oiling system, which also supplies the shaft seals in hydrogen-cooled units.

To function properly, rotating machines must have a magnetic circuit, usually involving both rotor and stator, and one or more insulated electrical circuits which interlink the magnetic circuit. To afford a low-reluctance magnetic path, the rotor and stator are separated only by a small clearance, called the air gap.

The windings are insulated electrically with materials such as enamel, cotton, varnished cambric, mica, asbestos, dacron, and glass fabric. The most common impregnants are shellac, asphaltum base varnish, and epoxy, polyester, or phenolic resins. External partially conducting varnish is sometimes applied to high-voltage coils for corona shielding.

Electrical-mechanical energy conversion. The force F in newtons produced on a conductor located at right angles to a magnetic field is $F = BIL$ newtons, where B is the flux density in webers per square meter in the vicinity of the conductor, I is the conductor current in amperes, and L is the length of the conductor, in meters, exposed to the flux. In a motor the magnetic field created by one member exerts a force on the current-carrying conductors of the other, producing a mechanical torque which drives the load. In a generator the changing magnetic field induces voltage in the armature windings when the rotor is driven by a source of mechanical power. Little power is required at no-load, but as the load current builds up, the prime mover must supply the torque to overcome the forces in the equation between field and conductors.

Ventilation. Rotating machinery must be ventilated to avoid overheating from internal losses. The principal cooling medium, usually air or hydrogen, is circulated by fans or blowers mounted on the rotor or separately driven. The illustration shows axial-flow fans at each end of the rotor, with arrows indicating the path taken by the gas. With conventional cooling, the cooling medium is blown over exposed surfaces of the insulated windings and core. In conductor cooling, the cooling medium

flows in ducts within the major insulation wall.

In large machines the superior effectiveness of conductor cooling is essential. In addition to hydrogen at pressures up to several atmospheres for cooling the rotor and the stator core iron, hydrogen at far greater pressure or a liquid such as oil or water is circulated through the stator conductors in the largest ratings. Generators having liquid-cooled rotor conductors are also being built. *See* ALTERNATING-CURRENT GENERATOR.

Losses. In all rotating machines, losses occur. Among them are I^2R losses, called copper losses, in the windings, connections, and brushes; stray load losses in windings, solid metal structures, and frame; core loss in the magnetic material and structural parts; windage and friction loss; and exciter and rheostat losses.

I^2R losses (in watts) in each path of the windings are equal to the square of the effective current in amperes times the resistance in ohms. Brush I^2R loss is the product of the potential drop in volts times the current in amperes. Stray load losses are caused mainly by eddy currents, due to variable magnetic fields (produced by the load current) within the conductors, pole surface, structural members, end shields, frame, and so forth.

Windage and friction losses are the result of circulation and turbulence of the cooling medium and friction of bearings, seals, and brushes. Windage loss is relatively large in air-cooled high-speed machines. In hydrogen the loss is only 7–15% of that in air within the operating range of purity. Bearing and seal friction losses are generally absorbed by the lubricating oil. To avoid excessive friction or overheating of bearings, an inlet oil temperature of 100–120°F is often recommended for large machines, with discharge at a temperature of about 150°F.

[LEON T. ROSENBERG]

Bibliography: D. G. Fink and J. M. Carroll (eds.), *Standard Handbook for Electrical Engineers*, 10th ed., 1968; D. G. Gehmlich and S. B. Hammond, *Electromechanical Systems*, 1967; E. Levi and M. Panzer, *Electromechanical Power*

Conversion, 1966; J. Rosenblatt and M. H. Friedman, *Direct and Alternating Current Machinery*, 1963; S. Seely, *Electromechanical Energy Conversion*, 1962; G. J. Thaler and M. L. Wilcox, *Electric Machines: Dynamics and Steady State*, 1966.

Electrical measurements

The measurement of any one of the many quantities by which the behavior of electricity is characterized. The knowledge of the quantitative behavior of electricity is essential to scientific and technical progress. Electrical measurements play a major role in industry, communications, and even in such unrelated fields as medicine.

Many electrical measurements can be made with direct-indicating instruments merely by connecting the instrument properly in the circuit. Thus a volt-meter provides a pointer which moves over a scale calibrated in volts, and an ammeter in the same way presents a reading of current in amperes. Other direct-reading instruments are wattmeters, frequency meters, power-factor or phase-angle meters, and ohmmeters. Many electrical quantities are measured both as instantaneous values and as values integrated over time. Some electrical measurements must be made with various specialized devices or systems requiring adjustment or balancing to obtain the measured value. Typical of these are potentiometers and bridges in many standard and specialized forms.

Because of differences in instruments and techniques, it is convenient to divide measurements into direct-current (dc) and alternating-current (ac) classes.

DC measurements. In dc circuits the measurement of voltage and current often suffices to define the operation of the circuit. The product of the two represents power. In the commercial sale of dc electricity the measurement of energy must be made with a dc watt-hour meter. Occasional use is made of a dc ampere-hour meter in battery-charging installations.

To measure high values of current, shunts are used to bypass all but a small fraction of the current around the measuring instrument. A newer technique employs a form of saturable reactor energized by alternating current to measure large direct currents. *See* ELECTRIC ENERGY MEASUREMENT; ELECTRIC POWER MEASUREMENT.

AC measurements. Alternating-current circuits involve more variables and hence more measurements than dc circuits. The most common measurements are voltage, current, and power; the last requires a wattmeter, as ac power cannot always be calculated directly from voltage and current. Also measured are frequency and power factor (or phase angle) and sometimes waveform or harmonic content. Energy is measured by means of the ac induction watt-hour meter. In general, ac instruments differ in principle and design from dc instruments, although many ac instruments may be used to measure dc quantities. Direct-current instruments do not respond to ac quantities, but some may be adapted by the addition of rectifiers to convert alternating current to direct. The thermocouple is another form of convertor by which a dc instrument may be made to read ac quantities.

If alternating voltages and currents above the normal ranges of self-contained instruments are to be measured, instrument transformers may be used to extend the ranges of those instruments. In the study of ac waveform a qualitative evaluation may be made with an oscillograph or a cathode-ray oscilloscope. Quantitative measurement of harmonic content requires the use of a harmonic analyzer.

Accuracy of measurements. Accuracy denotes the degree of compliance of the instrument reading with the true value of the measured quantity. It is common to describe the instrument's accuracy by stating the maximum allowable error. Thus an instrument with a maximum error of 2% is often described as having an accuracy of 2%. For many applications a small panel or miniature electric instrument with a maximum error of 2–5% of full-scale calibration will suffice. More refined instruments are available with maximum errors of 1, 0.5, or 0.25%. When measurements of higher accuracy than this are required, measurement systems, such as potentiometers and bridges, must be used. Direct current and voltage can readily be measured in this way to an accuracy of 0.01%. Alternating-current measurements can be readily made to an accuracy of 0.1% or better.

Laboratory measurements. In a laboratory, emphasis is normally placed on accuracy and on completeness of facilities to deal with all types of measurements. There is relatively little limitation on the size and complexity of equipment used. If standardizing service is a function of the laboratory, the equipment must include standard cells and precision standard resistors and also suitable potentiometers, bridges, shunts, and volt boxes (voltage-dividing resistors to extend the range of voltage measurements). With these, the calibration of dc instruments can be performed with high accuracy. Extension of calibration service to ac instruments requires transfer standards (instruments having negligible difference in performance when operated on alternating current or direct current).

Field measurements. This term is used to designate all measurements made outside a laboratory as in generating stations and substations, service shops, factory testing areas, ships, and aircraft. For these uses equipment is chosen to perform only specialized services. Accuracy well below that of laboratory measurements is usually permissible. Convenience, compactness, and often portability are prime considerations in choosing equipment. Electric instruments for this kind of service are often of the panel type, sometimes in miniature sizes. Multipurpose and multirange instruments, like the volt-ohm-milliammeter, are handy for service measurements where 2–5% error is permissible.

For field measurements of alternating current and voltage at power frequencies, the hook-on volt-ammeter provides readings within 2% maximum error. As a voltmeter it is connected to the circuit with spring clips; as an ammeter it operates on the current-transformer principle. The core, which is circular and extends outside the instrument case, has a hinged link that may be opened to slip around a conductor carrying current and then closed again. Thus there is no need to break the circuit under measurement; the conductor itself becomes a one-turn primary winding in the transformer measurement system. The measurement of power can also be made with a hook-on wattmeter.

On power systems, field measurements may be desired continuously over a period of time, sometimes at unattended locations. For this purpose recording instruments may be used or the reading may be telemetered to a manned station. Any instrument that is made in indicating form can be made in recording form, but the greater power necessary to drive a marking device over a chart may call for some kind of amplification.

The operation of a power system also calls for the recording of disturbances due to lightning strokes, insulation flashover, short circuits, and other transient phenomena. Recording oscillographs used for this purpose can be triggered by any condition that deviates from normal operation, thus making a record of the disturbance.

Frequency considerations. The measurement of voltage and current is commonly made over a frequency range from a few hertz (Hz) up through 2000 megahertz (MHz). The frequency at which measurements are to be made dictates the type of equipment needed, the precautions to be taken, and the degree of accuracy which may reasonably be expected. Alternating-current instruments of the moving-iron, fixed-coil type are intended primarily for 60 Hz applications but may be used with only moderate errors up to several hundred hertz. Electrodynamic (moving-coil, fixed-coil) instruments, which are generally of greater accuracy, may also be used in this frequency range. Errors in such instruments result mainly from reactance effects, which may be minimized by special design to permit operation to several kilohertz (kHz). Rectifier-type instruments possess only small frequency errors up to several kilohertz in relation to their overall accuracy rating, which is of the order of 2–5% error. Electronic voltmeters, which are of the same general accuracy, are especially suited for use over a wide range of frequencies.

Circuit loading. All electric instruments draw some power from the circuits to which they are connected, and ac instruments generally take more power than dc instruments. This circuit loading may alter appreciably the quantity being measured. For instance, an ac voltmeter rated 150 volts and having a resistance of 3000 ohms is perfectly suitable to measure voltage on a 120-volt house lighting circuit. However, if the same voltmeter is connected to the terminals of a small power amplifier with a maximum output of 200 milliamperes (ma), the voltmeter will load the source and seriously reduce its voltage. To avoid this error the measurement should be made with a rectifier voltmeter (about 150,000 ohms resistance) or an electronic voltmeter (above 0.5 megohm resistance).

In the measurement of current, consideration must be given to the voltage drop in the ammeter. If it is appreciable in relation to the source voltage, the current is not the same as that if the ammeter were not connected in the circuit. The magnitude of this error can usually be evaluated and minimized by proper choice of instruments.

[ISAAC F. KINNARD/EDWARD C. STEVENSON]

Time dependence. When voltage and current are variable functions of time, the measurement of frequency, wavelength, and waveform is of impor-

tance, in addition to phase-angle measurements. At frequencies where the physical dimensions of the electric circuit are small compared to the wavelength of the voltage and current, the frequency is said to be low and various forms of frequency meters are employed, depending upon the range involved.

At higher frequencies it often becomes necessary to measure frequency and wavelength independently since their product is not always a constant under this condition.

The waveform of the electrical quantity being measured is of importance. Many indicating instruments are calibrated to give correct readings only for sine-wave inputs. If the waveform is nonsinusoidal, it is necessary to consider the principles of operation of the particular instrument to interpret the meter indication correctly. For example, an electronic voltmeter, which measures peak values but is calibrated in root-mean-square (rms) values based on sinusoidal wave shape, would not give correct rms values for a nonsinusoidal wave.

Measurement of parameters. The parameters of any electric circuit are the resistances, inductances, and capacitances along and between the conducting branches of the circuit, including any ground plane that may be near or surrounding the circuit. The measurement of these parameters may be classified according to the apparent disposition of the parameters, which is a function of frequency. For any given circuit there is some frequency below which the circuit can be treated as having lumped parameters or circuit elements; above this frequency the parameters must be considered as being distributed throughout the circuit.

Lumped parameters. The measurement of lumped parameters may be subdivided according to the measurement accuracy desired. If errors of several percent are permissible, direct-reading instruments, which indicate the value of the parameter directly on a calibrated meter scale, are available. Inductance and capacitance measurements are made at some convenient frequency. Of this class of instruments the ohmmeter, used for measuring resistance at zero frequency (direct current), is the only one in common use.

For greater accuracies bridge measurements are preferred. Direct-current measurements are made with the Wheatstone bridge with maximum errors on the order of 0.01%. Resistances of less than a few ohms can be satisfactorily measured only with a bridge, regardless of the degree of accuracy desired.

Most circuit designers prefer to measure inductance and capacitance in the particular operating frequency range under consideration, and the ac bridge is commonly used. Bridge measurements provide numerous advantages over other methods, including high accuracy and the ability to compare the unknown to a known standard. Bridges are designed to operate in various frequency ranges from direct current to several hundred kilohertz, from several kilohertz to several megahertz, from 1 or 2 to several hundred megahertz, and from several hundred to several thousand megahertz. At the higher frequencies the application of the bridge becomes more complicated, and considerable caution and planning are necessary if reliable results are to be obtained.

A unique instrument known as a Q meter is available for measuring inductance or capacitance and effective resistance at radio frequencies.

[ROBERT L. RAMEY]

Distributed parameters. Electrical systems can be completely described by their associated electric and magnetic fields, the properties of the materials involved, the physical dimensions, and the velocity of light. When the dimensions are small compared with the wavelength, however, it is more convenient to treat them as circuits composed of lumped parameters.

At low frequencies lumped inductance and capacitance can be used, although they are rigorously derived only for nonvarying currents and voltages, respectively. At high frequencies the finite propagation velocity of electromagnetic waves cannot be neglected, and the derivations break down. If only one system dimension is comparable with the wavelength of the electrical disturbance, restricted conditions permit a rigorous definition of distributed inductance and capacitance. These distributed parameters combine with the resistance of a pair of conductors and the conductance between them to define the behavior of a transmission line for plane-wave propagation and to relate the voltage between conductors at any point on the line to the voltage at any other point. *See* TRANSMISSION LINES.

The concept of distributed parameters is also useful at low frequencies when it must be recognized that a circuit component, nominally representable by a single parameter, is actually modified by the presence of residual parameters. Thus a coil has not only inductance, but capacitance and resistance as well. This capacitance is definable by low-frequency analysis, since the dimensions are small, but it cannot be localized and represented as a unique lumped parameter because the winding is not an equipotential surface.

For example, a coil mounted over a ground plane has one terminal grounded. The voltage between winding and ground increases from zero at the grounded terminal to maximum at the other. Capacitance near the grounded terminal is therefore less effective than capacitance near the other. The resultant effective terminal capacitance is, in consequence, only one-third of the total capacitance for uniformly distributed capacitance. For other conditions of grounding, the ratio of effective capacitance to total distributed capacitance will again be different.

Values of distributed parameters are inferred from the behavior of the system that they define. For transmission lines, measurements may involve observing the voltage distribution along the line under different terminal conditions. For circuit elements, impedance may be measured at different frequencies or under different conditions of adjustment of some known lumped parameter.

[DONALD B. SINCLAIR]

Bibliography: D. Bartholomew, *Electrical Measurements and Instrumentation*, 1963; R. F. Field and D. B. Sinclair, A method for determining the residual inductance and resistance of a variable air condenser at radio frequencies. *Proc. IRE*, 24(2), 1936; E. L. Ginzton, *Microwave Measurements*, 1957; F. K. Harris, *Electrical Measurements*, 1952; F. A. Laws, *Electrical Measurements*, 1938; G. R. Partridge, *Principles of Electronic Instruments*, 1958; F. E. Terman and J. M. Pettit, *Electronic Measurements*, 1952.

Electrical units and standards

The standard in terms of which electrical quantities are evaluated, the quantities so adopted being known as units. The ohm, for example, is a unit of electrical resistance. The electrical units in practical use today, and also in extensive theoretical use, were designated by the Eleventh General Conference of Weights and Measures in 1960 as members of the International System of Units (Système International d'Unités, abbreviated SI in all languages). This action by the General Conference was the culmination of an effort initiated by A. Giorgi at the beginning of this century to bring the practical electrical units into a coherent system with appropriate mechanical units of the metric system.

ELECTRICAL UNITS

To accomplish the above objective, the base units for mechanical quantities were arbitrarily selected: the meter for the unit of length, the kilogram for the unit of mass, and the second for the unit of time. Units for other mechanical quantities are derived from these units in accordance with physical laws and concepts such as the unit of speed, the meter per second, and the unit for acceleration, the meter per second per second.

This system was originally called the mks system to distinguish it from the cgs system (based on the centimeter, gram, and second).

Meter-kilogram-second system. Acting under authority given it by the Eighth General Conference of Weights and Measures, the International Committee of Weights and Measures in 1937 proceeded to define a unit for force (now called the newton, N) and units for energy and power in mechanical terms. The theoretical magnitudes of these units are given below.

Unit of force. The force which gives to a mass of 1 kilogram an acceleration of 1 meter per second per second.

Joule (J). The work done when the point of application of the mks unit of force is displaced a distance of 1 meter in the direction of the force.

Watt (W). The power which gives rise to the production of energy at the rate of 1 joule per second.

The Committee then proceeded to define electric and magnetic units in terms of these mechanical units. The revised units were to replace the definitions which had been in effect for many years such as the "mercury ohm" and the "silver ampere." The revised definitions of electrical and magnetic units which have been accepted since 1948 were given by the Committee as follows.

Ampere (A). The constant current which, if maintained in two straight parallel conductors of infinite length, of negligible circular sections, and placed 1 meter apart in a vacuum, would produce between these conductors a force equal to 2×10^{-7} mks unit of force per meter of length.

Volt (V). The difference of electric potential between two points of a conducting wire carrying a constant current of 1 ampere, when the power dissipated between these points is equal to 1 watt.

Ohm (Ω). The electric resistance between two points of a conductor when a constant difference of potential of 1 volt, applied between these two points, produces in the conductor a current of 1 ampere, the conductor not being the seat of any electromotive force.

Coulomb (C). The quantity of electricity transported in 1 second by a current of 1 ampere.

Farad (F). The capacitance of a capacitor between the plates of which there appears a difference of potential of 1 volt when it is charged by a quantity of electricity equal to 1 coulomb.

Henry (H). The inductance of a closed circuit in which an electromotive force of 1 volt is produced when the electric current in the circuit varies uniformly at a rate of 1 ampere per second.

Weber (Wb). The magnetic flux which, linking a circuit of 1 turn, produces in it an electromotive force of 1 volt as it is reduced to zero at a uniform rate in 1 second.

The revised definitions were intended solely to fix the magnitudes of the units and not the methods to be followed for their practical realization. This realization is effected in accord with the well-known laws of electromagnetism. For example, the definition of the ampere represents only a particular case of the general formula expressing the forces which are developed between conductors carrying electric currents, chosen for the simplicity of its verbal expression. It serves to fix the constants in the general formula which has to be used for the realization of the unit.

A special name was added to the list by the Eleventh General Conference of Weights and Measures in 1960, the tesla (T) for the unit of magnetic flux density (one weber per square meter).

Centimeter-gram-second systems. Two systems of electric and magnetic units have been in use in scientific circles for a long time but both are rapidly giving way to the International System. They are the electrostatic system of units (esu) and the electromagnetic system of units (emu).

The electrostatic system defines a unit charge as that charge which exerts 1 cgs unit of force (1 dyne, which is equivalent to 10^{-5} newton) on another unit charge when separated from it by a distance of 1 centimeter in a vacuum. All other units of the system are derived from this definition by assigning unit coefficients in equations relating electric and magnetic quantities to each other. The units so derived are often referred to in terms of the SI units with the prefix "stat," for example, statvolts, statohms, and statamperes.

The electromagnetic system defines a unit magnetic pole (a highly fictitious concept) as that pole which exerts 1 cgs unit of force on another unit pole when separated from it by a distance of 1 centimeter in a vacuum. All other units of the system are derived from this definition in accord with the principles set forth above for the electrostatic system. Units so derived are often referred to in terms of the SI units with the prefix "ab," for example, abvolts, abohms, and abamperes. Special names are given to some magnetic units of the emu system such as the maxwell and the gauss, which correspond, respectively, to the weber and the tesla of SI, although differing from them in magnitude.

The magnitudes of corresponding units of the electrostatic and electromagnetic systems differ from each other by a factor theoretically equal to the speed of light, c, (3×10^{10} centimeters per second, approximately) or its square. Thus, 1 abampere is equal to 3×10^{10} statamperes; 1 statvolt is equal to 3×10^{10} abvolts; and hence 1 statohm is equal to 9×10^{20} abohms.

The esu system is found convenient for handling purely electrostatic problems. A combination of the two systems in which electrostatic quantities are expressed in esu and magnetic and electromagnetic quantities in emu, with appropriate use of the conversion constant c between the two systems, is called the Gaussian system. All of these systems are rapidly giving way to use of SI units in treatment of electric, magnetic, and electromagnetic phenomena.

ELECTRICAL STANDARDS

Electrical standards are the physical embodiments by means of which the electrical units are realized and maintained. The ampere is unique in this system, since an arbitrary constant, other than unity is employed in its definition to bring the entire system into agreement with the mechanical units while still adhering substantially to the old value for the unit.

Not all standards for electrical quantities are maintained in the national standards laboratories, such as the National Bureau of Standards; the only standards maintained are those for the volt, the ohm, the farad, and sometimes the henry, since very stable standards for these quantities can be produced. The other electrical quantities are determined from suitable combinations of these standards.

Determination of the ampere. Since the ampere is defined in terms of the force between two current-carrying wires, the conventional means of determining the ampere is by some kind of current balance in which the force between the current-carrying elements is compared with the force of

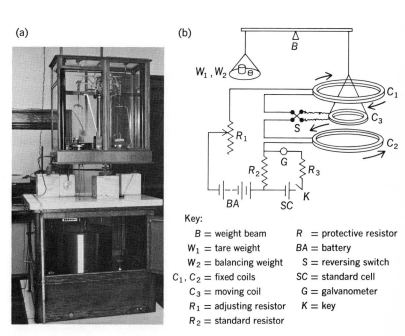

(a)   (b)

Key:

B	= weight beam	R	= protective resistor
W_1	= tare weight	BA	= battery
W_2	= balancing weight	S	= reversing switch
C_1, C_2	= fixed coils	SC	= standard cell
C_3	= moving coil	G	= galvanometer
R_1	= adjusting resistor	K	= key
R_2	= standard resistor		

Fig. 1. (*a*) Current balance. (*b*) Schematic of current balance.

Fig. 2. Electrodynamometer, the current balance used in absolute determination of the ampere at the National Bureau of Standards.

gravity on a known mass. One form of current balance consisting of two fixed coils and a movable coil supported by one arm of an equal arm balance is shown in Fig. 1. An electric current, supplied by a battery and controlled by a rheostat, is sent through the fixed and movable coils. The current flows through the fixed coils in opposite directions so that the magnetic fields produced by them are in opposition in the region of the movable coil. Here the magnetic field is directed horizontally and radially with respect to the axis of the coil system. The direction of the current in the movable coil is controlled by a reversing switch. When the current flowing through the movable coil is in one direction, the force exerted on it by the currents in the other coils is downward; but when the current is reversed, the force is upward. The force of gravity on the mass of the movable coil is balanced by a tare weight on the other arm of the balance.

The currents and the balancing weights are adjusted so that when the force due to the current is upward on the coil, the tare weight just balances the coil; but when the current is reversed, the additional weight must be added to achieve balance.

When balance is achieved, the current through the coil is given by Eq. (1), in which C is the con-

$$2 Ci^2 = mg \qquad (1)$$

stant computed from the dimensions of the coil assembly, i is the current in amperes, m is the mass of the small weight, and g is the acceleration

of gravity at the place where the experiment is performed.

Current circulating in the coil system is passed through a standard resistor having a resistance of 1 ohm, approximately, but its value must be accurately known.

A standard cell, that is, an electrochemical cell which produces a constant emf, with a protective resistor and a galvanometer and key in series with it, is placed across the known resistor. If no deflection of the galvanometer is observed when the key is closed in this circuit, then the emf of the standard cell is given by Eq. (2), where r is the re-

$$ir = v \qquad (2)$$

sistance of the known resistor. Thus the experiment for the determination of the ampere is actually an experiment for determining the emf of a standard cell.

Several such standard cells are calibrated by means of the current balance, and they in turn preserve the unit of voltage when the current balance is not in use. A current of 1 ampere may thus be established at anytime by sending a current of such strength through a 1-ohm resistor that it gives rise to exactly 1-volt drop across it.

In modern versions of the current balance, both the fixed coils and the movable coils are wound in a single layer on cylinders of marble, pyrex, or fused silica, in which accurate grooves have been lapped to maintain the wires in fixed positions so that the dimensions of the coils can be accurately measured.

Another form of the current balance, the Pellat balance, is called an electrodynamometer (Fig. 2). It consists of a long solenoid within which another cylindrical coil is balanced on knife edges so that it is free to rotate. The axis of the inner coil is at right angles to the axis of the solenoid. When the electric current is sent through both coils, a torque is exerted on the movable coil which is balanced by a weight suspended at the end of an arm extending out horizontally from the rotatable coil. When balance is attained, the current through the coil system is given by Eq. (3), where C is the computed

$$Ci^2 = mgl \qquad (3)$$

constant of the coil system, and l the length of the lever arm on which the weight is supported. When the current in the rotatable coil is reversed, the torque is reversed and the balancing weight is moved to an arm on the opposite side of the coil. The electrical circuit used with the Pellat balance is essentially identical with that used with the other type. Experiments at the National Bureau of Standards with the two forms of current balance gave agreement to within about 1 part in a million.

Similar experiments in the national standardizing laboratories of other countries have served to establish the value for the unit of voltage for those countries, but there were known differences in values assigned to the standards in various countries, as demonstrated by international comparisons. Increased accuracy with which such experiments have been conducted since World War II has permitted assignment of values to the electrical standards in much closer conformity with the theoretically defined values of the units. By international agreement, decision was made that electrical standards in use throughout the world should be referred to a uniform basis after Jan. 1, 1969, bringing them into closer agreement with the units they embody. The new volt at the National Bureau of Standards is 8.4 parts in a million smaller than the old volt. Thus, a standard cell which was assigned a value of, say, 1.0183000 volts before the changeover would now be assigned a value of 1.0183086 volts.

Determination of the ohm. Several methods have been employed for the determination of the ohm. The impedance of ohms of an electric circuit containing only resistive elements for direct or alternating current is given by Eq. (4), where r is the

$$z = r \qquad (4)$$

resistance in ohms. But for a circuit containing only inductive or capacitive elements the impedance for alternating current is, respectively, given by Eqs. (5) and (6), where L is the inductance in

$$z = \omega L \qquad (5)$$
$$z = 1/\omega C \qquad (6)$$

henries, C is the capacitance in farads, and ω is the angular frequency of the alternating current in radians per second.

The inductance of an arrangement of current-carrying elements or the capacitance of an arrangement of charge-bearing elements can be calculated readily from electromagnetic principles and the geometry of the arrangement in units of inductance (henries) or in units of capacitance (farads). The impedance in ohms of either of these arrangements may be calculated for given alternating current frequencies from Eqs. (5) or (6). If one of these calculated inductors or capacitors is placed in one arm of an alternating-current bridge, its impedance can be compared with that of a resistor in another arm of the bridge, thus establishing the value of the resistor in ohms (Fig. 3).

For many years the values of resistors were obtained from the values of computable inductors in most cases. The discovery of the Thompson-Lampard theorem in electrostatics in 1956 led to a great improvement in the art. In comprehending this theorem, a new form of capacitor should be visualized, consisting of a long metal tube divided into four segments by longitudinal cuts coplanar with the axis of the tube. If C_1 is the capacitance per unit length between one opposite pair of segments when the other pair is grounded, and C_2 is the capacitance per unit length between the other pair of opposite segments when the first pair is grounded, then Eq. (7) holds, where ϵ_0 is the electric constant.

$$e^{-\Pi C_1/\epsilon_0} + e^{-\Pi C_2/\epsilon_0} = 1 \qquad (7)$$

If C_1 and C_2 are nearly equal, the tube length is reduced as shown in Eq. (8), where $\Delta C = C_1 - C_2$.

$$\overline{C} = \frac{C_1 + C_2}{2} = \epsilon_0 \frac{\ln 2}{\Pi}\left(1 + 0.087\frac{\Delta C^2}{\overline{C}^2}\right) \qquad (8)$$

The tube need not be of cylindrical form. It is possible to replicate this arrangment with carefully machined parts such as cylindrical gage blocks insulated from each other, each block corresponding to one segment of the tube. With this arrangement C_1 and C_2 can be made nearly equal and ΔC

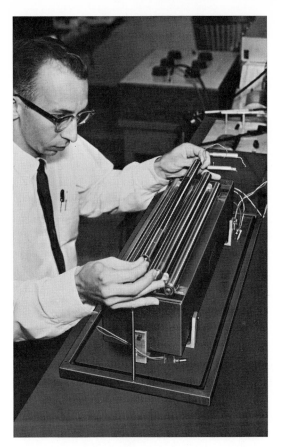

Fig. 3. Computable capacitance standard used in absolute determination of ohm.

becomes vanishingly small.

Since the length of cylindrical gage blocks may be measured with very great accuracy, the capacitance of this type of capacitor can be calculated from the length measurement with correspondingly great accuracy. The greatest uncertainty in calculation of the capacitance arises from uncertainty of the knowledge of the speed of light because the constant ϵ_0 is implicitly defined in the International System by Eq. (9), where c is the speed of light and μ_0 has the arbitrarily assigned numerical value $4\Pi \times 10^{-7}$.

$$c^2 = 1/\epsilon_0\mu_0 \qquad (9)$$

The impedance of a practical-size capacitor of this type is very great, about 10^8 ohms. However, its impedance can be compared with that of a 1-ohm resistor in successive steps so that nearly the full accuracy of the calculation can be realized.

[ALVIN G. MC NISH]

Bibliography: F. L. Hermech and R. F. Dziuba, *Precision Measurement and Calibration: Electrical*, Nat. Bur. Stand. Spec. Publ. no. 300, vol. 3, 1968.

Electrical utility industry

A modest upturn in sales of electrical energy during 1975, and record rate increases granted by regulatory commissions belatedly responsive to the utilities' financial plight, enabled the United States electrical utility industry to finish the year in somewhat better health than at the close of 1974. Auto-

United States electric power industry statistics for 1975*

Parameter	Amount	Increase or decrease compared with 1974, %
Generating capability, kW (×10³)		
Total	513,137	8.1
Conventional hydro	56,414	3.8
Pumped-storage hydro	9,449	7.1
Fossil-fueled steam	358,511	6.4
Nuclear steam	40,338	33.4
Combustion-turbine and internal combustion	48,425	9.4
Energy production, kW-hr (×10⁶)	1,901,700	1.3
Energy sales, kW-hr (×10⁶)		
Total	1,719,700	1.1
Residential	586,900	5.7
Commercial	416,300	6.0
Industrial	650,700	−5.6
Miscellaneous	65,800	3.3
Revenues, total (×10⁶)	$49,500	26.5
Capital expenditures, total (×10⁶)	$18,948	−7.8
Customers (×10³)		
Residential	72,300	1.9
Total	81,500	1.8
Residential usage, kW-hr (average)	8,210	3.8
Residential bill, ¢/kW-hr (average)	3.35	18.4

*From *Elec. World*, Sept. 15, 1975, and extrapolations from Edison Electric Institute monthly data.

matic fuel cost pass-alongs authorized for utilities supplying about 70% of the nation's electricity, and specific allowances for fuel cost hikes in rate increases granted to other utilities, compensated for most of the staggering rise in fuel prices that was sparked by the Arab oil embargo of November 1973. And the 1.1% upturn in energy sales, versus a 0.15% slump in 1974, provided a little load for the additional generating capability added to the utility systems during 1974 and 1975 to supply energy demands that had been projected, before the oil embargo, to grow at the historical 7–8% annual rate. The table shows electrical industry statistics for 1975.

Growth rate declines. In large measure, this 2-year period of near-zero load growth can be attributed to the United States government's early response to the oil embargo. Through its speedily created Federal Energy Office, it pressed for a 10% reduction in the use of electricity. And the 0.15% slump in 1974 sales of electricity instead of the expected 7–8% increase would indicate that most sectors of the economy cooperated effectively. But the loss of revenue to the electrical utility industry sorely strained its financial resources and spurred prompt demands for rate increases. As a result, average rates for electrical service jumped nearly 24% in 1974 and an additional 28% in 1975, a total over the 2 years of about 57% above the 1973 level, which added substantial economic incentive for reducing the use of electricity. And this incentive is expected to persist during the years ahead, making for a sharply lower growth rate than the industry had anticipated before the oil embargo.

After 2 years of essentially zero growth in energy sales, and only 3.6% growth in summer-peak demand, the industry finds itself with nearly 17% more generating capability than it needed to carry the 1975 peak load with a reasonable reserve mar-

gin. Moreover, generation additions nearing completion will carry this surplus into 1976. Such a surplus is costly in that it represents construction expenditures approaching $15,000,000,000, on which the annual charges (interest on investment, taxes, and so on) exceed $2,500,000,000, or about 0.15¢ per kW-hr sold.

New construction deferred. By early 1976, however, the sharp cutback of construction programs instituted by the electrical utilities, after lagging energy sales during the early months of 1974 indicated a slowing of the historical growth rate, will begin to slow their system growth. This cutback came too late to have much effect on 1974 expenditures, most of which were linked to projects requiring several years to complete. This is particularly true for generating stations, which generally require expenditures over periods ranging from 4 to 8 years before they are ready to operate. Hence this cutback had little effect on major system additions for 1975, because they were too near to completion. But additions scheduled for completion in 1976 have been shaved to match closely the expected increase in energy sales and peak demand, both of which are based on a slower growth rate than had prevailed before the oil embargo. And sharper cuts during the remainder of the decade will enable the industry to load up much of the surplus generating capability put on line during 1974 and 1975.

This surplus generating capability, despite its drain on the industry's financial resources, does afford some economies in fuel costs. Nearly 26% of the capability added is in nuclear stations, which burn fuel costing about a fourth as much as fuel oil. In addition, they reduce the industry's oil consumption at the rate of about 10,000 bbl (1590 m³) of oil per year per 1000 kW of generating capability. Hence the 19,268,000 kW of nuclear-fueled generation added since the oil embargo are saving nearly 200,000,000 bbl (10⁶ bbl = 158, 987 m³) of oil per year.

Somewhat more than 44,000,000 kW of fossil-fueled steam stations were completed during 1974 and 1975, with upwards of 40% of the capability designed to burn coal or lignite. Their operation also tends to reduce the consumption of fuel oil by nearly as great an amount as the nuclear stations. Thus the electrical utility industry moves toward the nation's objective of energy self-sufficiency by the early 1980s (see illustration).

Conversion to coal. Still to be resolved, however, is the effect of the recent selection by the Federal Energy Administration (FEA) of nearly 80 generating units to be converted from oil burning to coal burning. Most of them were built originally to burn coal—and presumably still have the facilities for handling it—but were converted to oil as the best way to satisfy Environmental Protection Agency (EPA) rulings on sulfur dioxide and other stack emissions. Hence the reconversion to coal will require either the costly retrofitting of additional antipollution facilities or some relaxation of EPA restrictions. In addition, it may call for burning more coal than the coal-mining industry is able to produce without expanding its facilities—an effort that also faces environmental pressures.

Nuclear stations. Nuclear power plants are the long-range key to easing the pressures of fuel shortages and inflated fuel prices. According to a survey of 27 modern steam stations completed by *Electrical World* late in 1973, three nuclear stations, constituting nearly 20% of the total capability covered, had total power costs averaging 10.4% below those of the fossil-fueled stations. The difference was attributable entirely to savings in fuel costs, which averaged 34% lower for the nuclear stations than for the fossil-fueled stations. The investment cost per kilowatt for these three nuclear stations was very nearly the same as that of coal-burning stations which had some of the costly antipollution facilities prescribed for burning coal.

The cost of fossil fuels has risen far more than that of nuclear fuel since the conclusion of the *Electrical World* survey. Unfortunately, however, only about 7% of the total generating capability of United States utilities during 1975 was nuclear, and some of that was out of service for modifications and repairs. The approximately 34,000,000 kW of nuclear generation that remained in service should be credited with saving the equivalent of nearly 340,000,000 bbl of fuel oil, which is a very important step toward achieving self-sufficiency in fuels.

Generating additions. The United States electrical utility industry entered 1975 with 364,433,000 kW of generating additions under construction or planned. Of this total, 41,044,000 kW were scheduled for completion in 1975, at a cost of about $13,000,000,000. They were to serve an expected gain of 25,900,000 kW in summer-peak demand and provide a modest margin for retiring obsolete units (over 40 years old) and for maintaining the reserve for reliability. But the conservation program slashed the increase in peak demand to 8,760,000 kW. Consequently the reserve margin soared to 37.0%, far higher than could possibly be supported financially by revenues that had been cut by the conservation program.

The tight austerity invoked by many utilities in 1974 became even tighter in 1975, and more planned generation additions were postponed or canceled. Savings during the remainder of the decade from the 1974 austerity actions, estimated at 27.5%, were expanded moderately in 1975 and extended into the early 1980s. These savings, of course, are based on the premise that the energy conservation ethic will continue into the early 1980s, or until energy self-sufficiency is attained, with the additional incentive of the sharply higher cost of electricity.

Nuclear stations. Some 10,072,000 kW of the generation additions during 1975 were nuclear, nearly 25% of the total additions, and 9,196,000 kW were completed in 1974. Thus the additions during just 2 years nearly equaled the total nuclear capability on line at the end of 1973. By 1980, nuclear generation is scheduled to reach a total of 92,062,000 kW, 14.2% of the 644,742,000 kW total generation in service. And this nuclear generation should supply upwards of 28% of all the nation's electricity while saving more than 900,000,000 bbl of fuel oil in 1980 alone and holding down the price of electricity by about 10%.

Fossil-fueled stations. During 1975 some 24,040,000 kW of fossil-fueled generation was completed and put in service. Of this, about 2,389,000 kW replaced 40-year-old stations that were too obsolescent for continued economical operation or too costly to retrofit with antipollution devices.

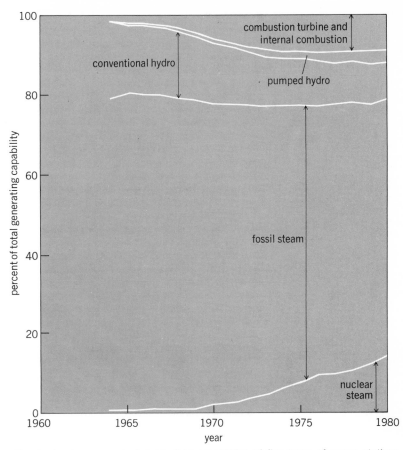

Graph showing percent of generating capabilities of five types of power stations from 1960 to 1980.

Generating units in these new stations are designed to burn fuels as follows: coal, 19 units; oil, 17; and natural gas (in gas-producing areas), 7. Most of them are prepared to burn alternative fuels if necessary. Additions nearing completion for 1976 total 16,581,000 kW, of which 1,923,000 kW will replace obsolescent generation. Of these 1976 additions, 20 units will burn coal, 11, oil, and only 2, natural gas. Thus the trend away from gas as boiler fuel will have become almost complete, as far as new electrical generating stations are concerned.

After 1976, fossil-fueled generation additions are slated to decline further until, at 6,172,000 kW (after retirements) in 1980, they will be less than nuclear. Moreover, as nuclear capability becomes more ample, the fossil-fueled generation will be used increasingly to supply the additional energy required on weekdays, and be shut down nights and weekends; this type of operation, commonly known as cycling, consumes considerably less fuel than continuous, or base-load, operation. Both trends reflect the electrical utility industry's effort to conserve fuels that are expected to remain in short supply.

Pollution regulations. The major problem for fossil-fueled stations today is compliance with air- and water-pollution regulations. The only fuels that seem likely to remain in ample supply are coal and, in some areas, lignite. Thus, essentially all new installations, and many of the existing ones, require high-efficiency electrostatic precipitators to capture 99% or more of the particulates in the combustion gases. And unless the coal is extremely low in sulfur content, they could be required by the EPA to add sulfur-removal equipment in the gas stream. These facilities add as much as 45% to the cost of the station and consume up to 5% of its electrical output. In addition, EPA has decreed that most steam stations have cooling towers for the water used to condense the exhaust steam; if these towers are of the evaporative type, they add another 3–4% to the station investment and further reduce its net output.

Combustion-turbine stations. Combustion, or gas, turbines dropped to a 7-year low of 3,995,000 kW in 1975 additions, even including those that are linked by heat recovery boilers to steam turbines in an arrangement known as combined cycle. This arrangement, by using part of the heat that is discharged in the exhaust of conventional combustion turbines, achieves a substantial improvement in fuel economy for a reasonable increase in investment cost. The greatest disadvantage is that present designs require either gas or distillate oil for fuel, both of which are likely to remain in short supply. Efforts today are directed toward developing simple gasifiers to produce suitable combustion-turbine fuel from coal or residual (heavy) oil. Until such devices become available, however, the construction of combustion-turbine stations is expected to decline to 3,561,000 kW in 1976 and to even lower levels thereafter. *See* COMBUSTION TURBINE.

Hydroelectric capability. Conventional hydroelectric additions climbed to 2,127,000 kW in 1975, but are scheduled to return to a more normal 1,880,000 kW in 1976 as additional 600,000-kW units go on line in Grand Coulee. With these additions, conventional-hydro capability will rise to 58,443,000 kW, some 10.7% of the total generating capability of United States electrical utilities. But it should be recognized that most of the added capability stems from converting existing stations to part-time, or cycling, operation or, as at Grand Coulee, to harness additional water flows made available by new upstream storage. Actually, very few sites remain where it would be feasible to build new hydro stations, and the share of the nation's electricity produced by falling water will decline continuously in the years ahead. *See* HYDROELECTRIC POWER.

Pumped-storage plants. Pumped-storage hydroelectric stations are a very desirable complement to nuclear generation in that they use the surplus of low-fuel-cost generation late at night to pump water into high-level reservoirs. The stored water is then used during heavy-load periods of the day to generate additional electricity in a hydroelectric station, thus functioning much like a giant storage battery. This generating cycle recovers about 65% of the energy used for pumping, and at a time when the energy is needed to serve electrical customers. Some 630,000 kW of such additions were completed in 1975, an amount considerably less than the 1,175,000 kW activated in 1974. But the total pumped-storage capability in service by 1980 is slated at 17,878,000 kW, nearly 2.8% of the industry's generating capability. *See* PUMPED STORAGE.

Transmission circuits. A heavy construction program during 1975 added upwards of 10,000 mi (15,200 km) of transmission circuits to the grid. These were needed to connect the 41,044,000 kW of new generating capability to load areas and to strengthen interconnections among load areas. About 33% of this added mileage was in overhead circuits for operation at upwards of 345,000 V, hence with carrying capabilities of 600,000 kW or more per circuit. Nearly 29% of these extra-high-voltage additions were in the heavily industrialized East North Central Region of the United States. But the West North Central Region had 20%, the Mountain Region 13%, the Pacific Region, 13%, and the remaining 25% was spread among all five other regions.

Lower-voltage circuits, mostly in the 115,000 to 230,000-V range, accounted for 67% of the total mileage added but only about one-fourth of the total carrying capability. Of these, nearly 23% of the mileage was built in the South Atlantic Region. But substantial additions were built in the West South Central Region (18%) and in the Mountain Region (16%), with the remaining 43% spread among the other six regions.

Only 0.7% of the transmission mileage added during 1975 was underground, most of it in the metropolitan areas of the Middle Atlantic Region and East North Central Region. Of the total underground mileage, only 9% was for operation at 230,000 V or higher.

Transformer capability. Substation capability for stepping up generating station outputs to transmission voltage levels, and for stepping the voltage back down at load centers, required the addition of 126,000,000 kVA in transformer capability during 1975. Of this total, about 37% was at new generating additions, and 43% at major substations to interconnect transmission systems operating at different voltage levels. The remaining 20% were placed in load-area industrial and distribution substations; this portion was far below normal, however, because of the slow growth of industrial and distribution loads attributable to the energy conservation effort, higher rates, and the generally low level of industrial activity.

Usage. Largely because of energy conservation and higher rates, distribution load growth lagged behind the rate of recent years, with the average residential usage, 8210 kW-hr, only slightly above the 1973 level. In addition, 25% fewer customers were connected during 1975 than in 1973, the latest substantially normal year. Consequently, the total kW-hr sales to residential customers gained only 5.7% in 1975 versus 0.14% in 1974 and 7.9% in 1973, and commercial consumption, largely served from distribution facilities, gained only 6.0%. Both factors contributed to a 21% drop in distribution expenditures below 1974 levels, and the drop approaches 27% if 1974 spending is adjusted for a year's inflation. Expenditures during 1974, however, were boosted by payments for material that was stored for use in 1975. *See the feature article* ENERGY CONSUMPTION.

Industrial sales dropped nearly 6% in 1975, reflecting the depressed economy. Aluminum production, which normally accounts for about 12% of the total industrial consumption, was down 13%. Other manufacturing usage dropped about 5.5%. And the only substantial gain, 21%, was for enriching the uranium used to fuel the nuclear generating stations; this usage alone accounted for nearly

7% of the total sales to industrial customers during 1975.

Combining these three major sales categories with sales to a miscellaneous group, which includes street and highway lighting and electric traction, brought the total for 1975 to 1,719,700,000,000 kW-hr, up only 1.1% over the 1,700,770,000,000 kW-hr sold during 1974.

Expenditures. As a consequence of this continuing near-zero growth, the electrical utility industry has revised its construction programs drastically. It had already invoked strict austerity during 1974 to hold its capital spending under budget, and it had delayed the completion of additions planned for service in 1975 and later years to match a slower probable growth rate. These moves deferred up to 30% of the expenditures that had been scheduled for later years. Of the $78,484,000,000 (in 1975 dollars) now projected for the remainder of the decade, 56.3% will go for generation additions, 11.1% for transmission, 27.6% for distribution, and 5.0% for such items as control centers, load control systems, operating headquarters, and a wide variety of construction and transportation devices.

Rates. Rates charged for electricity climbed steeply in 1975, reflecting an increase averaging about 0.30¢/kW-hr in fuel-cost adjustments and 0.12¢/kW-hr from a record $3,500,000,000 per year of rate increases authorized by various regulatory agencies. These increases boosted the average cost for residential service during 1975 to 3.35¢/kW-hr, commercial service to 3.34¢/kW-hr, industrial service to 2.15¢/kW-hr, and the average for all ultimate customers to 2.90¢/kW-hr. With these increases, the industry's total revenues for 1975 climbed to $49,500,000,000, up 26.5% over the 1974 level. Even this sharp increase, however, leaves the industry short of funds for paying still rising fuel costs and for financing continuing expansion that cost $18,948,000,000 in 1975 and is projected at $19,938,000,000 (1975 dollars) for 1976. *See* ELECTRIC POWER GENERATION; ELECTRIC POWER SYSTEMS; TRANSMISSION LINES.

[LEONARD M. OLMSTED]

Bibliography: Edison Electric Institute. *Statistical Yearbook of the Electric Utility Industry*, 1975; 18th Steam Station Cost Survey, *Elec. World*, 180(9):39–54, Nov. 1, 1973; Federal Power Commission, *The 1970 National Power Survey*, pts. 1 and 4, 1972; 1975 Annual Statistical Report, *Elec. World*, 183(6):43–74, Mar. 15, 1975; 26th Annual Electrical Industry Forecast, *Elec. World*, 184(6):35–50, Sept. 15, 1975.

Electricity

Electricity comprises those physical phenomena involving electric charges and their effects when at rest and when in motion. Electricity is manifested as a force of attraction, independent of gravitational and short-range nuclear attraction, when two oppositely charged bodies are brought close to one another. It is now known that the elementary (nondivisible) electric charges are possessed by electrons and protons. The charge of the electron is equal in magnitude to that of the proton, but is electrically opposite. The electron's charge is arbitrarily termed negative, and that of the proton, positive. Magnetism, those physical phenomena involving magnetic fields and their effects upon materials, manifests itself in the presence of moving electric charge. For this reason, magnetism was originally considered to be a part of electricity. *See* CHARGE, ELECTRIC.

Historical development. The earliest observations of electric effects were made on naturally occurring substances. Magnetism was observed in the attraction of metallic iron by the iron ore magnetite. The natural resin amber was found to become electrified when rubbed (triboelectrification) and to attract lightweight objects. Both of these phenomena were known to Thales of Miletus (640–546 B.C.). Jerome Cardan in 1551 first clearly distinguished the difference between the attractive properties of amber and magnetite, thus presaging the division of electric and magnetic effects. He also envisioned electricity as a type of fluid, a viewpoint that was developed more extensively in the late 18th and early 19th centuries. In 1600 W. Gilbert observed variations in the amounts of electrification of various substances. He divided substances into two classes, according to whether they did or did not electrify by rubbing. The division actually is into poor and good conductors, respectively. A two-fluid theory was first proposed by C. F. duFay in 1733. A one-fluid theory of electricity was propounded in 1747 by Benjamin Franklin, who called an excess of the fluid positive electrification, and a deficiency of fluid negative electrification. This theory fell into disrepute, but the choice of positive and negative remains. Although fluid theories of electricity were superseded at the end of the 19th century, the concept of electricity as a substance persists.

The quantitative development of electricity began late in the 18th century. J. B. Priestley in 1767 and C. A. Coulomb in 1785 discovered independently the inverse-square law for stationary charges. This law serves as a foundation for electrostatics.

In 1800 A. Volta constructed and experimented with the voltaic pile, the predecessor of modern batteries. It provided the first continuous source of electricity. In 1820 H. C. Oersted demonstrated magnetic effects arising from electric currents. The production of induced electric currents by changing magnetic fields was demonstrated by M. Faraday in 1831. In 1851 he also proposed giving physical reality to the concept of lines of force. This was the first step in the direction of shifting the emphasis away from the charges and onto the associated fields. *See* ELECTROMAGNETISM; INDUCTION, ELECTROMAGNETIC.

In 1865 J. C. Maxwell presented his mathematical theory of the electromagnetic field. This theory proposed a continuous electric fluid. It remains valid today in the large realm of electromagnetic phenomena where atomic effects can be neglected. Its most radical prediction, the propagation of electromagnetic radiation, was convincingly demonstrated by H. Hertz in 1887. Thus Maxwell's theory not only synthesized a unified theory of electricity and magnetism, but also showed optics to be a branch of electromagnetism.

The developments of theories about electricity subsequent to Maxwell have all been concerned with the microscopic realm. Faraday's experiments on electrolysis in 1833 had indicated a natural unit of electric charge, thus pointing toward a discrete rather than continuous charge. Thus, the

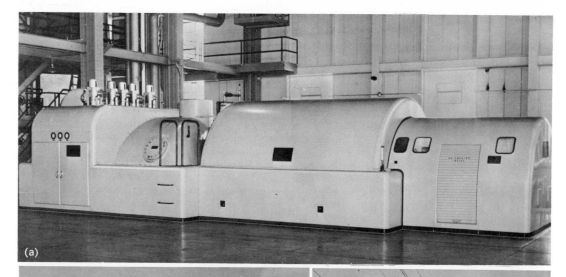

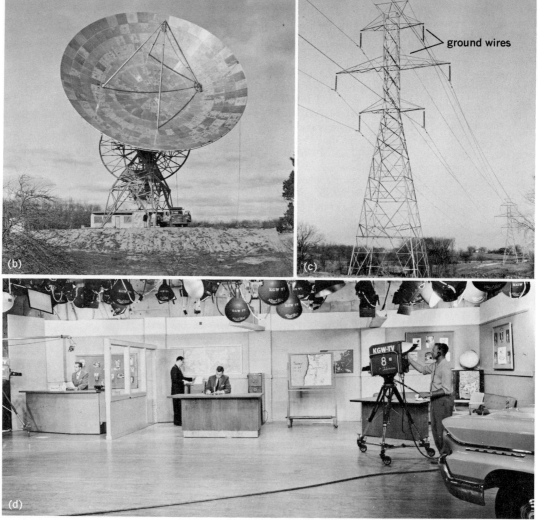

Some examples of large-scale electrical equipment. (a) Turbogenerator (*Allis-Chalmers*). (b) Radio telescope (*U.S. Office of Naval Research and University of Michigan*). (c) Transmission line (*Indiana and Michigan Electric Co.*). (d) General-purpose television studio equipment (*RCA Corporation*)

groundwork for exceptions to Maxwell's theory of electromagnetism was laid even before the theory was developed. H. A. Lorentz began the attempt to reconcile these viewpoints with his electron theory in 1895. He postulated discrete charges, called electrons. The interactions between the elec-trons were to be determined by the fields as given by Maxwell's equations. The existence of electrons, negatively charged particles, was dem-onstrated by J. J. Thomson in 1897 using a Crookes tube. The existence of positively charged particles (protons) was shown shortly afterward (1898) by

W. Wien, who observed the deflection of canal rays. Since that time, many particles have been found having charges numerically equal to that of the electron. The question of the fundamental (or elementary) nature of these particles remained unsolved in 1968, but the concept of a single elementary charge unit was apparently still valid. Of these many particles only two, the electron and the proton, exist in a stable condition on Earth.

A second departure from classical Maxwell theory was brought on by M. Planck's studies of the electromagnetic radiation emitted by "black" bodies. These studies led Planck to postulate that electromagnetic radiation was emitted in discrete amounts, called quanta. This quantum hypothesis ultimately led to the formulation of modern quantum mechanics. The most satisfactory fusion of electromagnetic theory and quantum mechanics was achieved in 1948 with the work of J. Schwinger and R. Feynman in quantum electrodynamics, which suppressed the particle aspect and emphasized the field. *See* HEAT RADIATION.

Sources. The sources of electricity in modern technology depend strongly on the application for which they are intended.

The principal use of static electricity today is in the production of high electric fields. Such fields are used in industry for testing the ability of components such as insulators and condensers to withstand high voltages, and as accelerating fields for charged-particle accelerators. The principal source of such fields today is the Van de Graaff generator.

The major use of electricity today arises in devices using electric currents alternating at low or zero frequency. The use of alternating current, introduced by S. Z. de Ferranti in 1885–1890, allows power transmission over long distances at very high voltages with a resulting low percentage power loss followed by highly efficient conversion to lower voltages for the consumer through the use of transformers. Large amounts of zero-frequency current, that is, direct current, are used in the electrodeposition of metals, both in plating and in metal production, for example, in the reduction of aluminum ore. To avoid power transmission difficulties, such facilities are frequently located near sources of abundant power. *See* ALTERNATING CURRENT; CURRENT (ELECTRICITY); DIRECT CURRENT.

The principal sources of low-frequency electricity are rotary generators whose operation is based on the Faraday induction principle. The force to drive such generators derives from the flow of water or the expansion of gases, as in steam and internal combustion engines. The primary heat has been derived principally from fossil fuels. Economic considerations, particularly the cost of natural gas and oil and the need to conserve these for petrochemical purposes, are leading to increased reliance on nuclear reactors as the heat source. In addition, the use of coal is reemerging, and intensive efforts are being made for the discovery and development of geothermal sources. Other sources, such as fusion, solar, and oceanic, appear several decades away from significant application.

A more direct method of using fission or fusion reactors is the direct conversion of the energy released in the nuclear process into electricity. This has been achieved on a laboratory scale in the case of fission reactors. *See* FUSION, NUCLEAR; GENERATOR, ELECTRIC; REACTOR, NUCLEAR.

Many high-frequency devices, such as communications equipment, television, and radar, involve the consumption of only moderate amounts of power, generally derived from low-frequency sources (see illustration). If the power requirements are moderate and portability is needed, the use of ordinary chemical batteries is possible. Ionpermeable membrane batteries are a later development in this line. Fuel cells, particularly hydrogen-oxygen systems, are being developed. They have already found extensive application in Earth satellite and other space systems. The successful use of thermoelectric generators based on the Seebeck effect in semiconductors has been reported in the Soviet Union and in the United States. In a particularly compact low-power device constructed in the United States, the heat needed for the operation of such a generator has been supplied by the energy release in the radioactive decay of suitably encapsulated isotopes produced in fission reactors. *See* BATTERY, ELECTRIC; THERMOELECTRICITY.

The Bell solar battery, also a semiconductor device, has been used to provide charging current for storage batteries in telephone service and in communications equipment in artificial satellites. *See* SOLAR BATTERY.

There are a number of other effects which might also serve to convert various forms of energy into electrical energy, but they do not seem generally practicable.

The changing magnetic flux required for the Faraday induction may be produced by an oscillating (rather than rotary) mechanism or by varying the temperature of a magnetic circuit whose components are made of a substance with a highly temperature-dependent permeability. It has been proposed to extract the energy of the fission (or possibly the fusion) reaction directly by inducing currents in external circuits by the changing magnetic field of bursts of ions from the reaction.

Direct conversion of mechanical energy into electrical energy is possible by utilizing the phenomena of piezoelectricity and magnetostriction. These have some application in acoustics and stress measurements. Pyroelectricity is a thermodynamic corollary of piezoelectricity.

Some other sources of electricity are those in which charged particles are released with some energy and collected in some manner. Charged particles are suitably released in radioactive decay, in the photoelectric effect, and in thermionic emission, among other ways. The photovoltaic effect may also be in this group.

The differences of work functions of various materials can be used for energy conversion. The contact potential difference may be used to convert heat directly to electricity or to provide improved collection for currents arising from some other source such as radioactivity. *See* WORK FUNCTION (THERMODYNAMICS).

Other possible sources of electricity arise from the existence of electrokinetic potentials in flowing fluids and of phase-transition potentials such as occur in the Workman-Reynolds effect. The possibilities of combining several effects also exist as exemplified in thermogalvanic potentials. It also appears that organic materials (as distinguished from the inorganic materials for which most of the work already described was done) merit investigation. A primitive type of organic solar battery has been developed. *See* Circuit (electricity); Conduction (electricity); Electric power measurement; Electrical units and standards.

[Walter Aron]

Bibliography: P. H. Abelson (ed.), *Energy: Use, Conservation and Supply*, American Association for the Advancement of Science, 1974; R. P. Feynman et al., *Feynman Lectures on Physics*, vol. 2, 1964; G. P. Harnwell, *Principles of Electricity and Electromagnetism*, 2d ed., 1949; A. F. Kip, *Fundamentals of Electricity and Magnetism*, 2d ed., 1969.

Electrodynamics

The study of the relations between electrical, magnetic, and mechanical phenomena. This includes considerations of the magnetic fields produced by currents, the electromotive forces induced by changing magnetic fields, the forces on currents in magnetic fields, the propagation of electromagnetic waves, and the behavior of charged particles in electric and magnetic fields. Classical electrodynamics deals with fields and charged particles in the manner first systematically described by J. C. Maxwell, whereas quantum electrodynamics applies the principles of quantum mechanics to electrical and magnetic phenomena. Relativistic electrodynamics is concerned with the behavior of charged particles and fields when the velocities of the particles approach that of light. Cosmic electrodynamics is concerned with electromagnetic phenomena occurring on celestial bodies and in space. *See* Electromagnetism.

[John W. Stewart]

Electromagnet

A soft-iron core that is magnetized by passing a current through a coil or wire wound on the core. Electromagnets are used to lift heavy masses of magnetic material and to attract movable magnetic parts of electric devices, such as solenoids, relays, and clutches.

The difference between cores of an electromagnet and a permanent magnet is in the retentivity of the material used. Permanent magnets, initially magnetized by placing them in a coil through which current is passed, are made of retentive (magnetically "hard") materials which maintain the magnetic properties for a long period of time. Electromagnets are meant to be devices in which the magnetism in the cores can be turned on or off. Therefore, the core material is nonretentive (magnetically "soft") material which maintains the magnetic properties only while current flows in the coil. All magnetic materials have some retentivity, called residual magnetism; the difference is one of degree.

A magnet, when brought near other susceptible material, induces magnetic poles in the susceptible material and so attracts it. A force will be developed in the susceptible material that will tend to move it in a direction to minimize the reluctance of the flux path of the magnet. The reluctance force may be expressed quantitatively in terms of the rate of change of reluctance with respect to distance.

In an engineering sense the word electromagnet does not refer to the electromagnetic forces incidentally set up in all devices in which an electric current exists, but only to those devices in which the current is primarily designed to produce this force, as in solenoids, relay coils, electromagnetic brakes and clutches, and in tractive and lifting or holding magnets and magnetic chucks.

Electromagnets may be divided into two classes: traction magnets, in which the pull is to be exerted over a distance and work is done by reducing the air gap; and lifting or holding magnets, in which the material is initially placed in contact with the magnet.

Examples of the latter type are magnetic chucks and circular lifting magnets. The illustration shows a cross-sectional view of a typical circular lifting magnet. The outer rim makes up one pole and the inner area is the opposing pole. Manganese steel, used as a protective cover plate for the coil, is nonmagnetic and does not provide a low reluctance shunt path for the flux.

The mechanical force between two parallel surfaces is given by Maxwell's equation, shown below, where B is the flux density (lines/in.^{2}) and A is

$$F = B^2 A / 72.13 \times 10^6 \quad \text{(lb)}$$

the cross-sectional area (square inches) through which the flux passes. When two poles are active, the force produced by each is calculated to find the total force. An interesting result of this relation is that the force is not simply the result of the total flux (BA) but also of the flux density. Thus if the same flux can be forced through one-half the cross-sectional area, the net pull will be doubled. In practice it is difficult, if not impossible, to calculate the actual lifting capacity of the magnet by using Maxwell's equation since the capacity varies with the shape and kind of material lifted, how the material is stacked, and other factors. Therefore, lifting magnets are usually rated on their all-day average lifting capacity.

Since current are large (10–20 amp) and the circuit is highly inductive, control of a lifting magnet is a problem. If the line switch were opened, a destructive arc would result. Therefore, the controller employed with a lifting magnet usually does the following things automatically: (1) reduces magnet current after initial high value to reduce heating of the magnet, (2) introduces a shunt discharge resistor across the magnet before allowing the line to be

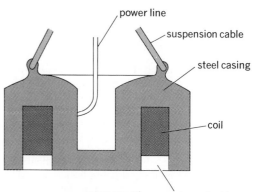

power line
suspension cable
steel casing
coil
nonmagnetic manganese steel bumper

Cross section of circular lifting electromagnet.

opened when the operator turns the magnet off, and (3) causes a reduced current of reverse polarity to flow in the magnet coil for a short time after the operator turns the switch off. Thus the residual magnetism is canceled and scraps and small chunks that might have continued clinging to the magnet are released.   [JOHN E. GIBSON]

Bibliography: D. G. Fink and J. M. Carroll (eds.), *Standard Handbook for Electrical Engineers*, 10th ed., 1968; E. Molley, M. G. Say, and R. C. Walker (eds.), *Electrical Engineer Reference Book*, 3d ed., 1948; H. Pender et al. (eds.), *Electrical Engineers' Handbook*, vol. 1, 4th ed., 1949, and vol. 2, 1950; E. M. Purcell, *Electricity and Magnetism*, vol. 2, 1965; R. M. Whitmer, *Electromagnets*, 2d ed.,

Electromagnetic propulsion

Motive power for flight vehicles produced by high-speed discharge of a plasma fluid. Together with electrostatic (ion) propulsion, electromagnetic propulsion collectively designates several mechanisms capable of attaining specific impulses, exceeding by a considerable margin those of thermal propulsion devices (Table 1). To heat fluids beyond the level of chemical reactions, a separate power source must be provided. In propulsion systems, group II in Table 1, a nuclear or solar power source heats the propellant directly or through electric discharges. Much more power can be transferred to a body of matter by electrical means than by heating, and material limitations impose a principal barrier to the rate of power transfer (propellant heating) in thermal propulsion systems. By increasing the power transfer enough to ionize the propellant material, it is possible to apply electric and electromagnetic driving forces, thereby circumventing the thermal barrier. However, different problems are encountered in the resulting electromagnetic and electrostatic drives.

Electromagnetic propulsion uses highly ionized gases (plasma), whose behavior is determined by interaction between electric currents in the plasma and magnetic fields from the vehicle. The discharged plasma is electrically neutral. Discharge density is, therefore, not limited by electrostatic

Table 1. Comparison of specific impulses and thrust densities*

Group†	Engine	Specific impulse	Thrust density
I	Chemical	≦475 sec	10–200
II	Nuclear (solid-core reactor with hydrogen)	700–900	5–100
	Arcjet (thermal adiabatic)	800–2000	0.5–5
III	MHD arcjet	1500–2500	2–12
	MPD arcjet	2000–10,000	0.8–1.2
	Electropulse	3000–5000	$5 \cdot 10^{-4} – 5 \cdot 10^{-3}$
IV	Colloid	600–2000	about 10^{-3}
	Ion	2500–20,000	about $2 \cdot 10^{-4}$

*Specific impulse is the thrust force obtained per unit weight of fluid discharged per second (in consistent units), or briefly, exhaust velocity divided by $g = 981$ cm/sec² $= 32.2$ ft/sec²; thrust density is the propulsive force per unit discharge area; above data is in pounds per square inch.

†Group I: energy stored in propellant; group II: separately powered thermal propulsion; group III: separately powered electromagnetic propulsion; group IV: separately powered electrostatic propulsion.

thrust density, and longest operating life.

Arcjet engine. The arcjet engine applies intense arc heating and partial ionization of a fraction of the total discharge fluid (Fig. 1). The current flowing through the plasma between the electrodes induces a magnetic field, which magnetically contains the plasma. As the plasma leaves the arc region, it expands rapidly in the direction of the decreasing magnetic field and decreasing pressure, transmitting its excess energy to the rest of the gas. The process is followed by thermal expansion.

High gas density is desired to establish high thrust density. This requirement intensifies problems of heat transfer to the nozzle wall. A high degree of dissociation (high temperature and low pressure) reduces the gas molecular weight and yields higher specific impulse; however, it also increases engine weight and therefore the thrust acceleration level of the system.

Magnetohydrodynamic (MHD) arcjet. In the arc MHD engine overall gas flow is transformed into a weakly ionized plasma (Fig. 2). Thermodynamic forces (space charge) present in an ion beam. Consequently, electromagnetic propulsion devices offer promise of higher thrust per unit discharge area. *See* MAGNETOGAS DYNAMICS; MAGNETOHYDRODYNAMICS.

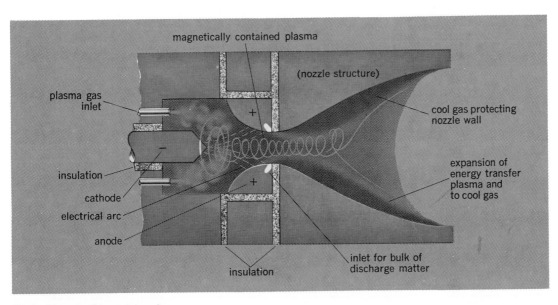

Fig. 1. Diagram of an arcjet engine.

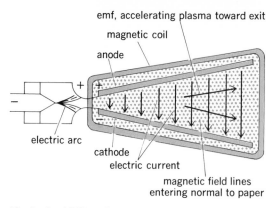

emf, accelerating plasma toward exit

magnetic coil

anode

electric arc

cathode

electric current

magnetic field lines entering normal to paper

Fig. 2. Arc MHD engine.

Electromagnetic propulsion is adaptable to a wide range of specific impulses. The optimization of a propulsion system within the framework of an overall vehicle system and a space mission calls for a compromise between acceleration and payload capability that best serves the particular purpose.

The accelerator provides direction and speed to the plasma flow and thus represents the thrust-producing mechanism. Electromagnetic propulsion devices can be divided, on the basis of accelerating mechanism, into steady-flow systems and pulsed systems. A survey is presented in Table 2. Of the presently recognized electromagnetic drives, the magnetoplasmadynamic (MPD) arcjet shows the greatest potential because it offers the widest range of high specific impulses, highest expansion of the neutral gas is augmented by plasma acceleration through the electromotive force (emf) in crossed electric and magnetic fields. To achieve plasma stability, the electron cyclotron frequency (the product of electron charge e times magnetic flux density B divided by the product of 2π times electron mass m_e or $eB/2\pi m_e$) should be much larger than the particle collision frequency. Thus, either low fluid density or strong magnetic fields are required. Steady fields of $10^4 - 10^5$ gauss, desirable in a gas pressure region of $10^{-5} - 10^{-3}$

atm, are required over extended periods of time.

Arc-heating methods suffer from limitations imposed by electrode erosion. Erosion is avoided by radio-frequency heating of the plasma. However, the efficiency of heating the gas tends to decrease as the plasma conductivity increases. The relatively massive equipment needed to generate the strong magnetic fields adds appreciably to the mass of the propulsion system, lowering the thrust-to-engine weight ratio to the low 10^{-3}-g regime. So low a ratio means that even a relatively small spacecraft powered by this engine probably will not attain thrust accelerations much above 10^{-4} g. The specific impulse expected of this drive is marginally low at such accelerations for purposes other than station keeping. For this reason and because of its comparative complexity, this drive is no longer attractive.

Magnetoplasmadynamic (MPD) arcjet. In the MPD arcjet the mechanism of electromagnetic propulsion is fully developed (Fig. 3). The MPD arcjet is the outgrowth of experimental work to improve the performance of conventional thermal arcjets. It was found that by reducing the propellant (hydrogen flow, and hence the chamber pressure, the arc was caused to diffuse over the cathode surface and to extend far into the exhaust jet behind the nozzle, causing a luminous exhaust plume to expand into a ball-shaped configuration. The arc current can be increased from hundreds to 3000 and more amperes without damaging throat and exhaust nozzle. The increase raises specific impulse from 3000 to 10,000 sec and improves corresponding overall efficiency (the product of mass utilization and power utilization efficiency) from about 15 to about 45%.

Experiments led to the tentative conclusion that the principal driving force (thrust contribution) is produced by the force of magnetic fields generated by the currents in the electric arc. This thrust component varies with the square of the arc current and becomes significant for currents in excess of 10^3 amp. The electron current flowing from cathode to anode (Fig. 3) produces a magnetic field of flux density B which generates a magnetic pressure $B^2/2\mu$ (μ being the magnetic susceptibility).

Table 2. Survey of electromagnetic propulsion systems

Engine	Operation	Acceleration	Propellant (working fluid)	Plasma formation	Problems and remarks
Arcjet (thermal adiabatic)	Steady flow	Adiabatic expansion of exhaust gas	Hydrogen; ammonia	Some plasma formation by electric arc heating	Electrode erosion and burnout of engine walls; losses due to ionization, dissociation, radiation
MHD arcjet	Steady flow	Adiabatic expansion, augmented by driving force of crossed electric and magnetic fields	Weakly ionized hydrogen plasma seeded by alkali metal (such as cesium)	Electric arc heating	Electrode erosion; massive auxiliary equipment for seeding propellant with readily ionized material and for generating the required magnetic field
MPD arcjet	Steady flow	Plasma is driven by magnetic pressure generated by the electric arc current	Strongly ionized hydrogen or other materials	Electric arc heating	Electrode erosion and wall burnout avoided or greatly reduced by magnetic confinement of plasma, keeping it from close contact with walls; has highest potential among electromagnetic drives
Electropulse	Nonsteady flow	Magnetic pressure	Plasma formed from a variety of materials	Electric discharge	Electrode erosion; system has not yet demonstrated its theoretical performance capability

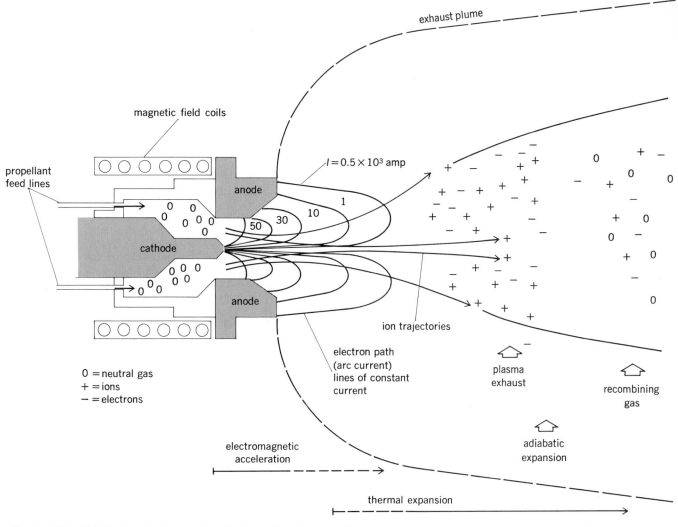

Fig. 3. MPD arcjet thrust device. Values given for lines of constant current are examples.

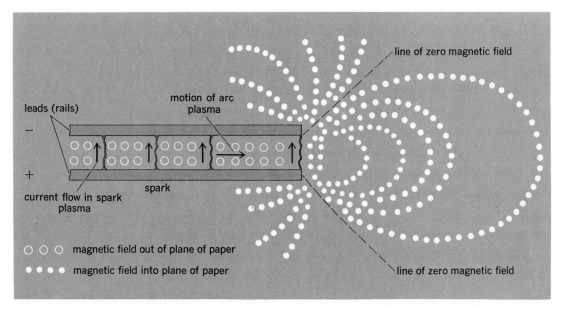

Fig. 4. Diagram of a spark accelerator.

This magnetic pressure is exerted against the electrically conductive plasma because a magnetic field cannot permeate a conductor and therefore exerts a mechanical force on it. In the MPD arcjet propellant is turned into plasma by flowing through the arc and subsequently expelled by the magnetic pressure. If an external magnetic field is superimposed the arc current interacts with it, producing an azimuthal, or ring, current which yields axial pressure when interacting with the external field and which also causes an azimuthal velocity component of the plasma. This motion is converted into axial velocity upon expansion in the nozzle. Because of the self-induced magnetic field, the plasma is kept away from the cathode, protecting the cathode from excessive damage due to sputtering and heating which would result from massive ion impact. This constraint also implies that, in a sense, the MPD arcjet uses a magnetic nozzle (a nozzle formed by magnetic lines of force) rather than the material nozzle walls, thereby protecting the latter from excessive ablative and thermal damage. At the same time, the plasma gas expands to a degree consistent with the acting electromagnetic forces, like a conventional thermal arcjet gas, thereby adding to the thrust force through conversion of enthalpy to kinetic energy of the exhaust flow.

Large MPD arcjet thrust units of the future may employ arc currents well above 10^4 amp. Their electric power consumption may be in the tens of megawatts regime. Thrust F in kilograms, produced as a function of electric power consumed P_e and specific impulse generated I_{sp}, is given by Eq. (1), where 20,855, in units of $(kg\text{-}m/sec)(sec^2/m)$

$$F = 20{,}855 \eta\, P_e / I_{sp} \qquad (1)$$

(1/mw), is a conversion factor; η is overall efficiency of the propulsion system (defined before); P_e is in megawatts; and I_{sp} is in seconds. Equation (2) gives the weight W_p in kilograms of

$$W_p = \alpha P_e \qquad (2)$$

this propulsion system, where α is the power specific weight of the system in kg/kw. Thus, the propulsion system thrust-to-weight ratio (in Earth-g units) is as in Eq. (3). Assuming that $I_{sp} = 6000$ sec

$$F/W_p = 20{,}855\, \eta / \alpha\, I_{sp} \qquad (3)$$

and that the corresponding $\eta = 0.25$, it is seen that $F/W_p = 10^{-4}$ or 2×10^{-4} g for α, about 9 or 4.5 kg/kw. The complete spacecraft, therefore, might have an initial acceleration of one-half to one-third of this value. At least for the lower value of α, this thrust places the MPD arcjet in the same specific impulse and acceleration category as ion propulsion. Advantages of the MPD arcjet drive are a simpler electric power conditioning system (lower voltages required); simpler operation (no space charge problem, because the ejection mass contains a mixture of ions and electrons); less structural damage to the thrust unit, and hence potentially longer operating life (due to reduced or zero sputtering); and greater compactness of the thrust units (because thrust density is about 1000 times higher for the MPD arcjet than for electrostatic thrusters). Poor overall efficiency of either system results in large radiator surfaces, and hence intrinsic vulnerability to inclement space environment.

Electropulse engine. Electropulsed methods are based on the use of spark discharges through which intense electric and magnetic fields are established for periods of microseconds up to a few milliseconds. A high-voltage storage capacitor is discharged through a spark gap between two leads (rail electrodes). The intense current in the arc plasma is at right angle to the magnetic field established by the current flow in the leads to the spark gap (Fig. 4). The resulting emf drives the plasma along the leads and away from the spark gap.

Propellants. Propellant fluids used in experimental plasma engines include water, air, argon, helium, hydrogen, deuterium, and metallic vapor. Potentially practical working fluids are lithium, some metal hydrides, methane, ammonia, and hydrogen. In principle, plasma engines have the widest choice of propellants (at least from among the nonoxidizing fluids) of all propulsion systems. This may eventually enable plasma-driven space ships to draw their propellant from many celestial bodies, rather than from Earth alone.

[KRAFFT A. EHRICKE]

Bibliography: G. Cann, A. Ducati, and V. Blackman, Experimental studies on the thrust from a continuous plasma jet, *Proceedings of USAF-OSR Rocketdyne Advanced Propulsion Symposium*, 1957; A. C. Ducati, G. M. Giannini, and E. Muehleberger, Experimental results in high specific-impulse thermionic acceleration, *AIAA J.*, 2:1452–1454, 1964; A. C. Ducati, G. M. Giannini, and E. Muehleberger, *Recent Progress in High Specific Impulse Thermo-Ionic Acceleration*, AIAA Pap. no. 65-96, 1965; R. G. John, *Physics of Electric Propulsion*, 1967; R. E. Jones, Results of large vacuum facility test of an MPD arc Thruster, *AIAA J.*, 4:1455–1456, 1966; R. J. Rosa, Application of magnetohydrodynamics to propulsion, *Proceedings of USAF-OSR Rocketdyne Advanced Propulsion Symposium*, 1957; E. Stuhlinger (ed.), *Progress in Astronautical and Aeronautical Electric Propulsion Development*, vol. 9, 1963.

Electromagnetic radiation

Energy transmitted through space or through a material medium in the form of electromagnetic waves. The term can also refer to the emission and propagation of such energy. Whenever an electric charge oscillates or is accelerated, a disturbance characterized by the existence of electric and magnetic fields propagates outward from it. This disturbance is called an electromagnetic wave. The frequency range of such waves is tremendous, as is shown by the electromagnetic spectrum in the table. The sources given are typical, but not mutually exclusive, as is shown by the fact that the atomic interstellar hydrogen radiation whose wavelength is 0.210614 m falls in the radar region. The other monochromatic radiation listed is that from positron-electron annihilation whose wavelength is 2.42626×10^{-12} m.

Detection of radiation. In theory, any electromagnetic radiation can be detected by its heating effect. This method has actually been used over the range from x-rays to radio. Ionization effects measured by cloud chambers, photographic emulsions, ionization chambers, and Geiger counters have been used in the γ- and x-ray regions. Direct photography can be used from the γ-ray to the infrared region. Fluorescence is effective in the x-ray and ultraviolet ranges. Bolometers, thermocouples, and other heat-measuring devices are used chiefly in the infrared and microwave regions. Crystal detectors, vacuum tubes, and transistors cover the microwave and radio frequency ranges.

Free-space waves. A charge in simple harmonic (linear sinusoidal) motion in a vacuum generates a simple wave which becomes spherical at distances from the source much larger than the amplitude of the motion and so great that many oscillations have occurred before the disturbance arrives. The wave is plane when the dimensions of the area observed are very small compared with the radius of spherical curvature. In this case the choice of the rectangular coordinates x and z as the directions of the oscillation and of the observation or field point, respectively, permits the electric intensity $\mathbf{E}$ and the magnetic flux density $\mathbf{B}$ to be written as Eq. (1). The field amplitude E_0 is con-

$$E_x = vB_y = E_0 \cos \left[\omega(t - v^{-1}z)\right] \qquad (1)$$

stant over the specified area and not dependent on z if the z-range is small compared with the source distance, as in stellar radiation. The angular frequency of the source is ω radians per second, which is the frequency ν in hertz multiplied by 2π. The velocity of the wave is v, the direction of propagation z, and the time t. The wavelength λ is $2\pi v/\omega$. If t is in seconds and z in meters, then v is in meters per second, and λ in meters. It is found that in a lossless, isotropic, homogeneous medium Eq. (2) holds; here μ is the permeability, and ϵ the ca-

$$v = (\mu\epsilon)^{-1/2} \qquad (2)$$

pacitivity, or dielectric constant. This wave is transverse because $\mathbf{E}$ and $\mathbf{B}$ are normal to z. It is plane-polarized because E_x and B_y are parallel to fixed axes. The plane of polarization is taken as that defined by the electric vector and the direction of propagation.

Plane waves. An electromagnetic disturbance is a plane wave when the instantaneous values of any field element such as $\mathbf{E}$ and $\mathbf{B}$ are constant in phase over any plane parallel to a fixed plane. These planes are called wavefronts. In empty unbounded space, $\mathbf{E}$ and $\mathbf{B}$ lie in the wavefront normal to each other; if the wave is unpolarized, their direction fluctuates in this plane in random fashion. If the plane waves are bounded, as on transmission lines and in wave guides, the amplitudes may vary over the wavefront, and in the case of wave guides and crystals some of the elements will not in general lie in the wavefront. The equation for an undamped plane wave whose front is normal to z is Eq. (3), where F is one of the field ele-

$$F = \Phi_1(x,y) f_1(z - vt) + \Phi_2(x,y) f_2(z + vt) \qquad (3)$$

ments such as $\mathbf{E}$ or $\mathbf{B}$. Note that if an observer sees a certain value of $\Phi_1(x,y)$ at z and then jumps instantaneously in the z direction to a point $z + \Delta z$, he will, after waiting a time $\Delta z/v$, see the same value $\Phi_1(x,y)$ because Eq. (4) is valid. Thus, the first term

$$f(z - vt) = f[z + \Delta z - v(t + \Delta z/v)] \qquad (4)$$

represents a wave moving in the z direction with a velocity v. The second term represents a wave in the negative z direction. The form of $\Phi_1(x,y)$ and $\Phi_2(x,y)$ depends on the boundary conditions.

Spherical waves. A wave is spherical when the instantaneous value of any field element such as

Electromagnetic spectrum

Frequency, Hz	Wavelength, m	Nomenclature	Typical source
10^{23}	3×10^{-15}	Cosmic photons	Astronomical
10^{22}	3×10^{-14}	γ-rays	Radioactive nuclei
10^{21}	3×10^{-13}	γ-rays, x-rays	
10^{20}	3×10^{-12}	x-rays	Atomic inner shell
		Positron-electron annihilation	
10^{19}	3×10^{-11}	Soft x-rays	Electron impact on a solid
10^{18}	3×10^{-10}	Ultraviolet, x-rays	Atoms in sparks
10^{17}	3×10^{-9}	Ultraviolet	Atoms in sparks and arcs
10^{16}	3×10^{-8}	Ultraviolet	Atoms in sparks and arcs
10^{15}	3×10^{-7}	Visible spectrum	Atoms, hot bodies, molecules
10^{14}	3×10^{-6}	Infrared	Hot bodies, molecules
10^{13}	3×10^{-5}	Infrared	Hot bodies, molecules
10^{12}	3×10^{-4}	Far-infrared	Hot bodies, molecules
10^{11}	3×10^{-3}	Microwaves	Electronic devices
10^{10}	3×10^{-2}	Microwaves, radar	Electronic devices
10^{9}	3×10^{-1}	Radar	Electronic devices
		Interstellar hydrogen	
10^{8}	3	Television, FM radio	Electronic devices
10^{7}	30	Short-wave radio	Electronic devices
10^{6}	300	AM radio	Electronic devices
10^{5}	3000	Long-wave radio	Electronic devices
10^{4}	3×10^{4}	Induction heating	Electronic devices
10^{3}	3×10^{5}		Electronic devices
100	3×10^{6}	Power	Rotating machinery
10	3×10^{7}	Power	Rotating machinery
1	3×10^{8}		Commutated direct current
0	Infinity	Direct current	Batteries

$\mathbf{E}$ or $\mathbf{B}$ is constant in phase over a sphere. The radiation from any source of finite dimensions becomes spherical at great distances in an unbounded, isotropic, homogeneous medium. The equation for an undamped spherical wave is Eq. (5). The first term represents a diverging and the

$$F = r^{-1}\Phi_1(\theta,\varphi) f(r - vt) + r^{-1}\Phi_2(\theta,\varphi) f(r + vt) \qquad (5)$$

second a converging wave. Again, the form of $\Phi_1(\theta,\varphi)$ and $\Phi_2(\theta,\varphi)$ depends on the nature of the source and other boundary conditions.

Damped waves. If there are energy losses which are proportional to the square of the amplitude, as in the case of a medium of conductivity γ which obeys Ohm's law, then the wave is exponentially damped, and Eq. (1) becomes Eq. (6). The symbol

$$E_x = E_0 e^{-\alpha z} \cos (\omega t - \beta z) \qquad (6)$$

α is called the attenuation constant, and β the wave number or phase constant which equals ω/v', where v' is the damped-wave velocity. The electric wave amplitude at the origin has been taken as E_0. The ratio of E_0 to B_0, as well as that of α to β, depends on the permeability μ, the capacitivity ϵ, and the conductivity γ of the medium. In terms of the phasor $\check{E}_x$, Eq. (6) may be written as the real part of Eq. (7). This is exactly the form for the cur-

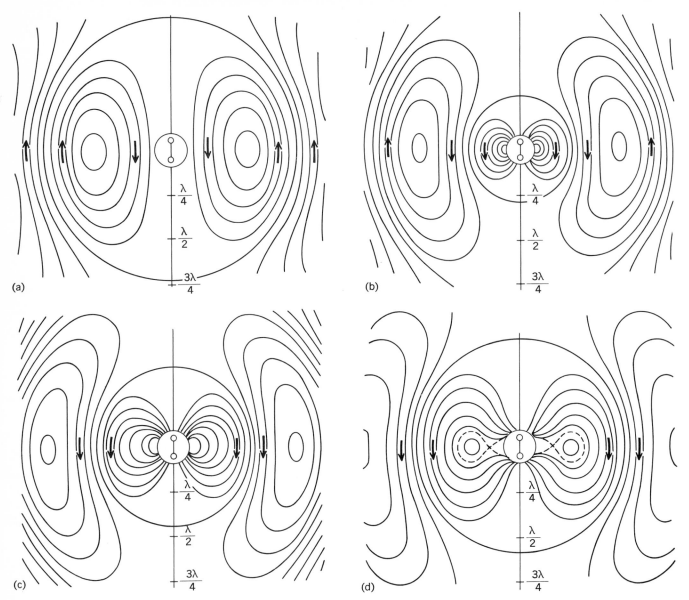

Diagrams of electric dipole. The outward moving electric field lines generated by the Hertzian oscillator are shown at successive eighth-period intervals. (a) $t=0$; (b) $t= T/8$; (c) $t= T/4$; (d) $t=3T/8$.

rent on a transmission line. (Phasors are complex numbers of form such that, when multiplied by

$$E_x = \check{E}_x e^{j\omega t} = E_0 e^{-(\alpha + j\beta)z} e^{j\omega t} \qquad (7)$$

$e^{j\omega t}$, the real part of the product gives the amplitude, phase, and time dependence.) *See* DAMPING.

Wave impedance. Those trained in transmission line theory find it useful to apply the same techniques to wave theory. Consider an isolated tubular section of the wave in Eq. (1) bounded by $x=0, x=1$, and $y=0, y=1$ as a transmission line. The potential across the line between $x=0$ and $x=1$ is **E**. The line integral of **B** around the $x=0$ boundary from $y=0$ to $y=1$ is μI by Ampère's law and equals **B** because **B** is zero on the negative side. Thus, the impedance of the line is, making use of Eqs. (1) and (2), given by Eq. (8). This depends only on the properties of the medium and is known as the wave impedance. In transmission line theory the ratio μ/ϵ would be replaced by the

ratio of the series impedance $\check{Z}_L = j\omega L$ to the shunt admittance $\check{Y} = j\omega C$, where L is the inductance per unit length, and C the capacitance per unit length

$$\check{Z}_k = \frac{V}{I} = \frac{\mu E}{B} = \frac{E}{H} = \left(\frac{\mu}{\epsilon}\right)^{1/2} = \eta \qquad (8)$$

across the line. If there is a resistance R per unit length across the line, then $1/R$ must be added to Y. This resistance is $1/\gamma$ for the tubular section. Thus, for a conducting medium, Eq. (8) becomes Eq. (9). The last term is a common transmission

$$\check{Z}_k = \left(\frac{j\omega\mu}{\gamma + j\omega\epsilon}\right)^{1/2} = \frac{j\omega\mu}{\alpha + j\beta} \qquad (9)$$

line form. The reflection and refraction of plane waves at plane boundaries separating different mediums may be calculated by transmission line formulas with the aid of Eqs. (8) and (9).

Electric dipole. A charge undergoing simple harmonic motion in free space is a dipole source when the amplitude of the motion is small compared with the wavelength. The term is loosely applied to the Hertzian oscillator, usually pictured as a dumbbell-shaped conductor in which the electrons oscillate from one end to the other, leaving the opposite end periodically positive. An electric dipole of moment M is defined as the product qa when two large, equal and opposite charges, $+q$ and $-q$, are placed a small distance a apart. A dipole is oscillating when M is periodic in time and is the simplest source of spherical waves. Much can be learned by a study of H. Hertz's picture of the outward moving electric field lines at successive time intervals of one-eighth period in a plane which passes through the Hertzian oscillator axis, shown in the figure. The most striking feature of the pictures is that, after breaking loose from the dipole, all electric field lines are closed, which means that the divergence of $\mathbf{E}$ is zero. This is true of all unbounded waves. It is also noteworthy that the waves become truly spherical with a fixed wavelength λ only in a direction perpendicular to the dipole and at a distance which greatly exceeds the dipole dimensions. This distance is beyond the edges of the picture. Lengths $\lambda/4$, $\lambda/2$, and $3\lambda/4$ are marked off on the axis for comparison. The magnetic field lines are circles coaxial with the oscillator, so they intersect the plane of the diagram normally. They are most dense where the electric lines are closely spaced. Radiant energy from atoms and molecules is essentially of the dipole type. *See* TRANSMISSION LINES.   [WILLIAM R. SMYTHE]

Bibliography: M. Born and E. Wolf, *Principles of Optics*, 1959; A. H. Compton and S. K. Allison, *X-Rays in Theory and Experiment*, 1935; G. P. Harnwell, *Principles of Electricity and Electromagnetism*, 2d ed., 1949; F. A. Jenkins and H. E. White, *Fundamental of Optics*, 3d ed., 1957; L. Page and N. I. Adams, *Principles of Electricity*, 3d ed., 1958; S. Ramo and J. R. Whinnery, *Fields and Waves in Modern Radio*, 1953; S. A. Schelkunoff, *Electromagnetic Waves*, 1943; W. R. Smythe, *Static and Dynamic Electricity*, 1968.

Electromagnetic wave

A disturbance, produced by the acceleration or oscillation of an electric charge, which has the characteristic time and spatial relations associated with progressive wave motion. A system of electric and magnetic fields moves outward from a region where electric charges are accelerated, such as an oscillating circuit or the target of an x-ray tube. The wide wavelength range over which such waves are observed is shown by the electromagnetic spectrum. The term electric wave, or Hertzian wave, is often applied to electromagnetic waves in the radar and radio range. Electromagnetic waves may be confined in tubes, such as wave guides, or guided by transmission lines. They were predicted by J. C. Maxwell in 1864 and verified experimentally by H. Hertz in 1884. *See* ELECTROMAGNETIC RADIATION.   [WILLIAM R. SMYTHE]

Electromagnetic waves, transmission of

The transmission of electrical energy by wires, the broadcasting of radio signals, and the phenomenon of visible light are all examples of the propagation of electromagnetic energy. Electromagnetic energy travels in the form of a wave. Its speed of travel is approximately 3×10^8 m/sec (186,000 mi/sec) in a vacuum and is somewhat slower than this in liquid and solid insulators. An electromagnetic wave does not penetrate far into an electrical conductor, and a wave that is incident on the surface of a good conductor is largely reflected.

Electromagnetic waves originate from accelerated electric charges. For example, a radio wave originates from the oscillatory acceleration of electrons in the transmitting antenna. The light produced within a laser originates when electrons fall from a higher to a lower energy level. *See* LASER.

The waves emitted from a source are oscillatory in character and are described in terms of their frequency of oscillation. Local telephone lines (not using carrier systems) carry electromagnetic waves with frequencies of about 200–4000 Hz. Medium-wave radio uses frequencies of the order of 10^6 Hz, radar uses frequencies of the order of 10^{10} Hz, and a ruby laser emits light with a frequency of 4.32×10^{14} Hz. The method of generating an electromagnetic wave depends on the frequency used, as do the techniques of transmitting the energy to another location and of utilizing it when it has been received.

The communication of information to a distant point is generally accomplished through the use of electromagnetic energy as a carrier. A familiar example is the telephone, in which sound waves in the range of frequencies from a few hundred to a few thousand hertz are converted into corresponding electromagnetic waves, which are then guided to their destination by a pair of wires. Another familiar example is radio, in which the signals are caused to modify an identifiable characteristic, such as the amplitude or frequency, of an electromagnetic carrier wave. The electromagnetic wave, thus modified, or modulated, is radiated from an antenna and can be received over a considerable region.

Features of electromagnetic waves. Figure 1 illustrates schematically some of the essential features of an electromagnetic wave. Shown in Fig. 1 are vectors that represent the electric field intensity E and the magnetic field intensity H at various points along a straight line taken in the direction of propagation of the wave. The electric field is in a vertical plane and the wave is said to be vertically polarized. The magnitude of the field, at a given instant, varies as a sinusoidal function of distance along the direction of propagation. The magnetic field intensity H lies in a plane normal to that of E and, at each point, is proportional in magnitude to E, as shown in Eq. (1), where H is the magnetic

$$\frac{E}{H} = \sqrt{\frac{\mu}{\epsilon}} \qquad (1)$$

field intensity in amp/m, E is the electric field intensity in volts/m, ϵ is the permittivity, or absolute dielectric constant, of the medium, and μ is the absolute permeability of the medium. For a vacuum, $\epsilon = 8.854 \times 10^{-12}$ farad/m and $\mu = 4\pi \times 10^{-7}$ henry/m; therefore for a vacuum the ratio E/H is approximately 377 ohms. This ratio is termed the wave impedance of the medium.

The E and H waves travel along a straight line, as suggested in Fig. 1. Of the two possible direc-

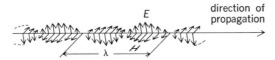

Fig. 1. Representation of an electromagnetic wave at a particular instant of time.

tions along this line, the actual direction of travel can be determined by imagining a screw with a right-hand thread placed along the axis and turned from E toward H; then the longitudinal direction of travel of the screw is the direction of propagation of the energy.

The velocity of travel of the wave is shown in Eq. (2). In a vacuum this is approximately 3×10^8

$$v = 1/\sqrt{\mu\epsilon} \qquad (2)$$

m/sec. The velocity in air is only slightly smaller.

The wavelength is the distance between two successive similar points on the wave, measured along the direction of propagation. The wavelength is denoted by λ in Fig. 1.

As the wave travels past a stationary point, the values of E and H at the point vary sinusoidally with time. The time required for one cycle of this variation is termed the period, T seconds. The number of hertz is the frequency f; and $f = 1/T$. In one cycle the wave, traveling at the velocity v, moves one wavelength along the axis of propagation. Therefore, $\lambda = vT$, or may be calculated by Eq. (3). Assuming a velocity of 3×10^8 m/sec, an

$$\lambda = v/f \qquad (3)$$

electromagnetic wave having a frequency of 60 Hz has a wavelength of 5×10^6 m, or approximately 3100 mi. At a frequency of 3 MHz (3×10^6 Hz), λ is 100 m, and at 3000 MHz, λ is 10 cm. Visible light has a frequency of the order of 5×10^{14} Hz and a wavelength of approximately 6×10^{-5} cm.

The density of energy in an electric field is $\epsilon E^2/2$ joules/m^3, and that in a magnetic field is $\mu H^2/2$ joules/m^3. With the aid of Eq. (1), the relationships become those shown in Eq. (4). Therefore the elec-

$$\mu H^2/2 = \mu(\sqrt{\epsilon/\mu}\, E)^2/2 = .\epsilon E^2/2 \qquad (4)$$

tric and magnetic fields carry equal energies in the electromagnetic wave. The total energy density at any point is equal to ϵE^2 joules/m^3. Since this is transported with a velocity equal to $1/\sqrt{\mu\epsilon}$, the rate of flow of energy per square meter normal to the direction of propagation is $\epsilon E^2 v$ or $E^2\sqrt{\epsilon/\mu}$ watts/m^2. In radio broadcasting a field of 50 mv/m is considered to be strong. An electromagnetic wave with this intensity has an average energy density of 2.2×10^{-14} joule/m^3, and the average rate of energy flow per square meter is 6.6×10^{-6} watt/m^2.

Radiation from an antenna. Figure 2 illustrates the configuration of the electric and magnetic fields about a short vertical antenna in which flows a sinusoidal current of the form $i = I_{max} \sin 2\pi ft$ amp. The picture applies either to an antenna in free space (in which case the illustration shows only the upper half of the fields), or to an antenna projecting above the surface of a highly conducting plane surface. In the latter case the conducting plane represents to a first approximation the sur-

face of the Earth. The fields have symmetry about the axis through the antenna. For pictorial simplicity only selected portions of the fields are shown in Fig. 2. The magnetic field is circular about the antenna, is perpendicular at every point to the direction of the electric field, and is proportional in intensity to the magnitude of the electric field, as indicated by Eq. (1). All parts of the wave travel radially outward from the antenna with the velocity given by Eq. (2); the wave is described as spherical, with the antenna located at the center of the wave. The wavelength of the radiation is given by Eq. (3).

If a short antenna projecting above a highly conducting plane surface carries a current of $i = I_{max} \sin 2\pi ft$ that is uniform throughout the length of the antenna, the intensity of the electric field in the radiated wave is that shown in Eq. (5), where l is

$$E_{max} = \sqrt{\frac{\mu}{\epsilon}} \frac{I_{max}}{r} \frac{l}{\lambda} \cos\ \theta \text{ volts/m} \qquad (5)$$

the length of the antenna, r is the radial distance from the antenna, and θ is the angle measured from the horizontal. The radiated field intensity is zero directly above the antenna, is greatest along the conducting plane where θ is 0, and varies inversely with the distance from the antenna.

If the rate of flow of energy per unit area $E^2\sqrt{\epsilon/\mu}$ is integrated over an imaginary spherical surface about the antenna, the average power radiated is that shown by Eq. (6). The factor $(l/\lambda)^2$ is of particu-

$$P_{av} = \frac{2\pi}{3} \sqrt{\frac{\mu}{\epsilon}} I^2_{max} \left(\frac{l}{\lambda}\right)^2 \quad \text{watts} \qquad (6)$$

lar importance, for it indicates that a longer antenna is required at the longer wavelengths (lower frequencies). The radiation of appreciable energy at a very low frequency requires an impracticably long antenna.

The foregoing relations assume a uniform current throughout the length of the antenna. An approximation to this can be achieved in practice by connecting a long horizontal conductor to the top of the antenna. Where some such construction is not utilized, the current will not be uniform in the

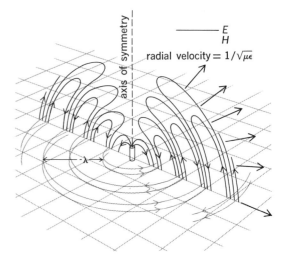

Fig. 2. Configuration of electric and magnetic fields about a short vertical antenna.

antenna and will, in fact, be zero at the tip. The results given by Eqs. (4) and (5) must be modified, but the qualitative features of the radiation remain as shown in Fig. 2.

Often it is desired to concentrate the radiated energy into a narrow beam. This can be done either by the addition of more antenna elements or by placing a large reflector, generally parabolic in shape, behind the antenna. The production of a narrow beam requires an antenna array, or alternatively a reflector, that is large in width and height compared with a wavelength. The very narrow and concentrated beam that can be achieved by a laser is made possible by the extremely short wavelength of the radiation as compared with the cross-sectional dimensions of the radiating system.

Propagation over the Earth. The foregoing discussion shows some of the important features of the radiation of electromagnetic energy from an antenna, but is oversimplified insofar as communication to and from positions on or near the Earth is concerned. The ground is a reasonably good, but not perfect, conductor; hence, the actual propagation over the surface of the Earth will show a more rapid decrease of field strength than that indicated by the factor of $1/r$ in Eq. (5). Irregularities and obstructions may interfere. In long-range transmission the spherical shape of the Earth is important. Inhomogeneities in the atmosphere refract the wave somewhat. For long-range transmission, the ionized region high in the atmosphere known as the Kennelly-Heaviside layer, or ionosphere, can act as a reflector. The electric field of the wave produces oscillation of the charged particles of the region, and this causes the refractive index of the layer to be smaller than that of the atmosphere below. The result is that, if the angle of incidence is not too near the normal and if the frequency of the wave is not too high, the wave may be refracted back toward the Earth. Successive reflections between ionosphere and Earth can provide communication for long distances around the periphery of the Earth.

Hollow waveguides. When an electromagnetic wave is introduced into the interior of a hollow metallic pipe of suitably large cross-sectional dimensions, the energy is guided along the interior of the pipe with comparatively little loss. The most common cross-sectional shapes are the rectangle and the circle. The cross-sectional dimensions of the tube must be greater than a certain fraction of the wavelength; otherwise the wave will not propagate in the tube. For this reason hollow waveguides are commonly used only at wavelengths of 10 cm or less (frequencies of 3000 MHz or higher).

A single wave of the type in Fig. 1 cannot propagate longitudinally inside a tubular conductor since, at some portions of the inner surface of the conducting tube, the E vector of the wave necessarily would have a component tangential to the surface. This is impossible because an electric field cannot be established along a good conductor, such as the wall of the tube. An electromagnetic wave can propagate along the interior of the tube only by reflecting back and forth between the walls of the tube. This reflection is a comparatively simple one between the plane surfaces of a rectangular tube, but is a complex reflection in tubes of other cross-sectional shapes.

A dielectric rod can also be used as a waveguide. Such a rod, if of insufficient cross-sectional dimensions, can contain the electromagnetic wave by the phenomenon of total reflection at the surface.

A hollow metallic waveguide of rectangular cross section is shown in Fig. 3a. The simplest mode of propagation is indicated in Fig. 3b. The entire space is filled with a plane electromagnetic wave which moves obliquely to the left in the direction shown by the solid arrows. This wave has its E vector normal to the paper and its H vector in the plane of the paper. Any plane normal to the direction of propagation is a plane of uniform phase (thus the name plane wave), and one such plane is indicated in the illustration by a broken line. The wave strikes the wall at an angle θ from the normal and is reflected at an equal angle. As the wave is reflected, the direction of its E vector reverses so as to make the tangential component of the electric field equal to zero at the conducting wall. The wave incident on the left wall thus is reflected to the right, where it is again reflected and moves to the left. By successive reflections the energy propagates longitudinally along the interior of the guide. As the wave incident upon the wall reflects and reverses the direction of its E vector, electric currents are caused to flow in the conducting wall. Since the wall is not a perfect conductor, some of the energy of the wave is transformed into heat. Consequently the amplitude of the wave diminishes exponentially as it passes down the guide; this phenomenon is termed attenuation. For an electromagnetic wave with a frequency of 3000 MHz (wavelength of 10 cm) propagating down the interior of a rectangular copper waveguide with cross-sectional dimensions of 4 cm by 8 cm, half the power is lost in a distance of approximately 150 m. Hollow waveguides are used chiefly for short-distance transmission, as from a transmitter to an antenna.

The requirements on the reflection of the wave, as outlined above, restrict the wavelength that can be propagated in a hollow guide. Consider the ray ABC in Fig. 3. The wave propagates from A to B, where it is reflected with reversal of the E vector; thereupon it propagates from B to C, where it is again reflected with another reversal of the E vector. But AC is a line of equal phase, and so the wave emerging from C must have the same phase as that at A. Thus the distance ABC must be an integral multiple of a wavelength, or $n\lambda$, where n is a positive integer. The distance ABC is $2b \cos \theta$ where b is the breadth of the guide; hence, $n\lambda = 2b \cos \theta$. The condition for propagation down the axis of the guide is that $\theta > 0$; hence $\cos \theta < 1$, and the restriction on wavelength is $\lambda < 2b/n$. The greatest ratio of wavelength to breadth of guide is obtained when $n = 1$, whence $\lambda/b < 2$. Thus, the breadth of the guide must be somewhat greater than $\lambda/2$.

In the simple mode of propagation described above, the fields are independent of distance in the direction of the dimension a, and this dimension has no influence on the propagation. The net electric vector caused by the sum of the two waves is everywhere transverse to the longitudinal axis of the guide, and so the mode is described as transverse electric (TE).

If the wavelength of the radiation is small

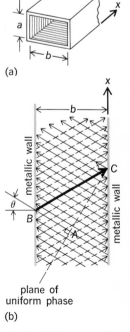

(a)

plane of uniform phase

(b)

Fig. 3. A hollow metallic waveguide of rectangular cross section. (a) The guide. (b) Paths of electromagnetic energy in the simplest mode of propagation.

enough in comparison with the cross-sectional dimensions of the guide, more complex modes of propagation are possible, in which the wave reflects obliquely against a side wall, proceeds to the top of the guide, and reflects from there to the other side wall, then to the bottom wall, and so on. With this type of reflection it is possible to have both transverse-electric and transverse-magnetic (TM) modes. In the latter the net H vector is everywhere transverse to the axis of the guide.

When the dimensions of the guide are such that complex modes are possible, so also are the simple ones. The transmission of energy by a combination of modes introduces complications in abstracting the energy from the guide at the receiving end. Propagation in only the simplest mode is ensured by selecting the dimension b to be greater than $\lambda/2$ but not as large as λ, and also by restricting the dimension a so as to render complex modes impossible.

Waveguides of circular cross section are sometimes used. Analysis of these shows that the first TE mode is propagated if the diameter of the guide is greater than 0.586λ, and that the first TM mode is propagated if the radius is greater than 0.766λ.

Two-conductor transmission lines. Electromagnetic energy can be propagated in a simple mode along two parallel conductors. Such a waveguiding system is termed a transmission line. Three common forms are shown in Fig. 4. If the spacing between conductors is a small fraction of the wavelength of the transmitted energy, only one mode of propagation is possible. This corresponds to the wave of Fig. 1, with the direction of propagation taken longitudinally along the line. The E and H vectors are in the plane of the cross section, and the mode is termed transverse electromagnetic (TEM). The E vector must be at right angles to a highly conducting surface, and the oscillating H vector must be parallel with such a surface. With two separated conductors, there is for each geometrical arrangement of conductors one and only one cross-sectional field configuration which will satisfy the boundary conditions at the metal surface. The field configurations for coaxial and two-wire lines are shown in Fig. 4. At each point the

ratio of E to H is as given by Eq. (1), and the velocity of propagation of the wave is given by Eq. (2). Half of the propagated energy is contained in the electric field and half in the magnetic field. This mode of propagation is in contrast with the more complex modes required in a hollow metal pipe, where the conditions required at the boundaries can be satisfied only by means of reflections at the metal walls. As a result, the two-conductor transmission line does not have the upper limit on wavelength that was imposed on the hollow waveguide by the requirement of reflections; in fact, the two-conductor line operates completely normally at zero frequency (direct current).

At wavelengths that are small enough to be comparable with the cross-sectional dimensions of the line, more complex modes, involving reflections from the surfaces of the conductors, become possible. High-frequency energy can thus be propagated in several modes simultaneously. In a coaxial cable a rough criterion for the elimination of higher modes is that the wavelength should be greater than the average of the circumferences of the inner and outer conductors.

As the wave propagates along the line, it is accompanied by currents which flow longitudinally in the conductors. These currents can be regarded as satisfying the boundary condition for the tangential H field at the surface of the conductor. The conductors have a finite conductivity, and so these currents cause a transformation of electrical energy into heat. The energy lost comes from the stored energy of the wave, and so the wave, as it progresses, diminishes in amplitude. The conductors are necessarily supported by insulators which are imperfect and cause additional attenuation of the wave. In a typical open-wire telephone line operating at voice frequencies, half the energy is lost in a distance of perhaps 60 mi. The losses increase with frequency, and for a typical air-insulated coaxial line operating at 5 MHz, half the energy is lost in a distance of less than 1 mi. At a frequency of 3000 MHz ($\lambda = 10$ cm), typical distances in which half the energy is lost are, for air-insulated coaxial cable, 25 m; for flexible coaxial cable insulated with polyethylene, 10 m.

Noise. In a transmission line intended for the transmission of large amounts of power, such as the cross-country lines joining electrical generating stations to centers of population, the loss of an appreciable proportion of the power en route is a serious matter. In a communication system, however, the average rate of flow of energy is rather small, and the intrinsic value of the energy itself is not of prime importance. The important characteristic of such a system is the accurate transmission of information, and the limiting factor is noise. Noise is always present in a transmission channel. Two common causes are thermal agitation and nearby electrical discharges. In a transmission system conveying information by an electromagnetic wave, the loss of energy in transmission becomes a serious matter if the wave is attenuated to the point where it is not large enough to override the noise. Amplifiers must be inserted in the transmission system at sufficiently close intervals so that the signal never falls into the noise level, from which it could not be recovered and interpreted accurately.

Circuit analysis of transmission lines. Because

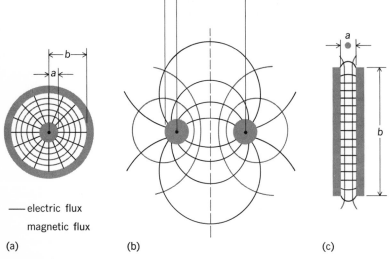

— electric flux

   magnetic flux

(a)   (b)   (c)

Fig. 4. Cross sections of common two-conductor transmission lines. (a) Coaxial cable. (b) Two-wire line. (c) Parallel-strip line.

the conductors of a transmission line are almost always spaced much closer together than a quarter wavelength of the electromagnetic energy that they are guiding, it is possible to analyze their performance quantitatively by circuit theory. It is then possible to deal with the voltages between the conductors and the currents flowing along the conductors, instead of with the electric and magnetic fields that exist in the insulating medium.

The waveguiding properties of the transmission line can be examined most conveniently if losses of energy are ignored. If L is defined as the inductance of the pair of conductors per unit length and C the capacitance between the conductors per unit length, field theory shows that L is μF_g and C is ϵ/F_g, where F_g is a geometrical factor that depends on the cross-sectional configuration of the conductors. For a coaxial line (Fig. 4a), $F_g = (1/2\pi)$ $\log_e (b/a)$. For a two-wire line (Fig. 4b), $F_g = (1/\pi)$ $\log_e (D/a)$. For a parallel-strip line (Fig. 4c), neglecting edge effects, $F_g = a/b$.

In the circuit analysis of a transmission line, the line can be visualized as being composed of a cascaded set of sections, each of short length Δx, as shown in Fig. 5b. The partial differential equations which describe the voltage e and the current i are shown by Eqs. (7) and (8).

$$\frac{\partial e}{\partial x} = -L \frac{\partial i}{\partial t} \qquad (7)$$

$$\frac{\partial i}{\partial x} = -C \frac{\partial e}{\partial t} \qquad (8)$$

The solution of these equations is shown by Eqs. (9) and (10), where f_1 and f_2 are any finite, single-

$$e = f_1(x - t/\sqrt{LC}) + f_2(x + t/\sqrt{LC}) \qquad (9)$$

$$i = \frac{1}{\sqrt{L/C}} [f_1(x - t/\sqrt{LC}) - f_2(x + t/\sqrt{LC})] \qquad (10)$$

valued functions of the arguments $x - t/\sqrt{LC}$ and $x + t/\sqrt{LC}$, respectively. These are interpreted physically as traveling waves, the first traveling in the positive x direction with the speed $1/\sqrt{LC}$ and the second traveling in the negative x direction at the same speed. Substitution of the values for L and C for any configuration of conductors yields the velocity $1/\sqrt{LC}$, which equals $1/\sqrt{\mu\epsilon}$.

The quantity $\sqrt{L/C}$ has the dimensions of ohms, termed the characteristic impedance Z_0 of the line, as shown by Eq. (11). Thus, Z_0 is a real quantity

$$Z_0 = \sqrt{L/C} = \sqrt{\mu/\epsilon}\, F_g \qquad (11)$$

(a resistance) and is equal to the wave impedance of the insulating medium, $\sqrt{\mu/\epsilon}$, multiplied by the geometrical factor, F_g, characteristic of the particular configuration of conductors. For the traveling waves of voltage and current, Eqs. (9) and (10), the ratio of voltage to current of the forward-traveling wave is Z_0; that of the backward-traveling wave is $-Z_0$.

In Fig. 5a, a source of electrical energy is connected at one end of a transmission line and an electrical load is connected to the other. Electromagnetic energy is propagated from the sending end to the receiving end, and a portion of the energy is reflected back toward the sending end if the load impedance Z_R is different from the characteristic impedance Z_0 of the line. If Z_R equals Z_0, there is no reflection of energy at the load, and in

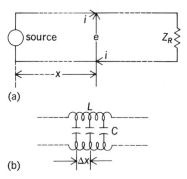

Fig. 5. Schematic representation of a transmission line. (a) Circuit diagram. (b) Visualization of L and C.

Eqs. (9) and (10), the function f_2, representing leftward-traveling energy, is absent. This is the condition desired when the purpose of the line is to deliver energy from the source of the load. The sending-end impedance of the line is then equal to Z_0.

In addition to impedance matching to reduce reflections (echoes) along a transmission line, it is also necessary to minimize signal distortion, which consists of amplitude and phase (delay) distortion. If the line attenuation is frequency-dependent, then a signal consisting of a group of different-frequency components will undergo amplitude distortion due to the unequal attenuation of each component of the signal. Similarly, if the velocity of propagation along the line is frequency-dependent, then a delay in phase of each component will result in associated phase distortion of the signal.

Signal distortion can be minimized by the use of line loading, which is the addition of series impedances along the line and which is used to adjust the line parameters to obtain the so-called distortionless condition. Under distortionless operation the attenuation and velocity of propagation are independent of frequency. For a discussion of the distortionless line see TRANSMISSION LINES.

Instead of loading a line, one may employ equalizing circuits to compensate for the phase distortion along the line.

Short sections of transmission line are sometimes used to provide low-loss reactive impedances and resonant circuits at high frequencies. This is done by open-circuiting or short-circuiting the receiving end of the line to provide complete reflection of the incident energy. A short-circuited low-loss line provides the sending-end impedance shown by Eq. (12), measured in ohms. When l

$$Z_s = jZ_0 \tan(2\pi fl/v) \qquad (12)$$

equals $v/4f$, the line is a quarter wavelength long, the argument $(2\pi fl/v)$ of the tangent function in Eq. (12) is $\pi/2$, and Z_s approaches an infinite value. In actual practice, losses keep Z_s to a finite value. However, at high frequencies the quarter wavelength is short and the losses are small, and such a short-circuited quarter-wave section can be used successfully as a low-loss insulator. Such a section is a resonant one and can be used as a substitute for a parallel-resonant LC circuit, for example, a tank circuit for a high-frequency oscillator. At low frequencies the required quarter wavelength is so large that losses impair the performance; also, the length becomes inconveniently great. At a frequency lower than $v/4l$, the sending-end impedance of the short-circuited line is inductive, and at frequencies between $v/4l$ and $v/2l$ the impedance is capacitive. This provides the possibility of using sections of short-circuited line as reactive elements in circuits.   [WALTER C. JOHNSON]

Bibliography: W. C. Johnson, *Transmission Lines and Networks*, 1950; E. C. Jordan, *Electromagnetic Waves and Radiating Systems*, 1950; J. J. Karakash, *Transmission Lines and Filter Networks*, 1950; J. D. Kraus, *Electromagnetics*, 1953; R. K. Moore, *Traveling-Wave Engineering*, 1960; S. Ramo, J. R. Whinnery, and T. Van Duzer, *Fields and Waves in Communication Electronics*, 1965; H. H. Skilling, *Electric Transmission Lines*, 1951; J. L. Stewart, *Circuit Analysis of Transmission Lines*, 1958.

Electromagnetism

The branch of science dealing with the observations and laws relating electricity to magnetism. Electromagnetism is based upon the fundamental observations that a moving electric charge produces a magnetic field and that a charge moving in a magnetic field will experience a force.

The magnetic field produced by a current is related to the current, the shape of the conductor, and the magnetic properties of the medium around it by Ampère's law.

The magnetic field at any point is described in terms of the force that it exerts upon a moving charge at that point. The electrical and magnetic units are defined in terms of the ampere, which in turn is defined from the force of one current upon another. *See* AMPERE.

The association of electricity and magnetism is also shown by electromagnetic induction, in which a changing magnetic field sets up an electric field within a conductor and causes the charges to move in the conductor. *See* ELECTRICITY; ELECTROMAGNET; INDUCTION, ELECTROMAGNETIC; INDUCTION, MAGNETIC; MAGNETIC CIRCUITS; MAGNETIC FIELD; MAGNETOMOTIVE FORCE.

[KENNETH V. MANNING]

Emissivity

The ratio of the radiation intensity of a nonblackbody to the radiation intensity of a blackbody. This ratio, which is usually designated by the Greek letter ϵ, is always less than or just equal to one. The emissivity characterizes the radiation or absorption quality of nonblack bodies. Published values are readily available for most substances. Emissivities vary with temperature and also vary throughout the spectrum. For an extended discussion of blackbody radiation and related information *see* HEAT RADIATION. *See also* BLACKBODY.

There are several methods by means of which the emissivity can be determined. The one most commonly used is the cavity method. In this technique a fine hole is provided in a radiating surface, and the ratio of the radiation intensity from the surface to the radiation intensity from the hole yields the emissivity directly. This method is quite accurate. One can also use an optical pyrometer to determine the emissivity from the brightness temperatures of the hole and the surface in conjunction with Wien's law of radiation.

The total emissivity when introduced into the Stefan-Boltzmann law gives the total radiated energy W in joules per square centimeter of the real heat radiator as $W = \epsilon\sigma T^4$. Here T represents the absolute temperature and σ, the radiation constant, has the value 5.67×10^{-12} joule cm^{-2} $°K^{-4}$. This energy is always smaller than the energy radiated by the blackbody, since ϵ is less than 1. For example, the total emissivity for tungsten is 0.32 at 2500°C, which means that at the same temperature tungsten radiates approximately one-third the energy of a blackbody.

The spectral emissivity ϵ_λ (the subscript λ denotes the wavelength) provides information on the energy distribution. Any spectral emissivity value is valid only for a narrow wavelength interval. The wavelength at which ϵ_λ has been determined is indicated by a subscript, for instance, $\epsilon_{0.655}$. A spectral emissivity of zero means that the heat radiator emits no radiation at this wavelength. Strongly selective radiators, such as insulators or ceramics, have spectral emissivities close to 1 in some parts of the spectrum, and close to zero in other parts. Carbon has a high spectral emissivity throughout the visible and infrared spectrum, exceeding 0.90 in certain portions; thus carbon is a good blackbody radiator. Tantalum is the only metal with a spectral emissivity greater than 0.5 in the visible spectrum. All other metals have a lower spectral emissivity. Tungsten is a relatively good emitter, with a spectral emissivity of $0.43 - 0.47$ within the visible region of the spectrum.

[HEINZ G. SELL; PETER J. WALSH]

Energy

The capacity for doing work. Energy is possessed, for example, by a body that is in motion, for stopping it provides work; by a compressed or stretched spring, for it is capable of doing work in returning to its ordinary configuration; by gunpowder or a bomb, because of the work it can do in exploding; by a charged electrical capacitor, for it can do work while being discharged. Energy, like work, is a scalar quantity. Its units are the same as those of work and include the foot-pound, foot-poundal, erg, joule, and kilowatt-hour. *See* WORK.

Because a system may possess an enormous store of energy that is not available for doing work, energy is better defined as that property of a system which diminishes when the system does work on any other system by an amount equal to the work so done. Although energy may be exchanged among various bodies or may undergo transformation from one form to another, it has the tremendously important property that it cannot be created or destroyed. *See* CONSERVATION OF ENERGY.

Energy occurs in several well-defined forms, as internal energy, kinetic energy, and potential energy.

Internal energy. This is the energy present within a body or system, such as a fuel, steam, or compressed gas, by virtue of the motions of, and forces between, the molecules and atoms of the body or system. Internal energy is sometimes erroneously referred to as the "heat energy" of a body. It is a property of any given state of a system, is evidenced by certain other properties of the system, notably temperature, and is to be distinguished from any kinetic or potential energy possessed by the system as a whole in its relation to other systems. According to the first law of thermodynamics, the gain of internal energy in any given process is equal to the difference between the heat gained by the system and the work done by the system on other systems external to it. *See* INTERNAL ENERGY; THERMODYNAMIC PRINCIPLES.

Kinetic energy. This is the term applied to the capacity for doing work that matter possesses because it is in motion. As everyday experience shows, the more massive a body is and the higher its speed, the more work it will do upon striking and being slowed down by an obstacle, and hence the larger is its kinetic energy. Specifically, for a body of mass m moving with a speed v, the kinetic energy E_k is given by Eq. (1).

$$E_k = \tfrac{1}{2}mv^2 \tag{1}$$

Thus if a car of mass 60 slugs (about 1900 lb) is moving with a speed of 90 ft/sec (about 60 mi/hr), its kinetic energy relative to the ground is 243,000 slug-ft²/sec², or 243,000 ft-lb. Now the change in the kinetic energy of a body during a given displacement is equal to the work done by the net, or resultant, force applied to the body during this displacement, a statement that is usually referred to as the work–kinetic energy theorem. Thus, to bring the car in the example to a stop in, say, 100 ft would require an average retarding force of 2430 lb, or about 1.2 tons of force. It will be noted that the value of the kinetic energy is always dependent on the body's speed relative to some chosen reference body. Thus the kinetic energy of a car relative to a man sitting in it is zero; only if the car changes speed relative to the man can it do work on him because of its motion.

To derive Eq. (1), suppose that a body of mass m and moving with speed v is brought to rest within a distance s by applying to it a constant force of magnitude f. The work done is fs and, by the work–kinetic energy theorem, is equal to the change in the body's kinetic energy E_k. By Newton's second law of motion, $f = ma$; and since the acceleration a is assumed here to be constant, $s = v^2/2a$. Therefore $E_k = fs = ma \cdot v^2/2a = \tfrac{1}{2}mv^2$. This expression for the kinetic energy is valid for all speeds except those comparable to the speed of light.

The kinetic energy of rotation of a body that is turning about a fixed axis with angular velocity of magnitude ω (radians/sec) is given by $E_k = \tfrac{1}{2}I\omega^2$, where I is the body's moment of inertia with respect to the axis in question.

Potential energy. This form of energy, as contrasted with kinetic energy, is the capacity to do work that a body or system has by virtue of its position or configuration. Thus elastic potential energy is possessed by a coiled spring that is compressed or stretched (Fig. 1); indeed, frictional forces in a spring are often so small that more than 99% of the work of deformation is recovered when the spring is released. Gravitational potential energy is possessed by a body that has been raised above the Earth's surface, for the body can do work in falling to the ground; it can strike and drive a nail or a pile, or can compress a spring. Electrical, magnetic, chemical, and nuclear systems may also possess potential energy.

In general, the potential energy of one configuration of a system relative to another configuration

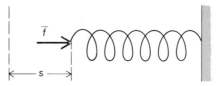

Fig. 1. A compressed spring possesses potential energy. Here s is the distance that the spring has been compressed from its normal length, and $\bar{f}$ is the average force of compression. It can be shown from Hooke's law that $f = \tfrac{1}{2}ks$, where k is a constant known as the stiffness coefficient of the spring. It follows that the elastic potential energy of the compressed spring is $\tfrac{1}{2}ks^2$.

tion of it may be defined as equal to the work done against the conservative forces of the system when its parts change from the one configuration to the other. Conservative forces are forces such as those of gravity or the force exerted by a spring, where the work is recoverable, that is, the net work done in a round trip is zero. As this definition implies, the potential energy of a system when in a particular configuration must always be computed with respect to some other arbitrarily selected configuration or position of the system; moreover, its value is a function only of the initial and final positions, and not of the paths followed by the parts in changing position. *See* FORCE.

Gravitational potential energy. Suppose that a body of mass m is at a height h above the ground and that h is small in comparison with the Earth's radius so that the body's weight mg does not change appreciably with the height. The gravitational potential energy E_p of the system body-Earth will then be mgh, for this is the work done against the weight mg in lifting the body to the height h and, in the absence of air resistance, is the work the body can do in returning to the ground. For example, if a 1-kg object is 1000 m above sea level, E_p with respect to sea level is $mgh = 1.0$ kg $\times 9.8$ (m/sec²) $\times 1000$ m $= 9800$ joules. But relative to a land surface of altitude, say, 500 m above sea level, E_p is only half as much, or 4900 joules, this being the work the body could do in dropping to this reference level rather than to the sea (Fig. 2).

If a body is at a great distance from the Earth, as when a missile is fired to a very high altitude, account must be taken of the change in the body's weight with distance r from the center of the Earth as given by the Newtonian law of gravitation, $f_g = Gmm'/r^2$, where f_g is the gravitational force of attraction, G is the gravitational constant, and m and m' are the masses of the body and Earth, respec-

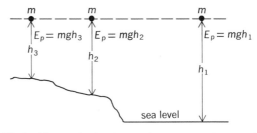

Fig. 2. The gravitational potential energy E_p of a body of weight mg for different reference levels.

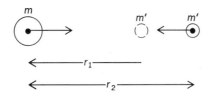

Fig. 3. Finding E_p for two particles of masses m and m'.

tively. If the distance between the body and the Earth's center is increased from r_1 to r_2 (Fig. 3), the work W done against gravitational attraction, and therefore the increase ΔE_p in the gravitational potential energy of the system body-Earth, is shown by Eq. (2). Equation (2) holds for any two bodies

$$W = \Delta E_p = \int_{r_1}^{r_2} f_g \, dr = E_{p2} - E_{p1}$$
$$= Gmm' \left(\frac{1}{r_1} - \frac{1}{r_2} \right) \qquad (2)$$

that can be treated as particles and therefore is applicable to the Earth as one of the attracting bodies to the extent that the Earth can be regarded as spherical and made up of concentric shells, each of which is homogeneous as regards density. In the theory of gravitation the results are often simplified when the reference distance r_1 between the two attracting bodies is infinitely great, that is, the gravitational potential energy of a body is assumed to be zero when it is far removed from all other bodies. Then, from Eq. (2), $E_{p2} = -Gmm'/r_2$, and since r_2 may be any distance, the subscripts can be dropped, giving $E_p = -Gmm'/r$. The negative sign means simply that E_p at any finite distance of separation r is smaller than it is at infinity.

Electrical potential energy. The electrical potential energy of two particles in vacuum having charges q and q' can similarly be shown, with the help of Coulomb's electrostatic law, to be $E_p = \pm k_0 qq'/r$, where k_0 is a constant; the algebraic sign is negative or positive, according as the charges have unlike or like signs. *See* ENERGY SOURCES.

[DUANE E. ROLLER/LEO NEDELSKY]

Bibliography: R. Benumof, *Concepts in Physics*, 1965; G. P. Harnwell and G. J. F. Legge, *Physics: Matter, Energy and the Universe*, 1967; F. A. Kaempffer, *The Elements of Physics: A New Approach*, 1967; E. M. Rogers, *Physics for the Inquiring Mind*, 1960.

Energy conversion

The process of changing energy from one form to another. There are many conversion processes that appear as routine phenomena in nature, such as the evaporation of water by solar energy or the storage of solar energy in fossil fuels. In the world of technology the term is more generally applied to man-made operations in which the energy is made more usable, for instance, the burning of coal in power plants to convert chemical energy into electricity, the burning of gasoline in automobile engines to convert chemical energy into propulsive energy of a moving vehicle or the burning of a propellant for ion rockets and plasma jets.

There are well-established principles in science which define the conditions and limits under which energy conversions can be effected, for example,

the law of the conservation of energy, the second law of thermodynamics, the Bernoulli principle, and the Gibbs free-energy relation. Recognizable forms of energy which allow varying degrees of conversion include chemical, atomic, electrical, mechanical, light, potential, pressure, kinetic, and heat energy. In some conversion operations the transformation of energy from one form to another, more desirable form may approach 100% efficiency, whereas with others even a "perfect" device or system may have a theoretical limiting efficiency far below 100%.

The conventional electric generator, where solid metallic conductors are rotated in a magnetic field, actually converts 95–99% of the mechanical energy input to the rotor shaft into electric energy at the generator terminals. On the other hand, an automobile engine might operate at its best point with only 20% efficiency, and even if it could be made perfect, might not exceed 60% for the ideal thermal cycle. Wherever there is a cycle which involves heat phases, the all-pervading limitation of the Carnot criterion precludes 100% conversion efficiency, and for customary temperature conditions the ideal thermal efficiency frequently cannot exceed 50 or 60%. *See* CARNOT CYCLE.

In the prevalent method of producing electric energy in steam power plants, there are many energy-conversion steps between the raw energy of fuel and the electricity delivered from the plant, for example, chemical energy of fuel to heat energy of combustion; heat energy so released to heat energy of steam; heat energy of steam to kinetic energy of steam jets; jet energy to kinetic energy of rotor; and mechanical energy of rotor to electric energy at generator terminals. This is a typical, elaborate, and burdensome series of conversion processes. Many efforts have been made over the years to eliminate some or many of these steps for objectives such as improved efficiency, reduced weight, less bulk, lower maintenance, greater reliability, longer life, and lower costs. For a discussion of major technological energy converters *see* POWER PLANT. *See also* ELECTRIC POWER GENERATION.

Efforts to eliminate some of these steps have been stimulated by needs of astronautics and of satellite and missile technology and need for new and superseding devices for conventional stationary and transportation services. Space and missile systems require more compact, efficient, self-contained power systems which can utilize energy sources such as solar and nuclear. With conventional services the emphasis is on reducing weight, space, and atmospheric contamination, on improving efficiency, and on lowering costs. The predominant objective of energy conversion systems is to take raw energy from sources such as fossil fuels, nuclear fuels, solar energy, wind, waves, tides, and terrestrial heat and convert it into electric energy. The scientific categories which are recognized within this specification are electromagnetism, electrochemistry (fuel cells), thermoelectricity, thermionics, magnetohydrodynamics, electrostatics, piezoelectricity, photoelectricity, magnetostriction, ferroelectricity, atmospheric electricity, terrestrial currents, and contact potential. *See* ENERGY SOURCES.

The electromagnetism principle today dominates the field. Electric batteries are an accepted form of electrochemical device of small capacity, for example, 1 kw in automobile service. Other categories are in various stages of development. Extensive efforts and funds are being given to some fields with attractive prospects of practical adaptation. *See* Battery, electric; Electric rotating machinery; Fuel cell; Magnetohydrodynamic power generator; Solar battery; Thermionic power generator; Thermoelectric power generator.

[THEODORE BAUMEISTER]

Bibliography: T. Baumeister (ed.), *Standard Handbook for Mechanical Engineers*, 7th ed., 1967; S. S. L. Chang, *Energy Conversion*, 1963; J. Kaye and J. H. Welsh, *Direct Conversion of Heat to Electricity*, 1960; W. Mitchell, Jr. (ed.), *Fuel Cells*, 1963.

Energy flow

Energy flows through the United States economy from coal, oil, natural gas, falling water, uranium, and other primary sources to provide heat, light, and power. Some primary energy is used directly as fuel: most of the heat is generated by burning coal, oil, or natural gas in stoves and furnaces, and most of the power for automobiles, trucks, airplanes, ships, and trains is generated by burning oil products in internal combustion engines. Other primary energy is converted to electric power before it is used; most lighting has been electrified, as has most of the power for machine tools, refrigeration plants, elevators, appliances, television, and other stationary machines and equipment. And some energy is lost in the processes of conversion and use. *See* Energy con-

version; Energy sources.

Transforming energy for use. Energy is used for three main purposes—heat, light, and mechanical power. Heat and light are easy to generate by means of fire, and people have done so for thousands of years, using wood and other organic materials as fuels. Mechanical power is more difficult to generate, but people have done so for thousands of years, using waterpower, windmills, sails, work animals, and their own muscles.

The ability to generate mechanical power from heat is relatively recent. On a practical scale, the transformation of heat to mechanical power began with steam engines in England in the 1770s, marking the beginning of the industrial revolution. Over a period of 2 centuries, steam engines and other heat engines replaced wind power and work animals throughout the societies that are now called the developed nations. This gradual replacement has been completed only recently. In the early 1900s the United States still obtained much of its mechanical power from work animals and windmills. Complete reliance on heat engines and waterpower came only since the 1950s. *See* Engine; Waterpower.

The transformation of mechanical power into electricity, which can be transmitted long distances over wires and converted to light or to mechanical power again at a distant location, is the most recent development in energy conversion technology. It began on a practical scale in the 1880s, and now nearly all illumination and all mechanical power except for farm work and transportation have been electrified. *See* Electric power generation.

Heat is still obtained from combustion, but fossil

ENERGY SOURCES   ENERGY CONVERSION FACILITIES   USEFUL ENERGY APPLICATIONS

Flow of energy through the United States economy in the mid-1970s, from major energy sources, through conversion facilities, to useful applications, with some resultant unavailable energy. The width of each channel is proportional to the amount of energy that flows along it each year.

fuels—coal, oil, and natural gas—are burned instead of organic materials. Some mechanical power is still obtained from falling water, but now in the form of electric power. Mechanical power for farm work and transportation comes from internal combustion engines burning oil products, and most other mechanical power comes in the form of electric power from steam engines whose boilers are fueled by coal, natural gas, or oil. Nuclear fuels have only just begun to be utilized as heat sources for stream engine boilers. *See* COMBUSTION; INTERNAL COMBUSTION ENGINE; NUCLEAR FUELS.

U.S. energy flow chart. The relative proportions of the major energy flows for the United States in the mid-1970s are shown in the chart. At the left are the nation's energy sources—fossil fuels and uranium as sources of high-temperature heat, and falling water as a source of mechanical power. In the middle are the nation's energy conversion facilities. On the right are the flows of energy into major useful applications: light, power, and heat. At the bottom of the figure are the flows of unavailable energy: low-temperature heat lost up stacks and chimneys and low-temperature heat lost in the conversion of high-temperature heat to mechanical power. *See* WASTE HEAT MANAGEMENT.

Most of the power from heat engines and from falling water is used for light and power applications. A small amount is used for electric heating applications, primarily for high-temperature industrial heat that cannot be achieved economically by direct combustion, for residential cooking and water heating where convenience and cleanliness are of importance, and for a limited amount of space heating where convenience, cleanliness, and low installation cost of heating equipment are important.

Efficiency considerations. The laws of nature limit the efficiency with which heat engines can convert heat into mechanical power or work. High-temperature heat input is converted partly to mechanical work output (the desired end-product) and partly to low-temperature heat output (the unavailable residue). The higher the temperature of the input heat and the lower the temperature of the output heat, the larger the proportion of work output that a heat engine can achieve. Because electric power plants are able to reject their unavailable heat at so low a temperature—only slightly warmer than the outdoor temperature—they are more efficient than internal combustion engines in converting input heat into output power. As a result, most stationary power applications have been electrified. *See* EFFICIENCY; POWER PLANT.

Illumination has been electrified, partly for greater convenience, cleanliness, and safety, but also for greater overall efficiency and lower cost. More light can be obtained from a gallon of oil burned in a power plant to make electricity that operates an electric lamp than from a gallon of oil burned directly in an oil lamp.

In future years nuclear energy sources are likely to grow in significance. Improvements in the efficiency of heat engines may increase the proportion of useful power relative to unavailable energy, and electricity may be more widely used for heating applications. Overall energy use may grow as the population and the general level of prosperity in-

crease. *See the feature articles* ENERGY CONSUMPTION; EXPLORING ENERGY CHOICES.

[JOHN C. FISHER]

Bibliography: John C. Fisher, *Energy Crises in Perspective*, 1974; National Academy of Engineering, *U.S. Energy Prospects: An Engineering Viewpoint*, 1974.

Energy sources

The sources from which energy can be obtained to provide heat, light, and power. Energy sources have progressed from human and animal power to fossil fuels and radioactive elements, and water, wind, and solar power. Industrial society was based on the substitution of fossil fuels for human and animal power. Future generations will have to increasingly use solar energy and nuclear power as the finite reserves of fossil fuels become exhausted. The principal fossil fuels are coal, lignite, petroleum, and natural gas—all of which were formed millions of years ago. Fossil fuels which have potential for future use are oil shale and tar sands. Nonfuel sources of energy include wastes, water, wind, geothermal deposits, biomass, and solar heat. At the present time the nonfuel sources contribute very little energy, but as the fossil fuels become depleted, the nonfuel sources and fission and fusion sources will become of greater importance since they are renewable. Nuclear power based on the fission of uranium, thorium, and plutonium, and fusion power based on the forcing together of the nuclei of two light atoms, such as deuterium, tritium, or helium 3, could become principal sources of energy in the 21st century.

As the world has become more industrialized, the consumption of fuels has increased at a rapid rate. World energy demand amounted to 132×10^{15} Btu (10^{15} Btu $= 1.05504 \times 10^{15}$ megajoules) in 1961 and increased to 238×10^{15} Btu in 1972 for an average annual growth of 5.5%. During the same period, population and real Gross National Product increased at rates of 1.9 and 5.1%, respectively.

The United States consumed a total of 73×10^{15} Btu in 1974, or 30% of the world energy consumption; the sources of energy in the United States in 1974 in absolute quantities and in Btu are shown in Table 1. The predictions for total United States energy demand by 1985 range from 103 to 125×10^{15} Btu, depending upon the average annual percent gain in energy requirements, as given in Table 2 (also see illustration). The Federal Energy Administration base case, which predicts a total 1985 energy demand in the United States of 109.0×10^{15} Btu, is based on the sources of fuel and energy

Table 1. United States consumption of fuels and energy in 1974

Fuel	Quantity	10^{15} Btu	%
Bituminous coal and lignite	551×10^6 short tons	13.037	17.8
Anthracite	5×10^6 short tons	0.132	0.2
Petroleum products			
From crude oil	4.4×10^9 bbl	24.861	34.0
From other sources	1.0×10^9 bbl	6.058	8.3
Natural gas, dry	21.8×10^{12} ft^3	22.237	30.4
Natural gas, liquid	638×10^6 bbl	2.571	3.5
Electricity, water power	294×10^9 kWhr	3.052	4.2
Electricity, nuclear power	110×10^9 kWhr	1.173	1.6
Total		73.121	100.0

Table 2. Predicted total United States energy demand by 1985*

Source of projection	Projected energy demand for 1985, 10^{15} Btu	Compound annual growth rate 1972–1985, %
FEA – oil @ $4/bbl	118.3	3.8
FEA – oil @ $7/bbl; base case	109.0	3.2
FEA – oil @ $11/bbl	102.9	2.7
Ford Foundation	115.0	3.6
National Petroleum Council	124.9	4.2

*Data from Federal Energy Administration, *Project Independence Report*, November 1974.

given in Table 3. Electric power generated in the United States in 1974 required 27% of the total fuel and energy consumed in that year; the sources of fuel and energy for electric power in 1974 are listed in Table 4. It is forecast that the fuels shown in Table 5 will be required for United States electric power generation by 1985. *See the feature article* ENERGY CONSUMPTION.

Petroleum. The production of crude petroleum in the United States in 1974 was 3.2×10^9 bbl (10^9 bbl $= 0.159 \times 10^9$ m³). Crude petroleum production in the United States peaked at 3.5×10^9 bbl in 1970 and has been declining since then. The consumption of petroleum products totaled 5.4×10^9 bbl in 1974, which entailed importing 2.2×10^9 bbl of crude oil and refined products. Oil imports accounted for over 40% of total oil consumption in 1974, and imports of oil will continue to grow as domestic oil production declines and demands for oil continue to increase.

The United States proved reserves of petroleum declined to a level of 35×10^9 bbl at the end of 1974, a life rate of 11 years at current levels of production. The decline in petroleum production in the United States was the result of progressively less drilling since about 1955. This decline in drilling activity was caused by government actions which made it uneconomical to drill for oil and gas in the United States. These actions included wellhead ceiling prices on oil and gas, the repeal of depletion allowances for the oil and gas industry, and a generally unfavorable climate for the oil industry in the United States. As a result, the oil companies sent their drilling rigs to foreign countries, particularly in the Middle East and Africa, where the cost of drilling was less and the expectation and probability of finding oil in large quantities were greater. Once it was established that very large reserves of oil were present in these areas and that the oil could be produced for 10–15 cents per barrel rather than several dollars and higher in the United States, the future of world oil production, with the predominance of the Middle East, became a reality.

World production of crude oil was over 20×10^9 bbl in 1974. While the United States produced only 3.2×10^9 bbl, or 16% of the total world production, it consumed 5.4×10^9 bbl, or 27% of total world consumption.

World proved recoverable reserves of crude oil

Table 3. Forecast of United States fuel and energy consumption by source in 1985

Fuel source	10^{15} Btu		%		Units
Coal		19.9		18.3	905×10^6 tons
Oil – Domestic	(23.1)		(21.2)	(3.7)	$\times 10^9$ bbl
Imports	(24.8)		(22.8)	(4.0)	$\times 10^9$ bbl
Total oil		47.9		44.0	7.7×10^9 bbl
Natural gas		23.9		21.9	23.4×10^{12} ft³
Hydro power		4.8		4.4	
Nuclear power		12.5		11.4	
Total		109.0		100.0	

Table 4. Fuels used for United States electrical power production in 1974

Fuel	Units	10^{15} Btu	%
Coal	389×10^6 tons	8.668	44.15
Natural gas	3.2×10^{12} ft³	3.328	16.95
Oil products	554×10^6 bbl	3.448	17.56
Nuclear power	110×10^9 kWhr	1.173	5.97
Hydro power	294×10^9 kWhr	3.018	15.37
Total		19.635	100.00

Table 5. Forecast of fuels for United States electric power production in 1985

Fuel	Units	10^{15} Btu	%
Coal	827×10^6 tons	18.240	52.06
Natural gas	1.42×10^{12} ft³	1.500	4.28
Oil products	836×10^6 bbl	5.200	14.84
Nuclear power	647×10^9 kWhr	6.900	19.69
Hydro power	312×10^9 kWhr	3.200	9.13
Total		35.040	100.00

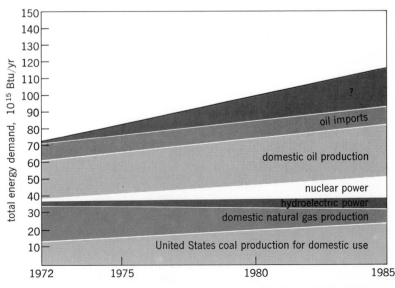

Projected total United States energy demand by types of fuel from 1972 to 1985.

as of January 1974 amounted to 85×10^9 metric tons, or 628×10^9 bbl. The United States proved recoverable reserves of crude oil as of January 1974 were 35×10^9 bbl, or 5.5% of the world's total reserves. *See* PETROLEUM.

Natural gas. At the end of 1974, the United States proved natural gas reserves amounted to 237×10^{12} ft³ (10^{12} ft³ $= 28.317 \times 10^9$ m³). The annual production of natural gas in the United States peaked in 1973 at 22.648×10^{12} ft³. Since 1973, natural gas production in the United States has declined at the rate of over 6% per year. By June 1975, production had declined to a level of about 20×10^{12} ft³ per year. These are indications that if there is no additional economic incentive to drill for oil and gas in the United States, the current rate of decline in natural gas production will reach 14×10^{12} ft³ by 1980 and 10×10^{12} ft³ by 1985.

The natural gas shortage first appeared in 1970 when curtailments amounted to only 0.1×10^{12} ft³, or less than 1% of demand. In 1974, curtailments totaled 2×10^{12} ft³, or 10% of demand. For 1975, the shortage was estimated at 2.9×10^{12} ft³, or 15% of demand. *See* NATURAL GAS.

Coal. Coal production in the United States ranged between $500-600 \times 10^6$ short tons (10^6 tons $= 0.9 \times 10^6$ metric tons) per year from 1965 to 1975. Forecasts of future United States coal production by 1985 range from 800 to 1200×10^6 short tons per year. In order to attain this level of production, it will be necessary to develop hundreds of new deep mines in the eastern United States and hundreds of new surface mines both east and west. The capital requirements for equipment and mine development will range up to $25,000,000,000. Failure to reach the 1985 goal of 1200×10^6 tons of coal production will result in an increasingly undesirable dependence upon imported oil, with more and more oil coming from the Middle East.

The U.S. Bureau of Mines estimated that as of January 1974 the demonstrated coal reserve base in the United States was 434×10^9 tons, of which 297×10^9 tons were underground coal reserves and 137×10^9 tons were mineable by surface methods. The world's total proved and currently recoverable reserve of coal in place amounted to 2376×10^9 short tons, and on this basis the United States contains 18% of the world's proved and currently recoverable reserves. The consumption of coal in the United States in 1974 totaled 612×10^6 short tons. The consumption by groups is shown in Table 6. *See* COAL; COAL MINING.

Nuclear energy. By 1975, nuclear energy in the United States was supplying 7% of the electric power, or nearly 2% of the total energy in the nation. A total of 56 nuclear reactors with a capacity of 38,000 MWe were operating in the United States as of September 1975 (see Table 7). By 1985, at least 200,000 MWe of nuclear plant capacity should be completed and operating, supplying 20% of the total United States electric power production. Nuclear power is expected to supply over 11% of the total United States energy by 1985.

The present United States nuclear program is based principally on light-water reactors (LWR) and, to a lesser extent, on the high-temperature gas-cooled reactor (HTGR). However, the HTGRs are expected to play a larger role in the United States. The cutback orders for power plants in

1975 by electric utilities resulted in the construction of a number of both LWRs and HTGRs being delayed or canceled.

Reserves of uranium are deemed to be inadequate to support LWR and HTGR nuclear power plants in the numbers planned by the year 2000, since these reactors make use of only small percentages of the energy which is potentially available in nuclear fuels. This makes it imperative that the breeder reactor program in the United States be accelerated and brought to a commercial stage faster than presently planned. *See* NUCLEAR POWER; REACTOR, NUCLEAR.

Fusion power. Fusion power is expected to be a major source of energy in the long-range future, but commercialization is not expected until the early part of the 21st century. The fusion process is based on the principle that when the nuclei of two light atoms are forced together to form one or two nuclei with smaller mass, energy is released. The principal fuels to be used in fusion reactors are the gases deuterium, tritium, or helium 3. Deuterium, a heavy isotope of hydrogen, is found in ordinary sea water. It can be extracted relatively cheaply as deuterium oxide, or heavy water. When used in a fusion reactor, the deuterium in sea water will provide an inexhaustible supply of energy for the world. *See* FUSION, NUCLEAR.

Oil shale. Oil shale deposits have been found in many areas of the United States, but the only deposits of sufficient potential oil content considered as near-term potential resources are those of the Green River Formation in Colorado, Wyoming, and Utah. The United States Geological Survey estimates that about 80×10^9 bbl of oil are recoverable under present economic conditions. When marginal and submarginal reserves of oil shale are included, it is estimated that reserves of 600×10^9 bbl of oil equivalent are present in economically recoverable oil shale deposits in the United States. *See* OIL SHALE.

Tar sands. Tar sands represent the largest known world supply of liquid hydrocarbons. Extensive resources are located throughout the world, but primarily in the Western Hemisphere. The to-

Table 6. United States consumption of coal in 1974

Category	10^6 short tons	%
Electric utilities	390	63
Coking coal	90	15
Industry	64	10
Retail	9	2
Total domestic consumption	553	90
Exports to Canada	14	2
Exports to other countries	46	8
Total	613	100
versus total production	601	

Table 7. United States nuclear reactor status as of September 1975

Number of reactors	Status	Capacity, MWe
56	Reactors with operating licenses	37,995
64	Reactors with construction permits	65,742
99	Reactors on order	111,493
17	Letter of intent/options	19,132
236		234,362

tal reserves in 1971 were estimated to contain 2000×10^9 bbl.

The best-known deposit is the Athabasca tar sands in northeastern Alberta, Canada. The oil in place is estimated to amount to 95×10^9 metric tons (over 700×10^9 bbl). Commercial operations have been underway for a number of years at a production level of 50,000 bbl per day. This operation is being expanded, and another commercial plant is contemplated.

The principal United States reserves are located in Utah. It is estimated that the resources in place amount to $18-28 \times 10^9$ bbl in five major formations. *See* OIL SAND.

Solar energy. Solar energy has always been a potential source of limitless, clean energy. However, the commercial development of solar energy has been slow because of storage requirements and high capital-cost requirements. The availability in the past of low-cost fossil fuels resulted in solar energy being considered uneconomical. With the sudden massive increase in world oil and gas prices, solar energy economics, particularly for low-level heating and cooling, are reaching the stage where commercial applications are possible.

The Federal Energy Administration predicts that by 1985, solar energy could provide between 0.3 and 0.6×10^{15} Btu of space heating. This would still amount to only one-fourth to one-half percent of the energy required by the United States in 1985. It is well to keep the use of commercial solar energy in perspective and recognize that the conventional fossil fuels—coal, oil, and gas—will still be providing the United States with most of its energy even in 1985. *See* SOLAR ENERGY.

Geothermal energy. The heat content of the Earth is immense and would appear to many to be an inexhaustible and plentiful supply for much of the United States energy needs. However, for a variety of reasons, it appears highly unlikely that the amount of geothermal energy will ever supply more than 10% of United States energy requirements. Even in California, the only area in the United States where geothermal energy has been commercialized, it is doubtful that geothermal energy will be able to supply more than 10% of the state's electrical generating requirements. It is important, nevertheless, to develop all available economical geothermal resources since they represent important sources of renewable energy. *See* GEOTHERMAL POWER.

Refuse, winds, tides, and biomass. Refuse in the form of residential, commercial, industrial, and agricultural wastes has been underutilized in the past. However, as the cost of conventional fossil fuels increases, the use of refuse as an alternative fuel becomes economically attractive. A number of plants utilizing wastes as fuel have been built to produce steam and electric energy, and more plants are being considered.

Total available combustible solid wastes expected in the United States annually amount to 800×10^9 lb (10^9 lb $= 0.4536 \times 10^9$ kg) or 400×10^6 tons. At 5000 Btu/lb heating value, the potential fuel available is 4×10^{15} Btu/year. For an annual energy consumption in the United States at about 75×10^{15} Btu, refuse is about 5% of the total.

Wind power's energy potential is very great. However, wind power is not expected to add significantly to the United States energy capability by 1990. *See* WIND POWER.

Tidal power, although potentially a large energy source, has not become a commercial reality except in a few areas of the world such as France, which has developed one site to 240 MWe of capacity. *See* TIDAL POWER.

Biomass production for energy is a set of technically feasible and potentially economic operations. The use of biomass on a significant scale would not require a large research or demonstration effort or excessively large capital and manpower investments. *See the feature articles* EXPLORING ENERGY CHOICES; OUTLOOK FOR FUEL RESERVES.

[GERARD C. GAMBS]

Bibliography: Atomic Industrial Forum, *Status Report: Energy Resources and Technology*, March 1975; Federal Energy Administration, *Project Independence Report*, November 1974; Federal Power Commission, *National Gas Survey*, 1975; Federal Power Commission, *U.S. Natural Gas Supply Staff Report*, December 1974; National Electric Reliability Council, *Forecast of Fuel Requirements*, July 1975; National Petroleum Council, *U.S. Energy Outlook*, December 1972; 9th World Energy Conference, *Survey of Energy Resources*, 1974; 9th World Energy Conference, Division 3, *Energy Resource Recovery*, 1974; Power Magazine, *The 1975 Energy Management Guidebook*, August 1975; U.S. Bureau of Mines, *Mineral Industry Surveys*, 1975; U.S. Bureau of Mines, *Minerals in the U.S. Economy*, July 1975; U.S. Bureau of Mines, *Reserves of Crude Oil, Natural Gas Liquids and Natural Gas in the U.S. and Canada as of December 31, 1974*, 1975.

Energy storage

Methods which permit energy generated when the supply exceeds the demand to be stored for later use. The most common energy storage device is the electric battery. Some electric utilities store energy in the form of water in large pumped-storage reservoirs. A wide variety of new storage systems are under investigation for possible use in motor vehicles and to store highly variable energy from renewable resources. These new systems include better batteries, fuel cells, compressed-air storage in undergound caverns, rapidly rotating flywheels, magnetic energy storage, and the "hydrogen economy." *See* BATTERY, ELECTRIC.

Fuels. The fossil fuels contain stored chemical energy from the sunlight of past geological eras in a form that is compact and convenient. Unfortunately, these fuels are finite in quantity, and the only really common one, coal, is physically and environmentally unsuitable for many purposes, such as powering automobiles or heating homes directly. *See* FUEL, FOSSIL.

Nuclear fuel, like the fossil fuels, can be used to generate electricity when needed, but the typical demand load of the utilities varies so much from day to day and from hour to hour that the capacity of the electric power system needs to be much larger (usually at least twice as large) as the average capacity utilized. As a result, expensive equipment that is rarely used must be available.

Renewable resources. Renewable resources, such as the sun or wind, generally have the problem of variability. The sun shines only part of the

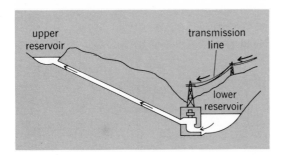

Fig. 1. Operation of a typical pumped-storage plant. The arrows show the energy flow in the pumping cycle, during off-peak power loads. In the generating cycle, during peak power loads, the direction of flow is reversed.

time; sometimes it is night and sometimes it is cloudy. The wind also varies in intensity in a manner not easily predicted. Practical storage systems could affect these shortcomings.

Pumped storage. Several utilities are using their unneeded capacity at night or other times to pump water uphill into a large pumped-storage reservoir (Fig. 1). When the utility experiences high demands, water from the reservoir flows downhill through large hydroelectric turbines to provide added capacity. Only about two-thirds of the energy needed to pump the water to the reservoir is recovered when the water flows back down. Nevertheless, the cost of electricity generated by the pumped-storage facility can be much less than bringing old steam plants or gas turbine peaking units into service at times of peak demand. The disadvantages of pumped storage are that it can be used only in parts of the country with suitable sites, and that its fluctuating water level can spoil the natural environment. *See* PUMPED STORAGE.

Batteries. At present, utilities do not have any other economical storage methods. Batteries, while convenient and inexpensive for many purposes, are much too costly for storing very large amounts of electricity. A battery is an energy storage device that contains chemical energy which can be converted into electricity simply by con-

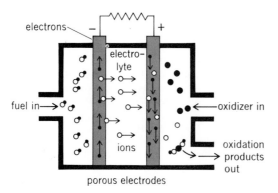

porous electrodes

Fig. 2. Diagram of fuel cell in which electrons are removed from the fuel atoms at the anode and are picked up by the oxidizer atoms on the right, after traveling through the external load circuit. In the hydrogen-oxygen fuel cell, hydrogen is the fuel and oxygen the oxidizer. (*From J. M. Fowler, Energy and the Environment, Mc-Graw-Hill, p. 307, 1975*)

necting a load to it. Primary batteries are constructed of materials which can furnish the energy directly. Secondary, or storage, batteries must first be charged, but can then be discharged and recharged many times.

Primary cell. A primary cell (battery unit) consists of an anode (negative terminal) which loses electrons, a cathode (positive terminal) which gains electrons, and an electrolyte between them. The simplest cell is a wet cell consisting of two different metals placed in a solution that can conduct electricity. A current will flow in a wire connected to the two metal electrodes. The common 1.5-V dry cell has a zinc can as the anode, a carbon rod as the cathode, and a paste of manganese dioxide and other compounds between the electrodes. The 1.5-V alkaline battery has a zinc anode and a manganese dioxide cathode. The 1.3-V mercury cell has a cathode of mercuric oxide and graphite and an anode of zinc and mercury.

Lead-acid battery. The most common storage battery is the lead-acid battery used in automobiles. Its basic cell consists of two lead plates placed in a sulfuric acid solution. When it is charged by direct current, the negative plate becomes spongy lead and the positive plate becomes lead peroxide. As the battery is discharged, both plates become lead sulfate. The 12-V car battery has six lead-acid cells, each providing 2 V, connected in series.

Edison and Ni-Cd cells. The 1.2-V Edison cell has a cathode of nickel peroxide, negative plates of powdered iron in steel tubes, and a potassium hydroxide solution as electrolyte. The 1.2-V nickel-cadmium cell has a cathode of nickel hydroxide, an anode of cadmium hydroxide, and an electrolyte of potassium hydroxide solution. It is rugged and can be recharged many times; a sealed type is commonly used in pocket calculators and other small electronic devices.

Other types. Several new types of batteries are being tested. Existing types store too little energy (compared with their mass, volume, or cost) and are short-lived, making them unsuitable for such uses as powering automobiles. Two storage batteries which operate at high temperatures (several hundred degrees Celsius) are receiving the most attention: the sodium-sulfur battery and the lithium-sulfur battery.

New systems. Several new energy storage systems are being developed in hopes of finding some that are both technically and economically feasible. One method for utilities is to use excess energy capacity to compress air into large underground caverns from which it can be extracted and used in gas turbines during times of peak demands. Utilities are also considering storing energy in magnetic fields, especially with superconducting magnets. In the future, utilities and vehicles may also use rapidly rotating flywheels, which can be quickly recharged.

Fuel cell. Fuel cells convert chemical energy directly into electricity (Fig. 2). One such cell uses hydrogen and oxygen gases which are pumped into hollow, porous carbon electrodes in a solution of potassium hydroxide. At the anode, the hydrogen atoms react with OH^- ions in the electrolyte to form water, while at the cathode the oxygen atoms react with water to form OH^- ions. *See* FUEL CELL.

Solar energy. Solar energy can be stored thermally by heating water, rocks, salts, or other substances. Solar energy might even be stored all summer, for use months later, in a solar pond, which differs from an ordinary pond in that the bottom layers of the water are kept very salty; it can trap heat at the bottom, producing water that is almost boiling, because the salt concentration gradient prevents convection currents from carrying the heat to the top. *See* SOLAR ENERGY.

Hydrogen. A great deal of attention has been paid to the concept of a "hydrogen economy." Hydrogen would be about as good as an energy storage method as gaseous fossil fuel. It could be piped around the country like natural gas, liquefied for use in vehicles, or converted into electricity by fuel cells or steam-electric plants. The hydrogen could be produced from water by electrolysis or thermal decomposition. Solar and wind energy could be used to produce hydrogen for a multitude of uses, freeing the consumer from dependence on the availability of sunlight and wind. *See* HYDROGEN-FUELED TECHNOLOGY.

Outlook. The future of energy sources depends not only on what energy resources are available and how they can be used, but to an ever-increasing degree on how they can be efficiently stored. Research into energy storage is as essential as research into energy supplied or energy demands. *See the feature article* EXPLORING ENERGY CHOICES.   [LAURENT HODGES]

Bibliography: L. R. Ember, Hydrogen, a future energy mediator?, *Environ. Sci. Technol.*, 9: 102–103, 1975; J. M. Fowler, *Energy and the Environment*, pp. 310–316, 457–459, 1975; A. L. Robinson, Energy storage (I): Using electricity more efficiently and Energy storage (II): Developing advanced technology, *Science*, 184:785–787, 884–887, 1974; W. M. Senger, Silent cars in Amsterdam, *Environment*, 16(8):14–17, 1974.

Energy transmission

The movement of energy from the point at which it is generated to the point at which it is used. The term is usually restricted to the transmission of electrical energy and does not encompass the transportation of coal, oil, gas, or other fuels. In the United States, electric energy is transmitted mainly as alternating current by overhead lines, although underground cables are occasionally used. Energy transmission technology is an active field of research because of the interest in developing transmission methods that are less expensive, carry greater amounts of power, and have fewer energy losses. Methods being developed today include transmission lines insulated with compressed gas and lines that are cooled to low temperatures.

Lines for ac or dc. The ordinary overhead transmission line has several copper or aluminum wires suspended in the air by glass or porcelain insulators. Alternating-current (ac) lines are generally used because of the simplicity of ac generators, motors, and transformers. Direct-current (dc) lines are less costly per mile than ac lines but must have expensive equipment at the ends of the line to convert back and forth between dc and ac. Lower conversion costs and longer transmission lines would favor dc lines. The Bonneville Power Ad-

ministration already has an 850-mi-long (1360 km) overhead dc transmission line from Oregon to southern California. *See* ALTERNATING CURRENT; DIRECT CURRENT.

Voltage ratings. The power losses in a transmission line (or other electric circuit) are the product, I^2R, of the square of the current I and the resistance R of the line. To minimize these losses, it is desirable to have a low resistance and low current. However, the power delivered by a line is the product, IV, of the current I and the voltage V, so a line carrying a low current but high power must have a high voltage. This explains why transmission lines have high voltage. The maximum voltages have increased from 287,000 V in 1950 to 500,000 V in 1960 and 765,000 V today, with even higher-voltage lines being designed. Transformers are used to raise the voltages from the 2000 to 30,000 V at the generating station. The transmission lines end at substations where transformers reduce the voltages to 10,000–15,000 V. Distribution lines carry the electricity to homes and other buildings, where the voltage is further reduced to 110 or 220 V.

Distribution. Transmission line mileage has increased rapidly along with the growth of electric power production. In 1970 there were 300,000 mi (480,000 km) of lines of 69,000 V or more, and by 1990 there are expected to be 500,000 mi (800,000 km).

The land used for transmission line rights-of-way is large, presently about 8,000,000 acres (32,370 km^2) in the United States. A 138,000-V line requires a path 100 ft (30 m) wide (12 acres per mile of line, or 3034 m^2 per kilometer) and a 765,000-V line a path 200 ft (60 m) wide (24 acres per mile). These rights-of-way must be kept cleared of vegetation, usually with herbicides. The rights-of-way are typically obtained by purchase or through condemnation proceedings, in which the utilities exercise their right of eminent domain.

Environmental effects. Transmission lines have certain environmental effects which are usually more significant with the higher-voltage lines. The high voltage can induce alternating currents in the ground or other objects in the vicinity. An ordinary fluorescent bulb will light up near a high-voltage line without batteries or electric cords, even several hundred feet away. There is medical evidence that the alternating electromagnetic fields have significant adverse health effects, but the hazards are not fully known.

The electromagnetic fields near the transmission lines can also interfere with radio and television reception. Loud static is commonly heard on the radio while driving under a transmission line. In some cases, utilities have built special antennas to improve the reception for families living near high-voltage lines.

Corona discharge. Transmission lines also have the problem of corona discharge, a blue glow which occurs in regions of high electric fields, such as near scratches or pieces of dust on the line. This occurs more prominently during bad weather since moist air is a better conductor than dry air. Corona discharge can be decreased by using larger wires and spacing them further apart, but these steps naturally increase the cost of the line.

In addition to the glow, the corona discharge can produce crackling and humming noises, especially loud during stormy weather. The energy dissipated into the air produces ozone, which has an unpleasant odor and irritates the eyes.

Towers. The towers used for transmission lines are generally made of steel or wood. The high-voltage lines require very large towers. A 765,000-V line may have several towers per mile, each 130 ft (39 m) high and 90 ft (27 m) across.

Underground transmission. Underground lines avoid many of these disadvantages. They have cables insulated with oil and paper operating at ambient temperatures. However, having earth near the cables makes both electrical insulation and heat dissipation more severe problems, so underground lines are typically 10 to 40 times as costly per mile as overhead lines. They are also more costly to repair. With cables close together, underground lines produce high heat losses which limit their usable length to about 20 mi (3.2 km). They are practical only in urban areas where the distances involved are not too great and the land costs are usually high (which works to the disadvantage of overhead lines). Only about 1% of the transmission lines in the United States are underground.

Compressed gas cable. One new type of underground transmission line, the compressed gas cable, is already being tested. The cable is insulated by sulfur hexafluoride gas and produces much less heat. These cables are presently rather expensive, in part because they require a trench several feet wide. The main value of such a line is not the saving in energy but the ability to carry safely much larger amounts of power.

Cooled lines. The same reasoning applies to the cooled lines. Cryoresistive underground transmission lines would be cooled down to liquid nitrogen temperature (77 K, −320°F). This would lower the electrical resistance, and the heat production might be only 10% as much as in an uncooled line. However, power would be needed to operate the refrigerating system.

Superconducting cables. The advantages of cooling could be multiplied in a superconducting line. At very low temperatures (generally less than 20 K), some metals and alloys lose all their electrical resistance and become superconducting. Superconducting cables would have low power losses with ac and no losses with dc. Very little cooling would be necessary once the line was superconducting. The major advantage of such a line would not be whatever power savings resulted but the fact that a very compact cable could carry very large amounts of power. A superconducting cable 60 cm (2 ft) in diameter could carry all the power for New York City.

Successful development of some of these concepts is expected to reduce considerably the cost of underground transmission but to leave the cost several times as high as ordinary overhead lines. The continued growth of electrical power will have substantial transmission costs associated with it as long as the electricity is produced at large central power stations. Alternatives are more on-site power production (for example, by solar photovoltaic cells or wind generators) or the transport of energy is some other form (such as piped hydrogen for use in fuel cells). *See* ELECTRIC DISTRIBUTION SYSTEMS; FUEL CELL; SOLAR ENERGY; WIND POWER.

[LAURENT HODGES]

Bibliography: J. M. Fowler, *Energy and the Environment*, pp. 114–119, 1975; A. L. Hammond, W. D. Metz, and T. H. Maugh II, *Energy and the Future*, American Association for the Advancement of Science, chs. 15 and 16, 1973; L. B. Young and H. P. Young, Pollution by electrical transmission, *Bull. At. Sci.*, 30(10):34–38, 1974.

Engine

A machine designed for the conversion of energy into useful mechanical motion. The principal characteristic of an engine is its capacity to deliver appreciable mechanical power, as contrasted to a mechanism such as a clock or analog computer whose significant output is motion. By usage an engine is usually a machine that burns or otherwise consumes a fuel, as differentiated from an electric machine that produces mechanical power without altering the composition of matter. Similarly, a spring-driven mechanism is said to be powered by a spring motor; a flywheel acts as an inertia motor. By this definition a hydraulic turbine is not an engine, although it competes with the engine as a prime source of mechanical power. *See* HYDRAULIC TURBINE; MOTOR, ELECTRIC; WATERPOWER.

Applications. A fuel-burning engine may be stationary, as a donkey engine used to lift cargo between wharf and ship, or it may be mobile, like the engine in an aircraft or automobile. Such an engine may be used for both fixed service and mobile operation, although accessory modifications that adapt the engine to its particular purpose are preferable. For example, the fan that draws air through the radiator of a water-cooled fixed engine is large and fitted in a baffle, whereas the fan of a similar but mobile engine can be small and unbaffled because considerable air is driven through the radiator by means of ram action as the engine propels itself along. *See* AIRCRAFT ENGINE; AUTOMOTIVE ENGINE.

Some types of engine can be designed for economic efficiencies in fixed service but not in mobile operation. Thus, the steam engine is widely used in central electric generator stations but is obsolete in mobile service. This is chiefly because, in a large ground installation, the furnace and boiler can be fitted with means for using most of the available heat. The engine proper can be a reciprocating (piston) or a rotating (turbine) type. Because shaft rotation is by far the most used form of mechanical motion, the turbine is the more common form of modern steam engine. In the operation of railroads the steam engine has given place to diesel and gasoline internal combustion engines and to electric motors. *See* POWER PLANT; STEAM ENGINE; STEAM TURBINE; STEAM-GENERATING UNIT.

Types. Traditionally, engines are classed as external or internal combustion. External combustion engines consume their fuel or other energy source in a separate furnace or reactor. *See* REACTOR, NUCLEAR.

Strictly, the furnace or reactor releases chemical or nuclear energy into thermal energy, and the engine proper converts the heat into mechanical work. The principal means for the conversion of heat to work is a gas or vapor, termed the working fluid. By extension, an engine which derives its heat energy from the Sun by solar radiation to working fluid in a boiler can be considered an external combustion type. To avoid loss of or contamination from nuclear fuel, the reactor and boiler are separated from (and may also be shielded from) the engine.

The working fluid takes on energy in the form of heat in the boiler and gives up energy in the engine, the engine proper being a thermodynamic device. The device may be a turbine for stationary power generation on a nozzle for long-range vehicular propulsion. *See* NUCLEAR POWER.

In an engine used for propulsion, the rearward velocity with which the working fluid is ejected and, thus, the forward acceleration imparted to the vehicle depend on the temperature of the fluid. For practical purposes, temperature is limited by the engine materials that serve to contain the chemical combustion or nuclear reaction. To achieve higher exhaust velocity, the working fluid may be contained by nonmaterial means such as electric and magnetic fields, in which case the fluid must be electrically conductive. *See* ELECTROMAGNETIC PROPULSION; MAGNETOGAS DYNAMICS.

The engine proper is then a magnetohydrodynamic device receiving electric energy from a separate fuel-consuming source such as a gas turbine and electric generator or a nuclear reactor and electric generator, or possibly from direct electric conversion of nuclear or solar radiation. *See* THERMOELECTRICITY.

A further basis of classification concerns the working fluid. If the working fluid is recirculated, the engine operates on a closed cycle. If the working fluid is discharged after one pass through boiler and engine, the engine operates on an open cycle. Closed-cycle operation assures the purity of the working fluid and avoids the discharge of harmful wastes. The open cycle is simpler. Thus the commonest types of engine use atmospheric air in open cycles as the principal constituent of the working fluids and as oxidizer for the fuels.

If open-cycle operation is used, the next modification is to heat the working fluid directly by burning fuel in the fluid; the engine becomes its own furnace. Because this internal combustion type engine uses the products of combustion as part of the working fluid, the fuel must be capable of combustion under the operating conditions in the engine and must produce a noncorrosive and nonerosive working fluid. Such engines are the common reciprocating gasoline and diesel units. *See* DIESEL ENGINE; INTERNAL COMBUSTION ENGINE; STIRLING ENGINE.

At low speeds the combustion process is carried out intermittently in a cylinder to drive a reciprocating piston. At high speed, however, friction between piston and cylinder walls and between other moving parts dissipates an appreciable portion of the developed power. Thus, where high power is developed at high speed, performance is improved by continuous combustion to drive a turbine wheel. *See* BRAYTON CYCLE; CARNOT CYCLE; DIESEL CYCLE; GAS TURBINE; OTTO CYCLE.

Engine shaft rotation may be used in the same way as in a reciprocating engine. However, for high-velocity vehicular propulsion, the energy of the working fluid may be converted into thrust more directly by expulsion through a nozzle. Once the vehicle is in motion, the turbine can be omitted. Alternatively, instead of drawing atmospheric oxygen into the combustion chamber, the engine may draw both oxidizer and fuel from storage tanks which are contained within the vehicle, or the combustion chamber may contain the full supply of fuel and oxidizer. *See* RAMJET; TURBOJET; TURBOPROP.

Despite all the variation in structure, mode of operation, and working fluid—whether of moving parts, moving fields, or only moving working fluid—these machines are basically means for converting heat energy to mechanical energy. *See* THERMODYNAMIC PROCESSES.

[FRANK H. ROCKETT]
Bibliography: D. H. Marter, *Engines*, 1962.

Engineering geology

The application of education and experience in geology and other geosciences to solve geological problems posed by civil engineering structures. The branches of the geosciences most applicable are (1) surficial geology, (2) petrofabrics; (3) rock and soil mechanics; (4) geohydrology; and (5) geophysics, particularly exploration geophysics and earthquake seismology. This article discusses some of the practical aspects of engineering geology.

Geotechnics is the combination of pertinent geoscience elements with civil engineering elements to formulate the civil engineering system that has the optimal interaction with the natural environment.

Engineering properties of rock. The civil engineer and the engineering geologist consider most hard and compact natural materials of the earth crust as rock, and their derivatives, formed mostly by weathering processes, as soil. A number of useful soil classification systems exist. Because of the lack of a rock classification system suitable for civil engineering purposes, most engineering geology reports use generic classification systems modified by appropriate rock-property adjectives.

Rock sampling. The properties of a rock element

Fig. 1. Rotary or core drill on damsite investigation.

can be determined by tests on cores obtained from boreholes. These holes are made by one or a combination of the following basic types of drills: the rotary or core drill (Fig. 1), the cable-tool or churn drill, and the auger. The rotary type generally is used to obtain rock cores. The rotary rig has a motor or engine (gasoline, diesel, electric, or compressed air) that drives a drill head that rotates a drill rod (a thick-walled hollow pipe) fastened to a core barrel with a bit at its end. Downward pressure on the bit is created by hydraulic pressure in the drill head. Water or air is used to remove the rock that is comminuted (chipped or ground) by the diamonds or hard-metal alloy used to face the bit. The core barrel may be in one piece or have an inner metal tube to facilitate recovery of soft or badly broken rock ("double-tube" core barrel). The churn-type drill may be used to extend the hole through the soil overlying the rock, to chop through boulders, or occasionally to deepen a hole in rock where core is not desired. When the rock is too broken to support itself, casing (steel pipe) is driven or drilled through the broken zone. Drill rigs range in size from those mounted on the rear of large multiwheel trucks to small, portable ones that can be packed to the investigation site on a man's back or parachuted from a small plane. *See* BORING AND DRILLING (MINERAL).

The rock properties most useful to the engineering geologist are compressive and triaxial shear strengths, permeability, Young's modulus of elasticity, erodability under water action, and density (in pounds per cubic foot, or pcf).

Compressive strength. The compressive (crushing) strength of rock generally is measured in pounds per square inch (psi) or kilograms per square centimeter (kg/cm²). It is the amount of stress required to fracture a sample unconfined on the sides and loaded on the ends (Fig. 2). If the load P of 40,000 lb is applied to a sample with a diameter of 2 in. (3.14 in.²), the compressive stress is $40,000 \div 3.14 = 12,738$ psi. If this load breaks the sample, the ultimate compressive strength equals the compressive stress acting at the moment of failure, in this case 12,738 psi. The test samples generally are cylindrical rock cores that have a length-to-diameter ratio (L/D) of about 2. The wide variety of classification systems used for rock results in a wide variation in compressive strengths for rocks having the same geologic name. The table gives a statistical evaluation of the compressive strengths of several rocks commonly encountered in engineering geology.

Most laboratory tests show that an increase in moisture in rock causes a decrease in its compressive strength and elastic modulus; what is not generally known, however, is that the reverse situation shown in Fig. 3 has been encountered in certain types of volcanic rocks. In sedimentary rock the compressive strength is strongly dependent upon the quality of the cement that bonds the mineral grains together (for example, clay cement gives low strength) and upon the quantity of cement (a rock may have only a small amount of cement, and despite a strong bond between the grains, the strength is directly related to the inherent strength of the grains). Strength test results are adversely affected by microfractures that may be present in the sample prior to testing, particularly if the mi-

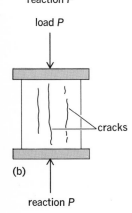

ENGINEERING GEOLOGY

load *P*

Q Q

O

cracks

S S

(a)

reaction *P*

load *P*

cracks

(b)

reaction *P*

Fig. 2. Unconfined compression test. (a) Shear failure showing failure planes *QS*. (b) Tension failure. *(From D. P. Krynine and W. R. Judd, Principles of Engineering Geology and Geotechnics, McGraw-Hill, 1957)*

Compressive strength of rocks*

Type of rock	No. of tests	Weighted mean, psi	Minimum and maximum values, psi	
Igneous				
Basalt	22	21,159	2,400	52,000
Diabase	14	31,680	6,000	46,600
Diorite	14	29,530	12,200	48,300
Granite	81	22,420	5,100	51,200
Porphyry	72	17,790	5,000	63,000
Sedimentary				
Dolomite	62	12,740	900	52,000
Limestone	211	10,760	200	37,800
Marlstone	14	15,620	8,100	28,200
Sandstone	257	9,200	300	47,600
Shale	67	9,660	100	33,500
Siltstone	14	15,740	500	45,800
Metamorphic				
Amphibolite	11	47,550	22,300	61,700
Gneiss	23	25,470	5,200	44,200
Greenstone	11	28,930	16,600	45,500
Marble	31	14,590	4,400	39,700
Quartzite	25	42,400	3,700	91,200
Schist	16	7,290	1,000	23,500
Mineral				
Hematite	18	35,050	16,200	99,900
Iron Ore	18	4,933	2,300	7,800

*Based on data used by W. R. Judd, *Correlating rock properties by statistical methods*, Purdue Research Foundation Report, 1969.

crofractures are oriented parallel to the potential failure planes.

The value of compressive strength to be used in an engineering design must be related to the direction of the structure's load and the orientation of the bedding in the foundation rock. This relationship is important because the highest compressive strength usually is obtained when the compressive stress is normal to the bedding. Conversely, the highest Young's modulus of elasticity (E) usually results when the compressive stress parallels the bedding.

When these strength and elastic properties apparently are not affected by the direction of the applied load, the rock is described as isotropic; if load applied parallel to the bedding provides physical property data that are significantly different than those obtained when the load is applied normal (perpendicular) to the bedding, the rock is anisotropic or aeolotropic. If the physical components of the rock element or rock system have equal dimensions and equal fabric relationships, the rock is homogeneous; significant variance in these relationships results in a heterogeneous rock. Most rocks encountered in foundations are anisotropic and heterogeneous.

Shear in rocks. Shearing stresses tend to separate portions of the rock (or soil) mass. Faults and folds are examples of shear failures in nature. In engineering structures, every compression is accompanied by shear stresses. For example, an arch dam compresses the abutment rock and, if the latter is intersected by fissures or weak zones, it may fail in shear with a resulting tensile stress in the dam concrete that may rupture the concrete. The application of loads over long periods of time on most rocks will cause them to creep or even to flow like a dense fluid (plastic flow).

Ambient stress. This type of stress in a rock system is actually potential energy, probably created by ancient natural forces, recent seismic activity, or nearby man-made disturbances. Ambient (residual, stored, or primary) stress may remain in rock long after the disturbance is removed. An excavation, such as a tunnel or quarry, will relieve the ambient stress by providing room for displacement of the rock, and thus the potential energy is converted to kinetic energy. In tunnels and quarries, the release of this energy can cause spalling, the slow outward separation of rock slabs from the rock massif; when this movement is rapid or explosive, a rock burst occurs. The latter is a different phenomenon from a rock bump, which is a rapid upward movement of a large portion of a rock system and, in a tunnel, can have sufficient force to flatten a steel mine car against the roof or break the legs of a man standing on the floor when the bump occurs. *See* MINING, UNDERGROUND.

One of two fundamental principles generally is used to predict the possible rock load on a tunnel roof, steel or timber supports, or a concrete or steel lining: (1) The weight of the burden (the rock and soil mass between the roof and the ground surface) and its shear strength control the load, and therefore the resultant stresses are depth-dependent, or (2) the shear strength of the rock system and the ambient stresses control the stress distribution, so the resultant loading is only indirectly dependent upon depth. The excavation process can cause rapid redistribution of these stresses to produce high loads upon supports some distance from the newly excavated face in the tunnel. Lined tunnels can be designed so that the reinforced concrete or steel lining will have to carry only a portion of the ambient or burden stresses.

Construction material. Rock as a construction material is used in the form of dimension, crushed, or broken stone. Broken stone is placed as riprap on slopes of earth dams, canals, and riverbanks to protect them against water action. Also, it is used as the core and armor stone for breakwater structures. For all such uses, the stone should have high density (± 165 pcf), be insoluble in water, and be relatively nonporous to resist cavitation. Dimension stone (granite, limestone, sandstone, and some basalts) is quarried and sawed into blocks of a shape and size suitable for facing buildings or for interior decorative panels. For exterior use dimension stone preferably should be isotropic (in physical properties), have a low coefficient of expansion when subjected to temperature changes, and be resistant to deleterious chemicals in the atmosphere (such as sulfuric acid). Crushed stone (primarily limestone but also some basalt, granite, sandstone, and quartzite) is used as aggregate in concrete and in bituminous surfaces for highways, as a base course or embankment material for highways, and for railroad ballast (to support the ties). When used in highway construction, the crushed stone should be resistant to abrasion as fine stone dust reduces the permeability of the stone layer; the roadway then is more susceptible to settling and heaving caused by freezing and thawing of water in the embankment. Concrete aggregate must be free of deleterious material such as opal and chalcedony; volcanic rocks containing glass, devitrified glass, and tridymite; quartz with microfractures; phyllites containing hydromica (illites); and other rocks containing free silica (SiO_2). These materials will react chemically with the cement in concrete and release sodium and potassium oxides (alkalies) or silica gels. Preliminary petrographic analyses of the aggregate and chemical analysis of the cement can indicate the possibility of alkali

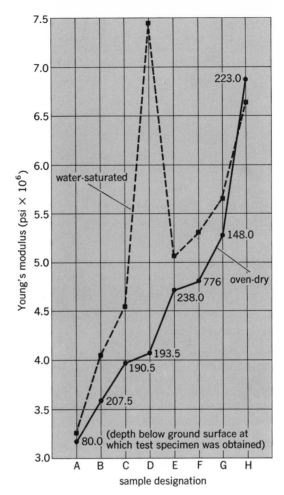

Fig. 3. Increase in Young's modulus caused by saturation of dacite porphyry. (*After J. R. Ege and R. B. Johnson, Consolidated tables of physical properties of rock samples from area 401, Nevada Test Site, U. S. Geol. Surv. Tech. Letter Pluto-21, 1962*)

reactions and thus prevent construction difficulties such as expansion, cracking, or a strength decrease of the concrete.

Geotechnical significance of soils. Glacial and alluvial deposits contain heterogeneous mixtures of pervious (sand and gravel) and impervious (clay, silt, and rock flour) soil materials (Fig. 4). The pervious materials can be used for highway subgrade, concrete aggregate, and filters and pervious zones in earth embankments. Dam reservoirs may be endangered by the presence of stratified or lenticular bodies of pervious materials or ancient buried river channels filled with pervious material. Deep alluvial deposits in or close to river deltas may contain very soft materials such as organic silt or mud. An unsuitable soil that has been found in dam foundations is open-work gravel. This material may have a good bearing strength because of a natural cement bond between grains, but it is highly pervious because of the almost complete lack of fine soil to fill the voids between the gravel pebbles.

Concrete or earth dams can be built safely on sand foundations if the latter receive special treatment. One requirement is to minimize seepage

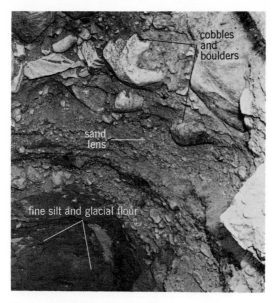

Fig. 4. Glacial deposit in test pit.

losses by the construction of cutoff walls (of concrete, compacted clay, or interlocking-steel-sheet piling) or by use of mixed-in-place piles 3 ft or more in diameter. The latter are constructed by augering to the required depth but not removing any of the sand. At the desired depth, cement grout is pumped through the hollow stem of the auger, which is slowly withdrawn while still rotating; this mixes the grout and the sand into a relatively impervious concrete pile. The cutoff is created by overlapping these augered holes. Some sand foundations may incur excessive consolidation when loaded and then saturated, particularly if there is a vibratory load from heavy machinery or high-velocity water in a spillway. This problem is minimized prior to loading by using a vibrating probe inserted into the sand or vibratory rollers on the sand surface or by removing the sand and then replacing it under vibratory compaction and water sluicing. *See* DAM.

Aeolian (windblown) deposits. Loess is a rela-

Fig. 5. Prewetting loess foundation for earth dam. (*U.S. Bureau of Reclamation*)

tively low-density soil composed primarily of silt grains cemented by clay or calcium carbonate. It has a vertical permeability considerably greater than the horizontal. When a loaded loess deposit is wetted, it rapidly consolidates, and the overlying structure settles. When permanent open excavations ("cuts") are required for highways or canals through loess, the sides of the cut should be as near vertical as possible: Sloping cuts in loesses will rapidly erode and slide because of the high vertical permeability. To avoid undesirable settlement of earth embankments, the loess is "prewetted" prior to construction by building ponds on the foundation surface (Fig. 5). Permanently dry loess is a relatively strong bearing material. Aeolian sand deposits present the problem of stabilization for the continually moving sand. This can be done by planting such vegetation as heather or young pine or by treating it with crude oil. Cuts are traps for moving sand and should be avoided.

Organic deposits. Excessive settlement will occur in structures founded on muskeg terrain. Embankments can be stabilized by good drainage, the avoidance of cuts, and the removal of the organic soil and replacement by sand and gravel or, when removal is uneconomical, displacement of it by the continuous dumping of embankment material upon it. Structures imposing concentrated loads are supported by piling driven through the soft layers into layers with sufficient bearing power.

Residual soils. These soils are derived from the in-place deterioration of the underlying bedrock. The granular material caused by the in-place disintegration of granite generally is sufficiently thin to cause only nominal problems. However, there are regions (such as California, Australia, and Brazil) where the disintegrated granite (locally termed "DG") may be hundreds of feet thick; although it may be competent to support moderate loads, it is unstable in open excavations and is pervious. A thickness of about 200 ft of DG and weathered gneiss on the sides of a narrow canyon was a major cause for construction of the Tumut-1 Power Plant (New South Wales) in hard rock some 1200 ft underground. Laterite (a red clayey soil) derived from the in-place disintegration of limestone in tropical to semitropical climates is another critical residual soil. It is unstable in open cuts on moderately steep slopes, compressible under load, and when wet produces a slick surface that is unsatisfactory for vehicular traffic. This soil frequently is encountered in the southeastern United States and southeastern Asia, including Indonesia and Vietnam.

Clays supporting structures may consolidate slowly over a long period of time and cause structural damage. When clay containing montmorillonite is constantly dried and rewetted by climatic or drainage processes, it alternately contracts and expands. During the drying cycle, extensive networks of fissures are formed that facilitate the rapid introduction of water during a rainfall. This cyclic volume change of the clay can produce uplift forces on structures placed upon the clay or compressive and uplift forces on walls of structures placed within the clay. These forces have been known to rupture concrete walls containing 3/4-in.-diameter steel reinforcement bars. A thixotropic or "quick" clay has a unique lattice structure that causes the clay to become fluid when subjected to vibratory forces. Various techniques are used to improve the foundation characteristics of critical types of clay: (1) electroosmosis that uses electricity to force redistribution of water molecules and subsequent hardening of the clay around the anodes inserted in the foundation; (2) provision of adequate space beneath a foundation slab or beam so the clay can expand upward and not lift the structure; (3) belling, or increasing in size, of the diameter of the lower end of concrete piling so the pile will withstand uplift forces imposed by clay layers around the upper part of the pile; (4) treatment of the pile surface with a frictionless coating (such as Teflon or a loose wrapping of asphalt-impregnated paper) so the upward-moving clay cannot adhere to the pile; (5) sufficient drainage around the structure to prevent moisture from contacting the clay; and (6) replacement of the clay by a satisfactory foundation material. Where none of these solutions are feasible, the structure then must be relocated to a satisfactory site or designed so it can withstand uplift or compressive forces without expensive damage.

Silt may settle rapidly under a load or offer a "quick" condition when saturated. For supporting some structures (such as residences), the bearing capacity of silts and fine sands can be improved by intermixing them with certain chemicals that will cause the mixture to "set" or harden when exposed to air or moisture; some of the chemicals used are sodium silicate with the later application of calcium chloride, bituminous compounds, phenolic resins, or special cements (to form "soil cement"). The last mixture has been used for surfacing secondary roads, for jungle runways in Vietnam, and as a substitute for riprap of earth dams. Some types of silt foundations can be improved by pumping into them soil-cement or clay mixtures under sufficient pressures to create large bulbs of compacted silt around the pumped area.

Geotechnical investigation. For engineering projects these investigations may include preliminary studies, preconstruction or design investigations, consultation during construction, and the maintenance of the completed structure.

Preliminary studies. These are made to select the best location for a project and to aid in formulating the preliminary designs for the structures. The first step in the study is a search for pertinent published material in libraries, state and Federal agencies, private companies, and university theses. Regional, and occasionally detailed reports on local geology, including geologic maps, are available in publications of the U.S. Geological Survey; topographic maps are available from that agency and from the U.S. Army Map Service. Oil companies occasionally will release the geologic logs of any drilling they may have done in a project area. Sources of geologic information are listed in the *Directory of Geological Material in North America* by J. A. Howell and A. I. Levorsen (NAS-NRC Publ. 556, 1957). Air photos should be used to supplement map information (or may be the only surficial information readily available). The U.S. Geological Survey maintains a current index map of the air-photo coverage of the United States. The photos are available from that agency, the U.S. Forest Service, the Soil Conservation Service, and

commercial air-photo companies; for some projects the military agencies will provide air-photo coverage. The topographic maps and air photos can be used to study rock outcrop and drainage patterns, landforms, geologic structures, the nature of soil and vegetation, moisture conditions, and land use by man (cultural features).

Field reconnaissance may include the collection of rock and soil specimens; inspection of road cuts and other excavations; inspection of the condition of nearby engineering structures such as bridges, pavements, and buildings; and location of sources of construction material. Aerial reconnaissance is essential at this stage and can be performed best in helicopters and second-best in slow-flying small planes.

Preconstruction. Surface and subsurface investigations are required prior to design and construction. Surface studies include the preparation of a detailed map of surficial geology, hydrologic features, and well-defined landforms. For dam projects, a small-scale geologic map (for example, 1 in. = 200 or 500 ft) is made of the reservoir area and any adjacent areas that may be directly influenced by the project; in addition, a large-scale geologic map (for example, 1 in. = 50 ft) is required of the specific sites of the main structures (the dam, spillway, power plant, tunnels, and so on). These maps can be compiled by a combination of field survey methods and aerial mapping procedures. They should have a grid system (coordinates) and show the proposed locations for subsurface investigations.

Subsurface investigations are required to confirm and amplify the surficial geologic data. These may include test pits, trenches, short tunnels (drifts or adits), and the drilling of vertical, horizontal, or oblique (angle) boreholes (Fig. 6). Geologic data obtained by these direct methods can be supplemented by indirect or interpreted data obtained by geophysical methods on the surface or in subsurface holes and by installation of special instruments to measure strain or deformation in a borehole or tunnel.

The geology disclosed by subsurface investigations is "logged" on appropriate forms. Tunnel logs display visual measurements of fractures and joint orientations (strike and dip); rock names and a description of their estimated engineering properties; alteration, layering, and other geologic de-

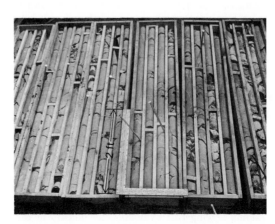

Fig. 6. Drill core obtained in dam foundation.

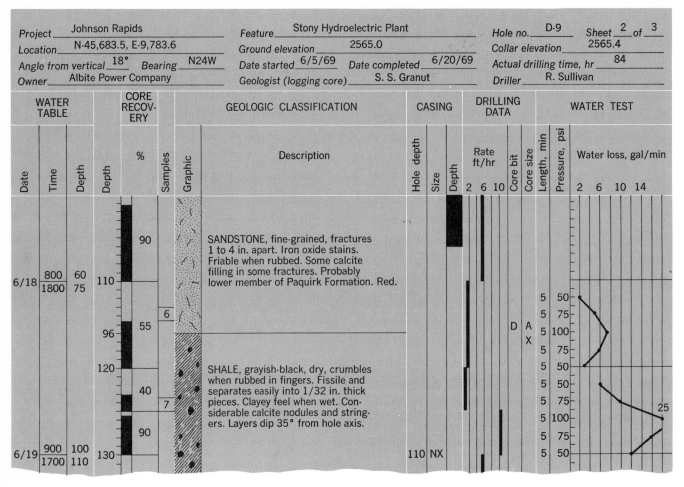

WATER TABLE			CORE RECOVERY			GEOLOGIC CLASSIFICATION		CASING			DRILLING DATA					WATER TEST	

Project Johnson Rapids
Location N-45,683.5, E-9,783.6
Angle from vertical 18° **Bearing** N24W
Owner Albite Power Company

Feature Stony Hydroelectric Plant
Ground elevation 2565.0
Date started 6/5/69 **Date completed** 6/20/69
Geologist (logging core) S. S. Granut

Hole no. D-9 **Sheet** 2 **of** 3
Collar elevation 2565.4
Actual drilling time, hr 84
Driller R. Sullivan

Date	Time	Depth	Depth	%	Samples	Graphic	Description	Hole depth	Size	Depth	Rate ft/hr 2 6 10	Core bit	Core size	Length, min	Pressure, psi	Water loss, gal/min 2 6 10 14
				90			SANDSTONE, fine-grained, fractures 1 to 4 in. apart. Iron oxide stains. Friable when rubbed. Some calcite filling in some fractures. Probably lower member of Paquirk Formation. Red.									
6/18	800 1800	60 75	110													
			96	55	6							D	A X	5 5 5 5	50 75 100 75	
			120				SHALE, grayish-black, dry, crumbles when rubbed in fingers. Fissile and separates easily into 1/32 in. thick pieces. Clayey feel when wet. Considerable calcite nodules and stringers. Layers dip 35° from hole axis.							5 5	50 50	
				40	7									5 5	75 100	25
				90										5	75	
6/19	900 1700	100 110	130					110	NX					5	50	

Fig. 7. Log recording of geological information from a borehole.

fects; the location and amount of water or gas inflow; the size and shape of blocks caused by fracturing or jointing and the width of separation or the filling material between blocks; and the irregularities in the shape of the tunnel caused by the displacement of blocks during or after excavation (rock falls, rock bursts, chimneying, and overbreakage). Geophysical seismic methods may be used to define the thickness of loosened rock around the tunnel; geoacoustical techniques that detect increases in microseismic noise during tunneling may be used to determine if the excavation is causing excessive loosening in the tunnel rock. This detection of "subaudible rock noise" occasionally is used to detect the potential movement of rock slopes in open excavations.

The borehole data can be logged on a form such as shown in Fig. 7. These data can be obtained by direct examination of the core, by visual inspection of the interior of the borehole using a borehole camera (a specially made television camera) or a stratoscope (a periscopelike device), or by geophysical techniques. Direct viewing of the interior of the hole is the only positive method of determining the in-place orientation and characteristics of separations and of layering in the rock system. The geophysical techniques include use of gamma-gamma logging that evaluates the density of the rock surrounding the borehole or at depths as great

as 150 ft beneath the gamma probe; neutron logging to determine the moisture content of the rock system by measuring the depth of penetration of the neutrons; traversing the borehole with a sonic logger that, by calibration, measures differences in the velocity of wave propagation in different strata (and thus can determine in place Young's modulus of elasticity and the thickness of each stratum encountered by the borehole); and electric logging that uses differences in the electrical resistivity of different strata to define their porosity, moisture content, and thickness.

Occasionally a hole is drilled through a talus deposit containing the same type of rock as the underlying rock in place (bedrock). Because of the similarity in rock types, the talus-bedrock contact sometimes is best identified by determining the orientation of the remnant magnetism in the core: The magnetic lines in the core will have a regular orientation, but the talus magnetism will have random directions. This method is useful only in rocks that contain appreciable remnant magnetism such as some basalts.

Geophysical seismic or electrical resistivity methods also can be used on the ground surface to define the approximate depth of bedrock or various rock layers. The results require verification by occasional boreholes, but this is an inexpensive and satisfactory technique for planning and design investigations. The seismic methods are not useful when it is necessary to locate soft strata (wherein the seismic waves travel at relatively low velocity) that are overlain by hard strata (that have higher wave velocity); the latter conceal and block the signal from the soft strata. Also, difficulties may occur when the strata to be located are overlain by soil containing numerous large boulders composed of rock having higher velocities than the surrounding soil, or when the soil is very compact (such as glacial till) because its velocity characteristics may resemble those of the underlying bedrock. Another problem is that the seismic method seldom can identify narrow and steep declivities in the underlying hard rock (because of improper reflection of the waves).

Construction. Geotechnical supervision is desirable during construction in or on earth media. The engineering geologist must give advice and keep a record of all geotechnical difficulties encountered during the construction and of all geological features disclosed by excavations. During the operation and maintenance of a completed project, the services of the engineering geologist often are required to determine causes and assist in the preparation of corrective measures for cracks in linings of water tunnels, excessive settlement of structures, undesirable seepage in the foundations of dams, slides in canal and other open excavations, overturning of steel transmission-line towers owing to a foundation failure, and rock falls onto hydroelectric power plants at the base of steep canyon walls (Fig. 8).

Legal aspects. An important consideration for the engineering geologist is the possibility of a contractor making legal claims for damages, purportedly because of unforeseen geologic conditions he encounters during construction. Legal support for such claims can be diminished if the engineering geologist supplies accurate and detailed geologic information in the specifications and drawings used for bidding purposes. These documents should not contain assumptions about the geological conditions (for a proposed structure), but they should show all tangible geologic data obtained during the investigation for the project: for example, an accurate log of all boreholes and drifts and a drawing showing the boundaries of the outcrops of all geological formations in the project area. The engineering geologist should have sufficient experience with design and construction procedures to formulate an investigation program that results in a minimum of subsequent uncertainties by a contractor. Numerous uncertainties about geologic conditions can result not only in increased claims but also may cause a contractor to submit a higher bid (to minimize his risks) than if detailed geologic information were available to him.

Special geotechnical problems. In arctic zones, structures built on permafrosted soils may be heaved or may cause thawing and subsequent disastrous settlement (Fig. 9). The growth of permafrost upward into earth dams seriously affects their stability and permeability characteristics. Obtaining natural construction materials in permafrosted areas requires thawing of the borrow area to permit efficient excavation; once excavated, the material must be protected against refreezing prior to placement in the structures. Perma-

Fig. 8. Kortes Dam, Wyo. Arrows show diverter walls protecting power plant against rock falls.

frost in rock seldom will cause foundation difficulties. In planning reservoirs, it is essential to evaluate their watertightness, particularly in areas containing carbonate or sulfate rock formations or lava flows. These formations frequently contain extensive systems of caverns and channels that may or may not be filled with claylike material or water. Sedimentation studies are required for the design of efficient harbors or reservoirs because soil carried by the moving water will settle and block or fill these structures. In areas with known earthquake activity, aseismic (earthquakeproof) design requires knowledge of the intensity and magnitude of earthquake forces. The prevention and rehabilitation of slides (landslides) in steep natural slopes and in excavations are important considerations in many construction projects and are particularly important in planning reservoirs, as was disastrously proved by the Vaiont Dam catastrophe in 1963.

Geohydrologic problems. In the foundation material under a structure, water can occur in the form of pore water locked into the interstices or pores of the soil or rock, as free water that is moving through openings in the earth media, or as included water that is a constituent or chemically bound part of the soil or rock. When the structure

Fig. 9. Door-frame distortion and floor settlement which occurred as a result of permafrost thaw and heave. (*U.S. Geological Survey*)

load compresses the foundation material, the resulting compressive forces on the pore water can produce undesirable uplift pressures on the base of the structure. Free water is indicative of the permeability of the foundation material and possible excessive water loss (from a reservoir, canal, or tunnel); uplift on the structure because of an increase in hydrostatic head (caused by a reservoir or the like); or piping, which is the removal of particles of the natural material by flowing water with a consequent unfilled opening that weakens the foundation and increases seepage losses.

The possibility of excessive seepage or piping can be learned by appropriate tests during the boring program. For example, water pressure can be placed on each 5-ft section of a borehole, after the core is removed, and any resulting water loss can be measured. The water pressure is maintained within the 5-ft section by placing an expandable rubber ring (packer) around the drill pipe at the top of the test section and then sealing off the section by using mechanical or hydraulic pressure on the pipe to force expansion of the packer. When only one packer is used, because it is desired to test only the section of hole beneath it, it is a "single-packer" test. In a double-packer test a segment of hole is isolated for pressure testing by placing packers at the top and the bottom of the test section. The best information on the permeability characteristics of the rock can be obtained by the use of three or more increments of increasing and then decreasing water pressure for each tested length of hole. The permeability K of the rock in feet per minute is $(Q\ln_e L/r)/2\pi LH$, where Q is the rate of flow loss in gallons per minute (gpm), r is the radius of the test section, L is the length of hole being tested and should be $\geq 10r$, and H (in feet) is

the height of the water column between the ground surface and the center of the test section plus 2.3 times the psi pump pressure. If the water loss continues to increase when the pressure is decreased, piping of the rock or filling material in fractures may be occurring or fractures are widening or forming. The water-pressure test can be supplemented by a groutability test in the same borehole. This test is performed in the same way as the water test except, instead of water, a mixture of cement, sand, and water (cement grout) or a phenolic resin (chemical grout) is pumped under pressure into the test section. The resulting information is used to design cutoff walls and grout curtains for dams. The pressures used in water-pressure or grouting tests should not exceed the pressure exerted by the weight of the burden between the ground surface and the top of the test section. Excessive test pressure can cause uplift in the rock, and the resulting test data will be misleading.

Included or pore water generally is determined by laboratory tests on cores; these are shipped from the borehole to the laboratory in relatively impervious containers that resist loss of moisture from the core. The cores with their natural moisture content are weighed when received and then dried in a vacuum oven at about 110°F until their dry weight stabilizes. The percentage of pore water (by dry weight) is (wet weight − dry weight) × 100 ÷ dry weight. Temperatures up to 200°F can be used for more rapid drying, provided the dried specimens are not to be used for strength or elastic property determinations. (High temperatures can significantly affect the strength because the heat apparently causes internal stresses that disturb the rock fabric or change the chemical composition of the rock by evaporation of the included moisture.)

Protective construction. Civilian and military structures may be designed to minimize the effects of nuclear explosions. The most effective protection is to place the facility in a hardened underground excavation. A hardened facility, including the excavation and its contents, is able to withstand the effects generated by a specified size of nuclear weapon. The degree of hardening commonly is indicated by a numerical "psi" value; for example, a facility having 100-psi hardness is designed to maintain operations despite nearby nuclear explosions that generate 100-psi air-blast overpressure on the surface directly above the structure. However, this type of designation does not necessarily indicate the capability of the facility to withstand all direct-induced effects from the nuclear blast, that is, the action on the structure that is induced by shock waves propagating through the earth media surrounding the structure. Direct-induced effects include the amount of acceleration, particle velocity, and displacement that can occur in earth media and the structure.

Desirable depths and configurations for hardened facilities are highly dependent upon the shock-wave propagation characteristics of the surrounding earth media. These characteristics are influenced by the type of rock (Fig. 10), discontinuities in the rock system, free water, and geologic structure. Therefore, prior to the design and construction of such facilities, extensive geotechnical field and laboratory tests are performed. In addi-

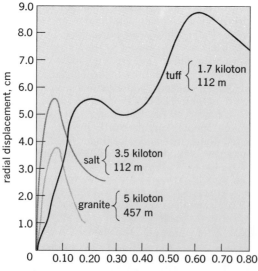

Fig. 10. Effect of rock type on the propagation of nuclear shock waves. The measuring point was 457 m from explosion source; all displacement measurements are extrapolated to the movement of the particle this distance from the source of a 5-kiloton explosion. (*Data from G. C. Werth and R. C. Herbst, Comparison of amplitudes of seismic waves from nuclear explosions in four media, University of California Lawrence Radiation Laboratory Report, UCRL-6962, June, 1962*)

tion to conventional physical properties, the laboratory tests also are directed toward determination of the appropriate Hugoniot curve for the media; this is done by exposing small elements of the rock to high pressures generated in a shock tube. The field tests may include small-scale nuclear explosions or chemical explosions designed to produce shock effects that simulate the nuclear explosion, and measurement of the shock effects by special instruments installed in the earth environment surrounding the blast point ("ground zero") and in any structures that are to be tested. Accurate interpretation of the explosion effects and adequate planning for the explosion require an accurate geologic map of the surface and of the underground environment that will be affected by the explosion. The map should show the precise location and orientation of all geologic defects that would influence the wave path, such as joints, fractures, and layers of alternately hard and soft rock. The importance of discontinuities was illustrated by the Rainier test (a 1.7-kiloton nuclear burst in tuff, 899 ft underground, in 1957); the shock wave caused a displacement of several inches along a fault that intersected the tunnel used for access to the shot point. *See* EXPLOSION AND EXPLOSIVE.

Application of nuclear energy. The use of nuclear energy for the efficient construction of civil engineering projects is under investigation in the Plowshare Program. Examples are the proposed use of nuclear explosions for the rapid excavation of a new route for the Panama Canal; the Gas Buggy nuclear shot in 1968 to test the possibility of increasing the production of natural gas by opening fractures in the reservoir rock; and a proposed underground nuclear explosion that might expedite the production of low-grade copper ore by causing extensive fracturing and possible concentration of the ore. All such uses of nuclear explosives must be preceded by extensive geotechnical investigations of more than usual accuracy: Sufficient geologic detail must be obtained to minimize the possible venting of the underground explosion; otherwise, unpredicted continuous fractures or joints would allow the radiation products from the blast to be released into the atmosphere in the vicinity of the shot. Geotechnical conditions also will influence the size and shape of the cavity that will be caused by the explosion (Fig. 11).

Waste disposal. Another geotechnical nuclear problem occurs in the use of nuclear energy to generate power and to produce radioisotopes: safe disposal of the radioactive waste products. These products can be mixed with concrete and buried in the ground, but geohydrologic conditions or the chemical nature of the earth environment must not be allowed to cause excessive deterioration of the concrete. Dangerous waste products from nuclear reactors or from the production of agents for chemical or bacterial warfare also may be disposed of in deep wells, but then it is necessary to determine if the groundwater is likely to transport the waste products to areas where the same water may be used by man. Another potential problem from this method became known when disposal of chemical agents in a deep well in the Denver, Colo., area between 1962 and 1965 apparently disturbed the ambient stress regime sufficiently to trigger a succession of small earthquakes. Dangerous wastes

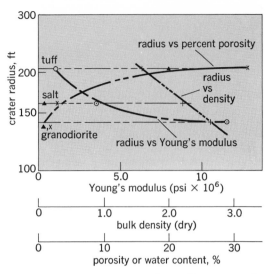

Fig. 11. Effect of rock properties and type on cavity radii for a 100-kiloton nuclear explosion buried 2000 ft. Solid triangles represent percentage moisture; curve not plotted. (*After C. R. Boardman et al., Characteristic effects of contained nuclear explosions for evaluation of mining applications, University of California Lawrence Radiation Laboratory Report, UCRL-7350, 1963*)

also are disposed of by sealing them in large concrete blocks that are dropped into the deep parts of the ocean. *See* NUCLEAR POWER; REACTOR, NUCLEAR. [WILLIAM R. JUDD]

Bibliography: E. C. Eckel (ed.), *Landslides and Engineering Practice*, National Academy of Sciences, Highway Research Board Special Report 29, 1958 (also HRB Reports 216, 1959, and 236, 1960); T. Fluhr and R. F. Legget (eds.), *Reviews in Engineering Geology*, vol. 1, Geological Society of America, 1962; Geological Society of America, *Engineering Geology Case Histories*, vols. 1–7, 1957–1968; N. V. Glazov and A. N. Glazov, *New Instruments and Methods of Engineering Geology*, 1959; M. E. Harr, *Groundwater and Seepage*, 1962; Idaho State University Department of Geology, *Proceedings of the 5th Annual Engineering Geology and Soil Engineering Symposium*, 1967; W. R. Judd, Geological factors in choosing underground sites, in J. J. O'Sullivan (ed.) *Protective Construction in a Nuclear Age*, 2 vols., 1961; D. P. Krynine and W. R. Judd, *Principles of Engineering Geology and Geotechnics*, 1957; U. Langefors and B. Kihlström, *Rock Blasting*, 1963; R. F. Legget, *Geology and Engineering*, 2d ed., 1962; D. S. Parasins, *Mining Geophysics*, 1966; U.S. Coast and Geodetic Survey, *Earthquake Investigations in the U.S.*, Special Publication No. 282, 1965; U.S. Department of the Army Office of the Chief of Research and Development, *Scientific and Technical Applications Forecast—1964—Excavation*, AD-607077 and AD-611555 (bibliography), 1963; R. C. S. Walters, *Dam Geology*, 1962.

Enthalpy

For any system, that is, the part of space under discussion, enthalpy is the sum of the internal energy of the system plus the system's volume multiplied by the pressure exerted on the system by its surroundings. This may be expressed as $U + PV = H$, where U is the system's internal energy, P the pressure of the surroundings on the system, V the system's volume, and H the enthalpy of the system. The sum of $U + PV$ is given the special symbol H primarily as a matter of convenience because this sum appears repeatedly in thermodynamic discussion. Consistent units must, of course, be used in expressing the terms in the above equation. Although each of the terms U and PV has the units of energy, enthalpy is not the energy of the system, but rather the sum of these two terms. Previously, enthalpy was referred to as total heat or heat content, but these terms are misleading and should be avoided. Enthalpy is, from the viewpoint of mathematics, a point function, as contrasted with heat and work, which are path functions. Point functions are usually much more readily handled in thermodynamic analysis than are path functions because the former depend only on the initial and final states of the system undergoing a change; they are independent of the paths or character of the change. Mathematically, the differential of a point function is a complete or perfect differential.

Because the absolute value of internal energy of even a simple system is usually unknown, recorded values of enthalpy are relative values measured above some convenient but arbitrarily chosen datum. Thus in the steam tables of Keenan and Keyes, the datum is liquid water at 32°F and under its own vapor pressure. At this state water is assumed to have an enthalpy equal to zero. Under this assumption the internal energy of water in this state is a negative quantity equal to PV. No complication is introduced by this fact, although visualization of negative energies of this kind may be disturbing to some. There is limited utility for absolute enthalpies because usually much changes in enthalpy are of most interest. It is instructive to examine the utility of the enthalpy function in terms of some simple but important thermodynamic processes.

The first law of thermodynamics is merely a statement of the law of conservation of energy. The first law alone indicates that:

1. For a chemical reaction carried out at constant pressure and temperature and in the absence of changes in kinetic, potential, and similar energies, and with no work performed except that resulting from keeping the pressure constant as the volume changes, the change in enthalpy of the system (the material taking part in the chemical reaction) is numerically equal to the heat that must be transferred to maintain the above-mentioned conditions. This heat is often loosely referred to as the heat of reaction. More properly, it is the enthalpy change for the reaction.

2. So-called heat balances on heat exchangers, furnaces, and similar industrial equipment that operate under steady flow conditions are really enthalpy balances.

3. The work developed in a steadily running adiabatic engine or turbine is equivalent to the enthalpy change of the steam or other fluid passing through the engine.

4. The adiabatic, irreversible, steady flow of a stream of materials through a porous plug or a partially opened valve under circumstances where difference in kinetic energy between entering and leaving streams is negligible (a Joule-Thomson process) results in no change in enthalpy of the flowing stream. Although no change in enthalpy results from this process, there is a loss in the energy available for doing work as a result of the Joule-Thomson flow. For change in enthalpy with pressure or temperature see THERMODYNAMIC PRINCIPLES. See also ENTROPY; THERMODYNAMIC PROCESSES.   [HAROLD C. WEBER]

Entropy

A function first introduced in classical thermodynamics to provide a quantitative basis for the common observation that naturally occurring processes have a particular direction. Subsequently, in statistical thermodynamics, entropy was shown to be a measure of the number of microstates a system could assume. Finally, in communication theory, entropy is a measure of information. Each of these aspects will be considered in turn. Before the entropy function is introduced, it is necessary to discuss reversible processes.

Reversible processes. Any system under constant external conditions is observed to change in such a way as to approach a particularly simple final state called an equilibrium state. For example, two bodies initially at different temperatures are connected by a metal wire. Heat flows from the hot to the cold body until the temperatures of both bodies are the same. As another example, a vessel containing a gas is connected through a stopcock to an evacuated vessel. When the stopcock is opened, the gas expands to fill the whole of the available space uniformly. It is common experience that the reverse processes never occur if the systems are left to themselves; that is, heat is never observed to flow from the cold to the hot body, nor will the gas compress itself into one of the vessels. Max Planck classified all elementary processes into three categories: natural, unnatural, and reversible.

Natural processes do occur, and proceed in a direction toward equilibrium. Unnatural processes move away from equilibrium and never occur. If A → B is a natural process between states A and B, then B → A is an unnatural process. A reversible process is an idealized natural process that passes through a continuous sequence of equilibrium states. Consider the evaporation of a liquid in the presence of its vapor at a pressure P. Let the equilibrium vapor pressure of the liquid be p. If $P < p$, liquid evaporates as a natural process. If $P > p$, evaporation is an unnatural process and will not occur; indeed, the opposite process — condensation — will take place. Finally, if $P = p$, both processes of condensation and evaporation are reversible and can be initiated by a very slight increase or decrease in the external pressure P.

A useful idea is that a reversible process may be exactly reversed by an infinitesimal change in the external conditions. If a hot object is placed adjacent to a much colder object, the heat-flow direction cannot be reversed by small changes in the temperature of either object. In reversible processes, work is accomplished through small pressure differences, and heat transfer occurs through small temperature differences.

Entropy function. The state function entropy S puts the foregoing discussion on a quantitative basis. The function is not derived in this article; but, rather, some of its properties are stated, and its implications are discussed mainly by example. Entropy is related to q, the heat flowing into the system from its surroundings and to T, the absolute temperature of the system. The important properties for this discussion are:

1. $dS > q/T$ for a natural change.

$dS = q/T$ for a reversible change.

It is necessary to introduce both S and T together. A formal derivation would show T^{-1} as an integrating factor leading to the complete differential dS.

2. The entropy of the system S is made up of the sum of all the parts of the system so that $S = S_1 + S_2 + S_3 \cdots$. See HEAT; TEMPERATURE; THERMODYNAMIC PRINCIPLES.

Heat flow. Consider two bodies, α and β, at different temperatures separated by an adiabatic (no heat transfer) wall. If the two bodies are connected by a fine wire that allows a small heat flow q from α to β, then $dS_\alpha = -q/T_\alpha$ and $dS_\beta = q/T_\beta$.

For the whole system, Eq. (1) holds. If $T_{-\alpha} > T_{-\beta}$,

$$dS = dS_\alpha + dS_\beta = q\left(\frac{1}{T_\beta} - \frac{1}{T_\alpha}\right) \qquad (1)$$

$dS > 0$, and heat flows from α to β as a natural process. The process could be continued until $T_{-\alpha} = T_{-\beta}$ and $dS = 0$.

Once the constraint of the adiabatic wall is abrogated, the entropy increases to a maximum value, and T_α becomes equal to T_β. This is a special case of the most important notion in thermodynamics; that is, the system will assume that equilibrium state which maximizes the entropy at constant energy, consistent with the constraints.

Nonconservation of entropy. In his study of the first law of thermodynamics, J. P. Joule caused work to be expended by rubbing metal blocks together in a large mass of water. By this and similar experiments, he established numerical relationships between heat and work. When the experiment was completed, the apparatus remained unchanged except for a slight increase in the water temperature. Work had been converted into heat with 100% efficiency. Provided the process was carried out slowly, the temperature difference between the blocks and the water would be small, and heat transfer could be considered a reversible process. The entropy increase of the water at its temperature T is $\Delta S = (Q/T) = (W/T)$.

Since everything but the water is unchanged, this equation also represents the total entropy increase. The entropy has been created from the work input, and this process could be continued indefinitely, creating more and more entropy. Unlike energy, entropy is not conserved. See CONSERVATION OF ENERGY.

Although the heat transfer is considered to be reversible in order to calculate the entropy increase, the overall process of converting work into heat is irreversible. The frictional process that converts kinetic energy into the heat of the metal blocks is a natural process. In fact, the impossibility of the reverse process is Lord Kelvin's statement of the second law of thermodynamics. Heat cannot be completely converted into work without other changes occurring in the surroundings. For example, a gas in a cylinder can be expanded reversibly by extracting heat from a large constant-temperature bath. All of the heat extracted from the bath is converted into work, but eventually the pressure of the gas system would be reduced to an unusable level. The system has changed, and the process cannot continue indefinitely. If one tries to convert heat into work through a system undergoing a cycle so that the system will return to its initial state, one finds that only a portion of the heat input does work and that the remainder must be rejected to a lower temperature; this is just the process which takes place in a heat engine. See THERMODYNAMIC CYCLE; THERMODYNAMIC PROCESSES.

Degradation of energy. Energy is never destroyed. But in the Joule friction experiment and in heat transfer between bodies, as in any natural process, something is lost. In the Joule experiment, the energy expended in work now resides in the water bath. But if this energy is reused, less useful work is obtained than was originally put in. The original energy input has been degraded to a less useful form. The energy transferred from a high-temperature body to a lower-temperature body is also in a less useful form. If another system is used to restore this degraded energy to its original form, it is found that the restoring system has degraded the energy even more than the original system had. Thus, every process occurring in the world results in an overall increase in entropy and a corresponding degradation in energy. R. Clausius stated the first two laws of thermodynamics as: "The energy of the world is constant. The entropy of the world tends toward a maximum."

Increasing entropy and mixing. Once the atomic theory of matter is accepted, the entropy concept can be made much clearer. It is then found through statistical thermodynamics that the increase of entropy toward its maximum value at equilibrium corresponds to the change of the system toward its most probable state consistent with the constraints. The most probable state represents the most mixed or most random state. Mixing must be given a broad interpretation which includes particle or configurational mixing and spreading of energy over the particles or thermal mixing. Diffusion of one gas into another represents obvious configurational mixing and increased entropy. Irreversible expansion of a gas represents configurational mixing of the molecules over the available space. Heat flow represents spreading of the kinetic energy between the particles. Friction spreads the kinetic energy of the body over the constituent particles. Sometimes the energy-spread entropy increase and the configurational entropy increase are not compatible, and a compromise is struck. A subcooled liquid adiabatically crystallizes to a lower configurational entropy but gains even more entropy through the additional energy levels made available. The same sort of behavior occurs in partially miscible liquids — some configurational entropy is sacrificed in order to gain a large amount of energy-spread entropy.

Absolute entropy. The third law of thermodynamics (Nernst's heat theorem) refers to the vanishing of entropy at zero temperature. In 1912 Planck proposed that the theorem applied to pure crystalline solids. However, the theorem is now known to be applicable to gases and, by all reasonable expectation, is applicable to any system. Thus, any substance at finite temperatures has an absolute entropy, the value of which can be determined from either calorimetric or spectroscopic data. Absolute entropies, together with thermochemical data, are very useful in the calculation of equilibrium compositions of reaction systems.

The statistical viewpoint is that a thermodynamic state at finite temperatures corresponds to many microstates. During an observation the microstates of a system undergo continuous rapid transitions. Since entropy is proportional to the logarithm of the number of available microstates, the Nernst theorem implies that the thermodynamic state at zero temperature corresponds to a single microstate. Thus, at zero temperature, even a ferromagnetic material should exist in a single state, fully magnetized in a direction determined by its inevitable interactions with the environment.   [WILLIAM F. JAEP]

Measure of information. The probability characteristic of entropy leads to its use in communication theory as a measure of information. The absence of information about a situation is equivalent to an uncertainty associated with the nature of the situation. This uncertainty, designated H, is the entropy of the information about the particular situation, Eq. (2), where $p_1, p_2, \ldots, p_n$ are the proba-

$$H(p_1, p_2, \ldots, p_n) = -\sum_{k=1}^{n} p_k \log p_k \qquad (2)$$

bilities of mutually exclusive events, the logarithms are taken to an arbitrary but fixed base, and $p_k \log p_k$ always equals zero if $p_k = 0$. For example, if $p_1 = 1$ and all others ps are zero, the situation is completely predictable beforehand; there is no uncertainty and so the entropy is zero. In all other cases the entropy is positive.

In introducing entropy of an information space, C. E. Shannon described a source of information by its entropy H in bits per symbol. The ratio of the entropy of a source to the maximum rate of signaling that it could achieve with the same symbols is its relative entropy. One minus relative entropy is the redundancy of the source.
                    [FRANK H. ROCKETT]

Bibliography: J. G. Aston and J. J. Fritz, *Thermodynamics and Statistical Thermodynamics*, 1959; H. B. Callen, *Thermodynamics*, 1960; K. G. Denbigh, *Principles of Chemical Equilibrium*, 2d ed., 1968; J. D. Fast, *Entropy: The Significance of the Concept of Entropy and Its Applications in Science and Technology*, 1962; E. A. Guggenheim, *Thermodynamics*, 1967; A. I. Khinchin, *Mathematical Foundations of Information Theory*, 1957; C. E. Shannon, A mathematical theory of communication, *Bell Syst. Tech. J.*, 27(3):379–423, 27(4):623–656, 1948; R. C. Tolman, *Statistical Mechanics, with Application to Physics and Chemistry*, Amer. Chem. Soc. Monogr., 1927.

Environmental engineering

The discipline which evaluates the effects of humans on the environment and develops controls to minimize environmental degradation.

In the 1960s the United States became acutely aware of the deterioration of its air, water, and land. The roots of the problem lie in the rapid growth of the national population and the industrial development of natural resources which has given Americans the highest standard of living in the world. Since 1964 there has been enactment of national, state, and local legislation directed toward the preservation of these resources.

The technology which provided society with all the necessities and luxuries of life is now expected to continue providing these services without degradation to the environment. How well this goal is attained will depend in great measure on the environmental engineers, who must cope with the enormous challenges presented by society.

It is the feeling of industry and governmental agencies that the ultimate goals should be the design of processes and systems which need minimal treatment for pollution control and the ultimate recycling of all wastes for reuse. This philosophy is both logical and necessary in a society that is rapidly depleting its nonrenewable natural resources.

Governmental policy. State and local governments have the prime responsibility for the adoption, administration, and enforcement of meaningful air, water, and land use standards. Legislation has been enacted by the state and Federal governments.

Federal legislation has included the Wilderness Act of 1964; the Clean Air and Water Act of 1965; the Clean Water Restoration Act of 1966; the Air Quality Act of 1967; the National Environmental Policy Act of 1969; Executive Order No. 11574 of December 1970 restating the Refuse Act of the Rivers and Harbors Act of 1899; the Clean Air Act Amendments of 1970; the Water Pollution Control Act Amendments, the Noise Control Act, and the Coastal Zone Management Act, all of 1972.

Environmental legislation enacted from 1970 to 1975 has had an unmistakable impact upon the community. Standards and limitations governing the release of pollutants to the environment have placed severe restrictions on industrial operations. The sweeping Federal legislation program for environmental protection has also led to the establishment of complementary legislation, regulations, and requirements at the state level. All states now have extensive programs for environmental control and protection. The confrontation between government and industry as a result of the legislation must be eliminated and replaced by a cooperative effort toward developing the technology necessary for environmental improvement in the best interests of all concerned.

The National Environmental Policy Act is an example of the all-inclusiveness of government regulation. This act makes it mandatory that an in-depth study of environmental impacts be made in connection with any new industrial or government activity that may involve the Federal government, directly or indirectly.

These requirements extend to such a wide range of activities as:

Construction of electric power plants and transmission lines, gas pipelines, railroads, highways, bridges, nuclear facilities, and airports.

Any industrial facilty releasing effluents to navigable waters and their tributaries.

Rights-of-way, drilling permits, mining leases, and other uses of Federal lands.

Applications to the Interstate Commerce Commission for the approval of transportation rate schedules.

In December 1970, by presidential order, the Environmental Protection Agency (EPA) was formed. Under Public Law 91-604, which extended the Clean Air and Air Quality acts, this agency now embodies under one administrator the responsibility for setting standards and a compliance timetable for air and water qualities improvement. This agency will also administer grants to state and local governments for construction of wastewater treatment facilities and for air- and water-pollution control.

Industrial policy. Through its interdisciplinary environmental teams, industry is directing large amounts of capital and technological resources both to define and resolve environmental challenges. The solution of the myriad complex environmental problems requires the skills and experience of persons knowledgeable in health, sanitation, physics, chemistry, biology, meteorology, engineering, and many other fields.

Each air and water problem has its own unique approach and solution. Restrictive standards necessitate high retention efficiencies for all control equipment. Off-the-shelf items, which were applicable in the past, no longer suffice. Controls must now be specifically tailored to each installation. Liquid wastes can generally be treated by chemical or physical means, or by a combination of the two, for removal of contaminants with the expectation that the liquid can be recycled. Air or gaseous contaminants can be removed by scrubbing, filtration, absorption, or adsorption and the clean gas discharged into the atmosphere. The removed contaminants, either dry or in solution, must be handled wisely, or a new water- or air-pollution problem may result.

Industries that extract natural resources from the earth, and in so doing disturb the surface, are being called on to reclaim and restore the land to a condition and contour that is equal to or better than the original state. Thirty-seven states now have reclamation laws; most require an approved restoration plan and bonding to assure that restoration is accomplished. Federal legislation, directly related to land disturbed by mining is being actively considered.

Air quality management. The air contaminants which pervade the environment are many and emanate from multiple sources. A sizable portion of these contaminants are produced by nature, as witnessed by dust that is carried by high winds across desert areas, pollens and spores from vegetation, and gases such as sulfur dioxide and hydrogen sulfide from volcanic activity and the biological destruction of vegetation and animal matter.

The greatest burden of atmospheric pollutants resulting from human activity are carbon monoxide, hydrocarbons, nitrogen oxides, and particulates. Public and private transportation using the internal combustion engine is a major source of these contaminants. It has been conservatively estimated that 60% of all major pollutants in the United States come from the internal combustion engine; restrictive guidelines by EPA to control these emissions have been extended to the auto industry. A catalytic converter is designed for installation on automobile exhausts. New cars include such controls, and not as an option. A question remains as to whether or not this specific control might increase the atmospheric burden of particulate sulfates while reducing emissions of other contaminants.

Industry contributes approximately 17% of the total air pollutants, with utility power plants following closely with 14%. The major pollutants from these sources are sulfur dioxide, particulates, hydrocarbons, and nitrogen oxides.

All states have ambient air and emission standards directed primarily toward the control of industrial and utility power plant pollution sources. The general trend in gaseous and particulate control is to limit the emissions from a process stack to a specified weight per hour based on the total material weight processed to assure compliance with ambient air regulation. Process weights become extremely large in steel and cement plants and in large nonferrous smelters. The degree of control necessary in such plants can approach 100% of all particulate matter in the stack. Retention equipment can become massive both in physical size and in cost. The equipment may include high-energy venturi scrubbers, fabric arresters, and electrostatic precipitators. Each application must be evaluated so that the selected equipment will provide the retention efficiency desired. *See* POWER PLANT.

Sulfur oxide retention and control present the greatest challenges to industrial environmental engineers. Ambient air standards are extremely low, and the emission standards calculated to meet these limits are beyond the state of control art (as of 1975) for weak sulfur oxide gas stream. The states of Arizona, Montana, Nevada, and Washington originally adopted standards requiring nonferrous smelters to limit sulfur emission to 10% of the total sulfur present in the concentrate treated. The above states, along with others having nonferrous smelters, have modified their emission standard to ensure that the Federal ambient sulfur dioxide standards are attained. In some instances the standards are attainable with less than 90% removal of sulfur. Most copper smelters and all utility power plants have large-volume, weak sulfur oxide effluent gas streams. To scrub these large, weak gas volumes with limestone slurries or caustics is extremely expensive, requires prohibitively large equipment, and can present water and solid waste problems of enormous magnitude. However, gas streams containing high concentrations of sulfur dioxide gas can be treated more economically to obtain such by-products as elemental sulfur, liquid sulfur dioxide, sulfuric acid, calcium sulfate, and ammonium sulfate. *See* ATMOSPHERIC POLLUTION.

The task of upgrading weak smelter gas streams to produce products which have no existing market has led to extensive research into other methods of producing copper. A number of mining companies are now piloting hydrometallurgical methods to produce electrolytic-grade copper from ores by chemical means, thus eliminating the smelting step. Liquid ion exchange, followed by electrowinning, is also being used more extensively for the heap leaching of low-grade copper. This method produces a very pure grade of copper without the emission of sulfur dioxide to the atmosphere.

Water quality management. The Federal effort in water-pollution control has been assumed by the EPA and its office of water quality under the National Pollutant Discharge Elimination System (NPDES). The EPA is proceeding with discharge permits for wastes or treated effluents with ultimate management of the system to become the responsibility of the states. The essential function of NPDES is compliance with effluent standards now being promulgated for municipal and industrial waste dischargers. Many states already have a permit program in effect; however, permit conditions and compliance schedules are still subject to Federal approval. As can be expected, EPA guidelines call for universal standards and, under this philosophy, criteria can only be arbitrary and, in some cases, punitive without epidemiological investigation of receiving-water biota. Some mining industry environmental engineers can be expected to take issue with the philosophy of wasting a portion of the assimilative capacity of receiving waters by not utilizing part of this natural phenomenon and thus producing finished products at a lower cost to the consumer.

The state of the art of treatment of wastewaters containing metals has barely advanced beyond neutralization and chemical precipitation. As water quality standards continue to become more stringent, especially concerning dissolved solids, it is apparent that chemical treatment of metals wastes will be unacceptable. Chemical treatment in many instances substitutes a different molecular species for another with no reduction in total dissolved solids.

Some segments of the mining industry are considering physical treatment methods of desalination techniques such as evaporation, reverse osmosis, and electrodialysis. These techniques require vast quantities of energy and are prohibitive in both capital and operational costs. Continued experimentation will undoubtedly improve the benefit-cost ratio.

For the most part the mining industry tends toward a policy of complete recycle of water. Some metallurgical processes are amenable to reuse of process waters with only minimal treatment, such as removal of suspended and settleable solids. Other processes require higher-quality water with lower dissolved-solids content. Even this generally requires removal of only a portion of the solids from a waste stream in order to maintain an acceptable process water quality. Many of these solids can also be reclaimed, thus recovering values now lost. This approach is not only logical but economical. If receiving-water quality criteria are more stringent than process-water treatment requirements, it is impractical to comply with those standards and then waste the water to the nearest surface drainage. In the more arid parts of the world, recycle of water has become a practice by necessity; in water-rich areas it will become practice by governmental decree.

The Federal Clean Water Act requires best practicable treatment technology by 1977. Best available technology (BAT) is a statute requirement by 1983. No pollutant discharge (ND) is a goal of the act by 1985, essentially prohibiting discharge of any material at any concentration. Hopefully, in the case of both BAT and ND, definition and application of this requirement and goal can be tempered by demonstrated need and economic practicality. As stated before, a significant portion of the nation's energy could be needlessly wasted, especially in efforts to attain ND. *See the feature article* PROTECTING THE ENVIRONMENT; *see also* WATER POLLUTION.

Land reclamation. It has been stated that strategic mineral development is the most productive use of land because of the great values that are received by the nation from such small land areas. Open-pit and strip mining have recently been criticized for their impact on the surrounding ecosystems. Some preservationists and conservationists feel that the Earth's surface should not be disturbed by either exploration or mining activities. Mining companies must reverse the spoiler image that has been created in the past, especially in the large-scale stripping of coal in the eastern and mid-central states. Most states now have regulations that require that all stripped land be reclaimed. Strip mining of coal has often produced an attendant acid mine drainage problem. *See* MINING, OPEN-PIT; MINING, STRIP.

Ecosystem studies are now being conducted by mining companies during exploration and prior to the commencement of mining operations. These studies lead to the effective planning for the most desirable method of mining that will least disturb the environment and yet lend itself to later reclamation.

Much controversey erupted with the presidential veto of the Federal Strip Mining Restoration Act in 1975; however, 37 states had by that time enacted land reclamation laws. Most of the remaining states are considering such regulations, and undoubtedly all states will soon regulate the mining and restoration of land. There is much legislative reclamation activity in the western states where large amounts of low-sulfur bituminous and subbitumious coals have become important in a time of increasing energy demands.

Mining companies, too, have taken the initiative in planning operations so as to limit the adverse impact on the environment. Notable examples of this forward-looking and concerned approach are programs carried on by American Cyanamid Company, American Metal Climax, Inc., Anamax Company (formerly Anaconda), Bethlehem Steel Company, and Peabody Coal Company.

Typical of these industrial programs is that of Anamax which undertook the development of the Twin Buttes Copper Mine in the desert area near Tucson, AZ. In excess of 240,000,000 tons of alluvium were removed from this open-pit site before production could begin. This material and future waste will be used to form large dikes that will impound tailings. The dikes are terraced and planted with vegetation indigenous to the area and, when completed, will be 1000 ft wide (1 ft = 0.3 m) at the base and 250 ft wide at the apex, with a maximum height of 230 ft. These dikes take on the appearance of mesas and blend into the desert landscape.

Moreover, American Cyanamid has found that restoration of Florida's phosphate-mined lands can best be accomplished by reclaiming the major portions of those areas simultaneously during mining operations.

Land reclamation is made an integral part of mine planning by advanced consideration and decisions regarding what the area should look like upon completion of mining activity. By systematically forming eventual lakes and distributing stripped wastes into previously mined cuts, grading the area and restoring it to desirable land become relatively easy. Many of these reclaimed areas have become useful as parks, recreational areas, wildlife sanctuaries, and agricultural and residential development sites.

Again, if a reasonable and empirical approach to mined-land reclamation is allowed, following the dictates of physical, chemical, and biological laws, the very small area of the Earth's surface disturbed by mining can be restored to beneficial use at minimal cost to the consumer and optimum conservation of energy.

[LEWIS N. BLAIR; JOHN C. SPINDLER, WALTER H. UNGER]

Bibliography: *Air Pollution Engineering Manual*, Public Health Serv. Publ. no. 999-AP-40, 1967; J. E. McKee and H. W. Wolf (eds.), *1963 State of California Water Quality Criteria*, California Water Quality Board and U.S. Public Health Service, 2d ed., 1963; *SME Mining Engineering Handbook*, vol. 1, sect. 8: Surface and Facility Requirements, Pollution and Environment, Society of Mining Engineers of the American Institute of Mining, Metallurical and Petroleum Engineers, 1973; *Symposium Proceedings on the Rehabilitation of Drastically Disturbed Surface Mined Lands*, Macon, GA, Nov. 4–5, 1971.

Explosion and explosive

An explosion is a violent and noisy outburst or sometimes a peaceful burning that would be better described as flame or combustion. This discussion, however, is limited to events associated with the buildup and release of pressure. In this sense the term includes not only transient occurrences, such as the bursting of a shell or the burning of powder in a gun barrel, but also continuous, steady-state effects such as the expulsion of gases from a rocket nozzle.

Explosion is usually associated with rapid chemical reaction. The reaction must evolve heat, and it also must produce some gaseous products. The explosive may be a single compound, such as trinitrotoluene (TNT), or a mixture, such as gunpowder (potassium nitrate, charcoal, and sulfur). All explosives give simple end products, such as CO_2, CO, H_2O, and N_2; these are the most stable atomic combinations at ambient temperatures. Oxygen-deficient explosives also may yield uncombined hydrogen and carbon. Certain transitory molecular fragments, or free radicals, such as OH, H, and O, usually are present in the high heat of the explosive reaction but combine to form stable molecules when the gases cool.

A distinction usually can be made between the chemical events in the explosion and the external events that result from it. Different phenomena may cause similar, and even indistinguishable, external effects. Thus the bursting of a large, pressurized vessel, the explosion of a gas mixture in a building, or the detonation of a high-explosive bomb may cause virtually identical effects at a distance. A remote observer could not know the true nature of the explosion. He might even refer to all three events as detonations; but, in the scientific sense, only the bomb produces a detonation.

Blast wave. When pressurized gas suddenly is released in a fluid medium (air or water), a pressure pulse travels outward. The speed is closely related to the sound velocity, about 1050 ft/sec in air and 5000 ft/sec in water. With high explosives, the pressure buildup is so fast that it creates a steep-fronted pulse called a shock wave, or, in air, a blast wave. The shock-wave velocity depends on the strength, that is, on the excess pressure at the wavefront; it is always higher than the speed of sound but approaches the acoustic level as the shock strength falls to zero (Fig. 1).

In air the relationship between excess pressure at the wavefront p_s and wave velocity D is given approximately by the formula shown below, where

$$\frac{p_s}{p_0} = \frac{7}{6}\left(\frac{D}{c_0}\right)^2 - \frac{1}{6}$$

p_0 is the atmospheric pressure and c_0 is the sound speed. In water, because it is much less compressible, the shock pressure is 100–1000 times higher than the air blast pressure from a similar explosive charge at a comparable distance. The shock-wave velocity in water, however, deviates only slightly from sound speed.

Much of the explosion energy is concentrated in the blast wave. Part is in the form of compressional work, represented in mathematical terms by the quantity $\int p\,dv$. Some of the energy is heat, produced when the gas passes through the shock front; the rest is kinetic energy associated with the wind behind the shock. A large fraction of the blast-wave energy is transferred to successive layers of the air as the shock front advances and, because the frontal area increases by the square of the distance, the energy is spread ever more thinly in space. For this reason, pressure, temperature, and wind velocity drop off sharply with distance from the explosion, and the damaging power diminishes accordingly.

The characteristics of shock waves are studied by electronic and optical devices. A pressure-sensing instrument for air blast first records the sharp pressure rise of the shock front, followed by a region of decaying pressure in the so-called positive phase of the blast. When air enters the shock front, it is given a strong forward motion and thus is greatly densified. This compression largely accounts for pressure rise; therefore the wind velocity and the shock pressure are closely related. The positive pressure region behind the shock front lasts for only a few thousandths of a second in the case of high explosive bombs, but it remains for about 1 sec in a nuclear blast. Behind it there follows a suction phase, in which the pressure dips below atmospheric level and the wind reverses direction. The seeming paradox of windows being blown outward by an explosive blast is due to this suction phase.

Underwater explosions create similar effects. However, the pressures are much higher, the water movement is slower, and the suction phase is unimportant. Additionally, there is a second com-

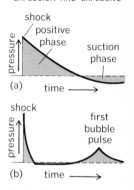

Fig. 1. Pressure versus time curve for (*a*) shock waves in air and (*b*) shock waves in water.

pression wave of lower pressure and longer duration caused by a contraction and second expansion of the explosion gas bubble; in fact, several weaker bubble pulses also are observed.

Blast damage. When an air-blast wave strikes a vertical wall, the air is brought to rest and piles up at the wall. Therefore, the instantaneous pressure on the wall is higher than the incident shock pressure. For acoustic waves of vanishing amplitude, the shock pressure is just doubled on impact, but for very strong shocks, the excess pressure in the reflected wave may be as much as eight times the incident pressure.

Under the force of an impacting blast wave, the wall begins to move. If it is massive and relatively flexible, that is, if the natural period of the wall is relatively long, the pressure may drop to zero before serious damage has been done. On the other hand, light and brittle members, such as windowpanes, may crack almost immediately. The destructive effect therefore depends on several factors, having to do both with the structure and with the blast wave.

For high-explosive bombs and industrial explosions, the damage-causing potential of the blast (aside from window breakage) depends on the total positive impulse, that is, on the area under the positive phase of the pressure-time curve. For nuclear explosions, the positive pressure phase lasts much longer than the natural period of most structural members. The damage caused by nuclear blast therefore depends only on the pressure level at the shock front. A similar criterion applies to window breakage from any but the smallest explosive sources.

Similarity. The laws of fluid motion impose a similarity restraint on the manner in which the shock wave variables, such as pressure and impulse, change with distance from the explosion. The similarity principle states that when the size of the charge is increased the scale of the shock-wave phenomenon increases in proportion. Thus, for a twofold increase in charge diameter, a given shock-wave pressure will be produced at just twice the distance from the charge. The time scale increases in a similar way. Because diameter is proportional to the cube root of charge weight w, the scaling factor is just $w^{1/3}$. The scaling law is highly significant because it means that if careful measurements are made for one size of charge the blast effects from any other charge of that explosive can be predicted (Fig. 2).

Strictly speaking, the similarity law operates only for charges of the same explosive composition and of similar shape, but it can be applied with considerable accuracy even to widely different explosives without shape restrictions. The starting point for comparison is the energy equivalence, the common basis being the equivalent weight of TNT. For example, the bomb dropped on Hiroshima was ascribed an energy equivalence of 20,000 tons of TNT. With this single figure, the Los Alamos scientists made reliable estimates of nuclear blast characteristics from measurements for TNT charges 1,000,000 times less powerful. The scale factor in this case was just 100.

Windowpanes commonly break under a blast pressure of about 1 psi. According to the similarity curve in Fig. 2, such a pressure level occurs at about 40 ft from the explosion of 1 lb of TNT and at about 14,000 ft, or 2.7 mi, from the explosion of a 20-kiloton nuclear bomb.

Energy release rate. Not all explosions produce a shock wave. For an unconfined gaseous explosion, the pressure rise is small and the energy is dissipated in a low-amplitude acoustic disturbance. It

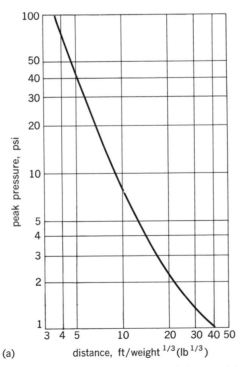

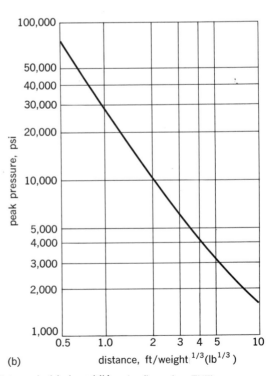

Fig. 2. Peak pressure as a function of charge weight and distance in (*a*) air and (*b*) water (based on TNT).

is a question of both reaction rate and scale; even a slow phenomenon on a sufficiently large scale can have all the destructiveness of a true detonation. For example, when the eruption of Krakatoa in 1883 spilled huge quantities of molten lava into the sea, the resulting steam caused a blast wave that shook the island of Batavia 96 mi away. A good measure of explosive violence might be the rate of total energy production divided by the surface area of the reacting medium.

Confinement also can be important. A burning gas mixture in the open air causes only a faintly audible whoosh, even when the flash of flame is quite spectacular. If the mixture is inside a large gasoline storage tank, however, the pressure builds up. Then, when the tank bursts and the hot gases erupt, a violent blast wave results. Here the final rate of energy release to the atmosphere is high.

Deflagration and detonation. The familiar burning mode is called deflagration or flame. Whether it produces an explosion depends on whether it is confined and causes the pressure to rise. Thus, when gasoline burns in an automobile cylinder, it is said to explode, whereas in a blowtorch it is said simply to burn. There is, however, little basic difference between the two phenomena. *See* COMBUSTION.

An important factor of deflagration is the burning rate. For the purpose of definition, the flame can be considered to be standing still, with the fuel mixture streaming into it and the burned products flowing away to the rear. The streaming velocity of the fuel mixture is called the burning velocity. It is characteristic, in the main, of mixture composition, but it also depends on temperature and pressure.

Burning velocities for gases ordinarily range from a few centimeters per second to about 1 m/sec. The flame advances by heat conduction and diffusion: Fast-moving and chemically active molecules, often free radicals such as H and OH, diffuse from the burning zone into the unburned layer and start reaction. The layer-to-layer speed depends on the molecular velocities in the flame. Light, high-speed molecules or molecular fragments such as hydrogen atoms favor a high flame speed, as does a high reaction temperature. The chemical reaction rate also affects the temperature gradients and controls the heat flow.

The advance of a flame in an open-ended glass tube filled with a combustible mixture can be easily followed with the eye. Ordinarily it makes little noise. Under certain circumstances, however, a new phenomenon occurs. Depending on the kind of gas, the dimensions of the tube, and other factors, a quiet, slow-moving flame may suddenly develop into an ear-shattering explosion that pulverizes the glass in a most violent fashion. This transformation can take place even when the ends of the tube are open to allow free expansion for the burning gases. The new effect is called detonation.

Detonation wave. To the eye and ear, a detonation is instantaneous, but with special instruments its speed can be measured. By comparison with a flame, the speed is incredible, being 1000 or more times as fast. For example, an ethylene-air mixture burns with a velocity of 63 cm/sec, and the detonation wave travels at a speed of 1734 m/sec.

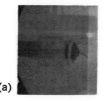

(a)   **(b)**

Fig. 3. Microsecond flash photographs of a detonating stick of solid explosive: (*a*) radiograph (x ray); (*b*) photograph. The radiograph shows the densification of the material immediately behind the detonation front and the rarefaction that sets in later as the gases expand. The outer cone of light in the photograph is the aerial shock wave, which is luminous at these very high pressures. (*From M. A. Cook, The Science of High Explosives, Amer. Chem. Soc. Monogr. no. 139, Reinhold, 1958*)

The enormous difference in speed signifies a basic difference in mode of propagation. The flame advances essentially by layer-to-layer heat conduction; the detonation is carried forward on the crest of a shock wave. Deflagration is a diffusion phenomenon similar to the spreading of an odor in still air; detonation is a wave phenomenon similar to the propagation of sound.

Detonation develops from flame by a complex series of events that are not always precisely repeated in similar experiments. The transition is associated with tubes and is not known to occur in free space. In a tube the expansion of the burned gas behind the flame can take place only by pushing away the gas in front (and also behind, if the tube is open at the rear; buildup to detonation is faster, however, in a tube closed at the rear). The gas flow caused by the expansion tends to become turbulent, and this speeds the burning rate. Small shock waves shoot out ahead of the flame, possibly because pockets of unburned gas are trapped in the turbulent wake. The final stage of transition to detonation is astonishingly sudden. The details of the process are complicated and little understood. In contrast, the final, fully developed detonation has a simplicity that sets it apart from other combustion effects (Fig. 3).

The impressive feature of detonation is its constant velocity. Once established, the speed does not change by more than 1 or 2 parts per thousand unless the mixture composition is nonuniform or the tube has irregularities. Changes in initial pressure and temperature have little effect. The detonation velocity is a characteristic constant for a combustible gas not unlike the velocity of sound for a noncombustible gas.

Detonation theory. To derive the detonation conditions for an explosive mixture, the wavefront can be considered to be standing still. Consider two planes, one in the unburned gas in front of the wave and the other in the burned gas at the rear (Fig. 4). Matter under conditions p_1, v_1, E_1, and ve-

Fig. 4. Physical model of detonation front.

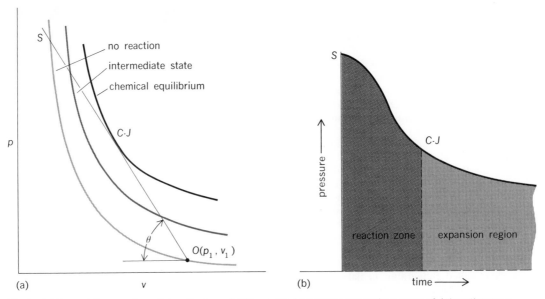

Fig. 5. (a) Hugoniot curves for a detonating gas. (b) Theoretical pressure versus time curve of detonation wave.

locity u_1 is streaming toward plane 1; and matter under conditions p_2, v_2, E_2, and velocity u_2 is streaming away from plane 2. The symbols p, v, and E represent pressure, specific volume, and specific energy, respectively. Three equations can express that (1) the mass flow rates, into 1 and out of 2, are equal (conservation of mass); (2) the difference between the total energy, internal and kinetic, flowing in at 1 and that flowing out at 2, is equal to the net work done by the gas (conservation of energy); and (3) the rate of momentum change between planes 1 and 2 is equal to the pressure difference (conservation of momentum). In addition, there is an equation of state that describes the particular properties of the detonating substance; this is a functional relationship, often not expressible as an explicit equation, between the final values p_2, v_2, and E_2 and the initial values p_1, v_1, and E_1. This equation involves the gas law and also reflects some assumption about the chemical state of the burned gas at 2. In all, since the initial gas conditions p_1, v_1, and E_1 are known, there are four equations with five unknowns: u_1, p_2, v_2, E_2, and u_2. Hence there is one degree of freedom, and an arbitrary choice of one of the unknowns will fix all five.

The solution to these four equations is usually represented by a functional relationship between p_2 and v_2, the pressure-volume curve for the burned gas. This is called the Rankine-Hugoniot curve or, simply, the Hugoniot.

Detonation wave velocity. The derivation outlined above is quite general. It applies equally to gases, liquids, and solids, not only to those that react chemically but also to those that do not so react. The Hugoniot is, in fact, the general equation for a shock wave. Its form depends on the equation of state and on an implicit assumption about the chemical state of the medium that emerges in plane 2. If there is no chemical reaction, the Hugoniot passes through the initial point p_1, v_1; if there is, the curve lies above it.

The final detonation state lies on the "equilib-

rium" Hugoniot in Fig. 5a. Any point on this curve could represent the final detonation conditions and meet the four physical requirements invoked to derive the Hugoniot. Thus a whole spectrum of detonation states seems to be possible, but only one state is actually observed. To fix the detonation state, a fifth condition must be found. This is the so-called Chapman-Jouguet hypothesis. The detonation state, according to this principle, is uniquely represented by the point C-J on the Hugoniot in Fig. 5a. This is the point of tangency of the chord drawn from $O(p_1, v_1)$. Although the choice of the C-J point was originally an ad hoc hypothesis, the C-J state can be shown to be the only one on the Hugoniot for which the wave is steady.

According to the derivation, the detonation velocity D is equal to $v_1\sqrt{\tan \theta}$. The C-J velocity can be shown to be equal to the mass flow rate plus the acoustic velocity in the burned gas. The stable detonation therefore has the same speed relative to fixed coordinates as a sound wave in the burned gas. Actually, transitory detonation states are possible, and are observed, that correspond to other points on the Hugoniot, but the C-J detonation is the only one that is stable.

Since, in computing the Hugoniot, any given chemical state can be postulated at plane 2, it is possible to erect a series of Hugoniots for all chemical states from no reaction to final equilibrium. A further extension of the theory shows that intermediate conditions in the chemical reaction zone lie on the straight line between S and C-J in Fig. 5a. The point S is the condition at the shock front (no reaction).

Detonation mechanism. In this emerging picture of the detonation wave, it appears that there is a shock front S followed by a region of decreasing pressure in which the reaction occurs. The reacting system comes to chemical equilibrium at C-J. The pressure here is about one-half that at S. The pressure contour of a detonation wave is depicted in Fig. 5b. In the reaction region, from S to C-J, the contour has a permanent form, but in the expansion region, beyond *C-J*, the contour changes and the pressure gradients become less steep with time.

The true nature of detonation is now revealed. The wave is fronted by a shock that quickly heats the medium to high temperature. A rapid chemical reaction follows. The heat released in the reaction maintains the shock front at constant strength and velocity. Detonation therefore can be described as a type of self-propelled shock wave.

The difference between detonation and deflagration does not lie in the chemistry, because the reaction for a given gas mixture may follow essentially the same kinetic mechanism in both cases; the difference, instead, lies in the manner in which successive layers of material are brought up to the high reaction temperature. A deflagration will transform into a detonation if conditions are favorable for the buildup of a shock wave. A high degree of lateral confinement, as in tubes, and a rapid chemical reaction are the main factors favoring such a changeover, but the geometry of the container and the sound velocity in the medium are also important.

Detonation calculations. The Chapman-Jouguet equations provide a framework, but many theoretical building blocks are needed to complete the structure. In order to apply the equations, thermal data up to 6000°C are needed, and these are quite inaccessible by direct experiment. However, with the aid of statistical mechanics, such data can be computed from spectral measurements, that is, from the wavelengths of light that molecules emit and absorb. A further problem arises in the fact that detonation pressures for condensed explosives lie in the region of 200,000 atm, far beyond the experimental range. To calculate the behavior of a gas at such pressures, it is necessary to rely on the theory of intermolecular forces. Indeed, the term gas as applied to the detonation state of a condensed explosive is only a formality, because the density is sometimes double that of liquid water. In spite of such difficulties, the theory has yielded reliable values of the parameters for all kinds of explosives. Where it is possible to check experimentally, the agreement is usually good, as indicated in the table.

The speed of the chemical reaction itself does not enter the Chapman-Jouguet theory. Indeed, it is possible to calculate detonation properties for materials that do not detonate at all. Presumably, the reaction is too slow. Other cases involve materials that detonate but, because the reaction is sluggish, the velocity is below theoretical. Ammonium nitrate, for example, has a calculated velocity of 3460 m/sec but, when it has been detonated (in a 9-in. diameter cartridge), the measured speed is only 1510 m/sec. Again, many explosives, such as nitroglycerine, exhibit a low-order detonation when poorly primed. The velocity in such cases may be only one-quarter to one-half of the *C-J* value. Thus an ideal detonation is in accord with the *C-J* theory, and a nonideal detonation, either because the reaction is slow or because it follows a different path, does not accord with theory.

With a good rationale available, development of new blasting compositions can be made on paper, with the elimination of much cut-and-try experimentation. A combination of theory and experi-

ment also is yielding basic information about the behavior of molecules at very high pressures and temperatures.

High explosives. In an explosive mixture such as gunpowder, the fuel and oxidizer are in separate compounds; but in the pure high explosives, groups that serve these functions are fused together into a single molecule. The oxidizer role in these explosives is played by the nitro group, $-NO_2$. In the aromatic nitro compounds, such as TNT, it is joined to carbon; in the nitramines, such as cyclotrimethylenetrinitramine (RDX), to nitrogen; and in the nitrate esters, such as nitroglycerine and pentaerythritol tetranitrate (PETN), to oxygen. Carbon and hydrogen supply the fuel for the reaction.

A fairly strict division can be made between explosives used for civil and military purposes. For many years following Alfred Nobel's epic discovery that nitroglycerine could be safely used and reliably detonated, nitroglycerine formed the base of the blasting dynamites. But its place has been increasingly taken over by ammonium nitrate, which is now the workhorse in a bewildering array of commercial explosives. Military explosives presently in use consist mostly of TNT, RDX, PETN, and tetryl. The structural formulas of various explosives are shown in Fig. 6.

Commercial explosives. Black powder was, from ancient times and until Nobel, the only explosive. It is not a high explosive, that is to say, a detonating explosive, because under most circumstances it simply burns (or deflagrates). Black powder is made from charcoal, saltpeter (KNO_3), and sulfur. Potassium nitrate is the oxidizer, and the sulfur and charcoal make up the fuel. The manufacture of good powder requires skill, but it will not be described because the importance of black powder as an explosive has almost vanished.

When Nobel discovered the use of nitroglycerine, the era of high explosives was opened. The shattering power of a detonating explosive when laid against rock or steel is in distinct contrast to the action of a deflagrating powder, which is completely ineffective unless confined. Nobel's discovery, therefore, created entirely new techniques of blasting and gave a great impetus to mining.

Calculated and observed detonation velocities

Composition	Calculated velocity, m/sec	Observed velocity, m/sec
Gases (by moles)		
$2H_2 + O_2$	2806	2819
$2H_2 + O_2 + 5N_2$	1820	1822–1840
$C_2H_2 + O_2$	2960	2920–2961
$C_2H_2 + O_2 + 4N_2$	2020	2015
C_2H_2	2070	2135–2160
Liquids		
Nitroglycerine	8060	8000
Nitroglycol	7630	8000
Solids (density in g/cm³)		
TNT (1.50)	6480	6700
TNT (1.00)	5060	4900
TNT (0.50)	3730	3200
Tetryl (1.50)	7550	7300
PETN (1.50)	8150	7600

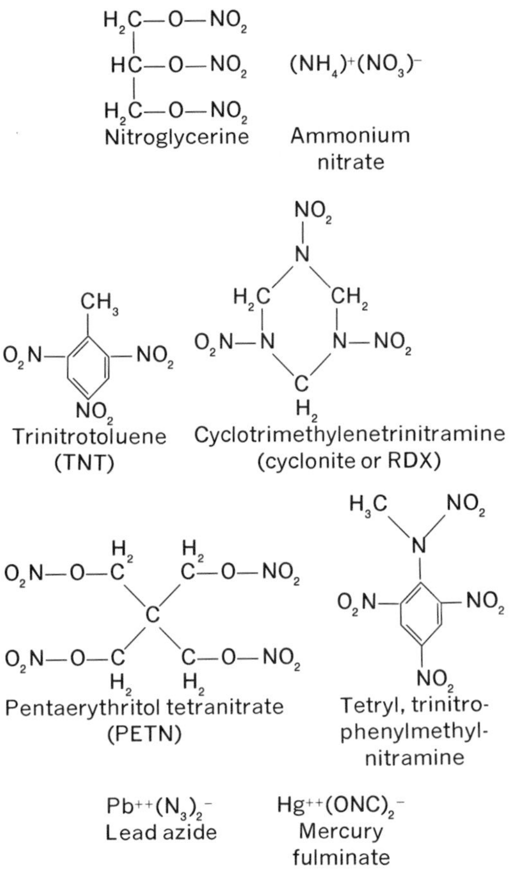

$$H_2C\!-\!O\!-\!NO_2$$
$$HC\!-\!O\!-\!NO_2 \qquad (NH_4)^+(NO_3)^-$$
$$H_2C\!-\!O\!-\!NO_2$$

Nitroglycerine   Ammonium nitrate

Trinitrotoluene (TNT)

Cyclotrimethylenetrinitramine (cyclonite or RDX)

Pentaerythritol tetranitrate (PETN)

Tetryl, trinitrophenylmethylnitramine

$Pb^{++}(N_3)_2{}^-$
Lead azide

$Hg^{++}(ONC)_2{}^-$
Mercury fulminate

Fig. 6. Structural formulas of various explosives.

Ammonium nitrate, NH_4NO_3, has slowly risen to the place of major importance for blasting. In one sense, the trend would seem to be a reversion to the black-powder type of explosive, with NH_4NO_3 replacing potassium nitrate, KNO_3. Ammonium nitrate explosives are similar to gunpowder. They contain the nitrate oxidizer and a fuel which, in this case, may be almost any carbonaceous material from oat hulls to the sweepings of breakfast-food factories. For various reasons, a bulky material is preferred. To make a dynamite, either nitroglycerine or TNT is added to sensitize this mixture. The result is a cap-sensitive high explosive, quite different from gunpowder.

In coal mining, regulations require that an explosive must not be capable of setting fire to coal gas (methane). In effect, this restriction places a limit on the temperature of the explosive products. Explosives that meet this requirement are called permissibles.

In open-pit mining an inexpensive explosive is gaining wide use. This material consists simply of prilled ammonium nitrate, an agricultural fertilizer, mixed with 5–6% furnace oil. The mixture is very insensitive, but in the wide and deep boreholes used for surface mining it can be detonated with a booster charge of dynamite. The earth-moving ability of this explosive is said to exceed that of dynamite.

Military explosives. TNT came into use as a military explosive during World War I and still retains an important place. It is an oxygen-deficient explosive and therefore can be improved and lowered in cost by the addition of ammonium nitrate to make a mixture called amatol. Although this mixture was important in World War I, it is not used today. TNT derives its important virtues from its relative insensitivity to shock and its convenient melting point (81°C). Because it can be melted with steam, it can be safely cast into shells and other armament. In this respect TNT is unique and, even though other explosives such as RDX are more powerful, TNT must still be used as the base for a castable mixture. The most common mixture is Composition B, which contains about 60% RDX, 40% TNT, and some wax as a desensitizer. Another mixture is Pentolite, a mixture of TNT and PETN.

RDX (cyclonite) rose to importance during World War II. In the early period of that conflict, a concentrated effort was made to develop a large-scale synthesis for RDX. This was successful, and the RDX plants soon began to turn out huge quantities. RDX is high-melting, fairly insensitive, and the most powerful compound in common use. Although slightly more energetic materials are known, RDX has almost the theoretical limit of stored energy in a molecule of this type. On decomposition it is balanced to CO, as may be seen by writing the decomposition reaction as in Eq. (1).

$$C_3H_6N_6O_6 \rightarrow 3CO + 3H_2O + 3N_2 \qquad (1)$$

If there were sufficient oxygen to oxidize the CO to CO_2, a slight gain in energy might be possible. RDX is used in Composition B and also, in conjunction with TNT, in the aluminized composition Torpex.

By adding powdered aluminum to explosives such as RDX, a further increase in explosive energy can be produced. The aluminum reacts with the explosion products to produce aluminum oxide, for example, as in Eq. (2). This reaction adds

$$Al + 3H_2O \rightarrow Al_2O_3 + 3H_2 \qquad (2)$$

more heat to that already liberated by the decomposition of the explosive. There is, of course, an optimum amount of aluminum for any given explosive, at which the energy output reaches a maximum.

PETN is almost as powerful as RDX and is used interchangeably with it in some instances. The main use is in the detonating fuse Primacord, a fabric-wrapped cord, which has a core of PETN and is widely used for commercial and military fuse trains.

Tetryl is a relatively sensitive explosive having excellent pressing properties. As a result, it is commonly used as the booster charge for military devices.

Detonators. The four explosives described above are called secondary explosives. They will ordinarily detonate only under the impetus of a powerful shock wave transmitted from a primary explosive. These latter materials, notably lead azide and mercury fulminate, are used in the form of detonator caps. When ignited by a fuse wire or by percussion, they go into detonation almost immediately. To further strengthen the shock wave from the detonator before it is transmitted to the main charge, a tetryl booster charge is normally employed.

Propellant burning in guns and rockets.

A gun or rocket propellant does not detonate; or, if it does, the results are disastrous. The combustion mechanism in this case is deflagration. It is often called cigarette-type burning to indicate that the propellant is steadily consumed by erosion at the surface. A flame stands at a short distance and heats the surface by radiation and conduction. The heating causes the surface material to evaporate and decompose. In this way a stream of gas is produced to feed the flame and a steady state is established.

The burning rate is reckoned as the rate at which the surface recedes. The rate depends on the gas pressure in the combustion chamber; in fact, for many propellants, the burning rate is just proportional to pressure. A representative speed is 2 cm/sec at 100 atm.

The function of the propellant, in both guns and rockets, is simply to produce gas. The rate of gas production is roughly proportional to the product of the surface area of the unconsumed propellant and the pressure in the chamber. Therefore, since in both rockets and guns it is desirable to maintain a steady rate of gas production, the propellant grains are shaped so as to maintain a constant surface area as they burn. To produce this result in gun propellants, a perforated cylindrical grain is commonly used (Fig. 7a). A grain of this shape, when it burns, both from the outside in and the inside out, tends to keep a constant area. In large, single-rocket grains, on the other hand, burning is inhibited on the outer surface and the grain burns only from the inside out. For this reason, a rocket grain usually has a star-shaped perforation to give it a large initial area (Fig. 7b).

The pressure in the gun breech rises rapidly when the powder is ignited. In a period of time t of about 1 or 2 msec (1 msec = 1/1000 sec), it reaches a maximum value of about 50,000 psi. The outward movement of the bullet then causes the pressure p to drop. Its muzzle velocity depends on the integral $\int p \, dt$. An appreciable fraction of the energy of the propellant is used in accelerating the hot gases, as well as the bullet; ballistic designers therefore seek propellants with light (low-molecular-weight) gaseous products.

The propellant system of a rocket is designed to operate at constant chamber pressure and, therefore, constant thrust. It is very undesirable to have a propellant whose burning rate changes rapidly with pressure; if it does, instability results and the pressure rises until the rocket explodes. Rockets therefore require special propellants with a low-pressure coefficient in the burning rate equation. As with gun propellants, light combustion products are preferred. A common figure of merit for the efficiency of a rocket propellant is the specific impulse I_{sp}; it is approximately proportional to $\sqrt{T/M}$, where T is the flame temperature and M is the average molecular weight of the propellant gases. See PROPELLANT.

Besides black powder, which is used mainly in sporting rifles, the common gun propellants are either nitrocellulose or a mixture of nitrocellulose and nitroglycerine (double-base propellant).

Rocket propellants are of two main types, liquid and solid. The liquid systems use an oxidizer such as liquid oxygen or fuming nitric acid and a fuel such as kerosine or hydrazine. Solid propellants may be of the nitrocellulose-nitroglycerine, double-base type (monopropellants) or mixtures (composite propellants) of an oxidizer, such as ammonium perchlorate, and a fuel, such as Thiokol.

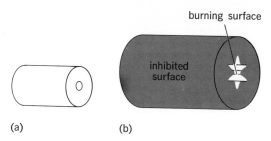

Fig. 7. Propellant grains. (a) A single powder grain, which may have a diameter, or a web, as it is called, from less than 1 mm up to 1/4 in. (b) Rocket grain, which may be several feet in diameter, depending on the size of the rocket. Only a single grain is used in the rocket chamber.

Gas and dust explosions.

Gas and dust explosions are serious industrial hazards, especially in coal mines. The severity of the effects depends on confinement, because gas explosions seldom result in detonation and dust explosions almost never do. The burning is usually so slow that the pressure would not rise, except in cases where the explosion gases cannot escape fast enough from the building or tunnel in which they are formed. See MINING SAFETY.

There are definite flammability limits for both combustible gases and dusts. Below a certain critical concentration of combustible, the air mixture becomes incapable of supporting a flame. In the case of gases, it is possible to provide warning of a rising concentration of combustible by means of a detector. See FLAME.

Coal mines, for example, make a regular practice of surveying the mine air for methane content. The minimum flammable concentration of methane is about 5%, but it can be readily detected at about 0.1%. The most common methane detector, which also is suitable for many other combustible gases, has a glowing platinum filament that catalyzes combustion of even slight traces of methane in the immediate vicinity of the wire. In a contaminated atmosphere the wire becomes heated above its normal temperature, and the temperature rise can be detected by the change in electrical resistance. Other methods of methane detection are used as well.

Prevention. For dusts there is no such convenient warning device. Measures can be taken, however, to prevent dust explosions. Besides a simple improvement in cleaning practices, these measures include the use of an inert dust, for example, stone dust, on ledges and other places where dust collects. If a sufficient quantity of the inert dust is present, it will stop the progress of flame in an otherwise combustible mixture. The action of the inert particles is to rob heat from the combustion and, in this way, to lower the temperature below the point at which the flame can spread.

Particle size is an extremely important factor in dust explosions. The minimum explosive concentration, that is, the minimum weight of dust in a cubic foot of air that will cause explosion, decreases sharply as the size of the particles is diminished. The total surface area of the dust particles is the underlying factor, and the area increases rapidly as particle diameter is reduced. More precisely, since the total weight is proportional to nd^3, where n is the number of particles and d is the average diameter, and since the total surface area is proportional to nd^2, the surface area for a given total weight of material will vary as $1/d$. As a result, a tenfold decrease in particle size leads to a tenfold increase in specific surface area.

Different gases vary in ease of ignition, and dusts vary in the ease with which the flame can be inhibited. For example, the most readily ignited hydrogen-air mixture requires only one-twentieth as much energy for ignition as the most sensitive methane-air mixture. For dusts it is easier to measure the amount of stone dust necessary to prevent inflammation. Enormous differences are found between dusts. Anthracite coal dust, for example, is barely flammable and requires very little stone dust to prevent burning. Cellulose-acetate resin, on the other hand, requires several times as much stone dust to stop a flame.

Dust explosions are very common. In some industries they occur almost routinely in the exhaust ducts, which are built to withstand the effects. The problem is aggravated by the fact that dusts generate static electricity and therefore are sometimes self-igniting. Furthermore, when an explosion starts, it can spread by raising the dust in clouds ahead of the flame. Some metal dusts, notably magnesium and zirconium, are extremely hazardous.

Extinguishers. Gas flammability also can be reduced by the presence of certain inhibitors. Most outstanding of these are the so-called alkyl halides. The best known compound of this type is carbon tetrachloride, a common fire extinguisher, but the group of fluorinated compounds known as Freons also is very effective. The action here is thought to be a specific chemical effect, possibly having to do with the halogen atoms that are broken away from the molecules by heat.

[WILLIAM E. GORDON]

Bibliography: M. A. Cook, *The Science of High Explosives*, Amer. Chem. Soc. Monogr. no. 139, 1958; B. Lewis and G. von Elbe, *Combustion, Flames and Explosions of Gases*, 1951; S. S. Penner and B. P. Mullins, *Explosions, Detonations, Flammability and Ignition*, 1959; J. Taylor, *The Detonation in Condensed Explosives*, 1952; U.S. Scientific Laboratory, *The Effects of Atomic Weapons*, 1950; T. Urbanski, *Chemistry and Technology of Explosives*, 2 vols., 1963, 1965.

Fire

A rapid but persistent chemical reaction accompanied by the emission of light and heat. The reaction is self-sustaining, unless extinguished, to the extent that it continues until the fuel concentration falls below a minimum value. Most commonly, it results from a rapid exothermic combination with oxygen by some combustible material. Flame and heat also may result from a reaction involving an agent other than oxygen. Thus, certain reactive metals such as zinc will burn in an atmosphere of chlorine.

Flame, the visible manifestation of fire, results from a heating to incandescence of minute particulate matter composed principally of incompletely burned fuel. The color of the flame depends upon the material undergoing reaction and the temperature. *See* COMBUSTION; FLAME; FUEL, FOSSIL.

[FRANCIS J. JOHNSTON]

Fischer-Tropsch process

The synthesis of hydrocarbons and, to a lesser extent, of aliphatic oxygenated compounds by the catalytic hydrogenation of carbon monoxide. The synthesis was discovered in 1923 by F. Fischer and H. Tropsch at the Kaiser Wilhelm Institute for Coal Research in Mülheim, Germany. The reaction is highly exothermic, and the reactor must be designed for adequate heat removal to control the temperature and avoid catalyst deterioration and carbon formation. The sulfur content of the synthesis gas must be extremely low to avoid poisoning the catalyst. The first commercial plant was built in Germany in 1935 with cobalt catalyst, and at the start of World War II there were six plants in Germany producing more than 4,000,000 bbl/year of primary products. Iron catalysts later replaced the cobalt.

Following World War II, considerable research was conducted in the United States on the iron catalysts. One commercial plant was erected at Brownsville, Tex., in 1948, which used a fluidized bed of mill scale promoted by potash. Because synthetic oil was not competitive with petroleum, the plant was shut down within a few years. A Fischer-Tropsch plant constructed in South Africa (SASOL) about the same time has continued to produce gasoline, waxes, and oxygenated aliphatics. The SASOL plant gasifies inexpensive coal in Lurgi generators at elevated pressure. After purification, the gas is sent to entrained and fixed-bed synthesis reactors containing iron catalysts. *See* COAL GASIFICATION.

Research at the U.S. Bureau of Mines Pittsburgh Coal Research Center has resulted in the development of active and low-cost iron catalysts of steel lathe turnings and flame-sprayed magnetite. With these catalysts, improved reactors have been tested by use of gas recycle to control the operating temperature while maintaining a low pressure drop.

Methanation. Since about 1960, interest has grown in the United States in catalytic methanation to produce high-Btu gas from coal. Synthesis gas containing three volumes of hydrogen to one volume of carbon monoxide is reacted principally to methane and water, as shown in Eq. (1). Nickel catalysts, first discovered by P. Sabatier and J. B. Senderens in 1902, are still the principal catalysts for methanation. Active precipitated catalysts were developed at the British Fuel Research Station, and Raney nickel catalysts were tested in fluid and fixed-bed reactors by the Bureau of Mines and the Institute of Gas Technology at the Illinois Institute of Technology. The Bureau of Mines has developed a technique for applying a thin layer of Raney nickel on plates and tubes by flame-spraying the powder. This has led to development of efficient gas recycle and tube-wall reactors. Commercial production of high-Btu gas from coal in the United States is expected to start about 1980 to supplement natural gas supplies.

Reactions. Typical Fischer-Tropsch reactions for the synthesis of paraffins, olefins, and alcohols are the pairs of equations labeled (1), (2), and (3), respectively.

$$(2n+1)H_2 + nCO \xrightarrow{\text{Co catalysts}} C_nH_{2n+2} + nH_2O \qquad (1)$$

$$(n+1)H_2 + 2nCO \xrightarrow{\text{Fe catalysts}} C_nH_{2n+2} + nCO_2$$

$$2nH_2 + nCO \xrightarrow{\text{Co catalysts}} C_nH_{2n} + nH_2O \qquad (2)$$

$$nH_2 + 2nCO \xrightarrow{\text{Fe catalysts}} C_nH_{2n} + nCO_2$$

$$2nH_2 + nCO \xrightarrow{\text{Co catalysts}} C_nH_{2n+1}OH + (n-1)H_2O \qquad (3)$$

$$(n+1)H_2 + (2n-1)CO \xrightarrow{\text{Fe catalysts}} C_nH_{2n+1}OH + (n-1)CO_2$$

The primary reaction on both the cobalt and the iron catalysts yields steam, which reacts further on iron catalysts with carbon monoxide to give hydrogen and carbon dioxide. On cobalt catalysts, at synthesis temperatures (about 200°C, or 392°F), the reaction $H_2O + CO = CO_2 + H_2$ is much slower than on iron catalysts at synthesis temperatures (250–320°C, or 482–617°F). All these synthesis reactions are exothermic, yielding 37–51 kcal/mole of carbon in the products or 4700–6100 Btu/lb of product.

The hydrocarbons formed in the presence of iron catalysts contain more olefins than those formed in cobalt catalyst systems. The products from both catalysts are largely straight-chain aliphatics; branching is about 10% for C_4, 19% for C_5, 21% for C_6, and 34% for C_7. Aromatics appear in small amounts in the C_7 and in larger amounts in the higher boiling fractions. Operating conditions and special catalysts required, such as nitrides and carbonitrides of iron, have been ascertained for the production of higher proportions of alcohols.

[JOSEPH H. FIELD]

Bibliography: H. A. Dirksen and H. R. Linden, *Inst. Gas Technol. Res. Bull.* no. 31, July, 1963; J. H. Field et al., *Ind. Eng. Chem. Prod. Res. Develop.*, 3:150–153, June, 1964; J. C. Hoogendoorn and J. M. Salomon, *Brit. Chem. Eng.*, pp. 238–243, May, 1957; L. A. Moignard and F. J. Dent, *Gas Times*, pp. 40–41, May 11, 1946; H. H. Storch, N. Golumbic, and R. B. Anderson, *The Fischer-Tropsch and Related Syntheses*, 1951.

Fission, nuclear

An extremely complex nuclear reaction representing a cataclysmic division of an atomic nucleus into two nuclei of comparable mass. This rearrangement or division of a heavy nucleus may take place naturally (spontaneous fission) or under bombardment with neutrons, charged particles, gamma rays, or other carriers of energy (induced fission). Although nuclei with mass number A of approximately 100 or greater are energetically unstable against division into two lighter nuclei, the fission process has a small probability of occurring, except with the very heavy elements. Even for these elements, in which the energy release is of the order of 200,000,000 electron volts (eV), the lifetimes against spontaneous fission are reason-

ably long. *See* NUCLEAR REACTION.

Liquid-drop model. The stability of a nucleus against fission is most readily interpreted when the nucleus is viewed as being analogous to an incompressible and charged liquid drop with a surface tension. Such a droplet is stable against small deformations when the dimensionless fissility parameter X in Eq. (1) is less than unity, where the

$$X = \frac{(\text{charge})^2}{10 \times \text{volume} \times \text{surface tension}} \qquad (1)$$

charge is in esu, the volume is in cm³, and the surface tension is in ergs/cm². The fissility parameter if given approximately, in terms of the charge number Z and mass number A, by the relation $X = Z^2/50A$.

Long-range Coulomb forces between the protons act to disrupt the nucleus, whereas short-range nuclear forces, idealized as a surface tension, act to stabilize it. The degree of stability is then the result of a delicate balance between the relatively weak electromagnetic forces and the strong nuclear forces. Although each of these forces results in potentials of several hundred million electron volts, the height of a typical barrier against fission for a heavy nucleus, because they are of opposite sign but do not quite cancel, is only 5,000,000 or 6,000,000 eV. Investigators have used this charged liquid-drop model with great success in describing the general features of nuclear fission and also in reproducing the total nuclear binding energies. *See* BINDING ENERGY, NUCLEAR.

Shell corrections. The general dependence of the potential energy on the fission coordinate representing nuclear elongation or deformation for a heavy nucleus such as ^{240}Pu is shown in Fig. 1. The expanded scale used in this figure shows the large decrease in energy of about 200 MeV as the fragments separate to infinity. It is known that ^{240}Pu is deformed in its ground state, which is represent-

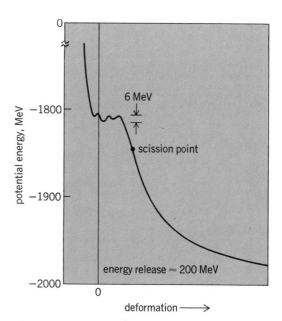

Fig. 1. Plot of the potential energy in MeV as a function of deformation for the nucleus ^{240}Pu. (*From M. Bolsteli et al., New calculations of fission barriers for heavy and superheavy nuclei, Phys. Rev., 5C:1050–1077, 1972*)

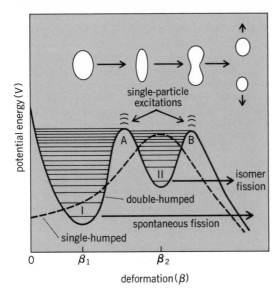

Fig. 2. Schematic plots of single-humped fission barrier of liquid-drop model and double-humped barrier introduced by shell corrections. *(From J. R. Huizenga, Nuclear fission revisited, Science, 168:1405–1413, 1970)*

ed by the lowest minimum of −1813 MeV near zero deformation. This energy represents the total nuclear binding energy when the zero of potential energy is the energy of the individual nucleons at a separation of infinity. The second minimum to the right of zero deformation illustrates structure introduced in the fission barrier by shell corrections, that is, corrections dependent upon microscopic behavior of the individual nucleons, to the liquid-drop mass. Although shell corrections introduce small wiggles in the potential-energy surface as a function of deformation, the gross features of the surface are reproduced by the liquid-drop model. Since the typical fission barrier is only a few million electron volts, the magnitude of the shell correction need only be small for irregularities to be introduced into the barrier. This structure is schematically illustrated for a heavy nucleus by the double-humped fission barrier in Fig. 2, which represents the region to the right of zero deformation in Fig. 1 on an expanded scale. The fission barrier has two maxima and a rather deep minimum in between. For purposes of comparison, the single-humped liquid-drop barrier is also schematically illustrated. The transition in the shape of the nucleus as a function of deformation is schematically represented in the upper part of the figure.

Double-humped barrier. The developments which led to the proposal of a double-humped fission barrier were triggered by the experimental discovery of spontaneously fissionable isomers by S. M. Polikanov and colleagues in the Soviet Union and by V. M. Strutinsky's pioneering theoretical work on the binding energy of nuclei as a function of both nucleon number and nuclear shape. The double-humped character of the nuclear potential energy as a function of deformation arises, within the framework of the Strutinsky shell-correction method, from the superposition of a macroscopic smooth liquid-drop energy and a shell-correction energy obtained from a microscopic single-particle

model. Oscillations occurring in this shell correction as a function of deformation lead to two minima in the potential energy, shown in Fig. 2, the normal ground-state minimum at a deformation of β_1 and a second minimum at a deformation of β_2. States in these wells are designated class I and class II states, respectively. Spontaneous fission of the ground state and isomeric state arises from the lowest-energy class I and class II states, respectively.

The calculation of the potential-energy curve illustrated in Fig. 1 may be summarized as follows. The smooth potential energy obtained from a macroscopic (liquid-drop) model is added to a fluctuating potential energy representing the shell corrections, and to the energy associated with the pairing of like nucleons (pairing energy), derived from a non-self-consistent microscopic model. The calculation of these corrections requires several steps, namely, (1) specification of the geometrical shape of the nucleus, (2) generation of a single-particle potential related to its shape, (3) solution of the Schrödinger equation, and (4) calculation from these single-particle energies of the shell and pairing energies.

The oscillatory character of the shell corrections as a function of deformation is caused by variations in the single-particle level density in the vicinity of the Fermi energy. For example, the single-particle levels of a pure harmonic oscillator potential arrange themselves in bunches of highly degenerate shells at any deformation for which the ratio of the major and minor axes of the spheroidal equipotential surfaces is equal to the ratio of two small integers. Nuclei with a filled shell, that is, with a level density at the Fermi energy that is smaller than the average, will then have an increased binding energy compared to the average, because the nucleons occupy deeper and more bound states; conversely, a large level density is associated with a decreased binding energy. It is precisely this oscillatory behavior in the shell correction that is responsible for spherical or deformed gound states and for the secondary minima in fission barriers, as illutrated in Fig. 2.

More detailed theoretical calculations based on this macroscopic-microscopic method have revealed additional features of the fission barrier. In these calculations the potential energy is regarded as a function of several different modes of deformation. The outer barrier B (Fig. 2) is reduced in energy for shapes with pronounced left-right asymmetry (pear shapes), whereas the inner barrier A and deformations in the vicinity of the second minimum are stable against such mass asymmetric degrees of freedom. Similar calculations of potential-energy landscapes reveal the stability of the second minimum against gamma deformations, in which the two small axes of the spheroidal nucleus become unequal, that is, the spheroid becomes an ellipsoid.

Experimental consequences. The observable consequences of the double-humped barrier have been reported in numerous experimental studies. In the actinide region more than 30 spontaneously fissionable isomers have been discovered between uranium and berkelium, with half-lives ranging from 10^{-11} to 10^{-2} s. These decay rates are faster by 20 to 30 orders of magnitude than the fission

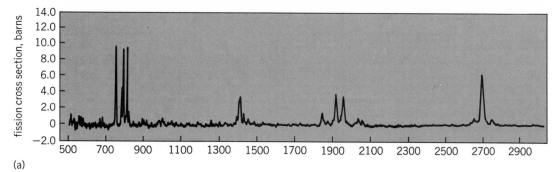

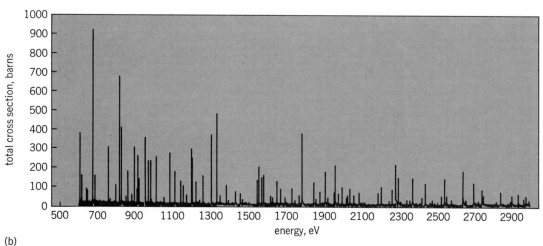

(a)

(b)

Fig. 3. Grouping of fission resonances demonstrated by (a) the neutron fission cross section of ^{240}Pu and (b) the total neutron cross section. (*From V. M. Strutinsky and* *H. C. Pauli, Shell-structure effects in the fissioning nucleus, Proceedings of the 2d IAEA Symposium on Physics and Chemistry of Fission, Vienna, pp. 155–177, 1969*)

half-lives of the ground states, because of the increased barrier tunneling probability (see Fig. 2). Several cases in which excited states in the second minimum decay by fission are also known. Normally these states decay within the well by gamma decay; however, if there is a hindrance in gamma decay due to spin, the state (known as a spin isomer) may undergo fission instead.

Qualitatively, the fission isomers are most stable in the vicinity of neutron numbers 146 to 148, a value in good agreement with macroscopic-microscopic theory. For elements above berkelium the half-lives become too short to be observable with available techniques; and for elements below uranium, the prominent decay is through barrier A into the first well, followed by gamma decay. It is difficult to detect this competing gamma decay of the ground state in the second well (called a shape isomeric state), but identification of the gamma branch of the 200-ns ^{238}U shape isomer has been reported.

Direct evidence of the second minimum in the potential-energy surface of the even-even nucleus ^{240}Pu has been obtained through observations of the E2 transitions within the rotational band built on the isomeric 0+ level. The rotational constant (which characterizes the spacing of the levels and is expected to be inversely proportional to the effective moment of inertia of the nucleus) found for this band is less than one-half that for the ground state and confirms that the shape isomers have a deformation β_2 much larger than the equilibrium

ground-state deformation β_1. From yields and angular distributions of fission fragments from the isomeric ground state and low-lying excited states some information has been derived on the quantum numbers of specific single-particle states of the deformed nucleus (Nilsson single-particle states) in the region of the second minimum.

At excitation energies in the vicinity of the two barrier tops, measurements of the subthreshold neutron fission cross sections of several nuclei have revealed groups of fissioning resonance states with wide energy intervals between each group where no fission occurs. Such a spectrum is illustrated in Fig. 3a, where the subthreshold fission cross section of ^{240}Pu is shown for neutron energies between 500 and 3000 eV. As shown in Fig. 3b, between the fissioning resonance states there are many other resonance states, known from data on the total neutron cross sections, which have negligible fission cross sections. Such structure is explainable in terms of the double-humped fission barrier and is ascribed to the coupling between the compound states of normal density in the first well to the much less dense states in the second well. This picture requires resonances of only one spin to appear within each intermediate structure group illustrated in Fig. 3a. In an experiment using polarized neutrons on a polarized ^{237}Np target, it was found that all nine fine-structure resonances of the 40-eV group have the same spin and parity: $I = 3 +$. Evidence has also been obtained for vibrational states in the second

Cross sections for neutrons of thermal energy to produce fission or undergo capture in the principal nuclear species, and neutron yields from these nuclei*

Nucleus	Cross section for fission, σ_f, 10^{-24} cm^2	σ_f plus cross section for radiative capture, σ_r	Ratio, $1 + \alpha$	Number of neutrons released per fission, ν	Number of neutrons released per slow neutron captured, $\eta = \nu/(1+\alpha)$
^{233}U	525 ± 2	573 ± 2	1.093 ± 0.003	2.50 ± 0.01	2.29 ± 0.01
^{235}U	577 ± 1	678 ± 2	1.175 ± 0.002	2.43 ± 0.01	2.08 ± 0.01
^{239}Pu	741 ± 4	1015 ± 4	1.370 ± 0.006	2.89 ± 0.01	2.12 ± 0.01
^{238}U	0	2.73 ± 0.04			0
Natural uranium	4.2	7.6	1.83	2.43 ± 0.01	1.33

*Data from *Brookhaven National Laboratory 325*, 2d ed., suppl. no. 2, vol. 3, 1965. The data presented are the recommended or least-squares values published in this reference for 0.0253-eV neutrons. All cross sections are in units of barns (1 barn $= 10^{-24}$ cm$^2 = 10^{-28}$ m^2).

well from neutron (n,f) and deuteron stripping (d,pf) reactions at energies below the barrier tops (f indicates fission of the nucleus).

A. Bohr suggested that the angular distributions of the fission fragments are explainable in terms of the transition-state theory, which describes a process in terms of the states present at the barrier deformation. The theory predicts that the cross section will have a steplike behavior for energies near the fission barrier, and that the angular distribution will be determined by the quantum numbers associated with each of the specific fission channels. The theoretical angular distribution of fission fragments is based on two assumptions. First, the two fission fragments are assumed to separate along the direction of the nuclear symmetry axis so that the angle θ between the direction of motion of the fission fragments and the direction of motion of the incident bombarding particle represents the angle between the body-fixed axis (the long axis of the spheroidal nucleus) and the space-fixed axis (some specified direction in the laboratory, in this case the direction of motion of the incident particle). Second, it is assumed that the transition from the saddle point (corresponding to the top of the barrier) to scission (the division of the nucleus into two fragments) is so fast that Coriolis forces do not change the value of K (where K is the projection of the total angular momentum I on the nuclear symmetry axis) established at the saddle point.

In several cases, low-energy photofission and neutron fission experiments have shown evidence of a double-humped barrier. In the case of two barriers, the question arises as to which of the two barriers A or B is responsible for the structure in the angular distributions. For light actinide nuclei like thorium, the indication is that barrier B is the higher one, whereas for the heavier actinide nuclei, the inner barrier A is the higher one. The heights of the two barriers themselves are most reliably determined by investigating the probability of induced fission over a range of several mega-electron volts in the threshold region. Many direct reactions have been used for this purpose, for example, (d,pf), (t,pf), and $(^3\text{He}, df)$. There is reasonably good agreement between the experimental and theoretical barriers. The theoretical barriers are calculated with realistic single-particle potentials and include the shell corrections.

Fission probability. The cross section for particle-induced fission $\sigma(y,f)$ represents the cross section for a projectile y to react with a nucleus and produce fission, as shown by Eq. (2). The quantities $\sigma_R(y)$, Γ_f, and Γ_t are the total reaction cross sections for the incident particle y, the fission width, and the total level width, respectively where $\Gamma_t = \Gamma_f + \Gamma_n + \Gamma_y + \cdots$ is the sum of all partial-level widths. All the quantities in Eq. (2)

$$\sigma(y,f) = \sigma_R(y)\,(\Gamma_f/\Gamma_t) \qquad (2)$$

are energy-dependent. Each of the partial widths for fission, neutron emission, radiation, and so on, is defined in terms of a mean lifetime τ for that particular process, for example, $\Gamma_f = \hbar/\tau_f$. Here $\hbar$, the action quantum, is Planck's constant divided by 2π and is numerically equal to 1.0546×10^{-34} J s $= 0.66 \times 10^{-15}$ eV s. The fission width can also be defined in terms of the energy separation D of successive levels in the compound nucleus and the number of open channels in the fission transition nucleus (paths whereby the nucleus can cross the barrier on the way to fission), as given by expression (3), where I is the angular momentum and i is

$$\Gamma_f(I) = \frac{D(I)}{2\pi} \sum_i N_{fi} \qquad (3)$$

an index labeling the open channels N_{fi}. The contribution of each fission channel to the fission width depends upon the barrier transmission coefficient, which, for a two-humped barrier (see Fig. 2), is strongly energy-dependent. This results in an energy-dependent fission cross section which is very different from the total cross section shown in Fig. 3 for ^{240}Pu.

When the incoming neutron has low energy, the likelihood of reaction is substantial only when the energy of the neutron is such as to form the compound nucleus in one or another of its resonance levels (see Fig. 3b). The requisite sharpness of the "tuning" of the energy is specified by the total level width Γ. The nuclei ^{233}U, ^{235}U, and ^{239}Pu have a very large cross section to take up a slow neutron and undergo fission (see table) because both their absorption cross section and their probability for decay by fission are large. The probability for fission decay is high because the binding energy of the incident neutron is sufficient to raise the energy of the compound nucleus above the fission barrier. The very large, slow neutron fission cross sections of these isotopes make them important fissile materials in a chain reactor. *See* CHAIN REACTION, NUCLEAR.

Scission. The scission configuration is defined in terms of the properties of the intermediate nucleus just prior to division into two fragments. In heavy nuclei the scission deformation is much larger than the saddle deformation at the barrier, and it is important to consider the dynamics of the descent from saddle to scission. One of the important questions in the passage from saddle to scission is the extent to which this process is adiabatic with respect to the particle degrees of freedom. As the nuclear shape changes, it is of interest to investigators to know the probability for the nucleons to remain in the lowest-energy orbitals. If the collective motion toward scission is very slow, the single-particle degrees of freedom continually readjust to each new deformation as the distortion proceeds. In this case, the adiabatic model is a good approximation, and the decrease in potential energy from saddle to scission appears in collective degrees of freedom at scission, primarily as kinetic energy associated with the relative motion of the nascent fragments.

On the other hand, if the collective motion between saddle and scission is so rapid that equilibrium is not attained, there will be a transfer of collective energy into nucleonic excitation energy. Such a nonadiabatic model, in which collective energy is transferred to single-particle degrees of freedom during the descent from saddle to scission, is usually referred to as the statistical theory of fission.

The experimental evidence indicates that the saddle to scission time is somewhat intermediate between these two extreme models. The dynamic descent of a heavy nucleus from saddle to scission depends upon the nuclear viscosity. A viscous nucleus is expected to have a smaller translational kinetic energy at scission and a more elongated scission configuration. Experimentally, the final translational kinetic energy of the fragments at infinity, which is related to the scission shape, is measured. Hence, in principle, it is possible to estimate the nuclear viscosity coefficient by comparing the calculated dependence upon viscosity of fission-fragment kinetic energies with experimental values. The viscosity of nuclei is an important nuclear parameter which also plays an important role in collisions of very heavy ions.

The mass distribution from the fission of heavy nuclei is predominantly asymmetric. For example, division into two fragments of equal mass is about 600 times less probable than division into the most probable choice of fragments when ^{235}U is irradiated with thermal neutrons. When the energy of the neutrons is increased, symmetric fission (Fig. 4) becomes more probable. In general, heavy nuclei fission asymmetrically to give a heavy fragment of approximately constant mean mass number 139 and a corresponding variable-mass light fragment (see Fig. 5). These experimental results have been difficult to explain theoretically. Calculations of potential-energy surfaces show that the second barrier (B in Fig. 2) is reduced in energy by up to 2 or 3 MeV, if octuple deformations (pear shapes) are included. Hence, the theoretical calculations show that mass asymmetry is favored at the outer barrier, although direct experimental evidence

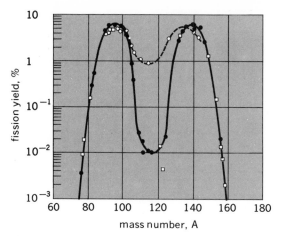

Fig. 4. Mass distribution of fission fragments formed by neutron-induced fission of ^{235}U $+ n = {}^{236}$U when neutrons have thermal energy, smooth curve (*Plutonium Project Report, Rev. Mod. Phys., 18:539, 1964*), and 14-MeV energy, dashed curve (*based on R. W. Spence, Brookhaven National Laboratory, AEC-BNL (C-9), 1949*). Quantity plotted is 100× (number of fission decay chains formed with given mass)/(number of fissions).

supporting the asymmetric shape of the second barrier is very limited. It is not known whether the mass asymmetric energy valley extends from the saddle to scission; and the effect of dynamics on mass asymmetry in the descent from saddle to scission has not been determined. Experimentally, as the mass of the fissioning nucleus approaches $A \approx 260$, the mass distribution approaches symmetry. This result is qualitatively in agreement with theory.

A nucleus at the scission configuration is highly elongated and has considerable deformation energy. The influence of nuclear shells on the scission shape introduces structure into the kinetic energy and neutron-emission yield as a function of frag-

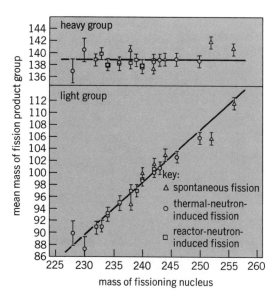

Fig. 5. Average masses of the light- and heavy-fission product groups as a function of the masses of the fissioning nucleus. Energy spectrum of reactor neutrons is that associated with fission. (*From K. F. Flynn et al., Distribution of mass in the spontaneous fission of ^{256}Fm, Phys. Rev., 5C:1725–1729, 1972*)

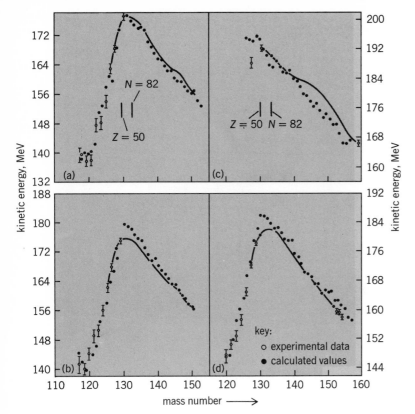

Fig. 6. Average total kinetic energy of fission fragments as a function of heavy fragment mass for fission of (a) ²³⁵U, (b) ²³³U, (c) ²⁵²Cf, and (d) ²³⁹Pu. Curves indicate experimental data. (*From J. C. D. Milton and J. S. Fraser, Time-of-flight fission studies on ²³³U, ²³⁵U and ²³⁹Pu, Can. J. Phys., 40:1626–1663, 1962*)

ment mass. The experimental kinetic energies for the neutron-induced fission of ²³³U, ²³⁵U, and ²³⁹Pu have a pronounced dip as symmetry is approached, as shown in Fig. 6. (This dip is slightly exaggerated in the figure because the data have not been corrected for fission fragment scattering.)

The variation in the neutron yield as a function of fragment mass for these same nuclei (Fig. 7) has a "saw-toothed" shape which is asymmetric about the mass of the symmetric fission fragment. Both these phenomena are reasonably well accounted for by the inclusion of closed-shell structure into the scission configuration.

A number of light charged particles (for example, isotopes of hydrogen, helium, and lithium) have been observed to occur, with low probability, in fission. These particles are believed to be emitted very near the time of scission. Available evidence also indicates that neutrons are emitted at or near scission with considerable frequency.

Postscission phenomena. After the fragments are separated at scission, they are further accelerated as the result of the large Coulomb repulsion. The initially deformed fragments collapse to their equilibrium shapes, and the excited primary fragments lose energy by evaporating neutrons. After neutron emission, the fragments lose the remainder of their energy by gamma radiation, with a lifetime of about 10^{-11} s. The kinetic energy and neutron yield as a function of mass are shown in Figs. 6 and 7. The variation of neutron yield with fragment mass is directly related to the fragment excitation energy. Minimum neutron yields are observed for nuclei near closed shells because of the resistance to deformation of nuclei with closed shells. Maximum neutron yields occur for fragments that are "soft" toward nuclear deformation. Hence, at the scission configuration, the fraction of the deformation energy stored in each fragment depends on the shell structure of the individual fragments. After scission, this deformation energy is converted to excitation energy, and, hence, the neutron yield is directly correlated with the fragment shell structure. This conclusion is further supported by the correlation between the neutron yield and the final kinetic energy. Closed shells result in a larger Coulomb energy at scission for fragments that have a smaller deformation energy

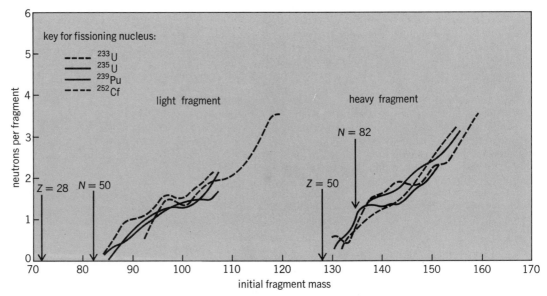

Fig. 7. Neutron yields as a function of fragment mass for four types of fission as determined from mass-yield data. Approximate initial fragment masses corresponding to various neutron and proton "magic numbers" N and Z are indicated. (*From J. Terrell, Neutron yields from individual fission fragments, Phys. Rev., 127:880–904, 1962*)

and a smaller number of evaporated neutrons.

After the emission of the prompt neutrons and gamma rays, the resulting fission products are unstable against β-decay. For example, in the case of thermal neutron fission of ^{235}U, each fragment undergoes on the average about three β-decays before it settles down to a stable nucleus. For selected fission products (for example, ^{87}Br and ^{137}I) β-decay leaves the daughter nucleus with excitation energy exceeding its neutron binding energy. The resulting delayed neutrons amount, for thermal neutron fission of ^{235}U, to about 0.7% of all the neutrons given off in fission. Though small in number, they are quite important in stabilizing nuclear chain reactions against sudden minor fluctuations in reactivity.

[JOHN R. HUIZENGA]

Bibliography: *Proceedings of the 3d IAEA Symposium on Physics and Chemistry of Fission*, Rochester, NY, 1973; R. Vandenbosch and J. R. Huizenga, *Nuclear Fission*, 1973.

Flame

A reaction front (or wave) in a gaseous medium into which the reactants flow and out of which the products flow. Solid or liquid particles may also be carried in by the gas and provide fuel for the flame, and sometimes the product gases contain solid particles, such as soot.

Flames produce heat and, usually, light. Hydrogen flames, for example, are almost invisible, although they do radiate in the infrared and ultraviolet regions of the spectrum. Nowadays, flames are used mostly for heat and very little for light. The greater part of the heat can be extracted only by direct conduction from the hot gases, because the radiant effect is small. For this reason the vigorous flames in a fireplace give very little warmth unless the air of the room is circulated through a duct in the firebox.

A seeming paradox is worthy of note, however. A faint, luminous phenomenon, known as a cool flame, is observed when, for example, a mixture of ether vapor and oxygen is slowly heated. Such flames are not really cool; they do produce some heat. But unlike ordinary flames that advance from layer to layer in the gas mainly by heat conduction, cool flames proceed by diffusion of reactive molecules, or free radicals, which initiate chemical processes as they go. The light from cool flames is a chemiluminescence that originates from excited molecules created by the reaction.

Combustion flames, in which oxygen or some other oxidizing substance combines with a fuel, are more common. But decomposition flames, in which a molecule such as ozone, O_3, breaks down into simple molecules, in this case oxygen, O_2, are also known. Not all combustion or decomposition reactions, however, produce a flame, even though they may evolve heat. For example, charcoal undergoes surface combustion and, although it glows, there is no flame. *See* COMBUSTION.

Premixed and diffusion flames. Gaseous fuels can be premixed with air or oxygen, in which case the mixture can be fed to a flame holder and burned in a very efficient manner. In a gas stove or a laboratory burner, an injection nozzle for fuel and ports for air are arranged so that air is drawn in by aspiration and mixed with fuel before it emerges at the flameholder.

Premixed flames can be either laminar or turbulent. The laminar flame front is smooth and sharply defined, and both the approach and exhaust flows are streamlined. The turbulent flame, on the other hand, may be described as bushy and is often quite noisy. The flame front fluctuates rapidly and is filled with swirls and eddies. Such turbulence results partly from the flow in the burner tube and partly from the flame itself. Turbulent flames are associated with high-speed flow; because of the large convoluted burning area, they are very efficient in terms of heat production per unit volume. Turbulent burning is therefore desirable in many practical applications.

Liquid and solid fuels burn in a so-called diffusion flame. A candle provides a good example. The fuel—wax, in this case—is melted and vaporized by the heat of the flame and emerges as a steady stream of vapor from the wick. Air, drawn in a convective stream toward the base of the flame, diffuses from the outside. The diffusion flame may therefore be pictured as an annulus, with fuel vapor diffusing from the inside and air from the outside.

Because certain parts are fuel-rich, hydrocarbon diffusion flames are yellow because of the presence of hot, incandescent carbon particles. If sufficient air is not drawn in to oxidize this carbon in later stages, the flame will be smoky. On the other hand, some diffusion flames, for example, the methyl alcohol flames, do not form carbon. Oil burners, coal furnaces, liquid-fueled rockets—in fact, the great majority of technical combustion devices—involve diffusion flames.

Bunsen flame. The bunsen burner has been extensively studied because it illustrates important concepts about flames. With a fuel of natural gas (methane, CH_4), or certain other hydrocarbons such as ethylene, C_2H_4, and with the air ports completely closed, a bright yellow flame is produced. This is the diffusion flame, in which the air is supplied entirely as secondary air that diffuses from the atmosphere. As the air ports are gradually opened, the yellow color diminishes and finally disappears. The flame then consists of three distinct regions.

The innermost region is dark. It is simply the cold, unburned gas. If a match head is suspended by a pin just above the mouth of the burner, it remains unlighted because it is kept cool by the flow of cold gas around it (Fig. 1).

The dark region of unburned gas is capped by a bright bluish-green, cone-shaped mantle, less than 1 mm thick. In this narrow layer, the fuel reacts with the premixed air. In the usual bunsen flame, however, the amount of premixed air is insufficient for complete combustion. Therefore, the gas emerging from the reaction zone still has fuel value. Analysis shows it to contain carbon monoxide, CO, and hydrogen, H_2, but none of the original methane. This gas can react with more oxygen to give carbon dioxide, CO_2, and water, H_2O.

The additional oxygen needed for combustion of gases emerging from the inner cone is supplied by secondary air from the atmosphere. Above the inner cone, therefore, is a pale bluish-violet diffusion flame. This is called the outer cone.

By means of two concentric tubes, called a Smithell's burner, the inner and outer cones can be separated. The gas between the cones in this burn-

FLAME

match

Fig. 1. Experiment showing that the flame cone is hollow. The match head will not light.

er emits no light. The light from a methane flame is emitted only by regions in which chemical reaction is taking place; none of the light comes from the hot combustion products. Spectroscopy shows that the flame radiation is produced by short-lived radicals such as C_2 and CH.

When the ports of the burner are opened to admit more primary air, the outer cone wanes and the inner cone becomes darker. Eventually, when enough primary air is available for complete combustion, the outer cone disappears.

Flameholder. A flame is stabilized on a burner tube because of heat flow to the cold rim. Drainage of heat lowers the flame speed at this point and, as a result, the front becomes anchored near the rim. If the front should momentarily move farther from the rim, less heat would drain, and the front would speed up and return toward its original position near the rim. On the other hand, if the front should shift toward the rim, it would lose more heat, the burning velocity would drop, and the gas stream would drive the flame back to its original place. Any solid surface may act in this way as a flameholder.

The height of the flame cone above the burner rim adjusts itself within wide limits to accommodate the gas flow. The surface of the cone is actually curved as a result of the parabolic flow distribution in the burner tube. With larger flow rates the cone height increases, and with smaller rates it decreases. However, above a certain limiting flow, the flame lifts off the burner and blows out; this is called blowoff. On the other hand, if the flow is sufficiently reduced, the cone becomes almost flat. A further decrease at this point will cause the flame to strike back into the burner tube; this is called flashback.

Between the flashback and blowoff limits, the cone assumes a height such that the volume rate of flow divided by the surface area of the cone retains a constant value, which is called the burning velocity and denoted by S_u.

Burning velocity. This is a basic property of a combustible gas mixture. It is affected by external factors only when heat is drained directly from the reaction zone near a flameholder. If, by special arrangements, a gas stream of uniform velocity over the cross section is produced, the flame front can be made almost perfectly flat. Under such a condition, the flame is stable only when the linear flow rate is exactly matched to the value of S_u.

Burning velocity always increases with rising gas temperature and usually shows a dependence on initial pressure also, but the behavior in this respect differs for various mixtures.

Burning velocities range through wide limits; methane-air, for example, has a velocity of only 35 cm/sec, whereas certain hydrogen-oxygen mixtures burn at a rate of 1200 cm/sec. No simple expression relates burning velocity and other properties of the mixture.

The theory of flames is, in fact, very complex. The properties influencing S_u are the heat conductivity, the diffusivity, the flame temperature, and the reaction rate. Because diffusivities of the various molecular species involved in the reaction differ and because the concentration of these species both affects, and is affected by, the rate of reaction, the mathematical analysis of what takes place in the burning zone is a formidable problem. The principles are well understood, but the interdependent factors are so complex that no general solution of the differential equations is possible. They have been solved in some cases by computers.

In spite of actual complications, there is no mystery in the mechanism of flame propagation in general. Both heat conduction and diffusion of reactive species serve to start reaction in successive layers of gas. When very mobile and reactive H atoms are involved, their diffusion from the burned to the unburned region dominates the process, but usually both diffusion and heat conduction are important. By considering only the latter and disregarding the effect of diffusion (which will never be strictly permissible), a good intuitive picture of the flame can be formed. Conduction of heat depends on the temperature gradient; and, by quite straightforward reasoning, it can be shown that the temperature distribution through the flame will always be S-shaped. This conclusion is indeed confirmed by actual thermocouple measurements (Fig. 2). The net rate at which heat is received by an element of gas depends on the curvature of this temperature profile. Where it is curving upward, at the lower part of the S, the gas is receiving more heat than it is losing; and where it is curving downward, in the upper portion of the curve, the gas is losing more heat than it is receiving. The first region may be considered as the preheat zone. Here, the gas is still not hot enough to react at an appreciable rate. In fact, most of the reaction occurs only near the region of highest temperature because the reaction rate, following the exponential Arrhenius law, increases very sharply as the temperature rises. Opposing the effect of temperature as the reaction nears completion, however, is the slowing effect resulting from the depletion of reactant concentrations. According to a much simplified picture, the flame propagates by the flow of heat from the reaction zone to the preheat zone. An element of gas passing through the flame is heated up in the preheat zone to a temperature at which it begins to react. Then, as it passes on through the reaction zone, it self-heats and, at the same time, hands back the extra heat it had previously received to the succeeding element of gas, now in the preheat zone. Thus the flame is maintained in a steady state. *See* EXPLOSION AND EXPLOSIVE; FUEL, FOSSIL.

[WILLIAM E. GORDON]

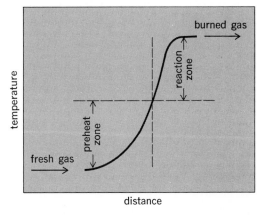

Fig. 2.   Temperature distribution through flame front.

Bibliography: R. M. Fristrom and A. A. Westenberg, *Flame Structure*, 1965; A. G. Gaydon and H. G. Wolfhard, *Flames: Their Structure, Radiation and Temperature*, 2d ed., 1960; B. Lewis and G. von Elbe, *Combustion, Flames and Explosions of Gases*, 1951; M. W. Thring, *The Science of Flames and Furnaces*, 2d ed., 1962; F. J. Weinberg, *Optics of Flames*, 1963.

Force

Force may be briefly described as that influence on a body which causes it to accelerate. In this way, force is defined through Newton's second law of motion.

This law states in part that the acceleration of a body is proportional to the resultant force exerted on the body and is inversely proportional to the mass of the body. An alternative procedure is to try to formulate a definition in terms of a standard force, for example, that necessary to stretch a particular spring a certain amount, or the gravitational attraction which the Earth exerts on a standard object. Even so, Newton's second law inextricably links mass and force.

Many elementary books in physics seem to expect the beginning student to bring to his study the same kind of intuitive notion concerning force which Isaac Newton possessed. One readily thinks of an object's weight, or of pushing it or pulling it, and from this one gains a "feeling" for force. Such intuition, while undeniably helpful, is hardly an adequate foundation for the quantitative science of mechanics.

Newton's dilemma in logic, which did not trouble him greatly, was that, in stating his second law as a relation between certain physical quantities, he presumably needed to begin with their definitions. But he did not actually have definitions of both mass and force which were independent of the second law. The procedure which today seems most free of pitfalls in logic is in fact to use Newton's second law as a defining relation.

First, one supposes length to be defined in terms of the distance between marks on a standard object, or perhaps in terms of the wavelength of a particular spectral line. Time can be supposed similarly related to the period of a standard motion (for example, the rotation of the Earth about the Sun, the oscillations of the balance wheel of a clock, or perhaps a particular vibration of a molecule). Although applying these definitions to actual measurements may be a practical matter requiring some effort, a reasonably logical definition of velocity and acceleration, as the first and second time derivatives of vector displacement, follows readily in principle.

Absolute standards. Having chosen a unit for length and a unit for time, one may then select a standard particle or object. At this juncture one may choose either the absolute or the gravitational approach. In the so-called absolute systems of units, it is said that the standard object has a mass of one unit. Then the second law of Newton defines unit force as that force which gives unit acceleration to the unit mass. Any other mass may in principle be compared with the standard mass (m) by subjecting it to unit force and measuring the acceleration ($\mathbf{a}$), with which it varies inversely. By suitable appeal to experiment, it is possible to conclude that masses are scalar quantities and that forces are vector quantities which may be superimposed or resolved by the rules of vector addition and resolution.

In the absolute scheme, then, Eq. (1) is written

$$\mathbf{F} = m\mathbf{a} \qquad (1)$$

for nonrelativistic mechanics; here boldface type denotes vector quantities. The quantities on the right of Eq. (1) are previously known, and this statement of the second law of Newton is in fact the definition of force. In the absolute system, mass is taken as a fundamental quantity and force is a derived unit of dimensions MLT^{-2} ($M =$ mass, $L =$ length, $T =$ time).

Gravitational standards. The gravitational system of units uses the attraction of the Earth for the standard object as the standard force. Newton's second law still couples force and mass, but since force is here taken as the fundamental quantity, mass becomes the derived factor of proportionality between force and the acceleration it produces. In particular, the standard force (the Earth's attraction for the standard object) produces in free fall what one measures as the gravitational acceleration, a vector quantity proportional to the standard force (weight) for any object. It follows from the use of Newton's second law as a defining relation that the mass of that object is $m = w/g$, g being the magnitude of the gravitational acceleration and w being the magnitude of the weight. The derived quantity mass has dimensions FT^2L^{-1}.

Because the gravitational acceleration varies slightly over the surface of the Earth, it may be objected that the force standard will also vary. This may be avoided by specifying a point on the Earth's surface at which the standard object has standard weight. In principle, then, the gravitational system becomes no less absolute than the so-called absolute system.

Composition of forces. By experiment one finds that two forces of, for example, 3 units and 4 units acting at right angles to one another at point 0 produce an acceleration of a particular object which is identical to that produced by a single 5-unit force inclined at arccos 0.6 to the 3-unit force, and arccos 0.8 to the 4-unit force (see figure). The laws of vector addition thus apply to the superposition of forces.

Conversely, a single force may be considered as equivalent to two or more forces whose vector sum equals the single force. In this way one may select the component of a particular force which may be especially relevant to the physical problem. An example of this would be the component of a railroad car's weight along the direction of the track on a hill.

Statics is the branch of mechanics which treats forces in nonaccelerated systems. Hence, the resultant of all forces is zero, and critical problems are the determination of the component forces on the object or its structural parts in static equilibrium. Practical questions concern the ability of structural members to support the forces or tensions.

Specially designated forces. If a force is defined for every point of a region and if this so-called vector field is irrotational, the force is designated conservative. Physically, it is shown in the development of mechanics that this property re-

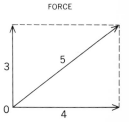

Vector addition of forces.

quires that the work done by this force field on a particle traversing a closed path is zero. Mathematically, such a force field can be shown to be expressible as the (conventionally negative) gradient of a scalar function of position V, as in Eq. (2).

$$\mathbf{F} = -\nabla V \qquad (2)$$

A force which extracts energy irreversibly from a mechanical system is called dissipative, or nonconservative. Familiar examples are frictional forces, including those of air resistance. Dissipative forces are of great practical interest, although they are often very difficult to take into account precisely in phenomena of mechanics.

The force which must be directed toward the center of curvature to cause a particle to move in a curved path is called centripetal force. For example, if one rotates a stone on the end of a string, the force with which the string pulls radially inward on the stone is centripetal force. The reaction to centripetal force (namely, the force of the stone on the string) is called centrifugal force.

Methods of measuring forces. Direct force measurements in mechanics usually reduce ultimately to a weight comparison. Even when the elastic distortion of a spring or of a torsion fiber is used, the calibration of the elastic property will often be through a balance which compares the pull of the spring with a calibrated weight or the torsion of the fiber with a torque arising from a calibrated weight on a moment arm.

In dynamic systems, any means of measuring acceleration—for example, through photographic methods or radar tracking—allows one to calculate the force acting on an object of known mass.

Units of force. In addition to use of the absolute or the gravitational approach, one must contend with two sets of standard objects and lengths, the British and the metric standards. All systems use the second as the unit of time. In the metric absolute system, the units of force are the newton and the dyne. The poundal is the force unit in the British absolute system, whereas the British gravitational system uses the pound. Metric gravitational systems are rarely used. Occasionally one encounters terms such as gram-force or kilogram-force, but no corresponding mass unit has been named.

[GEORGE E. PAKE]

Bibliography: M. L. Bullock, Systems of units in mechanics: A summary, *Amer. J. Phys.*, 22:291–299, 1954; C. Kittel, W. D. Knight, and M. A. Ruderman, *Mechanics*, 1965; R. B. Lindsay and H. Margenau, *Foundations of Physics*, rev. ed., 1957.

Forest resources

Forest resources consist of two separate but closely related parts: the forest land and the trees (timber) on that land. In the United States, forests cover one-third of the total land area of the 50 states, in total, about 754×10^6 acres (10^6 acres = 4046 km²). The fact that 1 of every 3 acres of the United States is tree-covered makes this land and its condition a matter of importance to every citizen. Recognizing this importance, Congress has charged the U.S. Forest Service with the responsibility of making periodic appraisals of the national timber situation. Most of the following data are

taken from the appraisal published in 1973.

About 500×10^6 acres, or about two-thirds, of the total forest area in the United States is classified as commercial forest land, that is, land suitable and available for growing commercial crops of wood. The remaining 254×10^6 acres is noncommercial forest consisting either of land with soils so poor or rocky that it is incapable of growing a commercially useful timber stand or of forest land being held for recreation or other nontimber purposes. Table 1 gives details of the distribution of forest land in the four main regions of the United States.

Most of the noncommercial forests are in public ownership, including approximately 17×10^6 productive forest acres legally withdrawn for such uses as national parks, state parks, and national forest wilderness areas. Another 2.7×10^6 acres, classed as "deferred," is under study for possible inclusion in the wilderness system. Of the remaining unproductive forest, about 113×10^6 acres are in Alaska (part of the Pacific Coast region). Most discussions of forest resources, however, concentrate upon the commercial areas, and these lands are the primary concern of the following discussion.

The South alone has about 38.5% of the total commercial area; 35.6% is in the North, and the remaining 25.9% in the West. Within this general pattern are very large differences: In Maine, for example, fully 85% of the land surface is covered with commercial forests; North Dakota and Nevada are at the opposite extreme, with less than 1% of their area similarly utilized.

For many years, changes in United States agriculture led to abandonment of marginal farms, which rapidly reverted to forest. This "new forest" was more than enough to offset those areas lost to highways, pipelines, urban development, and such. In fact, between 1943 and 1963 the total commercial forest area increased by about 31.6×10^6 acres. Between 1963 and 1970, however, the total area of commercial forest land declined about 8.4×10^6 acres, mostly in the South and in the Rocky Mountains. Some of this reduction, particularly in the West, came about as the result of shifts of public forest land to reserved or deferred status to satisfy public demand for recreation uses. In the South, timberland was cleared for crop production and for pasture. In all regions, substantial areas of forest land have been taken over by suburban development, highways, reserves, and other nontimber uses.

Forest types. There are literally hundreds of tree species used for commercial purposes in the United States. The most general distinction made is that between softwoods (the conifers, or cone-bearing trees, such as pine, fir, and spruce) and hardwoods (the broad-leafed trees, such as maple, birch, oak, hickory, and aspen). Viewed nationally, about 52% of United States commercial forest land is occupied by eastern hardwood types. Softwoods of various kinds make up 42%, western hardwoods only 3%; and 4% of the area is unstocked. Oak-hickory stands cover the largest area, accounting for 23% of all commercial timberland in 1970. The oak-pine type (14% of the eastern hardwood area) is mostly in the South and is primarily the residual resulting from the cutting of merchantable pine

Table 1. Distribution of forest land in the United States*

Type of forest land	Area, 10^6 acres†				
	Total	North	South	Rocky Mountain	Pacific Coast
Commercial forest	499.7	177.9	192.5	61.6	67.6
Noncommercial forest					
Unproductive	233.9	4.2	17.6	66.5	145.6
Productive reserves	17.2	4.3	1.7	7.9	3.3
Deferred	2.7	0	0	2.3	0.432
Total	253.8	8.6	19.3	76.6	149.3
Total forest land	753.5	186.5	211.9	138.2	216.9

*From U.S. Forest Service, *The Outlook for Timber in the United States*, Forest Resour. Rep. no. 20, USDA, October 1973.

†10^6 acres = 4046 km².

from mixed pine-hardwood forests. Increasing during the last few decades these stands have been converted to pine by the killing or cutting of hardwoods and, often, by the planting of pine.

Of the eastern hardwood forests, 44% are oak-hickory types, containing a large number of species but characterized by the presence of one or more species of oak or hickory. Other important eastern types are the maple-birch-beech (found throughout the New England, Middle Atlantic, and Lake states regions), the oak-gum-cypress forests (primarily in the Mississippi Delta and other southern river bottoms), and the aspen-birch type of the Lake states (relatively short-lived species that followed logging and fires). The bottomland hardwoods (oak-gum-cypress type) were reduced about 20% between 1962 and 1970, primarily by the clearing of forests for agriculture. For many years these forests have supplied much of the quality hardwoods in the United States.

Softwood types dominate the western forests, altogether occupying 85% of the region's commercial forest area. Douglas fir and ponderosa pine, the principal types, together constitute 47% of the region's commercial timberland. The western softwood types are the principal sources of lumber and plywood in the United States. Nearly all the commercial forest area of coastal Alaska is of the hemlock—Sitka spruce type. Hardwoods, mostly in Washington and Oregon, occupy only 10% of the West's commercial forest area, but have increased almost 2×10^6 acres since 1962 as the Douglas fir forests have been cut.

Growth. Growth on these areas has been more than enough to match the harvest since the mid-1960s, but the fact that growth and drain are in approximate balance provides no assurance that all is well. Much of the growth is by low-quality hardwoods in the East, yet about two-thirds of United States demand for the raw materials of the forest is met by the softwood production of the West and South. Potentially the most productive forest land is that of the West Coast states, where it is estimated that 24×10^6 acres are capable of growing more than 120 ft³ (10 ft³ = 0.28 m³) per acre annually, and that another 16×10^6 acres could produce more than 85 ft³/acre per year. The next most productive area is the South, with only 13×10^6 acres of highest-quality forest (more than 120 ft³/acre annually) but more than 53×10^6 acres

of good quality land (85–120 ft³/acre annually), and 90×10^6 acres producing 50 ft³/acre per year or more. The North is considerably less well off as far as forest growth rates are concerned.

Stocking. Second to the potential productivity of the soil itself, the most important factors in the growth of timber are probably stocking (the number of trees per acre) and the age of the trees. Gradually, as fire protection and forest management have been extended to larger and larger areas, and as the practice of forestry has won more widespread acceptance by industry and by the general public, the stocking of forest lands has been increased and is still increasing. There is still a long way to go, however, before commercial forest areas can be classified as well stocked. In 1970 only one-fifth of all hardwood forest land was estimated to support trees of desirable species, form, and size, and, in 21×10^6 acres, less than 10% of the land was tree-covered.

The commercial forests contained a truly vast amount of sound wood—about $714,545 \times 10^6$ ft³ ($20,229 \times 10^6$ m³)—at the beginning of 1970. Only 9% of this timber consisted of trees of poor or diseased condition or otherwise useless for harvesting, or of dead trees that might still be utilized. And almost two-thirds of the total was in trees sufficiently large to yield at least one sawlog; such

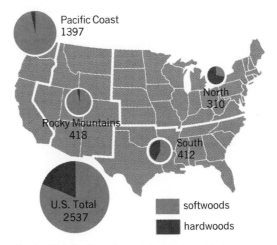

Fig. 1. Distribution of sawtimber in 10^9 bd ft for all species as of Jan. 1, 1963. (*U.S. Department of Agriculture*)

Table 2. Summary of net annual growth and removals of growing stock and sawtimber, by species group and region*

Section and species group	Growing stock, 10^9 ft^3†			Sawtimber, 10^9 bd ft‡		
	Net growth	Removals	Growth-removal ratio	Net growth	Removals	Growth-removal ratio
North						
Softwoods	1.4	0.6	2.2	3.6	2.1	1.7
Hardwoods	4.2	1.8	2.3	10.1	6.8	1.5
South						
Softwoods	5.4	4.0	1.4	20.1	15.0	1.3
Hardwoods	3.2	2.5	1.3	7.9	7.8	1.0
Rocky Mountains						
Softwoods	1.3	0.9	1.4	4.9	5.4	0.9
Hardwoods	0.1	—	26.2	0.1	—	11.7
Pacific Coast						
Softwoods	2.6	4.1	0.6	11.6	25.2	0.5
Hardwoods	0.5	0.1	4.1	1.5	0.4	4.0
United States						
Softwoods	10.7	9.6	1.1	40.3	47.7	0.8
Hardwoods	7.9	4.4	1.8	19.7	15.0	1.3

*From U.S. Forest Service, *The Outlook for Timber in the United States*, Forest Resour. Rep. no. 20, USDA, October 1973.

†10^9 ft$^3 = 28{,}316{,}846$ m^3.

‡10^9 bd ft $= 2{,}359{,}737$ m^3, as nominal recovered lumber.

trees are called, therefore, sawtimber (Fig. 1). Nearly 75% of this sawtimber inventory was in conifers, with the remainder in hardwoods, a fact that highlights an important facet of timber distribution in the United States: The western states have only 26% of the commercial forest land of the United States, but on this land 51% of the growing stock and 66% of the sawtimber are found. Further emphasizing the importance of the West to timber supplies is the fact that about 22% of all United States sawtimber is Douglas fir. Oaks are the principal hardwood sawtimber species.

Timber age, growth, and removal. The other factor affecting timber growth, that of age, must be considered before forest growth can be understood. The very heavy stands of virgin timber in the West are growing relatively slowly. Only as these old stands are replaced with young vigorous trees will growth of western forests begin to balance the cut they support. In 1970 growth of western softwood sawtimber averaged only 51% of the removals, with much of the imbalance in the highly important Douglas fir and ponderosa pine types. With southern pines, on the other hand, total

growth was 1.4 times removals, and in the important sawtimber size classes, the ratio was 1.3. The hardwood growth-removal ratio for the entire nation, 1.8, appears to be quite satisfactory; the sawtimber growth-removal ratio is 1.3. Net annual growth and removal data are summarized in Table 2.

Even with reasonably accurate data on growth and removals, there is room for misinterpretation of the forest situation. Total growth exceeds removals in most regions, but much of that growth is occurring on trees of low quality. Although more than a third of the hardwood sawtimber volume in 1970 was in preferred species, when quality or appearance is important (hard maple, yellow birch, black walnut or black cherry, select white oak or red oak, and so on), only 12% of the total growing stock volume was in trees 19 in. (48 cm) or more in diameter at breast height. Since timber size is closely related to quality, the sizes now available severely limit the supply of high-quality hardwood. Moreover, a substantial portion of the growth is being made by trees too widely scattered to be commercially harvestable, or by trees that are lo-

Table 3. Area of commercial timberland in the United States by type of ownership and section, Jan. 1, 1970*

	Total United States		North, 10^3 acres	South, 10^3 acres	Rocky Mountains, 10^3 acres	Pacific Coast, 10^3 acres
	Area, 10^3 acres†	Proportion, %				
Federal						
National forests	91,924	18	10,458	10,764	39,787	30,915
Bureau of Land Management	4,762	1	75	11	2,024	2,652
Bureau of Indian Affairs	5,888	1	815	220	2,809	2,044
Other	4,534	1	963	3,282	78	211
Total	107,109	21	12,311	14,277	44,699	35,822
State	21,423	4	13,076	2,321	2,198	3,828
County and municipal	7,589	2	6,525	681	71	312
Forest industry	57,341	14	17,563	35,325	2,234	12,219
Farm	131,135	26	51,017	65,137	8,379	6,602
Miscellaneous private	165,101	33	77,409	74,801	4,051	8,840
All ownerships	499,697	100	177,901	192,542	61,632	67,622

*From U.S. Forest Service, *The Outlook for Timber in the United States*, Forest Resour. Rep. no. 20, USDA, October 1973.

†10^3 acres $= 4.05$ km^2.

cated in areas of difficult accessibility. It is one thing—and a very good thing—to have timber cut in the United States balanced by forest growth; it is quite another to have that cut balanced by quality growth of species that are demanded by forest industries and located so that the timber can be logged profitably.

Ownership. The future condition of forest land in the United States is dependent to a very large extent on the decisions of the people who own these areas. The key factors in understanding the forest situation, therefore, are the forest land ownership pattern and the attitude these owners have toward forest management and hence toward the future of the forest lands they hold.

About 73% of the commercial forest is held in private ownership (363.6×10^6 acres). Of the remaining 27% held by various public owners, about 18% is in national forests, about 3% in other Federal ownership, 4% in state holdings, and 2% under county and municipal control. Among private owners, three major classes can be distinguished: industrial, farm, and a very heterogeneous group labeled miscellaneous. Table 3 gives details about forest ownership, as well as the distribution of privately owned forests, in the four major regions of the United States.

Private owners. About 14% of the privately owned commercial forest area is in the hands of industry, with the remainder divided between farmers (26%) and the miscellaneous private owners (33%).

Some of the most productive forest land of the United States is in private industrial holdings. Pulp and paper companies lead the forest-based industries, with much of their forest land concentrated in the southern states. More and more, however, as forest industries become integrated, it is increasingly difficult to make distinctions between pulp and paper, lumber, and plywood companies, and certainly the distinctions are much less meaningful than in former years.

The miscellaneous private category, which holds 33% of the privately owned forest land, consists of a tremendous variety of individuals and groups, ranging from housewives to mining companies. Relatively few such owners are holding this land for commercial timber production. Some owners, such as railroad companies or oil corporations, may indeed be interested in producing timber while holding the subsurface mineral rights, but most miscellaneous private owners are interested in other, nontimber, objectives.

Farmers own what is potentially the best forest land in the country, but it is also the most poorly managed and the most poorly stocked of all ownership categories. Most farm forests have been cut over several times; and, because farmers are primarily interested in the production of other kinds of crops, they are seldom concerned with the condition of their woodlands. The relatively poor condition of farm-owned woodlands has been a source of much disappointment, discussion, and considerable action on the part of public conservation agencies. Whether this situation is of crucial significance to the forest reserves of the United States is a matter of dispute. In past years some professional foresters have argued that these farm woodlands

should somehow be made to contribute their full share to the timber supply of the nation. Others, of a more economic persuasion, have argued that, as long as farmers have opportunities for investing time and money in ways that will yield a greater return, they should not be expected to worry about timber production. Farm owners are often unfamiliar with forest practices, usually lack the capital required for long-term investments, and in many cases are simply not interested in growing trees.

Public owners. Public agencies of several kinds hold large forest areas, the most important being the national forests, which contain 91.9×10^6 acres (10^6 acres $= 4050$ km^2) of commercial forest land and which are managed and administered by the U.S. Forest Service, a bureau of the U.S. Department of Agriculture. The Bureau of Land Management oversees about 4.76×10^6 acres, and the Bureau of Indian Affairs, another 5.89×10^6 acres. Various other agencies, especially those of the armed services, administer the remaining 4.53×10^6 acres under Federal supervision. State ownerships total 21.4×10^6 acres, and counties and municipalities control another 7.59×10^6 acres. Most of the public holdings are managed under multiple-use principles. As on most forest land, wood has always been the principal product of the national forests, but in recent years increasing attention has been given to recreation, wildlife, forage, and water. *See* CONSERVATION OF RESOURCES.

Public holdings now contain 58% of the softwood growing stock but only 17% of the hardwoods. Sixty-three percent of all softwood sawtimber is publicly owned, and most is in national forests. The high concentration of sawtimber in these areas makes many wood-based industries in the United States highly dependent upon government-owned raw material supplies.

World forest resources. In terms of forest land, the United States is far better off than many nations, but not as rich as others. For example, more than 50% of the land surface in South America is forested; in Great Britain, on the other hand, only a little more than 7% of the land surface is so utilized. Actually, about one-third of the world's land surface is forested—the same average percentage as in the United States as a whole. The Soviet Union has a greater forest area than all of North America and Europe combined. In fact, the Soviet Union possesses about one-third of the productive forests of the world. A substantial proportion of these forests, however, is probably unsuitable for commercial use, because of low productivity, long distance from markets, or rugged terrain.

In many countries, and especially in tropical areas, forests have not been well explored, nor have their boundaries been defined. In such circumstances, only approximate areas and general conditions can be given. Moreover, since forests reflect differences in soil, climate, situation, and past land use, merely listing the forest types of the world would take several pages. Yet there is value in distinguishing very generalized forest types by broad locational patterns (Fig. 2).

Coniferous and temperate mixed forests. In Europe, the Soviet Union, North America, and Japan, the forests are predominantly coniferous, a fact

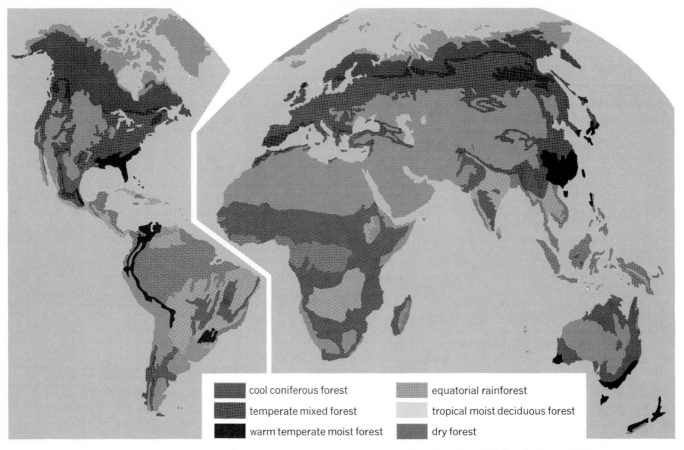

Fig. 2.   The world's forests. *(From Economic Atlas of the World, 3d ed., Oxford University Press, 1965)*

that has been of great significance in shaping the pattern and nature of wood use in the industrialized part of the world. Closely associated with the industrialized nations are the temperate mixed forests, which normally are composed of a high proportion of conifers along with a few broad-leafed species.

Most of the temperate mixed forests are in use, for these heavily populated areas have well-developed transportation systems. Growth rates in both the coniferous and temperate mixed forests of the North Temperate Zone are similar to those already given for the United States, which is an excellent example of the general region.

Tropical rain forests. These forests are made up exclusively of broad-leafed species and include the bulk of the volume of the world's broad-leafed woods. They are concentrated in and around the Amazon Basin in South America, in western and west-central Africa, and in Southeast Asia. Generally characterized by sparse population and only slight industrial development, these forests have been little used. The fact that they typically contain many species within a small area (as many as a hundred species per acre) has also served to limit their use, even though tropical rain forests include some of the most valuable of all woods, such as mahogany, cedar, and greenheart (South America); okoume, obeche, lima, and African mahogany (Africa); and rosewood and teak (Asia).

The rate of growth in the tropics can be very high, and someday it may be possible to obtain a major part of the world's wood fiber needs from the more than 2×10^9 acres of these forests. Today, however, so little is known about many of the species, and so few areas are under any form of systematic management, that little reliance can be placed on these vast areas to satisfy the future wood needs of humanity.

Savanna. Most of the other forests in the tropics and subtropics are dry, open woodlands, or savannas. These forests contain low volumes per acre, mostly in small sizes, and only a few of the species are of commercial value. Much of Africa (excluding western and west-central Africa) is of this low-yielding type.

Management. In recent years intensive afforestation has received considerable attention. The use of measures such as careful soil preparation prior to planting, application of fertilizer, and even irrigation to adapt the environment to high-yielding species can result in an enormous increase in returns from forests. For example, yields of 400–500 ft³/acre are common with eucalyptus in both South America and Africa, or with poplars in southern Europe. Only slightly smaller yields are obtained from the fast-growing pines under similarly intensive management. The high-yield potential of cultivated forests makes them much more important than their area might at first indicate. According to reports, most nations intend to have these forests make an even greater contribution in the future.

[G. R. GREGORY]

Bibliography: Food and Agriculture Organization of the United Nations, *Wood: World Trends and Prospects*, Freedom from Hunger Basic Study no. 16, 1967; U.S. Forest Service, *The Outlook for Timber in the United States*, Forest Resour. Rep. no. 20, USDA, October 1973.

Free energy

A term in thermodynamics which in different treatments may designate either of two functions defined in terms of the internal energy E or enthalpy H, and the temperature-entropy product TS.

The function $(E - TS)$ is the Helmholtz free energy and is the function ordinarily meant by free energy in European references. The Gibbs free energy is the function $(H - TS)$. For the Lewis and Randall school of American chemical thermodynamics, this is the function meant by the free energy F. To avoid confusion with the symbol F as applied elsewhere to the Helmholtz free energy, the symbol G has also been used. A recent development was the introduction of the name free enthalpy, with symbol G, for the Gibbs function. *See* WORK FUNCTION (THERMODYNAMICS).

Theory. For a closed system (no transfer of matter across its boundaries), the work which can be done in a reversible isothermal process is given by the series shown in Eq. (1). For these conditions,

$$W_{rev} = -\Delta A = -\Delta(E - TS) = -(\Delta E - T \Delta S) \quad (1)$$

$T \Delta S$ represents the heat given up to the surroundings. Should the process be exothermal, $T\Delta S < 0$, then actual work done on the surroundings is less than the decrease in the internal energy of the system. The quantity $(\Delta E - T \Delta S)$ can then be thought of as a change in free energy, that is, as that part of the internal energy change which can be converted into work under the specified conditions. This then is the origin of the name free energy. Such an interpretation of thermodynamic quantities can be misleading, however; for the case in which $T \Delta S$ is positive, Eq. (1) shows that the decrease in "free" energy is greater than the decrease in internal energy.

For constant temperature and pressure in a reversible process the decrease in the Gibbs function G for the system again corresponds to a free-energy change in the above sense, since it is equal to the work which can be done by the closed system other than that associated with its change in volume ΔV under the given constant pressure P. The relations shown in Eq. (2) can be formed since $\Delta H = \Delta E + P \Delta V$.

$$\Delta G = -(\Delta H - T \Delta S) = W_{net} = W_{rev} - P \Delta V \quad (2)$$

Each of these free-energy functions is an extensive property of the state of the thermodynamic system. For a specified change in state, both ΔA and ΔG are independent of the path by which the change is accomplished. Only changes in these functions can be measured, not values for a single state.

The thermodynamic criteria for reversibility, irreversibility, and equilibrium for processes in closed systems at constant temperature and pressure are expressed naturally in terms of the function G. For any infinitesimal process at constant temperature and pressure, $-dG \geq \delta w_{net}$. If δw_{net} is never negative, that is, if the surroundings do no net work on the system, then the change dG must be negative or zero. For a reversible differential process, $-dG > \delta w_{net}$; for an irreversible process, $-dG > \delta w_{net}$. The free energy G thus decreases to a minimum value characteristic of the equilibrium state at the given temperature and pressure. At equilibrium, $dG = 0$ for any differential process taking place, for example, an infinitesimal change in the degree of completion of a chemical reaction. A parallel role is played by the work function A for conditions of constant temperature and volume. Because temperature and pressure constitute more convenient working variables than temperature and volume, it is the Gibbs free energy which is the more commonly used in thermodynamics.

Partial molal quantities. For a particular homogeneous phase in the absence of surface, gravitational, and magnetic forces, the free energy G depends on the numbers of moles of the constituents present, the temperature T, and the pressure P. Let Ω represent the total number of constituents, n_i the number of moles of typical constituent i, and designate by subscript n constant composition, by subscript n_j constancy of the number of moles of all constituents except n_i; then Eq. (3) is formed.

$$dG(T,P,n_1, \ldots ,n_\Omega)$$
$$= \left(\frac{\partial G}{\partial T}\right)_{P,n} dt + \left(\frac{\partial G}{\partial p}\right)_{T,n} + \sum_{i=1}^{\Omega} \left(\frac{\partial G}{\partial n_i}\right)_{T,P,n_j} dn_i \quad (3)$$

In Eq. (3) the term $\left(\frac{\partial G}{\partial n_i}\right)_{T,P,n_i}$ is the chemical potential μ_i of the ith constituent. It is identical to the partial molal free energy $\overline{G}_i$ of Lewis and Randall. Therefore the relationship expressed as Eq. (4) holds.

$$dG = -S\, dT + V\, dP + \sum_{i=1}^{\Omega} \mu_i\, dn_i \quad (4)$$

Because the chemical potentials at constant T,P are intensive variables whose values are fixed, like that of the density, by the relative number of moles of the various constituents present, and are independent of the total mass of the phase, this equation can be integrated for constant T,P and relative composition starting from $n_i = 0$ to obtain Eq. (5).

$$G(T,P,n_1, \ldots ,n_\Omega) = \sum_{i=1}^{\Omega} n_i \mu_i \quad (5)$$

This yields Eq. (6). Consistency with the expression for dG in Eq. (4) requires that Eq. (7) hold.

$$dG = \sum_{i=1}^{\Omega} \mu_i\, dn_i + \sum_{i=1}^{\Omega} n_i\, d\mu_i \quad (6)$$

$$S\, dT - V\, dP + \sum_{i=1}^{\Omega} n_i\, d\mu_i = 0 \quad (7)$$

This is the Gibbs-Duhem equation. For constant temperature and pressure, this relation imposes a condition on the composition variation of the set of chemical potentials.

Heterogeneous systems. The free energy of a closed, heterogeneous system is the sum of the free energies of its various phases. In the absence of such a constraint as provided by the subdivision of the system by a rigid, semipermeable membrane, the general thermodynamic criterion of equilibrium requires that the temperature and pressure be uniform throughout the system and that the chemical potential of each constituent have a common value for all phases in which it is present. Further, if any of the constituents can be formed from others, the chemical potentials of the reactants and products are related in accordance with the stoichiometry of the reaction equation. Thus, for the reaction in Eq. (8), at equilibrium Eq.

$$A + 2B \rightleftharpoons 3C + 4D \qquad (8)$$

(9) can be formed. Expressing each chemical po-

$$\mu_A + 2\mu_B = 3\mu_C + 4\mu_D \qquad (9)$$

tential μ_i in terms of the standard value μ_i^0 and its associated activity term $RT \ln a_i$ results in Eq. (10).

$$RT \ln \left(\frac{a_C{}^3 a_D{}^4}{a_A a_B{}^2} \right)_{equil} = -(3\mu_C{}^0 + 4\mu_D{}^0 - \mu_A{}^0 - 2\mu_B{}^0)$$
$$= -\Delta G^0 \qquad (10)$$

In Eq. (10) ΔG^0 is called the standard free-energy change for the reaction. Its value depends on the standard states chosen, but for a given temperature and pressure, it is a constant characteristic of the reaction involved. A true equilibrium constant K then results as shown by Eqs. (11) and (12). If the

$$K = \left(\frac{a_C{}^3 a_D{}^4}{a_A a_B{}^2} \right)_{equil} \qquad (11)$$

$$RT \ln K = -\Delta G^0 \qquad (12)$$

pressure for each standard state is fixed and independent of the pressure of the reaction system, ΔG^0 and hence K are functions of temperature only. This is the conventional approach in treating gas-phase equilibria, but not ordinarily for condensed phases.

Since the activities can be correlated with partial pressures or concentrations through fugacity coefficients or activity coefficients, this thermodynamic approach eliminates the uncertainties otherwise associated with equilibrium calculations based on the law of mass action.

The prediction of an equilibrium constant then requires the calculation of ΔG^0 for the reaction. The so-called third-law method involves calculation for the reaction at 25°C of the value of ΔH^0, the standard heat of reaction, from tabulated standard heat of formation data and of ΔS^0 from tabulated third-law entropies. These are combined in the sense of $\Delta G^0 = \Delta H^0 - T \Delta S^0$ to permit calculation of the equilibrium constant for 25°C. This in turn is used for evaluation of the integration constant in the integration of the relation in Eq. (13).

$$\frac{d \ln K}{dT} = \frac{\Delta H^0}{RT^2} \qquad (13)$$

The integration requires expression of ΔH^0 as a function of temperature, which necessitates a knowledge of the heat capacities $C^0_{P(i)}$ for the various reactants over the temperature range involved.

Alternatively, if values of the free-energy function $(G^0 - H^0{}_{298}/T)$ are available, either from experimental measurement or from statistical thermodynamical computations, they can be combined with the standard heat of reaction at 25°C to give the desired result, Eq. (14).

$$\Delta G^0 = \Delta \left(\frac{G^0 - H^0{}_{298}}{T} \right) + \frac{\Delta H^0{}_{298}}{T} \qquad (14)$$

See ENTROPY; HEAT CAPACITY.

[PAUL BENDER]

Bibliography: K. G. Denbigh, *Principles of Chemical Equilibrium*, 2d ed., 1968; E. Fermi, *Thermodynamics*, 1956; J. W. Gibbs, *Collected Works*, vol. 1, 1948; K. S. Pitzer and L. Brewer, *Thermodynamics*, rev. ed., 1961; F. T. Wall, *Chemical Thermodynamics*, 2d ed., 1965.

Fuel, fossil

A material whose combustion is used to supply heat for any purpose. Fossil fuels are carbon-containing materials that are burned with air or with oxygen derived from the air.

Classification. Fuels are characterized (1) by their physical form at normal temperatures, whether they are solids, liquids, or gases; (2) by their heating value, that is, by the amount of heat given off when a unit weight or volume of the material is burned under standard conditions; and (3) by their combustion characteristics, which cover such points as ease of ignition, rate of combustion, flame temperature, and flame luminosity.

Fuels may be used to supply heat directly, as in a furnace. They may also be used to supply heat to perform some other function, such as to produce steam in a boiler, the steam then being used to produce power in a turbine or engine. The heat may be used to raise the temperature and pressure of the products of the combustion to provide a driving force, as in the internal combustion engine or gas turbine. *See* ENGINE; TURBINE.

Solid fuels today comprise primarily the various ranks of coal (anthracite, bituminous, subbituminous, and lignite) and the coke or char derived from them. Wood, charcoal, peat, and various other plant products, while important under special circumstances, represent only a small fraction of total fuel consumed. The important factors governing choice of coal for a given use are cost per unit of heating value, size range, moisture and ash content, and amount of smoke evolved. *See* COAL; PEAT.

Liquid fuels are derived almost entirely from petroleum. Historically, coal oil (a kerosine derived from coal tar) was used extensively for heat and light, but it is now largely supplanted by petroleum products. In general, the liquid fuels suitable for internal combustion engines command a premium, and the fuels designed for external combustion are derived from those portions of the petroleum that have the least potential for gasoline and diesel fuel manufacture. Factors in choice of liquid fuels are heating value, fluidity, boiling range, impurities such as sulfur, and freedom from water and sediment. For fuels for internal combustion

Approximate heating values of representative fuels

Fuel	Heating value
Anthracite coal	13,500 Btu/lb
Bituminous coal (low-volatile)	14,500 Btu/lb
Lignite	7,200 Btu/lb
Fuel oil	19,000 Btu/lb
Natural gas	1,100 Btu/ft³
Producer gas	129 Btu/ft³
Butane	3,200 Btu/ft³

engines *see* DIESEL FUEL; FUEL, LIQUID; FUEL OIL; GASOLINE; OIL SAND; OIL SHALE; PETROLEUM PRODUCTS.

Of the gaseous fuels, the most important are those derived from natural gas or petroleum, chiefly methane. Availability of these has increased sharply since World War II, and they have largely supplanted manufactured fuel gas for general distribution by public utilities. Manufactured gas was usually coke-oven gas or carbureted water gas made from coke, steam, and petroleum oil. During the era of gas illumination, the carbureting step yielded a gas which burned with a luminous flame. Other gaseous fuels are those that are byproducts of some manufacturing process and therefore available at a cost that makes them competitive with natural gas. These include coke-oven gas, blast-furnace gas, and producer gas. *See* FUEL GAS; NATURAL GAS.

Heating value. The heating value of gaseous fuels varies widely with composition. For this reason it is generally important to have an accurate and continuous determination of the heating value, and much effort has been devoted to the development of an apparatus for this purpose. Carefully metered streams of gas and combustion air are burned under constant temperature conditions in a submerged chamber. The temperature rise of a metered stream of cooling air is used as a measure of the heating value and is usually recorded continuously on a chart. This value in the United States is usually expressed as British thermal units (Btu) per cubic foot of gas (see table). *See* BRITISH THERMAL UNIT (BTU).

Heating values of solid and liquid fuels are determined by oxidation of a weighed sample in a system whose heat capacity is known and whose temperature rise can be measured. The oxidant may be gaseous oxygen or an oxidizing material such as sodium peroxide.

Heating values are reported as gross (or higher) heating value when the water formed in combustion is condensed, and as net (or lower) heating value when the water from combustion leaves the system as a vapor and the heat of condensation is not included. *See* COMBUSTION; ENERGY SOURCES; HEATING VALUE.

[HOWARD R. BATCHELDER]

Bibliography: G. N. Critchley (ed.), *Future of Fuel Technology*, 1964; W. Francis, *Fuels and Fuel Technology*, 1965; D. A. Williams and G. Jones, *Liquid Fuels*, 1963.

Fuel, liquid

Any of the liquids burned to produce usable energy in the form of heat or light. Ease of ignition, clean burning, and adaptability to transportation and storage have all been favorable features of the hot flame and the bright light from liquid fuels. Animal oils, vegetable oils, and petroleum have been used historically, but of all the liquid fuels used, petroleum has become the dominant basis for extensive industrial development and private convenience. This may be the century history will designate as "the age of petroleum." In the United States, beginning with production of petroleum from wells in Pennsylvania (1859), the extensive production of the 20th century has allowed escape from dependence upon animal oils. A naturally occurring hydrocarbon, petroleum ascended quickly to its principal role as an energy source as a result of the development of internal combustion engines and the emergence of machines of transportation. *See* INTERNAL COMBUSTION ENGINE.

Characteristics of petroleum. Petroleum is chemically a very complex mixture of carbon and hydrogen compounds. Minor impurities of oxygen, nitrogen, and sulfur vary in different crude oils. The term "crude oil" is in common usage to distinguish between the natural oil derived from rocks and the refined lubricants available from the neighborhood service station. All crude oil is lighter than water, immiscible with water, and soluble in ether, naptha, or benzine. Variations from solid black gilsonite to viscous black and brown asphalts, tar, and pitch, to light green and yellow crude oils are to be found in different localities where oil is produced in the United States. Terms such as "hydrocarbon" and "bitumen" are used interchangeably with the word "petroleum." However, petroleum (from the Latin *petra* meaning rock or stone and *oleum* meaning oil) is the most common usage. Petroleum occurs at the surface as springs and seepages, and in subsurface rocks where it is in the openings between the grains and in cracks in the rock. *See* PETROLEUM.

Origin of petroleum. Although it is generally agreed that petroleum is derived from organic matter mixed with sediments that later form the sedimentary rocks of the Earth's crust, there is no common agreement as to whether that organic matter was derived from animals or plants. Whether the organic matter was accumulated exclusively in marine sedimentary environments or in brackish-water continental-marginal sediments is also in contention. Conversion of organic matter into petroleum is associated with the heat, pressure, and fluids involved in converting sediments into sedimentary rock. Solid organic matter is dispersed through many sedimentary rocks, but the intermediate steps of conversion to liquid petroleum have not been observed.

Petroleum geology. Wide variation between crude oils makes a common origin doubtful, but there are recognizable requisites for the accumulation of usable quantities of petroleum in the crustal rocks of the Earth. Permeable rocks must exist which allow the passage of fluids through openings between the particles of the rock or through cracks and cavities that exist. For the accumulation of oil pools, rock porosity and permeability, then, are vital to the mechanism of accumulation. The rock layer or rock body in which an accumulation has occurred is referred to as reservoir rock. Equally important in the entrapment of petroleum is an impervious capping which impedes or stops the upward movement of fluids in the rock mass.

Petroleum is driven into an entrapped pool by the movement of water and gas in the subsurface. Oil migration and entrapment is most common within layers of sedimentary rocks and, hence, the principal production of the world is from sandstones, limestones, conglomerates, and other common sedimentary layers. Folded and faulted sedimentary layers provide potential sites for traps of oil. Subtle changes of porosity and permeability within sedimentary rock layers will also impede fluid movements and result in entrapment. These so-called stratigraphic traps are some of the most difficult sites to discover from surface investigations, yet many pools, still to be discovered, will be of this type.

Alternates to petroleum. There is disturbing evidence that over half of the oil that is to be discovered in the United States has already been produced. Since 1965, an increasing proportion of society's needs were met by importation of crude oil. Over a third of the crude oil used in the United States is imported. Along with the natural petroleum upon which the culture has become dependent, society must now accept the costly technology and availability of synthetic liquid fuels. Oils distilled from the kerogens of oil shale are an alternative to meet part of the demand for energy from liquid fuels. Crude oils and refined products from conversion of coal represent an energy resource within reach of modern-day technology. Liquefied natural gas, primarily methane, together with liquefied petroleum gas, propane and butane, must also be made available as a supplement for domestic oil supplies. The technology of producing liquefied natural gas, although not entirely efficient, is being perfected under the support of massive Federal and industrial research grants. As domestic production of petroleum gradually diminishes and prices escalate, it is expected that the substitution of these synthetic liquid fuels will become necessary. *See* COAL GASIFICATION; COAL LIQUEFACTION; FUEL, SYNTHETIC; LIQUEFIED NATURAL GAS (LNG); LIQUEFIED PETROLEUM GAS (LPG); OIL SHALE. [ORLO E. CHILDS]

Bibliography: M. King Hubbert, *The Environmental and Ecological Forum 1970–1971*, 1972; K. Landes, *Petroleum Geology*, 2d ed., 1975; A. I. Levorsen, *Geology of Petroleum*, 1954; F. Park, Jr., *Earthbound*, 1975.

Fuel, synthetic

A fuel which does not exist in nature, but which must be manufactured or synthesized. Generally, synthetic fuels are derived from other forms of fossil fuels which are less convenient for consumer use. Thus, substitute natural gas (SNG) is manufactured from coal, from shale, and from oil fractions, Synthetic liquid fuels are produced from coal, from shale, and from tar sands. *See* COAL GASIFICATION; COAL LIQUEFACTION; FUEL, FOSSIL; FUEL GAS; OIL SAND; PETROLEUM PROCESSING.

Gaseous and liquid fuels, the so-called fluid fuels, are the most desirable form of fossil fuels because they are conveniently stored, transmitted, and controlled in use. They are also convenient raw materials for the manufacture of chemicals and plastics. Accordingly, where an abundant supply of gaseous and liquid fuels is readily available at low cost, industry flourishes and a high standard of living generally ensues.

Where gaseous and liquid fuels are in short supply, interest is stimulated in the manufacture of substitute or synthetic fuels. Historically, the usual raw material has been coal. Coal was the source of kerosine and illuminating gas in the United States prior to the 1940s. It was the major source of synthetic liquid fuels used by Germany in World War II. Coal is the raw material for the only existing commercial liquefaction plant, located in South Africa.

Since 1970, United States reserves of natural gas and petroleum have declined significantly, requiring major imports of oil from other countries. As a result, a major program, called Project Independence, has been initiated under the sponsorship of the U.S. Energy Research and Development Administration to create new methods for converting coal into liquid fuels and SNG. A parallel effort to develop oil shale conversion technology has been carried out in part by the U.S. Bureau of Mines and in part by private industry. *See* NATURAL GAS; OIL SHALE.

In general, the objective of these programs is to find improvements in the existing processes so that the cost of the synthetic fuels can be reduced. A major investment is needed for equipment in which to carry out the synthesis of fuels. In addition, some energy is inevitably wasted during the process. As a result, synthetic fuels are quite expensive. SNG may cost more than $3.00 per thousand cubic feet (tcf; 28.317 m^3) whereas natural gas formerly sold for 30¢/tcf. Synthetic crude oil may cost more than $15/bbl whereas natural petroleum formerly sold for $5/bbl (1 bbl = 0.159 m^3).

[JOHN A. PHINNEY]

Bibliography: Federal Energy Administration, *Project Independence Report*, November 1974; H. C. Hottel and J. B. Howard, *New Energy Technology: Some Facts and Assessments*, 1971; H. H. Lowry, *Chemistry of Coal Utilization*, vols. 1, 2, and suppl. vol., 1945, 1963.

Fuel cell

An electric cell that converts the chemical energy of a fuel directly into electric energy in a continuous process. The efficiency of this conversion can be made much greater than that obtainable by thermal-power conversion. In the latter the chemical reaction is made to produce heat by combustion. The heat is then transformed partially into mechanical energy by a heat engine, which drives a generator to produce electric energy. Further loss is involved if the direct current generated is converted into alternating current.

Although, in principle, the nature of the reactants is not limited, the fuel-cell reaction almost always involves the combination of hydrogen with oxygen, as shown by Eq. (1). At 25°C and 1 atm

$$H_2(g) + \tfrac{1}{2}O_2(g) \rightarrow H_2O(l). \qquad (1)$$

pressure, that is, standard temperature and pressure (STP), the reaction takes place with a free energy change (ΔG) of $\Delta G = -56.69$ kcal/mole, that is, 237,000 joules/mole water.

If the reaction is harnessed in a galvanic cell working at 100% efficiency, a cell voltage of 1.23 volts results. In actual service such cells have shown steady-state potentials in the range 0.9–1.1 volts, with reported coulombic efficiencies of the order 73–90%.

Fuel cells are of 200–500 watts capacity and 50–100 ma/cm^2 current density. Larger prototypes have been produced, some as large as 15 kw capacity, while a system under study is expected to provide 100 kw.

In the present stage of development, it is difficult to make a classification of the fuel-cell types. The most popular and successful type remains the classical H_2-O_2 fuel cell of the direct or

Table 1. Theoretical cell potentials at various temperatures

Reaction	Cell potential, volts					
	25°C	100°C	250°C	500°C	750°C	1000°C
$C + O_2 \rightarrow CO_2$	1.02	1.02	1.02	1.02	1.02	1.01
$2C + O_2 \rightarrow 2CO$	0.71	0.75	0.82	0.93	1.04	1.15
$2CO + O_2 \rightarrow 2CO_2$	1.33	1.30	1.23	1.11	1.00	0.88
$2H_2 + O_2 \rightarrow 2H_2O$	1.23	1.18	1.12	1.05	0.97	0.90

indirect type. In the direct type, hydrogen and oxygen are used as such, the fuel being produced in independent installations. The indirect type, employs a hydrogen-generating unit which can use as raw material a wide variety of fuel. The reaction taking place at the anode is as in Eq. (2), and at the cathode as in Eq. (3).

$$2H_2 + 4OH^- \rightarrow 4H_2O + 4e^- \qquad (2)$$

$$O_2 + 2H_2O + 4e^- \rightarrow 4OH^- \qquad (3)$$

Because of the low solubility of H_2 and O_2 in electrolytes, the reactions take place at the interface electrode-electrolyte, requiring a large area of contact. This is obtained with porous materials called upon to fulfill the following main duties: The materials must provide contact between electrolyte and gas over a large area, catalyze the reaction, maintain the electrolyte in a very thin layer on the surface of the electrode, and act as leads for the transmission of electrons.

The porosity is obtained by the Raney technique or by sintering. When flooding of the pores is feared, the electrode is made with double porosity, fine at the electrolyte side and coarse at the gas side. The catalytic effect is obtained with noble metals, mainly platinum, silver, nickel, cobalt, and palladium.

The thickness of the electrolyte layer, on which depends the internal resistance of the cell, is controlled by pore size, wetting properties, and pressure of the fuel gas. When pressure is used, care must be taken not to increase it to the extent that gas is allowed to bubble through the electrolyte because of the danger of forming an explosive hydrogen-oxygen mixture.

The fuel cells may work with acid or alkaline electrolytes.

The acid electrolytes require costly corrosion-resistant construction materials but are not sensitive to CO and CO_2 in the fuel, which may lead to the buildup of carbonates. Some models using phosphoric acid proved quite successful. The alkaline electrolytes are more practical, and they are found in most fuel cells produced industrially at present.

Some fuel cells are designed to work with mol-ten carbonates as electrolyte, at temperatures as high as 800°C. These cells are attractive because they can use reformed hydrocarbon fuels, require a small investment, and can be made as large units. They are insensitive to carbon oxides but, at least in the present situation, are affected by important shortcomings. Mainly, these shortcomings are excessive size, rapid corrosion of metallic parts, and long periods of heating required before useful service.

The cells may use any alkali metal carbonate or eutectic mixtures of the same. The reactions taking place are those known for the systems involving hydrogen and oxygen or carbon monoxide and oxygen. In the case of carbon monoxide, the reaction step at the anode is as in Eq. (4) and at the cathode as in Eq. (5).

$$CO + CO_3^{--} \rightarrow 2CO_2 + 2e^- \qquad (4)$$

$$CO_2 + \tfrac{1}{2}O_2 + 2e^- \rightarrow CO_3^{--} \qquad (5)$$

The total cell reaction is given by Eq. (6), with a

$$2CO + O_2 \rightarrow 2CO_2 \qquad (6)$$

free energy change of $\Delta G = -61.45$ kcal/mole of CO. At 100% efficiency, this gives a theoretical cell voltage of 1.34 volts at STP.

Principal fuel-cell reactions. The principal overall reactions which have been employed in fuel-cell work are summarized in Tables 1 and 2.

The direct anodic use of carbon has been practically abandoned in modern fuel-cell work. Carbon potentials seem entirely due either to carbon monoxide, CO, or to hydrogen, H_2, formed at high temperature by direct reaction between the carbon and the electrolyte. For example, in the Jacques cell, which consists of carbon electrodes and iron (air) electrodes in molten sodium hydroxide, H_2 is liberated at the carbon by reaction with the electrolyte. It is this H_2 which is responsible for the observed potential.

Modern fuel cells use gaseous fuels, either H_2 or CO or mixtures of these gases. The oxidizer is normally oxygen or air. Hydrocarbons have not been made to function anodically. Where potentials have been measured, they are attributed to decomposition of the hydrocarbon to liberate H_2. For example, methane decomposes at high temperatures, as shown by Eq. (7).

$$2CH_4 \rightarrow 2C + 4H_2 \qquad (7)$$

For technical reasons, it is simpler to use the carbon or hydrocarbon fuel in a chemical reactor to produce the active gases, H_2 and CO, than to attempt to operate a cell under the conditions best suited for the chemical reaction. Typical chemical production of the active gases might be as in Eq. (8).

$$2C + O_2 \rightarrow 2CO \qquad (8)$$

Theoretically, CO has a requirement of 1.15 lb/kwhr when reacted anodically against an oxygen cathode. To produce 1.15 lb CO, 0.493 lb carbon is needed. Hence a perfect process, starting with 0.493 lb carbon, would yield 1 kwhr. A mixture of H_2 and CO can also be produced by reacting carbon with steam, as shown by Eq. (9).

$$C + H_2O \rightarrow CO + H_2 \qquad (9)$$

The complete engineering design of the chemi-

Table 2. Theoretical material consumption

Reaction	Temperature, °C	Consumption, lb/kwhr		
		Anode	Cathode	Total
$C + O_2 \rightarrow CO_2$	750	0.344	0.918	1.262
$2C + O_2 \rightarrow 2CO$	750	0.474	0.632	1.106
$2CO + O_2 \rightarrow 2CO_2$	750	1.15	0.656	1.816
$2H_2 + O_2 \rightarrow 2H_2O$	100	0.070	0.56	0.63
	750	0.085	0.68	0.763

cal reactor in conjunction with the fuel cell has been extensively studied. The main difficulties in the past were due to the large concentration of CO in the reformed fuel, which poisons the Pt catalyst often used in the fuel cell proper. At present, various new catalysts have been developed, such as Pt-Rh, Pt-Ir, and Pt-Ru, well capable of processing H_2-CO fuel mixtures.

The present fuel generators are capable of covering most of the needs of established fuel cells from almost pure hydrogen to 1:1 mixtures of H_2 and CO. One project, sponsored by the Office of Coal Research, U.S. Department of the Interior, has as its object a 100-kw coal reactor—solid electrolyte fuel-cell system.

Hydrogen-oxygen fuel cell. Work with H_2 has established that it can operate efficiently at moderate temperatures, with polarization decreasing as the temperature is increased. This permits the use of aqueous solutions. One has been reported to have, at 25°C, the following characteristics:

Current density, amp/ft²	0	1	10	50
Cell voltage	1.12	1.01	0.95	0.70

From the point of view of the working principle, three well-developed systems should be mentioned: those of General Electric, Allis-Chalmers, and Bacon. Only the General Electric system had been tested in service in 1968, but semiindustrial tests conducted on the Allis-Chalmers and Bacon systems show that these also can be considered as ready for practical use. Some of their general characteristics are listed in Table 3.

In the same category should be mentioned the hydrazine-air fuel cell under investigation by Monsanto Research Corp. and Allis-Chalmers, together with some research centers of the U.S. Army.

This cell is based on the reaction shown in Eq. (10). It is a medium-size system intended to

$$N_2H_4 + O_2 \rightarrow N_2 + 2H_2O \qquad (10)$$

produce 60–300 watts for the Monsanto project and up to 3 kw for the Allis-Chalmers project. The unit cell voltage is 0.6–0.7 volt, and its greatest advantage is that it uses a condensed fuel, convenient in some applications. A different concept is used in the alkaline metal—oxygen fuel cells developed by the M. W. Kellogg Co. In an actual unit the electrochemical process involves oxidation of Na with oxygen from air, with NaOH as electrolyte. The reaction taking place at the anode is as in Eq. (11) and at the cathode as in Eq. (12), with the cell reaction given as Eq. (13).

$$4Na \rightarrow 4Na^+ + 4e^- \qquad (11)$$

$$O_2 + 4e^- + 2H_2O \rightarrow 4OH^- \qquad (12)$$

$$4Na + O_2 + 2H_2O \rightarrow 4NaOH \qquad (13)$$

Because Na as such is too reactive, the cell uses a sodium amalgam which is quite stable in concentrated NaOH. The sodium amalgam—oxygen fuel cell possesses some remarkable features. It provides almost 1.5 volts at steady state and very high current densities of the order of 200 ma/cm², is insensitive to water quality, and requires a small gas consumption. It is expected to play an important role in some applications.

Ion-exchange membrane types. An interesting fuel cell in the laboratory stage is the one which uses inorganic ion-exchange membranes. This type of cell offers two potential advantages: It tolerates higher temperature operation and has a higher ionic conductivity because of the higher density of ion-exchange sites.

An advanced laboratory fuel-cell model developed by Armour Research Foundation uses zirconyl phosphate as an ion-exchange electrolyte, hydrogen and oxygen as fuels, and platinum black as catalyst. The fuel cell is capable of delivering almost 1 volt, but the current density reached so far is too small. The inorganic ion-exchange membrane requires water for its operation.

In the cells using solid electrolytes of the ionic conductive type, the need for water is eliminated. The electrolyte currently studied by Westinghouse Electric is a mixture of zirconium oxide and calcium oxide (0.85:0.15) and is expected to work at about 1000°C. Another promising electrolyte is the mixture zirconium oxide and yttrium oxide (0.9:0.1).

Closed cycle types. The last fuel-cell type to be considered is the closed-cycle type, in the frame of which the reactants are recovered by an auxiliary process. Some systems, such as those developed by Electro-Optical Systems Division of Xerox, are simply made of two converse cycles operating one at a time. The regeneration consists in producing hydrogen and oxygen by electrolysis when electrical energy is available and shifting to cell performance on stored fuel when the production of electrical energy becomes necessary. In the regenerative system developed by the United Aircraft, the electrolysis current is provided by a solar-energy converter made of regular silicon solar cells.

Another possibility is that of radiochemical regeneration. A model under development by Union Carbide is that proposed by J. A. Ghormley, using ferrous sulfate irradiated by gamma radiation. The electrical efficiency reported to the gamma radiation absorbed had been evaluated at 3%. The unit cell voltage is about 0.6 volt, and the module containing 6 units should deliver 5 watts over a period of 2 years.

Problem areas. In the development of fuel cells, there are some general difficulties which must be solved before they become mature industrial prop-

Table 3. Characteristics of three fuel-cell systems

Type	Principle	Temperature, °C	Power, watts	Application
General Electric	Ion-exchange membrane	25–35	100–1000	Gemini
Allis-Chalmers	Porous Ni electrodes and porous electrolyte vehicle	90–100	2000	Space flight
Bacon (Pratt and Whitney)	Porous Ni electrodes	200–220	500–1500	Apollo

ositions. The main problems are as follows.

Catalyst. Its importance increases while the temperature decreases. With few exceptions the catalyst is a very expensive constituent of the cell, and not only will its price increase when industrial production requires increased amounts, but it may even become unavailable at any price. This suggests that future models will tend to work at higher temperatures and pressures (like Bacon's fuel cell) at which cheaper catalysts (Ni, NiO, and so on) can be used.

Capital cost. At an estimated $1000/kw investment the fuel cell is not commercially attractive. By various standard improvements, its capital cost should be reduced to one-tenth this value before becoming competitive with other electrical energy sources.

Heat transfer. In most fuel cells the reaction product is water in gaseous or liquid form. At 100% efficiency there are 421 g of water per kilowatt-hour to dispose of. In addition, about 30% of the heat of reaction, that is, about 260 kcal/kw, must be disposed of in order to maintain the working temperature at a level at which the electrolyte is not decomposed and the material does not become too susceptible to corrosion.

High-temperature fuel cells. The use of carbon monoxide has been limited to high-temperature cells. One carbon monoxide cell using air as the cathodic material operated at about 700°C to yield 0.75–0.85 volt at 32 amp/ft². It has been demonstrated that molten-salt electrolyte cells can operate at 550–800°C on inexpensive hydrocarbons if steam is admitted to prevent carbon deposition. *See* DRY CELL.

[J. DAVIS; L. ROZEANU]

Bibliography: D. H. Archer and R. L. Zahradnik, The design of a 100-kilowatt, coal-burning fuel cell system, *Chem. Eng. Progr.*, 63:55, 1967; N. P. Chopey, What you should know about fuel-cells, *Chem. Eng.*, no. 125, May 25, 1964; E. Findl and M. Klein, Electrolytic regenerative hydrogen-oxygen fuel-cell battery, *Proceedings of the 20th Annual Power Sources Conferences*, 1966; M. I. Gillibrand and G. B. Lomax, Factors affecting the life of fuel cells, *Proceedings of the 20th Annual Power Sources Conferences*, 1966; D. W. McKee and A. G. Scarpellino, Electrocatalysts for hydrogen/carbon monoxide fuel cell anodes, *Electrochem. Technol.*, 6:101, 1968; E. Yeager and W. Mitchell, Jr. (eds.), *Fuel Cells*, 1963.

Fuel gas

A fuel in the gaseous state whose potential heat energy can be readily transmitted and distributed through pipes from the point of origin directly to the place of consumption. The development and use of fuel gases is closely associated with the progress of civilization. As shown in the illustration, such gases have become especially prominent in the industrial development period since 1900. Natural gas provides 30% of the energy needs of the United States, with LP (liquefied petroleum) fuel gases providing nearly another 2%.

The types of fuel gases are natural gas, LP gas, refinery gas, coke oven gas, and blast-furnace gas. The last two are used in steel mill complexes. Typical analyses of several fuel gases are presented in the table. Since these analyses are based on dry gases, the heating value of gases saturated with water vapor would be slightly lower than the values shown. *See* HEATING VALUE.

Most fuel gases are composed in whole or in part of the combustibles hydrogen, carbon monoxide, methane, ethane, propane, butane, and oil vapors and, sometimes, of mixtures containing the inerts nitrogen, carbon dioxide, and water vapor.

Natural gas. The generic term "natural gas" applies to gases commonly associated with petroliferous geologic formations. As ordinarily found, these gases are combustible, but nonflammable components such as carbon dioxide, nitrogen, and helium are often present. Natural gas is generally high in methane. Some of the higher paraffins may be found in small quantities.

The olefin hydrocarbons, carbon monoxide, and hydrogen are not present in American natural gases. The term "dry natural gas" indicates less than 0.1 gal (1 gal, U.S. = 0.003785 m³) of gasoline vapor occurs per 1000 ft³ (1 ft³ = 0.028 m³); "wet natural gas" indicates more than 0.1 gal/1000 ft³. "Sweet" and "sour" are terms that indicate the absence or presence of hydrogen sulfide.

There is no single composition which might be termed typical natural gas. Methane and ethane constitute the bulk of the combustible components; and CO_2 and nitrogen, the inerts. The net heating value of natural gas served by a utility company is often 1000–1100 Btu/ft³ (1 Btu = 1055 joule).

Natural gas is an ideal fuel for heating because of its cleanliness, ease of transporation, high heat content, and the high flame temperature. *See* LIQUEFIED NATURAL GAS (LNG): NATURAL GAS.

LP gas. This term is applied to certain specific hydrocarbons, such as propane, butane, and pentane, which are gaseous under normal atmospheric conditions but can be liquefied under moderate pressure at normal temperatures. *See* LIQUEFIED PETROLEUM GAS (LPG).

Oil gas. This term encompasses a group of gases derived from oils by exposure of such oils to elevated temperatures. Refinery oil gases are those obtained as by-products during the thermal processing of the oil in the refinery. They are used primarily for heating equipment in the refinery. Gas made by thermal cracking of oil was formerly very important as an urban fuel gas, but has been almost

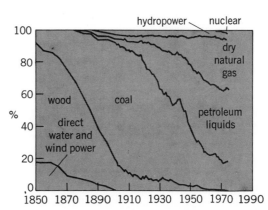

Contribution of various sources to primary energy consumption in the United States, 1850–1974.

Typical gas analyses*

Type	Analysis, % vol								Specific Gravity	Btu/ft³	
	CO₂	O₂	CO	H₂	CH₄	C₂H₆	C₃H₈ and C₄H₁₀	N₂		Gross	Net
Dry natural gas	0.2				99.2			0.6	0.56	1007	906
Propane (LP)						2.6	97.3	0.1	1.55	2558	2358
Refinery oil gas		0.2	1.2	6.1	4.4	72.5	15.0	0.6	1.00	1650	1524
Coke oven gas	2.0	0.3	5.5	51.9	32.3		3.2	4.8	0.40	569	509
Blast-furnace gas	11.5		27.5	1.0				60.0	1.02	92	92
Producer gas	8.0	0.1	23.2	17.7	1.0			50.0	0.86	143	133

*From L. Shnidman (ed.), *Gaseous Fuels*, 2d ed., American Gas Association, 1954.

totally displaced by natural gas, and the equipment for its production dismantled. A typical oil gas consists of saturated and unsaturated hydrocarbons and has a heating value of 1300–2000 Btu/ft³. Methane, ethane, propane, butane, ethylene, and propylene are the main constituents.

Coal gases. Until about 1940, gas produced from coal was an important part of the energy mix in the United States. These gases were rapidly replaced by natural gas in the distribution systems of gas utilities serving the residential, commercial, and industrial markets. During 1971 it became clear that the supply of natural gas would probably be insufficient to meet the ever-increasing demand. Therefore, as of the late 1970s, it is expected that gas will once again be produced from coal. In this period, the older processes will be replaced by modern methods capable of very high production rates at improved efficiency and reduced cost.

Coal gasification can be accomplished in a large number of ways, including pyrolysis or partial oxidation with air or oxygen and steam. Various processes operate as fixed beds or fluidized beds or with the coal entrained. The pressure may vary from near atmospheric to 1000 psig (1 psi = 6895 Pa) or more. All operate at high temperatures (1200–3000°F; 650–1645°C). The direct products of gasification vary in heating value from 120 to 150 Btu/ft³ (low-Btu gas), to 300 Btu/ft³ (medium-Btu gas or synthesis gas), to as high as 600 Btu/ft³. Low-Btu gas can be used as fuel for industrial processes or for production of electrical power by electric utilities. Medium-Btu gas, which consists principally of carbon monoxide and hydrogen, can be used directly as a fuel, or it can be upgraded by catalytic methanation to essentially pure methane, which is, for all practical purposes, identical to natural gas. Such a product gas is commonly called pipeline-quality gas or substitute natural gas (SNG). *See* COAL GASIFICATION.

Low-Btu gas. Producer gas is made by oxygen-deficient combustion of coal or coke, in which process a mixture of air and steam is blown upward through a thick hot bed of coal or coke. The gas is high in nitrogen introduced in the air. Its heating value is low, its specific gravity is high, and the percentage of inerts is high. Producer gas, which contains 23–27% carbon monoxide, is used as it comes from the generators after some preliminary purification. It was once the cheapest form of industrial gas, and could again become important; it may be possible to produce it in modernized equipment on a large scale.

Blast-furnace gas is a by-product from the manufacture of pig iron. Like producer gas, it is derived from the partial combustion of coke. Some of the combustibles in the gas are used to reduce the iron ore; thus the final gas contains about 27% carbon monoxide and more than 70% of inert gases (CO₂ and N₂), giving it the lowest heating value, less than 100 Btu/ft³, of any usable fuel gas. It is used for the operation of gas engines, heating by-product coke ovens, steel plant heating, steam raising, and crude heating. *See* FUEL, FOSSIL.

Medium-Btu gases. Coke oven gas is the only important gas that has an intermediate heating value. The gas is made by destructive distillation of a packed bed of coal out of contact with air. The process results in the formation of coke, which is used in the blast furnace. The gas is utilized totally within the steel-making complex.

The combination of the 1974 oil embargo, the energy crisis, and restrictions on air pollution resulted in the proposal of a large number of modern processes. Nearly all these processes use a mixture of steam and oxygen to combust coal. Less oxygen than is required for complete combustion is used; therefore, the products are primarily carbon monoxide and hydrogen. Depending on the end use of the gas, moderate amounts of methane and liquid products are produced.

SNG. Gas which is to be used within the modern natural gas transmission and distribution system must be essentially methane. The medium-Btu gas processes can be used as precursors to produce substitute natural gas (SNG), as noted above. Other new processes which operate at very high pressure (1000 psig) are being developd to produce very large quantities of gas (250,000,000 ft³/day of 1000 Btu/ft³ gas) to supplement the declining supplies of natural gas. Some of these processes use the partial combustion of oxygen as a direct source of heat, and others use air combustion in a variety of indirect modes. *See* SUBSTITUTE NATURAL GAS (SNG).

Nonfossil sources. In view of the immediate shortage of natural gas and the longer-range (A.D. 2030–2060) forecast that all fossil fuels will be consumed, there are several projects which may ensure a perpetual supply of fuel gas. These include conversion of waste materials to gas by pyrolysis or by anaerobic digestion. Large efforts have been initiated to convert solar energy to fuel gas. This would be accomplished by growing plants on land or sea and converting the harvest to gas in processes similar to those used on wastes. Finally, there has been a large effort to develop a "hydrogen economy." Hydrogen would be produced by thermochemical processes using heat from various sources such as atomic energy reactors.

[JACK HUEBLER]

Bibliography: 5th Synthetic Pipeline Gas Symposium, American Gas Association cat. no. L51173, October 1973; Hammond et al., *Energy and the Future*, American Association for the Advancement of Science, 1973.

Fuel oil

Any of the petroleum products which are less volatile than gasoline and are burned in furnaces, boilers, or other types of heaters. The two primary classes of fuel oils are distillate and residual. Distillate fuel oils are composed entirely of material which has been vaporized in a refinery distillation tower. Consequently, they are clean, free of sediment, relatively low in viscosity, and free of inorganic ash. Residual fuel oils contain fractions which cannot be vaporized by heating. These fractions are black and viscous and include any inorganic ash components which are in the crude. In some cases, whole crude is used as a residual fuel.

Uses. Distillate fuel oils are used primarily in applications where ease of handling and cleanliness of combustion are more important than fuel price. The most important use is for home heating. They are used in about 35% of United States homes with central heating systems, the remainder being heated with gas or electricity. Distillate fuel oils are also used in certain industrial applications where low sulfur or freedom from ash is important. Certain types of ceramics manufacture are examples. Increasing amounts of distillate fuel oils have been burned in gas turbines used for electricity generation. *See* ELECTRIC POWER GENERATION; GAS TURBINE.

Residual fuel oils are used where fuel cost is an important enough economic factor to justify additional investment to overcome the handling problems they pose. They are particularly attractive where large volumes of fuel are used, as in electric power generation, industrial steam generation, process heating, and steamship operation.

Combustion. Fuel oils are burned efficiently by atomizing them into fine droplets, about 50–100 μm in diameter, and injecting the droplet spray into a combusion chamber with a stream of combustion air. Small home heating units use a pressure atomizing nozzle. Larger units use pressure, steam, or air atomizing nozzles, or else a spinning cup atomizer. Older units employ natural draft to induce combustion air, but new units usually use forced draft, which gives better control of air-to-fuel ratio and increased turbulence in the combustion chamber. This permits more efficient operation and clean, smoke-free combustion. *See* OIL FURNACE.

Tests and specifications. Fuel oils are blended to meet certain tests and specifications which ensure that they can be handled safely and easily, that they will burn properly, and that they will meet air-pollution regulations. These tests include (1) flash point, which determines that the fuel can be stored and handled without danger of explosion; (2) pour point, which determines that it will not solidify under normal handling conditions; (3) carbon residue, which determines that it will not coke during handling or form undue amounts of carbon particulates during combustion; (4) sediment, which determines that it will not clog pumps and nozzles; (5) sulfur content, which determines how much sulfur dioxide will be emitted during combustion; (6) ash content; and (7) viscosity or resistance to flow.

Viscosity. Viscosity is an important factor in determining the grade of a fuel oil. Distillate fuels and light grades of residuals have a low viscosity so that they can be handled and atomized without heating. Heavier grades have high viscosity and must be heated in order to be pumped and atomized. The grade classification of the fuel oil determines what viscosity range it is blended to, and therefore the temperature level required in the preheat system.

Ash content. Fuel ash content may be important in residual fuels, even though ash content is negligible compared with coal, because certain ash components may cause slagging and corrosion problems. Vanadium and sodium are frequently present in oil ash. During combustion, they are converted to materials which tend to accumulate on boiler and superheater tubes. Under certain conditions, they may cause corrosion of these tubes.

Sulfur content. Sulfur content of fuel oil is carefully controlled in order to meet air-pollution regulations. This has required a drastic reduction in sulfur content of residual fuel oils in many areas since 1969. To achieve the reduction, extensive processing and changes in blending procedure have been required, which have caused major increases in the price of fuel oil.

In spite of the higher price, fuel oil demand has increased rapidly since 1970 because alternate fuels, coal and gas, either cannot meet the regulations or are not available in sufficient quantities to meet the demand. *See* COAL; NATURAL GAS; OIL ANALYSIS; PETROLEUM PRODUCTS.

A detailed list of specifications for fuel oils is given in the *Annual Book of ASTM Standards*, published by the American Society for Testing and Materials, Philadelphia.　[C. W. SIEGMUND]

Fusion, nuclear

One of the primary nuclear reactions, the name usually designating an energy-releasing rearrangement collision which can occur between various isotopes of low atomic number. *See* NUCLEAR REACTION.

Interest in the nuclear fusion reaction arises from the expectation that it may someday be used to produce useful power, from its role in energy generation in stars, and from its use in the fusion bomb. Since a primary fusion fuel, deuterium, occurs naturally and is therefore obtainable in virtually inexhaustible supply (by separation of heavy hydrogen from water, 1 atom of deuterium occurring per 6000 atoms of hydrogen), solution of the fusion power problem would permanently solve the problem of the present rapid depletion of chemically valuable fossil fuels. As a power source, the lack of radioactive waste products from the fusion reaction is another argument in its favor as opposed to the fission of uranium.

In a nuclear fusion reaction the close collision of two energy-rich nuclei results in a mutual rearrangement of their nucleons (protons and neutrons) to produce two or more reaction products, together with a release of energy. The energy usually appears in the form of kinetic energy of the reaction products, although when energetically allowed, part may be taken up as energy of an excited state of a product nucleus. In contrast to neutron-produced nuclear reactions, colliding nuclei, because they are positively charged, require a substantial initial relative kinetic energy to overcome their mutual electrostatic repulsion so that reaction can occur. This required relative energy increases with the nuclear charge Z, so that reactions between low-Z nuclei are the easiest to produce. The best known of these are the reactions between the heavy isotopes of hydrogen, deuterium and tritium.

Fusion reactions were discovered in the 1920s when low-Z elements were used as targets and bombarded by beams of energetic protons or deuterons. But the nuclear energy released in such bombardments is always microscopic compared with the energy of the impinging beam. This is because most of the energy of the beam particle is dissipated uselessly by ionization and single-particle collisions in the target; only a small fraction of the impinging particles actually produce reactions.

Nuclear fusion reactions can be self-sustaining, however, if they are carried out at a very high temperature. That is to say, if the fusion fuel exists in the form of a very hot ionized gas of stripped nuclei and free electrons termed a plasma, the agitation energy of the nuclei can overcome their mutual repulsion, causing reactions to occur. This is the mechanism of energy generation in the stars and in the fusion bomb. It is also the method envisaged for the controlled generation of fusion energy.

The cross sections (effective collisional areas) for many of the simple nuclear fusion reactions have been measured with high precision. It is found that the cross sections generally show broad maxima as a function of energy and have peak values in the general range of 0.01 barn (1 barn = 10^{-24} cm²) to a maximum value of 5 barns, for the deuterium-tritium (D-T) reaction. The energy releases of these reactions can be readily calculated from the mass differences between the initial and final nuclei or determined by direct measurement.

Simple reactions. Some of the important simple fusion reactions, their reaction products, and their energy releases in millions of electron volts (MeV) are given by Eqs. (1).

$$
\begin{aligned}
\text{D} + \text{D} &\to \text{He}^3 + n + \ \ 3.25 \text{ MeV} \\
\text{D} + \text{D} &\to \text{T} + p + \ \ 4.0 \text{ MeV} \\
\text{T} + \text{D} &\to \text{He}^4 + n + 17.6 \text{ MeV} \\
\text{He}^3 + \text{D} &\to \text{He}^4 + p + 18.3 \text{ MeV} \\
\text{Li}^6 + \text{D} &\to 2\text{He}^4 \ \ \ \ + 22.4 \text{ MeV} \\
\text{Li}^7 + p &\to 2\text{He}^4 \ \ \ \ + 17.3 \text{ MeV}
\end{aligned}
\qquad (1)
$$

If it is remembered that the energy release in the chemical reaction in which hydrogen and oxygen combine to produce a water molecule is about 1 eV per reaction, it will be seen that, gram for gram, fusion fuel releases more than 1,000,000 times as much energy as typical chemical fuels.

The two alternative D-D reactions listed occur with about equal probability for the same relative particle energies. Note that the heavy reaction products, tritium and helium-3, may also react, with the release of a large amount of energy. Thus it is possible to visualize a reaction chain in which six deuterons are converted to two helium-4 nuclei, two protons, and two neutrons, with an overall energy release of 43 MeV—about 10^5 kilowatthours (kWhr) of energy per gram of deuterium. This energy release is several times that released per gram in the fission of uranium, and several million times that released per gram by the combustion of gasoline.

Cross sections. Figure 1 shows the measured values of cross sections as a function of bombarding energy up to 100 keV for the total D-D reaction (both D-D,n and D-D,p), the D-T reaction, and the D-He³ reaction. The most striking feature of these curves is their extremely rapid falloff with energy as bombarding energies drop to a few kilovolts. This effect arises from the mutual electrostatic repulsion of the nuclei, which prevents them from approaching closely if their relative energy is small.

The fact that reactions can occur at all at these energies is attributable to the finite range of nuclear interaction forces. In effect, the boundary of the nucleus is not precisely defined by its classical diameter. The role of quantum mechanical effects in nuclear fusion reactions has been treated by G. Gamow and others. It is predicted that the cross sections should obey an exponential law at low energies. This is well borne out in energy regions reasonably far removed from resonances (for example, below about 30 keV for the D-T reaction). Over a wide energy range at low energies, the data for the D-D reaction can be accurately fitted by a Gamow curve, the result for the cross section being given by Eq. (2), where the bombarding energy W is in kilo-electron-volts.

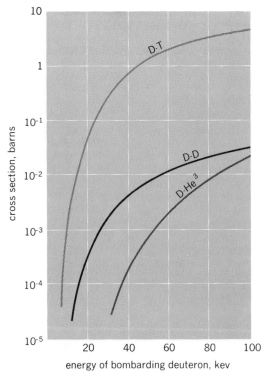

Fig. 1. Cross sections versus bombarding energy for three simple fusion reactions. (*From R. F. Post, Fusion power, Sci. Amer., 197(6):73–84, 1957*)

$$\sigma_{\text{D-D}} = \frac{288}{W} e^{-45.8W^{-1/2}} \times 10^{-24} \text{ cm}^2 \quad (2)$$

The extreme energy dependence of this expression can be appreciated by the fact that, between 1 and 10 keV, the predicted cross section varies by about 13 powers of 10, that is, from 3×10^{-42} to 1.5×10^{-29} cm^2.

Energy division. The kinematics of the fusion reaction stipulates that the reaction can occur only if two or more reaction products result. This is because both mass energy and momentum balance must be preserved. When there are only two reaction products (which is the case in all of the important reactions), the division of energy between the reaction products is uniquely determined, the lion's share always going to the lighter particle. The energy division (disregarding the initial bombarding energy) is as in Eq. (3). If Eq. (3) holds,

$$A_1 + A_2 \rightarrow A'_1 + A'_2 + Q \quad (3)$$

with the As representing the atomic masses of the particles and Q the total energy released, then Eqs. (4) are valid, where $W(A'_1)$ and $W(A'_2)$ are the kinetic energies of the reaction products.

$$W(A'_1) + W(A'_2) = Q$$

$$W(A'_1) = Q\left(\frac{A'_2}{A'_1 + A'_2}\right) \quad (4)$$

$$W(A'_2) = Q\left(\frac{A'_1}{A'_1 + A'_2}\right)$$

Thus in the D-T reaction, for example, A'_1, the mass of the α-particle, is four times A'_2, the mass of the neutron, so that the neutron carries off four-fifths of the reaction energy, or 14 MeV.

Reaction rates. When nuclear fusion reactions occur in a high-temperature plasma, the reaction rate per unit volume depends on the particle density n of the reacting fuel particles and on an average of their mutual reaction cross sections σ and relative velocity v over the particle velocity distributions. *See* THERMONUCLEAR REACTION.

For dissimilar reacting nuclei (such as D and T), the reaction rate is given by Eq. (5).

$$R_{12} = n_1 n_2 \langle \sigma v \rangle_{12} \quad \text{reactions/(cm}^3\text{)(s)} \quad (5)$$

For similar reacting nuclei (for example, D and D), the reaction rate is given by Eq. (6).

$$R_{11} = \frac{1}{2} n^2 \langle \sigma v \rangle \quad (6)$$

Note that both expressions vary as the square of the total particle density (for a given fuel composition).

If the particle velocity distributions are known, $\langle \sigma v \rangle$ can be determined as a function of energy by numerical integration, using the known reaction cross sections. It is customary to assume a maxwellian particle velocity distribution, toward which all others tend in equilibrium. The values of $\langle \sigma v \rangle$ for the D-D and D-T reactions are shown in Fig. 2. In this plot the kinetic temperature is given in units of kilo-electron-volts; 1 keV kinetic temperature $= 1.16 \times 10^7$ K. Just as in the case of the cross sections themselves, the most striking feature of these curves is their extremely rapid falloff with temperature at low temperatures. For example, although at 100 keV for all reactions $\langle \sigma v \rangle$

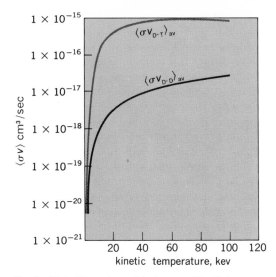

Fig. 2. Plot of the values of $\langle \sigma v \rangle$ versus kinetic temperature for the D-D and D-T reactions.

is only weakly dependent on temperature, at 1 keV it varies as $T^{6.3}$ and at 0.1 keV as T^{133}! Also, at the lowest temperatures it can be shown that only the particles in the "tail" of the distribution, which have energies large compared with the average, will make appreciable contributions to the reaction rate, the energy dependence of σ being so extreme.

Critical temperatures. The nuclear fusion reaction can obviously be self-sustaining only if the rate of loss of energy from the reacting fuel is not greater than the rate of energy generation by fusion reactions. The simplest consequence of this fact is that there will exist critical or ideal ignition temperatures below which a reaction could not sustain itself, even under idealized conditions. In a fusion reactor, ideal or minimum critical temperatures are determined by the unavoidable escape of radiation from the plasma. A minimum value for the radiation emitted from any plasma is that emitted by a pure hydrogenic plasma in the form of x-rays or bremsstrahlung. Thus plasmas composed only of isotopes of hydrogen and their one-for-one accompanying electrons might be expected to possess the lowest ideal ignition temperatures. This is indeed the case: It can be shown by comparison of the nuclear energy release rates with the radiation losses that the critical temperature for the D-T reaction is about 4×10^7 K. For the D-D reaction it is about 10 times higher. Since both radiation rate and nuclear power vary with the square of the particle density, these critical temperatures are independent of density over the density ranges of interest. The concept of the critical temperature is a highly idealized one, however, since in any real cases additional losses must be expected to occur which will modify the situation, increasing the required temperature.

Fusion reactor. Intense interest in nuclear fusion arises from its promise as a safe and inexhaustible source of energy for the future. Fusion reactors do not yet exist, but studies of the physics and technology that will be needed to construct such reactors have been underway since the 1950s.

The two key problems in achieving net power

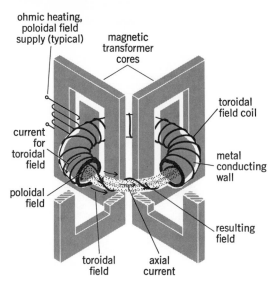

Fig. 3. Schematic illustration of a tokamak device. *(From R. F. Post and F. L. Ribe, Fusion reactors as future energy sources, Science, 186:397–407, 1974)*

from a fusion reactor are, first, to heat the fusion fuel charge to its required high temperature, and second, to confine the heated fuel for a long enough time for the fusion energy released to exceed the energy required to heat the fuel to its conbustion temperature, including all relevant losses. *See* LAWSON CRITERION.

The problem of achieving fusion power is in fact dominated by the quantitative requirements associated with the fusion process. The plasma heating technique employed must be capable of raising the fusion fuel charge to kinetic temperatures of order 100,000,000° or higher. The confinement system must be capable of satisfying stringent requirements on confinement time (which could be as long as seconds in some cases). At the same time it must be capable of sustaining the strong outward gas pressure exerted by the fuel charge. Furthermore, since the rate at which fusion power is generated varies as the square of the fuel density, for any continuously operating fusion reactor engineering limits on heat transfer must be taken into account, thus limiting the fuel density for such systems to a small fraction of the particle density of atmospheric air. At higher density it is only possible to conceive of pulsed operation—basically microexplosions. The quantitative requirements of fusion therefore strongly limit the possible approaches to fusion.

Two generically different approaches have emerged as constituting the most promising avenues to the eventual achievement of net fusion power, namely, magnetic confinement and pellet fusion.

Magnetic confinement relies on the fact that at fusion temperatures the fusion fuel charge will be completely ionized, that is, it will exist solely in the plasma state. Charged particles can be held trapped by a properly shaped magnetic field, and are thereby isolated from the reactor chamber walls, physical contact with which would instantly cool the plasma.

Pellet fusion aims at the same objective, but by an entirely different route. Here the idea to rapidly heat and compress a tiny fuel pellet, carrying out the entire operation so quickly that fusion can take place before the pellet flies apart—that is, the confinement is properly called "inertial." The major technical effort on pellet fusion is centered on the use of high-powered lasers to accomplish the heating and compression; substantial activity is also being devoted to pellet fusion induced by bombardment of the pellet with very-high-intensity electron beams; and use of heavy ions as the ignition probe is now receiving serious study.

Magnetic confinement. Magnetic confinement of a fusion plasma depends on the nature of the plasma state. The plasma may be viewed as an electrically conducting gas that exerts an outward pressure, or as a collection of free positive and negative charges. The pressure exerted by the plasma can be resisted by the electromagnetic stresses associated with a strong magnetic field; the individual charged particles can at the same time be guided by a properly shaped magnetic field that forces these particles to execute orbits that remain within the vacuum chamber surrounding the plasma without contact with the walls.

Adequate stability of the confined plasma is a prime requirement for effective magnetic confinement; otherwise, particles can escape prematurely, before having a sufficient probability to fuse. Thus, finding means for suppressing the inherent tendency for confined plasma to become unstable has been one of the central goals of nuclear fusion research since its inception.

Three general systems have been studies in large-scale experiments as candidates for fusion reactors based on magnetic confinement:

(1) The tokamak (Fig. 3), a closed or toroidal doughnut-shaped confinement system. The tokamak utilizes confining fields that are a combination of a strong external field produced by coils wound around the confinement chamber, and a weaker internal self-generated field arising from a circulating current induced in the plasma by transformer action. Here both the main field and the circulating current are toroidally directed. The resultant field forms a helical pattern of field lines. In the tokamak the circulating current also performs the important function of initial heating (ohmic heating) of the plasma.

(2) The theta pinch, a device that may either take the form of a skinny torus or of a long tube. The theta-pinch approach is directed in large part at the problem of heating a high-density plasma to fusion temperatures. In the theta pinch, heating is traditionally accomplished by pulsing on the confinement field in a time interval on the order of a microsecond, thereby both shock-heating and compressing the plasma to produce the high temperature (Fig. 4a). Following this heating phase, the magnetic field is sustained in order to maintain the plasma confinement (Fig. 4b).

(3) The mirror machine (Fig. 5), an "open-ended" system in which the plasma is held trapped by repeated reflections of its particles between magnetic mirrors (regions of intensified magnetic field at each end of the confinement chamber). It thus utilizes the repelling force that is exerted on a charged particle as it spirals along magnetic field lines that are converging, that is, by a field of increasing strength. Particles which spiral with sufficiently steep pitch angles will be repelled strongly

enough to become trapped, providing thereby a means for confinement of a plasma. However, mirror systems cannot trap an isotropic plasma; only particles whose pitch angles lie outside of the loss-cone defined by the strength of the mirrors will be contained. These systems must therefore rely on the injection of intense beams of energetic neutral atoms to maintain the plasma temperature and density in competition with particle leakage through the mirrors.

In theory, magnetic confinement should represent an almost ideal solution to the fusion reactor problem. In practice, it has been found difficult to achieve performance that comes close to the theoretical predictions. Nevertheless, major progress has been made toward perfecting the technique, to the point that scientists generally accept that it will be possible to demonstrate the confinement of plasmas at the temperatures and densities believed needed in a fusion reactor, sustained for times that will then be approaching those required in a full-scale reactor.

The problems that have to be solved to achieve successful magnetic confinement are both scientific and technological in nature. The scientific problem is to find those particular configurations of magnetic fields, and values of plasma parameters which, when scaled up to fusion reactor size, would ensure a viable net power yield from the reactor. Technologically, the problems are how to create the required high-intensity magnetic fields, how to heat the plasma toward fusion temperatures, at the same time protecting it from contamination by heavier atom impurities (which would quench the reaction).

A rough, and sometimes misleading, index of progress toward fusion requirements is the $n\tau$ confinement factor used in the Lawson criterion, where n is the particle density and τ is the confinement time. This parameter and the plasma ion temperature T_i at which the confinement is achieved together provide a useful index of progress toward achievement of the goal of thermonuclear power.

The table lists values of $n\tau$ and T_i achieved in some experiments taken from each of the three major approaches discussed above. For compari-

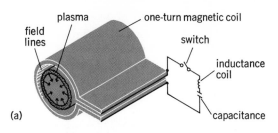

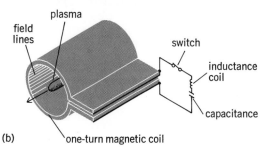

Fig. 4. Schematic illustration of theta-pinch system. (a) Dynamic phase; arrows indicate compression of plasma. (b) Quiescent phase; arrow indicates leakage of plasma along field lines. (From R. F. Post and F. L. Ribe, Fusion reactors as future energy sources, Science, 186: 397–407, 1974)

son, nominal values estimated to be required for a reactor based on the approach listed are also shown.

Pellet fusion. The basic ideas behind pellet fusion involve the rapid implosion of a high-density fuel pellet to produce a heated core that will fuse before it can fly apart. As usually conceived, the implosion would result from the rapid ablation of the surface of the pellet, giving rise to an inward-acting reaction force that compresses the core. But for this process to be successful, very high compression factors, of order 10,000, are required. That this should be the case can be seen from simple considerations: Consider the consequences of having to satisfy Lawson's $n\tau$ criterion while relying entirely on inertial effects to provide the confinement. As matter is compressed spherically, its density (n in the $n\tau$ product) increases as the cube

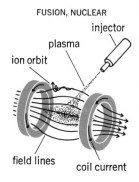

FUSION, NUCLEAR

Fig. 5. Schematic illustration of mirror machine using simple mirrors. (From R. F. Post and F. L. Ribe, Fusion reactors as future energy sources, Science, 186: 397–407, 1974)

Plasma parameters attained by magnetic confinement experiments compared with typical values believed needed for a fusion reactor based on the same approach

| Approach | Device | Experimental status | | | | | Reactor requirements | |
		Plasma diameter, d (cm)	Particle density, n (cm^{-3})	Ion temperature, T_i (keV)	Confinement time, τ (ms)	Lawson product, $n\tau_{\text{exper.}}$ (s/cm^3)	Ion temperature, T_i (keV)	Lawson product, $n\tau_{\text{reactor}}$ (s/cm^3)
Tokamak	T-4 (1970)	30	3×10^{13}	0.4	10	3×10^{11}	15	$10^{14} - 10^{15}$
	ST (1974)	25	4×10^{13}	0.6	10	4×10^{11}		
	TFR (1974)	40	4×10^{13}	0.8	15	6×10^{11}		
	Alcator (1975)	18	2×10^4	0.8	10	2×10^{12}		
Toroidal theta pinch	Scyllac	2	2×10^{16}	0.8	0.01	2×10^{12}	10	10^{15}
Mirror machine (magnetic-well type)	PR6 (1971)	10	2×10^{12}	0.5	0.15	3×10^8	150	10^{13}
	2XII (1975)	20	2×10^{13}	10.0	5.0	$\sim 10^{11}$		
	Baseball II (1972)	20	2×10^9	2.0	1000	2×10^9		
Early experiments for comparison								
Stellerator	Model C (1966)	10	10^{13}	0.06	0.2	2×10^9		
Mirror machine (simple mirrors)	Toy Top (1961)	5	10^{13}	3.0	0.05	5×10^8		

Fig. 6. One of the large laser amplifiers which are to be used in the 10,000-joule multiple-beam laser facility located at Lawrence Livermore Laboratory, shown in the process of construction.

of the radial compression factor. But the confinement time τ, which is here directly related to the time of flight of a particle out of the confinement region, decreases only as the first power of the radius. Thus $n\tau$ increases with the square of the radial compression factor. However, it is necessary to use tiny pellets to make the process technically accessible from the standpoint of engineering limits on heat transfer and energy release. This fact, taken together with the Lawson requirement, leads to the necessity for very large density compression factors, in turn leading to a requirement for very high compressing forces [of order 10^{12} atm (1 atm = 101,325 Pa); greater than those at the center of the Sun].

The problems thereby posed are twofold. First, very high pulsed powers are needed, (of order 10^{15} W/cm²), delivered within such a short time that an almost instantaneous ablation of the outer surface of the pellet occurs before an appreciable amount of heat can flow into the interior of the pellets; premature heat flow would prevent reaching the needed compression factors. In principle, this problem can be solved through the use of a sufficient number of converging laser beams from an array of ultra-high-powered lasers (Fig. 6) or alternately converging charged-particle beams from appropriate accelerator arrays. Second, although

simple in principle, the compression process is difficult in practice. If the compression action is not performed with high uniformity, it will go askew; small errors in uniformity can lead to a major reduction in the achievable compression. Sophisticated techniques must be used, and any instabilities occuring during compression must not be allowed to reach such a magnitude as to spoil the process. By 1976, experiments had successfully reached compression factors of order 100. The critical tests were therefore yet to come. These awaited the construction of sufficiently powerful laser systems to permit attempts at achieving the next orders of magnitude of compression. Charged-particle and ignition techniques are still in an early stage of development. *See* LASER-INDUCED FUSION.

Reactor studies. Preliminary studies have been made of the form that fusion reactors using the D-T fuel cycle might take if based on the various approaches that have been outlined. These studies cannot of course be definitive, but they help to indicate the sizes, costs, and special engineering problems that are likely to characterize fusion reactors, insofar as they can now be visualized.

The types of fusion reactors that have been studied encompass both pulsed and steady-state reactors, operating in either a driven mode or one in which plasma ignition is contemplated. A driven fusion reactor is one in which the plasma temperature is maintained primarily by a continuous input of energy—for example, by the injection of high-intensity beams of energetic neutral fusion fuel atoms—thereby maintaining the required temperature and density of the reacting plasma. The power input required to produce the beams is derived by recirculating a portion of the electrical energy yielded by the reactor. A positive power balance is here achieved when the recirculated power is less than the power output as recovered from the plasma, including not only reaction products but the kinetic energy content of the unreacted part (electrons and heated fuel ions). By contrast, a reactor operated in an ignition mode is one where it is required that the energy deposited directly within the plasma by charged reactor products (3.5-MeV alpha particles in the case of the D-T reaction; that is, only 20% of the total fusion energy release) is sufficient to maintain the plasma temperature, including the requirement for heating up any new cold fuel particles introduced to maintain the fuel plasma density as burning proceeds. The ignition mode is therefore more demanding with respect to confinement time than is a driven mode, but in principle could be technically simpler, since it does not impose strict requirements on the efficiencies of the energy recovery and plasma heating systems.

By taking advantage of the improvement in confinement associated with scale-up in size, large toroidal reactors might be able to operate in an ignition mode. Mirror systems, on the other hand, would have to be driven reactors, as now visualized, because of the smaller margin between fusion energy release and heating requirements.

Considering the magnetic confinement approach, reactor studies have lead to some important conclusions: Fusion reactors will be relatively large, thus suitable for central power stations; depending on the approach, electrical power outputs of 500 to several thousand megawatts are likely to be typical. Factors which determine the minimum output include the scaling laws for the confinement. For example, a toroidal reactor if too small in size could not produce a positive power balance, owing to the scaling of confinement time with size, whereas in mirror systems confinement time is determined by other factors, so that smaller systems could be contemplated. Another general result of the studies is to show the importance of choice of materials and heat-transfer characteristics of the inner wall of the reactor chamber. The flux of 14-MeV neutrons through this wall coming from DT reactions in the plasma will cause localized heating, radiation damage, and induced radioactivity. Thus the design of this portion of the reactor can be expected to be critical for any reactor based on magnetic confinement. Another critical factor is that of the generation of the confining magnetic field. Here the criterion would be to achieve the highest possible field at the least capital cost and for the least expenditure of energy to maintain the field. Fortunately, the development of practical high-current-density, high-field superconductors appears to provide an excellent solution to this problem.

[RICHARD F. POST]

Bibliography: J. Emmett, J. Nuckolls, and L. Woods, Fusion power by laser implosion, *Sci. Amer.*, 230(6):24–37, 1974; W. C. Gough and B. J. Eastland, Prospects of fusion power, *Sci. Amer.*, 224(2):50–64, 1971; B. B. Kadomtsev and T. K. Fowler, Fusion reactors, *Phys. Today*, 28(11):36–43, 1975; R. F. Post, Controlled fusion research and high temperature plasmas, *Annu. Rev. Nucl. Sci.*, 20:509–588, 1970; R. F. Post and F. L. Ribe, Fusion reactors as future energy sources, *Science*, 186:397–407, November 1974; Prospects for fusion power, *Phys. Today*, 26(4):30–39, 1973; D. J. Rose and M. Clark, *Plasmas and Controlled Fusion*, 1961; World Survey of Major Facilities in Controlled Fusion Research, *Nuclear Fusion*, 1969.

Gas dynamics

The study of the motion of gases which takes into account thermal effects generated by the motion. Gas dynamics combines fluid mechanics and thermodynamics and differs from gas statistics in that there is motion and from gas kinematics in that the forces exerted on or by the gas are considered.

Scope of subject. Several other names are used to define the subject. The most important ones are aerothermodynamics, aerothermochemistry, fluid dynamics, compressible fluid flow, and supersonic aerodynamics. This terminology reflects the fact that in each particular case different aspects of gas dynamics are emphasized. Magnetogasdynamics, which includes effects due to magnetism and electricity, has applications in rocket reentry, propulsion, and astrophysics. *See* MAGNETOGAS DYNAMICS.

Gas dynamics deals with the motion of a continuous medium and is not concerned with the behavior of individual atoms or molecules which constitute the gas. At low pressures, however, such as may be encountered at very high altitudes, the particle mean free path is very large and continuum considerations may not be applicable. The Knudsen number, kn, which is defined as the ratio of the mean free path to a characteristic length of interest, characterizes the regimes of continuum and free molecule flow, respectively. Thus, if $kn = 0.01$, the continuum gas dynamic equations are usually valid. On the other hand, when kn is 10 or greater, the flow must be described by the Boltzmann equation. The transition region when $kn = 1$ is referred to as slip flow.

Fundamental relations. The fundamental conservation principles of mechanics and thermodynamics constitute the theoretical basis of gas dynamics. The conservation laws can be derived in Lagrangian form with respect to a specific mass of flowing gas or in Eulerian form with respect to the rate at which gas enters and leaves a fixed control volume in space. The Eulerian conservation laws are expressed by (1) the continuity equation (conservation of mass); (2) the Navier-Stokes equations (conservation of momentum); and (3) the energy equation (conservation of energy).

A list follows of the principal notations used in the field.

A = cross-sectional area
a = acoustic velocity
c = speed of light
c_p = specific heat at constant pressure

c_v = specific heat at constant volume
f = friction factor, as in Eq. (53)
G = mass velocity, $G = \rho V$
G' = mass flow per unit time
h = enthalpy
k = thermal conductivity
L = characteristic length
m = molecular mass
n_i = number of particles of component i per unit volume
p = pressure
p_0 = total or stagnation pressure
Q_m = heat
R, R' = gas constant
r_h = hydraulic radius
T = temperature
T_0 = total or stagnation temperature
t = time
u = internal energy
V = average gas velocity
$\mathbf{V}$ = gas vector velocity
$\mathbf{v}_i$ = particle vector velocity
w_i = rate of production of species i
x = fractional dissociation
α = Mach angle
γ = specific heat ratio, c_p/c_v
ν = kinematic viscosity
ρ = density
∇ = gradient operator

Super and subscripts
$( )^*$ = critical values
$( )_o$ = reservoir condition

Continuity equation. Two types of continuity equation can be written: global and species conservation. The global continuity equation is expressed by Eq. (1), where the first term defines the

$$\nabla \cdot (\rho \mathbf{V}) + \frac{\partial \rho}{\partial t} = 0 \qquad (1)$$

rate of change of the mass flow with respect to the space coordinates, whereas the second term, usually called the source term, indicates changes with respect to time within the control volume.

If the flow is steady and there are no sources present, the continuity equation reduces to Eq. (2).

$$\nabla \cdot (\rho \mathbf{V}) = 0 \qquad (2)$$

For incompressible flow, it is given simply by Eq. (3).

$$\nabla \cdot \mathbf{V} = 0 \qquad (3)$$

It is often convenient in steady irrotational flow to introduce a function φ satisfying Eq. (4).

$$\nabla \varphi = \mathbf{V} \qquad (4)$$

The incompressible continuity equation, Eq. (3), becomes Eq. (5). The identity $\nabla \times (\nabla \varphi) = 0$ shows that the flow is irrotational. Equation (5) is known

$$\nabla^2 \varphi = 0 \qquad (5)$$

as Laplace's equation. Solutions to Laplace's equation which satisfy the appropriate boundary conditions are solutions of gas kinematics since they describe a flow without regard to the forces maintaining it.

If the gas undergoes chemical changes, the concentration of species is altered, which affects the spatial distribution of energy since each species

has its particular velocity $\mathbf{v}_i$. Thus, in order to properly keep track of the distribution of energy, one must account for the rate of change of each specie or component by a continuity equation of the form shown in Eq. (6). The term on the right-hand side

$$\frac{\partial n_i}{\partial t} + \nabla \cdot (n_i \mathbf{v}_i) = w_i \qquad (6)$$

represents the rate of specie production. When Eq. (6) is summed for all species, one arrives at the continuity equation, Eq. (1).

Frequently a one-dimensional approach is followed in gas dynamics, in which case the properties are assumed to vary mainly in the flow direction. The integral of the steady global continuity equation is then given by Eq. (7).

$$\rho A V = \text{constant} = G' \qquad (7)$$

Momentum equation. The momentum equation expresses the conservation of momentum. It must take into consideration the effects of friction and of external body forces such as gravity, magnetism, and possibly others. Written in vector form, it is given by Eq. (8). The term D/Dt represents the

$$D\mathbf{V}/Dt = -\frac{1}{\rho} \nabla p + \nu \nabla^2 \mathbf{V}$$
$$+ (\tfrac{1}{3})\nu \nabla (\nabla \cdot \mathbf{V}) + \sum_{l=1}^{N} \mathbf{F}_l \qquad (8)$$

mobile operator which is defined by Eq. (9). The

$$D/Dt = (\partial/\partial t) + \mathbf{V} \cdot \nabla \qquad (9)$$

first term on the right-hand side of Eq. (8) represents the forces exerted on the fluid at the control volume boundaries; the second and third terms on the right-hand side are the viscous stresses; and F_l stands for any body force such as gravity, electric, and magnetic forces. Equation (8) in its vectorial form is quite general and applies to classical gas dynamics and aerothermochemistry as well as to magnetohydrodynamics. The specialization to particular coordinate systems, for example, cartesian coordinates, is obtained by standard vector manipulation methods.

It may not be always necessary, however, to consider all terms. For example, if electromagnetic effects are not present but viscous effects are included, one obtains the Navier-Stokes equation.

A general solution for the Navier-Stokes equation has not been found, and only a few particular solutions exist. When the momentum equation is simplified to exclude viscous effects as well as forces on the body, one obtains Euler's equation, shown as Eq. (10). Its one-dimensional form is given by Eq. (11).

$$D\mathbf{V}/Dt = -\frac{1}{\rho} \nabla p \qquad (10)$$

$$\frac{dp}{\rho} + V\,dV = 0 \qquad (11)$$

Equation (11) can be integrated for incompressible flow (that is, $\rho = \text{constant}$) to yield Bernoulli's equation, shown as Eq. (12).

$$\frac{p}{\rho} + \frac{V^2}{2} = \text{constant} \qquad (12)$$

For compressible flow, one needs to consider the thermodynamics of the flow, which is done through the energy equation, discussed below.

Energy equation. The energy equation expresses the principle of conservation of energy (first law of thermodynamics) as applied to a flowing gas. There are a large number of equivalent forms in which this equation can be written. The apparent differences arise from the use of such subsidiary thermodynamic relations as, for example, the equation of state $p = \rho RT$ for a perfect gas, or $p/\rho^\gamma = $ constant for an isentropic process, to express one state variable in terms of another. A fairly common form of the energy equation is given in Eq. (13). The left-hand side of Eq. (13) gives the

$$\frac{\rho D(u + \frac{1}{2}V^2)}{Dt} = \nabla \cdot p\mathbf{V}$$
$$\text{(I)}$$
$$+ \rho \sum_{l=1}^{N} \mathbf{F}_l \cdot \mathbf{V} + \nabla \cdot k\nabla T + \sum_{m=1}^{M} Q_m \quad (13)$$
$$\text{(II)} \qquad\qquad \text{(III)} \qquad \text{(IV)}$$

change in internal and kinetic energy, which is balanced on the right-hand side by the rate at which work is done by (I) the pressure forces and (II) the body forces and (III) by the heat conducted across the boundary. Term (IV) accounts for any other energy-transfer mechanism, such as the transfer of heat generated by the dissipative action of viscosity and electrical conductivity or radiative heat transfer or transfer of electromagnetic energy.

If the flow is steady, one obtains Eq. (14). Furthermore,

$$\rho\mathbf{V} \cdot \nabla \left(u + \frac{1}{2}V^2 + \frac{p}{\rho} \right)$$
$$= \nabla \cdot k\nabla T + \rho\Sigma\mathbf{F} \cdot \mathbf{V} + \Sigma Q \quad (14)$$

thermore, if the flow is adiabatic, then the work done by the shear forces Φ does not leave the system but simply raises the internal energy of the gas and is, therefore, accounted for by u. In the absence of body forces, all terms on the right-hand side of Eq. (14) are zero, and the energy equation can be integrated to yield Eq. (15). Moreover, if the gas is perfect, one obtains Eq. (16).

$$u + \frac{1}{2}V^2 + \frac{p}{\rho} = \text{constant} \quad (15)$$

$$\frac{\gamma}{\gamma - 1}\frac{p}{\rho} + \frac{V^2}{2} = c_p T + \frac{V^2}{2} = \text{constant} \quad (16)$$

The idealizations introduced by the concept of a perfect gas are approximately satisfied at moderate temperatures and pressures. At very high pressures or temperatures or both, the real nature of the gas molecules must be taken into consideration. In gas dynamics, one is not ordinarily concerned with very high pressure; but high temperatures, and the attendant effects on the gas, are regularly encountered in missile flight, combustion, nuclear reactors, and many other technological applications. A very simple correction to the perfect equation of state can be made if the gas molecules dissociate. The equation of state then can be written as Eq. (17), where $R'/m = R$ and x is

$$p = R'Tn(1 + x) \quad (17)$$

the fraction of the gas which is dissociated—the bookkeeping is done by the conservation of species equations.

At sufficiently high temperatures, the electrons may become so energetic as to leave their orbits and become free electrons. When this happens, the gas is said to be ionized and is called a plasma.

In a fairly complex situation where the gas can consist of molecules, atoms, ions, and electrons, the equation of state may have to be modified to account for more than just one dissociating species, as in Eq. (18). Another consequence of the

$$p = R'Tn(1 + \Sigma x_i) \quad (18)$$

nonideal nature of the gas is the fact that it takes time for a molecule to dissociate, to ionize, and to combine with another species. The time that it takes to effect such a change is called relaxation time. When the relaxation time is short compared to a characteristic flow time, which might be the time it takes a fluid element to pass through a shock, one speaks of thermodynamic equilibrium; this state is characterized by Damköhler's first ratio, given by Eq. (19).

$$\text{Da}_\text{I} = \frac{t_\text{transit}}{t_\text{relaxation}} \gg 1 \quad (19)$$

At equilibrium, the degree of ionization X is given by the Saha equation, Eq. (20), where E_0 is a characteristic parameter of the gas.

$$\frac{x^2}{1 - x^2} = \text{constant}\frac{T^{5/2}}{p}\exp\ -(E_0/RT)] \quad (20)$$

For the case $\text{Da}_\text{I} \ll 1$, one speaks of frozen flow. Here the relaxation time is so large that the gas behaves nearly like a perfect gas.

When relaxation and transit times are comparable, the necessary accounting of all the possible species depends on a detailed knowledge of all possible chemical processes and the controlling rates; these rates have been established for a large number of reactions. However, the details of even a relatively simple phenomenon such as the burning of a Bunsen burner are still not completely understood.

Real gases deviate in other important aspects from a perfect gas. In the latter, the specific heat at constant pressure c_p, and at constant volume c_v are constants. In a real gas, the manner in which the molecules store, so to say, the heat must be considered. A gas that obeys the perfect gas thermal equation of state $p = \rho RT$ has the caloric equation of state given by Eq. (21), where c_{p0} and γ are

$$\frac{c_p}{c_{p0}} = 1 + \frac{\gamma - 1}{\gamma}\left[\left(\frac{\theta}{T}\right)^2 \frac{e^{\theta/T}}{(e^{\theta/T} - 1)^2} \right] \quad (21)$$

constant reference values. For $\gamma = 1.4$, the correction at 3000°K is approximately 26%. θ refers to molecular vibrational-energy constant. For air between 300 and 2500°K, the value of $\theta \approx 2800$°K.

Dimensionless parameters. Of the large number of dimensionless parameters that can be formed, there are a few that are particularly useful in effecting simplifications in the equations of motion. The simplifications are generally the consequence of one or more parameters being very large or very small.

The Knudsen number and Damköhler's ratio have been mentioned already. The Reynolds number $Re = VL/\nu$ is the dominating parameter when the effects of viscosity and the inertia of the fluid both contribute to the gas motion. The phenomena which are peculiar to gas dynamics, however, can best be categorized in terms of the Mach number. The Mach number is the ratio of the flow speed to the speed of sound. The speed of sound is the propagation velocity not only of audible sound but of any weak pressure disturbance.

Gas-dynamic flow regimes can then be classified as follows.

$$M < 1 \quad \text{subsonic flow}$$
$$M = 1 \quad \text{sonic flow}$$
$$0.9 < M < 1.1 \quad \text{transonic flow}$$
$$M > 1 \quad \text{supersonic flow}$$
$$M > 5 \quad \text{hypersonic flow}$$

Speed of sound. Consider a tube with insulated walls filled with a compressible nonviscous gas. Neglecting all external forces, Euler's equation in one dimension, Eq. (22), and continuity, Eq. (23), describe the motion.

$$\frac{\partial V}{\partial t} + \frac{V \partial V}{\partial x} = -\frac{1}{\rho} \frac{\partial p}{\partial x} \tag{22}$$

$$\frac{\partial \rho}{\partial t} + \frac{\partial (\rho V)}{\partial x} = 0 \tag{23}$$

For the case of small perturbations, the equation of state is given by Eq. (24), where ρ_0 is a reference density.

$$\rho = \rho_0 (1 + \epsilon) \tag{24}$$

Substituting Eq. (24) into Eqs. (22) and (23), one obtains Eqs. (25) and (26).

$$\frac{\partial V}{\partial t} = -\frac{\rho}{\rho_0} \frac{dp}{d\rho} \frac{\partial \epsilon}{\partial x} \tag{25}$$

$$\frac{\partial \epsilon}{\partial t} = -\frac{\partial V}{\partial x} \tag{26}$$

Eliminating ϵ between Eqs. (25) and (26), one obtains the one-dimensional wave equation, Eq. (27), where Eq. (28) defines the velocity of propaga-

$$\frac{\partial^2 V}{\partial t^2} = a^2 \frac{\partial^2 V}{\partial x^2} \tag{27}$$

$$a^2 = \frac{dp}{d\rho} \bigg|_{\rho = \rho_0} \tag{28}$$

tion of weak disturbances in general and sound in particular. If one assumes the transformation to be isentropic, which is quite reasonable, one obtains Eq. (29).

$$a^2 = \gamma R T \tag{29}$$

The derivation of the speed of sound is illustrative of a whole class of problems that can be treated by the simple wave equation. The term waves is used in gas dynamics not only in the classical sense but also to denote wavefronts.

Consider a pulsating pressure source moving with velocity V. This source starts pulsating at time $t = 0$. In a time interval t_1, the source travels a distance $V t_1$, while the signal reaches the surface of a sphere of radius $a t_1$ (the surface of this sphere constitutes the wavefront at time t_1). At a later time t_2, the source point is at $V t_2$, while the signal front is at $a t_2$. Thus, as long as $V/a = M < 1$, the source point is always inside the outermost wavefront (Fig. 1).

Still another use of the term wave refers to a

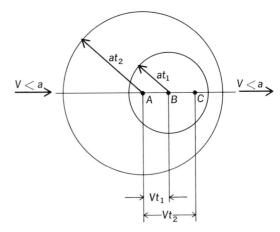

Fig. 1. Wavefronts produced by a point source moving at subsonic velocity. (*A. B. Cambel and B. H. Jennings, Gas Dynamics, McGraw-Hill, 1958*)

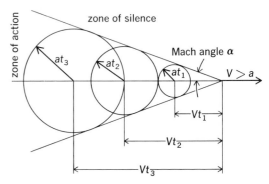

Fig. 2. Rule of forbidden signals from a point source moving at supersonic velocity. (*A. B. Cambel and B. H. Jennings, Gas Dynamics, McGraw-Hill, 1958*)

wave envelope. If the point source moves so fast that $V > a$, the wavefront spheres will no longer contain the source. The condition is shown in Fig. 2. The envelope to this family of spheres is a cone, known as the Mach cone, and the Mach angle is such that $\sin \alpha = 1/M$.

The Mach line constitutes a demarcation. In Fig. 2 the fluid outside the Mach line will receive no signal from the source. T. von Kármán has appropriately called this phenomenon the rule of forbidden signals and designated the region ahead of the Mach line the zone of silence and the region inside the Mach line the zone of action.

Shock waves. In the same manner in which a Mach wave is the envelope of infinitesimal disturbances, a shock wave is the envelope of finite disturbances. The steady conditions on either side of a standing shock wave can be obtained by applying the conservation laws (Fig. 3) expressed by Eqs. (30), (31), and (32), where $h = u + p/\rho$ is termed

$$\rho_1 V_1 = \rho_2 V_2 \qquad \text{(continuity)} \tag{30}$$

$$p_1 + \rho_1 V_1^2 = p_2 + \rho_2 V_2^2 \qquad \text{(momentum)} \tag{31}$$

$$h_1 + \frac{V_1^2}{2} = h_2 + \frac{V_2^2}{2} \qquad \text{(energy)} \tag{32}$$

enthalpy, which for a perfect gas is given by $c_p T$. By a simple rearrangement of these equations, one obtains for a perfect gas the approximate expressions for the shock Mach number, Eq. (33) for

$$M_s \approx 1 + \frac{\gamma+1}{4\gamma}\frac{p_2-p_1}{p_1} \qquad (33)$$

weak shocks and Eq. (34) for strong shocks.

$$M_s \approx \left(\frac{\gamma+1}{2\gamma}\frac{p_2}{p_1}\right)^{1/2} \qquad (34)$$

In a sound wave, $p_2 - p_1 \cong 0$ and, therefore, $M_s = 1$; for $p_2/p_1 = 4$, the shock speed is approximately equal to twice the speed of sound, as shown in Eq. (34).

Detonation and deflagration waves. Other interesting gas dynamic waves are characterized by the same continuity and momentum equations. The energy equation, however, is modified to include a term which accounts for chemical heat release; such waves are either detonations or deflagrations.

Eliminating the kinetic energy from the energy equation by the use of the momentum equation yields Eq. (35).

$$h_1 - h_2 + Q = \tfrac{1}{2}\,(p_1 - p_2)\left(\frac{1}{\rho_2} + \frac{1}{\rho_1}\right) \qquad (35)$$

For a given Q, zero or nonzero, and given p_1 and ρ_1 (which through the equation of state gives h_1) and one additional variable behind the wave, for example, V_2 or h_2, the locus of all possible combinations of ρ_2 and p_2 can be plotted on a so-called Hugoniot diagram (Fig. 4). The lines OJ and OK are tangent to the Hugoniot curve; O is the point $p_1, 1/\rho_1$. Points J and K separate strong and weak waves. Flows corresponding to J and K are characterized by the fact that all the thermodynamic and fluid-mechanic variables have an extremum. Transitions from B to C or vice versa involve a decrease in entropy and are therefore forbidden. The slope of the tangent is connected through the momentum equation to the flow velocity by Eq. (36).

$$\frac{p_2 - p_1}{\frac{1}{\rho_1} - \frac{1}{\rho_2}} = (\rho_2 V_2)^2 = \frac{\Delta p}{\Delta\left(\frac{1}{\rho}\right)}\bigg|_2 \qquad (36)$$

The derivative at constant entropy (since the entropy is stationary near J) is given by Eq. (37).

$$\rho^2\frac{dp}{d\rho}\bigg|_2 = \rho_2^2 a_2^2 \qquad (37)$$

In other words, the wave has a Mach number of unity with respect to the gas behind it. The points are appropriately named Chapman-Jouget points after the men who discovered their unique properties. The significance of these points lies in the fact that a stable detonation will eventually reach the point J and a deflagration point K.

Hydromagnetic (Alfvén) waves. Illustrative of the interaction of an electromagnetic field with a flowing plasma is the hydromagnetic, or Alfvén, wave. It is assumed that the velocity has only one component, for example, V_y, and the applied magnetic field B_x is perpendicular to it; the electric field E_z is perpendicular to B_x, and the current density j_z is parallel to the electric field. In order to simplify matters, it is assumed that the medium is

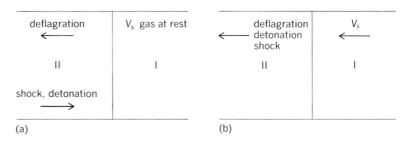

(a)   (b)

Fig. 3. Gas dynamic discontinuity. (a) Moving discontinuity, (b) Stationary discontinuity. (A. B. Cambel and B. H. Jennings, Gas Dynamics, McGraw-Hill, 1958)

incompressible so that the energy equation need not be considered. The fluid, moreover, is assumed to be inviscid and possesses infinite electrical conductivity. The equations of motion are then given by Eqs. (38) and (39).

Substituting these equations into Maxwell's equations yields a wave, Eq. (40), with the propaga-

$$\rho\frac{\partial V_y}{\partial t} = j_z B_x \qquad (38)$$

$$E_z - V_y B_x = 0 \qquad (39)$$

$$\frac{\partial^2 E_z}{\partial x^2} = \left(1 + \frac{4\pi\rho c^2}{B_x^2}\right)\frac{1}{c^2}\frac{\partial^2 E_z}{\partial t^2} \qquad (40)$$

tion velocity given by expression (41), called the Alfvén speed, where c is the speed of light.

$$\frac{c}{(1 + 4\pi\rho c^2/B_x^2)^{1/2}} \qquad (41)$$

Mach number functions. In many applications it is reasonable to assume that the gas is perfect — both thermally and calorically — and that the flow is

	M_1	M_2	p_2/p_1	V_2/V_1	ρ_2/ρ_1	T_2/T_1
detonation	>1	≤1	>1	<1	>1	>1
deflagration	<1	<1	<1	>1	<1	>1

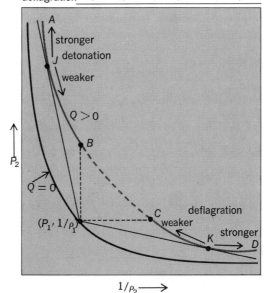

Fig. 4. Hugoniot diagram and Chapman-Jouget conditions. (A. B. Cambel and B. H. Jennings, Gas Dynamics, McGraw-Hill, 1958)

adiabatic. It then becomes very useful to express all the dependent variables in terms of the Mach number.

One usually starts with the energy equation and defines a stagnation temperature T_0 by expression (42). Physically, T_0 represents the temperature the gas would have if all its kinetic energy were transformed into thermodynamic enthalpy. The stagnation enthalpy can be measured by a thermometer immersed into a gas stream.

From Eq. (42), one simply obtains Eq. (43),

$$\frac{V^2}{2} + c_p T = c_p T_0 \qquad (42)$$

$$\frac{T_0}{T} = 1 + \frac{\gamma - 1}{2} M^2 \qquad (43)$$

where T is now called the static temperature. This temperature is measured with a thermometer at rest with respect to the gas. The stagnation temperature is constant in any adiabatic flow, even through a shock, and thus provides an excellent reference parameter. By using the isentropic relation $p = \rho^\gamma$ and the perfect gas law, one can define a reference stagnation pressure and density by means of Eqs. (44) and (45).

$$\frac{p_0}{p} = \left(1 + \frac{\gamma - 1}{2} M^2\right)^{\gamma/\gamma - 1} \qquad (44)$$

$$\frac{\rho_0}{\rho} = \left(1 + \frac{\gamma - 1}{2} M^2\right)^{1/\gamma - 1} \qquad (45)$$

When $M^2 \ll 1$, one may expand Eq. (44) by the binomial theorem to obtain Eq. (46).

$$p_0 = p + \frac{1}{2}\rho V^2 \left(1 + \frac{M^2}{4} + \ldots\right) \qquad (46)$$

The deviation from Bernoulli's equation, Eq. (12), due to compressibility at $M = 0.5$ is only 6%.

Flows can be classified as internal flow and external flow. Internal flow refers to the cases where the gas is constrained by a duct of some sort. Characteristically external flow is flow over an airplane or missile.

Internal one-dimensional flow. Internal flows are conveniently characterized by (1) the shape of the duct and its variation, (2) the heat transfer through the walls of the duct and internal heat sources, and (3) frictional effects. By varying one of these characteristics at a time, the essential features of internal flow can be discussed most simply.

Variable area flow. A device to accelerate the flow of a gas or liquid is termed a nozzle. In most engineering applications the contour of the nozzle

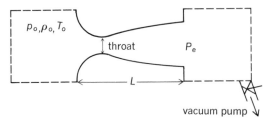

Fig. 5. Convergent-divergent nozzle. (*A. B. Cambel and B. H. Jennings*, Gas Dynamics, *McGraw-Hill, 1958*)

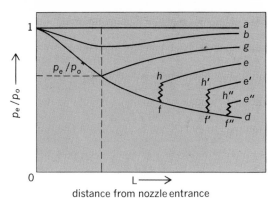

distance from nozzle entrance

Fig. 6. Pressure distribution in convergent-divergent nozzle between two reservoirs. (*A. B. Cambel and B. H. Jennings*, Gas Dynamics, *McGraw-Hill, 1958*)

is first converging and then diverging; it therefore has a minimum cross section, which is called a throat.

For isentropic flow in a convergent-divergent nozzle in which the flow is supersonic in the divergent section, the velocity at the throat is sonic, that is, $M^* = 1$. The throat pressure is then said to be critical p^* and is given by Eq. (47). Velocity and

$$p^* = p_0 \left(\frac{2}{\gamma + 1}\right)^{\gamma/(\gamma - 1)} \qquad (47)$$

pressure are related in this case by Eq. (48).

$$V = \frac{2\gamma R}{\gamma - 1} T_0 \left[1 - (p/p_0)^{(\gamma - 1)/\gamma}\right] \qquad (48)$$

Consider a convergent-divergent deLaval nozzle inserted between two reservoirs as in Fig. 5. There will be no flow if the ratio of exit pressure p_e to reservoir pressure p_0 is $p_e/p_0 = 1$ (case a in Fig. 6). If p_e is reduced so that it is slightly less than the entrance pressure, the nozzle will act like a conventional venturi, as represented by curve b in Fig. 6. For this case, the flow is always subsonic and resembles incompressible flow. When the exit pressure is reduced further, the critical pressure can be reached at the throat, as curve g shows. In this case, the velocity is sonic at the throat but is never supersonic within the nozzle, even though the pressure at the throat corresponds to the critical. The minimum pressure which can exist in the nozzle outlet is depicted by point d, for which the pressure at the throat will be the critical. Here the velocity in the converging section is subsonic, in the diverging section it is supersonic, and at the throat it is sonic. For the range of exit pressures from p_d to p_g, the rate-of-flow curve is the same, for a given reservoir pressure p_0, and is plotted in Fig. 7. The flow rate reaches a maximum value and remains at this value over this wide range of exit pressures.

Even in the absence of friction, isentropic flow can exist only for the range of exhaust pressures from p_a to p_g and at the pressure reached along curve d, but not at intermediate pressures. The pressures p_g and p_d are the significant design pressures for a given nozzle. For exhaust pressures in the range between p_d and p_g, shocks will occur in the nozzle, raising the pressure from f to h (or f' to h'), followed by a pressure rise after the shock points to an exit pressure such as p_e. If p_e is less than p_d, the jet leaving the nozzle is said to be underexpanded and will drop in pressure after leaving the mouth of the nozzle. The velocity V_1 in front of a shock is supersonic, and the velocity V_2 behind a normal in contrast to an oblique shock is always subsonic. For a normal shock $V_1V_2/a^{*2}=1$, where a^* is the critical speed of sound corresponding to $M=1$. Thus for $V_1 > V_2$, $V_1/a^* > 1$ and therefore $V_2/a^* < 1$.

Diabatic flow. Heat exchangers and combustion chambers are devices in which heat transfer occurs. The equations describing nonadiabatic or diabatic processes are complicated; consequently, certain limiting assumptions are usually required to make possible analytical solutions of the equations.

These assumptions are that (1) the flow takes place in a constant-area section, (2) there is no friction, (3) the gas is perfect and has constant specific heats, (4) the composition of the gas does not change, (5) there are no devices in the system which deliver or receive mechanical work, and (6) the flow is steady.

Equations which conform to these requirements are called Rayleigh equations, and the associated flow is designated as Rayleigh flow. Designating by $Q_{1\to2}$ the quantity of heat introduced between stations 1 and 2, one obtains (for the energy equation) Eq. (49).

If the stagnation enthalpy is introduced, then $Q_{1\to2}$ is given by Eq. (50), which can be expressed in terms of stagnation temperatures by Eq. (51).

$$Q_{1\to2} = c_p(T_2 - T_1) + (V_2^2 - V_1^2)/2$$
$$= h_2 - h_1 + (V_2^2 - V_1^2)/2 \quad (49)$$
$$Q_{1\to2} = h_{02} - h_{01} \quad (50)$$
$$Q_{1\to2} = c_p(T_{02} - T_{01}) \quad (51)$$

Because $Q_{1\to2} \neq 0$ for diabatic flow and because $c_p > 0$ always, it follows that $T_{02} \neq T_{01}$. This inequality states that in diabatic flow the stagnation temperature is not solely determined by the reservoir conditions, as is the case with adiabatic flow. Heating raises the stagnation temperature; cooling lowers it. *See* ENTHALPY.

The locus of points of properties during a

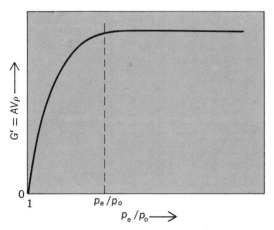

Fig. 7.   Flow rate for given p_0, through convergent-divergent nozzle between two reservoirs. (*A. B. Cambel and B. H. Jennings, Gas Dynamics, McGraw-Hill, 1958*)

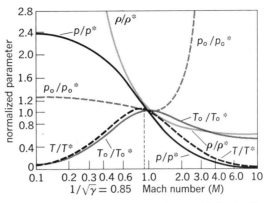

Fig. 8.   Diabatic flow parameters for specific heat ratio of 1.4. Asterisk where $M=1$. (*A. B. Cambel and B. H. Jennings, Gas Dynamics, McGraw-Hill, 1958*)

constant-area, frictionless flow with heat exchange is called the Rayleigh line. By definition, along the Rayleigh line the continuity equation and the momentum equation must apply. Equation (30) applies to steady flow in a constant-area duct. Mass velocity by definition is $G=\rho V$ and, from Eq. (31), the momentum relation is $p+\rho V^2=C$, where C is a constant. Consequently, one obtains Eq. (52), which is one of the many Rayleigh-line equations.

$$p + \frac{G^2}{\rho} = C \quad (52)$$

The variations of pressure, temperature, and density with Mach number for Rayleigh flow are plotted in Fig. 8. The fact that the curve for T_0/T_0^* in Fig. 8 reaches a maximum at a Mach number of unity indicates that it is impossible to pass from one flow domain into the other by the same heat-transfer process. Thus, if heat is added to a subsonic flow, the flow can be accelerated only until its Mach number becomes unity. Further addition of heat will not further accelerate the gas but will result in choking of the flow. As a consequence, the flow must readjust itself, which it will do by lowering its initial Mach number. Table 1 summarizes some of the Rayleigh flow phenomena.

Flow with friction. In long pipes the effects of friction may result in a significant pressure drop. Over a length dx, this pressure drop dp is given by the Fanning equation, Eq. (53), where f is a friction

$$dp = -f\frac{\rho V^2}{2r_h}dx \quad (53)$$

factor that must be determined experimentally. To solve friction-flow problems analytically, certain

Table 1. Variation of flow properties for Rayleigh flow

Property	Heating		Cooling	
	$M>1$	$M<1$	$M>1$	$M<1$
T_0	Increases	Increases	Decreases	Decreases
p	Increases	Decreases	Decreases	Increases
p_0	Decreases	Decreases	Increases	Increases
V	Decreases	Increases	Increases	Decreases
T	Increases	Increases when $M<1/\sqrt{\gamma}$ Decreases when $M>1/\sqrt{\gamma}$	Decreases	Decreases when $M<1/\sqrt{\gamma}$ Decreases when $M>1/\sqrt{\gamma}$

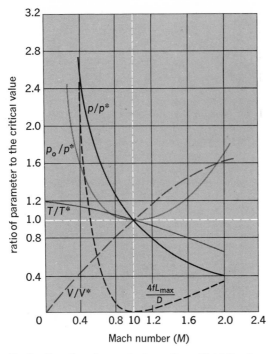

Fig. 9. Functions for constant-area flow with friction ($k=$ 1.4). Asterisk where $M=1$. (*A. B. Cambel and B. H. Jennings, Gas Dynamics, McGraw-Hill, 1958*)

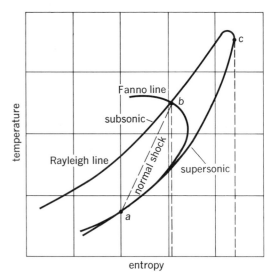

Fig. 10. Rayleigh and Fanno lines. (*A. B. Cambel and B. H. Jennings, Gas Dynamics, McGraw-Hill, 1958*)

simplifying assumptions are made and the resulting hypothetical flow is called Fanno flow. The Fanno flow assumptions are the same as those for Rayleigh flow except that the assumption that there is no friction is replaced by the requirement that the flow be adiabatic. Numerous Fanno flow equations may be written by combining the energy and the continuity equations in accordance with these assumptions. In Fig. 9 may be seen the variation of properties during Fanno flow. Table 2 summarizes the trends of the most important properties during subsonic and supersonic flow.

When the Fanno and Rayleigh lines are plotted for the same constant mass velocity $G=\rho V$, the curves appear as in Fig. 10. The Rayleigh and Fanno lines have two points of intersection, denoted by a and b; a normal shock connects these two points. The flow through a shock wave is irreversible; thus, associated with it is an increase in entropy. Point b, thus, always lies to the right of point a.

There is another interesting point about the Fanno curve. If frictional flow continues along the subsonic portion of the Fanno line, the Mach number tends to increase toward unity, whereas if it continues along the supersonic portion, the Mach number decreases toward unity. As in the case of Rayleigh flow, it is impossible, by virtue of the second law of thermodynamics, to pass from one flow

Table 2. Fanno flow phenomena

Property	Initial flow is subsonic	Initial flow is supersonic
M	Increases	Decreases
V	Increases	Decreases
p	Decreases	Increases
T	Decreases	Increases
ρ	Decreases	Increases

regime to the other (subsonic into supersonic or conversely) unless the mass velocity is readjusted.

External flow. Boundary layers and wakes are the centers of interest in external flows. Here the effects of compressibility are substantially more difficult to analyze than in internal flows, if for no other reason than the inapplicability of a one-dimensional approach.

Boundary layers. Ballistic missiles and space vehicles enter the Earth's atmsophere with velocities typically of 6–7 km/sec. The corresponding Mach number, depending on the altitude, is of the order of 20. The energy that maintains the bow shock and the work done to overcome the viscous shear stresses, Φ, reduce the kinetic energy of the vehicle, and it decelerates. This loss of kinetic energy, which is the work of the drag forces, reappears in part as the increased enthalpy and temperature of the fluid near the vehicle surface. The rate of heat transfer driven by the large enthalpy gradient $(h_g - h_s)/\Delta$, where h_g is the gas enthalpy at some suitably defined distance Δ from the surface and h_s is the enthalpy right at the surface, is so large that special protection must be afforded the vehicle. To this purpose, the vehicle can be covered with a material designed to char, melt, or gasify and in so doing absorb much of the heat that would otherwise penetrate into the structure. This "ablation" process, as it is called, introduces large amounts of material (some of it chemically active, some of it ionized, and some of it radiating) into what ordinarily would be called a boundary layer. Unfortunately, many of the assumptions that permit one to introduce the boundary-layer simplifications are violated here. Very complex computing-machine programs have been developed, however, that yield reasonably accurate estimates of these effects.

Wakes. Hypersonic wakes were observed long before the space age. The tails of "shooting stars" are the luminous wakes of meteors as they enter the atmosphere and burn up. Meteor velocities range between 20 and 70 km/sec, and temperatures as high as 3500°K have been estimated to be necessary for vaporization. Since meteor trails, as well as reentry-vehicle wakes, contain electrons which scatter electromagnetic energy, the trails can be observed with radar. The estimation of the decay rate of these electrons provides a simple example of wake chemistry.

The probability of capture per second of an electron by an ion is $\beta_i n_e$, where β_i is the recombination coefficient of the particular ion and n_e is the electron density. The capture rate dn_e/dt between an electron and an ion is then given by Eq. (54) for a single ionized species.

$$\frac{dn_e}{dt} = n_e \beta_i n_i \qquad (54)$$

In a neutral plasma $n_e = n_i$, and Eq. (54) can be integrated to yield Eq. (55), where n_0 is the ion

$$n_e = \left(\frac{1}{n_0} + \beta_i t\right)^{-1} \qquad (55)$$

density at time zero.

[JOSHUA MENKES; ALI B. CAMBEL]

Bibliography: A. B. Cambel and B. H. Jennings, *Gas Dynamics*, 1958; W. H. Dorrance, *Viscous Hypersonic Flow*, 1962; H. W. Liepmann and A. Roshko, *Elements of Gasdynamics*, 1957; A. H. Shapiro, *The Dynamics and Thermodynamics of Compressible Fluid Flow*, 2 vols., 1953 and 1954.

Gas field and gas well

Petroleum gas, one form of naturally occurring hydrocarbons of petroleum, is produced from wells that penetrate subterranean petroleum reservoirs of several kinds. Oil and gas production are commonly intimately related, and about one-third of gross gas production is reported as derived from wells classed as oil wells. If gas is produced without oil, production is generally simplified, in part at least because the gas flows naturally without lifting, and also because of fewer complications in reservoir problems. As for all petroleum hydrocarbons, the term field designates an area underlain with little interruption by one or more reservoirs of commercially valuable gas. *See* NATURAL GAS; OIL AND GAS FIELD EXPLOITATION; OIL AND GAS WELL DRILLING; PETROLEUM; PETROLEUM GEOLOGY; PETROLEUM RESERVOIR ENGINEERING.

[CHARLES V. CRITTENDEN]

Gas turbine

A heat engine that converts some of the energy of fuel into work by using gas as the working medium and that commonly delivers its mechanical output through a rotating shaft.

Cycle. In the usual gas turbine, the sequence of thermodynamic processes consists basically of compression, addition of heat in a combustor, and expansion through a turbine. The flow of gas during these thermodynamic changes is continuous in the basic, simple, open-cycle arrangement (Fig. 1a).

This basic open cycle can be modified through the addition of heat exchangers and multiple components for reasons of efficiency, power output, and operating characteristics (Fig. 1b−d). A regenerator recovers exhaust heat and returns it to the cycle by heating the air after compression and before it enters the combustor. An intercooler reduces the work of compression by removing some of the heat of compression. A second stage of heating can be added between sections of the turbine,

called an afterburner in aircraft turbines. In various combinations, the auxiliary features provide means for meeting a wide range of operating needs.

Types. The various gas-turbine cycle arrangements can be operated as open, closed, or semiclosed types.

Open cycle. In the open-cycle gas turbine, there is no recirculation of working medium within the structural confines of the power plant, the inlet and exhaust being open to the atmosphere (Fig. 1). This cycle offers the advantage of a simple control and sealing system. It also can be designed for high power-to-weight ratios (aircraft units) and for operation without cooling water. Most gas turbine plants are of this type.

Closed cycle. In the closed-cycle gas turbine, essentially all the working medium (except for seal leakage, bleed loss, and any addition or extraction of working medium for control purposes) is continuously recycled (Fig. 2a). Heat from a source such as fossil fuel (or, possibly, nuclear reaction) is transferred through the walls of a closed heater to the cycle. The closed cycle can be charged with gases other than air such as helium, carbon dioxide, or nitrogen. This is a particular advantage with a nuclear heat source. Other advantages of the closed cycle are (1) clean working fluid; (2) control of the pressure and composition of the working fluid; (3) high absolute pressure and density of the working fluid; and (4) constant efficiency over wide load range. A precooler is required to reduce the temperature of the working fluid before recompression. The higher densities of the working fluid increase the horsepower capacity of a plant of given volume. Changing the absolute pressure level at the compressor inlet changes the weight of working fluid circulated without changing the compression ratio or the temperatures, which results in relatively constant efficiency over a wide load range. The major disadvantage of the closed-cycle gas-turbine plant is the cost and size of the required high-temperature heater.

Semiclosed cycle. In the semiclosed cycle gas turbine, a portion of the working fluid is recirculated (Fig. 2b). This type requires a precooler for the recirculated gas, and a charging compressor to provide the necessary air for combustion. The semiclosed cycle can operate at high densities. The major disadvantages of this cycle are the corrosion and fouling which occur with the recirculation of the products of combustion, particularly when the fuels used have high sulfur or ash content.

Generalized performance. The overall performance of a given gas-turbine power-plant cycle depends basically on component efficiencies, pressure and leakage losses, pressure ratio (ratio of the highest to the lowest pressure), and temperature level.

For the simple cycle, the influence of pressure ratio on performance is illustrated in Fig. 3a. The curves are based on ambient inlet conditions of 80°F and 1000 ft altitude, 85% compressor efficiency, 90% turbine efficiency, 95% combustion efficiency, and 5% combustor pressure loss.

The effect of pressure ratio on plant efficiency for the four basic gas-turbine cycle arrangements illustrated in Fig. 1 is shown in Fig. 3b; the effect of inlet temperature on cycle efficiency for the four

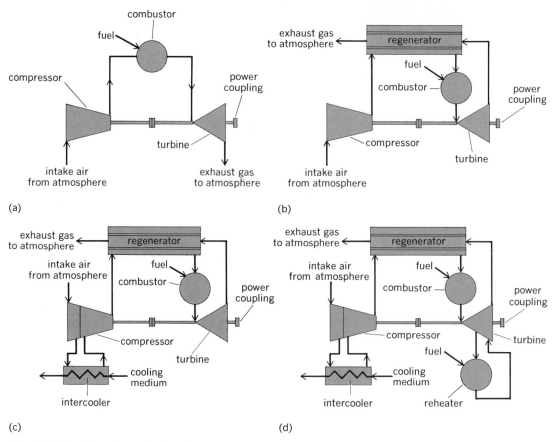

Fig. 1. Series-flow, single-shaft, gas-turbine power plant. (*a*) Simple open cycle. (*b*) Open cycle with regenerator. (*c*) Open cycle with intercooler and regenerator. (*d*) Open cycle with intercooler, regenerator, and reheater.

basic cycles is shown in Fig. 3*c*. The curves are drawn for the optimum pressure ratio for each cycle arrangement at the various turbine inlet temperatures.

Components. To achieve such overall performance, each process is carried out in the engine by a specialized component (Fig. 4). Air for the combustion chamber is forced into the engine by a compressor. In an aircraft, the intake may advance into the air fast enough to ram air into the engine. Fuel is mixed with the compressed air and burned in combustors. The heat energy thus released is converted by the turbine proper into rotary energy. Because of the high initial temperature of the combustion products, excess air is used to cool the combustion products to the allowable turbine inlet design temperature. To improve efficiency, heat exchangers can be added on the gas turbine exhaust to recover heat energy and to return it to the working medium after compression and prior to its combustion.

Compressors. Two basic types of compressors are used in gas turbines: axial and centrifugal. In a few special cases a combination type known as a mixed wheel, which is partially centrifugal and partially axial, has been used. The axial-flow compressor is the most widely used because of its ability to handle large volumes of air at high efficiency. For small gas turbines in the range of 500 hp and less, the centrifugal replaces the axial because it has comparable efficiency when handling reduced volume flow, and is smaller and more compact.

Axial and centrifugal compressors both have a stall or pumping limit where flow reverses. This limit is usually encountered on starting with stationary power units, and at high altitude and high speed with aircraft units. This stall limit is shifted out of the operating range of the gas turbine by the use of compressor bleed, variable compressor vanes, or water injection. All three methods are applied to reduce the aerodynamic loading on the stalled compressor stages.

Ram effect. Aircraft gas turbines moving at high speeds obtain a pressure rise from the ram effect in addition to the compressor pressure rise. The ram effect is the recovery of part of the air velocity, due to the forward motion of the plane, and conversion of this velocity energy to pressure.

Combustors. Combustors, sometimes referred to as combustion chambers, for gas turbines take a wide variety of shapes and forms. All contain fuel nozzles to introduce and meter the fuel to the gas stream and to atomize or break up the fuel stream for efficient combustion (Fig. 5). All are designed for a recirculating flow condition in the region of the nozzle to develop a self-sustaining flame front in which the gas speed is lower than the flame propagation speed. In addition to being designed to burn the fuel efficiently they also uniformly mix excess air with the products of combustion to maintain a uniform turbine-inlet temperature.

The combustor must bring the gas to a controlled, uniform temperature with a minimum of impurities and a minimum loss of pressure. In the

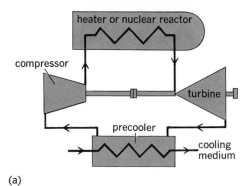

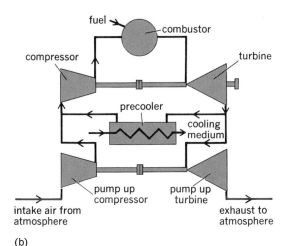

Fig. 2. Gas-turbine power plant. (*a*) Closed cycle with precooler, series flow, single shaft. (*b*) Semiclosed cycle with precooler, series flow, single shaft.

open-type gas turbine, the large excess of air must be controlled to avoid chilling the flame before complete combustion has taken place. Extremely high rates of combustion are common, the heat release being 1,000,000–5,000,000 Btu/ft³/hr/atm or 5–20 times that of high-output, steam-boiler furnaces. Refractory linings are not suitable because they cannot withstand the vibrations and velocities. Metal liners and baffles cooled by the incoming air are commonly used.

Turbine wheels. Two types of gas turbine wheels are used: radial-inflow and axial-flow. Small gas turbines use the radial-flow wheel. For large volume flows, axial turbine wheels are used almost exclusively. Although some of the turbines used in the small gas turbine plants are of the simple impulse type, most high performance turbines are neither pure impulse nor pure reaction. The high-performance turbines are normally designed for varying amounts of reaction and impulse to give optimum performance. This usually results in a blade which is largely impulse at the hub diameter and almost pure reaction at the tip. *See* TURBINE.

Turbine cooling. Gas turbines all employ cooling to various extents and use a liquid or gas coolant to reduce the temperature of the metal parts. The cooling systems vary from the simplest form where only first-stage disk cooling is involved to the more complex systems where the complete turbine (rotor, stator, and blading) is cooled. At high temperatures of operation the turbine material is designed to have a predetermined life, which can vary from a few minutes for missile applications to 100,000 hr or more for industrial units.

Heat exchangers. Two basic types of heat exchangers are used in gas turbines: gas-to-gas and gas-to-liquid. An example of the gas-to-gas type is the regenerator, which transfers heat from the turbine exhaust to the air leaving the compressor. The regenerator must withstand rapid large temperature changes and must have low pressure drop. Regenerators of both the shell-and-tube and the extended-surface type are used (Fig. 6). Rotary regenerators having high performance and reduced weight are under development. Intercoolers, which are used between stages of compression, are air-to-liquid units. They reduce the work of compression and the final compressor discharge

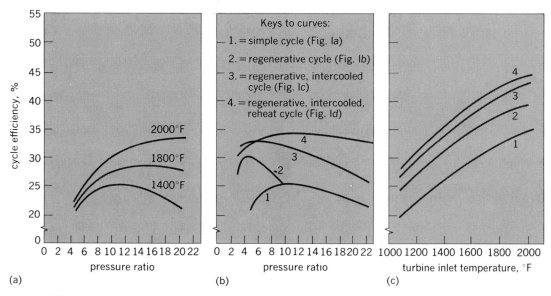

Fig. 3. Effects of pressure ratio on thermal efficiency. (*a*) Simple gas-turbine cycle at various turbine-inlet temperatures. (*b*) Various gas-turbine cycles at constant turbine-inlet temperature. (*c*) Effect of turbine-inlet temperature on thermal efficiency for various gas-turbine cycles at optimum pressure ratios.

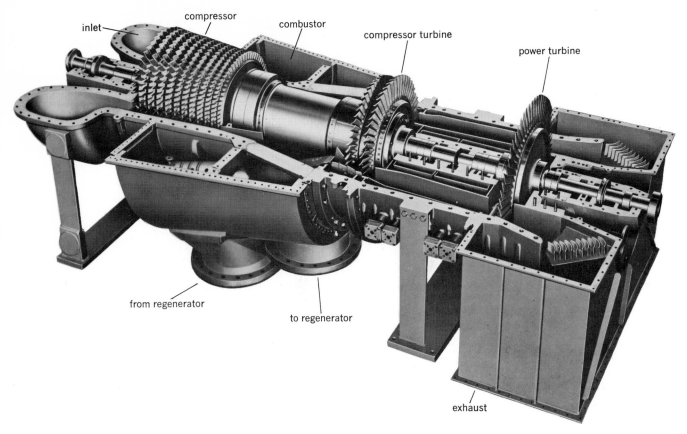

inlet

compressor

combustor

compressor turbine

power turbine

from regenerator

to regenerator

exhaust

Fig. 4. Dual-shaft open-cycle 5000-hp gas turbine with regenerator.

temperature. When used with a regenerator, they increase both the capacity and efficiency of a gas turbine power plant of a given size.

Controls and fuel regulation. The primary function of the subsystem that supplies and controls the fuel is to provide clean fuel, free of vapor, at a rate appropriate to engine operating conditions. These conditions may vary rapidly and over a wide range. As a consequence, fuel controls for gas turbines are, in effect, special-purpose computers employing mechanical, hydraulic, or electronic means, frequently all three in combination.

In aircraft the pilot controls the turbine engine

through a throttle lever, which usually establishes engine speed for each position of the lever. To achieve this relation between throttle position and engine speed under the design envelope range of flight conditions, the fuel control senses numerous engine conditions in addition to engine speed in revolutions per minute. Among these conditions are compressor inlet pressure and ambient air temperature. From these inputs, together with the position of the pilot's throttle, the fuel control continuously adjusts the rate of fuel flow to the engine. When the pilot moves the throttle to change engine power, the control senses the rate of engine acceleration or decleration and provides for the required rate of change in fuel rate.

The fuel control, as well as maintaining the engine at the speed called for by the throttle, should also prevent overspeeding, excessive temperature, or loss of the fire. Under some conditions, the rate at which a turbine engine can accelerate may be limited by the rate at which heat can be released by fuel combustion: Excessive internal engine temperatures could damage such engine parts as the turbine wheels. Under other conditions, the rate at which an engine can decelerate may be limited by the necessity of maintaining sufficient fuel flow to hold the flames; too low a fuel rate for the prevailing gas velocity in the combustors could blow out the flame.

Arrangements. Gas turbines can be constructed for both single-shaft or multiple-shaft arrangements, and can be arranged to supply power, high-pressure air, or hot-exhaust gases, singly or in combination.

transition to turbine inlet

combustion section

fuel nozzle

Fig. 5. Gas-turbine combustor components.

A single-shaft unit, consisting of a compressor, combustor, and turbine, is a compact, lightweight power plant. It is capable of rapid starting and loading and has no standby losses. It can be arranged to use little or no cooling water, which makes it particularly attractive for powering transportion equipment and as a mobile standby and emergency power plant. This type of plant can compete efficiently with small steam plants. Its simplicity involves a minimum of station operating personnel; some of these plants are arranged for completely remote operation.

To improve efficiency. the energy in the exhaust gases can be used either in a waste-heat boiler, or in combination with other processes. For example, a unit arranged to supply process air at 35 psig for operation of blast furnaces can use blast-furnace gas for its fuel (Fig. 7a).

Gas turbines might also be arranged to use a nuclear heat source. For aircraft application, the cycle is open because of weight and space considerations. The combustor is replaced with a nuclear reactor (Fig. 2a). For shipboard and stationary units, the cycle is closed so that fission products can be contained within the power plant in the event of a nuclear accident or fuel element failure. A closed-cycle, gas-turbine power plant for marine service could operate directly with a high-temperature, gas-cooled reactor (Fig. 7b).

Fuels. In the open-cycle plant, products of combustion come in direct contact with the turbine blading and heat-exchanger surfaces. This requires a fuel in which the products of combustion are relatively free of corrosive ash and of residual solids that could erode or deposit on the engine surfaces. Natural gas, refinery gas, blast-furnace gas, and distillate oil have proved to be ideal fuels for open-cycle gas turbines. Combustion chamber and fuel nozzle maintenance are negligible with these types of fuel. Residual fuel oil treated to avoid hot ash corrosion and deposition is also satisfactory although it requires frequent cleaning of fuel nozzles and higher combustor maintenance. Vanadium pentoxide and sodium sulfate are the principle ashes that have been found to cause corrosion and deposition in the 1250–1600°F temperature ranges of modern-day gas turbines. The treatment of residual oil to make it suitable consists of washing with water to remove sodium and introducing additives to raise the fusion temperature of the ash.

In closed-cycle turbines, gaseous fuel, distillate oils, and coal are satisfactory. Residual fuels also must be treated to avoid corrosion in the heater parts. Ash erosion is not a major problem because of the lower velocity of the combustion gases over the heater surfaces.

Fuels for semiclosed-cycle turbines must be selected to minimize both erosion and corrosion. Because the products of combustion are circulated through the entire cycle, the fuel must be selected to avoid corrosion at all pressures and temperatures of the cycle. These requirements limit the use of the semiclosed plant to fuels of low ash and sulfur content.

Applications. Gas-turbine power plants have been successfully applied in the following industries, where their characteristics have proved superior to competitive power plants.

Fig. 6. Air-to-gas regenerator, partly cut away.

Aviation. The gas turbine finds its most important application in the field of aviation. In the propulsion of aircraft, the gas-turbine power plant as a turboprop or turbojet engine has replaced the reciprocating engine in large, high-speed airplanes. This change is due primarily to its high power-to-weight ratio and its ability to be built in large horsepower sizes with high ratio of thrust per frontal area (Fig. 8). It makes use of the ram effect on the compressor inlet to give almost constant thrust at all aircraft speeds.

Gas pipeline transmission. Gas turbines have been installed along piplines to drive centrifugal compressors. Turbine sizes range from 1800–14,000 hp. Gas fuel is normally used, though units have been arranged for dual fuel firing using either natural gas or distillate oil.

Petroleum. In the petroleum industry the gas turbine generates power, drives air and gas compressors in refineries, supplies extraction air or exhaust gases for process, and in the oil fields drives gas compressors to maintain well pressure.

Steel. In the steel industry gas turbines drive air compressors to furnish extraction air for processes such as blast-furnace operation, as mentioned above, and drive electric generators for various plant services. The primary fuel in these applica-

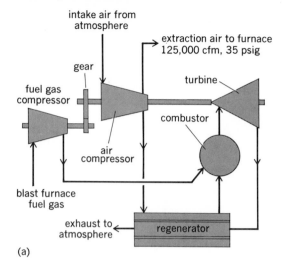

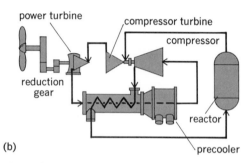

Fig. 7. Cycle arrangements (a) for a 125,000-ft³/min extraction unit and (b) for a closed-cycle gas turbine utilizing a gas-cooled reactor.

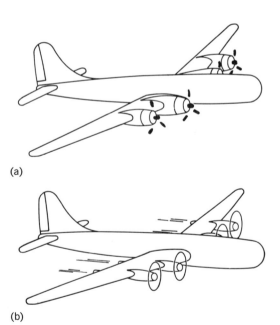

Fig. 8. Types of aircraft engine. (a) Reciprocating engine has greater frontal area for a given horsepower than (b) internal-combustion gas-turbine engine.

tions is blast-furnace gas with a number of units arranged for dual fuel firing.

Marine. Prototype gas turbines are being tested for marine use in ratings up to 6000 ship horsepower (shp). These units operate with residual oil as

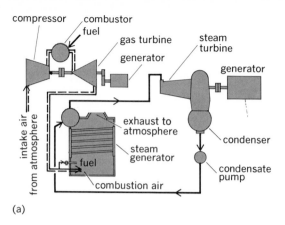

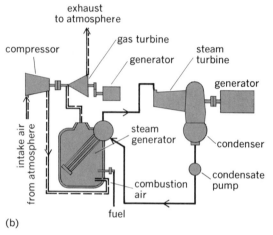

Fig. 9. Cycle arrangements (a) for combined steam- and gas-turbine cycle with gas-turbine, exhaust-fired steam generator and (b) for combined steam- and gas-turbine cycle utilizing a pressurized-steam generator.

fuel processed through fuel treatment equipment. A regenerative-cycle gas turbine with a separate power turbine provides the wide speed range required by such applications as marine service.

Marine installations of gas turbines range from small units used for propulsion of short-range, high-speed craft, emergency generator drive, minesweeper boat propulsion and generator drive, deicing, smoke generation, fire-pump drive, and pneumatic power applications, to large boost engines used in combination with steam turbines. *See* MARINE ENGINE.

Electric utilities. Gas turbines perform a variety of functions in electric-power generation. Peaking service uses simple single-shaft units in sizes up to 25,000 kw. Gas turbines have been installed as hydro-standby. Aircraft propulsion elements can be applied as gas generators, driving electric generator units of up to 300,000 kw. Rail-mounted mobile power plants provide 5000–6000 kw for emergency service. The compactness and simplicity of the gas turbine make it possible to house these units in a single cab. Units from 5000 to 30,000 kw carry base load. These units have been simple cycle machines in low fuel cost areas; units with regenerative, intercooled cycles are used in high fuel cost areas.

Combined steam and gas plants. Efficiency of a power plant is improved by combining a gas turbine with a steam turbine. Various combinations are possible; gas-turbine exhaust can heat feed water for the steam turbine, fuel-fired boilers can use gas turbine exhaust as combustion air (Fig. 9a), or in supercharged boiler plants the high-pressure boiler can serve as the pressurized combustor for the gas turbine (Fig. 9b). The gain in efficiency by using turbine exhaust (Fig. 9a) depends on efficiency of the overall steam plant to which the combined cycle is compared. For the most efficient plant using the best steam conditions and largest number of feedwater heaters, a gain of 2% is possible, while for the less efficient plant, a gain of 5% is possible.

Where the steam generator is pressurized with air from the gas turbine compressor and the hot gases from the boiler are then expanded through the gas turbine (Fig. 9b), a gain in overall efficiency of 4–8% is realized, depending on the plant used for comparison. These improvements can be made to all steam plant cycles, including the most efficient in operation, which employ the superpressure cycle.   [THOMAS J. PUTZ/JAY A. BOLT]

Bibliography: T. Baumeister (ed.), *Standard Handbook for Mechanical Engineers*, 7th ed., 1967; D. G. Fink and J. M. Carroll (eds.), *Standard Handbook for Electrical Engineers*, 10th ed., 1968; W. R. Hawthorne and W. T. Olson (eds.), *Designs and Performance of Gas Turbine Power Plants*, 1960; A. W. Judge, *Small Gas Turbines and Free Piston Engines*, 1960; D. G. Shepherd, *Introduction to the Gas Turbine*, 2d ed., 1960; C. W. Smith, *Aircraft Gas Turbines*, 1956; H. A. Sorensen, *Gas Turbines*, 1951.

Gasoline

A mixture of hydrocarbons used to fuel the spark-ignited internal combustion engine. This mixture consists of more than 100 different hydrocarbons. The lightest, that is, lowest-boiling, is isobutane, while the heaviest hydrocarbons consist of a variety of substituted naphthalenes. Figure 1 shows a drawing of a gas chromatographic trace of a typical gasoline; in practice, there are three to four times as many peaks, shoulders, and bumps. The area of each peak relates directly to the amount of that particular compound present in the mixture. The subsequent discussion of gasoline emphasizes its end use and considers its various properties from the point of view of how they affect its performance in an engine.

All the hydrocarbons in gasoline are the products of a limited number of refining processes designed to increase the yield or quality of gasoline

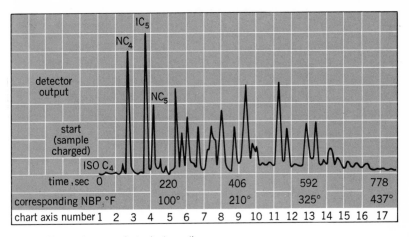

Fig. 1. Chromatogram of a typical gasoline.

components (naphthas). These processes are listed in Table 1 in order of decreasing volume output in United States refining; in addition, Table 1 briefly describes the primary chemical transformation and octane quality resulting from the process. *See* PETROLEUM PROCESSING.

With these various components at hand, the refiner chooses a combination which will meet the quality requirements of the particular market being supplied. The American Society for Testing and Materials (ASTM) specification on motor gasoline, D-439, provides general guidelines for gasoline quality. The qualities imparted to current motor gasoline can be grouped in three categories: volatility, octane, and additive derived.

Volatility. The volatility of a modern gasoline is controlled to provide a balance between a lack of vehicle performance due to insufficient vaporization and poor performance due to an excess of vapor. Figure 2 shows the summer-to-winter variation between gasoline distillation curves and identifies the critical upper and lower limits along the whole curve. In order for the spark-ignited internal combustion engine to function, there must be a gaseous mixture of fuel vapor and air that sustains combustion when ignited by the spark discharge. That is, in cold weather there must be enough low-boiling hydrocarbons to form a flammable mixture in the cold cylinder. *See* INTERNAL COMBUSTION ENGINE.

Shortly after the engine is running, it is expected to produce power and propel the vehicle. As power is required, greater quantities of vapor (up through the middle of the boiling range) are required. How-

Table 1. Common gasoline components

Naphtha	Chemical change	Major characteristic
Catalytic cracked	Heavy fractions are cracked to naphtha	Olefins and aromatics Good octane
Catalytic reformed	Paraffins are cyclized and dehydrogenated	Aromatics Very good octane
Straight run	Distillation from crude	Paraffins Low octane
Alkylate	Isobutane is added to a light olefin	Isoparaffin Good octane
Hydrocracked	Heavy fractions are cracked with hydrogen	Isoparaffins Good octane
Catalytic isomerized	Straight chains are branched	Isoparaffins Good octane

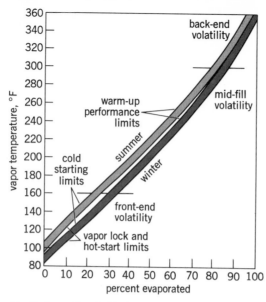

Fig. 2. A gasoline distillation versus performance.

ever, once the engine is fully warmed up, this same gasoline which was volatile enough to start the cold engine must not generate so much vapor as to create another set of operating problems.

A vehicle's carburetor meters liquid fuel as a function of the engine's air demand. For best operation, considering performance, fuel economy, and exhaust emissions, the ratio of air to fuel (A/F ratio) must be kept within narrow limits. If the gasoline in use creates excessive vapors in the warmed-up fuel system (these are not metered by the carburetor), the engine will operate with an overly rich fuel-to-air mixture. The result of a mild form can be poor fuel economy, high exhaust emissions, and roughness at idle. In the extreme (vapor lock), the engine will cease to operate (vapor lock).

Satisfactory vehicle performance as the engine warms to full operating temperature is enhanced by good midrange volatility, which can be characterized as the volume percent of a gasoline which has distilled at 212°F (100°C). This feature is particularly important to later-model vehicles with quick-acting chokes and generally lean carburetion. In the extreme, a high volume percent distilled at 212°F can limit the amount of aromatics that can be used in gasoline. Only benzene, the lightest of the aromatics (use limited because of toxicity), boils below 212°F.

The last portion of a gasoline to vaporize consists primarily of substituted aromatics. These compounds improve the total fuel energy content and thus its fuel economy. They do this because they are denser and have higher heat contents

Table 2. Conditions of laboratory knock methods

Condition	Research method	Motor method
Engine speed, rpm	600	900
Mixture temperature, °F	100 (approx.)	300
Spark timing, °BTC	13	Varied with compression ratio

than the total fuel. Again, there is a trade-off; these heavier aromatics contribute a disproportionately large share to the buildup of combustion chamber deposits. Many states have adopted a legal maximum final boiling point of 437°F (225°C).

The volatility measurements referred to so far are all based on the Engler distillation, ASTM D-86. This is a simple distillation from a specified flask at an imprecisely defined rate of heat input. That is, the Engler distillation is a poorly defined analytical technique. There is increasing interest in a gas chromatographic technique for a more precise definition of the vaporization characteristics of complex mixtures of hydrocarbons. It is likely that the simple, undefined distillation of gasoline will be replaced by a carefully controlled gas chromatographic technique, which will more precisely define the vaporization of complex mixtures of hydrocarbons.

Octane. A gasoline's octane number is an indication of the fuel's ability to prevent the occurrence of spark knock in an engine. Spark knock is an audible sound resulting from the explosion of a portion of the air-fuel charge prior to the arrival of the flame front; that is, the timed spark ignites a flame which propagates smoothly through the charge, except when spark knock occurs. Spark knock takes place when chain-branching reactions lead to an increasingly rapid buildup of hydroperoxide radicals (HO_2) in the region ahead of the advancing flame front. Tetraethyllead is an effective antiknock agent, because its decomposition gives rise to a nonstoichiometric lead oxide which acts as a radical trap.

The octane quality of a gasoline can be measured by two laboratory procedures which employ a single cylinder, CFR (Coordinating Fuels Research) knock rating engine (Table 2). The numerical octane scale used is based on blends of n-heptane and isooctane (2,2,4-trimethylpentane). Isooctane is assigned a value of 100, while the n-heptane value is designated 0; blends of the two standards are used to match the knock intensity of the test fuel.

For commercial gasolines, research octane numbers (RON) are higher than motor octane numbers (MON). According to the Energy Resources Development Agency (ERDA), premium gasoline averages 98.9 RON and 91.4 MON. Unfortunately, the average vehicle on the road does not operate under the narrowly specified conditions employed in the knock testing engines. Thus it is not surprising that neither RON nor MON alone is adequate to define antiknock quality for cars on the road; both are required.

Research octane number generally provides an indication of low-speed performance at full throttle, while motor octane number relates to part throttle (cruising and moderate acceleration) operation. A combination of both methods is required to define octane quality for vehicles equipped with automatic transmissions. Such a combination of laboratory octane numbers is said to define the road octane requirement of a vehicle, as shown in the equation below. The road octane number of a

$$\text{Road ON} = a\,(\text{RON}) + b\,(\text{MON}) + k$$

gasoline is the octane number of the primary reference fuel (n-heptane/isooctane) which produces the same knock intensity as the test fuel in a vehicle.

When refinery components are blended together, the octane number of the blend may be greater than, equal to, or less than that calculated from the volumetric average of the octane numbers of the individual blend components. Blending deviations of several octane numbers between experimentally observed and calculated antiknock rating can occur when blending refinery fuel components. Numerous calculational procedures have been developed over the years for predicting both research and motor octane numbers of multicomponent motor fuel blends. Some are based upon only the research and motor octane numbers of the components; others additionally consider composition data, such as olefins and aromatic contents, or detailed gas chromatographic analyses. Octane blending is nonlinear by nature, and blend deviations from linearity can be positive or negative, or both, over the range of blend compositions. These deviations are attributed to the effect of hydrocarbon type and concentration and their interactions on the chain-branching reactions occurring ahead of the flame front. *See* OCTANE NUMBER.

Additive derived properties. A host of properties exhibited by current gasolines results from the use of additives. Gasolines are colored with oil-soluble dyes to differentiate between the various grades marketed. Gasolines are made less susceptible to oxidation; this allows the gasoline to be stored for many weeks without excessive "gum" formation. In the absence of an oxidation inhibitor, the chemically reactive hydrocarbons and the trace quantities of oxygenated compounds undergo free-radical-catalyzed reactions which produce minor amounts (0.005–0.05%) of partially oxidized high-molecular-weight hydrocarbons. The "gums," now only partially soluble in gasoline, can deposit on critical parts of the carburetor and cause various parts to stick and thus fail to operate properly. Oxidation inhbitors, materials which are preferentially oxidized, are used to inhibit gum formation. The inhbitors when oxidized form stable compounds which remain gasoline-soluble. Thus the reactive hydrocarbons do not have the chance to oxidize and eventually form gasoline-insoluble gums.

Despite efforts to maintain a dry gasoline distribution system, small amounts of water (0.005–0.1%) are normally present in gasoline as a result of processing and condensation. Accordingly, rusting is an ever-present problem of the gasoline distribution system and ultimately in the end-use vehicle. Aside from shortening the useful life of the various containers involved, the corrosion products become finely dispersed and are transported by the fuel. This particulate matter can lodge in small passages in a carburetor and cause engine malfunction.

Numerous organic compounds with an affinity for polar surfaces are used as rust inhibitors. These materials used in concentrations of a few parts per million reduce corrosion throughout the system and thus reduce the particulate content of the fuel. This means longer life for the vehicle's fuel filter and less likelihood that critical carburetor passages will become blocked.

In addition to deposits that can result from the autoxidation of relatively unstable fuel components, a vehicle's carburetor can become deposited with a mixture of airborne dirt, engine compartment fumes, and in heavy traffic the exhaust products of other vehicles. These deposits build up in the throttle area of the carburetor. As these deposits accumulate, they change the metering characteristics of the carburetor. Generally, they restrict airflow and thus cause the engine to run overly fuel-rich. The positive crankcase ventilation (PCV) systems required on all vehicles sold in the United States since 1963 further aggravate induction system deposits by discharging the crankcase blowby in the intake manifold. In the absence of a carburetor cleaner or detergent, these deposits result in rough idle and poorer fuel economy and higher exhaust emissions. Carburetor detergents encompass a variety of chemical compound types, all of which can be generally described as having hydrocarbon tails for solubility and nonhydrocarbon heads containing polar oxygen and nitrogen functions which associate with the deposits. The combination of polar and nonpolar groups in a single molecule account for their ability to orient at surfaces. As these molecules concentrate at the surface, they loosen the deposits which are swept away to be burned in the engine.

Gasoline marketed in cold weather must have high volatility in order to provide adequate starting and drivability prior to the engine's becoming warmed up. During the first 10–15 minutes of operation on such a volatile fuel, ice can form in and around the carburetor throat. As the gasoline evaporates, it removes heat from the incoming airstream. With a volatile fuel the incoming air temperature can drop from 40°F (4.5°C) to 20°F (−6.7°C). Under these conditions most of the water vapor being carried by the 40°F (4.5°C) air condenses in the carburetor throat to liquid water at a temperature significantly below its freezing point of 32°F (0°C). As the supercooled water freezes, the ice buildup in carburetor or throat begins to restrict airflow to the engine. This becomes very apparent at idle when the engine may no longer operate.

Many of the compounds which function as rust inhibitors and carburetor detergents also help in this latter instance. The surfactant modifies ice-crystal growth, fostering the growth of small crystals which are more readily sloughed off the throttle plates and throats into the airstream.

Most of the desirable features of current gasolines require the deliberate application of technology during the refining and marketing of gasoline. The particular balance of properties arrived at will not be the same in all instances. Technical capability and economic considerations both play a part in determining which properties are imparted and emphasized. [HUGH F. SHANNON]

Bibliography: American Society for Testing and Materials, *Book of Standards*, pts. 23 and 47, 1975; Energy Resources Development Agency, Office of Public Affairs, Technical Information Center, *Motor Gasolines*, summer 1975.

Generator, electric

Any machine by which mechanical power is transformed into electric power. Generators fall into two main groups, alternating-current (ac) and

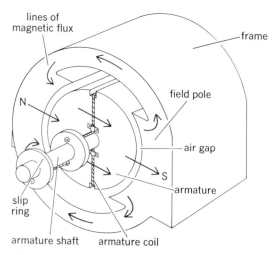

lines of
magnetic flux

frame

N

field pole

air gap

S

armature

slip
ring

armature shaft   armature coil

Fig. 1. Elementary generator.

direct-current (dc). They may be further classified by their source of mechanical power, called the prime mover. Generators are usually driven by steam turbines, hydraulic turbines, engines, gas turbines, or motors. Small generators are sometimes powered from windmills or through gears, belts, friction, or direct drive from parts of vehicles or other machines.

Theory of operation. The theory of operation of most electric generators is based upon Faraday's law. When the number of webers of magnetic flux linking a coil of wire is caused to change, an electromotive force proportional to the product of the number of turns times the rate of change of flux is generated in the coil. The instantaneous induced voltage is given in Eq. (1), where n is the

$$e = -n \, (d\phi/dt) \text{ volt} \qquad (1)$$

number of turns, ϕ is the flux in webers, and t is time in seconds. The minus sign indicates that the induced voltage opposes the effect which pro-

duced it. Voltage is induced in the windings of a generator by mechanically driving one member relative to the other, thereby causing the magnetic flux linking one set of coils, called the armature windings, to vary, pulsate, or alternate. The magnetic flux may originate from a permanent magnet, a dc field winding, or an ac source. See INDUCTION, MAGNETIC.

Figure 1 shows an elementary generator having a stationary field and a single, rotating, armature coil. It is apparent that the magnetic flux threading the coil reverses direction twice per revolution, thereby generating one cycle of voltage in the armature coil for each revolution. If this variation in flux is expressed as a function of time, the voltage generated in the coil would be given by Eq. (1). For example, let the flux vary as given by Eq. (2),

$$\phi = \Phi_m \cos 2\pi ft \qquad (2)$$

where Φ_m is the maximum flux, f is the frequency, and t is time. By differentiating Eq. (2) with respect to time, and substituting in Eq. (1), the instantaneous voltage in the coil is found to be Eq. (3). This

$$e = 2\pi fn\Phi_m \sin 2\pi ft \qquad (3)$$

ac voltage can be taken from the armature by brushes on the slip rings shown. If the coil terminals were brought to a two-segment commutator instead of slip rings, a pulsating dc voltage would appear at the brushes. See ALTERNATING-CURRENT GENERATOR; DIRECT-CURRENT GENERATOR.

Construction. In practice, permanent-magnet fields are used only in small generators. Large generators, except induction generators, are equipped with dc field windings. The field coils are wound on the stators of most dc generators to permit mounting the armature coils and commutator on the rotor. On ac generators, the field coils are normally located on the rotors. Field coils require only low voltage and power, and only two lead wires. They are more easily insulated and supported against rotational forces and are better

Fig. 2. Turbogenerator unit with direct-connected exciter, rated 22,000 kw. (*Allis-Chalmers*)

suited to sliding contacts than the relatively high-voltage armature windings, which often have six leads brought out.

Any part of the magnetic circuit not subject to changing flux may be of solid steel. This includes the field poles of dc machines and portions or all of the rotating field structure of some ac generators. In machines with small air gaps the poles are frequently of laminated steel, even though their flux may be substantially constant. Laminations help minimize pole-face losses arising from tooth-frequency pulsations. The armature core is almost always composed of thin sheets of high-grade electrical steel to reduce core loss. *See* CORE LOSS.

The windings are insulated from the magnetic structure, and are either embedded in slots distributed around the periphery or mounted to encircle the field poles. The terminals from the stator windings and from the brush holders are usually brought to a convenient terminal block for external wiring connections.

Turbogenerators. Generators driven by steam or gas turbines are sometimes called turbogenerators. Although in small sizes these may be gear-driven, and some may be dc generators, the term turbogenerator generally means an ac generator driven directly from the shaft of a steam turbine. A typical turbogenerator unit is shown in Fig. 2.

In order to achieve maximum efficiency, the steam turbine must operate at high speed. Consequently, direct-connected turbogenerators are seldom built to operate below 1500 rpm. To minimize windage loss and to keep rotational stresses down to a safe level, turbogenerator rotors are usually long and slender, in some cases five to six times the diameter in length of active iron. Long rotors operate above their first critical speed, and in some cases near or above their second, thereby introducing mechanical problems in balancing and resonance. To shorten the length of turbogenerators, conductor cooling has proved effective. *See* ELECTRIC POWER GENERATION; ELECTRIC ROTATING MACHINERY.

[LEON T. ROSENBERG]

Bibliography: L. V. Bewley, *Alternating Current Machinery*, 1949; D. L. Carr, *Electrotechnology*, vol. 6, 1962; D. G. Fink and J. M. Carroll, *Standard Handbook for Electrical Engineers*, 10th ed., 1968.

Geochemical prospecting

The use of geochemical and biogeochemical principles and data in the search for economic deposits of minerals, petroleum, and natural gases. In modern exploration programs, geochemical prospecting surveys are generally carried out in conjunction with geological and geophysical surveys. *See* GEOPHYSICAL EXPLORATION; PROSPECTING.

HISTORY

Both chemistry and geology stretch far back into antiquity and, as might be expected, suggestions as to how chemistry can be applied in the search for mineral deposits have been advanced since early times. Georgius Agricola in *De re metallica*, published in 1556, makes frequent reference to the use of springs, natural waters, and various chemical phenomena in prospecting for veins, and one can also find many interesting statements in the writings of the early Chinese and medieval European writers about the early lore of geochemical prospecting.

The techniques of modern geochemical prospecting had their origin mainly in the Soviet Union and in the Scandinavian countries, where much research on methods was carried out in the late 1930s. After World War II, the various methods were introduced into the United States, Canada, Great Britain, and other countries, where they have since been used extensively in mineral- and petroleum-exploration programs by both mining companies and government agencies.

The modern methods of geochemical prospecting owe their rapid development in the 20th century to the following:

1. Recognition of the nature of primary and secondary dispersion halos, trains, and fans that are associated with all mineral deposits and accumulations of hydrocarbons.

2. Development of accurate, rapid, inexpensive analytical methods utilizing the optical spectrograph, x-ray fluorescence spectrograph, atomic absorption spectrometer, and the various specific sensitive colorimetric reagents, especially dithizone (diphenylthiocarbazone).

3. Development of polyethylene laboratory ware of all types and the development of resins. These permit greater freedom of analysis in the field and reduce the incidence of contamination.

4. Development of rapid and precise methods of analyses of various volatile elements (such as mercury and radon) in rocks, soil gases, waters, and the atmosphere.

5. Development of airborne gamma-ray spectrometry for geochemical analysis of potassium, uranium, and thorium on a broad-scale reconnaissance basis.

6. Development of rapid and precise methods of analysis of various types of both organic and inorganic particulates in the atmosphere.

7. Development of gas chromatography and other precise methods of trace analysis of hydrocarbon compounds and various gaseous inorganic substances.

8. Refinement of field techniques for carrying out reconnaissance and detailed geochemical surveys of all types, especially those based on stream sediments, lake bottom sediments, heavy minerals, stream and lake waters, groundwaters, springs and their precipitates, and biological materials. The use of helicopters has revolutionized sample collection in reconnaissance surveys in practically all terrains.

9. Development and refinement of methods of detailed geochemical prospecting using overburden drilling techniques, especially in glacial terrains.

10. Development and refinement of methods using ore boulder and heavy mineral trains for prospecting in glacial terrains.

11. Research and development of efficient methods for the processing and assessment of geochemical prospecting data by statistical and computer techniques.

Geochemical prospecting is now being employed in all parts of the world, from the tundra to the tropical belts. Among its successes are a number of large low-grade deposits that yield gold, copper, nickel, lead, zinc, and other metals. Some oil fields

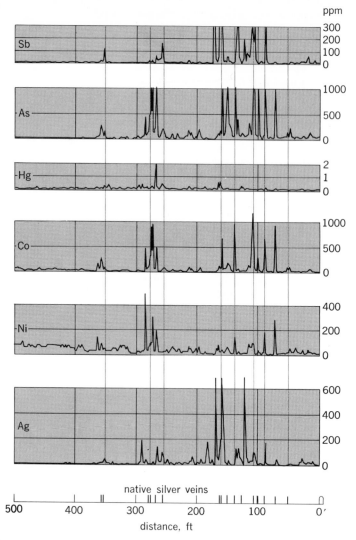

ppm

Fig. 1. Trace-element distribution in wall rocks, Silverfields Mine, Cobalt, Ontario. (*After Boyle et al., Geol. Surv. Can. Pap., 67-35, 1969*)

temporaneously or nearly so with their enclosing rocks. Examples are diamond pipes, certain sedimentary copper and uranium ores, and gypsum and salt deposits. Other deposits are epigenetic; that is, they were introduced into fractures, faults, porous zones, and structural traps generally long after their host rocks were formed. Examples are veins and lodes of metallic minerals in fractures and faults and accumulations of petroleum and natural gas in anticlinal traps and porous reef structures. *See* PETROLEUM GEOLOGY.

Primary and leakage halos. Most mineral deposits and accumulations of hydrocarbons are characterized by a central core or layer in which the valuable elements or minerals are concentrated in percentage quantities. Surrounding this core or layer, the valuable elements generally progressively diminish in quantity to amounts measured in parts per million (ppm) or parts per billion (ppb) which constitute the normal content or background of the enclosing rocks. The region through which this diminution of valuable elements takes place is called the primary halo of the deposit. The term "primary" refers to the fact that the elements in the halo were dispersed into the enclosing rocks at the same time, or nearly so, as those in the central core or layer. Another term, "leakage halo," refers to the dispersion of elements along channels and paths followed by mineralizing solutions leading into and away from the central focus of mineralization. Primary halos show infinite variety. Some are recognizable only a few inches from the central focus of mineralization; others can be detected over distances of hundreds, and in places thousands, of feet from deposits. Most are controlled by microfractures in the rock and by porosity and permeability considerations. Leakage halos are also controlled by the geometry of the available channels, such as fractures, faults, and shear zones, and by the porosity and permeability of the rocks. Leakage halos have been identified in fault and fracture systems up to 500 ft (150 m) and more from mineral deposits. Gaseous leakage halos, such as those related to oil and gas pools, are often detected several thousands of feet above or lateral to the economic concentrations of hydrocarbons.

Secondary halos, or dispersion trains. Mineral deposits exposed at the surface undergo extensive oxidation, during which the ore and gangue minerals are disintegrated, and some of their elemental constituents go into solution in the groundwaters. As the deposits are weathered down, the disintegrated particles of the ore and gangue minerals, and some of the elements in solution in the ground waters, are dispersed into the soil and weathered debris overlying the deposits. This dispersion produces a secondary halo, or dispersion train, in the soil. Plants growing in this soil take up elements generally in excess of that required for their physiological processes, thus producing an anomalous elemental halo or dispersion in the vegetation. The shape of the dispersion trains in the soil and vegetation are variable, depending on the topography and a host of other factors. Some are essentially halos that surround the locus of mineralization below. Others are fan-shaped with their apexes at or near the mineral deposits. These are generally referred to as dispersion fans.

have been discovered by detailed studies of hydrocarbons in groundwaters and overlying soils and glacial deposits. Even the hydrocarbon content of the sediments of the sea has been utilized in the search for submarine oil pools.

GENERAL PRINCIPLES

The earth is divisible into five spheres: lithosphere (rocks), pedosphere (soils, glacial till, and so forth), hydrosphere (natural waters), atmosphere (gases), and biosphere (living organisms and their products). The chemistry of these spheres is termed, respectively, lithochemistry, pedochemistry, hydrochemistry, atmospheric chemistry, and biochemistry. Geochemical prospecting methods have been developed that utilize analyses of the materials from each of these spheres, namely: lithogeochemical methods, pedogeochemical methods, hydrogeochemical methods, and so on.

Mineral deposits and accumulations of hydrocarbons represent anomalous concentrations of specific elements, usually within a relatively confined volume of the Earth's crust. Some of these deposits are syngenetic, that is, formed con-

Additional dispersion trains are produced by groundwaters that have dissolved some of the ore and gangue minerals. These waters ultimately appear at the surface as springs that feed the streams and rivers of an area. Since these waters frequently have higher-than-normal contents of ore and gangue minerals, derived from the deposits, they produce a hydrogeochemical train whose elemental content is markedly high at the spring orifices and decreases in intensity downstream as dilution by surface waters and other factors come into play. Finally, a dispersion train in the stream and river sediments may be associated with mineral deposits. This train results principally from adsorption of the elemental ions in stream water on the fine stream silt or from complex coprecipitation processes in the stream bed. In addition, there may also be a contribution of fine fragmented particles of ore and gangue minerals which reach the streams by mechanical processes from the deposits. The anomalous values in the trains of ore and gangue minerals in stream sediments are generally highest near the source of the dispersion (that is, the deposits), and they fall off progressively with distance.

Summarizing briefly, most mineral deposits have associated geochemical halos and trains. Those associated with the primary mineralization processes are called primary halos and leakage halos; those associated with weathering processes are called secondary dispersion trains and include those developed in the soil, in the vegetation, in the groundwater system, in the springs, streams, and rivers, and in the stream, river, and lake sediments. All of these halos and trains provide means of tracing and locating the source from which ore and gangue minerals are dispersed, namely, economic deposits. They provide the geochemical anomalies for which all geochemical prospectors search.

Background. Before anomalous conditions can be recognized, however, it is necessary to establish a background against which the anomalies can be compared. In any given area this is generally done by analyzing a relatively large number of samples of the materials to be used in the geochemical survey (rock, soil, vegetation, stream sediment, and so forth), excluding as far as possible any mineralized material. The values obtained may show a fairly large scatter, but the most frequently recurring values tend to fall within a relatively restricted range about a mode. This modal value is generally considered to represent the normal abundance or background of the area for the particular material sampled. Samples which contain amounts of elements twice background or more are generally assumed to be anomalous. When extensive analytical data are available for sampling materials in a surveyed area, a variety of statistical methods are commonly employed to evaluate the background, threshold, and anomalous values.

Interpretation of anomalies. Where anomalous samples are geometrically grouped in a fairly definite pattern, as in a train or halo, a geochemical anomaly or dispersion pattern is present. Figures 1 to 4 represent such anomalies. Figure 1 shows the distribution of a number of elements, expressed as parts per million, in the wall rocks of some silver veins at Cobalt, Ontario. Figure 2

shows the arsenic and silver contents in the soil horizons over another silver vein at Cobalt, Ontario. Figure 3 shows anomalous conditions in the water and sediments of a stream in the Keno Hill area, Yukon. Figure 4 represents a geochemical anomaly in lead in the ash of the twigs of trees growing in soil overlying a lead deposit in Nova Scotia.

When a geochemical anomaly is discovered, a decision has to be made as to its origin. Three possibilities exist: that the anomaly is genetically related to a mineral deposit or concentration of hydrocarbons, that the anomaly is related to subeconomic deposits, or that the anomaly is due to a concentration of elements resulting from a combination of chemical, topographic, and other factors. This is the most difficult part of geochemical prospecting. Every available piece of information must be brought to bear on the problem, such as a knowledge of the climatic, topographic, and geologic conditions, groundwater and surface-water movement, glacial movement, type and distribution of vegetation and humic deposits, and certain physicochemical parameters, such as the Eh, pH, and organic activity, which control the mobility and migration of the elements. These physicochemical parameters affect each element differently, and hence a basic knowledge of the

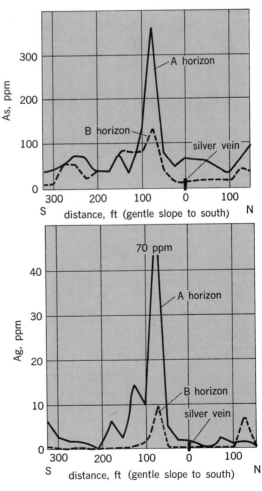

Fig. 2. Arsenic and silver contents on traverse across O'Brien No. 6 vein, Cobalt, Ontario. (*After Boyle and Dass, Econ. Geol., 62:274–276, 1967*)

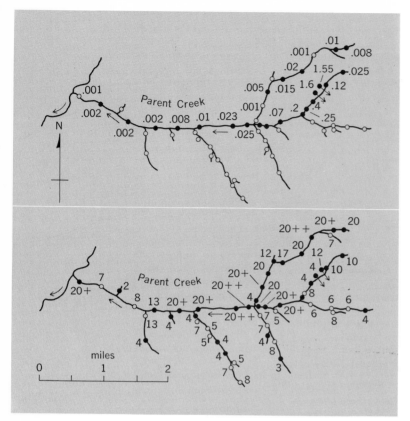

Fig. 3. Heavy-metal content (mainly zinc) in stream and spring (*a*) waters and (*b*) sediments, Parent Creek, Keno Hill area, Yukon Territory. All values in parts per million. (*After Gleeson et al., Geol. Surv. Can. Maps nos. 20–1964 and 21–1964*)

geochemistry of the elements is imperative if the geochemical prospector is to sort out those anomalies which are related to deposits from those that are not. After all interpretations have been made and ancillary data, such as those from geophysical surveys, have been integrated into the pattern, diamond drilling or other exploratory methods, such as trenching, are used to investigate those anomalies that are deemed worthy of detailed investigation.

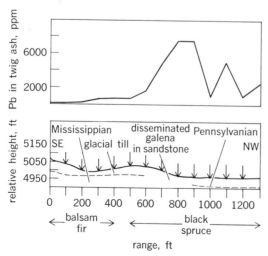

Fig. 4. Lead content in twigs, Silver Mine area, Cape Breton Island, Nova Scotia. (*After M. Carter, M.Sc. Thesis, Carleton University, Department of Geology, 1965*)

Indicator elements. In geochemical surveys the key elements in the deposits may be used to trace out the anomalies, for example, lead and zinc for lead-zinc deposits and copper for copper deposits. It often happens, however, that certain economic elements are present in amounts too small for the analytical methods employed or do not give good dispersion patterns because of poor mobility or other factors, for example, gold. In such cases elements occurring in larger amounts or with more favorable dispersion characteristics are employed. Such elements are generally referred to as indicator or pathfinder elements. Examples are arsenic for gold deposits; cobalt, nickel, or arsenic for certain types of native silver deposits; and molybdenum for certain types of copper deposits.

Sampling techniques. All geochemical prospecting surveys must be based on adequate sampling of the material or materials used in the survey. These materials may be rocks, waters, gases, soils, stream sediments, lake sediments, or vegetation. Where possible, sampling should be confined to one type of material for any particular survey, such as one rock types in lithogeochemical surveys, soil from one horizon in pedogeochemical surveys, or twigs from one species of tree. When diverse materials are sampled because of the heterogenous nature of the geology or vegetation, correlative factors must be applied to the resultant data before the geochemical patterns can be interpreted.

It should be emphasized that the sampling techniques used in geochemical surveys must be such that the sample obtained is representative of the rock, water, gas, soil, vegetation, or other types of organisms at the point of sampling. Failure to consider carefully this aspect of a survey leads only to spurious results and difficulties in interpretation. The spacing and distribution of the sampling points are controlled by the estimated size of the deposit, the mobility of the dispersed elements, and the type of dispersion pattern expected.

Chemical analysis. The analytical techniques used to determine the quantity of elements in geological and biological materials are varied. Some techniques have been devised for use at sampling sites in the field, whereas others have been developed for analysis in well-equipped mobile field laboratories or central laboratories operated by commercial firms, or government agencies. The analytical results are generally expressed in parts per million or parts per billion.

Analytical techniques for on-site sampling of water, soil, stream sediment, lake sediment and so on are usually based on colorimetry and use a specific selective organic reagent, such as dithizone. The loosely bonded or exchangeable metal in the geological material is extracted by means of a citrate, acetate, acid, or other extractant and then determined by the colorimetric reagent. When it is desirable to determine the total metal or other element content of a sample, various analytical methods are employed in mobile or central laboratories. These include optical spectrography, x-ray fluorescence spectrography, atomic absorption, and colorimetric, chromatographic, and polarographic methods. For hydrocarbons, gas chromatography and a variety of other methods employed in analytical organic chemistry have found wide usage.

Analytical techniques used for geochemical prospecting should be rapid, permitting a large number of samples to be done in a short time, and inexpensive, yet precise to within about 20%. Portability of equipment and reagents is the prime consideration in the case of methods used in the field at the sample sites.

GEOCHEMICAL PROSPECTING SURVEYS

Geochemical prospecting surveys are generally classified according to the type of materials sampled. These include lithogeochemical surveys (rocks), pedogeochemical surveys (soil, till), hydrogeochemical surveys (water, sediments), surveys based on volatiles and airborne particulates, biogeochemical surveys (plants, animals), remote-sensing surveys, and isotope surveys.

Geochemical prospecting surveys may be either reconnaissance surveys or detailed surveys. In reconnaissance surveys analyses of materials are carried out over a broad region, over a mineral or oil concession, or over a large number of claims with the express purpose of defining mineral belts, zones of mineralization, or sites of accumulation of hydrocarbons. Detailed surveys are carried out on a local basis with the purpose of locating individual deposits, oil pools, or favorable structures where these might occur. The amount of detail varies with the geological situation but may go down to sampling at 5-ft (1.5 m) intervals or less where primary halos associated with veins are sought.

Lithogeochemical surveys. These have not been used as extensively as other types of geochemical prospecting surveys, mainly because the sampling techniques and interpretation of the surveys have not been investigated sufficiently. Both reconnaissance and detailed lithogeochemical surveys may be carried out.

Most reconnaissance surveys are carried out on a grid or on traverses across a geological terrane, samples being taken of all available rock outcrops or at some specific interval. One or several rock types may be selected for sampling and analyzed for various elements. The distribution of the volatiles such as Cl, F, H_2O, S, and CO_2 in intrusives with associated mineralization has received some attention as indicators. Geochemical maps are compiled from the analyses, and contours of equal elemental values are drawn. These are then interpreted, often by using statistical methods, in the light of the geological and geochemical parameters. Under favorable conditions, mineralized zones or belts may be outlined in which more detailed work can be subsequently concentrated. If the reconnaissance survey is carried out over a large expanse of territory, geochemical provinces may be outlined.

Detailed lithogeochemical surveys are generally carried out on a local basis and have as their purpose the discovery or definition of primary or leakage halos associated with mineral deposits or petroleum accumulations. Chip samples of rocks on a definite grid are used in some cases, and samples from drill cores in others. All of the analytical data are plotted on plans and sections and compared with the geological situation. Frequently, primary halos can be discovered by this method (Fig. 1), and these and leakage halos can be traced to the focus of mineralization or to petroleum accumulations if sufficient work is done.

In detailed drilling and development work the use of ratios obtained from analyses of rocks along traverses or diamond drill holes can frequently be employed to estimate proximity to mineralized loci. Particularly useful ratios include K_2O/Na_2O, SiO_2/CO_2, and SiO_2/total volatiles; the volatiles commonly include H_2O, CO_2, S, As, and B. Many types of mineral deposits are characterized by a consistent increase in the ratio K_2O/Na_2O, which is essentially a manifestation of increasing potassic alteration. Similarly, a number of mineral deposits, particularly those enriched in gold and silver, are marked by a consistent decrease in the ratio SiO_2/CO_2 as ore is approached. The ratio SiO_2/total volatiles exhibits considerable variation among the various types of mineral deposits; in most cases, skarns excepted, there is a consistent decrease in the ratio as mineralization is approached.

Pedogeochemical surveys. Soil and glacial-till surveys have been used extensively in geochemical prospecting and have resulted in the discovery of a number of orebodies. Similar surveys have been used in searching for petroleum pools. Generally, soil and glacial-till surveys are of a detailed nature and are run over a closely spaced grid. One of the soil horizons is chosen for sampling, generally the B horizon, and a plan of the metal or hydrocarbon contents is plotted and contoured. Under favorable conditions, the highest values are centered over a deposit, but more generally the dispersion pattern is a train or a fan that requires careful interpretation to locate its source.

Certain geological conditions may require that all horizons of the soil be sampled, including the A, B, and C horizons. Frequently, sampling of the A (organic) horizons is effective in some areas (Fig. 2), whereas in others the B or C horizons are more rewarding. In some places deep sampling of the C horizon by drilling is the only satisfactory method for soil and glacial-till surveys. This technique has proven most effective in many of the heavily overburdened, glaciated terrains of Canada, in the permafrost regions of Canada and the Soviet Union, and in the deeply weathered lateritic regions of the tropics.

Heavy- and resistate-mineral surveys of soil, till, and weathered debris have found increasing usefulness in recent years. In these surveys the geological materials are panned, and the heavy and resistate minerals obtained. These are then examined microscopically for ore minerals or analyzed for ore or indicator elements, and the results are plotted on maps. Heavy- and resistate-mineral maps, prepared on the same grid as those for soil surveys, provide valuable ancillary data and often aid in the interpretation of the elemental dispersion patterns.

In soil analyses the fine fraction (minus 80 mesh) is generally analyzed for the chemically dispersed elements, whereas for heavy and resistate minerals a coarser fraction is used. *See* SOIL.

In glaciated terrains, as well as in certain other terrains, heavy- and light-clast (mineral fragments, stones and boulders) tracing has proven effective in the discovery of certain types of mineral deposits. Examples are quartz boulders or fragments as indicators of gold-quartz veins, and galena boulders or fragments as indicators of lead-zinc deposits. Surveys of this type are generally carried out on grids, the abundance of light or heavy clasts being visually noted and plotted at each sampling point or wherever they occur along the grid lines. In other surveys, large samples of the till or overburden are obtained at each sampling point, and the light- and heavy-clast indicators are counted in the light and heavy concentrates obtained from the samples. When all of the data from clast surveys are plotted, fans or trains are commonly outlined whose apexes or starting points often mark the sites of underlying mineralization.

Hydrogeochemical surveys. These include water surveys, sediment surveys, and heavy-mineral surveys. The last two are not strictly hydrogeochemical in nature, although they are generally carried out along the drainage systems of an area.

The background metal content of most natural waters is only a few parts per billion, rising to a few parts per million for certain elements in the vicinity of mineral deposits. Seasonal and diurnal fluctuations often occur in these values, features that have to be considered in water surveys. Accurate and sensitive analytical methods are a requirement in water surveys, and on-site analysis is recommended if possible to avoid problems introduced by transportation and contamination of water samples. Most water analyses are done by sensitive colorimetric methods or by atomic absorption spectrometry.

Either surface waters or groundwaters may be tested in water surveys, depending on local conditions. Surface waters are sampled at regular intervals along the drainage net, and a map of the values is prepared (Fig. 3). An increase in the metal content of the water upstream may indicate approach to a mineralized zone. Surveys based on groundwaters can be done only where there is a good distribution of wells, springs, or diamond drill holes. The metal content of the water in these is plotted on a map and contoured. Higher-than-normal contents in the water system may indicate sites or zones of mineralization. Extensive surveys of this type are being carried out in the Canadian Shield, Turkey, and elsewhere in the search for uranium and other metallic deposits.

Sediment and heavy-mineral surveys are carried out to determine the migration path of dispersed elements and minerals along the surface-drainage channels of an area. Samples are collected from the fresh sediment in the bottoms of streams and also from old sediment on the terraces and floodplains. For chemically dispersed elements, the fine fraction (minus 80 mesh) is generally used for analysis; for mechanically dispersed heavy minerals, a coarser fraction is panned from the sediment. Sampling points are located at intervals along the length of the drainage system. The results of the chemical analyses of the stream sediment are plotted on a map of the drainage (Fig. 3). An increase in the metal content of the stream sediment upstream may indicate approach to a mineralized zone. The heavy-mineral fraction may be examined microscopically for ore minerals or accompanying gangue minerals, or the fraction may be analyzed for ore elements or indicator elements. Both results are plotted on the drainage map and interpreted in the same way as the stream-sediment data.

Volatiles and airborne particulates. These surveys are based on gases, such as hydrocarbons, H_2S, and SO_2, and on volatile elements, such as mercury. Airborne particulates, of both an organic and inorganic nature, can also form the basis of surveys. These are usually carried out by sophisticated airborne equipment.

Despite its obvious importance, geochemical prospecting for accumulations of petroleum and natural gas has not been extensively employed in North America, mainly because of a lack of research into methods. It is evident that many oil and gas fields are marked by macroseeps (leakage halos) along faults and porous zones. It is logical to suppose that there are also microseeps, which should be detectable by modern methods of hydrocarbon analysis, especially gas chromatography.

The methods investigated using hydrocarbon analysis have been based mainly on soils. The technique is similar to that used for metalliferous deposits. A grid of soil samples is analyzed for hydrocarbons over suspected areas; the results are plotted and contoured and then interpreted. Techniques similar to those using rocks and groundwaters for the discovery of mineral deposits, but employing hydrocarbons as the indicators, should also be effective in locating accumulations of petroleum and natural gas.

Gases, such as SO_2 from oxidizing sulfide orebodies, have been suggested as good indicators in soil and water analyses. Proof of their effectiveness, however, requires much more research. Mercury, a volatile element, has been used effectively as an indicator element in soil and rock surveys for locating sulfide deposits.

Surveys based on the analysis of organic and inorganic particulates in the near-surface atmosphere have been conducted in Canada, Australia, and South Africa. The methods used are essentially at the research stage and require much more work to prove their usefulness in localizing mineralization.

Biogeochemical surveys. These surveys are of two types. One type utilizes the trace-element content of plants to outline dispersion halos, trains, and fans related to mineralization; the other uses specific plants or the deleterious effects of an excess of elements in soils on plants as indicators of mineralization. The latter type of survey is often referred to as a geobotanical survey.

In the first type of survey the trace-element content of selected plant material is determined on a grid over an area. Generally, samples are collected from parts of individuals of the same species of vegetation, such as twigs, needles, leaves, and seeds. These are dry- or wet-ashed, the trace-element content is determined, and the results are plotted for interpretation with respect to mineralized zones or deposits (Fig. 4). Interpretation of vegetation surveys is frequently difficult, since a number of factors enter into the uptake of elements by plants, some of which are unrelated to mineralization. These include the content and nature of exchangeable elements in the soil, drainage conditions, soil pH, and growth factors in the vegetation.

Geobotanical surveys are carried out by mapping the distribution of indicator-plant species, or plant symptoms diagnostic of high metal-bearing soils, such as chlorosis and dwarfing. Where suitable indicator plants are present, this method is rapid and inexpensive, but unfortunately such plants are seldom consistent in distribution from one area to another. Geobotanical techniques require careful preliminary orientation surveys carried out by trained personnel.

Some biogeochemical surveys of a research nature have been conducted in Canada and the Soviet Union utilizing various animals as the sampling media. The animals, or parts thereof, used have been fish (livers), mollusks (soft parts), and insects (whole organisms). The results of these surveys show that these animals commonly reflect the presence of mineralization in regions in which they occur by having higher-than-normal amounts of various elements.

An interesting development in petroleum prospecting is based on population counts of microflora and microfauna which oxidize hydrocarbons, particularly propane, during their metabolic processes. Where soils, rocks, and groundwaters are enriched in hydrocarbons, certain strains of bacteria and other similar forms of life flourish, and their density of population is apparently proportional to the content of hydrocarbons present. By utilizing specialized bacterial counting techniques, it is possible, as a number of geochemists in the Soviet Union have indicated, to plot contour maps showing the distribution of the bacteria and, hence, the hydrocarbons. Some of these contour maps show peaks and halos that mark accumulations of oil and gas at depth.

Dogs can locate mineral deposits by sniffing out boulders of ore occurring in the dispersion trains and fans of sulfide deposits. They can be trained to become quite sensitive to SO_2 and other gases associated with oxidizing sulfides and are said to be quite effective in the Scandinavian countries and the Soviet Union.

Remote sensing surveys. These surveys utilize various techniques such as airborne detection of radioactivity, infrared detection of anomalies (such as sulfide zones undergoing oxidation) in the geological terrane, and reflection studies. This is a large subject, and only a few examples will be given here.

Airborne gamma-ray spectrometers utilizing large detection crystals have been used extensively in many countries to outline positive radioactive belts and zones mineralized with uranium and thorium. Similar equipment can be utilized in demarcating zones rich in potassium (K^{40}) associated with gold and other types of mineralization. Negative radioactive zones, that is, those from which uranium and thorium have been leached during alteration processes often associated with certain types of gold, silver, lead, zinc, and copper mineralization, can also be located and outlined by these surveys.

Geochemists of the U.S. Geological Survey have observed that over certain types of ore deposits the reflectivity of leaves differs measurably from that of leaves on trees over barren ground. Efforts are being made to apply this method in low-lying jungle areas, where accessibility and deep weathering restricts the use of more conventional geochemical methods.

Isotope surveys. These surveys employ isotopic ratios such as those for lead (Pb^{204}, Pb^{206}, Pb^{207}, Pb^{208}) and sulfur (S^{32}, S^{34}). Only elements with two or more isotopes can be used.

Lead isotopes. Most surveys use the isotopic ratios of minerals in "fingerprinting" or indicating certain types of deposits. For instance, lead minerals in uranium deposits tend to have a high proportion of (radiogenic) uranium-lead, that is, they are enriched in Pb^{206} and Pb^{207}, the derivatives of U^{238} and U^{235}; in thorium-rich deposits, a high enrichment of (radiogenic) thorium-lead, Pb^{208}, can be expected in the lead minerals. In ordinary lead-bearing deposits, the lead minerals have a component of original (primal) lead isotopes (Pb^{204}, Pb^{206}, Pb^{207}, Pb^{208}) plus varying amounts of radiogenic lead (Pb^{206}, Pb^{207}, Pb^{208}), depending, among other factors, on the age of the deposit. Furthermore, there are deposits in which the lead minerals have isotopic ratios that are unusual or anomalous (the J-lead of the Mississippi Valley deposits and also of such deposits as Keno Hill, Yukon). The reasons for the variable isotopic composition of lead in deposits are extremely complex, to say the least, and certainly not understood as yet. Scientists need to understand the processes involved in the migration and concentration of lead isotopes before their theories can be placed on a firm basis. These problems notwithstanding, it is possible in certain cases to fingerprint lead deposits and minerals derived from lead-bearing deposits by means of their isotopic ratios. Thus, during geochemical prospecting surveys where the lead in soils, stream sediments, or in particles of galena in heavy concentrates exhibits high concentrations of radiogenic lead, uranium or thorium deposits should be suspected in the area. Similarly, galena particles exhibiting the so-called J-lead characteristics may indicate the presence of Mississippi-Valley-type deposits. Finally, the lead in rocks (feldspars, apatite, and so on), when isotopically analyzed, may give a clue to the type of lead deposit to be expected or may indicate the presence of uranium or thorium deposits in the terrane.

Sulfur isotopes. Sulfur isotopes commonly fingerprint certain types of deposits in a region, and hence isotopic analyses of the sulfides in heavy concentrates from soils and stream sediments, of sulfur in the ground and surface waters, and of sulfur in stream and lake sediments can be used in a general way to decide what types of deposits occur within a terrane. Thus, the sulfur in waters and stream sediments in regions containing abundant barite or evaporites is generally greatly enriched in the heavy isotope S^{34}; on the other hand, waters leaching sulfide deposits in most regions often have sulfur that is relatively enriched in the lighter isotope S^{32}. Sulfur isotopic ratios also vary systematically with distance from deposits; in some cases, the lighter isotope is enriched as ore is approached; in others, the heavier isotope is enriched. This information may be useful in detailed drilling work and in the interpretation of the ratios K_2O/Na_2O, SiO_2/CO_2, and so forth, obtained during lithogeochemical work.

Sulfur isotopes, as well as those of hydrogen and carbon, should find greater use in hydrocarbon prospecting, particularly in diffusion studies and in determining the migration characteristics of hydrocarbons from source to reservoirs.

EXPLORATION PROGRAM

A full-scale geochemical prospecting program for metals would include the following stages.

1. Preliminary evaluation of areas, selected on the basis of available geological data, by sampling and testing intrusive, metamorphic, and sedimentary rocks and by noting the presence of mineralized zones, faults, fractures, layers, and so forth associated with these rocks. In this way, a metallogenetic province can be identified.

2. Primary reconnaissance and orientation surveys, based on sampling major drainage basins, using water, stream sediment, lake sediment, and heavy-mineral surveys.

3. Secondary reconnaissance surveys based on detailed testing of drainage basins containing anomalous values. Poorly drained areas can be tested by widely spaced sampling of soil and groundwaters.

4. Followup surveys along dispersion trains or fans to determine the cutoff points and the extent of dispersion patterns. These surveys are normally a combination of stream-sediment, heavy-mineral, water, and soil testing, but biogeochemical surveys may also be useful. Priority for followup surveys should be based on the presence of favorable rocks and geological structures, favorable geophysical indications, and intensity of the geochemical anomaly.

5. Detailed surveys carried out in the vicinity of the suspected metalliferous source by soil or vegetation sampling at closely spaced intervals. Interpretation of the results at this stage generally suggests sites for trenching, sinking of shallow shafts, or drilling to locate the precise source of the body giving rise to the geochemical anomaly.

For hydrocarbons the stages in the geochemical exploration program would include the following stages.

1. Preliminary evaluation of areas, selected on the basis of available geological data and known or suspected to contain hydrocarbons. Normally, these will be underlain by relatively unmetamorphosed sediments, younger than Precambrian, and usually containing a high organic (kerogen) content.

2. Primary reconnaissance surveys based on relatively widely spaced sampling of rocks, waters, soils, glacial materials, lake sediments, or oceanic sediments for hydrocarbons, particularly those with molecular weights higher than CH_4.

3. Secondary reconnaissance surveys based on detailed sampling of anomalous areas indicated in stage 2. More closely spaced sampling of the materials in stage 2 should be carried out in addition to sampling of groundwater and stratal water by means of test holes. The cores from the test holes should also be analyzed for hydrocarbons.

4. Interpretation of hydrocarbon anomalies in conjunction with analysis of favorable geophysical indications followed by drilling of the most favorable anomalies.

[R. W. BOYLE]

Bibliography: R. W. Boyle and J. I. McGerrigle (eds.), *Geochemical Exploration*, Can. Inst. Min. Met. Spec. Vol. II, 1971; R. R. Brooks, *Geobotany and Biogeochemistry in Mineral Exploration*, 1972; J. B. Davis, *Petroleum Microbiology*, 1967; I. L. Elliott and W. K. Fletcher (eds.), *Geochemical Exploration*, 1974; I. I. Ginzburg, *Principles of Geochemical Prospecting*, 1960; H. E. Hawkes and J. S. Webb, *Geochemistry in Mineral Exploration*, 1962; M. J. Jones (ed.), *Geochemical Exploration*, 1972, Inst. Min. Met. London, 1973; A. A. Kartsev et al., *Geochemical Methods of Prospecting and Exploration for Petroleum and Natural Gas*, 1959; A. A. Levinson, *Introduction to Exploration Geochemistry*, 1974.

Geophysical exploration

Making, processing, and interpreting measurements of the physical properties of the earth with the objective of practical application of the findings. Most exploration geophysics is conducted to find commercial accumulations of oil, gas, or other minerals, but geophysical investigations are also employed with engineering objectives, in studies aimed at predicting the nature of the earth for the foundations of roads, buildings, dams, tunnels, nuclear power plants, and other structures, and in the search for geothermal areas, water resources, archaeological ruins, and so on.

Geophysical exploration is commonly called applied geophysics or geophysical prospecting. The physical properties and effects of subsurface rocks and minerals that can be measured at a distance include density, electrical conductivity, thermal conductivity, magnetism, radioactivity, and elasticity. Perhaps other properties not used now could also be measured. Occasionally, prospective features can be mapped directly, such as iron deposits by their magnetic effects, but most features are studied indirectly by measuring the properties or the geometry of rocks that are commonly associated with certain mineral deposits. Exploration geophysics is often divided into subsidiary fields according to the property being measured, such as magnetic, gravity, seismic, electrical, thermal, or radioactive properties. This article will first discuss principles applicable to most of the methods and then in subsequent sections the respective methods, the physical principles involved, instrumentation and field techniques employed, data processing, and geologic interpretation.

General principles. A number of principles apply to most of the different types of geophysical exploration. Ordinarily, one looks for an anomaly, that is, a departure from the uniform geologic characterics of a portion of the earth (Fig. 1). The primary objective of a survey is usually to determine the location of such departures. If the source of an anomaly is deep in the earth, the anomaly is spread over a wide area, and the anomaly magnitude is small at any given location. Sometimes, areas of anomalous data are obvious, but more often they are elusive because the anomaly magnitude is small compared to the background noise or because of the interference of the effects of different features. A variety of averaging and filtering techniques are used to accentuate the anomalous regions of change. An anomaly usually seems smaller as the distance between the anomalous source and the location of a measurement increases (Fig. 2). Hence, a nearby source usually produces a sharp anomaly detectable only over a limited region, although possibly of large magnitude in this region. The detail of measurement required to locate anomalies must be compatible with the depth of the sources of interesting anomalies. As the depth of the anomaly increases, more sensitive instruments are needed because the effects become much smaller. Hence, the depth of the feature sought governs both amount of detail and precision required in measurements. Many of the differences in geophysical methods derive from the different depths of interest. Engineering, mineral, and groundwater objectives are usually shallow, whereas petroleum and natural-gas accumulations are usually quite deep — 1 to 6 km.

Geophysical data usually are dominated by effects that are of no interest, and such effects must be either removed or ignored to detect and analyze the anomalous effects being sought. Noise caused by near-surface variations is especially apt to be large. The averaging of readings is the most common way of attenuating such noise.

The interpretation of geophysical data is almost always ambiguous. Since many different configurations of properties in the earth can give rise to the same data, it is necessary to select from among many possible explanations those that are most probable, and to select from among the probable explanations the few that are most optimistic from the point of view of achieving set objectives. Optimistic interpretations are desired in order to prevent overlooking a prospect. Additional measurements can test an interpretation hypothesis so that overoptimism will not be misleading in the final analysis. It is better to be wrong in interpreting a prospect than to miss one.

Geologic features affect the various types of measurements differently, hence more can be learned from several types of measurements than from any one alone. Combinations of methods are particularly useful in mining exploration. In petroleum surveys, magnetic, gravity, and seismic explorations are apt to be performed in that sequence, which is the order of relative cost. First, cheaper methods are used to narrow down the region to be explored by more expensive methods.

Usually the properties which geophysicists are able to measure are not directly related to objectives of interest, hence one must rely on some association between the measured properties and features of interest. Interpretation thus involves much inferential reasoning. For example, a serpentine dike often produces a magnetic anomaly, so that an ore that is often associated with a dike might be found by looking near magnetic anomalies, while realizing that most magnetic anomalies do not have associated ore bodies. Similarly, one might infer that the same factors that produced a particular structural feature also affected sedimentation, so locating such a structural feature may lead to the discovery of a stratigraphic accumulation.

A defect of relying on chains of inference is that one is apt to be wrong, although one is even more apt to be wrong using no inference. Geophysical exploration is thus justified by lessening the risk. The cost of geophysical work rarely exceeds a few percent of exploration costs. Cost effectiveness is a continuing concern. Different geophysical methods usually compete with each other along with nongeophysical methods, such as random drilling, for exploration funds.

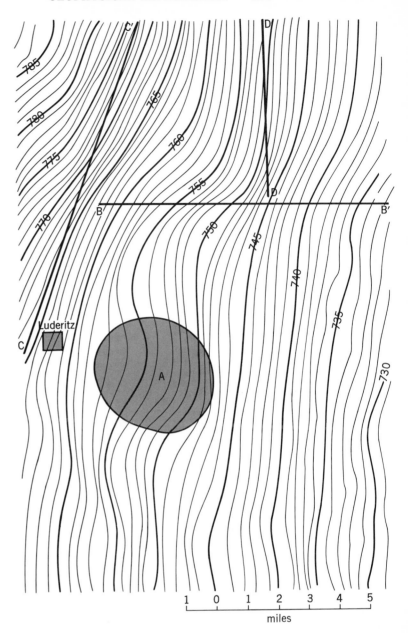

Fig. 1. Bouguer gravity of a portion of the Perth Basin, Western Australia. Contour interval is 1 mGal; data are arbitrary. Departures from the regularity are called anomalies. The bulge A results from an uplift area, the contour offset along BB′ is an east-west trending fault, downthrown to the south. Other faults (CC′ and DD′) are indicated by a tightening of the contours, both are downthrown to the east (*Western Australian Petroleum Pty. Ltd.*)

Magnetic exploration. Rocks and ores containing magnetic minerals become magnetized by induction in the Earth's magnetic field so that their induced field adds to the Earth's field. Magnetic exploration involves mapping variations in the Earth's field with the objective to determining the location, size, and shape of such bodies.

The magnetic susceptibility of sedimentary rock is generally orders of magnitude less than that of igneous or metamorphic rock. Consequently, the major magnetic anomalies observed in surveys of sedimentary basins usually result from the underlying basement rocks. Determining the depths of the tops of magnetic bodies is thus a way of estimating the thickness of the sediments.

Except for magnetite and a very few other min-

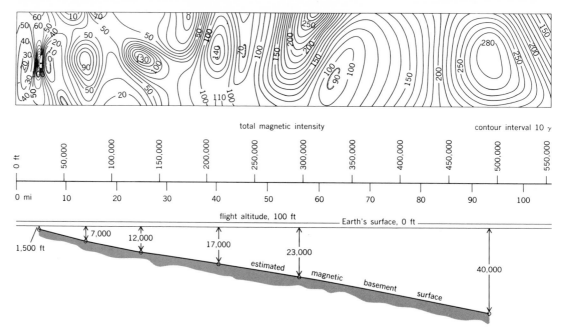

total magnetic intensity   contour interval 10 γ

Fig. 2. Portion of a magnetic map (top) and interpretation (below). On the map, sharper features at the left indicate basement rocks are shallow, whereas broad anomalies to the right indicate deep basement. (*After S. L. Hammer, 4th World Petroleum Congress Proceedings, Rome, Sec. 1, 1955*)

erals, mineral ores are only slightly magnetic. However, they are often associated with bodies such as dikes that have magnetic expression so that magnetic anomalies may be associated with minerals empirically. For example, placer gold is often concentrated in stream channels where magnetite is also concentrated.

Instrumentation. Several types of instruments are used for measuring variations in the Earth's magnetic field. Because the magnetic field is a vector quantity, both its magnitude and direction can be measured or, alternatively, components of the field can be measured in different directions. Often, however, only the magnitude of the total field is measured, and airborne and marine measurements usually are of the total field.

Optically pumped proton and fluxgate magnetometers are used extensively in magnetic exploration. Magnetometers used in exploration typically are accurate to 1 to 10 nanotesla (1 to 10 γ), an accuracy compatible with uncertainties in elevation for airborne measurements or noise background for measurements on land.

Field methods. Most magnetic surveys are made by aircraft, because a large area can be surveyed in a short time, and thus the cost per unit of area is kept very low. Hence, aeromagnetic surveying is especially adapted to reconnaissance, to locate those portions of large, unknown areas that contain the best exploration prospects so that future efforts can be concentrated there.

The spacing of measurements must be finer than the size of the anomaly of interest. Petroleum exploration is usually interested only in large anomalies, hence a survey for such objectives usually involves flying a series of parallel lines spaced 1 to 3 km apart, with tie lines every 10 to 15 km to assure that the data on adjacent lines can be related properly. The flight elevation is usually 300 to 1000 m. In mineral exploration, on the other hand, lines are usually located much closer—sometimes less than 100 m apart—and the flight elevation is as low as safety permits. Helicopters are sometimes used for mineral exploration.

The immediate product of aeromagnetic surveys is a graph of the magnetic field strength along lines of traverse. After adjustment, the data are usually compiled into maps on which magnetism is shown by contours (isogams, which connect points of equal magnetic field strength).

Sometimes, two magnetometers are towed by an aircraft so that they are displaced with respect to each other either horizontally or vertically or both. Thus, they detect magnetic anomalies at different times or in different magnitudes, whereas they simultaneously see variations in the Earth's magnetic field caused by diurnal effects and magnetic storms. This duplication increases confidence that anomalies represent local magnetic concentrations.

In ground mineral exploration, the magnetic field is measured at closely spaced stations. The effects of near-surface magnetic bodies is accentuated over measurements made in the air.

Magnetic surveying is often done in conjunction with other geophysical measurements, because it adds only a small increment to the cost and the added information often helps in resolving interpretational ambiguities. A magnetometer towed behind a marine seismic ship, for example, would distinguish between a volcanic plug and a salt diapir, features which might look nearly alike in the seismic data.

Data reduction and interpretation. The reduction of magnetic data is usually simple. Often, measurement conditions vary so little that the data can be interpreted directly, or else only network adjustments (to minimize differences at line intersections) are necessary because of location uncertainties or instrumental drift. In surveys of large areas, the variations in the Earth's overall magnetic field may be removed (magnetic latitude correction). In exceptional cases, such as in land surveys made over very irregular terrain, as in bottoms of canyons where some of the magnetic sources may be located above the instrument, reduction of the data can become very difficult.

Ths sharpness of a magnetic anomaly depends on the distance to the magnetic body responsible for the anomaly. Inasmuch as the depth of the magnetic body is often the information being sought, the shape of an anomaly is the most important aspect. Modeling is used to determine the magnetic field that would result from bodies of certain shapes and depths. The model anomalies are examined for a parameter of shape that is proportional to the depth (Fig. 3.) The shape parameter is measured on real anomalies and scaled to indicate how deep the body responsible for the anomaly lies. Such estimates are typically accurate to 10–20%, sometimes better.

Iterative modeling techniques are used in more detailed studies. The field indicated by the model is subtracted from the observed field to give an error field. Then the model is changed to obtain a new error field. This process is repeated until the error field is made sufficiently small. Inasmuch as many models can give the same magnetic field, additional constraints must be imposed to make the process meaningful. The significance of the conclusion depends greatly on how realistic the contraints are.

Gravity exploration. Gravity exploration is based on the law of universal gravitation: the gravitational force between two bodies varies in direct proportion to product of their masses and in inverse proportion to the square of the distance between them.

Because the Earth's density varies from one location to another, the force of gravity varies from place to place. Gravity exploration is concerned with measuring these variations to deduce something about rock masses in the immediate vicinity. The variations attributable to such local causes are superimposed on the much larger field determined by the mass, size, and shape of the Earth as a whole.

Vertical density changes affect all stations the same and so do not produce easily measured effects. Gravity field variations are produced by lateral changes in density. Absolute density values are not involved, only changes in density. The product of the volume of a body and the difference between the density of the body and of the horizontally adjacent rocks is called the anomalous mass.

Gravity surveys are used more extensively for petroleum exploration than for metallic mineral prospecting. The size of ore bodies is generally small; therefore, the gravity effects are quite small and local despite the fact that there may be large density differences between the ore and its surroundings. Hence, gravity surveys to detect ore bodies have to be very accurate and very detailed. In petroleum prospecting, on the other hand, the greater dimensions of the features more than offset the fact that density differences are usually smaller.

Instrumentation. The most common gravity instrument in use is the gravity meter or gravimeter.

The gravimeter basically consists of a mass suspended by springs comprising a balance scale. The gravimeter can be balanced at a given location. When the gravimeter is moved to another location, the minute changes in gravitational force, which require the instrument to be adjusted, can be measured. Hence the gravimeter measures differences in a gravity field from one location to another rather than the gravity field as a whole.

A gravimeter is essentially a very sensitive accelerometer, and extraneous accelerations affect the meter in the same way as the acceleration of gravity affects it. Typically, gravimeters read to an accuracy of 0.01 milligalileo, which amounts to 1/100,000,000 of the Earth's gravitational field. A mGal is 10^{-5} newton/kg. Anomalies of interest in petroleum exploration are often of the magnitude of 1/2 to 5 mGal.

Early gravity measurements were made with a torsion balance, a device that responds to the gradient of the gravity field. Absolute gravity measurements are made with the pendulum whose period varies with the gravity field. Occasionally two gravimeters are used, or one gravimeter is read successively at different elevations to measure the vertical gradient of gravity, an arrangement called a gradiometer. Other types of instruments for measuring gravity or gravity gradients are sometimes used.

Field surveys. Almost all gravity measurements are relative measurements; differences between locations are measured although the absolute values remain unknown. Ordinarily, the distance between the stations should be smaller than one-half the depth of the structures being studied.

Gravity surveys on land usually involve measurements at discrete station locations. Such stations are spaced as close as a few meters apart in some mining or archeological surveys, about 1/2 km for petroleum exploration, and even 10 to 20 kms for some regional geology studies. While one would like to have gravity values on a uniform grid, often this is not convenient, and so stations are located on traverses around loops. For petroleum exploration, the gravimeter might be read every 1/2 km around loops of about 60 k.

The gravity field is very sensitive to elevation. An elevation difference of 3 m represents a differ-

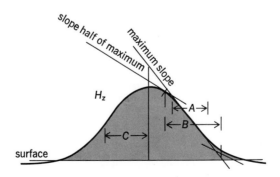

Fig. 3. Variation of measurements across an anomaly. Among the shape factors sometimes measured are A, the distance over which the slope is maximum; B, the distance between points where the slope is half of the maximum slope; C, half the width of the anomaly at half the peak magnitude. Such shape measurements multiplied by index factors give estimates of the depth of the body responsible for the anomaly.

ence in gravity of about 1 mGal. Hence, elevation has to be known very accurately, and the most critical part of a gravity survey often is determining elevations to sufficient accuracy.

Gravity measurements can be made by ships at sea. Usually the instrument is located on a gyro-stabilized platform which holds the meter as nearly level as possible. The limiting factor in shipboard gravity data is usually the uncertain velocity of the meter, especially east to west, since the ship is moving. The velocity of a ship traveling east adds to the velocity because of the rotation of the Earth. Consequently, centrifugal force on the meter increases and the observed gravity value decreases (Eötvos effect).

Gravity measurements are also sometimes made by lowering a gravimeter to the ocean floor and balancing and reading the meter remotely. Gravity measurements have been made by aircraft using techniques like those used at sea, but are not sufficiently accurate to be useful for exploration.

Specialized gravimeters are used to make measurements in boreholes. The main difference between gravity readings at two depths in a borehole is produced by the mass of the slab of earth between the two depths; this mass pulls downward on the meter at the upper level and upward at the lower level. Thus the difference in readings depends on the density of this slab. In sedimentary rocks, the borehole gravimeter is used primarily for measuring porosity.

Data reduction. Gravity measurements have to be corrected for factors other than the distribution of the Earth's mass. Meters drift or change their reading gradually because of various reasons. The Sun and Moon pull on the meter in different directions during the course of a day. The gravitational force varies with the elevation of the gravimeter both because at greater elevations the distance from the Earth's center increases (free-air correction) and because mass exists between the meter and the reference elevation, which is usually mean sea level (Bouguer correction). Gravity varies with latitude because the Earth's equatorial radius exceeds its polar radius and because centrifugal force resulting from the Earth's rotation varies with latitude. Nearby terrain affects a gravimeter; mountains exert an upward pull, valleys cause a deficit of downward pull.

Gravity measurements that have been corrected for all of these effects are called Bouguer anomaly values. They therefore represent the effects of masses within the earth, that is, effects for which corrections have not been made. Most gravity maps display contours (isogals) of Bouguer anomaly values. Sometimes the correction for mass intervening between the meter elevation and the reference elevation is not made; the results are then called free-air anomaly values.

Data interpretation. The most important part of gravity interpretation is locating anomalies that can be attributed to mass concentrations being sought isolating these from other effects. Separating the main part of the gravitational field, which is not of interest (the "regional"), from the parts attributed to local masses, the "residuals," is called residualizing (Fig. 1).

Many techniques for gravity data analysis are similar to those used in analyzing magnetic data. Shape parameters are used extensively to determine the depth of the mass's center. Another widely used technique is model fitting: a model of an assumed feature is made, its gravity effects are calculated, and the model is compared with measurements of the mass.

Continuation is a process by which one calculates from measurements of a field over one surface what values the field would have over another surface. A field can be continued if there is no anomalous mass between the surfaces. Continuing the field to a lower surface produces sharper anomalies as the anomalous mass is approached. However, if one carries the process too far, instability occurs when the anomalous mass is reached. The technique, however, is very sensitive to measurement uncertainties and often is not practical with real data.

The anomalous mass can also be calculated by integrating the residual anomaly. Anomalous mass calculation is useful in mining exploration in determining how much ore is present when the anomalous mass either is the ore or is related to the ore quantitatively. The accuracy of the calculation depends on the accuracy with which the residual has been defined. This calculation is not usually valuable in petroleum exploration.

Seismic exploration. Seismic exploration is the predominant geophysical activity. Seismic waves are generated by one of several types of energy sources and detected by arrays of sensitive devices called geophones or hydrophones. The most common measurement made is of the travel times of seismic waves, although attention is being directed increasingly to the amplitude of seismic waves or changes in their frequency content or wave shape.

Seismic exploration is divided into two major classes, refraction and reflection. Classification depends on whether the predominant portion of wave travel is horizontal or vertical, respectively.

Principles of seismic waves. A change in mechanical stress produces a strain wave that radiates outward as a seismic wave, because of elastic relationships. The radiating seismic waves are like those that result from earthquakes, though much weaker. Most seismic exploration involves the analysis of compressional waves in which particles move in the direction of wave travel, analogous to sound waves in air. Shear waves are occasionally studied, but most exploration sources do not generate very much shear energy. Surface waves, especially Rayleigh waves, are also generated, but these are mainly a nuisance because they do not penetrate far enough into the earth to carry much useful information. Recording techniques are designed to discriminate against them.

The amplitude of a seismic wave reflected at an interface depends on the elastic properties, often expressed in terms of seismic velocity and density on either side of the interface. Normally (when the direction in which the wave is traveling is perpendicular to the interface), the ratio of the amplitudes of reflected and incident seismic waves is given by the reflection coefficient R, as shown in Eq. (1), where $\Delta(\rho V)$ is the change in the product of velocity and density and $\overline{(\rho V)}$ is the average of the product of velocity and density on opposite sides of the interface. The ratio of the energy of the reflected to incident waves is R^2.

$$R = \Delta(\rho V)/2(\overline{\rho V}) \qquad (1)$$

Seismic waves are bent when they pass through interfaces, and Snell's law holds, shown in Eq. (2), where σ_i is the angle between a wavefront and the interface in the ith medium where the velocity is V_i. Because velocity ordinarily increases with depth, seismic-ray paths become curved with concave-upward curvature.

$$(\sin \sigma_1)/V_1 = (\sin \sigma_2)/V_2 \qquad (2)$$

The resolving power with seismic waves depends inversely on their wavelength λ and is often thought of as of the order of $\lambda/4$. The wavelength is often expressed in terms of the wave's velocity and frequency f: $\lambda = V/f$. Most seismic work involves frequencies from 20 to 50 Hz, and most rocks have velocities from 1500 to 6000 m/s, so that wavelengths range from 30 to 300 m. Usually, the frequency becomes lower and the velocity higher as depth in the earth increases, so that wavelength increases and resolving power decreases. Very shallow, high-resolution work involves frequencies high than those cited above, and long-distance refraction (and earthquakes) involve lower frequencies.

Refraction exploration. Refraction seismic exploration involves rocks characterized by high seismic velocity. Wavefronts are bent at interfaces (Fig. 4) so that appreciable energy travels in high-velocity members and arrives at detectors some distance from the source before energy that has traveled in overlying lower-velocity members. Differences in arrival time at different distances from the energy source yield information on the velocity and attitude (dip) of the high-velocity member.

A variant of refraction seismic exploration is the search for high-velocity masses in an otherwise low-velocity section, by looking for regions where seismic waves arrive earlier than expected. Such arrivals, called leads, were especially useful in locating salt domes in Louisiana, Texas, Mexico, and Germany in the late 1920s and 1930s.

Refraction seismic techniques are used in engineering geophysics and mining and groundwater studies to map bedrock under unconsolidated overburden, with objectives such as foundation information or locating buried stream channels in which heavy minerals might be concentrated or where water might accumulate. Refraction techniques are also used in petroleum exploration and for crustal studies.

Reflection exploration. Seismic-wave energy partially reflects from interfaces where velocity or density changes. The measurement of the arrival times of reflected waves (Fig. 5) thus permits mapping the interfaces that form the boundaries between different kinds of rock. This, the predominant geophysical exploration method, can be thought of as similar to echo sounding. In Fig. 5 a, a seismic source S generates seismic energy which is received at detectors located at intervals from A to B. The distance to the reflector RR' can be obtained from the arrival time of the reflection if the

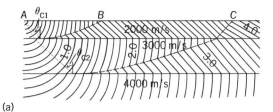

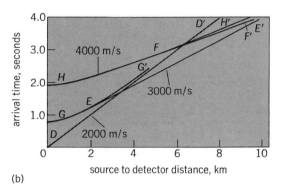

Fig. 4. Refractive seismic exploration data. Section through model of layered earth. (a) Curves (wavefronts) indicate location of seismic energy at successive times after a shot at A. Beyond B, energy traveling in the second layer arrives first; beyond C, that in the third layer. (b) Arrival time as a function of source-to-detector distance. DD', direct wave; EE', refracted wave (head wave) in the second layer; FF', refracted wave in the third layer; GG', reflection from interface between first and second layers; HH', reflection from that between second and third layers.

velocity is known. If the reflector dips as shown, the reflection will arrive sooner at B than at A; the difference in arrival times is a measure of the amount of dip. The angles between ray paths and perpendiculars to the reflector (i, i) are equal at any reflecting point. The image point I is used as an aid in constructing the diagram.

For a flat reflector (Fig. 5b) and constant velocity V, the arrival time at a detector (C or D) at the source S is $t_0 = 2Z/V$. From the Pythagorean theorem for the triangle CSI, Eqs. (3) and (4) are obtained. Similar relationships can be used for non-flat reflectors or nonconstant velocity to yield velocity information.

$$(Vt_c)^2 = (Vt_0)^2 + X^2 \qquad (3)$$

$$V = X(t_c^2 - t_0^2) \qquad (4)$$

Usually a number of detector groups are used, and the arrival of reflected waves is characterized by coherency between the outputs. Thus, if all of the detectors in a line move in a systematic way, a seismic wave probably passed. Multiple detectors make it possible to detect coherent waves in the presence of a high noise level and also to measure distinguishing features of the waves.

Instrumentation. Detectors of seismic energy on land (geophones) are predominantly electromechanical devices. A coil moving in a uniform magnetic field generates a voltage proportional to the velocity of the motion. Often the coil is an inertial element that tends to remain at rest as the case

GEOPHYSICAL EXPLORATION

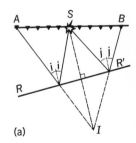

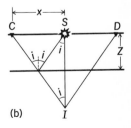

Fig. 5. Measuring reflected seismic wave energy. (a) Dipping reflector. (b) Flat reflector.

GEOPHYSICAL EXPLORATION

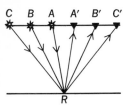

Fig. 6. Reflection of seismic energy obtained from reflection point R (common-depth point) by positioning a source at A and a detector at A'. The same point is involved with a source at B and detector at B', or C and C'. The redundancy in measuring reflections from the same point many times permits sorting out different kinds of waves.

and magnetic field move in response to a seismic wave. Usually the coil has only one degree of freedom and is used so it will be sensitive to vertical motion only. Three mutually perpendicular elements in a three-component detector are sometimes used to determine the direction from which waves come or to distinguish the type of waves (compressional, shear, Rayleigh, and such).

Detectors in water (hydrophones) are usually piezoelectric, and the changes in pressure produced as a seismic wave passes distort a ceramic element and induce a voltage between its surfaces. Such detectors are not directionally sensitive.

Detectors are usually arranged in groups (arrays) spread over a distance and connected electrically so that, in effect, the entire group acts as a single large detector. Such an arrangement discriminates against seismic waves traveling in certain directions. Thus a wave traveling horizontally reaches different detectors in the group at different times so that wave peaks and troughs tend to cancel, whereas a wave traveling vertically affects each detector at the same time so that the effects add.

The signal from the detectors is transmitted to recording equipment over a cable or streamer and is amplified and recorded. The output level from a geophone varies tremendously during a recording. Seismic recording systems are linear over ranges of 100 dB or more. Seismic amplifiers employ various schemes to compress the range of seismic signals without losing amplitude information. They also incorporate adjustable filters to permit discriminating on the basis of frequency.

Most recording systems are multichannel. In 1975, 24 or 48 channels were most common in petroleum exploration, but some 1024 channel systems were in use. Most engineering, mining, and groundwater applications employ one to six channels. The signals are usually digitized and recorded on magnetic tape so that computer analysis can be carried out subsequently. The signals are also displayed as a function of arrival time, on recorded paper. Sometimes signals are displayed on small cathode-ray tubes, and the time from source to first-energy arrival may be timed automatically and displayed on a counter. The latter type is used especially in engineering, groundwater, and mineral-refraction work.

Many sources are used to generate seismic energy. The classical energy source is an explosion in a borehole drilled for the purpose, and solid explosives continue to be used extensively for work on land and in marshes. The explosion of a gas mixture in a closed chamber, a dropped weight, a hammer striking a steel plate, and other sources of impulsive energy are used in land work. An air gun, which introduces a pocket of high-pressure air into the water, is the most common energy source in marine work. Other marine energy sources involve the explosion of gases in a closed chamber, a pocket of high-pressure steam introduced into the water, the discharge of an electrical arc, and the sudden mechanical separation of two plates (imploder).

An oscillatory mechanical source is also used, especially on land. Such a source introduces a long wavetrain so that individual reflection events cannot be resolved without subsequent processing (correlation with the input wavetrain), which, in effect, compresses the long wave train and produces essentially the same result as an impulsive source.

Field techniques. Most petroleum exploration geophysics is carried out by survey methods. These surveys are often run parallel to each other at right angles to the geological strike with occasional perpendicular tie lines, often run on a regular grid. Long lines, placed many kilometers apart, are sometimes run for regional information, but lines are often concentrated in regions in which anomalies have been detected by previous geophysical work. Most seismic work has the objective of mapping interfaces continuously along the seismic lines to map the geological structure.

Geophone groups are spaced 50 to 100 m apart with 24 to 48 adjacent groups of 6 to 24 geophones each being used for each recording. The source is sometimes located at the center of the active groups ("split-spread"), sometimes at one end ("end-on–spread"), and sometimes at a different location.

Following a recording, the layouts and sources are advanced down the line by half the distance over which the geophone groups are disposed (the "spread length") for continuous coverage or by some smaller multiple of the group interval for redundant coverage (Fig. 6). The active geophone groups are advanced usually by electrical switching rather than physical movement of the geophones and cables. Sometimes some geophone groups are laid out perpendicular to the seismic line on a cross-line to measure components of dip perpendicular to the line.

With surface sources such as vibrators, gas ex-

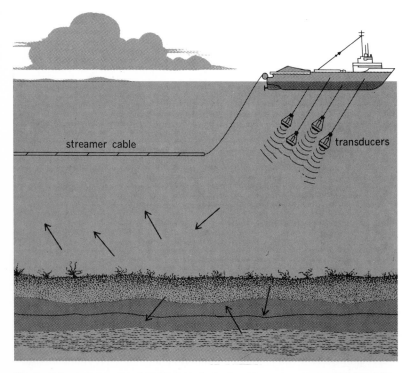

streamer cable

transducers

Fig. 7. Seismic surveying at sea is done by towing a long streamer containing detectors that sense seismic energy. Seismic energy is generated by several transducers that are also towed from the ship. (*After F. Levin, International Science and Technology, Conover-Mast Publications, 1969*)

arrival time, seconds

Fig. 8. Seismic record section in East Texas after processing. The reflection event at *A* is attributed to the Edwards Reef, which formed a barrier at the time the adjacent sediments were deposited. Downdip to the left of the reef formations the Woodbine sands can be seen pinching out in the updip direction. These are productive of oil and gas in this area. (*Seiscom-Delta, Inc.*)

ploders, and weight droppers, source trucks stop to deliver energy into the ground for a recording. They then move forward a few meters and repeat, and so on down the line. Several source trucks located a few meters apart are often used simultaneously; they are synchronized by radio from the recording unit. Often several individual recordings are summed together to make one field record. A typical seismic crew can survey from 1 to 10 km per day.

Small marine operations, often called profiling, consist of an energy source and a short streamer containing a number of hydrophones and feeding a single-channel recorder. Larger marine operations (Fig. 7) involve ships 60 m or more in length towing a streamer 2 to 3 km long with 24 or 48 groups of hydrophones spaced along the streamer. The streamer is typically towed at a depth of 10 to 15 m. An energy source is towed near the ship. Recordings are made as the ship is continuously underway at a speed of about 6 knots (3 m/s).

Marine operations require precise knowledge of position. Observations of navigation satellites are made whenever such a satellite passes overhead, yielding fixes every couple of hours with an uncertainty of the order of 50 m. Locations between satellite fixes are obtained by such techniques as Doppler-sonar navigation, various radio-location methods, and inertial navigation. Most marine seismic operations incorporate an integrated navigation system in which a number of sensors feed a computer that uses their data synergetically to determine the most probable location.

Data reduction and processing. Seismic data are corrected for elevation and near-surface variations on the basis of survey data and observations of the travel time of the first energy from the source to reach the detectors, which usually involve either travel in a direct path or in shallow refractors.

If data are not processed by computers, the arrivals of energy are timed and plotted on graphs for use in interpretation. It is essential to make certain that the same point on the same reflection or refraction is always picked, because interpretation depends much more heavily on differences between travel times than on the magnitude of travel times.

If data are processed by computers, the first step is often editing, wherein data are merged with identifying data, rearranged, checked for being either dead or wild (with bad values sometimes replaced with interpolated values), time shifted in accordance with elevation and near-surface corrections that have been determined in the field, scaled, and so on.

Following the editing, different processing sequences may follow, including (1) filtering (deconvolution) to remove undesired natural filter effects, trace-to-trace variations, variations in the strength or wave shape of the source, and so on; (2) grouping according to common-depth point (Fig. 6) or some other arrangement; (3) analyzing to see what velocity values will maximize coherency as a function of source-to-detector distance; (4) statistically analyzing to see what trace shifts will maximize coherency; (5) trace-shifting according to the results of steps 3 or 4; (6) stacking by adding together a number of individual traces; (7) automigrating by rearranging and combining data elements in order to position reflection events more nearly under the surface locations where the appropriate reflecting surface is located; (8) another filtering; and (9) displaying of the data.

A number of data outputs and displays are made during the processing sequence. These are analyzed for control and to determine parameters for subsequent processing steps.

Data interpretation. The travel times of seismic reflections are usually measured from record section displays (Fig. 8), which result from processing. Appropriate allowance ("migration") must be made because reflections from dipping reflectors appear at locations downdip from the reflecting points. Allowance must also be made for variations

North Sea

1 mi

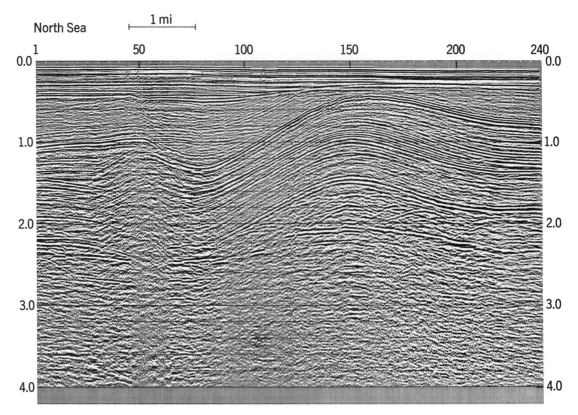

Fig. 9. Seismic record section from the North Sea. The layer from about 2 to 2.3 s at the right edge is composed primarily of salt that has undergone plastic flow as part of the folding of the rock formations. The upfolding (an anticline) is not productive here, but the formations at the left just below 1 s are believed to contain gas. This is a trap bounded at the right by a down-to-the right fault. (*Seiscom-Delta, Inc.*)

in seismic velocity, both vertically and horizontally. Seismic events other than reflections must be identified and explained.

In petroleum exploration the objective is usually to find traps, places in which porous formations are high relative to their surroundings and in which the overlying formation is impermeable. If oil or gas, which are lighter than water, were present, they would float on top of the water and accumulate in the pores in the rock at the trap. Seismic exploration determines the geometry, hence where traps might be located. However, one cannot usually tell from seismic data whether oil or gas was ever generated, whether the rocks have porosity, whether overlying rocks are impermeable, or whether oil or gas might have escaped or been destroyed, even if they were present at one time.

To reconstruct the geologic history, often several reflections at different depths are mapped, and attempts are made to reconstruct where the deeper reflectors were at the time of deposition of shallower rocks.

After some experience has been developed in an area, features in the seismic data that distinguish certain reflectors or certain types of structure often can be recognized. Seismic velocity measurements are helpful here.

In relatively unconsolidated sediments and in some other circumstances, gas and oil containing considerable gas may lower the seismic velocity or rock density or both sufficiently to produce a distinctive reflection, usually evidenced by strong amplitude (a "bright spot") and other distinguish-ing features (Fig. 9). This is often called direct detection, although such anomalies do not necessarily correspond to accumulations of commercial importance. Coal and peat beds are also characterized by reflections of strong amplitude.

Electrical and electromagnetic exploration. Variations in the conductivity or capacitance of rocks form the basis of a variety of electrical and electromagnetic exploration methods, which are used primarily in metallic mineral prospecting. Both natural and induced electrical currents are measured. Direct currents and low-frequency alternating currents are measured in ground surveys, and ground and airborne electromagnetic surveys involving the lower radio frequencies are made.

Some mineral deposits give rise to spontaneous earth currents. The attendant "self-potential" field can often be mapped so that the source of the self-potential can be found.

Natural currents in the earth, called telluric currents, affect large areas. They are believed to be related to ion-current circulations in the upper atmosphere. The current density of telluric currents varies with rock conductivity. Comparisons are made between readings at various locations, and the readings are observed simultaneously at a reference location.

Changes in electrical current flows give rise to associated magnetic fields, and the converse is also true, according to Maxwell's equations. Natural currents are somewhat periodic. Magnetotellurics involves the simultaneous measurement of natural electrical and magnetic variations from which the variation of conductivity with depth can be determined.

Certain mineral ores store energy as a result of current flow and, after the current is stopped, transient electrical currents flow. This phenomenon is called induced polarization. The storage mechanism is not only capacitative, but the exact mechanism is not understood. Observations of the rate of decay of these transient currents are studied in time-domain methods.

Alternating currents tend to flow along the surface of conductors rather than in their interior. The thickness which contains most of the current is called the skin depth. The skin depth, in meters, is given by $(2/\sigma\mu\omega)^{1/2}$, where σ is the conductivity in mhos per meter, μ is permeability in henrys per meter and ω is angular frequency in radians per second.

Since the skin depth becomes greater as frequency becomes lower, measurements at different frequencies give information on the variation of conductivity with depth. Methods in which apparent resistivity is determined as a function of frequency are called frequency-domain methods.

Instrumentation and field techniques. Direct-current and low-frequency alternating-current ground surveys are carried out with a pair of current electrodes, by which electrical current is introduced into the ground, and a pair of potential electrodes across which the voltage is measured. The equipment is often simple, consisting essentially of a source of electrical power (battery or generator), electrodes and connecting wires, ammeter, and voltmeter. A key problem here is providing equipment that will generate enough electrical or electromagnetic energy in the ground but is reasonably portable. The voltages are usually measured in a potentiometric arrangement so that current flow in the measuring circuit does not produce distortions. Various arrangements of electrodes are used. The depth of current penetration depends on the geometry of disposition of instruments, on the frequency used, and on the conductivity distribution. There are two basic types of measurement: (1) electrical sounding or electrical drilling, wherein apparent resistivity is measured as dimensions between the electrodes are increased—these measurements depend mainly on the variation of electrical properties with depth; (2) electrical profiling, in which variations are measured as the electrode array is moved from location to location.

Electromagnetic methods generally involve a transmitting coil, which is excited at a suitable frequency, and a receiving coil, which measures one or more elements of the electromagnetic field at a number of observation points. The receiving coil is usually oriented in a way that minimizes its direct coupling to the transmitter, and the residual effects are then caused by the currents that have been induced in the ground. A multitude of configurations of transmitting and receiving antennae are used in electromagnetic methods, both in ground surface and airborne surveys. Among the methods used are long linear antennae, large rectangular loops, and small portable loops, which are varied both in orientation and location. In airborne surveys, the transmitting and receiving coils and all

associated gear are carried in an aircraft, which normally flies as close to the ground as is safe. Airborne surveys often include multisensors, which may record simultaneously electromagnetic, magnetic, and radioactivity data along with altitude and photographic data. Sometimes several types of electromagnetic configurations or frequencies are used.

The effective penetration of most of the electromagnetic methods into the earth is not exceptionally great, but they are used extensively in searching for mineral ores within about 100 m of the surface. Electrical methods are effective in exploring for groundwater and in mapping bedrock, for example, at dam sites. Electrical methods are used also for detecting the position of buried pipelines and in land-mine detection and other military operations.

Radioactivity exploration. Natural radiation from the earth, especially of gamma rays, is measured both in land surveys and airborne surveys. Natural types of radiation are usually absorbed by a few feet of soil cover, so that the observation is often of diffuse equilibrium radiation. The principal radioactive elements are uranium, thorium, and potassium; radioactive exploration has been used primarily in the search for uranium and other ores, such as columbium, which are often associated with them. Radioactive methods are sometimes used in the search for potash deposits. The Geiger counter and scintillation counter are instruments generally used to detect and measure the radiation.

Remote sensing. Measurements of natural and induced electromagnetic radiation made from high-flying aircraft and earth satellites are referred to collectively as remote sensing. This comprises both the observation of natural radiation in various spectral bands, including both visible and infrared radiation, such as by photography and measurements of the reflectivity of infrared and radar radiation.

Well logging. A variety of types of geophysical measurements are made in boreholes, including self-potential, electrical conductivity, velocity of seismic waves, natural and induced radioactivity, and temperature variations. Borehole logging is used extensively in petroleum exploration to determine the characteristics of the rocks which the borehole has penetrated, and to a lesser extent in mineral exploration.

Measurements in boreholes are sometimes used in combination with surface methods, as by putting some electrodes in the borehole and some on the surface in electrical exploration, or by putting a seismic detector in the borehole and the energy source on the surface. *See* WELL LOGGING (OIL AND GAS).   [R. E. SHERIFF]

Bibliography: M. B. Dobrin, *Introduction to Geophysical Prospecting,* 3d ed., 1976; F. S. Grant and G. F. West, *Interpretation Theory in Applied Geophysics,* 1965; D. H. Griffiths and R. F. King, *Applied Geophysics for Engineers and Geologists,* 1965; L. L. Nettleton, *Gravity and Magnetic Prospecting,* 1975; D. S. Parasnis *Principles of Applied Geophysics,* 2d ed., 1972; R. E. Sheriff, *Encyclopedic Dictionary of Exploration Geophysics,* 1973; W. M. Telford et al., *Applied Geophysics,* 1975.

Geothermal power

Generation of electrical power from the stored heat in the Earth's crust. This generation is now possible in two forms, natural and drilled (or artificial). In the natural form, existing emissions of steam or hot water, such as fumaroles, geysers, or even quiescent underground hot reservoirs, are harnessed to generate high-pressure steam, which in turn drives a conventional turbine generator. In the case of drilled geothermal power, the deep heat in the Earth's crust is used directly by drilling a deep hole, sending cold water down, and bringing back hot water or steam through an insulated center pipe or through a parallel pipe, after permeating the intervening state.

Natural geothermal power generation is generally limited to volcanic regions of the Earth, which rarely coincide with regions of large power demand. The drilled geothermal power concept, on the other hand, seeks to tap the almost unlimited body of high-temperature heat that is everywhere no farther than $10-15$ mi ($16-24$ km) beneath the surface and therefore equally accessible to every center of power demand.

Natural geothermal power. Additions are planned or in process at principal power sites in Italy, Japan, Mexico, New Zealand, and California, as shown in Table 1. Plans or evaluations to develop the many other potential sites listed are also in progress, and worldwide exploration is going on for new sites for either power or process heat. Figure 1 shows a typical cross section of a geothermal natural hot-water generating system before installation of aboveground equipment to convert the heat to electrical power.

Although the planned capacity in each case listed in Table 1 is only a small percentage of the total national demand, the potential capacity is many times greater. In the United States, for instance, the 600 MW installed at The Geysers in California represents 0.1% of the nation's total, but geologists estimate its ultimate capacity at 5000 MW. The Salton Sea area has recently been measured to

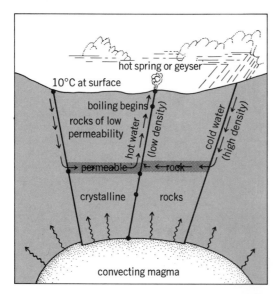

Fig. 1. Model of a high-temperature hot-water geothermal system.

Table 1. Natural geothermal power developments, 1975

| Site | First installation | Total capacity, MW | |
		Installed	Planned
Ecuador			*
Ethiopia (Danakil)			*
Hungary			*
Italy (Larderello)	1904	400	~200
Iceland	1969	5	50
Indonesia			*
Japan (Honshu)	1966	50	150
Kenya			*
Kinshasa (Kiabukwu)		~1	*
Mexico			
Cerro Prieto	1970	75	200
Hidalgo	1959	~5	*
New Zealand			
Wairekai	1959	200	500
Kawerau	1955	10	*
St. Lucia			*
Salvador		~1	*
United States†			
The Geysers, Calif.	1960	600	1200
Mojave, Calif.			2 sites
Inyo County, Calif.			4 sites
Salton Sea, Calif.		3	~100
Hawaii			
Eureka County, Nev.			*
Alaska			*
Soviet Union			*
Kamchatka	1967	25	~400
50 other basins			*

*Sites with large known potential, still in planning stage.
†Also potential sites in New Mexico, Arizona, Oregon, Washington, Nevada, and Wyoming.

contain 6×10^9 acre-feet (7.4×10^{12} m³) of water under high pressure at temperatures above 500°F, a resource sufficient to produce 20,000 MW for southern California. In fact, known total geothermal resources in the West are now estimated to embrace 1.35×10^6 acres (5500 km²), and it is entirely possible that natural geothermal power could ultimately supply 25% of the western demand.

The power-generating equipment employed to harness these natural emissions is now fairly standard. Deep wells are drilled, and live steam is piped directly to a steam turbine. The optimum depth of the well is determined by the quantity of steam available and its superheat at a given pressure. In the most efficient cases, the steam leaves the turbine at very low pressure by virtue of direct condensation with cooling water, thus greatly increasing turbine power. Where a local source of cooling water is not available, as at The Geysers, the condensate is recirculated through cooling towers. The small percentage of noncondensable gases (CO_2, CH_4, and H_2S) in the geothermal steam is removed by steam jet ejectors. These gases, plus other corrosive impurities in the steam, require that piping and turbine blading be of stainless steel. Nevertheless the capital cost per kilowatt is less than $150/kWe, and the absence of any fuel cost makes geothermal energy the most inexpensive source for electric power generation. *See* STEAM TURBINE.

Approximate data for the three principal installations in the world today are shown in Table 2.

Drilled geothermal power. Drilled geothermal power may ultimately become an equally attractive concept. It permits the harnessing of the Earth's

Table 2. Comparison of approximate data for three natural geothermal installations

	The Geysers, Calif.	Larderello, Italy	Wairekai, New Zealand
Installed capacity, MW	300	390	340
Well depth, ft	800–5000	1000–3000	574–3200
Steam pressure, psia			
With no flow	175	400	235
At full power	115	80	195
Steam temperature, °F	348	266–446	315–395
Steam condition	SH	SH	Wet
Condenser pressure, in. Hg	4	4–30	2
Steam flow, M lb/hr	12,000	~9000	~6000
Noncondensable gases, %	0.7	5	0.5

heat at any site without the need of any natural steam emission, thus avoiding the problems of corrosion and geographical location. However, this method requires the development of ultradeep drilling and casing of greater holes than 24 in. (60 cm) in diameter, to a depth of 50,000 ft (15,000 m) or more, at a cost less than $10,000,000 in order to prove economical. This goal is considered possible in the light of recent achievements in drilling technology. *See* BORING AND DRILLING (MINERAL); WELL LOGGING (OIL AND GAS).

Drilled geothermal power is based on the temperature gradient of about 1.5°F per 100 ft (2.75°C per 100 m) in the Earth's crust, so that, at a depth of 50,000 ft (15,000 m), rock temperatures of about 800°F (420°C) can be expected. A 24-in. (60-cm) cased hole drilled to this depth would have an effective heat-transfer surface in excess of 300,000 ft² (about 30,000 m²). Water sent down such a well would drain heat from the surrounding rock as it descended, with the temperature and pressure rising to 800°F (420°C) and 16,500 psi (114 × 10⁶ N/m²) at the bottom. With an insulated inner pipe in the well, this water would then return to the top at 700°F (370°C), and the density difference of the two columns of water would be sufficient to drive the loop without pumps. By keeping the emerging water at about 4000 psi (28 × 10⁶ N/m²), its heat could be transferred in heat exchangers to generate superheated steam at lower pressure for conventional turbine generators. The spent water, still at about 400°F (200°C) and high pressure, could generate further power in a hydraulic turbine and deliver substantial heat for municipal central heating and other process industries before being returned to the wellhead. The entire process is shown schematically in Fig. 2.

A 24-in.-diameter (60 cm) well 50,000 ft (15,000 m) deep would generate up to 20,000 kW when operated cyclically as a peak-load unit 3 hr per day. It would require little attention and would incur no surface contamination or air pollution. In addition, it would furnish some 300 × 10⁶ Btu (75 × 10⁶ kg-cal) per day of process heat between 100 and 400°F (40 and 200°C). At greater depths and diameters, both efficiency and power output would increase. A typical operating cycle is shown in Fig. 3.

Economical generation of continuous (that is, noncyclic) power from drilled holes in dry strata requires either extremely deep holes, or some means of breaking up the lower strata far out from the hole and flooding them with a high-heat transfer fluid or percolating water through the fractured rock. This latter method is being investigated, using hydraulic fractionation of the rock, at Los Alamos, NM, and in the Maryville, MT, batholith, where an anomalous high-temperature zone is nearer the surface than normal. The percolating flow passes from a downflow well to the bottom of a parallel "riser" well, which must be insulated.

Such drilled wells, whether using percolation between two holes, or counterflow in a single hole, are to be thought of in economic terms as "mining" the stored heat of the Earth. This concept is analogous to mining coal or oil, with the one important difference that the stored heat replenishes itself from an essentially inexhaustible reservoir deeper in the Earth.

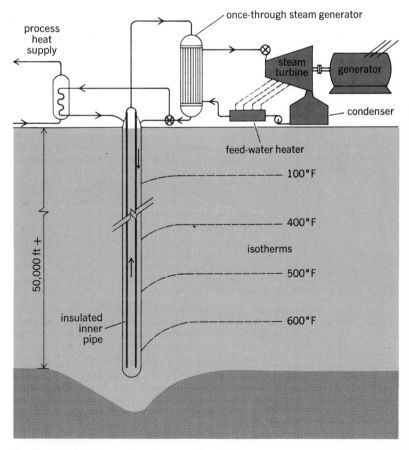

Fig. 2. Drilled geothermal power-generating system.

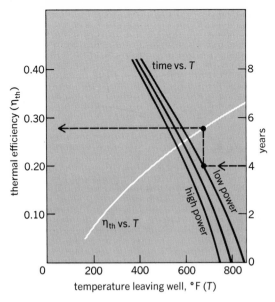

Fig. 3. Typical performance curves (given the well depth and the diameter) for a drilled geothermal power system.

The economic problem, evident in Fig. 3, is to determine the optimum rate of removal of the heat, followed by the optimum dormant period while the heat returns. A high rate of removal gives a short-term high return on the capital investment, with a rapid fall in efficiency as the well cools down. A low rate of removal gives a longer useful life (between replenishment periods) but at a small return. If the heat is removed at such a low rate that the depression of the Earth isotherms away from the well reach steady state, the power generation would appear to be uneconomically low. Hence cyclical operation would appear to be optimum.

The objective in optimizing the design of such a plant is to determine the flow rate, and therefore the heat removal rate, for a given well depth and diameter, at which long-term power generation would be cheapest. The average properties of the Earth's crust now available for such analysis are a density of 170 lb (specific gravity of 2.73), a specific heat of 0.17 B/lb-°F (0.17 cal/g-°C), and a thermal diffusivity of 0.05 ft²/hr (0.013 cm²/sec).

The concept is sufficiently sound and the economic and environmental advantages are so significant that serious research and development are now being considered. *See* ELECTRIC POWER GENERATION.

[ROBERT L. WHITELAW]

Bibliography: H. C. H. Armstead (ed.), *Geothermal Energy: Review of Research and Development*, UNESCO, Paris, 1973; J. B. Koenig, *Worldwide Status of Geothermal Exploration and Development*, Society of Petroleum Engineers, SPE-4179, November 1972; P. Kruger and C. Otte, *Geothermal Power*, 1973; F. D. Stacey, *Physics of the Earth*, 1969; U.S. House of Representatives Committee on Science and Astronautics, *Geothermal Energy*, H. R. 8628 and 9658, Sept. 11, 13, 18, 1973; G. A. Waring, *Thermal Springs of the U.S. and Other Countries of the World*, Geol. Prof. Pap. no. 492, 1965.

Graphite

A low-pressure polymorph of carbon, the common high-pressure polymorph being diamond. Several other rare polymorphs have been synthesized or discovered in meteorites. The contrast in physical properties between these two polymorphs is remarkable: Graphite is metallic in appearance and very soft, whereas diamond is transparent and one of the hardest substances known.

Graphite is hexagonal, space group $P\ 6_3/m\ mc$, $a = 2.48$ A, $c = 6.80$ A, with 4C in the unit cell. Its atomic arrangement consists of sheets of carbon atoms at the vertices of a planar network of hexagons (Fig. 1). Thus, each carbon atom has three nearest-neighbor carbon atoms. The layer distance between the sheets is $c/2 = 3.40$ A. Diamond, on the other hand, is a three-dimensional framework structure with the carbon atoms in tetrahedral (fourfold) coordination. Crystals of graphite are infrequently encountered since the mineral usually occurs as earthy, foliated, or columnar aggregates often mixed with iron oxide, quartz, and other minerals (Fig. 2).

Properties. The sheetlike character of the graphite atomic arrangement results in distinctive physical properties. The mineral is very soft, with hardness $1\frac{1}{2}$; it soils the fingers and leaves a black streak on paper, hence its use in pencils. The specific gravity is 2.23, often less because of the presence of pore spaces and impurities. The color is black in earthy material to steel-gray in plates, and thin flakes are deep blue in transmitted light. One perfect cleavage is parallel to the hexagonal sheets, allowing the mineral to be split into thin flexible but nonelastic folia. Graphite is a conductor of electricity, distinguishing it from amorphous carbon (lampblack).

Graphite closely resembles molybdenite, MoS_2, a mineral with similar crystal structure, but the two can be distinguished by the greenish streak and much higher (4.70) specific gravity of the latter mineral. Early confusion with the brittle gray lead sulfide, galena, resulted in the synonymous trade

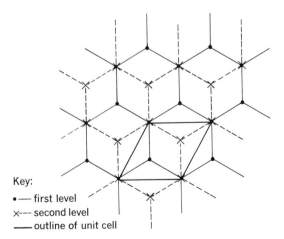

Key:
•— first level
×--- second level
—— outline of unit cell

Fig. 1. The graphite atomic arrangement down the *c*-axis. There are two distinct layers of carbon atoms, related by symmetry. The first level consists of carbon atoms (solid circles) at the vertices of a network of hexagons. The second level, also a hexagonal network with carbon atoms at the vertices (crosses), is shifted relative to the first level.

names "plumbago" and "black lead" for graphite. In a similar manner, the misleading term "lead pencil" has persisted and is still in common usage.

Occurrence. Graphite arises from the thermal and regional metamorphism of rocks such as sandstones, shales, coals, and limestones which contained organic products not exposed to an oxidizing environment. It also can form in a strongly reducing environment, such as in serpentinites and limestones where hydrogen gas may reduce carbon dioxide. Platy graphite showing crude crystal surfaces often occurs speckled in coarsely crystallized marbles. The major sources of graphite are in gneisses and schists, where the mineral occurs in foliated masses mixed with quartz, mica, and so on. Noteworthy localities include the Adirondack region of New York, Korea, and Ceylon. In Sonora, Mexico, graphite occurs as a product of metamorphosed coal beds. Graphite is also observed in meteorites, where the mineral was formed under strongly reducing conditions, usually in association with metallic iron.

[PAUL B. MOORE]

Synthetic graphite. Graphite has a highly developed crystalline structure, and its softness, high thermal and electrical conductivity, and self-lubricating qualities differentiate it from other forms of carbon.

Carbon in graphitic form has both metallic and nonmetallic properties. Commercially produced synthetic graphite is a mixture of crystalline graphite and cross-linking intercrystalline carbon. Its physical properties are the result of contributions from both sources. Thus, among engineering materials, synthetic graphite is unusual because a wide variation in measurable properties can occur without significant change in chemical composition.

At room temperature the thermal conductivity of synthetic graphite is comparable to that of aluminum or brass. An unusual property of graphite is its increased strength at high temperature. The crushing strength is about 20% higher at 1600°C and the tensile strength is 50–100% higher at 2500°C than at room temperature.

Graphite is resistant to thermal shock because of its high thermal conductivity and low elastic modulus. It is one of the most inert materials with respect to chemical reaction with other elements and compounds. It is subject only to oxidation, reaction with and solution in some metals, and formation of lamellar compounds with certain alkali metals and metal halides.

Uses. Graphite has many uses in the electrical, chemical, metallurgical, nuclear, and rocket fields: electrodes in electric furnaces producing carbon steel, alloy steel, and ferroalloys; anodes for electrolytic production of chlorine, caustic and chlorates, magnesium, and sodium; motor and generator brushes; sleeve-type bearings; seal rings; electronic tube anodes and grids; nuclear reactor moderators, reflectors, and thermal columns (Fig. 3); rocket motor nozzles; missile rudder vanes; metallurgical molds and crucibles; linings for chemical reaction vessels; and in the resin-impregnated impervious form, for heat exchangers, pumps, piping, valves, and other process equipment.

Preparation. Synthetic graphite can be made from almost any organic material that leaves a high carbon residue on heating to 2500–3200°C. In commercial operations, raw materials are carefully selected because not all substances with high carbon content undergo a suitably complete transformation to graphite at these temperatures. Petroleum coke is raw material for the most commonly used production process. After calcining and sizing, the coke is mixed with coal tar pitch, heated to about 165°C, and formed by extrusion or molding to "green" shapes. Baking to 750–1400°C in gas- or oil-fired kilns follows the forming operation. Graphite is produced by heating the baked shapes to 2600–3000°C by passing electricity amounting to 1.6–3.0 kw per pound of graphite through the bed of a furnace made of the shapes laid in granular coke (Fig. 4). The whole bed is covered by an insulating blanket of silicon carbide, coke, and sand. Higher-density synthetic graphite can be obtained by impregnating the baked carbon with pitch prior to graphitization. Graphite with total ash content less than 20 parts per million is needed for a number of nuclear and electrolytic uses, and is obtained by heating the graphite shapes electrically to about 2500°C while bathing them in a purifying gas. *See* COKING IN PETROLEUM REFINING.

Highly ordered crystalline graphite can be produced up to about 1/4-in. thickness by pyrolyzing organic gases under controlled conditions at 1400–2000°C. Pyrolytic graphite exhibits a high degree of anisotropy (varying properties in different directions). Parallel to the thickness the thermal conductivity is comparable to copper, but perpendicular to the thickness conductivity is about 1/200 the conductivity of copper. Tensile strength and thermal expansion also vary greatly with orientation. The room temperature density of pyrolytic graphite reaches up to 2.22, about 98% of the 2.26 density of the graphite single crystal, whereas the density of graphite electrodes ranges from 1.5 to 1.7. Pyrolytic graphite has been formed into rocket nose cones.

Graphite can be recrystallized to increase its density and improve other related and significant properties. Starting with an initial level of 1.7 for a conventional grade, bulk densities as high as 2.18 at room temperature have been obtained on a laboratory basis. Recrystallized graphite was introduced in 1960 and is now commerically available at a density range of 1.92–1.97. The permeability of recrystallized graphite is comparable to brass

GRAPHITE

(a) |— 25mm —|

(b)

Fig. 2. Graphite. (a) Earthy aggregate (*American Museum of Natural History* specimen). (b) Hexagonal crystal of graphite with triangular markings on face (*from C. Palache, H. Berman, and C. Frondel, Dana's System of Mineralogy, vol. 1, 7th ed., Wiley, 1944*).

Fig. 3. Graphite reflector for a test reactor.

Table 1. Graphite cloth oxidation in still air

Temperature, °C	Hr to lose 1% weight
266	10,000
360	100
493	1.0
699	0.01

and is $10^5 - 10^7$ less than electrode-grade graphite. Since the degree of anisotropy (always less than pyrolytic graphite) can be controlled, the usefulness of recrystallized graphite is not limited (or restricted by production considerations) to thin sections in the presence of thermal stresses. Initial use of this form of graphite has been in nozzles of rocket motors.

Graphite and carbon fibers. Commercial carbon and graphite fibers are made from any carbonaceous fibrous raw material that pyrolyzes to a char, does not melt, and leaves a high carbon residue. The most commonly used precursor materials are rayon, polyacrylonitrile, and pitch. Working in the laboratory near the triple point of carbon (92 atm, 3650°C), submicron-diameter graphite whiskers of near theoretical strength—100 to 140×10^6 psi modulus—are found inside boules. Vapor-deposited pyrolytic graphite fibers have been made on a small scale.

There are two distinct families of carbonaceous fibers. Rigid classification depends on x-ray diffraction measurements, but fibrous carbon and graphite may be loosely distinguished from each other by the highest temperature reached during their manufacture. Fibers and fabrics heated to less than 2000°C are designated carbon. Graphite fiber products are furnaced above 2500°C.

Fibrous carbon is prepared by pyrolyzing spun, felted, or woven raw materials to 700–1800°C. Prior to the heat treatment, chemical or other treatments are employed in some processes. Graphite articles are made by further processing the carbonized material to 2500–3000°C.

In comparison with natural and synthetic fibers, carbon and graphite filaments have a high modulus of elasticity, over 4,000,000 psi, and low elongation at break. In these respects, they resemble glass filaments and, in general, can be handled on equipment designed for glass yarns.

Graphite cloth and yarn oxidize in air relatively slowly, considering the large surface-area-to-weight ratio, because their relatively high purity excludes most of the oxidation-promoting catalysts found in most synthetic graphite. Table 1 tabulates the results of a typical oxidation experiment.

It can be said that graphite cloth oxidizes in air at about the same rate as common graphite rods and plates. The threshold of oxidation of graphite in steam is about 700°C and in carbon dioxide is 900°C. Depending on its carbon assay, fibrous carbon has less oxidation resistance than graphite.

As with all forms of carbon and graphite, these fibers do not melt at ordinary pressures, and temperatures above 3600°C are required to volatilize them. The strength of general-purpose carbon and graphite cloth and yarn increases with temperature up to about 2000°C, and then declines. High-performance fibers have usable strength at 3000°C.

Cloth. One of the major uses of fibrous carbon products is for ablative composites for rocket motors in the form of rayon cloth converted to carbon and graphite cloth. The most widely used fabrics weigh 7–8 oz/yd² and have filament diameters of 0.0003–0.0004 in. (Table 2).

Both carbon and graphite cloths as reinforcement in resin composites are used in a variety of applications which require short-term high-temperature strengths, good ablation characteristics, and insulation with low density. Since graphite is somewhat more resistant than carbon and physically stable at elevated temperatures, graphite reinforcements are preferred in rocket nozzle throats and ablation chambers. Carbon cloths are used where greater strength and lower thermal conductivity is required, such as in reentry vehicles and rocket nozzle entrance sections and exit cones of spacecraft.

Continuous filament yarns and strands. General-purpose and high-performance continuous filament yarns have yields from 300 yd/lb up to 4100 yd/lb, depending on construction (Table 3). They can be woven, knitted, or braided into fabrics, and are well suited for filament winding or unidirectional tape-winding equipment. Carbon yarns are characterized by higher surface area, generally

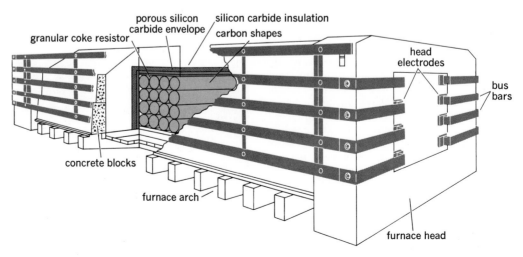

Fig. 4. Graphitizing furnace, for making synthetic graphite.

Table 2. Typical properties of carbon and graphite fabrics and felts at 20°C*

Weave:	Carbon fabric			Graphite fabric			Carbon felt Needled	Graphite felt Needled
	5 H.S.	8 H.S.	Plain	5 H.S.	8 H.S.	Plain		
Characteristics								
Width, in.	42–44	33–46	32–42	42.5–45.0	33.0–44	41.5–44.6	44	43.5
Weight, oz/yd²	8	8–9	6–8	8.0	7.2	7.4	15.7	16.6
Gage, in.	0.018	0.020	0.017–0.038	0.018	0.023	0.022	0.2–0.3	0.2
Count, yarns/in. {Warp	40	50–52	27–34	40	51	27	—	—
{Fill	36	49–51	23–32	36	49	21	—	—
Filaments/yarn, bundle	960	720	1440	960	720	1440	—	—
Filaments, diam., in.	0.0004	0.0005	0.0004	0.0003	0.0004	0.0004	0.0005	0.0005
Tensile strength, {Warp	38.0	21–59	10–100	38	85	42	2	1
lb/in. {Fill	32.0	11–30	15–90	34	75	25	2	1
Surface area, m²g	6	1	250	3	3	3	20–200	1–2
Density (He), g/cm³	1.47	1.50	1.4–1.8	1.50	1.50	1.42	.08	.08
Carbon assay, w/o	99	86–95	86–99	99.9	99.9	99.9	91–98	99.6
Ash, w/o	0.02	0.9	0.5–10	0.009	0.009	0.009	0.5	0.04
pH	7.0	2.2	7.0–10			7.9	—	—
Resistivity, 70°F, {Warp	0.40	280	—	0.41	0.46	0.38	—	0.35
ohm/yd² {Fill	0.48	270	0.5–1.6	0.42	0.50	0.50	—	0.55

*H.S. = harness satin; w/o = weight percent.

over 100 m²/g, lower thermal and electrical conductivity, and lower carbon assay than graphite yarns. They have significant moisture regain and are readily wet with resins to form low-thermal-conductivity composites. Graphite yarns have low surface area, do not absorb moisture, and are more oxidation resistant. Applications for these refractory yarns are in reentry systems and ablative structures requiring greater strength in woven form than that possible in conventional, carbonized broad goods.

Graphite yarns braided into packing for pumps and valves have the leak-free, low-maintenance performance of radial seal rings at stuffing box cost. Carbon- and graphite-fiber-filled materials are being developed for nonlubricated sleeve-bearing applications.

Carbon yarns are many times stronger than metallic wire of the same electrical resistance per unit length. The volume resistivity of carbon yarns at 20°C is about .005 ohm-cm, approximately 60 times that of nickel alloy resistance wire, and about 3500 times that of copper. Carbon yarns have a slight negative temperature coefficient of resistance; thus no current surge occurs in heating elements at start-up. An important advantage of

Table 3. Typical carbon and graphite filament and yarn properties at 20°C

Property	General-purpose carbon	General-purpose graphite	High-modulus graphite*	
Filament				
Tensile strength, lb/in.² (1-in. gage)	120,000	90,000	285,000	360,000
Tenacity, g/denier (1-in. gage)	6.2	5.3	13.7	—
Elongation at break, %	2.0	1.5	0.6	1.0
Elastic recovery, %	100	100	100	100
Modulus of elasticity, lb/in.² (1-in. gage)	6,000,000	6,000,000	50,000,000	30,000,000
Stiffness, g/denier (1-in. gage)	310	350	2300	—
Density, Hg immersion, g/cm³	1.53	1.32	1.63	1.76
Equivalent diameter, μ	9.5	8.9	6.6	6.9
Carbon assay, w/o†	90.0	98.8	99.9	92.0
Volume electrical resistivity, ohm-cm × 10⁶	6000	3500	—	—
Surface area, m²/g	130	4	1	1
Thermal conductivity, Btu/hr/ft²/°F/ft	13	22	68	12
Specific heat, 70°F, Btu/lb/°F	0.17	0.17	0.17	0.17
(mean 70 to 2700°F)	0.40	0.40	0.40	0.40
Construction	*2-ply yarn‡*	*2-ply yarn‡*	*2-ply yarn‡*	*Strand*
Yield, yd/lb	3100	4100	6355	2600
Breaking strength, lb (average, 5-in. gage)	12	6	6.4	13.5
Tenacity, g/denier (5-in. gage)	3.8	2.3	4.7	—
Elongation at break, %	2.4	1.3	0.53	1.0
Yarn diameter, in. (approx.)	0.03	0.03	0.015	0.017
pH, ASTM D-1512-60	7.0	6.0	7.5	—

*10,000-, 40,000- and 160,000-filament tows are also commercially available.

†w/o = weight percent.

‡Single-, 5-, 10-, and 30-ply yarn constructions have been made commercially.

carbon yarns is their high electrical resistance, which facilitates design of heating elements connected in parallel in which loss of a single yarn will not affect overall performance significantly.

High-modulus carbon and graphite fibers are available in multifilament yarns, strands, and tows (Table 3). Depending on precursor and processing conditions, a wide range of modulus and tensile properties can be provided. By applying stress to a rayon precursor yarn during pyrolysis, fibers with modulus from 50 to 75×10^6 psi and tensile strength from 285 to 350,000 psi can be produced. Because of the relatively high cost of these products, applications today are primarily in aerospace reentry components. Polyacrylonitrile precursors process through pyrolysis to carbon and graphite fibers with modulus from 30 to 70×10^6 psi. The 30×10^6 psi modulus fiber with tensile strength of 360,000 offers higher elongation and a balance of tension and compression in resin matrix composites. This product has found wide acceptance in aircraft and space structures, sporting goods, and industrial and construction markets. The high specific strength and modulus of the composites provide cost-effective performance in competition with metals and other fibrous composites.

Pitch is a very-low-cost precursor material. Ordinary petroleum or coal tar pitches can now be converted to high-modulus continuous fibers by a process that eliminates hot stretching of the fibers. This technical breakthrough offers the opportunity for very-low-cost high-modulus fibers to exploit volume commercial markets. High-modulus pitch-based fibers are also available in a blown filament product in mat form. They can be chopped and made into high-modulus carbon paper or used as a modulus modifier in polymeric molding compounds. These electrically conductive fibers can be used in resistance-heating applications.

Felts. Carbon and graphite felts are used as insulation and heat shields in high-temperature induction and resistance furnaces (Table 2). Graphite-tape electrical heating elements are used in temperatures up to 2500°C in metalworking furnaces. Protective atmospheres or high vacuum are necessary to prevent oxidation. These felts are better insulators than carbon black. Felt density is uniform and from 1/4 to 1/10 that of tamped carbon black. The low density of felt speeds furnace thermal response and reduces energy requirements. The felts do not develop voids characteristic of powders, thus increasing furnace life.

[R. B. FORSYTH; W. M. GAYLORD]

Bibliography: R. Bacon, *J. Appl. Phys.*, 31(2): 283–290, 1960; R. Bacon and W. H. Smith, Tensile behavior of carbonized rayon filaments at elevated temperature, *Society of Chemical Industries Second Conference on Industrial Carbon and Graphite, London, April, 1965*, 1966; E. N. Cameron and P. L. Weis, *Strategic Graphite: A Survey*, Geol. Surv. Bull. no. 1082-E, 1960; L. M. Currie et al., The production and properties of graphite for reactors, *Proc. Int. Conf. Peaceful Uses At. Energy*, 8:451–473, 1956; R. B. Forsyth, *Low cost continuous fiber from a pitch precursor*, 20th National SAMPE Symposium, 1975; J. E. Hove, Graphite as a high temperature material, *Met. Soc. AIME Trans.*, 212(1):7–13, 1958; N. J. Johnson, V. J. Nolan, and J. W. Shea, *Carbon and Graphite in the Metallurgical Industries*, 1961; J. K. Lancaster, *The Effect of Carbon Fibre Reinforcement on the Function and Wear of Polymers*, Roy. Aircraft Estab. Tech. Rep. no. 66378, 1966; National Carbon Co., *The Industrial Graphite Handbook*, 1959.

Graybody

An energy radiator which has a blackbody energy distribution, reduced by constant factor, throughout the radiation spectrum or within a certain wavelength interval. The designation "gray" has no relation to the visual appearance of a body but only to its similarity in energy distribution to a blackbody. Most metals, for example, have a constant emissivity within the visible region of the spectrum and thus are graybodies in that region. The graybody concept allows the calculation of the total radiation intensity of certain substances by multiplying the total radiated energy (as given by the Stefan-Boltzmann law) by the emissivity. The concept is also quite useful in determining the true temperatures of bodies by measuring the color temperature. For a discussion of the Stefan-Boltzmann law and color temperature *see* HEAT RADIATION. See also BLACKBODY.

[HEINZ G. SELL; PETER J. WALSH]

Heat

For the purposes of thermodynamics, it is convenient to define all energy while in transit, but unassociated with matter, as either heat or work. Heat is that form of energy in transit due to a temperature difference between the source from which the energy is coming and the sink toward which the energy is going. The energy is not called heat before it starts to flow or after it has ceased to flow. A hot object does contain energy, but calling this energy heat as it resides in the hot object can lead to widespread confusion. *See* INTERNAL ENERGY.

Heat is a result of a potential difference between the source and sink called temperature. Work is energy in transit as a result of a difference in any other potential such as height. Work may be thought of as that which can be completely used for lifting weights. Heat differs from work, the other type of energy in transit, in that its conversion to work is limited by the fundamental second law of thermodynamics, or Carnot efficiency. This natural law is that the fraction of the heat Q convertible to work is determined by the relation $dW = Q \, (dt/T)$ for processes where the source and sink are but differentially different in temperature, or by the relation $dW = dQ(T_1 - T_2)/T_1$ where the source (at T_1) and the sink (at T_2) differ by a finite temperature interval. *See* BRITISH THERMAL UNIT (BTU).

For the above relations to be valid, temperature must be expressed on a thermodynamic temperature scale. Conversely, any temperature scale for which the above relations are valid, irrespective of the substance or material under investigation, is a thermodynamic temperature scale. The perfect gas law temperature scales (degrees Kelvin equals degrees centigrade plus 273.15°, and degrees Rankine equals degrees Fahrenheit plus 456.67°) are thermodynamic temperature scales to the extent to which the constants 273.15° and 456.67° are known. The true nature of heat continues to be the subject of widespread speculation from a philosophical viewpoint.   [HAROLD C. WEBER]

Bibliography: C. O. Bennett and J. E. Myers, *Momentum, Heat, and Mass Transfer*, 1962; P. W. Bridgman, *The Nature of Thermodynamics*, 1941; H. C. Weber and H. P. Meissner, *Thermodynamics for Chemical Engineers*, 2d ed., 1957.

Heat balance

A particular form of an energy balance. The heat balance is generally useful in science but of special importance in process engineering. It is basic to the design and analysis of operating equipment since it provides a relationship between the energy terms in a process. As a simple example, the heat loss from a pipe carrying a hot fluid is difficult to measure directly but is easily calculated by a heat balance if the fluid properties at the ends of the pipe are known.

The conservation of energy requires that for any system the accumulation of energy within must equal the difference between energy entering and leaving. Energy terms are conveniently classified as: the heat Q transferred into the system across the boundary, the work W put into the system, and the energy of the material streams flowing to and from the system. The flowing streams carry internal energy U, potential energy Z by virtue of their height, kinetic energy $u^2/2g_c$, where u is velocity and g_c is the gravitational constant, and a flow work term pV which arises from forcing a volume of material V into or out of the system under the restraint of the pressure p.

Frequently, the kinetic energy, potential energy, and work terms are negligible or cancel out so that the energy balance simplifies to Eq. (1), where

$$U_1 + p_1 V_1 + Q = U_2 + p_2 V_2 + \Delta E \qquad (1)$$

subscripts 1 and 2 refer to input and output streams, respectively, and ΔE represents accumulation of energy in the system. This simplified energy balance is generally referred to as a heat balance.

The term $U + pV$ occurs in balance equations frequently and is given the name enthalpy, designated by H, so that the heat balance may be written as Eq. (2).

$$Q = H_2 - H_1 + \Delta E \qquad (2)$$

The balance in this form is termed an enthalpy balance and is especially convenient since values of H are tabulated for many of the common fluids and are easily calculated for other materials. For a steady-state flow process such as usually occurs in the continuous operation of boilers, reactors, and so forth, ΔE is zero and the heat exchange is simply equal to the enthalpy difference of the streams. *See* ENTHALPY.   [WILLIAM F. JAEP]

Heat capacity

The quantity of heat required to raise a unit mass of homogeneous material one unit in temperature along a specified path, provided that during the process no phase or chemical changes occur is known as the heat capacity of the material in question. The unit mass may be 1 g, 1 lb, or 1 gram-molecular weight (1 mole). Moreover, the path is so restricted that the only work effects are those necessarily done on the surroundings to cause the

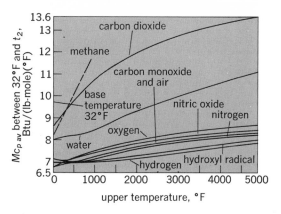

Fig. 1. Variation of average molar heat capacity with temperature for several common gases. (*Based on data of H. C. Hottel, MIT*)

change to conform to the specified path. The path, except as noted later, is at either constant pressure or constant volume. This definition conforms to an average heat capacity for the chosen unit change in temperature.

Instantaneous heat capacity at a particular temperature is defined as the rate of heat addition relative to the temperature change at the temperature in question; that is, on a plot of heat addition Q as a function of temperature T, instantaneous heat capacity is given by the slope of the curve at the temperature in question. Units of heat capacity are energy units per unit mass of material per unit change in temperature.

In accordance with the first law of thermodynamics, heat capacity at constant pressure C_p is equal to the rate of change of enthalpy with temperature at constant pressure $(\partial H/\partial T)_p$. Heat capacity at constant volume C_v is the rate of change of internal energy with temperature at constant volume $(\partial U/\partial T)_v$. Moreover, for any material, the first law yields the relation in Eq. (1). *See* ENTHALPY; INTERNAL ENERGY.

$$C_p - C_v = \left[P + \left(\frac{\partial U}{\partial V} \right)_T \right] \left(\frac{\partial U}{\partial T} \right)_P \qquad (1)$$

Gases. For one mole of a perfect gas, the preceding relation becomes $MC_p - MC_v = R$, where M

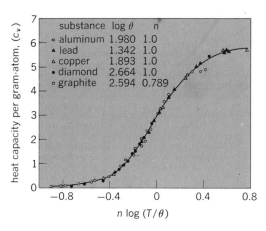

Fig. 2. Atomic heat capacity as a function of temperature (°K). (*Based on data of G. N. Lewis and G. E. Gibson, J. Amer. Chem. Soc., 39:2554–2581, 1917*)

Heat capacities of some elements

Element	Heat capacity
All heavy elements	6.4
Boron	2.7
Carbon	1.8
Fluorine	5.0
Hydrogen	2.3
Oxygen	4.0
Phosphorus and sulfur	5.4
Silicon	3.5

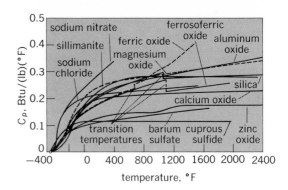

Fig. 4. Change in heat capacity of compounds with temperature. (*Based on data of K. K. Kelly, U.S. Bur. Mines Bull., no. 371, 1934*)

is the molecular weight of the gas under discussion and R is the perfect gas law constant. For gases the ratio C_p/C_v is usually designated by the symbol K.

For monatomic gases at moderate pressures, MC_v is about 3, K is about 1.67, and the heat capacity changes but little with temperature. For diatomic gases, MC_v is approximately 5 at 20°C and moderate pressures. Change of heat capacity with temperature is usually small. The value of K is between 1.40 and 1.42. For triatomic gases at moderate pressures, MC_v varies from 6 to 7 and changes rapidly with temperature. The value of K varies but is always smaller than that for the less complex molecules at the same conditions of pressure and temperature.

For gases with more than three atoms per molecule, no generalizations are reliable. However, as molecular complexity increases, heat capacity increases, the influence of temperature on heat capacity increases, and K decreases. Figure 1 shows average MC_p for several common gases. Up to pressures of a few atmospheres, the effect of pressure on heat capacity of gases is small and is usually neglected.

Solids. For solids, the atomic heat capacity (heat capacity when the unit mass under discussion is 1 at.wt) may be closely approximated by an equation of type (2), where $n=1$ for elements of simple crys-

$$C_v = J\left(\frac{T}{\theta}\right)^n \qquad (2)$$

talline form, but has a smaller value for those of more complex structures; θ is characteristic of each element; J is a function that is the same for all substances; T is absolute temperature. Figure

2 compares measured with calculated values.

For all solid elements at room temperature, C_v is about 6.4 calories per gram atom per degree Celsius. This approximation may be used when no experimental data are available, but errors may be considerable, particularly for elements with atomic weights less than 39. Kopp's law states that for solids the molal heat capacity of a compound at room temperature and pressure approximately equals the sum of heat capacities of the elements in the compound. Errors are considerable but may be reduced by judicious choice of atomic heat capacities for the lighter elements. Recommended values for some of these are given in the table of constants for Kopp's law. Use of Kopp's law is justified only when no experimental data are available.

Figures 3 and 4 give instantaneous heat capacities for some industrially important solids.

Liquids. For liquids and solutions no useful generally applicable approximations are available. For aqueous solutions of inorganic salts the approximate heat capacity of the solution may be estimated by assuming the dissolved salt to have negligible heat capacity. Thus, in a 20% by weight solution of any salt in water 0.8 would be the estimated heat capacity.

Effect of pressure on heat capacities at any temperature may be calculated by the relations in Eqs. (3a) and (3b).

$$\left(\frac{\partial MC_p}{\partial P}\right)_T = -T\left(\frac{\partial^2 V}{\partial T^2}\right)_P \qquad (3a)$$

$$\left(\frac{\partial MC_v}{\partial V}\right)_T = T\left(\frac{\partial^2 P}{\partial T^2}\right)_V \qquad (3b)$$

Constant temperature. Not so familiar as C_p and C_v are the heat necessary to cause unit change in pressure in a unit mass of material at constant temperature and the heat required to cause unit change in volume at constant temperature. These are designated $\partial Q_p/\partial P$ and $\partial Q_T/\partial V$. Similarly, $\partial Q_v/\partial P$ and $\partial Q_p/\partial V$ may be called heat capacities. See THERMODYNAMIC PRINCIPLES.

[HAROLD C. WEBER]

Bibliography: W. H. Brown, *Thermodynamics and Heat Engines*, 1964; H. C. Weber and H. P. Meissner, *Thermodynamics for Chemical Engineers*, 2d ed., 1957.

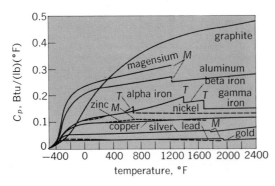

Fig. 3. Change in heat capacity of some industrially important solids with temperature. *M*= melting point; *T*= transition temperature. (*Based on data of K. K. Kelly, U.S. Bur. Mines Bull., no. 371, 1934*)

Heat exchanger

A device used to transfer heat from a fluid flowing on one side of a barrier to another fluid (or fluids) flowing on the other side of the barrier.

When used to accomplish simultaneous heat transfer and mass transfer, heat exchangers become special equipment types, often known by other names. When fired directly by a combustion process, they become furnaces, boilers, heaters, tube-still heaters, and engines. If there is a change in phase in one of the flowing fluids—condensation of steam to water, for example—the equipment may be called a chiller, evaporator, sublimator, distillation-column reboiler, still, condenser, or cooler-condenser.

Heat exchangers may be so designed that chemical reactions or energy-generation processes can be carried out within them. The exchanger then becomes an integral part of the reaction system and may be known, for example, as a nuclear reactor, catalytic reactor, or polymerizer.

Heat exchangers are normally used only for the transfer and useful elimination or recovery of heat without an accompanying phase change. The fluids on either side of the barrier are usually liquids, but they may also be gases such as steam, air, or hydrocarbon vapors; or they may be liquid metals such as sodium or mercury. Fused salts are also used as heat-exchanger fluids in some applications.

With the development and commercial adoption of large, air-cooled heat exchangers, the simplest example of a heat exhanger would now be a tube within which a hot fluid flows and outside of which air is made to flow for the purpose of cooling. By similar reasoning, it might be argued that any container of a fluid immersed in any fluid could serve as a heat exchanger if the flow paths were properly connected, or that any container of a fluid exposed to air becomes a heat exchanger when a temperature differential exists. However, engineers will insist that the true heat exchanger serve some useful purpose, that the heat recovery be meaningful or profitable.

Most often the barrier between the fluids is a metal wall such as that of a tube or pipe. However, it can be fabricated from flat metal plate or from graphite, plastic, or other corrosion-resistant materials of construction. If, as is often the case, the barrier wall is that of a seamless or welded tube, several tubes may be tied together into a tube bundle (see diagram) through which one of the fluids flows distributed within the tubes. The other fluid (or fluids) is directed in its flow in the space outside the tubes through various arrangements of passes. This fluid is contained by the heat-exchanger shell. Discharge from the tube bundle is to the head (heads) and channel of the exchanger. Separation of tube-side and shell-side fluids is accomplished by using a tube sheet (tube sheets).

Applications. Heat exhangers find wide application in the chemical process industries, including petroleum refining and petrochemical processing; in the food industry, for example, for pasteurization of milk and canning of processed foods; in the generation of steam for production of power and electricity; in nuclear reaction systems; in aircraft and space vehicles; and in the field of cryogenics for the low-temperature separation of gases. Heat exchangers are the workhorses of the entire field of heating, ventilating, air-conditioning, and refrigeration.

Classifications. The exchanger type described in general terms above and illustrated by the diagram is the well-known shell-and-tube heat exchanger. Shell-and-tube exchangers are the most numerous, but constitute only one of many types. Exchangers in use range from the simple pipe within a pipe—with a few square feet of heat-transfer surface—up to the complex-surface exchangers that provide thousands of square feet of heat-transfer area.

In between these extremes is a broad field of shell-and-tube exchangers often specifically named by distinguishing design features; for example, U tube, fin tube, fixed tube sheet, floating head, lantern-ring packed floating head, socket-and-gland packed floating head, split-ring internal floating head, pull-through floating head, nonremovable bundle with floating head or U-tube construction, and bayonet type.

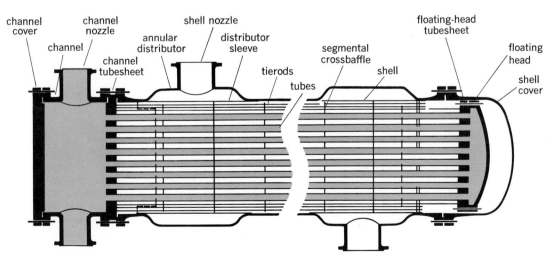

Schematic diagram of heat exchanger.

Also, varying pass arrangements and baffle-and-shell alignments add to the multiplicity of available designs. Either the shell-side or tube-side fluids, or both, may be designed to pass through the exchanger several times in concurrent, countercurrent, or cross flow to the other fluids.

The concentric pipe within a pipe (double pipe) serves as a simple but efficient heat exchanger. One fluid flows inside the smaller-diameter pipe, and the other flows, either concurrently or countercurrently, in the annular space between the two pipes, with the wall of the larger-diameter pipe serving as the shell of the exchanger.

To solve new processing problems and to find more economical ways of solving old ones, new types of heat exchangers are being developed. There has been much emphasis on cramming more heat-transfer surface into less and less volume. Extended-surface exchangers, such as those built with fin tubes, are finding wide application.

Water shortage has added a new dimension to heat-exchanger design. In its four newest refineries (all outside the United States) a large oil company has not used a single water-cooler exchanger. All are air-cooled. On the other hand, in the United States, two new plants on opposite banks of the same river disagreed on the merits of air-cooling. One chose water-cooling and the other air-cooling.

Plate-type heat exchangers, long used in the milk industry for pasteurization and skimming, are moving into the chemical and petroleum industries. Coiled tubular exchangers and coiled-plate heat exchangers are winning new assignments. Spiral exchangers offer short cylindrical shells with flat heads, carrying inlets and outlets leading to internal spiral passages. These passages may be made with spiral plates or with spiral banks of tubes. Exchangers with mechanically scraped surfaces are finding favor for use with very viscous and pastelike materials.

A somewhat unusual type of plate heat exchanger is one in which sheets of 16-gage metal, seam- and spot-welded together, are embossed to form transverse internal channels which carry the heat-transfer medium. This type is often used for immersion heating in electroplating and pickling.

Materials of construction. Every metal seems to be a possible candidate as a material of construction in fabrication of heat exchangers. Most often, carbon steels and alloy steels are used because of the strength they offer, especially when the exchanger is to be operated as a pressure vessel. Because of excellent heat conductance, brass and copper find wide use in exchanger manufacture.

Corrosion plays a key role in the selection of exchanger construction materials. Often, a high-priced material will be selected to contain a corrosive tube-side fluid, with a cheaper material being used on the less corrosive shell side.

For special corrosion problems, exchangers are built from graphite, ceramics, glass, bimetallic tubes, tantalum, aluminum, nickel, bronze, silver, and gold.

Problems of use. Each of the fluids and the barrier walls between them offers a resistance to heat transfer. However, another major resistance that must be considered in design is the formation of dirt and scale deposited on either side of the barrier wall. This resistance may become so great that the exchanger will have to be removed from service periodically for cleaning.

Chemical and mechanical methods may be used to remove the dirt and scale. For mechanical cleaning, the exchanger is removed from service and opened up. Perhaps the entire tube bundle is pulled from the exchanger shell if the plant layout has provided space for this to be done. If the deposit is on the inside of straight tubes, cleaning may be accomplished merely by forcing a long worm or wire brush through each tube.

More labor is required to remove deposits on the shell side. After removal of the tube bundle, special cleaning methods such as sandblasting may be necessary.

Much engineering effort has gone into the design of heat exchangers to allow for fouling. However, D. Q. Kern has suggested that methods are available to design heat exchangers that, by accommodating a certain amount of dirt in a thermal design, will allow heat exchangers to run forever without shutdown for cleaning. Commercial units designed in this fashion are operating today.

Another operating problem is allowance for differential thermal expansion of metallic parts. Most operating difficulties arise during the startup or shutdown of equipment. Therefore, M. S. Peters suggests the following general rules:

1. Startup. Always introduce the cooler fluid first. Add the hotter fluid slowly until the unit is up to operating conditions. Be sure the entire unit is filled with fluid and there are no pockets or trapped inert gases. Use a bleed valve to remove trapped gases.

2. Shutdown. Shut off the hot fluid first, but do not allow the unit to cool too rapidly. Drain any materials which might freeze of solidify as the exchanger cools.

3. Steam condensate. Always drain any steam condensate from heat exchangers when starting up or shutting down. This reduces the possibility of water hammer caused by steam forcing the trapped water through the lines at high velocities.

Standardization. Users have requested that heat exchangers be made available at lower prices through the standardization of designs. Organizations active in this work are the Tubular Exchangers Manufacturers' Association, American Petroleum Institute, American Standards Association, and the American Institute of Chemical Engineers. See COOLING TOWER; HEAT RADIATION; HEAT TRANSFER.

[RAYMOND F. FREMED]

Bibliography: W. E. Glausser and J. A. Cortright, How to specify heat exchangers, *Chem. Eng.*, 62(12):203–206, 1955; W. M. Kays and A. L. London, *Compact Heat Exchangers*, 2d ed., 1964; D. Q. Kern, Speculative process design, *Chem. Eng.*, 66(20):127–142, 1959; M. S. Peters, *Elementary Chemical Engineering*, 1954; J. C. Smith, Trends in heat exchangers, *Chem. Eng.*, 61(6):232–238, 1953.

Heat pump

The thermodynamic counterpart of the heat engine. A heat pump raises the temperature level of heat by means of work input. In its usual form a compressor takes refrigerant vapor from a low-pressure, low-temperature evaporator and delivers it at high pressure and temperature to a condenser (Fig. 1). The pump cycle is identical with the customary vapor-compression refrigeration system. *See* REFRIGERATION CYCLE.

Application to comfort control. For air-conditioning in the comfort heating and cooling of space, a heat pump uses the same equipment to cool the conditioned space in summer and to heat it in winter, maintaining a generally comfortable temperature at all times (Fig. 2).

This dual purpose is accomplished, in effect, by placing the low-temperature evaporator in the conditioned space during the summer and the high-temperature condenser in the same space during the winter (Fig. 3). Thus, if 70°F is to be maintained in the conditioned space regardless of the season, this would be the theoretic temperature of the evaporating coil in summer and of the condensing coil in winter. The actual temperatures on the refrigerant side of these coils would need to be below 70°F in summer and above 70°F in winter to permit the necessary transfer of heat through the coil surfaces.

If the average outside temperatures are 100°F in summer and 40°F in winter, the heat pump serves to raise or lower the temperature 30° and to deliver the heat or cold as required. The ultimate ideal cycle for estimating performance is the same Carnot cycle as that for heat engines. The coefficient of performance cp_c as a cooling machine is given in Eq. (1), and the coefficient cp_w as a warming machine is given in Eq. (2), where T is temperature in

$$cp_c = \frac{\text{refrigeration}}{\text{work}} = \frac{T_c}{T_h - T_c} \quad (1)$$

chine is given in Eq. (2), where T is temperature in

$$cp_w = \frac{\text{heat delivered}}{\text{work}} = \frac{T_h}{T_h - T_c} \quad (2)$$

degrees absolute and the subscripts c and h refer to the cold and hot temperatures, respectively.

For the data cited, the theoretical coefficients of performance are as in Eqs. (3). The significance of

$$cp_c = \frac{460 + 70}{(460 + 100) - (460 + 70)} = 17.7$$

$$cp_w = \frac{460 + 70}{(460 + 70) - (460 + 40)} = 17.7 \quad (3)$$

these coefficients is that ideally for 1 kilowatt-hour (kwhr) of electric energy input to the compressor there will be delivered $3413 \times 17.7 = 60,000$ Btu/hr as refrigeration or heating effect as required. This is a great improvement over the alternative use of resistance heating, typically, where 1 kwhr of electric energy would deliver only 3413 Btu. The heat pump uses the second law of thermodynamics to give a much more substantial return for each kilowatt-hour of electric energy input. The electric energy serves to move heat already present to a desired location.

Effect of seasonal loads. With the prevalent acceptance of summer comfort cooling of space, it is entirely possible and practical to use the same compressor equipment and coils for winter heating and for summer cooling and to dispense with the need for direct-fired apparatus using oil, gas, coal, or wood fuel.

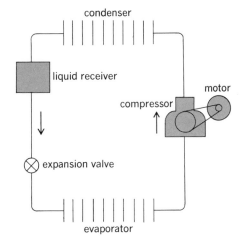

Fig. 1. Basic flow diagram of heat pump with motor-driven compressor. For summer cooling, condenser is outdoors and evaporator indoors; for winter heating, condenser is indoors and evaporator outdoors.

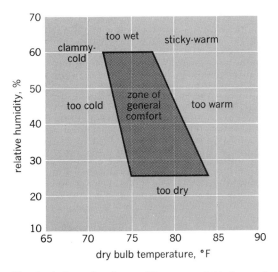

Fig. 2. Indoor climatic conditions acceptable to most people when doing desk work; continuous air motion with 5–8 air changes per hour.

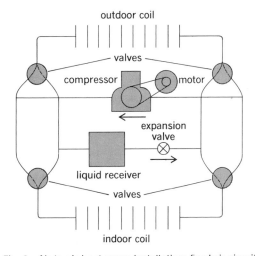

Fig. 3. Air-to-air heat pump installation; fixed air circuit with valves in the summer positions (the broken lines show the winter positions).

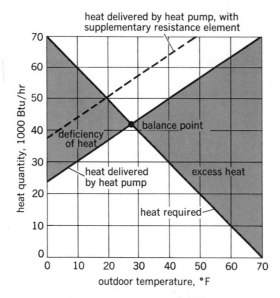

Fig. 4. Performance of air-to-air, self-contained, domestic heat pump on the heating cycle.

For an economical installation, equipment must be of correct size for both the summer cooling and winter heating loads. Climatic conditions have a significant influence and can lead to imbalance on sizing. If the heating and cooling loads are equal, the equipment can be selected with minimum investment. However, generally the loads are not balanced; in the temperate zone the heating load is usually greater than the cooling load. This necessitates (1) a large, high horsepower compressor fitted to the heating demand, (2) a supplementary heating system (electrical resistance or fuel), or (3) a heat-storage system.

If well water or the ground serves as the heat source, the imbalance is less severe than when atmospheric air is the source. However, the uncertain heat transfer rates with ground coil, the impurities, quality, quantity, and disposal of water, and the corrosion problems mitigate the use of these sources.

Atmospheric air as a heat source is preferable, particularly with smaller domestic units. A self-contained, packaged unit of this type offers maximum dependability and minimal total investment. The performance of such a unit for the heating cycle is illustrated in Fig. 4. Curves of heat required and heat available show the limitations on capacity. The heat delivered by the pump is less than the heat required at low temperatures, so that there is a deficiency of heat when the outside temperature goes below, in this case, about 28°F. The intersection of the two curves is the balance point. There is an area of excess heat to the right and of deficiency of heat to the left of the balance point. Many devices and methods are offered to correct this situation, such as storage systems, supplementary heaters, and compressors operating alternatively in series or in parallel.

In temperate regions heat-pump installations achieve coefficients of performance in the order of 3 on heating loads when all requirements for power, including auxiliary pumps, fans, resistance heaters, defrosters, and controls are taken into account. Automatic defrosting systems, when air

is the heat source, are essential for best performance, with the defrost cycle occurring twice a day.

Heat pumps are uneconomical if used for the sole purpose of comfort heating. The direct firing of fuels is generally more attractive from an overall financial viewpoint. The investment in heat-pump equipment is higher than that for the conventional heating system. Unless the price of electric energy is sufficiently low or the price of fuels very high, the heat pump cannot be justified solely as a heating device. However, if there is also need for comfort cooling of the same space in summer, the heat pump, to do both the cooling and heating, becomes attractive. This use of the heat pump will probably increase with the broad acceptance of summer air conditioning represented by the sale of more than 1,000,000 units a year in the United States.

One can also expect that the heat pump will find increasing acceptance for operation in conjunction with solar heating systems. The efficiency of a solar collector rises rapidly with lower collection temperatures so that an electrically driven heat pump can use such a heat source with a high overall coefficient of performance. *See* SOLAR COLLECTORS.

The heat pump is also used for a wide assortment of industrial and process applications such as low-temperature heating, evaporation, concentration, and distillation. [THEODORE BAUMEISTER]

Bibliography: E. R. Ambrose, *Heat Pumps and Electric Heating*, 1966; T. Baumeister (ed.), *Standard Handbook for Mechanical Engineers*, 7th ed., 1967; W. H. Carrier et al., *Modern Air-Conditioning, Heating, and Ventilating*, 3d ed., 1959; C. W. MacPhee (ed.), *Guide and Data Book: Fundamentals and Equipment, Applications*, American Society of Heating, Refrigerating, and Air-Conditioning Engineers, 1966; C. Strock and R. L. Koral (eds.), *Handbook of Heating, Air-Conditioning and Ventilation*, 2d ed., 1965.

Heat radiation

The energy radiated by solids, liquids, and gases as a result of their temperature. Such radiant energy is in the form of electromagnetic waves and covers the entire electromagnetic spectrum, extending from the radio-wave portion of the spectrum through the infrared, visible, ultraviolet, x-ray, and γ-ray portions. From most hot bodies on Earth this radiant energy lies largely in the infrared region. *See* ELECTROMAGNETIC RADIATION; INFRARED RADIATION.

Radiation is one of the three basic methods of heat transfer, the other two methods being conduction and convection. *See* HEAT TRANSFER.

A hot plate at 400°K (261°F) may show no visible glow; but a hand which is held over it senses the warming rays emitted by the plate. A temperature of more than 1000°K is required to produce a perceptible amount of visible light. At this temperature a hot plate glows red and the sensation of warmth increases considerably, demonstrating that the higher the temperature of the hot plate the greater the amount of radiated energy. Part of this energy is visible radiation, and the amount of this visible radiation increases with increasing temperature. A steel furnace at 1800°K shows a strong yellow glow. If a tungsten wire (used as the filament in incandescent lamps) is raised by resistance heating to a temperature of 2800°K, it emits a bright white light. As the temperature of a substance increases, additional colors of the visible portion of the spectrum appear, the sequence being first red, then yellow, green, blue, and finally violet. The violet radiation is of shorter wavelength than the red radiation, and it is also of higher quantum energy.

In order to produce strong violet radiation, a temperature of almost 3000°K is required. Ultraviolet radiation necessitates even higher temperatures, and there is no solid on Earth which can withstand such temperatures without melting. The Sun emits considerable ultraviolet radiation, as evidenced by the sunburn it produces. The spectral distribution of the Sun's radiation has been measured, and the temperature of the Sun's surface has been determined from Wien's displacement law and corresponds to about 6000°K (Wien's law is discussed later). Such temperatures have been produced on Earth in gases ionized by electrical discharges. The mercury-vapor lamp used on highways, the fluorescent lamp used in offices, and the xenon compact-arc lamp used in searchlights are good examples of such gas discharges. They emit large amounts of ultraviolet radiation. Temperatures up to 20,000°K, however, are still much too low to produce x-rays or γ-radiation. Approaches to the utilization of nuclear energy have made use of the fusion of deuterons in magnetically constricted arcs at extremely high currents. By this means, temperatures above 1×10^6 °K have been obtained for small fractions of a second. These devices require enormous amounts of energy to produce such high temperatures. A gas maintained at such temperatures emits x-rays and γ-rays. *See* FUSION, NUCLEAR.

Theory. The emission of radiation is explained in terms of excited atoms and nuclei. For example, electrons in an atom can be ejected from their normal orbits around the atom into those farther from the nucleus. When this happens the atom is said to be in an excited state. This occurs when energy, supplied from outside a substance, is converted into thermal motion and finally into excitation. A short time after excitation, the electrons return to their normal orbits and give off their excess energy ΔE in the form of radiation of a particular frequency ν. This wavelength may be determined by the relation $\Delta E = h\nu$, where h is Planck's constant.

In a gas, the thermal motion consists of substantially unhindered movement of the individual particles with different velocities. In a solid, on the other hand, the thermal motion is an oscillating movement of the particles, with varying displacements, about their fixed positions. The extent of the thermal motion depends upon the temperature. The hotter the substance, the greater the thermal motion and the higher the intensity and energy of the radiation. An energy distribution of the radiation intensity results, for example, from the distribution of velocities of the particles in a gas or from the distribution of displacements of the particles about their positions in a solid.

Further, the maximum available energy (excitation energy) depends upon temperature, and this explains why the energy of emitted radiation shifts to shorter wavelength (that is, higher

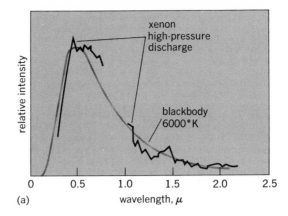

(a)

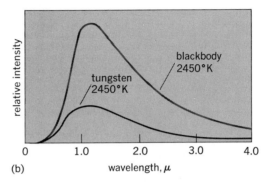

(b)

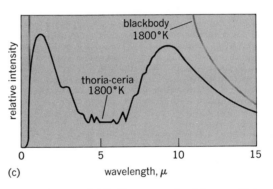

(c)

Fig. 1. Energy distribution curves for (a) xenon high-pressure discharge (from W. Meyer, ed., *Technischwissenschaftliche Abhandlungen aus dem Osramkonzern*, vol. 6, Springer, 1953); (b) tungsten (from *Amer. Inst. Mining Met. Eng., Pyrometry: The Papers and Discussions of a Symposium on Pyrometry, 1920*); (c) a typical ceramic (from R. W. Pohl, *Einführung in die Optik*, Springer, 1948).

energy) as the temperature is increased. For instance, a temperature of 1000°K produces just enough excitation energy for the dark red glow which contains the longest wavelengths within the visible portion of the spectrum. As explained before, higher temperatures or greater excitation energies are necessary to excite measurable quantities of the shorter wavelength regions. It is obvious that with decreasing temperatures, less excitation energy is available, and the amount of heat radiation decreases until finally at the absolute zero of temperature (0°K) substances radiate no energy because all atomic motion has ceased.

The radiated energy per second is commonly expressed in terms of joules per second, or watts. Other units often used are ergs per second or calories per second. These are related to each other as follows: 1 watt = 1 joule/sec = 10^7 erg/sec = 0.239 cal/sec. For instance, the Sun radiates onto 1 cm² of the Earth's surface 2 cal/min or $\frac{1}{30}$ cal/sec or about $\frac{1}{7}$ watt. The total energy radiated from 1 cm² of a tungsten wire in an incandescent lamp at 2800°K is 112 watts. The same wire at room temperature emits only 0.0015 watt.

Energy distribution curves. In order to evaluate the usefulness of a heat radiator, energy distribution curves are used. These are graphs of relative or absolute radiated energy versus the wavelength of radiation (expressed in microns, 10^{-6} m, or in angstroms, 10^{-10} m) or the frequency (velocity of light/wavelength) expressed in hertz (cycles per second).

Such graphs show how the energy radiated from a substance at a certain temperature is distributed over the various portions of the spectrum. The usefulness of these graphs lies in the fact that they provide information, for example, on the effectiveness of a radiator as a light source or as a heating element. Furthermore, the area under the energy distribution curve is equivalent to the total radiated energy. The energy distribution curves for tungsten metal, thoria plus 1% ceria (a ceramic), and the xenon high-pressure electrical gas discharge are illustrated in Fig. 1.

The energy distribution of various substances differs because of their internal properties and their surface condition. As a common rule, which holds well above 30,000 angstroms (A) or 3 microns (μ), substances with good electrical conductivity, especially metals, are poor emitters of radiation and are good reflectors (for example, silver or aluminum). Insulators radiate strongly in the infrared region of the spectrum and have gaps of low radiation intensity near the visible portion of the spectrum, as shown in Fig. 1c. These gaps are due to the electronic band structure of insulators. Roughening the surface of all radiators increases the emitted energy. This is true because tiny holes in the surface act as cavity radiators, radiating almost blackbody energy. (Cavity radiators and blackbody radiation are discussed later.)

Energy distribution curves are obtained by passing white (heterochromatic) light through a monochromator (quartz prism, grating, and the like) and measuring the spectral intensity of radiation with a phototube or a thermopile. The measured intensities at the various wavelengths are then plotted either as percent of the maximum intensity (relative energy) or as absolute intensity (absolute energy) versus the wavelength.

A radiator used in heating rooms should produce much infrared radiation (heat) and no light, whereas much visible light and little heat is desired from a light source. Unfortunately, an energy distribution curve gives a true picture of the radiator for one particular temperature only. If more information is needed, a set of such curves would have to be provided for the temperature range of interest.

Blackbody radiation. Because of the tedious experimental work involved in determining such curves, a different and more fruitful approach is generally taken. Two quantities characterize a heat radiator completely: the total emissivity and the spectral emissivity, which are designated by ϵ and ϵ_λ, respectively, where the subscript λ designates wavelength. Both emissivities, in conjunction with the radiation properties of a blackbody, describe fully the behavior of a real heat radiator. The radiation properties of a blackbody are completely stated by Planck's radiation law.

Planck's law and the concept of blackbody radiation are of utmost importance for the understanding of heat radiation. The blackbody signifies in the domain of heat radiation what any other standard, such as the standard meter, signifies in its own domain. A blackbody is defined as a body which emits the maximum amount of heat radiation. Although there exists no perfect blackbody radiator in nature, it is possible to construct one on the principle of cavity radiation. *See* BLACKBODY.

A cavity radiator is usually understood to be a heated enclosure with a small opening which allows some radiation to escape or enter. The escaping radiation from such a cavity has the same characteristics as blackbody radiation.

As can be seen from Fig. 2, radiation energy which enters the cavity is almost completely absorbed because of the multiple reflections it encounters. This follows because at each reflecting point some of the energy is absorbed by the walls. The absorptivity of the cavity hole is essentially unity, independent of the wall material. As a consequence of Kirchhoff's law, the emissivity of the cavity is also unity, and this fulfills the definition of a blackbody radiator. In practice, the cavity is approximated by a small hole or even a wedge cut into a surface.

Kirchhoff's law. This law correlates mathematically the heat radiation properties of materials at thermal equilibrium. It is often called the second law of thermodynamics for radiating systems.

Kirchhoff's law can be expressed as follows: The ratio of the emissivity of a heat radiator to the absorptivity of the same radiator is a function of frequency and temperature alone. This function is the same for all bodies, and it is equal to the emissivity of a blackbody. When ϵ is the emissivity of a real radiator, α its absorptivity, and $E = 1$ the emissivity of a blackbody, Kirchhoff's law takes the form of Eq. (1).

$$\epsilon/\alpha = E = 1 \qquad (1)$$

A substance, when brought without contact into an evacuated enclosure the walls of which are at a constant but higher temperature than the body, will assume the wall temperature after some time. However, it will not exceed it. Under these con-

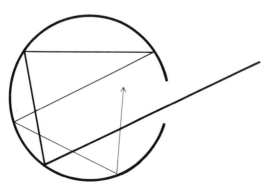

Fig. 2. Cavity radiator.

ditions, the exchange of energy can take place only by radiation. As the test body receives radiation from the walls it will absorb some of it, transforming it into motion of its elementary particles, and thereby raising its own temperature. Thermal equilibrium is obtained when the temperature of the walls and the test body is the same; in this case the test body must emit as much energy as it receives. If it absorbs all the impinging radiation, it is a blackbody. If it absorbs only a fraction of the impinging radiation, the other part must be reflected in order to maintain the equilibrium. These statements require that the absorptivity be equal to the emissivity. This is the form in which Kirchhoff's law is often stated. For opaque bodies, absorptivity plus reflectivity must be equal to unity, and therefore the emissivity and the absorptivity respectively must be unity minus the reflectivity. A consequence of Kirchhoff's law is the postulate that a blackbody has an emissivity which is greater than that of any other body.

Planck's radiation law. This celebrated law represents mathematically the energy distribution of the heat radiation from 1 cm² of surface area of a blackbody at any temperature. It is the only heat radiation law which is accurate throughout the entire spectrum. The basis of Planck's radiation law was experimental data obtained from measurements on cavity radiators.

Planck's radiation law has great importance. Formulated by Max Planck early in the 20th century, it laid the foundation for the advance of modern physics and the advent of quantum theory. In determining the heat radiation of hot bodies Planck's radiation law is a basic tool in research and development, both in science and industry. The radiation law can be used to predict light output of incandescent lamps, the cooling time of molten steel, heat dissipation of nuclear reactors, the energy radiated from the Sun, the temperature of the stars, and many other important applications.

Although Planck's law can be derived on theoretical grounds alone, it was deduced from experiment. Prior attempts to calculate the heat radiation of a blackbody had described the radiation as consisting of electromagnetic waves whose energy content could vary continuously. Those attempts did not match the experimental results. Planck replaced the concept of continuous energy with the idea that the energy existed in bundles; that is, the energy was quantized.

This concept was a drastic innovation at that time. However, upon calculating the radiation, Planck found that the expression in Eq. (2) de-

$$R_\lambda = 37{,}410/\lambda^5(e^{14.39/\lambda T} - 1) \qquad (2)$$

scribed the experimental results completely. This is the mathematical expression of Planck's radiation law, where R_λ is the total energy radiated from the body measured in watts per square centimeter per unit wavelength, at the wavelength λ. The wavelength in this formula is measured in microns. The quantity T is the temperature in degrees Kelvin, and e is the base of the natural logarithms. Figure 3 presents graphs of Planck's law for various temperatures and shows substances which attain these temperatures. Note that these substances will not radiate as predicted by Planck's law since they are not blackbodies themselves.

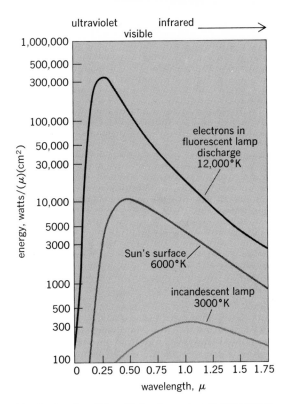

Fig. 3. Graphs of Planck's law for various temperatures.

As can be seen from Fig. 3, the radiation increases at every point of the energy spectrum as the temperature is increased. At all temperatures, the energy radiated at the extremes of the energy spectrum approaches zero and has a maximum at some place in between. The total area under any of the curves, when plotted as in Fig. 1, measures the total energy radiated by the body at the temperature represented by the curve.

Three important aspects of Planck's radiation law can be examined. First, the behavior of the law at the extremes of the energy spectrum leads to a discussion of Wien's radiation law, the Rayleigh-Jeans radiation law, and the so-called ultraviolet catastrophe. Second, the shift of the wavelength at which the maximum energy is radiated can be studied as the temperature is changed. This leads to Wien's displacement law. Finally, the total amount of energy radiated at any temperature can be investigated. This leads to the Stefan-Boltzmann law. The four laws mentioned were well known prior to the formulation of Planck's law. It is the fact that the Planck law so neatly sums up the four earlier laws and introduces the implication of energy quantization which made it of such importance in the development of modern physics during the 20th century.

Rayleigh-Jeans law. The heat radiation from a blackbody at long wavelengths is adequately described by the Rayleigh-Jeans radiation law. For larger values of λT, Planck's law simplifies to the Rayleigh-Jeans law, as shown in Eq. (3). This law

$$R_\lambda = 2560\,T/\lambda^4 \qquad (3)$$

states that the energy radiated at any temperature increases without limit as the wavelength decreases. As can be seen from Figs. 1 and 3, this law can

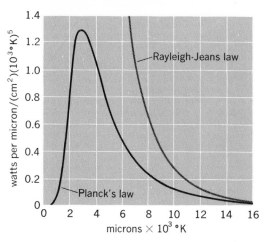

Fig. 4. Planck's law expressed as Wien's displacement law. Rayleigh-Jeans law is shown for comparison.

be accurate only for wavelengths much larger than that at which the maximum occurs. For wavelengths shorter than this maximum, the energy radiated from a blackbody actually decreases again. If a blackbody acted as predicted by the Rayleigh-Jeans law, then the energy radiated at very short wavelengths, in the ultraviolet region, would become extremely large and the total energy radiated would be infinite. This is known as the ultraviolet catastrophe, and would be valid at any temperature, no mater how low.

Wien's radiation law. This law is valid at short wavelengths and is obtained from Planck's law by taking λT as very small. Planck's formula then becomes Eq. (4). This law is accurate in the visible region of the spectrum below 3000°K.

$$R_\lambda = \frac{37{,}410}{\lambda^5} e^{-14.39/\lambda T} \qquad (4)$$

Wien's displacement law. This law is obtained from Planck's law by the process of differentiation. It describes the shift with temperature of the wavelength at which the maximum amount of radiation occurs by Eq. (5). Thus, the product of the

$$\lambda_{max} T = 2898 \qquad \text{(micron-degrees)} \qquad (5)$$

temperature of a blackbody and the wavelength at which the maximum amount of radiation occurs is a constant. Wien's law has wider significance than this, however. Dividing Planck's law by T^5 results in Eq. (6). On the right-hand side of Eq. (6) the

$$R_\lambda / T^5 = 37{,}410/(\lambda T)^5 (e^{14.39/\lambda T} - 1) \qquad (6)$$

wavelength and temperature always appear multiplying each other. This means that only one curve is needed to express Planck's law, for all temperatures, if a graph is used in which the radiation energy divided by the fifth power of the temperature is plotted versus the product of wavelength and temperature. This is illustrated in Fig. 4. It is helpful here to measure temperature in thousands of degrees. For comparison, the Rayleigh-Jeans law is also illustrated.

Wien's displacement law is helpful in determining the temperature of hot bodies. If λ_{max} can be measured, the temperature is immediately obtained from the displacement law. This is how an astronomer measures the temperature of a star. As an example, the radiation from the Sun's surface has a maximum in the green region of the energy spectrum in the vicinity of $\lambda = 0.5 \mu$ (Fig. 3). From Wien's law the surface temperature of the Sun must then be about 6000°K.

Stefan-Boltzmann law. This law states that the total energy radiated from a hot body increases with the fourth power of the temperature of the body. This law can be derived from Planck's law by the process of integration and is expressed mathematically as Eq. (7), where R_T is the total

$$R_T = 5.669 \times 10^{-12} T^4 \qquad (7)$$

amount of energy radiated from a blackbody in watts per square centimeter. When R_T is multiplied by the total emissivity, the total energy radiated from a real heat radiator is obtained.

The rapid increase in heat radiation with temperature is quite evident from the Stefan-Boltzmann law. If the absolute temperature is doubled, say from 273 to 546°K (32 to 524°F), then the energy radiated increases sixteenfold. Thus, the attainment of very high temperatures requires large amounts of energy to overcome the loss of energy by heat radiation. Temperatures greater than 3×10^7 °K are encountered in a hydrogen bomb explosion. Such temperatures are 100,000 times higher than room temperature. Therefore, the energy radiated by 1 cm² of a substance at this high temperature will be 1×10^{20} times as much as that radiated at room temperature by the same substance. This energy, if radiated by a blackbody, would boil 2×10^7 tons of ice water in 1 sec.

Temperature determination. The apparent temperature of a real heat radiator can be determined by comparison with a blackbody whose temperature is known. This is done in any one of three customary ways, based upon the radiation laws described.

1. The radiation temperature of a surface is the temperature of the blackbody which radiates the same total energy per unit area in 1 sec as does the surface. This temperature is based upon the Stefan-Boltzmann law.

2. The brightness temperature of a surface is the temperature of the blackbody which has the same brightness at a certain wavelength as has the surface. The wavelength is normally taken as 0.655 μ, and the brightness temperature thus measured can be used in conjunction with the spectral emissivity $\epsilon_{0.655}$ and Wien's radiation law to calculate the true temperature of the body.

3. The color temperature of a surface is the temperature of a blackbody whose radiation has the same or approximately the same energy distribution as the surface. This is illustrated in Fig. 1a, where it is seen that the xenon high-pressure discharge has a color temperature approximating 6000°K. For a graybody, the color temperature and true temperature are the same. *See* GRAYBODY.

[HEINZ G. SELL; PETER J. WALSH]

Bibliography: M. Born, *Atomic Physics,* 6th ed., 1957; S. Fluegge (ed.), *Handbuch der Physik,* vol. 26, 1958; G. Joos, *Theoretical Physics,* 3d ed., 1958; A. L. King, *Thermophysics,* 1962; R. L. Weber, *Heat and Temperature Measurement,* 1950; M. R. Wehr and J. A. Richards, Jr., *Physics of the Atom,* 1960.

Heat transfer

Heat, a form of kinetic energy, is transferred in three ways: conduction, convection, and radiation. Heat can be transferred only if a temperature difference exists, and then only in the direction of decreasing temperature. Beyond this, the mechanisms and laws governing each of these ways are quite different. This article gives introductory information on the three types of heat transfer (also called thermal transfer) and on important industrial devices called heat exchangers.

Conduction. Heat conduction involves the transfer of heat from one molecule to an adjacent one as an inelastic impact in the case of fluids, as oscillations in solid nonconductors of electricity, and as motions of electrons in conducting solids such as metals. Heat flows by conduction from the soldering iron to the work, through the brick wall of a furnace, through the wall of a house, or through the wall of a cooking utensil. Conduction is the only mechanism for the transfer of heat through an opaque solid. Some heat may be transferred through transparent solids, such as glass, quartz, and certain plastics, by radiation. In fluids, the conduction is supplemented by convection, and if the fluid is transparent, by radiation.

The conductivities of materials vary widely, being greatest for metals, less for nonmetals, still less for liquids, and least for gases. Any material which has a low conductivity may be considered to be an insulator. Solids which have a large conductivity may be used as insulators if they are distributed in the form of granules or powder, as fibers, or as a foam. This increases the length of path for heat flow and at the same time reduces the effective cross-sectional area, both of which decrease the heat flow. Mineral wool, glass fiber, diatomaceous earth, glass foam, Styrofoam, corkboard, Celotex, and magnesia are all examples of such materials.

Convection. Heat convection involves the transfer of heat by the mixing of molecules of a fluid with the body of the fluid after they have either gained or lost heat by intimate contact with a hot or cold surface. The transfer of heat at the hot or cold surface is by conduction. For this reason, heat transfer by convection cannot occur without conduction. The motion of the fluid to bring about mixing may be entirely due to differences in density resulting from temperature differences, as in natural convection, or it may be brought about by mechanical means, as in forced convection.

Most of the heat supplied to a room from a steam or hot-water radiator is transferred by convection. In fact, the heat from the fire in the furnace heating the hot water or steam is transferred to the boiler wall by convection, and the hot water or steam transfers heat from the boiler to the radiator by convection. Iced tea is cooled and soup heated by convection.

Radiation. Solid material, regardless of temperature, emits radiations in all directions. These radiations may be, to varying degrees, absorbed, reflected, or transmitted. The net energy transferred by radiation is equal to the difference between the radiations emitted and those absorbed.

The radiations from solids form a continuous spectrum of considerable width, increasing in intensity from a minimum at a short wavelength through a maximum and then decreasing to a minimum at a long wavelength. As the temperature of the object is increased, the entire emitted spectrum decreases in wavelength. As the temperature of an iron bar, for example, is raised to about 1000°F, the radiations become visible as a dark red glow. As the temperature is increased further, the intensity of the radiation increases and the color becomes more blue. This process is quite apparent in the filament of a light bulb. When the bulb is operated at less than normal voltage, the light appears quite red. As the voltage is increased, the filament temperature increases and the light progressively appears more blue.

Liquids and gases only partially absorb or emit these radiations, and do so in a selective fashion. Many liquids, especially organic liquids, have selective absorption bands in the infrared and ultraviolet regions.

Transfer of energy by radiation is unique in that no conducting substance is necessary, as with conduction and convection. It is this unique property that makes possible the transfer of large amounts of energy from the Sun to the Earth, or the transfer of heat from a radiant heater in the home. It is the ready transfer of heat by radiation from a California orange grove to outer space on a clear night that sometimes results in a frost. The presence of a shield of clouds will tend to prevent this loss of heat and often prevent the frost. By means of heat lamps and gold-plated reflectors, heat may be transferred deep into the layer of enamel on a car body, with resultant hardening of the enamel from the inside out. It is also the transfer over great distance of quantities of radiant energy that makes the atomic bomb so destructive.

Design considerations. By utilizing a knowledge of the principles governing the three methods of heat transfer and by a proper selection and fabrication of materials, the designer atttempts to obtain the heat flow required for his purposes. This may involve the flow of large amounts of heat to some point in a process or the reduction in flow in others. It is possible to employ all three methods of heat transfer in one process. In fact, all three methods operate in processes that are commonplace. In summer, the roof on a house becomes quite hot because of radiation from the Sun, even though the wind is carrying some of the heat away by convection. Conduction carries the heat through the roof where it is distributed to the attic by convection. The prudent householder attempts to reduce the heat that enters the rooms beneath by reducing the heat that is absorbed in the roof by painting the roof white. He may apply insulation to the underside of the roof to reduce the flow of heat through the roof. Further, heated air in the attic may be vented through louvers in the roof.

Heat transferred by convection may be transferred as heat of the convecting fluid or, if a phase change is involved, as latent heat of vaporization, solidification, sublimation, or crystallization. The human body can be cooled to less than ambient temperature by evaporation of sweat from the skin. Dry ice absorbs heat by sublimating the carbon dioxide. Heat extracted from the products of combustion in the boiler flows through the gas film and the metal tube wall and converts the water inside the tube to steam, all without greatly changing the temperature of the water.

Heat exchangers. In industry it is generally desired to extract heat from one fluid stream and add it to another. Devices used for this purpose have passages for each of the two streams separated by a heat-exchange surface in the form of plates or tubes and are known as heat exchangers. Needless to say, the automobile radiator, the hot-water heater, the steam or hot-water radiator in a house, the steam boiler, the condenser and evaporator on either the household refrigerator or air conditioner, and even the ordinary cooking utensils in everyday use are all heat exchangers. In power plants, oil refineries, and chemical plants, two commonly used heat exchangers are the tube-and-shell and the double-pipe exchangers. The first consists of a bundle of tubes inside a cylindrical shell. One fluid flows inside the tubes and the other between the tubes and the shell. The double-pipe type consists of one tube inside another, one fluid flowing inside the inner tube and the other flowing in the annular space between tubes. In both cases, the tube walls serve as the heat-exchange surface. Heat exchangers consisting of spaced flat plates with the hot and cold fluids flowing between alternate plates are also in use. Each of these exchangers essentially depends upon convection heat flow through a film on each side of the heat-exchange surface and conduction through the surface. Countless special modifications, often also utilizing radiation for heat transfer, are in use in industry.

In these exchangers, the fluid streams may flow parallel concurrently or in mixed flow. In most cases, the temperatures of the various streams remain essentially constant at a given point, and the process is said to be a steady-state process. As the streams move through the exchangers, unless there is a phase change, the fluids are continuously changing in temperature, and the temperature gradient from one stream to the other may be continuously varying. To determine the amount of surface needed for a given process, the designer must evaluate the effective temperature gradient for the particular condition and exchanger.

With extremely high temperatures, or with gas streams carrying suspended solids, the use of conventional heat exchangers becomes impractical. Under these conditions, the transfer of heat from one stream to another becomes more economical by the alternate heating and cooling of refractory solids or by checkerwork as in the blast-furnace hot stove, in the glass-furnace regenerator, or in the Royster stove. At lower temperatures, metal packing is frequently employed, as in the Ljungstrom preheaters or in regenerators for liquid-air production. In petroleum refining and in the metallurgical industry, exchangers are being employed in which one or more of the streams are fluidized beds of solids, the large area of the solids tending to produce very high rates of heat exchange. In some of these devices and also in nuclear power reactors, large quantities of heat are being generated in the exchangers. Here one of the principal problems involves the rapid removal of this heat before the temperature rises to the point where the equipment is damaged or destroyed.

Often the heating or cooling of a body is desired. In this case, the body representing the second stream does not remain at constant temperature, the heat being transferred representing a change in the heat content of the body. Such a process is known as an unsteady-state process. The heating or cooling of food and canned products in utensils, refrigerators, and sterilizers; the heating of steel billets in metallurgical furnaces; the burning of brick in a kiln; and the calcination of gypsum are examples of this type of process. *See* HEAT; HEAT EXCHANGER; HEAT RADIATION.

[RALPH H. LUEBBERS]

Bibliography: G. M. Dusinberre, *Numerical Analysis of Heat Flow*, 1949; E. R. G. Eckert, *Introduction to Heat and Mass Transfer*, 1963; H. C. Hottel and E. S. Cohen, Radiant heat exchange in gas filled enclosure, *A.I.Ch.E. J.*, 4(1):3–14, 1958; M. Jakob, *Heat Transfer*, vol. 1, 1949; D. Q. Kern, *Process Heat Transfer*, 1950; J. G. Knutson and D. L. Katz, *Fluid Dynamics and Heat Transfer*, 1958; B. Lubarsky and S. J. Kaufman, Review of experimental investigations of liquid-metal heat transfer, *Nat. Adv. Comm. Aeronaut. Tech. Notes*, 3336:115, 1955; W. H. McAdams, *Heat Transmission*, 3d ed., 1954.

Heating, comfort

The maintenance of the temperature in a closed volume, such as a home, office, or factory, at a comfortable level during periods of low outside temperature. Two principal factors determine the amount of heat required to maintain a comfortable inside temperature: the difference between inside and outside temperatures and the ease with which heat can flow out through the enclosure.

Heating load. The first step in planning a heating system is to estimate the heating requirements. This involves calculating heat loss from the space, which in turn depends upon the difference between outside and inside space temperatures and upon the heat transfer coefficients of the surrounding structural members.

Outside and inside design temperatures are first selected. Ideally, a heating system should maintain the desired inside temperature under the most severe weather conditions. Economically, however, the lowest outside temperature on record for a locality is seldom used. The design temperature selected depends upon the heat capacity of the structure, amount of insulation, wind exposure, proportion of heat loss due to infiltration or ventilation, nature and time of occupancy or use of the space, difference between daily maximum and minimum temperatures, and other factors. Usually the outside design temperature used is the median of extreme temperatures.

The selected inside design temperature depends upon the use and occupancy of the space. Generally it is between 66 and 75°F.

The total heat loss from a space consists of losses through windows and doors, walls or partitions, ceiling or roof, and floor, plus air leakage or ventilation. All items but the last are calculated from $H_l = UA(t_i - y_o)$, where heat loss H_l is in British thermal units per hour, U is overall coefficient of heat transmission from inside to outside air in Btu/(hr)(ft²)(°F), A is inside surface area in square feet, t_i is inside design temperature, and t_o is outside design temperature.

Values for U can be calculated from heat transfer coefficients of air films and heat conductivities for building materials or obtained directly for various materials and types of construction from heating guides and handbooks.

The heating engineer should work with the architect and building engineer on the economics of the completed structure. Consideration should be given to the use of double glass or storm sash in areas where outside design temperature is 10°F or lower. Heat loss through windows and doors can be more than halved and comfort considerably improved with double glazing. Insulation in exposed walls, ceilings, and around the edges of the ground slab can usually reduce local heat loss by 50–75%. Table 1 compares two typical dwellings. The 43% reduction in heat loss of the insulated house produces a worthwhile decrease in the cost of the heating plant and its operation. Building the house tight reduces the normally large heat loss due to infiltration of outside air.

Insulation and vapor barrier. Good insulating material has air cells or several reflective surfaces. A good vapor barrier should be used with or in addition to insulation, or serious trouble may result. Outdoor air or any air at subfreezing temperatures is comparatively dry, and the colder it is the drier it can be. Air inside a space in which moisture has been added from cooking, washing, drying, or humidifying has a much higher vapor pressure than cold outdoor air. Therefore, moisture in vapor form passes from the high vapor pressure space to the lower pressure space and will readily pass through most building materials. When this moisture reaches a subfreezing temperature in the structure, it may condense and freeze. When the structure is later warmed, this moisture will thaw and soak the building material, which may be harmful. For example, in a house that has 4 in. or more of mineral wool insulation in the attic floor, moisture can penetrate up through the second floor ceiling and freeze in the attic when the temperature there is below freezing. When a warm day comes, the ice will melt and can ruin the second floor ceiling. Ventilating the attic helps because the dry outdoor air readily absorbs the moisture before it condenses on the surfaces. Installing a vapor barrier in insulated outside walls is recommended, preferably on the room side of the insulation. Good vapor barriers include asphalt-impregnated paper, metal foil, and some plastic-coated papers. The joints should be sealed to be most effective. *See* INSULATION, HEAT.

Infiltration. In Table 1, the loss due to infiltration is large. It is the most difficult item to estimate accurately and depends upon how well the house is built. If a masonry or brick-veneer house is not well calked or if the windows are not tightly fitted and weather-stripped, this loss can be quite large. Sometimes, infiltration is estimated more accurately by measuring the length of crack around windows and doors.

Illustrative quantities of air leakage for various types of window construction are shown in Table 2. The figures given are in cubic feet of air per foot of crack per hour.

Design. Before a heating system can be designed, it is necessary to estimate the heating load for each room so that the proper amount of radiation or the proper size of supply air outlets can

Table 1. Effectiveness of double glass and insulation*

Heat-loss members	Area, ft²	With single-glass weather-stripped windows and doors	With double-glass windows, storm doors, and 2-in. wall insulation
		Heat loss, Btu/hr	
Windows and doors	439	39,600	15,800
Walls	1,952	32,800	14,100
Ceiling	900	5,800	5,800
Infiltration		20,800	20,800
Total heat loss		99,000	56,500
Duct loss in basement and walls (20% of total loss)		19,800	11,300
Total required furnace output		118,800	67,800

*Data are for two-story house with basement in St. Louis, Mo. Walls are frame with brick veneer and 25/32-in. insulation plus gypsum lath and plaster. Attic floor has 3-in. fibrous insulation or its equivalent. Infiltration of outside air is taken as a 1-hr air change in the 14,400 ft³ of heating space. Outside design temperature is −5°F; inside temperature is selected as 75°F.

be selected and the connecting pipe or duct work designed. *See* CENTRAL HEATING; STEAM HEATING; WARM-AIR HEATING SYSTEM.

Heat is released into the space by electric lights and equipment, by machines, and by people. Credit to these in reducing the size of the heating system can be given only to the extent that the equipment is in use continuously or if forced ventilation, which may be a big heat load factor, is not used when these items are not giving off heat, as in a factory. When these internal heat gain items are large, it may be advisable to estimate the heat requirements at different times during a design day under different load conditions to maintain inside temperatures at the desired level.

Cost of operation. Design and selection of a heating system should include operating costs. The quantity of fuel required for an average heating season may be calculated from

$$F = \frac{Q \times 24 \times DD}{(t_i - t_o) \times \text{Eff} \times H}$$

where F = annual fuel quantity, same units as H
Q = total heat loss, Btu/hr
t_i = inside design temperature, °F
t_o = outside design temperature, °F
Eff = efficiency of total heating system (not just the furnace) as a decimal
H = heating value of fuel
DD = degree-days for the locality for 65°F base, which is the sum of 65 minus each

Table 2. Infiltration loss with 15-mph outside wind

Building item	Infiltration, ft³/(ft)(hr)
Double-hung unlocked wood sash windows of average tightness, non-weather-stripped including wood frame leakage	39
Same window, weather-stripped	24
Same window poorly fitted, non-weather-stripped	111
Same window poorly fitted, weather-stripped	34
Double-hung metal windows unlocked, non-weather-stripped	74
Same window, weather-stripped	32
Residential metal casement, 1/64-in. crack	33
Residential metal casement, 1/32-in. crack	52

day's mean temperature for all the days of the year.

If a gas furnace is used for the insulated house of Table 1, the annual fuel consumption would be

$$F = \frac{56,500 \times 24 \times 4699}{[75 - (-5)] \times 0.80 \times 1050} = 94,800 \text{ ft}^3$$

For a 5°F, 6- to 8-hr night setback, this consumption would be reduced by about 5%.

[GAYLE B. PRIESTER]

Bibliography: American Society of Heating and Air Conditioning Engineers, *Handbook of Fundamentals*, 1967; C. Strock (ed.), *Handbook of Air Conditioning, Heating, and Ventilating*, 1966.

Heating, electric

Methods of converting electric energy to heat energy by resisting the free flow of electric current. Electric heating has no upper limit to the obtainable temperature except for the materials used. It has the advantage of electrical temperature control and provides uniform heating.

Any comparison of costs between electric heating and other methods should consider total costs and not just the cost of obtaining equal heat energies from electricity and fuels. The cost per unit of manufactured product is the true measure. This involves labor, quality, time, cleanliness, safety, and maintenance costs. Electric heat can usually excel in all of these.

Types of electric heaters. There are four major methods of electric heating.

1. Resistance heaters produce heat by passing an electric current through a resistance. Resistance heaters have an inherent efficiency of 100% in converting electric energy into heat. A high proportion of this heat can be transferred to the work material by conduction and by radiation.

2. Dielectric heaters use currents of high frequency which generate heat by dielectric hysteresis within the body of a nominally nonconducting material. The power factor varies but, due to inherent losses, can never approach unity.

3. Induction heaters produce heat by means of a periodically varying electromagnetic field within the body of a nominally conducting material. The power factor can never approach unity, because of inherent reactances. *See* INDUCTION HEATING.

4. Electric-arc heating is really a form of resistance heating in which a bridge of vapor and gas carries an electric current between electrodes. The arc has a property of resistance. Both electrodes may be of carbon, or one may be the conducting work material in a furnace.

Thermal problems. Electric heating differs radically from other methods that have a constant temperature. The heat-energy input of those methods depends on the temperature difference, which decreases as the work temperature rises. The heat input decreases as the work temperature increases.

Electric heating has a constant heat-energy input (for a given voltage), and the surface temperature of the heater rises to compensate for work temperature rise. Therefore, some control function is necessary to prevent overheating. To avoid damage, a suitable heat density per unit of heating surface must be maintained. With fluids, heat is distributed by natural thermal or forced convection currents. This helps heat transfer by wiping away surface films, and greater heat density can be tolerated. Heat capacity and mass play large parts in satisfactory heating. In large masses, such as tanks of oil or water, the temperature changes slowly and seldom presents a serious control problem. With small masses, uneven heat zones, or variable heating time, the control of the input heat becomes more critical. Heat control is required when heating such materials as gas vapor and paper. Moisture content also affects the heat requirements.

The heater rating can be found from the heat required for the process. This must include the heat required to heat the material and its container, any heat of fusion or vaporization in the process, and the heat losses. The heat required for the material and its container can be found from the weight, the specific heat, and the temperature rise of the material and container. The total heat required for the process is the sum of all the pertinent factors and can be put in units of kilowatt-hours. The number of kilowatt-hours divided by the hours allowed for the process will then give the required rating of the heater in kilowatts. The heater rating therefore depends on the time allowed for the process. A longer time will permit use of a heater of lower rating and will result in reduced costs of equipment and operation.

General design features. All electrical parts must be well protected from contact by operators, work materials, and moisture. Terminals must be enclosed within suitable boxes, away from the high heat zone, to protect the power supply cables. Repairs and replacements should be possible without tearing off heat insulations.

Resistance heaters are often enclosed in pipes or tubes suitable for immersion or for exposure to difficult external conditions. Indirect heating is done by circulating a heat transfer medium, such as special oil or Dowtherm (liquid or vapor), through jacketed vessels. This permits closer control of heating-surface temperature than is possible with direct heating.

Some conducting materials can be heated by passing electric current through them, as is done in the reduction of aluminum. Some conducting liquids can be heated by passing an electric current between immersed electrodes. Heat is produced by the electrical resistance of the liquid.

The supply of necessary electric power for large heating installations necessitates consultation with the utility company. The demand, the power factor of the load, and the load factor all affect the power rates. Large direct-current or single-phase alternating-current loads should be avoided. Polyphase power at 440–550 volts permits lower current and reduced costs.

[LEE P. HYNES]

Bibliography: J. J. Barton, *Space Heating Design and Practice*, 1963; Edison Electric Institute, *Power Sales Manual: Dielectric Heating*, 1959; Edison Electric Institute, *Power Sales Manual: Induction and High Frequency Resistance Heating*, 1961; L. P. Hynes, Industrial electric resistance heating, *AIEE Trans.*, 67:1359–1361, 1948; L. P. Hynes, Some unusual designs of electric resistance heating, *AIEE Trans.*, pap. no. 59-77; H. Pender and W. A. Del Mar (eds.), *Electrical Engineers' Handbook*, vol. 1, 4th ed., 1949.

Heating value

The energy per unit mass of material (fuel) that can be released from it as heat under specified conditions, where heat is simply energy in transit due to a temperature difference. In a sense, the heating value of a fuel can be thought of as energy which can potentially be released as heat if certain things are done to the fuel. For example, a typical bituminous coal has a heating value of about 7500 kcal/kg (13,500 Btu/lb). This means that 7500 kcal (13,500 Btu) of energy per kg (lb) of the coal can be released as heat. For a fossil fuel, such as coal, it is assumed that the energy is released upon complete combustion of the fuel at atmospheric pressure. On the other hand, 1 kg (2.2 lb) of nuclear fuel may have a heating value of 2.0×10^{10} kcal (7.92×10^{10} Btu), but this energy is released as heat upon fission, rather than combustion. *See* COMBUSTION; ENERGY; FISSION, NUCLEAR.

The units used in specifying heating value are defined as follows. One kilocalorie (kcal) is the quantity of heat necessary to raise the temperature of 1 kilogram of water 1 centigrade degree. One British thermal unit (Btu) is the quantity of heat necessary to raise the temperature of 1 pound (mass) of water 1 Fahrenheit degree. (It is the normal convention that the 1 C° is from 14.5 to 15.5°C, and the 1 F° from 59.5 to 60.5°F.) By direct conversion, 1 Btu/lb is equal to 0.555 kcal/kg. *See* BRITISH THERMAL UNIT (BTU).

Fossil fuels. The heating value of fossil fuels generally refers to the heat released upon complete combustion with oxygen at atmospheric pressure. This is called the heat of combustion at atmospheric pressure. It would be possible to react the fossil fuel with another material, such as chlorine, and this would release a different amount of heat. Strictly speaking, it would not be wrong to call this heat the heating value of the fossil fuel; however, convention dictates that the heating value of fossil fuels refers to combustion with oxygen. The reason for this convention is that fossil fuels such as coal are generally burned with oxygen in air, as in a home or industrial furnace at atmospheric pressure. *See* FUEL, FOSSIL.

Ideally, measurement of the heat of combustion of fossil fuels at atmospheric pressure would require a device that could maintain constant atmospheric pressure of the gases released when the fuel is combusted. Because of the difficulty of doing this, the heat of combustion is measured at constant volume instead of constant atmospheric pressure, and then the heat of combustion at atmospheric pressure (which is the practical value wanted) is calculated from the measurement using thermodynamics. *See* HEAT.

The heat of combustion at constant volume is measured using a bomb calorimeter as shown in the illustration. In this device, a measured quantity of the fossil fuel is placed in the bomb, which is a massive steel cylinder capable of withstanding the high pressures of the expanding gases released upon combustion. Enough oxygen is introduced to the bomb to ensure complete combustion of the fuel, and the combustion is started by heating the fuel with an electric current running through a fine wire in the bomb. The heat of combustion at constant volume (the volume of the bomb is constant) is calculated from the final temperature rise, after

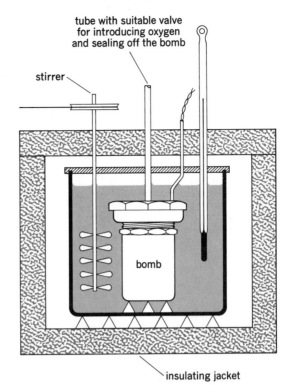

Diagram of a bomb calorimeter. (*From G. Shortley and D. Williams, Principles of College Physics, 2d ed., p. 305, Prentice-Hall, Inc., 1967*)

thermal equilibrium is reached. The result of this measurement is called the gross or higher heating value of the fossil fuel. In general, reported heating values of fossil fuels in the literature are the gross or higher heating values. Two corrections must be applied to obtain the net or lower heating value, which refers to the heat of combustion at atmospheric pressure. The first correction accounts for the latent heat of vaporization of the water in the combustion products. In the bomb calorimeter, water as steam is formed as one of the gaseous products. When the bomb reaches thermal equilibrium, the steam condenses and releases its heat of vaporization. This must be subtracted to obtain the net heating value of the fossil fuel. The second correction accounts for the fact that the measured value is at constant volume, while the desired value is for constant (atmospheric) pressure which occurs under practical use.

The heating value of gaseous fossil fuels is more

Table 1. Typical values* of heat of combustion of fossil fuels

Substance	kcal/kg	Btu/lb	kcal/SCM	Btu/SCF
Solid fuels				
Anthracite	7,900	14,220		
Bituminous coal	7,500	13,500		
Lignite	4,100	7,380		
Charcoal	7,000	12,600		
Liquid fuels				
Gasoline	11,100	19,980		
Kerosine	11,400	20,520		
No. 6 fuel oil	10,500	18,900		
Benzol	10,000	18,000		
Gaseous fuels				
Methane			9,016	1,013
Hydrogen			2,892	325

*Higher heating value.

Table 2. Range of some characteristics of coal classes

Coal class	Heating value*		Fixed carbon content, wt. %	Moisture content, wt. %
	kcal/kg	Btu/lb		
Anthracite	Up to 8900	Up to 16,020	86–98	1–3
Bituminous	6100–8300	10,980–14,940	50–86	3–12
Subbituminous	4400–6700	7920–12,060	40–60	20–30
Lignite	3100–4400	5580–7920	<40	Up to 40

*Higher heating value.

conveniently reported in energy per standard cubic meter (SCM) or foot (SCF), where SCM (SCF) is 1 cubic meter (foot) of the gas at standard pressure and temperature. For example, natural gas, which is largely methane, has a heating value of 8900 kcal/SCM (1000 Btu/SCF), and synthesis gas, which is largely carbon monoxide and hydrogen, may have a heating value of 3000 kcal/SCM (337 Btu/SCF). The actual heating value of synthesis gas depends upon the percentage of carbon monoxide and hydrogen.

Table 1 gives some typical values of the heat of combustion of some fossil solid, liquid, and gaseous fuels. As usual in reporting heating values, these are at constant volume and are the higher heating values.

The heating value of a fossil fuel such as coal may vary widely because of the wide variance in the proportion of fixed carbon it contains. This variance is illustrated for coal in Table 2. (Fixed carbon has a large heating value, while moisture has none.) Crude oil does not have as large a variance, because its composition does not vary much. Its heating value is about 36,000 kcal/gal (143,000 Btu/gal). Organic material such as coal can be converted in various ways (for example, by reacting it with steam and oxygen) to fossil fuel gases with heating values ranging from 1150 to 8900 kcal/SCM (129–1000 Btu/SCF). The variance in the heating values is due to the variance in the composition of the gases.

Other ways of obtaining heat. Fossil fuels are presently the greatest practical source of heat, but nuclear fission is being used more and more. In nuclear fission, the nucleus of an unstable heavy element such as uranium is broken up into lighter nuclei by absorbing a neutron. The energy released as heat comes from the conversion of mass into energy (that is, the difference between the mass of the reactants and that of the products becomes energy). One kg (2.2 lb) of fissionable material has about 2.0×10^{10} kcal (7.92×10^{10} Btu), which is the same as the energy obtained from burning 2930 metric tons of 7500 kcal/kg (13,500 Btu/lb) coal. In general, nuclear heat is used to generate steam, which can then be used to drive a turbogenerator to produce electricity.

There are other sources of energy which may be used for heat, but these have not been classically called fuels. Hot geothermal fluids may be used for home heating, as in Iceland, or to generate electrical power. Similarly, solar radiation brings an average of 3800 kcal/m² day (1400 Btu/ft² day) onto the surface of the United States. This can be used for direct heating by means of solar collectors or for electrical power by means of solar cells or focused solar collectors and a steam cycle. Any electrical power from geothermal energy, solar energy, nu-

clear energy, or wind energy can be used for resistance heating. There is also the energy available in the tides, which is a form of gravitational energy. This, too, could eventually be converted to heat. So far, no one has specified heating values for these sources, but they are real sources of heat and will become more and more important as fossil fuel resources are used up. *See the feature article* EXPLORING ENERGY CHOICES; *See also* GEOTHERMAL POWER; NUCLEAR POWER; SOLAR COLLECTORS; SOLAR RADIATION; TIDAL POWER; WIND POWER.

[ROBERT C. BINNING; RICHARD S. HOCKETT]

Bibliography: J. P. Holman, *Thermodynamics*, 2d ed., 1974; H. H. Lowry (ed.), *Chemistry of Coal Utilization*, Suppl. Vol., 1963; R. H. Perry and C. H. Chilton (eds.), *Chemical Engineers' Handbook*, 5th ed., 1973; G. Shortley and D. Williams, *Principles of College Physics*, 2d ed., 1967.

Horsepower

The unit of power in the British engineering system of units. One horsepower (hp) equals 550 ft-lb/sec, or 746 watts. *See* POWER.

The horsepower is a unit of convenient magnitude for measuring the power generated by machinery. As an example of the size of the horsepower, consider a 200-lb man walking up a stairway, the top of which is 10 ft higher than the bottom. In walking up the stairway, the man does 2000 ft-lb of work. If he climbs the stairs in 5 sec, his rate of work is 400 ft-lb/sec, or about 3/4 hp.

[PAUL W. SCHMIDT]

Hot-water heating system

A heating system for a building in which the heat-conveying medium is hot water. Heat transfer in British thermal units (Btu) equals pounds of water circulated times drop in temperature of water. For other liquids, the equation should be modified by the specific heats. The system may also be modified to provide cooling.

A hot-water heating system consists essentially of water-heating or -cooling means and of heat-emitting means such as radiators, convectors, baseboard radiators, or panel coils. A piping system connects the heat source to the various heat-emitting units and includes a method of establishing circulation of the water or other medium and an expansion tank to hold the excess volume of water as it is heated and expands. Radiators and convectors have such different response characteristics that they should not be used in the same system.

Types. In a one-pipe system (Fig. 1), radiation units are bypassed around a one-pipe loop. This type of system should only be used in small installations.

In a two-pipe system (Fig. 2), radiation units are connected to separate flow and return mains, which may run in parallel or preferably on a reverse return loop, with no limit on the size of the system.

In either type of system, circulation may be provided by gravity or pump. In gravity circulation each radiating unit establishes a feeble gravity circulation; hence such a system is slow to start, is unpredictable, and is not suitable for convectors, baseboard radiation, or panel coils because circu-

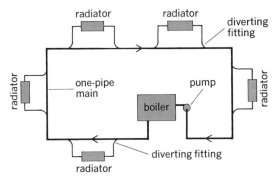

Fig. 1. One-pipe hot-water heating system.

lating head cannot be established, and circulation cannot be supplied to units below the mains. The pipes must be large in size. For these reasons gravity systems are no longer being used.

In forced circulation a pump is used as the source of motivation. Circulation is positive and units may be above or below the heat source. Smaller pipes are used.

Operation. For perfect operation it is imperative that the friction head from the heat source through each unit of radiation and back to the heat source be the same. To achieve this condition usually requires careful balancing after installation and during operation.

Expansion tanks may be open or closed. Open tanks are vented to the atmosphere and are used where the water temperature does not exceed 220°F (at sea level). They provide the safest operation, practically free from explosion hazards. Closed tanks, used for higher water temperatures, must be provided with safety devices to avoid possible explosions.

One outstanding advantage of hot-water systems is the ability to vary the water temperature according to requirements imposed by outdoor weather conditions, with consequent savings in fuel. Radiation units may be above or below water heaters, and piping may run in any direction as long as air is eliminated. The system is practically indestructible. Flue-gas temperatures are low, resulting in fuel savings. The absence of myriads of special steam fittings, which are costly to purchase and to maintain, is also an important advantage. Hot water is admirably adapted to extensive central heating where high temperatures and high pressures are used and also to low-temperature panel-heating and -cooling systems.

Circulating hot-water pumps must be carefully specified and selected. On medium-size installations, it is recommended that two identical pumps be used, each capable of handling the entire load. The pumps operate alternately but never together in parallel. On large installations three or more pumps may be used in parallel, provided they are identical and produced by the same manufacturer. The casings and runners must be cast from the same molds, and the metals and other features that affect their temperature characteristics must be identical. All machined finishes must be identical, and the pumps must be thoroughly shop-tested to operate with identical characteristics.

When the system is in operation, the pump can be disconnected from the boiler by throttling down the valve at the boiler return inlet; it should not be closed completely. This procedure permits the water in all boilers to be at the same temperature so that when a boiler is thrown back into service, the flue gases do not impinge on any cold surfaces, thus producing soot and smoke to further contaminate the outdoor atmosphere. *See* HEATING, COMFORT: OIL BURNER.

[ERWIN L. WEBER]

Bibliography: American Society of Heating, Refrigerating, and Air Conditioning Engineers, *Guide*, 1962, 1964, 1966, and 1967; American Society of Heating and Ventilating Engineers, *Heating, Ventilating, Air Conditioning Guide*, vol. 37, 1959; F. E. Giesecke, *Hot-water Heating and Radiant Heating and Radiant Cooling*, 1947.

Hydraulic turbine

A machine which converts the energy of an elevated water supply into mechanical energy of a rotating shaft. Most old-style waterwheels utilized the weight effect of the water directly, but all modern hydraulic turbines are a form of fluid dynamic machinery of the jet and vane type operating on the impulse or reaction principle and thus involving the conversion of pressure energy to kinetic energy. The shaft drives an electric generator, and speed must be of an acceptable synchronous value. *See* GENERATOR, ELECTRIC.

The impulse or Pelton unit has all available energy converted to the kinetic form in a few stationary nozzles and subsequent absorption by reversing buckets mounted on the rim of a wheel (Fig. 1). Reaction units of the Francis or the Kaplan types run full of water, submerged, with a draft tube and

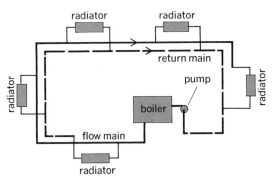

Fig. 2. Two-pipe reverse return system.

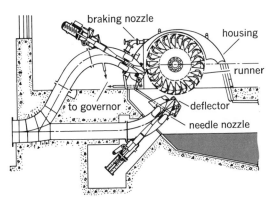

Fig. 1. Cross section of an impulse (Pelton) type of hydraulic-turbine installation.

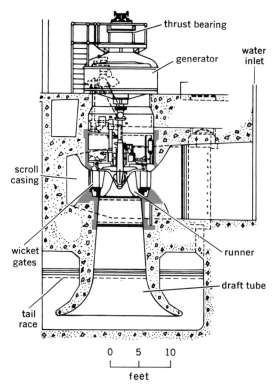

Fig. 2. Cross section of a reaction (Francis) type of hydraulic-turbine installation.

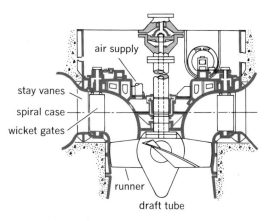

Fig. 3. Cross section of a propeller (Kaplan) type of hydraulic turbine installation.

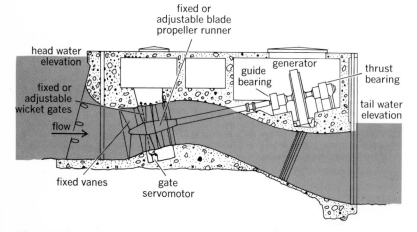

Fig. 4. Axial-flow tube-type hydraulic-turbine installation.

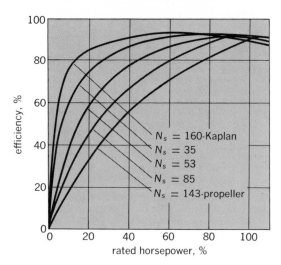

Fig. 5. The efficiency characteristics of some selected hydraulic-turbine types.

a continuous column of water from head race to tail race (Figs. 2 and 3). There is some fluid acceleration in a continuous ring of stationary nozzles with full peripheral admission to the moving nozzles of the runner in which there is further acceleration. The draft tube produces a negative pressure in the runner with the propeller or Kaplan units acting as suction runners; the Francis inward-flow units act as pressure runners. Mixed-flow units produce intermediate degrees of both rotor pressure drop and fluid acceleration. See IMPULSE TURBINE.

For many years reaction turbines have generally used vertical shafts for better accommodation of the draft tube whereas Pelton units have favored the horizontal shaft since they cannot use a draft tube. Vertical-shaft Pelton units have found increasing acceptance in large sizes because of multiple jets (for example, 4–6) on a single wheel; these provide reduced runner windage and friction losses and, consequently, higher efficiency. Axial-flow (Fig. 4) and diagonal-flow reaction turbines offer improved hydraulic performance and economic powerhouse structures for large-capacity low-head units. Kaplan units employ adjustable propeller blades as well as adjustable stationary nozzles in the gate ring for higher sustained efficiency (Fig. 5). Pelton units are preferred for high-head service (1000± ft), Francis runners for medium heads (200± ft), and propeller or Kaplan units for low heads (50± ft).

Hydraulic-turbine performance is rigorously defined by characteristic curves, such as the efficiency characteristic (Fig. 5). The proper selection of unit type and size is a technical and economic problem. The data of Fig. 6 are significant because they show synoptically the relationship between unit type and site head as the result of accumulated experience on some satisfactory turbine installations. Specific speed N_s is a criterion or coefficient which is uniquely applicable to a given turbine type and relates head, power, and speed, which are the basic performance data in the selection of any hydraulic turbine. Specific speed is defined in the equation shown, where rpm is revolutions per minute, shp is shaft horsepower, and head is head on unit in feet. Specific speed is usu-

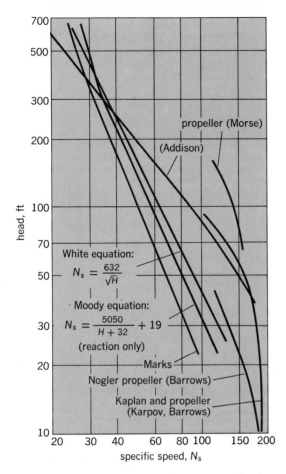

Fig. 6. Hydraulic-turbine experience curves, showing specific speed versus head.

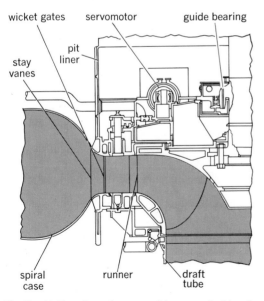

Fig. 7. Half-section of a reversible pump-turbine for pumped-storage application.

ally identified for a unit at the point of maximum

$$N_s = \frac{\text{rpm} \times (\text{shp})^{0.5}}{(\text{head})^{1.25}}$$

efficiency. Cavitation must also be carefully scrutinized in any practical selection.

The draft tube (Fig. 2) is a closed conduit which (1) permits the runner to be set safely above tail water level, yet to utilize the full head on the site, and (2) is limited by the atmospheric water column to a height substantially less than 30 ft. When made flaring in cross section, the draft tube will serve to recover velocity head and to utilize the full site head.

Efficiency of hydraulic turbine installations is always high, more than 85% after all allowances for hydraulic, shock, bearing, friction, generator, and mechanical losses. Material selection is not only a problem of machine design and stress loading from running speeds and hydraulic surges, but is also a matter of fabrication, maintenance, and resistance to erosion, corrosion, and cavitation pitting.

Governing problems are severe, primarily because of the large masses of water which are involved, their positive and negative acceleration without interruption of the fluid column continuity, and the consequent shock and water-hammer hazards.

Pumped-storage hydro plants have employed various types of equipment to pump water to an elevated storage reservoir during off-peak periods and to generate power during on-peak periods where the water flows from the elevated reservoir through hydraulic turbines. Although separate, single-purpose, motor-pump and turbine-generator sets give the best hydraulic performance, the economic burden of investment has led to the development of reversible pump-turbine units, as shown in Fig. 7. Components of the conventional turbine are retained, but the modified pump runner gives optimum performance when operating as a turbine. Compromises in hydraulic performance, with some sacrifice in efficiency, are more than offset by the investment savings with the dual-purpose machines. *See* WATERPOWER.

[THEODORE BAUMEISTER]

Bibliography: H. K. Barrows, *Water Power Engineering*, 3d ed., 1943; T. Baumeister (ed.), *Standard Handbook for Mechanical Engineers*, 7th ed., 1967; W. P. Creager and J. D. Justin, *Hydroelectric Handbook*, 2d ed., 1950; D. G. Fink and J. M. Carroll, *Standard Handbook for Electrical Engineers*, 10th ed., 1968.

Hydrocracking

A catalytic, high-pressure process flexible enough to produce either of the two major light fuels— high octane gasoline or aviation jet fuel. It proceeds by two main reactions: adding hydrogen to molecules too massive and complex for gasoline and then cracking them to the required fuels. The process is carried out by passing oil feed together with hydrogen at high pressure (1000–2500 psig) and moderate temperatures (500–750°F) into contact with a bifunctional catalyst, comprising an acidic solid and a hydrogenating metal component. Gasoline of high octane number is produced, both directly and through a subsequent step such as catalytic reforming; jet fuels may also be manufactured simply by changing conditions with the same catalysts. The process is characterized by a long catalyst life (2–4 years), though a slow decline in activity occurs, caused by the deposition of carbonaceous material on the catalyst. Regeneration at intervals by burning off these deposits restores the activity, but eventually the catalyst porosity is destroyed and it must be replaced.

Generally, the process is used as an adjunct to catalytic cracking. Oils, which are difficult to convert in the catalytic process because they are highly aromatic and cause rapid catalyst decline, can be easily handled in hydrocracking, because of the low cracking temperature and the high hydrogen pressure, which decreases catalyst fouling. Usually, these oils boil at 400–1000°F, but it is possible to process even higher-boiling feeds if very high hydrogen pressures are used. However, the most important components in any feed are the nitrogen-containing compounds, as these are severe poisons for hydrocracking catalysts and must be removed to a very low level.

Hydrocracking was carried out on a practical scale in Germany and England starting in the 1930s. In this early work, a common hydrocracking catalyst was tungsten disulfide on acid-treated clay; thus, both hydrogenation and acidic components were present. Generally, a light oil from coal or coking products was vaporized and passed over the catalyst at high pressure. After separation of gasoline from the products, the unconverted material was returned to the reactor with a fresh portion of feed. Because this catalyst was not very active, the process had to be carried out at very high pressures and temperatures (4000 psig; 750°). It was costly and the products were not of high quality.

Research in the United States concentrated on the development of much more active catalysts, a different mode of operation, and the use of heavier oil feeds. As a result, the reaction is carried out in two separate, consecutive stages; in each, oil and hydrogen at high pressure flow downward over fixed beds of catalyst pellets placed in large vertical cylindrical vessels.

First stage. In the first, or pretreating, stage the main purpose is conversion of nitrogen compounds in the feed to hydrocarbons and to ammonia by hydrogenation and mild hydrocracking. Typical conditions are 650–740°F, 150–2500 psig, and a catalyst contact time of 0.5–1.5 hr; up to 1.5 wt% hydrogen is absorbed, partly by conversion of the nitrogen compounds, but chiefly by aromatic compounds which are hydrogenated. It is most important to reduce the nitrogen content of the product oil to less than 0.001 wt% (10 parts per million). This stage is usually carried out with a bifunctional catalyst containing hydrogenation promotors, for example, nickel and tungsten or molybdenum sulfides, on an acidic support, such as silica-alumina. The metal sulfides hydrogenate aromatics and nitrogen compounds and prevent deposition of carbonaceous deposits; the acidic support accelerates nitrogen removal as ammonia by breaking carbon-nitrogen bonds. The catalyst is generally used as $1/8 \times 1/8$ in. or $1/16 \times 1/8$ in. pellets, formed by extrusion.

Second stage. Most of the hydrocracking is accomplished in the second stage, which resembles the first but uses a different catalyst. Ammonia and some gasoline are usually removed from the first-stage product, and then the remaining oil, which is low in nitrogen compounds, is passed over the second-stage catalyst. Again, typical conditions are 600–700°F, 1500–2500 psig hydrogen pressure, and 0.5–1.5 hr contact time; 1–1.5 wt% hydrogen may be absorbed. Conversion to gasoline or jet fuel is seldom complete in one contact with the catalyst, so the lighter oils are removed by distillation of the products and the heavier, high-boiling product combined with fresh feed and recycled over the catalyst until it is completely converted.

The catalyst for the second stage is also a bifunctional catalyst containing hydrogenating and acidic components. Metals such as nickel, molybdenum, tungsten, or palladium are used in various combinations, dispersed on solid acidic supports such as synthetic amorphous or crystalline silica-aluminas, such as zeolites. These supports contain strongly acidic sites and sometimes are enhanced by the incorporation of a small amount of fluorine. A long period (for example, 1 year) between regenerations is desirable; this is achieved by keeping a low nitrogen content in the feed and avoiding high temperatures, which lead to excess cracking with consequent deposition of coke on the catalyst. When activity of the catalyst does decrease, it can be restored by carefully controlled burning of the coke.

The catalyst is the key to the success of the hydrocracking process as now practiced, particularly the second-stage catalyst. Its two functions must be most carefully balanced for the product desired; that is, too much hydrogenation gives a poor gasoline but a good jet fuel. The oil feeds are composed of paraffins, other saturates, and aromatics—all complex molecules boiling well above the required gasoline or jet-fuel product. The catalyst starts the breakdown of these components by forming from them carbonium ions, that is, positively charged molecular fragments, via the protons (H^+) in the acidic function. These ions are so reactive that they change their internal molecular structure spontaneously and break down to smaller fragments having excellent gasoline qualities. The hydrogenating function aids in maintaining and controlling the ion reactions and protects the acid function by hydrogenating coke precursors off the catalyst surface, thus maintaining catalyst activity. Any olefins formed in the carbonium ion decomposition are also hydrogenated.

Products. The products from hydrocracking are composed of either saturated or aromatics compounds; no olefins are found. In making gasoline, the lower paraffins formed have high octane numbers; for example, the 5- and 6-carbon number fractions have leaded research octane numbers of 99–100. The remaining gasoline has excellent properties as a feed to catalytic reforming, producing a highly aromatic gasoline which, with added lead, easily attains 100 octane number. Both gasolines are suitable for premium-grade motor gasoline. Another attractive feature of hydrocracking is the low yield of gaseous components, such as methane, ethane, and propane, which are less desirable than gasoline. When making jet fuel, more hydrogenation activity of the catalysts is used, since jet fuel contains more saturates than gasoline. *See* GASOLINE.

The hydrocracking process is being applied in other areas, notably, to produce lubricating oils and to convert very asphaltic and high-boiling residues to lower-boiling fuels. Its use will certainly increase greatly in the future, since it accomplishes two needed functions in the petroleum-fuel economy: Large, unwieldly molecules are cracked, and the needed hydrogen is added to produce useful, high-quality fuels. *See* CRACKING; REFORMING IN PETROLEUM REFINING.

[CHARLES P. BREWER]

Bibliography: *Advan. Petrol. Chem. Refining*, 8:168–191, 1964; W. F. Bland and R. L. Davidson (eds.), *Petroleum Processing Handbook*, sec. 3, pp. 16–25, 1967; *Hydrocarbon Process.*, 47(9): 139–144, 1968; Hydroprocesses, *Kirk-Othmer Encyclopedia of Chemical Technology*, 2d ed., vol. 11, 1966.

Hydroelectric generator

An electric rotating machine that transforms mechanical power from a hydraulic turbine or water wheel into electric power. Hydroelectric generators may have horizontal or vertical shafts, depending upon the turbine. The most common type in large ratings is the vertical-shaft, synchronous generator. The hydroelectric generator may be arranged to operate as a motor during periods of low power demand. In such a case the turbine serves as a pump, raising water to a high elevation for reuse to generate power at a period of peak load. Such a unit is termed a pump turbine. *See* ELECTRIC POWER GENERATION; GENERATOR, ELECTRIC; HYDRAULIC TURBINE; TURBINE; WATERPOWER.   [LEON T. ROSENBERG]

Hydroelectric power

Electric power generated through the use of flowing water. Water stored by dams is released to turn modern waterwheels called turbines. Coupled to the turbines are generators which supply electricity.

Place in energy economy. Hydroelectric power developments in the United States, including pumped-storage, had a generating capability of more than 63×10^9 kW at the end of 1974. This accounts for more than 13% of the United States power industry's total capability.

Worldwide, about 1/5 of all energy utilized is in the form of electric power, and about 1/5 of this electric power is generated from hydroelectric power plants. Thus, while hydro power produced only about 1/25 of the energy utilized by humans, it is still an important power source with particular characteristics which assure its future development. *See* ELECTRIC POWER GENERATION.

Hydro power is the oldest known mechanical power source. It is usually the lowest cost form of bulk energy. It is in most cases the most efficient form of power. It is the least polluting form of power. It is the most responsive—easy to start and stop—of any electric generating source. And finally, it is the most compatible power where large river projects are being built for nonpower purposes such as flood control, irrigation, navigation, or water supply. Hydro power, in the form of pumped-storage, is to date the only practical means to store excess electric power for use at a later time.

Theory of hydro power. Hydro power is the opposite of a ship's propeller or a pump. A propeller or a pump uses power to move water; a hydro turbine uses moving water to produce mechanical power. This mechanical power is almost invariably used to rotate an electrical generator, which produces electricity. Physically, hydro turbines and propellers or pumps resemble each other closely, comprising flow-controlling blades mounted on a rotating shaft connected to a power-using device. In some cases, hydro turbines actually are designed for reverse-direction rotation use as pumps. The largest and highest lift pumps in the world are such "reversible pump-turbines." These machines, rated at millions of horsepower, can literally pump a river to the top of a mountain. *See* HYDRAULIC TURBINE.

Hydro power is a form of nondepleting, self-replenishing energy—actually a form of solar energy. In this view, the water-covered 7/8 of the Earth's surface receives a continuous inflow of solar radiant energy. Part of this inflow is radiated back out into space, and most of the rest is utilized to evaporate (and thus desalt!) sea water, which becomes clouds. These clouds pass over land masses, which receive rain from the clouds. The topography, and drainage pattern, of the continents concentrates this "solar energy in the form of rain" into narrow, powerful rivers. Hydro power plants placed strategically on these rivers can utilize a highly concentrated—and therefore both powerful and economical—form of solar energy. *See* SOLAR ENERGY.

History of waterpower. Wind and waterpower are the most ancient form of nonanimal mechanical (and motive) power used by man. Sail and ocean currents are prehistoric means of navigation, and the use of windmills and waterwheels is roughly coincident with the birth of civilization. Early civilizations such as the Chinese, Greek, Persian, and Roman were masters of wind or water mills, or both.

Where rivers were available, waterpower was more useful because of its inherent lower cost, greater reliability, and greater force or power. As the technological and industrial revolutions developed, wind power was progressively displaced by steam, internal combustion, or electrical power. But waterpower has kept pace with the advances of technology and is more used today than ever before. Thus, waterpower is both the most ancient and one of the most modern forms of power used for the good of humankind. *See* WATERPOWER; WIND POWER.

Modern waterpower had its beginnings in the early 19th century with the development of the hydraulic reaction turbine, and the modernizing of the impulse wheel. A radial outward-flow turbine was installed at Pont sur l'Ognon in France in 1827 after 4 years of experimentation by M. Fourneynon. James B. Francis in 1847 significantly improved an inward-flow turbine patented in 1836, and is considered by many as the originator of the American-type reaction turbine.

In 1853 Jearum Atkins of the United States began studying the scientific theory behind the ancient flutter wheel, which in plan resembled a wagon wheel and moved when a water jet from a spout hit the wheel's flat, vertical vanes near their ends. Practical development of this wheel is credited to Lester A. Pelton, who commenced his radical improvements in 1882.

The first hydroelectric central generating station in the United States was built at Appleton, WI, in 1882. It furnished enough power to light 250 electric lights.

Before 1897 nearly 300 hydroelectric plants were in operation in the world. By 1900 the first pumped-storage plant had been constructed in Europe. Since the turn of the century hydroelectric power has developed and advanced with the electric power industry. Giant plants — Hoover, Grand Coulee, Chief Joseph, Churchill Falls (Canada), Krasnoyarsk (Soviet Union), Guri (Venezuela), Aswan (Egypt) — have been built and are being expanded. Developing countries have found hydroelectric power the most economical means to bring industrial advancement to their people.

Typical hydroelectric plant. A typical hydroelectric project includes a water-diverting structure, such as a dam or canal, a conduit to transport water to the turbines, turbines and governors, generators, control and switching apparatus, a powerhouse to protect the equipment, transformers, and transmission lines to carry the generated power to distribution centers (see Fig. 1). In most cases a forebay or a vertical shaft is provided to store the surge of water which occurs when the plant is shut down or its power output is reduced, thus reducing water hammer. Water hammer in a hydro plant is similar to (but on a grander scale than) that com-

mon in home plumbing systems not provided with surge pipes near water faucets. If the power station is some distance from the water source, the forebay also serves as the initial supply of water for startup, or when increasing power output, while water in the canal or pipeline accelerates. An intake structure at the head of the water conduit houses trash racks to keep debris from the turbines, and also gates or valves to stop the flow of water for conduit inspection and maintenance. *See* DAM.

The water conduit may be very short, as when the powerhouse is an integral part of a concrete dam. Many conduits are longer, ranging from a few feet to several miles. Those made of steel are called penstocks. Generally, the steel-lined portions of water passages through a concrete dam or of tunnels are also referred to as penstocks.

Hydraulic turbines are classed as either impulse or reaction turbines. The impulse turbine is represented today by the Pelton-type waterwheel. In an impulse turbine the water is discharged through one or more nozzles as one or more free jets which act upon the runner (a series of buckets around the circumference of a hub). A housing prevents splash and guides the discharge, which falls freely into the tail water. *See* HYDRAULIC TURBINE; IMPULSE TURBINE; REACTION TURBINE.

In reaction turbines the entire flow from head water to tail water occurs in a closed conduit system. Whereas with a Pelton wheel all available head is converted into kinetic energy, with a reaction turbine only part of the available head is converted to kinetic energy at the entrance to the runner. A substantial part of the available head remains as pressure head which varies throughout the passage through the turbine.

Reaction turbines are represented by two types: the Francis turbine and the vertical-flow propeller turbine. Propeller turbines are subdivided into fixed-blade and adjustable-blade (Kaplan) types. Diagonal-flow turbines have also been developed. In the Francis turbine (also vertical-flow), flow passes from the penstock to a spiral case and then inwardly around the inside circumference through a series of pivoting gates (wicket gates) to the runner, which resembles the blade arrangement in a jet airplane engine. The curved runner vanes react to the impulse of the flowing water, causing the runner to rotate. The vanes also deflect the water to the tailrace. In the propeller turbine, flow also passes through wicket gates but then is deflected downward onto the propeller blades and through the blades to the draft tube (see Fig. 2). The draft tube is a diverging discharge passage connecting the runner with the tailrace. It is shaped to decelerate the flow of water to the tailrace, and at the same time accelerate the momentum of the turbine runner by converting the kinetic energy remaining in the discharge into suction head, thereby increasing the pressure difference on the runner. Draft tubes are not used with impulse wheels, so that the head from nozzle to tail water is not utilized.

The generator is located above the turbine except in plants having horizontal-flow turbines. Generator and turbine are connected by a common shaft. Thus when the turbine turns, the generator rotor turns. The rotor, passing through a magnetic

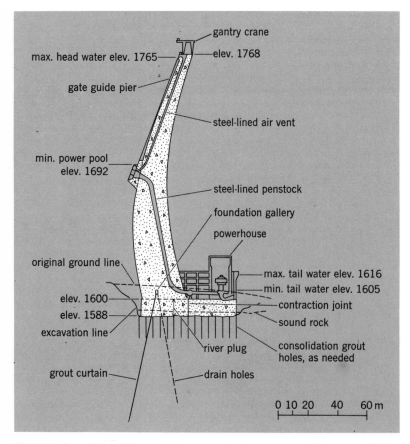

Fig. 1. Section of Karadj River arch dam and powerhouse, in Iran. Three Francis-type turbines by Harland of Scotland, each rated 49,700 hp (37,000 kW), 482 ft (147 m) net head. Three generators by Hitachi of Japan. Total plant capacity, 113,000 kW. (*From C. V. Davis and K. E. Sorensen, Handbook of Applied Hydraulics, 3d ed., McGraw-Hill, 1969*)

Fig. 2. Model of turbine and generator for Priest Rapids power station, Columbia River, Washington. Kaplan-type turbine rated 114,000 hp (85,000 kW); 78 ft (23.4 m) net head, 85.7 rpm, runner diameter 284 in. (7.2 m). Ten units by English Electric Company, Ltd. Total plant capacity, 790,000 kW. (*From C. V. Davis and K. E. Sorensen, Handbook of Applied Hydraulics, 3d ed., McGraw-Hill, 1969*)

field, generates electricity. This electricity is carried by cables to the control and switching apparatus, and thence to transformers from which it is transmitted over power lines to the distribution centers of the power system. *See* HYDROELECTRIC GENERATOR; TRANSMISSION LINES.

Hydro power classifications. Hydroelectric developments are classified as multipurpose or single-purpose, as base-load or peak-load, as run-of-river or storage, as high-head or low-head, as indoor or outdoor, as aboveground or underground, and as conventional hydro or pumped-storage or tidal. Any project may fit several of these classifications.

A multipurpose project is developed for a number of uses, including flood control, irrigation, navigation, power and silt removal. Grand Coulee Project on the Columbia River, the largest hydroelectric plant in the United States, with new construction bringing it to an installed capacity of 6×10^6 kW, was built for flood control, irrigation, and power. The Hoover Project, with a capacity of 1.34×10^6 kW, is used for silt removal as well as the uses cited for Grand Coulee. Guri Project on the Rio Caroni in Venezuela, which will exceed Grand Coulee in size (9×10^6 kW), is a single-purpose hydroelectric project built for power only.

A base-load plant supplies power almost continuously to satisfy the minimum constant power demand in a system. Several power plants may be required to furnish a system's base load. A peak-load plant generates power only during periods of peak demand, that is, when the power needed in a system exceeds that generated by the base-load plants. Fossil fuel and nuclear power projects are difficult to shut down and to start up and, therefore, are not efficient peaking plants. *See* POWER PLANT.

A run-of-river project utilizes the flow passing the plant for hydro power but does not alter the amount or intensity of flow to areas downstream. A storage project retains water so river flow downstream may be regulated. This assures farmers or other water users downstream they will have an adequate water supply during periods of low flow or drought.

Description of a project as high-head or low-head refers to the differences in water levels just upstream (head water) and downstream (tail water) of the plant. High-head plants have heads over 100 ft (1 ft = 0.3 m), while low-head plants have heads of 100 ft or less. The amount of head has a direct effect on the type of hydraulic turbine adopted. Francis turbines of the reaction type have been applied for heads up to 2205 ft (Rosshag power station, Austria), impulse turbines of the Pelton type up to 5800 ft (Reisseck plant, Austria), and propeller turbines up to 289 ft (Kaplan turbine at Nembia plant, Italy). Generally, for any one turbine type, the higher the head, the smaller the turbine needed to furnish the same power.

Designation of a plant as indoor or outdoor depends upon whether the powerhouse superstructure encloses the powerhouse overhead crane and generator housing. If the crane is outdoors but the generator housing is indoors, the project is referred to as semi-outdoor.

Aboveground plants are those where the powerhouse is at ground surface. Certain project sites make it less expensive to excavate a cavern to house the generating equipment than to build a surface structure. Sometimes a site, such as a narrow gorge, lacks space for the powerhouse aboveground, and belowground may be a good alternative, foundation conditions permitting. Generally, underground powerhouses are constructed in sound rock rather than in poor rock or soft ground.

Conventional hydroelectric power plants include all the hydro projects which fit the description of a typical hydro project given above. Pumped-storage projects, though almost as old as conventional hydro in Europe, have had in the United States the greatest development since 1960. They are designed to provide peaking power. To make power available from a pumped-storage plant during periods of high demand, water is pumped to an upper reservoir during off-peak hours, utilizing unneeded power generated by base-load plants, and then released as required to provide the energy necessary to turn the turbines. In Europe most pumped-storage plants have separate pumps and turbines. In the United States, however, reversible pump-turbines are used. Examples of such plants are the Kinzua or Seneca projects on the Allegheny River in western Pennsylvania, Muddy Run on the lower Susquehanna River in south-central Pennsylvania, and Ludding-ton on the eastern shore of Lake Michigan. *See* PUMPED STORAGE.

A tidal power station harnesses the energy available from water-level fluctuations. Two tidal power developments have been constructed: La Rance in the Bay of St. Malo on the north coast of Brittany, and Kislogubskaya in the Soviet Union. At La Rance, the amplitude of the tides in spring reach 44 ft, but average head is 27 ft. Considerable distance from areas needing power seems to be the deterrent to construction at most potential sites. *See* TIDAL POWER.

Future of hydroelectric power. Waterpower enjoys a great future for development. It is almost always the lowest-cost and least-polluting form of power. Advanced, industrialized countries have already utilized the "cream" of their waterpower potential, but modern economics of thermal and nuclear power are making the remaining waterpower potential projects — which were often marginal economically — now look much more economical and rewarding. And the largest rivers and therefore the greatest potential hydro projects are in the underdeveloped and developing countries — Brazil, Mozambique, Zaïre, Venezuela, Argentina, and Paraguay, to name a few. In these cases, construction is just starting, and the power developments will be built between 1975 and 2000. These will be vast, international, multibillion-dollar undertakings. Their construction will raise world hydroelectric generation from 1.3×10^9 MW-hr to 3.4×10^9 MW-hr by the year 2000, with a projected ultimate potential of 5×10^9 MW-hr. They will require decades, and absence of war, to complete. But the technology for development is in hand, economic need for such projects is evident, and initial projects are underway. *See the feature articles* EXPLORING ENERGY CHOICES; WORLD ENERGY ECONOMY.

[RICHARD D. HARZA; H. CLARK DEAN]

Bibliography: A. M. Angelini, The utilization of hydraulic resources still available in the world, 9th World Energy Conference, Detroit, September 1974, quoted in *Water Power Dam Constr.*, p. 47, February 1975; J. Cotillon, La Rance: Six years of operating a tidal power plant in France, *Water Power (London)*, 26(10):314–322, October 1974; C. V. Davis and K. E. Sorensen, *Handbook of Applied Hydraulics*, 3d ed., 1969; Federal Power Commission, *Hydroelectric Power Resources of the United States, Developed and Undeveloped*, FPC P-42, Jan. 1, 1972; International Commission on Large Dams, *The World's Highest Dams, Largest Earth and Rock Dams, Greatest Man-made Lakes, Largest Hydroelectric Plants*, compiled by T. W. Mermel, Scientific Affairs, Bureau of Reclamation, Department of the Interior, May 1973; 1975 Statistical Report, *Elec. World*, pp. 58–63, Mar. 15, 1975.

Hydrogen

The first chemical element in the periodic system. Under ordinary conditions it is a colorless, odorless, tasteless gas composed of diatomic molecules, H_2. The hydrogen atom, symbol H, consists of a nucleus of unit positive charge and a single electron. It has atomic number 1 and an atomic weight of 1.00797. The element is a major constituent of water and all organic matter, and is widely distributed not only on the Earth but throughout the universe. There are three isotopes of hydrogen: protium, mass 1, makes up 99.98% of the natural element; deuterium, mass 2, makes up about 0.02%; and tritium, mass 3, occurs in extremely small amounts in nature but may be produced artificially by various nuclear reactions.

Although hydrogen had been produced by earlier workers by the reaction of metals with acid, it was Henry Cavendish (1731–1810) who first distinguished it from other flammable gases.

Uses. The largest single use of hydrogen is in the synthesis of ammonia. Ammonia plants are often built adjacent to petroleum refineries or coking plants to utilize by-product hydrogen which might otherwise be wasted. A rapidly expanding use for hydrogen is in petroleum-refining operations, such as hydrocracking and hydrogen treatment for removal of sulfur. Large quantities of hydrogen are consumed in the catalytic hydrogenation of unsaturated liquid vegetable oils to make solid fats. Hydrogenation is used in the manufacture of organic chemicals, such as alcohols from esters and glycerides, amines from nitriles, and cycloparaffins from aromatic hydrocarbons. Methanol is produced commercially by reaction of hydrogen with carbon monoxide. Reaction of hydrogen with chlorine is a source of hydrochloric acid.

Large quantities of hydrogen are used in the United States space program. It is used both as a rocket fuel, in conjunction with oxygen or fluorine, and as a propellant for nuclear-powered rockets. These uses have led to rapid development of the technology of the production and handling of liquid hydrogen. Liquid hydrogen is also used in bubble chambers for studying high-energy particles from nuclear accelerators. Hydrogen gas is used in the oxyhydrogen torch and in the atomic hydrogen torch to produce high temperatures for welding and cutting metals. It is used in the metallurgical industries to reduce metal oxides, such as those of tungsten and molybdenum, to provide a reducing atmosphere in the heat treatment of metals, and in the manufacture of metal hydrides. It is used increasingly in the production of iron from its ores. Although once used extensively to inflate dirigible balloons, it has been largely superseded by helium because of a number of disastrous explosions. When not used directly at the site of manufacture, hydrogen is transported and stored as gas in steel cylinders at a pressure of 120–150 atm, or as liquid in large, well-insulated tanks.

Natural occurrence. Hydrogen, in the free state, is only a minor component of the Earth. It constitutes less than 1 ppm of the atmosphere. It is found in the gases from some volcanoes, oil wells, and coal mines. It may be liberated as a product of the decomposition of organic matter and has been observed in the intestinal gases of animals. In the combined state, hydrogen makes up 0.76% of the weight of the Earth's crust to make it the ninth most abundant element; 13.5% of the atoms of the Earth's crust are hydrogen, exceeded in number only by oxygen and silicon. Most of this hydrogen is present in sea water, of which it constitutes 10.82% by weight. Other important occurrences are in minerals, as hydrates, in the hydrocarbons of petroleum deposits, and in the organic constituents of all living organisms.

Hydrogen is believed to constitute approximately 90% of the atoms in the universe. The thermonuclear energy produced by fusion reactions of hydrogen nuclei is the source of most of the energy radiated by the Sun and other stars.

Physical properties. Ordinary hydrogen has an atomic weight of 1.00797, and a molecular weight of 2.01594. The gas has a density at 0°C and 1 atm of 0.08987 g/liter. Its specific gravity, compared to air, is 0.0695. The lightest substance known, it has a buoyancy in air of 1.203 g/liter. Some additional properties of hydrogen are listed in Table 1.

Hydrogen dissolves in water to the extent of 0.0214 volume per volume of water at 0°C, 0.018 volume at 20°C, and 0.016 volume at 50°C. It is somewhat more soluble in organic solvents, and 0.078 volume dissolves in 1 volume of ethanol at 25°C. Many metals adsorb hydrogen. Palladium is particularly notable in this respect, and dis-

Table 1. Properties of hydrogen

Property	Value
Melting point	−259.2°C
Boiling point at 1 atm	−252.8°C
Density of solid at −259.2°C	0.0866 g/cm³
Density of liquid at −252.8°C	0.0708 g/cm³
Critical temperature	−240.0°C
Critical pressure	13.0 atm
Critical density	0.0301 g/cm³
Specific heat at constant pressure	
Gas at 25°C	3.42 cal/(g)(°C)
Liquid at −256°C	1.93 cal/(g)(°C)
Solid at −259.8°C	0.63 cal/(g)(°C)
Heat of fusion at −259.2°C	14.0 cal/g
Heat of vaporization at −252.8°C	107 cal/g
Thermal conductivity at 25°C	0.000444 cal/(cm)(cm²)(sec)(°C)
Viscosity at 25°C	0.00892 centipoise

solves about 1000 times its volume of the gas. The adsorption of hydrogen in steel may cause "hydrogen embrittlement," which sometimes leads to the failure of chemical processing equipment.

The hydrogen atom has an ionization potential of 13.54 volts. The hydrogen nucleus (proton, mass 1) has a spin of $1/2\,\hbar$ and a magnetic moment of 2.79270 nuclear magnetons. Its absorption cross section for thermal neutrons is 0.332×10^{-24} cm^2.

The hydrogen molecule may exist in either of two forms, known as ortho- and parahydrogen. These distinct forms are possible because the nucleus of the hydrogen atom is spinning in a toplike manner and two atoms may combine with their nuclei spinning in the same direction (ortho) or in opposite directions (para). These spin isomers, as they are called, are ordinarily fairly stable but may be rapidly interconverted by a suitable catalyst, such as activated charcoal or platinized asbestos. The ratio of the two isomers in an equilibrium mixture varies markedly with the temperature: near the absolute zero, equilibrium hydrogen consists entirely of the para modification; at room temperature and above, parahydrogen constitutes 25% and orthohydrogen 75% of the mixture. Pure parahydrogen may be readily prepared by passing liquid hydrogen over activated charcoal. Pure orthohydrogen has been prepared only in small amounts by separating it from parahydrogen by processes such as thermal diffusion or gas chromatography. Physical properties of parahydrogen are measurably different from those of ordinary hydrogen. For example, parahydrogen melts at $-259.34°C$ and boils at $-252.90°C$, whereas ordinary hydrogen melts at $-259.21°C$ and boils at $-252.77°C$. The thermal conductivity of parahydrogen is markedly greater than that of the ortho form; the difference is used in analyzing mixtures of the two. When ordinary hydrogen is liquified, the heat evolved during the slow conversion of the equilibrium mixture to parahydrogen is responsible for the evaporation of large amounts of liquid hydrogen during storage. These losses may be averted by catalytically converting all of the orthohydrogen to the para form during the liquefaction. The orthoparahydrogen interconversion can be catalyzed by free hydrogen atoms. Measurement of the rate of interconversion permits the determination of hydrogenatom concentrations in chemical reactions.

Chemical properties. At ordinary temperatures hydrogen is a comparatively unreactive substance unless it has been activated in some manner, for example, by a suitable catalyst. At elevated temperatures it is highly reactive.

Although ordinarily diatomic, molecular hydrogen dissociates at high temperatures into free atoms according to Eq. (1). The heat of dissociation

$$H_2 \rightleftharpoons 2H \qquad (1)$$

at 25°C is 104.2 kcal/mole. The calculated percentage dissociation is 0.08 at 2000°K, 7.8 at 3000°K, 62.2 at 4000°K, and 95.5 at 5000°K. Atomic hydrogen is also produced when an electrical discharge is passed through hydrogen gas at low pressure, or when a mixture of hydrogen and mercury vapor is irradiated with light of wavelength 2537 A from a mercury arc.

Atomic hydrogen is a powerful reducing agent, even at room temperature. It reacts with the ox-

ides and chlorides of many metals, including silver, copper, lead, bismuth, and mercury, to produce the free metals. It reduces some salts, such as nitrates, nitrites, and cyanides of sodium and potassium, to the metallic state. It reacts with a number of elements, both metals and nonmetals, to yield hydrides such as NaH, KH, H_2S, and PH_3. With oxygen atomic hydrogen yields hydrogen peroxide, H_2O_2. With organic compounds atomic hydrogen reacts to produce a complex mixture of products. With ethylene, C_2H_4, for example, the products include ethane, C_2H_6, and butane, C_4H_{10}. The heat liberated when hydrogen atoms recombine to form hydrogen molecules is used to obtain very high temperatures in atomic hydrogen welding.

Hydrogen reacts with oxygen to form water, as shown by Eq. (2). The heat of reaction is 57.6

$$2H_2 + O_2 \rightarrow 2H_2O \qquad (2)$$

kcal/mole of hydrogen. At room temperature this reaction is immeasurably slow, but is accelerated by catalysts, such as platinum, or by an electric spark, and then may take place with explosive violence. In the absence of catalysts, reaction between hydrogen and oxygen becomes measurable at about 300°C and rapid above 500°C. The reaction is believed to take place in a series of steps constituting chain reaction (3). The intense heat of

$$\begin{aligned} OH + H_2 &\rightarrow H_2O + H \\ H + O_2 &\rightarrow O + OH \qquad (3) \\ O + H_2 &\rightarrow OH + H \end{aligned}$$

the reaction is utilized in the oxyhydrogen torch for cutting and welding metals.

Hydrogen reacts less vigorously with the other group VI elements. The reaction with sulfur, Eq. (4), is exothermic by only 5 kcal/mole; the corresponding reactions with selenium, Eq. (5), and tellurium, Eq. (6), are endothermic.

$$H_2 + S \rightarrow H_2S \qquad (4)$$

$$H_2 + Se \rightarrow H_2Se \qquad (5)$$

$$H_2 + Te \rightarrow H_2Te \qquad (6)$$

The reactivity of hydrogen with the halogens decreases in the order fluorine, chlorine, bromine, and iodine. The reaction with fluorine, Eq. (7), is

$$H_2 + F_2 \rightarrow 2HF \qquad (7)$$

violent, even in the dark at $-252°C$. Properly controlled, it has been used to produce the extremely high temperature of 4000°C. The reaction with chlorine, as shown by Eq. (8), is slow under ordi-

$$H_2 + Cl_2 \rightarrow 2HCl \qquad (8)$$

nary conditions, but may become explosive under the influence of light or heat. As in the case of oxygen, a chain reaction is involved, sequence (9).

$$\begin{aligned} Cl_2 + h\nu &\rightarrow Cl + Cl \\ Cl + H_2 &\rightarrow HCl + H \qquad (9) \\ H + Cl_2 &\rightarrow HCl + Cl, \text{ etc.} \end{aligned}$$

With nitrogen, hydrogen undergoes the important reaction, shown by Eq. (10), to give ammonia.

$$N_2 + 3H_2 \rightarrow 2NH_3 \qquad (10)$$

Hydrogen reacts at elevated temperatures with a number of metals, including lithium, sodium, potassium, calcium, strontium, and barium, to give hydrides. An example is given by Eq. (11).

$$H_2 + 2Li \rightleftharpoons 2LiH \qquad (11)$$

The oxides of many metals are reduced by hydrogen at elevated temperatures either to the free metal or to lower oxides. Metals which can be produced from their oxides in this way include copper, silver, bismuth, mercury, tungsten, iron, nickel, and cobalt. Similarly, the chlorides of silver, copper, nickel, and cobalt react with hydrogen at 300–750°C to give hydrogen chloride and the free metal. Metallic bromides and iodides are more easily reduced than the chlorides.

Hydrogen reacts at room temperature with the salts of the less electropositive metals, such as gold, silver, copper, or mercury, in aqueous solution, and reduces them to the metallic state. These reductions are ordinarily quite slow, but are markedly accelerated if the hydrogen is adsorbed on platinum or palladium.

In the presence of a suitable catalyst, such as platinum, palladium, or nickel, hydrogen reacts with unsaturated organic compounds and adds to the double bond. With ethylene, the reaction is as shown in Eq. (12). Aldehydes and ketones are similarly reduced to alcohols. Acetaldehyde, for example, is reduced to ethyl alcohol, as shown by Eq. (13).

$$H_2C{=}CH_2 + H_2 \rightarrow H_3C{-}CH_3 \qquad (12)$$

larly reduced to alcohols. Acetaldehyde, for example, is reduced to ethyl alcohol, as shown by Eq. (13).

$$CH_3CH{=}O + H_2 \rightarrow CH_3CH_2OH \qquad (13)$$

Principal compounds. Hydrogen is a constituent of a very large number of compounds containing one or more other elements. Such compounds include water, acids, bases, most organic compounds, and many minerals. Compounds in which hydrogen is combined with a single other element are commonly referred to as hydrides. These may be divided into three general classes: the ionic or saltlike hydrides, the covalent or molecular hydrides, and the transitional metal hydrides.

With the halogens, fluorine, chlorine, bromine, and iodine, hydrogen forms hydrides containing 1 atom of hydrogen per atom of halogen. These compounds are usually referred to as hydrogen halides, for example, hydrogen chloride, HCl, or as hydrohalic acids, for example, hydrochloric acid. The hydrogen halides are gases at room temperature. Hydrofluoric acid $(HF)_x$ is polymeric; the others are monomeric. Their physical properties are given in Table 2. The hydrogen halides dissolve in water to give strongly acid solutions. Hydrogen fluoride is commonly prepared from calcium fluoride and sulfuric acid, as shown in Eq. (14). Hydrogen chloride is similarly prepared from

$$CaF_2 + H_2SO_4 \rightarrow CaSO_4 + 2HF \qquad (14)$$

sodium chloride and sulfuric acid. Hydrogen bromide and hydrogen iodide may be prepared by direct union of the elements or by hydrolysis of the corresponding phosphorus trihalides, PBr_3 and PI_3.

Hydrogen forms two compounds with oxygen, water, H_2O, and hydrogen peroxide, H_2O_2. The physical properties of these compounds are com-

Table 2. Properties of hydrogen halides

Property	Hydrogen fluoride, $(HF)_x$	Hydrogen chloride, HCl	Hydrogen bromide, HBr	Hydrogen iodide, HI
Melting point, °C	−83.1	−114.8	−86.9	−50.7
Boiling point, °C	19.5	−84.9	−66.8	−35.4
Density of liquid, g/ml	0.991	1.194	2.77	2.85
At temperature °C	19.5	−86	−67	−47

pared in Table 3. Hydrogen peroxide may be prepared by the hydrolysis of a metal peroxide, as shown by Eq. (15), or by electrolysis reactions. It is

$$Na_2O_2 + 2H_2O \rightarrow H_2O_2 + 2NaOH \qquad (15)$$

unstable, and slowly evolves oxygen on standing. It is a powerful oxidizing agent, and reacts vigorously with many organic compounds.

With sulfur, hydrogen forms the compound hydrogen sulfide, H_2S, a colorless gas with the odor of rotten eggs. It may be prepared by the direct reaction of hydrogen with sulfur, or by treating a metal sulfide with an acid. A series of polysulfides of hydrogen, with the formulas H_2S_2, H_2S_3, H_2S_4, H_2S_5, and H_2S_6, are also known. With selenium and tellurium, hydrogen forms the compounds hydrogen selenide, H_2Se, and hydrogen telluride, H_2Te.

Boron forms a series of covalent hydrides, the best-known member of which is the gas, diborane, B_2H_6. Other members of the series are pentaborane, B_5H_9, a liquid, and decaborane, $B_{10}H_{14}$, a crystalline solid. Because of their high heats of combustion, these compounds and their derivatives have received considerable attention as possible rocket fuels.

The largest class of the covalent hydrides comprises the compounds of hydrogen with carbon, known as hydrocarbons. The first member of this series is methane, CH_4, a gas.

Among the hydrides of nitrogen are ammonia, NH_3, and hydrazine, N_2H_4. Ammonia is a colorless gas which melts at −78°C and boils at −33.3°C. Hydrazine is a liquid with a melting point of 1.4°C and a boiling point of 113.5°C.

Other covalent hydrides are silane, SiH_4; phosphine, PH_3; arsine, AsH_3; and stibine, SbH_3.

Preparation. A large number of methods may be used to prepare hydrogen gas. The choice of method is determined by such factors as the quantity of hydrogen desired, the purity required, and the availability and cost of raw materials. Among the processes frequently used are the reactions of metals with water or acids, the electrolysis of water, the reaction of steam with hydrocarbons or other organic materials, and the thermal decomposition of hydrocarbons.

Small amounts of hydrogen are readily prepared in the laboratory by the reaction of zinc with dilute

Table 3. Properties of hydrogen peroxide and water

Property	Water, H_2O	Hydrogen peroxide, H_2O_2
Melting point, °C	0	−0.9
Boiling point, °C	100	151.4
Density, g/ml	1.000 at 4°C	1.465 at 0°C

hydrochloric acid; this reaction is shown by Eq. (16). The gas may be purified by passing it through

$$Zn + 2HCl \rightarrow ZnCl_2 + H_2 \qquad (16)$$

an acidified solution of potassium permanganate or potassium dichromate and then through a solution of sodium hydroxide. It may be dried by passage through concentrated sulfuric acid or over silica gel. Other metals more electropositive than hydrogen will likewise liberate hydrogen gas upon treatment with water or acid solutions. Thus, metallic sodium reacts violently with water according to Eq. (17). The reaction may be moderated by amal-

$$2Na + 2H_2O \rightarrow 2NaOH + H_2 \qquad (17)$$

gamating the sodium with mercury. The dissolution of iron filings in dilute sulfuric acid has also been frequently used as a source of hydrogen.

For the preparation of somewhat larger quantities of hydrogen, for example, for use in filling balloons, a convenient method is the reaction of calcium hydride with water, as shown by Eq. (18). In

$$CaH_2 + 2H_2O \rightarrow Ca(OH)_2 + 2H_2 \qquad (18)$$

this reaction 1 kg of calcium hydride will produce about 1 m³ of hydrogen. Another process sometimes used for the production of moderate amounts of hydrogen is the dissolution of aluminum or of silicon in alkali, shown by Eqs. (19) and (20).

$$2Al + 2NaOH + 6H_2O \rightarrow 2NaAl(OH)_4 + 3H_2 \qquad (19)$$

$$Si + 4NaOH \rightarrow Na_4SiO_4 + 2H_2 \qquad (20)$$

Hydrogen gas is produced on a large scale industrially. The amount produced annually in the United States is over 100,000,000,000 ft³, excluding that consumed in the manufacture of ammonia and methanol or in petroleum refining. The principal raw materials for hydrogen production are now hydrocarbons, such as natural gas, oil refinery gas, gasoline, fuel oil, and crude oil. In catalytic steam-hydrocarbon reforming, which has become the dominant production process, volatile hydrocarbons are reacted with steam over a nickel catalyst at 700–1000°C to produce carbon oxides and hydrogen. The carbon monoxide formed, as indicated in Eq. (21), using propane as a typical hydrocarbon, is converted to carbon dioxide according to the reaction shown in Eq. (22). The latter

$$C_3H_8 + 3H_2O \rightarrow 3CO + 7H_2 \qquad (21)$$

drocarbon, is converted to carbon dioxide according to the reaction shown in Eq. (22). The latter

$$CO + H_2O \rightarrow CO_2 + H_2 \qquad (22)$$

conversion takes place at 350°C over an iron oxide catalyst.

The steam-hydrocarbon processes have largely replaced the once widely used steam–water gas process. Water gas, a mixture of carbon monoxide and hydrogen, is made by treating coke or coal with steam at a temperature of 1000°C or higher, as shown in Eq. (23). The carbon monoxide is con-

$$C + H_2O \rightarrow CO + H_2 \qquad (23)$$

verted to carbon dioxide as in the hydrocarbon processes. Hydrogen may be separated from carbon dioxide by scrubbing with an aqueous solution of monoethylamine.

Hydrogen is also produced industrially by the electrolysis of water, containing dissolved potas-

sium hydroxide. Although comparatively expensive, this process generates hydrogen of very high purity (over 99.9%). High purity oxygen is a by-product. The overall reaction is Eq. (24).

$$2H_2O \rightarrow 2H_2 + O_2 \qquad (24)$$

Other important industrial processes are the reaction of steam with iron at 800°C, Eq. (25), and

$$Fe + H_2O \rightarrow FeO + H_2 \qquad (25)$$

the thermal dissociation of ammonia, Eq. (26),

$$2NH_3 \rightarrow N_2 + 3H_2 \qquad (26)$$

used for the production of comparatively small amounts of hydrogen for metal treating or for catalytic hydrogenation.

Analytical methods. There are no fully satisfactory reagents for the direct absorption of hydrogen. Colloidal palladium absorbs large amounts of hydrogen, but the reagent is unstable, and it is necessary to remove carbon dioxide, olefins, and carbon monoxide before absorption of the hydrogen. Silver phosphate and silver borate have also been used to absorb hydrogen from a gas mixture.

The most commonly used methods for determining hydrogen in a gas mixture depend on burning the hydrogen to water and measuring the reduction in volume of the gas. The oxidation may be performed by adding excess oxygen and exploding the mixture either by passing an electrical spark between platinum electrodes immersed in the gas, or by passing the gas over a platinum wire heated to about 600°C. Since 1/2 mole of oxygen is required to burn 1 mole of hydrogen, the volume of hydrogen is equal to two-thirds the reduction in volume. Alternatively, the hydrogen may be oxidized by passing it through a tube filled with copper oxide heated to 290°C. The latter procedure is quite specific for hydrogen, because the only other common gas oxidized by copper oxide at this temperature is carbon monoxide, the oxidation of which results in no volume change. Because no oxygen gas is consumed in this procedure, the volume of hydrogen is equal to the measured reduction in volume.

A number of physical measurements may be used to determine hydrogen, especially in mixtures with a single other gas whose identity is known. The exceptionally high thermal conductivity and low density of hydrogen make the measurement of either of these properties a sensitive indicator of hydrogen content. Other physical methods useful for hydrogen analysis include measurement of the velocity of sound in the gas and measurement of the refractive index. The hydrogen content of quite complex gas mixtures can often be determined by mass spectrometry or by gas chromatography.

The determination of chemically bound hydrogen, particularly in organic compounds, is usually performed by burning the compound in a stream of oxygen at about 700°C. The combustion water is absorbed in a dehydrating agent, such as anhydrous magnesium perchlorate; the amount of hydrogen is calculated from the increase in weight of the absorption tube. Alternatively, the combustion water may be determined by gas chromatography, as is done in some automated systems for simultaneous determination of carbon, hydrogen, and nitrogen in organic compounds. Nuclear magnetic resonance (NMR) spectroscopy may be used, not only for the quantitative determination of total bound hydrogen, but also for measurement of hydrogen attached at specific locations in a molecule.   [LOUIS KAPLAN]

Bibliography: A. Farkas, *Orthohydrogen, Parahydrogen and Heavy Hydrogen*, 1935; D. T. Hurd, *An Introduction to the Chemistry of the Hydrides*, 1952; K. M. MacKay, *Hydrogen Compounds of the Metallic Elements*, 1966; J. W. Mellor, *A Comprehensive Treatise on Inorganic and Theoretical Chemistry*, vol. 1, 1922; P. W. Mullen, *Modern Gas Analysis*, 1955; H. Remy, *Treatise on Inorganic Chemistry*, vol. 1, 1956; R. B. Scott (ed.), *Technology and Uses of Liquid Hydrogen*, 1964; M. C. Sneed and R. C. Brasted, *Comprehensive Inorganic Chemistry*, vol. 6, 1957; F. G. A. Stone, *Hydrogen Compounds of the Group IV Elements*, 1962.

Hydrogen-fueled technology

Hydrogen may play a prominent role as a synthetic chemical fuel as fossil fuels become depleted. Future attractive energy sources (solar energy, nuclear fusion, and so on) are generally characterized by their immobility and large scale, so that their broad utility will be dependent on energy carriers, including electricity and synthetic chemical fuels such as hydrogen. The advantages and disadvantages of hydrogen as a synthetic fuel are explored below, and the current state of knowledge in its application to various energy-use sectors in society, especially transportation, is reviewed. The economics of hydrogen production and use are discussed, and a perspective on the place of hydrogen in an overall self-consistent, steady-state future energy metabolism is given. *See* FUEL, SYNTHETIC; HYDROGEN.

Properties of hydrogen. The physical and chemical properties of hydrogen are recalled in Table 1, where particular attention is given to hy-

Table 1. Properties of liquid hydrogen

Property	Value
Boiling point	20.4 K
Liquid density	0.0708 g/cm³
Latent heat of vaporization	108 cal/g
Energy release upon combustion	29,000 cal/g or 2050 cal/cm³ or 1.21×10^5 joule/g
Flame temperature	2483 K
Autoignition temperature	858 K

drogen as a cryogenic liquid. The advantages of hydrogen as a chemical fuel are that: (1) it has the greatest specific energy (energy per unit mass) of any chemical fuel (Fig. 1); (2) its combustion results in only water as a product, with CO, CO_2, and unburned hydrocarbons not present in principle; even oxides of nitrogen are reduced greatly, relative to those accompanying hydrocarbon combustion; and (3) it may be used in every application where hydrocarbons are now used, often with greater combustion efficiency, and lends itself to low-temperature catalytic combustion not practical with hydrocarbons. In addition, hydrogen can be burned with pure oxygen to provide a turbine power system with exceptional thermal efficiency and can be oxidized in fuel cells to produce electricity. Nonenergy uses of hydrogen include desulfurization of coal and petroleum and clean metal ore reduction. Currently, the largest industrial use of hydrogen is in ammonia synthesis (for fertilizer) and for hydrogenation of oils.

The chief disadvantage of hydrogen is its extremely low boiling point (20 K), requiring handling only in special cryogenic containers or dewar vessels as a liquid, or its storage as a metal hydride. Even as a liquid, its energy density per unit volume is about three times poorer than liquid hydrocar-

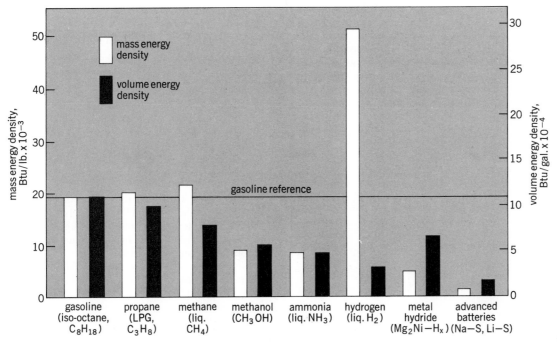

Fig. 1. A graph comparing the energy density characteristics of a number of fuels. ("Advanced batteries" includes adjustment for different energy conversion efficiencies.)

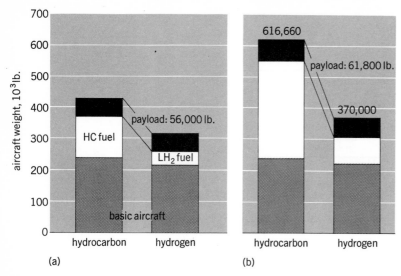

Fig. 2. Comparative aircraft weight breakdowns for subsonic and supersonic transport designs, based on Lockheed California Company estimates. (10^3 lb = 454 kg.) (a) Subsonic transport (modified present design). (b) Supersonic transport (new design).

bons, due to its very low density. A second disadvantage often cited is the fire and explosion hazard, and indeed, the tragedy of the 1937 burning of the German zeppelin *Hindenburg* still haunts the public memory. While the safety problems with hydrogen are different than with liquid hydrocarbons, they are not a priori more severe, and the safety record of liquid hydrogen as a rocket fuel in the United States space program has been excellent.

Hydrogen storage. Hydrogen may be stored as a compressed gas, as a liquid, or as a hydride. Compressed gas storage is the least attractive, except for very small systems, in view of the very low energy density of even highly compressed hydrogen gas and the great weight of the required pressure vessel.

Liquid hydrogen has been used extensively for over a decade, primarily in the space program. Double-walled vessels with vacuum (dewar vessels) or foam insulation are widely used in sizes from small laboratory vessels, through transport truck and rail freight cars of 50,000 to 100,000 liters, to large storage dewars of up to 3,000,000 liters. The loss from small dewars is 1–2% per day

Fig. 3. A Chevrolet modified to burn hydrogen by Billings Energy Research, Inc., of Provo, UT. Hydrogen is stored onboard in a 50-liter liquid dewar.

through evaporation (due to thermal leakage). Hydrogen liquefaction is currently carried out with at best about 40% of the Carnot (ideal) efficiency, and since the energy difference between hydrogen gas at ambient temperature and atmospheric pressure and liquid hydrogen is about 12% of the chemical energy content of the gas, the energy required for liquefaction is about 30% of the available chemical energy. In principle, part of this energy could be beneficially recovered in applications of liquid hydrogen as a fuel. *See* CARNOT CYCLE.

Hydrogen may also be combined with metals to form loosely bound hydrides, which may then be dissociated at elevated temperature. Such hydrides can store hydrogen at a volume density as great as or greater than liquid hydrogen, however at a weight characteristic of the metal. Attractive hydrides are based on Mg-Ni and Mg-Cu alloys. The dissociation temperatures at 1 atm (100,000 N/m^2) are about 250°C. This dissociation heat might be obtained from the exhaust of a hydrogen-fueled engine. Metal hydrides have practical problems, such as the required dissociation heat, and there is much less engineering experience with them than with liquid hydrogen storage. Nevertheless, they provide an alternative storage worthy of serious research.

Aircraft fuel. The most appealing application of hydrogen is as a replacement fuel for aircraft, in view of its excellent mass energy density. Several aircraft companies have considered airframe designs for subsonic commercial jet aircraft and concluded that the same payload and range could be achieved with a much lighter fuel load, hence a lighter overall aircraft, requiring shorter runways and resulting in overall fuel economy. The craft would be more bulky, requiring either attached wing-pod tanks or an expanded fuselage to accommodate the more bulky liquid hydrogen tanks. The relative sophistication of airport crews and the relative security of airports simplify the cryogenic handling and safety problems of hydrogen. Experiments have been made on hydrogen-fueled jet engines, both in static tests and in flight. Even greater advantages accrue in the use of hydrogen fuel in supersonic and hypersonic aircraft, as the cryogenic liquid may then be used to cool the leading-edge airfoil surfaces which are heated by the air friction (see Fig. 2). *See* AIRCRAFT FUEL.

Hydrogen-burning automobiles. The private automobile represents the most difficult application of hydrogen fuel because of the storage problem. There appear at this time two options, either cryogenic or metal hydride storage. In either case, the storage is bulky and, based on present technology, expensive. However, the performance of the hydrogen-burning automobile (Fig. 3) is excellent. The common, noxious pollutants that have come to be identified with the urban environment are absent altogether or (in the case of oxides of nitrogen) greatly reduced. Hydrogen burns well in a lean mixture (in a fuel-air ratio as low as half the stoichiometric value) and provides efficient power at reduced throttle and idling speeds. Because of this and because the freedom from pollution permits high compression ratios and more efficient exhaust systems, experience with hydrogen-powered automobiles points to a fuel energy efficiency as much as 50% greater than with gasoline for a typical driving cycle. Although gas carburetion has been most widely used in hydrogen experiments (as with methane combustion), fuel injection has also been used successfully.

The application of hydrogen fuel to fleet vehicles and to the surface transportation vehicles now powered by diesel engines (trucks, buses, construction and earthmoving vehicles, and railroad locomotives) is in every case simpler technologically and economically than the application to private automobiles. However, private autos are the single largest transportation consumer of energy, and their conversion to a clean fuel would have the greatest impact on air quality of any other single improvement in the energy use patterns of the United States. *See* ATMOSPHERIC POLLUTION; INTERNAL COMBUSTION ENGINE.

Energy storage. One interesting application of hydrogen fuel technology is in electric power generation for energy storage. The demand for electric power in a typical metropolitan area fluctuates with diurnal, weekly, and seasonal variations by over a factor of 2. Hence, the installed generating capacity must be about twice what would be needed for the minimum demand, or about 30% greater than the average. Electric power companies often make use of special, small gas-turbine-driven generators which can be started quickly to handle power demand peaks, and the main steam-generating boilers can be throttled back for power demand minima. Pumped hydroelectric storage is also used in those areas where the terrain is appropriate. The problem becomes more severe when nuclear fission power plants supply most of an electricity grid's power. Fission reactors are sensitive to thermal cycling and are less able to tolerate a "throttling back" than conventional boilers. *See* NUCLEAR POWER; PUMPED STORAGE.

Hydrogen could play a role if surplus power were used to electrolyze water and the resulting hydrogen and oxygen were stored. During demand peaks, these gases would be burned to power turbines to generate the required electric power. Water electrolysis is currently done on a large scale with commercially available units with an energy efficiency of 60−65%. The storage of hydrogen and oxygen might be in pressurized tanks but more probably underground in salt domes, depleted oil or gas wells, or limestone caverns. For power plants near or on large bodies of water, storage under water has been proposed wherein large plastic domes would be anchored to the bottom. By using hydrogen and oxygen rather than hydrogen and air, a combustion temperature of over 3500°C could be achieved; the actual temperature would be regulated by adding water to the combustion chamber. With an initial temperature of 2000°C, a hydrogen-oxygen-driven turbine could generate electricity with an energy efficiency of 60−70% (as compared with a conventional steam boiler-turbine generator efficiency of 40%). Hence, the overall efficiency of the water electrolysis and reburning of hydrogen and oxygen to generate electricity might exceed 40%.

Hydrogen may be combined with pure oxygen or with air in a fuel cell to provide energy with 50−60% efficiency, and indeed it is the fuel of choice for fuel cells. Although they are energy sources of low power density, and hence less attractive for transportation, fuel cells are very promising for electric power peaking systems with hydrogen energy storage. *See* FUEL CELL.

Hydrogen production. Hydrogen might be produced in one of several ways. Currently, the industrial production of about 20,000,000 tons (20×10^9 kg) of hydrogen per year is predominantly by steam reforming of natural gas or petroleum. Where hydrogen is required for its unique properties as in ammonia synthesis or as a rocket fuel, this method is efficient and economical. However, this is not a defensible means of producing hydrogen as a fuel to replace hydrocarbons; the hydrocarbons might better and more economically be used as fuels directly. The use of hydrogen fuel is primarily considered as a replacement for hydrocarbon fluid fuels following their depletion or exhaustion. Hydrogen can be readily produced from coal, however, and would be one of several alternative fuels possible as a result of coal gasification or liquefaction. Steam reacts with coal at about 950°C and high pressure to produce CO, CO_2, and H_2. Subsequent reactions result in only easily separated CO_2 and nearly pure hydrogen. *See* COAL GASIFICATION; COAL LIQUEFACTION.

Where electricity is the immediate product of a thermal cycle (the primary energy source may be nuclear fission, including the breeder reactor, or solar or geothermal energy), hydrogen may be produced by electrolysis with an energy efficiency of at least 60%. The energy efficiency of electrolysis may be increased, for example, by operating at elevated temperature and pressure, albeit with increased first costs. Nuclear fusion may evolve as a major energy source in the 21st century, and the fast neutron energy could then also supply heat for a thermal cycle. An interesting alternative proposed by KMS Fusion, Inc., is to employ the neutrons to directly dissociate water into hydrogen and oxygen. *See* FUSION, NUCLEAR.

The heat developed by a nuclear reactor, solar furnace, or such may also be used directly to drive a series of chemical reactions which result in the separation of water into hydrogen and oxygen. For example, one such sequence, proposed by G. Beni and C. Marchetti, is shown in the equations:

Table 2. Costs of hydrogen fuel production*

Production process	Cost range, per 10^6 Btu
Coal gasification	$1.00−1.70
Nuclear electric	$2.50−5.00
Liquefaction	$1.20−1.80
(Gasoline at $0.36/gal for comparison)	($2.89)

SOURCE: J. Hord (ed.), *Selected Topics on Hydrogen Fuel*, NBS1R 75-803, Cryogenics Division, Institute of Basic Standards, National Bureau of Standards, Boulder, CO, January 1975; J. E. Johnson, Union Carbide Corporation, Linde Division, *The Economics of Liquid Hydrogen Supply for Air Transportation*, Cryogenics Engineering Conference, Atlanta, August 1973; J. Michel (ed.), *Hydrogen and Other Synthetic Fuels*, a summary of work of the Synthetic Fuels Panel, prepared for the Federal Council on Science and Technology R & D Goals Study, USAEC, Rep. T1D 26136, U.S. Government Printing Office, September 1972.

*Costs include plant construction and operation plus input energy. The range of variation reflects both different authors and different prices of coal and nuclear electricity. Distribution, marketing, and taxes are not included.

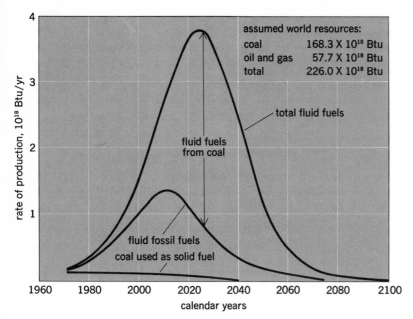

Fig. 4. Projection of rate of production of fluid fossil fuels with and without coal conversion to fluid fuels. (*From M. A. Elliot and N. C. Turner, Estimating the future rate of production of the world's fossil fuels, Texas Eastern Transmission Corporation*)

$$CaBr_2 + 2H_2O \rightarrow 2HBr + Ca(OH)_2 (730°C)$$
$$2HBr + Hg \rightarrow H_2 + HgBr_2 (250°C)$$
$$Ca(OH)_2 + HgBr_2 \rightarrow HgO + CaBr_2 + 2H_2O (200°C)$$
$$HgO \rightarrow Hg + \tfrac{1}{2}O_2 (600°C)$$

The overall energy efficiency of such processes might be 40–60%, significantly better than the overall efficiency of hydrogen production via electric power and electrolysis.

Hydrogen-fuel economy. Inevitably, a driving force of the hydrogen economy, or any other future energy scenario, will be the relative costs of different energy options. Table 2 is a summary of possible cost range estimates for various hydrogen production options. Hydrogen from natural gas or petroleum is currently cheaper than the options listed but is not included for the reasons given above. Hydrogen from coal, the plausible large-scale interim source until solar and fusion energy sources mature, is next in attractiveness. Hydrogen from water electrolysis is naturally closely tied to the cost of electric energy. Cost estimates for direct thermal dissociation are more speculative, as no pilot plant even has been built. The cost of liquefaction is noted as a separate item. *See the feature article* WORLD ENERGY ECONOMY.

Distribution of hydrogen would probably be by gas pipeline, with local liquefaction near the marketing or supply area. Gas pipeline distribution is about 15% more expensive than the corresponding distribution of natural gas, and much cheaper than the corresponding energy distribution costs of electricity. This observation has led to the suggestion that some energy users might be supplied only with pipeline gas hydrogen and might use it to generate electricity locally in a fuel cell. It is then a matter of economic balance whether this option is preferable to parallel distribution of hydrogen and centrally generated electricity.

The driving motivation for the hydrogen fuel economy is the search for a cyclic, indefinitely viable energy metabolism pattern to replace the current reliance on fossil fuels as these approach exhaustion and their recovery ceases to be economically attractive.

While coal is widely heralded as an economically attractive long-range replacement for fossil hydrocarbons (with synthetic hydrocarbon fluid fuels manufactured from coal), it is also limited, albeit at a more future date than petroleum (Fig. 4). Even here, however, caution is necessary in view of the possible buildup of CO_2 in the atmosphere to a level that would severely modify the global climate. Data collected over the past 20 years show that the atmospheric CO_2 is increasing at a rate of 0.4% per year, and increased by a total of 5% between 1950 and 1970. This corresponds to about half of the carbon burned as fuel remaining in the atmosphere as CO_2. Extrapolation to the end of the century indicates that CO_2 will increase another 15%. If the world's readily recoverable coal is all burned, the CO_2 in the atmosphere will increase by a factor of 6! It therefore seems most prudent both to endeavor to totally understand the consequences of carbon release on the global climate and meantime to seriously pursue alternative energy systems not reliant on carbon.

An energy pattern for the 21st century wherein nuclear fusion or solar power provides primary energy, and electricity and synthetic chemical fuels serve as energy carriers, seems logical and perhaps inevitable. Hydrogen appears in many respects to be the most attractive synthetic chemical fuel. *See the feature article* EXPLORING ENERGY CHOICES.   [LAWRENCE W. JONES]

Bibliography: D. P. Gregory, *Sci. Amer.*, 228:1, 13, 1973; J. Hord (ed.), *Selected Topics on Hydrogen Fuel*, NBSIR 75-803, Cryogenics Division, Institute of Basic Standards, National Bureau of Standards, Boulder, CO, January 1975; L. W. Jones, *Environ. Plann. Pollut. Contr.*, 1:12, 1973; L. W. Jones, *Science*, 174:367, 1971; J. Michel (ed.), *Hydrogen and Other Synthetic Fuels*, Rep. TID 26136, U.S. Government Printing Office, September 1972; T. Nejat Veziroglu (ed.), *Hydrogen Energy*, pts. A and B, 1975.

Hysteresis motor

A synchronous motor without salient poles and without dc excitation which makes use of the hysteresis and eddy-current losses induced in its hardened-steel rotor to produce rotor torque. The stator and stator windings are similar to those of an induction motor and may be polyphase, shaded-pole, or capacitor type. The rotor is usually made up of a number of hardened steel rings on a nonmagnetic arbor. The hysteresis motor develops constant torque up to synchronous speed. The motor can, therefore, synchronize any load it can accelerate. These motors are built in small sizes, for instance, for electric clocks. *See* INDUCTION MOTOR; SYNCHRONOUS MOTOR.

[LOYAL V. BEWLEY]

Impedance matching

The use of electric circuits and devices to establish the condition in which the impedance of a load is equal to the internal impedance of the source. This condition of impedance match provides for the maximum transfer of power from the source to the load. In a radio transmitter, for example, it is desired to deliver maximum power from the power amplifier to the antenna. In an audio amplifier, the requirement is to deliver maximum power to the loudspeaker.

The maximum power transfer theorem of electric network theory states that at any given frequency the maximum power is transferred from the source to the load when the load impedance is equal to the conjugate of the generator impedance. Thus, if the generator is a resistance, the load must be a resistance equal to the generator resistance for maximum power to be delivered from the generator to the load. When these conditions are satisfied, the power is delivered with 50% efficiency; that is, as much power is dissipated in the internal impedance of the generator as is delivered to the load.

Impedance matching network. In general, the load impedance will not be the proper value for maximum power transfer. A network composed of inductors and capacitors may be inserted between the load and the generator to present to the generator an impedance that is the conjugate of the generator impedance. Since the matching network is composed of elements which, in the ideal case of no resistance in the inductors and perfect capacitors, do not absorb power, all of the power delivered to the matching network is delivered to the load. An example of an L-section matching network is illustrated. Matching networks of this type are used in radio-frequency circuits. The values of inductance and capacitance are chosen to satisfy the requirements of the maximum power transfer theorem. The power dissipated in the matching network is a small fraction of that delivered to the load, because the elements used are close approximations of ideal components.

Transformers. The impedance measured at the terminals of one winding of an iron-cored transformer is approximately the value of the impedance connected across the other terminals multiplied by the square of the turns ratio. Thus, if the load and generator impedances are resistances, the turns ratio can be chosen to match the load resistance to the generator resistance for maximum power transfer. If the generator and load impedances contain reactances, the transformer cannot be used for matching because it cannot change the load impedance to the conjugate of the generator impedance (the L-section matching network can). The turns ratio can be chosen, however, to deliver maximum power under the given conditions, this maximum being less than the theoretical one.

Iron-cored transformers are used for impedance matching in the audio and supersonic frequency range. The power dissipated in the core increases with frequency because of hysteresis. Above the frequency range at which iron-cored transformers can be used, the air-core transformer or transformers with powdered-iron slugs can be used effectively. However, in these cases the turns-ratio-squared impedance-transforming property is no longer true. Since the transformer is usually part of a tuned circuit, other factors influence the design of the transformer.

The impedance-transforming property of an iron-cored transformer is not always used to give maximum power transfer. For example, in the de-

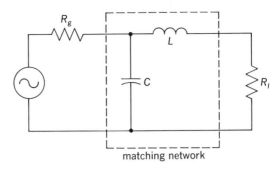

L-section impedance matching network.

sign of power-amplifier stages in audio amplifiers, the impedance presented to the transistor affects distortion. A study of a given circuit can often show that at a given output power level, usually the maximum expected, there is a value for the load resistance which will minimize a harmonic component in the harmonic distortion, such as the second or third harmonic. The transformer turns ratio is selected to present this resistance to the transistor. *See* TRANSFORMER.

Emitter follower. In electronic circuitry a signal source of large internal impedance must often be connected to a low-impedance load. If the source were connected directly to the load, attenuation of the signal would result. To reduce this attenuation, an emitter follower is connected between the source and the load. The input impedance of the emitter follower is high, more nearly matching the large source impedance, and the output impedance is low, more nearly matching the low load impedance. If the object were the delivery of maximum power to the load, it might be possible to design the emitter follower to have an output resistance equal to the load resistance, assuming that the load is a resistance. (Special audio amplifiers have been designed to use emitter followers, rather than a transformer, to connect the loudspeaker to the power amplifier.) In many cases, maximum power transfer is not the goal; the emitter follower is introduced primarily to reduce to a minimum the attenuation of the signal.

There exist a number of applications where the emitter follower is not useful as an impedance matching circuit. For example, if a very-low-impedance source must be matched to a high-impedance load, then a transistor is used in the common base configuration.

[CHRISTOS C. HALKIAS]

Bibliography: E. W. Kimbark, *Electrical Transmission of Power and Signals*, 1949; J. Millman and C. C. Halkias, *Electronic Devices and Circuits*, 1967; M. E. Van Valkenburg, *Network Analysis*, 1964.

Impulse turbine

A prime mover in which fluid (water, steam, or hot gas) under pressure enters a stationary nozzle where its pressure (potential) energy is converted to velocity (kinetic) energy. The accelerated fluid then impinges on the blades of a rotor, imparting its energy to the blades to produce rotation and overcome the connected rotor resistance. The impulse principle is basic to many turbines.

The impulse principle can be distinguished from

IMPULSE TURBINE

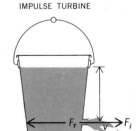

Fig. 1. Water escaping through a nozzle near the base of a bucket, free to swing, illustrating the impulse F_i and reaction F_r forces of the jet issuing with a velocity v.

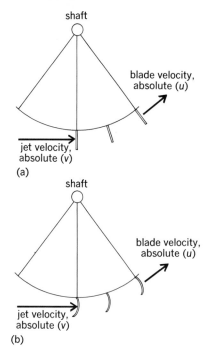

Fig. 2. Jet impinging on a series of (a) flat blades and (b) curved blades on periphery of a wheel.

the reaction principle by considering the flow of water from a hole near the bottom of a bucket (Fig. 1). The hole is a nozzle that serves to convert potential energy to the kinetic form ΔE; the impulse force F_i in the issuing jet is given by the expression $\Delta E = F_i v/2$, where v is the velocity of the jet.

If the jet is allowed to impinge on a series of vanes mounted on the periphery of a wheel, the impulse force can overcome the resistance connected to the shaft (Fig. 2). The efficiency of the device is ideally dependent upon the vane curvature and the absolute vane velocity. For flat blades (Fig. 2a), the efficiency cannot exceed 50%. With curved blades and complete reversal of the jet (Fig. 2b), the efficiency will be 100% when the vane velocity u is one-half the jet velocity v.

If the bucket in Fig. 1 is suspended and free to move, there will be, by Newton's third law of motion, a reaction force F_r equal and opposite to the impulse force F_i. For maximum efficiency (100%) the swinging bucket will have to move with an absolute tip velocity u equal to the jet velocity, v. This is the reaction principle and is demonstrated by the Barker's mill (Fig. 3).

The basic difference between the reaction principle and the impulse-turbine principle is determined by the presence or absence of moving nozzles. A nozzle is a throat section device in which there is a drop in pressure with consequent acceleration of the emerging fluid. An impulse turbine has stationary nozzles only. A reaction turbine must have moving nozzles but may have stationary nozzles also so that the fluid can reach the moving nozzle. This is the usual condition for any practical reaction turbine.

The idealized vector diagrams of Fig. 4 demonstrate distinguishing features for the construction which uses a row of blades mounted on the periphery of a wheel and for which flow is axially through the blade passages from one side of the wheel to the other. The theoretical condition further presupposes complete (180°) reversal of the jet and no friction losses. In Fig. 4, v is the absolute velocity of the fluid, u is the absolute velocity of the moving blade, and w is the relative velocity of the fluid with respect to the moving blade. Subscripts 1 and 2 apply to entrance and exit conditions, respectively. The vectors of the illustration demonstrate that maximum efficiency (zero residual absolute fluid velocity v_2) obtains (1) when the blade speed is half the jet speed for impulse turbines and (2) when the blade speed is equal to the jet speed for reaction turbines.

Figure 5 shows the situation more practically as complete reversal of the jet is not realistic. The

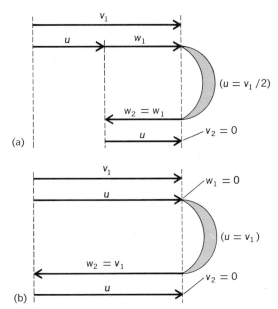

Fig. 4. Velocity vector diagrams for idealized axial fluid flow on (a) impulse-turbine blading and (b) reaction-turbine blading; 180° jet reversal. In b the fluid is accelerated to w across the moving blades (nozzles). In both cases the leaving absolute fluid velocity v is zero, giving maximum efficiency.

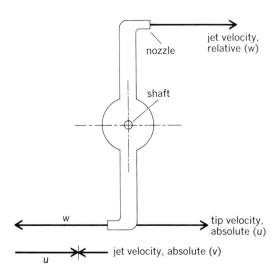

Fig. 3. Barker's mill, a reaction-type jet device, illustrating the application of moving nozzles and the resultant absolute and relative velocities.

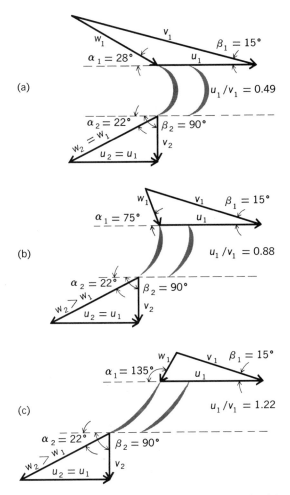

Fig. 5. Vector diagrams for idealized (frictionless) axial fluid flow on (a) impulse-turbine blading and (b, c) reaction-turbine blading; jet reversal less than 180°. Exit vector triangles are identical with consequent equal efficiencies. Resultant speed ratios (u_1/v_1) are 0.49, 0.88, and 1.22, respectively.

vector diagrams of Fig. 5 show speed ratios of 0.49, 0.88, and 1.22 for reasonable degrees of jet reversal and no friction losses. By varying the angle of entrance α_1 to the moving blades, a wide range of speed ratios is available to the designer. In each case shown in Fig. 5, the stationary nozzle entrance angle β_1 is fixed at 15°; the exit vector triangle from the moving blades is identical in all cases. The data in Fig. 5c demonstrate that it is entirely possible to have speed ratios greater than unity with the reaction principle. This condition is utilized in many designs of hydraulic turbines, for example, propeller and Kaplan units.

The choice of details is at the discretion of the turbine designer. He can determine the extent to which the basic principles of impulse and reaction should be applied for a practical, reliable, economic unit in the fields of hydraulic turbines, gas turbines, and steam turbines. *See* GAS TURBINE; HYDRAULIC TURBINE; REACTION TURBINE; STEAM TURBINE.

[THEODORE BAUMEISTER]

Bibliography: T. Baumeister (ed.), *Standard Handbook for Mechanical Engineers*, 7th ed., 1967; G. T. Csanady, *Theory of Turbomachines*, 1964; D. G. Shepherd, *Principles of Turbomachinery*, 1956; V. L. Streeter, *Handbook of Fluid Dynamics*, 1961.

Induction, electromagnetic

The production of an electromotive force either by motion of a conductor through a magnetic field in such a manner as to cut across the magnetic flux or by a change in the magnetic flux that threads a conductor.

Motional electromotive force. A charge moving perpendicular to a magnetic field experiences a force that is perpendicular to both the direction of the field and the direction of motion of the charge. In any metallic conductor, there are free electrons, electrons that have been temporarily detached from their parent atoms.

If a conducting bar (Fig. 1) moves through a magnetic field, each free electron experiences a force due to its motion through the field. If the direction of the motion is such that a component of the force on the electrons is parallel to the conductor, the electrons will move along the conductor. The electrons will move until the forces due to the motion of the conductor through the magnetic field are balanced by electrostatic forces that arise because electrons collect at one end of the conductor, leaving a deficit of electrons at the other. There is thus an electric field along the rod, and hence a potential difference between the ends of the rod while the motion continues. As soon as the motion stops, the electrostatic forces will cause the electrons to return to their normal distribution.

From the definition of magnetic induction (flux density) B, the force on a charge q due to the motion of the charge through a magnetic field is given by Eq. (1), where the force F is at right angles to a

$$F = Bqv \sin \theta \qquad (1)$$

plane determined by the direction of the field, and the component $v \sin \theta$ of the velocity is perpendicular to the field. When B is in webers/m², q is in coulombs, and v is in meters/sec, the force is in newtons. *See* INDUCTION, MAGNETIC.

The electric field intensity E due to this force is given in magnitude and direction by the force per unit positive charge. The electric field intensity is equal to the negative of the potential gradient along the rod. In motional electromotive force (emf), the charge being considered is negative. Thus, Eqs. (2) hold. Here l is the length of the conductor in a direction perpendicular to the field, and $v \sin \theta$ is the component of the velocity that is per-

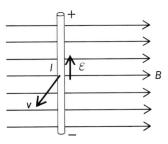

Fig. 1. Flux density B, motion v, and induced emf$_e$ when a conductor of length l moves in a uniform field. (*From R. L. Weber, M. W. White, and K. V. Manning, Physics for Science and Engineering, McGraw-Hill, 1957*)

pendicular to the field. If B is in webers/m², l is in meters, and v is in meters/sec, the emf $\mathscr{E}$ is in volts.

$$E = \frac{F}{-q} = -Bv \sin \theta = -\frac{\mathscr{E}}{l} \quad (2)$$

$$\mathscr{E} = Blv \sin \theta$$

This emf exists in the conductor as it moves through the field whether or not there is a closed circuit. A current would not be set up unless there were a closed circuit, and then only if the rest of the circuit does not move through the field in exactly the same manner as the rod. For example, if the rod slides along stationary tracks that are connected together, there will be a current in the closed circuit. However, if the two ends of the rod were connected by a wire that moved through the field with the rod, there would be an emf induced in the wire that would be equal to that in the rod and opposite in sense in the circuit. Therefore, the net emf in the circuit would be zero, and there would be no current.

Emf due to change of flux. When a coil is in a magnetic field, there will be a flux Φ threading the coil the magnitude of which will depend upon the area of the coil and its orientation in the field. The flux is given by $\Phi = BA \cos \theta$, where A is the area of the coil and θ is the angle between the normal to the plane of the coil and the magnetic field. Whenever there is a change in the flux threading the coil, there will be an induced emf in the coil while the change is taking place. The change in flux may be caused by a change in the magnetic induction of the field or by a motion of the coil. The magnitude of the induced emf, Eq. (3), depends upon the

$$\mathscr{E} = -N\frac{d\Phi}{dt} \quad (3)$$

number of turns of the coil N and upon the rate of change of flux. The negative sign in Eq. (3) refers to the direction of the emf in the coil; that is, it is always in such a direction as to oppose the change that causes it, as required by Lenz's law. If the change is an increase in flux, the emf would be in a direction to oppose the increase by causing a flux in a direction opposite to that of the increasing flux; if the flux is decreasing, the emf is in such a direction as to oppose the decrease, that is, to produce a flux that is in the same direction as the decreasing flux.

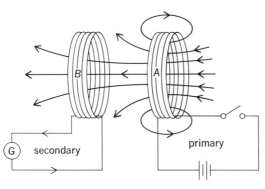

Fig. 2. Mutual induction. An emf is induced in the secondary when the current changes in the primary. (*R. L. Weber, M. W. White, K. V. Manning, Physics for Science and Engineering, McGraw-Hill, 1957*)

Consider the case of a flat coil of area A rotating with uniform angular velocity ω about an axis perpendicular to a uniform magnetic field of flux density B. For any position of the coil, the flux threading the coil is $\Phi = BA \cos \theta = BA \cos \omega t$, where the zero of time is taken when θ is zero and the normal to the plane of the coil is parallel to the field. Then the emf induced as the coil rotates is given by Eq. (4).

$$\mathscr{E} = -N\frac{d\Phi}{dt}$$

$$= -NBA\frac{d(\cos \theta)}{dt} = NBA\omega \sin \omega t \quad (4)$$

The induced emf is sinusoidal, varying from zero when the plane of the coil is perpendicular to the field to a maximum value when the plane of the coil is parallel to the field.

Self-induction. If the flux threading a coil is produced by a current in the coil, any change in that current will cause a change in flux, and thus there will be an induced emf while the current is changing. This process is called self-induction. The emf of self-induction is proportional to the rate of change of current. The ratio of the emf of induction to the rate of change of current in the coil is called the self-inductance of the coil.

Mutual induction. The process by which an emf is induced in one circuit by a change of current in a neighboring circuit is called mutual induction. Flux produced by a current in a circuit A (Fig. 2) threads or links circuit B. When there is a change of current in circuit A, there is a change in the flux linking coil B, and an emf is induced in circuit B while the change is taking place. Transformers operate on the principle of mutual induction. *See* TRANSFORMER.

The mutual inductance of two circuits is defined as the ratio of the emf induced in one circuit B to the rate of change of current in the other curcuit A.

Coupling coefficient. This refers to the fraction of the flux of one circuit that threads the second circuit. If two coils A and B having turns N_A and N_B, respectively, are so related that all the flux of either threads both coils, the respective self-inductances are given by Eqs. (5) and the mutual inductance of the pair is given by Eq. (6). Then Eqs. (7) hold.

$$L_A = \frac{N_A\Phi_A}{I_A} \qquad L_B = \frac{N_B\Phi_B}{I_B} \quad (5)$$

$$M = \frac{N_A\Phi_B}{I_B} = \frac{N_B\Phi_A}{I_A} \quad (6)$$

$$M^2 = \frac{N_AN_B\Phi_A\Phi_B}{I_AI_B} = \frac{N_A\Phi_A}{I_A}\frac{N_B\Phi_B}{I_B} = L_AL_B \quad (7)$$

$$M = \sqrt{L_AL_B}$$

In general, not all the flux from one circuit threads the second. The fraction of the flux from circuit A that threads circuit B depends upon the distance between the two circuits, their orientation with respect to each other, and the presence of a ferromagnetic material in the neighborhood, either as a core or as a shield. It follows that for the general case that Eq. (8) holds.

$$M \leqq \sqrt{L_AL_B} \quad (8)$$

The ratio of the mutual inductance of the pair to the square root of the product of the individual self-inductances is called the coefficient of coupling K, given by Eq. (9). The coupling coefficient has a

$$K = \frac{M}{\sqrt{L_A L_B}} \qquad (9)$$

maximum value of unity if all the flux threads both circuits, zero if none of the flux from one circuit threads the other. For all other conditions, K has a value between 0 and 1.

Applications. The phenomenon of electromagnetic induction has a great many important applications in modern technology. For example, see GENERATOR, ELECTRIC; INDUCTION HEATING; MOTOR, ELECTRIC.   [KENNETH V. MANNING]

Bibliography: S. S. Attwood, *Electric and Magnetic Fields,* 3d ed., 1949; L. Page and N. I. Adams, *Principles of Electricity and Magnetism,* 3d ed. 1958; E. M. Purcell, *Electricity and Magnetism,* vol. 2, 1965; F. W. Sears, *Principles of Physics,* vol. 2, 1951; R. P. Winch, *Electricity and Magnetism,* 1955.

Induction, magnetic

A vector quantity that is used as a quantitative measure of a magnetic field. It is defined in terms of the force on a charge moving in the field by Eq. (1), where B is the magnitude of the magnetic in-

$$B = \frac{F}{qv \sin \theta} \qquad (1)$$

duction and F is the force on the charge q, which is moving with speed v in a direction making an angle θ with the direction of the field. The direction of the vector quantity B is the direction in which the force on the moving charge is zero.

The magnetic induction may also be expressed in terms of the force F on a current element of length l and current I, as in Eq. (2). The meter-

$$B = \frac{F}{Il \sin \theta} \qquad (2)$$

kilogram-second (mks) unit of magnetic induction is derived from this equation by expressing the force in newtons, the current in amperes, and the length in meters. The unit of B is thus the newton/ampere-meter.

Magnetic flux density is the magnetic flux per unit area through a surface perpendicular to the magnetic induction. Magnetic flux density and magnetic induction are equivalent terms. *See* MAGNETIC FIELD.

The magnetic induction may be represented by lines that are drawn so that at every point in the field the tangent to the line is in the direction of the magnetic induction. To represent the magnetic induction qualitatively, as many such lines may be drawn as are necessary to portray the field. If, however, the lines are to represent the magnetic induction quantitatively, an arbitrary choice must be made for the number of lines to represent a given condition. One such choice is that in which the number of lines per square meter of a surface perpendicular to B is set equal to the value of B. These lines are called magnetic flux. One line of induction as here selected is called a flux of 1 weber. The corresponding unit of flux density is then the weber per square meter. From the manner of

defining flux used here, it follows that the weber per square meter is equivalent to the newton/ampere-meter.

Another unit of flux density is defined by using centimeter-gram-second (cgs) units in both the defining equation for magnetic induction and in the area in which there is unit flux. This cgs unit of flux density is called the gauss. The relationship between the gauss and the weber is given by 1 weber/m² = 10⁴ gauss.

[KENNETH V. MANNING]

Bibliography: D. S. Parasuis, *Magnetism,* 1961; E. R. Peck, *Electricity and Magnetism,* 1953; E. M. Purcell, *Electricity and Magnetism,* vol. 2, 1965; F. W. Sears, *Principles of Physics,* vol. 2, 1951.

Induction heating

The heating of a nominally electrical conducting material by currents induced by a varying electromagnetic field.

The principle of the induction heating process is similar to that of a transformer. In Fig. 1, the inductor coil can be considered the primary winding of a transformer, with the workpiece as a single-turn secondary. When an alternating current flows in the primary coil, secondary currents will be induced in the workpiece. These induced currents are called eddy currents. The current flowing in the workpiece can be considered as the summation of all of the eddy currents.

In the design of conventional electrical apparatus, the losses due to induced eddy currents are minimized because they reduce the overall efficiency. However, in induction heating, their maximum effect is desired. Therefore close spacing is used between the inductor coil and the workpiece, and high coil currents are used to obtain the maximum induced eddy currents and therefore high heating rates.

Applications. Induction heating is widely employed in the metal working industry to heat metals for soldering, brazing, annealing, hardening, and for induction melting and sintering.

As compared to other conventional processes, it has these inherent advantages:

1. Heating is induced directly into the material. It is therefore an extremely rapid method of heating. It is not limited by the relatively slow rate of heat diffusion in conventional processes using surface-contact or radiant heating methods.

2. Because of skin effect, the heating is localized and the heated area is easily controlled by the shape and size of the inductor coil.

3. Induction heating is easily controllable, resulting in uniform high quality of the product.

4. It lends itself to automation, in-line processing, and automatic process cycle control.

5. Start-up time is short, and standby losses are low or nonexistent.

6. Working conditions are better because of the absence of noise, fumes, and radiated heat.

Heating process. The induced currents in the workpiece flow roughly parallel to the current in the inductor coil turns.

Because of the skin effect, these induced currents concentrate near the surface of the workpiece. The effective depth of current penetration is greater for lower frequencies than for higher frequencies, and is greater for high-resistivity metals

INDUCTION HEATING

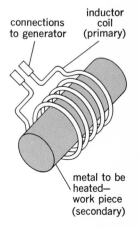

connections to generator

inductor coil (primary)

metal to be heated— work piece (secondary)

Fig. 1. Basic elements of induction heating.

air-setting refractory cement

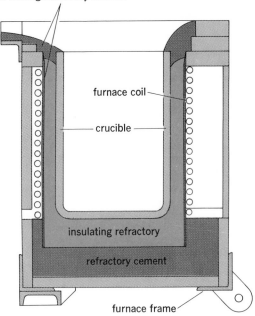

Fig. 2. Cross section of a typical induction melting furnace. (*Inductotherm Corp.*)

Fig. 3. Manually loaded induction heater for hardening of gears. Gear is loaded on spindle under coil in center. Spindle raises gear for heating and lowers it into oil quench for hardening. (*Westinghouse Electric Corp.*)

than for low-resistivity metals. In magnetic materials such as steel, the depth of current penetration is less below the Curie temperature (approximately 1350°F where the steel is magnetic) than it is above the Curie temperature.

For efficient heating, the frequency used must be high enough to make the depth of current penetration considerably less than the thickness of the workpiece, measured at right angles to the coil turns. The table shows the range of frequencies used for applications of the induction heating process.

In mass heating applications the lowest frequency consistent with efficient heating is used, because as a rule the initial cost of equipment goes up with frequency. When the workpieces are small, or when it is desired to concentrate the heating near the surface, as in surface hardening, it is necessary to use higher frequencies, even though efficient power transfer could be accomplished at some lower frequency. In large production use, such as forging, two separate frequencies, with separate inductor coils, are employed to maximize production at minimum equipment cost.

Power sources. The equipment used as power sources depends on the frequency range for the application. When line frequencies (generally 60 cycles) are used, suitable transformers, power factor correction capacitors, and control equipment are required.

For higher frequencies, up to 10,000 Hz, inductor-type alternators are used. These are usually driven by induction motors and are available in ratings from $7\frac{1}{2}$ to over 300 kw.

Converters are used for the 10,000–60,000 Hz range, principally for small-scale melting. These produce the desired frequency by repeatedly charging a large capacitor from the 60-Hz line and discharging it through an output circuit tuned to the desired frequency. The output is a train of damped oscillations.

For frequencies above 200 kHz, vacuum-tube oscillators are used. These are self-excited, and are complete with high-voltage rectifier, oscillator tank circuit, controls, and instrumentation. When operating from a three-phase supply, they put out a continuous wave of rf power.

Frequencies used in induction and dielectric heating

Frequency, Hz	Source of power	Uses
60 – 960	Rotating generators or converters	Mass induction heating for forging, forming, extrusion, or preheating
960 – 10,000	Motor-generator sets	Induction heating for melting, heat-treating, and hardening
10,000 – 60,000	Converters	Induction heating for small-scale melting and sintering
200,000 – 550,000	Vacuum-tube oscillators	Surface induction heating for brazing, soldering, hardening, and strip and wire heating
2,000,000 – 90,000,000	Vacuum-tube oscillators	Dielectric heating

Process use. Induction heating is used for many heat processes, as shown in the table. The construction of a typical melting furnace is shown in Fig. 2. An induction heater used for hardening is shown in Fig. 3.

Induction heating differs from other methods of heat treating in that it heats the metals very rapidly, and that holding time at hardening temperature approaches zero. A minimum of time is therefore available for metallurgical reactions, and this has a significant influence on the selection of steel to be used. For other electric heating methods *see* HEATING, ELECTRIC.

[CARL P. BERNHARDT]

Bibliography: D. W. Brown, *Induction Heating Practice*, 1956; G. H. Brown, C. N. Hoyler, and R. A. Bierwirth, *Theory and Application of Radio-Frequency Heating*, 1947; J. W. Cable, *Induction and Dielectric Heating*, 1954; P. G. Simpson, *Induction Heating: Coil and System Design*, 1960; C. A. Tudbury, *Basics of Induction Heating*, 1960.

Induction motor

An alternating-current motor in which the currents in the secondary winding (usually the rotor) are created solely by induction. These currents result from voltages induced in the secondary by the magnetic field of the primary winding (usually the stator). An induction motor operates slightly below synchronous speed and is sometimes called an asynchronous (meaning not synchronous) motor.

Induction motors are the most commonly used electric motors because of their simple construction, efficiency, good speed regulation, and low cost. Polyphase induction motors come in all sizes and find wide use where polyphase power is available. Single-phase induction motors are found mainly in fractional-horsepower sizes, and those up to 25 hp are used where only single-phase power is available.

POLYPHASE INDUCTION MOTORS

There are two principal types of polyphase induction motors: squirrel-cage and wound-rotor machines. The differences in these machines is in the construction of the rotor. The stator construction is the same and is also identical to the stator of a synchronous motor. Both squirrel-cage and wound-rotor machines can be designed for two- or three-phase current.

Stator. The stator of a polyphase induction motor produces a rotating magnetic field when supplied with balanced, polyphase voltages (equal in magnitude and 90 electrical degrees apart for two-phase motors, 120 electrical degrees apart for three-phase motors). These voltages are supplied to phase windings, which are identical in all respects. The currents resulting from these voltages produce a magnetomotive force (mmf) of constant magnitude which rotates at synchronous speed. The speed is proportional to the frequency of the supply voltage and inversely proportional to the number of poles constructed on the stator.

Figure 1 is a simplified diagram of a three-

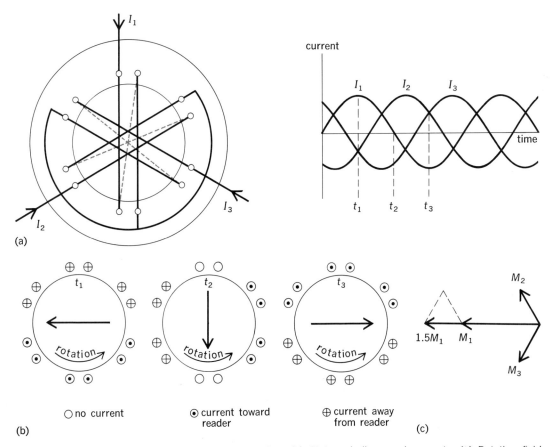

(a)

(b)

○ no current   ◉ current toward reader   ⊕ current away from reader   (c)

Fig. 1.   Three-phase, two-pole, Y-connected stator of induction motor supplied with currents I_1, I_2, and I_3.

(a) Stator windings and currents. (b) Rotating field. (c) Magnetomotive forces produced by stator winding.

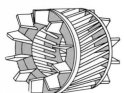

Fig. 2. Bars, end rings, and cooling fins of a squirrel-cage rotor.

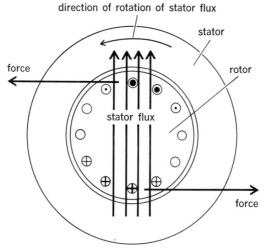

direction of rotation of stator flux

stator

force

rotor

stator flux

force

⦿ current toward reader

⊕ current away from reader

weight of ● or + indicates magnitude of current

Fig. 3. Forces on the rotor winding.

phase, two-pole, Y-connected stator supplied with currents I_1, I_2, and I_3. Each stator winding produces a pulsating mmf which varies sinusoidally with time. The resultant mmf of the three windings (Fig. 1c) is constant in magnitude and rotates at synchronous speed. Figure 1b shows the direction of the mmf in the stator for times t_1, t_2, and t_3 shown in Fig. 1a and shows how the resultant mmf rotates. The synchronous speed N_s is shown by Eq. (1), where f is the frequency in hertz and p is the

$$N_s = \frac{120f}{p} \quad \text{rpm} \quad (1)$$

number of stator poles. For any given frequency of operation, the synchronous speed is determined by the number of poles. For 60-Hz frequency, a two-pole motor has a synchronous speed of 3600 rpm; a four-pole motor, 1800 rpm; and so on.

Squirrel-cage rotor. Figure 2 shows the bars, end rings, and cooling fins of a squirrel-cage rotor. The bars are skewed or angled to prevent cogging (operating below synchronous speed) and to reduce noise. The end rings provide paths for currents that result from the voltages induced in the rotor bars by the stator flux. The number of poles on a squirrel-cage rotor is always equal to the number of poles created by the stator winding.

Figure 3 shows how the two motor elements interact. A counterclockwise rotation of the stator flux causes voltages to be induced in the top bars of the rotor in an outward direction and in the bottom bars in an inward direction. Currents will flow in these bars in the same direction. These currents interact with the stator flux and produce a force on the rotor bars in the direction of the rotation of the stator flux.

When not driving a load, the rotor approaches synchronous speed N_s. At this speed there is no motion of the flux with respect to the rotor conductors. As a result, there is no voltage induced in the rotor and no rotor current flows. As load is applied,

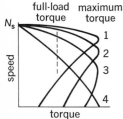

full-load torque   maximum torque

N_s

speed

torque

1
2
3
4

Fig. 4. Speed-torque characteristic of polyphase induction motor.

the rotor speed decreases slightly, causing an increase in rotor voltage and rotor current and a consequent increase in torque developed by the rotor. The reduction in speed is therefore sufficient to develop a torque equal and opposite to that of the load. Light loads require only slight reductions in speed; heavy loads require greater reduction. The difference between the synchronous speed N_s and the operating speed N is the slip speed. Slip s is conveniently expressed as a percentage of synchronous speed, as in Eq. (2).

$$s = \frac{N_s - N}{N_s} \times 100\% \quad (2)$$

When the rotor is stationary, a large voltage is induced in the rotor. The frequency of this rotor voltage is the same as that of the supply voltage. The frequency f_2 of rotor voltage at any speed is shown by Eq. (3), where f_1 is the frequency of the

$$f_2 = f_1 s \quad (3)$$

supply voltage and s is the slip expressed as a decimal. The voltage e_2 induced in the rotor at any speed is shown by Eq. (4), where e_{2s} is the rotor

$$e_2 = (e_{2s})s \quad (4)$$

voltage at standstill. The reactance x_2 of the rotor is a function of its standstill reactance x_{2s} and slip, as shown by Eq. (5). The impedance of the rotor at

$$x_2 = (x_{2s})s \quad (5)$$

any speed is determined by the reactance x_2 and the rotor resistance r_2. The rotor current i_2 is shown by Eq. (6). In the equation, for small

$$i_2 = \frac{e_2}{\sqrt{r_2^2 + x_2^2}}$$
$$= \frac{(e_{2s})s}{\sqrt{r_2^2 + (x_{2s})^2 s^2}} = \frac{e_{2s}}{\sqrt{\left(\frac{r_2}{s}\right)^2 + (x_{2s})^2}} \quad (6)$$

values of slip, the rotor current is small and possesses a high power factor. When slip becomes large, the r_2/s term becomes small, current increases, and the current lags the voltage by a large phase angle. Standstill (or starting) current is large and lags the voltage by 50–70°. Only in-phase, or unity powerfactor, rotor currents are in space phase with the air-gap flux and can therefore produce torque. The current i_2 contains both a unity power-factor component i_p and a reactive component i_r. The maximum value of i_p and therefore maximum torque are obtained when slip is of the correct value to make r_2/s equal to x_{2s}. If the value of r_2 is changed, the slip at which maximum torque is developed must also change. If r_2 is doubled and s is doubled, the current i_2 is not changed and the torque is unchanged.

This feature provides a means of changing the speed-torque characteristics of the motor. In Fig. 4, curve 1 shows a typical characteristic curve of an induction motor. If the resistance of the rotor bars were doubled without making any other changes in the motor, it would develop the characteristic of curve 2, which shows twice the slip of curve 1 for any given torque. Further increases in the rotor resistance could result in curve 3. When r_2 is made equal to x_{2s}, maximum torque will be

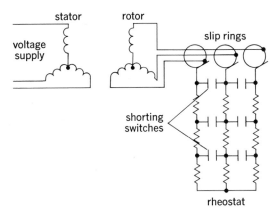

Fig. 5. Connections of wound-rotor induction motor.

developed at standstill, as in curve 4. These curves show that higher resistance rotors give higher starting torque. However, since the motor's normal operating range is on the upper portion of the curve, the curves also show that a higher-resistance rotor results in more variation in speed from no load to full load (or poorer speed regulation) than the low-resistance rotor. Higher-resistance rotors also reduce motor efficiency. Except for their characteristic low starting torque, low-resistance rotors would be desirable for most applications.

Wound rotor. A wound-rotor induction motor can provide both high starting torque and good speed regulation. This is accomplished by adding external resistance to the rotor circuit during starting and removing the resistance after speed is attained.

The wound rotor has a polyphase winding similar to the stator winding and must be wound for the same number of poles. Voltages are induced in these windings just as they are in the squirrel-cage rotor bars. The windings are connected to slip rings so that connections may be made to external impedances, usually resistors, to limit starting currents, improve power factor, or control speed.

Figure 5 shows the connection of a rheostat used to bring a wound-rotor motor up to speed. The rheostat limits the starting current drawn from the supply to a value less than that required by a squirrel-cage motor. The resistance is gradually reduced to bring the motor up to speed. By leaving various portions of the starting resistances in the circuit, some degree of speed control can be obtained, as in Fig. 4. However, this method of speed control is inherently inefficient and converts the motor into a variable-speed motor, rather than an essentially constant-speed motor. For other means of controlling speed of polyphase induction motors and for other types of ac motors see ALTERNATING-CURRENT MOTOR.

SINGLE-PHASE INDUCTION MOTORS

Single-phase induction motors display poorer operating characteristics than polyphase machines, but are used where polyphase voltages are not available. They are most common in small sizes (1/2 hp or less) in domestic and industrial applications. Their particular disadvantages are low power factor, low efficiency, and the need for special starting devices.

The rotor of a single-phase induction motor is of the squirrel-cage type. The stator has a main winding which produces a pulsating field. At standstill, the pulsating field cannot produce rotor currents that will act on the air-gap flux to produce rotor torque. However, once the rotor is turning, it produces a cross flux at right angles in both space and time with the main field and thereby produces a rotating field comparable to that produced by the stator of a two-phase motor.

An explanation of this is based on the concept that a pulsating field is the equivalent of two oppositely rotating fields of one-half the magnitude of the resultant pulsating field. In Fig. 6, ϕ_m is the maximum value of the stator flux ϕ, which is shown only by its two components ϕ_f and ϕ_b, which represent the two oppositely rotating fields of constant equal magnitudes of $\phi_m/2$. Each component ϕ_f and ϕ_b produces a torque T_f and T_b on the rotor. Figure 7 shows that the sum of these torques is zero when speed is zero. However, if started, the sum of the torques is not zero and rotation will be maintained by the resultant torque.

This machine has good performance at high speed. However, to make this motor useful, it must have some way of producing a starting torque. The method by which this starting torque is obtained designates the type of the single-phase induction motor.

Split-phase motor. This motor has two stator windings, the customary main winding and a starting winding located 90 electrical degrees from the main winding, as in Fig. 8a. The starting winding has fewer turns of smaller wire, to give a higher resistance-to-reactance ratio, than the main winding. Therefore their currents I_m (main winding) and

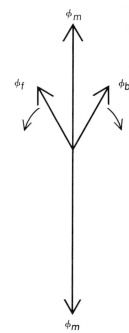

Fig. 6. Fluxes associated with the single-phase induction motor.

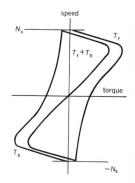

Fig. 7. Torques produced in the single-phase induction motor.

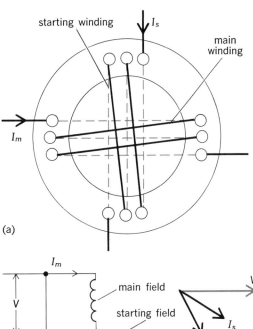

(a)

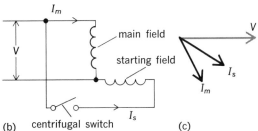

(b)   (c)

Fig. 8. Split-phase motor. (a) Windings. (b) Winding connections. (c) Vector diagram.

INDUCTION MOTOR

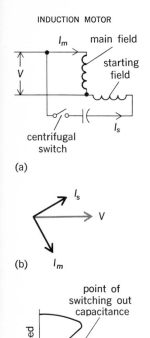

(a)

(b)

(c)

Fig. 9. Capacitor motor.
(a) Winding connections.
(b) Vector diagram.
(c) Characteristic.

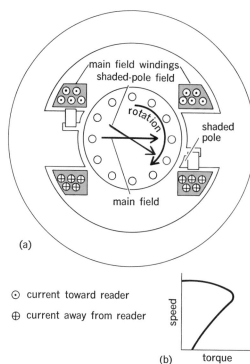

(a)

⊙ current toward reader

⊕ current away from reader

(b)

Fig. 10. Shaded-pole motor. (a) Cross-sectional view. (b) Typical characteristic.

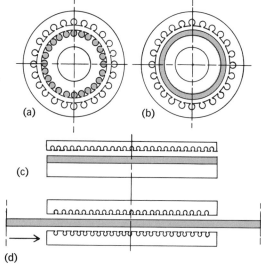

(a)   (b)

(c)

(d)

Fig. 11. Evolution of linear induction motor. (a) Polyphase squirrel-cage induction motor. (b) Conduction sheet motor. (c) One-sided long-secondary linear motor with short-flux return yoke. (d) Double-sided long-secondary linear motor.

I_s (starting winding) are out of time phase, as in Fig. 8c, when the windings are supplied by a common voltage V. These currents produce an elliptical field (equivalent to a uniform rotating field superimposed on a pulsating field) which causes a unidirectional torque at standstill. This torque will start the motor. When sufficient speed has been attained, the circuit of the starting winding can be opened by a centrifugal switch and the motor will operate with a characteristic illustrated by the broken-line curve of Fig. 7.

Capacitor motor. The stator windings of this motor are similar to the split-phase motor. However, the starting winding is connected to the supply through a capacitor (Fig. 9a). This results in a starting winding current which leads the applied voltage. The motor then has winding currents at standstill which are nearly 90° apart in time, as well as 90° apart in space. High starting torque and high power factor are therefore obtained. The starting winding circuit can be opened by a centrifugal switch when the motor comes up to speed. A typical characteristic is shown in Fig. 9c.

In some motors two capacitors are used. When the motor is first connected to the voltage supply, the two capacitors are used in parallel in the starting circuit. At higher speed one capacitor is removed by a centrifugal switch, leaving the other in series with the starting winding. This motor has high starting torque and good power factor.

Shaded-pole motor. This motor is used extensively where large power and large starting torque are not required, as in fans. A squirrel-cage rotor is used with a salient-pole stator excited by the ac supply. Each salient pole is slotted so that a portion of the pole face can be encircled by a short-circuited winding, or shading coil.

The main winding produces a field between the poles as in Fig. 10. The shading coils act to delay the flux passing through them, so that it lags the flux in the unshaded portions. This gives a sweeping magnetic action across the pole face, and consequently across the rotor bars opposite the pole face, and results in a torque on the rotor. This torque is much smaller than the torque of a split-phase motor, but it is adequate for many operations. A typical characteristic of the motor is shown in Fig. 10b.

For another type of single-phase alternating-current motor see REPULSION MOTOR. For synchronous motors built for single-phase see HYSTERESIS MOTOR; RELUCTANCE MOTOR.

Linear motor. Figure 11 illustrates the arrangements of the elements of the polyphase squirrel-cage induction motor. The squirrel cage (secondary) is embedded in the rotor in a manner to provide a close magnetic coupling with the stator winding (primary). This arrangement provides a small air gap between the stator and the rotor. If the squirrel cage is replaced by a conducting sheet as in Fig. 11b, motor action can be obtained. This machine, though inferior to that of Fig. 11a, will function as a motor. If the stator windings and iron are unrolled (rectangular laminations instead of circular laminations), the arrangement of the elements will take a form shown in Fig. 11c, and the field produced by polyphase excitation of the primary winding will travel in a linear direction instead of a circular direction. This field will produce a force on the conducting sheet that is in the plane of the sheet and at right angles with the stator conductors. A reversal of the phase rotation of the primary voltages will reverse the direction of motion of the air-gap flux and thereby reverse the force on the secondary sheet. No load on the motor corresponds to the condition when the secondary sheet is

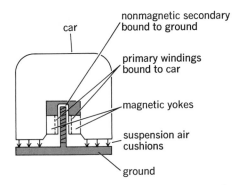

car

nonmagnetic secondary
bound to ground

primary windings
bound to car

magnetic yokes

suspension air
cushions

ground

Fig. 12. Drawing of linear induction-motor configuration. Magnetic interaction supplies the forces between the car and the ground. (*Le Moteur Lineaire and Société de l'Aerotrain*)

moving at the same speed as the field produced by the primary. For the arrangement of Fig. 11c, there is a magnetic attraction between the iron of the stator and the iron of the secondary sheet. For some applications, this can be a serious disadvantage of the one-sided motor of Fig. 11c. This disadvantage can be eliminated by use of the double-sided arrangement of Fig. 11d. In this arrangement the primary iron of the upper and lower sides is held together rigidly, and the forces that are normal to the plane of the sheet do not act on the sheet.

In conventional transportation systems, traction effort is dependent on the contact of the wheels with the ground. In some cases, locomotives must be provided with heavy weights to keep the wheels from sliding when under heavy loads. This disadvantage can be eliminated by use of the linear motor. Through use of the linear motor with air cushions, friction loss can be reduced and skidding can be eliminated. Figure 12 illustrates the application of the double-sided linear motor for high-speed transportation.

Conveying systems that are operated in limited space have been driven with linear motors. Some of these, ranging from 1/2 to 1 mi (805 – 1609 m) in length, have worked successfully.

Because of the large effective air gap of the linear motor, its magnetizing current is larger than that of the conventional motor. Its efficiency is somewhat lower, and its cost is high. The linear motor is largely in the experimental stage.

[ALBERT G. CONRAD]

Bibliography: P. D. Agarwal and T. C. Wang, Evaluation of fixed and moving primary linear induction motor systems, *Proc. IEEE*, pp. 631–637, May 1973; P. L. Ager, *The Nature of Induction Machines*, 1964; J. H. Dannon, R. N. Day, and G. P. Kalman, A Linear-induction motor propulsion system for high-speed ground vehicles, *Proc. IEEE*, pp. 621–630, May 1973; A. M. Dudley and S. F. Henderson, *Connecting Induction Motors*, 4th ed., 1960; E. R. Laithwaite, Linear electric machine: A personal view, *Proc. IEEE*, pp. 220–290, February 1975; A. F. Puchstein, T. C. Lloyd, and A. G. Conrad, *Alternating-current Machines*, 3d ed., 1954; David Schieber, Principles of operation of linear induction devices, *Proc. IEEE*, pp. 647–656, May 1973.

Infrared radiation

Electromagnetic radiation whose wavelengths lie in the range from about 0.8 μ to about 1000 μ (1 mm). The lower of these boundaries is set by the long-wavelength limit of the human eye's sensitivity to red light and the upper by the short-wavelength limit to radiation which can be generated and measured conveniently by microwave electronic devices. All solid bodies whose temperature is above absolute zero radiate some energy in the infrared, and if their effective temperature has a value up to about 3500°K, the radiation falls preponderantly in the infrared. Hence, infrared radiation is often called heat radiation or heat rays.

The infrared region was discovered around 1800 by William Herschel. He found that sunlight, dispersed into a spectrum by a glass prism, showed its greatest heating effect outside the visible part of the spectrum just beyond the red end. Herschel concluded that this effect was due to invisible radiation which was of the same nature as light except for its inability to affect the eye. In the next 50 years, experimental proof of this view slowly accumulated, and it has been generally accepted by scientists for the past 100 years.

Among the scientific, industrial, and military applications of infrared radiation are qualitative and quantitative chemical analyses by infrared spectroscopy; control of industrial processes; radiant heating; invisible signaling for burglar alarms, military messages, and the like; active and passive detection of military targets; and missile guidance.

Infrared spectrum. The infrared region is often subdivided, but there is no general agreement about the names and boundaries of the subdivisions. The table gives a subdivision which is based on instrumental characteristics but which also corresponds in a rough way to the subdivision of natural frequencies of molecules, as shown in the last column. The boundaries of the subdivisions are rather arbitrary and approximate.

As the table implies, near-infrared radiation can be detected by photoelectric cells, and it corresponds in frequency range to the lower energy levels of electrons in molecules and semiconductors (the higher electronic levels correspond to the frequencies of visible and ultraviolet radiation). The intermediate infrared region covers the frequency range of most molecular vibrations; only the overtones of the higher vibrational frequencies fall outside it. This region is sometimes called the prism infrared because spectrometers equipped with

Subdivisions of the infrared (IR) spectrum

Wavelength range, μ	Wave number range, cm^{-1}	Names used	Appropriate molecular motions
0.8–2.5	4000–12,500	Near IR; photoelectric IR	Low electronic levels; vibrational overtones
2.5–50	200–4000	Intermediate IR; prism IR	Molecular vibrations
50–1000	10–200	Far IR; grating IR	Molecular rotations

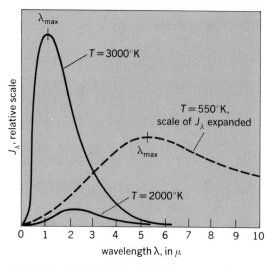

Fig. 1. Spectral distribution of J_λ, the power radiated by an incandescent blackbody (a theoretically perfect radiator) at temperature T. The ordinate J_λ is the radiant power in watts per unit area of radiating surface per unit solid angle per unit of wavelength at λ in the infinitesimal wavelength range $d\lambda$.

alkali-halide prisms are commonly used here. The far-infrared, where diffraction gratings must be used in spectrometers because no suitable prism materials are known, spans the region of very low molecular vibrational frequencies and most rotational frequencies, at least of the lighter molecules. Because radiation of about 1-mm wavelength, the upper limit set for the far infrared, can also be produced and detected by microwave radar devices, there is no gap between the infrared and microwave domains, and in fact the region on

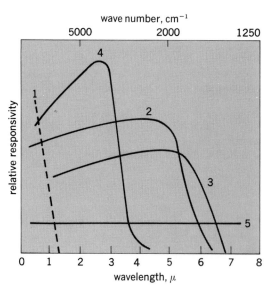

Fig. 2. Spectral sensitivity of infrared detectors. 1, infrared-sensitized photographic emulsion; 2, PbTe (lead telluride) photoconductive detector (refrigerated); 3, PbSe (lead selenide) photoconductive detector (refrigerated); 4, PbS (lead sulfide) photoconductive detector; 5, thermal detector. Responsivity scale is not the same for various detectors; approximate wavelengths at which detectors 1, 2, 3, and 4 become less responsive than a thermal detector are correctly shown.

either side of 1 mm is under investigation by both techniques.

Sources. The usual source of infrared radiation is an incandescent solid body. Such sources emit radiant power having a continuous broad range of wavelengths. The variation of this radiant power with wavelength ordinarily has the form shown in Fig. 1, with a pronounced maximum at some wavelength λ_{max}, a rather sharp decrease on the low-wavelength side, and a gentler slope at higher wavelengths.

Infrared radiation may be emitted by systems other than hot solid bodies, for example, by gases through which an electrical discharge is passed. Although such systems are relatively inefficient emitters, they are sometimes used for special purposes. The mercury-vapor arc has been found to be one of the best sources of radiant power in the far-infrared.

The development of the laser has resulted in sources of infrared radiation that are sharply monochromatic and of extremely high radiant power. The most powerful continuously operating laser in the infrared is the carbon-dioxide-gas laser, which can be made to emit several hundred watts of power at a wavelength of 10.6 μ. Other infrared gas lasers are the helium-neon laser (prominent wavelengths of 0.692, 1.19 and 3.39 μ), the water-vapor laser (27.97, 47.7, 78.46 and 118.6 μ), and the hydrogen-cyanide laser (311 and 337 μ). See LASER.

While the gas laser is sharply monochromatic, its wavelength spread or half-width being smaller than one-millionth of its wavelength, the wavelength cannot be varied continuously over a significant range. The solid-state laser of gallium arsenide produces near- and mid-infrared radiation whose wavelength can be adjusted, or tuned, by means of hydrostatic pressure. Apart from a tunable laser, the only way of producing monochromatic infrared radiation of any desired wavelength is to isolate it from an infrared source containing that wavelength and many others, such as a blackbody source (Fig. 1). The isolating device may be a selective optical filter or a spectrometer.

Detectors. Detectors of electromagnetic radiation may be classified broadly as quantum detectors, resonant detectors, and heat engines or thermal detectors. The first are devices which convert a quantum of the radiation in question into a proportionate signal by some process which is insensitive to quanta of less than a certain energy (for example, the mean energy of quanta emitted by a body at room temperature). Photographic emulsions, photoelectric cells, and Geiger counters are examples of quantum detectors. Resonant detectors are devices that are responsive only to radiation of the frequency to which they are tuned. Heat engines act as detectors by converting the radiation into heat and using the heat to operate a device that produces a signal which is proportionate to the amount of radiant energy received.

No resonant detectors have been constructed which can be tuned to infrared frequencies. For the photoelectric infrared (see table), quantum detectors in the form of specially sensitized photographic emulsions, photoemissive cells, and particularly photoconductive cells are usable. As can be seen from Fig. 2, the responsivity of such detec-

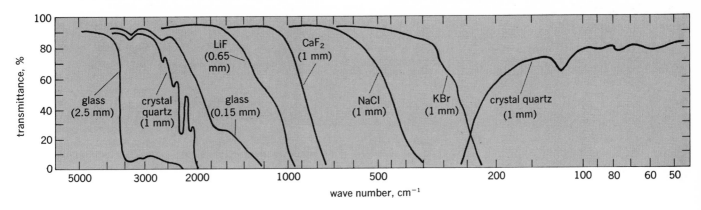

Fig. 3. Transmission curves of some widely used infrared optical materials.

tors varies considerably with frequency and drops to low values at wave numbers of about 2000–3000 cm⁻¹ (3.5–5 μ in wavelength). Photoconductive detectors are known having good responsivity below 2000 cm⁻¹ (above 5μ), but these must be operated at or near liquid helium temperatures (<10°K).

Therefore, in a large part of the infrared region, heat-engine detectors—whose responsivity is the same to all kinds of radiation provided the radiation is converted entirely to heat in the detector—are the only generally usable kind. Examples of heat-engine detectors are thermocouples and thermopiles, which produce an electromotive force when heated; bolometers, which change their electrical resistance when heated; and pneumatic radiometers, in which heat is detected by the increase in pressure of a heated gas. Because these devices are all subject to the laws of thermodynamics governing the conversion of heat into useful work (that is, into a signal), their ultimate responsivity is expected to be approximately the same, and this is found to be the case. Thermal detectors operating at room temperature have a lower limit of sensitivity of the order of 10⁻¹⁰ watt, with response times of the order of 0.1 sec. This limit can be reduced by a factor of 100 or more if the detector is capable of operation at temperatures near 1°K.

Propagation. The properties of various media for the transmission of infrared radiation are often quite different from those of light. For example, window glass is quite opaque to 5-μ radiation, whereas pure germanium crystals, which do not transmit visible radiation, are very transparent to this wavelength (apart from reflection losses which can be reduced by surface coatings). The alkali halide crystals, such as sodium chloride (common salt) and potassium bromide, are well known for their transparency in the near- and mid-infrared regions (Fig. 3).

The attenuation of infrared radiation by the atmosphere is of special interest. Nitrogen, oxygen, and the rare gases are transparent to all infrared wavelengths, but water vapor and carbon dioxide are strongly absorbing in certain regions. In the range 0.8–4.0 μ, there is irregular absorption due mainly to water vapor, with more or less open "windows" at about 1.0, 1.4, 1.6, 2.2, and 3.4–4.0 μ. From 4 to 8 μ water vapor and carbon dioxide together are strongly absorbing, but there is an

extensive window from 8 to 13.5 μ. This window is of great meteorological importance because the peak of the radiation curve of the Earth falls near 10 μ. Beyond 14 out to 600 μ, there is more or less continuous absorption by atmospheric water vapor, arising from transitions between the rotational energy levels of this molecule.

Liquid water is rather generally opaque in the infrared above 2 μ in path lengths larger than 1 mm. The transmission of infrared radiation through fog is little better than that of visible light, because of a combination of scattering and absorption by the water droplets. The popular misconception that fog is transparent to infrared radiation has perhaps arisen from the well-advertised effectiveness of red-sensitive photographic film for photography through atmospheric haze of dust particles. [RICHARD C. LORD]

Bibliography: R. S. Estey, Infra-red: New uses for an old technique, *Missiles and Rockets*, 3(7): 107–112, 1958; G. R. Harrison, R. C. Lord, and J. R. Loofbourow, *Practical Spectroscopy*, 1948; P. W. Kruse, L. D. McGlauchlin, and R. B. McQuistan, *Elements of Infrared Technology*, 1962; Ivan Simon, *Infrared Radiation*, 1966; R. A. Smith, F. E. Jones, and R. P. Chasmar, *Detection and Measurement of Infrared Radiation*, 2d ed., 1968.

Insulation, heat

Materials whose principal purpose is to retard the flow of heat. Thermal- or heat-insulation materials may be divided into two classes, bulk insulations and reflective insulations. The class and the material within a class to be used for a given application depend upon such factors as temperature of operation, ambient conditions, mechanical strength requirements, and economics.

Examples of bulk insulation include mineral wool, vegetable fibers and organic papers, foamed plastics, calcium silicates with asbestos, expanded vermiculite, expanded perlite, cellular glass, silica aerogel, and diatomite and insulating firebrick. They retard the flow of heat, breaking up the heat-flow path by the interposition of many air spaces and in most cases by their opacity to radiant heat.

Reflective insulations are usually aluminum foil or sheets, although occasionally a coated steel sheet, an aluminumized paper, or even gold or silver surfaces are used. Refractory metals, such as tantalum, may be used at higher temperatures.

Their effectiveness is due to their low emissivity (high reflectivity) of heat radiation. *See* EMISSIVITY.

Thermal insulations are regularly used at temperatures ranging from a few degrees above absolute zero, as in the storage of liquid hydrogen and helium, to above 3000°F in high-temperature furnaces. Temperatures of 4000–5000°F are encountered in the hotter portions of missiles, rockets, and aerospace vehicles. To withstand these temperatures during exposures lasting seconds or minutes, insulation systems are designed that employ radiative, ablative, or absorptive methods of heat dissipation.

Heat flow. The distinguishing property of bulk thermal insulation is low thermal conductivity. Under conditions of steady-state heat flow the empirical equation that describes the heat flow through a material is Eq. (1), where q = time rate of

$$\frac{q}{A} = -k\frac{(\theta_2 - \theta_1)}{l} \tag{1}$$

heat flow, A = area, θ_1 = temperature of warmer side, θ_2 = temperature of colder side, l = thickness or length of heat-flow path, and k = thermal conductivity, representative values being listed in the table. For a given thickness of material exposed to a given temperature difference, the rate of heat flow per unit area is directly proportional to the thermal conductivity of the material.

In the unsteady state, or transient heat flow, the density and specific heat of a material have a strong influence upon the rate of heat flow. In such cases, thermal diffusivity $\alpha = k/\rho\, C_p$ is the important property. Here ρ = density and C_p = specific heat at constant pressure. In the simple case of one-dimensional heat flow through a homogeneous material, the governing equation is Eq. (2), where

$$\frac{d\theta}{dt} = \alpha \frac{d^2\theta}{dx^2}\bigg|_0^l \tag{2}$$

t = time and x is measured along the heat-flow path from 0 to l.

Thermal conductivity. In general, thermal conductivity is not a constant for the material but varies with temperature. Generally, for metals and other crystalline materials, conductivity decreases with increasing temperature; for glasses and other amorphous materials, conductivity increases with temperature. Bulk insulation materials in general behave like amorphous materials and have a positive temperature coefficient of conductivity.

Thermal conductivity of bulk insulation depends upon the nature of the gas in the pores. The conductivities of two insulations, identical except for the gases filling the pore spaces, will differ by an amount approximately proportional to the difference in the conductivities of the two gases.

Increasing the pressure of the gas in the pores of a bulk insulation has little effect on the conductivity even with pressures of several atmospheres. Decreasing the pressure has little effect until the mean free path of the gas is in the order of magnitude of the dimensions of the pores. Below this pressure the conductivity decreases rapidly until it reaches a value determined by radiation and solid conduction. A few materials have such fine pores that at atmospheric pressure their dimensions are smaller than the mean free path of air. Such insulations may have conductivities less than the conductivity of still air. *See* HEAT RADIATION; HEAT TRANSFER.

[HARRY F. REMDE]

Bibliography: H. S. Carlslaw and J. C. Jaeger, *Conduction of Heat in Solids*, 2d ed., 1959; P. E. Glaser et al., *Investigation of Materials for Vacuum Insulators up to 4000F*, ASD TR-62-88, 1962; W. H. MacAdams, *Heat Transmission*, 3d ed., 1954; E. M. Sparrow and R. D. Cess, *Radiation Heat Transfer*, 1967; H. M. Strong, F. P. Bundy, and H. P. Bovenkerk, Flat panel vacuum thermal insulation, *J. Appl. Phys.*, 31:39, 1960; J. D. Vershoor and P. Greebler, Heat transfer by gas conduction and radiation in fibrous insulations, *Trans. ASME*, 74(6):961–968, 1952; G. B. Wilkes, *Heat Insulation*, 1950.

Internal combustion engine

A prime mover, the fuel for which is burned within the engine, as contrasted to a steam engine, for example, in which fuel is burned in a separate furnace. *See* ENGINE.

The most numerous of internal combustion engines are the gasoline piston engines used in passenger automobiles, outboard engines for motor boats, small units for lawn mowers, and other such equipment, as well as diesel engines used in trucks, tractors, earth-moving, and similar equipment. This article describes the gasoline piston and diesel types of engines. For other types of internal combustion engines *see* GAS TURBINE; TURBINE PROPULSION.

The aircraft piston engine is fundamentally the same as that used in automobiles but is engineered for light weight and is usually air cooled. *See* AIRCRAFT ENGINE, RECIPROCATING.

Thermal conductivities of selected solids*

Material	Density, lb/ft³		Conductivity, k† Btu/(in.) (hr) (ft²) (°F)
Asbestos cement board	120	75	4
Cotton fiber	0.8–2.0	75	0.26
Mineral wool, fibrous rock, slag, or glass	1.5–4.0	75	0.27
Insulating board, wood, or cane fiber	15	75	0.35
Foamed plastics	1.6	75	0.29
Glass			3.6–7.32
Hardwoods, typical	45	75	1.10
Softwoods, typical	32	75	0.80
Cellular glass	9	75	0.40
Fine sand (4% moisture content)	100	40	4.5
Silty clay loam (20% moisture content)	100	40	9.5
Gypsum or plaster board	50	75	1.1

*From American Society of Heating, Refrigerating, and Air Conditioning Engineers, *Heating, Ventilating and Air Conditioning Guide*, 1959.

†Typical; suitable for engineering calculations.

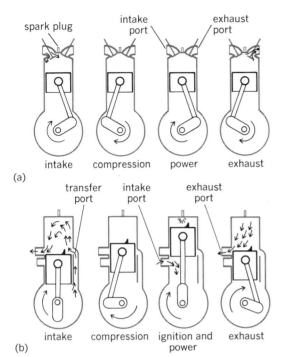

Fig. 1. (a) The four strokes of a modern four-stroke engine cycle. For intake stroke the intake valve (left) has opened and the piston is moving downward, drawing air and gasoline vapor into the cylinder. In compression stroke the intake valve has closed and the piston is moving upward, compressing the mixture. On power stroke the ignition system produces a spark that ignites the mixture. As it burns, high pressure is created, which pushes the piston downward. For exhaust stroke the exhaust valve (right) has opened and the piston is moving upward, forcing the burned gases from the cylinder. (b) The same action is accomplished without separate valves and in a single rotation of the crankshaft by a three-port two-cycle engine. (From M. L. Smith and K. W. Stinson, Fuels and Combustion, McGraw-Hill, 1952)

ENGINE TYPES

Characteristic features common to all commercially successful internal combustion engines include (1) the compression of air, (2) the raising of air temperature by the combustion of fuel in this air at its elevated pressure, (3) the extraction of work from the heated air by expansion to the initial pressure, and (4) exhaust. William Barnett first drew attention to the theoretical advantages of combustion under compression in 1838. In 1862 Beau de Rochas published a treatise that emphasized the value of combustion under pressure and a high ratio of expansion for fuel economy; he proposed the four-stroke engine cycle as a means of accomplishing these conditions in a piston engine (Fig. 1). The engine requires two revolutions of the crankshaft to complete one combustion cycle. The first engine to use this cycle successfully was built in 1876 by N. A. Otto. See OTTO CYCLE.

Two years later Sir Dougald Clerk developed the two-stroke engine cycle by which a similar combustion cycle required only one revolution of the crankshaft. In this cycle, exhaust ports in the cylinder were uncovered by the piston as it approached the end of its power stroke. A second cylinder then pumped a charge of air to the working cylinder through a check valve when the pump pressure exceeded that in the working cylinder.

In 1891 Joseph Day simplified the two-stroke engine cycle by using the crankcase to pump the required air. The compression stroke of the working piston draws the fresh combustible charge through a check valve into the crankcase, and the next power stroke of the piston compresses this charge. The piston uncovers the exhaust ports near the end of the power stroke and slightly later uncovers intake ports opposite them to admit the compressed charge from the crankcase. A baffle is usually provided on the piston head of small engines to deflect the charge up one side of the cylinder to scavenge the remaining burned gases down the other side and out the exhaust ports with as little mixing as possible.

Engines using this two-stroke cycle today have been further simplified by use of a third cylinder port which dispenses with the crankcase check valve used by Day. Such engines are in wide use for small units where fuel economy is not as important as mechanical simplicity and light weight. They do not need mechanically operated valves and develop one combustion cycle per crankshaft revolution. Nevertheless they do not develop twice the power of four-stroke cycle engines with the same size working cylinders at the same number of revolutions per minute (rpm). The principal reasons for this are (1) the reduction in effective cylinder volume due to the piston movement required to cover the exhaust ports, (2) the appreciable mixing of burned (exhaust) gases with the combustible mixture, and (3) the loss of some combustible mixture through the exhaust ports with the exhaust gases.

Otto's engine, like almost all internal combustion engines developed at that period, burned coal gas mixed in combustible proportions with air prior to being drawn into the cylinder. The engine load was generally controlled by throttling the quantity of charge taken into the cylinder. Ignition was accomplished by a device such as an external flame or an electric spark so that the timing was controllable. These are essential features of what has become known as the Otto or spark-ignition combustion cycle.

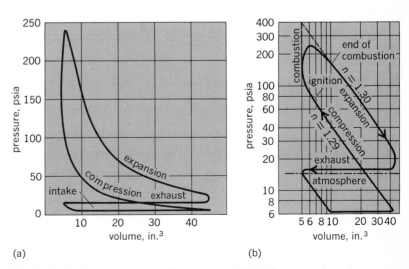

Fig. 2. Typical pressure-volume indicator card (a) plotted on rectangular coordinates and (b) plotted on logarithmic coordinates.

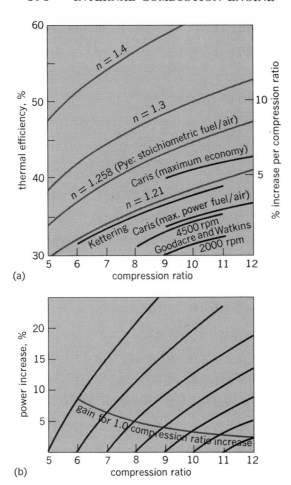

Fig. 3. (a) Effect of compression ratio on thermal efficiency as calculated with different values of n and compared with published experimental data. (b) Increase in power from raising compression ratio as calculated with $n=1.3$. Percentage values are but little altered by calculating with different values of n.

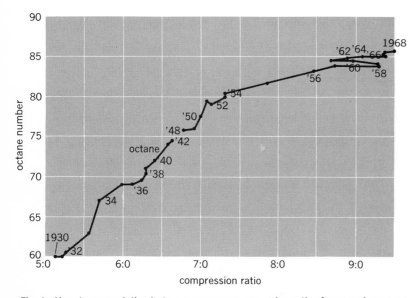

Fig. 4. Year-to-year relation between average compression ratio of cars and average octane number (ASTM research method) of regular-grade summer gasolines sold in the United States. (*Automotive Industries, March, 1954, 1969*)

Ideal and actual combustion. In the classical presentation of the four-stroke cycle, combustion is idealized as instantaneous and at constant volume. This simplifies thermodynamic analysis, but fortunately combustion takes time, for it is doubtful that an engine could run if the whole charge burned or detonated instantly.

Detonation of a small part of the charge in the cylinder, after most of the charge has burned progressively, causes the knock which limits the compression ratio of an engine with a given fuel. *See* COMBUSTION.

The gas pressure of an Otto combustion cycle using the four-stroke engine cycle varies with the piston position as shown by the typical indicator card in Fig. 2a. This is a conventional pressure-volume (PV) card for an 8.7:1 compression ratio. For simplicity in calculations of engine power, the average net pressure during the working stroke, called the mean effective pressure (mep), is frequently used. It may be obtained from the average net height of the card, which is found by measurement of the area with a planimeter and by division of this area by its length. Similar pressure-volume data may be plotted on logarithmic coordinates as in Fig. 2b, which develops expansion and compression relations as approximately straight lines, the slopes of which show the values of exponent n to use in equations for pressure-volume relationships.

The rounding of the plots at peak pressure, with the peak developing after the piston has started its power stroke, even with the spark occurring before the piston reaches the end of the compression stroke, is due to the time required for combustion. The actual time required is more or less under control of the engine designer, because he can alter the design to vary the violence of the turbulence of the charge in the compression space prior to and during combustion. The greater the turbulence the faster the combustion and the lower the antiknock or octane value required of the fuel, or the higher the compression ratio that may be used with a given fuel without knocking. On the other hand, a designer is limited as to the amount he can raise the turbulence by the increased rate of pressure rise, which increases engine roughness. Roughness must not exceed a level acceptable for automobile or other service.

Compression ratio. According to classical thermodynamic theory, thermal efficiency η of the Otto combustion cycle is given by Eq. (1), where

$$\eta = 1 - \frac{1}{r^{n-1}} \tag{1}$$

the compression ratio r_c and expansion ratio r_e are the same ($r_c = r_e = r$). When theory assumes atmospheric air in the cylinder for extreme simplicity, exponent n is 1.4. Efficiencies calculated on this basis are almost twice as high as measured efficiencies. Logarithmic diagrams from experimental data show that n is about 1.3. Even with this value, efficiencies achieved in practice are less than given by Eq. (1). This is not surprising considering the differences found in practice and assumed in theory, such as instantaneous combustion and 100% volumetric efficiency.

Attempts to adjust classical theory to practice by use of variable specific heats and consideration of dissociation of the burning gases at high temperatures have shown that this exponent should vary with the fuel-air mixture ratio, and to some extent with the compression ratio. G. A. Goodenough and J. B. Baker have shown that, for an 8:1 compression ratio, the exponent should vary from about 1.28 for a stoichiometric (chemically correct) mixture to about 1.31 for a lean mixture. Similar calculations by D. R. Pye showed that, at a compression ratio of 5:1, n should be 1.258 for the stoichiometric mixture, increasing with excess air (lean mixture) to about 1.3 for a 20% lean mixture and to 1.4 if extrapolated to 100% air. Actual practice gives thermal efficiencies still lower than these, which might well be expected because of the assumed instantaneous changes in cyclic pressure (during combustion and exhaust) and the disregard of heat losses to the cylinder walls. These theoretical relations between compression ratio and thermal efficiency, as well as some experimental results, are shown in Fig. 3a. The data published by C. F. Kettering and D. F. Caris are about 85% and 82%, respectively, of the theoretical relations for the corresponding fuel-air mixtures.

Figure 3b gives the theoretical percentage gain in indicated thermal efficiency or power from raising the compression of an engine from a given value. They were plotted from Eq. (1) with $n = 1.3$, but would differ only slightly if obtained from any of the curves shown in Fig. 3a. The dotted line crossing these curves shows the diminishing gain obtainable by raising the compression ratio one unit at the higher compression ratios.

Experimental data indicate that a change in compression ratio does not appreciably change the mechanical efficiency or the volumetric efficiency of the engine. Therefore, any increase in thermal efficiency resulting from an increase in compression ratio will be revealed by a corresponding increase in torque or mep; this is frequently of more practical importance to the engine designer than the actual efficiency increase, which becomes an added bonus.

Compression ratio and octane rating. For years compression ratios of automobile engines have been as high as designers considered possible without danger of too much customer annoyance from detonation or knock with the gasoline on the market at the time (Fig. 4). Engine designers continue to raise the compression ratios of their engines as suitable gasolines come on the market.

Little theoretical study has been given to the effect of engine load on indicated thermal efficiency. Experimental evidence reveals that it varies little, if at all, with load, provided that the fuel-air-ratio remains constant and that the ignition time is suitably advanced at reduced loads to compensate for the slower rate of burning which results from dilution of the combustible charge with the larger percentages of burned gases remaining in the combustion space and the reduced turbulence at lower speeds.

Ignition timing. Designers obtain high thermal efficiency from high compression ratios at part loads, where engines normally run at automobile cruising speeds, with optimum spark advance, but

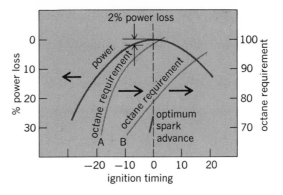

Fig. 5. Effects of advancing or retarding ignition timing from optimum on engine power and resulting octane requirement of fuel in an experimental engine with a combustion chamber having typical turbulence (A) and a highly turbulent design (B) with the same compression ratio. Retarding the spark 7° for a 2% power loss reduced octane requirement from 98 to 93 for design A.

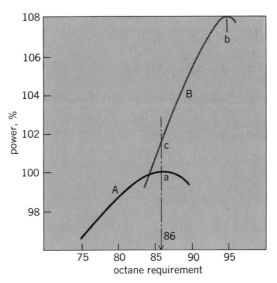

Fig. 6. Effect of raising compression ratio of an experimental engine on the power output and octane requirement at wide-open throttle. While an 86-octane fuel was required for optimum spark advance (maximum power) with the original compression ratio, the same gasoline would be knock-free at the higher compression ratio by suitably retarding the ignition timing.

avoid knock on available gasolines at wide-open throttle by use of a reduced or compromise spark advance. The tendency of an engine to knock at wide-open throttle is reduced appreciably when the spark timing is reduced 5–10° from optimum, as is shown in Fig. 5. Advancing or retarding the spark timing from optimum results in an increasing loss in mep for any normal engine as shown by the solid curve. The octane requirement falls rapidly as the spark timing is retarded, the actual rate depending on the nature of the gasoline as well as on the design of the combustion chamber. The broken-line curves A and B show the effects on a given gasoline of the use of moderate- and high-turbulence combustion chambers, respectively, with the same compression ratio. Because the mep curve is relatively flat near optimum spark advance, retarding the spark for a 1–2% loss is considered normally acceptable in practice because of the appreciable reduction in octane requirement.

In Fig. 6 similar data are shown by curve A for another engine with changes in mep plotted on a percentage basis against octane requirement as the spark timing was charged. Point a indicates optimum spark timing, where 85 octane was required of the gasoline to avoid knock. By raising the compression ratio, the power and octane requirement were also raised as shown by the broken-line curve B. Although optimum spark required 95 octane (point b), retarding the spark timing and thus reducing the octane requirement to 86 (point c) developed slightly more power than with the original compression ratio at its optimum spark advance. The gain may be negligible at wide-open throttle, but at lower loads where knock does not develop the spark timing may be advanced to optimum (point b), where appreciably more power may be developed by the same amount of fuel.

In addition to the advantages of the higher compression ratio at cruising loads with optimum spark advance, the compromise spark at full load may be advanced toward optimum as higher-octane fuels become available, and a corresponding increase in full-throttle mep enjoyed. Such compromise spark timings have had much to do with the adoption of compression ratios of 10:1 to 13:1.

Fuel-air ratio. A similar line of reasoning shows that a fuel-air mixture richer than that which develops maximum knock-free mep will permit use of higher compression ratios. However, the benefits derived from compromise or superrich mixtures vary so much with mixture temperature and the sensitivity of the octane value of the particular fuel to temperature that it is not generally practical to make much general use of this method. Nevertheless it has been the practice with piston-type aircraft engines to use fuel-air mixture ratios of 0.11 or even higher during takeoff, instead of about 0.08, which normally develops maximum mep in the absence of knock.

Compression-ignition engines. About 20 years after Otto first ran his engine, Rudolf Diesel successfully demonstrated an entirely different method of igniting fuel. Air is compressed to a pressure high enough for the adiabatic temperature to reach or exceed the ignition temperature of the fuel. Because this temperature is in the order of 1000°F, compression ratios of 12:1 to 20:1 are used commercially with compression pressures generally over 600 psi. This type of engine cycle requires the fuel to be injected after compression at a time and rate suitable to control the rate of combustion.

Conditions for high efficiency. The classical presentation of the diesel engine cycle assumes combustion at constant pressure. Like the Otto cycle, thermal efficiency increases with compression ratio, but in addition it varies with the amount of heat added (at the constant pressure) up to the cutoff point where the pressure begins to drop from adiabatic expansion. *See* DIESEL CYCLE; DIESEL ENGINE.

Practical attainments. Diesel engines were highly developed in Germany prior to World War I, and made an impressive performance in submarines. Large experimental single-cylinder engines were built in several European countries with cylinder diameters up to 1 m. As an example, the two-stroke Sulzer S100 single-acting engine with a bore of 1 m and a stroke of 1.1 m developed 2050 gross horsepower at 150 rpm. Multiple-cylinder engines developing 15,000 hp are in marine service. Small diesel engines are in wide use also.

Fuel injection. In early diesel engines, air injection of the fuel was used to develop extremely fine atomization and good distribution of the spray. However, the need for injection air at pressures in the order of 1500 psi necessitated the use of expensive and bulky multistage air compressors and intercoolers.

A simpler fuel-injection method was introduced by James McKechnie in 1910. He atomized the fuel as it entered the cylinder by use of high fuel pressure and suitable spray nozzles. After considerable development it became possible to atomize the fuel sufficiently to minimize the smoky exhaust which had been characteristic of the early solid-injection engines. By 1930 solid or airless injection had become the generally accepted method of injecting fuel in diesel engines.

Contrast between diesel and Otto engines. There are many characteristics of the diesel engine which are in direct contrast to those of the Otto engine. The higher the compression ratio of a diesel engine, the less the difficulties with ignition time lag. Too great an ignition lag results in a sudden and undesired pressure rise which causes an audible knock. In contrast to an Otto engine, knock in a diesel engine can be reduced by use of a fuel of higher cetane number, which is equivalent to a lower octane number. *See* OCTANE NUMBER.

The larger the cylinder diameter of a diesel engine, the simpler the development of good combustion. In contrast, the smaller the cylinder diameter of the Otto engine, the less the limitation from detonation of the fuel.

High intake-air temperature and density materially aid combustion in a diesel engine, especially of fuels having low volatility and high viscosity. Some engines have not performed properly on heavy fuel until provided with a super charger. The added compression of the supercharger raised the temperature and, what is more important, the density of the combustion air. For an Otto engine, an increase in either the air temperature or density increases the tendency of the engine to knock and therefore reduces the allowable compression ratio.

Diesel engines develop increasingly higher indicated thermal efficiency at reduced loads because of leaner fuel-air ratios and earlier cutoff. Such mixture ratios may be leaner than will ignite in an Otto engine. Furthermore, the reduction of load in an Otto engine requires throttling of the engine, which develops increasing pumping losses in the intake system.

TRENDS IN AUTOMOBILE ENGINES

Cylinder diameters of average American automobile engines prior to 1910 were over $4\frac{1}{4}$ in. By 1917 they had been reduced to only a little over $3\frac{1}{4}$ in., where they stabilized until after 1945. Since then the increased demand for more power, with the number of cylinders limited to eight for practical mechanical reasons, the diameters have been increased from year to year until they averaged 3.98 in. in 1969, with a maximum of 4.36 in.

Stroke-bore ratio. Experimental engines differing only in stroke-bore ratio show that this ratio has no appreciable effect on fuel economy and friction at corresponding piston speeds. Practical advantages which result from the short stroke include (1) the greater rigidity of crankshaft from the shorter crank cheeks, with crankpins sometimes overlapping main bearings, and (2) the narrower as well as lighter cylinder block which is possible. On the other hand, the higher rates of crankshaft rotation for an equivalent piston speed necessitate greater valve forces and require stronger valve springs. Also the smaller depth of the compression space for a given compression ratio increases the surface-to-volume ratio and the proportion of heat lost by radiation during combustion. Nevertheless, stroke-bore ratios have been decreasing for more than 25 years and in 1969 reached 0.9 for the average automobile in the United States.

Cylinder number and arrangement. Engine power may be raised by increasing the number of cylinders as well as the power per cylinder. The minimum number of cylinders has generally been four for four-cycle automobile engines, because this is the smallest number that provides a reasonable balance for the reciprocating pistons. Many early cars had four-cylinder engines. After 1912 six-cylinder in-line engines became popular. They have superior balance of reciprocating forces and more even torque impulses. By 1940 the eight-cylinder 90° V engine had risen in popularity until it about equaled the six-in-line. After 1954, the V-8 dominated the field for American automobile engines. There are several important reasons for this

besides the increased power. For example, the V-8 offers appreciably more rigid construction with less bearing deflection at high speeds, provides more uniform distribution of fuel to all cylinders from centrally located downdraft carburetors, and has a short, low engine that fits within the hood demanded by style trends. With the introduction of the smaller "compact" cars in 1959, where the power and cost of eight-cylinder V-type engines were not required, six-cylinder designs increased. By 1969 about 38% of all engine designs were of the six-cylinder in-line type. The evolution of cylinder arrangements included for a short period the V-12 and even a V-16 cylinder design, but experience showed that in their day there was too much practical difficulty in providing good manifold distribution of fuel, especially when starting cold, and too much difficulty in keeping all spark plugs firing.

Compression ratio. The considerable increase in power of the average automobile engine over the years is shown in Fig. 7 together with the compression ratios which have had much to do with the increased mep. Such ratios approach practical limits imposed by phenomena other than detonation, such as preignition, rumble, and other evidences of undesirable combustion.

The modern trend toward high compression ratios, with their small compression volumes, has dictated the universal use of overhead valves in all American engine designs. High compression ratios also tend to restrict cylinder diameters because the longer flame travel increases the tendency to knock (Fig. 8).

Improved breathing and exhaust. Added power output has been brought about by reducing the

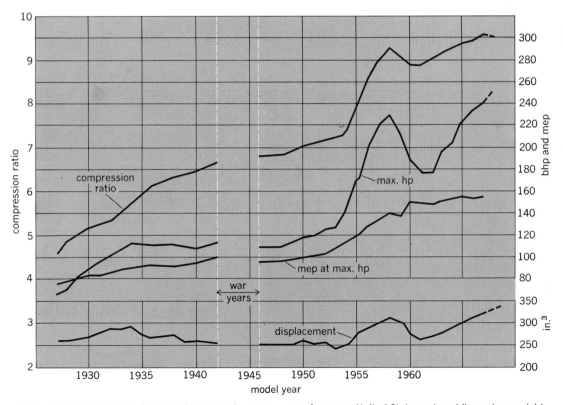

Fig. 7. Trend toward considerable increases in compression ratio, mean effective pressure (mep), and power of average United States automobile engines weighted for production volume. (*Ethyl Corp. data*)

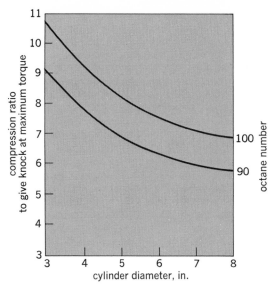

Fig. 8.   Relation between cylinder diameter and limiting compression ratio for engines of similar design, one using 90- and the other 100-octane gasoline. (*From L. L. Brower, unpublished SAE paper, 243, 1950*)

pressure drop in the intake system at high speeds and by reducing the back pressure of the exhaust systems (Fig. 9). These results were accomplished by larger valve areas and valve ports, by larger venturi areas, and by more streamlined manifolds. Larger valve areas were achieved by higher lift of the valves and by larger valves. Larger venturi areas in the carburetors were achieved by use of one or more two-stage carburetors; in these, sufficient air velocity was developed to meter the fuel on one venturi at low power; the second venturi was opened for high power. Better streamlining of the manifold passages between carburetors and valves, especially at the cylinder ports, and the use of dual exhausts and mufflers with reduced

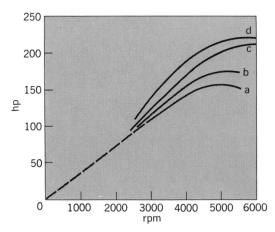

Fig. 9.   Increases in peak power of a six-cylinder engine (210 in.³ displacement) by successive reductions in resistance to airflow through it. Curve a, power developed by original engine with two 1¾ in. carburetors; curve b, higher valve-lift valves and dual exhaust systems; curve c, larger valves, smoother valve ports, and two 2-in. carburetors; curve d, three double-barrel carburetors. (*Adapted from W. M. Heynes, The Jaguar engine, Inst. Mech. Eng., Automot. Div. Trans., 1952–1953*)

back pressure have also improved engine breathing and exhaust.

Valve timing. The times of opening and closing of the valves of an engine in relation to the piston position are usually selected to develop maximum power over a desired speed range at wide-open throttle. For convenience the timing of these events is expressed as the number of degrees of crankshaft rotation before or after the piston reaches the end of one of its strokes. Because of the time required for the flow of the burned gas through the exhaust valve at the end of the power or expansion stroke of a piston, it is customary to start opening the valve considerably before the end of the stroke. If the valve should be opened when the piston is nearer the lower end of its stroke, power would be lost at high engine speeds because the piston on its return (exhaust) stroke would have to move against gas pressure remaining in the cylinder. On the other hand, if the valve were opened before necessary, the burned gas would be released while it is still at sufficient pressure to increase the work done on the piston. Thus for any engine there is a time for opening the exhaust valve which will develop the maximum power at some particular speed. Moreover, the power loss at other speeds does not increase rapidly. It is obvious that, when an engine is throttled at part load, there will be less gas to discharge through the exhaust valve and there will be less need for it to be opened as early as at wide-open throttle.

The timing of intake valve events is normally selected to trap the largest possible quantity of air or combustible mixture in the cylinder when the valve closes at some desired engine speed and at wide-open throttle. The intermittent nature of the flow through the valve subjects it to alternate accelerations and retardation which require time. During the suction stroke, the mass of air moving through the pipe leading to the intake valve is given velocity energy which may be converted to a little pressure at the valve when the air mass still in the pipe is stopped by its closure. Advantage of this phenomenon may be obtained at some engine speed to increase the air mass which enters the cylinder. The engine speed at which the maximum volumetric efficiency is developed varies with the relative valve area, closure time, and other factors, including the diameter and particularly the length of this pipe (Fig. 10). These curves reveal the characteristic falling off at high speeds from the inevitable throttling action as air flows at increased velocities through any restriction such as a valve or intake pipe or particularly the venturi of a carburetor.

The curves shown in Fig. 10 are smoothed averages drawn through data obtained from experiments, but if they had been drawn through data taken at many more speeds, they would have revealed a wavy nature (Fig. 11) due to resonant oscillations in the air column entering the cylinder; these oscillations are somewhat similar to those which develop in an organ pipe.

Volumetric efficiency has a direct effect on the mep developed in a cylinder, on the torque, and on the power that may be realized at a given speed. Since power is a product of speed and torque, the peak power of an engine occurs at a higher speed than for maximum torque, where the rate of torque loss with any further increase in speed will exceed the rate of speed increase. This may be seen in Fig. 12, where the torque and power curves for a typical six-cylinder engine have been plotted. This engine developed its maximum power at a speed about twice that for maximum torque. The average 1969 V-8 engine developed its maximum power at a speed about 60% higher than its maximum torque.

Table 1 shows that the engine speeds at which maximum torque and power are desired in practice require closure of the intake valve to be delayed until the piston has traveled almost half the length of the compression stroke. At engine speeds below those where maximum torque is developed by this valve timing, some of the combustible charge which has been drawn into the cylinder on the suction stroke will be driven back through the valve before it closes. This reduces the effective compression ratio at wide-open throttle—thus the increasing tendency of an engine to develop a fuel knock as the speed and the resulting gas turbulence are reduced.

Another result of this blowback into the intake pipe is the possible reversal of the flow of some of the fuel-air mixture through the carburetor, which will draw fuel each time it passes through the metering system, thereby producing a much richer mixture than if it had passed through the same carburetor only once. The increased fuel supplied to the air reaching the intake valve because of this reversed air flow may be over 100% greater for a single cylinder engine at wide-open throttle than if there were no reversal of flow. Throttling the engine reduces and almost eliminates the blowback through the metering system and the ratio of fuel to air approaches that which would be expected from air flow in one direction only. Manifolding more than one cylinder to the metering system of one carburetor also reduces the blowback through a carburetor because it averages the blowback from individual cylinders. When six cylinders are supplied by a single carburetor, there may be practically no blowback through the carburetor or enrichment of the combustible mixture by this phenomenon. However, when three carburetors are installed on the same six-cylinder engine, with two cylinders supplied by each, the enrichment may be 80–90%. When four cylinders are supplied by a single metering system, as with most V-type eight-cylinder engines, the enrichment may be 10–50% and varies with many factors besides the closing time of the intake valve, such as the exhaust valve events and exhaust back pressure, so that carburetor settings and compensation must usually be made on the engine which it is to supply, and preferably as installed in the car, if optimum power and economy are to be realized. Unfortunately, such settings can not be predicted by simple calculations of fuel flow through an orifice into an airstream flowing at constant velocity.

Intake manifolds. Intake manifolds for multicylinder engines should meet several requirements for the satisfactory performance of spark-ignition engines. They should (1) distribute fuel equally to all cylinders at temperatures where unvaporized fuel is present, as when starting a cold engine or during the warm-up period; (2) supply sufficient

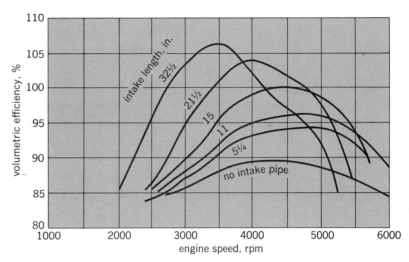

Fig. 10. Effect of intake-pipe length and engine speed on volumetric efficiency of one cylinder of a six-cylinder engine. (*From E. W. Downing, Proc. Automot. Div., Inst. Mech. Eng., no. 6, p. 170, 1957–1958*)

heat to vaporize the liquid fuel from the carburetor as soon after starting as possible; (3) distribute the vaporized fuel-air mixture evenly to all cylinders during normal operation and at low speeds; (4) offer minimum restriction to the mixture flow at high power; and (5) provide equal ram or dynamic boost to volumetric efficiency of all cylinders at some desired part of the engine speed range. This requires that each branch from the carburetor to the valve port should be equal in length, as may be inferred from Fig. 10. Accordingly, no cylinder port should be siamesed with another at the end of a leg of the manifold.

For the warming-up period with liquid fuel present, rectangular sections are desirable to impede spiraling of liquid fuel along the walls, and right-angle bends should be sharp, at least at their inner corner, so as to throw the liquid flowing along the inner wall back into the air stream, and there should be an equal number in each branch.

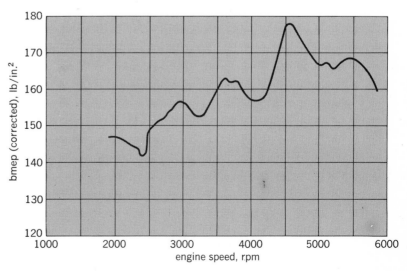

Fig. 11. Evidence of resonant oscillations in the intake pipe to a single cylinder shown by readings taken at small speed increments. (*From E. W. Downing, Proc. Automot. Div., Inst. Mech. Eng., no. 6, p. 170, 1957–1958*)

Table 1. Valve timing of 1969 American automobile engines in degrees crankshaft rotation*

	Intake valve		Exhaust valve	
Engine	Opens before top center	Closes after bottom center	Opens before bottom center	Closes after top center
Six-cylinder				
Average	13	58.5	50.3	16.6
Range	9–26	48–62	42–60	6–28
Eight-cylinder V				
Average	20	71	72	28
Range	10–31	50–114	44–90	10–58

*Adapted from *Automotive Industries*, March 15, 1969.

Manifold heat. Intake manifolds of most American automobile engines are heated to the temperature required to vaporize the fuel from the carburetor (120–140°F) by exhaust gas passing through a suitable passage in the manifold casting, particularly at the first T beyond the carburetor where the liquid fuel impinges before turning to side branches.

To speed the warm-up process, thermostatically operated valves are generally placed in the engine exhaust system so as to force most of the exhaust gases through the intake manifold heater passages when the engine is cold. After the intake manifold has reached the desired temperature, such valves are intended to open and permit only the necessary small portion of exhaust gases to continue passing through the heater. This is an important feature, for too much heat causes a loss of engine power and aggravates the tendency for the engine toward knock and vapor lock.

On some engines, the intake manifolds are heated by water jackets taking hot water from the engine cooling system. This gives uniform heating over a wide range of operating conditions without danger of the overheating that might result from exhaust gas heat if the thermostatic exhaust valve should fail to open. It has the disadvantage, however, of requiring more time to reach normal manifold temperature, even though the water supply from the cylinder heads is short-circuited through the manifold jacket by a suitable water thermostat during warm-up.

One of the advantages of the V-8 engine is the excellent intake manifold design permitted by the centrally located carburetor with but small differences in the lengths of the passage between the carburetor and each cylinder, and an equal number of right-angle bends in each, as in a typical intake manifold using a dual carburetor (Fig. 13). With the usual firing order shown in Fig. 20, the firing intervals for each of the lower branches (shown dotted) are evenly spaced 360 crankshaft degrees apart, but for each of the upper branches two cylinders fire 180° and then 540° apart.

Icing. Because gasoline has considerable latent heat of vaporization, it lowers the air temperature as it evaporates. This is true even at the low temperatures, where only a small part of it is vaporized. It is therefore possible for moisture which may be carried by the air to freeze under certain conditions. Ice is most likely to form when the atmosphere is almost saturated with moisture at temperatures slightly above freezing and up to about 40°F. When ice forms around or near the throttle, it can seriously interfere with the operation of an engine. For this reason small passages have been provided on some engines for jacket water, or exhaust gas from the heating supply for the intake manifold, to warm at least the flange of the carburetor. Here, again, too much heat would produce vapor lock and this would interfere with normal fuel metering. This is one of the reasons for designing some carburetors with separate casting for the throttle bodies which are heated by the manifold through only a thin gasket, while a thick gasket acting as a heat barrier is inserted between it and the float chamber containing the fuel metering systems.

FUEL CONSUMPTION AND SUPERCHARGING

Fuel consumption at loads throughout the operating range of an engine provide insight into such characteristics as friction loss within the engine. Volumetric efficiency of an engine can be increased by use of supercharging.

Part-load fuel economy. When the fuel consumption of a spark-ignition engine is plotted against brake horsepower, straight lines may generally be drawn through the test points at given speeds, as shown in Fig. 14, provided that the

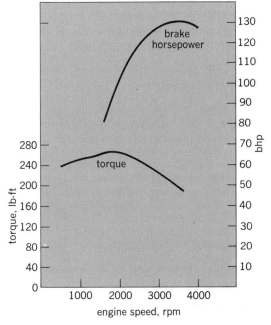

Fig. 12. Typical relation between engine speeds for maximum torque and maximum horsepower.

tests are run with optimum spark advances and at constant fuel-air ratios. Such lines are similar to the Willans lines long used for the steam consumption of steam engines.

For practical purposes the lines at various speeds may be considered parallel over a wide range of speeds. The assumption that the negative power indicated by extrapolating these lines to zero fuel consumption reveals the power absorbed by internal friction of the engine would be justified only when the thermal efficiency remains constant over the load range. On these coordinates, lines radiating from the origin represent constant ratios of fuel consumed to power developed and therefore constant specific fuel consumption (sfc). Several such lines are indicated in Fig. 14, from which the sfc at various loads may be read directly where they cross the performance lines at the various speeds.

Similar plots of even greater utility may be drawn on an indicated horsepower basis, as has been done in Fig. 15 for the same data. For many engines, a single performance line may be drawn through all test points at a given fuel-air ratio over a considerable range of speeds. When extrapolated, the performance line passes through the origin as it does for the engine shown in Fig. 15; the indicated sfc and thermal efficiency of the engine remain constant over the load range covered. Frequently a performance line for an engine passes a little to the left of the origin because of conditions causing a decrease in thermal efficiency as the load is reduced, such as insufficient turbulence or too low a manifold velocity. For a more complete picture of the fuel consumption performance of an engine, similar plots may be made on an mep basis. When this is done and both fuel consumption and horsepower are divided by the engine factor which converts horsepower to mep, the slope and nature of the fuel performance line remain unchanged. The fuel consumption scale then becomes equivalent to the product of mep and sfc. Such plots may be on the basis of either indicated mep (imep) or brake mep (bmep). Figure 16 shows the same data as Figs. 14 and 15 plotted on an imep basis for two different fuel-air ratios.

The fuel consumption performance of diesel engines at part loads may be shown on similar bases, but the plots should not be expected to be straight because the effective fuel-air ratio varies with load. This is illustrated in Fig. 17. It is characteristic of most diesel engines that the curvature of the plot generally flattens out at low loads so that it becomes tempting to extrapolate it to zero fuel consumption, and to consider the negative power intercept as friction. Such an intercept would represent friction only if the efficiency did not change with load, as for the engine characteristics shown in Fig. 14. If the thermal efficiency of a diesel engine improves as the load is reduced, as it would in theory for the classical diesel cycle, the zero fuel intercept for a curve such as shown in Fig. 17 would be to the right of the negative power representing engine friction. Although these fuel consumption performance plots are of considerable utility for recording such data for an engine, the fact that they are curved requires at least three or even more points to fix their location on the plot.

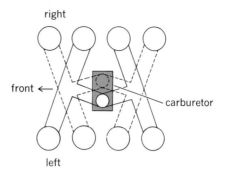

Fig. 13. Schematic of typical intake manifold with dual carburetor on a V-8 engine.

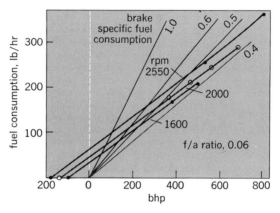

Fig. 14. Fuel consumption of an engine at part loads, plotted as typical Willans lines against brake horsepower. Data were taken with optimum spark advance and with fuel-air ratio adjusted at each test point.

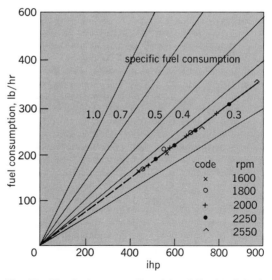

Fig. 15. The fuel consumption data of Fig. 14 plotted against the indicated horsepower, by which speed differences are neutralized.

Supercharging spark-ignition engines. Volumetric efficiency and thus the mep of a four-stroke spark-ignition engine may be increased over a part of or the whole speed range by supplying air to the engine intake at higher than atmospheric pressure.

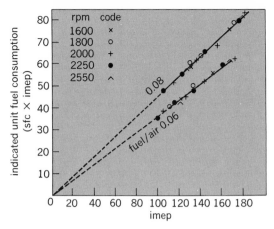

Fig. 16. The fuel consumption data of Fig. 15 plotted on the basis of imep by dividing the fuel scale and the power scale by the same factor ($k=$ ihp/imep). Line for 0.08 fuel-air ratio has been added to show effect of fuel-air ratio on slope of such plots.

This is usually accomplished by a centrifugal or rotary pump. The indicated power of an engine increases directly with the absolute pressure in the intake manifold. Because fuel consumption increases at the same rate, the indicated sfc is generally not altered appreciably by supercharging.

The three principal reasons for supercharging four-cycle spark-ignition engines are (1) to lessen the tapering off of mep at higher engine speed; (2) to prevent loss of power due to diminished atmospheric density, as when an airplane (with piston engines) climbs to high altitudes; and (3) to develop more torque at all speeds.

In a normal engine characteristic, torque rises as speed increases but falls off at higher speeds because of the throttling effects of such parts of the fuel intake system as valves and carburetors. If a supercharger is installed so as to maintain the volumetric efficiency at the higher speeds without increasing it in the middle-speed range, peak horsepower can be increased.

The rapid fall of atmospheric pressure at increased altitudes causes a corresponding decrease in the power of unsupercharged piston-type aircraft engines. For example, at 20,000 ft the air density, and thus the absolute manifold pressure

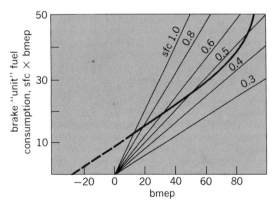

Fig. 17. Fuel consumption of a diesel engine at part loads showing the curvature, typical of such engines on these coordinates, caused by changing effective fuel-air ratios as the loads are increased.

and indicated torque of an aircraft engine, would be only about half as great as at sea level. The useful power developed would be still less because of the friction and other mechanical power losses which are not affected appreciably by volumetric efficiency. By the use of superchargers, which are usually of the centrifugal type, sea-level air density may be maintained in the intake manifold up to considerable altitudes. Some aircraft engines drive these superchargers through gearing which may be changed in flight, from about 6.5 to 8.5 times engine speed. The speed change avoids oversupercharging at medium altitudes with corresponding power loss. Supercharged aircraft engines must be throttled at sea level in order to avoid any damage from detonation or excessive overheating caused by the high mep which would otherwise be developed.

Normally an engine is designed with the highest compression ratio allowable without knock from the fuel expected to be used. This is desirable for the highest attainable mep and fuel economy from an atmospheric air supply. Any increase in the volumetric efficiency of such an engine would cause it to knock unless a fuel of higher octane number were used or the compression ratio were lowered. When the compression ratio is lowered, the knock-limited mep may be raised appreciably by supercharging but at the expense of lowered thermal efficiency. There are engine uses where power is more important than fuel economy, and supercharging becomes a solution. The principle involved is illustrated in Fig. 18 for a given engine. With no supercharge this engine, when using 93-octane fuel, developed an imep of 180 psi at the border line of knock at 8:1 compression ratio. If the compression ratio were lowered to 7:1, the mep could be raised by supercharging along the 7:1 curve to 275 imep before it would be knock-limited by the same fuel. With a 5:1 compression ratio it could be raised to 435 imep. Thus the imep could be raised until the cylinder became thermally limited by the temperatures of critical parts, particularly of the piston head.

Supercharged diesel engines. Combustion in a four-stroke diesel engine is materially improved by supercharging. In fact, fuels which would smoke badly and misfire at low loads will burn otherwise satisfactorily with supercharging. The imep rises directly with the supercharge pressure, until it is limited by the rate of heat flow from the metal parts surrounding the combustion chamber, and the resulting temperatures. A practical application of this limitation was made on a locomotive built by British Railways where the powers, and thus the heats developed, were held reasonably constant over a considerable speed range by driving the supercharger at constant speed by its own engine. In this way the supercharge pressure varied inversely with the speed of the main engine. The corresponding torque rise at reduced speed dispensed with much gear-shifting which would have been required during acceleration with a conventional engine.

When superchargers of either the centrifugal or positive-displacement type are driven mechanically by the engine, the power required becomes an additional loss to the engine output. Experience shows that there is a degree of supercharge for any

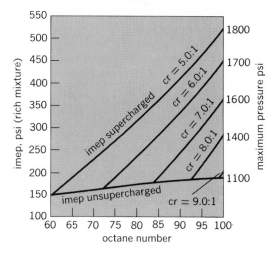

Fig. 18. Graph showing the relationship between compression ratio and knock-limited imep for given octane numbers, obtained by supercharging a laboratory engine. (*From H. R. Ricardo, The High-Speed Internal Combustion Engine, 4th ed., Blackie, 1953*)

engine which develops maximum efficiency; too high a supercharge absorbs more power in the supercharger than is gained by the engine, especially at low loads. Another means of driving the supercharger which is becoming quite general is by an exhaust turbine, which recovers some of the energy that would otherwise be wasted in the engine exhaust. This may be accomplished with so small an increase of back pressure that little power is lost by the engine. This type of drive results in an appreciable increase in efficiency at loads high enough to develop the necessary exhaust pressure.

Supercharging a two-cycle diesel engine requires some means of restricting or throttling the exhaust in order to build up cylinder pressure at the start of the compression stroke, and is used on a few large engines. Most medium and large two-cycle diesel engines are usually equipped with blowers to scavenge the cylinders after the working stroke and to supply the air required for the subsequent cycles. These blowers, in contrast to superchargers, do not build up appreciable pressure in the cylinder at the start of compression. If the capacity of such a blower is greater than the engine displacement, it will scavenge the cylinder of practically all exhaust products, even to the extent of blowing some air out through the exhaust ports. Such blowers, like superchargers, may be driven by the engine or by exhaust turbines.

Engine balance. Rotating masses such as crank pins and the lower half of a connecting rod may be counterbalanced by weights attached to the crankshaft. The vibration which would result from the reciprocating forces of the pistons and their associated masses is usually minimized or eliminated by the arrangement of cylinders in a multicylinder engine so that the reciprocating forces in one cylinder are neutralized by those in another. Where these forces are in different planes, a corresponding pair of cylinders is required to counteract the resulting rocking couple.

If piston motion were truly harmonic, which would require a connecting rod of infinite length, the reciprocating inertia force at each end of the

stroke would be as in Eq. (2), where W is the total

$$F = 0.000456 W N^2 s \qquad (2)$$

weight of the reciprocating parts in one cylinder, N is the rpm, and s is the stroke in inches. Both F and W are in pounds. But the piston motion is not simple harmonic because the connecting rod is not infinite in length, and the piston travels more than half its stroke when the crankpin turns 90° from firing dead center. This distortion of the true harmonic motion is due to the so-called angularity a of the connecting rod, shown by Eq. (3), where r is the

$$a = \frac{r}{l} = \frac{s}{2l} \qquad (3)$$

crank radius, s the stroke, and l the connecting rod length, all in inches.

Reciprocating inertia forces act in line with the cylinder axis and may be considered as combinations of a primary force—the true harmonic force from Eq. (2)—oscillating at the same frequency as the crankshaft rpm and a secondary force oscillating at twice this frequency having a value of Fa, which is added to the primary at firing dead center and subtracted from it at inner dead center. In reality there is an infinite but rapidly diminishing series of even harmonics at 4, 6, 8, . . . , times crankshaft speed, but above the second harmonic they are so small that they may generally be neg-

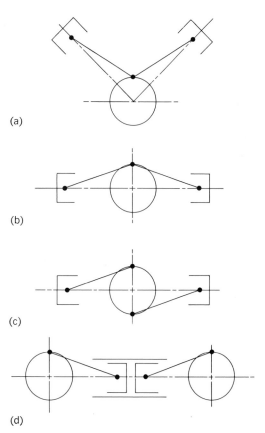

(a)

(b)

(c)

(d)

Fig. 19. Arrangements of two cylinders. (*a*) A 90° V formation with connecting rods operating on the same crankpin. (*b*) Opposed cylinders with connecting rods operating on the same crankpin. (*c*) Opposed cylinders with pistons operating on crankpins 180° apart. (*d*) Double-opposed pistons in the same cylinder but with pistons operating on separate crankshafts.

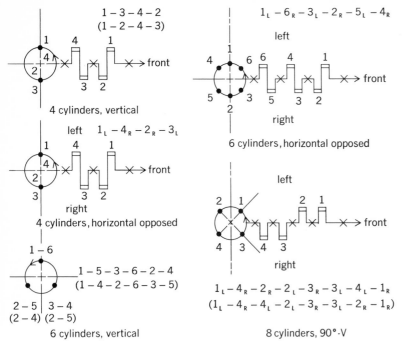

Fig. 20.   Typical cylinder arrangements and firing orders.

connecting rods operate on the same crankpin, as in Fig. 19b, the primary forces are added and are twice as great as for one piston, but the secondary forces cancel. If the pistons operate on two crankpins 180° apart, as in Fig. 19c, all reciprocating forces are balanced. However, as they will be in different planes, a rocking couple will develop unless compensated by an opposing couple from another pair of cylinders. Double-opposed piston pairs, operating in a single cylinder on two crankshafts as in Fig. 19d (with a cross shaft to maintain synchronism), are in perfect balance for primary and secondary reciprocating forces as well as for rotating masses and torque reactions.

In the conventional four-cylinder in-line engines with crankpins in the same plane, the primary reciprocating forces of the two inner pistons (2 and 3) cancel those of the two outer pistons (1 and 4), but the secondary forces from all pistons are added. They are thus equivalent to the force resulting from a weight about $4a$ times the weight of one piston and its share of the connecting rod, oscillating parallel to the piston movement, having the same stroke, but moving at twice the frequency. A large a for this type of engine is advantageous. Where the four cylinders are arranged alternately on each side of a similar crankshaft, and in the same plane, both primary and secondary forces are in balance. Six cylinders in line also balance both primary and secondary forces.

Early eight-cylinder 90° engines with four crankpins in the same plane, like those of the early four-cylinder engine, had unbalanced horizontal secondary forces acting through the crankshafts, which were four times as large as those from one pair of cylinders.

In 1927 Cadillac introduced a crank arrangement for its V-8 engines, with the crankpins in two planes 90° apart. Staggering the 1 and 2 crankpins 90° from each other equalizes secondary forces, but the forces are in different planes. The couple thus introduced is cancelled by an opposite couple from the pistons operating on crankpins 3 and 4. This arrangement of crankpins is now universally used on V-8 engines.

Torsion dampers. In addition to vibrational forces from rotating and reciprocating masses, vibration may develop in an engine from torsional

lected. Thus, for a connecting rod with the average angularity of the 1969 automobile engines, where $a = 0.291$, the inertia force caused by a piston at firing dead center is about 1.29 times the pure harmonic force, and at inner dead center it is about 0.71 times as large.

Where two pistons act on one crankpin, with the cylinders in 90° V arrangements, as in Fig. 19a, the resultant primary force is radial and of constant magnitude and rotates around the crankshaft with the crankpin. Therefore, it may be compensated for by an addition to the weight required to counterbalance the centrifugal force of the revolving crankpin and its associated masses. The resultant of the secondary force of the two pistons is 1.41 times as large as for one cylinder, and reciprocates in a horizontal plane through the crankshaft at twice crankshaft speed.

In engines with opposed cylinders, if the two

Table 2. Selected average dimensions (in inches) of 1969 engines*

Category	Six-cylinder	Eight-cylinder V
Cylinder bore	3.68	4.07
Cylinder stroke	3.32	3.68
Stroke-bore ratio	0.90	0.91
Displacement	212	385
Maximum hp/rpm	131/4100	294/4640
bmep at maximum hp	120	131
Maximum torque (lb/ft)/rpm	194/2000	409/2870
bmep at maximum torque	139	160
Connecting-rod length	5.69	6.32
Crank radius/rod length ratio	0.292	0.292
Crankpin diameter	2.08	2.27
Wrist-pin diameter	0.91	0.99
Main-bearing diameter	2.41	2.79
Intake-valve-head diameter (max)	1.67	1.96
Exhaust-valve-head diameter (max)	1.42	1.60
Piston-head area/intake-valve-head area ratio	4.85	4.31

*Adapted from *Automotive Industries*, March 15, 1969.

resonance of the crankshaft at various critical speeds. The longer the shaft for given bearing diameters, the lower the speeds at which these vibrations may develop. Such vibrations are dampened on most six- and eight-cylinder engines by a vibration damper which is similar to a small flywheel on the crankshaft at the end opposite to the main flywheel but coupled to the shaft only through rubber so arranged as to reduce the torsional resonances. Vibration dampers of this type are usually combined with the pulley that is driving the cooling fan and generator of the engine.

Even though the majority of American automobile engines are now dynamically balanced, it has been general practice for several years to mount them in the chassis frame on rubber blocks. This reduces the transmission of the small-amplitude high-frequency vibrations in torque reaction as well as small unbalance of reciprocating parts in individual cylinders, so that a low noise level is developed in the car from the operation of the engine.

Firing order. Cylinder arrangements are generally selected for even firing intervals and torque impulses, as well as for balance. As a result, the cylinder arrangements and firing orders shown in Fig. 20 may be found in automobile use. It is generally customary to identify cylinder banks as left or right as seen from the driver's seat, and to number the crankpins from front to rear. Manufacturers do not agree on methods of numbering individual cylinders of V-type engines. However, the arrangements and firing orders shown are in general use with the addition in parentheses of alternate arrangements only occasionally used.

Typical American automobile engine. American six- and eight-cylinder engines manufactured in 1969 had many features in common. All had overhead valves located in a removable cylinder head and cylinder blocks cast integral with the upper crankcase. The valves were operated from a chain-driven camshaft in the crankcase through push rods and rockers having arm lengths giving a valve lift about 50–75% greater than the cam lift. With the exception of one six-cylinder design, all silenced the valve action by use of hydraulic valve lifters.

Many designs provided means for rotating at least the exhaust valve to improve valve life. All cylinder barrels were completely surrounded by the jacket cooling water, and most designs extended it the full length of the bore. Main bearings supported the crankshaft between each crankpin. Almost all designs locked the wrist pin to the connecting rod; the others permitted it to "float" in both the piston and connecting rod. The use of three piston rings above the wrist pin had become general practice, two being narrow compression rings and the lowest an oil scraper. Compression rings were generally of cast iron about 0.078 in. wide and provided with a coating such as chromium or tin to prevent scuffing the surface during the wearing-in period of a new engine and when it is started cold with but little oil on the cylinder wall. Oil-scraper rings were about 3/16 in. wide, were provided with a nonscuffing surface, and had drain holes through the piston for the return of the excess oil scraped from the cylinder wall.

The largest six-cylinder engine developed a

Fig. 21. Cutaway view of a typical V-8 overhead valve automobile engine with air filter. (*Pontiac Div., General Motors Corp.*)

maximum of 155 hp, with a cylinder displacement of 250 in.³ Where higher power was required, the eight-cylinder V-type was installed. A typical engine of this type is shown in cross section in Fig. 21. The right and left banks of cylinders were staggered to enable connecting rods of opposing cylinders to be located side by side on the same crankpin. The V arrangement provides a short and very ridged structure, which is important for high engine speeds because of the minimized deflection of the main bearings. It also makes possible efficient intake-manifold designs and almost symmetrical and equal-length branches to each cylinder port from a centrally located downdraft carburetor. The short length and low height of these engines are also important features for car styling. Some of the principal dimensions and other statistics of these engines have been averaged in Table 2.

[NEIL MAC COULL]

Bibliography: T. Baumeister (ed.), *Standard Handbook for Mechanical Engineers*, 7th ed., 1967; W. H. Crouse, *Automotive Engines*, 2d ed., 1959; W. H. Crouse, *Automotive Mechanics*, 4th ed., 1960; L. C. Lichty, *Combustion Engine Processes*, 1967; E. F. Obert, *Internal Combustion Engines*, 3d ed., 1968; H. R. Ricardo, *The High-Speed Internal Combustion Engine*, 1931, 4th ed., 1953; C. F. Taylor and E. S. Taylor, *The Internal Combustion Engine*, rev. ed., 1948.

Internal energy

A characteristic property of the state of a thermodynamic system, introduced in the first law of thermodynamics. For a static, closed system (no bulk motion, no transfer of matter across its boundaries), the change ΔE in internal energy for a process is equal to the heat Q absorbed by the system from its surroundings minus the work w done by the system on its surroundings. Only a change in internal energy can be measured, not its value for any single state. For a given process, the change in internal energy is fixed by the initial and final states, and it is therefore independent of the path by which the change in state is accomplished.

The internal energy includes the intrinsic energies of the individual molecules of which the system is composed and contributions from the interactions among them. It does not include contributions from the potential energy or kinetic energy of the system as a whole; these changes must be accounted for explicitly in the treatment of flow systems. Because it is more convenient to use an independent variable (the pressure P for the system instead of its volume V), the working equations of practical thermodynamics are usually written in terms of such functions as the enthalpy $H = E + PV$, instead of the internal energy itself. *See* ENTHALPY.

[PAUL J. BENDER]

Jet fuel

Fuel blended from the light distillates fractionated from crude petroleum. There are two general types, a wide-cut heavy naphtha-kerosine blend used by the U.S. Air Force as JP-4 (or commercially as Jet B) and a kerosine used by the world's airlines as Jet A (or Jet A-1) or by the U.S. Navy as JP-5. Since 1970, commercial kerosine of 38°C flash point has grown from a small-volume household-heating fuel into a major product of commerce rivaling gasoline in importance as a source of transportation energy. During the 1960–1975 period, JP-4 diminished in relative importance, and commercial use of Jet B also declined to a small percent. *See* KEROSINE.

All jet fuels must meet the stringent performance requirements of aircraft turbine engines and fuel systems, which demand extreme cleanliness and freedom from oxidation deposits in high-temperature zones. Combustors require fuels that atomize and ignite at low temperatures, burn with adequate heat release and controlled radiation, and produce neither smoke nor attack of hot turbine parts. The operation of the aircraft in long-duration flights at high altitude imposes a special requirement of good low-temperature flow behavior; this need establishes Jet A-1 which has a freezing point of −50°C (wax) as an international flight fuel; Jet A which has a freezing point of −40°C (wax) can serve shorter domestic routes. *See* AIRCRAFT ENGINE.

Fuels pumped through long multiproduct pipelines or delivered by tanker are usually clay-filtered to ensure freedom from surfactants. Many stages of filters operate to ensure clean, dry product as the fuel moves into airport tanks, hydrant systems, and finally aircraft. Because high-speed filtration can generate static charges, fuels may contain an electrical conductivity additive to ensure rapid dissipation of charge.

[W. G. DUKEK]

Kerosine

A refined petroleum fraction used as a fuel for heating and cooking, jet engines, lamps, and weed burning, and as a base for insecticides. Kerosine, known also as lamp oil, is recovered from crude oil by distillation. It boils in the approximate range of 350–550°F. Most marketed grades, however, have narrower boiling ranges. The specific gravity is about 0.8. Determined by the Abel tester, the flash point is not below 73°F, but usually a higher flash point is specified. Down to a temperature of −25°F, kerosine remains in the liquid phase. Components are mainly paraffinic and naphthenic hydrocarbons in the C_{10}–C_{14} range. A low content of aromatics is desirable except when kerosine is used as tractor fuel.

Specifications are established for specific grades of kerosine by government agencies and by refiners. Since these specifications are developed from performance observations, they are adhered to rigidly to assure satisfactory operation. For use in lamps, for example, a highly paraffinic oil is desired because aromatics and naphthenes give a smoky flame; and for satisfactory wick feeding, a viscosity no greater than 2 centipoises is required in this application. Furthermore, the nonvolatile components must be kept low. In order to avoid atmospheric pollution, sulfur content must be low; a minimum flash point of 100°F is desirable to reduce explosion hazards.

Today, kerosine represents only a little over 4% of the total petroleum-products production in the United States, whereas it was the major product in the 1800s. Tractor fuel now represents an insignificant percentage of the total production. Use of kerosine as a jet fuel continues to increase as more jet planes are put into operation. Kerosine also is the principal hydrocarbon fuel for rockets.

The price of kerosine has followed the price of crude oil, and this cost continues to be the dominant price factor; however, changing use patterns and specifications may exert an additional upward force on the price of kerosine. *See* DISTILLATE FUEL; JET FUEL; PETROLEUM PRODUCTS.

[HAROLD C. RIES]

Laser-induced fusion

A process (now primarily theoretical) in which a laser would be used to heat a plasma to a sufficiently high temperature for a minimum critical length of time to cause an efficient thermonuclear fusion reaction to occur throughout the plasma. The energy given off in the resulting thermonuclear microexplosion could then be converted to useful energy as is done presently by nuclear fission reactors. The achievement of thermonuclear fusion by a technique known as magnetic confinement has been under extensive investigation as a potential energy source since about 1950, but has not reached the conditions whereby an efficient thermonuclear reaction could be obtained. With the development of high-power pulsed lasers, the concept of laser fusion has indicated enough promise to be considered by many as a possible alternative to the magnetic confinement process. *See* REACTOR, NUCLEAR.

Investigations of laster-induced fusion have taken two directions, both requiring plasma temperatures of the order of $10^8°C$. The most extensively studied technique involves the concepts of implosion and compression of the plasma to increase its density and temperature and of inertial confinement to provide enough time to allow the reaction to occur throughout the entire mass of the plasma. The other approach is to use the laser as an auxiliary heating source to heat a very long, magnetically confined plasma. This process involves some of the ideas from both the laser implosion technique and the magnetic confinement technique, along with some new concepts and problems of its own.

Deuterium, proposed as a fuel from which the plasma is to be created, is a low cost, relatively clean fuel that can be obtained from a virtually inexhaustible supply in the oceans. The higher energy yield and lower contamination from the fusion reaction, as compared to fission reactions, makes the difficult program involved in attempting to initiate such a reaction a worthwhile task.

Fission. Present-day nuclear power plants use a fission reaction as their source of energy. This reaction occurs spontaneously in a special isotope of uranium in which the heavy uranium nucleus breaks apart into two nuclei of lighter elements. During this process a large amount of extra energy is produced that results in the emission of radiation and the impartation of high velocities to the newly formed particles. When the rapidly moving particles are absorbed within the reaction chamber, heat is produced that can be used to generate useful power. The high velocities of the particles help speed up reactions of surrounding atoms, and if the mass density is high enough a thermonuclear explosion could result. It is thus essential to control the reaction rate by diluting the uranium mass in the reaction region. *See* FISSION, NUCLEAR.

Fusion. In the fusion reaction, rather than causing a heavy element to divide, two light elements are fused together to form a heavier element, again giving off energy in the form of radiation and of kinetic energy of the newly formed particles. The fusion reaction occurring with the fastest rate involves the fusion of a deuterium ion (hydrogen with one extra neutron) and a tritium ion (hydrogen with two extra neutrons) resulting in a helium atom and a neutron. The $10^8°C$ temperature required to initiate this reaction is essential in order to provide high enough impact velocities when the deuterium and tritium ions collide so as to overcome the strong repulsive forces caused by the positive electrical charges of both ions. When the nuclei fuse together to form a helium atom, most of the extra energy is given off to the emitted neutron. This neutron can transfer its energy to neighboring deuterium and tritium ions, causing the reaction to spread if the density is high enough and the confinement time is long enough. The Lawson criterion is a useful guide to determine the density-time product which must be satisfied in addition to the minimum temperature requirement. If n is the plasma density and τ is the confinement time of the plasma, $n\tau = 10^{14}$ cm^{-3}s is the Lawson criterion for the deuterium-tritium reaction. The energy re-

leased in a deuterium-tritium fusion reaction is approximately seven times that obtained in a uranium fission reaction.

Fusion reactions commonly occur in stars where the necessary temperatures are already present and the confinement time occurs as a result of gravitational pressure. These reactions were first produced by humans in the development of thermonuclear weapons. The ideas soon evolved as to how such a reaction could be controlled on a much smaller scale and the energy harnessed for peaceful uses. Since 1950 the extensive development and testing of magnetic confinement techniques, in which a plasma is created and magnetically confined within a "magnetic bottle" for times of the order of seconds, has not yet achieved the requirements for a sustained thermonuclear reaction. Thus the high-power laser has become available at an opportune time to allow new methods to be explored in achieving the goal of useful, efficient electrical power generation using thermonuclear energy. *See* FUSION, NUCLEAR.

Fusion power by laser implosion. The laser implosion technique, which is the most extensively studied laser technique, involves the creation of a miniature explosion uniformly over the surface of a very small pellet (50–100 μm diameter) consisting of a mixture of frozen deuterium and tritium. Some newer pellet designs incorporate hollow spheres or layered regions of other materials to improve the burning characteristics. In the explosion process, electrons in the surface region absorb energy from the laser pulse. This energy is then rapidly transferred by collisions to the heavier surrounding ions which then move outward with explosive force at very high velocities. The opposite, momentum-conserving reaction is an inward-moving shock wave that compresses and heats the remaining major portion of the pellet mass. Compressions of 10,000 times liquid density corresponding to a decrease in the pellet diameter to approximately 1/20 of its original value should yield densities of the order of 10^{26} cm^{-3} in the compressed region. From the Lawson criterion the minimum confinement time for such a density would have to be of the order of 10^{-11} to 10^{-12} s. Such a confinement time would inherently occur during the compression because the particle velocities of 10^8 cm/s would result in very little movement for such short time periods. This type of confinement is referred to as inertial confinement. When the high-temperature requirement and the Lawson criterion are simultaneously satisfied, the nuclear reaction or burn will occur throughout the entire pellet mass, releasing the maximum possible energy to the chamber walls. Although such a laser-induced reaction has never been successfully demonstrated in the laboratory, detailed mathematical models have been used to simulate the reaction with computers. Enough details are known from previous studies of nuclear reactions to place a high degree of confidence in these models.

One of the most crucial requirements indicated by the theoretical models is that of having the proper amount of laser energy arriving at the target at each instant of time. If too much energy arrives at the target too soon, the outer surface can be heated sufficiently rapidly that only part of the pellet will be "burned." If too little energy is available,

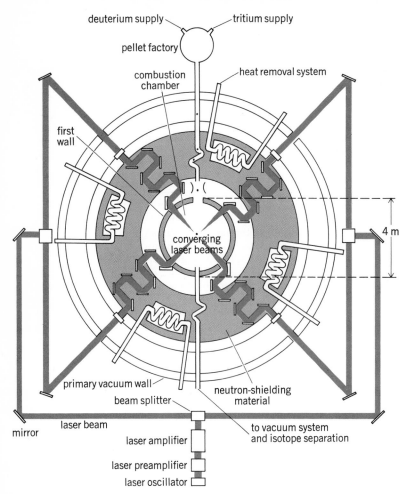

Fig. 1. Diagram of a laser fusion power plant. *(From J. L. Emmett, J. Nuckolls, and L. Wood, Fusion power by laser implosion, Sci. Amer., 230(6):24–37, 1974)*

German laboratory is also developing an atomic iodine laser operating at 1.31 μm for fusion experiments. In addition, several new kinds of molecular lasers operating in the near ultraviolet may prove to be useful lasers for fusion if the high energy storage and efficiency of current small laboratory-size lasers can be shown to scale to larger dimension. Problems of optical breakdown and plasma instabilities tend to favor a short-wavelength laser for laser fusion experiments, whereas two-photon absorption problems favor longer wavelengths. The ideal wavelength is thought to be somewhere between 0.3 and 0.8 μm (ranging from near ultraviolet to near infrared).

As for the progress in the laser implosion experiments, a group in the Soviet Union demonstrated the first production of high-energy neutrons from a laser irradiated target in 1968, and reported indications of implosion and compression in 1972. Later a United States laboratory demonstrated compression by a factor of 100 using a deuterium-tritium pellet irradiated from opposite sides by two neodymium-glass lasers, each having an energy of 50 J.

A simplified drawing of a possible design for a laser fusion power plant is shown in Fig. 1. The converging laser beams would arrive at the central target area via zigzag mirrored paths through a neutron shielding wall. The frozen deuterium-tritium pellets would be dropped from a pellet factory, guided electrostatically on a zigzag path through a neutron shielding wall, and timed to arrive at the target in the central region of the chamber simultaneously with the laser pulse. There they are symmetrically irradiated and imploded by converging laser beams. Compression and heating of the pellet to 10^8°C temperatures will cause a thermonuclear fusion reaction to occur throughout the pellet mass, releasing large amounts of energy. Energy would be extracted by a circulating liquid, such as liquid lithium, which could then be used for making steam to drive a turbogenerator. If 100 microexplosions could be produced every second, a 1000 MW power plant might be possible.

Fusion power by laser heating and magnetic confinement. The development of the high-power, long-pulse, CO_2 laser made possible the serious consideration of another technique for initiating laser-induced fusion. This technique involves the use of a long-wavelength laser as an auxiliary source for heating a magnetically confined plasma. Because the plasma would consist primarily of deuterium and tritium ions, the magnetic field would trap these charged particles for long periods of time, and thus the entire laser heating process could be performed much more slowly than in the pellet implosion scheme, thereby making lower densities possible. Much of the technology for laser heating with magnetic confinement can be obtained from existing knowledge available from the extensive studies of magnetic confinement devices. Such devices have been steadily improved upon for many years, but the plasmas produced have never quite reached the appropriate temperature and density for the necessary time (Lawson criterion) to obtain significant energy yields from the fusion reaction. The devices under consideration for laser heating are those that would use a pulsed magnetic field for confinement, which are referred to as laser-augmented pinches,

the ignition temperature will not be reached at all. It is thus essential to "tailor" the laser pulse shape to create the appropriate burn sequence. This places severe requirements upon the laser system. Extremely high energies, of the order of 5000 J, will probably be needed to achieve a gain of 1 (which means that as much energy is obtained from the fusion reaction as was put into the reaction by the laser). To achieve a gain of 75, calculations indicate that the laser must provide a 300,000-J pulse lasting for 1 nanosecond, with an efficiency of 10%. From such a laser system a 1000-MW power plant might be possible if 100 pellets are exploded every second.

There are no lasers available that can provide anywhere near the above-mentioned input-pulse energy or even the high efficiency for such short pulses. The neodymium-doped glass lasers operating at a wavelength of 1.06 μm (in the near infrared) have achieved energies of the order of 1000 J but with efficiencies of only up to 0.1 %. These lasers are being used by laboratories in the United States and the Soviet Union for experiments which are designed to test many of the ideas relating to pellet implosion. The carbon dioxide laser operating at 10.6 μm (middle infrared) has a potential short-pulse efficiency of somewhere between 5 and 10 %. A 10,000-J version of this laser is being developed at a United States laboratory. A West

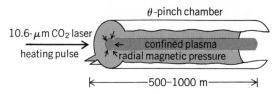

Fig. 2. Diagram of a laser-augmented pinch showing a long-wavelength laser pulse arriving from the left.

and those that use steady fields, which are termed laser-heated solenoids. In either case, the laser would be used to provide the necessary increase in heating that has not so far been possible with magnetic confinement alone.

A possible scheme is shown in Fig. 2, where the pulsed CO_2 laser arrives from the left and enters the already formed plasma within the θ-pinch. Plasma temperatures of the order of 10^{8}°C and densities of $10^{17}-10^{18}$/cm³ must be maintained for times ranging from 1 to 0.1 ms to satisfy the Lawson criterion. In this density range the absorption coefficient for 10.6 μm radiation would be relatively small and therefore a plasma length of 500–1000 m would be required in order to use the CO_2 laser energy efficiently. At 10^{8}°C a thermonuclear burn will occur, releasing high-energy neutrons that will heat the chamber wall. This heat energy can be used to produce steam to drive a turbogenerator. Also, for these densities, confinement could be achieved with available magnetic field technology. Plasma leaks from the ends of the cylinder would be minimized by the long plasma lengths and also possibly by incorporating magnetic mirrors at the ends of the cylinder.

One of the problems in this laser-heating scheme is that of keeping the laser beam from being refracted out of the magnetically confined region while it is being absorbed over the long heating path length. It was suggested that a lower plasma density on the cylinder axis might cause the beam to be guided and therefore trapped in the central region due to the refractive effects of the plasma. Energy losses would therefore be minimized as the laser beam was being efficiently absorbed over the entire plasma length. Experiments showing that such guiding occurs have been performed in United States and Canadian laboratories.

Once the ignition temperature is reached in this long plasma, the energy from the high-energy neutrons could be collected in the chamber walls in a similar fashion to that of the laser implosion scheme. The heat would be extracted to produce steam which could then drive a turbogenerator. *See the feature article* EXPLORING ENERGY CHOICES.   [WILLIAM T. SILFVAST]

Bibliography: J. L. Emmett, J. Nuckolls, and L. Wood, Fusion power by laser implosion, *Sci. Amer.*, 230(6):24–37, 1974; M. S. Feld, A. Javan, and N. A. Kurnit (eds.), *Fundamental and Applied Laser Physics*, 1973; W. C. Gough and B. J. Eastlund, The prospects of fusion power, *Sci. Amer.*, 224(2):50–64, 1971; S. Jacobs, M. Sargent III, and M. O. Scully (eds.), *High Energy Lasers and Their Applications*, 1974; Thermonuclear neutrons from laser implosion, *Phys. Today*, 27(8):17–19, 1974; C. Yamanaka (ed.), *Laser Interaction with Matter*, Japan Society for the Promotion of Science, 1973.

Lawson criterion

A necessary but not sufficient condition for the achievement of a net release of energy from nuclear fusion reactions in a fusion reactor. As originally formulated by J. D. Lawson, this condition simply stated that a minimum requirement for net energy release is that fusion fuel charge must combust for at least a long enough time for the recovered fusion energy release to equal the sum of energy invested in heating that charge to fusion temperatures, plus other energy losses that occur during the combustion period.

The result is usually stated in the form of a minimum value of $n\tau$ that must be achieved for energy break-even, where n is the fusion fuel particle density and τ is the confinement time. Lawson considered bremsstrahlung (x-ray) energy losses in his original definition. For many fusion reactor cases, this loss is small enough to be neglected compared to the heating energy. With this simplifying assumption, the basic equation from which the Lawson criterion is derived is obtained by balancing fusion energy release against heat input to the fuel plasma. Assuming hydrogenic isotopes, deuterium and tritium at densities n_D and n_T respectively, with accompanying electrons at density n_e, all at a maxwellian temperature T, one obtains Eq. (1),

$$n_D n_T \langle \sigma v \rangle Q \tau \eta_r \geq \left[\frac{3}{2} kT \left(n_D + n_T + n_e \right) \right] \frac{1}{\eta_h} \quad (1)$$

where the recovered fusion energy release is set equal to or greater than the energy input to heat fuel. Here $\langle \sigma v \rangle$ is the reaction cross section as averaged over the velocity distribution of the ions, Q is the fusion energy release, η_r is the efficiency of recovery of the fusion energy, η_h is the heating efficiency, and k is the Boltzmann constant.

For a fixed mixture of deuterium and tritium ions, Eq. (1) can be rearranged in the general form of Eq. (2). For a 50-50 mixture of deuterium and tri-

$$n\tau \geq F(n_r, n_h, Q) \left[\frac{T}{\langle \sigma v \rangle} \right] \quad (2)$$

tium the minimum value of $T/\langle \sigma v \rangle$ occurs at about 25 keV ion kinetic temperature (mean ion energies of about 38 keV). Depending on the assumed efficiencies of the heating and recovery processes, the lower limit values of $n\tau$ range typically between about 10^{14} and 10^{15} cm⁻³. These values therefore serve as a handy index of progress toward fusion, although their achievement does not alone guarantee success. Under special circumstances (unequal ion temperatures, unequal deuterium and tritium densities, and nonmaxwellian ion distributions), lower $n\tau$ values may be adequate for nominal break-even.

The discussion up to this point has been oriented mainly to situations in which the fusion reactor may be thought of as a driven system, that is, one in which a continuous input of energy from outside the reaction chamber is required to maintain the reaction. Provided the efficiencies of the external heating and energy recovery systems are high, a driven reactor generally would require the lowest $n\tau$ values to produce net power. An important alternative operating made for a reactor would be an ignition mode, that is, one in which, once the initial heating of the fuel charge is accomplished, energy directly deposited in the plasma by charged reaction products will thereafter sustain the reaction. For example, in the D-T reaction, approximately 20% of the total energy release is imparted to the alpha particle; in a magnetic confinement system, much of the kinetic energy carried by this charged nucleus may be directly deposited in the plasma, thereby heating it. Thus if the confinement time is adequate, the reaction may become self-sustaining without a further input of energy from external sources. Ignition, however, would generally require $n\tau$ products with a higher range of values, and is thus expected to be more difficult to achieve than the driven type of reaction. However, in all cases the Lawson criterion is to be thought of as only rule of thumb for measuring fusion progress; detailed evaluation of all energy disruptive and energy recovery processes is required in order properly to evaluate any specific system. *See* FUSION, NUCLEAR.

[RICHARD F. POST]

Lignite

Soft and porous carbonaceous material intermediate between peat and subbituminous coal, with a heat value less than 8300 Btu on a mineral-matter-free (mmf) basis. In the United States, lignitic coals are classified as lignite A or lignite B, depending upon whether they have calorific value of more or less than 6300 Btu. Elsewhere than in North America, lignites are called brown coals, two major classes being recognized, namely, hard and soft brown coals. The soft brown coals are described as either earthy, resembling peat, or as fragmentary; the hard brown coals are regarded as dull (matte) or bright (glance). The soft brown coals correspond in a general way with lignites of class B, and the hard brown coals with lignites of class A, but the two classifications should not be regarded as strictly interchangeable. Brown coals generally have a brown color, compared with the usual black color of bituminous coal. At the present time there is a vigorous effort being made by the International Committee of Coal Petrology that will make possible a satisfactory differentiation on a petrologic basis of bituminous and lignitic coals and provide a satisfactory classification of varieties of lignite on a basis other than calorific.

Reserves. The lignite of North America is mainly of lignite class A. The known reserves of lignite in 15 states have been estimated as 224,-000,000,000 short tons, as compared with an estimated total coal reserve of 828,000,000,000 tons. About 98% of the lignite reserves in the United States are in the northern Great Plains area, which includes the western half of North Dakota, eastern Montana, northeastern Wyoming, and northeastern South Dakota. A single lignite mine in California operates for the production of montan wax. The reserves in the northern Great Plains are believed to represent a total of 99% of the total reserves of lignite in the United States.

Uses. The main use of lignite in the United States is in residential and industrial heating and in the generation of power. Because of the competition with bituminous coal, the market area of lignite in the United States is relatively small. When used at any great distance from its source, lignite requires dehydration and even briquetting.

Some investigation is under way that may lead to the production of pipeline gas from lignite. The U.S. Bureau of Mines regards the total gasification of lignite a potential for a large tonnage outlet for this variety of coal. *See* COAL; COAL GASIFICATION.

[GILBERT H. CADY]

Bibliography: Mineral Facts and Problems, U.S. Bur. Mines Bull. no. 360, 1965.

Liquefied natural gas (LNG)

A product of natural gas which consists primarily of methane. Its properties are those of liquid methane, slightly modified by minor constituents. One property which differentiates liquefied natural gas (LNG) from liquefied petroleum gas (LPG), which is principally propane and butane, is the low critical temperature, about −100°F. This means that natural gas cannot be liquefied at ordinary temperatures simply by increasing the pressure, as is the case with LPG; instead, natural gas must be cooled to cryogenic temperatures to be liquefied and must be well insulated to be held in the liquid state.

Peak shaving system. The earliest commercial use for LNG was for storage to meet winter home-heating peak demands. In the construction of long-distance gas pipelines normal practice is to size the line somewhat larger than the average yearly demand rate. Capacity can be increased by 15−20% by operating all compression equipment at the maximum possible flow rate. However, during the coldest days of winter the demand for heating fuel is extremely high and some temporary additional supply of gas is required. Several different techniques are available to supply this temporary peak demand. One reasonable approach is to collect gas during the summer months when demand is low and to store it until winter when it can be withdrawn to meet demands. Since 600 ft³ of natural gas condenses to less than 1 ft³ of liquid, this form of storage is relatively convenient.

Such a peak shaving system was constructed in Cleveland, Ohio, in the early 1940s and operated successfully for several years until one of the storage tanks developed a leak and the escaping liquid caught fire, producing a very destructive conflagration. It was almost a generation before this approach to peak shaving was used again. In the many LNG peak shaving systems that have been built since then, careful attention has been given to separation of storage tanks and other equipment, so that even if there were complete destruction by fire of one storage tank, it would not be likely to result in damage in any other area of the plant.

The large insulated storage tanks constitute the main cost element in LNG peak shaving. Conventional double-walled metal or prestressed concrete tanks with perlite powder insulation between the walls have proved to be satisfactory for units of all sizes. In the development of LNG technology, a number of novel storage schemes have been attempted, but all have weaknesses that result in unsafe or uneconomical operations. The most radical of these attempts utilized frozen earth to form the tank lining for an excavated reservoir. Thermal insulation would theoretically be provided by the earth itself. With large enough tanks, the ratio of surface area to volume declines sufficiently so that the heat-conduction rate of natural soil results in a tolerably low evaporation rate of the liquefied gas.

Several attempts to exploit this scheme have failed because the intrusion of an underground stream into the frozen earth area created an unacceptable increase in heat input and liquid evaporation rate. A second weakness of this scheme is that the soil around the tank becomes saturated with hydrocarbon vapors, greatly increasing the potential for an uncontrollable conflagration should a fire start in the tank area.

Systems relying on reservoirs lined either with special concrete or with plastic films or foams have all failed. In these experiments the unequal contraction of the liner during cool-down to the cryogenic temperature of the LNG has caused cracks to open in the liner. Subsequent loss of LNG into the insulation space or into the soil around the tank has made the projects both unsafe and uneconomical.

A tragic accident occurred at an installation on Staten Island, NY, in 1973. Workers were attempting to repair the plastic film liner in the tank. The tank had been emptied and held out of service for a year before the repair work was started. Somehow, a fire originated in the equipment being used inside the tank. The fire spread, and eventually the roof collapsed; all of the men working in the tank were killed. The circumstances were unlike those of the earlier accident in Cleveland, for no LNG was in the tank or anywhere in the plant site at the time. No LNG facility has had a fire or other serious problem since the Cleveland disaster. With more than 100 LNG storage facilities operating around the world, this is an exemplary safety record.

Liquefaction equipment represents a relatively small portion of total project cost and has not produced much variation in design. One novel approach has been used successfully by the San Diego Gas and Electric Co. The peak shaving plant is located adjacent to a large electrical generating station, which is fueled by natural gas supplied at transmission-line pressure of about 600–800 psi. The gas is first processed by the liquefier through heat exchangers and expansion turbines. A fraction of the gas is liquefied, and the remainder returns through the heat exchangers to pass on to the electric generating station at low pressure. In this way the liquefaction plant uses the otherwise wasted pressure of the pipeline. The maximum amount of LNG which can be produced in any given location is limited to at most 20% of the quantity of gas which drops from pipeline pressure to low pressure.

Ocean transport. In Algeria, Libya, Alaska, and other oil-producing areas there is not much demand for the natural gas that is produced along with the oil. One economical way of transporting the surplus gas to markets in industrial centers is to liquefy the gas and ship it in specially designed insulated tankers.

The original transoceanic LNG system transported liquefied gas from Algeria to England and northern France. Since 1965 about 150,000,000 ft³ of gas has been liquefied daily for shipment aboard three specially insulated tank ships. Larger systems have been built to supply Italy and Spain from Libya, Japan from Alaska and Borneo, and southern France from Algeria. Smaller quantities have been supplied to cities in the eastern United States from Libya and Algeria. Liquefaction capacity is expanding in these source areas and in other areas rich in gas resources.

Other uses. As peak shaving and LNG tanker systems have developed, attention has been directed to the potential use of LNG as a fuel in different types of vehicles. LNG has a number of attractive characteristics as an engine fuel. It has an antiknock value well over 100, without any additives. It is extremely clean-burning, resulting in low maintenance costs and a minimum of air pollution from engine exhaust. In gas turbines it provides a large heat sink and burns with a relatively nonluminous flame, both of which assist in engine cooling. Its specific energy per pound of fuel is 15% higher than for gasoline or kerosine. In rocket engines it provides the highest specific impulse of any hydrocarbon fuel.

The primary limitation on the development of LNG as an engine fuel is the general shortage of natural gas in the United States, Europe, and Japan. LNG imported into these areas is intended to supplement the available supplies of natural gas for home heating and cooking. Studies on the use of LNG engine fuel for a variety of transport vehicles have shown its advantages. However, manufacturers are reluctant to undertake engine and vehicle redesign efforts since adequate supplies of LNG cannot be assured. *See* LIQUEFIED PETROLEUM GAS (LPG).

[ARTHUR W. FRANCIS]

Bibliography: R. Blakely, *Oil Gas J.*, 66(1):60–62, 1968; W. B. Emery, E. L. Sterrett, and C. F. Moore, *Oil Gas J.*, 66(1):55–59, 1968; C. H. Gatton, *Liquefied Natural Gas Technology and Economics*, 1967; M. Sittig, *Cryogenics: Research and Applications*, 1963; C. M. Sliepcevich and H. T. Hashemi, *Chem. Eng. Progr.*, 63(6):68–72, 1967; J. W. White and A. E. S. Neumann (eds.), *Proceedings of 1st International Conference on Liquefied Natural Gas*, Chicago, 1968.

Liquefied petroleum gas (LPG)

A product of petroleum gases, principally propane and butane, which must be stored under pressure to keep it in a liquid state. At atmospheric pressure and above freezing temperature, these substances would be gases. Large quantities of propane and butane are now available from the gas and petroleum industries. These are often employed as fuel for tractors, trucks, and buses and mainly as a domestic fuel in remote areas. Because of the low boiling point (−44 to 0°C) and high vapor pressure of these gases, their handling as liquids in pressure cylinders is necessary. Owing to demand from industry for butane derivations, LPG sold as fuel is made up largely of propane. On a gallonage basis, production of LPG in the United States exceeds that of kerosine and approaches that of diesel fuel.

Operating figures for gasoline, diesel, and LPG fuels show that LPG compares favorably in cost per mile. LPG has a high octane rating, making it useful in engines having compression ratios above 10:1.

Another factor of importance in internal combustion engines is that LPG leaves little or no engine deposit in the cylinders when it burns. Also, since it enters the engine as a vapor, it cannot wash down the cylinder walls, remove lubricant, and increase cylinder-wall, piston, and piston-ring wear. Nor does it cause crankcase dilution. All these factors reduce engine wear, increase engine life, and keep maintenance costs low. However, allowances must be made for the extra cost of LPG-handling equipment, including relatively heavy pressurized storage tanks, and special equipment to fill fuel tanks on the vehicles. *See* GAS TURBINE; INTERNAL COMBUSTION ENGINE; PETROLEUM PRODUCTS.   [MOTT SOUDERS]

Machine

A combination of rigid or resistant bodies having definite motions and capable of performing useful work. The term mechanism is closely related but applies only to the physical arrangement that provides for the definite motions of the parts of a machine. For example, a wristwatch is a mechanism, but it does no useful work and thus is not a machine.

Machines vary widely in appearance, function, and complexity from the simple hand-operated paper punch to the ocean liner, which is itself composed of many simple and complex machines. No matter how complicated in appearance, every machine may be broken down into smaller and smaller assemblies, until an analysis of the operation becomes dependent upon an understanding of a few basic concepts, most of which come from elementary physics.

[RICHARD M. PHELAN]

Bibliography: R. M. Phelan, *Fundamentals of Mechanical Design*, 2d ed., 1962; J. E. Shigley, *Theory of Machines*, 1961.

Magnetic circuits

Closed paths of magnetic flux; also a design method using such paths to compute the magnetic field of a core geometry that is often encountered, for instance, high-permeability flux-path segments (and their associated air gaps), each segment having reasonably definite length and area. Examples of magnetic circuits are transformer cores, relay frames, and iron parts of electrical machinery. The magnetic-field equations for these devices look so similar to dc circuit equations that they are called magnetic circuits.

Reluctance. If NI ampere-turns link one closed core, then Eqs. (1) hold, where l is length, A is

$$\Phi \times \frac{l}{\mu A} = NI$$

(Flux) × (reluctance)   (1)
        = magnetomotive force (mmf)

$$\Phi \times R = \text{mmf}$$

cross section, and μ is permeability large enough so that the flux Φ is nearly completely confined to the core (Fig. 1).

If the flux path is a sequence of dissimilar seg-

ments, its total reluctance is the sum of segment reluctances as given by Eq. (2).

$$R = \sum \frac{l_n}{\mu_n A_n}$$

(2)

If the (constant) flux divides among several parallel flux paths, it does so in inverse proportion to their reluctances.

These series and parallel reluctances have the same algebra as do series and parallel dc resistances; hence the fundamental equation is often called Ohm's law for magnetic circuits, even though mmf is not a force, Φ is not a flow, and the equation is not Ohm's but Hopkinson's.

Specific use of magnetic-circuit reluctance is nearly always qualitative; it is handy for explanation and discussion. How much flux change occurs when reluctance is altered is usually determined by a nonlinear calculation which does not compute a numerical value for reluctance itself. Accordingly, there is no common name for a reluctance unit.

Quantitative calculations. Three principles are applied:

1. Magnetic-flux lines are endless: Φ is the same at every point in the flux bundle, even though the lines are specially crowded at some places.

2. The relation between flux density B and field vector H at any point is a property of the matter at that point. From (experimental) charts for the material, either B or H can be found from the other (Fig. 2).

3. Ampère's line integral $\oint H \cdot dl = \Sigma I_{\text{linked}}$ is taken along the path followed by the flux bundle being analyzed.

Given total flux Φ, flux-density in the nth segment is $B_n = \Phi/A_n$. Each H_n is then found from its B_n by chart, and the Ampère integral is evaluated as shown in Eq. (3).

$$\oint H \cdot dl = \Sigma H_n l_n = NI$$

(3)

Units. Numerical evaluation of the Ampère integral requires consistent units. Modern practice uses the ones listed in the table.

Many published curves still use the older cgs units: Φ in maxwells, B in gauss, and H in oersteds, with $\mu_{\text{air}} = 1$. Conversion to one of the systems in the table is recommended; it avoids the absolute amperes and factors of 4π that occur in the cgs equations. This conversion, given below, is straightforward because the flux lines keep their identity.

Commonly employed units for magnetic circuits

Quantity	mks units	Engineering units
Flux density B	Webers/m^2	Kilolines/in.^{2}
Flux Φ	Webers	Kilolines
Field vector H	Amp-turns/m	Amp-turns/in.
$\mu_{\text{air}} = B/H$ in air	$4\pi \times 10^{-7} = \dfrac{1}{7.95 \times 10^5}$	$3.18 \times 10^{-3} = \dfrac{1}{313}$
Length	Meters	Inches

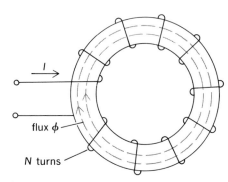

Fig. 1. Diagram of a toroidal magnetic circuit. (*A. E. Fitzgerald, D. E. Higginbotham, and A. Grabel, Basic Electrical Engineering, McGraw-Hill, 1967*)

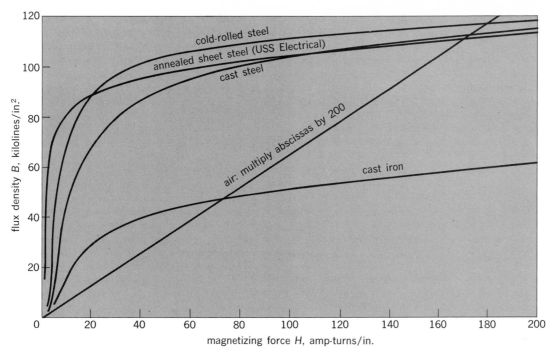

Fig. 2. Graph of the normal magnetization curves for some magnetic materials. (A. E. Fitzgerald, D. E. Higgin-botham, and A. Grabel, Basic Electrical Engineering, Mc-Graw-Hill, 1967)

One weber is 10^5 kilolines (10^8 lines or 10^8 max-wells).

One weber/m² is 64.5 kilolines/in.² (10^4 lines/cm² or 10^4 gauss).

One amp-turn/m is 2.51×10^{-2} amp-turns/in. and corresponds to $4\pi \times 10^{-7}$ weber/m² in air (or to $4\pi \times 10^{-3}$ gauss in air or $4\pi \times 10^{-3}$ oersteds).

Special situations. Some rules of thumb allow for leakage flux and for reduction of flux density when the lines spread in air gaps. Magnetic circuits that include permanent magnets or superconductors or plasmas need care in their analysis. For example, when a slice of permanent-magnet material in inserted as one of the flux-path segments, it supplies a magnetomotive force that requires calculation. Instead of puzzling over what H is in a permanent magnet, it is better to replace with a slice of normal material (that is, of the same dimensions and same incremental permeability) plus ampere-turns around its edge to make the same magnetic moment as the original magnet slice.

In each piece of superconducting material, two things happen: (1) Unless the superconductor is very thin, all the flux lines detour around the outside of it, and (2) any attempt to change Φ near a superconductor starts nondecaying eddy currents that hold Φ to (almost) its initial value. More accurate (quantum) descriptions of these phenomena have been published. The flux rejection of a superconductor can be used to channel the flux when this is desired; several devices employ very thin superconductors into which some flux can be driven in order to modify the onset of superconductivity.

The added ampere-turns from a few ions in the flux path can be found by eddy-current methods if time variation is not too violent. However, when there are enough of these ions to be called a plasma they interact by collision and the resulting currents and magnetomotive force are not easily found. *See* MAGNETOHYDRODYNAMICS.

[MARK G. FOSTER]

Bibliography: A. S. Bishop, *Project Sherwood*, 1958; A. E. Fitzgerald, D. E. Higginbotham, and A. Grabel, *Basic Electrical Engineering*, 1967; J. Hopkinson and E. Hopkinson, Dynamo-electric Machinery, *Phil. Trans. Roy. Soc.*, pt. 1, 1886; W. T. Hunt and R. Stein, *Static Electromagnetic Devices*, 1963; V. L. Newhouse, *Applied Superconductivity*, 1964; R. J. Smith, *Circuits, Devices and Systems*, 1966.

Magnetic field

A condition existing in the vicinity of a magnetic body (or a current-carrying medium) whereby the magnetic forces due to the body (or current) are detectable. In a magnetic field, there is a force on a moving charge in addition to the electrostatic (Coulomb) forces between charges, or there is a force on a magnetic pole. The description of the field may be in terms of magnetic induction (flux density) or it may be in terms of magnetic field strength.

The magnetic induction B of the field is defined from the force F on a moving charge q or current element of length l carrying a current I by Eq. (1).

$$B = \frac{F}{qv \sin \theta} = \frac{F}{Il \sin \theta} \qquad (1)$$

The direction of the magnetic induction is that direction in which the force on the moving charge is zero. The factor $v \sin \theta$ is then the component of the velocity of the charge in a direction perpendicular to B. The mks unit of B is the newton/ampere-meter or weber/meter².

The magnetic induction at a given point due to a current may be found from Ampère's law, Eq. (2), which may also be expressed as Eq. (2b). Equation (2a) gives the contribution of a current element Idl to the magnetic induction, where μ_0 is the permeability of empty space, r the distance of the point to the current element, and θ the angle between the

$$dB = \frac{\mu_0}{4\pi} \frac{Idl \sin \theta}{r^2} \qquad (2a)$$

$$B = \frac{\mu_0}{4\pi} \int \frac{Idl \sin \theta}{r^2} \quad \text{(vector sum)} \qquad (2b)$$

current element and the line joining the element to the point.

The direction of B, from Ampère's law, is perpendicular to the plane determined by a line tangent to dl and the line joining the current element to the point at which B is determined.

A second magnetic vector quantity, the magnetic field strength or intensity H, may be defined in part from Ampère's law. The part of the field strength that is due to currents is given by the vector sum from Eq. (3). Thus, H is computed in the

$$H = \frac{1}{4\pi} \int \frac{Idl \sin \theta}{r^2} \qquad (3)$$

same manner as the flux density B, but without the factor μ_0. The direction is specified in exactly the same manner as B. Thus, H depends only on the current present, and not upon the properties of the surrounding medium. This definition of H is only partial because it does not include contributions by magnetic poles if they are present in the neighborhood.

The mks unit of magnetic field strength appears from the defining equation, when current is in amperes and dl and r are in meters, as ampere per meter. Because many of the equations that are derived from the defining equation involve the number of turns N of a coil times the current, the ampere-turn per meter (amp-turn/m) is also used as an equivalent unit.

Magnetic poles. A body can be magnetized by bringing it into a magnetic field due to currents or magnets. Except in the case of a ring magnetized along its circumference, the field associated with a magnetized body extends to the region surrounding the body. The external effect usually appears in limited regions of the body called poles. A magnetized bar of iron has two poles, one at either end; and from the fact that the bar will set itself in an approximate north-south direction in the Earth's field, it appears that there are two kinds of poles. The pole that is at the north end of the bar is called a north-seeking pole; that at the south end is called a south-seeking pole. The two poles at the ends are merely indications of the continuous magnetization within the body. An indication of the validity of this statement is the fact that when a bar magnet is broken into two parts, two new poles appear at the break, and the orientation of the poles in each fragment is the same as it was in the original magnet.

It is observed that magnetic poles exert forces on each other and upon moving charges in the region near the poles. There is a field near the pole, and the pole may be considered as the cause of that field.

If the poles of a magnetized body are small enough that they may be considered point poles, the force that one pole exerts upon another is found to be proportional to the product of the pole strengths m and m' and inversely proportional to the square of the distance r between them. This statement is called Coulomb's law of magnetostatics and is written as Eq. (4). The proportionality

$$F = k' \frac{mm'}{r^2} \qquad (4)$$

factor k' depends upon the units used and upon the medium between the poles. For empty space in the mks system, k' is assigned the value $\frac{1}{4\pi\mu_0}$. The unit of pole strength associated with this choice is the weber.

The magnetic field strength or magnetic intensity H may be expressed as the force per unit north-seeking magnetic pole as in Eq. (5). Then,

$$H = \frac{F}{m'} \qquad (5)$$

from Coulomb's law, the contribution of a point pole of strength m to the magnetic field strength near the pole is given by Eq. (6). The direction of H

$$H = \frac{F}{m'} = \frac{1}{4\pi\mu_0} \frac{m}{r^2} \qquad (6)$$

is away from north-seeking poles and toward south-seeking poles. The contribution of several poles is the vector sum of the contributions of the individual poles as shown in Eq. (7). If the H due to

$$H = \frac{1}{4\pi\mu_0} \sum \frac{m}{r^2} \quad \text{(vector sum)} \qquad (7)$$

poles is to be in the same units as the H due to currents, the unit of pole strength must be chosen properly. If H is to be in amperes per meter when μ_0 is in webers per ampere-meter and r is in meters, then m must be in webers.

If the poles are distributed over surfaces or throughout volumes, the summation becomes an integral as shown in Eq. (8).

$$H = \frac{1}{4\pi\mu_0} \int \frac{dm}{r^2} \quad \text{(vector sum)} \qquad (8)$$

The general expression for the field strength due to both currents and poles is given by Eq. (9), where the integrals represent vector sums.

$$H = \frac{1}{4\pi} \int \frac{Idl \sin \theta}{r^2} + \frac{1}{4\pi\mu_0} \int \frac{dm}{r^2} \qquad (9)$$

If a toroidal coil has an iron core, the iron is magnetized by the current of the coil. The magnetic field strength within the core is given entirely by the first term of the equation for H, since there are no poles. For a long straight solenoid with an air core (illustration a), H is entirely due to the current and is found by integration to be $H_I = NI/l$, where N is the number of turns of the coil, and l is the length of the solenoid. When an iron core is inserted into the solenoid (illustration b), the iron becomes magnetized, and poles appear at each end. The contribution of the current to H remains the same as before, but the second term now contributes to H components that are opposite in direction to H_I and that vary along the bar because of the variation in distances from the poles and because of the variation of the permeability of the iron. The effect of the poles on H is greatest at the ends near the poles. The magnetization of the iron is not uniform. The effect of the poles is essentially demagnetizing; it is large for short magnets and negligible at the center of a long magnet.

Lines of force. As in the case of magnetic induction B, which can be represented by lines called magnetic flux, the vector quantity H may be represented by lines called lines of force. The number of lines of force per unit area of a surface perpendicular to H is made equal to the value of H. The direction of the lines of force is the direction of the field. The lines of force are closed curves, as are the lines of induction.

Energy of a magnetic field. Consider a Rowland ring (a ring-shaped sample of magnetic material) surrounded by a coil of N turns in which there is a current I. The field strength H within the ring is given from Ampère's law by Eq. (10), where l is the mean circumference of the ring.

$$H = \frac{NI}{l} \qquad (10)$$

In building up the current in the coil, energy must be supplied that becomes energy of the magnetic field. This energy is given by Eq. (11), where

$$W = \tfrac{1}{2}LI^2 \qquad (11)$$

L is the self-inductance of the coil.

But L is defined by Eq. (12), where Φ is the flux,

$$L = \frac{N\Phi}{I} \qquad (12)$$

and W by Eq. (13), where V is the volume of the

$$W = \frac{1}{2}\frac{N\Phi}{I}I^2 = \tfrac{1}{2}N\Phi I = \tfrac{1}{2}NBAI$$

$$= \frac{1}{2}\frac{NI}{l}BV = \tfrac{1}{2}HBV \qquad (13)$$

core, and A is the mean cross-sectional area of the ring. The energy per unit volume of the field is then given by Eq. (14). If μ is the permeability of

$$\frac{W}{V} = \tfrac{1}{2}HB \qquad (14)$$

the core, then $B = \mu H$ and the energy density of the field may be written down as shown below in Eq. (15).

$$\frac{W}{V} = \tfrac{1}{2}HB = \tfrac{1}{2}\mu H^2 = \frac{1}{2}\frac{B^2}{\mu} \qquad (15)$$

Magnetic potential. When a magnetic pole is in a magnetic field, there will be a force F acting on it. If it is moved a distance ds in the field, work dW done against the field is given by Eq. (16), where θ

$$dW = -F\cos\theta\,ds = -Hm\cos\theta\,ds \qquad (16)$$

is the angle between the positive sense of H and the positive direction of s. The magnetic potential difference V may be defined as the work done per unit pole in taking the pole from one point to the other as in Eq. (17a) or (17b).

$$dV = \frac{dW}{m} = -H\cos\theta\,ds \qquad (17a)$$

$$V = -\int_{s_1}^{s_2} H\cos\theta\,ds \qquad (17b)$$

The resulting equation for the magnetic potential does not include the concept of the magnetic pole, and can be used as a defining equation for the magnetic potential. The integral represents the sum along a path of the products of ds and the component of H in the direction of ds. Such an in-

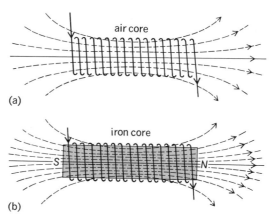

Flux in a solenoid (*a*) containing only air (essentially vacuum) and (*b*) containing an iron core.

tegral is called a line integral and is represented by the symbol $\oint$. The defining equation for magnetic potential may be written as Eq. (18). From this

$$V = -\oint H\cos\theta\,ds \qquad (18)$$

equation the relationship between magnetic potential and H may be written as Eq. (19). Thus the

$$H\cos\theta = -\frac{dV}{ds} \qquad (19)$$

component of field strength in any direction is the negative of the magnetic potential gradient in that direction. This statement is similar to the relation between electric field intensity and electric potential gradient.

The rise in magnetic potential around any closed path in a magnetic field may be deduced from consideration of the field about a long straight conductor.

For this special case, the lines of force are concentric circles about the conductor and at a distance a from the conductor and Eq. (20) holds. For

$$H = \frac{I}{2\pi a} \qquad (20)$$

a path that follows a circular line of force in a sense opposite to that of H, $\theta = 180°$ and $\cos\theta = -1$. Then the rise in magnetic potential around the closed path is given by Eq. (21). This result is inde-

$$V = -\oint H\cos\theta\,ds = \int_0^{2\pi a}\frac{I}{2\pi a}ds = I \qquad (21)$$

pendent of the radius of the circle followed and, in fact, is independent of the shape of the path, since any path may be resolved into components that are along circles and along radii. The contributions of the radial parts are zero, since they are perpendicular to H.

The result deduced from this special case may be generalized for a closed loop of any shape. The rise in magnetic potential around the path is equal to the line integral of $H\cos\theta$ around the path, and this in turn is equal to the current through the surface bounded by the path as shown in Eq. (22). If

$$\oint H\cos\theta\,ds = I \qquad (22)$$

the path taken includes no current, the integral is

zero. If there are N equal currents inside the path, the integral becomes NI.

By analogy to an electric circuit, in which the line integral of the electric intensity around the circuit is the electromotive force of the circuit, the rise in magnetic potential around the closed path, that is, the line integral of $H \cos \theta$ around the path, is called the magnetomotive force (mmf). Thus, the mmf of a coil of N turns in which there is a current I is NI. *See* MAGNETOMOTIVE FORCE.

[KENNETH V. MANNING]

Bibliography: S. S. Attwood, *Electric and Magnetic Fields*, 3d ed., 1949; W. B. Boast, *Vector Fields: A Vector Foundation of Electric and Magnetic Fields*, 1964; L. Page and N. I. Adams, *Principles of Electricity*, 3d ed., 1958; F. W. Sears, *Principles of Physics*, vol. 2, 1951; R. P. Winch, *Electricity and Magnetism*, 1955.

Magnetogas dynamics

The science of motion in a plasma under the influence of mechanical, electric, and magnetic forces; also termed hydromagnetics and magnetohydrodynamics. A plasma may be a partially or fully ionized gas, or a mixture of elementary particles such as protons, electrons, and neutrons. A plasma gas usually contains positive ions, electrons, and—if partially ionized—neutral atoms. When a gas is ionized, positive and negative charges are produced in equal numbers, so that the plasma is electrically neutral.

Electrons move freely in a plasma, thereby making it an electrical conductor. As such, its dynamic behavior responds to the presence of electric and magnetic fields. The effect of mechanical forces (such as those caused by a pressure gradient or by thermodynamic expansion) depends upon gas density and temperature, degree of ionization, and the strength of electric and magnetic fields.

Although magnetofluid dynamics includes the motion of any deformable, electrically conductive substance in the presence of electric and magnetic fields, the term is frequently applied specifically to liquids such as mercury.

Interactions of particles and fields. In magnetogas dynamics, individual particle motion, as well as the macroscopic dynamics of a plasma, must be considered. An electric field transfers energy to a charged particle, accelerating it parallel to the field toward the oppositely charged electrode (Fig. 1).

A magnetic field exerts a force only if the charged particle is in motion relative to the field.

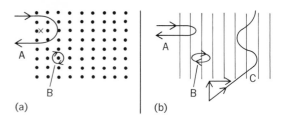

Fig. 3. (*a*) Magnetic field lines entering normal to paper. (*b*) Magnetic field lines parallel to paper. Particle A is reflected by a magnetic field; B is trapped in a magnetic field; C is in helical motion in a magnetic field.

Then the magnetic force is proportional to the field strength and to the velocity component which is normal to the field direction (field lines). The direction of force is normal to the field and normal to the velocity component perpendicular to the field, thereby deflecting the path of the particle (Fig. 2). In extreme cases its direction of motion may be reversed, the effect being termed a magnetic mirror (Fig. 3*a*). If the particle velocity is constant and normal to the field and if the field is constant in space and time, the particle will be trapped in the field, because continuous deflection will cause it to move in a circular path, the action being that of gyration. If the velocity of the particle is inclined to the field, the normal component will determine the radius of gyration, while the parallel velocity component is unaffected. Consequently, the particle describes a helical path (Fig. 3*b*). In contrast to the electrical field, the magnetic field, therefore, can merely change the direction of motion of a particle, not its kinetic energy. In the presence of crossed electric and magnetic fields, the particle is simultaneously accelerated and deflected, which causes it to drift in a direction normal to the magnetic as well as to the electric field. *See* MAGNETOHYDRODYNAMICS.

Interaction of whole plasma with fields. The two most important macroscopic aspects of magnetogas dynamics are the effect of electric or magnetic fields on the magnitude and direction of plasma motion. The use of crossed electric and magnetic fields in the manner shown in Fig. 4 causes high drift velocities. However, the macroscopic plasma motion tends to be more complex. In steady fields, complications arise because of collisions, neutralization, and re-ionization, which destroy the field pattern and dissipate energy in random velocities. In principle, these conditions can be avoided by the use of nonsteady electric and magnetic fields, such as pulsed electric and magnetic fields. In this manner, currents and fields of much higher peak intensity can be attained, resulting in more powerful electromotive forces and less time for energy dissipation in random velocities. *See* ELECTROMAGNETIC PROPULSION.

The macroscopic manifestation of charged-par-

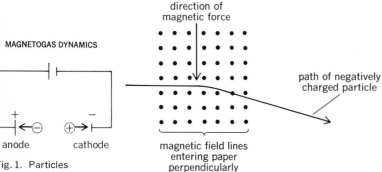

Fig. 1. Particles accelerated in an electric field.

Fig. 2. Particle deflected in a magnetic field.

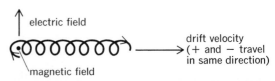

Fig. 4. Drift of particles in crossed electric and magnetic fields. Crossed fields cause high drift velocities.

ticle behavior in a magnetic field is that, being a conductor, the plasma is diamagnetic in that it resists penetration by a magnetic field. Field lines running parallel to the direction of plasma motion lend rigidity to the flow by resisting lateral (transverse) motion of the charges. By changing the field intensity in flow direction, the magnetic field lines converge (increasing intensity) or diverge, causing compression or expansion of the gas as in a convergent or divergent duct. By producing highly ionized plasma in an intense magnetic field, the hot gas, except for the neutral particles, can be kept away from solid walls (magnetic containment). A highly or completely ionized plasma flow represents a strong current, inducing a powerful surrounding magnetic field by means of which the plasma contains itself (pinch effect). In weaker plasma streams, the induced field has to be augmented by a field of external origin, induced by a current flowing through coils wrapped around the plasma-carrying duct. By closer coil winding, the field intensity is increased; conversely, the intensity can be decreased, corresponding to a convergent-divergent duct.

Magnetoaerodynamics. High-velocity vehicles ionize the air by causing intense shock waves. This occurs at flight velocities upward of about 18,000 ft/sec and is experienced by returning ballistic vehicles and spacecraft. Drag increase due to magnetic retardation of the environmental plasma flow (electromagnetic drag) increases the effective bluntness of the vehicle, causing the shock wave to be displaced forward and to become steeper (increased shock drag).

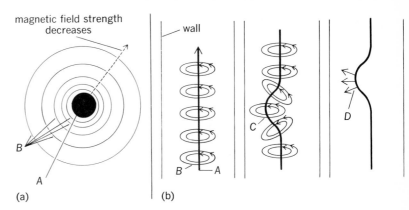

Fig. 5. Pinch device. (*a*) Plasma current A generates surrounding magnetic field B. (*b*) Typical kink instability is shown by C, direction of driving power by D.

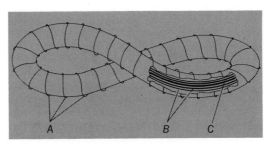

Fig. 6. Stellarator. External, closely spaced, current-carrying windings A produce a magnetic field whose lines of force B run parallel to the walls, confining plasma C in tube's center.

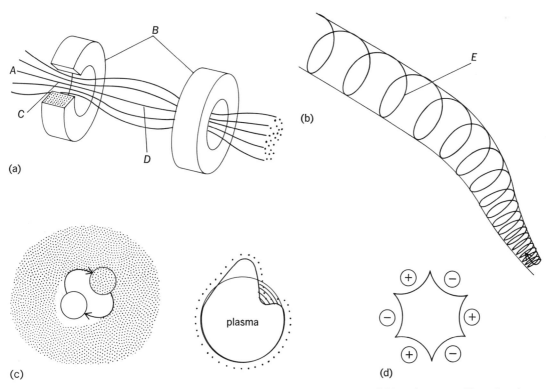

Fig. 7. Basic mirror device. (*a*) Axial magnetic field A is compressed at both ends by current-carrying mirror coils B; C and D designate regions of stable and unstable confinement. (*b*) Slow-down and reflection by mirror is shown by a particle's line of motion E. (*c*) In flute instability plasma and field exchange positions, forming a bulge in cross-sectional configuration of originally circular plasma tube. (*d*) Current-carrying longitudinal bars stabilize plasma in mirror device.

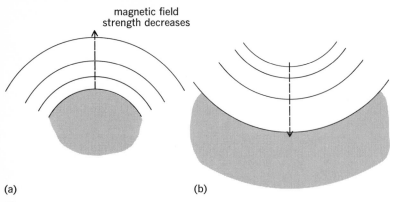

magnetic field
strength decreases

(a)   (b)

Fig. 8. Magnetic confinement. (a) Unstable, by a concavely oriented magnetic field. (b) Stable, by a convex field.

Formation of plasma sheaths around reentering spacecraft causes a communication blackout for several minutes, since heavily ionized gases resist penetration by electromagnetic signals.

Controlled fusion. Controlled thermonuclear reaction is probably the most important application of magnetogas dynamics for the future. In the fusion of hydrogen isotopes (deuterium, tritium) to helium, about 50,000,000 times as much energy is released per unit mass as in an oxygen-hydrogen reaction. The most critical problem in this process is the containment of the extremely hot gas. In stars, the reacting gas in the stellar core is confined by gravitational means (the pressure of vast masses of gas surrounding the core). Within the far more limited dimensions of engineering systems, confinement must be achieved through application of magnetogas dynamics. The reacting gas, 100,000,000 to 1,000,000,000°C, cannot be contained by solid walls, even though the gas density typically corresponds to only about one-ten-thousandth of an atmosphere. Being a fully ionized plasma, the gas can be confined by strong magnetic fields. The key ingredient, however, is stability, a problem of magnetogas dynamics.

Confinement by a magnetic field means that the outward pressure of the plasma must be balanced by the inward pressure of the surrounding magnetic field. The ratio of plasma pressure to magnetic field pressure is denoted by β. A low value of β means that the magnetic pressure is much larger than the plasma pressure.

The three principal confinement concepts are known as pinch, stellarator, and magnetic mirror. In the pinch concept (Fig. 5), internal currents generate a magnetic field encircling the current. In the stellarator concept (Fig. 6), the plasma is confined, by an externally imposed magnetic field, in a tube which closes on itself (as in figure-eight or racetrack configurations). The magnetic mirror concept (Fig. 7) is based on plasma confinement in a straight tube. The magnetic field is externally imposed and is particularly strong at the ends of the tube, reflecting the plasma back into the tube like a mirror.

Confined plasmas are subject to a wide variety of instabilities. They can be divided into two major categories. The gross instabilities involve bulk motion of large fractions of the plasma across magnetic lines of force. The microinstabilities

(also called wave-particle instabilities) involve resonant interactions between plasma waves and particle motion.

Instabilities are caused by drivers, that is, energies which power the instability motion. Gross instabilities occur when the plasma is not properly confined by the magnetic field, that is, when it is not in a magnetic well. In a proper magnetic well, the magnetic field strength increases with distance from the plasma. The plasma tends to move in the direction of decreasing field strength. The latter is the case if the magnetic field lines are concave with respect to the plasma (Fig. 8). This instability can cause a gross interchange in position between plasma and field lines whenever this results in a decrease in potential energy. This interchange is called flute-interchange instability (Fig. 7). The release of energy associated with the decrease in potential energy drives the instability, just as a stone rolling down the mountain is driven by the potential energy released as the stone seeks a lower energy level.

In a stable confinement, the lines of force are convex to the plasma (Fig. 8). If a plasma is trapped between magnetic mirrors at each end, the mirror region provides a stable confinement. In a longitudinal magnetic mirror, a charged particle follows an increasingly tight helical path during which its longitudinal motion is slowed down and eventually brought to a halt and the particle is reflected back into the central region of the mirror device. In the region between the mirrors, flute-interchange instabilities can occur.

The Soviet physicist A. F. Ioffe showed that by adding a set of current-carrying longitudinal bars, a true magnetic well can be produced in which the field strength increases in every direction outward from a minimum value along the axis.

Typical macroinstability in the pinch device is the kink instability (Fig. 5). As the kink develops, the plasma moves into regions of lower magnetic field intensity. This produces the driving effect which causes the kink to increase rapidly. Another pinch instability is the so-called sausage instability, causing a necking-down or local constriction of the plasma beam. As shown in Fig. 9, methods of stabilizing the pinch include encasement of the pinch in a conducting shell (A) outside an insulating inner linear (B), and insertion of an axial magnetic field (C) whose lines of force lend rigidity to the plasma column; magnetic pinch field (F) is produced by axial current D. Thus, kink and sausage instabilities can be suppressed.

Microinstabilities are driven primarily by local nonequilibrium conditions, such as density or temperature gradients or irregular velocity distribution (for instance, drift), in the plasma. The key to suppressing these instabilities lies in preventing gradients, thereby minimizing the driving energies. In mirror configurations, that is, in open-ended systems, this primarily means minimizing irregularities in non-Maxwellian velocity distributions as plasma is lost at the ends.

The ultimate objective of all attempts to achieve plasma stability is to prolong the residence time of the plasma in the region where thermal and pressure conditions make fusion possible. Plasma leakage means at least loss of energy, that is, inefficiency, and can readily bring the fusion

MAGNETOGAS DYNAMICS

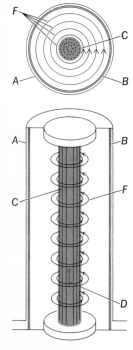

Fig. 9. Pinch device for plasma stabilization.

process to a halt altogether. Closed configurations, such as the stellarator, result from an attempt to eliminate open ends as a source of plasma losses. But the escape of plasma from its magnetic confinement to the wall, where it is immediately chilled down below temperatures at which fusion can continue, represents a loss as critical as open-end losses. Plasma stabilization within closed configurations offers greater difficulties than in open-end configurations.   [KRAFFT A. EHRICKE]

Bibliography: A. S. Bishop, *Project Sherwood: The U.S. Program in Controlled Fusion*, 1958; A. B. Cambel, *Plasma Physics and Magnetofluidmechanics*, 1963; T. G. Cowling, *Magnetohydrodynamics*, 1957; J. Drummond (ed.), *Plasma Physics*, 1960; D. J. Rose, *Plasma and Controlled Fusion*, 1961; A. Simon, *Thermonuclear Research*, 1959; L. Spitzer, *Physics of Fully Ionized Gases*, 1956.

Magnetohydrodynamic power generator

A system for the generation of electrical power through the interaction of a flowing, electrically conducting fluid with a magnetic field. As in a conventional electrical generator, the Faraday principle of motional induction is employed, but solid conductors are replaced by an electrically conducting fluid. The interactions between this conducting fluid and the electromagnetic field system through which electrical power is delivered to a circuit are determined by the magnetohydrodynamic (MHD) equations, while the properties of electrically conducting gases or plasmas are established from the appropriate relationships of plasma physics. Major emphasis has been placed on MHD systems utilizing an ionized gas, but an electrically conducting liquid or a two-phase flow can also be employed. *See* INDUCTION, ELECTROMAGNETIC; MAGNETOHYDRODYNAMICS.

Improvement of the overall thermal efficiency of central station power plants has been the continuing objective of power engineers. Conventional plants based on steam turbine technology are limited to about 40% efficiency, imposed by a combination of working-fluid properties and limits on the operating temperatures of materials. Application of the MHD interaction to electrical power generation removes the restrictions imposed by the blade structure of turbines and enables the working-fluid temperature to be increased substantially. This enables the working fluid to be rendered electrically conducting and yields a conversion process based on a body force of electromagnetic origin. *See* ELECTRIC POWER GENERATION; STEAM TURBINE.

Principle. Electrical conductivity in an MHD generator can be achieved in various ways. At the heat-source operating temperatures of MHD systems (1000–3000 K), the working fluids usually considered are gases derived from fossil fuel combustion, noble gases, and alkali metal vapors. For combustion gases, a seed material such as potassium carbonate is added in small amounts, typically about 1% of the total mass flow. The seed material is thermally ionized and yields the electron number density required for adequate electrical conductivity above about 2500 K. With monatomic gases, operation at temperatures down to about 1500 K is possible through the use of cesium as a seed material. In plasmas of this type, the electron temperature can be elevated above that of the gas (nonequilibrium ionization) to provide adequate electrical conductivity at lower temperatures than with thermal ionization. In so-called liquid metal MHD, electrical conductivity is obtained by injecting a liquid metal into a vapor or liquid stream to obtain a continuous liquid phase.

When combined with a steam turbine system to serve as the high-temperature or topping stage of a binary cycle, an MHD generator has the potential for increasing the overall plant thermal efficiency to around 50%, and values higher than 60% have been predicted for advanced systems. This follows from the increased efficiency made available by the higher source temperature. Thus, the MHD generator is a heat engine or electromagnetic turbine which converts thermal energy to a direct electrical output via the intermediate step of the kinetic energy of the flowing working fluid. *See* ENGINE; THERMODYNAMIC PROCESSES.

The MHD generator itself consists of a channel or duct in which a plasma flows at about the speed of sound through a magnetic field. The power output per unit volume W_e is given by $W_e = \sigma v^2 B^2 k (1-k)$, where σ is the electrical conductivity of the gas, v the velocity of the working fluid, B the magnetic flux density, and k the electrical loading factor (terminal voltage/induced emf). Typical values are $\sigma = 10-20$ mhos/m; $B = 5-6$ tesla; $v = 600-1000$ m/s; and $k = 0.7-0.8$. These values yield W_e in the range 25–150 MW/m^3. The high magnetic field strengths required are to be provided by a super-conducting magnet, and the development of suitable electrodes which will conduct current into and out of the gas is one of the major development problems.

Types. When this generator is embedded in an overall electrical power generation system, a number of alternatives are possible, depending on the heat source and working fluid selected. The temperature range required by MHD can be achieved through the combustion of fossil fuels with oxygen or compressed preheated air. The association of MHD with nuclear heat sources has also been considered, but in this case limitations on the temperature of nuclear fission heat sources with solid fuel elements has thus far precluded any practical scheme being developed where a plasma serves as the working fluid. The possibility of coupling MHD to a fusion reactor has been explored, and it is possible that 21st-century central station power systems will comprise a fusion reactor and an MHD energy conversion system. *See* FUSION, NUCLEAR; REACTOR, NUCLEAR.

Development efforts on MHD are focused on fossil-fired systems, the fossil fuel selected being determined by national energy considerations. In the United States, coal is the obvious candidate, whereas in the Soviet Union, where ample reserves of natural gas are available for electric power generation, this is the preferred fuel. In Japan, major emphasis is on the use of petroleum-based fuels.

MHD power systems are classified into open- or closed-cycle systems, depending respectively on whether the working fluid is utilized on a once-through basis or recirculated via a compressor. For fossil fuels, the open-cycle system (Fig. 2) offers the inherent advantage of interposing no solid heat exchange surface between the combustor and the MHD generator, thus avoiding any limitation being placed on the cycle by the temperature attainable over a long period of operation by construction materials in the heat exchanger. Closed-cycle systems were originally proposed for nuclear heat sources, and the working fluid can be either a seeded noble gas or a liquid metal – vapor mixture.

Features. The greatest development effort in MHD power generation has been applied to fossil-fired open-cycle systems, but sufficient progress has been made in closed-cycle systems to establish their potential and to identify the engineering problems which must be solved before they can be considered practical. The rest of this article discusses fossil-fired open-cycle systems.

In addition to offering increased power plant efficiencies, MHD power generation also has important potential environmental advantages. These are of special significance when coal is the primary fuel, for it appears that MHD systems can utilize coal directly without the cost and loss of efficiency resulting from the processing of coal into a clean fuel required by competing systems. The use of a seed material to obtain electrical conductivity in the working fluid also places the requirement on the MHD system that a high level of recovery be attained to avoid adverse environmental impact and also to ensure acceptable plant economics. The seed recovery system required by an MHD plant also serves to recover all particulate material in the plant effluent. A further consequence of the use of seed material is its demonstrated ability to remove sulfur from coal combustion products. This occurs because the seed material is completely dissociated in the combustor, and the recombination phenomena downstream of the MHD generator favor formation of potassium sulfate in the presence of sulfur. Accordingly, seed material acts as a built-in vehicle for removal of sulfur. Laboratory experiments have shown that the sulfur dioxide emissions can be reduced to levels below those experienced with natural-gas – fired plants. A further important consideration is the reduction of the emission of oxides of nitrogen through control of combustion and the design of component operating conditions. Laboratory scale work has demonstrated that these emissions can be controlled to the most exacting standards prescribed by the Environmental Protection Administration (EPA). *See* ATMOSPHERIC POLLUTION.

While not a property of the MHD system in itself, the potential of MHD to operate at higher thermal plant efficiencies has the consequence of substantial deduction in thermal waste discharge, following the relationship that the heat rejected per unit of electricity generated is given by $(1-\eta)/\eta$, where η is the plant efficiency. As the technology of MHD is developed along with that of advanced gas turbines, there also exists the possibility that MHD systems can dispense entirely with the need for large amounts of cooling water for steam condensation through the coupling of MHD generators with closed-cycle gas turbines. *See* WATER POLLUTION.

State of development. A number of test facilities and experimental installations have developed engineering information on the MHD process and demonstrated operation of the generator and other components of the complete system. The most technically advanced installation is located on the northern outskirts of Moscow. Known as the U-25 installation, it is of the open-cycle type and has delivered its rated power of 20.5 MW to the Moscow grid using natural gas as the fuel. Operation with direct coal firing has been successfully demonstrated in the United States, and a test generator has been operated in Japan with a 5-tesla superconducting magnet. [WILLIAM D. JACKSON]

Bibliography: International MHD Liaison Group, Nuclear Energy Agency, OECD, Paris, *1976 MHD Status Report*, 1976; W. D. Jackson and P. S. Zygielbaum, Open cycle MHD power generation: Status and engineering development approach, *Proc. Amer. Power Conf.*, 37:1058–1071, 1975; R. J. Rosa, *Magnetohydrodynamic Energy Conversion*, 1968.

Magnetohydrodynamics

The science that deals with the dynamics or motion of a fluid interacting with a magnetic field. The fluid must be a good conductor of electricity and hence can be a liquid metal or, more usually, an ionized gas or plasma. Magnetohydrodynamics is important in the development of controlled thermonuclear reactors. In these devices, the fusion reaction takes place in a high-temperature plasma composed of heavy hydrogen isotopes; the plasma is surrounded by a magnetic field which serves to confine the plasma and isolate it from the walls of the reaction chamber. Magnetohydrodynamics is employed to study the usefulness of different plasma and magnetic field configurations for this purpose. Other applications include simulation of hypersonic flight conditions, ionic thrust for outer-space propulsion, space-vehicle braking upon reentry to the atmosphere, high-energy particle accelerators, microwave generators, thermionic energy-conversion devices, application of thin metallic coatings, and study of cosmic and upper atmospheric phenomena. This science is also called hydromagnetics or magnetogas dynamics.

The conducting fluid and magnetic field interact through electric currents that flow in the fluid. The currents are induced as the conducting fluid and the magnetic field lines move across each other. In turn, the currents influence both the magnetic field and the motion of the fluid. Qualitatively, the magnetohydrodynamic interactions tend to link the fluid and the field lines so as to make them move together.

The generation of the currents and their subsequent effects are governed by the familiar laws of electricity and magnetism. The motion of a conductor across magnetic lines of force causes a voltage drop or electric field at right angles to the direction of the motion and the field lines; the induced voltage drop causes a current to flow as in the armature of a generator.

The currents surround themselves with magnetic field lines, heat the conductor, and give rise to mechanical ponderomotive forces when flowing across a magnetic field. (These are the forces which cause the armature of an electric motor to turn.) In a fluid, the ponderomotive forces combine with the pressure forces to determine the fluid motion. *See* ELECTRICITY; GENERATOR, ELECTRIC; MOTOR, ELECTRIC.

It has been suggested that magnetic disturbances which develop in the very hot and turbulent center of the Sun whip up to the surface along magnetic lines of force in this way to produce sunspots. Other theories relate the heating of the solar corona and the acceleration of cosmic rays to Alfvén waves. It has been possible to generate these waves in the laboratory by twisting a column of liquid sodium placed in a strong magnetic field.

Although some early work in magnetohydrodynamics has been concerned with liquid metals, a much wider interest has developed in phenomena which involve ionized gases or plasmas. A large electrical conductivity in a plasma requires a high density of high-energy electrons. This can occur for a plasma either in thermal equilibrium at relatively high temperatures, from a few electron volts upward (1 electron volt is equivalent to a temperature of 11,400°K), or in a nonequilibrium situation where the ions and molecules remain at a low temperature and the electrons are supplied with energy by an external source such as a microwave generator or ultraviolet radiation.

Plasmas are encountered in interstellar space, in hot stars, and in the upper atmosphere, as well as in man-made devices into which energy is fed from electrical, chemical, or nuclear sources. Strong shock waves forming ahead of a blunt object traveling with hypersonic velocities through low-density air may heat the air sufficiently to ionize it.

FUNDAMENTAL LAWS

Magnetohydrodynamic phenomena involve two well-known branches of physics, electrodynamics and hydrodynamics, with some modifications to account for their interplay. *See* ELECTRODYNAMICS.

The basic laws of electrodynamics as formulated by J. C. Maxwell apply without any change. However, Ohm's law, which relates the current flow to the induced voltage, has to be modified.

It is useful to consider first the extreme case of a fluid with a very large electrical conductivity σ. Maxwell's equations predict, according to H. Alfvén, that for a fluid of this kind the lines of the magnetic field **B** move with the material. The picture of moving lines of force is convenient but must be used with care because such a motion is not observable. It may be defined, however, in terms of observable consequences by either of the following statements: (1) a line moving with the fluid, which is initially a line of force, will remain one; or (2) the magnetic flux through a closed loop moving with the fluid remains unchanged.

If the conductivity is low, this is not true and the fluid and the field lines slip across each other. This is similar to a diffusion of two gases across one another and is governed by similar mathematical laws. Numerically, the distance the magnetic field will slip through the fluid in a time t is $\delta = \sqrt{t/\mu\sigma}$, where μ is the magnetic permeability (a constant depending upon the magnetic properties of the fluid). The condition that the conductivity be very large can now be stated more precisely: σ should be large enough so that the distance δ for the time of interest t is small compared to the dimension of the system L.

As in ordinary hydrodynamics, the dynamics of the fluid obeys theorems expressing the conservation of mass, momentum, and energy. These theorems treat the fluid as a continuum. This is justified if the mean free path λ of the individual particles is much shorter than the distances that characterize the structure of the flow. Although this assumption does not generally hold for plasmas, one can gain much insight into magnetohydrodynamics from the continuum approximation. The ordinary laws of hydrodynamics can then easily be extended to cover the effect of magnetic and electric fields on the fluid by adding a magnetic force to the momentum-conservation equation and electric heating and work to the energy-conservation equation.

The mathematical descriptions of electrodynamic and hydrodynamic phenomena — Maxwell's equations for the electromagnetic field and the equations of ordinary fluid dynamics — both involve a set of partial differential equations.

Maxwell's equations. Maxwell's equations, written in the rationalized mks system of units, are Eqs. (1)–(4). Equations (3) and (4) lead to the con-

$$\nabla \times \mathbf{E} + \frac{\partial \mathbf{B}}{\partial t} = 0 \tag{1}$$

$$\nabla \cdot \mathbf{B} = 0 \tag{2}$$

$$\nabla \times \mathbf{H} - \frac{\partial \mathbf{D}}{\partial t} = \mathbf{j} \tag{3}$$

$$\nabla \cdot \mathbf{D} = \rho_e \tag{4}$$

servation of charge density ρ_e defined in Eq. (5), where **j** is the current density.

$$\frac{\partial \rho_e}{\partial t} + \nabla \cdot \mathbf{j} = 0 \tag{5}$$

The electric and magnetic fields **E** and **B** are related to the electric displacement **D** and the magnetic induction **H** by the equations $\mathbf{B} = \mu\mathbf{H}$ and $\mathbf{D} = \epsilon\mathbf{E}$, where μ is the magnetic permeability and ϵ is the dielectric constant for the medium. For good electrical conductors, a very small local excess or deficiency of electrons compared to the positive charge carriers would be removed almost instantly by the resulting electric field. The charge equalization takes place in a time which is roughly the larger one of the two characteristic times $t_1 = \epsilon/\sigma$ and $t_2 = \sqrt{m\epsilon/ne^2}$, where n, m, and e are the number density, mass, and charge of the electrons. In a metallic conductor, t_1 is roughly 10^{-18} sec; in the ionosphere it drops from 10^{-9} sec at a height of 100 km (E layer) to 10^{-12} sec at 250 km (F layer). Usually t_2 is larger, and the corresponding values are 6×10^{-17} sec, 6×10^{-8} sec, and 4×10^{-8} sec.

With these high rates of charge neutralization, it is not practical to calculate the electric field from the charge density through Eq. (4). The electric field is related much more effectively to the current distribution through Ohm's law, Eq. (6), which

$$\mathbf{E} = -\mathbf{v} \times \mathbf{B} + \frac{1}{\sigma}\mathbf{j} + \frac{m}{ne^2}\frac{\partial \mathbf{j}}{\partial t} \tag{6}$$

has been reformulated to include an electric field induced by the velocity **v** of the fluid across a magnetic field and the effect of electron inertia. The latter, however, need be retained only for very rapid oscillation; otherwise the last term in Eq. (6) can be dropped.

The displacement current $\partial \mathbf{D}/\partial t$ in Eq. (3) is important only when currents can pile up electrical charges. Because of the high rate with which charges are neutralized in a good conductor, it is usually possible to drop the term. This brings about a considerable simplification of Maxwell's equations, because one can now, with the help of Ohm's law, eliminate the electric field altogether and arrive at the relation in Eq. (7).

$$\frac{\partial \mathbf{B}}{\partial t} = \mathbf{\nabla} \times (\mathbf{v} \times \mathbf{B}) + \frac{1}{\sigma\mu}\,\mathbf{\nabla}^2\mathbf{B} \qquad (7)$$

In this equation, the first term on the right makes the fluid and the field lines move together; the second term makes them slip across each other. The magnetic Reynolds number $R_m = vL\sigma\mu$ is a measure of the effect the motion of the fluid has on the magnetic field. When this number is very large, the second term can be dropped, producing a still simpler equation which is the mathematical basis for Alfvén's description.

Hydrodynamic equations. The conservation equations of hydrodynamics need additional terms, to take into account the interaction of the fluid with the electromagnetic field. The mass equation can remain unchanged. In the momentum equation one must add the force density $\mathbf{j} \times \mathbf{B}$. Just as the hydrodynamic force is expressed (when viscosity effects are included) as the divergence of the pressure tensor, so the magnetic force $\mathbf{j} \times \mathbf{B}$ can be expressed as the divergence of the magnetic part of Maxwell's stress tensor. This obvious analogy has led to regarding certain components of Maxwell's tensor as a magnetic pressure.

The energy equation requires the addition of a term, $\mathbf{j} \cdot \mathbf{E}$, to account for the transfer of energy from the electromagnetic field to the fluid. Thus, the three hydrodynamic conservation equations modified for magnetohydrodynamics are Eqs. (8)–(10). Gravitational forces can be added readily if necessary.

$$\frac{\partial \rho}{\partial t} + \mathbf{\nabla} \cdot (\rho\mathbf{v}) = 0 \qquad (8)$$

$$\rho\frac{d\mathbf{v}}{dt} = -\mathbf{\nabla} \cdot \mathbf{P} + \mathbf{j} \times \mathbf{B} \qquad (9)$$

$$\rho\frac{d}{dt}\left(E + \frac{v^2}{2}\right) = -\mathbf{\nabla} \cdot (\mathbf{P}\mathbf{v} + \mathbf{Q}) + (\mathbf{j} \cdot \mathbf{E}) \qquad (10)$$

If the mean free path $\lambda \ll L$, the fluid is everywhere nearly in a state of thermal equilibrium, and one can express all state variables in terms of the

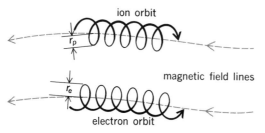

Fig. 1. Charged particles moving in a magnetic field describe helixes about field lines. The sense of rotation for positively charged ions is opposite that for negatively charged electrons. (*From Fusion/1958, Brochure for U.S. Fusion Research Exhibit, 2d International Conference on Peaceful Uses of Atomic Energy, Geneva, 1958*)

density ρ, the temperature T, and the velocity $\mathbf{v}$. The components of the pressure tensor $\mathbf{P}$ and of the heat flow vector $\mathbf{Q}$ are given in Eqs. (11) and (12), where $\delta_{ij} = 1$, if $i = j$; and $\delta_{ij} = 0$, if $i \neq j$. The

$$P_{ij} = \left[p + \tfrac{2}{3}\eta(\mathbf{\nabla} \cdot \mathbf{v})\right]\delta_{ij} - \eta\left(\frac{\partial v_i}{\partial x_j} + \frac{\partial v_j}{\partial x_i}\right) \qquad (11)$$

$$Q_i = -K\frac{\partial T}{\partial x_i} \qquad (12)$$

scalar pressure p, the viscosity η, the heat conductivity K, and the internal energy E per unit mass are functions of ρ and T which depend upon the fluid and which can be found experimentally or from kinetic theory.

Using Ohm's law in the form $\mathbf{j} = \sigma(\mathbf{E} + \mathbf{v} \times \mathbf{B})$, one can split $(\mathbf{j} \cdot \mathbf{E}) = \mathbf{v} \cdot (\mathbf{j} \times \mathbf{B}) + \mathbf{j}^2/\sigma$ into work done by the force $\mathbf{j} \times \mathbf{B}$ and the joule heating $\mathbf{j}^2/\sigma$. The three conservation equations lead to the entropy equation, Eq. (13). Here the rate of viscous

$$\rho T\frac{dS}{dt} = \rho\left(\frac{dE}{dt} + p\,\frac{d}{dt}\frac{1}{\rho}\right)$$
$$= \mathbf{\nabla} \cdot (K\mathbf{\nabla} T) + \eta\phi + \mathbf{j}^2/\sigma \qquad (13)$$

dissipation $\eta\phi$ is positive. In many problems the three dissipation terms on the right can be ignored and the entropy S then is conserved along a streamline.

Ohm's law. The formulation of Ohm's law given in Eq. (6) is still incomplete. Yet to be added are two effects due to the force $\mathbf{j} \times \mathbf{B}$ and the different rate of diffusion of electrons and ions in the presence of pressure and temperature gradients. The nature of these effects depends on the size of λ in relation to the radii $r_i = \sqrt{3m_iKT}/e\mathbf{B}$ of the spiral paths followed by particles with charge e and speed $\sqrt{3KT/m_i}$ in a field $\mathbf{B}$.

For the first time the detailed motion of the individual charged particles in the plasma must be considered. In a magnetic field, both positively and negatively charged particles describe helixes about the field lines (Fig. 1).

If a magnetic field $\mathbf{B}$ combines with an electric field $\mathbf{E}$, the center of the spiral drifts sideways in the direction of $\mathbf{E} \times \mathbf{B}$ rather than staying on a line of force. This drift is reduced roughly by a factor $1/[1 + (r_i/\lambda)^2]$ because of collisions. If $\mathbf{B}$ is small, the ion drift practically vanishes compared to the electron drift and the latter constitutes the Hall effect. If $\mathbf{B}$ is large, the two components drift nearly alike; this causes a mass flow but nearly cancels the Hall current. In a strong magnetic field the effective mean free path of a charged particle in a direction perpendicular to $\mathbf{B}$ is its radius of gyration. Thus, the diffusion current is also reduced significantly by a strong magnetic field. Two forms of a generalized Ohm's law can be used:

For small $\mathbf{B}$ Eq. (14) holds.

$$\mathbf{E} = -\mathbf{v} \times \mathbf{B} + \mathbf{j}/\sigma + (\mathbf{j} \times \mathbf{B} - \mathbf{\nabla}p_e)/ne \qquad (14)$$

For large $\mathbf{B}$ Eqs. (15) hold.

$$\mathbf{E}_\parallel = \mathbf{j}_\parallel/\sigma$$
$$\mathbf{E}_\perp = -\mathbf{v} \times \mathbf{B} + \left(\mathbf{j}_\perp + \frac{3n}{4\mathbf{B}^2}\mathbf{\nabla}kT \times \mathbf{B}\right)\Big/\sigma_\perp \qquad (15)$$

The symbols $\parallel$ and $\perp$ indicate the direction relative to $\mathbf{B}$, p_e stands for the partial pressure of the electron gas, and the conductivity $\sigma_\perp$ is approximately $\sigma/2$. For some applications it is desirable to solve Eq. (14) for $\mathbf{j}$ and this leads to Eq. (16), where $\mathbf{E}^1 = \mathbf{E} + \mathbf{v} \times \mathbf{B} + \nabla p_e/ne$, $\sigma_1 = \sigma/1 + \alpha^2$, $\sigma_2 = \alpha\sigma$, and $\alpha = \sigma B/ne$.

$$\mathbf{j} = \sigma\mathbf{E}^1_{\parallel} + \sigma_1\mathbf{E}_{\perp}^1 + \sigma_2\frac{\mathbf{B} \times \mathbf{E}^1}{B} \qquad (16)$$

The parameter α is of the same order as λ/r_e, so that Eq. (16) does not apply in the limit of large α.

The electrical conductivity σ can be found experimentally or from kinetic theory. The latter leads to the formula $\sigma = ne^2\tau/m$, where τ is the effective collision time of an electron, that is, the time in which collisions alone would bring the velocity of an electron into equilibrium with that of the surrounding ions.

A partially ionized gas can be regarded as a mixture of a completely ionized plasma and a neutral gas. One can consider separate densities ρ_p, ρ_n and velocities $\mathbf{v}_p, \mathbf{v}_n$ for these two components. The density and velocity of the mixtures are $\rho = \rho_p + \rho_n$ and $\mathbf{v} = (\rho_p\mathbf{v}_p + \rho_n\mathbf{v}_n)/\rho$. Ohm's law retains the form given by Eq. (14) if one replaces $\mathbf{v}$ by $\mathbf{v}_p$. If one writes τ_{rs} for the effective collision time of a particle of type r with all particles of type s, the effective collision time in the formula for σ can be expressed by the relation in Eq. (17). The indices e, i, and n refer to electrons, ions, and neutral atoms.

$$\frac{1}{\tau} = \frac{1}{\tau_{ei}} + \frac{1}{\tau_{en} + \tau_{in}} \qquad (17)$$

The motion of the plasma relative to the neutral component is called ambipolar diffusion. It gives rise to drag forces which dissipate energy in the form of heat. The part of the dissipation resulting from the drag between ions and neutrals is most often not included in (although it is often larger than) the joule heating.

MAGNETOHYDRODYNAMIC PHENOMENA

The combination of fluid motion and electromagnetic effects can lead to much more varied phenomena than either one alone. The analysis, however, can often be simplified by observing that the equations permit similar solutions which differ only by scaling factors. The results of a particular calculation or experiment can then be applied to an entire class of arrangements. Scaling laws can be most conveniently used by specifying the value of certain dimensionless parameters such as the magnetic Reynolds number R_m mentioned before, the ratio of magnetic to kinetic energy $S = B^2/\mu\rho v^2$, and others yet to be introduced.

Equilibrium. Before turning to the peculiar interplay of hydrodynamic and electromagnetic forces, the conditions for equilibrium in the absence of motion can be considered. The formulation of equilibrium conditions can be simplified by expressing the force density as the divergence of Maxwell's stress tensor, as in Eq. (18). The stress-

$$(\mathbf{j} \times \mathbf{B})_i = \frac{\partial}{\partial x_k}(B_iB_k/\mu - B^2\,\delta_{ik}/2\mu) \qquad (18)$$

es represented by this tensor are two-fold: a pressure $B^2/2\mu$ at right angles to the field and a tension $-B^2/2\mu$ along the field lines. The field lines thus tend to repel each other, and they also tend to contract like elastic strings. When the magnetic field lines are straight and parallel to

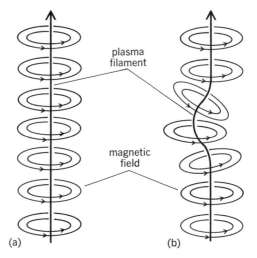

Fig. 2. Plasma-magnetic-field configuration for cylindrical pinch. (a) Equilibrium configuration of plasma filament and magnetic field generated by axial current flow through plasma. (b) Onset of kink instability.

each other, equilibrium is obtained for $p + B^2/2\mu =$ constant. When the magnetic field lines form circles around the axis of a cylinder (Fig. 2a), the equilibrium condition is given by Eq. (19), which

$$\frac{dp}{dr} = \frac{B}{\mu r}\frac{d}{dr}(rB) = 0 \qquad (19)$$

requires an additional relation for its integration. For example, B can be assumed to be proportional to r, and this leads to $p + B^2/\mu =$ constant. The magnetic field in this geometry can balance a pressure twice as large as in the first example. Intuitively this can be understood by considering that the tension along the field lines causes an inward force adding to the magnetic pressure gradient. The particular geometry of the second example arises in the theories of the filaments in the solar atmosphere and of the static pinch in controlled fusion.

In discussing equilibrium configurations, it is quite common to assume simple mathematical expressions for the field. Although one usually does not consider how to generate or maintain the electric currents which give rise to the magnetic field assumed in the mathematical model, one tries to use only plausible current distributions. For instance, the magnetic field in the cylindrical geometry above could be generated by a constant current density along the axis.

It is possible to maintain appropriate combinations of toroidal and polar fields in equilibrium with both pressure and gravitational forces in axially symmetric configurations, where the fields are confined to an interior region. Specific solutions are known for the sphere and the infinite cylinder. The latter is a possible model for explaining the relatively strong magnetic fields in the spiral arms of the Galaxy.

If $\mathbf{B}$ and $\mathbf{j}$ are parallel, no force is exerted on the fluid. Such force-free fields are in equilibrium in the absence of other forces. A case of special interest is encountered if $\nabla \times \mathbf{B} = \alpha\mathbf{B}$ with constant α. Whereas normally the decay of fields due to finite conductivity creates forces which are not balanced, this special type remains force-free. An

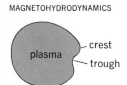

crest

plasma

trough

Fig. 3. Cross section
of plasma cylinder
with flute instability.
Magnetic field is
into the paper.

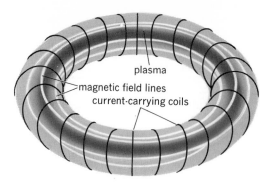

plasma

magnetic field lines

current-carrying coils

Fig. 4. Stellarator geometry. Plasma is confined in
magnetic-field torus generated by external windings.
Additional windings (not shown) cause field lines to rotate about chamber axis to improve stability.

example is the twisted field in a cylinder in which
the axial and tangential components of the field are
proportional to the Bessel functions $J_0(\alpha r)$ and
$J_1(\alpha r)$.

Stability. It is important to establish whether an
equilibrium configuration is stable. To determine
this, one can use an energy principle. One defines
a potential energy as the sum of internal energy E,
magnetic energy as in Eq. (20), and, if necessary,

$$E_m = \int (\mathbf{B}^2/2\mu)\,dv \qquad (20)$$

gravitational energy E_g. If an arbitrary deformation of the plasma always leads to an increased
potential energy, the equilibrium is stable. In applying this principle, the magnetic field lines are
considered to be attached to the plasma.

Another method of ascertaining stability considers small amplitude disturbances by linearizing
the equations of motion. One carries out an analysis for the normal modes of oscillation and the
equilibrium is unstable if any of the modes are.
This analysis furnishes the rate of growth of an
instability.

Generally, plasma contained on the concave
side of curved field lines is unstable. Important
types of instability are the formation of kinks and
flutes. The first is associated with cylindrical pinch
discharges. A minute kink in the plasma cylinder
will grow until it disrupts the discharge. Flute-
shaped ripples appear along the surface of a plasma which is confined in a flux tube (Fig. 3).

Such a disturbance can be regarded as an interchange of the field and plasma between the crest
and the trough. This interchange will result in a
decreased potential energy and, thus, instability of
the flute unless the value of $\int dl/B$ decreases toward the outside where the material pressure falls
off. In the stellarator the confining flux tube is bent
to form a torus (Fig. 4). The coils producing the field
can be arranged so that the lines (except for one
called the magnetic axis) do not close on themselves after a single turn around the torus. By this
means it is possible to prevent interchange and
thus interchange instability.

In a pinch discharge a magnetic field superimposed along the axis of the pinch stabilizes some
but not all modes of deviation from equilibrium.

An attempt to confine the plasma on the convex side of the field has led to the cusp geometry
(Fig. 5).

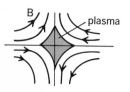

B

plasma

Fig. 5. Cusp geometry
of plasma confined
on convex side of
magnetic field.

A plasma configuration in equilibrium with
magnetic and gravitational forces will be unstable
if the magnetic field is too large. The virial theorem
leads to an upper limit $E_m < |E_g|$ for the magnet-
ic field energy. Actually, no definitely stable
configurations are known. However, in some
configurations the buildup rate of instabilities is so
slow they are practically stable. This is true for the
cylinders which are proposed as models for galac-
tic spiral arms.

Some interesting results appear if one enlarges
the class of gravitational equilibrium configurations to include internal motions. In a rotating
equilibrium configuration with axial symmetry,
all points along a line of force rotate with the
same angular velocity. This is known as the law
of isorotation. The inclusion of internal motions
has made possible the proof of the stability
of all axis-symmetric solutions whose motion is
$\mathbf{v} = \mathbf{B}/\sqrt{\rho\mu}$ and whose pressure p is given by $p/\rho +
v^2/2 + \phi =$ constant, where ϕ is the gravitational
potential.

Steady flow. A perfectly conducting fluid tends
to push the field lines out of the way. With a finite
conductivity, however, the field lines slip through
the fluid. Generally, a large magnetic Reynolds
number indicates a strong effect of the flow on the
magnetic fields in its path. This criterion does not
depend upon the size of the magnetic field. The
extent to which a magnetic field influences the
motion of the fluid, on the other hand, must clearly
increase with the magnitude of the field. If viscous
forces are negligible in the momentum balance,
the ratio of ponderomotive to inertial force $N_p =
B^2L\sigma/\rho v$ is a measure of this influence.

As an example consider the one-dimensional
flow of a fluid at supersonic speed coming from a
field-free region and passing through a region of
width L with a strong magnetic field at right angles
to the direction of the flow. The magnetic field exerts a drag on the fluid and an increase in N_p causes a reduction in the rate of flow. When a critical
value is passed, the flow will become subsonic.
However, the fluid cannot pass continuously from
supersonic to subsonic conditions. Instead, a
standing shock wave will form in the magnetic region effecting the transition to subsonic flow.

In the previous example the magnetic field region was bounded by parallel planes perpendicular
to the flow. Consider next an axial magnetic field
confined to a cylindrical region perpendicular to
the flow. In this case, a two-dimensional flow pattern will develop. The flow will be deflected sideways as it passes through the field because of the
Hall effect. A reaction force will push the magnetic
field region in the opposite direction. It has been
suggested that this lift may have aerodynamic applications. For strong fields the flow pattern goes
around the cylinder as if it were a solid obstacle.

When a fluid goes around an object which has a
magnetic field at right angles to its surface, it has
to cross magnetic field lines. This slows the flow
down and increases the size of the stagnation region (the region of zero velocity in front of the object) so that velocity and temperature gradients
near the stagnation point are reduced. The magnitude of these effects also depends upon the parameter N_p where L is a typical linear dimension of
the object.

Near a wall, viscous forces influence the flow and the viscosity also enters the scaling laws. The profile of a pressure-induced incompressible flow between parallel insulating walls (Poiseuille flow) with a magnetic field at right angles depends in the laminar region upon the Hartmann number $N_H = BL\sqrt{\sigma/\rho\nu}$, where $2L$ is the distance between the walls (Fig. 6). For $N_H = 0$ one has the parabolic profile of classical Poiseuille flow. For large N_H the velocity is nearly constant across the channel, decreasing only in the layers adjacent to the walls. The pressure gradient necessary to maintain the same average flow velocity increases by the factor in notation (21). The lines of force remain perpen-

$$N_H^2/3(N_H \coth N_H - 1) \approx 1 + N_H^2/15 \quad (21)$$

dicular at the walls but bulge in direction of the flow at the center line. The size of the bulge is proportional to R_m.

Frequently it is good approximation to assume that R_m is so small that the flow does not distort the magnetic field lines, and this approximation greatly simplifies the solution of otherwise difficult flow problems. One can, for example, determine the flow around a sphere in a field which is parallel to the axis of the flow by assuming that the field is constant throughout the region of the flow. The drag on such a sphere is increased by the factor in notation (22), over the classical Stokes value.

$$1 + \frac{3}{8}N_H + \frac{7}{960}N_H^2 + \cdots \quad (22)$$

In general, laminar flow through channels or past objects shows an increased drag with application of a magnetic field which has a component normal to the walls. The heat transfer to the wall, on the other hand, is generally reduced.

In the presence of a magnetic field a steady-state solution for the boundary layer flow along a semi-infinite flat plate can exist only when the Alfvén speed $v_a = B/(\mu\rho)^{1/2}$ is less than the speed of the flow or equivalently, $S < 1$. This conclusion is based upon a mathematical approximation procedure carried out for the case of a longitudinal magnetic field. Either $S = 0$ or $R_m/R = \mu\nu\sigma = 0$ reduces this theory to the classical case treated by H. Blasius. If $\mu\nu\sigma$ is small but not zero, the Blasius solution is not significantly modified unless S is very close to one. In the limit as $S \to 1$ the thickness of the boundary layer increases to infinity and the skin friction decreases to zero.

The case of a transverse field has been treated by an expansion in powers of $N_{px} = \sigma B^2 x/\rho v$, where x is the variable distance along the plate measured from the leading edge. This procedure does not, however, answer the question of whether there is a limit to the existence of steady flow. One can distinguish the cases in which the field is tied either to the motion of the plate or of the fluid. In the first case one finds a reduction of the skin friction coefficient at the wall of $(1 - 2.7N_{px} + 1.1N_{px}^2)$ and in the second case an increase of $(1 + 3.4N_{px} - 4.2N_{px}^2)$ compared to the Blasius solution.

Small-amplitude waves. One difficulty of the magnetohydrodynamic equations is their nonlinear character. By adopting a restriction to small-amplitude motions it is possible to linearize these equations. For Fourier analysis one can show the existence of a large variety of waves. Some of

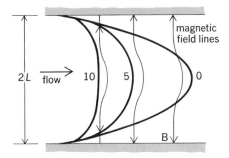

Fig. 6. Approximate profiles of pressure-induced incompressible flow between insulating walls for three values of N_H keeping the pressure gradient unchanged.

these are predominantly electromagnetic (with no fluid motion involved). In the presence of a dc magnetic field $\mathbf{B}_0$ there are, however, low-frequency wave modes in which there is a strong interaction between fluid motion and fields. These waves are similar to ordinary sound waves, but in contrast to the ordinary fluid situation there are three sound speeds. These speeds depend strongly upon the direction of propagation relative to $\mathbf{B}_0$.

The tendency of the fluid and the field lines to move together causes characteristic restoring forces which combine with the ordinary pressure forces in a number of different ways. This gives rise to the larger variety of wave types. In particular, it is possible to have shear waves with the fluid moving at right angles to the wave motion. The most simple shear wave, called the Alfvén wave, moves along the magnetic field lines with the Alfvén speed $v_a = B_0/\sqrt{\rho\mu}$. The speed of compression waves at right angles to the magnetic field is $\sqrt{v_a^2 + v_s^2}$, where v_s is the ordinary sound speed.

For an arbitrary angle θ between the direction of propagation and the direction of the magnetic field, one of the modes is a shear wave where the fluid is moving at right angles to the plane formed by the field $\mathbf{B}_0$ and the wave vector $\mathbf{k}$. The speed of this wave is $v_a \cos\theta$. The two other modes are in general hybrid forms with components of fluid motion both parallel and perpendicular to $\mathbf{k}$ and in the $(\mathbf{B}_0, \mathbf{k})$ plane. The two velocities in Eq. (23) are respectively faster and slower than the shear wave velocity.

$$v = \sqrt{\tfrac{1}{2}[v_a^2 + v_s^2 \pm \sqrt{(v_a^2 + v_s^2)^2 - 4v_a^2 v_s^2 \cos^2\theta}]} \quad (23)$$

In the two extreme cases where $\mathbf{k}$ is parallel or perpendicular to $\mathbf{B}_0$, the two hybrid modes are unscrambled into pure shear and compression modes. In the parallel case, $\mathbf{k}$ and $\mathbf{B}_0$ do not define a plane so that the two shear modes become undistinguishable. In the perpendicular case, the velocity of both shear modes formally goes to zero and only the compression mode exists.

The strong coupling between fluid motion and field disappears as the frequency of the wave approaches $\omega_i = eB/m_i$, the angular velocity of ions spiraling in the field B. Another limit is the frequency $V_a^2\mu\sigma$, at which the slipping of the fluid across the field lines destroys the wave structure.

Alfvén waves have been generated in the laboratory by twisting a column of liquid sodium in a strong magnetic field. In the design of this experiment, scaling laws have been very useful. The Lundquist number $N_L = R_m \sqrt{S} = BL\sigma \sqrt{\mu/\rho}$ seems to be natural to the description of this phenomenon. This number (where L stands for the radius of the column) must be large compared to 1, to permit the experiment. For mercury under normal laboratory conditions, say $L = 10$ cm and $B = 1000$ gauss, N_L is 10^{-1}. By using liquid sodium it is possible to increase this to $N_L = 5$. It is of interest to compare this to the interior of the Earth and the Sun, where $N_L \sim 10^3$ and $N_L \sim 10^7$ respectively.

Shock waves. The threefold structure of magnetohydrodynamic waves that follows from the linearized theory also shows up in the nonlinear case. One is led to three speeds depending on p, ρ, and B in the same manner as before. These speeds define the typical motion of certain disturbances.

Just as do ordinary hydrodynamic waves, magnetohydrodynamic waves tend to get steeper in front and develop into shocks, that is, surfaces across which the physical state changes discontinuously. These changes are restricted by the conservation laws, which relate the state variables on one side of the shock front with those on the other side and with the speed u of the front. If $\Delta F = F_{\text{ahead}} - F_{\text{behind}}$ is the difference between the values of a quantity F on the two sides of the shock front and $\mathbf{n}$ is the unit vector normal to the shock front pointing ahead, the shock conditions for a fluid with infinite conductivity are given by Eqs. (24)–(28).

$$\Delta \mathbf{B} \cdot \mathbf{n} = 0 \tag{24}$$

$$\Delta[(\mathbf{v} \cdot \mathbf{n} - u)\mathbf{B} - (\mathbf{B} \cdot \mathbf{n})\mathbf{v}] = 0 \tag{25}$$

$$\Delta[(\mathbf{v} \cdot \mathbf{n} - u)\rho\mathbf{v} + (p + \mathbf{B}^2/2\mu)\mathbf{n}$$
$$- (\mathbf{B} \cdot \mathbf{n})\mathbf{B}/\mu] = 0 \tag{26}$$

$$\Delta[(\mathbf{v} \cdot \mathbf{n} - u)\rho] = 0 \tag{27}$$

$$\Delta\left[(\mathbf{v} \cdot \mathbf{n} - u)\left(\frac{1}{2}\rho v^2 + \rho E + \mathbf{B}^2/2\mu\right)\right.$$
$$+ (\mathbf{v} \cdot \mathbf{n})(p + \mathbf{B}^2/2\mu)$$
$$\left. - (\mathbf{B} \cdot \mathbf{n})(\mathbf{B} \cdot \mathbf{v})/\mu\right] = 0 \tag{28}$$

These equations can be supplemented by the statement that a discontinuity of the tangential component of $\mathbf{B}$ implies the flow of a sheet current $I = (\mathbf{n} \times \mathbf{B})/\mu$ along the front.

One can again distinguish fast, slow, and intermediate shocks. Special cases of interest are parallel and perpendicular shocks where the names indicate the direction of $\mathbf{B}$ relative to $\mathbf{n}$. In the first of these the hydrodynamic motion is not coupled to the magnetic field and the shock proceeds as in ordinary hydrodynamics, so that the slow and intermediate types do not exist for this special case. Parallel as well as slow and fast perpendicular shocks can be considered in a frame in which the flow velocity is normal on both sides.

Shocks which are neither parallel nor perpendicular are called oblique. For these a frame of reference can be introduced in which the stream lines are parallel to $\mathbf{B}$ on both sides of the shock front but change their direction in passing through. In this frame the electric field vanishes on both sides.

In the three shock modes the change of the tangential component $\mathbf{B}_t$ of the magnetic field is distinctly different. Across a fast shock, $\mathbf{B}_t$ retains its direction and increases in magnitude; across a slow shock it retains or reverses its direction and decreases in magnitude; and across an intermediate shock it changes its direction and retains its magnitude.

In a fast shock it is possible for $\mathbf{B}_t$ to change from zero ahead of the shock front to a nonzero value behind; this is called a switch-on shock. In a slow shock it is similarly possible for $\mathbf{B}_t$ to change to zero behind the front; this is called a switch-off shock. Switch-on shocks exist only if $v_a > v_s$ ahead of the shock front and if the shock strength lies below a critical valve. Switch-off shocks always exist if behind the shock front $v_a < v_s$, and they exist for $v_a > v_s$ provided the shock is strong enough.

Transient flow. Transient flows can be started either by setting an object into motion or by switching on electrical circuits which create fields.

The impulsive motion of an infinite flat plate (Rayleigh problem) in a transverse magnetic field starts a transient flow which approaches the steady state in a time of the order $\rho/\sigma B^2$.

The flow which develops along the semi-infinite flat plate in a parallel field ($S < 1$) has a different character in different regions. Between the leading edge and a point $x = (1 - \sqrt{S})vt$ the flow approaches the steady-state solution discussed earlier. Beyond $x = (1 + \sqrt{S})vt$ the flow approaches the infinite plate solution. In between there is a transition region.

The sudden release of the energy stored in a large condenser and its conversion to mechanical and thermal energy of a plasma give rise to problems whose theoretical treatment requires fast computers. One can simplify the theory of a collapsing pinch discharge considerably by assuming that all material which has been swept up by the contracting magnetic field is piled up in a very thin layer which is snow plowed toward the axis. In this manner one can set up an ordinary differential equation for the radius of this layer whose numerical integration can be carried out with more modest equipment. The time of collapse of such a pinch is given by $Et \approx r(\rho\mu)^{1/4}$, where ρ and r are the initial density and radius of the plasma, and E is the electric field causing the discharge. *See* THERMONUCLEAR REACTION.

Flow instability. Laminar flow breaks down when R rises beyond a critical number. It has been demonstrated both theoretically and experimentally that a magnet field improves flow stability; the critical R increases in proportion to N_H. For flow between parallel flat plates with a transverse magnetic field the predicted onset of instability takes place at $R = 50,000N_H$. The measured suppression of turbulence takes place at the much lower value $R = 225N_H$. Between coaxial cylinders, where the inner one rotates faster than the outer one, an axial magnetic field raises the critical angular velocity by a factor which approaches $N_H/2$ for $N_H \geqq 20$ (the difference in radii is used to define N_H).

Turbulence. The mathematical structure of Eq. (7) for the field vector $\mathbf{B}$ is identical with that of Eq. (29) for the vorticity vector $\boldsymbol{\omega}$ in ordinary hydrody-

$$\frac{\partial \boldsymbol{\omega}}{\partial t} = \nabla \times (\mathbf{v} \times \boldsymbol{\omega}) + \nu \nabla^2 \boldsymbol{\omega} \qquad (29)$$

namics. The first term on the right-hand side of either equation tends to increase the mean square value of the respective field vector; the second term causes a decrease due to resistive and viscous losses. In ordinary turbulence the increase and decrease of $\overline{\omega^2}$ roughly balance each other. It has been suggested that small magnetic disturbances in a turbulent flow field will increase if resistive losses are less than viscous ones, that is, if $\sigma \mu \nu > 1$. In a Fourier expansion of the magnetic field energy, the spectrum within an uncertain range of wave numbers k tends toward a distribution $\sim k^{-5/3}$ (Kolmogoroff spectrum). Within that range the ratio of magnetic to kinetic energy is about 1.6.

Ordinary turbulence requires large values of R and magnetohydrodynamic turbulence requires even larger values of R_m. Such conditions are encountered only in geophysical and astrophysical situations.

TRANSPORT-EQUATION DESCRIPTION

The continuum approach to magnetohydrodynamics ceases to be valid when the mean free paths of the particles are of the order of or larger than the lengths characterizing the structure of the flow. A description of more general validity can be based on the functions $f_i(r,v,t)$ that give the densities of particles with velocity v at position r for the various particle species identified by the index i. These functions obey "transport equations," Eq. (30), where $\mathbf{E}$ and $\mathbf{B}$ are macroscopic fields that

$$\frac{\partial f_i}{\partial t} + v \cdot \nabla f_i + \frac{e_i}{m_i}(\mathbf{E} + v \times \mathbf{B})\, \nabla_v f_i = \left(\frac{df_i}{dt}\right)_{\text{coll}} \qquad (30)$$

satisfy Maxwell's equations. The rate of change due to collisions $(df_i/dt)_{\text{coll}}$ can be brought into manageable form by certain assumptions. One of these is that only two particles participate in any collision. This assumption yields the Boltzmann equation. Another assumption is that collisions produce predominantly small deflections. In this case one can make certain expansions leading to the Fokker-Planck equation. Both these forms can be used to derive continuum-type equations.

Although a charged particle in a plasma collides simultaneously with very many other particles, each collision has only a minute effect. For this situation, the mean free path can be defined as the distance a particle travels until its momentum or energy is appreciably changed by the random addition of minute changes. This effective mean free path increases as the square of the temperature and can become very large. In this case one can set $(df_i/dt)_{\text{coll}} = 0$.

In a magnetic field, charged particles spiral around the lines of force; this restricts their motion at right angles to the field. If the radii of the spirals are small compared to L, these gyrations will cause some degree of randomness of the particle motion even in the absence of collisions. Gyrations are not as effective in doing this as collisions, and in particular exclude any energy transfer between the components perpendicular and parallel to the magnetic field. Nevertheless, they may produce a nonthermal equilibrium in which a modified continuum approach is still possible.

One such modification replaces the scalar pressure by a tensor with two pressures $p_\parallel$ and $p_\perp$ parallel and perpendicular to $\mathbf{B}$. With no collisions, viscous and resistive losses are absent, and therefore the assumption of no heat flow in the energy equation of the ordinary continuum theory (Eq. 13) leads to constant entropy along stream lines. In the modified theory, no heat flow leads to the constancy of $p_\perp/\rho \mathbf{B}$ and of $p_\parallel \mathbf{B}^2/\rho^3$ along stream lines. Such constants of motion are used mainly in investigating the stability of equilibria by means of a variational principle. However, it is not obvious why the no-heat-flow assumption should hold.

Small changes from an equilibrium distribution as they occur in applying a variational principle can also be handled by a direct use of the transport equation. This theory requires the Debye length $\sqrt{\epsilon KT/ne^2}$ to be small compared to L, which is true for quite general conditions in the plasma. The stability obtained in this manner is stronger than that obtained using the constant-entropy approximation and weaker than that obtained from the two-pressure modification. *See* MAGNETOGAS DYNAMICS.

[ROLF LANDSHOFF]

Bibliography: H. Alfvén and C. G. Falthammer, *Cosmical Electrodynamics*, 2d ed., 1963; T. G. Cowling, *Magnetohydrodynamics*, 1957; R. K. M. Landshoff (ed.), *Magnetohydrodynamics*, 1957; J. A. Shercliff, *Magnetohydrodynamics*, 1964; L. Spitzer, Jr., *Physics of Fully Ionized Gases*, 1956; G. W. Sutton and A. Sherman, *Engineering Magnetohydrodynamics*, 1965.

Magnetomotive force

The magnetomotive force (mmf) around a magnetic circuit is the work per unit magnetic pole required to carry the pole once around the circuit. It is the analog of electromotive force. *See* MAGNETIC CIRCUITS.

It is expressed mathematically in Eq. (1), where

$$\text{mmf} = \oint H \cos \theta\, ds \qquad (1)$$

$H \cos \theta$ is the component of magnetic field strength in the direction of a length of path ds. The line integral is taken around any closed path in the field.

The magnetomotive force is the rise in magnetic potential around the path. For a discussion of magnetic potential *see* MAGNETIC FIELD.

For a path that encloses a current I, Eq. (2)

$$\oint H \cos \theta\, ds = I \qquad (2)$$

holds, and for a path that encloses N equal currents, for example, a path that loops through a coil of N turns, Eq. (3) is valid. If no current is enclosed by the path, the line integral is zero.

$$\oint H \cos \theta = NI \qquad (3)$$

The meter-kilogram-second (mks) unit of magnetomotive force is the ampere-turn.

[KENNETH V. MANNING]

Marine engine

An engine that propels a waterborne vessel. Even in small craft the marine engine must have the following characteristics: reliability, light weight, compactness, fuel economy, low maintenance, long life, relative simplicity for operating personnel, ability to reverse, and ability to operate steadily at low or cruising speed. The relative importance of these characteristics varies with the service performed by the vessel, but reliability is of prime importance.

Steam engines. Steam, used to drive the earliest powered vessels, is still a common type of propulsion for large ships. The diesel engine has gained wide acceptance in foreign merchant ships, but in the United States the majority of seagoing vessels use steam propulsion.

Reciprocating steam engines. Early engines commonly used steam flowing in series through as many as four cylinders whose pistons had the same stroke but were of increasing diameters. This system provided for an expansion or increase in steam volume which accompanied the decrease in pressure to the exhaust under a vacuum. The modern, multicylinder, uniflow marine steam engine, with complete expansion in each cylinder, shows better steam economy. Because it has the same diameter for all cylinders (two to six in number), it is preferable from a manufacturing viewpoint. Equal power is developed by each cylinder; units of four cylinders or more have good torque and balance characteristics. A steam rate of 10 lb/hp/hr with 275 psi at 240°F superheat is attained. Uniflow engines as large as 5000 hp have been used on shipboard. Normally, steam engines are double-acting; that is, steam acts on each side of the pistons. With superheated steam, piston-cylinder lubrication must be provided. Pure feedwater is required by modern, high-capacity boilers; therefore, an effective oil filter is installed where the condensate must be returned to the boiler.

Steam turbines. The marine steam turbine has the advantages of direct rotary motion, little or no rubbing contact of pressure-confining surfaces, and ability to use effectively both highly superheated steam and steam at low pressure, that is, at a high vacuum where specific volumes of over 400 ft³/lb are reached.

For good efficiency of steam turbines, high rotative speeds are required. This requirement led to the introduction of the reduction geared turbine and turboelectric drive. These systems give efficient turbine speeds and efficient propeller rpm. With geared turbines, for example, turbine rotor speeds range from 3000 to 10,000 rpm, while propeller rpm is reduced to the 80 to 400 range.

Steam is generally supplied to the turbine at 850 psi and 950°F (510°C) by a pair of oil-fired marine water-tube boilers, and the exhaust from the turbine is usually at 1.5 in. Hg absolute. Forced draft fans and other auxiliaries are usually motor-driven, except for the main feed pumps which are usually driven by an auxiliary turbine. Electric power is provided by a separate turbogenerator.

In low-powered geared turbines, steam completes its expansion in one rotor and casing. Such a design has been used in geared turbines of up to 8000 shaft horsepower (shp). However, series flow through two or even three casings is preferable in most steam turbines. This arrangement provides more flexibility in turbine design, allowing for different and optimum revolutions for high- and low-pressure rotors. Also, in a seagoing vessel in case of casualty to one turbine or its high-speed pinion, the vessel usually can make port with the remaining turbine in operation.

A steam turbine is made up of fixed blades, usually called nozzles, and rotating blades. A stage is generally one stationary row and one moving row. Impulse staging has all the steam pressure drop taking place in the fixed blades. The moving row then absorbs the kinetic energy produced. Reaction staging results when some of the pressure drop occurs in the moving blades, the degree of reaction depending on the design.

Modern marine practice favors impulse staging in the high-pressure end of the turbine because of the reduced parasitic losses with this type, but as the steam progresses toward the low-pressure end where the volume is much greater, the reaction stage is more efficient. The usual arrangment has a cross compound system with a high-pressure unit of 7 to 10 stages and a low-pressure unit of 6 to 8 stages, each driving a pinion of a reduction gear. *See* IMPULSE TURBINE; REACTION TURBINE.

A turbine is capable of operating in only one direction. In order to provide reverse power, a second turbine is installed on the shaft of the low-pressure, ahead turbine. The astern turbine is usually not more than three moving rows of blades, but it may be only two. It produces about 40% of the normal ahead horsepower. Since this unit is turning backward in normal ahead operation, it is located in the low-pressure end of the low-pressure turbine. The steam at this point has a very low density, and hence the astern turbine has a low windage loss.

The propeller is reversed by closing the steam valve to the ahead turbine and opening the valve to the astern turbine.

Gas turbines. The gas turbine is a relative newcomer to the marine field. It generally consists of an axial compressor discharging compressed air to a combustion chamber where fuel is burned, adding heat. The products of combustion at high temperature and pressure then pass through a gas turbine that drives the compressor and load. Generally, the term "gas turbine" is applied to the entire plant. If lower pressure ratios (final pressure leaving the compressor divided by initial pressure entering) are used, a large amount of heat is available in the exhaust gas which may be recovered by heating the compressed air before it enters the combustion chamber. This is done in a regenerator. With higher pressure ratios the expansion through the turbine is so great that the exhaust gas temperature is insufficient to heat the compressed air.

Two distinct types of gas turbines are appearing in the marine field: the aircraft-derived type and the industrial type. The aircraft-derived type uses a jet engine as a gas generator, which discharges to a gas turbine driving the load. This type of plant offers simplicity and light weight but must burn high-quality fuel. The industrial gas turbine is a more rugged machine designed for long life and is capable of using low grades of fuel, properly washed. This plant usually uses a regenerator. The gas turbine offers simplicity, ease of control, and efficiency, but requires special fuel or special treatment of the fuel. Large amounts of air and exhaust gas are used, and as a result up-

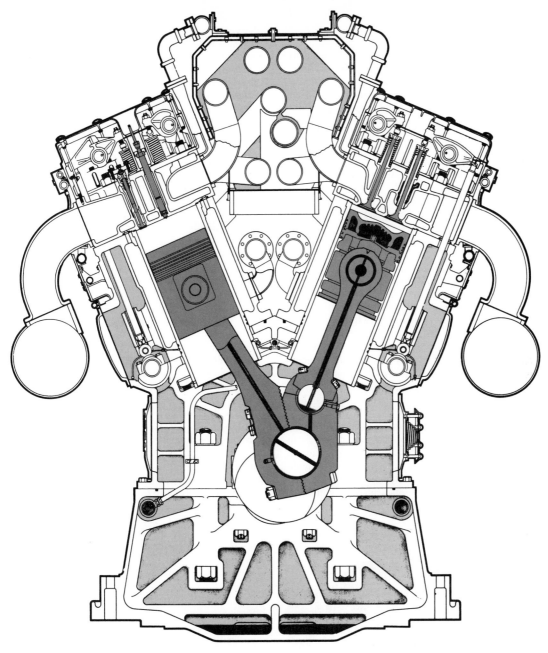

Fig. 1. Cross section of four-cycle V-type diesel.

takes and air supply are a special problem. The aircraft-derived gas turbine seems destined to drive a large number of naval combatant ships. Selection of this type for a new class of naval destroyers has been announced. The industrial type will be used for merchant vessels where its greater weight will be of little disadvantage. *See* GAS TURBINE.

Internal combustion engines. Both diesel and gasoline internal combustion engines are used in marine applications. Many moderate- and low-power marine installations use automotive or locomotive engines designed for variable load and intermittent service. High-power marine propulsion units normally are called on to operate continuously under load. Therefore, the brake horsepower (bhp) rating of units selected for marine service should be conservative.

The gasoline engine is the most common power plant for pleasure craft. It is inexpensive to buy and maintain. Because of its widespread use in automobiles, most parts are readily available. In most areas gasoline costs slightly more than diesel fuel, but the cost differential is usually insufficient to make up the difference between the cost of gasoline and diesel engines. Gasoline presents an explosion and fire hazard, which is its major disadvantage. *See* INTERNAL COMBUSTION ENGINE.

Direct-drive diesels. For typical commercial freight vessels, direct-drive diesels provide economical service. For good propeller efficiency, the propeller rpm should be under 120±. Such a top limit on engine revolutions results in a large, heavy, bulky, slow-rpm engine. However, the direct-drive diesels have a lower fuel oil consumption than do higher-rpm units, and with suitable fuel treatment they will operate on the better grades of the cheaper fuel oil burned in boilers. *See* DIESEL ENGINE.

Slow-speed, direct-drive diesels are favored by many European owners and shipbuilders. Turbocharged, two-cycle, single-acting diesel engines of 50,000 bhp are now available; such engines weigh more than 100 lb/bhp. For high horsepower the total machinery weight for diesels is more than the weight of geared turbine machinery, including boilers and auxiliaries.

Moderate-speed diesels. Diesel engines of 250–500 rpm are available in two- and four-cycle, single-acting types, generally with trunk pistons (Fig. 1). In some marine applications they are connected directly to the propeller and thus fitted only with reverse gear. However, they are also employed with geared diesel and diesel-electric drive. The weight of such engines runs about 35–70 lb/bhp.

High-speed diesels. Many high-speed diesel engines of 600 rmp and more (some types originally developed for truck and locomotive service) are available for marine propulsion. Opposed piston types have been developed; other manufacturers favor a V type to reduce weight. Such engines are of two- and four-cycle types and usually weigh 10–40 lb/bhp. Because of less efficient scavenging, lower powers, and other factors, their fuel and lubricating oil rates are higher than for large, low-speed diesels.

Except for direct drive in moderate or fairly high-speed craft, marine applications of diesel engines are fitted with either mechanical reduction gearing or with diesel-electric drive to provide good propeller efficiency. Because their pistons, valves, and other components are small, standard-ized, and carried in stock, repairs are readily made, with the result that engines of this type are popular for nonoceangoing services.

Oil consumption and starting. Lubricating oil consumption of diesel engines is high because of the cylinder-piston lubrication that must be provided and the contamination of the crankcase oil with residues blown by the piston rings. In large engines this contamination is avoided by using piston rod–crosshead construction so that the crankshaft, connecting rods, and crossheads operate in a closed casing separated from the working cylinder. These engines are started and maneuvered by pressure from one or more reservoirs filled with air at about 250 psi. To make it feasible to start and readily reverse, two-cycle, single-acting marine engines should have at least four cylinders; four-cycle engines should have five or more cylinders.

Nuclear power. Very successful installations of nuclear power have been made in submarines and a few surface ships. Operation of the first nuclear merchant ship, the *Savannah*, was successful technically but not commercially. (The operating crew required special training, and it would be difficult to replace with only one commercial nuclear ship in service.)

Mechanical reduction gears. Reduction gearing for diesel and gasoline engines allows the use of a relatively high engine speed and lower, more efficient propeller speed. Speed reduction ratios of 1.8:1 to 4:1 are common, preferably with helical teeth to give better wear and quieter performance. A reverse gear device often is incorporated in low-power gears for astern operation. Other methods for providing reverse rotation use a direct reversing engine or a controllable-pitch propeller.

One, two, three, or four engines may drive the same gear through individual pinions. The use of a friction, electromagnetic, pneumatic, or hydraulic coupling serves to disconnect any engine. By reversing one or more engines, ready maneuvering, including astern operation, is provided for by the use of the respective coupling.

The high rpm (3000–9000) of modern marine steam turbines and the low revolutions of an effective propeller (as low as 80 rpm) require the use of two-stage gearing. Gear teeth are of involute form, with the pinion teeth of harder material than the gear. The gear trains are of the double helical type to avoid heavy axial thrust (Fig. 2). Double reduction gears are constructed with flexible couplings between the high-speed train and the low-speed elements.

Mechanical reduction gears are carefully constructed to close tolerances. They have forced lubrication in sprays ahead of the meshing teeth, to the bearings, and to the flexible couplings. Tests have shown that bearings represent at least half the power loss of the entire gear set.

Turboelectric drive. This type of drive, comprising one or more steam turbine generators and ac propulsion motors, is also used for ship propulsion. It was installed in many United States tankers during World War II because of available manufacturing facilities. The synchronous motors are provided with an induction winding for starting and reversing. Relatively large changes in propeller revolutions are made by alteration of the turbogenerator speed.

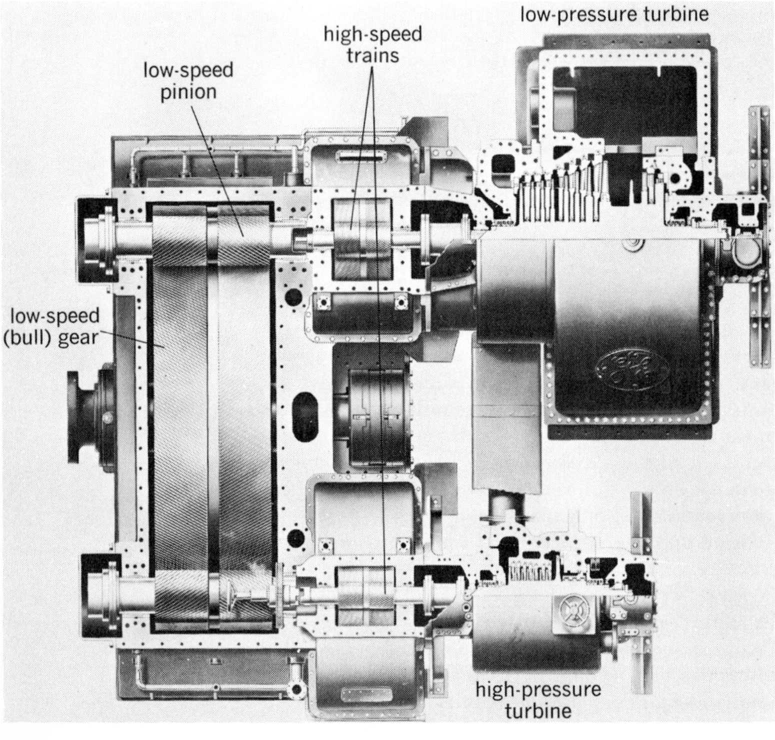

Fig. 2. Double reduction articulated gear design for use with modern high-rpm marine steam turbines. (*General Electric Co.*)

Motor, generator, exciter, and cooling equipment losses result in several percent lower efficiency than with geared steam turbines. Weights and costs are generally 25–30% higher than for the comparable turbine gear arrangement. Electric drive is not employed unless it offers significant operational or design advantages. These include flexibility of control and the independence of the location of the turbo generator relative to the propeller shaft or propulsion motor.

Diesel-electric drive. This type of drive, composed of one or more dc diesel generator sets and often a double-armature propulsion motor, is used in tugs, dredges, Coast Guard cutters, and icebreakers, where maneuvering and a wide range in propeller speed are necessary. For slow-speed operating during maneuvering, the engine speed is often reduced and the generator field excitation is altered to provide wide variation in the motor output and propeller speed.

Control arrangements. Bridge or pilothouse control of the ship propulsion unit, without action by the engineer on watch, is used with diesel-electric drive and for small, low-powered, direct-drive, and mechanically geared diesel installations. This is the customary arrangement for tugs and dredges.

Modern practice for large diesel and steamships is to provide pilothouse control of the main engines and no engine-room watch except day workers and when entering and leaving port. Propulsion plant monitoring is usually provided in a central control space, with chart recorders or a data storage system. All levels of automation are now being used, from simple manual surveillance of all systems in a fully manned engine space to a completely unmanned system recording data and monitoring trends to pinpoint possible trouble.

Governors. Above the operating rpm, ship propeller torque increases faster than engine or turbine torque, and thus ship propeller drive is inherently stable. Because of the ship's pitching, the propeller may lift partially out of the water and the engine may tend to race. To allow for this situation, or for propeller shafting failure, American Bureau of Shipping regulations require that a governor be fitted to limit overspeed to 15% above the rated speed.

A common type of governor uses oil pressure developed by small pumps incorporated with the main turbine rotors to activate the governor; low lubricating oil pressure will also shut off the steam supply.

With turboelectric and diesel-electric drive, there is no mechanical connection between the generator set and the propulsion motor and propeller. The operating governor holds generator speed at the set value by throttling turbogenerator steam or the amount of fuel injected in the diesel engine cylinders.

[JENS T. HOLM]

Bibliography: Babcock and Wilcox Co., *Steam*, 37th ed., 1963; T. Baumeister (ed.), *Marks' Mechanical Engineer's Handbook*, 7th ed., 1967; I. G. Broersma, *Marine Gears*, 1962; I. G. DeRooij, *Practical Shipbuilding*, 1961; R. L. Harrington, *Marine Engineering*, 1971; R. F. Latham, *Introduction to Marine Engineering*, 1958; C. N. Payne, *Descriptive Analyses of Naval Turbine Propulsion Plants*, 1958; C. E. Pounder (ed.), *Diesel Engine Principles and Practice*, 1955; P. de W. Smith, *Modern Marine Electricity and Electronics*, 1966.

Marine geology

The study of the portion of the Earth beneath the oceans. More than 70% of the Earth's surface is covered by marine waters. Of the oceanic area $(361 \times 10^6 \text{ km}^2)$ approximately $300 \times 10^6 \text{ km}^2$ is contributed by the deep-sea floor; the remaining $60 \times 10^6 \text{ km}^2$ represents the submerged margins of the continents. The distribution of elevations on the Earth is shown in Fig. 1.

Soundings. Soundings are measurements of ocean depth made from ships. Early soundings were made with a lead attached to a hemp line; about 1875 the hemp line was replaced by piano wire. Since about the middle of the 1920s, virtually all deep-sea soundings have been made by echo sounding. The echo-sounding machine sends out a sound pulse (10–20 Hz) and then times the interval from the sound pulse to the returning echo. The early sounders required manual operation, but since about 1935 automatic recording sounders, which plot a graph of depth versus time or distance, have been used almost exclusively. Since 1953 precision, high-resolution echo sounders have been used in increasing numbers.

Sounding corrections. Wire and hemp line soundings require a correction for wire angle, for stretch of the wire, and for calibration of the metering counters used. Echo soundings require a correction for sound velocity, since the average vertical velocity is not constant, and for slope of the bottom, as the point from which the first echo returns is not always directly beneath the ship. In addition, corrections for inaccuracies of timing and mechanical imperfections must be made for most sounders. The position of the sounding lines is generally determined by standard astronomical fixes and dead reckoning. Errors of a few miles are the rule in deep-sea sounding surveys.

PHYSIOGRAPHIC PROVINCES

The Earth relief lies at two dominant levels (Fig. 1); one, within a few hundred meters of sea level, represents the normal surface of the continental

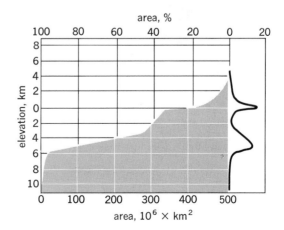

Fig. 1. Hypsographic curve showing area of Earth's solid surface above any given level of elevation or depth. Curve at the right shows frequency distribution of elevations and depths for 2-km intervals.

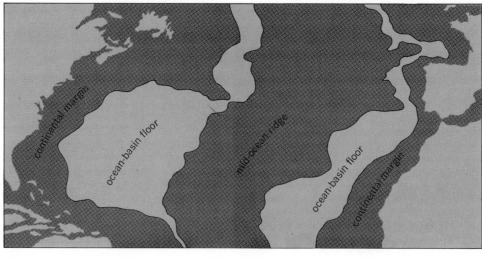

Fig. 2.   Major morphologic divisions of North Atlantic Ocean. The profile is from New England to Sahara.

blocks; the other, between 4000 and 5000 m below sea level, and comprising more than 50% of the Earth's surface, represents the ocean-basin floor. The topographic provinces beneath the sea can be included under three major morphologic divisions: continental margin, ocean-basin floor, and mid-oceanic ridge. These are indicated on a typical transoceanic profile taken from the North Atlantic in Fig. 2. Each of these major divisions can be further divided into categories of provinces and those into individual physiographic provinces (Fig. 3).

Continental margins.   The continental margin includes those provinces associated with the transition from continent to ocean floor. The continental margin in the Atlantic and Indian oceans is generally composed of continental shelf, continental slope, and continental rise. A typical profile off the northeastern United States is shown in Fig. 4.

Gradients on the continental shelf average 1:1000, while on the continental slope gradients range from 1:40 to 1:6, and occasionally local slopes approach the vertical. The continental rise lies at the base of the continental slope. Continental rise gradients average 1:300, but individual slope segments may be as low as 1:700 or as steep as 1:50. The continental slopes are cut by many submarine canyons. Some of the larger canyons such as the Hudson extend across the continental rise (Fig. 4). Submarine alluvial fans extend out from the seaward ends of the larger canyons.

The continental margin can be divided into three categories of provinces. Category I includes the continental shelf, marginal plateaus, and shallow epicontinental seas, all slightly submerged portions of the continental block. Category II includes the continental slope, marginal escarpments, and the landward slopes of marginal trenches, all expressions of the outer edge of the continental block. Category III includes the continental rise, the ridge-basin complex, and the ridge-trench complex. The continental slope of the

northeastern United States can be traced directly into the marginal escarpment (Blake Escarpment) off the southeastern United States (Fig. 5) and the landward slope of the Antilles marginal trench (Puerto Rico). The continental rise off New England can be traced into the Antilles Outer Ridge. Seismic refraction studies show that a trench filled with sediments and sedimentary rocks lies at the base of the continental slope off New England. Thus the main difference in morphology between the trenchless continental margins and continental margins with a marginal trench is that in the former the trench has been filled with sediments.

In the continental margins of the Atlantic, Indian, Arctic, and Antarctic oceans and the Mediterranean Sea, the continental rise generally represents the category III provinces. The Pacific, however, is bounded by an almost continuous line of marginal trenches. The high seismicity, vulcanism, and youthful relief of the Pacific borders suggest a very recent origin. In contrast, the nonseismic, nonvolcanic character, as well as the lower relief, of the Indo-Atlantic margins suggests a greater age. Thus on the old, stable, continental margins the deposition of sediment derived from the land has filled the marginal trench and produced the continental rise. The local relative relief on the continental margin rarely exceeds 20 fathoms, with the major exception of submarine canyons and occasional seamounts.

Submerged benches.   Submerged marine beach terraces have been identified throughout the world. Since the beaches seem to correlate well between areas of vastly different tectonic development, it has been concluded that those which are listed in the accompanying table represent submerged late Pleistocene beaches.

Structural benches.   Structural benches, the topographic expression of outcropping beds, have been identified on the continental slope. Near Cape Hatteras, N.C., the structural benches have

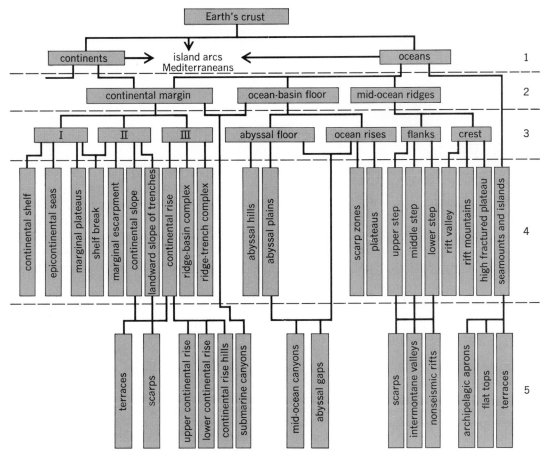

Fig. 3. Outline of submarine topography. Line 1, first-order features of the crust; line 2, major topographic features of the ocean; line 3, categories of provinces and superprovinces; line 4, provinces; line 5, subprovinces and other important features.

been dated by extrapolating data obtained in several test borings near the coastline (Fig. 6). Benches on Georges Bank have been dated from bottom samples obtained by dredging. Through the action of slumps, bottom currents, and turbidity currents, sediments are continually removed from the continental slope. Thus, there is no cover of recent sediments to obscure the outcrops of the ancient formations.

Ocean-basin floor. Excluding the marginal trenches and mid-oceanic ridges, the deepest portions of the ocean are included in this division. Approximately one-third of the Atlantic and three-fourths of the Pacific fall under this heading. The ocean-basin floor can be divided into three categories of provinces; the abyssal floor, oceanic rises, and seamounts and seamount groups.

Abyssal floor. The abyssal floor includes the broad, deep areas of the central portion of the ocean. In the Atlantic, Indian, and northeast Pacific oceans, abyssal plains occupy a large part of the abyssal floor. An abyssal plain is a smooth portion of the deep-sea floor where the gradient of the bottom does not exceed 1:1000. Abyssal plains adjoin all continental rises and can be distinguished from the continental rise by a distinct change in bottom gradient. At their seaward edge, most of the abyssal plains gradually give way to abyssal hills. Individual abyssal hills are 50–200 fathoms high and 2–6 mi wide. In the Atlantic, the abyssal hill provinces only locally exceed 50 mi in width. Abyssal plains in the same area range from 100 to 200 mi in width. Core samples of sediment obtained from the Atlantic abyssal plain invariably contain beds of sand, silt, and gray clay intercalated in the red or gray pelagic clay which is generally characteristic of the deep-oceanic environment. These deep-sea sands were transported by turbidity currents from the continental margin. Some of the currents probably descended along a broad front, while others certainly followed the submarine canyons and spread out fanwise from their submarine alluvial cones.

The abyssal hills are thought to represent tectonic or volcanic relief of a type identical with that buried beneath the abyssal plains. Abyssal plains are also found in the marginal trenches, marginal basins, and in epicontinental marginal seas. Features of exactly the same morphology and origin are found in some lakes. Of similar origin are archipelagic aprons, which spread out from the base of oceanic islands.

Oceanic rises. Oceanic rises are areas slightly elevated above the abyssal floor which do not belong to the continental margin or the mid-oceanic ridges. In the North Atlantic, the Bermuda Rise is the best-known example (Fig. 7). In contrast to the mid-oceanic ridges, oceanic rises are nonseismic; their relief is more subdued, and they are asymmetrical in cross section. The western and central

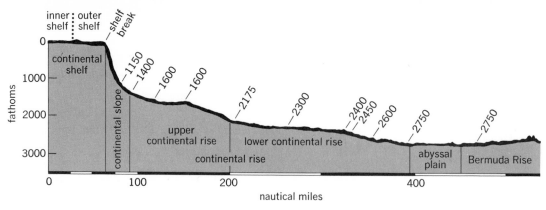

Fig. 4. Continental margin provinces: type profile off northeastern United States.

Bermuda rise is characterized by gentle, rolling relief. The average depth gradually decreases toward the east. In the eastern third, the rise is cut by a series of scarps, 500–1000 fathoms in height, from which the sea floor drops to the level of the abyssal plain on the east. The series of eastward-facing scarps suggest block faulting. Situated approximately in the center of the Bermuda Rise is the volcanic pedestal of Bermuda. A small archipelagic apron surrounds the pedestal. The turbidity-current origin of the smooth apron is supported by cores containing shallow-water carbonate clastic sediments in depths of 2300 fathoms. In the Pacific, extending for more than 3000 mi west of Cape Mendocino, Calif., is an asymmetrical rise with a southward-facing scarp, which has been named the Mendocino Fracture Zone. The Bermuda Rise is less than one-fourth as long as the Mendocino Rise, but otherwise the relief of both features is quite similar. Although the circum-Pacific seismic belt crosses its trend, the Mendocino Escarpment is nonseismic. Other nonseismic fracture zones, which probably can be classified as oceanic rises, have been reported from the eastern Pacific. The Rio Grande Rise of the South Atlantic and the Mascarene Ridge of the Indian Ocean are similar in form.

Seamounts and seamount groups. A seamount is any submerged peak more than 500 fathoms high. This discussion, however, is limited to the larger, more or less conical peaks more than 1000 fathoms in height. Seamounts are distributed through all the physiographic provinces of the oceans. Sea-

mounts sometimes occur randomly scattered but more often lie in linear rows. It seems safe to conclude that virtually all conical seamounts are extinct or active volcanoes. The Kelvin seamount group, a line of seamounts 800 mi long, stretches out from the vicinity of the Gulf of Maine toward the Mid-Atlantic Ridge. The Atlantis–Great Meteor seamount group extends for 400 mi along a north-south line, south of the Azores. In the southwestern Pacific, many lines of islands and seamounts crisscross the ocean. In the mid-Pacific, southwest of Hawaii, is a large area of seamounts whose flat summits range from 50 to 850 fathoms beneath sea level. These tablemounts have been termed guyots. From the flat summits, shallow-water fossils of Cretaceous age have been dredged. Such sunken islands are not limited to the Pacific. Several of the Kelvin seamounts are flat-topped at 650 fathoms, and the seamounts of the Atlantis–Great Meteor group have flat summits at 150–250 fathoms.

Mid-Oceanic Ridge. The middle third of the Atlantic, Indian, and South Pacific oceans is occupied by a broad, fractured swell known as the Mid-Oceanic Ridge. In the Atlantic, it is known as the Mid-Atlantic Ridge; in the southern Indian Ocean, the Mid-Indian Ridge; in the Arabian Sea, the Carlsberg Ridge and Murray Ridge; and in the South Pacific, the Easter Island Ridge.

The Mid-Atlantic Ridge can be divided into distinctive physiographic provinces which can be identified on most transatlantic profiles (Figs. 2 and 8).

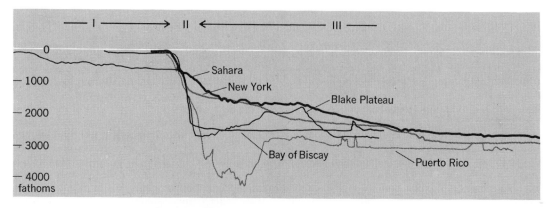

Fig. 5. Three categories of continental-margin provinces.

Depth in fathoms of prominent continental-shelf terraces*

Placentia Bay, Newfoundland	Norfolk, Va.	Charleston, S.C.	Bimini, B.W.I.	St. Vincent, Cape Verde Is.	Dakar, Senegal	San Pedro, Calif.
10		12	10	8	10	10
				15	15	15
20	18	20	20	24	20	20
	30	30	28	28	28	28
35	35	35		32		
40				38	38	38
42		45	42	42	45	45
	50					
55	58			54	55	55
				60		
68		68	65			
80	80	80	85	80	78	80

*Each column based on a single nonprecision echogram.

Crest provinces. The rift valley, rift mountains, and high-fractured plateau which constitute this category form a strip 50–200 mi wide. The rift valley is bounded by the inward-facing scarps of the rift mountains. The floor of the rift valley lies 500–1500 fathoms below the adjacent peaks of the rift mountains, which drop abruptly to the high fractured plateau, lying at depths of 1600–1800 fathoms on either side of the rift mountains. The topography of the crest provinces is the most rugged submarine relief. An earthquake belt accurately follows the rift valley through a distance of over 40,000 mi. Heat-flow measurements in the crest provinces give values several times greater than have been obtained in normal ocean or continental areas. A large, positive magnetic anomaly and a moderate (−20 mgal) negative gravity anomaly are associated with the rift valley. Seismic refraction measurements indicate a crust intermediate in composition between the oceanic crust and the mantle. The crest provinces of the Mid-Oceanic Ridge can be traced directly into the rift valleys, rift mountains, and high plateaus of Africa. These features of African geology are clearly the result of extensional forces in the Earth's crust. The Mid-Oceanic Ridge is probably similar in all essential characteristics, including origin, to the African rift valley complex.

Flank provinces. The flank provinces of the Mid-Oceanic Ridge can be divided into several steps or ramps, each bounded by scarps somewhat larger than those which characterize the entire area.

Parts of the flank provinces, particularly the Upper Step south of the Azores, are characterized by smooth-floored intermontane valleys. Photographs, cores, and dredging indicate that the crest of the Mid-Atlantic Ridge north of the Azores is being denuded of its sediments. As these sediments are eroded from the crest provinces and deposited on either side, they are gradually filling the intermontane basins and smoothing the relief of the flanks.   [BRUCE C. HEEZEN]

UNDERLYING STRUCTURE

Because approximately 70% of its surface is covered by the oceans, the typical structure of the Earth is found in the oceanic and not in the land areas. Statistical examination shows that most of the Earth's solid surface is either at the elevation of the ocean floors or at the elevation of the continents. The anomalous areas, those of extreme or of intermediate elevation, are long, narrow features—the mountain ranges, island arcs, deep-sea trenches, and continental margins. With the exception of a few intermediate areas, such as the Red Sea, the crustal structure of the ocean basins is distinctly different from the crustal structure of the continents, and it appears that this has been

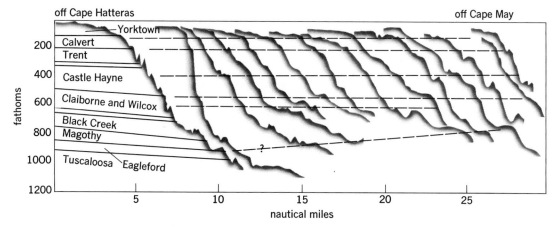

Fig. 6. Correlation of structural benches (outcropping beds) on the continental slope from Cape Hatteras, N.C., to Cape May, N.J. The soundings were taken by the U.S. Coast and Geodetic Survey.

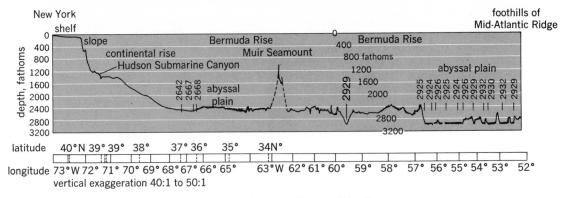

Fig. 7. Precision-depth-recorder profile between Mid-Atlantic Ridge and New York.

the case for most of the Earth's history.

To have a rough model of a section through the Earth, one draws a circle about 5 in. in diameter and a concentric one of about half that diameter. Inside the smaller circle is the core, probably of a nickel-iron composition. The part between the circles is the mantle, a crystalline, basic rock with density about 3.3 g/cm³. The line forming the outer circle, if made with an average pencil point, will include all of the crust of the Earth. The crust has a density of about 2.7 g/cm³. The oceanic crust is about 6 km thick compared to about 36 km for the continents. The boundary between the crust and the mantle is called the M-discontinuity (Mohorovičić discontinuity) by seismologists.

Instruments and techniques. Unlike the continental areas, the ocean floors cannot be studied directly; hence, most of the information about the structure of the Earth beneath the oceans comes from geophysical measurements, from samples dredged or cored from the ocean floor, and, to a smaller extent, from observations made from deep submersibles. Principally employed geophysical techniques are earthquake seismology, explosion seismology, and measurements of the variations in the Earth's gravitational and magnetic fields. Seismology is the study of the propagation of sound waves or elastic waves in the Earth. The source of these waves can be natural, which is the case in earthquake seismology, or man-made. Commonly used man-made sound sources include explosions, high-energy electrical sparkers, and pneumatic devices. The speed of sound waves traveling in the various layers, considered together with gravity and magnetic measurements, makes it possible to estimate certain physical properties such as density and elastic constants. These in

turn suggest the type of rock constituting each layer. The travel time of sound waves reflected or refracted by a particular layer provides a measure of the depth and thickness of the layer. In marine investigations, a relatively new device for measuring the thickness and structure of the sediments is the continuous seismic reflection profiler. A sound source, towed behind the ship, emits periodic pulses of acoustic energy. These pulses travel through the water and are reflected back to the surface from the interfaces between the layers of sediment and rock. The reflected pulses are received by a towed array of hydrophones, amplified, and printed by a scanning recorder. This technique produces a cross-sectional view of the sediments on the ocean floor and usually shows the topography of the basement beneath the sediments as well (Fig. 9).

The sea gravimeter is essentially a highly damped, extremely sensitive spring balance. Changes in the displacement of the springs reflect minute variations in gravitational force. Total magnetic field intensity is measured with a magnetometer towed well astern to minimize the magnetic effects of the ship. The gravitational and magnetic anomalies (variations) when combined with seismological data are indications of compositional changes, or structural features such as folds or faults, and have provided much information about the structure of the Earth.

Paleomagnetism is the study of the remanent magnetization in rocks. The direction of remanent magnetization in sedimentary rocks is parallel to the Earth's magnetic field at the time that the rock was deposited. The remanent magnetic vector in igneous rocks indicates the direction of the Earth's field at the time the rock cooled through the Curie

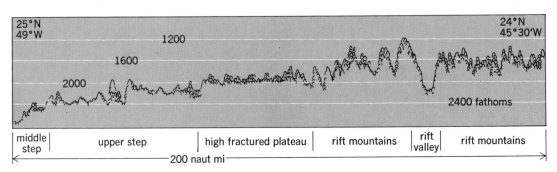

Fig. 8. Tracing of a precision-depth-recorder record showing crest and western flank of Mid-Atlantic Ridge.

temperature. Paleomagnetic studies of rocks of various ages and wide geographic distribution appear to indicate that the magnetic poles have shifted with respect to the geographic poles through geologic time or that the continents have shifted with respect to each other. Data have been accumulated that prove there have been many reversals of the Earth's magnetic field. These reversals have been used to explain linear magnetic anomaly patterns in some parts of the oceans and to support the hypothesis of continental drift by spreading of the sea floor away from the axes of the mid-ocean ridges.

Ocean basins and continental margins. Figure 10 is a structure section across an Atlantic type (rifted) margin based on an interpretation of geologic and geophysical data within the framework of plate tectonics and sea-floor spreading. On the continent and continental shelf, the rocks beneath the sedimentary layers are mainly of the acidic type, such as granite, gneiss, or schist. Beneath the rocks is an intermediate layer believed to be gabbroic or basaltic. The mantle is probably composed of ultramafic rocks, such as peridotite, enstatite, or eclogite. This is the most prominent layer in the Earth, extending from near the surface approximately halfway to the center. There are some variations in the upper parts of the mantle between continental and oceanic areas, but the most apparent difference is in the thickness and composition of the crust. The continental crust is six or seven times thicker than the oceanic crust and contains almost all of the acidic rocks, such as granites, whereas the oceanic crust is almost entirely composed of basic rock.

The average depth of the ocean basins is about 4.8 km. The topography of the ocean floor is rough in the majority of the explored parts, although there are broad areas, particularly in the Atlantic, where the bottom is almost completely flat. These abyssal plains are thought to have been formed by turbidity current deposition where sediments, set in motion during underwater landslides and thrown into suspension in the water, flow to the deepest parts of the basins.

Away from the abyssal plains and continental rises (see section above on physiographic provinces) the sediment layers are mainly composed of clay-sized particles (less than 2μ) and of the skeletons of plankton that lived and died in the waters above. Pelagic sediments of this type generally form an approximately uniform blanket over preexisting topography. In some areas the sea-floor topography is shaped by nonuniform deposition and erosion by bottom currents, particularly by the flow of the cold, dense water generated in the polar regions. The thickness of the unconsolidated sediments varies from tens of meters in areas which receive no turbidities and where the planktonic population is sparse to thousands of meters in other areas. The average is less than 1 km.

Layer 2, immediately below the layer-1 sediments, has a seismic velocity between 4 and 6 km/s. This range of velocities encompasses those appropriate to compacted or metamorphosed sediments or volcanic rocks. Shallow penetrations into layer 2 by deep-sea drilling indicate that its upper part consists of a mixture of pillow basalts, dikes, volcanic debris, and sediments. The lower

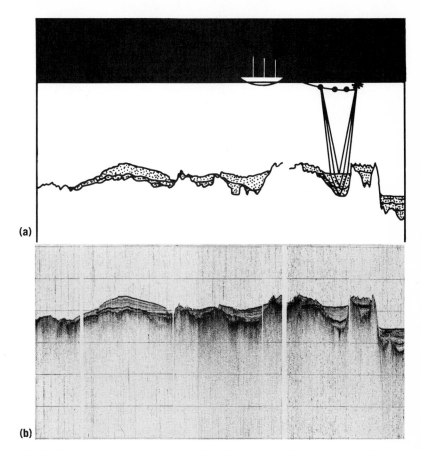

(a)

(b)

Fig. 9. Continuous seismic profiling: (a) technique and (b) typical record. Record shows approximately 200 km of traverse. The average thickness of these ocean-bottom sediments is about 500 m. Vertical exaggeration × 25.

part is probably massive basalt or metabasalt or both.

The principal layer of the oceanic crust, layer 3, shown by check marks in the structure sections (Figs. 10–12), has been found by most of the numerous seismic refraction measurements made in the Atlantic, Pacific, and Indian oceans. The average velocity of sound in this layer is near 6.7 km/s, and the thickness is about 5 km. Variations from these averages are often found in the neighborhood of anomalous areas such as seamounts, trenches, or continental margins. Laboratory sound-velocity measurements of various rock types dredged from sea-floor escarpments suggest that layer 3 is composed of gabbro and metagabbro. Whether or not a similar layer exists under the continents is in dispute. There is much evidence

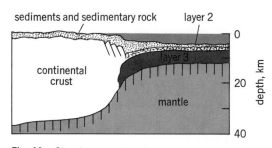

Fig. 10. Structure section across Atlantic-type (rifted) margin.

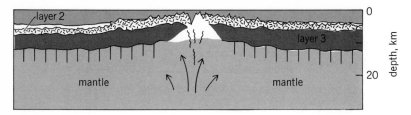

Fig. 11. Structure section across Mid-Atlantic Ridge, indicating crustal accretion and divergence.

that the velocity in the lower part of the continental crust is intermediate between that in the upper part and that in the mantle, and this could indicate the presence of material similar to that in the oceanic crust.

Continental shelves. These are the submerged borders of the continents. The water depth is on the order of a few hundred meters, and the width of the shelf varies from a few kilometers in some places to a few hundred kilometers in others. The thickness of the crust is intermediate between that of the continents and oceans, and its composition is continental. In some places, such as parts of the east coast of North America and South America, the continental shelf and continental slope are broad areas where erosion of the continental masses has resulted in the deposition of many thousands of meters of sediment during the past several million years. In other areas which do not receive much sediment, notably the west coast of the Americas, there is only a narrow continental shelf.

Submarine ridges. There are two types of oceanic ridges, seismic and aseismic. The mid-ocean ridge system is of the former type and is the largest single feature on the Earth. It is more than 80,000 km long and completely encircles the globe. In many places the ridge is offset by large fracture zones, along which seismic activity is generally higher than elsewhere on the ridge system. Over its entire length the ridge is characterized by a narrow zone of shallow-focus earthquake epicenters. Usually associated with this zone are a rift valley and a large positive magnetic anomaly. In addition, parallel linear bands of alternating positive and negative magnetic anomalies are found on the flanks in many areas. The sediment cover is typically thin and in some places in the crest region is entirely absent in a zone 100–500 km wide. On the flanks an appreciably thicker cover is found; in some places it covers the basement rock more or less uniformly, while in others it is collected in pockets separated by exposed basement peaks or ridges.

The tops and flanks of seismic ridges have been dredged and have yielded basaltic volcanic rocks with some inclusions of gabbroic or ultramafic materials. Figure 11 shows a structure section across the Mid-Atlantic Ridge based on seismic refraction and gravity measurements.

In addition to the mid-ocean ridge system, there are other oceanic ridges which are seismically inactive. Typical examples of these are the Walvis Ridge off southwestern Africa and the Hawaiian Ridge, Emperor Seamount Chain, and Line Island Chain in the Pacific. These ridges are thought to have been formed by extrusion of large volumes of volcanic material as the oceanic crust drifted over "hot spots" in the lower part of the mantle. Another class of aseismic ridge is exemplified by the Lomonosov Ridge in the Arctic Ocean and the Jan Mayen Ridge in the Norwegian Sea. These appear to be thin slivers of continental rocks separated from the larger continental masses by rifting associated with sea-floor spreading.

Deep-sea trenches. Deep-sea trenches are important structural features associated with some continental margins, island arcs, earthquake belts, and areas of volcanic activity. Most of them are confined to the margins of the Pacific Ocean, although there are trenches in the Atlantic and Indian oceans. The greatest depths in the oceans are found in these trenches, the deepest in the Pacific being about 10.7 km in the Marianas Trench, and the deepest in the Atlantic about 8.4 km in the Puerto Rico Trench. The trench is formed by the depression of the high-velocity crustal layer and the mantle by several kilometers. In the bottom are layers of sediments which are generally thickest landward of the trench axes. The great depth of the underlying dense layers causes a pronounced deficiency of gravity, a characteristic feature of all deep-sea trenches. Earthquake foci tend to lie on inclined planes dipping landward from the trenches.

Several hypotheses have been advanced to account for the existence of island arcs and the associated trenches and to relate them to a major processes going on in the Earth. The theory receiving much attention currently is that there are convection cells in the mantle which cause upwelling of new crustal material along the mid-ocean ridges and convergence at the trenches. The convergent flow causes the oceanic crust to be thrust underneath the continents or island arcs, creating the negative gravity anomalies, the earthquake foci along the shear zones, and the volcanic and orogenic activity landward of the trench itself. Figure 12 illustrates the concept of trench structure at a convergent margin.

Continental drift and sea-floor spreading. The parallel Atlantic borders of the Americas and Africa first led scientists to speculate that these continents were originally a single unit which was split apart by drifting. Alfred Wegener first popularized the idea in 1910. Throughout this century, the theory has been the subject of much debate, and although the early evidence was confined to the continents themselves, an impressive body of geological, climatological, and paleontological observations was brought to bear on the theory, including the existence of truncated mountain ranges and similar fossils and mineral assemblages

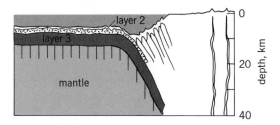

Fig. 12. Structure section across Pacific-type (convergent) margin.

found on opposite sides of the Atlantic. Paleomagnetic data also have been interpreted as indicating that magnetic poles have moved and that the continents have drifted relative to each other. Opposition to the theory has been due largely, but not entirely, to the difficulty in finding a suitable mechanism for drift that was not in conflict with concurrent theories on the composition of the Earth's interior.

Convection cells. A concept of continental drift by spreading of the sea floor has attracted much attention. An important concept in the sea-floor spreading hypothesis is that heat generated in the Earth causes convective overturn of mantle material in a number of convection cells. The mid-ocean ridges are thought to lie along the juncture of the upwelling limbs of separate cells. At the surface, the upwelling material diverges and flows away from the ridge axes, carrying the continental blocks along with it until they come to rest over or near a downflowing convergence. It is suggested that reversals of the magnetic field produce areas of reversely magnetized crustal rock, symmetrical with respect to the ridge axis. The times of reversals are known for the last 5,000,000 years, and on the basis of this hypothesis values of 1–4 cm/year have been deduced for spreading rates. The descending flow is thought to create the deep-sea trenches and earthquake fault planes mentioned previously, and the resulting compression to be responsible for the mountain building and volcanism associated with the trenches. This mechanism has also been offered as an explanation for the fact that no sediments or rocks older than Jurassic have yet been found in the ocean basins on the assumption that all older ones have been swept into the mantle at the trenches. Such a description is greatly simplified, and many complexities are encountered in attempting to fit the entire surface of the Earth into a plausible pattern and in visualizing convection cells with the great horizontal dimensions implied by this theory and the small vertical dimensions implied by mantle stratification. But there are many arguments in favor of it.

Mid-ocean ridge system. Modern investigations in submarine geology and geophysics have supplied many new facts with which to test the theory. Among these was the discovery of the mid-ocean ridge system and the fact that, particularly in the Atlantic Ocean, the ridge lies almost exactly equidistant between the continents on either side. In addition to the median position of the ridge in the Atlantic, several other features about it have been discovered which seem to support the concept of continental drift. The great majority of earthquake epicenters in the oceans are concentrated in the axial zone of the ridge. An axial rift valley characterizes much of its length and appears to be a tensional feature, such as would be produced if the crust were being pulled apart. In many places long fracture zones are found, perpendicular to the ridge axis, which may represent flowlines of the continents' movements. There are distinct concentrations of earthquake activity near the intersections of the fracture zones and the ridge axis. Directions of first motion for these earthquakes appear to be best explained by transform faults associated with crustal spreading away from the

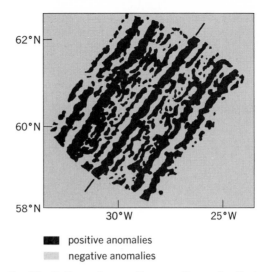

positive anomalies
negative anomalies

Fig. 13. Pattern of magnetic anomalies on the Reykjanes Ridge southwest of Iceland. (*After Heirtzler, Le Pichon, and Baron, 1966*)

ridge axis. Measurements of the thermal gradient in the bottom sediments have shown that the crest of the ridge is, in many places, characterized by abnormally high heat flow, which would be expected if convection cells were bringing new, hot material to the surface at this point.

Rock magnetism. Widely acclaimed evidence cited in support of sea-floor spreading has come from the observed bands of alternating positive and negative magnetic anomalies parallel to and symmetric about the ridge axis. Figure 13 shows an example of this magnetic pattern on the Mid-Atlantic Ridge southwest of Iceland. It has been proposed that the positive anomaly bands (including the central one) correspond to rocks that cooled through the Curie temperature with the Earth's magnetic field in its present polarity. The negative bands correspond to rocks magnetized by a reversed field. According to the sea-floor spreading hypothesis, the banded pattern was formed by upwelling and outward spreading of new material at the ridge crest, each newly generated strip of crust acquiring permanent magnetization in the direction of the ambient field. The spatial arrangement of the positive and negative bands near the ridge crest, normalized for spreading rate, is in agreement with that predicted by the history of field reversals.

Sediment distribution. Since the mid-1960s many hundreds of thousands of kilometers of seismic reflection measurements have shown the general pattern of sediment thickness in the oceans. On and near the crests of the mid-ocean ridges the sediment cover is very thin, becoming progressively thicker down the ridge flanks and into the basins. The total amount of sediment on the ridges varies from area to area because sediment productivity varies, but the increase of thickness with increasing distance from the ridge crest is a general characteristic of the pattern. The deep-sea sediment drilling program has now provided the additional evidence that the age of the sediment lying on the igneous basement (that deposited earliest) increases with increasing distance from the ridge crest. Thus, both the patterns of sediment thickness and sediment age are consistent with a young crust and progressively older flanks, concepts inherent in the sea-floor spreading hypothesis.

[MAURICE EWING; JOHN EWING]

Bibliography: C. A. Burk and C. L. Drake (eds.), *The Geology of Continental Margins*, 1974; A. E. Maxwell (ed.), *The Sea*, vol. 4, pts. I and II: *New Concepts of Sea Floor Evolution*, 1970; W. C. Pitman and J. R. Heirtzler, Magnetic anomalies over the Pacific-Antarctic Ridge, *Science*, 154:1164–1171, 1966; L. R. Sykes, Mechanism of earthquakes and nature of faulting on the mid-oceanic ridges, *J. Geophys. Res.*, 72:2131–2153, 1967; Symposium on Continental Drift, *Phil. Trans. Roy. Soc. London, Ser. A*, 258(1088):41–75, 1965; F. J. Vine, Spreading of the ocean floor; New evidence, *Science*, 154:1405–1415, 1966; T. J. Wilson, A new class of faults and their bearing on continental drift, *Nature*, 267:343–347, 1965.

Marine resources

The oceans cover 71% of the Earth's surface, to an average depth of 3795 m, with a total volume of 1.37×10^9 km^3. Their living and nonliving contents constitute the basis of several extractive industries. Many of the sea's resources, however, cannot be used profitably at the present stage of knowledge, but a moderate and reasonably certain advance in technology would make them valuable. The development of these latent resources is an important frontier of modern science. Both extractive and nonextractive resources of the oceans are discussed here.

Extractive resources. The extractive resources of the oceans include (1) nonrenewable resources, that is, those resources for which the rate of renewal is so slow as to be negligible, such as petroleum and natural gas under the sea floor; and (2) renewable resources, such as the living resources of the sea.

The continental shelves (the land submerged under less than 600 ft of water) under the margins of the seas extend over about 11,800,000 mi^2 and include some 30,000,000 mi^3 of possible oil-bearing sediments. By comparison with the petroleum content of such sediments on land, it is estimated that they contain about 400,000,000,000 barrels of recoverable crude oil, plus large amounts of natural gas.

Extensive geophysical and geological prospecting has located some of these deposits in the Gulf of Mexico, off the coast of California, in the Persian Gulf, and elsewhere. Successful drilling to recover them has been accomplished in depths of water up to 700 ft and at distances up to 100 mi from shore. Rapidly developing new techniques are extending the water depths and distances from shore in which drilling can be economically conducted. *See* OIL AND GAS, OFFSHORE.

Nonextractive resources. Other aspects of marine resources include the disposal of waste products and the use of the sea as a source of energy.

Waste disposal. Disposal of domestic sewage and industrial wastes is conveniently accomplished near coastal population centers by running them into the adjacent sea. There, the large volume and rapid mixing of the waters dilute the wastes, and the bacteria in the sea break down the organic constituents. It is necessary, however, to consider in some detail the local effects of tidal and wind currents, density stratification, rates and volumes of mixing, the character of the bottom sediments, and the rate of disappearance of human bacteria as a basis for planning such disposal without incurring a pollution hazard. *See* WATER POLLUTION.

In the coming era of large nuclear fission power plants, it may be necessary, particularly for countries with densely populated land areas and long sea coasts, to dispose of some fission waste products in the sea. Safe ocean disposal of radioactive wastes involves the selection of sites where rapid and profound dilution will occur, or where sufficient decay will take place before the radioactive waters and their contained organisms come into contact with human beings. Some low-level radioactive wastes are already being safely disposed of in coastal waters, but it is certainly not safe to introduce large quantities of high-level wastes there. Deep-ocean disposal may be possible. However, much more must be known about the deep-ocean circulation, and the transfer of elements between the deep sea and the surface waters (where man uses the sea) by physical and biological processes, before it can be stated with certainty where and under what circumstances specified quantities of radioactive waste products can safely be introduced.

Energy. Of the several forms of energy in the sea that are capable of being used to produce power, the most apparent is the ebb and flow of the tide. Several attempts have been made to harness it, but the only major tidal power plant in operation is near the mouth of the Rance River in France. *See* TIDAL POWER.   [MILNER B. SCHAEFER]

Bibliography: J. F. Brahtz (ed.), *Ocean Engineering*, 1968; H. W. Menard, *Marine Geology of the Pacific*, 1964; J. L. Mero, *Mineral Resources of the Sea*, 1964; J. E. G. Raymont, *Plankton and Productivity in the Oceans*, 1963; R. Revelle et al., *The Effects of Atomic Radiation on Oceanography and Fisheries*, Nat. Acad. Sci.–Nat. Res. Counc. Publ. no. 551, 1957; M. B. Schaefer and R. R. Revelle, *Natural Resources*, 1959; D. K. Tressler and J. M. Lemon, *Marine Products of Commerce*, 2d ed., 1951.

Mass defect

The difference between the mass of an atom and the sum of the masses of its individual components in the free (unbound) state. The mass of an atom is always less than the total mass of its constituent particles; this means, according to Albert Einstein's well-known formula, that an energy of $E = mc^2$ has been released in the process of combination, where m is the difference between the total mass of the constituent particles and the mass of the atom, and c is the velocity of light.

The mass defect, when expressed in energy units, is called the binding energy, a term which is perhaps more commonly used.

[W. W. WATSON]

Mercury battery

A primary dry-cell battery consisting of a zinc anode, a cathode of mercuric oxide (HgO) mixed with graphite, and an electrolyte of potassium hydroxide (KOH) saturated with zinc oxide (ZnO). With carefully purified materials and balanced amounts of ZnO and HgO, the cell has very low self-discharge and makes efficient use of the active materials.

In some cells which require long-term continuous drains, for example, in hearing aid use, MnO_2 is added to the HgO. In these cells, the open circuit voltage is slightly above 1.4 volts, compared with 1.35 volts for cells using 100% HgO as depolarizer.

Within the steel can, the active materials are separated by a porous material which prevents migration of conducting particles from the mercuric oxide pellet. Dense dialysis paper and porous polyvinyl chloride have been used for this purpose. The electrolyte is completely absorbed in the active materials, separator, and absorbent materials. The steel can serves as the contact to the HgO. A metal top with a concentric neoprene grommet closes off the top of the can and serves as the contact to the zinc.

The electrochemical system may be written as below. This does not involve the electrolyte. The

Anode: $Zn + 2OH^- \rightarrow ZnO + H_2O + 2e^-$
Cathode: $HgO + H_2O + 2e^- \rightarrow Hg + 2OH^-$
Overall: $Zn + HgO \rightarrow ZnO + Hg$

cell potential, therefore, does not change appreciably with different concentrations of alkali.

The cutaway view shown in Fig. 1 is of the flat pellet structure. Two other types are manufactured: the cylindrical structure, which also uses pressed amalgamated zinc powder but is made in sizes corresponding to the N, AA, and D Leclanche cells; and the wound anode flat structure, with large diameter and high surface area, giving superior performance at low temperatures.

When current flows, the ZnO formed in the cell reaction quickly saturates the small amount of electrolyte and then precipitates out. This maintains a constant composition of the electrolyte. Offsetting this is the transport of water away from the anode by the solvated potassium ions. The equilibrium under steady current flow results from

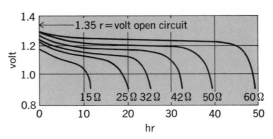

Fig. 2. Voltage-discharge characteristics of mercury cells under continuous load conditions at 70°F. At 1.25 volts, equivalent current drains for resistances are: 15 ohms, 83 ma; 25 ohms, 50 ma; 32 ohms, 40 ma; 42 ohms, 30 ma; 50 ohms, 25 ma; 60 ohms, 20 ma.

complex exchanges through the separator, making cell-voltage characteristics under load dependent on initial electrolyte composition.

The mercury cell has a theoretical output of 0.247 amp-hr/g of HgO. In practice, the cathode pellet contains about 95% HgO and 5% graphite, having a theoretical output of 0.234 amp-hr/g. The anode is 90% zinc and 10% mercury. This has a theoretical output of 0.738 amp-hr/g.

As built, the cells have slight excess of cathodic capacity. A discharged cell will then have no zinc left to react with the electrolyte and evolve hydrogen. Thus a cell with 12.5 g of cathodic material has 3.6 g of zinc amalgam, compared with 3.98 needed for exact balance.

The electrolyte used is about 1 ml/amp-hr. One composition is 100 g KOH, 100 ml H_2O, 16 g ZnO. The actual cell capacity is only slightly less than the theoretical. The overall cell output is approximately 0.36 amp-hr/lb and 5 amp-hr/in.[3]

The ampere-hour capacity of mercury cells is relatively unchanged with variation of discharge schedule and to some extent with variation of discharge current. The cells have a relatively flat discharge characteristic, as shown in Fig. 2. The outstanding features of the mercury cell include flat discharge curve, small variation in capacity with intermittent or continuous discharge, shelf life of several years, and good high-temperature characteristics. *See* DRY CELL; PRIMARY BATTERY; RESERVE BATTERY.   [JACK DAVIS]

Bibliography: R. R. Clune, Recent developments in the mercury cell, *Proc. Power Sources Conf.*, 14:117, 1960; R. R. Clune and D. Naylor, Alkaline zinc-mercuric oxide cells and batteries, *Proc. Int. Symp. Batteries*, 1958; M. Friedman and C. E. McCauley, The Ruben cell: A new alkaline primary dry cell battery, *Trans. Electrochem. Soc.*, 92:195, 1947; D. Naylor, Wound anodes for mercury cells, *Proc. Int. Symp. Batteries*, 1960; S. Ruben, Balanced alkaline dry cells, *Trans. Electrochem. Soc.*, 92:183, 1947; G. W. Vinal, *Primary Batteries*, 1950.

Metal-base fuel

A fuel containing a metal of high heat of combustion as a principal constituent. High propellant performance in either a rocket or an air-breathing engine is obtained when the heat of combustion of the fuel is high. Chemically, high heats of combustion are attained by the oxidation of the low-atomic-weight metals in the upper left-hand corner

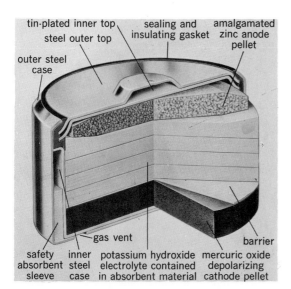

Fig. 1. Cutaway view of mercury cell.

tin-plated inner top
steel outer top
sealing and insulating gasket
amalgamated zinc anode pellet
outer steel case
gas vent
barrier
safety absorbent sleeve
inner steel case
potassium hydroxide electrolyte contained in absorbent material
mercuric oxide depolarizing cathode pellet

Heats of combustion of the lightweight metals compared with those of some other fuels

Fuel	Specific gravity	Product	Heat of combustion		
			Btu/ lb pdts	Btu/ lb fuel	Btu/ in.³ fuel
Lithium	0.53	Li_2O	8,570	18,500	360
Beryllium	1.85	BeO	10,450	29,100	1,950
Boron	2.35	B_2O_3	7,820	25,200	2,140
Carbon	2.00	CO_2	3,840	14,100	1,020
Magnesium	1.74	MgO	6,430	10,700	670
Aluminum	2.70	Al_2O_3	7,060	13,400	1,310
Gasoline	0.75	CO_2, H_2O	4,600	20,400	550
Hydrazine	1.00	N_2, H_2O	4,180	9,360	340
Hydrogen	0.07	H_2O	6,830	61,500	155

of the periodic table. The generally preferred candidates are lithium, beryllium, boron, carbon, magnesium, and aluminum.

Performance parameters. For the rocket where the total propellant containing both fuel and oxidizer is carried on board, the performance parameter of interest is the heat of combustion per unit mass of combustion product. In air-breathing engines, however, where only the fuel is carried on board and the oxidizer in the form of air is attained free, the figure of merit is the heat of combustion per unit mass of fuel. There is still a third figure of merit for the air-breathing engine in which most of the flight involves a cruise operation in which the thrust just balances the drag. In this case, it is desirable to minimize the cross-sectional area (consequently volume) and maximize the heat of combustion so that a given total energy output can be realized in a minimum-volume vehicle to present low drag. The figure of merit, in this case, is the heat of combustion per unit volume of fuel. *See* PROPULSION.

The table shows a listing of the heats of combustion of various candidate metals along with three nonmetallized high-performance fuels for comparison. The three different heats of combustion described above are figures of performance merit, depending upon the vehicle mission and type of propulsion system employed. In general terms, for an accelerating rocket, the important figure of merit is Btu/lb pdts (pound of products); for the accelerating air breather, it is Btu/lb fuel; and for the cruising air breather it is Btu/in.³ fuel. *See* PROPELLANT.

The metal with the highest heat release for all applications is beryllium, whereas boron shows

essentially equivalent performance for air-breathing applications. For rocket applications where Btu/lb pdts is important, lithium, boron, and aluminum are close to beryllium. As discussed later, however, because of inefficiencies in the combustion heat release as well as thermodynamic limitations, not all of these candidates are equivalent.

Method of use. The metallized additive can be used in either a liquid or solid propellant. When the pure metal is added to liquid fuels, an emulsifying or gelling agent is employed which maintains the particles in uniform suspension. Such gelling agents are added in small concentrations and so do not negate the beneficial effect of the fuel. Particle sizes in such suspensions are generally in the range $0.1-50 \mu$, and stable suspensions have been made which do not separate for long periods of time. When a metallic compound is employed, the compounding group is chosen so that the additive is soluble in the fuel. Thus, if gasoline is the fuel base, the suitable soluble boron compound is ethyldecaborane and, for aluminum, triethylaluminum may be employed. In both cases, the metallic additive contains a hydrocarbon constituent which provides solubility in the hydrocarbon base fuel.

When used in composite solid rocket propellants, the metal powder is usually mixed with the oxidizer and unpolymerized fuel, and the propellant is then processed in the usual way. If a compound of the metal additive is employed, it is convenient to dissolve it in the fuel, and then to process the fuel containing the metallic compound and oxidizer according to normal procedures. Homogeneous solid propellants also employ metallizing constituents, both as the pure metal and as a compound.

Compounds of interest. Early compounding of metallized propellants employed the free metal itself. As a result of extensive research in metalloorganic compounds, however, several classes of metallic compounds have been employed. Two major reasons for the use of such compounds are that the solubility of the metal in the fuel can be realized, resulting in a homogeneous propellant, and performance higher than that for the pure metal can be obtained in some cases.

The major classes of metallic compounds of interest as high-performance propellants include the hydrides, amides, and hydrocarbons. Additional classes include mixtures either of two metals or of two chemical groups, such as an amine hydrocarbon. Some examples of the hydrides include LiH, BeH_2, B_2H_6, B_5H_9, $B_{10}H_{14}$, MgH_2, and AlH_3. These compounds form the more exotic class of metallics because they are capable of yielding the highest performances, exceeding those of the pure metals. The presence of hydrogen provides the low-molecular-weight gases needed for high thermodynamic performance. Examples of the propellant amides include $LiNH_2$ and $B(NH_2)_2BNHCH_3$. The largest class is the hydrocarbons, which includes LiC_2H_5, $Be(CH_3)_2$, $B(CH_3)_3$, and $Al(C_2H_5)_3$. The major advantage of this class is the potential solubility in conventional hydrocarbon fuels.

Operational limitations. Most metallic fuels are costly and many, because of particle-size requirements or synthesis in a specific compound, are in limited supply. The combustion gases all produce smoky exhausts which may be objectionable in

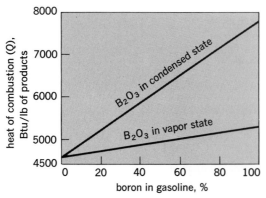

Total heat of combustion of gasoline-boron mixtures.

use. Engine development problems are also increased because of the appearance of smoke and deposits in the engine. These engine deposits can sometimes occur on the injector face, nozzle, and turbine blades of turbojet engines. Careful adjustment of operating conditions is required in such cases to minimize these difficulties with resultant added development costs.

Combustion losses. In rocket engines, where the flame temperatures are high, vaporization and dissociation are particularly pronounced. In the tabular listing of total heats of combustion, the heat value is given as measured in a laboratory calorimeter, the end product being the condensed oxide. At high flame temperatures, however, the oxides vaporize, producing a lower effective heat of combustion. In the case of boron oxide, B_2O_3, for example, the normal boiling point is about 2500°K (4000°F), whereas the flame temperature in most high-performing rocket engines exceeds 3000°K. These effects are illustrated in the graph, which compares the total heat of combustion Q of gasoline-boron mixtures when the B_2O_3 is solid or liquid to the heat when it is a vapor. At still higher flame temperatures, in addition to vaporization, the oxides undergo endothermic dissociation such as shown in Eqs. (1), (2), and (3).

$$B_2O_3(g) = B_2O_2(g) + 1/2\,O_2 - 108\text{ kcal} \qquad (1)$$

$$B_2O_2(g) = 2BO(g) - 126\text{ kcal} \qquad (2)$$

$$Al_2O_3(l) = Al_2O(g) + O_2 - 330\text{ kcal} \qquad (3)$$

The net result of these reactions is a further reduction in flame temperature to an extent that sensitively depends on the temperature itself. At temperatures below about 2500°K, as in air-breathing engines operating below Mach 3, vaporization and dissociation for the oxides of beryllium, aluminum, boron, and magnesium occur hardly at all, and these limitations do not apply.

In actual engines, a certain fraction of the useful energy which is lost by vaporization and dissociation is recovered by condensation and recombination in nozzles. If thermodynamic equilibrium prevails and if the nozzle area ratio is infinite, all of this energy loss is recovered. In actual cases, however, because of the finite area ratio and finite reaction and condensation rates, only a part of the energy is recovered for propulsion.

Two-phase flow losses. A second type of propulsive efficiency loss occurs as a result of the appearance of a condensed phase. Because the acceleration of the gases in the nozzle results from pressure-volume work, a reduction in gas content reduces the thermodynamic efficiency. The combustion gas behaves as if the molecular weight were increased. A second consideration is involved with the momentum exchange between condensed phase and gas. If the particles are large and lag appreciably behind the gas in velocity, impulse is lost. This effect can be appreciable when the percentage of condensed phase is high.

The net result of the thermodynamic limitations is to partially reduce in some cases the improvement in performance expected on the basis of heat of combustion alone. As additives, metals are most effective when mixed with the lower-performing propellants because the lower temperatures minimize the vaporization and dissociation losses. For

this reason, the greatest use of metal additives has been in solid propellants, which develop 10–20% lower specific impulses than do liquid propellants. The result of the use of such additives has been to close the gap in performance between the solid and liquid propellants. The second attractive application is in air-breathing engines, where flame temperatures are relatively low.   [DAVID ALTMAN]

Bibliography: C. duP. Donaldson, J. V. Charyk, and M. Summerfield (eds.), *High Speed Aerodynamics and Jet Propulsion*, vol. 2, 1956; M. Gilbert, L. Davis, and D. Altman, Velocity lag of particles in linearly accelerated combustion gases, *Jet Propul.*, 25(1):26–30, 1955; F. E. Marble, The role of approximate analytical results in the study of two-phase flow in nozzles, *Proceedings of the A. F. Rocket Propulsion Laboratory Two-Phase Flow Conference*, 1967; W. R. Maxwell, W. Dickinson, and C. F. Caldin, Adiabatic expansion of a gas stream containing solid particles, *Aircraft Eng.*, 18:350–351, 1946.

Methane

A member of the alkane or paraffin series of hydrocarbons with the formula shown below.

$$
\begin{array}{c}
\text{H} \\
| \\
\text{H}-\text{C}-\text{H} \\
| \\
\text{H}
\end{array}
$$

Methane is called marsh gas because it forms by anaerobic bacterial decomposition of vegetable matter in swampy land. Coal miners know it as firedamp because mixtures with air are combustible. It is a major constituent of natural gas (50–90%) and of coal gas. It forms in large amounts in the activated-sludge process of sewage disposal. As a liquid it freezes at −182.6°C and boils at −161.6°C.

In addition to its use as a fuel, methane is important as a source of organic chemicals and of hydrogen. Its reaction with steam at high temperatures in the presence of catalysts yields carbon monoxide and hydrogen (synthesis gas), which can be catalytically converted to liquid alkanes (Fischer-Tropsch process) or to methanol and other alcohols. The catalytic reaction of the synthesis gas with olefins yields higher alcohols (the Oxo process), and reaction with steam produces additional hydrogen and carbon dioxide.

The incomplete combustion of methane with air produces finely divided carbon, called carbon black, hundreds of millions of pounds of which are used annually as a reinforcing and filling agent in compounding rubber and as a pigment in printing ink.

Chlorination of methane yields methyl chloride, methylene chloride, chloroform, and carbon tetrachloride. See FISCHER-TROPSCH PROCESS.

[LOUIS SCHMERLING]

Mineral resources conservation

The effort to ensure to society the maximum present and future benefit from the use of mineral resources. It stresses maximum use of a commodity for the benefit of the largest numbers of people. Conservation is influenced by many economic and political factors. As costs increase, people tend to use less; they retain materials for more essential uses, thus practicing conservation. Governments influence conservation by imposing taxes, by regulating imports and exports and by controlling production and prices, particularly of gas and oil. Many governments control both the price and the sale of all raw materials. A strong sense of awareness of the needs and the values of minerals was created throughout the world by the embargo on petroleum products enforced by the Organization of Petroleum Exporting Countries (OPEC) in 1973–1974. Many underdeveloped countries from which the industrial nations obtain their needed supplies maintain close controls over their mines and oilfields, primarily to regulate production and obtain greater revenues. The worldwide result has been promotion of conservation and curtailment of waste.

Reserves and resources. Reserves represent that fraction of a commodity that can be recovered economically; resources include all of the commodity that exists in the Earth. For example, in 1972 the National Petroleum Council estimated the total oil-shale resources in the Piceance Basin, CO, to be 1.2×10^{12} bbl (192×10^9 m³), but the reserves (the recoverable fraction) made up only 50×10^9 bbl (8×10^9 m³). Traces of copper and iron exist in most rocks, but their recovery is inconceivable. They are resources, but not reserves.

Reserves are classed as proved, probable, possible, and potential. When a company publishes reserve figures, it is usually referring to proved and probable reserves. Proved reserves are those known without doubt; probable reserves are those that are nearly certain but about which a slight doubt exists; possible reserves are those with an even greater degree of uncertainty but about which some favorable information is known; potential reserves are based upon geological reasoning; they represent an educated guess.

At some time in the future, certain resources may become reserves. A change of this type can be brought about by improvements in recovery techniques, either mining or metallurgical, or by improved secondary recovery methods whereby oil and gas are forced to a well and can be pumped to the surface. New methods of extraction and recovery may result in lowering costs enough to make a deposit economic. New uses may also be found for a commodity, and the increased demand may result in an increase in price; or a large deposit may become exhausted, thus forcing production from a lower-grade and higher-cost ore body.

A knowledge of reserves and resources is essential to an understanding of the problems of conservation.

Classification of raw materials. For convenience, raw materials are classified as fuels, metals (including ferrous metals, nonferrous metals, light metals, and precious metals), and industrial minerals or nonmetallics. Some metals and minerals, and their major uses, are listed in Table 1.

Many minerals are lost during use and cannot be recovered and recycled. Petroleum in gasoline and natural gas and oil in fuels are dissipated, as are lead and other additives in gasoline. It is not economic to recover the tiny amounts of silver on some types of photographic film. Other metals, for example, iron, steel, copper, brass, and aluminum,

Table 1. Classification of selected materials and some of their major uses

Materials	Uses
Fuels	
Bituminous and anthracite coal	Direct fuel, electricity, gas, chemicals
Lignite	Electricity, gas, chemicals
Petroleum	Gasoline, heating, chemicals, plastics
Natural gas	Fuel, chemicals
Uranium	Nuclear power, explosives
Metals	
Ferrous metals	
Chromium	Alloys, stainless steels, refractories, chemicals
Cobalt	Alloys, permanent magnets, carbides
Columbium (niobium)	Alloys, stainless steels
Iron	Steels, cast iron
Manganese	Scavenger in steelmaking, batteries
Molybdenum	Alloys
Nickel	Alloys, stainless steels, coinage
Tungsten	Alloys
Vanadium	Alloys
Nonferrous metals	
Copper	Electrical conductors, coinage
Lead	Batteries, gasoline, construction
Tin	Tinplate, solder
Zinc	Galvanizing, die casting, chemicals
Light metals	
Aluminum	Transportation, rockets, building materials
Beryllium	Copper alloys, atomic energy field
Magnesium	Building materials, refractories
Titanium	Pigments, construction, acid-resistant plumbing
Zirconium	Alloys, chemicals, refractories
Precious metals	
Gold	Monetary, jewelry, dental, electronics
Platinum metals	Chemistry, catalysts, automotive
Silver	Photography, electronics, jewelry
Industrial minerals	
Asbestos	Insulation, textiles
Boron	Glass, ceramics, propellants
Clays	Ceramics, filters, absorbents
Corundum	Abrasives
Feldspar	Ceramics, fluxes
Fluorspar	Fluxes, refrigerants, acid
Phosphates	Fertilizers, chemicals
Potassium salts	Fertilizers, chemicals
Salt	Chemicals, foods, glass, metallurgy
Sulfur	Fertilizers, acid, metallurgy, paper, foods, textiles

are recoverable and may be reused many times. The gold of the ancient Egyptian and Incan treasuries probably still reposes in the vaults of some banks; certainly, it was never destroyed.

More metals are now being recovered and reused than ever before. As the prices of primary materials increase, scrap becomes increasingly valuable, and conservation is encouraged. Table 2 lists the primary metals and scrap used in the United States in 1960 and 1970.

Supplies of minerals. Most of the Earth's surface has been closely examined in the search for useful minerals. Future supplies will depend in large part upon techniques that permit exploration below the surface and in remote and rather inaccessible areas. Exploration is thus becoming increasingly costly and unrewarding. Petroleum is being found in the waters of the continental shelves, at depths and under conditions never before tested, and in remote parts of the Arctic. New techniques are being developed and used. Likewise, greater amounts of hard minerals are being obtained from undeveloped and politically unstable lands. The locations of mineral deposits are fixed by nature; their positions cannot be changed. Competition to obtain the materials needed to sustain industry is growing rapidly; demands are overtaking supplies, and shortages of several commodities can develop within a few years. In order to maintain their economies, the industrial nations must encourage development and conservation of all their resources; they must find substitutes for some resources and must increase recycling. As prices of commodities increase and shortages develop, standards of living are certain to fall everywhere.

The United States imports at least part of 88 of the 100 most commonly used mineral products. For example, 100% of such essential materials as the ores of chromium (used in the manufacture of stainless steels), manganese (without which sound steel cannot be made), tin, columbium-tantalum, industrial diamonds, rutile (for the manufacture of titanium metal), platinum metals, block mica, and amorphous graphite is imported. The United States imports about 90% of the ores of aluminum, cobalt, nickel, antimony, bismuth, and beryl, and more than one-third of the petroleum and iron ores that are used. Only with a few commodities—coal, sulfur, phosphates, molybdenum, and boron—is a surplus available for export. Imports of minerals in recent years have been growing at the rate of about 1% a year.

After World War II the United States established national stockpiles of minerals in an effort to assure the country of adequate supplies during times of emergency. Over the years the needs of the country have changed, and, likewise, materials in the stockpiles have changed. Efforts to use stockpiles for purposes other than the original purpose have been made, and advocates of price regulation still attempt to buy for the stockpile when prices are low and to sell when they are high. Handling of the national stockpiles clearly influences conservation policies.

Government conservation policies. Many policies of the United States government directly influence conservation and availability of energy and other minerals. The effects of government activities are most clearly evident insofar as oil and gas are concerned. Many state governments have laws that regulate the spacing of oil wells and prevent overpumping. They establish the most effective rate of recovery and ensure maximum life of a field. On the other hand, the Federal government has established price controls that discourage oil exploration in areas where costs are high. The government maintains price controls on natural gas at so low a level that drilling for gas is uneconomic. Companies are reluctant to pipe gas across state borders, for it then becomes subject to Federal price controls. Thus gas can be obtained in the states where it is produced but is becoming increasingly difficult to buy elsewhere.

Periodically the government offers blocks of public land for auction and lease. These blocks, thought to contain oil and gas, are bid on by companies or, more commonly, by groups of companies. Legislation passed by the U.S. Congress prohibits large oil companies from making joint bids except with small companies. Bids commonly total $1,000,000,000 or more, and the government also receives royalty payments from the oil and gas produced.

The depletion allowance is a tax-exemption policy based upon the rate of depletion of a natural resource and is intended to equate tax policy in the extractive industries with the depreciation allowance permitted manufacturing industries. Producers are allowed to deduct a certain percentage of their gross revenues from the part of their income subject to Federal income tax. Early in 1975 Congress canceled the depletion allowance for the large oil companies, although permitted retention of the allowance for smaller companies and the mining industry. Removal of the depletion allowance has reduced capital available for exploration and has resulted in an upward pressure on the price of petroleum products.

The political history and legal climate in a country strongly influence private investment in mines and oil fields. In the past few years many nations have expropriated raw materials properties, in some places without compensation, and in others with compensation less than the former owners thought reasonable. Such actions greatly discourage private investment, and, as a result, many governments now must finance and operate their own mines and oil fields.

Government policies also influence petroleum and mineral exploration; there is a myriad of regulations and laws concerning environment, right of access, land restoration, and other items. Open-pit or strip mining coal is cheaper, has a much higher recovery, and is safer than underground mining, which results in more than 50% of the coal being left underground. In spite of these facts, government continues to discourage strip mining. The many government bureaus that issue regulations concerning exploration and development are also restrictive. Especially in the energy fields, unslightly strip mines, oil spills offshore, and worries about radioactive leakage have resulted in rigorously enforced regulations. The United States has

Table 2. Consumption of primary metals and scrap, in short tons (except for silver in ounces and tin in long tons), in the United States*

	Primary metal		Scrap	
	1960	1970	1960	1970
Pig iron	66,626,336	90,126,000	66,468,708	85,559,000
Copper	1,349,896	2,070,000	1,208,434	869,400
Lead	1,102,172	1,350,000	469,903	597,390
Zinc	1,158,938	1,198,000	265,820	339,527
Silver	151,007,000	250,000,000	49,007,000	70,000,000
Tin	51,530	58,027	29,030	20,802

SOURCE: U.S. Bureau of Mines.
*1 short ton = 0.9 metric ton, 1 troy ounce = 3.11 grams, 1 long ton = 1016 kilograms.

also withdrawn from exploration extensive wilderness areas and other tracts of public lands. The country badly needs an energy and minerals policy that recognizes the requirements for raw materials and establishes guidelines to assure their availability and conservation.

Technological changes. Developments in oil well drilling, in secondary recovery methods, and in mining, milling, and smelting methods all have profound impacts on the availability of raw materials. Oil well drilling is being carried out in the open oceans at depths of 900 ft (270 m) or more, and techniques are available that will enable wells to be drilled to considerably greater depths. New shutoff devices and other safety devices greatly reduce the dangers of oil spills and wellhead fires. *See* OIL AND GAS WELL DRILLING.

The development of large machines for handling huge tonnages of rock and the use of refined explosives have improved mining methods. Open-pit and strip mines are larger and more efficient than ever before, and lower grades of ore are being recovered. It is common for a mine to produce 50,000 tons (45,000 metric tons) of ore a day, and some yield more than 100,000 tons (90,000 metric tons). Industry is taking advantage of economies of size. Methods of transportation, concentration, and smelting are also undergoing changes that permit less costly handling of large tonnages and improve recoveries. Many older smelting methods are being replaced by methods that are cleaner, that have as good or better recovery records, and that may prove to be less expensive to operate. *See* COAL MINING.

Demand and supply problems. A serious problem facing the raw materials industry is the cyclical nature of demands for its products. When economic times are good, the markets are strong; but consumption is greatly curtailed during recessions, and excessive inventories are created. As a result, prices are cyclical and, with some commodities, for example, mercury, are highly volatile. In order to stabilize prices, many countries favor the establishment of cartels and associations. Probably the most effective association of this type is the International Tin Commission, an organization that includes both consumers and producers. A ceiling and a floor price are established: If the market price goes below the floor, the Commission buys tin; if the price goes above the ceiling, the Commission sells. In this way, the price and production of tin are maintained at reasonably stable rates.

Governments that have copper, aluminum, and iron for export have established cartels in these commodities and are making efforts to consolidate controls over rates and prices. Most such cartels aim to increase the price of the commodity and to improve the financial status of the producer at the expense of the consumer. As industrial needs increase and minerals become difficult to purchase, cartels will probably be more successful, but whether or not such associations succeed, the prices of raw materials will increase with demand. As a national policy conservation is not only desirable, it is a necessity. *See* CONSERVATION OF RESOURCES.   [CHARLES F. PARK, JR.]

Bibliography: P. A. Bailly, *Conversion of Resources to Reserves*, 1975; D. A. Brobst and W. P.

Pratt, *United States Mineral Resources*, U.S. Geol. Surv. Prof. Pap. no. 820, 1973; C. F. Park, Jr., *Earthbound: Minerals, Energy, and Man's Future*, 1975; A. Sutulov, *Minerals in World Affairs*, 1972; U.S. Bureau of Mines, *U.S. Minerals Yearbook*, annual.

Mining

The taking of minerals from the earth, including production from surface waters and from wells. Usually the oil and gas industries are regarded as separate from the mining industry. The term mining industry commonly includes such functions as exploration, mineral separation, hydrometallurgy, electrolytic reduction, and smelting and refining, even though these are not actually mining operations.

The use of mineral materials, and thus mining, dates back to the earliest stages of man's history, as shown by artifacts of stone and pottery and gold ornaments. The products of mining are not only basic to communal living as construction, mechanical, and raw materials, but salt is necessary to life itself, and the fertilizer minerals are required to feed a populous world.

Methods. Mining is broadly divided into three basic methods: opencast, underground, and fluid mining. Opencast, or surface mining, is done either from pits or gouged-out slopes or by strip mining, which involves extraction from a series of successive parallel trenches. Dredging is a type of strip mining, with digging done from barges. Hydraulic mining uses jets of water to excavate material.

Underground mining involves extraction from beneath the surface, from depths as great as 10,000 ft, by any of several methods.

Fluid mining is extraction from natural brines, lakes, oceans, or underground waters; from solutions made by dissolving underground materials and pumping to the surface; from underground oil or gas pools; by melting underground material with hot water and pumping to the surface; or by driving material from well to well by gas drive, water drive, or combustion. Most fluid mining is done by wells. One type of well mining is to wash insoluble material loose by underground jets and pump the slurry to the surface. *See* MINING, OPEN-PIT; MINING, STRIP; MINING, UNDERGROUND; PETROLEUM ENGINEERING.

The activities of the mining industry begin with exploration which, since mankind can no longer depend on accidental discoveries or surficially exposed deposits, has become a complicated, expensive, and highly technical task. After suitable deposits have been found and their worth proved, development, or preparation for mining, is necessary. For opencast mining this involves stripping off overburden; and for underground mining the sinking of shafts, driving of adits and various other underground openings, and providing for drainage and ventilation. For mining by wells drilling must be done. For all these cases equipment must be provided for such purposes as blasthole drilling, blasting, loading, transporting, hoisting, power transmission, pumping, ventilation, storage, or casing and connecting wells. Mines may ship their crude products directly to reduction plants, refiners, or consumers, but commonly concentrating mills are provided to separate useful from useless (gangue) minerals.

Economics. Mining is done by hand in places where labor is cheap, but in the more industrialized countries it is a highly mechanized operation. Some surface mines use the largest and most expensive machines ever developed—unless a large ship can be called a machine—including giant bucket wheels that can mine 12,000 tons/hr, shovels and draglines with 200-yd³ buckets, and 250-ton capacity trucks. The $25,000 capital outlay per worker in mining is five times the average for domestic industries.

There are many small- and medium-sized mines but also a growing number of large ones. The trend to larger mines and particularly to large opencast operations is due to the great demand for mineral products, depletion of high-grade reserves, technological progress, and the need for economies of scale in mining low-grade deposits. The largest open-pit coal mine moves 300,000 tons of material per day and the largest metal mine, 300,000 tons. The largest underground coal mine produces 17,000 tons of coal per day, and the largest metal mine 50,000 tons of ore.

The quality of deposits which can be mined economically depends on the market value of the contained valuable minerals, on the costs of mining and treatment, and on location. Alluvial gold gravels may run as low as 1/350 oz of gold/yd³ and gold from lodes as low as 1/10 oz/ton. Uranium ore containing 1.5 lb of uranium oxide per ton is mined underground, and copper-molybdenum ore as low grade as 0.35% copper and 0.05% molybdenum is taken from open pits. On the other hand, iron ore is rarely mined below 25% iron, aluminum ore below 30% aluminum, and coal less than 90% pure. Crude petroleum is usually over 95% pure.

Use of natural resources. A unique feature of mining is the circumstance that mineral deposits undergoing extraction are "wasting assets," meaning that they are not renewable as are other natural resources. This depletability of mineral deposits not only requires that mining companies must periodically find new deposits and constantly improve their technology in order to stay in business, but calls for conservational, industrial, and political policies to serve the public interest. Depletion means that the supplies of any particular mineral, except those derived from oceanic brine, must be drawn from ever-lower-grade sources. Consciousness of depletion causes many countries to be possessive about their mineral resources and jealous of their exploitation by foreigners. Depletion also accounts for some controversial attitudes toward conservation. Some observers would reduce the scale of domestic production and increase imports in order to extend the lives of domestic deposits. Their opponents argue that encouragement of mining through tariff protection, subsidies, or import quotas is desirable, on the grounds that only a dynamic industry can develop the means of mining low-grade deposits or meet the needs of a national emergency. They point out that protection encourages the extraction of marginal resources that would otherwise be condemned through abandonment. *See the feature articles* EXPLORING ENERGY CHOICES; U.S. POLICIES AND POLITICS.

Despite its essential nature, mining today is being constrained as to where it can operate by wilderness lovers and by rapidly expanding urbanization. Concern over pollution, some of which is caused by mine water, mining wastes, and smelter effluents, is growing rapidly. More and more objection is developing to defacement of landscapes by surface mining, and many states now require restoration of the surface to a cultivable or forested condition after strip mining. People in residential areas usually resist the development of industries near their homes, particularly if they employ blasting, produce smoke or fumes, or cause traffic by large vehicles. *See the feature article* PROTECTING THE ENVIRONMENT.   [EVAN JUST]

Bibliography: Eng. Mining J., Centennial Issue, June, 1966; D. Frasche, *Mineral Resources*, NAS-NRC Publ. no. 1000, pt. C, 1962; M. K. Hubbert, *Energy Resources*, NAS-NRC Publ. no. 1000, pt. D, 1962; R. B. Lewis and G. B. Clark, *Elements of Mining*, 3d ed., 1964; R. Peele et al., *Mining Engineer's Handbook*, 1941; E. Pfleider et al., *Surface Mining*, 1969; T. A. Rickard, *Man and Metals*, 1932; T. A. Rickard, *The Romance of Mining*, 1944; E. H. Robie et al., *Economics of the Mineral Industries*, 2d ed., 1964; U.S. Bureau of Mines, *Mineral Facts and Problems*, Bull. no. 630, 1965; W. V. Van Royer and O. Bowles, *Atlas of the World's Resources*, vol. 2: *Mineral Resources*, 1952.

Mining, open-pit

The extraction of ores of metals and minerals by surface excavations. This method of mining is applicable for deposits which have large tonnage reserves, which can be exploited at a high rate of production and which do not have a ratio of overburden (waste material that must be moved) to ore that would make the operation uneconomic. Where these criteria are met, large scale earth moving equipment can be used to give low unit mining costs. For a discussion of some of the other methods of surface mining *see* COAL MINING; MINING, STRIP.

Fig. 1. Roads and rounds that have developed open cuts on Cerro Bolivar in Venezuela. (*U.S. Steel Corp.*)

Fig. 2.   Rotary drill used for drilling blast holes in large open-pit mine. (*Kennecott Copper Corp.*)

Most open-pit mines are developed in the form of an inverted cone with the base of the cone on surface. Exceptions are open-pit mines developed in hills or mountains (Fig. 1). The walls of all open pits are terraced with benches to permit shovels to excavate the rock and to provide haul roads for trucks to transport the rock out of the pit. The final depth of the pits ranges from less than 100 ft to nearly 3000 ft and is dependent on the depth and value of the ore and the cost of mining. The cost of mining usually increases with the depth of the pit because of the increased distance the ore must be hauled to the surface and the increased amount of

Fig. 3.   Crawler-mounted drilling machine. Horizontal toe holes are drilled by percussion. (*Kennecott Copper Corp.*)

overburden, or waste material, that must be removed to expose the ore. Numerous benches or terraces in the pit permit a number of areas to be worked at one time and are necessary to give a slope to the sides of the pit. Using multiple benches also permits a balanced operation in which much of the overburden is removed from upper benches at the same time ore is being mined from lower benches. It is undesirable to remove all of the overburden from an ore deposit before mining the ore because the initial cost of developing the ore is then extremely high.

The principal operations in mining are (1) drilling, (2) blasting, (3) loading, and (4) hauling. These operations are usually required for both ore and overburden. Occasionally, the ore or the overburden is soft enough that drilling and blasting is not required.

Rock drilling. Drilling and blasting are interrelated operations. The primary purpose of drilling is to provide an opening in the rock for the placing of explosives. If the ore or overburden is soft enough to be easily excavated without blasting by explosives, it is not necessary for it to be drilled. The cuttings from the hole are often used to fill the hole after the explosive charge has been placed at the bottom. Blast holes drilled in ore serve a secondary function in that the cuttings can be sampled and assayed to determine the mineral content of the ore. Sampling of drill holes is frequently done for ores not readily identified by visual means.

The basic methods of drilling rock are (1) rotary, (2) percussion, (3) jet piercing, and (4) churn drilling. The drill holes range in size from $1\frac{1}{2}$ in. to 15 in. in diameter and are drilled to varying depths and spacings as required for the particular type of rock being mined. The drill hole depth ranges from 20 to 50 ft depending on the bench height used in the mine. The drill hole spacing (distance between holes) is governed by the depth of the holes and the hardness of the rock, and generally ranges from 12 to 20 ft. For short holes or hard rock, the hole spacing must be close; and for long holes and soft rock, the spacing may be increased. *See* BORING AND DRILLING (MINERAL).

Rotary drilling. Drilling is accomplished by rotating a bit under pressure. Compressed air is forced down a small hole in the drill shank or steel, and is allowed to escape through small holes in the bit for cooling the bit and blowing the rock cuttings out of the hole. Rotary drill holes range from a minimum diameter of 4 in. to a maximum diameter of 15 in. and are drilled by machines mounted on trucks or a crawler frame for mobility (Figs. 2 and 3). The weight of these machines ranges from 30,000 to 200,000 lb and, in general, the heavier machines are required for the larger-diameter holes. The rotary method is the most common type of drilling used in open-pit mines because it is the cheapest method of drilling; however, it is restricted to soft and medium-hard rock.

Percussion drilling. Percussion drilling is accomplished with a star-shaped bit which is rotated while being struck with an air hammer operated by high-pressure air. Air is also forced down the drill steel to cool the bit and to blow the cuttings out of the hole. Small amounts of water are frequently added with the air to reduce the dust. The hole diameters range from $1\frac{1}{2}$ to 9 in. depending upon the type of drill. The drilling machine used for the smaller-diameter holes ($1\frac{1}{2}$–$4\frac{1}{2}$ in.) is small and usually mounted on a lightweight crawler or rubber-tired frame weighing a few thousand pounds. The air hammer and rotating device are mounted on a boom on the machine and the hammer blows are transferred by the drill steel to the drill bit. For the larger-diameter and deeper holes, the air hammer is attached directly to the bit and is lowered down the hole because the impact of the hammer blows is dissipated in the larger and longer columns of drill steel. Percussion drilling is most applicable to brittle rock in the medium-hard to hard range. The depth of the smaller-diameter holes is limited to about 25 or 30 ft, but the larger holes can be drilled to 40 or 50 ft without significant loss of efficiency.

Jet piercing. The jet piercing method of drilling was developed to drill the very hard iron ores (taconite). In this method, a hole is drilled by applying a high-temperature flame produced by fuel oil and oxygen to the rock. The holes range in size from 6 to 18 in., tend to be irregular in diameter, and require careful control of the flame to prevent overenlargement. The drilling machines are integrated units which control the fuel oil and oxygen mixture and lower the jet piercing bit down the hole. The jet piercing method is limited to very hard rock for which other types of drilling are more costly.

Churn drilling. Churn drilling was one of the first methods of drilling used in open-pit mining and was common until the 1950s; however, it has been almost entirely supplanted by rotary drilling. Churn drilling is accomplished by repeatedly raising a heavy bit weighing several tons and dropping it to shatter the rock in the bottom of the hole. Water is used in the hole to keep the cuttings in a muddy suspension; and after several feet of hole is drilled, it is necessary to remove the mud with a bailing device. Churn drilling is relatively slow and not economically competitive with other methods of drilling.

Blasting. The type and quantity of explosive are governed by the resistance of the rock to breaking. The primary blasting agents are dynamite and ammonium nitrate, which are detonated by either electric caps or a fuselike detonator called primacord.

Dynamite is available in varying strengths for use with varying rock conditions. It is commonly used either in cartridge form or as a pulverized, free-flowing material packed in bags. It can be used for almost any blasting application, including the detonation of ammonium nitrate, which cannot be detonated by the conventional blasting cap.

Commercial, or fertilizer-grade, ammonium nitrate has become a popular blasting medium because of its low cost. It is also safer to handle, store, and transport than most explosives. Granular or prill-size ammonium nitrate is commonly packed in paper, textile, or polyethylene bags. The carbon necessary for the proper detonation of ammonium nitrate is usually provided by the addition of fuel oil. Granular or prilled ammonium nitrate is highly soluble and becomes insensitive to detonation if placed in water, and therefore its use is restricted to water-free holes. The primary advantages of ammonium nitrate over dynamite are

low cost (less than one-third that of dynamite), safety, and ease of handling. Ammonium nitrate is not, however, quite as powerful as dynamite, and a higher ratio of powder to rock is required with ammonium nitrate than with dynamite.

During the early 1960s ammonium nitrate slurries were developed by mixing calculated amounts of ammonium nitrate, water, and other ingredients such as TNT and aluminum. These slurries have several advantages when compared to dry ammonium nitrate: They are more powerful because of higher densities and added ingredients, and they can be used in wet holes. Further, the slurries are generally safer and less costly than dynamite and are being used extensively in mines which are wet or have rock difficult to blast with the less powerful ammonium nitrate. Both ammonium nitrate and slurries are detonated by a small charge of dynamite. *See* EXPLOSION AND EXPLOSIVE.

Mechanical loading. Ore- and waste-loading equipment in common use includes power shovels for medium to large pits and tractor-type front-end loaders for the smaller pits. The loading unit must be selected to fit the transportation system, but because the rock will usually be broken to the largest size that can be handled by the crushing plant, the size and weight of the broken material will also have a significant influence on the type of loading machine. Of equal importance in determining the type of loading equipment is the required production or loading rate, the available working room, and the required operational mobility of the loading equipment.

Power shovels. Shovels in open-pit mines range from small machines equipped with 2-yd³ buckets to large machines with 25-yd³ buckets. A 6-yd³ shovel will, under average conditions, load about 6000 tons per shift, while a 12-yd³ shovel will load about 12,000 tons per shift. However, the nature of the material loaded will have a significant effect on productivity. If the material is soft or finely broken, the shovel productivity will be high; but if the material is hard or poorly broken with a high percentage of large boulders, the shovel production will be adversely affected. Power may be derived from diesel or gasoline engines or diesel-electric or

Fig. 4. An 8-yd³ shovel loading a 75-ton haulage truck. (*Kennecott Copper Corp.*)

Fig. 5. Typical open-pit loading. Electric shovel loads ore into railroad cars. (*Kennecott Copper Corp.*)

electric motors. The use of diesel or gasoline engines for power shovels is usually limited to shovels of up to 4-yd³ capacity, while diesel-electric drives are not common in shovels of more than 6-yd³ capacity. Electric drives are used in shovels ranging from 4-yd³ to 25-yd³ capacity and are the most widely used power sources in the open-pit mining operations (Figs. 4 and 5).

Draglines. A dragline is similar to a power shovel but uses a much longer boom. A bucket is suspended by a steel cable over a sheave at the end of the boom. The bucket is cast out toward the end of the boom and is pulled back by a hoist to gather a load of material which is deposited in an ore haulage unit or on a waste pile. Draglines range in size from machines with buckets of a few yards capacity to machines with buckets of 150-yd³ capacity. Draglines are extensively used in the phosphate fields of Florida and North Carolina. The overburden and the ore (called matrix in phosphate operations) are quite soft and no blasting is required. The phosphate ore is deposited in large slurrying pits where it is mixed with water and transported hydraulically to the concentrating plant.

Front-end loaders. Front-end loaders are tractors, both rubber-tired and track-type, equipped with a bucket for excavating and loading material (Fig. 6). The buckets are usually operated hydraulically and range in capacity from 1 to 15 yd³. Front-

Fig. 6. A 3-yd³ front-end loader loading a haulage truck in a small open-pit mine. (*Eaton Yale and Towne, Inc.*)

end loaders are usually powered by diesel engines and are much more mobile and less costly than power shovels of equal capacity. On the other hand, they are not generally as durable as a power shovel nor can they efficiently excavate hard or poorly broken material. A front-end loader has about one-half the productivity of a power shovel of equal bucket capacity; that is, a 10-yd³ front-end loader will load about the same tonnage per shift as a 5-yd³ shovel. However, the loader is gaining in popularity where mobility is desirable and where the digging is relatively easy.

Mechanical haulage. The common modes of transporting ore and waste from open-pit mines are trucks, railroads, inclined skip hoists, and belt conveyors. The application of these methods or combination of these methods depends upon the size and depth of the pit, the production rate, the length of haul to the crusher or dumping place, the maximum size of the material, and the type of loading equipment used.

Truck haulage. Truck haulage is the most common means of transporting ore and waste from open-pit mines because trucks provide a more versatile haulage system at a lower cost than rail, skip, or conveyor systems in most pits. Further, trucks are frequently used in conjunction with rail and conveyor systems where the haulage distance is greater than 2 or 3 mi. In this case trucks haul from the pit to a permanent loading point on the surface from which the material is transferred to one of the other systems. Trucks range in size from 20-ton to 200-ton capacity and are powered by diesel engines or a combination of diesel-electric units. Diesel drives are used almost exclusively in trucks up to 75-ton capacity. In the larger trucks, the diesel-electric drive, in which a diesel engine drives a generator to provide power for electric motors mounted in the hubs of the wheels, is most common. The diesel engines in the conventional drives range from 175 to 1000 hp, while the engines in the diesel-electric units range from 700 to 2000 hp. Most haulage trucks can ascend road grades of 8–12% fully loaded and are equipped with various braking devices, including dynamic electrical braking, to permit safe descent on equally steep roads. Tire cost is a major item in the operating cost of large trucks, and the roads must be well designed and maintained to enable the trucks to operate efficiently at high speed.

The size of truck used is primarily dependent upon the size of the loading equipment, but the required production rate and the length of haul are also factors of consideration. Usually 20–40-ton capacity trucks are used with 2–4-yd³ shovels, while the 70–100-ton trucks are used with 8–10-yd³ shovels.

In the early 1960s major advances were made in truck design which put the truck haulage system in a favorable competitive position compared to the other methods of transportation. Trucks are used almost exclusively in small and medium-size pits and are being used in conjunction with rail haulage in the larger pits. Trucks have the advantage of versatility, mobility, and low cost when used on short-haul distances.

Rail haulage. When mine rock must be transported more than 1½ or 2 mi, rail haulage is generally employed. Since rail haulage requires a larger capital outlay for equipment than other systems, only a large ore reserve justifies the investment. As a rough rule of thumb, the reserve should be large enough to support for 25 years a production rate of 30,000 tons of ore per day and an equivalent or greater tonnage of stripping. Adverse grades should be limited to a maximum of 3% on the main lines and 4% for short distances on switchbacks. Good track maintenance requires the use of auxiliary equipment such as mechanical tie tampers and track shifters. The latter are required for relocating track on the pit benches as mining progresses and on the waste dumps as the disposed material builds up adjacent to the track. Ground movement in the pit resulting from disturbance of the Earth's crust or settling of the waste dumps makes track maintenance a large part of mining cost.

Locomotives in use range from 50 to 125 tons in weight, with the largest sizes coming into increased use for steeper grades and larger loads. Most mines operate either all-electric or diesel-electric models. The use of all-electric locomotives creates the problem of electrical distribution in the pit and on the waste dumps, and requires the installation of trolley lines adjacent to all tracks.

Mine cars range in capacity from 50 to 100 tons of ore, and to 40 yd³ of waste. Ore is transported in various types of cars: solid-bottom, side-dump, or bottom-dump. The solid-bottom car is cheapest to maintain, but requires emptying by a rotary dumper. Waste is mostly handled by the side-dump car. Truck haulage has replaced much of the rail haulage because of advances in truck design and reduction in truck haulage costs.

Inclined skip hoist. Skip hoists consist of two counterbalanced cars of 20–40-ton capacity which operate on steeply inclined rails placed on a pit slope or wall. The skips are hoisted by a cable from the hoist house at the top of the pit. The skip system was developed to eliminate the need to haul ore and waste on long spiraling roads or tracks from the bottom of the pit. Either truck or rail haulage can be used to haul the rock from the shovel to the skip. Skips were very popular in the 1950s, but major advances in truck design and efficiency made skips all but obsolete by 1960.

Conveyors. Rubber-belt conveyors may be used to transport crushed material from the pit at slope angles up to 20°. Conveyors are especially useful for transporting large tonnages over rugged terrain and out of pits where ground conditions preclude building of good haulage roads. Improved belt design is permitting greater loading, higher speeds, and the substitution of single-flight for multiple-flight installations. The chief disadvantage of this transport system is that, to protect the belt from damage by large lumps, waste as well as ore must be crushed in the pit before it is loaded on the belt.

Waste-disposal problems. To keep costs at a minimum, the dump site must be located as near the pit as possible. However, care must be taken to prevent location of waste dumps above possible future ore reserves. In the case of copper mines, where the waste contains quantities of the metal which can be recovered by leaching, the ground on which such waste is deposited must be impervious to leach water. Where the creation of dumps is necessary, problems of possible stream pollution

and the effect on farms and on real-estate and land values must all be considered.

Slope stability and bench patterns. In open-cut and pit mining, the material ranges from unconsolidated surface debris to competent rock. The slope angle, that is, the angle at which the benches progress from bottom to top, is limited by the strength and characteristics of the material. Faults, joints, bedding planes, and especially groundwater behind the slopes are known to decrease the effective strength of the material and contribute to slides. In practice, slope angles vary from 22 to 60° and under normal conditions are about 45°. Steeper slopes have a greater tendency to fail but may be economically desirable because they lower the quantity of waste material that must be removed to provide access to the ore.

Recently developed technology permits an engineering approach to the design of slopes in keeping with measured rock and water conditions. Quantitative estimates of the factor of safety of a given design are now possible. Precise instrumentation has been developed to detect the boundaries of moving rock masses and the rate of their movement. Instrumentation to give warning of impending failure is being used in some pits. *See* ENGINEERING GEOLOGY.

Communication. Efficiency of mining operations, especially loading and hauling, is being improved by the use of communication equipment. Two-way, high-frequency radiophones are proving useful for communicating with haulage and repair crews and shovel operators. [DONALD P. BELLUM]

Mining, strip

A surface method of mining by removing the material overlying the bed and loading the uncovered mineral, usually coal. It is safer than underground mining because neither the workers nor the equipment is subjected to such hazards as roof falls and explosions caused by gas or dust ignitions. Coal near the outcrop or at shallow depth can be stripped not only more cheaply but more completely than by deep mining, and the need for leaving pillars of coal to support the mine roof is elimi-

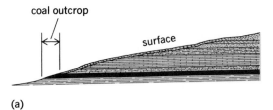

(a)

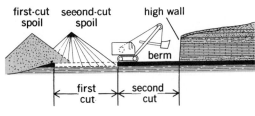

(b)

Fig. 1. Representative cross-section profile diagrams of contour strip mining of coal. (*a*) Section before stripping. (*b*) Section after second cut.

Fig. 2. Large electric shovel of the type used in prairie regions, high wall (right), and spoil bank (left) at Hanna Coal Co.'s Georgetown mine, eastern Ohio. (*U.S. Bureau of Mines and Marion Power Shovel Co.*)

nated. The roof over coal at shallow depth is weak and difficult to support in underground workings by conventional methods, yet this same weakness, of cover and of coal seam, makes stripping less difficult.

Stripping techniques. Power shovels, draglines, bulldozers, and other types of earth-moving equipment slice a cut through the overburden down to the coal. The cut ranges from 40 to 150 ft wide, depending on the type and size of equipment used. The stripped overburden (spoil) is stacked in a long ridge (spoil bank) parallel with the cut and as far as possible from undisturbed overburden (high wall). The slope of a spoil bank is approximately 1.4:1 and that of a high wall under average conditions is 0.3:1. The uncovered coal (berm) is then fragmented, loaded, and transported from the pit. Spoil from each succeeding cut is stacked overlapping and parallel with the previous ridge and also fills the space left by the coal removed (Fig. 1).

Techniques of stripping methods are similar, but the size of equipment used depends on whether the mine is in prairie or hill country. In prairie areas the thickness of overburden is nearly uniform, the coal bed is extensive, and equipment can be used for years at one mine without dismantling and moving to another location. Large-capacity shovels requiring many months to erect on the site are used at prairie mines. A unit of this type is the 60-yd³ rig shown in Fig. 2.

Most coal underlying hills is mined by underground methods, but where the working approaches the outcrop and the overburden is thin, the roof becomes difficult and expensive to support. The coal between the actual or potential underground workings and the outcrop is then more suitable for stripping. Usually, only two or three cuts 40–50 ft wide can be made on the contour of the coal bed, after which the shovel has to be moved to another site. Thus, in contour stripping, mobility of shovels up to 5-yd³ capacity is more important than those of larger capacity.

Large draglines are used instead of shovels to strip pitching beds of anthracite to depths surpassing 400 ft; however, this use of large draglines could more properly be classed as open pit. *See* MINING, OPEN-PIT.

Removal of unconsolidated overburden by hydraulic monitoring is a technique used especially in Alaska. Water under a high-pressure head is directed through a nozzle against the overburden to wash it into deep valleys where swift streams carry it away.

Although the character of the overburden determines the thickness of overburden that can be stripped, the maximum for shovels up to 5-yd³ capacity is about 50 ft and for the largest equipment about 110 ft. To reach these goals it frequently is necessary to use a dragline, carryall, or bulldozer on the high wall or a dragline in tandem with the shovel on the berm to strip the upper few feet of overburden. *See* MINING EXCAVATION.

Digging equipment known as the wheel can be used ahead of a large power shovel to remove the upper 20–40 ft of unconsolidated soil, clay, or weathered strata. This spoil is discharged onto a belt conveyor, then onto a stacker, and finally deposited several hundred feet from the high wall. Overburden thus removed improves the shovel productivity rate materially. Also, coal reserves can be mined that would have been too deep for the power shovel alone to handle.

Rocks overlying coal beds present some diversity of conditions for removal. Materials generally comprise shale, sandstone, and limestone with shale predominating. Proper fragmentation before stripping may be necessary to produce sizes that are smaller than the shovel dipper. Probably more research has been done on overburden drilling and blasting than on any other phase of stripping. The diameter, depth, and spacing of drill holes, the type of drill (whether vertical or horizontal), and the amount and type of explosive for each blast hole are the variables that must be determined for optimum production. Truck-mounted rotary drills have replaced churn drills for drilling vertical holes, and for horizontal drilling, auger drills are used. Package explosives, Airdox, Cardox, and commercial ammonium nitrate mixed with diesel fuel are used for blasting. The ammonium nitrate–diesel fuel mixture is one of the cheaper explosives and has gained favor rapidly. Equipment for mixing this explosive and automatically injecting it through a plastic tube into horizontal drill holes is being tested. If perfected, it will mechanize the only manual operation remaining in the stripping cycle.

Before coal is loaded, spoil remaining on the berm is removed by bulldozer, grader and rotary brooms, or at small mines by hand brooms. Small-diameter vertical holes are augered into the bed and blasted with small charges of explosive to crack the coal. A ripper pulled by a bulldozer is effective in replacing coal drilling and blasting. Broken coal is loaded by $\frac{1}{2}$–5-yd^3 capacity shovels into trucks ot 5–80-ton capacity and transported from the stripping.

Land reclamation. The disfiguration of the Earth's surface and scars left by high walls and spoil piles of early stripping activity have been well publicized. Over the years, the states in which strip mining was done have enacted legislation to compel the restoration of strip-mined land to some usable form. Responsible operators have been involved in such reclamation work to preserve good public relations even before being required to do so by law. Water was impounded for recreation lakes; spoil banks were partially graded and trees planted; and some areas were graded and planted in grasses for livestock grazing.

Most states require a bond from the operator as assurance that the stripped land will be reclaimed as specified by law. If the restoration is not completed satisfactorily, the bond is forfeited and the state uses that money to do the work.

Some state strip mine reclamation laws are more demanding than others, and in 1967 the U.S. Department of the Interior published guides for restoring stripped areas. Legal requirements for reclamation have become so detailed in some states that strip mine operators usually integrate the work as part of the mining cycle rather than waiting until the stripping is completed. By so doing, some stripping equipment is operated continuously rather than intermittently, as was done in the past. Furthermore, prompt restoration lessens silting of waterways and can eliminate the formation of acid water that would otherwise contaminate them. *See the feature article* PROTECTING THE ENVIRONMENT; *see also* COAL MINING.

[JAMES J. DOWD]

Bibliography: *Mining Guidebook and Buying Directory*, annual; R. Peele, *Mining Engineers' Handbook*, 3d ed., 1948; A. L. Toenges et al., *Some Aspects of Strip Mining of Bituminous Coal in Central and South Central States*, U.S. Bur. Mines, Inform. Circ. no. 6959, 1937; U.S. Department of the Interior, *Surface Mining and Our Environment*, 1967.

Mining, underground

An underground mine is a system of underground workings for the removal of ore from its place of occurrence to the surface, and involves the deployment of men and services.

There are several basic physical elements in an underground mining system. The passageways (openings) in a mine are called drifts if they are parallel to the geological structure, and cross-cuts if they cut across it. They range in size about 60–200 ft^2 in cross section, depending on their functions. The workings on a level (horizontal plane) are joined with those on another level by passageways of similiar cross section, called raises if they are driven upward and winzes if driven downward.

The passageways give access to, and provide transportation routes from, the stopes, which are the excavations where the ore is mined. The stopes are between levels. There may be rooms on the level, such as pump rooms, service shops, and lunchrooms. This article discusses exploration of a mine site, methods of removing ore material, and design of underground openings for ore removal and mine facilities. For other aspects *see* COAL MINING; MINING EXCAVATION; MINING MACHINERY.

EXPLORATION

A mine is designed and the mine openings specified after the exploration phase of a mine's history. Exploration in this context is not to be confused with prospecting. Exploration is the process of finding the characteristics of the mineralized rocks and the environmental rocks that make up the mine site. These attributes are absolute and unchanging, but they can only be predicted from sampling so there is always the risk that the predictions may be wrong. Some of the attributes of a mineral deposit and their limits of variation that are found by sampling and measurement are: shape, tabular or curvilinear; attitude, flat or vertical; dimensions, thick or thin, uniform or variable, long or short, or shallow or deep; physical character, hard or soft, strong or weak, laminar, jointed, or massive; mineralization, massive, globular, or disseminated, intense or sparse, or chemically stable or unstable; and surface and overlying formations, expendable or not expendable.

During the exploration there is a feedback from predictions of the revenue and expense that would result from operation. The end result of the exploration phase is a forecast of the grade (amount of valuable mineral per ton) and tonnage that can be mined at a specified rate. The ore grade acceptable could be different at another mining rate.

Parts of the risk involved are (1) a change in the mineralization or the environmental rocks, or both, as mining progresses, (2) drastic changes in the exchange value of the production relative to wages, equipment, and supplies, (3) the availability of new or better equipment, and (4) a change in the governmental attitude, such as in taxation. Any of these will affect the grade and the tonnage for the mine site.

MINE OPERATION

A mine is designed, developed, and worked in blocks of levels and stopes. The size of the blocks may be determined by the amount of ore that has been sufficiently explored, by geological boundaries, or by the need for effective supervision. The design must meet ventilation requirements and the openings must be maintained as long as they are needed. When mining in a block is completed it may, if expedient, be cut off from the ventilating system and the workings may be allowed to collapse. This can be done only if the failure will not disrupt other operations, for example, by causing rock bursts.

There are two basic plans of attack. The choice of plan depends on whether or not the surface, or the rocks overlying the ore, may be disturbed, and on stress redistribution problems.

Longwall. The principle of longwall mining is to advance in line all the stopes and pillars being mined in a block. No remnants are left either to support the back (overlying rocks) or to constitute stress concentrators. The line of attack may advance toward the shaft or other entrance or retreat from it. In the latter case, or if there are blocks beyond the current mining to be mined later, passageways must be maintained. Longwall mining is the method generally used to mine flat-lying deposits such as coal. Rooms with uniform dimensions separated by pillars with uniform dimensions are mined. The pillars support the backs. If the preservation of the backs or the surface is not a factor, the pillars may be systematically mined (robbed) in a longwall retreat or advance.

The result of mining with rooms and pillars is a cellular pattern. Most mining methods available to the mine designer involve the creation of a cellular pattern made up of stopes and pillars. Generally the ore from the stopes is won with less expense than is involved in recovering the pillars. Often the pillars are not mined because they are worth more as pillars than they would be as ore. They are stronger than any material that could economically be used to replace them, and they fit better.

Fill. There are few situations in which it is possible to recover a worthwhile amount of pillar ore unless the adjacent stopes or rooms have been filled. No filling will support the overlying rocks as well as the ore or pillars can, but if there is some settlement onto the fill, it will be limited because the rock is dilated and occupies more volume than the solid rock does. The swell will finally support the back.

The fill may be any incombustible available material—waste rock, sand or gravel, or mill tailings. The tailings from the mill are treated to meet the mine specifications for settling and percolation rates by removing some of the fine sizes. They are then transferred from the mixing plant to the stopes as a slurry by a pipeline and distributed in the stopes through hoses. Fine-grained natural sand may be placed that way as well. The water must be taken out by decantation or by percolation, or both. Some operators add portland cement to form a weak concrete. A mixture of smelter slag and sulfide-bearing mill tailings has been used to form a weak rock by chemical action.

When fill is placed in a stope alongside a pillar that is to be recovered, a partition, usually of light timber, is installed between the fill and the pillar. Some operators use an enriched mixture of cement at the interface to form a concrete.

MINE DESIGN

Fundamentally, mining is materials handling, and a mine is designed accordingly. Three functions are involved: breaking the ore from the face, delivering it to the surface, and delivering supplies to where they are needed. Ancillary functions are getting men and services, air, power, and water to where they are needed. Power may be electrical or compressed air.

Ore breaking and transporting together constitute 30–55% of the total cost of mining. If the ore has to be broken by explosives, the drilling and blasting cost is 10–20% of the total mine cost and the transportation portion is 20–35%.

Primary breaking cost varies inversely, and the transportation cost varies directly, with the size of the broken ore. There is an optimum size for the product of blasting. Each stage of the transportation phase has a limiting size that can be accommodated, so expense saved in the breaking phase may be exceeded by that of secondary breaking between stages in the transportation system.

A mine is designed around the method of primary breaking that is chosen, and the choice is governed by the forecasts from exploration. Stopes may be open or filled if the ore has to be broken by explosives. Caving may be used if conditions are favorable, and the ore will be broken by natural forces that make up the stress field in the ore and in the environmental rocks. Mining methods will be described under major headings, but since each ore body is unique and operators are ingenious, there are variations and hybrid methods.

Open stopes. There is a further qualification to this type of stope—with or without delayed filling. Using open stopes without delayed filling may be expedient, but if there are other extensive workings, the open stopes may redistribute the ground stresses in a manner which will interfere with subsequent work.

The mining method is chosen according to the thickness and inclination of the deposit. The breaking point between thick and thin in a tabular deposit is about 16 ft. That is about the maximum length of a stull (round timber) that can be handled conveniently in an open stope to give casual support where it is thought that loose rock might develop, or to provide a working platform for the miners. The breaking point between steep and flat is about 40° from horizontal, the limit at which rock will move by gravity on a rough surface.

The stope may be worked either overhand or underhand. If the ore will move by gravity to the drawpoints, the overhand attack is by advancing a breast (face) parallel to the level and so breaking out a slice. Because a platform to work off must be constructed with stulls and lagging or plank for each drilling site, the method is limited to thin deposits. The underhand attack is started from the top of the stope, usually from the access raise, and a block is broken out by drilling and blasting down holes. The bench must be cleaned off after each blast before drilling is resumed. This method has an advantage in gold mines, where coarse gold may lodge on the footwall and have to be swept out. Once cleaned, an area will not have to be cleaned again.

If the deposit is steep enough to deliver the broken ore to the drawpoints by gravity and thick enough to prevent the broken ore from arching over the opening, it probably should not be mined off platforms in an open stope.

Shrinkage stoping. An open stope requires successive platforms for the miners to work off. If the broken ore is drawn off just enough to give working room for the miners and they can work off the broken ore, the stope is called a shrinkage stope. The length of the stope is established by two raises, one at each end, which serve as manways and service entrances for air and water lines. By cribbing up with timber, they are maintained through the broken ore. The draw is from one-third to one-half of the break. The rest of the ore is retained in the stope until the stope is complete, and then the stope may be drawn empty through chutes which are installed during stope preparation.

The broken ore, which moves when drawn, has little ground-support capability. Shrinkage stoping may be used only up to the limit where the back is not self-supporting, or the walls will slab off and give unacceptable dilution. The limit may be extended by using casual timber support, or rock bolts for the back, and by rock-bolting potential slabs to the walls. Once a series of shrinkage stopes has been established and some of the stopes completed, the storage available provides flexibility in production beyond that of any other method.

If the ore body is too wide for the span of the back to be self-supporting or to be cheaply supported when mined parallel to the long dimension of the ore, it may be worked with transverse shrinkage stopes and intervening pillars. This involves delayed filling to permit mining the pillars.

Sublevel stoping. A deposit that is wide enough (about 40 ft) and has walls and ore sufficiently strong to permit shrinkage stoping may be worked by sublevel stopes (Fig. 1).

A vertical face is maintained. At vertical intervals, spaced to accommodate the drilling method to be used for breaking, sublevels are driven in the long direction of the stope from the entry raise. Two procedures are available after the initial slot has been made, which is generally done by widening a raise to the width of the stope. Holes may be drilled radially from the sublevels to give an acceptable distribution of explosive behind the face or slice to be broken off, or a cut may be slashed out across the face from each sublevel and drilled with vertical holes, quarry fashion, to give a better explosive distribution. The choice depends in part on the equipment available and in part on the control wanted at the sides. When the stope is wide, or for better control of the sides (walls), the radial drilling may be done from each of two sublevels at the same level which have been driven on either side of the stope.

The stope may be longitudinal with respect to the long dimension of the ore body, or it may be transverse and separated from the adjacent stope by a pillar. It follows that it will be a delayed-fill stope if the pillar is to be recovered. Even delayed-fill stopes are not satisfactory if the pillar material is sufficiently valuable to require complete and clean recovery.

Filled stopes. When the wall rock or ore is not sufficiently strong to permit the use of one of the open-stope methods, or if clean pillar recovery is important, methods have been devised to mine with only a small area of unsupported rock or ore exposed.

Cut-and-fill stoping. The breaking phase of this method is not different from that which has been described for the overhand shrinkage stope, but the broken ore is removed after each breast (slice) is completed and a layer of fill is placed. The cycle is: breaking, removing the broken ore, picking up the floor, raising the fill level, and replacing the floor (Figs. 2 and 3). Before the new fill is placed, the manways and the ore chute are raised with cribbing. The cribbing is covered with burlap if hydraulic fill is used. The manway at one end of the stope is built with two compartments. One compartment is used as the ore chute, called a mill hole in filled stopes.

The back must be strong enough to be self-supporting over the span, but the method is sufficiently lower in cost than the alternative so that the back may be supported with casual timber

Fig. 1. Sublevel stoping.

or rock bolts, which are broken out with the new breast.

Square-set stopes. When the backs, and perhaps the walls, are not strong enough to permit cut-and-fill stoping, temporary support may be provided by carefully framed and placed timbers called sets. The sets are filled when no longer needed, at a rate that depends on the rate at which they take weight and might collapse (Fig. 4).

A set is made up of posts 8 or 9 ft long, and caps and girts about 6 ft long, cut to exact lengths and framed to give a good fit at the corners. The timber is usually about 8 in. square, though round timber may be used.

A set is installed in an opening just large enough to accommodate it, and then blocked against the surrounding unstable rock or ore. Little blasting is needed because of the characteristics of the rock that make square-setting the best choice. The

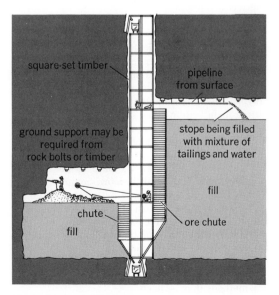

Fig. 2. Hydraulic-fill stope.

Fig. 3. A cut-and-fill stope. (a) Ready for hydraulically placed sand fill, except for the burlap lining. (b) Placing the floor over the fill.

problem is to hold back the broken rock until the set can be placed. This may be accomplished by extending boom timbers out over the caps.

Some of the ore may be moved manually. Generally, however, by retaining open sets in the fill and fitting them with inclined slides, provision is made for gathering the ore for scraping to the mill hole.

The sets may be installed either overhand or underhand, depending on the problem, or, if the choice is not critical, on the skill of the miners. The overhand technique is more common but in some camps the miners work better underhand.

Square-setting is the usual method for removing pillars if clean, complete extraction is needed. It is a flexible method and can be used to recover ore in offsets from the main deposit.

Caving. When the surface is expendable and other characteristics are favorable, one of the caving methods may be used.

Top slicing. In some respects this is like square-

setting but it is less expensive after it is underway. It is used when the surface is expendable and the ore is too weak to stay in place over a useful span.

The mining block is developed by driving a two-compartment raise through the ore to serve as an access manway and a mill hole, and by driving a longitudinal drift from the raise, at the top of the ore, to the extremity of the ore or the end of the proposed stope.

The initial unit of mining is a timbered crosscut driven each way from the drift to the edge of the block to be mined. Subsequent units are crosscuts driven adjacent to each preceding crosscut to take out a slice of ore. As the face of the slice is retreated toward the mill hole, the timbers in abandoned workings (several sets back) are permitted to fail, or forced to fail by blasting. The routine is continued until the slice is completed to the raise. The overlying formations collapse onto the broken timbers. As the routine is continued by taking successive slices, the broken timbers form a feltlike mat that has some tensile strength, and little timber support is needed in the crosscuts. Several slices are mined concurrently, step fashion. The overlying caving formations must follow the mat. No caverns can be left which could collapse and create an air blast.

Sublevel caving. If the ore is sufficiently strong, and after a timber mat has been developed, one or more slices may be omitted. The cantilever shelf formed when the next slice is taken will collapse under the load of caved material and its own weight. The broken ore is moved to the mill hole as in top slicing. Several slices are advanced simultaneously as in top slicing.

An adaption has been used in which the slices are taken out as sublevels in open stoping are taken, and the over lying formations are caved against the face. The sublevel slices are advanced in steps as in top slicing (Fig. 5). There is no mat and some ore is lost into the cave material. However, it is low-grade and the overall low cost of mining makes up for the loss.

Block-caving. If the ore texture (blockiness) and strength are suitable, and if the primitive stress field is favorable, an entire block 150–250 ft on a side and several hundred feet high may be induced to cave after it is undercut. The broken ore is drawn off through bell-shaped drawpoints (Fig. 6).

The drawing cycle is critical. It must keep the undersurface of the block unstable and continuing to fail. The lateral dimensions of the block are controlled by weakening the perimeter with raises and lateral workings, or even short shrinkage stopes. No large cavities are permitted to develop. In the final phase, when caving has reached the overlying formation, care must be taken to avoid drawing it with the ore. There is no primary breaking expense but the cost of secondary breaking for transportation may be high.

Ground control. Mine openings must be kept open as long as they are needed. Mining engineers recognize that rock is not necessarily solid or inert. The study of the behavior of rocks when subjected to force is called rock mechanics. It is a comparatively new field, although knowledge of the phenomena under study has been utilized for years without formal analyses of what was going on.

The observations that rocks around a mine

opening do not always behave in a manner that would be predictable by classical mechanics imply that there are other than gravitational forces involved, and that there is strong lateral component of strain energy. The source and reservoir of the strain energy have been less obscure since geologists have measured the rate of spreading of the North Atlantic Ocean floor and associated continental land masses (average 6 cm. per year since Carboniferous times). The resultant force vector from the combination of gravitational force and tectonic force is referred to as primitive stress in this description of underground mining. The rock mass is in equilibrium until a mine opening is made; that is, it is in equilibrium for a relatively short time involved in the mine operation. The mine opening accepts no force and the force is diverted to around the opening.

Ground support. The ideal support is a pillar of appropriate size, but the use of pillars is not always feasible.

Timber. Traditionally timber has been the usual support for the perimeters of mine openings. It is usually supplied as stulls or as lagging, depending on the slenderness ratio, diameter to length, and to some extent on the use. If it is slender, it is lagging. If a log (stull) is placed vertically, it is called a post. If it is placed nearly horizontally, it is usually called a stull, whether it is acting as a beam or as a column. Both posts and stulls are installed with lower ends in hitches in the rock and upper ends loosely fitted to the back or wall, depending on the location. The final fit is achieved by driving wooden wedges between the end of the timber and the rock. The hitch may be chiseled into the rock, but in hard rock a natural recess is generally used.

When a lateral working requires timbered support, the stulls are usually framed to give a neat fit at the corners, and flatted on two sides to save space in the working. Sawn square timber is often used. The unit is two posts and a cap (stull), usually with a sill on the floor. Whether or not the sill is used depends on the expected loading. The posts and the cap are wedged tightly to the walls and the back at the corners. Lagging or plank is laid over the caps to provide overhead protection and placed behind the posts if a loose wall is expected. Raises and winzes are similarly protected unless they are to be used for hoisting and more precise timbering is needed. Steel is frequently used in the same manner as timber.

Concrete. Openings are frequently lined with concrete if permanence or added strength is needed, if the ventilation friction factor must be reduced, or if the operator does not trust timber because of the fire hazard. Generally the opening is made round or ovaloid so the concrete will be in compression. Forms and poured concrete are commonly used, generally with reinforcing bars. Circumferential steel reinforcement is not effective if the concrete is loaded in circumferential compression.

Concrete may be blown onto the rock surface with a cement gun (guniting). Sand and small-sized aggregate are mixed dry and blown through a hose to the face to be coated. Water is added as the mixture passes the nozzle. The low-moisture mixture hits the face and a portion of the aggregate falls out. A tight bond is formed at the concrete-

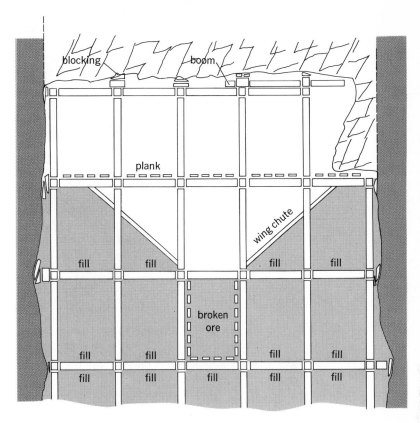

Fig. 4. Square-set stoping.

rock interface. It is thought that the peining action of the aggregate helps to make the bond and to produce a dense concrete. In treacherous rock quite large rooms, such as underground hoist rooms 30 ft or more across, have been successfully secured in this way. The angle of impingement for the application is critical. The thickness of the

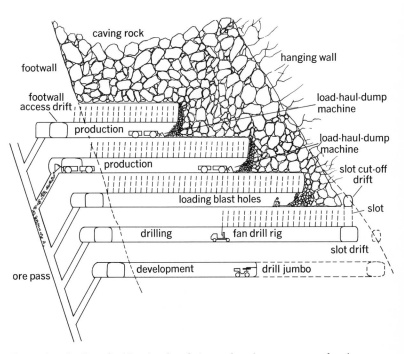

Fig. 5. An adaption of sublevel caving. Cutaway view shows progress of caving.

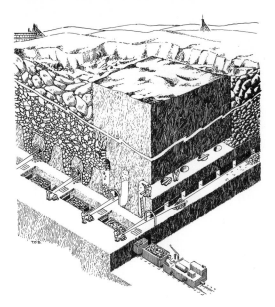

Fig. 6. Block-caving.

Fig. 7. The reinforcement of a mine opening with concrete. (*a*) A drift reinforced with gunite. (*b*) The same drift beyond the gunite.

coating is not more than a few inches (say 3 or 4) over the depressions in the rock surface and thinner over the bumps (Fig. 7).

Rock bolts. The systematic use of rock bolts for rock reinforcement has increased rapidly. These are steel bolts about 3/4 in. in diameter and generally 3–5 ft long, anchored at the bottoms of holes drilled at a right angle to the rock surface, and tensioned by a nut over a small plate at the rock surface. The anchorage is generally a split shell forced against the wall of the bore hole by a wedge as tension is applied at the bolt end. Some suppliers offer a method to anchor the bolt in an epoxy resin, and some others supply a bolt that is to be embedded in concrete or cement for its entire length.

There is no consensus as to the reason that rock bolts are effective, but it is agreed that they should be installed as soon as possible after an opening is made, and should be under high tension (Fig. 8). The mechanism offered most frequently for the effectiveness of bolts in a bedded formation is that a compound beam is built up by binding several laminar beds together to act as a single thick beam. For massive rocks that have no bedding it is commonly accepted that the bolts must extend into the compression arch that is postulated to be formed above the opening, and for that reason long bolts are often specified.

Actually, the abutments of the arch are restrained from moving outward, and the tendency is to move inward, especially if there is a high primitive stress lateral component. In narrow openings there is compression close to the skin of the opening and the function of the bolt in tension is to reduce the tendency for failure in oblique shear by preventing the thickening of the rock. At some width, as an opening is enlarged from narrow to wide, the compressive primitive stress is neutralized and the back goes into tension. Rock has little tensile strength because of discontinuities. If the rock is blocky, the blocks may be held together by bolts and form a flat, or nearly flat, voussoir arch.

Transportation. Gravity is used wherever it can be effective in the movement of ore toward the surface. Ore from open stopes that are steep enough for gravity flow is loaded through chutes directly into the level transportation units. When a stope is wider than about 25 ft, bell-shaped openings (drawpoints) are driven into the floor of the stope. If more than one row of them is needed, these drawpoints are driven on about 25-ft centers on a regular pattern so one crosscut can serve the outlets of several drawpoints. When the ore is loaded into the haulage equipment, it is taken to an ore pass and moved by gravity either to a loading pocket at the shaft or to a crushing plant, and thence to the shaftpocket. An inclined ore pass will give considerable lateral movement. A mine will also have a system of waste passes.

A drawpoint may discharge through a chute into

a haulage vehicle or into a short branch off the haulage line and be loaded into the main-line vehicle mechanically. If the stope is wide, the drawpoints may discharge onto the floor of a scraper drift at a higher elevation than the back of the haulage level. The ore is then delivered to the main-line vehicle by scraping (Fig. 6).

When a lode is too flat to permit the use of gravity, ore is moved to a central gathering point by a scraper, either in one stage or two. If the lode is flat enough to permit the use of wheeled or crawler-tracked vehicles, the broken ore may be loaded into a gathering vehicle and taken either to the ore pass or to the main-line transportation unit. An alternative is to use a load-haul-dump vehicle. Smaller versions of this type of machine are being introduced into large stopes. They are displacing the scraper, which in turn had displaced the small railcar and the wheelbarrow (Fig. 5).

Equipment is designed to do a specific job and its value in use beyond that job decreases rapidly. The primary gathering equipment is designed for a short haul.

Entry from surface. When the topography of the area has low relief, the entry will be by a shaft or a ramp. Sometimes both means are employed, the ramp being used for moving heavy, large equipment within the mine. If the relief is high, an adit (tunnel) may be used.

A shaft is usually located in the footwall far enough from the mine workings to avoid ground movement. It is designed for specific functions which determine the area (cross section) and shape, if the shape is not modified to accommodate ground stresses and sinking problems. It may be vertical or inclined, though the vertical shaft is the more common. Functionally either kind should be rectangular to accommodate the equipment used in it.

Many vertical shafts are circular or elliptical, but a rectangular framework is fitted in them to guide the shaft vehicles. There is an exception, not common in North America, when rope (steel-cable) guides are hung in the shaft to guide the vehicles. On the other hand, a round or elliptical shaft is better for ventilation because it may be smooth-lined and offers less air resistance.

A shaft is designed after its functions have been decided and the rock conditions have been forecast. It may be multipurpose and the cross section (plan) must include space for each of the functions, as well as a ladderway for an emergency exit. A shaft may be specialized, that is, designed exclusively for ore hoisting, for services, or for ventilation. A mine must have two shafts to provide alternate routes to the surface in case one shaft is out of commission in an emergency.   [A. V. CORLETT]

Bibliography: R. S. Lewis and G. B. Clark, *Elements of Mining*, 3d ed., 1964; R. Peele (ed.), *Mining Engineers' Handbook*, vol. 1, 3d ed., 1941.

Mining excavation

In mining for coal, metallic, and nonmetallic minerals, the process of removing minerals from the Earth. Excavation consists of fragmentation (or in special cases solution) of minerals from their solid state, loading them, and transporting them to the surface. Fragmentation is accomplished by the use of explosives or mechanical means, the former being most commonly applied. Excavation is also involved in establishing mine entries and other development workings in waste rock for access to the minerals. Mechanical fragmentation by means of boring machines (Figs. 1–3) is being introduced at some mines for this purpose.

Fig. 8. Testing the tension on a rock bolt by using a simple instrument to measure torque.

Fig. 1. Raise boring machine (drills pilot holes between levels) in Idaho silver-lead mine. (*Hecla Mining Co.*)

Fig. 2.   Raise boring machine set to ream pilot hole to finished dimension of raise. (*Homestake Mining Co.*)

Hard rock is generally broken with explosives to attain fragmentation. Some moderately soft deposits, such as coal, potash, and borax, are fragmented mechanically by machines without the use of explosives. When fragmentation is by machine, the loading device is commonly an integral part of the machine. In special cases (sulfur, salt, and potash) the mineral may be excavated by solution. When solution mining is used, the mineral-bearing solution is pumped to the surface; thus loading and transport become integral parts of the process.

Explosives. Two general classes of explosives are used for mining, black powder and high explosives or dynamites. Black powder is not used underground, where commercial high explosives

Fig. 3.   Tunnel boring machine used in mining. (*James S. Robbins and Associates, Inc.*)

commonly known as dynamites are preferred. A special class of explosive, designated "permissible" after being tested and passed by the United States Bureau of Mines, is required in gaseous and dusty coal mines. Permissible explosives are especially designed to produce a flame of small volume, short duration, and low temperature. Dynamites are typed according to properties and further subdivided into grades based on strength. The principal properties are strength, density, sensitiveness, velocity, water resistance, freezing resistance, and fume products of detonation. Strength refers to the energy content of an explosive, and is based on the percent by weight of nitroglycerin or equivalent energy when other strength-imparting ingredients are substituted for nitroglycerin.

High explosives are detonated with blasting caps, small metal tubes closed at one end and charged with a highly heat-sensitive explosive.

Blasting agents not technically classified as explosives are increasingly being used for blasting, both in surface mines and underground. They are non-cap-sensitive materials or mixtures in which none of the ingredients are classed as explosives but which can be detonated by a high-explosive primer for blasting purposes. A commonly used blasting agent of this type is a mixture of fertilizer-grade ammonium nitrate and fuel oil (ANFO) that has found wide acceptance because of its safety features and low cost. *See* EXPLOSION AND EXPLOSIVE.

Fragmentation. Fragmentation is the process of breaking ground with explosives and machines.

Explosives fragmentation. The object of explosives fragmentation is to break the minerals and produce fragments of a size best suited for handling. It is cheaper and generally desirable to break waste rock in coarse sizes. Customarily, coal has been produced as lumps, but now the trend is toward using fine sizes for more economical handling and combustion. Fine sizes are desirable for most metallic and some nonmetallic ores that are processed after mining. Some nonmetallic minerals can be marketed more profitably in medium coarse sizes (1/4 to 2 in.), so fine material is wasted and must be minimized. The size of mineral fragments can be controlled partly by the amount and strength of the explosive used in blasting and somewhat by the spacing between shot holes and by their depth.

An assemblage of shot holes drilled into the face of a stope, drift, crosscut, shaft, raise, winze, adit, or tunnel and blasted at one time is a "round" (Fig. 4). The pattern of the round contributes to the effectiveness of a blast. A wide variety of patterns is used, depending upon the character of the material to be broken, the size and shape of the desired opening, and the desired size of fragments. These rounds have been given names such as pyramid, triangle, V, and fan, suggestive of the pattern of the drill holes that are blasted to give the initial cut in the center of the face. Others, such as Norwegian and Michigan, have been named according to place of origin. Michigan is also called burned cut because the holes which are drilled close together in a group near the center of the face, when blasted, pulverize the rock and discharge it at a high velocity. *See* BORING AND DRILLING (MINERAL).

To break a round, each hole is charged with ex-

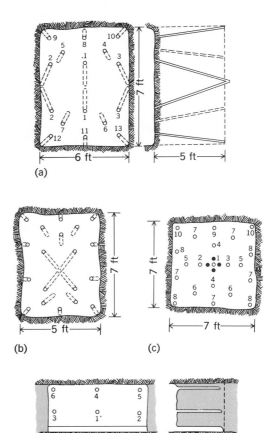

Fig. 4. Typical rounds for explosives fragmentation. (a) Horizontal V cut. (b) Four-hole pyramid cut. (c) Five-hole burn cut. (d) Undercut coal face.

plosives. The explosives are in the form of cylindrical cartridges, or in the case of blasting agents, may be in bulk. Cartridges are tamped into place in the drill hole and explosives in bulk form are blown into the hole pneumatically. A detonating cap with attached safety fuse or electric wire is inserted into a cartridge and placed in each hole to form a primer. The round is then blasted by igniting the fuse or sending an electric current through the wiring. The sequence in which holes of a round are blasted is important. Usually a group of holes near the center or along one side is blasted first. Then successive groups of holes are blasted in series. Crude timing between successive blasts is attained by trimming fuses to different lengths. Precision timing is possible with electric blasting caps constructed to vary in detonation by a few thousandths of a second.

Mechanical fragmentation. This method is used primarily in coal mining. The principal machines are the continuous miner, auger, coal plow, and coal cutter. The continuous miner is manufactured in several sizes by both domestic and foreign companies. There are many variations in detail to meet varying conditions in coal seams.

One representative model is shown in Fig. 5. The coal is fragmented by a front-end cutting head comprising a number of continuously revolving chains upon which are mounted hard-metal-tipped cutters or picks. The chains are mounted on a bar which can be rotated horizontally to cut an entry 12–20 ft wide and vertically to cut from 6 in. below the bottom of the machine to 7 or 8 ft above. The cutter-bar frame is mounted on a carriage which in turn is mounted on caterpillar treads. The broken coal is collected by a conveyor which transports it from the cutting head to the transportation system at the rear of the machine.

Augers up to 5 ft in diameter (Fig. 6) are used under high banks of abandoned strip mines to recover coal to a boring depth of about 200 ft. The auger is made in sections, the first one having a hard-metal-faced cutting tool that does the cutting as the auger is rotated. The force providing thrust and rotation to the auger is supplied by a diesel engine at the surface. Some machines are self-propelling and handle and store the auger sections mechanically.

The coal plow or planer was developed in Germany to mine coal by the longwall system. The plow is drawn back and forth along the face and cuts or slabs off the coal in a series of slices. The planer unit comprises (1) an armored double-chain (Panzer) conveyor which rests on the mine floor and extends the entire length of the longwall face, (2) an electric driving unit at each end of the conveyor, (3) the planer assembly, and (4) pneumatic conveyor shifters. The electric motors move the plow by means of a rotating drum and chain attached to the planer. Broken coal falls to the conveyor, which moves it to entries adjacent to the mining areas. The pneumatic shifters move the conveyor and the plow toward the face as mining advances.

Where coal is fragmented by explosives, it is first undercut by a coal-cutting machine along the bottom of the seam. It either falls by gravity (undercutting and mass falling), or is broken by permissible explosives or other substitutes, shot

holes are placed to take advantage of such natural features as the presence of hard and soft bands and the direction of cleavage planes in the coal.

Three classes of machines are used for cutting coal: disk, bar, and chain, classified by the method employed for holding the cutting tools or cutter picks.

The chain coal cutter (Fig. 7) is in most common use and is manufactured in various sizes. These electrically powered machines are designed to cut

Fig. 5. A continuous miner, finishing cut as it rips overlapping paths in coal seam 18 in. deep by 42 in. wide by 90 in. high. Coal is discharged into waiting shuttle. (*U.S. Bureau of Mines, and Joy Manufacturing*)

Fig. 6. A view from the spoil bank of a strip mine showing the operation of auger mining and coal being loaded into a truck in the Pittsburgh coal bed, West Virginia. (*U.S. Bureau of Mines*)

in an arc of 180°. The smaller units are maneuvered by power-actuated drums and cable attached to stationary anchors. The larger self-propelled units are mounted on track or use pneumatic tires. The cutting jib can be rotated while cutting without moving the machine. It can also be rotated to cut either vertically or horizontally.

Loading. Since most loading is mechanized, hand shoveling is limited. Underground loaders include the continuous loader, scraper loader, and revolving and overcast shovels. Clamshell loaders are used in sinking shafts. The continuous loader (Fig. 8) is used extensively in underground coal mining and less extensively in other mines. Its essential parts comprise a gathering headframe, or scoop, on the front end that is crowded into the broken material by the machine, and a set of gathering arms to rake the material onto a bar chain conveyor that transfers it to the haulage system. The scraper loader consists of a hoe-type scraper, double-drum electric hoist, and ramp. Fragmented material is scraped up the ramp and discharged

Fig. 7. Electrically powered chain coal cutter making shear cut in coal face of West Virginia mine. (*U.S. Bureau of Mines, and Joy Manufacturing*)

into the transportation system. The scraper loader is used in many places in mining to scrape ore and waste short distances up to about 200 ft. Revolving power shovels are used for excavation in large underground openings. The overcast shovel loader, which runs on rail or caterpillar tracks, is best adapted for use in confined areas. Loading is accomplished by crowding the unit into the broken rock; when the shovel is full, it casts back over the unit into cars or trucks.

Transportation. Moving or hauling of men, supplies, and broken minerals is one of the most complicated operations in mining. Transportation includes (1) transfer of broken material through mine openings by gravity and scraping, (2) rail haulage on surface and underground, (3) trackless, wheeled haulage on surface and underground, (4) hoisting and cable haulage from open pits and underground mines, (5) movement of broken material by numerous types of conveyors, and (6) pumping broken ore through pipelines.

Gravity and scraping methods are used to gather small quantities of material into larger volume for transportation by some other method. Material may fall directly into raises and ore passes where it flows by gravity to a chute or ore pocket.

Underground rail haulage has been adapted from railroad practice, but on a much smaller scale. It is used mostly for collecting broken material from chutes (Fig. 9) and various loaders, and transporting it to the hoist or directly to the surface. Track gage ranges from 18 to 48 in. Trains are made up of cars ranging from 1 ton to about 20 tons capacity. Cars are equipped for manual dumping or with some dumping device. Small units are drawn by storage battery locomotives and larger units by electric trolley locomotives.

Trackless, wheeled haulage underground usually is by electric cable reel and diesel-electric shuttle cars with self-contained discharging conve-

Fig. 8. Caterpillar-mounted electric machine loading broken limestone into a diesel-powered haulage truck in a deep limestone mine in Ohio. (*U.S. Bureau of Mines*)

Fig. 10. Loader and a trackless, diesel-electric shuttle car mounted on pneumatic tires in an underground Missouri lead-zinc mine. (*U.S. Bureau of Mines*)

yors (Fig. 10). They are mounted on pneumatic tires and have capacities ranging to 20 tons. Dump trucks are used underground to a limited extent. When so used they are equipped with diesel engines for safety. Most trackless underground haulage equipment negotiates grades up to 10%.

Electric hoisting is employed in deep underground mines. Either vertical or inclined shafts are divided into three to six parallel shaftways—one for a manway, pipe, or power lines, and a minimum of two for hoisting. In large deep shafts, two compartments are commonly used for hoisting broken material, and two others for handling supplies, equipment, and personnel.

The hoisting layout comprises a headframe, usually of timber, concrete, or steel, erected over the collar of the shaft. A sheave wheel is mounted at the top of the headframe for each hoisting compartment. In addition, the headframe contains bins into which hoisted material is discharged.

Most hoists are electrically driven and have two winding drums that can be operated in balance or separately. When four compartments are employed for hoisting, a larger, double-drum unit hoists rock and a smaller one hoists men and supplies. Each drum is wound with a steel wire cable long enough to reach the sump (bottom) of the shaft. Cables range up to about $1\frac{1}{2}$ in. in diameter. Rock is hoisted from various levels of the mine

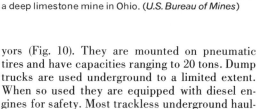

Fig. 9. Loading 60 ft³-capacity, Granby-type mine car at uranium ore-pass station on haulage level, San Juan County, Utah. Underswing arc gate is operated by a compressed-air cylinder controlled by motorman helper on right. (*U.S. Bureau of Mines*)

in skips (buckets, baskets, or open cars) ranging up to 10 tons capacity. Men and supplies and, at some smaller mines, loaded cars are hoisted in cages with one to three decks.

An important safety feature of the skip and the cage is a device that stops them if the hoisting cable fails. When failure occurs, a spring releases a set of safety catches that engage the wooden guides along each side of the skip or cage, thereby stopping it.

Several different types of conveyors are used in transporting coal in underground mines. The most important of these are belt conveyors. In some mines virtually all the coal is transported by a system of belt conveyors from the mining face through the various entries to surface. *See* MINING, UNDERGROUND. [JAMES E. HILL]

Mining machinery

Apparatus used in removing and transporting valuable solid minerals from their place of natural origin to a more accessible location for further processing or transportation. Many of the machines are identical to, or minor adaptations of, those used for excavating in the construction industry. In a wider sense mining machinery could also include all equipment used in finding (exploring and prospecting), removing (mining: developing and exploiting), and improving (processing: ore dressing, milling, concentrating or beneficiating, and refining) valuable minerals; it could even include metallurgical (smelting) and chemical processing equipment used in extracting or purifying the final product for industry. The term is also applied to special equipment for recovery of minerals from beneath the sea. In usual context the term does not include apparatus used principally in the petroleum industry. Perhaps those machines most often considered as uniquely mining machinery are drills, mechanical miners, and specially adapted materials-handling equipment for use in mining underground or on surface (where a large proportion of mines are located). In addition, some unique auxiliary equipment and processing equipment are used in the mining industry.

Design and construction. All components of mining machinery—including primary mechanism, controls, means of powering, and frame—require the following features to a much greater extent than do other machines (with the possible exception of some military, construction, oil well, and marine units).

Ruggedness. Equipment is handled roughly, frequently receiving severe and sudden shock from dropping, striking, and blasting vibration; overloading is common, and long life is demanded by the economics of mining.

Weather resistance. Operations extend over a wide range of climate and altitude.

Abrasion resistance. Minerals include some of the hardest substances known. Dust and fine particles are always present.

Water and corrosion resistance. Moisture and water, often acidic, are common in mining operations.

Infrequent and simple maintenance. Equipment is often widely scattered and in locations with restricted access. Trained mechanics and repair parts are generally limited in availability because of the remoteness of operations.

Easy disassembly and reassembly. Access to the machinery at the site of operation is frequently limited. Also, the working space near it and mechanical aids to moving or lifting it may be limited or nonexistent.

Safety. Mining has had a poor safety record. As a result, most governments have testing bureaus, inspection agencies, and enforcement laws for the approval of mining equipment. Simplicity of operation, low initial cost, and low operating costs are also desirable features of mining machinery.

Exploration machinery (vehicles, drills, and accessory equipment) is subjected to the same operating conditions as other mining equipment. Mineral-beneficiating equipment must have, above all, abrasion resistance. Smelting equipment has the requirement of heat resistance. Chemical refining process equipment must be highly resistant to corrosion. Reliability of all processing equipment is critical because slurries are commonly handled, and they can cause considerable difficulty in restarting after shutdowns.

Underground requirements. Machinery operated underground must meet special design requirements.

Low-ventilation demand. Quantity and geometry of passageways for air are rather rigidly fixed, so that high air consumption is a problem and noxious gases cannot be readily dispersed. Heat removal is a problem in deep mines.

Compactness. Space is at a premium, especially height, particularly in bituminous coal mines.

Easy visibility. Most operating areas are lighted only by individual cap-mounted or hand-held lamps.

Hand portability. Units or components must frequently be hand-carried into an operating area.

Absence of spark and flame. Equipment is often used in or near explosives, timber supports, and natural or man-made combustible gas or dust. In the presence of hydrocarbons, as well as of certain metal ores such as some sulfides, complete absence of open sparks or flames is a major requirement.

Power source. Mining machinery is very commonly powered by compressed air, but electricity is also widely used and is often the basic source. Compressed air has the advantages of simplicity of transmission and safety under wet conditions. It is especially advantageous underground as an aid to ventilation. Machines powered by compressed air can be easily designed to accommodate overloads or jamming, which is desirable on surface as well as underground. Large central compressors and extensive pipeline distribution systems are common, especially at underground mines.

Electric power, purchased from public sources or locally generated at a large central station, is common in open-pit and strip mines and dredging operations. Underground coal and saline-mineral mines often use electric-powered production machinery, but in other underground mines electricity is normally used only for pumps and transportation systems in relatively dry or permanent locations. Direct-current devices are dominant because of simplicity of speed and power control, but alternating-current apparatus is becoming common. Mobile equipment is often either battery of cable-reel (having a spring-loaded reel of extension power cable mounted on the machine) type. Processing machinery units are almost exclusively powered by individual electric motors.

Diesel engines are popular for generating small quantities of electric power in remote areas and for transportation units. Underground, abundant ventilation is essential, as well as wet scrubbers, chemical oxidizers, and other accessories to aid the removal of noxious and irritant exhaust gases. Hydraulic (oil) control and driving mechanisms are widely used. Transfer of power by wire rope is common, especially for main vertical transportation.

Drills. Drills make openings, of relatively small cross section and long length, which are used to obtain samples of minerals during exploration, to emplace blasting explosives, and to extract natural or artificial solutions or melts of minerals. Exploration holes are generally vertical or inclined steeply downward, less than 6 in. in diameter and up to 10,000 ft long. Blastholes range from 3/4 to 12 in. in diameter and usually are under 50 ft long in any direction, with the larger usually downward. Solution wells normally are vertical and 6 to 12 in. in diameter, sometimes reach depths of several thousand feet, and are equipped with several concentric strings of pipe.

Rock drills. Percussion, rotary, or a combination action of a steel rod or pipe, tipped with a harder metal chisel or rolling gearlike bit, chips out holes up to 12 in. (or more) in diameter by 125 ft (or more) long from the surface, and 1–3 in. by 5–200 ft from underground. Crawler or wheeled carriers are used, and the smaller drills are often attached to hydraulically maneuvered booms. Air or liquids flush out the chips.

Diamond drills. Rotation of a pipe tipped with a diamond-studded bit is used in exploration to penetrate the hardest rocks. Large units make holes up to 3 in. in diameter and 5000 ft or more deep; at the other extreme are units so small that they can be pack-carried. A cylindrical core is usually recovered.

Water-jet drills. For exploration and blasting in loose or weakly bonded materials, a water jet washes out a hole as a wall-supporting pipe is inserted.

Jet flame drills. For economical surface blast holes in hard abrasive quartzitic rock, a high-velocity flame is used to spall out a hole.

Mechanical miners. There are many machines designed to excavate the valuable mineral or the access openings by relatively continuous dislodgement of material without resorting to the more common practice of intermittent blasting in drill holes. These units also frequently transport the mineral a short distance, and when designed for weakly bonded minerals, they often become primarily materials-handling equipment.

Continuous miners. For horizontal openings in coal and saline deposits, toothlike lugs on moving chains, or rotating drums or disks rip material from the face of the opening as the assembly crawls ahead.

Plows or planers. In coal and other mineral deposits of medium hardness, bladelike devices continuously break off a 6-in. layer as they are pulled by various mechanisms along a wall several hundred feet long and 3–5 ft high.

Augers. Coal and soft sediments are mechanically mined by augers up to 5 ft in diameter and 100 ft long, usually used horizontally.

Shaft and raise drills and borers. For vertical and inclined openings up to 8 ft and even larger in diameter, various rotary coring or fullface boring equipment is used, both in an upward direction (raising) for several hundred feet or downward (sinking) for several thousand feet. These units usually use many rigid teeth or rolling gearlike bits to chip out the mineral.

Tunneling machines. There are similar rotary boring units for horizontal, or nearly so, openings of any length and up to 35 ft in diameter (in soft rock).

Rock saws. To remove large blocks of material, narrow slots or channels are cut by the action of a moving steel band or blade and a slurry of abrasive particles (sometimes diamonds) rather than teeth. Small flame jets are also used.

Hydraulic monitors. Water jets of medium to high pressure (some to 5000 psi) are used to excavate weakly cemented surface material and brittle hydrocarbons both on the surface and underground.

Special materials-handling equipment. Loose material (muck) is picked up (mucked or loaded) and transported (hauled or hoisted) by a wide variety of equipment.

Excavator loaders. For confined places underground there are various unique grab-bucket shaft muckers and overcasting shovel tunnel muckers, and also gathering-head loading-conveyor units having eccentric arms, lugged chains, and screws or oscillating pans for handling muck in horizontal openings.

Dragline scrapers. Scrapers (slushers) with a flat plowlike blade or partially open bucket pulled by a wire rope are commonly used to move muck up to a couple hundred feet, especially in underground mines.

Dipper shovels and dragline cranes. In surface mining single-bucket loads can handle up to 100 yd³ of material. Many of the intermediate size (20–40 yd³) units move on unique walking shoes.

Bucketline and bucketwheel excavators. These are for surface use and can dig up to 5000 yd³/hr, using a series of buckets on a moving chain or a rotating wheel supported on crawlers, railcars, or floating hulls (dredges).

Suction dredges and pipelines. On surface, up to 3700 yd³/hr of moderately loose mineral up to several inches in diameter can be picked up and moved as a slurry (mix of water and solid material) by pumps usually mounted on floating hulls.

Trucks. Diesel and electric shuttle cars (short-haul trucks) of unusual design, often having very low profile and conveyor bottoms, are used underground. At surface mines there are diesel-electric- and electric-trolley-type dump trucks of over 100-ton capacity.

Railroads. Underground locomotives range from 1/2 to 80 tons in weight, with electric (storage battery, cable-reel, or trolley), diesel, and sometimes compressed-air power units. Cars are usually of special design.

Conveyors. Unique movable, self-propelled, sectional and extensible conveyors are used in underground mining.

Wire rope hoists. Hoists or winders of up to 6000 hp are used in shafts for vertical or steeply inclined transportation in single lifts of as much as 6000 ft. Of various particular designs, there are two basic types: drum, simply a powered reel of rope; and friction, in which the rope is draped over a powered wheel and a counterweight is attached to one end and a conveyance to the other.

Auxiliary equipment. Drainage pumps handling hundreds of gallons of water per minute at heads of 1000 ft or more are used underground. For underground roof support, primarily in coal, there are mechanically moved jacks of 100-ton capacity. Ventilation fans are capable of moving several hundred thousand cubic feet of air per minute. Crushers can handle pieces of hard rock several feet across in two dimensions. In processing, minerals are sorted by size or density, or both, by a variety of screens, classifiers, and special concentrators, using vibration, fluid flow, centrifugal force, and other principles. Froth flotation and magnetic and electrostatic equipment take advantage of other special properties of minerals.

[LLOYD E. ANTONIDES]

Bibliography: Engineering and Mining Journal, *Mining Guidebook*, annual; R. S. Lewis, *Elements of Mining*, 3d ed., 1964; R. Peele (ed.), *Mining Engineers' Handbook*, 2 vols., 3d ed., 1941.

Mining safety

The prevention of mine worker injury by precautionary practices in mining operations. This article discusses records of mine safety, methods of fire prevention, government regulations, training programs for mine workers, and mine environment problems causing disease. For other aspects of mine operation *see* COAL MINING; MINING, UNDERGROUND.

Safety records. Vigilance is the price of safety in a mine, as it is in any other activity. Fortunately, the presentation of mining hazards in motion pictures and television bears little relation to fact. The Mine Safety Appliances Co. in the late 1930s originated the John T. Ryan trophies, to be awarded annually to the Canadian mines with the best safety records each year. These awards include two national trophies, one for metalliferous mining and one for coal mining, and six regional trophies, two for coal mining and four for metalliferous mining. The awards are based on the number of lost-time accidents during 1,000,000 man-hours of exposure for metalliferous mines, and for 120,000 man-hours of exposure in coal mines.

A lost-time accident is defined as one that results in a compensation payment for a total or partial disability that lasts more than 3 days. In Canada, before Aug. 1, 1968, a 3-day waiting period was required in some provinces before an injury became compensable. The records of the workmen's compensation boards are used to compile safety records, and all locations of employment are included, such as mine, mill, offices, and shops, but smelters and open-pit or strip-mine operations are not included.

The Canada John T. Ryan trophy for metalliferous mines is usually won by a mine with no lost-time accidents as defined. A mine with four or five accidents is likely to be disqualified from the countrywide competition. A fatal accident eliminates a mine. The coal mining record is not as good as that of the metalliferous mines; for one thing, there are fewer coal mines. The low incidence of accidents is not attained without effort. Supervisors, like traffic police, spend a lot of their time trying to keep people from acting foolishly. In North America governments have regulations regarding the safety of workmen in mines and inspectors to see that they are observed.

Fire hazards. Government regulations for mine operation and the safety of workmen in mines were established in North America during the 19th century, patterned on those of Europe and Great Britain, where there was a longer history of coal mining than in North America. The regulations were first directed to coal mining because it presents the hazard of combustible gas from bituminous coal and flammable dust. Coal is made up of volatile material and fixed carbon. If the ratio

$$\frac{\text{volatile}}{\text{volatile} + \text{fixed carbon}}$$

exceeds 0.12, the dust can burn explosively when mixed with air in dangerous proportions, and the range of dangerous proportions is quite wide. All bituminous coals have a ratio in excess of 0.12. Combustible gas is especially dangerous because it is mobile and if it is ignited the ensuing fire or explosion may affect men remote from the source of trouble. Other accidents, a fall of rock, for instance, do not affect men working outside the location of the accident.

Methane gas, the major component of the combustible gas, is dangerous in two ways. It dilutes the mine air and so reduces the amount of life-sustaining oxygen. Coal mines are classified as gassy, slightly gassy, and nongassy. A nongassy mine has less than 0.05% methane in the air. If methane is in a concentration of 5–13.9%, it may be ignited and will burn explosively. The most violent explosive mixture is 9.4% methane. At some point, as the location of the emission is approached, the explosive ratio will be reached; a single spark can set off an explosion. The resultant turbulence in the air will stir up the coal dust and the fire can spread with explosive violence. *See* METHANE.

A mine fire has side effects. Men may be caught in the actual fire, but the burning produces carbon monoxide and carbon dioxide which will spread far beyond the fire area. If the roof is sulfurous shale, it may be brought down by the heat.

Prevention of ignition. The propagation of a fire in the dust stirred up by the air turbulence accompanying an explosion may be inhibited by dusting.

An incombustible rock dust is spread throughout the workings. When stirred up, it dilutes the mine dust and absorbs heat, so combustion cannot be maintained. The initiation of combustion must be prevented.

Equipment. In the United States only equipment certified by the U.S.B.M. (United States Bureau of Mines) as permissible may be used where gas or combustible dust may be present. The restriction applies to electrical equipment, including cap lamps and machinery. The certification is based on the no-sparking qualities of switches and adequate current-carrying capacities for electrical equipment; and on ensurance that machinery has no operating parts that can heat to the ignition temperature for gas or dust. In Canada a British certification is also accepted.

Explosives. Only permitted explosives may be used. All explosives involve combustion and high temperatures. The U.S.B.M. has certified certain explosives as permissible for use in coal mines. They are compounded to produce a lower temperature and volume of flame than ordinary explosives, and certain salts are included in the formulas to quench the flame rapidly. Permitted explosives must be used as prescribed for prescribed conditions. Special attention is required for stemming to prevent blowouts. *See* EXPLOSION AND EXPLOSIVE.

Safety lamp. The biggest single advance in coal mine safety was probably the invention of the Davy safety lamp in 1815. A mantle of wire gauze around the flame dissipated the heat from the flame to below the ignition temperature of methane, which is about 650°C. It provided light and the miner no longer had to work with an open flame. The height and color of the flame gave a measure of the amount of methane in the air. Improved models are still in use, though they are being superseded by no-flame instruments that give a prompt reading or can monitor the methane content continually. The miner is no longer dependent on a flame for light since the electric cap lamp became available.

Ventilation. Coal mine operators have developed ventilation techniques, which until recently, surpassed those used in metal mines. Great care is taken to sweep working places with enough air to prevent dangerous accumulations of gas or of dust. The advent of mechanical coal-cutters and other mechanical equipment has made ventilation more difficult, but the problem has been overcome.

Other mine fires. Methane is not confined to coal mines. It is occasionally released from pockets in metalliferous mines. It is seldom troublesome unless it has accumulated in an unused unventilated working or in a sump at the bottom of a shaft.

Sulfide ores containing certain sulfide minerals may oxidize and generate sulfur dioxide. This may generate enough heat to ignite wood. Pyrrhotite should always be suspected as such an oxidizing sulfide.

Dry, timbered mines in which there is careless smoking are obvious fire hazards. Most metalliferous mines are damp and vigorous fires are not expected, but smoldering fires may be even more hazardous. They may burn for a long time without being detected and the incomplete combustion generates carbon monoxide which is heavier than air, odorless, and lethal. The fire is usually in rubbish accumulated in abandoned places. Good housekeeping is required. Many fires are started in old power cables when the insulation has broken down, and a surprising number start during locomotive battery charging. Sparks from acetylene torch burning are another common source of fire.

Mine rescue teams. The U.S.B.M. officials have accumulated a vast store of fire-fighting knowledge. Most mines have teams of men trained in mine rescue and in the use of U.S.B.M.-approved equipment in accordance with manuals prepared by the U.S.B.M. officials. They work under their own supervisors but U.S.B.M. officials are available for consultation.

Government regulations. Government regulations for safety in mine operations must be enforced. The higher the government authority and the less localized the enforcement agent, the more effective the enforcement will be. The basic enforcement unit in the United States is the state authority, and even obviously good regulations are often hard to enforce. In addition ot the human tendency to resist change, pressure may be brought to bear on the legislatures of the mining states. Some government inspectors may be appointed for political expediency; in the past the company safety inspector was too often an employee given a sinecure in lieu of a pension.

The U.S.B.M. has been assigned responsibility for the health and safety of United States miners. The period of 1907–1913 was formative. States are jealous of their prerogatives, and so it was 1941 before the U.S.B.M. officers had the right to enter a mine. They were chiefly involved by invitation after a disaster. They attained the right to order a coal mine closed for dangerous practices only in 1952. Legislation in 1969 resulted in a safety code to apply to all mines, coal and metalliferous. The initial responsibility continues to be with the state authority, and the U.S.B.M. will interfere only when the state regulations and enforcement do not meet the requirements of the Federal code.

Regulations in Canada are prepared and enforced by the provinces. The Northwest Territories, Yukon, and the Arctic Islands are not provinces and are under the authority of the federal Canadian government. The Ontario metal mining code is the most comprehensive and has been used as a model for the preparation of the codes in other provinces and countries. There is no coal mining in Ontario. The coal mining provinces based their regulations on British codes originally, and these regulations have since been modified in the light of experience in the United States.

Occurrence of accidents. The following discussion is based on data from the *Ontario Department of Mines Inspection Branch Annual Report for 1967*, which gives accident experience for an average of 17,461 men who put in 32,391,000 man-hours of work in underground mines and open pits of all sizes and degrees of organization. There were 1887 lost-time accidents. The experience did not vary much from that of the preceding 4- or 5-year period.

The time distribution for lost-time accidents was 1 per 17,100 man-hours worked or, using 1850 man-hours per man-year, 1 per 9.4 man-years. Whether or not a man has had an accident during any one year has no bearing on whether he could have another in the same year or any other year, so this is a Poisson distribution. A man would have a 0.33 probability of having no lost-time accidents in 9.4 years.

Actually, the mathematical probability is misleading because a man is more apt to have an accident in his earlier years of employment before he becomes mine-wise, or if he is employed in a developing mine before a safety program is well organized.

The 16 fatal accidents had an incidence of 1 per 2,020,000 man-hours worked, or 1 per 1090 man-years. Mathematically a group of 100 men working 10 years would have about 0.37 probability of having no fatal accident. That is subject to the same reservations as those set out above for the lost-time accidents.

Training. The necessity for training men to work safely cannot be overstated. Of the 1887 nonfatal accidents in Ontario in 1967 about 34% (641) were personal accidents—fall of persons, or strains while moving or lifting or handling material other than rock or ore. There is reason to suspect laxity in the use of safety equipment, such as gloves and safety belts and boots; ineffective instruction on how to lift; and insufficient emphasis on the importance of good housekeeping in working places. It must be emphasized that the data are from all sizes of mines and from mines that range from those operated under rather primitive conditions to those using modern equipment and management. Some of them would be close runners-up in the Canada John T. Ryan trophy competition, and one of them won the Regional trophy. Others were barely passing the government inspections. One runner-up for the Canada trophy had 2 accidents and the next had 3 accidents in 1967. The winner of the Ontario Regional trophy had 7 accidents per 1,000,000 man-hours.

The most lethal class of accident is the fall of rock or ore. About 10% (185) of all accidents, including fatal accidents, in Ontario in 1967, were falls of rock or ore, but 31% (5) of the fatal accidents were from that cause. The real cause was probably the failure to recognize an unsafe condition due to neglect or lack of experience. Either cause requires better training of the supervisors and the workmen. Nearly half of the accidents occurred during drilling or scaling of loose rock, clearly showing a lack of skill or the neglect of an unsafe condition.

The advent of mechanized mining in large deposits requires greater areas of exposure to provide room to maneuver equipment. Fortunately, it has been accompanied by the extended use of rock bolts as a means for rock reinforcement, and the incidence of falls of rock has been reduced. The illustration shows an experimental stope in a mine where U.S.B.M. engineers tested the effectiveness of rock bolting.

The ultimate in mine safety cannot be achieved without the complete support of top management, but the key person in the achievement is the workman, and in the chain of responsibility it is the supervisor at the lowest level of contact with the workman. Though he may have the best of intentions, the supervisor is handicapped because men work in scattered places, out of his sight except for short intervals each day. That condition has improved in mines that work larger deposits with me-

Rock bolting in a Michigan copper mine. (*Photograph by H. R. Rice*)

chanical equipment. There is a concentration of working places and the crews are under less intermittent surveillance.

Most mines have a staff safety engineer responsible only to top management. The safety department usually has no line authority. The most important duty that the safety engineer has is the education of supervisors and workmen by lectures and training sessions.

It must be recognized that any accident could be fatal, even a neglected scratch from a nail. Furthermore, an accident need not involve injury to a person. Any undesirable happening at any time or in any place that has not been foreseen is an accident. It may be only by chance that there is no personal injury.

Environmental problems. Instantly recognizable injuries to persons attract more attention, but for ages miners have been subject to pneumoconiosis. This includes all lung diseases, fibrotic or nonfibrotic, caused by breathing in a dusty atmosphere. It takes time to develop, and began to be recognized as something to be eliminated early in the 20th century. Most varieties are not fibrotic and will clear up when the person is out of the dust-laden atmosphere; silicosis and the effects of breathing radioactive dust or gas, however, will not.

Silicosis. South African mine doctors led in the diagnosis of silicosis. About 1920 mine operators in North America became aware of the magnitude of the problem and moved to do something about it. It is caused by inhaling silica dust, and possibly some other mineral dusts. It is not confined to mining. Any industry that produces a silica-dusty atmosphere is dangerous. Furthermore, a clear-looking atmosphere may still be dangerous be-

cause the particles that are harmful are less than 5 μ (0.001 mm) in size. At that size they settle slowly and do not show in a beam of light.

The small particles pass the filtering hairs and mucus in the nose and throat. They reach the lungs and by some action, mechanical or chemical, create fibrosis. The useful volume of the lung is reduced. The disease is seldom lethal in itself but the victim is susceptible to tuberculosis and pneumonia.

As soon as the cause was recognized, the operators and the government authorities increased ventilation requirements and insisted on water sprays, particularly after blasting, to knock down the dust. Wet drilling was well established at that time but there have been some improvements in the drills. The dust content of the mine air at working places and elsewhere was measured, especially in the fine sizes because the presence of coarse dust is easily detected. What are thought to be safe, or acceptable, working levels have been established.

All employees were examined by x-ray for fibrosis in the lungs, and for a proneness to develop silicosis. Some lung shapes are more likely to develop it than others are. Employees with developing fibrosis were given work in dust-free locations, or treatments and pensions. Employees are now examined at least yearly to detect the disease in the primary stage so that prompt action may be taken, and the incidence of silicosis is lessening. Aluminum dust sprayed into the air in mine change houses and underground has been found to have a prophylactic, and possibly therapeutic, effect. Ventilation is the only completely effective remedy.

Radioactivity. It was observed that in some mine areas silicosis seemed to be much more virulent than in others. This was thought to be due to some additive to the silica. It was found that coal dust speeded up the development of the disease.

When uranium mining got under way, it was found that the rocks emitted radon gas which broke down in the lungs and produced lung cancer. When the instruments developed for radioactivity research became available, it was learned that many rocks that had a low level of radioactivity emitted small amounts of radon gas, which in some cases were carried into the mine workings in the mine water and released under the reduced pressure. That provided one reason for the virulence of silicosis in some mining areas.

The government authorities moved quickly and established safe levels of radioactivity and safe exposure time in those levels. The regulations are enforced. The acceptable levels are attained by ventilation.

Other considerations. When diesel engines were introduced underground for motive power, they had to be certified for an acceptable level of carbon monoxide and oxides of nitrogen in the fumes. The required levels were attained by engine design and by scrubbers in the exhaust system. An engine must have a U.S.B.M. certificate of approval, and then it is only approved to travel certain routes for which the government inspectors consider the ventilation to be adequate.

[A. V. CORLETT]

Bibliography: R. S. Lewis and G. B. Clark, *Elements of Mining*, 3d ed., 1964; R. Peele (ed.), *Mining Engineers' Handbook*, 3d ed., vol. 2, 1941.

Motor, electric

An electric rotating machine which converts electric energy into mechanical energy. Because of its many advantages, the electric motor has largely replaced other motive power in industry, transportation, mines, business, farms, and homes. Electric motors are convenient, economical to operate, inexpensive to purchase, safe, free from smoke and odor, and comparatively quiet. They can meet a wide range of service requirements—starting, accelerating, running, braking, holding, and stopping a load. They are available in sizes from a small fraction of a horsepower to many thousands of horsepower, and in a wide range of speeds. The speed may be fixed (or synchronous), constant for given load conditions, adjustable, or variable. Many are self-starting and reversible. For uniformity and interchangeability, motors are standardized in sizes, types, and speeds. *See* ELECTRIC ROTATING MACHINERY.

Electric motors may be alternating-current (ac) or direct-current (dc). There are many types of each. Although ac motors are more common, dc motors are unexcelled for applications requiring simple, inexpensive speed control or sustained high torque under low-voltage conditions.

Motor classification. Motors are classified in many ways. The following classifications show some of the many available variations in types of motors.

1. Size: flea, fractional, or integral horsepower.
2. Application: general purpose, definite purpose, special purpose, or part-winding start. May be further classified as crane, elevator, pump, and so forth.

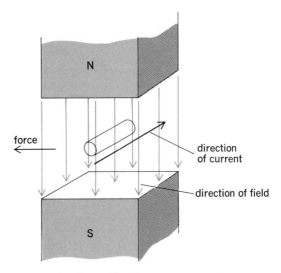

Relative directions of field flux, current, and force.

3. Electrical type: alternating-current induction, synchronous, or series; direct-current series, permanent magnet, shunt, or compound.
4. Mechanical protection and cooling: (*a*) open: dripproof, splashproof, semiguarded, fully guarded, externally ventilated, pipe ventilated, weather protected; (*b*) totally enclosed: nonventilated, fan cooled, explosionproof, dustproof, ignitionproof, waterproof, water cooled, water-air cooled, air-to-air cooled, pipe ventilated, fan cooled guarded.
5. Speed variability: constant speed, varying speed, adjustable speed, adjustable varying speed, multispeed.
6. Mounting: floor, wall, ceiling, face, flange, vertical shaft.

Characteristics. Each electrical type of motor has its own individual characteristics. Each motor is selected to meet the requirements of the job it must perform. For individual motor characteristics *see* DIRECT-CURRENT MOTOR; INDUCTION MOTOR; REPULSION MOTOR; SYNCHRONOUS MOTOR. For comparison of all ac motors *see* ALTERNATING-CURRENT MOTOR.

Principles of operation. When a conductor located in a magnetic field carries current, a mechanical force is exerted upon it (see illustration). This force has the value shown in Eq. (1), where i

$$F = Bil \qquad \text{newtons} \qquad (1)$$

is the current in amperes, B is the magnetic density in webers per square meter, and l is the conductor length in meters. The illustration shows the relative directions of current, field, and force. The force reverses with either current or field reversal, but not when both are reversed. The torque T is the product of this force and the rotor radius.

If the conductor moves in the direction of F, an emf e is generated which opposes the current (motor action). If the conductor is moved against F, this emf will assist the current (generator action). Its value is shown in Eq. (2), where $v =$ velocity of conductor across the flux in meters per second.

$$e = vBl \qquad \text{volts} \qquad (2)$$

The product ei represents the power converted,

in watts, shown in Eqs. (3) and (4), which are the

$$\text{Motor output} = ei - \text{rotative loss} \quad (3)$$
$$\text{Generator shaft input} = ei + \text{rotative loss} \quad (4)$$

bases for the emf and output formulas of dynamo machinery. Many machines will operate as either a motor or a generator, but they should be designed for the particular service.

Current may be fed into the field and armature by conduction, as in dc machines and ac series motors, or into the stator by conduction and the rotor by induction, as in ac induction and repulsion motors.

[ALBERT F. PUCHSTEIN]

Bibliography: A. E. Fitzgerald and C. Kingsley, *Electric Machinery*, 2d ed., 1961; H. Majmundar, *Electromechanical Energy Converters*, 1965; G. J. Thaler and M. L. Wilcox, *Electric Machines*, 1966.

Motor-generator set

A motor and one or more generators, with their shafts mechanically coupled, used to convert an available power source to another desired frequency or voltage. The motor of the set is selected to operate from the available power supply; the generators are designed to provide the desired output. Motor-generator sets are also employed to provide special control features for the output voltage.

The principal advantage of a motor-generator set over other conversion systems is the flexibility offered by the use of separate machines for each function. Assemblies of standard machines may often be employed with a minimum of engineering required. Since a double energy conversion is involved, electrical to mechanical and back to electrical, the efficiency is lower than in most other conversion methods. In a two-unit set the efficiency is the product of the efficiencies of the motor and of the generator.

Motor-generator sets are used for a variety of purposes, such as providing a precisely regulated dc current for a welding application, a high-frequency ac power for an induction-heating application, or a continuously and rapidly adjustable dc voltage to the armature of a dc motor employed in a position control system. *See* GENERATOR, ELECTRIC; MOTOR, ELECTRIC.   [ARTHUR R. ECKELS]

Natural gas

A combustible gas that occurs in porous rock of the Earth's crust and is found with or near accumulations of crude oil. Being in gaseous form, it may occur alone in separate reservoirs. More commonly it forms a gas cap, or mass of gas, entrapped between liquid petroleum and impervious capping rock layer in a petroleum reservoir. Under conditions of greater pressure it is intimately mixed with, or dissolved in, crude oil.

Composition. Typical natural gas consists of hydrocarbons having a very low boiling point. Methane, CH_4, the first member of the paraffin series, and with a boiling point of $-254°F$, makes up approximately 85% of the typical gas. Ethane, C_2H_6, with a boiling point of $-128°F$, may be present in amounts up to 10%; and propane, C_3H_8, with a boiling point of $-44°F$, up to 3%. Butane, C_4H_{10}; pentane, C_5H_{12}; hexane; heptane; and octane may also be present. Structural formulas of four of these compounds are given here.

Whereas normal hydrocarbons having 5–10

Methane · Ethane · Normal pentane · iso-Pentane

carbon atoms are liquids at ordinary temperatures, they have a definite vapor pressure and therefore may be present in the vapor form in natural gas. Carbon dioxide, nitrogen, helium, and hydrogen sulfide may also be present.

Types of natural gas vary according to composition and can be dry or lean (mostly methane) gas, wet gas (considerable amounts of so-called higher hydrocarbons), sour gas (much hydrogen sulfide), sweet gas (little hydrogen sulfide), residue gas (higher paraffins having been extracted), and casinghead gas (derived from an oil well by extraction at the surface). Natural gas has no distinct odor. Its main use is for fuel, but it is also used to make carbon black, natural gasoline, certain chemicals, and liquefied petroleum gas. Propane and butane are obtained in processing natural gas. *See* PETROLEUM PRODUCTS.

Distribution and reserves. Gas occurs on every continent (see table). Wherever oil has been found, a certain amount of natural gas is also present. In production and known reserves the United States stands first among nations. Six states account for more than 90% of the known reserves (Texas, Louisiana, New Mexico, Kansas, Oklahoma, and California). Among these states Texas has 43% of the total with 125×10^{12} ft³, and Louisiana ranks second with 29% of the total with 83×10^{12} ft³. The estimated known reserves in the United States at the end of 1967 were 293×10^{12} ft³. Consumption in the United States in 1967 was 18×10^{12} ft³. Since 1960 the annual rate of finding new reserves in the United States is about 20×10^{12} ft³. Natural gas is being discovered in western Canada at a rapid rate, and at the beginning of 1967 the proven reserves were 43×10^{12} ft³.

Long before supplies of natural gas run out or

World gas reserves, Dec. 31, 1967

Region	Volume	
	10^{12} m³	10^{12} ft³
North America	9.6	338
South and Central America	1.9	69
Western Europe	3.6	129
Africa	4.8	168
Eastern Europe, including China and Soviet Union	4.8	171
Near East	6.4	225
Middle and Far East, including Australia	1.1	38

become expensively scarce, it is expected that some process of coal gasification will produce a gas which is completely interchangeable with natural gas and at a competitive price. This is important because coal makes up a majority of the world's known fossil fuel reserves. But since energy consumers have indicated in the marketplace their preference for fluid and gaseous fuels over the solid forms, coal gasification research, already well under way, will be given additional impetus. *See* COAL GASIFICATION.

In estimating gas reserves, the volumetric method is preferred. The volume of the reservoir is determined by means of the thickness, porosity, and permeability of the producing zones. A study of many depleted fields suggests that about 85% of all gas in dry-gas reservoirs is recovered. Some engineers use the production versus pressure-decline method. They calculate future production by plotting past production against the decline in reservoir pressure.

In California 60% of the gas is associated with oil, but in western Texas the percentage is about 40%. The percentage figure for the United States as a whole is only 23%. This means that a large proportion of the reserves is stored in such dry-gas fields as the Hugoton-Panhandle area of Kansas, Oklahoma, and Texas; Monroe, Louisiana; Carthage, northeastern Texas; Bethany-Waskom, Texas; Big Sandy, Kentucky; Sligo, Louisiana; Blanco Mesaverde, New Mexico; Red Oak – Norris, Oklahoma; Long Lake, Texas; Katy, Texas; San Salvador, Texas; Joaquin-Logansport, Texas; Lake Arthur, Louisiana; Big Piney, Wyoming; Chocolate Bayou, Texas; and Kenai and Cook Inlet, Alaska; there are also about 20 other so-called giant fields in the United States. In addition there are at least 12 giant (10^{12} ft^3 or more) gas fields already discovered in offshore Louisiana in the Gulf of Mexico. Furthermore, prospects of offshore activity in other parts of the world, especially Africa, Asia, the North Sea, and South America, are tremendous for the development of future giant gas fields. This is made possible by the almost incredible technological advances in deep-water drilling, which has enabled explorers to drill in waters in excess of 10,000 ft deep as compared with some 200 ft in depth in the late 1950s. In western Canada some of the large gas fields are the Pincher Creek, the Waterton, and the Jumping Pound. The largest dissolved-gas area in the United States lies along the Gulf Coast of Texas and Louisiana and contains about 50% of the total known reserves of associated gas. Offshore drilling in the waters of the Gulf will add considerably to these reserves. In an average year slightly over 91% of the gas produced is marketed, while 7.7% is used for repressuring, and 1.4% is vented or wasted. In earlier years a much larger percentage was piped away from oil fields and burned. *See* OIL AND GAS, OFFSHORE.

Geological associations. Natural gas is present in every system of rocks down to the Cambrian. The first gas deposits found in the United States were those in the Eastern states. In New York and Pennsylvania 85% of the gas came from Devonian rocks. In West Virginia, Kentucky, and eastern Ohio, Devonian and Mississippian rocks rank nearly equally, but Silurian rocks are also important. In Indiana and Illinois, Pennsylvanian rocks

outrank the Mississippian. The Hugoton-Panhandle field is one of the largest in the world, and the Permian dolomites produce gas from five different levels. The fact that oil is found lower down in Pennsylvanian and older rocks proves the superior migratory capacity of gas. Up to the end of 1966 this field has produced about 38×10^{12} ft^3 of the original 70×10^{12} ft^3 initial reserves. This field has a high percentage of nitrogen (almost 15%).

Oklahoma and the western part of Texas have gas in many stratigraphic zones, from the Permian down to the Cambrian. Most of it is associated with crude oil, either in solution or in the form of gas-cap accumulations. The Carthage field is in northeastern Texas, where 10 different layers in the Trinity division of the Cretaceous system have been found productive. The cumulative total production to the end of 1966 was 311×10^{12} ft^3. During 1967 gas production in the whole state of Texas was about 7.1×10^{12} ft^3, of this amount 72% was gas-well gas. Cretaceous rocks are the principal reservoir rocks in northern Louisiana and in Mississippi. Throughout the Rocky Mountain states various layers in the Cretaceous system account for most of the gas. There are many dry-gas pools, of which the outstanding are the Blanco, northwestern New Mexico; the Baxter Basin, southwestern Wyoming; and the Cedar Creek, southwestern Montana. In California gas production is derived from various layers in the Tertiary system. Although about 60% of the gas is associated with oil reservoirs, there are a number of dry-gas fields. *See* PETROLEUM GEOLOGY.

Of much interest is the trend toward low-temperature transportation and storage of liquid petroleum gas and methane. The ability to store frozen gas (as a liquid) underground will enable pipeline companies to more efficiently meet the cyclical demands of the seasons. Storage of gas close to markets will do away with the necessity of having the large pipeline capacity, which is needed to take care of peak seasons but which lies more or less idle during periods of lessened demand. The ability to condense 600 ft^3 of gas into 1 ft^3 of liquid opens great possibilities for the movement of gas across oceans in tankships. Through such transportation remote areas of the world can become consumer areas, fuel-short consumer areas will have access to needed supplies, and producing areas will benefit from new revenues. *See* LIQUEFIED NATURAL GAS (LNG); OIL AND GAS STORAGE.

Helium, which has many industrial uses, is a by-product of natural gas and is present in some fields. The Rattlesnake field in New Mexico contains 7.5%, the highest percentage of helium to total gas content found up to 1968.

[MICHEL T. HALBOUTY]

Pipeline distribution. The analysis and design of natural-gas distribution network systems have undergone tremendous change in the 1970s. The Hardy Cross method continues to form the analytical basis for all steady-state network studies, but rather than the tedious trial-and-error hand calculations, more efficient and sophisticated computer techniques are used. Design procedures that are based upon the classical optimization theories such as linear programming and dynamic programming have been useful in isolated applications. The most exciting aspect in current transmission and distribution simulations is the inclusion of the unsteady behavior of the natural gas in the system. Along with this added degree of sophistication has come the realization that potential benefits exist, not only for system operation but also for improved design strategies that include the interaction of potential capital expansion and existing system operating characteristics.

Steady-state analysis. Given a piping network configuration with specified deliveries and supply points, the normal analysis problem consists of identifying the pressures and pipeline flows throughout the system. A large number of possible groupings of known and unknown pressures and flows are possible. However, in a specific analysis, the known variables are generally apparent so that a solution to the problem is feasible. A satisfactory solution is achieved when Kirchhoff's first law (continuity equation) is satisfied at each pipe junction or node, and for the system as a whole $\Sigma Q = 0$, where Q is the flow rate. Kirchhoff's second law for each loop in the system must also be satisfied, that is, $\Sigma \Delta p = 0$ along each closed path, where Δp is the pressure drop in each pipeline. When the physical laws governing flow through elements that connect nodes are satisfied and either of the Kirchhoff laws is satisfied, then the other is automatically satisfied and the system is balanced. Such an element relationship—for example, the basic pipeline—is described by an equation of the form $\Delta p^2 = RQ^n$, in which p is pressure, n a constant, and R a function of the pipe geometry, gas properties, and the flow rate Q. Either a loop-balancing procedure or a node-balancing scheme may be used for the analysis. For the fundamental pipeline analysis problem, as opposed to the problem that includes other elements such as compressors, R. Epp and A. G. Fowler provided adequate evidence that the loop-oriented scheme offers some advantage.

In networks that involve different source gases such as liquefied natural gas (LNG), substitute natural gas (SNG), and propane-air mixtures as well as pipeline gas, the specific gravity and the heating value of the components vary widely. M. A. Stoner and M. A. Karnitz described the use of a simulator which balances a system on a thermal or Btu basis and also uses the correct gravity for each branch or element in the system. This scheme enables the user to trace the ultimate destination in the network of the different source gases.

Design. Typical design problems may encompass innumerable variables, ranging from the obvious pressure levels, deliveries, and pipe sizing to conduit alignments, location of compressor stations, location of storage, compression horsepower, and even staging component installation to optimize total cost over the design-life of the system. In this article the discussion is limited to determinate design criteria in which feasible alignments, pressure constraints, and loads are provided. This restriction is desirable in order that a more specific problem may be addressed. Even within these limitations a direct procedure to produce the optimal natural gas distribution network design was not available in 1975.

In view of the vastly improved capabilities of current analysis programs, the design of complex network systems, and particularly of additions to existing systems, is entirely feasible through parametric studies. A design is assumed, analyzed, altered, and reanalyzed until the design objectives are satisfied. Computer software companies currently provide programs for use at remote computer consoles at which the designer is able to interact with an analysis program to improve the design of the system. As a side benefit to this type of parametric study, an improved understanding of the operating characteristics of the system is gained by the design engineer.

In 1972 Stoner borrowed a technological development in water distribution analysis and design and presented it along with additional theoretical advances as a convenient, practical, efficient tool for natural-gas network design. Included in the study, in addition to the standard pipeline system elements, are compressors, control valves, and storage fields. The mathematical model is constructed by writing the continuity equation at each node in the system. The flow equation for each element connected to the node is substituted to eliminate the element flow. A set of nonlinear simultaneous equations result, which are solved by the n-dimensional Newton-Raphson method. A very sparse coefficient matrix exists in this formulation, and Stoner's procedure takes advantage of this sparsity in both computer core storage savings and execution time efficiency. A further development was also proposed which involves a sensitivity study, that is, a convenient means of observing the interaction of system variables without independent solutions of the entire system for each new set of specified variables. Such questions as the approximate additional required horsepower at a remote station to provide an increased nearby nodal pressure of 10 psi may be answered very efficiently. This direct network element design procedure and sensitivity capability are necessary forerunners to packaged optimal design methods.

Unsteady flow. In the actual operation of a distribution system, the volume of gas stored in the pipelines is extremely important in meeting peak-flow demands. This line pack may properly be accounted for in a transient analysis of a system, whereas the potential benefits of the stored volume are ignored in the conventional steady-state study. Inasmuch as real piping systems rarely attain steady-state operation due to changes in demand, supply, failures, additions of equipment, and so forth, there is much motivation for the capability to analyze unsteady flows. H. H. Rachford presented a study of increased operating efficiency and fuel conservation by improved operating strategies in the face of time-varying sales. Operation times for accommodating sudden large loads imposed on a system may be accurately predicted with a transient simulator. In the design process, compressor stations may be optimally positioned for maximum utility. In the operational phase, a transient simulator can provide invaluable experience in the training of operators to handle new and different systems. One of the major difficulties in analyzing natural gas piping, either steady or unsteady, is the accurate evaluation of the frictional characteristics of the system. In 1974 Stoner and Karnitz presented an adaptation of an unsteady flow simulator to utilize real time-varying field data to ascertain the pipeline friction parameters.

In any transient simulation, the partial differential equations of motion and mass continuity and an equation of state are treated numerically to describe the unsteady behavior of the fluid in the pipeline. A total system response is available when boundary conditions are introduced at pipeline junctions and variable supply, or use rates are specified by compressor stations, well fields, or valve operations. In 1974 E. B. Wylie and co-workers presented an analysis procedure that utilizes the method of characteristics to numerically model the equations. Other procedures have also been used in the development of simulators of varying degrees of sophistication, as described by Rachford and T. Dupont in 1974.

The industry has been slow to adopt these simulation procedures. In view of current heavy demands on the limited sources of supply of natural gas, the transmission phase of total operation becomes increasingly important. Transient simulators are essential for leak and failure detection, for optimal system operation. They also represent an essential element in the implementation of automatically controlled systems. See PIPELINE.

[E. BENJAMIN WYLIE]

Bibliography: American Gas Association, Inc., American Petroleum Institute, and Canadian Petroleum Association, *Reserves of Crude Oil, Natural Gas Liquids, and Natural Gas in the United States and Canada*, vol. 22, 1968; American Petroleum Institute, *Petroleum Facts and Figures*, 1967; J. A. Clark, *The Chronological History of Petroleum and Natural Gas*, 1963; R. Epp and A. G. Fowler, *J. Hydraul. Div. ASCE*, vol. 96, no. HYl, 1970; A. M. Leeston, J. A. Chrichton, and J. C. Jacobs, *The Dynamic Natural Gas Industry*, 1963; E. J. Neuner, *The Natural Gas Industry*, 1960; H. H. Rachford, *Oil Gas J.*, pp. 93–96, July 16, 1973; H. H. Rachford and T. Dupont, *J. Petrol. Eng. AIME*, vol. 14, no. 2, 1974; M. A. Stoner, *J. Petrol. Eng. AIME*, vol. 12, no. 1, 1972; M. A. Stoner and M. A. Karnitz, *Oil Gas J.*, pp. 97–100, Dec. 10, 1973; M. A. Stoner and M. A. Karnitz, *Transp. Eng. J. ASCE*, vol. 100, no. TE3, 1974; E. B. Wylie, V. L. Streeter, and M. A. Stoner, *J. Petrol. Eng. AIME*, vol. 14, no. 1, 1974.

Natural gas and sulfur production

Hydrogen sulfide, or sour gas, is a highly toxic gas that can cause failure in certain materials and is associated with the deposition of sulfur in the Earth. Sulfur deposition has been recognized as a serious problem in petroleum reservoirs containing high content of hydrogen sulfide, sulfur, hydrogen, and carbon dioxide. Moreover, depending on pressure and temperature, elemental sulfur may precipitate from the fluid mixtures and cause serious plugging of the formation, tubing, and surface equipment. A practical scheme for producing this type of fluid from a deep reservoir has been proposed, and it may avoid the problem of impairment to the formation by sulfur deposition. This has been accomplished by analysis taking into account the flow of gas mixtures accompanying sulfur deposition in porous media.

Mechanisms of sulfur deposition. A survey of operations of sour gas wells in Canada and Europe was reported by J. B. Hyne. In 9 of 31 cases tabulated, deposition of sulfur in tubing was observed.

This survey revealed that sulfur precipitation almost always occurs in wells if the content of pentane-plus in the reservoir fluid is low and that the content of carbon dioxide is not an important factor. High temperatures at bottom hole and wellhead, and low pressure at wellhead, were also found to provide favorable conditions for sulfur deposition in tubing. These results may be accounted for by the solubility behavior of sulfur in a reservoir fluid. For example, J. G. Roof measured the solubility of sulfur in hydrogen sulfide at pressures ranging from 1020 to 4520 psia (1 psia equals 6890 N/m²) and temperatures varying from 110 to 230°F (43 to 110°C). His results illustrated that the solubility change per unit pressure drop is greater at higher temperature. Therefore, if the temperature in the tubing is high and the pressure at the wellhead is low (large difference between the bottom-hole and wellhead pressures), the solubility reduction between the bottom-hole and wellhead conditions is very significant. Consequently, a large quantity of solid sulfur can precipitate from the solution for a given production rate, and serious plugging in the tubing can occur. This sulfur deposition can be prevented, for instance, by circulating hot fluids under pressure from a specific location in the tubing.

The foregoing evidence confirms that for a given fluid containing dissolved sulfur, the elemental sulfur can precipitate from the solution because of solubility reduction. Furthermore, the solubility of sulfur in the solution is controlled mainly by pressure and temperature. Since the fluid flow in the formation can be considered isothermal under normal operating conditions, pressure remains the most important factor.

Prediction of performance. A mathematical model has been formulated by C. H. Kuo to describe the flow of a fluid accompanied by solid precipitation in porous media. In such a treatment, a mixture of hydrogen sulfide, hydrogen, sulfur, and carbon dioxide can be considered as a homogeneous solution contained in a formation of porosity ϕ_i at its initial condition. As the reservoir, bounded by impermeable cap and base rocks, is depleted, the contained fluid flows radially toward the production well, which is located at the center of the reservoir. Some of the initially dissolved component may precipitate from the solution as the pressure declines. In view of the experimental evidence discussed earlier, it can be assumed that the solid deposits formed are incapable of flowing and accumulate in the voids.

The velocity of the isothermal, horizontal fluid flow through this porous medium is assumed to be governed by the one-dimensional form of Darcy's law, and the vertical velocity component and gravity effects are unimportant. Thus, the equations of continuity for the mixture (solution) and the sulfur component can be written as Eqs. (1) and (2).

$$\frac{1}{r}\frac{\partial}{\partial r}\left(\rho\frac{k}{\mu}r\frac{\partial p}{\partial r}\right) = \frac{\partial}{\partial t}[\rho\phi + \rho^*(\phi_i - \phi)] \quad (1)$$

$$\frac{1}{r}\frac{\partial}{\partial r}\left(R\frac{k}{\mu}r\frac{\partial p}{\partial r}\right) = \frac{\partial}{\partial t}[R\phi + \rho^*(\phi_i - \phi)] \quad (2)$$

In Eqs. (1) and (2) the density of the solution ρ and the solubility of sulfur in the solution R are

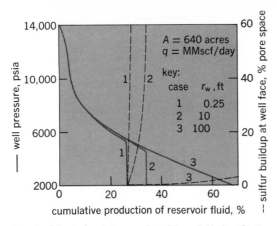

Fig. 1. Effect of well-bore radius. (*From C. H. Kuo, On the production of hydrogen sulfide–sulfur mixture from deep formations, J. Petrol. Technol., 24(9):1142–1146, 1972*)

considered functions of pressure. The density of the deposited sulfur ρ^* is assumed constant, since there is little information in the literature regarding the influence of pressure on this density. The permeability k and the viscosity μ of the fluid are treated as functions of porosity and pressure respectively.

These partial-differential equations can be solved by numerical methods to predict the pressure p and the porosity ϕ as functions of time t and radial distance r. If the fluid is initially undersaturated with the sulfur, it is necessary to solve only Eq. (1) in the early production period to obtain the pressure distribution ($\phi = \phi_i$). Once the reservoir is depleted below the saturation pressure, however, Eqs. (1) and (2) must be solved simultaneously to predict $p(r,t)$ and $\phi(r,t)$, since the pore space is partially filled by deposited sulfur. The fraction of the pore space occupied by the deposited sulfur is obtained, then, as $1 - \phi/\phi_i$.

Optimum production scheme. As an example, theoretical performances of a reservoir similar to that discovered in southern Mississippi will be examined. The gas mixture was found to exist

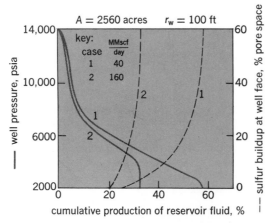

Fig. 2. Effect of production rate. (*From C. H. Kuo, On the production of hydrogen sulfide–sulfur mixture from deep formations, J. Petrol. Technol., 24(9):1142–1146, 1972*)

in a formation 20,000 ft deep (1 ft is about 0.3 m), with a net pay of 300 ft, at 14,000 psi and 390°F (200°C). A flow test indicated that the gas contains 78% hydrogen sulfide, 20% carbon dioxide, and 2% other compounds. In addition, it was estimated that elemental sulfur is dissolved in the mixture at 7500 lb/MMscf (1 lb is 0.45 kg; MMscf represents million standard cubic feet). Assuming an initial porosity of 0.1, the sour gas initially in place is calculated to be 600 MMscf/acre (1 acre is 4047 m²), or 380 and 1530 Bscf, billion standard cubic feet, for 640 and 2560 acres respectively. In the study by Kuo using estimated reservoir and fluid properties, it was assumed that the well can be produced at a constant mass rate q varying from 10 to 160 MMscf/day with 640- or 2560-acre well spacing before the pressure at the well face drops to a minimum value of 2000 psi. Thereafter, the rate decreases, and the minimum pressure is maintained at the well face. The well is shut in when the rate has declined to less than a certain minimum production rate.

The predicted histories of pressure and of sulfur buildup at the well face are plotted against the cumulative production of the reservoir fluid in Figs. 1 and 2. As illustrated in Fig. 1, assuming an original well-bore radius r_w of 0.25 ft (case 1), 25% of the reservoir fluid is produced before the well pressure declines to the estimated saturation pressure, 5500 psia. The well pressure then drops rapidly, and the pore space in the vicinity of the well bore is filled with deposited sulfur before any significant amount of the reservoir fluid is produced during this period. This implies that the formations must be cleaned frequently to remove deposited sulfur. If the formation can be fractured to create a large, effective well-bore radius, however, sulfur plugging may be alleviated, as indicated by the results for cases 2 and 3.

The effect of well spacing can be investigated by comparing the results predicted for case 3 in Fig. 1 and case 1 in Fig. 2. Although the design for case 3 is preferable to alleviate impairment due to sulfur deposition, the high cost of drilling in deep formations may dictate the choice of case 1, designed with wider spacing. Figures 1 and 2 also demonstrate that the effects of well-bore radius and production rate on the well pressure history are unimportant during early production. Therefore, with an effective well-bore radius equal to that obtained in the drilling of the well, it should be feasible to produce the reservoir fluid at a fairly high flow rate for a certain period. Once the reservoir is depleted to the saturation pressure, the formation should be fractured to create a large, effective well-bore radius. This could be the most desirable scheme because fracturing is relatively inexpensive, and, furthermore, a large well-bore radius makes it possible to produce the reservoir fluid for a long period without seriously impairing the formation. *See* PETROLEUM RESERVOIR ENGINEERING.

[C. H. KUO]

Bibliography: T. W. Hamby and J. R. Smith, *J. Petrol. Technol.*, 24(3):347–356, 1972; J. B. Hyne, *Oil Gas J.*, pp. 107–113, Nov. 25, 1968; C. H. Kuo, *J. Petrol. Technol.*, 24(9):1142–1146, 1972; J. G. Roof, *Soc. Petrol. Eng. J.*, 11(3):272–276, 1971.

Nuclear battery

A battery that converts the energy of particles emitted from atomic nuclei into electric energy. Two basic types have been developed: (1) A high-voltage type, in which a beta-emitting isotope is separated from a collecting electrode by a vacuum or a solid dielectric, provides thousands of volts but the current is measured in micromicroamperes ($\mu\mu$a); (2) a low-voltage type gives about 1 volt with current in microamperes (μa).

High-voltage nuclear battery. In the high-voltage type, a radioactive source is attached to one electrode, emitting charged particles. The source might be strontium-90, krypton-85, or hydrogen-3 (tritium), all of which are pure beta emitters. An adjacent electrode collects the emitted particles. A vacuum or solid dielectric separates the source and the collector electrodes.

One high-voltage model, shown in Fig. 1, employs tritium gas sorbed in a thin layer of zirconium metal as the radioactive source. This source is looped around and spot-welded to the center tube of a glass-insulated terminal. A thin coating of carbon applied to the inside of a nickel enclosure acts as an efficient collector having low secondary emission. The glass-insulated terminal is sealed to the nickel enclosure. The enclosure is evacuated through the center tube, which is then pinched off and sealed.

The Radiation Research Corporation model R-1A is 3/8 in. in diameter and 0.531 in. in height. It weighs 0.2 oz and occupies 0.05 in.³ It delivers about 500 volts at 160 $\mu\mu$a. Future batteries are expected to deliver 1 μa at 2000 volts, with a volume of 64 in.³

Earlier models employed strontium-90. This isotope has the highest toxicity in the human body of the three mentioned. Tritium has only one one-thousandth the toxicity of strontium-90. Both strontium-90 and krypton-85 require shielding to reduce external radiation to safe levels. Tritium produces no external radiation through a wall that is thick enough for any structural purpose. Tritium was selected on the basis of these advantages.

The principal use of the high-voltage battery is to maintain the voltage of a charged capacitor. The current output of the radioactive source is sufficient for this purpose.

This type of battery may be considered as a constant-current generator. The voltage is proportional to the load resistance. The current is determined by the number of emissions per second captured by the collector and does not depend on ambient conditions or the load. As the isotope ages, the current declines. For tritium, the inten-

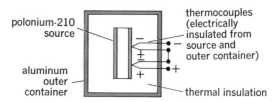
Fig. 2. Thermoelectric nuclear battery.

sity drops 50% in a 12-year interval. For strontium-90, the intensity drops 50% in a 25-year interval.

Low-voltage nuclear battery. Three different concepts have been employed in the low-voltage type of nuclear batteries: (1) a thermopile, (2) the use of an ionized gas between two dissimilar metals, and (3) the two-step conversion of beta energy into light by a phosphor and the conversion of light into electric energy by a photocell.

Thermoelectric-type nuclear battery. This low-voltage type, employing a thermopile, depends on the heat produced by radioactivity (Fig. 2). It has been calculated that a sphere of polonium-210 of 0.1 in. diameter, which would contain about 350 curies, if suspended in a vacuum, would have an equilibrium surface temperature of 2200°C, assuming an emissivity of 0.25. For use as a heat source, it would have to be hermetically sealed in a strong, dense capsule. Its surface temperature, therefore, would be lower than 2200°C.

To complete the thermoelectric battery, the heat source must be thermally connected to a series of thermocouples which are alternately connected thermally, but not electrically, to the heat source and to the outer surface of the battery. After a short time, a steady-state temperature differential will be set up between the junctions at the heat source and the junctions at the outer surface. This creates a voltage proportional to the temperature drop across the thermocouples. The battery voltage decreases as the age of the heat source increases. With polonium-210 (half-life, 138 days) the voltage drops about 0.5% per day. The drop for strontium-90 is about 0.01% per day (20-year half-life).

A battery containing 57 curies of polonium-210 sealed in a sphere 0.4 in. in diameter and 7 chromelconstantan thermocouples delivered a maximum power of 1.8 milliwatts. It had an open-circuit voltage of 42 millivolts with a 78°C temperature differential. Over a 138-day period, the total electrical output would be about 1.5×10^4 joules (watt-sec).

Total weight of the battery was 34 g. This makes the energy output per pound equal to

$$\frac{1.5 \times 10^4}{3600} \text{ watt-hours (whr)} \times \frac{1}{34} \times 454 = 55.6$$

This is the same magnitude as with conventional electric cells using chemical energy. This nuclear energy, however, is being dissipated whether or not the electric energy is being used.

The choice of isotope for a thermoelectric nuclear battery is somewhat restricted. Those with a half-life of less than 100 days would have a short useful life, and those with a half-life of over 100 years would give too little heat to be useful. This

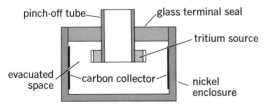
Fig. 1. Tritium battery in cross section.

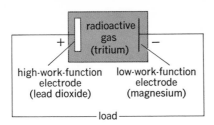

Fig. 3. Gas-ionization nuclear battery (schematic).

leaves 137 possible isotopes. This number is further reduced by the consideration of shielding.

The trend is to use plutonium-238 (unable to support a chain reaction) and strontium-90. The most frequently used thermocouple material is a doped lead-telluride alloy that performs between 200 and 480°C. Another material under investigation is silicon-germanium alloy, which can operate at 800°C (hot junction).

The thermoelectric systems developed so far are in the small power range (5–60 watts), and work with a maximum efficiency of 7%.

When shielding is not a critical problem, as in unmanned satellites, they are extremely convenient, being reliable and light. SNAP-3 and SNAP-9A (Systems for Nuclear Auxiliary Power) have weights in the 1–2 lb/watt range.

The above discussion applies only to a portable power source. Drastic revision would be required if a thermoelectric-type nuclear battery were to be designed for central-station power.

Gas-ionization nuclear battery. In this battery a beta-emitting isotope ionizes a gas situated in an electric field (Fig. 3). Each beta particle produces about 200 current carriers (ions), so that a considerable current multiplication occurs compared with the rate of emission of the source. The electric field is obtained by the contact potential difference of a pair of electrodes, such as lead dioxide (high work function) and magnesium (low work function). The ions produced in the gas move under the influence of the electric field to produce a current.

A cell containing argon gas at 2 atmospheres, electrodes of lead dioxide and magnesium, and a radioactive source consisting of 1.5 millicuries of tritium has a volume of 0.01 in.^{3} and an active plate area of 0.2 in.^{2}, and gives a maximum current of 1.6×10^{-9} amp. The open-circuit voltage per cell depends on the contact potential of the electrode couple. A practical value appears to be about 1.5 volts. Voltage of any value may be achieved by a series assembly of cells.

The ion generation is exploited also in a purely thermal device, the thermionic generator. It consists of an emitting electrode heated to 1200–1800°C and a collecting electrode kept at 500–900°C. The hot electrode is made of tungsten or rhenium, metals which melt at more than 3000°C and have low vapor pressures at the working temperature. The collecting electrode is made out of a metal with a low work function, such as molybdenum, nickel, or niobium. The electrons emitted by the hot electrode traverse a small gap of 1–2 mm to the collector, and recover the emitter by an external circuit.

The ionization space is filled with cesium vapor which acts in two ways. First, it covers the surface of the two electrodes with adsorbed cesium atoms and thus reduces the work function to the desired level. Second, it creates an ionized atmosphere, thus controlling the electron space charge.

The heating of the emitting electrode can be obtained by any of the known means: concentrated solar energy, radioisotopes, or conventional nuclear reactors. While reaching the high temperature was a problem easy to solve, the cooling of the collecting electrode appeared for a long time to be a critical technical problem. Eventually this problem was solved by the development of what is known as the "heat pipe."

In principle the heat pipe is an empty cylinder absorbing heat at one end by vaporization of a liquid and releasing heat at the other end by condensation of the vapor; the liquid returns to the heat-absorbing end by capillarity through a capillary structure covering the internal face of the cylinder. The heat transfer of a heat pipe may be 10,000 times or more higher than that of copper or silver.

The prototype developed under the sponsorship of NASA (NASA/RCA Type A1279) represents an advanced design capable of supplying 185 watts at 0.73 volt with an efficiency of 16.5%.

Since no major limitations are foreseen in the development of thermionic fuel cells, it is expected that they will provide the most convenient technique for using nuclear energy in the large-scale production of electricity.

Scintillator-photocell nuclear battery. This type of cell is based on a two-step conversion process (Fig. 4). Beta-particle energy is converted into light energy; then the light energy is converted into electric energy. To accomplish these conversions, the battery has two basic components, a light source and photocells.

The light source consists of a mixture of finely divided phosphor and promethium oxide (Pm_2O_3) sealed in a transparent container of radiation-resistant plastic. The light source is in the form of a thin disk. The photocells are placed on both faces of the light source. These cells are modified solar cells of the diffused-silicon type.

Since the photocells are damaged by beta radiation, the transparent container of the light source must be designed to absorb any beta radiation not captured in the phosphor. Polystyrene makes an excellent light-source container because of its resistance to radiation.

The light source must emit light in the range at which the photocell is most efficient. A suitable phosphor for the silicon photocell is cadmium sulfide or a mixture of cadmium and zinc sulfide.

In a prototype battery, the light source consisted of 50 milligrams (mg) of phosphor and about 5 mg of the isotope promethium-147. This isotope is a pure beta-emitter with a half-life of 2.6 years. It is deposited as a coating of hydroxide on the phosphor particles, which are then dried to give the oxide. For use with the low light level (about 0.001 times sunlight) of the light source, special treatment is necessary to make the equivalent shunt resistance of the cell not less than 100,000 ohms. For a description of the photocell *see* SOLAR BATTERY.

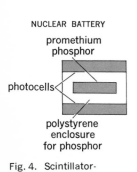

NUCLEAR BATTERY

promethium phosphor

photocells

polystyrene enclosure for phosphor

Fig. 4. Scintillator-photocell battery.

The phototype battery, when new, delivers 20×10^{-6} amp at 1 volt. In 2.6 years (half-life) the current drops about 50% but the voltage drops only about 5%.

The power output improves with decreasing temperature, as a result of improved photocell diode characteristics which more than compensate for a decrease in short-circuit current. At $-100°F$, the power output is 1.7 times as great as at room temperature. At 144°F, the power output is only 0.6 times as great as at room temperature.

The battery requires shielding to reduce the weak gamma radiation to less than 9 milliroentgen per hour (mr/hr), which is the tolerance for continuous exposure of human extremities. The unshielded battery has a radiation level of 90 mr/hr. By enclosing the cell in a case of tungsten alloy, density 16.5, the external radiation becomes less than 9 mr/hr.

The unshielded battery has a volume of 0.014 in.³ and a weight of 0.016 oz. Over a 2.5 year period, the total output would be 0.32 whr (whether or not used). This gives a unit output of 320 whr/lb, which is about six times as great as chemical-battery output. The shielded battery, however, has a volume of 0.07 in.³ and a weight of 0.6 oz. This reduces the unit output to 8.5 whr/lb. The cell can undergo prolonged storage at temperatures of 200°F.   [JACK DAVIS; L. ROZEANU]

Bibliography: B. I. Leefer, Nuclear-thermionic energy converter, *Preceedings of the 20th Annual Power Sources Conferences*, p. 172, 1966; L. I. Shure and H. J. Schwartz, Survey of electric power plants for space applications, *Chem. Eng. Progr.*, 63:99, 1967.

Nuclear engineering

That branch of engineering that deals with the production and use of nuclear energy. It is concerned with the development, design, construction, and operation of power plants which convert energy produced by fission or fusion to other useful forms such as heat or electrical energy. Development of these unique sources of energy requires novel solutions to difficult mechanical, electrical, and materials problems. Because many of the components and systems operate in the presence of intense high-energy radiation, special problems that are generated by the interaction of radiation with various materials are encountered. Such problems are unique to nuclear engineering. Training of nuclear engineers places special emphasis on this area. *See* FISSION, NUCLEAR; FUSION, NUCLEAR; NUCLEAR FUELS; NUCLEAR POWER; REACTOR, NUCLEAR.

Radioactive materials are used in a wide variety of industrial processes and equipment, ranging from nondestructive testing of welds to low-temperature sterilization of pharmaceuticals. Handling and storage of the large quantities of radioactive substances used as reactor fuel, produced as by-product material, or generated as waste introduce problems in the protection of personnel, equipment, and the environment. Such protection from high-energy radiation requires the design and construction of a variety of radiation shields and of shipping and handling equipment. *See* DECONTAMINATION OF RADIOACTIVE MATERIALS; NUCLEAR FUELS REPROCESSING; RADIOACTIVE WASTE MANAGEMENT.

Nuclear explosives are being investigated for use in large-scale excavation and for stimulation of the production of natural gas. Nuclear reactors are now used for propulsion of a variety of naval vessels. Serious consideration is being given to the use of nuclear power for propulsion of commercial ships.

More than 60 colleges and universities in the United States offer educational programs in nuclear engineering. Undergraduate curricula emphasize design and analysis of fission reactor power plants, industrial applications of radiation and radioactive isotopes, and radiation protection. Graduate programs typically place emphasis on research in fission reactor fuels management, reactor safety, effects of radiation on materials, generation and control of magnetically confined high-temperature plasmas, laser-generated fusion, design of fusion power plants, radiation measuring devices and systems, and medical applications of radiation. Many nuclear engineering departments operate research or training reactors which are used for laboratory instruction and as intense sources of neutron and gamma radiation for research.

[WILLIAM KERR]

Bibliography: J. J. Duderstadt and L. J. Hamilton, *Nuclear Reactor Analysis*, 1976; A. R. Foster and R. L. Wright, *Basic Nuclear Engineering*, 2d ed., 1973; J. R. Lamarsh, *Introduction to Nuclear Engineering*, 1975; R. L. Murray, *Nuclear Energy*, 1975.

Nuclear fuels

The fissionable and fertile elements and isotopes used as the sources of energy in nuclear reactors. Although many heavy elements can be made to fission by bombardment with high-energy α-particles, protons, deuterons, or neutrons, only neutrons can provide a self-sustaining reaction.

The number of neutrons ν released in the fission process varies from one per many fissions for elements just beyond the fission point (silver) to two or more per fission for the heavier elements, such as thorium and uranium. Even in such elements, neutron capture by the nucleus accompanied by the release of excess energy in the form of a γ-ray occurs in many cases, rather than nuclear fission. This reduces the number of neutrons available for further fission. The ratio of neutron capture to neutron fission varies from nucleus to nucleus and changes with the energy of the bombarding neutrons. Only a few isotopes of the heavy elements have a higher probability of fission than capture. These fissionable isotopes, U^{233}, U^{235}, and Pu^{239}, are the only materials that can sustain the fission reaction and are therefore called nuclear fuels. *See* REACTOR, NUCLEAR.

Of these isotopes, only U^{235} occurs in nature as 1 part in 140 of natural uranium, the remainder being U^{238}. The other two fissionable isotopes must be produced artificially, U^{233} by neutron capture in Th^{232} and Pu^{239} by neutron capture in U^{238}. The isotopes Th^{232} and U^{238} are called fertile materials.

By using a mixture of both fissionable and fertile isotopes in a nuclear reactor, it is possible to re-

Table 1. Nuclear reactions in a thermal-neutron spectrum

Reaction number	Equation	Cross section, barns*	Half-life
(1)	$U^{235} + n \rightarrow U^{236} + \gamma$	107	
(2)	$U^{235} + n \rightarrow$ Fission $+ 2.47\,n$	582	
(3)	$U^{238} + n \rightarrow U^{239} + \gamma$	2.74	
(4)	$U^{239} \rightarrow Np^{239} + \beta^-$		23.5 m
(5)	$Np^{239} \rightarrow Pu^{239} + \beta^-$		2.33 d
(6)	$Pu^{239} + n \rightarrow Pu^{240} + \gamma$	277	
(7)	$Pu^{239} + n \rightarrow$ Fission $+ 2.88\,n$	748	
(8)	$Pu^{240} + n \rightarrow Pu^{241} + \gamma$	250	
(9)	$Pu^{241} + n \rightarrow Pu^{242} + \gamma$	390	
(10)	$Pu^{241} + n \rightarrow$ Fission $+ 3.06\,n$	1025	
(11)	$Pu^{242} + n \rightarrow Pu^{243} + \gamma$	19	
(12)	$Pu^{243} \rightarrow Am^{243} + \beta^-$		4.98 h
(13)	$Th^{232} + n \rightarrow Th^{233} + \gamma$	7.3	
(14)	$Th^{233} \rightarrow Pa^{233} + \beta^-$		23.3 m
(15)	$Pa^{233} \rightarrow U^{233} + \beta^-$		27.4 d
(16)	$U^{233} + n \rightarrow U^{234} + \gamma$	52	
(17)	$U^{233} + n \rightarrow$ Fission $+ 2.51\,n$	527	
(18)	$U^{234} + n \rightarrow U^{235} + \gamma$	90	

*Accepted values for monoenergetic thermal neutrons at 2200 m/sec (0.0252 ev); 1 barn $= 10^{-24}$ cm².

duce the rate of depletion of the nuclear fuel, because capture of excess neutrons by the fertile materials replenishes the fissionable material. Thus, U^{235} can be burned (fissioned) and the surplus neutrons used to produce plutonium from U^{238} or U^{233} from thorium. Such nuclear reactors are called converter reactors.

The efficiency of production of new nuclear fuel depends on the extent of neutron losses due to undesirable neutron absorptions in the reactor or to neutron leakage. In some cases these losses can be kept small enough so that more nuclear fuel is produced than burned. Moreover, the U^{233} and Pu^{239} can be subsequently used as fuel in place of the original U^{235}, and by this means a large fraction of fertile material can be gradually converted into fissionable material. Reactors that burn U^{233} and Pu^{239} and produce as much fuel as is consumed, or more, are called breeders.

The total energy that can be produced from the fissionable U^{235} in known resources of high-grade uranium ores corresponds to less than 5% of that from economically recoverable fossil fuels. Thus, atomic energy will not become an important source of power unless the breeding and conversion fuel cycles are utilized.

Breeding and conversion. The nuclear reactions governing the consumption and production of nuclear fuel in a reactor are listed in Table 1. Also shown are values for the thermal-neutron (0.025-ev neutron) cross section (probability that the reaction will take place) and the half-life for radioactive decay of the relatively unstable isotopes.

In a mixture of U^{235} and U^{238}, three competing

reactions take place with thermal neutrons: (1) U^{235} capture, (2) U^{235} fission, and (3) U^{238} capture (numbers in parentheses refer to reactions in Table 1). Reaction (3) leads to the production of Pu^{239} by successive decay of U^{239} and Np^{239}, as shown by reactions (4) and (5). The conversion ratio (relative production and consumption of nuclear fuel) of the system is given by the relative probability that reaction (3) will take place as compared to reactions (1) and (2).

Similarly, in a mixture of Pu^{239} and U^{238} the conversion ratio (breeding ratio) is given by the relative probability of reaction (3) as compared to reactions (6) and (7). In this case, however, the higher isotopes of Pu^{239} that are formed have a long half-life and start to absorb neutrons as their concentration builds up by means of reactions (6), (8), and (9). The Pu^{243} formed by reaction (11) decays rapidly to americium, as shown, to end the chain effectively. Thus, after long exposure to thermal neutrons in a reactor, a mixture of U^{235} and U^{238} will contain appreciable concentrations of U^{236}, Pu^{239}, Pu^{240}, Pu^{241}, and Pu^{242}, all of which must be taken into consideration in determining the overall conversion ratio.

Reactions (1), (2), and (13)–(18) represent the reactions taking place in a mixture of U^{235} and thorium. In this fuel cycle secondary isotopes of importance to the conversion ratio are U^{233}, U^{234}, U^{236}, and Pa^{233}. For most efficient neutron utilization (capture in thorium), it is important to minimize losses due to neutron absorption in Pa^{233} by keeping the average neutron flux low.

Maximizing conversion ratio. When it is desired to maximize the neutron-conversion ratio in a reactor, neutron losses are held to a minimum by suitable selection of the materials making up the reactor system, their arrangement in the reactor, and its operating conditions. For example, neutron leakage is reduced if the reactor is made large; fission-product poisons (neutron absorbers) can be lowered by frequent processing of fuel; and nonfission neutron capture by fuel can be minimized by designing the reactor so that the average energy of the neutrons is optimum for causing fission. However, the extent to which these methods of improving neutron utilization can be applied is limited by economic considerations. Thus for any given nuclear power application, there will be an optimum reactor size and configuration and an optimum fuel-processing cycle.

The control of neutron losses due to parasitic capture in fuel by varying the relative amounts of neutron-scattering material (moderator) and fuel to give the proper neutron energy is the most important factor in achieving a high conversion ratio. The effect of neutron energy ν on α (the ratio of neutrons lost by parasitic capture in fuel to those leading to fission) and the number of neutrons emitted per neutron absorbed in fuel, as seen in the equation below, are shown in Table 2.

$$\eta = \frac{\nu}{1 + \alpha}$$

Table 2 indicates that the theoretical maximum conversion ratio (given by $\eta - 1$) is above 1.0 for all three fissionable materials as long as the average energy of the neutrons causing fission is either

Table 2. Capture-to-fission ratio (α) and neutron yield (η) as functions of energy

Neutron energy, ev	U^{233}		U^{235}		Pu^{239}	
	α	η	α	η	α	η
0.025	0.102	2.28	0.190	2.07	0.380	2.09
0.10	0.08	2.33	0.17	2.11	0.59	1.81
0.30	0.15	2.19	0.25	1.97	0.70	1.70
10^2			0.52	1.62	0.72	1.67
10^5			0.18	2.09	0.60	1.80
10^6	0.03	2.44	0.08	2.28	0.10	2.62

Table 3. Effect of moderator temperature on the nuclear properties of U²³⁵ and Pu²³⁹

Average moderator temperature, °C	Average neutron energy (kT), ev	Fast neutrons produced per thermal neutron absorbed in:	
		U²³⁵	Pu²³⁹
75	0.030	2.083	2.006
200	0.041	2.094	1.936
350	0.054	2.102	1.875
600	0.075	2.103	1.871

very high (~1 Mev) or very low (~0.025 ev). In a practical reactor design, however, both these neutron energy conditions are difficult to achieve, because for any given mixture of fuel and moderator there will exist neutrons moving at all energies, ranging from those for fission neutrons (fast or high-energy neutrons) down to those moving at approximately the same velocities as the moderator atoms. Even in a highly thermalized reactor (high ratio of moderator to fuel), the neutron energy will vary considerably from the mean that is established by the moderator temperature. Because of this, the conversion ratio is affected by the moderator temperature. This is especially true in the case of the U²³⁵, U²³⁸, Pu²³⁹ fuel cycle, as shown in Table 3.

In nuclear power reactors that operate with high moderator and coolant temperatures to achieve high thermal efficiencies, it is difficult to get a high conversion ratio because of the effect just described. One solution to this problem is to insulate the moderator thermally from the coolant and to maintain the moderator at a lower temperature. Such a technique cannot be applied in graphite-moderated reactors because of the necessity for keeping the graphite hot to minimize its expansion due to radiation damage and to minimize the buildup of stored energy.

Fast reactors. By eliminating the moderator, it is possible to raise the average neutron energy in a reactor to a value close to that of the fission neutrons (2.0 Mev, average). Coolants, fertile material, and structural material in the core, however, tend to degrade the energy so that the average is normally 0.6–0.2 Mev. Under these conditions, the ratio of parasitic fuel captures to fuel fissions varies from 0.12 to 0.25. Corresponding breeding ratios range from 1.96 to 1.40 for Pu²³⁹-fueled fast (unmoderated) reactors and from 1.34 to 1.08 for U²³⁵-fueled reactors. In all cases the neutron yield from fissions in fuel is increased by fast-neutron fissions in fertile material (U²³⁸ or Th²³²) resulting in a higher breeding ratio than that given simply by $\eta - 1$.

It is evident from the foregoing that considerably higher conversion or breeding ratios are possible in a U²³⁵- or Pu²³⁹-fueled fast reactor than in a thermal reactor. Fast reactors, therefore, provide a means of utilizing a far greater proportion of natural uranium than would be otherwise possible.

In the case of thorium utilization by means of the U²³³-thorium cycle, breeding is possible with both fast and thermal neutrons. Here, the difference in breeding ratio between thermal and fast reactors is not as great as for the U²³⁵-pluto-

nium cycle, and the choice depends upon other considerations, such as the amount of fissionable material required for criticality in each case.

In addition to achieving a high conversion ratio in a nuclear power reactor, it is also desirable to have a high thermal efficiency and high material economy (heat output per unit weight of fuel and fertile material). Unfortunately, in most cases these three characteristics cannot be maximized simultaneously. For example, in a boiling water reactor, which generates steam inside the reactor core for power production, an increase in the rate of steam generation increases the neutron losses and decreases the neutron economy. Therefore, the optimum design of a boiler reactor and most other reactor types involves a compromise between high power density and high neutron economy.

Fuel requirements. To estimate future fuel requirements for the United States nuclear industry, it is necessary to predict the industry's growth rate, the probable types of nuclear reactors, and the amount of uranium or thorium needed for each type of reactor. Because of uncertainties in these predictions, it is obvious that the resulting estimate of future nuclear fuel requirements must be very approximate. Nevertheless, it is important to make an approximate estimate and to compare it with estimated resources of low-cost uranium to evaluate whether such resources are sufficient to meet long-term needs. Some facts concerning the present nuclear industry help in such an evaluation. It is clear that up to 1980 the industry will consist primarily of light-water reactors, probably half pressurized water and half boiling water. The optimum fueling characteristics of each type of reactor can be calculated fairly accurately based on known and anticipated fuel cycle economic conditions which will prevail during the 1970s. Expressing these fuel requirements in terms of the amount of natural uranium fed to the isotope separation plant per megawatt of installed electrical capacity and assuming equal numbers of boiling water and pressurized water reactors, it is found that such reactors will require 500 kg of U₃O₈ for the initial fuel inventory plus a net annual requirement of 110 kg to make up for fissile consumption. The word net implies that the plutonium produced during fuel irradiation is used for enrichment in subsequent fuel cycles. The consensus of most recent projections of the growth of the United States nuclear industry indicates an installed nuclear capacity in 1980 of 150,000 MwE. On this basis the cumulative uranium requirement will amount to about 250,000 short tons by 1980. If in addition an 8-year forward reserve were set aside to supply reactors in operation at that time, the total need would increase to 650,000 short tons of

Table 4. Estimated United States resources of natural uranium (U₃O₈)

Short tons	Price, $/lb
500,000	7–10
500,000	10–15
600,000	15–30
8,400,000	30–50

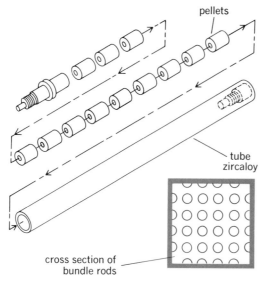

pellets

tube
zircaloy

cross section of
bundle rods

Fig. 1. Cylindrical pellets of UO_2 are pressed to exacting specifications for size and weight. After finishing, pellets are inserted into stainless steel or Zircaloy tubes. Tubes are sealed and welded, then assembled into bundles to form the rod-type element. (*From Nuclear Fuel Elements, General Electric Co.*).

natural uranium up to 1980. Comparing this figure with estimated United States reserves as a function of cost, as shown in Table 4, indicates that most of the low-cost uranium will be used or committed by 1980. It is evident from Table 4 that continued construction of large numbers of light-water reactors after 1980 would soon lead to uranium prices ranging $30–50/lb and add considerably to the cost of nuclear power. Under such conditions nuclear power could not compete with conventional fuels and would never become a long-range source of electricity. *See* NUCLEAR POWER.

Fortunately, the development of breeder reactors, which produce more fissile fuel than they consume, is being actively pursued in the United States and elsewhere, and such breeders should become commercially available in the early 1980s.

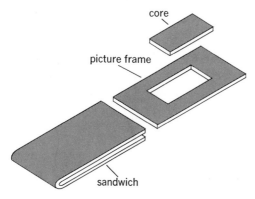

core

picture frame

sandwich

Fig. 2. A fuel plate is assembled with the core, or uranium alloy piece, fitting into a picture frame of aluminum plate. Aluminum plate is placed on either side, and the entire sandwich is hot-rolled to effect bonding. After centering the core by x-ray, the plates are trimmed to size, assembled, and mechanically bonded to the side plates. (*From Nuclear Fuel Elements, General Electric Co.*)

The initial inventory requirements per MwE of such breeders amount to 3–5 kg Pu^{239} (fast breeders) or 1 kg U^{233} (thermal breeders). These breeders, however, annually produce about 0.4 kg Pu^{239} MwE and 0.1 kg U^{233} MwE, respectively, when operating 7000 hr per year. Thus they provide an inventory for subsequent generations of reactors. This sequence continues until the nuclear industry ultimately becomes self-sustaining on bred fuel and independent of mined U^{235}. Under such conditions the high cost of natural uranium would no longer be an important economic factor.

The situation regarding nuclear fuel requirements for other countries is analogous to that in the United States; however, the dates at which low-cost resources will be used up are not likely to be as early as 1980. Nevertheless, the development of breeders is just as essential to the long-term future of nuclear power in other countries as it is in the United States.

Preparation of uranium fuel. Starting with ore, four steps are required in the preparation of a natural uranium fuel: (1) recovery of uranium from ore (concentration), (2) purification of crude concentrate, (3) conversion of oxide to metal, and (4) fabrication of the fuel element. To enrich the fuel (increase the U^{235}/U^{238} ratio), the steps following step 2 are (3) conversion of oxide to UF_6, (4) isotope separation by gaseous diffusion, (5) reduction of enriched UF_6 to metal, alloy, or compound, and (6) fabrication of the fuel element. These steps are described as follows.

Concentration. Because of the variety of natural sources of uranium, no one concentration method is uniquely suited to all ores. Concentration by gravity methods, for example, is applicable for pitchblende but not for carnotite or autunite, from which uranium is extracted almost exclusively by leaching with acid or alkali carbonate. This is followed by a precipitation process (or by ion exchange or solvent extraction) to recover the uranium from the leach solutions.

Purification. To make natural uranium most suitable for use in a nuclear reactor, it is desirable to reduce the concentration of neutron-absorbing impurities such as boron, cadmium, and the rare earths to levels of 0.1–10 parts per million. This is accomplished either by selective extraction of uranyl nitrate from aqueous solutions by certain oxygenated organic solvents, notably diethyl ether, methyl isobutyl ketone, or tributyl phosphate in kerosine, or by quantitative precipitation of uranium peroxide, $UO_4 \cdot 2H_2O$, from weakly acid solutions of uranyl salts.

Conversion. Conversion of the purified uranyl nitrate or UO_4 to UF_6 or U metal is carried out by first calcining the salt to produce UO_3. This is then reduced to UO_2, which is treated with HF to produce green salt, UF_4. Uranium metal is produced from green salt by reduction with calcium or magnesium metal, and UF_6 gas is produced from UF_4 by reaction with fluorine.

Isotope separation. Separation of the uranium isotopes, U^{235} and U^{238}, depends upon the physical differences arising from the difference in their atomic weights. Gaseous diffusion is the method now used to take advantage of this difference in atomic weights.

UF_6 reduction and fabrication. The UF_6 product

from the diffusion plant must be reduced to uranium oxide or uranium metal for incorporation in fuel elements. For most reactor applications, these fuel elements consist of plates or rods, protected by a cladding of aluminum, stainless steel, or zirconium, and assembled into a unit (Figs. 1 and 2). This cladding must be in intimate contact with the uranium-bearing material for good heat removal. It must also be chemically compatible with the material and absolutely leak-tight to prevent the release of radioactive fission products and chemical reaction of the uranium with coolant. Typical fuel elements include aluminum-clad uranium metal rods for plutonium production reactors; zirconium-clad plates containing U-Zr alloy or stainless-clad UO_2 dispersed in stainless steel for propulsion reactors; and zirconium- or stainless-steel-clad UO_2 pellets for central station power reactors (Fig. 1). *See* FISSION, NUCLEAR; NUCLEAR POWER; PLUTONIUM; RADIOACTIVITY; THERMONUCLEAR REACTION; URANIUM.   [JAMES A. LANE]

Bibliography: M. Benedict and T. H. Pigford, *Nuclear Chemical Engineering*, 1957; F. R. Bruce, J. M. Fletcher, and H. H. Hyman (eds.), *Process Chemistry*, in *Progress in Nuclear Energy*, ser. 3, vol. 1, 1958; H. Etherington (ed.), *Nuclear Engineering Handbook*, 1958; S. Glasstone, *Sourcebook on Atomic Energy*, 2d ed., 1958; J. A. Lane, *Economics of nuclear power*, *Annu. Rev. Nucl. Sci.*, vol. 16, 1966; R. Stephenson, *Introduction to Nuclear Engineering*, 2d ed., 1959; S. M. Stoller and R. B. Richards (eds.), *Reactor Handbook*, vol. 2: *Fuel Reprocessing*, 1961; C. R. Tipton, *Reactor Handbook*, vol. 1: *Materials*, 1960; U.S. Atomic Energy Commission, *1967 Supplement to the 1962 Report on Civilian Nuclear Power*, 1967; W. D. Wilkinson and W. F. Murphy, *Nuclear Reactor Metallurgy*, 1958.

Nuclear fuels reprocessing

The treatment of spent reactor fuel elements to recover fissionable and fertile material. Spent fuel is usually discharged from reactors because of chemical, physical, and nuclear changes that make the fuel no longer efficient for the production of heat, rather than because of the complete depletion of fissionable material. Therefore, discharged fuel usually contains fissionable material in sufficient amounts to make its recovery attractive. In the case of breeder reactors, reprocessing of fuel must be done to recover the fissile material bred into part of the fuel in order to take advantage of the characteristic of the breeder reactor to produce more fissile material than is used. If fertile material is also contained in the fuel, it is ordinarily recovered and purified during fuel reprocessing. Purification of the valuable constituents consists of the removal of fission products and extraneous structural material present in the fuel. *See* NUCLEAR FUELS; REACTOR, NUCLEAR.

Because of the frequency of fuel discharge and because of the high value of fissionable materials, it is important that the degree of recovery approach 100% as closely as practicable. It is often necessary to reduce the fission product impurity content of discharged fuel by a factor of 10^6 to 10^7 in order to make the recovered material safe to handle during refabrication into new fuel for reuse.

There are several basic steps involved in fuel reprocessing. After fuel has been discharged from a nuclear reactor, it is common practice to store the fuel submerged in 15–20 ft (4.6–6.1 m) of water (for cooling and radiation-shielding purposes) for a period of 50–200 days; this allows the short-lived fission products to decay radioactively. During this period the radioactivity of the fuel decreases rapidly and substantially, so that when reprocessing is commenced, shielding requirements are reduced to practical thicknesses, heat evolution of the spent fuel assemblies is reduced to more easily managed levels, and radiation damage to chemicals or special structural materials in the reprocessing plant can be held to tolerable magnitudes. Following the cooling period, the fuel is mechanically cut or disassembled into convenient sizes. At this point the fuel is ready for chemical reprocessing to enable recovery and purification of the desired materials.

The specific steps next undertaken depend upon the particular reprocessing method employed to separate the desired products from each other, from fission products, and from extraneous structural materials. Although many separation methods exist, the one based upon solvent extraction principles is most frequently used for fuel reprocessing. Therefore, the discussion of the sequence of steps for recovery and purification will be based on the use of the solvent extraction method. Further details of this process are given below.

Dissolution of spent fuel. The fuel assembly has been cut into pieces so that the fuel, normally held in metal tubes called cladding, is exposed. Fuel from reactors other than those using UO_2 contained in Zircaloy tubes is disassembled to allow access of chemical reagents to the fuel or fuel assembly components. The cut-up or disassembled fuel is charged, along with an appropriate aqueous dissolution medium, generally nitric acid, into a vessel. Here the solid fuel is dissolved. Except for a few fission products which are volatilized during dissolution, all the constituents initially in the fuel are retained in the dissolver solution as soluble salts or, in the case of minor constituents, as an insoluble sludge. Generally, the resulting solution must be treated by various means to accommodate its use as a feed solution to the solvent extraction process. Important reasons for such pretreatment are the adjustments of oxidation states, removal of sludge, and the adjustment of concentrations of solution constituents for optimum recovery and purification performance in the solvent extraction process.

Solvent extraction. In the solvent extraction steps, the actual recovery and purification are performed by the action of special organic solvents. By far the most frequently used solvent is a mixture of tributyl phosphate (TBP) and purified kerosine. The separation process using TBP with nitric acid as the aqueous medium and salting agent is called the Purex process. This is the process in general use for power reactor fuels. The basic requirements for separation during solvent extraction are immiscibility of the organic solvent with the aqueous solution of irradiated fuel, and appreciable differences with which components, initially present in the aqueous fuel solution, distribute or partition themselves between the organic and the aqueous solutions when the organic solution is first thoroughly mixed with the aqueous solution and is later separated from it.

If a quantity of suitably prepared solution of irradiated fuels is mixed with a similar quantity of TBP-kerosene solution and is allowed to stand, the following results. The TBP-kerosene mixture floats, essentially quantitatively, on the aqueous solution because the organic mixture is not miscible with, and is less dense than, the aqueous solution. Analyses of the separated liquids show that a large fraction of the uranium and plutonium (and thorium if it is also present) transfers to the organic mixture, but only a minute fraction of the fission products and other impurities transfers.

To enhance the transfer of uranium and plutonium into the organic mixture without appreciably influencing the transfer of fission products, it is customary to have large concentrations of certain chemicals called salting agents in the aqueous solution. If the organic solvent containing the uranium and plutonium is now brought into contact with an aqueous solution wherein salting agents are absent, the uranium and plutonium will retransfer almost quantitatively to the new aqueous solution. The solvent can then be reused. Because only a minute fraction of the fission products was initially transferred to the organic mixture, the new aqueous solution contains recovered uranium and plutonium well separated from fission products. It is also possible to separate the uranium and plutonium from each other. The same organic solvent can be used, taking advantage of the fact that under certain conditions (reduced oxidation state of plutonium) plutonium extraction by the solvent is very small. Thus separation of the two heavy elements is achieved in much the same way that the impurities (for example, fission products) were initially removed from uranium and plutonium.

Extraction of uranium and plutonium can be done by repeated contact of the fuel solution with quantities of fresh solvent. This is the batch extraction mode and, while useful in the laboratory, it is seldom employed in large-scale practice. Operation of extraction equipment in the continuous countercurrent mode is normally more convenient, efficient, and economic for the purification of large amounts of fuel. The basic principles of separation are the same in the continuous countercurrent mode as in the batch extraction mode. The continuous operation provides repeated mixing and separation of the organic solvent and the aqueous solution of irradiated fuel from which it is desired to remove all of the valuable products freed of impurities. The continuous nature of operation is also applied in the step wherein the purified products are retransferred to a solution free of salting agents. Continuous separation of plutonium and uranium (or thorium and uranium in the case of thorium–uranium-fueled reactors) can also be performed. The principal advantages of continuous over batch operations are more uniform product quality, greater ease of instrumentation, and economy.

The countercurrent aspects of the operation are derived from having the organic solvent flow in equipment in a direction opposite to that of the aqueous solution. This allows maximum loading of the organic solution with the components to be extracted because fresh solvent encounters initially low concentrations of these components. Progressively higher concentrations occur as the solvent moves toward the point at which the aqueous solution is introduced. The aqueous solution moves counter to the flow of organic extractant and is thereby depleted to a high degree of the valuable products (for example, uranium and plutonium). Similarly, reextraction (stripping) of the products from the organic phase into a clean aqueous phase, when done in a countercurrent mode, reduces the product losses to the organic solvent. In this way a minimum of solvent is required to achieve maximum recovery of desired materials with solvent extraction equipment of a given efficiency.

With proper process conditions and suitable equipment, the solvent extraction operation yields nearly complete recovery and purification. Products are usually obtained in the form of dilute aqueous solutions. These solutions are subsequently further processed to give the form of plutonium, uranium, or thorium which is suitable for reuse in nuclear reactors. In the case of uranium which has been depleted in its ^{235}U content, the processing may include ^{235}U isotope reenrichment in gaseous diffusion plants. The fission products and other impurities initially present in the fuel are also obtained in the form of aqueous solution waste. The waste is concentrated by evaporation and then may be introduced into underground tanks for interim storage. *See* RADIOACTIVE WASTE MANAGEMENT.

Processing plants. Plants for fuel reprocessing are large and expensive. They can be a few hundred yards in length, are normally built above and below ground level, and may cost hundreds of millions of dollars. Modern reprocessing plants are usually integrated into facilities that also provide other fuel cycle services such as conversion of uranium to UF_6, solidification of solutions of high-level waste, and conversion of plutonium product solutions to a solid suitable for shipment.

There are many factors that contribute to the high cost of these plants. Some of the more important factors are the large amounts of massive shielding (up to 7 ft or 2.1 m of high-density concrete) used to separate the process equipment from normally occupied work areas; the stringent design criteria for resistance of critical parts of the structure and equipment against natural forces such as earthquakes and tornadoes; the high integrity and rigid manufacturing standards of process equipment; the extensive systems to confine particulate radioactivity; and the barriers interposed between stored wastes and the environment.

Operation of the plants, including sampling for process control, is conducted by remote means. In some plants, even repairs and modifications of equipment in high-radiation zones are made by remote techniques. The additional cost of this type of maintenance is large. For those plants in which maintenance is performed by direct methods, the initial capital cost is reduced. This saving may be offset to some extent by increased operating costs when decontamination is difficult and permissible working time of maintenance personnel is limited. Because of the difficulty and cost of maintenance by either remote or direct methods, more spare equipment and higher standards of design, construction, and installation are necessary in fuel-reprocessing plants than in conventional chemical plants. Special precautions, which also contribute to increased capital and operating costs, must be taken in fuel-reprocessing plants to avoid nuclear accidents from inadvertent accumulation of fissionable materials. This is particularly important when highly enriched fuels are reprocessed.

Thus the gross capital and operating costs are high for a fuel-reprocessing plant. The unit cost of recovered products is also very large because the output of moderately large plants is relatively small. They produce only a few tons of uranium per day, for uranium containing 3% or less of the ^{235}U isotope, or as little as 10–20 lb/day (4.5–9 kg/day) when uranium containing greater than 90% of the ^{235}U isotope (highly enriched uranium) is processed. Unit cost can be substantially reduced, however, by increased capacity, because total capital and operating costs do not increase proportionally.

The operating experience to date with fuel-reprocessing plants has shown them to be relatively safe in spite of hazards from radiation and nuclear criticality, as well as other hazards of a more conventional nature.

[MARTIN J. STEINDLER]

Bibliography: M. Benedict and T. H. Pigford, *Nuclear Chemical Engineering*, 1957; F. R. Bruce, J. M. Fletcher, and H. H. Hyman, *Process Chemistry*, ser. 3, vol. 3, 1961; H. Etherington (ed.), *Nuclear Engineering Handbook*, 1958; J. T. Long, *Engineering for Nuclear Fuel Reprocessing*, 1967; C. E. Stevenson, A. T. Gresky, and E. A. Mason (eds.), *Process Chemistry*, ser. 3, vol. 4, 1970; S. M. Stoller and R. B. Richards (eds.), *Reactor Handbook*, vol. 2: *Fuel Reprocessing*, 2d ed., 1961.

Nuclear materials safeguards

A system of controls designed to assure that nuclear materials and the means of their production are devoted to peaceful ends in keeping with national policy and international commitments. In this context, nuclear materials consist of both source and special nuclear materials. Source materials are the natural elements uranium and thorium; special nuclear materials are uranium enriched in the isotopes U^{233} and U^{235} and plutonium, derived from the source materials through irradiation in nuclear reactors or through the separation of the rare U^{235} isotope from the predominant U^{238}.

Safeguard coverage. The control system must cover the complete cycle of refining, purification, isotope separation, reactor use, chemical recovery, and recycling and the flow of materials between these phases. The cycle begins with the processing of uranium and thorium ores and proceeds through the production of pure metals, alloys, or compounds. A gaseous uranium compound, UF$_6$, is the form used for U^{235} isotopic enrichment and subsequent conversion to other compounds or to metal. Other compounds or alloys, including mixtures with thorium, are used as fuels for nuclear reactors to produce heat and neutrons, which in turn produce the fissionable plutonium or U^{233}. These fissionable materials can remain in the fuel to contribute to the chain reaction, or they may be separated chemically from the fuel materials and used for enhancing new reactor fuel, or in essentially pure form they may be used for building nuclear explosives. Worldwide production in power reactors is expected to approach quantities of plutonium sufficient for tens of nuclear weapons per day.

A special difficulty arises in the need to estimate the content and quality of fissionable material in finely machined products and in nonhomogeneous scrap resulting from mechanical and chemical processes. This estimation often can be met only by the use of reliable methods for measuring the radioactivity emitted by nuclear material isotopes. Some methods may be passive, measuring selected radiations continuously emitted by the material. Another course is to bombard the material with radiations, x-rays or neutrons, and measure selected reaction emissions induced therefrom.

Whenever feasible, of course, more precise measurements are made by sampling and chemical and mass spectrometric analytical techniques. These definitive techniques are also used to validate the nondestructive techniques mentioned above.

Accounting system. The measurement activities form the basis for recording in an accounting system the location and responsibility for all nuclear materials. For the United States, such a system is maintained in the form of an automated Nuclear Materials Information System (NMIS).

NMIS receives reports of all production, transfers, and other significant inventory changes from all nuclear material custodians, regardless of ownership of the material. The significant inventory changes include fabrication discards, consumption, and growth of fissionable isotopes in nuclear reactors.

Custodians are required to establish and maintain administrative and physical controls over their facilities and operations to protect nuclear materials from theft or misuse.

Special controls that include prescheduling, monitoring, tracing, and reporting govern shipments of materials, since this is the phase in which material is most vulnerable to theft or loss.

The effectiveness of the government-prescribed accounting and physical controls is determined by a program of surveillance by government inspectors and auditors. These staffs review custodian systems for applying the controls and verify that the systems are effective, are adhered to, and are modified to meet changing conditions. These reviews are planned also to include examination of construction design of facilities and equipment to assure that location and quality of materials in process can be verified.

Improving measurements. The program outlined above is supported by continuing research, development evaluation, and testing activities. Primarily these activities are needed to improve measurement capabilities in terms of accuracy and promptness.

One approach to active interrogation is the use of accelerators as neutron generators which provide a copious source of neutrons of well-defined energy that can be varied. Californium-252, a spontaneous fission source, also offers a practical maintenance-free neutron source in extremely compact form. Mobile nondestructive assay laboratories, embodying such systems, are being evaluated in field use of nondestructive assay techniques to verify nuclear materials inventories.

Passive gamma or neutron assay techniques are useful for items of a size and shape that allow natural radiation to emerge from the item. The development of gamma detectors with high energy resolution makes it feasible to apply this technique in many situations. Significant developments in fission neutron-counting assemblies permit wider applications in the assay of plutonium which undergoes spontaneous fission.

Calorimetry is another nondestructive method which provides improved accuracy when the isotopic composition of heat-emitting isotopes is known. By using sensitive calorimeters to measure the total power output of a sample, the mass of each isotope can be calculated. Work with PuO_2-UO_2 mixtures (heat output 0.08 to 2.5 watts) attains an accuracy of a calorimetric measurement of ± 0.1 to 0.3%.

Systems studies. A somewhat broader research and development effort is devoted to safeguards systems studies. Systems studies encompass the evaluation of material balances and the limits of error relating to material balances and the measurements on which they are based. They involve the development, testing, and implementation of statistical techniques to determine and analyze measurement uncertainties and error propagation. These studies are also designed to improve understanding of where losses occur in the nuclear fuel cycle, their causes, and the practical limits for reducing them. In addition, the studies should help identify those portions of the nuclear fuel cycle where diversions are most likely to be difficult to detect because of the inherent loss mechanisms at work. *See* REACTOR, NUCLEAR.

[DELMAR L. CROWSON]

Bibliography: W. C. Bartels and W. A. Higinbotham, in *Safeguards Techniques*, vol. 1, International Atomic Energy Agency, Vienna, 1970; D. L. Crowson, in *Safeguards Techniques*, vol. 1, International Atomic Energy Agency, Vienna, 1970; W. A. Higinbotham, *Phys. Today*, 22(11):33–37, 1969; *Safeguards Glossary*, WASH-1162, U.S. Atomic Energy Commission, 1970; B. W. Sharpe, *Phys. Today*, 22(11):40–44, 1969.

Nuclear physics

The discipline involving the structure of atomic nuclei and their interactions with each other, with their constituent particles, and with the whole spectrum of elementary particles that is provided by very large accelerators. The nuclear domain occupies a central position between the atomic range of forces and sizes and elementary-particle physics, characteristically within the nucleons themselves. As the only system in which all the known natural forces can be studied simultaneously, it provides a natural laboratory for the testing and extending of many of the fundamental symmetries and laws of nature.

Containing a reasonably large, yet manageable number of strongly interacting components, the nucleus also occupies a central position in the universal many-body problem of physics, falling between the few-body problems, characteristic of elementary-particle interactions, and the extreme many-body situations of plasma physics and condensed matter, in which statistical approaches dominate; it provides the scientist with a rich range of phenomena to investigate—with the hope of understanding these phenomena at a microscopic level.

Activity in the field centers on three broad and interdependent subareas. The first is referred to as classical nuclear physics, wherein the structural and dynamic aspects of nuclear behavior are probed in numerous laboratories, and in many nuclear systems, with the use of a broad range of experimental and theoretical techniques. Second is higher-energy nuclear physics (referred to as medium-energy physics in the United States), which emphasizes the nuclear interior and nuclear interactions with mesonic probes. Third is heavy-ion physics, internationally the most rapidly growing subfield, wherein the accelerated beams of nuclei spanning the periodic table are used to study nuclear phenomena which were previously inaccessible.

Nuclear physics is unique in the extent to which it merges the most fundamental and the most applied topics. Its instrumentation has found broad applicability throughout science, technology, and medicine; nuclear engineering and nuclear medicine are two very important areas of applied specialization. *See* NUCLEAR ENGINEERING.

Nuclear chemistry, certain aspects of condensed matter and materials science, and nuclear physics together constitute the broad field of nuclear science; outside the United States and Canada elementary particle physics is frequently included in this more general classification. *See* FISSION, NUCLEAR; FUSION, NUCLEAR; NUCLEAR REACTION; REACTOR, NUCLEAR.

[D. ALLAN BROMLEY]

Nuclear power

Power derived from fission or fusion nuclear reactions. More conventionally, nuclear power is interpreted as the utilization of the fission reactions in a nuclear power reactor to produce steam for electrical power production, for ship propulsion, or for process heat. Fission reactions involve the breakup of the nucleus of heavy-weight atoms and yield energy release which is more than a millionfold greater than that obtained from chemical reactions involving the burning of a fuel. Successful control of the nuclear fission reactions provides for the utilization of this intensive source of energy, and with the availability of ample resources of uranium deposits, significantly cheaper fuel costs for electrical power generation are attainable. Safe, clean, economic nuclear power has been the objective both of the Federal government and of industry's programs for research, development, and demonstration. Critics of nuclear power seek a complete ban or at least a moratorium on construction of new commercial plants. *See* FISSION, NUCLEAR; FUSION, NUCLEAR.

Considerations. Fission reactions provide intensive sources of energy. For example, the fissioning of an atom of uranium yields about 200 MeV, whereas the oxidation of an atom of carbon releas-

Table 1. United States uranium resources*

Production cost, $/lb U_3O_8	Proven reserves	Potential reserves			Total reserves
		Probable	Possible	Speculative	
8	200,000	300,000	200,000	30,000	730,000
15	420,000	680,000	640,000	210,000	1,950,000
30	600,000	1,140,000	1,340,000	410,000	3,490,000

SOURCE: Atomic Industrial Forum, Inc.
*Tons of U_3O_8 as of Jan. 1, 1975 (1 short ton = 0.9 metric ton).

es only 4 eV. On a weight basis, the 50,000,000 energy ratio becomes about 2,500,000. Only 0.7% of the uranium found in nature is uranium-235, which is the fissile fuel used. Even with these considerations, including the need to enrich the fuel to several percent uranium-235, the fission reactions are attractive energy sources when coupled with abundant and relatively cheap uranium ores. Although resources of low-cost uranium ores are extensive (see Tables 1 and 2), more explorations in the United States are required to better establish the reserves. In 1975 the distribution of the United States' uranium resources at $10 per pound of U_3O_8 was: New Mexico 53%, Wyoming 32%, Colorado 3%, Utah 2%, and other states 10%. Major foreign sources of uranium are Australia, Canada, South Africa, and southwest Africa; smaller contributions come from France, Niger, and Gabon; other sources include Sweden, Spain, Argentina, Brazil, Denmark, Finland, India, Italy, Japan, Mexico, Portugal, Turkey, Yugoslavia, and Zaire.

Projected use of nuclear power for the 236 reactors identified in 1975 as operable, being built, or planned, representing 234,900 MW electrical capability, would have required about 940,000 tons (846,000 metric tons) of uranium oxide by the year 2000, with additional needs of about 300,000 tons (270,000 metric tons) to provide for 30 years' operation for all the reactors. (For example, under this assumption, reactors starting operation in 1988 would continue to operate to 2018.) These projected fuel requirements are based upon the assumption that the enrichment plants produce tails with 0.3% uranium-235 and that no plutonium is recovered from the irradiated fuel and recycled back into the fuel. *See* URANIUM.

Table 3 illustrates the progression of orders for nuclear power reactors from 1965 to 1975, indicating that, based upon capacity, there were more orders for nuclear than for fossil plants. By the mid-1970s, economic slowdowns, rising construction costs, prolongation of schedules, more rigid regulatory practices, energy conservation practices, and public intervention and criticisms contributed in various degrees to reevaluations of the nuclear power expansion programs, as well as fossil plants. In 1975, for the 226 fossil units committed with a generating capacity of 132,000 MWe (megawatts of electrical generating capacity), and 180 nuclear units committed with 196,000 MWe, 4% of the units had been canceled for both fossil and nuclear, and 20% and 47% deferred for fossil and nuclear, respectively. At the end of 1975, 56 plants representing 38,000 MWe were licensed for

Table 2. Foreign resources of uranium*

Production cost, $/lb U_3O_8	Reasonably assured	Estimated additional	Total resources
10	1,040,000	440,000	1,480,000
15	1,780,000	960,000	2,740,000
30	2,200,000	1,900,000	4,100,000

*Tons of U_3O_8 (1 short ton = 0.9 metric ton).

operation; 64 plants with 65,700 MWe had construction permits; 97 plants with 110,000 MWe were on order; and 17 plants with 19,100 MWe were firmly planned.

Government administration and regulation. The development and promotion of the peaceful uses of nuclear power in the United States was under the direction of the Atomic Energy Commission (AEC), which was created by the Atomic Energy Act of 1946 and functioned through 1974. The Atomic Energy Act of 1954, as amended, provided direction and support for the development of commercial nuclear power. Congressional hearings established the need to assure that the public would have the availability of funds to satisfy liability claims in the unlikely event of a serious nuclear accident, and that the emerging nuclear industry should be protected from the threat of unlimited liability claims. In 1957 the Price-Anderson Act was passed to provide a combination of private

Table 3. Summary of commerical nuclear power reactors licensed for operation and orders placed.

Period	Orders	Percent of total station capacity	Licensed for operation†
1965	20*	15	5
1966	24	43	
1967	30	47	8
1968	14	36	
1969	7	22	12
1970	16	30	16
1971	26	50	21
1972	33	63	27
1973	36	59	39
1974	27		53
1975	5		54

*Includes years previous to 1965.
†Totals do not include the Shippingport and N Reactor, and eight reactors that have been decommissioned.

insurance and governmental indemnity to a maximum of $560,000,000 for public liability claims. The act was extended in 1965 and again in 1975, each time following congressional hearings which probed the need for such protection and the merits of having nuclear power. The Federal Energy Reorganization Act of 1974 separated the promotional and regulatory functions of the AEC, with the creation of a separate Nuclear Regulatory Commission (NRC) and the formation of the Energy Research and Development Administration (ERDA).

Safety measures. The AEC, overseen by the Joint Committee on Atomic Energy (JCAE), a statutory committee of United States senators and representatives, had sought to encourage the development and use of nuclear power while still maintaining the strong regulatory powers to ensure that the public health and safety were protected. The inherent dangers associated with nuclear power which involves unprecedented quantities of radioactive materials, including possible wide-scale use of plutonium, were recognized, and extensive programs for safety, ecological, and biomedical studies, research, and testing have been integral with the advancement of the engineering of nuclear power. Safety policies and implementation reflect the premises that any radiation may be harmful and that exposures should be reduced to "as low as reasonably achievable" (ALARA); that neither humans nor their creations are perfect and that suitable allowances should be made for failures of components and systems and human error; and that human knowledge is incomplete and thus designs and operations should be conservatively carried out. The AEC regulations, inspections, and enforcements sought to develop criteria, guides, and improved codes and standards which would enhance safety, starting with design, specification, construction, operation, and maintenance of the nuclear power operations; would separate control and safety functions; would provide redundant and diverse systems for prevention of accidents; and would provide to a reasonable extent for engineered safety features to mitigate the consequences of postulated accidents.

Criteria for siting a nuclear power station involve thorough investigation of the region's geology, seismology, hydrology, meteorology, demography, and nearby industrial, transportation, and military facilities. Also included are emergency plans to cope with fires or explosions, and radiation accidents arising from operational malfunctions, natural disasters, and civil disturbances. The AEC Directorates of Licensing, Regulatory Operations, and Regulatory Standards thus functioned to achieve an extraordinary program of safety measures which would be commensurate with the extraordinary risks involved with the use of nuclear power.

Starting about 1970, regulatory safety measures have been significantly augmented in response to the introduction of nuclear power reactors with larger powers and higher specific power ratings; to improved technology, experiences, and more sophisticated analytical methods; to the National Environmental Policy Act of 1969 (NEPA) and the interpretations of its implementations; and to public participation and criticism. A variety of special assessments, studies, and hearings have been undertaken by such parties as congressional committees other than JCAE, the U.S. General Accounting Office (GAO), the American Physical Society, the National Research Council representing the National Academy of Sciences and the National Academy of Engineering, and by organizations representing public interests related to the environmental impacts of the continued use of nuclear power.

Nuclear Regulatory Commission. The independent NRC is charged solely with the regulation of nuclear activities to protect the public health and safety and the environmental quality, to safeguard nuclear materials and facilities, and to ensure conformity with antitrust laws. The scope of the activities include, in addition to the regulation of the nuclear power plant, most of the steps in the nuclear fuel cycle; milling of source materials; conversion, fabrication, use, reprocessing, and transportation of fuel; and transportation and management of wastes. Not included in the activities of the Commission are uranium mining and operation of the government enrichment facilities.

Public issues. Public issues of nuclear power have covered many facets and have undergone some changes in response to changes being effect-

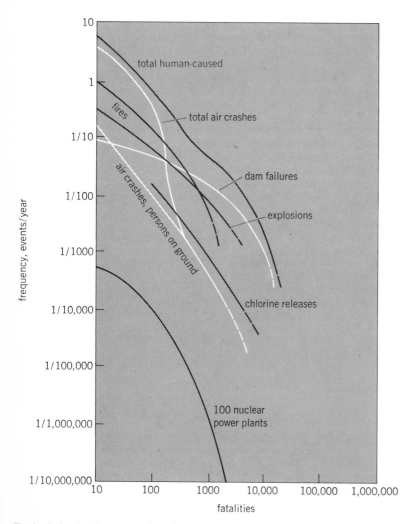

Fig. 1. Estimated frequency of fatalities due to human-caused events.

ed. Key issues include possible theft of plutonium, with threatening consequences; management of radioactive wastes; whether, under present escalating costs, nuclear power is economic and reliable; and protection of the nuclear industry from unlimited indemnity for catastrophic nuclear accidents.

Special nuclear materials. Guidance for improving industrial security and safeguarding special nuclear materials has been initiated. Scenarios studied include possible action by terrorist groups, and evaluations have been undertaken of effective methods for preventing or deterring thefts and for recovering stolen materials. Loss of plutonium by theft and diversionary tactics could pose serious dangers through threats to disperse toxic plutonium oxide particles in populated areas or to make and use nuclear bombs. The more critical segments in the safeguard program for the commercial nuclear fuel cycle would be at the chemical reprocessing plants where the high-level radioactive wastes are separated from the uranium and the plutonium, the shipment of the plutonium oxide to the fuel manufacturing plant, and the fuel manufacturing plant where plutonium oxide would be incorporated in the uranium oxide fuel. Plutonium is produced in the normal operation of a nuclear power reactor through the conversion of uranium-238. For each gram of uranium-235 fissioned, about 0.5–0.6 g of plutonium-239 is formed, and about half of this amount is fissioned to contribute to the operation of the power reactor. Reactor operations require refueling at yearly intervals, with about one-fourth to one-third of the irradiated fuel being replaced by new fuel. In the chemical reprocessing, most of the uranium-238 initially present in the fuel would be recovered, and about one-fourth of the uranium-235 initially charged remains to be recovered along with an almost equal amount of plutonium-239.

The only commercial chemical reprocessing plant, the Nuclear Fuel Services, Inc., facility in West Valley, NY, recovered uranium and plutonium as nitrates, and stored the high-level wastes in large, underground tanks. Nuclear Fuel Services operated from 1966 to 1972. An expansion and modernization program has been planned, with operations not expected until the 1980s. The Midwest Fuel Recovery Plant at Morris, IL, was to have begun operations in 1974, but functional pretests revealed that major modifications would have to be undertaken before initiating commercial operations. In the interim, maximum use of the facility has been made to accommodate storage of irradiated fuel. A third plant, the Allied General Nuclear Services Barnwell Nuclear Fuel Plant at Barnwell, SC, has been under construction since 1971. Thus, in the mid-1970s no commercial reprocessing plant was in operation, and there was only very limited use of test fuel assemblies containing mixed oxides of plutonium and uranium. *See* NUCLEAR FUELS REPROCESSING.

The NRC has developed a system of reviews, including public participation through hearings and through comments received on draft regulations, to determine whether recycling of plutonium is to be licensed in a generic manner. Consideration has also been given to the possible colocation of chemical reprocessing and fuel manufacturing

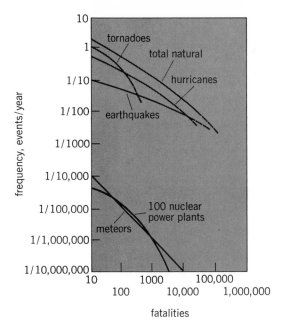

Fig. 2. Estimated frequency of fatalities due to natural events.

plants, and whether there are net gains achieved through the concentration of nuclear power reactors and fuel facilities in energy parks.

Public participation in licensing. The procedures for licensing a nuclear facility for construction and for operation provide opportunities for meaningful public participation. Unique procedures have evolved from the Atomic Energy Act of 1954, as amended, which are responsive to public and congressional inquiry. The applicant is required to submit to the NRC a set of documents called the Preliminary Safety Analysis Report (PSAR), which must conform to a prescribed and detailed format. In addition, an environmental report is prepared. The docketed materials are available to the public, and with the Freedom of Information Act and the Federal Advisory Com-

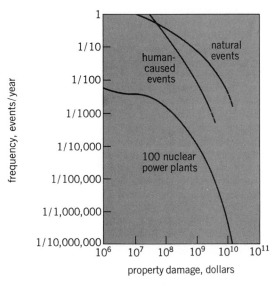

Fig. 3. Estimated frequency of property damage due to natural and human-caused events.

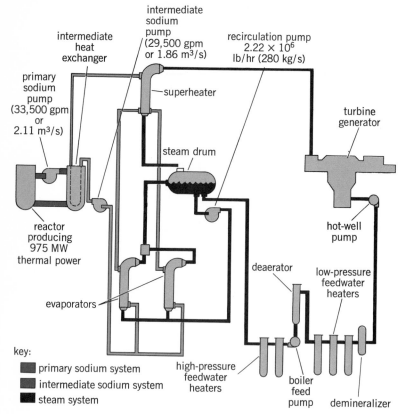

Fig. 4. Heat transport and power generation system for the Clinch River Breeder Reactor.

its contractors and consultants. Early in the review process, the NRC attempts to identify problems to be resolved, including concerns from citizens in the region involved with the siting of the plant. Formal questions are submitted to the applicant, and the replies are included as amendments to the PSAR.

Environmental Impact Statement. Major Federal actions that significantly affect the quality of human environment require the preparation of an Environmental Impact Statement (EIS) in accordance with the provisions of NEPA. The EIS presents (1) the environmental impact of the proposed action; (2) any adverse environmental effects which cannot be avoided should the proposal be implemented; (3) alternatives to the proposed action; (4) relationships between short-term uses of the environment and the maintenance and enhancement of long-term productivity; and (5) any irreversible and irretrievable commitments of resources which would be involved in the proposed action should it be implemented. To better achieve these objectives, the NRC staff supplements the applicant's submittal with its own investigations and analyses and issues a draft EIS so as to gain the benefit of comments from Federal, state, and local governmental agencies, and from all interested parties. A final EIS is prepared which reflects consideration of all comments.

Safety Evaluation Report. A second major report issued by the NRC staff is the Safety Evaluation Report (SER), which contains the staff's conclusions on the many detailed safety items, including discussions on site characteristics; design criteria for structures, systems, and components; design of the reactor, fuel, and coolant system; engineered safety features; instrumentation and control; both off-site and on-site power systems; auxiliary systems, including fuel storage and han-

mittee Act, even more public access to information is available. The NRC carries out an intensive review of the PSAR, extending over a period of about a year, involving meetings with the applicant and

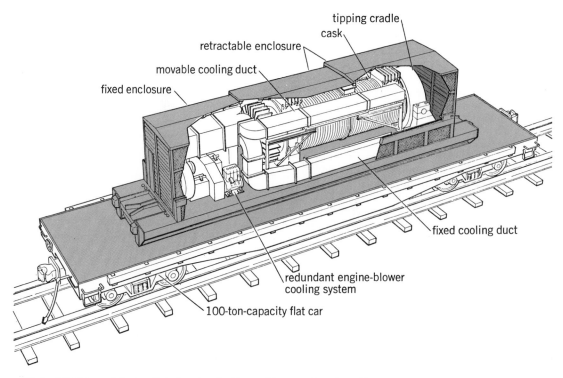

Fig. 5. Shipping cask for irradiated nuclear fuel in a rail transport configuration.

dling, water systems, and fire protection system; radioactive waste management; radiation protection for employees; qualifications of applicant and contractors; training programs; review and audit; industrial security; emergency planning; accident analyses; and quality assurance.

Independent review. Two additional steps are required before the decision is made regarding the construction license. An independent review on the radiological safety items is made and reported by the Advisory Committee on Reactor Safeguards (ACRS), and a public hearing is held by the Atomic Safety and Licensing Board (ASLB). The ACRS is a statutory committee consisting of a maximum of 15 members, covering a variety of disciplines and expertises. Appointments are made for this part-time activity by the NRC. Members are selected from universities, national laboratories and institutes, and industry, including experienced engineers and scientists who have retired, and, in each case, any possible conflicts-of-interest are carefully evaluated. The ACRS has a full-time staff and has access to more than 90 consultants. The ACRS conducts an independent review on nuclear safety issues and prepares a letter to the chairman of the NRC. Both subcommittee and full committee meetings are held to review the documents available and to discuss the applicant's and NRC staff's views on specific and generic issues.

Public hearing and appeal. Public participation is a major objective in the public hearing conducted by the ASLB. The ASLB is a three-member board, chaired by a lawyer, with usually two technical experts. For each application, a board is chosen from among the members of the Atomic Safety and Licensing Panel. Most members of the panel are part-time, and all members are appointed by the NRC. Prehearing conferences are held by the ASLB to identify parties who may wish to qualify and participate in the public hearing. Attempts are made to improve the understanding of the contentions, to see which contentions can be settled before the hearing, and to agree on the issues to be contested. The hearing may probe the need for additional electrical power, the suitability of the particular site chosen over possible alternative sites, the justification of the choice of nuclear power over alternate energy sources, and special issues regarding environmental impact and safety. The ASLB makes a decision on the construction application and may prescribe conditions to be followed. The decision is reviewed by and may be appealed to the Atomic Safety and Licensing Appeal Board. The appeal board is chosen from a panel completely separate from the ASLB panel. The NRC retains the authority to accept, reject, or modify the decisions rendered. Parties not satisfied by the review process and the decisions rendered can take their case to the courts. In several cases, resort to the U.S. Supreme Court has been utilized.

Authorization. A construction permit license is not issued until the NRC, ACRS, and ASLB reviews have been completed and the application has been approved, including conditions to be met during the construction review phase. Depending upon the justification of need, a Limited Work Authorization may be granted for limited construction activities following satisfactory review of the EIS, but prior to completion of the public hearings.

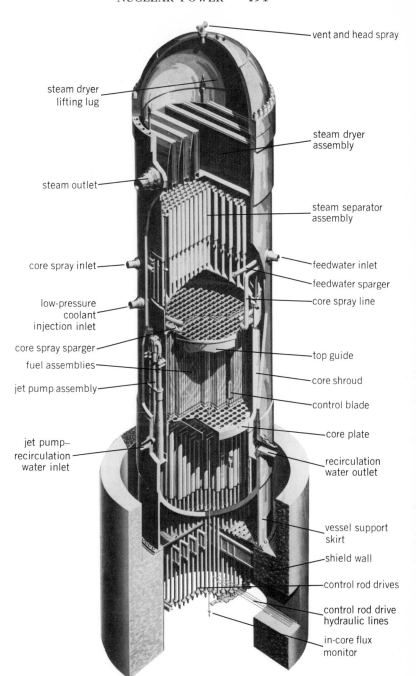

Fig. 6. Cutaway view of pressure vessel of the BWR-6. (*General Electric Co.*)

Construction of a nuclear power plant may take 5 to 6 years or more, during which time the NRC Office of Inspection and Enforcement is involved in monitoring and inspection programs. Several years prior to the completion of construction, a Final Safety Analysis Report (FSAR) is submitted by the applicant, and again an intensive review is undertaken by the NRC staff, and later by the ACRS. A detailed Safety Evaluation Report is prepared by the NRC, and a letter is prepared by the ACRS. If all items have been satisfactorily resolved and the construction and preoperational tests have been completed, a license for operation up to the full power is granted by the NRC. Technical specifications accompany the operating

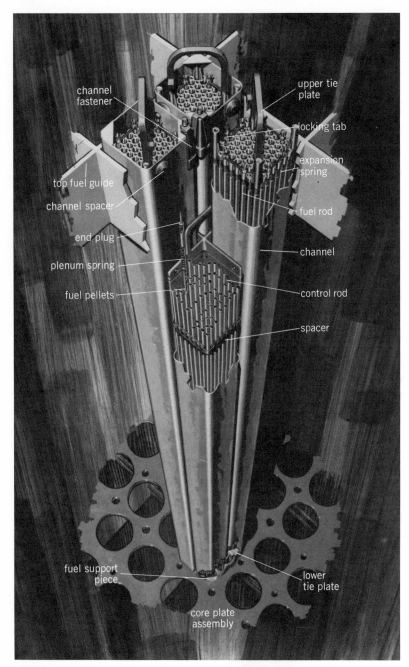

channel
fastener

upper tie
plate

locking tab

expansion
spring

top fuel guide

channel spacer

fuel rod

end plug

channel

plenum spring

fuel pellets

control rod

spacer

fuel support
piece

lower
tie plate

core plate
assembly

Fig. 7. BWR-6 fuel assemblies and control rod module. (*General Electric Co.*)

ably twice as much additional material in supporting exhibits. EISs on major activities, such as the liquid metal fast breeder reactor (LMFBR) program and the management of commercial high-level and transuranium-contaminated radioactive waste, have provided a process for public interactions and influence, and final decisions had not been made in early 1976.

Reactor Safety Study. A detailed, quantitative assessment of accident risks and consequences in United States commercial nuclear power plants has been carried out (WASH-1400, October 1975). The final report has had the benefit of comments and criticisms on a previous draft from governmental agencies, environmental groups, industry, professional societies, and a broad spectrum of other interested parties. Although the study was initiated by the AEC and continued by the NRC, the ad hoc group directed by N. C. Rasmussen of the Massachusetts Institute of Technology carried out an independent assessment. Aside from the very significant technical advancements made in the risk assessment methodology, the "Reactor Safety Study" represents an approach to deal with one phase of the controversial impact of technology upon society. The study presents estimated risks from accidents with nuclear power reactors and compares them with risks that society faces from both natural events and nonnuclear accidents caused by people. The judgment as to what level of risk may be acceptable for the nuclear risks still remains to be made.

Figures 1 and 2 illustrate that the frequency of human-caused nonnuclear accidents and natural events is about 10,000 times more likely to produce large numbers of fatalities than accidents at nuclear plants. The study examined two representative types of nuclear power reactors from the 50 operating reactors, has considered that extrapolation to a base for 100 reactors is a reasonable representation. Improvements in design, construction, operation, and maintenance would be expected to reduce the risks for later expansion in nuclear reactor operations. The fatalities shown in Figs. 1 and 2 do not include potential injuries and longer-term health effects from either the nonnuclear or nuclear accidents. For the nuclear accidents, early illness would be about 10 times the fatalities in comparison to about 8,000,000 injuries caused annually by other accidents. The long-term health effects

license and provide for detailed limits on how the plant may be operated. In some situations where additional information is sought, less than full power is authorized. A public hearing at the operation license stage is not mandatory and would be held only if an intervenor justifies sufficient cause.

Hearings on general matters. Public hearings have also been held on generic matters to establish rules for operation. The two rulemaking hearings conducted by the AEC that have attracted much attention were the Emergency Core Cooling Systems (ECCS) and the "As Low As Practicable" (ALAP) hearings. The hearing on the criteria and conditions for evaluating the effectiveness of the ECCS for a postulated loss of coolant accident lasted from January 1972 to July 1973, and provided more than 22,000 pages of transcript, with prob-

Table 4. Average risk of fatality by various causes

Accident type	Total number	Individual chance per yr
Motor vehicle	55,791	1 in 4,000
Falls	17,827	1 in 10,000
Fires and hot substances	7,451	1 in 25,000
Drowning	6,181	1 in 30,000
Firearms	2,309	1 in 100,000
Air travel	1,778	1 in 100,000
Falling objects	1,271	1 in 160,000
Electrocution	1,148	1 in 160,000
Lightning	160	1 in 2,000,000
Tornadoes	91	1 in 2,500,000
Hurricanes	93	1 in 2,500,000
All accidents	111,992	1 in 1,600
Nuclear reactor accidents (100 plants)	—	1 in 5,000,000,000

such as cancer and genetic effects are predicted to be smaller than the normal incidence rates, with increases in incidence difficult to detect even for large accidents. Thyroid illnesses, which rarely lead to serious consequences, would begin to approach the normal incidence rates only for large accidents.

The likelihood and dollar value of property damage arising from nuclear and nonnuclear accidents are illustrated in Fig. 3. Both natural events (tornadoes, hurricanes, earthquakes) and human-caused events (air crashes, fires, dam failures, explosions, hazardous chemicals) might result in property damages in billions of dollars at frequencies up to 1000 times greater than that for accidents arising from the operation of 100 nuclear power plants.

Figures 1, 2, and 3 represent overall risk information. Risk to individuals being fatally injured through various causes is summarized in Table 4. The results of the study indicate that the predicted nuclear accidents are very small compared to other possible causes of fatal injuries.

The probability of an accident leading to the melting of the fuel core was estimated to be one chance per 20,000 reactor-years of operation, or for 100 operating reactors, one chance in 200 per year. The consequences of a core melt depend upon a number of subsequent factors, including addition-

al failures leading to release of radioactivity, type of weather conditions, and population distribution at the particular site. The factors would have to occur in their worst conditions to produce severe consequences. Table 5 illustrates the progression of consequences and the likelihood of occurrence.

Enrichment facilities. Only 0.7% of the uranium that is found in nature is the isotope uranium-235,

Table 5. Approximate values of early illness and latent effects for 100 reactors

Chance per yr	Early illness	Latent Cancer Fatalities* per year	Thyroid illness* per year	Genetic effects† per year
1 in 200‡	< 1.0	< 1.0	4	< 1.0
1 in 10,000	300	170	1,400	25
1 in 100,000	3,000	460	3,500	60
1 in 1,000,000	14,000	860	6,000	110
1 in 10,000,000	45,000	1500	8,000	170
Normal incidence per yr	4×10^5	17,000	8,000	8,000

SOURCE: Nuclear Regulatory Commission, *Reactor Safety Study*, WASH-1400, 1975.
*This rate would occur approximately in the 10–40-yr period after a potential accident.
†This rate would apply to the first generation born after the accident. Subsequent generations would experience effects at decreasing rates.
‡This is the predicted chance per yr of core melt for 100 reactors.

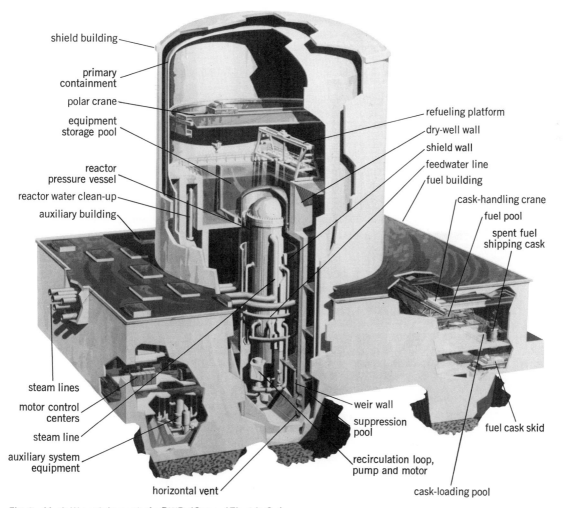

Fig. 8. Mark III containment of a BWR. (*General Electric Co.*)

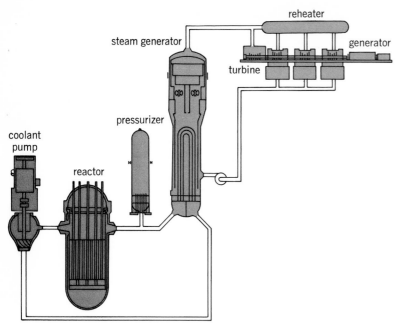

Fig. 9. Closed-cycle PWR. (*Westinghouse Electric Corp.*)

which is used for the fuel in the nuclear power reactors. An enrichment of several percent is needed, and the Federal government (ERDA) owns the enrichment facilities. Expansion of the enrichment facilities is needed to meet expected demands. In early 1976 Congress had not decided whether the commercial sector should participate in the ownership and expansion of such facilities. *See* NUCLEAR FUELS.

Breeder reactor program. In a breeder reactor, more fissile fuel is generated than is consumed.

Fig. 10. Transport of a pressure vessel for a PWR. (*Babcock and Wilcox*)

For example, in the fissioning of uranium-235, the neutrons released by fission are used both to continue the neutron chain reaction which produces the fission and to react with uranium-238 to produce uranium-239. The uranium-239 in turn decays to neptunium-239 and then to plutonium-239. Uranium-238 is called a fertile fuel, and uranium-235, as well as plutonium-239, is a fissile fuel and can be used in nuclear power reactors. The reactions noted can be used to convert most of the uranium-238 to plutonium-239 and thus provide about a 60-fold extension in the available uranium energy source. Breeder power reactors can be used to generate electrical power and to produce more fissile fuel. Breeding is also possible using the fertile fuel thorium-233, which in turn is converted to the fissile fuel uranium-233. An almost inexhaustible energy source becomes possible with breeder reactors. The breeder reactors would decrease the long-range needs for enrichment and for mining more uranium.

Development. The development of the breeder nuclear power reactors had been initiated by the AEC, and the experimental breeder reactor I (EBR-I) was the first reactor to demonstrate production of electrical energy from nuclear energy (Dec. 20, 1951) and to prove the feasibility of breeding utilizing a fast reactor and a liquid metal coolant. In a special test in 1955, an operator error led to a substantial melting of the fuel, but with no off-site effects. The construction on EBR-II was begun in 1958, and it has been operating since 1963. The EBR-II installation is an integrated nuclear reactor power, breeding, and fuel cycle facility. The first commercial breeder licensed for operation, in 1965, was the Enrico Fermi Fast Breeder Reactor. In 1966, while the reactor was operating at a low power, partial coolant blockage occurred, leading to some melting of fuel and release of some radioactivity within the containment building. Although the operation of the reactor was resumed in 1970, the project was discontinued in 1972, primarily for economic considerations.

Environmental impact studies. The AEC had established that the priority breeder program would utilize the LMFBR concept. In response to a 1973 Court of Appeals ruling requiring an environment impact statement for the total LMFBR research and development program, the AEC issued a seven-volume *Proposed Final Environmental Statement* in December 1974 (WASH-1535), covering environmental, economic, social, and other impacts; alternative technology options; mitigation of adverse environmental impacts; unavoidable adverse environmental impacts; short- and long-term losses; irreversible and irretrievable commitments of resources; cost-benefit analysis; and responses to the many critical comments received during the development of the environmental statements in previous drafts. With due regard to the inherent hazards and the need to carefully manage plutonium, including a vigorous program to strengthen and improve safeguards, the conclusion reached was: "The LMFBR can be developed as a safe, clean, reliable and economic electric power generation system and the advantages of developing the LMFBR as an alternative energy option for the Nation's use far outweigh the attendant disadvantages."

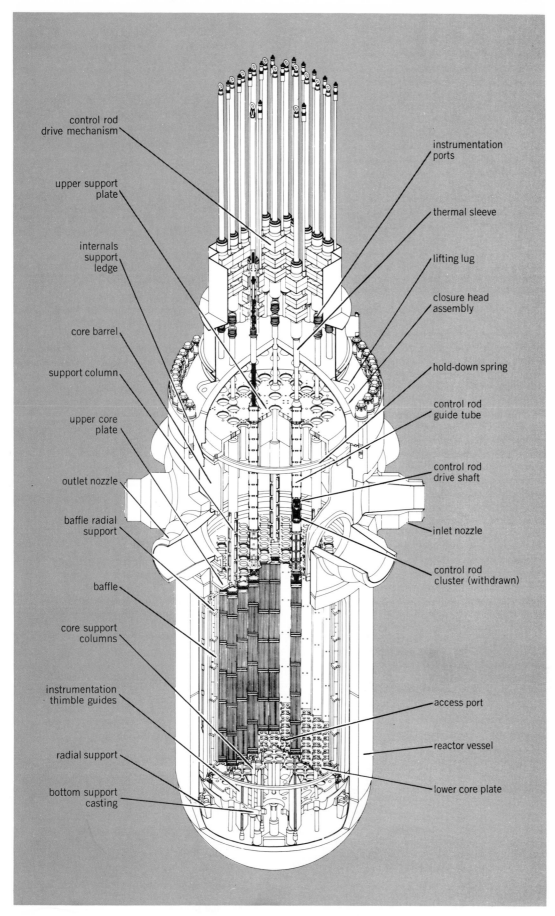

control rod
drive mechanism

upper support
plate

internals
support
ledge

core barrel

support column

upper core
plate

outlet nozzle

baffle radial
support

baffle

core support
columns

instrumentation
thimble guides

radial support

bottom support
casting

instrumentation
ports

thermal sleeve

lifting lug

closure head
assembly

hold-down spring

control rod
guide tube

control rod
drive shaft

inlet nozzle

control rod
cluster (withdrawn)

access port

reactor vessel

lower core plate

Fig. 11.   Internal structure of pressure vessel of a PWR. (*Westinghouse Electric Corp.*)

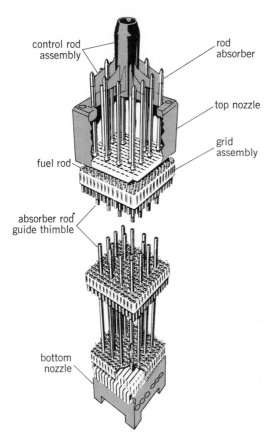

control rod
assembly

rod
absorber

top nozzle

grid
assembly

fuel rod

absorber rod
guide thimble

bottom
nozzle

Fig. 12. Cutaway of typical rod cluster control assembly for a PWR. (*Westinghouse Electric Corp.*)

A reexamination of the LMFBR program was undertaken by ERDA with its issuance of the 10-volume *Final Environmental Statement* (ERDA-1535) to cover additional and supporting information for ERDA's findings and responses to the comment letters on WASH-1535. The possible impact of a new technology upon society has never

Fig. 13. Oconee Nuclear Power Station containment structures.

been so thoroughly questioned. The *Final Environmental Statement* addressed major uncertainties in nuclear reactor safety, fuel cycle performance, safeguards, waste management, health effects, and uranium resource availability. The environmental acceptability, technical feasibility, and economic advantages of the LMFBR cannot be ascertained until additional research, development, and demonstration programs are undertaken. The ERDA decision was to proceed with a plan to continue the research and supporting programs so as to provide sufficient data by 1986 for ERDA to make the decision on commercialization of the LMFBR technology. The plan contemplates the licensing, construction, and operation of the Clinch River Breeder Reactor; the design, procurement, component fabrication and testing, and the licensing for construction of the prototype large breeder reactor; and the planning of a commercial breeder reactor (CBR-I). The heat transport and power generation system for the Clinch River Breeder Reactor is illustrated in Fig. 4.

Radioactive waste management. The responsibility for the development of a proposed Federal repository for high-level radioactive wastes was transferred from the AEC to ERDA in 1975. The AEC had established regulations that the high-level radioactive wastes from the chemical reprocessing of irradiated fuel must be converted to a stable solid within 5 years of its generation at the reprocessing plant, and that the solids be sealed in high-integrity steel canisters and delivered to the AEC for subsequent management within 10 years of its generation at the reprocessing plant. Initially the AEC had sought to develop a salt mine near Lyons, KS, as the repository, but effective public intervention disclosed deficiencies for the site selected and the proposal was withdrawn. Subsequently, a more extensive study was undertaken to review permanent, safe, geologic formations and to locate possible sites. Stable, deep-lying formations could serve to isolate the wastes and dissipate their heat generation without need for maintenance or monitoring. During the period of time that might be used for the demonstration of a repository on a pilot plant scale, a retrievable surface storage facility (RSSF) would be placed in operation. The RSSF would require maintenance and monitoring, and could be engineered using known technology.

Critics of nuclear power consider the radioactive wastes generated by the nuclear industry to be too great a burden for society to bear. Since the high-level wastes will contain highly toxic materials with long half-lives, such as about a few tenths of one percent of plutonium that was in the irradiated fuel, the safekeeping of these materials must be assured for time periods longer than social orders have existed in the past. Nuclear proponents answer that the proposed use of bedded salts, for example, found in geologic formations that have prevented access of water and have been undisturbed for millions of years provide assurance that safe storage can be engineered. A relatively small area of several square miles would be needed for disposal of projected wastes. As of the beginning of 1976, decisions for the repository, requiring environmental impact statements and public hearings, had not been made.

Research and development has been underway

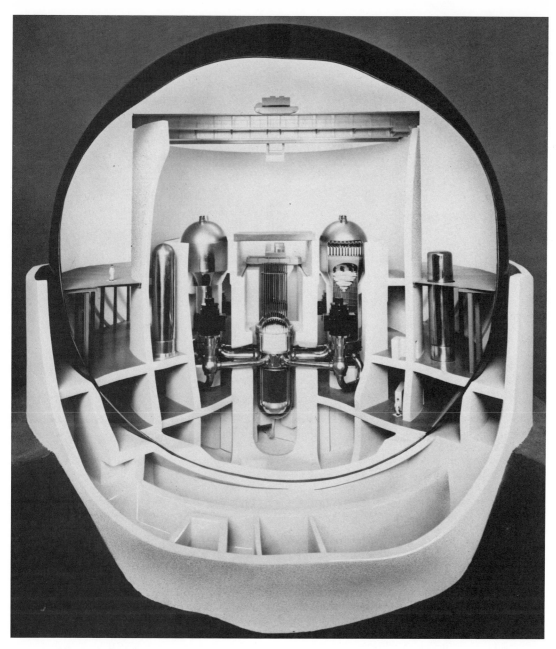

Fig. 14. Spherical containment design for a PWR. (*Combustion Engineering, Inc.*)

since the 1950s on methods for solidifying the wastes; however, neither ERDA nor the NRC has provided the detailed criteria and guides needed by industry to carry forth the design of the waste solidification portions for the chemical reprocessing plants. Until the chemical reprocessing plants begin operation, irradiated fuel will be retained at the reactor sites and at the chemical reprocessing plants in appropriate water storage pools. Though the need for either retrievable storage or permanent storage extends to about the later 1980s, as of early 1976 the necessary decisions by ERDA, NRC, and other Federal agencies such as the Environmental Protection Agency (EPA) on waste management policies, requiring the input of public hearings to be held by Federal agencies and congressional committees, had not been made.

Transportation. Transportation of nuclear wastes has received special attention. With increasing truck and train shipments and increased probabilities for accidents, the protection of the public from radioactive hazards is achieved through regulations and implementation which seek to provide transport packages with multiple barriers to withstand major accidents. For example, the cask used to transport irradiated fuel is designed to withstand severe drop, puncture, fire, and immersion tests. Figure 5 illustrates a shipping cask in a rail transport configuration.

Low-level wastes. Management of low-level wastes generated by the nuclear energy industry requires use of burial sites for isolation of the wastes and decay to innocuous levels. Operation of the commercial burial sites is subject to regulations by Federal and state agencies.

Routine operations of nuclear power stations

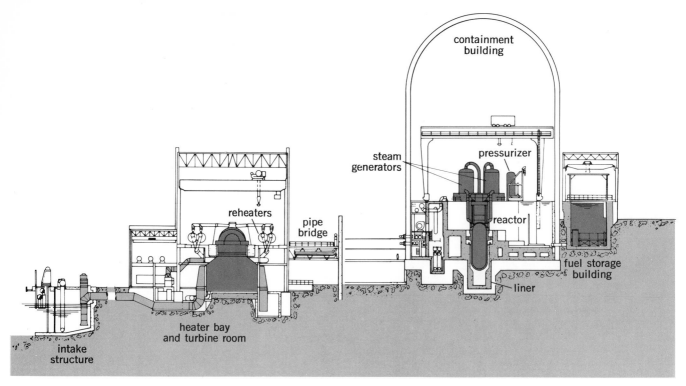

Fig. 15. Cross section of a typical PWR plant. (*Westinghouse Electric Corp.*)

result in very small releases of radioactivity in the gaseous and water effluents. The NRC has adopted the principle that all releases should conform to the ALARA standard. Extension of ALARA guidance to other portions of the nuclear fuel cycle has been undertaken. *See* RADIOACTIVE WASTE MANAGEMENT.

Economics. During the first half of 1975, nuclear power plants in the United States produced more than 76,000,000,000 net kilowatt-hours, representing about 8.3% of all the electricity generated. The average capacity factor was 64.2% versus 45.9% for the average of all nonnuclear units, and the average availability factor was 71.3% for nuclear and 78.5% for nonnuclear units. (Capacity is the percentage of the energy which could be achieved if the plant were running at full power for the period of comparison; availability is the percentage of the time period that the plant is available for use, regardless of power level of operation.) The Atomic Industrial Forum estimated that nuclear operations had total costs of 11.41 mills/kWhr in comparison to 32.73 mills for oil and 14.71 mills for coal.

Capital costs. Economic projections for new plants must reflect the escalation of capital costs for both nuclear and fossil plants. As of August 1975, one evaluation for a nuclear station to have a 1200-MWe nuclear unit achieving commercial operation by January 1985, and a second unit 2 years later, used a basis of $1005/kWe for the capital costs. By using a 15% fixed-charge rate and a 75% capacity factor, capital costs were estimated to be 23 mills/kWhr. Critics of nuclear power believe that capital costs for nuclear plants might escalate significantly and that capacity factors might be significantly lower, leading to capital costs 50–100% higher than indicated. For such

cases, savings in nuclear fuel costs would not compensate for increased capital costs. The detailed study of the nuclear station, a midwestern station using western coal and no SO_2 removal and an eastern station using eastern coal and SO_2 removal, yielded the results shown in Table 6 for a 10-year levelized period, 1985–1995. Over the 30-year lifetime of the plant, the 35 mills/kWhr would increase to 67 for the nuclear plant, but the coal plant totals of 59 and 38 would increase to 149 and 96 mills/kWhr. For this presentation, the economic advantage of nuclear over fossil fuel is still maintained. Both nuclear and coal energy sources will be needed to meet projected increases in electrical power utilization.

Phase-out of governmental indemnity. In 1954 the Atomic Energy Act of 1946 was revised, making possible the possession and use of fissionable materials for industrial uses. As noted above, the Price-Anderson Act of 1957, extended in 1965 and again in 1975, protects the nuclear industry against

Table 6. Comparison of nuclear and coal plant costs, mills/kWhr*

Cost	Eastern coal station	Midwestern coal station	Nuclear station
Fuel	36	20	10
Operation and maintenance	2	2	2
Capital	21	16	23
Total	59	38	35

*From L. F. C. Reichle, Ebasco Services, Inc., presentation to the New York Society of Security Analysts, Aug. 27, 1975.

unlimited liability in the unlikely event of a catastrophic nuclear accident. The total amount was set at $560,000,000, with private liability insurance starting at $60,000,000 in 1957 and increasing to $125,000,000 by 1975, and the government indemnity commensurately decreasing to $435,000,000. Up to early 1976, there had not been a single claim arising from operations of nuclear power reactors. About 67% of the premiums paid to the private sector are placed in a reserve fund, and after a 10-year period, approximately 97–98% of this reserve fund has been returned. The smaller premiums paid to the Federal government are not returned. For a 1000-MW electrical power plant, the annual private insurance premium was about $250,000, and the government premium about $90,000. A 1966 amendment to the Price-Anderson Act provided features for no-fault liability and provisions for accelerated payment of claims. The 1975 extension of the act to 1987 provides for phasing out government indemnification, permits the $560,000,000 limit to float upward, and extends coverage to certain nuclear incidents that may occur outside the territorial limits of the United States. Each licensee is to be assessed a deferred payment, with a maximum level to be set per reactor. For example, by using $3,000,000 per reactor and with 100 licensed operating reactors, $300,000,000 would be available in addition to the base level of $125, leaving $135,000,000 for the government indemnity. As the number of operating reactors increases, the government indemnity would phase out and would permit increases of total indemnity to exceed $560,000,000.

In addition to the indemnity insurance, private pools have provided property damage up to $175,000,000. New nuclear power stations would have capital costs in excess of $1,000,000,000.

Types of power reactors. There are five commercial nuclear power reactor suppliers in the United States: three for pressurized-water reactors, Combustion Engineering, Inc., Babcock and Wilcox, and Westinghouse Electric Corporation; one for boiling-water reactors, General Electric Company; and one for high-temperature gas-cooled reactors, General Atomic Company. Approximately two-thirds of the orders for nuclear power reactors are shared by General Electric and Westinghouse, and the remaining one-third by Combustion and Babcock and Wilcox. One high-temperature gas-cooled reactor (HTGR), Fort St. Vrain Nuclear, with 330 MW net electrical power, has been licensed for operation, but orders placed for the higher-power units (with up to 1160 MW net electrical power) have been canceled or deferred indefinitely.

Boiling-water reactor (BWR). A BWR assembly for a design to produce about 3580 MW thermal and 1220 MW net electrical power is illustrated in Fig. 6. The reactor vessel is 238 in. (6.05 m) in inside diameter, 5.7 in. (14.5 cm) thick, and about 71 ft (21.6 m) in height. The active height of the core containing the fuel assemblies is 148 in. (3.76 m). Each fuel assembly contains 63 fuel rods, and 732 fuel assemblies are used. The diameter of the fuel rod is 0.493 in. (12.5 mm). The reactor is controlled by the cruciform-shape control rods moving up from the bottom of the reactor in spaces between the fuel assemblies (177 control rods are provided).

The fuel assemblies and control rod module are shown in Fig. 7. The water coolant is circulated up through the fuel assemblies by 20 jet pumps at about 70 atm (7 MPa), and boiling occurs within the core. The steam is fed through four 26-in. diameter (66 cm) steam lines to the turbine. About one-third of the energy released by fission is converted into electrical energy, and the remaining heat is removed in the condenser. The condenser operations are typical of both fossil and nuclear power plants, with heat being removed by the condenser having to be dissipated to the environment. Some limited use of the low-temperature heat source from the condenser is possible. The steam produced in the nuclear system will be at lower temperatures and pressures than that from fossil plants, and thus the efficiency of the nuclear plant in producing electrical power is less, leading to proportionately greater heat rejection to the environment.

Shielding is provided to reduce radiation levels, and pools of water are used for fuel storage and when access to the core is necessary for fuel transfers. Among the engineered safety features to minimize the consequences of reactor accidents is the containment; its general arrangements are indicat-

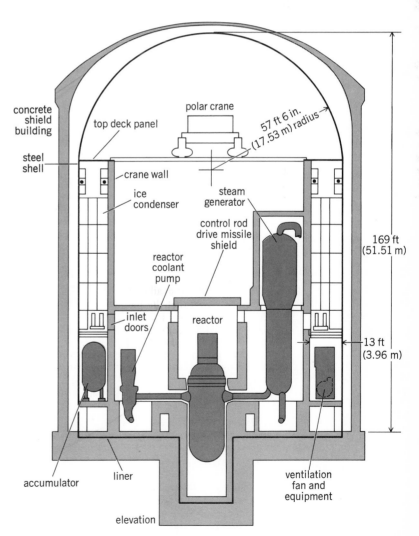

Fig. 16. Cross section of ice condenser containment for a PWR. (*Westinghouse Electric Corp.*)

Fig. 17.   Training simulator for a PWR plant. (*Combustion Engineering, Inc.*)

ed in Fig. 8 for the Mark III version. The function of the containment is to cope with the energy released by depressurization of the coolant should a failure occur in the primary piping, and to provide a secure enclosure to minimize leakage of radioactive material to the surroundings. The BWR utilizes a pool of water to condense the steam produced by the depressurization of the primary water coolant. Various arrangements have been used for the suppression pool. Other engineered safety features include the emergency core-cooling systems.

Pressurized-water reactor (PWR). Whereas in the BWR a direct cycle is used in which steam from the reactor is fed to the turbine, the PWR employs a closed system, as shown in Fig. 9. The water coolant in the primary system is pumped through the reactor vessel, transports the heat to a steam generator, and is recirculated in a closed primary system. A separate secondary water system is used on the shell side of the steam generator to generate steam, which is fed to the turbine, condensed, and recycled back to the steam generator. A pressurizer is used in the primary system to maintain about 150 atm (15 MPa) pressure to suppress boiling in the primary coolant. One loop is illustrated in Fig. 9, but up to four loops have been used.

The reactor pressure vessel is about 44 ft (13.4 m) in height, about 14.5 ft (4.4 m) in inside diameter, and has wall thickness in the core region at least 81/2 in. (22 cm). Figure 10 shows the transport of a reactor vessel enroute to its site; Fig. 11 illustrates the reactor vessel internals. The active length of the fuel assemblies may range from 12 to

14 ft (3.7 to 4.3 m), different configurations are used by the manufacturers. For example, one type of fuel assembly contains 264 fuel rods, and 193 fuel assemblies are used for the 3411-MW-thermal, four-loop plant. The outside diameter of the fuel rods is 0.374 in. (9.5 mm). For this arrangement, the control rods are grouped in clusters of 24 rods per cluster, with 61 clusters provided. In the PWR the control rods enter from the top of the core; an arrangement is illustrated in Fig. 12. Control of the reactor operations is carried out by using both the control rods and a system to increase or decrease the boric acid content of the primary coolant.

Figure 13 is an external view of the Oconee Nuclear Power Station with three reactors, each housed in a separate containment building. The prestressed concrete containment buildings are designed for a 4-atm (400 kPa) rise in pressure and have inside dimensions of about 116 ft (35.4 m) in diameter and 208.5 ft (63.6 m) in height. The walls are 45 in. (1.14 m) thick. The containments have cooling and radioactive absorption systems as part of the engineered safety features. Figure 14 is a view of a design for a 3800-MW-thermal nuclear power reactor utilizing a two-loop system placed in a spherical steel containment (about 200, or 60 m, in diameter), surrounded by a reinforced-concrete shield building. The cross section of a typical PWR plant is given in Fig. 15. The ice condenser containment, shown in Fig. 16, provides an annular array of ice "cubes" held in baskets for the condensation of the steam released by the depressurization of the primary coolant in the unlikely event of a break in the primary piping. The ice condenser design, which permits the use of smaller containments, is proposed for use in the offshore floating-barge nuclear power stations.

The instrumentation and control for a nuclear power station involves separation of control and protection systems, redundant and diverse features to enhance the safety of the operations, and ex-core and in-core monitoring systems to ensure safe, reliable, and efficient operations. Figure 17 shows a training facility arrangement of the instrumentation and controls for a reactor plant. *See* ELECTRIC POWER GENERATION; ENERGY SOURCES; REACTOR, NUCLEAR.

[H. S. ISBIN]

Bibliography: Atomic Energy Commission, *Proposed Final Environmental Statement: Liquid Metal Fast Breeder Reactor Program*, WASH-1535, 1974; Energy Research and Development Administration, *Final Environmental Statement*, ERDA-1535 (including WASH-1535), 1975; H. S. Isbin (coordinator), *Public Issues of Nuclear Power*, 1975; Nuclear Regulatory Commission, *Operating Units Status Reports, such as Licensed Operating Reactors*, NUREG-75/020-12, December 1975; Nuclear Regulatory Commission, *Reactor Safety Study*, WASH-1400, 1975; Subcommittee on Energy and the Environment of the Committee on Interior and Insular Affairs, House of Representatives, 94th Congress, first session, pt. 1: *Overview of the Major Issues*, pt. 2: *Nuclear Breeder Development Program*, 1975.

Nuclear radiation

A term used to denote all the particles and radiations which emanate from the atomic nucleus as a result of radioactive decay and nuclear reactions. The term was originally used to denote only the ionizing radiations observed from naturally radioactive materials. These were α-rays (high-speed helium nuclei), β-rays (negative electrons or negatrons), and γ-rays (electromagnetic radiation of much shorter wavelength than visible light).

The distinction between nuclear radiations and others with similar physical properties lies in whether a nuclear process is involved in their production. Thus, although γ-rays and x-rays are both electromagnetic radiations and, for the same wavelength, are not distinguishable physically, one (γ-radiation) is emitted as a result of a rearrangement of protons and neutrons within the nucleus, while the other (x-rays) results from rearrangement of electrons outside the nucleus.

In addition to α-, β-, and γ-rays, other commonly encountered nuclear radiations are positively charged electrons (positrons), protons, and neutrons. Another radioactive decay product is the neutrino; it is not ordinarily considered as nuclear radiation, since its interaction with matter is very slight. Nuclear reactors are excellent sources of neutrinos. *See* NUCLEAR REACTION; REACTOR NUCLEAR.   [WILLIAM W. BUECHNER]

Nuclear reaction

A reaction which is produced as a result of interactions between atomic nuclei when the interacting particles approach each other to within distances of the order of nuclear dimensions (10^{-12} cm). In the usual experimental situation, one of the interacting particles, the target nucleus, is essentially at rest, and the reaction is initiated by bombarding it with nuclear projectiles of some type.

Means of producing reactions. Because of the intense electrostatic field produced by the nuclear charge, positively charged bombarding particles must have a large kinetic energy in order to overcome the electrostatic (Coulomb) repulsion and reach the target nucleus. Whereas for the lightest target nuclei, protons with kinetic energies of a few hundred thousand electron volts are sufficient to cause certain reactions, energies of many hundreds of millions of electron volts (MeV) are required to initiate reactions between the heavier nuclei. Beams of such energetic charged particles are provided by particle accelerators of various types (Van de Graaff generators, cyclotrons, linear accelerators, and so forth).

Since neutrons are uncharged, they are not repelled by the electrostatic field of the target nucleus, and neutron energies of only a fraction of an electron volt are sufficient to initiate nuclear reactions. Neutrons for reaction studies may be obtained from nuclear reactors or from various nuclear reactions. The interaction of electromagnetic radiation with nuclei may also lead to nuclear reactions. So-called photodisintegration may take place if the radiation has sufficient energy to cause the target nucleus to break up into two or more fragments. In a similar manner, high-energy electrons may also cause nuclear disintegrations. Electromagnetic radiation and electrons, however, interact strongly with the atomic electrons surrounding the target nucleus and are relatively less effective in causing nuclear reactions than are nuclear particles such as protons and neutrons.

Typical reactions. The most common and most extensively studied reactions are those which result in two products, one of which, the residual nucleus, is of nearly the same mass number and charge as the target nucleus, while the other product, the emitted particle, is either a single nucleon (a proton or a neutron) or a small assembly of nucleons, such as an α-particle. Examples are the reactions initiated when a layer of carbon of atomic mass 12 is bombarded with deuterons of a few MeV of kinetic energy. Deuterons, protons, neutrons, and α-particles are emitted, reactions (1)–(4) being responsible.

$$^{12}_{6}C + ^{2}_{1}H \rightarrow ^{2}_{1}H + ^{12}_{6}C \tag{1}$$

$$^{12}_{6}C + ^{2}_{1}H \rightarrow ^{1}_{1}H + ^{13}_{6}C \tag{2}$$

$$^{12}_{6}C + ^{2}_{1}H \rightarrow ^{1}_{0}n + ^{13}_{7}N \tag{3}$$

$$^{12}_{6}C + ^{2}_{1}H \rightarrow ^{4}_{2}He + ^{10}_{5}B \tag{4}$$

In these equations the subscript is the atomic number (nuclear charge) of the nucleus indicated by the usual chemical symbol, while the superscript is the mass number of the particular isotope involved. These reactions are conventionally written as $^{12}C(dd)^{12}C$, $^{12}C(dp)^{13}C$, $^{12}C(dn)^{13}N$, and $^{12}C(dd)^{10}B$, respectively. In each case, the interaction of the incident particle with the target nucleus results in the formation of a residual nucleus and an emitted particle. In the (dd) reaction, in which the residual nucleus is the same as the target nucleus, the process is referred to as scattering, either elastic or inelastic, depending upon whether the residual nucleus is left in its ground state or in one of its various excited states. The other three reactions lead to the production of a residual nucleus different from the target nucleus and are examples of nuclear disintegrations or transmutations. In these cases, also, the residual nucleus may be formed in its ground state or in one of its excited states. If the latter situation occurs, the residual nucleus will subsequently emit the excitation energy in the form of γ-radiation or, occasionally, electrons. The residual nucleus may also be a radioactive species, as in the case of ^{13}N formed in the $^{12}C(dn)$ reaction. In this case, the residual nucleus undergoes further transformations in accordance with its characteristic radioactive decay scheme.

Q value. In a nuclear reaction, the sum of the kinetic energies of the products may be greater than, equal to, or less than the sum of the kinetic energies before the reaction. The difference between these sums is the Q value for the particular reaction. Through Eq. (5), it is also equal to the

$$m_1c^2 + T_1 + m_2c^2 + T_2 = m_3c^2 + T_3 + m_4c^2 + T_4 \quad (5)$$

difference between the rest (proper) energies of the products and the rest energies of the initial nuclei. Reactions with a positive Q value are called exoergic or exothermic, while those with a negative Q are endoergic or endothermic. In reactions (1)–(4), for those cases in which the residual nucleus is formed in its ground state, the Q values are: $^{12}C(dd)^{12}C$, $Q = 0$; $^{12}C(dp)^{13}C$, $Q = 2.72$ MeV; $^{12}C(dn)^{13}N$, $Q = -0.28$ MeV; $^{12}C(d\alpha)^{10}B$, $Q = -1.39$ MeV. For reactions with a negative Q, a definite minimum energy, or threshold energy, is necessary for the reaction to take place. While there is no threshold energy for positive Q reactions, the yields of those reactions involving charged incident particles are quite low unless the bombarding energy is high enough to enable the incident particles to overcome the repulsive electric field from the charge on the target nucleus. A nuclear reaction and its inverse are reversible in the sense that their Q values are numerically equal but have opposite signs. Thus, the Q for the $^{10}B(\alpha d)^{12}C$ reaction is +1.39 MeV.

Conservation laws. The probability that a particular reaction will take place when an individual target nucleus interacts with an incident particle is a function of the bombarding energy, and the factors which determine it are not completely understood. However, it has been found experimentally that certain physical quantities are conserved in all nuclear reactions, and these conservation laws restrict the reactions which may take place. Those quantities which are conserved are described in the following paragraphs.

Charge. The total electric charge is always conserved. Except for high-energy reactions involving meson production, the total number of protons is also conserved. In the $^{12}C(d\alpha)^{10}B$ reaction, for example, there are seven protons involved in both the initial components and the final products.

Mass number. The total number of nucleons is always the same both before and after the reaction. For each of the four reactions listed, 14 nucleons are involved. Since, except for reactions which result in meson production, the number of protons is conserved, the number of neutrons is

also constant at each stage in nuclear reactions.

Energy. The total energy is conserved in all nuclear reactions. The energy of each particle involved is $mc^2 + T$, where m is its rest mass, c the velocity of light, and T its kinetic energy. For two-particle reactions, such as those of Eqs. (1)–(4), the conservation of total energy is expressed in Eq. (5). In this equation the subscripts 1, 2, 3, and 4 refer to the incident particle, the target nucleus, the residual nucleus, and the emitted particle, respectively. In the common experimental situation, T_2 is so small as to be negligible. In this equation, the kinetic energies and the rest masses are usually expressed in units of millions of electron volts, the conversion factor between the two being 1 atomic mass unit (amu) $= 931.502$ MeV.

Linear momentum. The total linear momentum is the same before and after any nuclear reaction. A consequence of this conservation law is that the threshold energy necessary to initiate an endoergic reaction is not numerically equal to the negative Q value but is higher by the amount required to enable the final products to have a combined linear momentum equal to that brought into the reaction by the incident particle. The threshold energy of the $^{12}C(dn)^{13}N$ reaction, for example, is 0.33 MeV.

Angular momentum. The total angular momentum in nuclear reactions is the sum of the angular momentum associated with the relative motion of the reaction components and their intrinsic angular momentum, or spin. This total is always conserved.

Parity. Experimental evidence shows that, in most nuclear reactions, the total parity is the same before and after the interaction. Since the parity associated with the wave function describing the motion of a particle is determined by the angular momentum quantum number l (the parity is even if l is even, and odd if l is odd), and since every nucleus in any one of its allowed states has either even or odd parity, this conservation law, together with that for angular momentum, acts to restrict excited states of the residual nucleus which can be formed by an incident particle of given angular momentum.

Statistics. Since the total number of nucleons is conserved during a nuclear reaction, the statistics which govern the system are the same before, during, and after the interaction; Fermi-Dirac statistics are obeyed if the total number of nucleons is odd, and Bose-Einstein if the total number is even.

Reaction mechanisms. A number of mechanisms have been proposed to account for the observed features of nuclear reactions. Although none has been completely successful, they provide means for correlating and at least partially understanding many of the experimental facts. The most generally used models for nuclear reactions involve either compound nucleus formation or a direct interaction.

Compound nucleus formation. According to this point of view, originally proposed by N. Bohr, a nuclear reaction is visualized as proceeding in two distinct steps. The incident nucleus and the target nucleus are assumed to combine to form a compound nucleus, which exists for a time (of the order of 10^{-16} sec) which is much longer than the approximately 10^{-22} sec that would be required for the incident particle to pass through the target nucleus. The compound nucleus is always in a highly excited, unstable state and can subsequently decay into a number of different products, or through a number of so-called exit channels. In the four examples cited earlier, $^{14}_7N$ is the compound nucleus formed by the amalgamation of a deuteron and $^{12}_6C$, and four possible decay modes or exit channels are indicated. Two essential features of this hypothesis are that, during its relatively long lifetime, the compound nucleus "forgets" the particular way in which it was formed, and that the energy brought in by the incident particle is shared by all the nuclear constituents. The probability that a particular reaction will occur is, then, the product of the probability of forming the compound nucleus and the probability that it will decay through a particular exit channel. Experiments indicate that, for a given energy of excitation in the compound nucleus, this latter factor (probability of decay through a particular exit channel) is independent of the manner in which the compound nucleus is formed. In the case of ^{14}N, it can be formed by $^{13}C + p$ or $^{10}B + \alpha$, as well as by $^{12}C + d$. While certain features of various types of interactions cannot be completely explained on the compound nucleus hypothesis, it appears that this mechanism plays some role in nearly all nuclear reactions.

Direct interactions. Some reactions have probabilities or other properties which conflict with the predictions of the compound nucleus hypothesis, and many are better explained on the assumption that the incident nucleus does not combine with the target nucleus as a whole, but rather that it, or some component, interacts only with the surface or with some individual constituent. The entire process is completed in the time required for the bombarding particle to traverse the diameter of the target nucleus.

Many direct reactions can be classified as "pickup" reactions or as "stripping" reactions. A pickup reaction is visualized as involving the acquisition by the bombarding nucleus of some component, such as a proton, neutron, or alpha particle, from the target nucleus. In a stripping reaction, some component from the bombarding nucleus is "stripped off" and acquired by the target nucleus. The (dp) reaction is a simple example of a stripping reaction, whereas the (dt) process is a reaction which involves pickup.

Coulomb excitation. It is observed that, in the bombardment of nuclei with charged particles, gamma rays characteristic of transitions between excited states of the target nucleus are produced at bombarding energies so low that the probability of either direct interaction or compound nucleus formation is negligible. This process is well explained by the assumption that the nuclei interact with the rapidly changing electric field caused by the passage of the charged bombarding particle.

Elastic scattering. This process leaves the quantum state of the scatterer unchanged. For charged bombarding particles with low energies, the elastic nuclear scattering is accurately described in terms of the inverse-square force law between electric charges. In this case, the process is known as Rutherford scattering. For higher bombarding energies, where the particle can come within the range of the various nuclear forces (approximately

10^{-13} cm), the scattering deviates from predictions based on the inverse-square law. In the case of neutrons, the elastic scattering is entirely due to the nuclear forces.

Nuclear cross sections. The cross section for a nuclear reaction is a measure of its probability. Consider a reaction initiated by a beam of particles bombarding a region which contains N atoms per unit area (uniformly distributed) and where I particles per second striking the area result in R reactions of a particular type per second. This result can be expressed in terms of the fraction of the bombarded region which is effective in producing reaction products R/I. If this is divided by the number of nuclei per unit area, the effective area or cross section per target nucleus is obtained. The cross section $\sigma = R/IN$. This is referred to as the total cross section, since it involves all the disintegration products of the reaction. The dimensions are those of an area, and total cross sections are expressed in either square centimeters or in barns (1 barn $= 10^{-24}$ cm^2). The differential cross section refers to the probability that a particular reaction product will be observed at a given angle with respect to the beam direction. Its dimensions are those of an area per unit solid angle.

Types of reactions. Aside from elastic and inelastic scattering, the most common interactions initiated by the usual bombarding particles are discussed in the following paragraphs.

Proton-induced reactions. Capture reactions, in which the proton combines with the target nucleus to form a compound nucleus in an excited state, occur over a wide range of proton energies. If the compound nucleus decays to its ground state by the emission of a γ-ray, the process is known as a (p,γ) reaction. With higher proton energies, a (p,n) reaction is possible. This always has a negative Q value and leads to a radioactive residual nucleus. For many target nuclei, the (p,α) reaction has a high positive Q, but the yields are low, except at high proton energies, because of the difficulty of the doubly charged α-particle in penetrating the nuclear barrier.

Deuteron-induced reactions. The (d,p), (d,n), and (d,α) reactions usually have positive Q values. Except for light nuclei, in which the nuclear potential barrier is low, the (d,α) reactions have low probabilities. The (d,n) reactions of deuterium, tritium, and beryllium are important as sources of neutrons. Both the (d,p) and (d,n) reactions often lead to radioactive residual nuclei that are useful in various fields of investigation. Deuteron-induced reactions among the very light nuclei, for example, $^3H(dn)^4He$ ($Q = 17.6$ MeV), are important thermonuclear processes. *See* FUSION, NUCLEAR; THERMONUCLEAR REACTION.

Neutron-induced reactions. Neutron capture leading to an (n,γ) reaction is important for all stable nuclei and occurs even with very low-energy neutrons. With a given target nucleus, it yields the same final product as the (d,p) reaction. The capture γ-rays usually have maximum energies of about 8 Mev. This reaction is the source of many of the radioactive isotopes produced by nuclear reactors. For high-energy neutrons, the (n,p) and (n,α) reactions are also observed. In very heavy nuclei, neutron capture may lead to disintegration of the compound nucleus into two massive fragments, with the release of large amounts of kinetic energy and several additional neutrons. For a discussion of this phenomenon *see* FISSION, NUCLEAR.

Alpha-particle-induced reactions. The (α,p) reactions of various light nuclei using the α-particles from naturally occurring radioactive substances were the first examples of artificially produced nuclear disintegrations. High α-particle energies are required for other than light nuclei because of the Coulomb barrier of the nucleus. At sufficiently high energies (about 30 MeV), (α,p) and (α,n) reactions are observed, even in the bombardment of heavy nuclei.

Heavy-ion reactions. Heavy-ion reactions are those produced by nuclei having masses greater than those of the helium isotopes. High-energy beams of nuclei such as ^{12}C, ^{32}S, and ^{40}Ca can produce reactions of much greater variety and can involve much higher angular momentum transfers than those induced by light ions, and are of great value in nuclear spectroscopy. There is a possibility that the interaction of such heavy ions with other heavy nuclei may lead to the production of new elements with atomic numbers even higher than the known transuranic atoms.

[WILLIAM W. BUECHNER]

Bibliography: J. Cerny (ed.), *Nuclear Spectroscopy and Reactions*, 1974; J. de Boer and H. J. Mang (eds.), *Proceedings of the International Conference on Nuclear Physics*, Munich, 1973; P. E. Hodgson, *Nuclear Reactions and Nuclear Spectroscopy*, 1971; M. Jean and R. A. Ricci (eds.), *Nuclear Structure and Nuclear Reactions*, 1969; R. L. Robinson et al. (eds.), *Proceedings of the International Conference on Reactions between Complex Nuclei*, Gatlinburg, 1974.

Nucleonics

The technology based on phenomena of the atomic nucleus. These phenomena include radioactivity, fission, and fusion. Thus, nucleonics embraces such devices and fields as nuclear reactors, radioisotope applications, radiation-producing machines (such as cyclotrons and Van de Graaff accelerators), the application of radiation for biological sterilization and for the induction of chemical reactions, and radiation-detection devices. Nucleonics makes use of and serves virtually all other technologies and scientific disciplines. *See* NUCLEAR ENGINEERING; NUCLEAR PHYSICS.

That part of the industry concerned with nuclear reactors involves a cross section of the entire industrial complex. The chemical industry is concerned with uranium ore refining, fuel and moderator preparation, and fuel reprocessing; the light and heavy metals industry, with fuel fabrication, special component fabrication to withstand environmental conditions including radiation, and containment materials; the machinery industry with control rods, fuel charge and discharge devices, and manipulators; and the instrument industry with control systems. The many applications of nuclear reactors and isotopes also bring the industries making use of them into the field, so that electrical generation, marine propulsion, process heat, special industrial devices, and agriculture, to name a few, are industries participating to some degree in nucleonics.

A number of service activities such as reactor-design consultation, film-badge reading, special shipping and disposal of radioactive nuclear materials and wastes, and analytical services by such techniques as low-level counting and activation are included in the nucleonics industry. The unique radiation hazards and benefits associated with nuclear technology have also engendered special legal, political, and mercantile aspects.

[BERNARD I. SPINRAD]

Nucleus, atomic

The central region of an atom. Atoms are composed of negatively charged electrons, positively charged protons, and electrically neutral neutrons. The protons and neutrons (collectively known as nucleons) are located in a small central region known as the nucleus. The electrons move in orbits which are large in comparison with the dimensions of the nucleus itself. Protons and neutrons possess approximately equal masses, each roughly 1840 times that of an electron. The number of nucleons in a nucleus is given by the mass number A and the number of protons by the atomic number Z. Nuclear radii r are given approximately by $r = 1.4 \times 10^{-13} A^{1/3}$ cm.

[HENRY E. DUCKWORTH]

Octane number

A standard laboratory measure of a fuel's ability to resist knock during combustion in a spark-ignition engine. A single-cylinder four-stroke engine of standardized design is used to determine the knock resistance of a given fuel by comparing it with that of primary reference fuels composed of varying proportions of two pure hydrocarbons, one very high in knock resistance and the other very low. A highly knock-resistant isooctane (2,2,4-trimethylpentane, C_8H_{18}) is assigned a rating of 100 on the octane scale, and normal heptane (C_7H_{16}), with very poor knock resistance, represents zero on the scale. Octane number is defined as the percentage of isooctane required in a blend with normal heptane to match the knocking behavior of the gasoline being tested.

The CFR (Cooperative Fuel Research) knock-test engine used to determine octane number has a compression ratio that can be varied at will and a knockmeter to register knock intensity. In the classic method of knock rating, the engine is run on the fuel to be tested, and its compression ratio is adjusted to give a standard level of knock intensity. Without changing the compression ratio, this knock level is then bracketed by running the engine on two primary reference fuel blends, one of which knocks a little more than the test fuel, and the other a little less. The octane number of the fuel being rated is then determined by interpolation from the knockmeter readings of the bracketing reference fuels.

Alternatively, the engine's compression ratio can be adjusted to close to the limit for the fuel being tested, and then, while the engine is run on each of two closely bracketing test fuels, the ratio can be readjusted to a standard intensity reading on the knockmeter. Finally, the engine is again run on the test fuel, and its compression ratio is adjusted to give the same knockmeter reading. The octane number of the test fuel is then interpolated from the compression ratio settings.

Knock tests are performed under one of two sets of engine operating conditions. Results of tests using the so-called Motor (M) method correlated well with the fuels and automobile engines of the 1930s, when the method was developed. The Research (R) method was developed later when improved refining processes and engines gave gasolines better road performance than their M ratings indicated. Today, the arithmetic average of a gasoline's R and M ratings usually is a good indicator of its performance in a typical car on the road. Therefore, that average, $(R + M)/2$, is posted at service stations to show the antiknock quality of a particular fuel.

For fuels with a rating higher than 100 octane, the rating is usually obtained by determining the amount of tetraethyllead compound that needs to be added to pure isooctane to match the knock resistance of the test fuel. For example, if the amount is 1.3 ml, the fuel's rating is expressed as $100 + 1.3$ or extrapolated above 100 (in this case, about 110 octane) by means of a correlation curve.

[JOHN C. LANE]

Bibliography: *ASTM Manual for Rating Motor, Diesel, and Aviation Fuels*, 4th ed., 1974–1975.

Oil analysis

Analysis of petroleum, or crude oil, to determine its value in modern refinery operations. The analysis, or assay, must provide the refinery planner with the data needed to predict yields, qualities, and operating costs for a bewildering variety of refinery operating conditions and product demands.

In a refinery, crude oil is distilled and separated into products according to the boiling points of the crude oil components (see table). *See* PETROLEUM PRODUCTS.

Procedure. A crude assay follows much the same procedure. The oil is distilled and separated into up to 40 narrow-boiling-range cuts. Cuts with normal boiling points higher than about 190°C are distilled at greatly reduced pressure. This allows the oil to vaporize at temperatures below 190°C and avoids thermal reactions which would alter the chemical properties of the oil. Each cut is then subjected to a variety of tests sufficient to characterize it for the products the cut could be included in. The refinery planners may then calculate yield and qualities of any product by blending the yields and qualities of the cuts that are included in the product.

Petroleum consists primarily of compounds of carbon and hydrogen containing from 1 to about 60 carbon atoms. Carbon atoms in natural petroleum occur in straight and branched chains (paraffins), in single or multiple saturated rings (cycloparaffins or naphthenes), and in cyclic structures of the aromatic type such as benzene, naphthalene, and phenanthrene. Cyclic structures may have attached to them side chains of paraffinic carbons. In lubricating oil it is usual to have naphthene rings built onto the aromatic rings and side chains attached. In products produced by cracking in the refinery, olefins or compounds with carbon-carbon double bonds not in aromatic rings are also found. The high-boiling fractions of petroleum contain increasing amounts of oxygen, nitrogen, and sulfur compounds, as well as traces of organic compounds of metals such as vanadium, nickel, and iron.

Three types of tests are used to characterize a crude oil and its narrow boiling range cuts: (1) tests for physical properties such as specific gravity, refractive index, freeze temperature, vapor pressure, octane number, and viscosity; (2) tests for specific chemical species such as sulfur, nitrogen, metals, and total paraffins, naphthenes, and aromatics; (3) test for determining actual chemical composition.

Physical properties. The physical properties are used to predict how petroleum products will perform. Octane number and vapor pressure are two important qualities of gasolines. Jet fuels must be formulated so that they remain fluid, and therefore pumpable, at the low temperatures encountered by high-flying airplanes. Lubricating oils are characterized by their viscosity level together with the change of viscosity with temperature. The many different tests used for physical properties are relatively simple. Methods have been standardized and published by the American Society for Testing and Materials (ASTM).

Chemical species. The amounts of specific chemical species are also important in evaluating petroleum products. In these days of environmental concern and emission controls, the sulfur content often determines whether or not a fuel can be used. Aromatic components are desirable in gasoline and undesirable in diesel fuels. Fractions with vanadium, nickel, and iron compounds should not be used in catalytic processes since these metals permanently deactivate most catalysts. Tests are available to measure all the chemical species of interest. Many of the tests involve controlled burning of an oil sample, followed by analysis of the combustion products.

Chemical composition. The analysis of petroleum in terms of chemical composition can be done with varying degrees of thoroughness. Anything approaching the complete analysis of a crude oil is so time-consuming and expensive that in the whole world only one sample of crude oil is being analyzed thoroughly. This project, known as Project No. 6 of the American Petroleum Institute, was carried on from 1928 to 1968, when the remaining work was transferred to another project. Obviously, such a detailed analysis will never be done on a routine basis.

Since 1965, gas chromatography (GC), mass spectroscopy (MS), and nuclear magnetic resonance (NMR) methods have been developed that are capable of economically determining the de-

Typical products derived from crude oil

Product	Carbon atom range	Boiling point range, °C
Gas	C_1 and C_2	
Liquefied petroleum gas	C_3 and C_4	
Gasolines	C_4 to C_{10}	15 to 190
Kerosines	C_9 to C_{15}	150 to 280
Middle distillates	C_{12} to C_{20}	200 to 340
Gas oils	C_{20} to C_{45}	340 to 560
Bottoms	Unvaporized residua	

tailed chemical composition of petroleum fractions through the C_9 or C_{10} boiling range plus the distribution of hydrocarbon types up to about C_{40}. The hydrocarbon-type analysis includes distinguishing between one-, two-, three-, and four-ring naphthenes and aromatics.

In a GC analysis a small sample is introduced into a column packed with an adsorbent material such as silica gel. The sample is swept along the column by an inert carrier gas such as nitrogen. The carrier gas forces all components in the sample to move along the column, but each component moves at a unique rate depending on how strongly it is adsorbed by the silica gel. Each component leaves the column as a discrete packet, or peak, of noncarrier gas. The time of each peak identifies the component, and amount in the peak gives the composition. The operation of the GC column and the calculation of results have been automated so that very little human effort is required. However, setting up a GC column is a tedious business since each column seems to behave differently and so must be calibrated for all components it will be used to measure. Also, the method has trouble distinguishing between similar isomers that give overlapping peaks.

In an MS analysis the molecules of the sample being tested are bombarded with an electron beam. This breaks the molecules into ionized fragments whose weights are determined by measuring how much the ionized fragments bend as they move through a magnetic field. The fragment pattern obtained is then correlated against the known fragment patterns of pure compounds to find the composition of the sample. It is difficult to apply the method to compounds for which calibration patterns are not available. It is, however, not feasible to obtain calibration patterns for the many thousands of compounds which appear in petroleum. Nevertheless, even in the higher-boiling fractions where these uncertainties exist, the MS analysis is the best available method for hydrocarbon type analysis.

An NMR analyses in its simplest form determines the type of hydrogen present in a sample, that is, whether the hydrogen is attached to an aliphatic, naphthenic, or aromatic type carbon atom. This is determined from a measurement of the magnetic resonance spectrum of the diluted sample under the influence of a very strong magnetic field. Information of this type, coupled with other analytical information, makes possible the development of a rather detailed picture of molecular types and structures.

Oil analysis, then, depends on a wide variety of tests which have been developed over the years and which continue to be extended and improved. Each refiner selects those tests which best characterize the oil for the refinery operations contemplated. A good analysis is critical when evaluating a crude oil. See PETROLEUM.   [JACK REES]

Bibliography: American Society for Testing and Materials, *ASTM Standards*, annual; ASTM, *Composition of Petroleum Oils*, Spec. Tech. Publ. no. 224, 1958; ASTM, *Manual on Hydrocarbon Analysis*, Spec. Tech. Publ. no. 332A, 1968: ASTM, *Significance of ASTM Tests for Petroleum Products*, 3d ed., 1956; ASTM, *Symposium on Hydrocarbon Analysis*, Spec. Tech. Publ. no. 389, 1965; F. D. Rossini, B. J. Mair, and A. J. Streif, *Hydrocarbons from Petroleum*, 1953; K. Van Nes and H. A. Van Western, *Aspects of the Constitution of Mineral Oil*, 1951; H. I. Waterman, C. Boelhouwer, and J. Cornelissen, *Correlation between Physical Constants and Chemical Structure*, 1958.

Oil and gas, offshore

Oil and gas prospecting and exploitation on the continental shelves and slopes. Since Mobil Oil Co. drilled what is considered the first offshore well off the coast of Louisiana in 1945, exploration for petroleum and natural gas on the more than 8,000,000 mi^2 of the world's continental shelves and slopes, lying between the shore and 1000-ft water depth, has expanded rapidly to include exploration or drilling or both off the coasts of more than 75 nations.

More than 119 national and private petroleum companies have joined in the worldwide offshore search for oil and gas. Near the end of 1974, the mobile offshore drilling fleet consisted of about 265 rigs. In addition, 167 mobile rigs were under construction. The construction costs of these rigs rose tremendously due to inflation and the added sophistication of the equipment. The largest new semisubmersible rigs have cost between $40- and $50,000,000. In mid-1975, the investment in the offshore drilling fleet plus rigs under construction was approximately $6,000,000,000.

In 1977, with a total fleet of 432 units operating, the drilling costs were estimated to be about $7,000,000,000 per year. For this expenditure, the industry should be able to drill between 1200 and 1300 holes per year.

Until early in 1975, most of the offshore rigs were employed in drilling on the United States Outer Continental Shelf. After that time, this changed dramatically, with more than 75% of the offshore drilling being conducted outside the United States waters.

Paralleling the development of drilling vessels able to withstand the rigors of operation in the open sea have been remarkable technological developments in equipment and methods. More than 125 companies in the United States devote a large share of their efforts to the development and manufacture of material and devices in support of offshore oil and gas.

For many years petroleum companies stopped at the water's edge or sought and developed oil and gas accumulations only in the shallow seas bordering onshore producing areas. These activities were usually confined to water depths in which drilling and producing operations could be conducted from platforms, piers, or causeways built upon pilings driven into the sea floor. Major accumulations along the Gulf Coast of the United States, in Lake Maracaibo of Venezuela, in the Persian Gulf, and in the Baku fields of the Caspian Sea were developed from such fixed structures.

Exploration deeper under the sea did not begin in earnest until the world's burgeoning appetite for energy sources, coupled with a lessening return from land drilling, provided the incentives for the huge investments required for drilling in the open sea.

Geology and the sea. There is a sound geologic basis for the petroleum industry turning to the continental shelves and slopes in search of needed reserves. Favorable sediments and structures exist beneath the present seas of the world, in geologic settings that have proven highly productive onshore. In fact, the subsea geologic similarity—or in some cases superiority—to geologic conditions on land has been a vital factor in the rapid expansion of the free world's investment in offshore exploration and production. More than $12,000,000,000 had been invested in the free world offshore effort by 1969. World offshore expenditures for the year 1972 were approximately $4,000,000,000.

Subsea geologic basins, having sediments considered favorable for petroleum deposits, total approximately 6,000,000 mi^2 out to a water depth of 1000 ft, or about 57% of the world's total continental shelf. This 6,000,000-mi^2 area is equivalent to one-third of the 18,000,000 mi^2 of geologic basins on land.

As of January 1974, world proven reserves were estimated at 640×10^9 bbl of oil and 550×10^{12} ft^3 of natural gas. Of these reserves, 18% of the oil and 9.5% of the gas were located offshore. The cumulative world oil and gas production as of the same date was approximately 270×10^9 bbl of oil and 550×10^{12} ft^3 of natural gas, with the offshore contributing about 6% and 4.5%, respectively. During 1973, about 3.6×10^9 bbl of oil and 53×10^{12} ft^3 of natural gas were produced offshore, 18% and 14% of world production, respectively. When these figures are contrasted with total offshore production as of Jan. 1, 1974, the rapid growth of offshore production through the years is seen.

Free world proven oil reserves are estimated to total 375×10^9 bbl, of which 71×10^9 or 19%, are located offshore. Estimates of future offshore oil reserves remaining to be discovered provided that an economic incentive is present go as high as 700×10^9 bbl.

Furthermore, this production rate has been achieved despite the fact that only 166,000 mi^2, or about 2%, of the world's continental shelves out to 1000-ft water depths has been tested by the drill. Estimates indicate that only a small percentage of the seismic work necessary to evaluate the continental shelves has been accomplished, and that even by pursuing a work rate nearly double the average for the 1960s, it would take another 127 years to complete the seismic survey of all those areas judged to be prospective. Surveying of the world's total favorable continental shelf area will take an estimated 68,400 seismic crew months, with 8300 months actually accomplished.

Virtually all of the world's continental shelf has received some geological study, with active seismic or drilling exploration of some sort either planned or in effect. Offshore exploration planned or underway includes action in the North Sea; in the English Channel off the coast of France; in the Red Sea bordering Egypt; in the seas to the north, west, and south of Australia; off Sumatra; along the Gulf Coast and east, west, and northwest coast of the continental United States; in the Cook Inlet and off the west coast of Alaska; in the Persian Gulf; off Mexico; off the east and west coasts of Central America and South America; and off Nigeria and North Africa. In addition, offshore exploration has been undertaken in the Caspian and Baltic seas off the coasts of East Germany, Poland, Latvia, and Lithuania; the China seas; Gulf of Thailand; Irish Sea; Arctic Ocean; and Arctic islands. *See* PETROLEUM GEOLOGY.

Successes at sea. The price of success in the offshore drilling is staggering in the amount of capital required, the risks involved, and the time required to achieve a break-even point.

Offshore Louisiana, long the center of much of the world's offshore drilling, has yielded a large number of oil and gas fields. By the end of 1973, the United States government received revenues amounting to $11,100,000,000 from the industry activities in the Gulf of Mexico. At the same time, the value of the products produced by industry yielded a gross revenue of $15,400,000,000. A total of 768 platforms were set by the end of 1973 in the Gulf of Mexico. It is obvious that industry has a long way to go before the capital investment in production platforms and facilities and exploration costs are recovered. *See* PETROLEUM PRODUCTS.

Major oil and gas strikes have been made in the North Sea, and oil and gas production has been established in the Bass Straits off Australia. These have the potential for changing historical energy sources and the economies of the countries involved. Discoveries made since 1964 in the Cook Inlet of Alaska are making a major impact on the West Coast market of the United States; although they involve between 1.5×10^9 to 2×10^9 bbl of oil, the extremely high exploration and development costs will cause the break-even point to be somewhere near 10 years. The most active offshore areas outside the United States are the North Sea, Southeast Asian seas, Nigeria, and the Persian Gulf.

The Okan field of Nigeria reached 45,000 barrels per day (bpd) production in 1966. Total production from Nigeria, where fields other than Okan are in swamps at the edge, reached 325,000 bpd in 1966.

The Persian Gulf area, where the world's largest

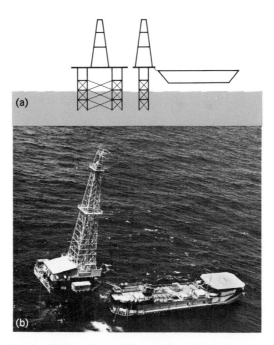

Fig. 1. Offshore fixed drilling platform. (*a*) Underwater design (*World Petroleum*). (*b*) Rig on a drilling site (*Marathon Oil Co., Findlay, Ohio*).

fields on land have been found, has also produced the world's largest offshore field, Safaniya, in Saudia Arabia. This field is capable of producing more than 1,000,000 bpd.

Mobile drilling platforms. The underwater search has been made possible only by vast improvements in offshore technology. Drillers first took to the sea with land rigs mounted on barges towed to location and anchored or with fixed platforms accompanied by a tender ship (Fig. 1). A wide variety of rig platforms has since evolved, some designed to cope with specific hazards of the sea and others for more general work. All new types stress characteristics of mobility and the capability for work in even deeper water.

The world's mobile platform fleet can be divided into four main groupings: self-elevating platforms, submersibles, semisubmersibles, and floating drill ships.

The most widely used mobile platform is the self-elevating, or jack-up, unit (Fig. 2). It is towed to location, where the legs are lowered to the sea floor, and the platform is jacked up above wave height. These self-contained platforms are especially suited to wildcat and delineation drilling. They are best in firmer sea bottoms with a depth limit out to 300 ft of water.

The submersible platforms have been developed from earlier submersible barges which were used in shallow inlet drilling along the United States Gulf Coast. The platforms are towed to location and then submerged to the sea bottom. They are very stable and can operate in areas with soft sea floors. Difficulty in towing is a disadvantage, but this is partially offset by the rapidity with which they can be raised or lowered, once on location.

Semisubmersibles (Fig. 3) are a version of submersibles. They can work as bottom-supported units or in deep water as floaters. Their key virtue

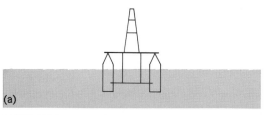

Fig. 3.   Offshore semisubmersible drilling platform. (a) Underwater design (*World Petroleum*). (b) Santa Fe Marine's Blue Water no. 3 drilling rig on a drilling site (*Marathon Oil Co., Findlay, Ohio*).

is the wide range of water depths in which they can operate, plus the fact that, when working as floaters, their primary buoyancy lies below the action of the waves, thus providing great stability. The "semis" are the most recent of the rig-type platforms.

Floating drill ships (Fig. 4) are capable of drilling in 60-ft to abyssal depths. They are built as self-propelled ships or with a ship configuration that requires towing. Several twin-hulled versions have been constructed to give a stable catamaran design. Floating drill ships use anchoring or ingenious dynamic positioning systems to stablize their position, the latter being necessary in deeper waters. Floaters cannot be used in waters much shallower than 70 ft because of the special equipment required for drilling from the vessel subject to vertical movement from waves and tidal changes, as well as minor horizontal shifts due to stretch and play in anchor lines. Exploration in deeper waters necessitates building more semisubmersible and floating drill ships. A conventional exploratory hole has been drilled 50 mi off the coast of Gabon in 2150 ft of water by Shell Deep Water Drilling Company using the Sedco 445 drill ship. The *Glomar Challenger* has drilled stratigraphic holes in the sea floor to a depth of 3334 ft in water depths of 20,000 ft.

Production and well completion technology. The move of exploration into the open hostile sea has required not only the development of drilling vessels but a host of auxiliary equipment and techniques. A whole new industrial complex has developed to serve the offshore industry.

Of particular interest is the development of diving techniques and submersible equipment to aid in exploration and the completion of wells. A platform was constructed for installation in 850 ft of

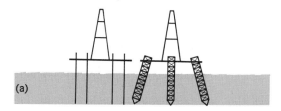

Fig. 2.   Offshore self-elevating drilling platform. (a) Underwater design (*World Petroleum*). (b) Self-elevating drilling platform (*Marathon Oil Co., Findlay, Ohio*).

water in the Santa Barbara Channel. Tentative plans for platforms in 1000 ft of water have been announced for the Gulf of Mexico. Economics will soon force sea-bottom completions which require men or robots or both to make the necessary pipe and well connections. Such work is necessary even in the water depths now being developed.

A robot device has been developed which operates from the surface and uses sonar and television for viewing; it can excavate ditches for pipelines and make simple pipe connections and well hookups in water depths to 2500 ft. Limitations of robot devices are such that the more complex needs of well completions and service require the actions of men. To fill this need, diving specialists have been used to sandbag platform bases, recover conductor pipe, survey and remove wreckage, and make pipeline connections and well hookups. Diving depths have been increased to the point where useful work has been performed at depths in excess of 600 ft. Pressure chambers to take divers to the bottom and to return them to the surface to be decompressed are operational. One has been operating routinely in 425 ft of water for Esso Exploration, Norway. This deep diving is made possible by the development of saturation diving, which uses a mixture of oxygen, helium, nitrogen, and argon. This technique has allowed divers to remain below 200 ft for 6 days while doing salvage work on a platform. It has also allowed prolonged submergence at 600 ft in preparation for actual work on wells at this depth.

Miniature submarines have taken their place in exploration and completion work allowing the viewing of conditions, the gathering of samples, and simple mechanical tasks. Their depth range is for all practical purposes unlimited.

Technical groups are experimenting with the design of drilling and production units that would be totally enclosed and be set on the sea bottom. Living in and working from these units, personnel would be able to carry out all the necessary oil field operations. In effect, such units would resemble a miniature city on the sea floor, from which a man would need to return to the surface only when his tour of work was completed. *See* OIL AND GAS FIELD EXPLOITATION.

Concomitant with the progress of the petroleum industry in its venture into the open sea has been a vast increase in the knowledge of the sea and its contained wealth. Mining of the sea floor using some of petroleum's technology has started in several areas of the world, and actual farming or ranching of the life in the sea is being planned.

Hazards at sea. As the petroleum industry pushed farther into the hostile environment of the sea, it sustained a series of disasters, reflected by the doubling of offshore insurance rates in April 1966. Between 1955 and 1968, for example, 23 offshore units were destroyed by blowout and 6 by hurricane and breakup and collapse at sea. The United States Gulf Coast, where a large percentage of the world's offshore drilling has taken place, was severely hit in 1964 and 1965 by hurricanes, which claimed over $7,500,000 in tow and service vessels, $21,000,000 in fixed platforms, and $28,000,000 in mobile platforms.

These figures do not include expenses sustained from loss of wells, removal of wrecked equipment, and loss of production. Such liabilities have raised insurance rates on a $5,000,000 platform to as much as $500,000 or more per year, depending on platform type and location. Much current design work is aimed at engineering better safety features for the benefit of both the crews and structures.

Despite the hazards and monumental cost involved in extracting oil and gas from beneath the sea, the world's population explosion and its ever increasing demand for petroleum energy will force the search for new reserves into even deeper waters and more remote corners of the world. In truth, the search is only just beginning.

[G. R. SCHOONMAKER]

Bibliography: Ocean engineering, *Petrol. Eng.*, vol. 39, no. 11, 1967; *OCS Statistics*, United States Geological Survey, 1973; Offshore news, *Offshore*, vol. 34, no. 10 and 13, 1974; Offshore 1973, *World Oil*, vol. 177, no. 1, 1973; Offshore report and offshore rigs, *World Petrol.*, vol. 36, no. 5, 1965; Rig insurance, *Oil Gas J.*, vol. 64, no. 19, 1966; Unmanned subsea device, *World Oil*, vol. 165, no. 7, 1968; L. G. Weeks, Petroleum resources potential of continental margins, in C. A. Burk and C. L. Drake (eds.), *The Geology of Continental Margins*, 1974.

Oil and gas field exploitation

In the petroleum industry, a field is an area underlain without substantial interruption by one or more reservoirs of commercially valuable oil or gas, or both. A single reservoir (or group of reservoirs which cannot be separately produced) is a pool. Several pools separated from one another by barren, impermeable rock may be superimposed one above another within the same field. Pools have variable areal extent. Any sufficiently deep well located within the field should produce from one or more pools. However, each well cannot produce from every pool, because different pools have different areal limits.

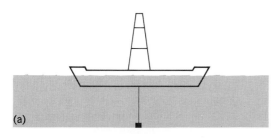

Fig. 4. Floating drill ship. Such ships can drill in depths from 60 to 1000 ft or more. (*a*) Underwater design (*World Petroleum*). (*b*) Floating drill ship on a drilling site (*Marathon Oil Co., Findlay, Ohio*).

DEVELOPMENT

Development of a field includes the location, drilling, completion, and equipment of wells necessary to produce the commercially recoverable oil and gas in the field.

Related oil field conditions. Petroleum is a generic term which, in its broadest meaning, includes all naturally occurring hydrocarbons, whether gaseous, liquid, or solid. By variation of the temperature or pressure, or both, of any hydrocarbon, it becomes gaseous, liquid, or solid. Temperatures in producing horizons vary from approximately 60° to more than 300°F, depending chiefly upon the depth of the horizon. A rough approximation is that temperature in the reservoir sand, or pay, equals 60°F, plus 0.017°F/ft of depth below surface. Pressure on the hydrocarbons varies from atmospheric to more than 11,000 psi. Normal pressure is considered as 0.465 psi/ft of depth. Temperatures and pressure vary widely from these average figures. Hydrocarbons, because of wide variations in pressure and temperature and because of mutual solubility in one another, do not necessarily exist underground in the same phases in which they appear at the surface.

Petroleum occurs underground in porous rocks of wide variety. The pore spaces range from microscopic size to rare holes 1 in. or more in diameter. The containing rock is commonly called the sand or the pay, regardless of whether the pay is actually sandstone, limestone, dolomite, unconsolidated sand, or fracture openings in relatively impermeable rock.

Development of field. After discovery of a field containing oil or gas, or both, in commercial quantities, the field must be explored to determine its vertical and horizontal limits and the mechanisms under which the field will produce. Development and exploitation of the field proceed simultaneously. Usually the original development program is repeatedly modified by geologic knowledge acquired during the early stages of development and exploitation of the field. See PETROLEUM GEOLOGY.

Ideally, tests should be drilled to the lowest possible producing horizon in order to determine the number of pools existing in the field. Testing and geologic analysis of the first wells sometimes indicates the producing mechanisms, and thus the best development program. Very early in the history of the field, step-out wells will be drilled to determine the areal extent of the pool or pools. Step-out wells give further information regarding the volumes of oil and gas available, the producing mechanisms, and the desirable spacing of wells.

The operator of an oil and gas field endeavors to select a development program which will produce the largest volume of oil and gas at a profit. The program adopted is always a compromise between conflicting objectives. The operator desires (1) to drill the fewest wells which will efficiently produce the recoverable oil and gas; (2) to drill, complete, and equip the wells at the lowest possible cost; (3) to complete production in the shortest practical time to reduce both capital and operating charges; (4) to operate the wells at the lowest possible cost; and (5) to recover the largest possible volume of oil and gas.

Selecting the number of wells. Oil pools are pro-duced by four mechanisms: dissolved gas expansion, gas-cap drive, water drive, and gravity drainage. Commonly, two or more mechanisms operate in a single pool. The type of producing mechanism in each pool influences the decision as to the number of wells to be drilled. Theoretically, a single, perfectly located well in a water-drive pool is capable of producing all of the commercially recoverable oil and gas from that pool. Practically, more than one well is necessary if a pool of more than 80 acres is to be depleted in a reasonable time. If a pool produces under either gas expansion or gas-cap drive, oil production from the pool will be independent of the number of wells up to a spacing of at least 80 acres per well (1866 ft between wells). Gas wells often are spaced a mile or more apart. The operator accordingly selects the widest spacing permitted by field conditions and legal requirements, as discussed later. See PETROLEUM RESERVOIR ENGINEERING.

Major components of cost. Costs of drilling, completing, and equipping the wells influence development plans. Having determined the number and depths of producing horizons and the producing mechanisms in each horizon, the operator must decide whether he will drill a well at each location to each horizon or whether a single well can produce from two or more horizons at the same location. Clearly, the cost of drilling the field can be sharply reduced if a well can drain two, three, or more horizons. The cost of drilling a well will be higher if several horizons are simultaneously produced, because the dual or triple completion of a well usually requires larger casing. Further, completion and operating costs are higher. However, the increased cost of drilling a well of larger diameter and completing the well in two or more horizons is 20–40% less than the cost of drilling and completing two wells to produce separately from two horizons.

In some cases, the operator may reduce the number of wells by drilling a well to the lowest producible horizon and taking production from that level until the horizon there is commercially exhausted. The well is then plugged back to produce from a higher horizon. Selection of the plan for producing the various horizons obviously affects the cost of drilling and completing individual wells, as well as the number of wells which the operator will drill. If two wells are drilled at approximately the same location, they are referred to as twins, three wells at the same location are triplets, and so on. See OIL AND GAS WELL COMPLETION; OIL AND GAS WELL DRILLING.

Costs and duration of production. The operator wishes to produce as rapidly as possible because the net income from sale of hydrocarbons is obviously reduced as the life of the well is extended. The successful operator must recover from his productive wells the costs of drilling and operating those wells, and in addition he must recover all costs involved in geological and geophysical exploration, leasing, scouting, and drilling of dry holes, and occasionally other operations. If profits from production are not sufficient to recover all exploration and production costs and yield a profit in excess of the rate of interest which the operator could secure from a different type of investment, he is discouraged from further exploration. See

GEOPHYSICAL EXPLORATION.

Most wells cannot operate at full capacity because unlimited production results in physical waste and sharp reduction in ultimate recovery. In many areas, conservation restrictions are enforced to make certain that the operator does not produce in excess of the maximum efficient rate. For example, if an oil well produces at its highest possible rate, a zone promptly develops around the well where production is occurring under gas-expansion drive, the most inefficient producing mechanism. Slower production may permit the petroleum to be produced under gas-cap drive or water drive, in which case ultimate production of oil will be two to four times as great as it would be under gas-expansion drive. Accordingly, the most rapid rate of production generally is not the most efficient rate.

Similarly, the initial exploration of the field may indicate that one or more gas-condensate pools exist, and recycling of gas may be necessary to secure maximim recovery of both condensate and of gas. The decision to recycle will affect the number of wells, the locations of the wells, and the completion methods adopted in the development program.

Further, as soon as the operator determines that secondary oil-recovery methods are desired and expects to inject water, gas, steam, or, rarely, air to provide additional energy to flush or displace oil from the pay, the number and location of wells may be modified to permit the most effective secondary recovery procedures. *See* PETROLEUM SECONDARY RECOVERY.

Legal and practical restrictions. The preceding discussion has assumed control of an entire field under single ownership by a single operator. In the United States, a single operator rarely controls a large field, and this field is almost never under a single lease. Usually, the field is covered by separate leases owned and operated by different producers. The development program must then be modified in consideration of the lease boundaries and the practices of the other operators who are in the field.

Oil and gas know no lease boundaries. They move freely underground from areas of high pressure toward lower-pressure situations. The operator of a lease is obligated to locate his wells in such a way as to prevent drainage of his lease by wells on adjoining leases, even though the operator may own the adjoining leases. In the absence of conservation restrictions, an operator must produce petroleum from his wells as rapidly as it is produced from wells on adjoining leases. Slow production on one lease results in migration of oil and gas to nearby leases which are more rapidly produced.

The operator's development program must provide for offset wells located as close to the boundary of his lease as are wells on adjoining leases. Further, the operator must equip his wells to produce as rapidly as the offset produces and must produce from the same horizons which are being produced in offset wells. The lessor who sold the lease to the operator is entitled to his share of the recoverable petroleum underlying his land. Negligence by the operator in permitting drainage of a lease makes the operator liable to suit for damages or cancellation of the lease.

A development program acceptable to all operators in the field permits simultaneous development of leases, prevents drainage, and results in maximum ultimate production from the field. Difficulties may arise in agreement upon the best development program for a field. Most states have enacted statutes and have appointed regulatory bodies under which judicial determination can be made of the permissible spacing of the wells, the rates of production, and the application of secondary recovery methods.

Drilling unit. Commonly, small leases or portions of two or more leases are combined to form a drilling unit in whose center a well will be drilled. Unitization may be voluntary, by agreement between the operator or operators and the interested royalty owners, with provision for sharing production from the well between the parties in proportion to their acreage interests. In many states the regulatory body has authority to require unitization of drilling units, which eliminates unnecessary offset wells and protects the interests of a landowner whose acreage holding may be too small to justify the drilling of a single well on his property alone.

Pool unitization. When recycling or some types of secondary recovery are planned, further unitization is adopted. Since oil and gas move freely across lease boundaries, it would be wasteful for an operator to repressure, recycle, or water-drive a lease if the adjoining leases were not similarly operated. Usually an entire pool must be unitized for efficient recycling of secondary recovery operations. Pool unitization may be accomplished by agreement between operators and royalty owners. In many cases, difference of opinion or ignorance on the part of some parties prevents voluntary pool unitization. Many states authorize the regulatory body to unitize a pool compulsorily on application by a specified percentage of interests of operators and royalty owners. Such compulsory unitization is planned to provide each operator and each royalty owner his fair share of the petroleum products produced from the field regardless of the location of the well or wells through which these products actually reach the surface.

EXPLOITATION—GENERAL CONSIDERATIONS

Oil and gas production necessarily are intimately related, since approximately one-third of the gross gas production in the United States is produced from wells that are classified as oil wells. However, the naturally occurring hydrocarbons of petroleum are not only liquid and gaseous but may even be found in a solid state, such as asphaltite and some asphalts.

Where gas is produced without oil, the production problems are simplified because the product flows naturally throughout the life of the well and does not have to be lifted to the surface. However, there are sometimes problems of water accumulations in gas wells, and it is necessary to pump the water from the wells to maintain maximum, or economical, gas production. The line of demarcation between oil wells and gas wells is not definitely established since oil wells may have gas-oil ratios ranging from a few cubic feet per barrel to many thousand cubic feet of gas per barrel of oil. Most gas wells produce quantities of condensable vapors, such as propane and butane, that may be liquefied and marketed for fuel, and the more stable liquids produced with gas can be utilized as natural gasoline. *See* PETROLEUM.

Factors of method selection. The method selected for recovering oil from a producing formation depends on many factors, including well depth, well-casing size, oil viscosity, density, water production, gas-oil ratio, porosity and permeability of the producing formation, formation pressure, water content of producing formation, and whether the force driving the oil into the well from the formation is primarily gas pressure, water pressure, or a combination of the two. Other factors, such as paraffin content and difficulty expected from paraffin deposits, sand production, and corrosivity of the well fluids, also have a decided influence on the most economical method of production.

Special techniques utilized to increase productivity of oil and gas wells include acidizing, hydraulic fracturing of the formation, the setting of screens, and gravel packing or sand packing to increase permeability around the well bore.

Aspects of production rate. Productive rates per well may vary from a few barrels per day to several thousand barrels per day, and it may be necessary to produce a large percentage of water along with the oil.

Field and reservoir conditions. In some cases reservoir conditions are such that some of the wells flow naturally throughout the entire economical life of the oil field. However, in the great majority of cases it is necessary to resort to artificial lifting methods at some time during the life of the field, and often it is necessary to apply artificial lifting means immediately after the well is drilled.

Market and regulatory factors. In some oil-producing states of the United States there are state bodies authorized to regulate oil production from the various oil fields. The allowable production per well is based on various factors, including the market for the particular type of oil available, but very often the allowable production is based on an engineering study of the reservoir to determine the optimum rate of production.

Useful terminology. A few definitions of terms used in petroleum production technology are listed below to assist in an understanding of some of the problems involved.

Porosity. The percentage porosity is defined as the percentage volume of voids per unit total volume. This, of course, represents the total possible volume available for accumulation of fluids in a formation, but only a fraction of this volume may be effective for practical purposes because of possible discontinuities between the individual pores. The smallest pores generally contain water held by capillary forces.

Permeability. Permeability is a measure of the resistance to flow through a porous medium under the influence of a pressure gradient. The unit of permeability commonly employed in petroleum production technology is the darcy. A porous structure has a permeability of 1 darcy if, for a fluid of 1 centipoise (cp) viscosity, the volume flow is 1 cm³/(sec)(cm²) under a pressure gradient of 1 atm/cm.

Productivity index. The productivity index is a measure of the capacity of the reservoir to deliver oil to the well bore through the productive formation and any other obstacles that may exist around the well bore. In petroleum production technology, the productivity index is defined as production in barrels per day per pound drop in bottom-hole pressure. For example, if a well is closed in at the casinghead, the bottom-hole pressure will equal the formation pressure when equilibrium conditions are established. However, if fluid is removed from the well, either by flowing or pumping, the bottom-hole pressure will drop as a result of the resistance to flow of fluid into the well from the formation to replace the fluid removed from the well. If the closed-in bottom-hole pressure should be 1000 psi, for example, and if this pressure should drop to 900 psi when producing at a rate of 100 bbl/day (a drop of 100 psi), the well in question would have a productivity index of one.

Barrel. The standard barrel used in the petroleum industry is 42 U.S. gal.

API gravity. The American Petroleum Institute (API) scale that is in common use for indicating specific gravity, or a rough indication of quality of crude petroleum oils, differs slightly from the Baume scale commonly used for other liquids lighter than water. The table shows the relationship between degrees API and specific gravity referred to water at 60°F for specific gravities ranging from 0.60 to 1.0.

Viscosity range. Viscosity of crude oils currently produced varies from approximately 1 cp to values above 1000 cp at temperatures existing at the bottom of the well. In some areas it is necessary to supply heat artificially down the wells or circulate lighter oils to mix with the produced fluid for maintenance of a relatively low viscosity throughout the temperature range to which the product is subjected.

In addition to wells that are classified as gas wells or oil wells, the term gas-condensate well has come into general use to designate a well that produces large volumes of gas with appreciable quantities of light, volatile hydrocarbon fluids. Some of these fluids are liquid at atmospheric pressure and

Degrees API corresponding to specific gravities of crude oil at 60°/60°F

Specific gravity, in tenths	Specific gravity, in hundredths									
	.00	.01	.02	.03	.04	.05	.06	.07	.08	.09
0.60	104.33	100.47	96.73	93.10	89.59	86.19	82.89	79.69	76.59	73.57
0.70	70.64	67.80	65.03	62.34	59.72	57.17	54.68	52.27	49.91	47.61
0.80	45.38	43.19	44.06	38.98	36.95	34.97	33.03	31.14	29.30	27.49
0.90	25.72	23.99	22.30	20.65	19.03	17.45	15.90	14.38	12.89	11.43
1.00	10.00									

temperature; others, such as propane and butane, are readily condensed under relatively low pressures in gas separators for use as liquid petroleum gas (LPG) fuels or for other uses. The liquid components of the production from gas-condensate wells generally arrive at the surface in the form of small droplets entrained in the high-velocity gas stream and are separated from the gas in a high-pressure gas separator.

PRODUCTION METHODS IN PRODUCING WELLS

The common methods of producing oil wells are (1) natural flow; (2) pumping with sucker rods to actuate a pump located in the well fluid; (3) gas lift; (4) hydraulic subsurface pumps, in which the subsurface unit is a reciprocating hydraulic motor joined to a reciprocating pump and driven by circulating oil or other power fluid down the well to actuate the hydraulic engine; (5) electrically driven centrifugal well pumps; and (6) swabbing.

Numerous other methods, including jet pumps and sonic pumps, have been tried and are used to slight extent. The sonic pump is a development in which the tubing is vibrated longitudinally by a mechanism at the surface and acts as a high-speed pump with an extremely short stroke.

A brief discussion of production methods, in the approximate order of their relative importance and popularity, follows.

Natural flow. Natural flow is the most economical method of production and generally is utilized as long as the desired production rate can be maintained by this method. It utilizes the formation energy, which may consist of gas in solution in the oil in the formation; free gas under pressure acting against the liquid and gas-liquid phase to force it toward the well bore; water pressure acting against the oil; or a combination of these three energy sources. In some areas the casinghead pressure may be of the order of 10,000 psi, so it is necessary to provide fittings adequate to withstand such pressures. Adjustable throttle values, or chokes, are utilized to regulate the flow rate to a desired and safe value. With such a high-pressure drop across a throttle valve the life of the valve is likely to be very short. Several such valves are arranged in parallel in the tubing head "Christmas tree" with positive shutoff valves between the chokes and the tubing head so that the wearing parts of the throttle valve, or the entire valve, can be replaced while flow continues through another similar valve.

An additional safeguard that is often used in connection with high-pressure flowing wells is a bottom-hole choke or a bottom-hole flow control valve that limits the rate of flow to a reasonable value, or stops it completely, in case of failure of surface controls. Figure 1 shows a schematic outline of a simple flowing well hookup. The packer is not essential but is often used to reduce the free gas volume in the casing.

Flow rates for United States wells seldom exceed a few hundred barrels per day because of enforced or voluntary restrictions to regulate production rates and to obtain most efficient and economical ultimate recovery. However, in some countries, especially in the Middle East, it is not uncommon for natural flow rates to exceed 10,000 bpd/well.

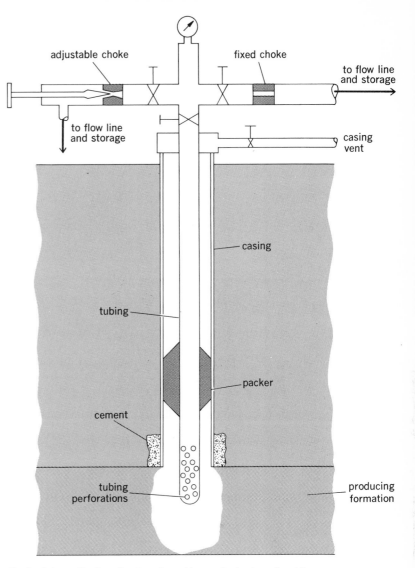

Fig. 1. Schematic view of well equipped for producing by natural flow.

Lifting. Most wells are not self-flowing. The common types of lifting are outlined here.

Pumping with sucker rods. Approximately 90% of the wells made to produce by some artificial lift method in the United States are equipped with sucker-rod–type pumps. In these the pump is installed at the lower end of the tubing string and is actuated by a string of sucker rods extending from the surface to the subsurface pump. The sucker rods are attached to a polished rod at the surface. The polished rod extends through a stuffing box and is attached to the pumping unit, which produces the necessary reciprocating motion to actuate the sucker rods and the subsurface pump. Figure 2 shows a simplified schematic section through a pumping well. The two common variations are mechanical and hydraulic long-stroke pumping.

1. Mechanical pumping. The great majority of pumping units are of the mechanical type, consisting of a suitable reduction gear, and crank and pitman arrangement to drive a walking beam to produce the necessary reciprocating motion. A counterbalance is provided to equalize the load on the upstroke and downstroke. Mechanical pump-

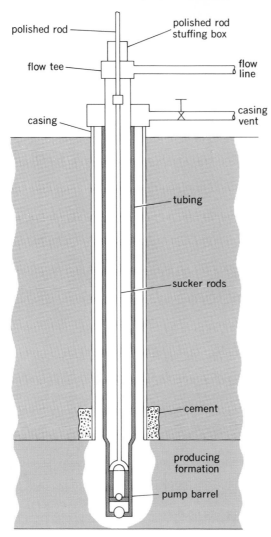

Fig. 2. A schematic view of a well which is equipped for pumping with sucker rods.

Fig. 3. Pumping unit with adjustable rotary counterbalance. (*Oil Well Supply Division, U.S. Steel Corp.*)

ing units of this type vary in load-carrying capacity from about 2000 to about 43,000 lb, and the torque rating of the low-speed gear which drives the crank ranges from 6400 in.-lb in the smallest API standard unit to about 1,500,000 in.-lb for the largest units now in use. Stroke length varies from about 18 to 192 in. Usual operating speeds are from about 6 to 20 strokes/min. However, both lower and higher rates of speed are sometimes used. Figure 3 shows a modern pumping unit in operation.

Production rates with sucker-rod–type pumps vary from a fraction of 1 bpd in some areas, with part-time pumping, to approximately 3000 bpd for the largest installations in relatively shallow wells.

2. Hydraulic long-stroke pumping. For this the units consist of a hydraulic lifting cylinder mounted directly over the well head and are designed to produce stroke lengths of as much as 30 ft. Such long-stroke hydraulic units are usually equipped with a pneumatic counterbalance arrangement which equalizes the power requirement on the upstroke and downstroke.

Hydraulic pumping units also are made without any provision for counterbalance. However, these units are generally limited to relatively small wells, and they are relatively inefficient.

Gas lift. Gas lift in its simplest form consists of initiating or stimulating well flow by injecting gas at some point below the fluid level in the well. With large-volume gas-lift operations the well may be produced through either the casing or the tubing. In the former case, gas is conducted through the tubing to the point of injection; in the latter, gas may be conducted to the point of injection through the casing or through an auxiliary string of tubing. When gas is injected into the oil column, the weight of the column above the point of injection is reduced as a result of the space occupied by the relatively low density of the gas. This lightening of the fluid column is sufficient to permit the formation pressure to initiate flow up the tubing to the surface. Gas injection is often utilized to increase the flow from wells that will flow naturally but will not produce the desired amount by natural flow.

There are many factors determining the advisability of adopting gas lift as a means of production. One of the more important factors is the availability of an adequate supply of gas at suitable pressure and reasonable cost. In a majority of cases gas lift cannot be used economically to produce a reservoir to depletion because the well may be relatively productive with a low back pressure maintained on the formation but will produce very little, if anything, with the back pressure required for gas-lift operation. Therefore, it generally is necessary to resort to some mechanical means of pumping before the well is abandoned, and it may be more economical to adopt the mechanical means initially than to install the gas-lift system while conditions are favorable and later replace it.

This discussion of gas lift has dealt primarily with the simple injection of gas, which may be continuous or intermittent. There are numerous modifications of gas-lift installations, including various designs for flow valves which may be installed in the tubing string to open and admit gas to the tubing from the casing at a predetermined pressure differential between the tubing and casing. When the valve opens, gas is injected into the tubing to initiate and maintain flow until the tubing pressure drops to a predetermined value; and the valve closes before the input gas-oil ratio becomes excessive. This represents an intermittent-flow–type valve. Other types are designed to maintain continuous flow, proper pressure differential, and proper gas injection rate for efficient operation. In some cases several such flow valves are spaced up the tubing string to permit flow to be initiated from various levels as required.

Other modifications of gas lift involve the utilization of displacement chambers. These are installed on the lower end of the well tubing where oil may accumulate, and the oil is displaced up the tubing with gas injection controlled by automatic or mechanical valves.

Hydraulic subsurface pumps. The hydraulic subsurface pump has come into fairly prominent use. The subsurface pump is operated by means of a hydraulic reciprocating motor attached to the pump and installed in the well as a single unit. The hydraulic motor is driven by a supply of hydraulic fluid under pressure that is circulated down a string of tubing and through the motor. Generally the hydraulic fluid consists of crude oil which is discharged into the return line and returns to the surface along with the produced crude oil.

Hydraulically operated subsurface pumps are also arranged for separating the hydraulic power fluid from the produced well fluid. This arrangement is especially desirable where the fluid being produced is corrosive or is contaminated with considerable quantities of sand or other solids that are difficult to separate to condition the fluid for use as satisfactory power oil. This method also permits the use of water or other nonflammable liquids as hydraulic power fluid to minimize the fire hazard in case of a failure of the hydraulic power line at the surface.

Centrifugal well pumps. Electrically driven centrifugal pumps have been used to some extent, especially in large-volume wells of shallow or moderate depths. Both the pump and the motor are restricted in diameter to run down the well casing, leaving sufficient clearance for the flow of fluid around the pump housing. With the restricted diameter of the impellers the discharge head necessary for pumping a relatively deep well can be obtained only by using a large number of stages and operating at a relatively high speed. The usual rotating speed for such units is 3600 rpm, and it is not uncommon for such units to have 50 or more pump stages. The direct-connected electric motor must be provided with a suitable seal to prevent well fluid from entering the motor housing, and electrical leads must be run down the well casing to supply power to the motor.

Swabs. Swabs have been used for lifting oil almost since the beginning of the petroleum industry. They usually consist of a steel tubular body equipped with a check valve which permits oil to flow through the tube as it is lowered down the well with a wire line. The exterior of the steel body is generally fitted with flexible cup-type soft packing that will fall freely but will expand and form a seal with the tubing when pulled upward with a head of fluid above the swab. Swabs are run into the well on a wire line to a point considerably below the fluid level and then lifted back to the surface to deliver the volume of oil above the swab. They are often used for determining the productivity of a well that will not flow naturally and for assisting in cleaning paraffin from well tubing. In some cases swabs are used to stimulate wells to flow by lifting, from the upper portion of the tubing, the relatively dead oil from which most of the gas has separated.

Bailers. Bailers are used to remove fluids from wells and for cleaning out solid material. They are run into the wells on wire lines as in swabbing, but differ from swabs in that they generally are run only in the casing when there is no tubing in the well. The capacity of the bailer itself represents the volume of fluid lifted each time since the bailer does not form a seal with the casing. The bailer is simply a tubular vessel with a check valve in the bottom. This check valve generally is arranged so that it is forced open when the bailer touches bottom in order to assist in picking up solid material for cleaning out a well.

Jet pumps. A jet pump for use in oil wells operates on exactly the same principle as a water-well jet pump. Advantage is taken of the Bernoulli effect to reduce pressure by means of a high-velocity fluid jet. Thus oil is entrained from the well with this high-velocity jet in a venturi tube to accelerate the fluid and assist in lifting it to the surface, along with any assistance from the formation pressure. The application of jet pumps to oil wells has been insignificant.

Sonic pumps. Sonic pumps essentially consist of a string of tubing equipped with a check valve at each joint and mechanical means on the surface to vibrate the tubing string longitudinally. This creates a harmonic condition that will result in several hundred strokes per minute, with the strokes being a small fraction of 1 in. in length. Some of these pumps are in use in relatively shallow wells, but it appears that their field of application is very limited.

Lease tanks and gas separators. Figure 4 shows a typical lease tank battery consisting of four 1000-bbl tanks and two gas separators. Such equipment is used for handling production from wells produced by natural flow, gas lift, or pumping. In some pumping wells the gas content may be too low to justify the cost of separators for saving the gas.

Natural gasoline production. An important phase of oil and gas production in many areas is the production of natural gasoline from gas taken from the casinghead of oil wells or separated from the oil and conducted to the natural gasoline plant. The plant consists of facilities for compressing and extracting the liquid components from the gas. The natural gasoline generally is collected by cooling and condensing the vapors after compression or by absorbing in organic liquids having high boiling points from which the volatile liquids are distilled. Many natural gasoline plants utilize a combination of condensing and absorbing techniques.

Fig. 4. Lease tank battery with four tanks and two gas separators. *(Gulf Oil Corp.)*

Fig. 5. Modern natural gasoline plant in western Texas. (*Gulf Oil Corp.*)

Figure 5 shows an overall view of a natural gasoline plant operating in western Texas.

PRODUCTION PROBLEMS AND INSTRUMENTS

To maintain production, various problems must be overcome. Numerous instruments have been developed to monitor production and to control production problems.

Corrosion. In many areas the corrosion of production equipment is a major factor in the cost of petroleum production. The following comments on the oil field corrosion problem are taken largely from *Corrosion of Oil- and Gas-Well Equipment* and reproduced by permission of NACE-API.

For practical consideration, corrosion in oil and gas-well production can be classified into four main types.

1. Sweet corrosion occurs as a result of the presence of carbon dioxide and fatty acids. Oxygen and hydrogen sulfide are not present. This type of corrosion occurs in both gas-condensate and oil wells. It is most frequently encountered in the United States in southern Louisiana and Texas, and other scattered areas. At least 20% of all sweet oil production and 45% of condensate production are considered corrosive.

2. Sour corrosion is designated as corrosion in oil and gas wells producing even trace quantities of hydrogen sulfide. These wells may also contain oxygen, carbon dioxide, or organic acids. Sour corrosion occurs in the United States primarily throughout Arbuckle production in Kansas and in the Permian basin of western Texas and New Mexico. About 12% of all sour production is considered corrosive.

3. Oxygen corrosion occurs wherever equipment is exposed to atmospheric oxygen. It occurs most frequently in offshore installations, brine-handling and injection systems, and in shallow producing wells where air is allowed to enter the casing.

4. Electrochemical corrosion is designated as that which occurs when corrosion currents can be readily measured or when corrosion can be mitigated by the application of current, as in soil corrosion.

Corrosion inhibitors are used extensively in both oil and gas wells to reduce corrosion damage to subsurface equipment. Most of the inhibitors used in the oil field are of the so-called polar organic type. All of the major inhibitor suppliers can furnish effective inhibitors for the prevention of sweet corrosion as encountered in most fields. These can be purchased in oil-soluble, water-dispersible, or water-soluble form.

Paraffin deposits. In many crude-oil–producing areas paraffin deposits in tubing and flow lines and on sucker rods are a source of considerable trouble and expense. Such deposits build up until the tubing or flow line is partially or completely plugged. It is necessary to remove these deposits to maintain production rates. A variety of methods are used to remove paraffin from the tubing, including the application of heated oil through tubular sucker rods to mix with and transfer heat to the oil being produced and raise the temperature to a point at which the deposited paraffin will be dissolved or melted. Paraffin solvents may also be applied in this manner without the necessity of applying heat.

Mechanical means often are used in which a scraping tool is run on a wire line and paraffin is scraped from the tubing wall as the tool is pulled back to the surface. Mechanical scrapers that attach to sucker rods also are in use. Various types of automatic scrapers have been used in connection with flowing wells. These consist of a form of piston that will drop freely to the bottom when flow is stopped but will rise back to the surface when flow is resumed. Electrical heating methods have been used rather extensively in some areas. The tubing is insulated from the casing and from the flow line, and electric current is transmitted through the tubing for the time necessary to heat the tubing sufficiently to cause the paraffin deposits to melt or go into solution in the oil in the tubing. Plastic coatings have been utilized inside tubing and flow lines to minimize or prevent paraffin deposits. Paraffin does not deposit readily on certain plastic coatings.

A common method for removing paraffin from flow lines is to disconnect the line at the well head and at the tank battery and force live steam through the line to melt the paraffin deposits and flow them out. Various designs of flow-line scrapers have also been used rather extensively and fairly successfully. Paraffin deposits in flow lines are minimized by insulating the lines or by burying the lines to maintain a higher average temperature.

Emulsions. A large percentage of oil wells produce various quantities of salt water along with the oil, and numerous wells are being pumped in which the salt-water production is 90% or more of the total fluid lifted. Turbulence resulting from production methods results in the formation of emulsions of water in oil or oil in water; the commoner type is oil in water. Emulsions are treated with a variety of demulsifying chemicals, with the application of heat, and with a combination of these two treatments. Another method for breaking emulsions is the electrostatic or electrical precipitator type of emulsion treatment. In this method the emulsion to be broken is circulated between electrodes subjected to a high potential difference. The resulting concentrated electric field tends to rupture the oil-water interface and thus breaks the emulsion and permits the water to settle out. Figure 6 shows two pumping wells with a tank battery in the background. This tank battery is equipped with a wash tank, or gun barrel, and a gas-fired heater for emulsion treating and water separation before the oil is admitted to the lease tanks.

Gas conservation. If the quantity of gas produced with crude oil is appreciably greater than that which can be efficiently utilized or marketed, it is necessary to provide facilities for returning the excess gas to the producing formation. Formerly, large quantities of excess gas were disposed of by burning or simply by venting to the atmosphere. This practice is now unlawful. Returning excess gas to the formation not only conserves the gas for future use but also results in greater ultimate recovery of oil from the formation.

Salt-water disposal. The large volumes of salt water produced with the oil in some areas present serious disposal problems. The salt water is generally pumped back to the formation through wells drilled for this purpose. Such salt-water disposal wells are located in areas where the formation already contains water. Thus this practice helps to maintain the formation pressure as well as the productivity of the producing wells.

Offshore production. Offshore wells present additional production problems since the wells must be serviced from barges or boats. Wells of reasonable depth on land locations are seldom equipped with derricks for servicing because it is more economical to set up a portable mast for pulling and installing rods, tubing, and other equipment. However, the use of portable masts is not practical on offshore locations, and a derrick is generally left standing over such wells throughout their productive life to facilitate servicing. There are a considerable number of offshore wells along the Gulf Coast and the Pacific Coast of the United States, but by far the greatest number of offshore wells in a particular region is in Lake Maracaibo in Venezuela. Figure 7 shows a considerable number of derricks in Lake Maracaibo with pumping wells in the foreground. These wells are pumped by electric power through cables laid on the lake bottom to conduct electricity from power-generating stations onshore. An overwater tank battery is visible at the extreme right. All offshore installations, such as tank batteries, pump stations, and the derricks and pumping equipment, are supported on pilings in water up to 100 ft or more in depth. There are approximately 2300 oil derricks in Lake Maracaibo. A growing number of semipermanent platform rigs and even some bottom storage facilities are now being used in Gulf of Mexico waters at depths of more than 100 ft. *See* OIL AND GAS, OFFSHORE.

Instruments. The commoner and more important instruments required in petroleum production operations are included in the following discussion.

1. Gas meters, which are generally of the orifice type, are designed to record the differential pressure across the orifice, and the static pressure.

Fig. 6. Two pumping wells with tank battery. (*Oil Well Supply Division, U.S. Steel Corp.*)

2. Recording subsurface pressure gages small enough to run down 2-in. ID (inside diameter) tubing are used extensively for measuring pressure gradients down the tubing of flowing wells, recording pressure buildup when the well is closed in, and measuring equilibrium bottom-hole pressures.

3. Subsurface samplers designed to sample well fluids at various levels in the tubing are used to determine physical properties, such as viscosity, gas content, free gas, and dissolved gas at various levels. These instruments may also include a recording thermometer or a maximum reading thermometer, depending upon the information required.

4. Oil meters of various types are utilized to meter crude oil flowing to or from storage.

5. Dynamometers are used to measure polished-rod loads. These instruments are sometimes known as well weighers since they are used to record the polished-rod load throughout a pumping cycle of a sucker-rod—type pump. They are used to determine maximum load on polished rods as well as load variations, to permit accurate counterbalancing of pumping wells, and to assure that pumping units or sucker-rod strings are not seriously overloaded.

6. Liquid-level gages and controllers are used. They are similar to those used in other industries, but with special designs for closed lease tanks.

A wide variety of scientific instruments find application in petroleum production problems. The above outline gives an indication of a few specialized instruments used in this branch of the industry, and there are many more. Special instruments developed by service companies are valued for a wide variety of purposes and include calipers to

Fig. 7. Numerous offshore wells located in Lake Maracaibo, Venezuela. (*Creole Petroleum Corp.*)

detect and measure corrosion pits inside tubing and casing and magnetic instruments to detect microscopic cracks in sucker rods. *See* OIL AND GAS STORAGE.   [ROY L. CHENAULT]

Bibliography: American Petroleum Institute, *History of Petroleum Engineering*, 1961; E. L. DeGolyer (ed.), *Elements of the Petroleum Industry*, 1940; T. C. Frick (ed.), *Petroleum Production Handbook*, vol. 1: *Mathematics and Production Equipment*, 1962; T. C. Frick (ed.), *Petroleum Production Handbook*, vol. 2: *Reservoir Engineering*, 1962; V. B. Guthrie, *Petroleum Products Handbook*, 1960; W. F. Lovejoy and P. T. Homan, *Economic Aspects of Oil Conservation Regulation*, 1967; W. F. Lovejoy and P. T. Homan, *Methods of Estimating Reserves of Crude Oil, Natural Gas, and Natural Gas Liquids*, 1965; B. W. Murphy (ed.), *Conservation of Oil and Gas: A Legal Story*, 1949; M. Muskat, *Physical Principles of Oil Production*, 1949; S. J. Pirson, *Oil Reservoir Engineering*, 2d ed., 1958; L. C. Uren, *Petroleum Production Engineering: Oil Field Development*, 4th ed., 1956; L. C. Uren, *Petroleum Production Engineering: Oil Field Exploitation*, 3d ed., 1953.

Oil and gas storage

Crude oil and natural gas, after being produced from their natural reservoirs, are stored in great quantities. Large amounts of refined products are stored as well. Storage is necessary to meet seasonal and other fluctuations in demand and for efficient operation of producing equipment, pipelines, tankers, and refineries. Storage also provides ready reserves for emergency use. According to the U.S. Bureau of Mines, 265×10^6 bbl of crude oil were in storage in the United States at the end of 1974. In addition, 808×10^6 bbl in the form of refined products, natural gasoline, plant condensate, and unfinished oils were in storage. The American Gas Association reported 3.969×10^{12} ft^3 of natural gas stored in underground reservoirs in the United States at the end of 1974.

Crude oil and refined products. Oil from producing wells is first collected in welded-steel, bolted-steel, or wooden tanks of 100 bbl or greater capacity located on individual leases. These tanks, upright cylinders with low-pitched conical roofs, provide temporary storage while the oil is awaiting shipment. Several tanks grouped together are called a tank battery. Assemblages of large steel tanks, known as tank farms, are used for more permanent storage at pipeline pump stations, points where tankers are loaded and unloaded, and refineries.

With the trend toward giant tankers, accelerated by the closing of the Suez Canal in 1967, large storage facilities are needed at both the loading and unloading ends of the tanker runs. Some tanks with capacities of 1×10^6 bbl are now in use. Large-capacity excavated reservoirs with concrete linings have been used for many years in California to store both crude and fuel oil. One such reservoir with a fixed roof and elliptical in form, is 780 ft long, 467 ft wide, and 23 ft deep. It covers $9\frac{1}{4}$ acres and provides storage for more than 1×10^6 bbl. Another reservoir has a capacity of 4×10^6 bbl and covers 16 acres.

Offshore storage. For offshore producing fields a number of unique storage systems have been designed. In several instances old tankers have been adapted for storage, and barges have been constructed especially for offshore storage use. One underwater installation consists essentially of three giant inverted steel funnels. Each unit is 270 ft in diameter and 205 ft high, weighs 28×10^6 lb, and has a capacity of 0.5×10^6 bbl. The bottom is open, and the unit is anchored to the sea floor by 95-ft pilings. A reinforced concrete installation features a nine-module storage unit with 1×10^6 bbl capacity surrounded by a perforated wall 302 ft in diameter that serves as a breakwater. The outer wall is about 270 ft high and extends about 40 ft above the water surface. A submerged floating storage tank 96 ft in diamter and 305 ft high is held in place by six anchor lines and has a capacity of 300,000 bbl. One relatively small unit consists of a platform with four vertical legs, each holding 4700 bbl, and four horizontal tanks at the bottom holding 1850 bbl each, for a total capacity of 26,200 bbl. A second small unit, utilizing bottom tanks as an anchor, holds 2400 bbl underwater and 600 bbl in a spherical tank above the surface. A third small unit consists of a sea-floor base, connected by a universal joint to a large-diameter vertical cylinder, about 350 ft high, which extends above the surface of the water. One proposed design includes an excavated cavern beneath the sea floor, and nuclear cavities have also been suggested.

Volatility problems. To minimize vaporization losses, lease tanks are sometimes equipped to hold several ounces pressure. At large-capacity storage sites, special tanks are generally used. Tanks with lifter or floating roofs are used to store crude oil, motor gasoline, and less volatile natural gasoline. Motor and natural gasolines are also stored in spheroid containers. Spherical containers are used for more volatile liquids, such as butane. Horizontal cylindrical containers are used for propane and butane storage. Refrigerated insulated tank systems enabling propane to be stored at a lower pressure are also in use. One tank has a capacity of 900,000 bbl.

Underground storage. Large quantities of volatile liquid-petroleum products, including propane and butane, are stored in underground caverns dissolved in salt formations and in mined caverns, gas reservoirs, and water sands. In 1973 the underground storage capacity for liquid-petroleum prod-

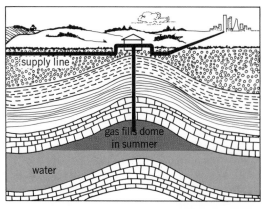

Schematic cross section of typical aquifer gas storage field showing injected gas displacing water. (*Natural Gas Pipeline Company of America*)

supply line

gas fills dome in summer

water

Growth of underground gas storage in United States, 1916–1974*

Year	Number of reservoirs	Number of states	Total reservoir capacity, $\times 10^9$ ft³
1916	1	1	
1919	2	2	5
1920	3	3	5
1925	4	4	7
1927	5	5	7
1928	7	5	15
1929	8	6	18
1930	9	6	18
1931	11	6	25
1933	12	7	27
1934	15	7	39
1935	15	7	38
1936	22	8	66
1937	31	8	146
1938	35	8	198
1939	39	9	205
1940	44	9	232
1941	51	10	283
1942	59	12	289
1943	66	12	301
1944	67	13	251
1945	75	14	416
1946	78	14	424
1947	91	15	560
1948	102	15	584
1949	118	15	625
1950	125	15	770
1951	142	15	911
1952	151	16	1285
1953	167	17	1726
1954	172	17	1849
1955	178	18	2084
1956	187	19	2389
1957	199	19	2589
1958	205	19	2703
1959	209	20	2507
1960	217	20	2854
1961	229	21	3201
1962	258	21	3485
1963	278	23	3674
1964	286	24	3942
1965	293	24	4086
1966	303	25	4421
1967	308	25	4520
1968	315	26	4783
1969	320	26	4927
1970	325	26	5178
1971	333	26	5575
1972	348	26	6040
1973	360	26	6279
1974	367	26	6360

*American Gas Association, *Underground Storage of Gas in the United States and Canada*, 1975.

ucts in the United States was 255.231×10^6 bbl. Underground storage capacity in Canada was 14.996×10^6 bbl. Liquid-petroleum products are also being stored underground in Belgium, France, Germany, Italy, and the United Kingdom. Caverns are also used for storing crude oil in Sweden, Germany, and France. In Pennsylvania an abandoned quarry with a capacity of 2×10^6 bbl has been equipped with a floating roof for storing fuel oil. Refrigerated propane is also being stored in excavations in frozen earth and in underground concrete tanks. To provide security in the event of

another oil embargo, the National Petroleum Council has proposed developing salt cavern storage for 500×10^6 bbl of crude oil.

Natural gas. Natural gas is stored in low-pressure surface holders, buried high-pressure pipe batteries and bottles, depleted or partially depleted oil and gas reservoirs, water sands, and several types of containers at extremely low temperature ($-258°F$, or $-161°C$) after liquefaction.

Low-pressure holders, which store relatively small volumes of gas, basically use either a water or a dry seal, and variations of each type exist. With the displacement of manufactured gas by natural gas in the United States, the need for surface holders has greatly diminished and they have disappeared almost entirely.

Underground storage. In the United States gas pipeline and utility companies store large quantities of natural gas in underground reservoirs. In most cases these reservoirs are located near market areas and are used to supplement pipeline supplies during the winter months when the gas demand for residential heating is very high. Since gas can be stored in the summer when the gas demand is low, underground storage permits greatly increased pipeline utilization, resulting in lower transportation costs and reduced gas cost to the consumer. Underground storage is the only economical method of storing large enough quantities of gas to meet the seasonal fluctuations in pipeline loads, and has enabled gas companies to meet market requirements which otherwise could not be satisfied.

Gas was first stored underground in 1915 in a partially depleted gas field in Ontario, Canada. The following year gas was injected into a depleted gas field near Buffalo, NY. The table, prepared by the American Gas Association, shows the growth in underground storage capacity in the United States. At the end of 1974 gas was being stored in 367 reservoirs, of which 311 were depleted gas and oil fields, 51 were water sands, 3 were salt caverns, and 1 was an abandoned coal mine. These reservoirs are located in 26 states and are operated by 86 different companies. They have a total capacity of 6.360×10^{12} ft³ and, at the end of 1974, held 3.969×10^{12} ft³ of stored gas plus 0.993×10^{12} ft³ of native gas. The maximum volume of gas in storage during 1974 was 4.505×10^{12} ft³, excluding native gas. During 1974 a total of 1.705×10^{12} ft³ was withdrawn from storage through 13,005 wells. The total maximum daily output from all of these reservoirs was 28.3×10^9 ft³. Canada has 17 storage reservoirs, 14 in gas and oil reservoirs and 3 in salt caverns, with a total capacity of 325×10^9 ft³. At the end of 1974 these reservoirs held 136×10^9 ft³ plus 90×10^9 ft³ of native gas. The maximum volume in storage in 1974 was 174×10^9 ft³, excluding native gas. During 1974 a total of 124×10^9 ft³ was withdrawn from storage through 163 wells. The total maximum daily output of these reservoirs was approximately 1.5×10^9 ft³. Gas is also being stored in underground reservoirs in France, East and West Germany, Austria, Italy, Poland, Rumania, Czechoslovakia, and the Soviet Union. In the United Kingdom a salt cavern is being used.

Gas storage in water sands was first undertaken in 1952, and this method of storage has steadily increased, especially in areas where no gas or oil fields are available. In the United States at the end of 1974, aquifer-type storages numbered 51 and these reservoirs were operated by 25 companies in 10 states. They had a total capacity of 1.408×10^{12} ft³ and, at the end of 1974, held 849×10^9 ft³. During 1974 the maximum volume in storage was 899×10^9 ft³. The total maximum daily output from these reservoirs was 4.7×10^9 ft³. A cross section of a typical aquifer storage field is shown in the illustration. A geologic trap having adequate structural closure and a suitable caprock is needed. The storage sand must be porous and thick enough and under sufficient hydrostatic pressure to hold large quantities of gas. The sand must also be sufficiently permeable and continuous over a wide enough area so that water can be pushed back readily to make room for the stored gas. In some cases, water removal wells are also utilized.

Reservoir pressure. In operating storage reservoirs only a portion of the stored gas, called working gas, is normally withdrawn. The remaining gas, called cushion gas, stays in the reservoir to provide the necessary pressure to produce the storage wells at desired rates. In aquifer storages some water returns to help maintain the reservoir pressure. The percentage of cushion gas varies considerably among reservoirs. Based on American Gas Association figures, cushion gas amounted to 59% of the maximum gas in storage, including native gas, in the United States in 1974. In some instances, the original reservoir pressure of the oil and gas field is exceeded in storage operations. This has resulted in storage volumes greater than the original content and has substantially increased well deliverabilities. In aquifer storages the original hydrostatic pressure must be exceeded in order to push the water back.

In the Soviet Union aquifer gas storage is being undertaken in one area where no appreciable structure. In an inconclusive field test, air was injected into a center well with control of the lateral spread of the air bubble attempted by injecting water into surrounding wells. Storage of gas in cavities created by nuclear explosions has been proposed and seriously considered.

Liquefied gas. Storage of liquefied natural gas has rapidly increased throughout the world. Storage is in connection with shipment of liquefied natural gas by tanker, and is located at the loading and unloading ends of the tanker runs as well as at peak sharing facilities operated by gas pipeline and local utility companies. In 1974 the United States and Canada had more than 100 liquefied natural-gas storage installations either operational or under construction. These had a total storage capacity of 22.8×10^6 bbl, or 78.5×10^9 ft³. England, France, the Netherlands, West Germany, Italy, Spain, Algeria, Libya, and Japan also have installations. Storage is in insulated metal tanks, buried concrete tanks, or frozen earth excavations. In two projects using frozen earth excavations, excessive boil-off of the liquefied gas has led to replacement with insulated metal tanks. *See* LIQUEFIED NATURAL GAS (LNG); OIL AND GAS FIELD EXPLOITATION; PETROLEUM PROCESSING; PIPELINE.   [PETER G. BURNETT]

Bibliography: American Gas Association, *The Underground Storage of Gas in the United States and Canada*, 24th Annual Report on Statistics, 1975; D. C. Bond, *Underground Storage of Natural Gas*, Illinois State Geological Survey, 1975; T. C. Frick, *Petroleum Production Handbook*, 1962; D. L. Katz et al., *Handbook of Natural Gas Engineering*, 1959; D. L. Katz and P. A. Witherspoon, *Underground and Other Storage of Oil Products and Gas*, Proceedings of the 8th World Petroleum Congress, 1971. E. L. Katz and K. H. Coats, *Underground Storage of Fluids*, 1968; Stone and Webster Engineering, *Gas Storage at the Point of Use*, Amer. Gas Ass. Proj. PL-56, 1965.

Oil and gas well completion

The operations that prepare for production a well drilled to an oil or gas reservoir. Various problems of well casing during and at the end of drilling are related to modes of completing connection between the proper reservoirs and the surface. Tubing inside the casing and valves and a pumping unit at the surface must deliver reservoir products to the surface at a controlled rate. Variations in reservoir and overlying formations may require special techniques to keep out water or sand or to increase the production rate. *See* OIL AND GAS FIELD EXPLOITATION; OIL AND GAS WELL DRILLING.

Casing. Oil and gas wells are walled with steel tubing which is cemented in place.

Tubing. Steel tubing is manufactured in various diameters, wall thicknesses, lengths, and steel alloys, selected to satisfy specific needs. These lengths, called joints, are threaded and coupled so that they may be joined together in a continuous string in the well bore. Properly placed and cemented in the hole, casing protects fresh-water

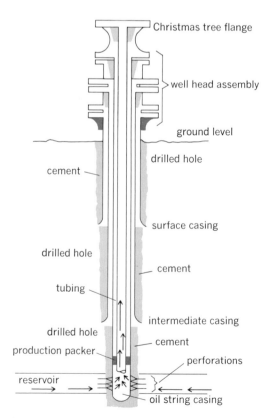

Fig. 1. Casing detail; casing strings in an oil well.

reservoirs from contamination, supports unconsolidated rock formations, maintains natural separation of formations, aids in the prevention of blowouts and waste of reservoir energy, and acts as a conduit for receiving pipe of smaller diameters through which the well effluent may be brought to the surface under controlled conditions.

It may be necessary to set many strings of casing in one hole before reaching the objective (Fig. 1). The determining factors are many, such as depth of hole, loss of circulation, high-pressure formations, hole sloughing, and wearing out a string of casing while rotating drill pipe through it over a long period of time.

Cementing casing. Basically, ordinary portland cement is used in cementing casing. In order to obtain the protection and fulfil the purposes, it is imperative that each string of casing be securely sealed to the walls of the hole for at least some distance up from the bottom of the casing string. After casing is in place, the cement is pumped down the inside and up the outside to a predetermined height to occupy the space between the casing and the walls of the hole, thereby effecting the desired seal. In the pumping and measuring process, plugs are used to separate the cement from other fluids to eliminate contamination; also, the cement inside the casing is displaced with fluid. Oil or gas well cementing is not performed with the drilling equipment; an outside service company equipped with mobile, high-pressure mixing and pumping equipment and accessories operated by trained personnel is employed (Fig. 2).

Well hole–reservoir connection. Reservoir conditions, known to exist or later defined, determine the type of completion technique to be followed: (1) barefoot completion, (2) preperforated liner, or (3) casing set through and perforated.

Barefoot completion. This type of completion is frequently used when the character of the producing rock is such that it does not require supplemental support or screening, for example, in formations such as limestone, dolomite, or hard sandstone. With this method, the production casing is seated above the producing section in the conventional manner. The casing is cleaned out by insertion of fluid-separating plugs and drilled out through the casing and into the producing formation below. The formation contents enter the bore hole from the bare or unlined producing stratum or strata, hence the term barefoot (Fig. 3).

Liner-type completion. This type of completion is similar to a barefoot completion except that the open portion of the hole is cased with a preperforated section of casing called a liner (Fig. 4). This liner is smaller in diameter than the casing previously set in the hole and is usually suspended from the upper casing near the bottom from a liner hanger. The hanger is attached to the top of the liner, and when it is set, it effects a seal between the liner and the casing. The purpose of the liner is to permit gas and liquids to enter the hole and screen out formation particles.

Gun perforating. Gun perforating is a method of forming holes through the casing and into a formation from within a well bore. The two more popular methods are bullet perforating, as with a rifle, and jet perforating, as with a torch.

The gun is fitted around the outside with barrels

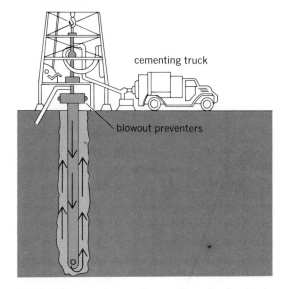

Fig. 2. Diagram of cementing process, showing truck, equipment, and well job.

containing the perforating medium. Each barrel is wired to fire by remote control from the surface. The gun is run into the hole on a wire line from a service company's shooting truck. The wire line serves to lower and raise the gun in and out of the hole and, when the gun is in position to be fired, the operator sends an electric impulse down the line to trigger it. The hole is thus formed through the steel casing, the cement sheath, and some inches into the reservoir rock, creating an entry for the reservoir content into the well bore. Guns of the bullet type are retrieved and reloaded (Fig. 5). Jet-type guns are expendable and disintegrate (Fig. 6).

Production-flow control. A steel tube, the same as casing except that it is smaller in diameter, serves as a production flow line within the well. The tubing is run inside the casing and is either suspended in the hole or set on a production pack-

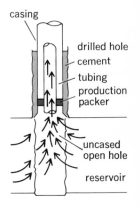

Fig. 3. Diagram of barefoot completion.

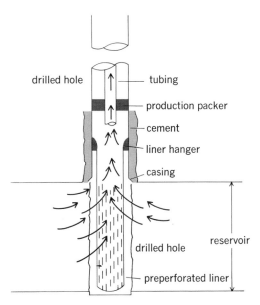

Fig. 4. Liner-type completion; preperforated liner.

trol the rate of daily production from the reservoir and direct it into a pipeline, generally at reduced pressures, to the oil-gathering station.

Pumping unit. Relatively few oil wells flow from natural pressures. Most require secondary means of removing the reservoir product. The most common of several methods is the pumping unit. A walking beam is operated like a seesaw, raising and lowering a plunger-type pump set near the bottom of the hole. The rods between the walking bean and pump are called sucker rods (Fig. 8).

Multiple completion. An oil or gas well from which several separate horizons are individually, separately, and simultaneously produced is called a multiple completion. Such a completion is accomplished by the use of multiple-zone packers and separate tubing strings. The producing zones are separated one from another by proper placement of packers in the well bore. An individual string of tubing is attached to each packer and extended to the surface where each is interconnected with the Christmas tree or flow assembly (Fig. 9).

The several advantages of such a completion from a single well bore are increase in daily production, more efficient and economic utilization of a well bore with multiple reservoirs, increase in ultimate recovery, accurate measurement of product withdrawal from each reservoir, and elimination of mixing of products of different gravities and basic sediment and water content.

Water problems. The production of water in quantity from an oil or gas well renders it uneconomical; means are provided for water exclusion.

Water-exclusion methods. Water exclusion may be effected by the application of cements or various types of plastic. If it is determined that water is entering from the lower portion of a producing sand in a relatively shallow, low-pressure well, a cement plug may be placed in the bottom of the hole so that it will cover the oil-water interface of the reservoir. This technique is called laying in a plug and may be accomplished by placing the cement with a dump-bottom bailer on a wire line or by pumping cement down the drill pipe or tubing. For deeper, higher-pressure, or more troublesome wells, a squeeze method is used. Squeeze cementing is the process of applying hydraulic pressure to force a cementing material into permeable space of an exposed formation or through openings in the casing or liner. In many conditions cement, plastic, or diesel-oil cement may be squeezed into water-, oil-, or gas-bearing portions of a producing zone to eliminate excessive water without sealing off the gas or oil. A few of the applications are: repair of casing leaks; isolation of producing zones prior to perforating for production; remedial or secondary cementing to correct a defective condition, such as channeling or insufficient cement on a primary cement job; sealing off a low-pressure formation that engulfs oil and gas or drilling fluids; and abandonment of depleted producing zones to prevent migration of formation effluent and to reduce possibilities of contaminating other zones or wells.

The squeeze-method tool is a packer-type device designed to isolate the point of entry between or below packing elements. The tool is run into the hole on drill pipe or tubing, and the cementing material is squeezed out between or below these

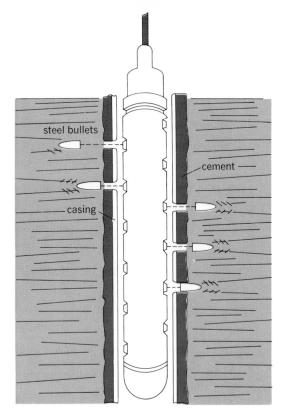

Fig. 5. Gun perforator bullets from bullet-type gun.

er at or near the producing interval. The top of the tubing string terminates at the surface in a sealing element in the wellhead assembly to which the so-called Christmas tree is attached. A manifold constructed of steel valves and fittings, placed on top of the casings protruding above the surface, is called a Christmas tree (Fig. 7). Its purpose is to maintain the well under proper control, to receive the formation products under pressure, and to con-

OIL AND GAS WELL
COMPLETION

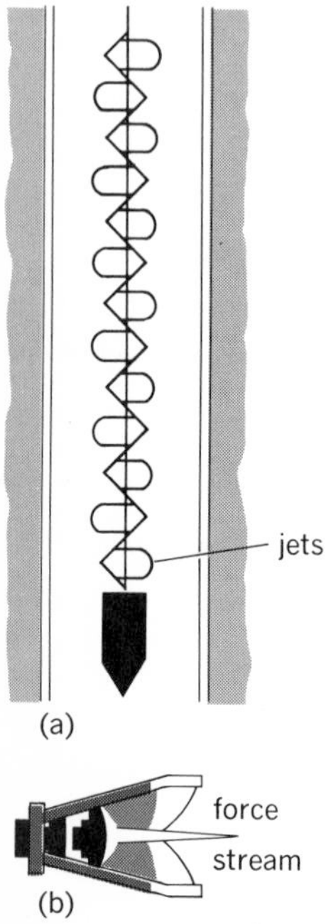

(a)

force

stream

(b)

Fig. 6. Jet-type perforator. (*a*) Charges in firing position. (*b*) Side view of a shaped charge.

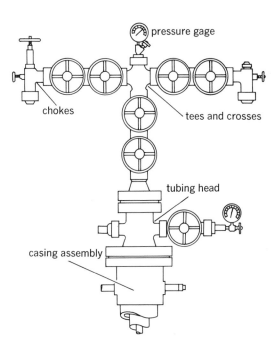

Fig. 7. Typical layout of a Christmas tree.

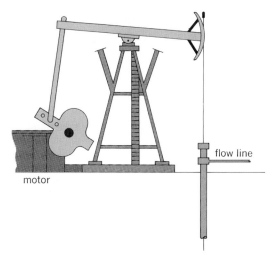

Fig. 8. Pumping unit diagram. These are most common if natural pressure is lacking for well flow.

confining elements into the problem area. The well is then recompleted. It may be necessary to drill the cement out of the hole and reperforate, depending upon the outcome of the job performed in the squeeze process.

Water-exclusion plug back. Simple water shutoff jobs in shallow, deep, or high-pressure wells may also be performed in multizone wells in which the lower producing interval is depleted or the remaining recoverable reserves do not justify rehabilitation.

Here, water may be excluded by placing a packer-type plug (cork) above the interval, then producing formations that are already open or perforating additional intervals that may be present higher up the hole (Fig. 10).

Production-stimulation techniques. The initial testing or production history often indicates subnormal production rates, signifying the necessity for remedial action. Any method designed to increase the production rate from a reservoir is defined as production stimulation. Three of the methods used are acidizing, fracturing, and employing explosives.

Acidizing. Varied volumes of hydrochloric acid are used in limestone and dolomite or other acid-soluble formations to dissolve the existing flow-channel walls and enlarge them (Fig. 11). High-pressure equipment, pumps, and wellheads are necessary for satisfactory performance. Fast pumping speeds and acid inhibitors are used to alleviate corrosion of the well equipment.

Fracturing. Formation fracturing is a hydraulic process aimed at the parting of a desired section of formation. Selected grades of sand or particles of other materials are added to the fracturing fluid in varied quantities. These particles pack and fill the fracture, acting as a propping agent to hold it open when the applied pressure is released (Fig. 12). Such fractures increase the flow channels in size and number, improving the fluid-flow characteristics of the reservoir rock. The particle-carrying agent (fluid) is of considerable importance and is varied to fit particular demands. Some of the fluids which are used in this process are crude oil (sand oil fracturing), special refined oils (sand oil frac-

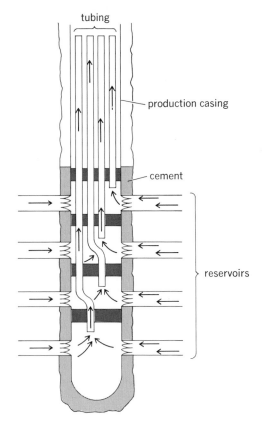

Fig. 9. A representative multiple completion diagram.

turing), water (river fracturing), acid (acid fracturing), and oil, water, and chemical emulsion (emulsifracturing).

Explosives. The idea of stimulating production by use of explosives was first used in a well in Pennsylvania on Jan. 21, 1865. The first torpedo consisted of 8 lb of gunpowder, contained in a metal tube, which was lowered into the well and detonated. Its more important function, in this shallow well, was to clear away paraffin, which was accomplished, but a decided increase in production was also accomplished. The method has since

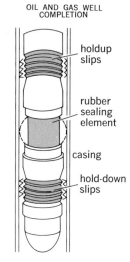

OIL AND GAS WELL COMPLETION

Fig. 10. Bridge plug.

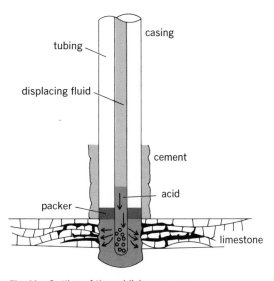

Fig. 11. Outline of the acidizing process.

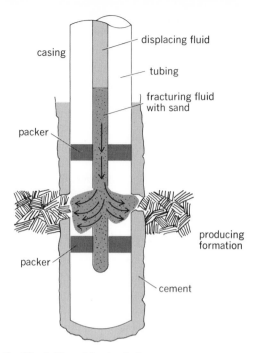

Fig. 12.   Outline of the fracturing process.

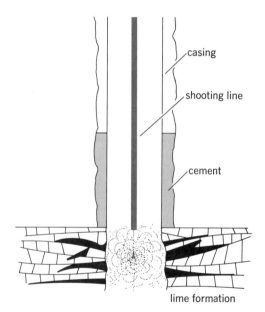

Fig. 13.   Stimulating production by use of explosives.

OIL AND GAS WELL
COMPLETION

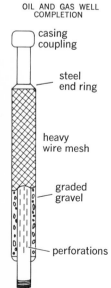

Fig. 14.   Prepacked
gravel liner.

preslotted pipe wrapped with wire screen designed to screen out or retain outside the bore hole all except the fine particles that may be brought to the surface with the reservoir fluid. The original design is such that a calculated snug fit is obtained between the liner and bore hole. The coarse screened-out particles form a secondary gravel pack between the liner and the hole, additionally supporting the formation and reducing sand incursion.

Sand consolidation. Sand consolidation is the result of successfully placing a binding material in the producing sand. In effect this glues the sand grains together without completely destroying the porosity and permeability of the sand. The binding material, generally a form of plastic, is forced into the sand through perforations in the casing. The purpose is to consolidate the sand around the well bore to eliminate sloughing and sand incursion.

Prepack gravel liners. This type of liner (Fig. 14) is made by using a perforated section of steel pipe, over which has been fitted a tubular sleeve formed of an inner and outer screen of heavy wire mesh or perforated sheet steel. The pipe section is held in concentric relationship by spacers with gravel of proper size packed between the screens and sealed at both ends to retain the gravel. Set in the hole, it serves as a screen liner. *See* PETROLEUM RESERVOIR ENGINEERING.

[HARRY S. BRIGHAM]

Oil and gas well drilling

The drilling of holes for exploration and extraction of crude oil and natural gas. Deep holes and high pressures are characteristics of petroleum drilling not commonly associated with other types of drilling. In general, it becomes more difficult to control the direction of the drilled hole as the depth increases, and additionally, the cost per foot of hole drilled increases rapidly with the depth of the hole. Drilling-fluid pressure must be sufficiently high to prevent blowouts but not high enough to cause fracturing of the bore hole. Formation-fluid pressures are commonly controlled by the use of a high-density clay-water slurry, called drilling mud. The chemicals used in drilling mud can be expensive, but the primary disadvantage in the use of drilling muds is the relatively low drilling rate which normally accompanies high bottom-hole pressure. Drilling rates can often be increased by using water to circulate the cuttings from the hole; when feasible, the use of gas as a drilling fluid can lead to drilling rates as much as 10 times those attained with mud. Drilling research has the objectives of improving the utilization of current drilling technology and the development of improved drilling techniques and tools.

Hole direction. The hole direction must be controlled within permissible limits in order to reach a desired target at depths as great as 25,000 ft. Inclined layers of rocks with different hardness tend to cause the direction of drilling to deviate; consequently, deep holes are rarely truly straight and vertical. The drilling rate generally increases as additional drill-collar weight is applied to the bit by adjusting the pipe tension at the surface. However, crooked-hole tendency also increases with higher weight-on-bit. In recent years a so-called packed-hole technique has been used to reduce the tendency to hole deviation. One version of this

been improved with new explosives, firing mechanisms, and procedures, but the basic idea is the same, that is, to remove reservoir-blocking material from the reservoir face and to create fractures in the rock to increase production (Fig. 13).

Sand exclusion. Some reservoir rock is of an unconsolidated nature similar to beach sands; after having been penetrated with a bore hole, it will slough, if unsupported. This rock also has a tendency to flow with its formation fluids, resulting in plugging of the well bore and restriction or elimination of the entry of formation fluids. Several measures can be used to combat this condition, but no single measure can be used universally.

Screen liner. This type of liner is a segment of

technique makes use of square drill collars which nearly fill the hole on the diagonals but permit fluid and cuttings to circulate around the sides. This procedure reduces the rate at which the hole direction can change.

Cost factors. Drilling costs depend on the costs of such items as the drilling rig, the bits, and the drilling fluid, as well as on the drilling rate, the time required to replace a worn bit, and bit life. At depths below 15,000 ft, for example, the cost per foot of hole drilled can exceed $100. Operating costs of a large land rig required for deep holes are about $2000 per day, whereas comparable costs for an offshore drilling platform or floating drilling vessel can vary from $500 to $30,000 per day. Although weak rock at shallow depth can be drilled at rates exceeding 100 ft/hr, drilling rates often average about 5 ft/hr in deep holes. Conventional rotary bits have an operating life of 10–20 hr, and 10–20 hr are required for removing the drill pipe, replacing the worn bit, and lowering the pipe back into the hole. Diamond bits may drill for as long as 50–200 hr, but the drilling rate is relatively low and the bits are expensive. For more economical drilling it is desirable to increase both bit life and drilling rate simultaneously.

Drilling fluids. The increased formation or pore-fluid pressures existing at great depths in the Earth's crust adversely affects drilling. Gushers, blowouts, or other uncontrolled pressure conditions are no longer tolerated. High-density drilling fluids maintain control of well pressures. A normal fluid gradient for salt water is about 0.5 psi/ft of depth, and the total stress due to the weight of the overburden increases approximately 1 psi/ft. Under most drilling conditions in permeable formations, the well-bore pressure must be kept between these two limiting values. If the mud pressure is too low, the formation fluid can force the mud from the hole, resulting in a blowout; whereas if the mud pressure becomes too high, the rock adjacent to the well may be fractured, resulting in lost circulation. In this latter case the mud and cuttings are lost into the fractured formation.

High drilling-fluid pressure at the bottom of a bore hole impedes the drilling action of the bit. Rock failure strength increases, and the failure becomes more ductile as the pressure acting on the rock is increased. Ideally, cuttings are cleaned from beneath the bit by the drilling-fluid stream; however, relatively low mud pressure tends to hold cuttings in place. In this case mechanical action of the bit is often necessary to dislodge the chips. Regrinding of fractured rock greatly decreases drilling efficiency by lowering the drilling rate and increasing bit wear.

Drilling efficiency can be increased under circumstances where mud can be replaced by water as the drilling fluid. This might be permissible, for example, in a well in which no high-pressure gas zones are present. Hole cleaning is improved with water drilling fluid because the downhole pressure is lower and no clay filter cake is formed on permeable rock surfaces. So-called fast-drilling fluids provide a time delay for filter cake buildup. This delay permits rapid drilling with no filter cake at the bottom of the hole and, at the same time, prevents excessive loss of fluid into permeable zones above.

In portions of wells where no water zones occur, it is frequently possible to drill using air or natural gas to remove the cuttings. Drilling rates with a gas drilling fluid are often 10 times those obtained with mud under similar conditions. Sometimes a detergent foam is used to remove water in order to permit gas drilling in the presence of limited water inflow. In other instances porous formations can be plugged with plastic to permit continued gas drilling. However, in many cases it is necessary to revert to either water or mud drilling when a water zone is encountered. Another possibility is the cementing of steel casing through the zone containing water and then proceeding with gas drilling.

Research. Drilling research includes the study of drilling fluids, the evaluation of rock properties, laboratory simulation of field drilling conditions, and the development of new drilling techniques and tools. Fast-drilling fluids have been developed by selecting drilling-fluid additives which plug the pore spaces very slowly, thereby providing the desired time delay for filter cake buildup. Water-shutoff chemicals have been formulated which can be injected into a porous water-bearing formation in liquid form and then, within a few hours, set to become solid plastics. Well-logging techniques can warn of high-pressure permeable zones so that the change from gas or water drilling fluid to mud can be made before the drill enters the high-pressure zone. Downhole instrumentation can lead to improved drilling operations by providing information for improved bit design and also by feeding information to a computer for optimum control of bit weight and rotary speed. Computer programs can also utilize information from nearby wells to determine the best program for optimum safety and economy in new wells.

Since rock is a very hard, strong, abrasive material, there is a challenge to provide drills which can penetrate rock more efficiently. A better understanding of rock failure can lead to improved use of present equipment and to the development of better tools. Measurements of physical properties of rocks are beginning to be correlated with methods for the theoretical analysis of rock failure by a drill bit. A small $1\frac{1}{4}$-in.-diameter bit, called a microbit, has been used in scale-model drilling experiments in which independent control is provided for bore-hole, formation-fluid, and overburden pressures. These tests permit separation of the effects of the various pressures on drilling rates.

Novel drilling methods which are being explored include studies of rock failure by mechanical, thermal, fusion and vaporization, and chemical means. Some of the new techniques currently have limited practical application while others are still in the experimental stage. For example, jet piercing is widely used for drilling very hard, spallable rocks, such as taconite. Other methods include the use of electric arc, laser, plasma, spark, and ultrasonic drills. Better materials and improved tools can be expected in the future for drilling oil and gas wells to greater depths more economically and with greater safety. *See* BORING AND DRILLING (MINERAL); CABLE-TOOL DRILL; OIL AND GAS WELL COMPLETION; PETROLEUM GEOLOGY; ROTARY TOOL DRILL.

[J. B. CHEATHAM, JR.]

Bibliography: C. Fairhurst, *Failure and Breakage of Rock*, 1967; L. W. Ledgerwood, Efforts to develop improved oil-well drilling methods, *J. Petrol. Tech.*, 219:61–74, 1960; W. C. Maurer, *Novel Drilling Techniques*, 1968.

Oil burner

A device for converting fuel oil from a liquid state into a combustible mixture. A number of different types of oil burners are in use for domestic heating. These include sleeve burners, natural-draft pot burners, forced-draft pot burners, rotary wall flame burners, and air-atomizing and pressure-at-omizing gun burners. The most common and modern type that handles 80% of the burners used to heat United States homes is the pressure-atomizing—type burner shown in Fig. 1.

Characteristics. The sleeve burner, commonly known as a range burner because of its use in kitchen ranges, is the simplest form of vaporizing burner. The natural-draft pot burner relies on the draft developed by the chimney to support combustion. The forced-draft pot burner is a modification of the natural-draft pot burner, since the only significant difference between the two types is the means of supplying combustion air. The forced-draft pot burner supplies its own air for combustion and does not rely totally on the chimney. The rotary wall flame burners have mechanically assisted vaporization. The gun-type burner uses a nozzle to atomize the fuel so that it becomes a vapor, and burns easily when mixed with air.

The most important feature of a high-pressure atomizing gun burner is the method of delivering the air. The most efficient burner is the one which completely burns the oil with the smallest quantity of air. The function of the oil burner is to properly proportion and mix the atomized oil and air required for combustion (Fig. 2).

Efficiency. If a large quantity of excess air is used to attempt to burn the oil, there is a direct loss of usable heat. This air absorbs heat in the heating unit which is then carried away through the stack with the combustion gases. This preheated air causes high stack temperatures which lower the efficiency of the combustion. The higher the

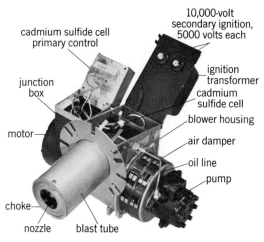

Fig. 1. An oil burner of the pressure-atomizing type. (*Automatic Burner Corp.*)

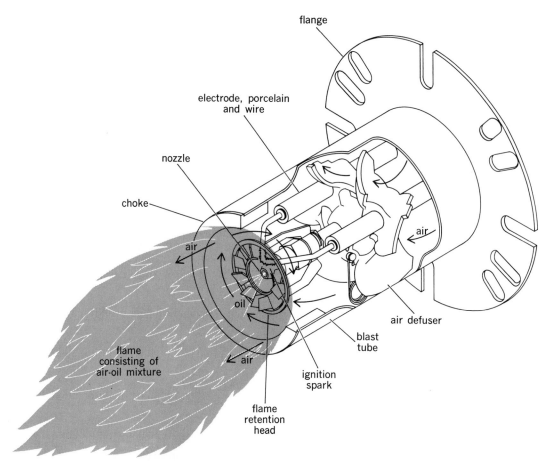

Fig. 2. Flame-retention–type burner, an example of a high-efficiency burner.

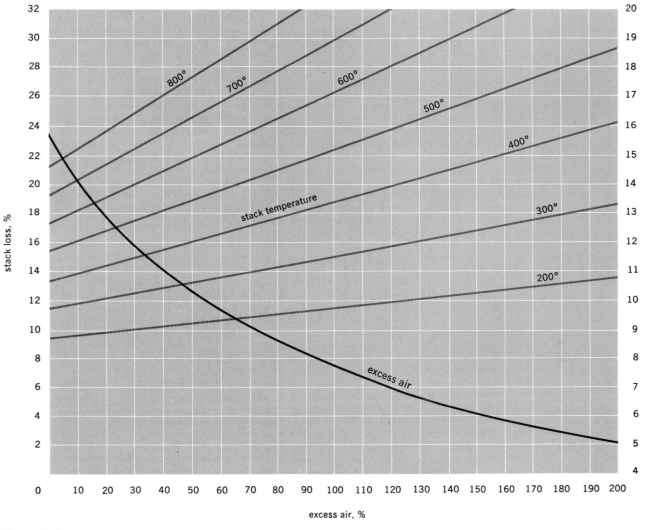

Fig. 3. Typical stack loss chart. To determine stack loss and efficiency, start with correct CO₂ and follow horizontal line to excess air curve, then vertical line to stack temperature, and finally horizontal line to stack loss. Overall efficiency percent = 100 − stack loss.

CO_2 (carbon dioxide), the less excess air. Overall efficiencies or stack loss can be estimated by the use of the stack loss chart (Fig. 3).

Increased efficiency can be obtained through the use of devices located near the flame end of the burner. CO_2 is not the ultimate factor in efficiency. Burners producing high CO_2 can also produce high smoke readings. Accumulations of 1/8-in. soot layers on the heating unit surface can increase fuel consumption as much as 8%. An oil burner is always adjusting to start smoothly with the highest CO_2 and not more than a number two smoke on a Bacharach smoke scale. For designs in high-efficiency burners commonly known as flame-retention–type burners see Fig. 2.

Nozzle. The nozzle is made up of two essential parts: the inner body, called the distributor, and the outer body, which contains the orifice that the oil sprays through. Under this high pump pressure of 100–300 psi the oil is swirled through the distributor and discharged from the orifice as a spray (Fig. 4). The spray is ignited by the spark and combustion is self-sustaining, provided the proper amount of air is supplied by the squirrel cage blower. Air is delivered through the blast tube and

moved in such a manner that it mixes well with the oil spray (Fig. 2). This air is controlled by a damper either located on the intake side of the fan or the discharge side of the fan (Fig. 1).

The essential parts of the pressure gun burner are the electric motor, squirrel-cage–type blower, housing, fuel pump, electric ignition, atomizing nozzle, and a primary control (Figs. 1 and 5).

The fuel oil is pumped from the tank through the pump gears, and oil pressure is regulated by an internal valve that develops 100–300 psi at the burner nozzle. The oil is then atomized, ignited, and burned.

The fine droplets of oil that are discharged from the nozzle are electrically ignited by a transformer which raises the voltage from 115 to 10,000 volts. The recently developed electric ignition system called the capacitor discharge system incorporates a capacitor that discharges voltage into a booster transformer developing as much as 14,000 volts. The electric spark is developed at the electrode gap located near the nozzle spray (Figs. 2 and 5).

The high velocity of air produced by the squirrel cage fan helps develop the ignition spark to a point where it will reach out and ignite the oil with-

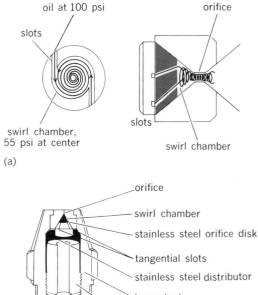

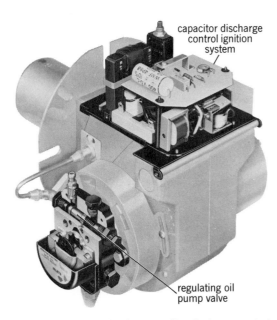

Fig. 4. Nozzle. (*a*) Operation of a nozzle. (*b*) Cutaway of nozzle. (*Delavan Manufacturing Co.*)

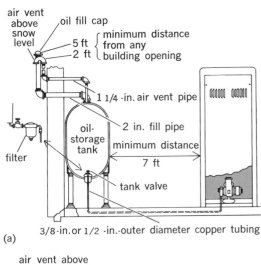

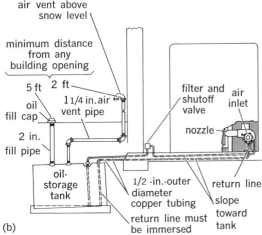

Fig. 6. Two typical installation methods for oil storage tank and oil burner. (*a*) Inside basement. (*b*) Outside underground storage tank.

Fig. 5. Oil burner, showing capacitor discharge control ignition system and regulating valve for fuel pump. (*Automatic Burner Corp.*)

out the electrode tips actually being in the oil spray. Figure 2 shows a typical gun-type burner inner assembly.

Installations. Fuel oils for domestic burners are distilled from the crude oil after the lighter products have been taken off. Consequently, the oil is nonexplosive at ordinary storage temperatures. Domestic fuel oils are divided into two grades, number one and number two, according to the

ASTM specification. There are several different methods of installing an oil tank system with an oil burner (Fig. 6). *See* FUEL OIL.

One of the most important safety items of an automatic oil burner is the primary control. This device will stop the operation when any part of the burner or the heating equipment does not function properly. This control protects against such occurrences as incorrect primary air adjustments, dirt in the atomizing nozzle, inadequate oil supply, and improper combustion. The most modern type of primary control is called the cadmium cell control. The cadmium sulfide flame detection cell is located in a position where it directly views the flame. If any of the above functional problems should occur, the electrical resistance across the face of the cell would increase, causing the primary control to shut the burner off in 70 sec or less.

Draft caused by a chimney or technical means is a very important factor in the operation of domestic oil burners. The majority of burners are designed to fire in a heating unit that has a minus 0.02 in. water column draft over the fire. Because the burner develops such low static pressures in the draft tube, over fire drafts are an important factor in satisfactory operation.

If the heating unit is designed to create pressure

over the fire, a burner developing high static pressures in the blast tube must be used. These burners develop high static by means of higher rpm on the motor and fan. Burners of this nature are available and are being produced. Flame control is very important. Firing heads similar to Fig. 2 are used on applications of this nature.

The oil burner is also used for a wide assortment of heating, air conditioning, and processing applications. Oil burners heat commercial buildings such as hospitals, schools, and factories. Air conditioners using the absorbtion refrigeration system have been developed and fired with oil burners. Oil burners are used to produce CO_2 in greenhouses to accelerate plant growth. They produce hot water for many commercial and industrial applications. *See* AIR COOLING; HEATING, COMFORT; HOT-WATER HEATING SYSTEM; OIL FURNACE.

[ROBERT A. KAPLAN]

Bibliography: American Society of Heating, Refrigerating, and Air-conditioning Engineers (ASHRAE), *Guide and Data Book: Systems and Equipment*, 1967; ASHRAE, *Handbook of Fundamentals*, 1967; ASHRAE, *Heating and Ventilating Engineering Data Book*, 1948; C. H. Burkhardt, *Domestic and Commercial Oil Burners*, 2d ed., 1961; D. Graham, *Audel's Oil Burner Guide*, 1963; W. F. B. Shaw, *Domestic Heating*, 1960.

Oil field model

A small-scale and commonly simplified replica of subsurface conditions of interest and value in petroleum prospecting and oil-field development. The term model has been applied by geologists to a simplified diagram and by mathematical physicists to a formal analysis with special boundary conditions and related attributes. These and somewhat more complex physical models have value in transmitting concepts and relationships to the nonspecialist, but they are generally designed by geologists and petroleum engineers to aid investigation of a particular problem. To do so, the model is either scaled to size, shape, and like attributes, or according to forces upon it, relative to its archetype.

Size-scaled models. These are commonly models of ore and mineral bodies and of those blocks on which surface topography and surface geology are shown on top and geologic cross sections on the sides, also to scale. Peg models are constructed for oil fields with wells shown as pegs or wires. Cutout block diagrams are also used. Models of this kind are found in many geological museums and in mining companies' offices.

Forces and movement models. This type contains mobile material to simulate by its movement a movement that has taken place or will take place in the prototype. Transparent solid plastics have been incorporated in mobile parts of models for enhanced internal visibility and to obtain photoelastic data. Such models are also used to make graphic representation of changes recorded from observations with geophysical instruments, for example, from microinstruments used to survey physical response of a small mass in an artificial field. Force and movement models have come into use for study of movements of earth materials. By this, past changes resulting in present features can be made as dimensionally credible experiments in periods tremendously short as compared with geologic time. Models with mobility are of three principal types.

Fluid or fluids moving in porous medium. The models for study of fluid movement through a geometrically stable medium have important applications to hydrologic and petroleum engineering. The hydrologic case is mostly of an interface underground between fresh water and salt water caused by extraction or by injection of fresh water through wells. Where the interface owes its position to fresh-water–salt-water patterns found in many coastal zones and along many shorelines, the setting is difficult to describe without a model. Such demonstrations have added to the understanding of water conditions in California's Santa Clara Valley and other critical water areas.

The petroleum engineering application is to movement of fluid in respect to another kind of interface underground. This is between petroleum and natural gas, or both, and salt water, and especially when petroleum is being extracted through wells and salt water simultaneously injected through other wells into the same porous stratum, so that the important flow is radial, that is, two-dimensional.

The model used involves scaling down from the actual distance between wells in the oil field under study, and also the device of an analog for the fluids, whereby the pressure distribution in steady-state porous flow is exactly the same as the potential distribution in an electrical conducting medium.

Uniform high-viscosity materials. The second geologic realm where models are used extensively is that of the movement of large segments of the earth made of materials of essentially uniform high viscosity, comprising nearly all rock species in the Earth's crust. This has little direct bearing on oil-field models.

Adjacent materials of diverse viscosities. Among the movements of the earth which are of interest to geologists are those involving what may be called soft layers between relatively hard ones. This is the realm where diapir structures are inferred to develop and also their subform, the salt dome. Because of the economic importance of salt domes and their associated oil traps, investigations with models have become widespread.

L. L. Nettleton produced a model for salt-dome formation using two viscous fluids of different density to emphasize the role of the lower density of salt to that of adjacent sediments in the process of salt-dome movements. M. B. Dobrin made a parallel mathematical analysis. T. J. Parker and A. N. MacDowell extended salt-dome studies by using materials with higher yield points than Nettleson's first model, and their model showed fractures above and around the "salt" analog in their model.

Also in both the prototype and the model, boundary conditions modify vectors nearby. This seems particularly hard to handle in cases of hydrologic model study, such as at the University of California.

Models versus mathematical analysis. There is a possible alternative to the use of physical models—mathematical analysis, which is more attractive now that computing machines are available.

Dobrin used this method for the case of Nettleton's model investigation of salt dome. However, mathematical analysis is as yet a subject of controversy among physicists in many simple geologic settings; physical models are presently available and their value is acknowledged.

Physical models are probably more helpful in educating the nonspecialist in geology; they present the earth features in a readily visible fashion easier to grasp than mathematical analysis. For a discussion of mathematical models *see* PETROLEUM RESERVOIR MODELS.   [PAUL WEAVER]

Bibliography: J. W. Amyx, D. M. Bass, and R. L. Whiting, *Petroleum Reservoir Engineering*, vol. 1: *Physical Properties*, 1960; M. B. Dobrin, Some quantitative experiments on a fluid salt-dome model and their geological implications, *Trans. Amer. Geophys. Union*, 22:528–542, 1941; H. E. McKinstry et al., *Mining Geology*, 1948; L. L. Nettleton, Recent experimental and geophysical evidence of mechanics of salt-dome formation, *Bull. Amer. Ass. Petrol. Geol.*, 27(1):51–63, 1943; A. E. Scheidegger, *Principles of Geodynamics*, vol. 1, 1958; D. K. Todd, *Ground Water Hydrology*, 1959.

Oil field waters

Waters of varying mineral content which are found associated with petroleum and natural gas or have been encountered in the search for oil and gas. They are also called oil field brines, or brines. They include a variety of underground waters, usually deeply buried, and have a relatively high content of dissolved mineral matter. These waters may be (1) present in the pore space of the reservoir rock with the oil or gas, (2) separated by gravity from the oil or gas and thus lying below it, (3) at the edge of the oil or gas accumulation, or (4) in rock formations which are barren of oil and gas. Brines are commonly defined as water containing high concentrations of dissolved salts. Potable or fresh waters usually are not considered oil field waters but may be encountered, generally at shallow depths, in areas where oil and gas are produced.

Oil field waters or oil field brines differ widely in composition and concentration. They may differ from one geologic province to another, from one formation to another within a given geologic province, or from one part of a specific geologic horizon to another. They range from slightly salty water with 1000–3000 parts of dissolved substances in 1,000,000 parts of solution to very nearly saturated brines with dissolved mineral content of more than 270,000 parts per million (ppm).

The most common and abundant mineral found in oil field waters is sodium chloride, or common table salt. Calcium chloride is next in order of abundance. Carbonates, bicarbonates, sulfates, and the chlorides of magnesium and potassium are present in lesser quantities. In addition to the above mentioned salts, salts of bromine and iodine are also found. Traces of strontium, boron, copper, manganese, silver, tin, vanadium, and iron have been reported. Barium has been reported in many of the Paleozoic brines of the Appalachian region. The commercial value of a brine depends upon the concentration of salts, purity of the products to be recovered, and value and practicability of by-product recovery. Concentrations less than 200,000 ppm are seldom of commercial interest.

Slightly salty waters, while not suitable for human consumption, may be used in some industrial processes or may be amenable to beneficiation for municipal supplies in areas lacking fresh waters.

Classified genetically, oil field waters are generally considered connate; that is, they are sea waters which (presumably) originally filled the pore spaces of the rock in which they are now confined. However, few analyses of these waters correspond to present-day sea water, thus indicating some mixing and modification since confinement. Dilute solutions suggest that rainwater has percolated into the rocks along bedding planes, fractures, faults, and other permeable zones. Presence of carbonates, bicarbonates, and sulfates in an oil field water further suggests that at least some of the water had its origin at the surface. Concentrations of dissolved solids greater than that of modern sea water suggest partial evaporation of the water or addition of soluble salts from the adjacent or enclosing rocks.

Waters in most sedimentary rocks increase in mineral concentration with depth. This increase may be due to the fact that, since salt water is heavier than fresh water, the more dense solution will eventually find a position as low as possible in the aquifer. An additional factor would be the longer exposure of the deeper waters to the mineral-bearing rocks. Exceptions have been noted and probably are due to the presence of larger quantities of soluble salts in some geological formations than in others.

Probably the most important geological use of oil field water analyses is their application to the quantitative interpretation of electrical and neutron well logs, particularly micrologs. In order to compute the connate water saturation of a formation in a quantitative manner from electrical data, it is necessary to know with accuracy the connate water resistivity.

Naturally mineralized waters are frequently the only waters available for water-flooding operations. Water analyses are useful in predicting the effect of the water on minerals in the reservoir rock and on the mechanical equipment employed on the project. Waters which exert a corrosive action on the lines and pumps or which tend to plug up the pay zone are not suitable for water-flooding operations.

Oil field water composition may be an important factor in the determination of the source of water in oil wells which have leaky casings or improper completions with resulting communication between wells, and in identifying and correlating reservoirs in multipay oil pools, particularly in those containing lenticular sand bodies.

Industrial wastes, including mineralized water produced with oil, may be disposed of in underground reservoirs. Between the zone of potable water and the horizon of commercial brines, there commonly are rock formations, the waters of which contain chemicals in amounts sufficient to make the waters unsuitable for domestic, municipal, industrial, and livestock consumption, but not in sufficient quantity to be considered as a source for recovery of chemicals. Provided there is sufficient porosity and permeability, these rock

Oil furnace

A combustion chamber in which oil is the heat-producing fuel. Fuel oils, having from 18,000 to 20,000 Btu/lb, which is equivalent to 140,000 to 155,000 Btu/gal, are supplied commercially. The lower flash-point grades are used primarily in domestic and other furnaces without preheating. Grades having higher flash points are fired in burners equipped with preheaters.

The ease with which oil is transported, stored, handled, and fired gives it a special advantage in small installations. The fuel burns almost completely so that, especially in a large furnace, combustible losses are negligible. *See* FUEL OIL; OIL BURNER.

Domestic oil furnaces with automatic thermostat control usually operate intermittently, being either off or operating at maximum capacity. The heat-absorbing surfaces, especially the convection surface, should therefore be based more on maximum capacity than on average capacity if furnace efficiency is to be high. The combustion chamber should provide at least 1 ft³ for each 1.5–2 lb of fuel burned per hour. Gas velocity should be below 40 ft/sec. The shape of the chamber should follow the outline of the flame.   [FRANK H. ROCKETT]

Oil mining

The surface or subsurface excavation of petroleum-bearing sediments for subsequent removal of the oil by washing, flotation, or retorting treatments. Oil mining also includes recovery of oil by drainage from reservoir beds to mine shafts or other openings driven into the oil rock, or by drainage from the reservoir rock into mine openings driven outside the oil sand but connected with it by bore holes or mine wells.

Surface mining consists of strip or open-pit mining. It has been used primarily for the removal of oil shale or bituminous sands lying at or near the surface. Strip mining of shale is practiced in Sweden, Manchuria, and South Africa. Strip mining of bituminous sand is conducted in Canada.

Subsurface mining is used for the removal of oil sediments, oil shale, and Gilsonite. It is practiced in several European countries and in the United States. Some authorities consider subsurface mining the best method to recover oil when oil sediments are involved, because virtually all of the oil is recovered.

European experience. Subsurface oil mining was used in the Pechelbronn oil field in Alsace, France, as early as 1735. This early mining involved the sinking of shafts to the reservoir rock, only 100–200 ft below the surface, and the excavation of the oil sand in short drifts driven from the shafts. These oil sands were hoisted to the surface and washed with boiling water to release the oil. The drifts were extended as far as natural ventilation permitted. When these limits were reached, the pillars were removed and the openings filled with waste.

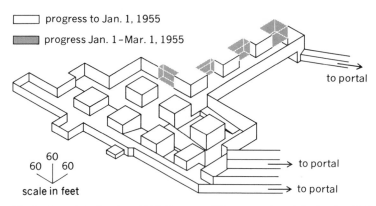

☐ progress to Jan. 1, 1955
▨ progress Jan. 1–Mar. 1, 1955

to portal
to portal
to portal

60
60   60
scale in feet

Fig. 1. Isometric drawing of U.S. Bureau of Mines experimental mine for oil shale at Rifle, Colo. Developed by 39-ft top or advanced heading, a 34-ft following bench had just been initiated when a roof fall made it necessary to conclude operations. (*U.S. Bur. Mines Rep. no. 5237, 1956*)

This type of mining continued at Pechelbronn until 1866, when it was found that oil could be recovered from deeper, more prolific sands by letting it drain in place through mine openings, without removing the sands to the surface for treatment.

Subsurface mining of oil shale also goes back to the mid-19th century in Scotland and France. It is not so widely practiced now because of its high cost as compared with that of usual oil production, particularly in the prolific fields of the Middle East.

United States oil shale mining. The U.S. Bureau of Mines carried out an experimental mining and processing program at Rifle, Colo., between 1944 and 1955 in an effort to find economically feasible methods of producing oil shale.

One of the more important phases of this experimental program was a large-scale mine dug into what is known as the Mahagony Ledge, a rich oil shale stratum that is flat and strong, making it favorable for mining. This stratum lies under an average of about 1000 ft of overburden and is 70–90 ft thick.

The Bureau of Mines adopted the room-and-pillar system of mining, advancing into the 70-ft ledge face in two benches. The mine roof was supported by 60-ft pillars staggered at 60-ft intervals and supplemented by iron roof bolts 6 ft long (Fig. 1).

Fig. 2. Mine locomotive and cars removing shale from U.S. Bureau of Mines shale mine at Rifle, Colo. At left, in middle ground, is the Colorado River, nearly 3000 ft below. (*After R. Fleming, U.S. Bureau of Mines*)

Fig. 3. Close-up view of Union Oil Co. of California shale-oil retort near Grand Valley, Colo. Right part of structure is portion of the system for removing oil vapors that would otherwise escape in the gas stream.

Multiple rotary drills mounted on trucks made holes in which dynamite was placed to shatter the shale; the shale was then removed from the mine by electric locomotive and cars (Fig. 2).

The experimental mining program ended in February, 1955, when a roof fall occurred. Despite this occurrence, however, the Bureau is convinced that the room-and-pillar method used in coal, salt, and limestone mines is feasible for shale oil mining in Colorado.

Oil shale does not contain oil, as such. Draining methods, therefore, are not applicable. It does, however, contain an organic substance known as kerogen. This substance decomposes and gives off a heavy, oily vapor when it is heated above 700°F in retorts. When condensed, this vapor becomes a viscous black liquid called shale oil, which resembles ordinary crude but has several significant differences.

Since 1955, several companies have conducted experimental efforts to produce shale oil. Those continuing to move toward commercialization (not expected before 1978) include companies which plan to use traditional mining techniques combined with some form of surface retorting. One company is testing an on-site process in which both mining and retorting are done underground, and another firm is well along in development work using a method involving a gas-combustion retort similar to that used by the Bureau of Mines in various pilot plants operated during the 1950s and 1960s.

Another possibility is a modified open-cast surface method, which proponents claim has a 95% recovery rate of the minable reserves, compared with a maximum 40% rate by room-and-pillar underground mining and 20% by on-site mining and retorting.

Colorado's Mahagony Ledge yields an average of about 30 gal of oil per ton. This means that large amounts of oil shale must be mined, transported, retorted, and discarded for production of commer-cial quantities of oil. Various types of retort also have been tested in Colorado, but none is in commercial use (Fig. 3). *See* OIL SHALE.

Gilsonite. Gilsonite is a trade name, registered by the American Gilsonite Co., for a solid hydrocarbon found in the Uinta Basin of eastern Utah and western Colorado. The American Gilsonite Co. uses a subsurface wet-mining technique to extract about 700 tons of Gilsonite daily from its mine at Bonanza, Utah.

Conventional mining methods were found unsuitable for mass output of Gilsonite because it is friable and produces fine dust when so mined. This dust can be highly explosive. In the system now being used, tunnels are driven from the main shaft by means of water jetted through a 1/4-in. nozzle under pressure of 2000 psi. The stream of water penetrates tiny fissures and the ore falls to the bottom of the drift. The drifts are cut on a rising grade of about 2.5°. The ore is washed down to the main shaft where it is screened. Particles of sizes smaller than 3/4 in. are pumped to the surface in a water stream; larger pieces are hoisted in buckets. A long rotary drill with carbide-tipped teeth is used to remove ore that cannot be broken with water jets.

Gilsonite is moved through a pipeline in slurry form to a refinery, where it is dried and melted and then heated to about 450°F. The melted oil is fed to a coker and other processing units to make gasoline and other petroleum products.

[ADE L. PONIKVAR]

North American tar sands. The world's only tar sands mining operation is being conducted at the 50,000 barrels per day (bpd) synthetic crude complex of Great Canadian Oil Sands (GCOS), 21 mi north of Fort McMurray, Alberta. A second operation, the 125,000-bpd project of Syncrude Canada Ltd., located approximately 5 mi away, was scheduled to begin production in 1978. Both projects are situated in the minable area of the gigantic Athabasca Tar Sands deposit, the world's largest oil reservoir. The minable area, with an overburden thickness of less than 150 ft, embraces about 10% of the estimated 624×10^9 bbl of total in-place bitumen reserves. *See* OIL SAND.

In both leases about half of the terrain is covered with muskeg, an organic soil resembling peat moss, which ranges anywhere from a few inches to 20 ft in depth. The major part of the overburden, however, consists of Pleistocene glacial drift and Clearwater formation sands and shales. The total overburden varies from 20 to 120 ft in thickness. Underlying this is the McMurray Formation (Lower Cretaceous) in which the oil-impregnated sands reside. The tar sands strata, in the region of the GCOS and Syncrude operations, are also variable in thickness; they average about 150 ft although typically strata of 20–30-ft thickness have a bitumen content below the economic cutoff grade of 6 wt %.

Composition. The bitumen content of tar sand can rage from 0 to 20 wt %, but feed-grade material normally runs between 10 and 12 wt %. The balance of the tar sand is composed, on the average, of 5 wt % water and 84 wt % sand and clay. The bitumen, which has an API gravity of 8°, is heavier than water and very viscous. Tar sand is a competent material, but it can be readily dug in the summer months; in the winter months, however, which see the temperature plunging to −50°F, tar sand assumes the consistency of concrete. To maintain an acceptable digging rate in the winter, mining must proceed faster than the rate of frost penetration; if not, supplemental measures, such as blasting, are required.

Overburden removal. For muskeg removal, a series of ditches are dug two or three years in advance of stripping to permit as much of the water as possible to drain. Despite this, the spongy nature of the muskeg persists and removal is best accomplished after freeze-up.

Great Canadian has tried several overburden removal methods, including shovel and trucks, scrapers, and front-end loaders and trucks. In 1974 the overburden removal equipment consisted of 5 Caterpillar D9G bulldozers for ripping and dozing, 7 Marathon LeToureau L700 15-yd³ front-end loaders, and a fleet of 21 WABCO 150-ton-capacity trucks. Additional equipment is used for maintaining the haul roads and for spreading and compacting the spoiled material.

Bucket wheel excavators. Mining of tar sand is performed mainly by two large bucket wheel excavators, of German manufacture, each operating on a separate bench (Fig. 4). These units, weighing 1800 tons, have a 33-ft-diameter digging wheel on the end of a long boom. Each wheel has a theoretical capacity of 9500 tons/hr, but the average output when digging is about 5000 tons/hr. Because the availability of these machines is normally 55–60%, the extraction plant has been designed to accept a widely fluctuating feed rate. To facilitate digging of the highly abrasive tar sand and to achieve a reasonable bucket and tooth life, GCOS routinely preblasts the tar sand on a year-round basis. Tar sand at the rate of 140,000 tons/day is transferred from the mine to the plant by a system of 60-in. face conveyor belts and 72-in. trunk conveyors, operating at 1080 ft/min. Following extraction by a process using hot water, the bitumen is upgraded by coking and hydrogenation to a high-quality synthetic crude.

Dragline scheme. Syncrude has opted for an even more capital-intensive mining scheme. Four large draglines, each equipped with an 80-yd³

Fig. 4.   Krupp wheel excavator removing bituminous sand near Mildred Lake, Athabaska region, Alberta, Canada. (*Cities Service Co.*)

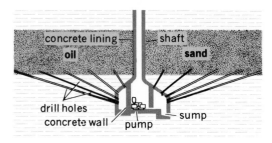

Fig. 5.   Sketch of Wright system for draining oil sands by a series of bore holes drilled from a mine shaft. (*After L. C. Uren, Petroleum Production Engineering: Oil Field Exploitation, 3d ed., McGraw-Hill, 1953*)

bucket at the end of a 360-ft boom, will be employed to dig both the overburden, which will be free-cast into the mining pit, and the tar sand, which will be piled in windrows behind the machines. Four bucket-wheel reclaimers, quite similar to the GCOS excavators except larger to handle the additional capacity, will load the tar sand from the windrows onto conveyor belts which will transfer it to the plant. With a peak tar sand mining rate of 336,000 tons/day, the Syncrude project will be the largest mining operation in the world.

The advantages of the dragline scheme lie in its ability to handle overburden at a lower cost and its greater selectivity in rejecting lenses of low-grade tar sand and barren material. The disadvantages include the necessity of rehandling the tar sand and the probability of an increased percentage of lumps in the plant feed, which can damage conveyor belting.

Several years of comparative operation will be required to firmly establish whether one of the schemes will enjoy an economic advantage. *See* PETROLEUM GEOLOGY.

[G. RONALD GRAY]

Bibliography: A. R. Allen and E. R. Sanford, *The Great Canadian Oil Sands Operation*, Canadian Society of Petroleum Geology, Oil Sands Symposium, Calgary, Alberta, 1973; F. L. Hartley and C. S. Brinegar, Oil shale and bituminous sand, *Energy Resources Conference*, 1956; J. B. Jones, Jr. (ed.), *Hydrocarbons from Oil Shale, Oil Sands and Coal*, 1964; *Synthetic Liquid Fuels: Annual Report of the Secretary of the Interior for 1955*, pt. 11: *Oil From Shale*, U.S. Bur. Mines Rep. no. 5237, 1956; L. C. Uren, *Petroleum Production Engineering: Oil Field Exploitation*, 3d ed., 1953.

Oil sand

A loose sand or a semiconsolidated sandstone impregnated with a heavy asphaltic crude oil too viscous to be processed by conventional methods; also known as tar sand or bituminous sand.

Distribution. Oil sands are distributed throughout the world but occur primarily in Canada, Venezuela, the United States, Madagascar, Albania, Trinidad, Rumania, and the Soviet Union. The known world reserves of heavy hydrocarbons approximate 1.7×10^9 bbl (210×10^9 m³). Estimates of the amount that will ultimately prove recoverable—generally of the order of 10 to 33%—are highly speculative and depend on the development of successful technologies at competitive costs.

The Alberta Oil Sands, comprising four enor-

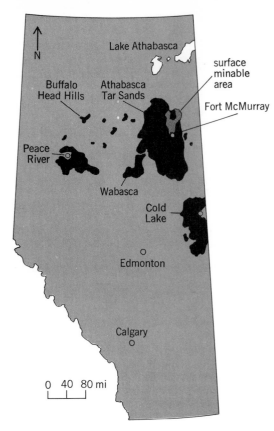

Tar sand deposits in Alberta.

mous deposits—including the famous Athabasca Tar Sands, location of the only existing commercial oil sand mining development—and a number of lesser deposits, contain the largest and best-known of the oil sand reservoirs. Established reserves exceed 900×10^9 bbl (143×10^9 m³).

Ranking closely behind the Alberta Oil Sands in total reserves volume are the tar sands and heavy oil deposits of the Orinoco Petroleum Belt of Venezuela. Here an estimated 700×10^9 bbl (111×10^9 m³) of 8–15° API oil occur in reserves stretching along the northern bank of the Orinoco River for a distance of 375 mi (600 km). In 1975 about 80,000 bbl (12,700 m³) per day were being produced by injecting light oil into the formation to dilute the heavy crude. Numerous in-place, recovery methods have been applied in mainly unsuccessful attempts to boost the sagging production.

In the United States, oil sands are found in Utah, California, Texas, Kentucky, Missouri, and Kansas. Over 90% of the total reserves, or some 28×10^9 bbl (1.3×10^9 m³) are situated in 24 Utah tar sand deposits. Most of these are located in the Uinta Basin of northeast Utah. Thickness of the reservoirs varies between 0 and 300 ft (100 m), and depth ranges from 0 to 2000 ft (670 m).

One interesting characteristic distinguishes much of the heavy oil in Utah. The sulfur content of less than 0.5% is about one-tenth that of the Canadian and Venezuelan tars.

Far back in second place is California, which contains an estimated 200×10^6 bbl (31.8×10^6 m³) of oil sand reserves. The Edna deposit, located midway between Los Angeles and San Francisco, is the largest. It is a rarity among oil sand deposits

in that it is considered a marine facies. Virtually all of the other major oil sand deposits in the world occur in fresh-water fluviatile and deltaic environments. The Edna formation is fossiliferous and consists largely of diatomaceous sandstone.

Though there are claims of a large deposit in the Olenek anticline in northeastern Siberia, the only other substantial oil sands reservoir about which much is known is the Bemolanga deposit in western Madagascar. It covers approximately 150 mi² (388 km²) and contains reserves estimated at 1.75×10^9 bbl (278×10^6 m³).

Alberta deposits. While the energy crisis is focusing attention on all oil sand deposits, much of the experimental and developmental activity currently is directed toward the Alberta Oil Sands. The deposits range across the northern part of the province (see illustration).

The Athabasca deposit is the largest known oil field in the world. The aerial extent of this deposit is about 13,000 mi² (33,670 km²). The McMurray formation, in which the oil-impregnated sands reside, belongs to the Lower Cretaceous age. The formation outcrops along the Athabasca River, north of Fort McMurray. Elsewhere, the deposit is buried under a variable layer of overburden which reaches up to 1700 ft (almost 600 m) in thickness. As a fortunate circumstance, some of the richest parts of the deposit are covered by the thinnest overburden.

It is in this area that Great Canadian Oil Sands (GCOS), a subsidiary of Sun Oil Company, was producing 50,000 bbl (7950 m³) per day of synthetic crude from mined tar sands in 1975. On an adjacent lease, the immense Syncrude Canada Ltd. mining project was scheduled to start production in 1978. This 125,000 barrels-per-day (19,860 m³) project was severely affected by an unprecedented worldwide escalation in process-industry construction costs, and the original capital cost estimates of $1,000,000,000 doubled by 1975.

Surface mining. Both the GCOS and Syncrude projects employ open-pit mining methods to remove the overburden and underlying tar sands, hot water extraction units to recover 90% or more of the tarry bitumen, and upgrading units to convert it to high-quality synthetic crude. Differences in the two projects, aside from the size—the Syncrude project with a peak mining rate of 336,000 tons (302,400 metric tons) of tar sand per day will be the largest mine in the world—lie mainly in the mining and upgrading methods employed. GCOS uses large bucket-wheel excavators to mine the abrasive tar sand, whereas Syncrude has purchased large draglines for this purpose; GCOS converts the bitumen to lighter products in a delayed coking unit, whereas Syncrude is building two large fluid coking units. In each case, the high-sulfur-content coker distillates are upgraded by high-pressure hydrogenation to "sweet" 34–38° API gravity synthetic crudes. The economically recoverable, or "proved," minable reserves in the Athabasca deposit have been calculated at 38×10^9 bbl (6×10^9 m³) of bitumen, corresponding to 26.5×10^9 bbl (4.2×10^9 m³) of synthetic crude. *See* OIL MINING.

Recovery of deep deposits. Many experimental programs have been undertaken in an attempt to devise a commercially feasible method of recovering the 90% of the bitumen which is buried too deeply for surface mining. Amoco Canada refers to its process as COFCAW, for "combination of forward combustion and waterflood." Most of the other companies have utilized some form of steam stimulation. Imperial Oil was producing 4000 bbl (637 m³) per day at its Cold Lake pilot project in 1975. The heavy oil at Cold Lake is 2–3° lighter than the 8° API Athabasca oil, and this has a significantly beneficial effect on the ease of recovery. To date, however, recovery efficiencies of all in-place schemes have been too low to justify a commercial project.

[G. RONALD GRAY]

Bibliography: M. A. Carrigy (ed.), *Guide to the Athabasca Oil Sands Area*, Alberta Research, 1973; G. R. Gray, *Conversion of Athabasca Bitumen*, American Society of Chemical Engineers Conference, 1971; L. V. Hills (ed.), *Oil Sands: Fuel of the Future*, Canadian Society of Petroleum Geologists, Calgary, Alberta, 1974; *Properties of Utah Tar Sands*, U.S. Bur. Mines Rep. no. 7923, 1974; *Reserves of Crude Oil, Gas, Natural Gas Liquids and Sulfur*, Province of Alberta, ERCB Rep. 75–18, 1974.

Oil shale

A sedimentary rock containing solid, combustible organic matter in a mineral matrix. The organic matter, often called kerogen, is largely insoluble in petroleum solvents, but decomposes to yield oil when heated. Although "oil shale" is used as a lithologic term, it is actually an economic term referring to the rock's ability to yield oil. No real minimum oil yield or content of organic matter can be established to distinguish oil shale from other sedimentary rocks. Additional names given to oil shales include black shale, bituminous shale, carbonaceous shale, coaly shale, cannel shale, cannel coal, lignitic shale, torbanite, tasmanite, gas shale, organic shale, kerosine shale, coorongite, maharahu, kukersite, kerogen shale, and algal shale.

Origin and mineral composition. Oil shale is lithified from lacustrine or marine sediments relatively rich in organic matter. Most sedimentary rocks contain small amounts of organic matter, but oil shales usually contain substantially more. Specific geochemical conditions are required to accumulate and preserve organic matter, and these were present in the lakes and oceans whose sediments became oil shale. R. M. Garrels and C. L. Christ define these conditions in terms of oxidation-reduction potential (Eh) and acid-base condition (pH) of the water in and around the sediment. Organic matter accumulates under the strongly reducing conditions and neutral or basic pH present in euxinic marine environments and organic-rich saline waters. The organic-rich sediments which became oil shale accumulated slowly in water isolated from the atmosphere, a condition relatively rare in natural waters. This isolation was achieved by stagnation or stratification of the water body and the accompanying protection of its sediments.

Quartz, illite, and pyrite (sometimes with marcasite and pyrrhotite) occur in virtually every oil shale. Feldspars and other clays, particularly montmorillonite, are found in many oil shales. Most oil shale deposits contain small amounts of carbonate minerals, but some, notably the Green River Formation in Colorado, Utah, and Wyoming, contain large amounts of dolomite and calcite. The oil shale minerals were probably formed in the sediment by chemical processes related, at least in part, to the presence of organic matter.

Some oil shales, particularly those called black shales because of the coallike color of their organic matter, have tended to become enriched in trace metals. The reducing conditions necessary to preserve organic matter were conducive to precipitating available trace metals, frequently as sulfides. The Kupferschiefer of Mansfield, Germany, contains an unusually high content of copper, and the Swedish Alum shale has been exploited for its uranium content. The Devonian Chattanooga Shale of Tennessee and neighboring states contains an average of close to 0.006 wt % uranium and has been extensively studied as a potential low-grade source of this element. Vanadium in potentially commercial amounts occurs in the Permian Phosphoria Formation of Wyoming and Idaho. Enrichment of As, Sb, Mo, Ni, Cd, Ag, Au, Se, and Zn has also been noted in black oil shales.

Physical properties. Oil shales are fine-grained rocks generally with low porosity and permeability. Many are thinly laminated and fissile. On outcrop, some oil shales weather to form stacks of thin organic-rich layers called paper shale. The colors of oil shales range from black to light tan and are produced or altered by organic matter.

The physical properties of oil shale are strongly influenced by the proportion of organic matter in the rock. Its decrease in density with increasing organic content illustrates this most graphically. The mineral components have densities of about 2.6–2.8 g/cm³ for silicates and carbonates and 5 g/cm³ for pyrite, but the density of organic matter is near 1 g/cm³. Larger fractions of organic matter produce rocks with appreciably lower density. The equation below quantifies this relationship. Here,

$$D_T = \frac{D_A D_B}{A(D_B - D_A) + D_A}$$

A = weight fraction of organic matter, B = weight fraction of mineral matter, and $A + B = 1$. The organic fraction has an average density D_A, and the mineral fraction an average density D_B. This relationship is applicable to any relatively uniform oil shale deposit when appropriate values for D_A and D_B are known. For Green River Formation oil shales, D_A is 1.06–1.07 g/cm³ and D_B is about 2.7 g/cm³. By incorporating a factor for conversion of organic matter to oil, the equation yields a relationship between oil yield and oil shale density which is particularly useful in calculating resources and reserves in an oil shale deposit.

The volume of organic matter in an oil shale rock affects its physical strength properties and its crushability. The rapid increase in volume of organic matter in the shale with increasing organic content weight fraction can be demonstrated by calculations based on the equation above. For example, in the Green River Formation, in oil shale containing 4 wt % organic matter (a very lean shale yielding 2.6 wt % oil), the organic matter makes up 10 vol % of the rock. In richer oil shales, the organic matter is the largest volume component of the

Table 1. Relationship between organic carbon–organic hydrogen ratio and conversion of oil shale organic matter to oil by heating

Deposit sampled	Carbon-hydrogen value	Organic carbon recovered, wt %
Pictou County, Nova Scotia, Canada	12.8	13
Top Seam, Glen Davis, Australia	11.5	26
New Albany Shale, Kentucky, United States	11.1	33
Ermelo, Transvaal, South Africa	9.8	53
Cannel Seam, Glen Davis, Australia	8.4	60
Garfield County, Colorado, United States	7.8	69

rock, and the physical properties of the organic matter predominate. In richer Green River Formation oil shales the organic matter makes the rock tough and resilient; under load, the shales will deform plastically rather than break.

Organic composition and oil production. The organic matter in oil shales and other sedimentary rocks has been extensively studied by organic geochemists, but a specific description of it has not been produced. Although some oil shales contain recognizable organic fragments like spores or algae, most do not, because the basic reducing conditions associated with oil shale development digested and homogenized the organic debris. The resulting organic matter (kerogen) is best described as a high-molecular-weight organic mineraloid of indefinite composition. This composition varies from deposit to deposit and is influenced by the depositional conditions and the nature of the organic debris. Variations in the hydrogen content of this organic matter are significant, however, because the fraction of organic matter converted to oil on heating increases as the amount of hydrogen available in the organic matter increases. To illustrate this relationship, Table 1 compares the proportion of organic carbon recovered as oil during Fischer assay with the weight ratio of organic carbon to organic hydrogen in several oil shales. For petroleum, the carbon-hydrogen values range from 6.2 to about 7.5; for coal, they range upward from 13. The carbon-hydrogen values for organic matter in oil shales range from near petroleum to near coal.

Analytical determination of the elemental composition of the organic matter has been difficult because of the heterogeneous nature of oil shales.

Carbon, hydrogen, sulfur, oxygen, and nitrogen are the major elements of the organic matter; but (except for nitrogen) they also occur in the mineral material for oil shales. The organic matter and the mineral matter in oil shales are difficult to separate either physically or chemically. Analytical techniques designed to distinguish between organic and mineral forms of elements, specialized organic matter enrichment techniques, and specialized evaluation techniques have been and are still being developed to aid in the study of oil shales.

The Fischer assay is the best known of the specialized analytical procedures. It was developed by the U.S. Bureau of Mines for oil shale evaluation. The method, employing a modified Fischer retort, determines the quantities of liquid oil and other products recoverable from an oil shale sample heated under prescribed conditions. Although the procedure does not measure the total amount of organic matter in the sample, it approximates the oil available by commercial operations. This simple procedure has proved to be suitable for most oil shale evaluation purposes. Resource information for United States oil shales is based on Fischer assay oil-yield data accumulated by the Laramie Energy Research Center of the U.S. Energy Research and Development Administration.

World oil shale resources. The world's organic-rich shale deposits represent a vast store of fossil energy. They occur on every continent in sediments ranging in age from Cambrian to Tertiary. D. C. Duncan and V. E. Swanson estimated the shale oil represented by the world's oil shale deposits. Their evaluations are summarized in Table 2. The values given for known resources refer only to evaluated resources. To these values Duncan and Swanson added possible extensions of known resources and geologically based estimates of undiscovered and unappraised resources to obtain their estimate of order-of-magnitude values for the total in-place oil resource in the world's oil shale deposits.

A barrel of oil, the 42-gal volume unit used in Western petroleum commerce, has no direct equivalent in metric countries. A barrel of oil represents 0.159 kiloliter. Specification of oil density, a variable, is necessary to convert the barrels into the tonne, the metric ton. A conversion factor agreed on at the World Energy Conference in 1974 defines a barrel of oil as 0.145 tonne, ignoring the

Table 2. Shale-oil resources of the world

Continents	Known resources*			Order of magnitude of total resources*		
Range in grade (oil yield in gal/ton):	25–100	10–25	5–10	25–100	10–25	5–10
Africa	100	Small	Small	4,000	80,000	450,000
Asia	90	14	†	5,500	110,000	590,000
Australia and New Zealand	Small	1	†	1,000	20,000	100,000
Europe	70	6	†	1,400	26,000	140,000
North America	600	1,600	2,200	3,000	50,000	260,000
South America	50	750	†	2,000	40,000	210,000
Totals	910	2,400	2,200	17,000	325,000	1,750,000

*In 10^9 bbl. †Not estimated.

density conversion. The Western shale grade unit, gallons per ton, is also a volume unit equivalent to 4.172 liters per metric ton (tonne). The relationship 1 gal/ton × 0.29 = 1 kg/tonne is an approximation agreed on at the 1975 World Energy Conference.

The size of the potential shale-oil resource is staggering. The richest part of the world's evaluated oil shale resource [9×10^{11} bbl (1.3×10^{11} tonne); see Table 2] alone is equivalent to the world's crude oil reserves in 1975 (7×10^{11} bbl, or 1×10^{11} tonne). These petroleum reserves represent only 4% of the projected total resource of rich oil shale (25–100 gpt, or gallons per ton). Since the 1965 estimates shown in Table 2 were made, some deposits have been moving from unknown to known resource classification.

The resource estimates in Table 2 are separated into three grades according to oil yield, recognizing that the richest deposits are more amenable to economic development. Since many factors besides richness affect the economics of development, the grade designations in Table 2 have limited significance.

World oil shale developments. Although the oil potential of the world's oil shales is great, commercial production of this oil has generally been considered uneconomic. Oil shales are lean ores, producing only limited amounts of oil which historically has been low in price. Mining and heating 1 ton of 25 gal/ton (104 liters/tonne) oil shale produces only 0.6/bbl (0.087 tonne) of oil.

In special situations when other fuels were in short or uncertain supply, or when energy transportation was difficult, energy development from oil shales has been carried out commercially. The 1694 English patent granted for a process "to distill oyle from a kind of stone" is the earliest such record, although medicinal oils were apparently produced from oil shales earlier. A French operation initiated in 1838 is probably the earliest energy development recorded; and Scotland, Canada, and Australia produced shale oil commercially before 1870. From 1875 to 1960, energy equivalent to about 250,000,000 bbl (31×10^6 tonne) of oil was produced from Europe's oil shale deposits. Most of this production was derived from deposits in Estonia, S.S.R., and Scotland. Low-priced oil from the Near East and improving oil transport systems stopped most oil shale developments.

World War II caused sharp increases in petroleum demand and disrupted both petroleum production and petroleum distribution, reactivating interest in oil shale development. Oil shale production operations during and since World War II have been conducted in Germany, France, Spain, Manchuria (China), Estonia and other areas of the Soviet Union, Sweden, Scotland, South Africa, Australia, and Brazil.

Two modern developments, the Manchurian and the Estonian, are relatively large. The Manchurian shale development is near the city of Fushun. The Oligocene oil-shale deposit averages about 450 ft (150 m) of shale, yielding approximately 15 gpt (63 liters/tonne). The deposit overlies a thick coal seam. Removal of the oil shale deposit to enable the coal to be mined by open-pit methods has resulted in the development of the world's largest oil shale industry. Production information has been difficult to obtain, but a daily output of 40,000 bbl of oil (5800 tonne) has been reported. Successful development at Fushun appears to have generated other oil shale developments in the area.

Broad Soviet areas in Estonia and the adjacent Leningrad region are underlain by Ordovician oil shale (kukersite) beds at shallow depths. These shales, reaching 10 aggregate feet (3.3 m) of 50 gpt (210 liters/tonne) shale, are being used to generate electricity and large quantities of low-heating-value gas for domestic and industrial purposes in Leningrad and Tallin. Production exceeded 30,000,000 tons (27×10^6 tonne) of shale in 1973. Most of this was burned directly to generate electricity.

Several smaller-scale oil shale operations have been conducted during and since World War II. Australia operated an oil shale plant at Glen Davis, New South Wales, during World War II. Problems associated with mining thin seams caused this plant to close about 1950. Brazil, always short of domestic petroleum, has intensively investigated two major deposits for shale-oil production. Petrobras, a corporation partly sponsored by the Brazilian government, collaborated with a United States firm to develop and apply the Petrosix retort to the Permian Irati shale. France pioneered destructive distillation of oil shale at Autun, and operated plants on three other deposits after World War II. All had ceased to operate by 1957. Germany operated oil shale plants on the Jurassic deposits in Württemberg during World War II, but these developments did not survive postwar economics. Scotland, an oil shale pioneer, continued to produce shale oil by mining and processing the Carboniferous Lothians oil shale deposits until 1963. In South Africa, the South African Torbanite Mining and Refining Company, Ltd., began operations on a deposit near Ermelo, Transvaal, in 1935. This grew into a large-scale operation which exhausted the 20,000,000–30,000,000-ton (tonne) oil shale deposit. An integrated company operating at Puertollano in Spain's Ciudad Real Province has produced gasoline, diesel and fuel oils, lubricants, and other by-products on a small scale since about 1922. An enlarged company created in 1942 by the National Institute of Spain built a new installation which incorporated a low-temperature hydrogenation plant to upgrade shale oil. The Swedish government built a large plant at Kvarntorp in Närke Province during World War II to produce oil from the Alum black shale. This plant was in full production by 1947 but closed in the 1950s. In conjunction with this plant, Sweden tested underground gasification of oil shale by electrical heating. In this procedure, known as the Ljungstrom method, hydroelectric power available during times of low demand was used. Its proponents claimed that the calorific value of the oil and gas vaporized was about three times that of the energy used to produce them. Developments being investigated in other areas have been continually reported. Like the more extensive efforts, these came to an end because of the economic pressures caused by the abundant and inexpensive petroleum supplies from the Near East.

Two major laboratories working primarily on oil shales exist, both government-sponsored. The Oil Shale Institute at Kohtla-Javre, Estonia, S.S.R.,

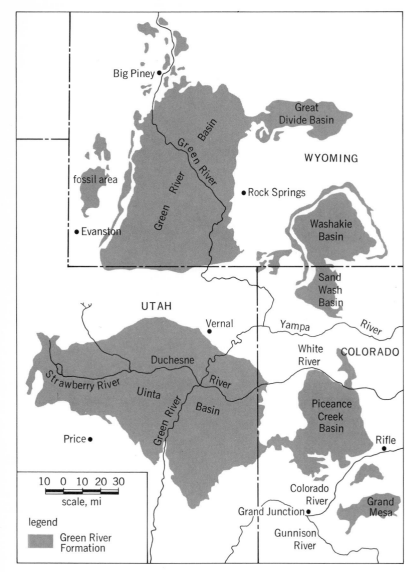

Fig. 1.   Location of Green River Formation oil shale deposits.

and from Alabama to the Canadian border. This vast area, estimated at 250,000 mi² (65,000,000 hectares), is underlain by a time-transgressive continuum of black shale marine sediments occurring in formations locally referred to by names such as Chattanooga, New Albany Shale, Antrim, Ohio, Sunbury, Marcellus, Middlesex, Rhinestreet, Genessee, and Woodford. *U.S. Geological Survey Bulletin no. 523* indicates that the combined deposits offer about 1×10^{12} bbl (145×10^9 tonne) of oil from shale yielding 10 gpt (42 liters/tonne). Approximately 20% of this total resource is classed as known.

Organic matter in these Devonian-Mississippian black shales tends to be low in hydrogen, yielding only a small fraction (from zero to about one-third) of its weight as oil on heating (see Table 2). The heating value of the organic matter ranges from 8500 to 8900 cal/g. Although the black shales underlie a large area and the organic matter disseminated in them represents a huge amount of fossil energy, the deposits are low-grade and the resource in any area is relatively small. In Kentucky, where the deposit seems richest, the resource reaches 50×10^6 bbl of oil per square mile (28,000 tonne oil/hectare) in a 100-ft-thick (30 m) section, with an average oil yield of 10 gal/ton (42 liters/tonne). Organic matter in the New Albany Shale in this area is 10 wt %, representing 22×10^6 tons of organic matter per square mile (32,000 tonne/hectare).

The world's largest oil shale resource is the Eocene Green River Formation in Colorado, Utah, and Wyoming (Fig. 1). The oil potential of oil shales in this 16,500-mi² (1 mi² = 259 hectares) deposit exceeds 2×10^{12} bbl (290×10^9 tonne). Of this, at least 600×10^9 bbl (87×10^9 tonne) occur in deposits yielding 25 gal or more of oil per ton of shale (104 liters/tonne) in continuous sections of oil shale at least 10 ft (3 m) thick. In the Piceance Creek Basin of Colorado, the oil shale beds reach a thickness of 2100 ft (630 m) and contain about 500×10^6 tons of organic matter per square mile (700,000 tonne/hectare). In Utah and Wyoming, the shale beds are not as thick and are sometimes separated by lean or barren rock.

Minerals ubiquitous in the Green River Formation oil shales are dolomite, quartz, sodium and potassium feldspars, illite, and pyrite. In some locations, the Green River Formation also contains large amounts of sodium carbonate minerals, including trona ($Na_2CO_3 \cdot NaHCO_3 \cdot 2H_2O$), nahcolite ($NaHCO_3$), dawsonite [$NaAl(OH)_2CO_3$], and shortite ($Na_2CO_3 \cdot 2CaCO_3$). More than 60% of the soda ash supply in the United States was produced from Green River Formation trona in 1975. Nahcolite and dawsonite occur in some rich Colorado oil shales. Nahcolite can yield soda ash, and dawsonite can yield alumina; consequently, production of soda ash and alumina together with shale oil is being investigated. The mineral shortite, unique to the Green River Formation, has no known commercial value but occurs in huge amounts in Wyoming and Utah oil shales.

The entire Green River Formation is characterized by a remarkable lateral homogeneity, showing only very gradual geographic changes in its organic and mineral composition. Vertically, however, the formation is extremely variable, most notably

was founded in 1950 to investigate Estonian and other oil shales. Two additional Soviet laboratories devote part of their effort to other organic shales of the country. The Laramie Energy Research Center in Laramie, WY, began work on oil shale in 1944 under the U.S. Department of Interior's Bureau of Mines. The center's primary interests are the Green River Formation and other United States oil shales. In 1975 the center became part of the U.S. Energy Research and Development Administration.

United States oil-shale resources. Organic-rich sedimentary deposits underlie about 20% of the United States land area. They range in age from Cambrian to late Tertiary, and most have not been evaluated or have shown only limited shale-oil potential. The Cretaceous Niobrara Formation in Wyoming, Colorado, Nebraska, and South Dakota; the Tertiary Humboldt Formation in Nevada; and several Alaskan occurrences are examples of unevaluated, organic-rich deposits.

In the United States, the largest deposit in terms of area is the Devonian-Mississippian black shale composite, which extends from Texas to New York

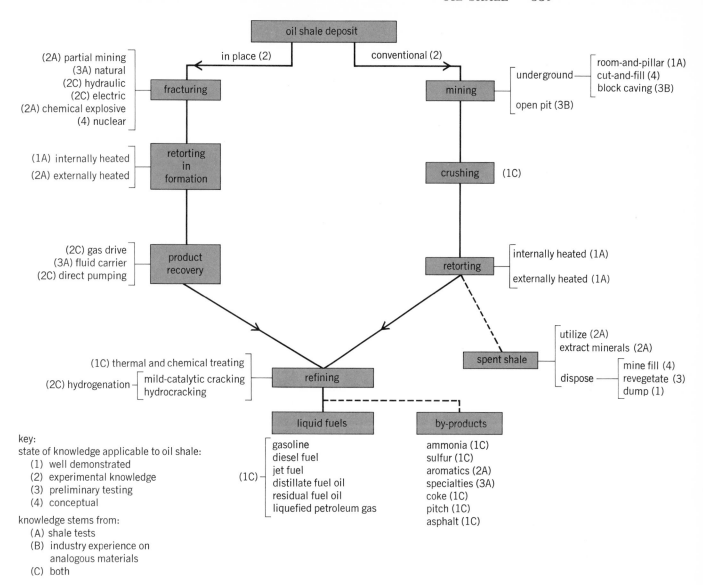

Fig. 2. Oil shale technology routes and states of knowledge.

in its organic content. Although the oil shale may be vertically continuous, its oil yields range from a few gallons per ton to nearly 100 gpt (417 liters/tonne).

The Mahogany Zone, a particularly organic-rich bed in the Green River Formation of Colorado and Utah, has been investigated intensively as a source of fossil energy. Organic matter in the Mahogany Zone has the following average elemental composition, in wt %: C, 80.5; H, 10.3; N, 2.4; S, 1.0; and O, 5.8, with a gross heating value of 9500–9600 cal/g. The high hydrogen content of this organic matter correlates with the large fraction of oil recovered when the shale is heated (Table 1). Hydrogen-rich organic matter is characteristic of all the Green River Formation oil shales.

Shale oil. Shale oil is produced from the organic matter in oil shale when the shale rock is heated in the absence of oxygen (destructive distillation). This heating process is called retorting, and the equipment that is used to do the heating is called a retort. The rate at which the oil is produced depends upon the temperature at which the shale is retorted. For example, if Mahogany Zone oil shale is heated rapidly to 500°C and held at that temperature, it will take about 10 min for the reaction to reach completion. However, if the shale is heated rapidly to only 340°C, it will take more than 100 hr; when the shale is heated rapidly to a temperature of 660°C, the reaction takes only seconds. Most references report retorting temperatures as being about 500°C.

Retorting temperature affects the nature of the shale oil produced. Low retorting temperatures produce oils in which the paraffin content is greater than the olefin contents; intermediate temperatures produce oils that are more olefinic; and high temperatures produce oils that are nearly completely aromatic, with little olefin or saturate content.

In those retorting systems not capable of rapidly heating the oil shale to a constant temperature, the nature of the oil is determined by the rate at which the oil shale is heated. Thus at heating rates of about 1°C/min, the reaction is essentially completed by the time the temperature reaches

425°C, and the oil is principally paraffinic. At heating rates of about 100°C/min, the reaction is not complete until the temperature reaches more than 600°C, and the oil is more olefinic. As heating rates are increased above 100°C/min, both the paraffin and olefin contents of the oil decrease, and its aromatic content increases.

In general, shale oils can be refined to marketable products in modern petroleum refineries. There is no really typical shale oil produced from Green River oil shale, but the oils do have many properties in common. They usually have high pour points, 20–32°C; high nitrogen contents, 1.6–2.2 wt %; and moderate sulfur contents, about 0.5 wt %. High pour points make necessary some processing before the oils are amenable to pipeline transportation. The high nitrogen contents make hydrogenation necessary to reduce the nitrogen contents so that the oils can be processed into fuels. Hydrogenation also reduces the sulfur content to an acceptable level.

United States technology. The two general approaches to recovering shale oil from Green River Formation oil shales are (1) mining, crushing, and aboveground retorting, called conventional processing; and (2) in-place processing. The basic problems facing conventional processing are handling and heating huge amounts of low-grade ore and disposing of huge volumes of spent shale, the residue remaining after oil production. The in-place approach largely avoids the problems of handling and disposal, with its attendant environmental questions, but faces a different basic problem—the impermeability of the oil shale beds. Progress toward solving the basic problems of both approaches has been made. Figure 2 summarizes the state of knowledge of each step required with the two development approaches.

With the conventional approach, oil shale mining by the room-and-pillar technique developed by the U.S. Bureau of Mines appears capable of producing the huge amounts of ore necessary to operate a large production plant. The procedure has also been tested by industry in Mahogany Zone shales. Outputs on the order of 2500 tons of oil shale per manshift have been reported from the highly mechanized operation. Crushing technology is well demonstrated. Retorting must be done con-

tinuously in order to reach the throughput necessary for economic production of shale oil. Two general systems for heating a continuous stream of oil shale are outlined in Fig. 3. In the internally heated system the oil shale furnishes its own heat because part of its organic matter is burned inside the retort. The U.S. Bureau of Mines gas combustion retort and one form of the Paraho retort of Development Engineering, Inc., are examples of this system. In the externally heated systems, heat generated outside the retort is carried inside. The Oil Shale Corporation's TOSCO II retort, in which preheated ceramic balls heat the oil shale stream, is an example of an externally heated retort. Several retorting systems, including both internally and externally heated designs, have been tested on pilot or semiworks scales, but a full-scale retort capable of operating in a commercial oil shale development had not been built by 1975.

Spent shale disposal has been studied intensively. More than 80% of the mined oil shale remains as residue after oil production. An oil shale plant producing 50,000 bbl of oil (7250 tonne) per day from Mahogany Zone shale might mine 75,000 tons (68,000 tonne) of rock and dispose of 60,000 tons (54,000 tonne) of spent shale daily. In the vast and largely unpopulated areas of the Green River Formation, dumping these volumes of spent shale is not as large an environmental problem as it appears to be. The spent shale is virtually insoluble, and contoured dumping to control water flow will minimize leaching, already low in an arid region. Native vegetation will establish itself on spent shale dumps, and revegetation procedures can accelerate this process. Returning spent shale to the mine to furnish roof support may permit recovery of additional ore, but this approach has not been tested.

Research and development efforts toward in-place processing have concentrated on creating permeability in the impermeable oil shale (Fig. 2). In-place processing may be accomplished by two means: (1) a borehole technique in which oil shale is first fractured undergound and heat is applied, and (2) a process in which some rock is first removed by mining, then the remaining oil shale is fragmented into the voids created by mining, and finally heat is applied. These two methods are referred to as in-place (no mining) and modified in-place (some mining) processing. Several investigators have tested in-place methods, and Occidental Oil Shale Corporation has tested the vertical-burn type of modified in-place production. Both of the heating methods outlined for retorts (Fig. 3) have also been applied to in-place processing.

Less than one-third of the total Green River Formation oil shale resource is considered minable for conventional processing. In-place processing may make the remainder of the resource available to production. Conventional processing offers process control advantages, including ready adaptation of many existing industrial procedures, but it is capital-intensive, requiring huge investments before production begins. In-place processing is less easily controlled and evaluated but requires less capital outlay before production begins.

Prospects for United States oil shale development. By 1975, there was still no commercial production of shale oil. Oil shale is a major fossil fuel resource. The balance between energy supply and

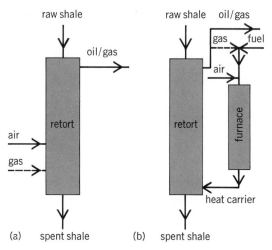

Fig. 3. Oil shale retorting systems. (a) Internally heated. (b) Externally heated.

energy demand controls the position of oil shale in the United States energy market. Shale oil can compete in the petroleum market when the price of oil exceeds the cost of producing shale oil. The Middle East oil embargo in the winter of 1973–1974, together with subsequent petroleum price increases by OPEC members, intensified interest in oil shale as a domestic supply of fossil energy. Consumption of domestic petroleum reserves and the ever-increasing domestic demand for petroleum products are additional forces acting toward bringing oil shale into the energy market.

Federal land ownership is one of several factors affecting oil shale development. More than 80% of the Green River Formation oil shale reserves, including the richest and thickest deposits, are federally owned. President Herbert Hoover "temporarily" withdrew oil shale from leasing pending evaluation of the resource. The National Petroleum Council believes that oil shale development would be unlikely without leasing of Federal lands. Four tracts for commercial shale-oil production were leased in 1974 by bonus bidding, and leasing of two more tracts restricted to in-place techniques is planned. The current Federal attitude is that no more oil shale land will be leased until these prototype tracts are in production. Of the privately held land, two-thirds is held by five major oil companies. Although the total resource is large enough to support any reasonable level of shale-oil production, its legal availability may be a major deterrent to development.

Environmental and socioeconomic impacts of oil shale operations also complicate oil shale development. It has been estimated that 7 years is required between the time a lease is offered and the time mature production is reached. Much of this time is spent preparing detailed environmental-impact statements, obtaining required Federal, state, and local licenses and permits, and preparing to meet housing and service requirements.

Water availability has been frequently cited as a problem in oil shale development. The Federal Energy Administration Project Independence oil shale report concludes that water is available for normal development, and that the water requirement of a sharply accelerated development could be met by building additional storage facilities.

Many forms of government support for developing synthetic fuel industries, including an oil shale industry, have been proposed to augment the energy supply available in the United States. Implementation of these programs is being evaluated. However, the continually increasing demand for petroleum is the largest spur toward development of oil shale.

[JOHN WARD SMITH; HOWARD B. JENSEN]

Bibliography: D. C. Duncan and V. E. Swanson, *U.S. Geological Survey Circular no. 523*, 1965; R. M. Garrels and C. L. Christ, *Solutions, Minerals, and Equilibria*, 1965; T. A. Hendrickson, *Synthetic Fuels Data Handbook*, 1975; D. K. Murray, *Energy Resources of the Piceance Creek Basin*, Rocky Mountain Association of Geology, 1974; *Oil Shale: Prospects and Constraints*, Federal Energy Administration Project Independence Blueprint, Govt. Print. Office stock no. 4118-00016, 1974; M. P. Rogers, *Bibliography of Oil Shale and Shale Oil*, O.S.R.D. 59, Laramie Energy Research Center, 1974.

Otto cycle

The basic thermodynamic cycle for the prevalent automotive type of internal combustion engine. The engine uses a volatile liquid fuel (gasoline) or a gaseous fuel to carry out the theoretic cycle illustrated in Fig. 1. The cycle consists of two isentropic (reversible adiabatic) phases interspersed between two constant-volume phases. The theoretic cycle should not be confused with the actual engine built for such service as automobiles, motor boats, aircraft, lawn mowers, and other small (generally <300± hp) self-contained power plants.

The thermodynamic working fluid in the cycle is subjected to isentropic compression, phase 1–2; constant-volume heat addition, phase 2–3; isentropic expansion, phase 3–4; and constant-volume heat rejection (cooling), phase 4–1. The ideal performance of this cycle, predicated on the use of a perfect gas, Eqs. (1), is summarized by Eqs. (2) and (3) for thermal efficiency and power output.

$$V_3/V_2 = V_4/V_1 \qquad T_3/T_2 = T_4/T_1$$

$$\frac{T_2}{T_1} = \frac{T_3}{T_4} = \left(\frac{V_1}{V_2}\right)^{k-1} = \left(\frac{V_4}{V_3}\right)^{k-1} = \left(\frac{P_2}{P_1}\right)^{\frac{k-1}{k}} \qquad (1)$$

$$\text{Thermal eff} = \frac{\text{net work of cycle}}{\text{heat added}}$$

$$= [1 - (T_1/T_2)] = \left[1 - \left(\frac{1}{r^{k-1}}\right)\right] \qquad (2)$$

Net work of cycle = heat added − heat rejected
$$= \text{heat added} \times \text{thermal eff}$$
$$= \text{heat added} [1 - (T_1/T_2)] \qquad (3)$$
$$= \text{heat added} [1 - (1/r^{k-1})]$$

In Eqs. (2) and (3) V is the volume in cubic feet; P is the pressure in pounds per square inch; T is the absolute temperature in degrees Rankine; k is the ratio of specific heats at constant pressure and constant volume, C_p/C_v; and r is the ratio of compression, V_1/V_2.

The most convenient application of Eq. (3) to the positive displacement type of reciprocating engine uses the mean effective pressure and the horsepower equation, Eq. (4), where hp is horsepower;

$$\text{hp} = \frac{\text{mep } Lan}{33,000} \qquad (4)$$

mep is mean effective pressure in pounds per square foot; L is stroke in feet; a is piston area in square inches; and n is the number of cycles completed per minute. The mep is derived from Eq. (3) by Eq. (5), where 778 is the mechanical equivalent

$$\text{mep} = \frac{\text{net work of cycle} \times 778}{144 \ (V_1 - V_2)} \qquad (5)$$

of heat in foot-pounds per Btu; 144 is the number of square inches in 1 ft²; and $(V_1 - V_2)$ is the volume swept out (displacement) by the piston per stroke in cubic feet.

Air standard. In the evaluation of theoretical and actual performance of internal combustion engines, it is customary to apply the above equations to the idealized conditions of the air-standard cycle. The working fluid is considered to be a perfect gas with such properties of air as volume at 14.7 psia and 492°R equal to 12.4 ft³/lb, and the ratio of specific heats k as equal to 1:4. Figure 2

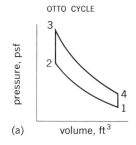

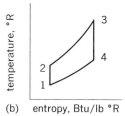

Fig. 1. (*a*) Pressure-volume and (*b*) temperature-entropy diagrams for Otto cycle: phase 1–2, isentropic compression; phase 2–3, constant-volume heat addition; phase 3–4, isentropic expansion; phase 4–1, constant volume heat rejection.

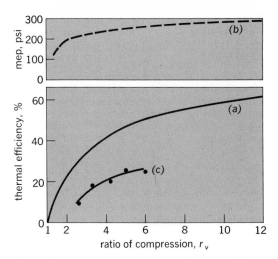

Fig. 2. Effect of compression ratio on thermal efficiency and mean effective pressure of Otto cycle. Curve *a* shows thermal efficiency, air-standard cycle; curve *b* mean effective pressure, air-standard cycle; and curve *c* thermal efficiency of an actual engine.

shows the thermal efficiency for this air-standard cycle as a function of the ratio of compression r, and the mep for a heat addition of 1000 Btu/lb of working gases. These curves demonstrate the intrinsic worth of high compression in this thermodynamic cycle.

The table gives a comparison of the important gas-power cycles on the ideal air-standard base for the case of compression ratio = 10 and 1000 Btu added per pound of working gases. The Otto, Brayton, and Carnot cycles show the same thermal efficiency of 60%. The mean effective pressures, however, show that the physical dimensions of the engines will be a minimum with the Otto cycle but hopelessly large with the Carnot cycle. The Brayton cycle ideal mep is only about one-fifth that of the Otto cycle, and it is accordingly at a distinct disadvantage when applied to a positive displacement mechanism. This disadvantage can be offset by use of a free-expansion, gas-turbine mechanism for the Brayton cycle. The Diesel cycle offers a lower thermal efficiency than the Otto cycle for the same conditions, for example, 42 versus 60%, and some sacrifice of mep, 200 versus 290 psi. As opposed to the Otto engine, the diesel can utilize a much higher compression ratio without preignition troubles and without excessive peak pressures in the cycle and mechanism. Efficiency and engine weight are thus nicely compromised in the Otto and diesel cycles. *See* GAS TURBINE.

Thermal efficiency, mean effective pressure, and peak pressure of air-standard gas-power cycles*

Cycle	Efficiency	Mep	Peak pressures
Otto	60	290	2100
Diesel	42	200	370
Brayton	60	61	370
Carnot	60	Impossibly small	Impossibly high

*Ratio of compression = 10; heat added = 1000 Btu per pound working gases.

Actual engine process. This reasoning demonstrates some of the valid conclusions that can be drawn from analyses utilizing the ideal air-standard cycles. Those ideals, however, require the modifications of reality for the best design of internal combustion engines. The actual processes of an internal combustion engine depart widely from the air-standard cycle. The actual Otto cycle uses a mixture of air and a complex chemical fuel which is either a volatile liquid or a gas. The rate of the combustion process and the intermediate steps through which it proceeds must be established. The combustion process shifts the analysis of the working gases from one set of chemicals, constituting the incoming explosive mixture, to a new set representing the burned products of combustion. Determination of temperatures and pressures at each point of the periodic sequence of phases (Fig. 1) requires information on such factors as variable specific heats, dissociation, chemical equilibrium, and heat transfer to and from the engine parts.

N. A. Otto (1832–1891) built a highly successful engine that used the sequence of engine operations proposed by Beau de Rochas in 1862. Today the Otto cycle is represented in many millions of engines utilizing either the four-stroke principle or the two-stroke principle. *See* INTERNAL COMBUSTION ENGINE.

The actual Otto engine performance is substantially poorer than the values determined by the theoretic air-standard cycle. An actual engine performance curve *c* is added in Fig. 2, in which the trends are similar and show improved efficiency with higher compression ratios. There is, however, a case of diminishing return if the compression ratio is carried too far. Evidence indicates that actual Otto engines offer peak efficiencies (25 ± %) at compression ratios of 15 ±. Above this ratio, efficiency falls. The most probable explanation is that the extreme pressures associated with high compression cause increasing amounts of dissociation of the combustion products. This dissociation, near the beginning of the expansion stroke, exerts a more deleterious effect on efficiency than the corresponding gain from increasing compression ratio. *See* BRAYTON CYCLE; CARNOT CYCLE; DIESEL CYCLE; THERMODYNAMIC CYCLE.

[THEODORE BAUMEISTER]

Bibliography: T. Baumeister (ed.), *Standard Handbook for Mechanical Engineers*, 7th ed., 1967; J. B. Jones and G. A. Hawkins, *Engineering Thermodynamics*, 1960; J. H. Keenan, *Thermodynamics*, 1941; L. C. Lichty, *Combustion Engine Processes*, 1967; E. F. Obert, *Internal Combustion Engines*, 3d ed., 1968.

Oxygen

A gaseous chemical element, O, atomic number 8, and atomic weight 15.9994. Oxygen is of great interest because it is the essential element both in the respiration process in most living cells and in combustion processes. It is the most abundant element in the Earth's crust. About one-fifth (by volume) of the air is oxygen.

Oxygen is separated from air by liquefaction and fractional distillation. The chief uses of oxygen in order of their importance are (1) smelting, refining, and fabrication of steel and other metals; (2) manufacture of chemical products by controlled oxida-

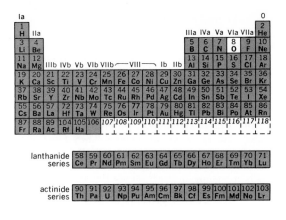

tion; (3) rocket propulsion; (4) biological life support and medicine; and (5) mining, production, and fabrication of stone and glass products.

Uncombined gaseous oxygen usually exists in the form of diatomic molecules, O_2, but oxygen also exists in a unique triatomic form, O_3, called ozone.

In 1774, Joseph Priestley, an English clergyman who later immigrated to the United States and settled in Northumberland, Pa., observed that mercuric oxide, on heating, yielded a gas that vigorously supported the combustion of a candle. Priestley found that the gas would support respiration and called the gas dephlogisticated air. The name oxygen, meaning acid-former, was given the gas by a group of French chemists in 1787 in recognition of the ability of some oxides, such as the oxides of sulfur, to form acids.

Use in industry. All fuel combustion processes utilize oxygen. In most circumstances the 21% concentration in air is sufficient. However, in an increasing number of applications the use of higher concentrations of oxygen improves the process sufficiently to justify the expense of separating oxygen from air.

Metallurgical uses. Oxygen is a component in the metallurgical processes of smelting, refining, welding, cutting, and surface conditioning.

Smelting. Smelting of ore in the blast furnace involves the combustion of about 1 ton of oxygen for each ton of metal produced. When air is used, $3\frac{1}{2}$ tons of nitrogen accompany each ton of oxygen and must be compressed, heated, and blown into the furnace. A large amount of heat is lost with the exhaust gases, which also carry powdered ore and coke away as dust and limit the capacity of the furnace. By removing some or all of the nitrogen, the furnace capacity can be increased, less expensive fuels can be used in place of some of the coke, and fuels can be used more efficiently.

Metal refining. In refining copper and in making steel from pig iron various impurities such as carbon, sulfur, and phosphorus must be removed from the metal by oxidation. If air is blown through the molten metal, as in the Bessemer converter, nitrogen is picked up, limiting the product quality. Nitrogen also carries away a great deal of the heat produced by the oxidation process. Better-quality steel and copper can be produced by injecting pure oxygen into the molten metal until the impurities are completely removed. Oxygen injection can be utilized in the open hearth or electric furnaces. However, new steelmaking equipment has been

developed which depends entirely on high-purity oxygen. All the heat for the furnace operation is supplied by oxidation of carbon and other impurities. The technique is called the basic oxygen process. The most common form is known as the L-D process, named after the Austrian cities of Linz and Donawitz, where the procedure was first used in 1951.

In 1967 more than 6,000,000 tons of oxygen were used in producing 127,000,000 tons of steel.

Welding, cutting, and surface conditioning. The high-temperature flame of the oxyacetylene torch can be used in welding steel, although most welding is now done by the electric arc process.

In cutting, the point of the steel at which the cutting is to start is first heated by an oxygen-acetylene flame. A powerful jet of oxygen is then turned on. The oxygen burns some of the iron in the steel to iron oxide, and the heat of this combustion melts more iron; the molten iron is blown out of the kerf by the force of the jet. By feeding powdered iron into the oxygen stream this cutting process can be extended to alloys, such as stainless steel, which are not readily cut by oxygen alone and to completely noncombustible materials such as concrete.

Steel ingots normally have oxide inclusions and other defects at the outer surface. After preliminary rolling, the steel in slab or billet form has the surface skin removed to eliminate these defects. This can be most easily accomplished by scarfing. Streams of oxygen from many nozzles are played on all sides of the billet at once. The oxygen burns off the surface defects and some of the steel in a spectacular shower of sparks. The billet is then ready for further rolling. Oxygen scarfing (skinning) is now standard practice in most steel mills.

Chemical synthesis. Several syntheses in the chemical industry involve oxygen. These processes are outlined.

Partial oxidation of hydrocarbons. When natural gas or fuel oil is burned, the heat of combustion first cracks the hydrocarbon molecules into fragments. These fragments usually encounter oxygen molecules within a few hundredths of a second and are oxidized to water and carbon dioxide. However, if the supply of oxygen is carefully controlled and the passage of material through the combustion zone is very rapid, it is possible to freeze the reaction at various stages of completion.

In this manner natural gas (mostly methane, CH_4) can be converted to acetylene (C_2H_2), ethylene (C_2H_4), or propylene (C_3H_6). Ethylene (C_2H_4), in turn, can be partially oxidized to ethylene oxide (CH_2CH_2O).

Syngas production. Reaction of carbon or hydrocarbons with oxygen and steam yields a mixture of carbon monoxide (CO) and hydrogen (H_2), that is, syngas. By use of suitable catalysts, syngas can be recombined to form various organic compounds such as methanol (CH_3OH), octane (C_8H_{16}), and many others. In the presence of other catalysts, carbon monoxide can combine with steam to form more hydrogen and carbon dioxide. After removal of the carbon dioxide, the hydrogen can be used for chemical reactions, such as the manufacture of ammonia (NH_3), hydrogenation of fats, and hydrocracking of petroleum. *See* COAL LIQUEFACTION; CRACKING.

Liquid fuel rockets. In rocket engines liquid oxygen is used as an oxidizer either with kerosine or liquid hydrogen fuels. While fluorine could theoretically provide somewhat improved performance in terms of specific impulse, oxygen is very nearly as good, and is also much cheaper and is easier to handle.

Solid-fueled rockets, based on hydrocarbon polymers that contain sufficient oxidizer to effect self-combustion, dominate the short-range military uses. Liquid-fueled rockets are expected to remain dominant in space work until the full development of nuclear propulsion. The Saturn-Apollo launch vehicle has a fully loaded weight of about 3000 tons of which more than 2000 tons are liquid oxygen. Most of the liquid oxygen consumed by the aerospace industry has been used in the development and proof testing of rocket engines mounted in static test stands. The usage of oxygen in this testing has been in excess of 1000 tons per day. *See* PROPELLANT.

Occurrence. About 49.5% by weight of the Earth's crust, including the oceans and atmosphere, is oxygen. Most of this oxygen is combined in the form of silicates, oxides, and water. Water is composed of 88.81% oxygen by weight.

Oxygen also exists outside the atmosphere of the Earth, but since more than 98% of the matter in the visible universe (stars, nebulae, and interstellar space) is composed of hydrogen and helium, the cosmic concentration of oxygen is relatively low.

Dry air contains 20.946% oxygen by volume, and this concentration has been found to be the same at any level between the surface of the Earth and a height of 40 mi. The atoms in atmospheric oxygen consist of three isotopes in the following atomic proportions: 99.759%, oxygen-16; 0.037%, oxygen-17; and 0.204%, oxygen-18. The molecules of oxygen in the air, each of which has two atoms, consist of the statistically expected proportion of the possible combinations of these isotopes, the most abundant molecules being $O^{16}O^{16}$, $O^{16}O^{18}$, and $O^{16}O^{17}$. The isotopic composition of the oxygen in water is slightly different from that in air and varies slightly in samples from different bodies of water (lakes, oceans, and seas).

Even though large quantities of oxygen from the air are continuously being used in respiration, combustion, and other oxidation processes, the concentration of oxygen in the atmosphere remains very nearly constant, chiefly because oxygen is liberated in the process of photosynthesis. In this process carbohydrates are produced by green plants from carbon dioxide and water. The primary source of the free oxygen in the atmosphere is believed by some authorities to have been the decomposition of water vapor by ultraviolet radiation in the upper atmosphere. Almost all the hydrogen formed in this way escaped from the Earth's gravitational field, but the oxygen molecules were too heavy to escape. They remained, therefore, in the atmosphere. This photochemical decomposition of water vapor to produce oxygen gas is still going on.

The following radioactive isotopes of oxygen are known: O^{14}, O^{15}, and O^{19}. These isotopes may be formed in particle accelerators, such as the cyclotron, or by neutron bombardment of the appropriate atomic species; for example, O^{19} is formed when the nucleus of an atom of stable O^{18} absorbs a neutron. All three of the radioactive istopes of oxygen are very short-lived, the one with the longest half-life, that of about 120 sec, being O^{15}.

Physical properties. Under ordinary conditions oxygen is a colorless, odorless, and tasteless gas. It condenses to a pale blue liquid, in contrast to nitrogen, which is quite colorless in the liquid state. Oxygen is one of a small group of slightly paramagnetic gases, and it is the most paramagnetic of the group. Liquid oxygen is also slightly paramagnetic. Some data on oxygen and some properties of its ordinary form, O_2, are listed in the table.

Before the mass spectrometer was invented and when nothing was known about isotopes, the average weight of the oxygen atoms in oxygen obtained from water was selected by chemists as the standard of weight for the atoms of all elements. This weight was assigned the value 16.0000. It is now known that isotopes exist and that the isotopic composition of many elements is subject to considerable variation. Consequently, there is no longer a good theoretical basis for the system of chemical atomic weights, based on the mixture of oxygen isotopes as they happen to occur in the Earth's atmosphere. Chemists concluded that the single isotope $C^{12} = 12.0000$ should be taken as the standard.

Chemical properties. Practically all chemical elements except the inert gases form compounds with oxygen. Most elements form oxides when heated in an atmosphere containing oxygen gas. Many elements form more than one oxide; for example, sulfur forms sulfur dioxide (SO_2) and sulfur trioxide (SO_3). Among the most abundant binary compounds of oxygen are water, H_2O, and silica, SiO_2, the latter being the chief ingredient of sand. Among compounds containing more than two elements, the most abundant are the silicates, which constitute most of the rocks and soil. Other widely occurring compounds are calcium carbonate (limestone and marble), calcium sulfate (gypsum), aluminum oxide (bauxite), and the various oxides of iron which are mined as a source of iron. Several other metals are also mined in the form of their oxides. Hydrogen peroxide, H_2O_2, is an interesting compound used extensively for bleaching.

Oxygen production. Oxygen is produced on a large scale by the liquefaction and fractional distillation of air. A little oxygen is also made by the electrolysis of water, but oxygen produced in this

Properties of oxygen

Property	Value
Atomic number	8
Atomic weight	15.9994
Triple point (solid, liquid, and gas in equilibrium)	$-218.80°C$ ($54.35°K$)
Boiling point at 1 atm pressure	$-182.97°C$ ($90.18°K$)
Gas density at 0°C and 1 atm pressure, g/liter	1.4290
Liquid density at the normal boiling point, g/ml	1.142
Solubility in water at 20°C, ml oxygen (STP) per 1000 g water at 1 atm partial pressure of oxygen	30

way is more expensive than oxygen distilled from air. Electrolysis of water is not used, therefore, unless there is some special reason, such as a need for the hydrogen that is also produced during electrolysis.

The traditional methods of preparing oxygen in school chemistry courses are (1) heating potassium chlorate with or without addition of a little manganese dioxide or other catalyst; (2) heating mercuric oxide (Priestley's original method); and (3) electrolysis of water to which an electrolyte has been added. When oxygen is needed in the laboratory, however, it is usually obtained from a cylinder of compressed oxygen.

Distribution. Oxygen is commonly distributed in three ways: (1) Most oxygen is piped directly to users; (2) about 10% is liquefied for transportation and storage in insulated tanks; and (3) about 1% is compressed to high pressure (more than 2000 psi) for transport in steel cylinders or tube bundles.

Oxygen pipelines are usually short since the raw material for air separation is readily available. In industrial areas a single large plant may supply a dozen consumers through a network of pipelines. One such system operating in the heavy industrial area along the south shore of Lake Michigan supplies more than 5000 tons of oxygen daily.

For smaller or intermittent uses or for rocket engines oxygen is produced and distributed as a liquid. In liquid form oxygen is about one-third heavier than water. So long as it is kept at low temperature, the liquid can be stored, transported, pumped, or handled much as any other liquid. To keep heat away from this very cold liquid, the storage and transport tanks use the best possible insulating techniques. Two concentric tanks are constructed. The space between the tanks is filled with a powdered material of low thermal conductivity which is also opaque to radiant heat. The powder-filled space is then evacuated. The combination of vacuum and insulating powder minimizes heat transfer by convection, conduction, and radiation. The result is a container in which liquid oxygen can be transported hundreds of miles with little or no loss. Large liquid oxygen tanks have been mounted on trucks, trailers, and railroad cars. Smaller tanks can be wheeled around by hand. Special lightweight liquid oxygen tanks have been manufactured to permit sick people to carry several hours' supply in a pack about the size of a binocular case.

Generally speaking, liquid transport is preferred to high-pressure gas containers because much more product can be carried per pound of total weight. However, for some applications high gas pressures are desired. For others the use is so intermittent that liquid supply would involve excessive losses during idle periods. For these applications oxygen may be transported at high pressure in steel cylinders. Ordinary cylinders are about 9 in. in diameter, 4 ft in height, and about 150 lb in weight when filled with 240 ft (20 lb) of oxygen. Individual cylinders may be clustered and longer tubes may be mounted on trailers to achieve greater capacity.

Detection and quantitative analysis. The traditional laboratory test for oxygen gas is that it will cause a glowing wooden splinter to burst into flame; this test does not distinguish between oxygen and nitrous oxide.

In laboratory gas-analysis apparatus oxygen is usually determined by absorption in an alkaline solution of pyrogallol or in an ammoniacal solution of copper (I) chloride. The concentration of oxygen in oxygen tents and gas streams is readily determined with oxygen meters that measure the content of the oxygen by its paramagnetism. Oxygen in a mixture of gases may be determined in a gas chromatograph. There are a number of colorimetric tests for traces of oxygen.

[ARTHUR W. FRANCIS]

Bibliography: J. E. Browning, New processes focus interest on oxygen, *Chem. Eng.*, 75(5):88–92, Feb. 26, 1968; J. W. Giachino et al., *Welding Skills and Practices*, 2d ed., American Technical Society, 1965; M. Sittig, *Combining Hydrocarbons and Oxygen for Profit*, 1962; M. Sittig, Oxygen chemical raw material, *Chem. Eng. News*, 39(49): 92–102, Nov. 27, 1961.

Panel heating and cooling

A system in which the heat-emitting and heat-absorbing means is the surface of the ceiling, floor, or wall panels of the space which is to be environmentally conditioned. The heating or cooling medium may be air, water, or other fluid circulated in air spaces, conduits, or pipes within or attached to the panel structure. For heating only, electric current may flow through resistors in or on the panels. *See* HEATING, ELECTRIC.

Warm or cold water is circulated in pipes embedded in concrete floors or ceilings or plaster ceilings or attached to metal ceiling panels. The coefficient of linear expansion of concrete is 0.000095; for steel it is 0.000081, or 15% less than for concrete. For copper it is 0.000112, or 20% more than for concrete, and for aluminum it is 0.000154, or 60% more than for concrete. Since the warmest or coolest water is carried on the inside of the pipes and the heat is transmitted to the concrete, only steel pipe should be used for panel heating and cooling systems, except when metal panels are used. Cracks are bound to develop in the concrete or plaster, breaking the bonds between the pipes and the concrete or plaster. The pipes move freely, causing scraping noises. An insulating layer of air is formed between the concrete or plaster and this markedly reduces the coefficient of conductivity between the liquid heating medium and the active radiant surfaces.

Heat transfer. Heat energy is transmitted from a warmer to a cooler mass by conduction, convection, and radiation. Radiant heat rays are emitted from all bodies at temperatures above absolute zero. These rays pass through air without appreciably warming it, but are absorbed by liquid or solid masses and increase their sensible temperature and heat content. *See* HEAT TRANSFER.

The output from heating surfaces comprises both radiation and convection components in varying proportions. In panel heating systems, especially the ceiling type, the radiation component predominates. Heat interchange follows the Stefan-Boltzmann laws of radiation; that is, heat transfer by radiation between surfaces visible to each other varies as the difference between the fourth power of the absolute temperatures of the two surfaces, and is transferred from the surface with the higher temperature to the surface with the lower temperature.

The skin surface temperature of the human body under normal conditions varies from 87 to 95°F and is modified by clothing and rate of metabolism. The presence of radiating surfaces above these temperatures heats the body, whereas those below produce a cooling effect. *See* RADIANT HEATING.

Cooling. When a panel system is used for cooling, the dew-point temperature of the ambient air must remain below the surface temperature of the heat-absorbing panels to avoid condensation of moisture on the panels. In regions where the maximum dew point temperature does not exceed 60°F, or possibly 65°F, as in the Pacific Northwest and the semiarid areas between the Cascade and Rocky mountains, ordinary city water provides radiant comfort cooling. Where higher dew points prevail, it is necessary to dehumidify the ambient air. Panel cooling effectively prevents the disagreeable feeling of cold air blown against the body and minimizes the occurrence of summer colds.

Fuel consumption records show that panel heating systems save 30–50% of the fuel costs of ordinary heating systems. Lower ambient air temperatures produce comfort and air temperatures within the room are practically uniform and not considerably higher at the ceiling, as in radiator- and convector-heated interiors. *See* HEATING, COMFORT; HOT-WATER HEATING SYSTEMS.

[ERWIN L. WEBER/RICHARD KORAL]

Bibliography: American Society of Heating, Refrigerating, and Air Conditioning Engineers, *Guide*, 1962, 1964, 1966, 1967; American Society of Heating and Ventilating Engineers, *Heating, Ventilating, Air Conditioning Guide*, vol. 37, 1959; Chase Brass and Copper Co., *Chase Radiant Heating Manual*, 1945; F. E. Giesecke, *Hot Water Heating and Radiant Heating and Radiant Cooling*, 1947; Keeney Publishing Co., *Panel Heating and Cooling*.

Peat

A dark-brown or black residuum produced by the partial decomposition and disintegration of mosses, sedges, trees, and other plants that grow in marshes and other wet places. Forest-type peat, when buried and subjected to geological influences of pressure and heat, is the natural forerunner of most coal.

Peat may accumulate in depressions such as the coastal and tidal swamps in the Atlantic and Gulf Coast states, in abandoned oxbow lakes where sediments transported from a distance are deposited, and in depressions of glacial origin. Moor peat is formed in relatively elevated, poorly drained moss-covered areas, as in parts of Northern Europe. *See* COAL.

In the United States, where the principal use of peat is for soil improvement, the estimated reserve on an air-dried basis is 13,827,000 short tons. In Ireland and Sweden peat is used for domestic and even industrial fuel. In Germany peat is the source of low-grade montan wax.   [GILBERT H. CADY]

Petroleum

A naturally occurring, oily, flammable liquid composed principally of hydrocarbons, and occasionally found in springs or pools but usually obtained from beneath the Earth's surface by drilling wells. Formerly called rock oil, unrefined petroleum is now usually termed crude oil.

Petroleum is separated by distillation into fractions designed as (1) straight-run gasoline, boiling at up to about 200°C; (2) middle distillate, boiling at about 185–345°C, from which are obtained kerosine, heating oils, and diesel, jet, rocket, and gas turbine fuels; (3) wide-cut gas oil, which boils at about 345–540°C, and from which are obtained waxes, lubricating oils, and feed stock for catalytic cracking to gasoline; and (4) residual oil, which may be asphaltic.

The physical properties and chemical composition of petroleum vary markedly, depending on its source. As it comes from the earth, it ranges from an occasional nearly colorless liquid consisting chiefly of gasoline to a heavy black tarry material high in asphalt content. Although most crudes are black, many are amber, red, or brown by transmitted light and show a greenish fluorescence by reflected light. Their specific gravity is usually in the range about 0.82–0.95.

Hydrocarbons constitute 50–98% of petroleum, and the remainder is composed chiefly of organic compounds containing oxygen, nitrogen, or sulfur, and trace amounts of organometallic compounds. Pennsylvania crude oils contain 97–98% hydrocarbons; some California oils contain only 50%.

Hydrocarbon types. The hydrocarbon types found in petroleum are paraffins (alkanes), cycloparaffins (naphthenes or cycloalkanes), and aromatics. Olefins (alkenes) and other unsaturated hydrocarbons are usually absent.

Paraffins. The paraffins range from methane (found together with ethane, propane, and the butanes in the natural gas which accompanies petroleum) to *n*-hexacontane ($C_{60}H_{122}$, a microcrystalline wax) and compounds of even higher molecular weight. Both straight-chain and branched-chain paraffins are present. The former usually predominate, particularly in the higher-boiling fractions. Commercial paraffin wax ordinarily consists chiefly of straight-chain paraffins of from about 22–30 carbon atoms isolated from the wide-cut, gas-oil fraction.

Cycloparaffins. The cycloparaffins are chiefly those having five or six carbon atoms in the ring. These include not only the monocyclic compounds (cyclopentane, cyclohexane, alkylcyclopentanes, and alkylcyclohexanes) but also polycyclic hydrocarbons, such as the bicycloparaffins (*trans*-decahydronaphthalene and *cis*-bicyclo[3·3·0]octane) as well as tri- and higher cycloparaffins.

Aromatics. Aromatic hydrocarbons are usually present in smaller amounts than the paraffins and cycloparaffins. The aromatic compounds boiling in the gasoline range are chiefly alkylbenzenes (such as toluene, the xylenes, and *p*-cymene). Higher-boiling fractions contain polynuclear aromatics of both fused-ring (alkylnaphthalenes) and linked-ring (biphenyl) types. The fused-ring polycyclics usually predominate. Mono- and polynuclear aromatic rings fused to one or more cycloparaffin rings (as in indan and 1,2,3,4-tetrahydronaphthalene) are also present.

Petroleum fractions. The number of carbon atoms in hydrocarbons of a given boiling range depends on the hydrocarbon type. In general, gasoline will include hydrocarbons having 4–12 carbon atoms; kerosine, 10–14; middle distillate, 12–

20; and wide-cut gas oil, 20–36.

A study of the gasoline fractions from representative petroleums from seven different areas in the United States has permitted some interesting conclusions. Five main classes of compounds are present in the gasoline fraction: straight-chain paraffins, branched-chain paraffins, alkylcyclopentanes, alkylcyclohexanes, and alkylbenzenes. Although the relative amounts of the classes vary from petroleum to petroleum, the relative amounts of the individual compounds within a given class are of the same magnitude for the different petroleums. Hence, the gasoline fraction of different crudes is characterized by specifying the relative amounts of the five main classes of compounds in the fraction.

Petroleums may be classified in accordance with their composition. Thus Pennsylvania and Michigan crude oils are largely paraffinic and contain little or no asphalt. Some Texas and California oils are rich in naphthenes, whereas others are unusually high in aromatics; most contain much asphalt.

Asphalt is a dark-brown to black solid or semisolid consisting of carbon, hydrogen, oxygen, sulfur, and sometimes nitrogen. It is made up of three components: (1) asphaltene, a hard, friable, infusible powder; (2) resin, a semisolid to solid ductile and adhesive material; and (3) oil, which is structurally similar to the lubricating oil fraction from which it is derived. The asphalts are almost completely soluble in carbon disulfide, carbon tetrachloride, and pyridine, but are only partly soluble in low-boiling paraffins, which dissolve the oils and resins and precipitate the asphaltenes.

Components other than hydrocarbons. The total oxygen content of crude oils is generally low but may be as high as 2%. The oxygen-containing compounds consist principally of phenols and carboxylic acids. The phenols comprise cresols and higher-boiling alkylphenols. The acids include straight-chain and branched-chain acids, such as hexanoic acid and 3-methylpentanoic acid, and cyclopentane and cyclohexane derivatives, such as cyclopentaneacetic acid and cyclohexanecarboxylic acid. There is also some indication of the presence of acids containing aromatic rings (mono- and dinuclear). Hence the name naphthenic acids, which has been applied to the carboxylic acids derived from petroleum, is a misnomer; petroleum acids is a preferable term.

The nitrogen content of crude oils ranges from less than 0.05 to about 0.8%. Up to about one-half is in the form of basic pyridine and quinoline compounds, the latter predominating. The nonbasic nitrogen compounds or complexes include pyrroles, indoles, and carbazoles.

The sulfur content varies over a wide range, from traces to more than 5%. Pennsylvania and midcontinent crudes usually contain less than 0.25% by weight of sulfur, whereas some California and Texas stocks contain over 2%. Part of the sulfur may be in the form of elemental sulfur and hydrogen sulfide. Most is present as mercaptans (thiols), aliphatic sulfides, and cyclic sulfides. The mercaptans and sulfides exist as both straight-chain compounds, such as n-propyl mercaptan and methyl ethyl sulfide, and branched-chain compounds, such as tert-butyl mercaptan and methyl isopropyl sulfide. The cyclic sulfides consist of five- and six-membered ring compounds, such as thiacyclopentanes and thiacyclohexanes.

A number of metals have been identified in the ash (about 0.01–0.05% by weight) obtained by burning crude petroleum. These include sodium, magnesium, calcium, strontium, copper, silver, gold, aluminum, tin, lead, vanadium, chromium, manganese, iron, cobalt, nickel, platinum, and uranium. Boron, silicon, and phosphorus have also been detected. It is quite probable that the sodium and strontium are present chiefly in the form of aqueous solutions of salts that are finely dispersed in the oil. Most of the other metals are present as oil-soluble salts or organometallic compounds. For example, nickel and vanadium, which are the most abundant of these, occurring in 5–40 ppm in many crude oils of the United States, are probably present as porphyrin complexes. See PETROLEUM ENGINEERING; PETROLEUM GEOLOGY; PETROLEUM PROCESSING; PETROLEUM PRODUCTS.

[LOUIS SCHMERLING]

Bibliography: B. T. Brooks et al. (eds.), *The Chemistry of Petroleum Hydrocarbons*, vol. 1, 1954; A. E. Dunstan et al. (eds.), *The Science of Petroleum*, vol. 2, 1938, and vol. 5, pt. 1, 1950; R. F. Goldstein and A. L. Waddams, *The Petroleum Chemicals Industry*, 3d ed., 1967; W. A. Gruse and D. R. Stevens, *Chemical Technology of Petroleum*, 3d ed., 1960; H. L. Lochte and E. R. Littmann, *The Petroleum Acids and Bases*, 1955; F. D. Rossini, B. J. Mair, and A. J. Streiff, *Hydrocarbons from Petroleum*, 1953; H. Steiner (ed.), *Introduction to Petroleum Chemicals*, 1961.

Petroleum engineering

The application of almost all types of engineering to the drilling for and production of oil, gas, and liquefiable hydrocarbons. Petroleum operations depend on a region's geology, economic conditions, and government policies as well as on related technology. However, this article discusses only the engineering aspects. Technologies in petroleum engineering include mechanical engineering of drilling equipment and surface installations, civil engineering of structures for offshore drilling, and chemical engineering of fluid flow in petroleum reservoirs. For a discussion of the refining of petroleum into fuels, lubricants, and chemicals see PETROLEUM PROCESSING. *See also the feature article* OUTLOOK FOR FUEL RESERVES.

Drilling. Through advanced drilling engineering, depths of 20,000 ft are becoming common, allowing exploration in new geologic basins. Much research, design, and development have been devoted to drilling wells in open deep water. New structural engineering and oceanographic technologies have been developed. It is becoming possible to do exploratory drilling in 400 ft of water, and techniques are being developed to produce from this depth.

Drilling deeper than 3 mi into the earth is still very costly. New technology is needed. The lack of efficient transmission of drilling energy to the bottom of the hole is a major problem because fluid and mechanical friction take up much of the energy. One new method being studied is electrical transmission of energy. Between 1955 and 1970 improved drilling technology reduced by more than 33% the cost of drilling wells to about 3 mi in depth.

The type of drilling fluid greatly affects the drilling rate. The fluid, synthetic mud, is pumped down inside the drill pipe and back up the formation face. The fluid prevents loss of mud to the formation by forming a filter cake, and prevents a blowout by keeping formation fluids from forcing their way into the hole. Drilling fluids have been highly developed. A high-pressure differential between the mud and the cutting face in the bottom of the hole, caused by instantaneous filter cake buildup, greatly reduces drilling speed. Drilling muds are being developed to reduce this hold-down pressure, termed chip hold-down.

Significant new methods have been developed to complete the well and put it on production. The type of completion is determined by the individual reservoir and fluid properties. Many formations are fractured hydraulically to improve flow into the well, or they are perforated by shaped charge or jet perforation. Some wells are completed with as much as 16,000 psi pressure at the surface well head. See OIL AND GAS WELL COMPLETION; OIL AND GAS WELL DRILLING; ROTARY TOOL DRILL.

Reservoir exploitation. Reliable estimates of recovery depend on specific properties of both the hydrocarbon and the reservoir rock. Three levels of phenomena are involved: (1) the reservoir mechanism as a whole; (2) the properties of unit cubes of the reservoir media; and (3) the physical chemistry of the individual pore.

Once the hydrocarbon deposit has been located by an exploratory well, successive wells are placed to obtain the maximum geologic information commensurate with predicted production efficiency. The interpretation of the geologic history and the prediction of the present distribution of petroleum in the internal rock fabric is termed production geology. Subsurface interpretation is evolved step by step as drilling progresses but it is rarely free from speculation, particularly for complex geologic structures. Speculation has been reduced by a development enabling prediction of the original environment of deposition of sediments making up the petroleum structure. Models are used which combine previous geologic studies of surface outcrops with well-documented examples of actual fully developed petroleum structures. See PETROLEUM GEOLOGY.

The quality of a reservoir is also affected by changes in porosity through geologic time. Predictions of porosity trends can often be made by examining the geologic history regarding compaction, leaching, and cementing. Dolomite is the most common reservoir rock in carbonate formations. An important discovery concerning the origin of large masses of dolomite has greatly improved the predictability of porosity trends in a reservoir.

All basic physical, chemical, and geologic properties of the reservoir must be obtained from the bore hole. Many properties are determined by electrical, acoustical, and radioactive measurements. The complexity of obtaining and analyzing these data has led to petrophysical engineering. Rock fluid relationships with electrical, radioactive, and other physicochemical properties have been discovered. Ingenious devices lowered into the hole to obtain these data have been developed, making it possible to determine geologic and fluid conditions of the formations penetrated by the drill. See WELL LOGGING (MINERAL).

Many fundamental features of reservoir performance as a whole have been established from research, theoretical analysis, and study of actual reservoirs through reservoir engineering. Predictions of reserves and producing life are fundamental to the proper management of a petroleum deposit, the financing of operations, and further exploration for new deposits.

Reservoir rocks and hydrocarbon fluids have been characterized. Material and volumetric balance equations have been developed for various reservoir conditions. A mathematical theory of transient flow has been developed as well as models for determining sweep efficiency of natural water and gas or of water injected to force petroleum to the producing wells. See PETROLEUM RESERVOIR ENGINEERING; PETROLEUM SECONDARY RECOVERY.

Models. Sophisticated mathematical and physical models have been devised to aid in predicting reservoir performance and in determining optimum operating policies. Models can account for up to three-phase flow in two or three space dimensions. The effects of capillary and gravity forces, compressibility of fluids, and variations in pressure and oil saturation can be taken into account for a certain reservoir geometry.

The more rigorous methods of approach include: (1) an iterative method for interface movement which involves direct solution of Laplace's equation, using boundary conditions at each time step; (2) a continuous saturation approach which involves simultaneous solution of flow equations for each fluid over the entire problem domain; and (3) the method of characteristics used for multidimensional displacement problems.

The more approximate numerical approaches include the generalized volumetric balance models. In these models a system of volumetric balances on oil, gas, and water around small segments (cells) of reservoir is formulated. Physical laboratory models can be used with improved accuracy because scaling rules have been developed. See PETROLEUM RESERVOIR MODELS.

Increasing recovery. Two promising methods of increasing recovery are being field-tested. One is the injection of small percentages of chemicals in aqueous solution into the formation to reduce capillary forces and increase sweep efficiency. Experimental work with chemicals has been done for years, but the cost of chemicals lost to adsorption on the enormous surface area of porous rock has been far too great. However, economically acceptable processes appear to be available based on new knowledge in chemistry and research with fluid flow in natural porous rocks. Chemical processes could greatly increase recovery of known reserves. The second method of increasing recovery now being field-tested raises the permeability of a low-permeability formation by fracturing with nuclear explosives. Intended originally for recovery of gas, this method may also be applicable to oil shale. See ENGINEERING GEOLOGY; OIL AND GAS FIELD EXPLOITATION. [G. E. ARCHIE]

Bibliography: National Petroleum Council, *Impact of New Technology on U.S. Petroleum Exploration and Production 1946–1965*, 1967.

Petroleum geology

The application of geological principles in the discovery and development of oil and gas pools. The geology of petroleum includes the origin, migration, and accumulation of petroleum; the structural and stratigraphic relations of oil and gas pools; and the lithologic and paleontologic characteristics of geological formations and producing horizons. Petroleum geology is strongly influenced by the economic aspects of petroleum exploration.

Geological aspects treated here include the occurrence of petroleum, the character of reservoir rocks, typical reservoir traps, and the general nature of reservoir fluids. For further treatment of the physical and chemical properties of petroleum, the origin and migration of petroleum, reservoir and production mechanics, and geological and geophysical methods of exploration *see* PETROLEUM; PETROLEUM ENGINEERING.

OCCURRENCE OF PETROLEUM

Petroleum deposits may be classified as surface occurrences and subsurface occurrences.

Surface occurrences. These occurrences may be thought of as currently active or "live" occurrences, such as seepages, springs, exudates, mud volcanoes, and mud flows. Others may be termed fossil or "dead" occurrences, such as bitumen-impregnated sediments, inspissated deposits, and dike and vein fillings of solid bitumens.

Seepages, springs, and exudates. Petroleum that exudes in any of these forms may reach the surface along fractures, joints, fault planes, unconformities, or bedding planes or through the connected porous openings of the rocks. Most seepages (or springs) are formed by the slow escape of petroleum from accumulations that are close to the surface or have been tapped by faults and fractures. Many oil and gas pools and producing regions have been discovered by drilling near petroleum seepages.

Exudates of asphaltic oils issuing at the surface are likely to be changed to asphalt, partly by the escape of volatile fractions, but mainly by chemical changes such as combination with oxygen or sulfur. The asphalt is black and varies in consistency from a sticky liquid to a substance hard enough to walk on. Outcrops of asphaltic oils are sometimes marked by small pools, some of which contain the bones of animals caught in the sticky material. Asphalt particles may be transported by water, and some deposits consist largely of transported material.

Mud volcanoes and mud flows. These are high-pressure gas seepages that carry with them water, mud, sand, fragments of rock, and occasionally oil. Mud volcanoes are usually confined to regions underlain by incompetent softer shales, boulder and submarine landslide deposits, clays, sands, and unconsolidated sediments. The surface of a mud volcano is often a conical mound or hill, with an opening or crater at the top through which issue mud and water which is usually salty. Only the type which emits gas, with or without oil, in addition to the mud and water, should be considered a surface indication of oil or gas. Mud volcanoes occur chiefly in areas of Cenozoic rocks that have been strongly deformed.

Solid and semisolid deposits. Tar, asphalt, wax, and hard brittle bitumen (any of the flammable, viscid, liquid or solid hydrocarbon mixtures soluble in carbon disulfide) are popularly regarded as solid although, strictly speaking, some of them are highly viscous liquids. Outcrops of solid petroleum are found in the form of disseminated deposits and as veins or dikelike deposits filling cracks and fissures.

Disseminated deposits are sediments containing petroleum in the form of asphalts, bitumen, pitch, or thick heavy oil, disseminated through the pore spaces of rock either as a matrix or as the bonding material. They are commonly called bituminous sands or bituminous limestones, depending on the nature of the host rock. Two different types of disseminated occurrences are found, inspissated deposits and primary mixtures of rock and bitumen.

Inspissated or dried-up deposits occur in place and were probably once a pool in liquid and gaseous form. They now consist of only the more resistant and heavier residues, the lighter fractions having been lost. An inspissated deposit may be thought of as a fossil oil field. As erosion gradually removed the overburden, bringing the surface closer to the pool, the decreased pressure permitted gases and lighter oil fractions to come out of solution and expand, leaving the heavier hydrocarbon fractions behind. As the pool approached the zone of weathering, the opening of incipient fractures allowed the gases to escape more readily. Oxidizing agents aided in solidifying the heavier oils that remained behind.

Primary mixtures are those in which the sediments were mixed with the oil, asphalt, or tar during their deposition, the whole deposit having later been buried by younger sediments and then exposed by erosion. The Athabasca oil sands in Alberta, Canada, are thought by many to be such a deposit. One theory is that oil seeped up from the underlying and then outcropping Devonian organic limestone and was redeposited during Cretaceous time together with Cretaceous sands in lagoons and barred basins along the shore. These oil sands may be considered a primary disseminated deposit in which the sand was deposited in or with the oil. *See* OIL SAND.

Dike and vein fillings may be regarded as fossil or dead seepages from which the gaseous and liquid fractions have been removed, leaving only the solid residues behind. In inspissated deposits the separation of the lighter constituents occurred in place in the rock. In the primary deposits the separation of the gas from the liquid took place before the contemporaneous deposition of the oil and asphalt with the enclosing sediments. In the solid vein and dike fillings the loss of the gaseous and liquid fractions probably occurred while the petroleum was filling the opening.

Oil shale is rock which yields oil upon being heated. Deposits of oil shale which are suitable for commercial exploitation yield 5–15 gal or more of oil per ton. The organic material, termed kerogen, in oil shale is in solid form prior to being heated. Some of the kerogen decomposes into gaseous and liquid petroleum hydrocarbons when heated to 350°C or more. So many different meanings have been given to the word kerogen that it is not in

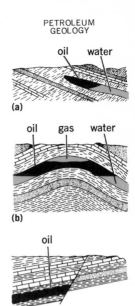

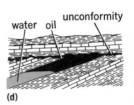

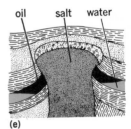

Typical petroleum traps. (*a*) Lens. (*b*) Anticline. (*c*) Fault. (*d*) Unconformity. (*e*) Salt dome.

good scientific standing today. *See* OIL SHALE.

Asphalt is a black, plastic to fairly hard substance, easily fusible and soluble in carbon disulfide. It occurs in nature, but it is also obtained as the residue from the refining of certain petroleums; then it is known as artificial asphalt. Asphalt melts between 150 and 200°F. It may occur as seepages, surface accumulations, and impregnations, and also in large lakes such as the Rancho La Brea deposit in Los Angeles.

Asphaltites are harder solid hydrocarbons which differ from asphalt in being strictly of an intrusive nature. They are found in veins or dikes cutting across the sediments. Asphaltites are fusible, but melt at somewhat higher temperatures and are harder and heavier than the asphalts.

Naturally occurring mineral waxes are solid hydrocarbons believed to result from the drying out of a paraffin-base oil. One example is ozokerite, a plastic waxlike paraffin vein material which is found in Utah and near Boryslaw, Poland.

Subsurface occurrences. Underground occurrences of petroleum may be classified as pools, fields, and provinces.

Pools. Underground accumulations of petroleum characterized by a single and separate natural reservoir (usually a porous sandstone or limestone) and a single natural pressure system are called pools. The production of petroleum from one part of a pool affects the reservoir pressure throughout its extent. A pool is bounded by geological barriers in all directions, such as rock structure, impermeable strata, and water in the formations, so that the pool is effectively separated from any other pools that may be present in the same district or on the same geological structure.

Fields. An oil field may be a single pool, or it may consist of two or more pools which are commonly but not necessarily related to the same geological structure. Where more than one pool is present in the same field, the different pools are separated from one another. The different pools may occur at several stratigraphic horizons separated by impermeable strata, and they may partially or completely coincide in their horizontal distribution. Geological features that influence the accumulation of petroleum include salt domes, anticlinally folded strata, and combinations of faulting, folding, and stratigraphic variations.

Provinces. A petroleum province is a region in which a number of oil and gas pools and fields occur in a similar or related geological environment. The term is used to indicate the larger producing regions, such as the mid-continent region of the United States.

RESERVOIR ROCKS

Reservoir rocks are rocks with sufficient porosity and permeability to allow oil and gas to accumulate and be produced in commercial quantities. There are three requisites for a reservoir rock: (1) It must be porous, that is, have enough pore space to contain oil or gas; (2) it must be permeable to allow fluids, including oil and gas, to move through it; and (3) there must be a trap which prevents escape of the oil and gas. Any rock with these characteristics may become a reservoir for oil and gas provided that hydrocarbons are available to migrate into the rock.

The reservoir character of a rock may be an original feature (intergranular porosity of sandstones) or a secondary feature resulting from chemical changes (solution porosity of limestones), or it may be the result of physical changes (fracturing of brittle rock).

Types of reservoir rocks. Reservoir rocks may be classified as fragmental or clastic (broken), chemical and biochemical, or miscellaneous. They may also be classified as marine and nonmarine reservoir rocks.

Fragmental type. Some reservoir rocks are aggregates of particles, that is, fragments of older rocks. They are also called clastic or detrital rocks because they consist of mineral and rock particles derived from eroded areas. The constituent particles of fragmental rocks may range in size from colloidal particles up to pebbles and cobbles. The most common of the fragmental reservoir rocks are sandstones, conglomerates, arkoses, graywackes, and siltstones. Many, however, are carbonate rocks, such as oolitic rocks, calcarenites, and coquinas, which are made up of oolites and skeletal fragments that have been cemented and in some cases recrystallized.

Some sandstone reservoir rocks consist either entirely or in part of loose, uncemented sand grains. The grains tend to be brought to the surface in large quantities along with oil during production. The sand grains in most sandstones, however, are held together by various kinds of cementing material, mostly carbonates, silica, or clays. Some of the cementing materials are primary, having been deposited along with the sand grains. Other cementing materials are secondary, having been precipitated from solutions that entered the rock after it was deposited.

Clastic limestones and dolomites consist of particles of calcite and dolomite that have been transported and deposited in much the same manner as quartz and other mineral grains. Rocks thus formed are commonly recemented with calcite and, if fine grained, may resemble chemically deposited limestones or dolomites. Carbonate rocks thus may form good oil reservoirs because of high porosity and permeability, particularly if the original pore spaces are not completely filled with cement.

Chemical types. These rocks are made up chiefly of chemical or biochemical precipitates. They are composed of mineral matter that was precipitated at the place where the rocks were formed (in contrast to the transported grains in clastic carbonates). The most important chemical reservoir rocks are limestones and dolomites. Some chemically precipitated rocks consist entirely or almost entirely of silica in the form of chert or novaculite, but such rocks provide few reservoirs. The porosity of carbonate rocks is largely the result of solution leaching by percolating groundwaters.

Miscellaneous types. Other reservoir rocks include igneous and metamorphic rocks and mixtures of both. Any porous and permeable igneous rock in close association with sedimentary rock may become a reservoir rock when saturated by oil derived from the sediments. Igneous and metamorphic rocks are only a minor source of oil and gas because, generally, they are not permeable enough, and when they are they are not often associated with suitable source rocks.

Marine and nonmarine types. A distinction may be made between reservoir rocks which were deposited in ancient seas and those deposited in fresh water. Most petroleum occurs in rocks deposited under marine conditions but some occurs in sediments of nonmarine origin. The occurrence of oil in nonmarine sediments is sometimes explained as the result of migration of oil along faults, fractures, or bedding planes from adjacent marine sediments, but some petroleum probably has formed in nonmarine rocks.

Properties of reservoir rocks. The porosity and permeability of reservoir rocks, as well as the nature of the traps, are all factors which regulate the accumulation of petroleum. Porosity is the total space in the rock (pores, voids, interstices) not occupied by solid material. It is expressed as a percentage. Factors which influence porosity are the size of the rock particles, arrangement, sorting, shape, and cementing material. Most oil-producing rocks have porosities above 10% and thicknesses greater than 10 ft.

Total pore space is not the sole determinant of a petroleum reservoir. A reservoir must also have permeability; that is, it must allow fluids to flow through it with relative ease. Pumice, for instance, has a large amount of pore space, but the pores are not connected and it has very low permeability. A rock that is permeable necessarily has interconnected pores or fractures that are greater than capillary in size.

RESERVOIR TRAPS

Reservoir traps contain the accumulated oil or gas so that it cannot escape. The upper boundary of a reservoir trap is called the roof or cap rock; the lower boundary is the oil-water or gas-water contact.

Roof or cap rock is an impermeable layer of rock forming the roof of an oil trap. The connecting pores in the reservoir rock, which are individually minute, are as a rule saturated with water. Since oil and gas are lighter than water, the petroleum rises through the water until it is stopped by the roof rock. If the roof rock is concave (domed, arched, folded, peaked, or roof-shaped), it acts as a trap, keeping the oil and gas from escaping upward or laterally.

The oil-water contact or gas-oil contact generally forms the lower boundary of the accumulation. The water is the water that normally fills the pores of the reservoir rock. The water supports the pool of oil and gas, and the forces of buoyancy impel the petroleum upward against the bounding surfaces of the trap, holding it in place, as shown in the illustration.

Geological structure. The anticlinal theory is a successful theory of petroleum geology. Most of the oil in the major oil fields of the world occurs in anticlines. The fact that oil and gas commonly occur on anticlinal axes was first noted by W. E. Logan in 1842. He observed that oil seeps occurred in the vicinity of anticlinal axes near the mouth of the St. Lawrence River. Although the term anticlinal theory has fallen into disuse, the fundamental principle on which it is based remains generally valid—oil and gas tend to accumulate in the highest places within the reservoir. It is rec-

ognized today, however, that other factors affect the accumulation of oil in many pools, and that the anticlinal theory by no means provides an explanation for all oil accumulations.

Classification of traps. Three basic types of traps generally are recognized: structural traps, stratigraphic traps, and combination traps.

Structural traps. A trap whose upper boundary has been made concave by folding or faulting, or both, of the reservoir rock is known as a structural trap. The edges of a pool occurring in a structural trap are determined by the intersection of the underlying water table with the enclosing roof or cap rock. Structural traps include closed anticlines or domes, faulted anticlines with closure, closure against faults, anticlines on downdip sides of faults, and oil and gas accumulations in fractures produced by structural deformation.

Stratigraphic traps. Also known as varying permeability traps, stratigraphic traps are those in which the chief trap-making element is some variation in the stratigraphy or lithology, or both, of the reservoir rock. These include facies change, variable local porosity and permeability, and any upstructure termination of the reservoir rock. Stratigraphic traps include sandstone lenses, channels, bars, and reefs and porosity lenses. Some of the most common stratigraphic traps are strandline pools, shoestring sand traps, biostromes, and bioherms.

Shoestring sand traps are long, narrow, discontinuous sandstone deposits which are commonly a few miles wide or less and which may be many miles in extent. Except at their terminal ends they are generally surrounded by impermeable shales and clays. Some sand traps of this nature are believed to be channel fillings and others offshore sandbars.

Two general classes of primary stratigraphic traps occur in rocks of chemical origin, almost all of them carbonate rocks. These are biostromes and bioherms. Biostromes include porous lenses, enclosed by impermeable shales, limestones, or dolomites. Bioherms, or organic reefs, are porous, domelike, moundlike, or otherwise circumscribed masses, built largely by lime-secreting organisms such as corals, algae, brachiopods, mollusks, or crinoids and enclosed in strata of different lithologic character.

Combination traps. These traps result from both structural and stratigraphic conditions. An example of such a trap is the salt dome. Salt domes are cylindrical or steeply conical masses of salt which has flowed plastically under pressure. These masses, called plugs or domes, originate at depths of 20,000 ft or more and pierce the overlying sedimentary strata. Three kinds of traps are associated with salt plugs: cap rock, flanking sands, and supercap sands. Cap rock consists of limestone. gypsum, and anhydrite and occurs as a capping over the tops of the salt plugs. Flanking sands are strata abutting upon and cut off by the salt plug. Supercap sands are sandy strata that arch over the tops of the plugs in the form of structural domes. In many salt domes recurrent movement of the dome during deposition of overlying strata caused variations in the thickness and lithology of the strata, creating traps that are in part stratigraphic traps.

RESERVOIR FLUIDS

Fluids fill the voids or pore spaces in all reservoir rocks. The fluid may be water, water and oil, water and gas, or a mixture of water, oil, and gas. The fluid content of a gas pool consists of water and gas; that of an oil pool consists of gas, oil, and water. Gas is almost invariably present in solution in oil and, in addition, free gas may be present.

The distribution of gas, oil, and water in the reservoir depends upon relative buoyancy, relative saturation of pore space with each fluid, and capillary and displacement pressures, as well as the porosity, permeability, and composition of the reservoir rock. In traps that contain oil, water, and free gas, the fluids occur in distinct zones. Gas, being the lightest, occurs at the top of the trap. Below the gas, oil occurs, and below that, water. Where there is gas but no oil, the gas is immediately underlain by water and the contact is the gas-water table. Interstitial water (adsorbed water or wetting water which lines the pore walls or occurs on the surfaces of mineral grains) is generally present throughout the reservoir, occupying 10–30% of the pore space.

Oil field waters are waters associated with oil and gas pools. They may be classified as meteoric waters, connate water, and mixed water. Most oil field waters are saline. See OIL FIELD WATERS.

Oil saturation is the amount of oil contained in a petroleum reservoir. It is measured as a percentage of the effective pore space.

Gas volume, or natural-gas content, of a petroleum reservoir may range from small quantities dissolved in oil up to 100% of petroleum content. The natural gas in a reservoir may occur as free gas, as gas dissolved in oil, and as gas dissolved in water. See NATURAL GAS; PETROLEUM RESERVOIR ENGINEERING.

[JOHN W. HARBAUGH]

Bibliography: G. D. Hobson, *Some Fundamentals of Petroleum Geology*, 1954; K. K. Landes, *Petroleum Geology*, 2d ed., 1959; A. I. Levorsen, *Geology of Petroleum*, 2d ed., 1967; W. I. Russell, *Principles of Petroleum Geology*, 2d ed., 1960; E. N. Tiratsoo, *Petroleum Geology*, 1952.

Petroleum processing

The recovery and processing of various usable fractions from the complex crude oils. The usable fractions include gasoline, jet fuel, kerosine, fuel oil, asphalt, lubricating oils, and many others.

The petroleum refining industry is one of the largest manufacturing industries. The distribution of refining capacity among regions of the world is shown in Table 1. Almost $5,000,000,000 was spent in 1973 for materials and labor to place new facilities within these refineries. An additional $4,000,000,000 was spent to maintain and modernize existing facilities.

Refineries in the United States produced 7,450,000 barrels per day (bpd) of gasoline in 1973. In addition, the following products were produced in the quantities shown: middle distillate (including jet fuel, diesel oil fuels, and others), 5,100,000 bpd; residual fuel oil (for heating purposes), 3,290,000 bpd; all others (such as waxes, lubricating oils, asphalt, coke), 3,790,000 bpd.

Crude oil is a mixture of many different hydro-

Table 1. Refining capacity in the world, Jan. 1, 1973

Region	Crude oil charge, 10^6 bpcd*
United States	13.46
Other Western Hemisphere countries	8.07
Western Europe	17.42
Soviet Union and Eastern Europe	7.79
Africa	0.86
Middle East	2.70
Asia and Far East	7.88
World total	58.18

SOURCE: United Nations, *Statistical Yearbook, 1973*, pp. 273–275, 1974.

*Barrels per calendar day (bpcd) equals reported metric tons per year multiplied by 0.02. Estimates are given for missing data so that the total here (equivalent 2.91×10^9 metric tons) is greater than the source total (2.56×10^9 metric tons).

carbon compounds of the paraffin type (wax compounds) and of the naphthene type (asphalt compounds), making the chemistry of petroleum refining extremely complex. The refining processes can be grouped under three main headings: (1) separating the crude oils to isolate the desired products; (2) breaking the remaining large chemical compounds into smaller chemical compounds by cracking; (3) building desired product properties by chemical reactions, such as reforming, alkylation, and isomerization. The capacities for some of these downstream processes are compared to the total crude distillation for United States refining in Table 2.

Refinery products, such as gasoline, kerosine, diesel oil, and others, are not pure chemical compounds but mixtures of chemical compounds. Some of the hydrocarbon compounds contained in gasoline are shown in Table 3, along with the individual specific gravities, molecular weights, and normal boiling points.

A simplified flow sheet of refinery operations is shown in Fig. 1. By means of distillation a typical crude oil may be separated quite easily into many fractions of raw products. Some of these are shown in Table 4.

A more complex flow sheet of a refinery for light oils is shown in Fig. 2. Here are included the cracking equipment, reforming equipment, extrac-

Table 2. United States processing capacity, Jan. 1, 1973

Process type	Charge capacity, 10^6 bpcd
Crude distillation	13.46
Catalytic cracking	4.50
Hydrocracking	0.78
Thermal cracking	0.45
Coking	0.99
Catalytic reforming	3.23
Alkylation*	0.79

SOURCE: U.S. Bureau of Mines, *Petroleum refineries in the United States and Puerto Rico, January 1, 1973*, July 24, 1973.

*Alkylation given in product capacity.

Table 3. Some chemical compounds found in gasoline*

Name	Formula	Molecular weight	API gravity	Normal boiling point, °F	Research blending octane number
n-Pentane	C$_5$H$_{12}$	72	92.7	97	62
n-Hexane	C$_6$H$_{14}$	86	81.6	158	19
n-Heptane	C$_7$H$_{16}$	100	74.1	209	0
n-Octane	C$_8$H$_{18}$	114	68.7	260	−18
n-Nonane	C$_9$H$_{20}$	128	64.6	310	−18
n-Decane	C$_{10}$H$_{22}$	142	61.3	343	−41
n-Endecane	C$_{11}$H$_{24}$	156	58.0	387	−55

*Only the straight-chain paraffin hydrocarbons are shown here to indicate the range. Actually the gasoline contains also branched-chain paraffins, alkenes, naphthenes, aromatics, and other compounds with higher octane numbers.

tion units, polymerization units, and other facilities. Figure 3 is a schematic diagram of a refinery for producing lubricating oils.

Separating the crude oil. There are two principal separating procedures: topping of crude oil, and lubricating oil processing. Both of these procedures include combinations of several operations, such as distillation, centrifuging, filtration, and treating processes.

Topping, or distilling, the crude oil. The crude oil is desalted and dehydrated, then passed through heaters where the temperature is raised to about 650°F, at which temperature all of the gas, gasoline, jet fuel, and light fuel oil fractions are in the vapor phase. This vapor and liquid mixture enters a large distillation tower, about one-third the distance up from the bottom (Figs. 1 and 2). Into the bottom of the tower about 1 – 2 lb of steam per gallon of crude oil is usually introduced to

Table 4. Some fractions obtained from crude oil

Fraction	Carbon atoms	Molecular weight	API gravity	Boiling range, °F
Gas	1 – 4	16 – 58		−259 – 31
Gasoline	5 – 12	72 – 170	58 – 62	31 – 400
Jet fuel	10 – 16	156 – 226	40 – 46	356 – 525
Gas oil	15 – 22	212 – 294	34 – 38	500 – 700
Lube oil	19 – 35	268 – 492	24 – 30	640 – 875
Residuum	36 – 90	492 – 1262	8 – 18	875+

make the separation easier. From the top of the tower some gases are evolved and sent to units which process light ends. The next higher-boiling fraction is the gasoline, followed successively by the jet fuel, the gas oil, the cracking stock, and the lubricating distillate. Below the feed entrance

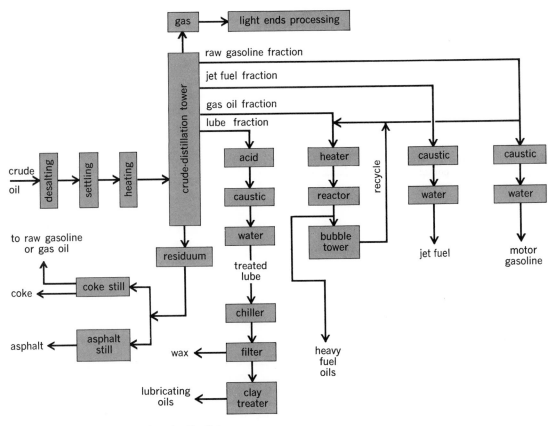

Fig. 1. Simplified flow sheet of crude oil refining.

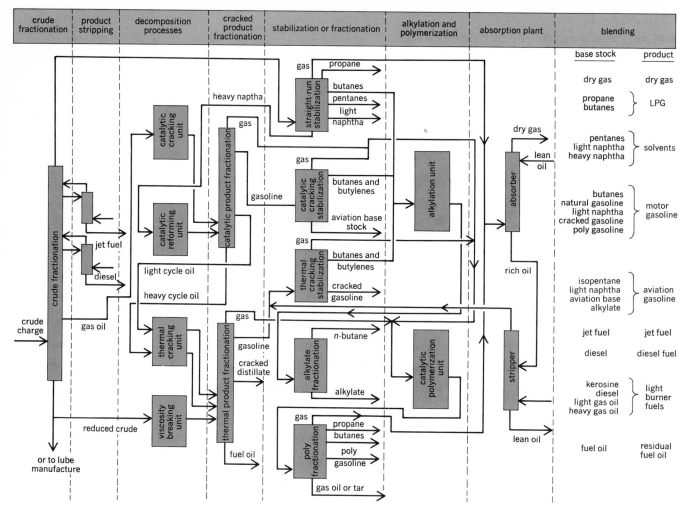

Fig. 2. Refinery for light oils (mainly gasolines, kerosine, and distillates).

a fraction called the residuum is removed.

The temperature of the feed to the tower depends considerably upon the ultimate plans for the residual oil. If this residual oil is to be processed further for the manufacture of lubricating oils, the feed is not heated to so high a temperature.

Each of the streams from the distillation unit must be treated further before it can be sold. The gasoline fraction is treated, then blended with other stocks. Finally, certain chemicals called additives are added to the stream to improve its properties.

Lubricating oil processing. The most important property of lubricating oil is its viscosity. The lube fraction produced in the vacuum distillation column contains some hydrocarbons that give the oil a poor viscosity-temperature characteristic. In addition, the lube oil fraction has poor oxidation resistance and contains wax and other impurities which must be removed. Consequently, the lubricating oil fraction must be treated to remove or to reduce the concentrations of the following: free-carbon—forming material, low viscosity-index materials, wax, unstable compounds which may decompose to form asphaltic substances or coke, and chemicals that affect the color of the lube oil products.

The flow sheet shown in Fig. 3 describes a

process for the production of lubricating oils. Not only the lubricating fractions but also a portion of the residuum fraction is used to make the lubricating oils. In this case, the residuum is treated with a solvent to remove the asphaltic material. The deasphalted residuum is further extracted along with the other lubricating oil fractions, dewaxed, acidtreated, clay-treated, blended with additives, and then sent to storage.

The solvents used for extraction include furfural, cresylic acid, phenol, sulfur dioxide, chlorex, nitrobenzene, propane, benzene, and many others. Since the solvent must be removed from the oils after the extraction, elaborate distillation equipment is required. The oils are freed from the final traces of solvent by steam stripping or vacuum flashing. Quite often the solvent is more expensive than the oil being treated so that, from the standpoint of economy alone, all of the solvent must be recovered. More important, however, the solvent itself may have properties which are detrimental to the finished oil when they are present in trace amounts.

Distillation. All distillation processes are essentially the same. The factors to be considered for different types of distillation processes include the sensitivity of liquid with respect to heat, the specifications of the product, and the boiling range

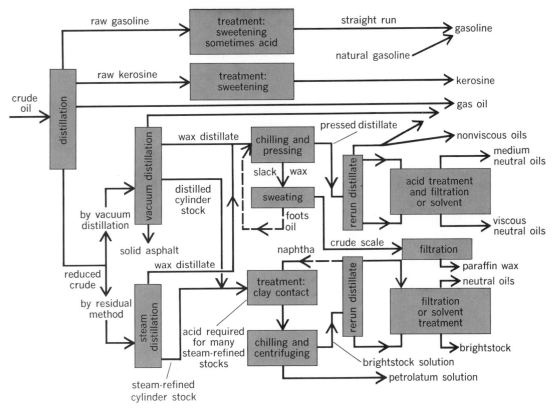

Fig. 3. Refinery for lubricating oils. (*From W. L. Nelson, Petroleum Refinery Engineering, 4th ed., McGraw-Hill, 1958*)

of the feed.

In topping or skimming procedures, the crude is heated to a certain temperature and fed to a distillation tower where the product fractions are removed at various heights along the column. Figure 1 includes a schematic diagram of such a separation.

Stabilization is the distillation process that removes the lighter hydrocarbons (usually the dissolved gaseous hydrocarbons) from the particular fraction being processed. Here the feed is heated and sent to a fractionation column, where gases are removed overhead and the stabilized product at the bottom. In natural gasoline stabilizers, 40–60 plates are required in the distillation column to remove the dissolved propane and lighter hydrocarbons.

Steam distillation is used to increase the amount of distilled products obtainable at a fixed feed temperature. The feed stock is heated to approximately 550–660°F in the presence of a large amount of steam. The effect of steam is to reduce the boiling point by partial pressure effects. The boiling point of a material can be reduced either by reducing the total pressure or by adding an inert gas such that the same total pressure will be partially due to the inert gas.

Vacuum distillation is used for the redistillation of the pressure distillate, lube stock, topped crudes, and other fractions. Lubricating oil, for example, is thermally sensitive and partially decomposes if exposed to high temperatures. Therefore the distillation is done under a high vacuum to take advantage of the lower temperatures required at the lower operating pressures. Sometimes high

vacuum is not sufficient; it is then necessary to combine vacuum distillation with steam distillation in a combination unit. In this case, steam is added to the distillation column operating under the vacuum. The amount of steam required will vary, of course, but may be as high as 1–2 lb/gal of oil processed. The dry vacuum distillation processes have the advantages that smaller towers and smaller condensing equipment are required for a given throughput.

Filtration. This is also an important operation in the refining of petroleum. Regular gravity-type settlers are used wherever possible, but occasionally the solids are too finely divided to settle. There are many types of filters which are used for the removal of finely divided clay from treated stocks in the clay-contact process. These filters are classified as filter presses, leaf filters, rotary filters, and others.

Breaking the large molecules. The major product from the refinery is motor fuel (gasoline). Of course kerosine, diesel oils, jet fuels (mostly kerosine fraction), and others are extremely important also. However, each barrel of oil charged to the distillation tower has a given fraction of gasoline. This varies, but on the average is not over 20% of the total volume of crude. If more gasoline than this 20% obtainable by distillation is desired, and it almost always is, it is necessary to resort to other means than straight separation to obtain it. This can be done either by recombining the gaseous, or lighter, molecules (polymerization), or by breaking down the heavier molecules (cracking).

Table 3 shows that gasoline molecules seldom contain more than 11–12 atoms of carbon. The crude oil, however, contains many molecules consisting of more than 50–60 atoms of carbon. The heavy naphtha fraction and the jet fuel and gas oil fractions, for example, all contain large molecules compared with the gasoline fraction (Table 4). In order to use these fractions for gasoline production, the long or large molecules must be broken into smaller ones of the gasoline type. This process is called cracking.

The cracking may be done either by thermal means (maintaining the heavy fractions at high temperatures) or by catalytic means. In thermal cracking the charge stocks are usually light and heavy gas oils, residual oils, or any of the topping column fractions heavier than the gasoline fraction. The resulting gasoline yields depend upon the composition of the charge stock but ranges 15–40% by volume of gasoline (100–400°F boiling range).

In catalytic cracking the fraction to be cracked is contacted with a catalyst under lower pressure conditions than in thermal cracking, although the temperatures are still almost as high. Catalytic cracking gives much better yields of gasoline, lower carbon formation, and a gasoline of much higher octane number. The use of a newer zeolitic-type catalyst gives even better yields of gasoline.

About 85% of the cracking capacity in use in the United States today is of the catalytic type. *See* CRACKING; HYDROCRACKING.

Rebuilding desired chemical compounds. The saturated straight-chain paraffins shown in Table 3 have very low octane numbers. These compounds can be altered, however, by chemical reaction to yield a different kind of molecule with much higher octane characteristics. In general, the straight paraffin compounds have the lowest octane rating and the aromatic compounds (benzene family) have the highest. The olefins and the naphthenes have intermediate octane numbers.

Some of the forms of a six-carbon-atom hydrocarbon and their research octane numbers (RON) are shown below. All these forms, and many others, are found in the gasoline fraction. It is possible to convert hexane (RON = 24.8) into benzene, which has an octane number of over 100.

Among the many processes that are used for altering the chemical structure of the molecules are the following:

1. Hydrogenation is used mostly for producing

n-Hexane, C_6H_{14} (straight-chain paraffin), RON = 24.8

2-Methylpentane or isohexane, C_6H_{14} (branched paraffin or isoparaffin), RON = 73

1-Hexene, C_6H_{12} (olefin or alkene), RON = 80

Benzene, C_6H_6 (aromatic), RON = over 100

Cyclohexane, C_6H_{12} (naphthene), RON = 83

saturated hydrocarbons from unsaturated ones. During World War II, this process was used for making isooctane from isooctene. Later this process was used almost exclusively for desulfurization processes.

2. Dehydrogenation is the removal of hydrogen from a molecule. For example, 1-hexene may be made from *n*-hexane by removing hydrogen. This reaction often results in increased octane number.

3. Aromatization yields aromatic type hydrocarbons from other types, as benzene from hexane or cyclohexane. Aromatization and isomerization predominate in the reforming operation.

4. Cyclization is the transformation of a hydrocarbon of the chain type to one of the ring type; for example, making cyclohexane from *n*-hexane.

5. Isomerization is the rearrangement of the atoms in a molecule, such as *n*-hexane, to form an isomer, such as isohexane.

6. Polymerization involves two or more molecules in a building process. For example, as shown by Eq. (1), propene and the butenes, which are

$$CH_3{=}CHCH_2 + CH_3CH_2CH{=}CH_2 \xrightarrow[\text{catalyst}]{\text{Heat or}}$$

$$CH_2{=}CHCH_2CH_2\overset{\overset{\displaystyle CH_3}{|}}{C}HCH_3 \qquad (1)$$

5-Methyl-1-hexene

present in the gases of thermal- or catalytic-cracking operations, are polymerized to form a larger liquid molecule with a high octane number. The catalytic polymerization capacity for making gasoline in the United States is dropping so that the total stands at 100,000 bpd (1968).

7. Alkylation also makes use of two or more molecules in the reaction. This process uses an isoparaffin, such as isobutane, and an olefin, such as ethylene, to yield a larger molecule with a high octane number, 2,2-dimethylbutane, as shown by Eq. (2). This reaction uses only one-half as many of

$$CH_2{=}CH_2 + CH_3\underset{\underset{CH_3}{|}}{CHCH_3} \xrightarrow[\text{catalyst}]{\text{Heat or}} CH_3\underset{\underset{CH_3}{|}}{\overset{\overset{CH_3}{|}}{C}}CH_2CH_3 \quad (2)$$

Ethylene Isobutane   2,2-Dimethyl-
                                    butane

the expensive olefin molecules as the polymerization process. By the beginning of 1968 the alkylation capacity in the United States was about 790,000 bpd. This indicates that alkylation is replacing polymerization as a means of making gasoline from gaseous feedstocks. *See* OCTANE NUMBER; REFORMING IN PETROLEUM REFINING.

Treating processes. Both the crude oil and the petroleum products must, on occasion, be treated to remove undesirable impurities or to improve the properties of the product. The important treating processes are desalting and dehydrating, sweetening and desulfurization, acid treatment, clay-contact adsorption treatment, vapor-phase treatment, and solvent treatment. Some of these processes are used both on the crude oil and on the products, whereas others are used on only one or the other.

Desalting and dehydration of the crude oil. The salt content of the crude oil which enters the refinery may be as high as 4 or 5%, and the water content may be much higher than the equilibrium amount because water is present as an emulsion.

Because of the high temperatures of the heater tubes, the introduction of the wet crude into the heaters would be dangerous. In addition, the salt would precipitate onto the tube walls, reducing the rate of heat transmission and thereby the efficiency of the heaters.

Many processes are available for the removal of both the salt and the water from the crude oil. These are grouped into four types as shown in Table 5.

The crude oil (containing the salt and the oil) is heated, an emulsion breaker is added, and the resultant mass is settled, or even filtered, to remove the salt and water phase from the oil phase.

Sweetening and desulfurization. Since the original crude oils contain some sulfur compounds, the resulting gasolines and other products also contain sulfur compounds, including hydrogen sulfide, mercaptans, sulfides, disulfides, and thiophenes.

The processes used to sweeten, or desulfurize, the products depend upon the type of sulfur compounds present and the specifications of the finished gasoline or other stocks.

Mercaptans are removed or converted into less undesirable disulfides in the following ways:

1. Mercaptan removal: (*a*) unisol process using alkaline solution of methyl alcohol; (*b*) solutizer processes using sodium hydroxide along with minute amounts of sodium isobutyrate; and (*c*) mercapsol process using an alkaline solution of naphthenic acids and phenols. These are regenerative solution processes.

2. Mercaptan conversion (oxidation to disulfides): (*a*) lead sulfide doctor sweetening; (*b*) copper chloride–oxygen sweetening; (*c*) sodium hypochlorite sweetening; (*d*) oxygen sweetening with chelated cobalt catalyst in either a caustic solution or fixed bed.

3. Hydrogen sulfide removal by regenerative solution processes using aqueous solutions of (*a*) sodium hydroxide, (*b*) calcium hydroxide, (*c*) trisodium phosphate, and (*d*) sodium carbonate.

Hydrotreating is the most widely practiced treating process for all types of petroleum products, whether fuels or lubricants, which is in use in the world today. Total hydrotreating capacity for all products in the United States was in excess of 3,000,000 bpd in 1974. The process, through the selection of the appropriate catalysts and operating conditions, is used to achieve desulfurization, eliminate other undesirable impurities such as nitrogen and oxygen, decolorize and stabilize products, correct odor problems, and improve many other product deficiencies. Fuel products treated range from naphthas to heavy burner fuels. Specialty products such as lubes and waxes and solvents are also treated to improve various characteristics.

Solvent treatment in petroleum refining. Undesired constituents may also be removed by selective solvent extraction. In this case a liquid that will selectively dissolve the undesired constituents is added to the oil. The solvent processes may be divided into two main categories, solvent extraction and solvent dewaxing.

The solvents used in the extraction processes include propane and cresylic acids, 2,2′-dichlorodiethyl ether, phenol, furfural, sulfur dioxide, benzene, and nitrobenzene.

In the dewaxing process, the principal solvents are benzene, methyl ethyl ketone, methyl isobutyl ketone, propane, petroleum naphtha, ethylene dichloride, methylene chloride, and sulfur dioxide.

Before the solvent-extraction processes were developed, only a few types of crudes were considered to be good lubricating oil crudes. By using these solvent processes, the original properties of the crudes can be changed so greatly that almost any crude will make good lubricating oils.

The early developments of solvent processing were concerned with the lubricating oil end of the crude. Solvent-extraction processes are being

Table 5. Desalting and dehydrating methods

Method	Temperature, °F	Type of treatment
Chemical separation	140–210	0.05–4% solution of soap in water 0.5–5% solution of soda ash in water
Electrical separation	150–200	10,000–20,000 volts
Gravity separation	180–200	Up to 40% water added
Centrifugal separation	180–200	Up to 20% water added (sometimes no water added)

applied to many useful separations in the purifications of gasoline, kerosine, diesel fuel, and others. In addition, solvent extraction may replace fractionation in many separation processes in the refinery. For example, propane deasphalting has replaced, to some extent, vacuum distillation as a means of removing asphalt from reduced crudes. *See* OIL ANALYSIS; PETROLEUM; PETROLEUM PRODUCTS.

[JOHN J. MC KETTA; HAROLD L. HOFFMAN]
Bibliography: *Hydrocarbon Processing Process Developments*, published September, odd-numbered years; *Hydrocarbon Processing Process Handbook*, published September, even-numbered years; W. L. Nelson, *Petroleum Refinery Engineering*, 4th ed., 1958.

Petroleum products

Crude petroleum is the starting material not only for the fuels used for transportation and energy production but also for petrochemical feedstocks, solvents, lubricants, asphalts, and many other specialities (see table). The more complex the family of petroleum products, the more energy is needed to refine the crude. A simple refinery consumes only 5% of the energy in the crude; to make the quality and types of products to satisfy today's market, however, only about 90% of the input energy emerges in products.

Fuels for transportation. Liquid products such as gasoline, jet fuel, diesel fuel, and marine fuel, which together account for more than half the volume of output, serve the transportation industry. Hydrocarbons from butane to C_{12} appear in gasoline blends, which still constitute the most important product from petroleum, about 40% of crude in the United States. Most specialized energy-consuming process units in refineries, such as catalytic reformers and cracking units, exist to convert crude fractions to gasoline of a quality that meets the needs of the modern automobile engine. Added to the usual engine requirements for antiknock performance and volatility of fuel are environmental demands for emission standards in exhaust. Concern about air pollution has required the introduction of unleaded fuel for cars equipped with catalytic mufflers and, in turn, has made necessary the increased use of aromatic hydrocarbons for antiknock performance. *See* ATMOSPHERIC POLLUTION; CRACKING; GASOLINE.

Aviation gasoline, once the leading military and air-transport fuel, is now serving general aviation. Several grades of various antiknock and lead levels are blended from alkylate, pentane, aromatics, and naphthas to the exacting requirements of high-performance aircraft piston engines. Higher-boiling-fraction C_8 to C_{15} hydrocarbons are the blend stocks for jet fuels for the air-transport industry. The U.S. Air Force uses JP-4, a wide-cut naphtha-kerosine blend, and the U.S. Navy uses a kerosine called JP-5. Kerosine used as a jet fuel for the world's airlines accounts for about 6% of crude. *See* JET FUEL.

Diesel fuels are blended from both distilled and cracked fractions—up to C_{22}-type hydrocarbons. They range in volatility from kerosine for lightweight automotive diesel engines to gas oil for large, lower-speed industrial or marine diesel engines, and account for 10% of the crude. Diesel fuels must have a controlled viscosity and must exhibit good combustion performance. Low-temperature properties are important for diesel fuels used in railroads, trucks, and buses that are operated in climatic extremes. *See* DIESEL FUEL.

Marine fuels are made from the highest-boiling crude fraction—the residuum from crude distillation blended with heavy cracked gas oils. In a ship equipped with steam turbines, marine fuel is burned in a boiler to generate steam, just as in an electric power generating station. But the same fuel can be burned directly in a ship's gas turbine engine, provided it meets certain requirements in viscosity and trace metals.

Fuels for energy production. Petroleum fuels heat homes, factories, and offices and also generate electric power. Home-heating oils are similar to diesel fuels in boiling range but must exhibit storage stability and good atomization and combustion performance in small oil burners. Like diesel fuels, they must have a high flash point to ensure safety in handling, and must also display good low-temperature flow characteristics. Their viscosity ranges from that of kerosines, for simple vaporizing-type heating units, to heavy gas oils, for large burners used in the heating plants of large buildings. But the residual oils delivered to the public utilities to generate electricity constitute the largest volume of energy-producing fuels. In this application, petroleum competes with gas, coal, and nuclear power. Heavy fuel oil must be heated to reduce its viscosity and to atomize it successfully in the fireboxes of the boilers that generate steam. Today this heavy fuel must frequently conform to strict limits on sulfur content in order for the power station to meet air-pollution standards on stack gases. The energy-producing fuels account for about 30% of the crude. *See* FUEL OIL.

Nonfuel products. From each of the fractions distilled from crude or its cracked or processed products, valuable and indispensable nonfuel materials are made. Some of the major products are:

Ethylene, propylene, butylene, and other reactive gases for the petrochemical industry's output of polymers, rubbers, chemicals, textiles, and films.

Liquefied petroleum gases for heating, cooking, and drying. *See* LIQUEFIED PETROLEUM GAS (LPG).

Solvents for the paint and dry-cleaning industries and as vehicles for aerosol products such as insecticides.

Lubricants of all types, from light spindle oils to heavy turbine oils and greases. Engine and machine lubricants are compounded in a wide range of viscosities from both paraffin- and napthene-base stocks.

Specialty oils for hydraulic fluids, transformers, emulsions, cutting fluids, pharmaceuticals, inks, preservatives, and so on.

Wax for a host of applications, from candles, medicines, coatings, and compounding to petrochemical feedstocks.

The residues from processing crude can also yield such familiar products as asphalt, the major road-building material, and coke, burned as fuel or made into electrodes. Useful petroleum products, made directly or indirectly, number in the thousands. No part of the petroleum barrel is wasted, and great efforts are made to recover spills and slops, not only to minimize water pollution but to recover valuable fuel energy.

Crude petroleum and some of its products*

```
Hydrocarbon
gases
    Liquefied gases───── Metal cutting gas, illumination gas
    Petroleum ether
    Polymers─────────── Antiknock fuels, lubricating oils
    Alcohols, esters,─── Solvents
        ketones          Aldehydes────────── Resins
    Acetylene            Acetic acid───────── Esters
                         Synthetic rubber
                         Acetylene black───── Batteries
    Gas black─────────── Rubber tires, inks, paints
    Fuel gas
    Light naphthas────── Light naphthas────── Gas machine gasoline
                                              Pentane, hexane

Light
distillates
    Naphthas                                  Aviation gasoline
                                              Motor gasoline              Rubber solvent
                         Intermediate naphthas── Commercial solvents── Fatty oil solvent (extraction)
                                              Blending naphtha            Lacquer diluents
                                              Varnish-makers and painters naphtha
                         Heavy naphthas────── Dyers and cleaners naphtha
                                              Turpentine substitutes
    Refined oils         Refined kerosine──── Stove fuel, lamp fuel, tractor fuel
                         Signal oil────────── Railroad signal oil, lighthouse oil
                         Mineral seal oil──── Coach and ship illuminants, gas absorption oils

Intermediate
distillates                                  Water gas carburetion oils
    Gas oil                                   Metallurgical fuels
                                              Cracking stock for gasoline manufacture
                                              Household heating fuels
                                              Light industrial fuels
                                              Diesel fuel oils
    Absorber oil─────── Gasoline recovery oil, benzol recovery oil

Heavy
distillates                                                      Tree spray oils
                                                                 Bakers machinery oil, fruit packers oil
                                              Technical────────── Candymakers oil
                         White oils                              Egg packers oil
    Technical oils                                               Slab oil
                                              Medicinal────────── Internal lubricant, salves, creams, oint-
                                                                      ments
                         Saturating oils──── Wood oils, leather oils, twine oils
                         Emulsifying oils─── Cutting oils, textile oils, paper oils, leather oils
                         Electrical oils──── Switch oils, transformer oils, Metal recovery oils
                         Flotation oils
                         Candymakers and chewing gum wax
                         Candle wax, laundry wax, sealing wax, etchers wax
    Paraffin wax         Saturating wax, insulation wax────── Match wax, cardboard wax, paper wax
                         Medicinal wax
                         Canning wax
                         Paraflow
                         Fatty acids──────── Grease, soap, lubricant
                         Fatty alcohols
                            and sulfates──── Rubber compounding, detergents, wetting agents

                         Light spindle oils        Steam cylinder oils
                         Transformer oils          Valve oils
                         Household lubricating oils Turbine oils
                         Compressor oils           Dust-laying oils
                         Ice machine oils          Tempering oils
                         Meter oils                Transmission oils
    Lubricating oils     Journal oils              Railroad oils
                         Motor oils                Printing ink oils
                         Diesel oils               Black oils
                         Engine oils               Lubricating greases

                                              Medicinal────────── Salves, creams, and ointments
    Petroleum grease── Petrolatum                                  Petroleum jelly
                                                                   Rust-preventing compounds
                                              Technical────────── Lubricants
                                                                   Cable-coating compound

Residues
    Residual fuel oil    Wood preservative oils    Gas manufacture oils
                         Boiler fuel               Metallurgical oils
    Still wax─────────── Roofing material
                         Liquid asphalts
                         Binders────────────── Roofing saturants, road oils, emulsion bases
                         Fluxes
                                                              Briqueting asphalts
                                                              Paving asphalts
    Asphalts             Steam-reduced asphalts── Shingle saturants
                                                              Paint bases
                                                              Flooring saturants
                                                              Roof coatings
                         Oxidized asphalts──────── Waterproofing asphalts
                                                              Rubber substitutes
                                                              Insulating asphalts

Refinery
sludges
    Coke────────────── Carbon electrode coke, carbon brush coke, fuel coke
    Acid coke────────── Fuel
                         Saponification agents
    Sulfonic acid────── Demulsifying agents
                         Emulsifiers
    Heavy fuel oils──── Refinery fuel
    Sulfuric acid────── Fertilizers
```

CRUDE PETROLEUM

---

*From P. Albert Washer, Texas A. and M. College Extension Division (First Session).

Crudes differ substantially in their characteristics as sources of valuable nonfuel petroleum products. Some are rich in asphaltenes and are segregated for asphalt manufacture. Others yield excellent lubricating oil feedstocks or high-quality wax. Manufacturing these nonfuel products from the best feedstocks requires an assortment of specialized processing methods such as asphalt oxidation, phenol extraction, hydrogenation, acid treating, dewaxing, and grease blending, and careful blending with additives such as antioxidants, viscosity improvers, rust inhibitors, dispersants, and other specific agents. The distinction between natural and synthetic petroleum products becomes blurred; for example, many synthetic lubricants originate from petrochemical-derived base stocks. Other petroleum products are fortified with chemical agents or compounded with synthetic materials to achieve the proper balance of properties. The amount of petroleum which is not consumed but is turned into valuable and indispensable nonfuel products is about 5% of the supply of crude. See PETROLEUM; PETROLEUM PROCESSING.

[W. G. DUKEK]

Petroleum reservoir engineering

The applied science concerned with the development and operation of reservoirs for maximum economic recovery of oil or gas, or both. It is a composite technology requiring coordinated application of many special scientific disciplines, such as physics, geology, chemistry, and mathematics, as well as other engineering sciences, in the study of the complex reservoir systems.

The gross measures of a reservoir as an entity of commercial interest are (1) the amount of oil or gas, or both, initially present in the reservoir; (2) the rates at which the hydrocarbons can be withdrawn; and (3) the fraction of the original hydrocarbons in place which can be economically recovered.

Oil or gas in place. The amount initially in place can be determined simply as the volume of reservoir rock containing the hydrocarbons of interest times the content per unit volume of rock. The former can be obtained by multiplying the thickness of the productive formation at wells drilled for its development, and the total productive area defined and outlined by these wells. The oil or gas content per unit volume of rock is essentially given by the measured or calculated porosity, reduced by the amount of water saturation; from the reservoir volume of the hydrocarbons the volume at the surface can be calculated. The two volumes differ because of shrinkage of crude oil on evolution of its solution gas or expansion of free reservoir gas as its pressure declines to atmospheric. The formation thickness, porosity, and connate or interstitial water saturation should be considered as locally variable, to the extent that such variations can be determined from information obtained at the individual wells.

If the total measured oil-bearing or productive area in acres is indicated by A; the average formation thickness of the productive zone, excluding nonproductive members such as shales, in feet by h; the average porosity by φ; the average water saturation by S_w; and the formation volume factor of the oil by B_o; the initial surface (stock tank) oil content in place N is given by Eq. (1).

$$N = 7758.4\, Ah\varphi\, (1 - S_w)/B_o \text{ barrels} \qquad (1)$$

For the gas cap of an overlying oil reservoir, or a nonassociated gas reservoir, the gas content is determined by the same equation, provided the term B_o is replaced by B_g, the volume at reservoir conditions of a unit volume of gas at the surface. The gas content in cubic feet is obtained by multiplying the calculated volume in barrels by the factor 5.6146. For a complete accounting of the gas content, that dissolved in the oil (the solution gas) is calculated as the oil content times the gas solubility at the initial conditions, expressed as cubic feet per barrel. See PETROLEUM GEOLOGY.

Material balance equation. The initial oil in place can also be inferred from observations on the pressure behavior within the reservoir as oil and gas are produced. Making a material balance for the gas and oil by interrelating the volumes produced with the amounts assumed to be initially present, it is found that N will be given by the material balance equation, Eq. (2).

$$N = \frac{N_p\left(\dfrac{R_p - R_s + B_o}{B_g}\right) - G_i - \left(\dfrac{1}{B_{gi}} - \dfrac{1}{B_g}\right)G - \dfrac{W_e}{B_g}}{R_{si} - R_s - (B_{oi} - B_o)/B_g} \qquad (2)$$

G = initial reservoir volume of free gas phase present
W_e = net water intrusion volume
N_p = cumulative oil production
R_p = cumulative gas-oil ratio (total gas produced divided by N_p)
R_s = gas solubility in oil
G_i = cumulative gas injection, if any; subscript i elsewhere indicates initial values

In Eq. (2), N and G are the basic unknown constants; N_p, R_p, and G_i are the actual quantities of production or injection; and R_s, B_o, and B_g, as well as their initial values, are functions of the pressure and can be determined by experiments with the oil and gas. W_e is, in principle, a variable unknown.

If it is known that the water intrusion term is not of an important magnitude and can be neglected, the material balance equation reduces to one with the two constant unknowns N and G. Application of the equation to two or more time periods for which the other terms are known will then permit its solution for N and G.

If G may be taken as zero, but the water intrusion term W_e is not an insignificant factor, calculations of N on ignoring W_e will show an increasing trend as production continues. This in itself will be strong evidence that water encroachment is playing a role in the pressure performance. Extrapolation of the calculated values of N to the time of initial production will often indicate reasonable values of the true magnitude of N. Conversely, if the latter is known or can be estimated independently the material balance equation can be inverted to calculate the volumes of water encroachment corresponding to the production performance.

In reservoirs producing highly volatile oils or

condensate, it may be necessary to refine the gross material balance analysis by imposing compositional material balance requirements. Here the well-fluid composition is translated into molar volumes at reservoir conditions by use of equation of state correlations. On the basis of a preliminary reservoir volume estimate and the assumption of an original single-phase condition, as oil or gas phase alone, the initial total molar content and composition are obtained. For several values of reservoir pressure during depletion the corresponding residual molar content is determined by subtracting the total produced well-stream molar content from the initial levels. By applying the appropriate equilibrium ratios, the distribution of these reservoir residual molar contents between a gas and liquid phase and their respective volumes are then computed. If the calculated combined gas and liquid phase volume equals that assumed for the initial total reservoir volume, and if this calculated total remains substantially constant during the depletion process, it can be considered as a good estimate, with the assumption that there was originally present only a single phase and that there has been no substantial water intrusion during the period studied. If the latter assumptions are actually valid, a few trial-and-error checks with different initial estimates of the total reservoir volume will yield satisfactory constancy in the calculated reservoir volumes and establish it as reliable. If, however, the initial reservoir fluids include both gas and liquid phases, or appreciable water intrusion occurs during the course of production, or both of these conditions occur, then constancy of the computed total reservoir volumes will not be found and supplementary assumptions will have to be introduced regarding the magnitudes of the initial phase volumes and the rates of water intrusion. As in the case of the gross material balance method, the analysis then becomes more complex and more limited in its power to fix the initial reservoir volumes or other characteristics to high degrees of precision.

Darcy's law; permeability. The rates at which the hydrocarbon fluids can be withdrawn from a reservoir depend on the number of wells draining the reservoir, the average thickness of the formation, and the inherent transmissibility of the reservoir rock for these fluids. The last factor is expressed by the term permeability. Its significance lies in that it is the coefficient of proportionality in the basic physical law governing the flow of fluids through porous materials, namely, Darcy's law. In its generalized form, applicable to flow in a direction s inclined to the horizontal by the angle θ, it may be expressed as Eq. (3), where u is the volu-

$$u = -\frac{k}{\mu}\left(\frac{\partial p}{\partial s} + \rho g \sin \theta\right) \tag{3}$$

metric rate of flow per unit area, μ the fluid viscosity, ρ the fluid density, g the acceleration of gravity, $\partial p/\partial s$ the pressure gradient, and k the permeability. If u is expressed in cm³/sec, μ in centipoises, $\partial p/\partial s$ in atmospheres/cm and ρg in atmospheres/cm, then k is in darcys. The permeability unit, darcy, may be defined as the permeability of a porous medium which will carry a flow of 1 ml/ (sec)(cm²) of a 1-centipoise (cp) viscosity fluid

under a pressure or hydraulic gradient of 1 atm/cm.

In most practical applications it is convenient to express the actual permeability in the unit of the millidarcy (md), or thousandth of a darcy. Consolidated producing sands generally have permeabilities in the range of a few to several hundred millidarcys. The permeabilities of unconsolidated sands and fractured or highly vugular limestones often range in the thousands of millidarcys. Tight productive limestones frequently have matrix permeabilities even lower than 1 md.

In the above differential forms of Darcy's law, the permeability k is to be considered as being variable from point to point in the medium, if the latter is not uniform throughout, even though the fluid itself persists as a single-phase liquid or gas. It may also have different values for different directions of flow. The primary variable is the pressure p. The validity of Darcy's law has been established by extensive experimentation, although, as most linear relationships do, it tends to break down if the fluid velocities are indefinitely increased. Within such limits, which encompass virtually all situations of practical importance, the flow is considered to be viscous.

The permeability defined above is independent of the nature of the fluid, provided it occupies the whole of the pore space, and depends only on the character of the porous medium. The viscosity and density alone suffice to discriminate between one fluid and another. At low pressures, however, permeabilities measured for gas flow have been found by L. J. Klinkenberg to be higher than permeabilities which have been determined for liquid flow, but this effect is of minor importance except in laboratory experimentation on low-permeability materials.

Perhaps the simplest application of Darcy's law to a problem simulating one of actual oil and gas production relates to the steady-state horizontal flow into a well bore. Assuming the flow is radially symmetrical about the well, it is readily found that the pressure will increase as the logarithm of the distance from the center of the well. The rate of liquid flow is then given by Eq. (4), where k is the

$$q = \frac{2\pi kh(p_e - p_w)}{\mu B \ln r_e/r_w} \tag{4}$$

permeability, h the formation thickness, p_w the pressure at the well of radius r_w, p_e the external pressure at radius r_e, μ the viscosity, and B the formation volume factor of the liquid.

In common practical units this becomes Eq. (5),

$$q = \frac{0.003076kh(p_e - p_w)}{\mu B \log_{10} r_e/r_w} \text{ barrels/day} \tag{5}$$

where k is expressed in millidarcys, h in feet, μ in centipoises, and p_e, p_w in psi. The external radius r_e, though not precisely defined, represents the area from which the liquid is being drained or that where the pressure may be assumed to be p_e.

For similar steady-state radial flow of gas into a well, the rate of flow q_g at 60°F and 1 atm is found to be given by Eq. (6), where Z represents the aver-

$$q_g = \frac{0.3056kh(p_e{}^2 - p_w{}^2)}{\mu Z T_r \log_{10} r_e/r_w} \text{ ft}^3/\text{day} \tag{6}$$

age supercompressibility or deviation factor of the gas in the reservoir as compared to an ideal gas, and T_r is the reservoir temperature in °R (Rankine).

Multiphase production. Actual systems always involve more than one fluid phase. Gas and oil and water and oil are the most common flow stream combinations. But even when gas and oil are flowing individually as single phases, the presence of the immobile connate water must be taken into account.

Experimental studies have shown that the multiphase flow through porous media can still be described by the basic Darcy type of equation, provided it is applied to each distinct phase separately and the associated permeabilities are considered to be functions of the fluid-phase saturations. Indicating the oil, gas, and water phases by the subscripts o, g, and w, their simultaneous flow in the direction s, at angle θ with the horizontal, will be governed by Eqs. (7). The velocities u_o, u_g, and u_w

$$u_o = -\frac{k_o}{\mu_o}\left(\frac{\partial p_o}{\partial s} + g\rho_o \sin\theta\right)$$

$$u_g = -\frac{k_g}{\mu_g}\left(\frac{\partial p_g}{\partial s} + g\rho_g \sin\theta\right) \qquad (7)$$

$$u_w = -\frac{k_w}{\mu_w}\left(\frac{\partial p_w}{\partial s} + g\rho_w \sin\theta\right)$$

represent the volumetric flux rates of the corresponding phases. The pressures p are expressed individually, since they will change discontinuously across the curved interfaces between the phases. These pressure differences are referred to as capillary pressures, and may be considered as functions of the phase saturations, to be measured experimentally, although they are determined directly by the interfacial curvatures and interfacial tensions. They are often ignored in the treatment of large-scale systems, but they may be of importance near fluid fronts and in regions of rapid change of the fluid saturations.

Effective and relative permeabilities. The permeabilities k_o, k_g, and k_w are termed effective permeabilities. When expressed as fractions of the permeability for a single fluid phase, the absolute permeability, they are called relative permeabilities. The latter are always less than 1, reflecting the interference of each phase with the flow of the others. The saturations, of which they are functions, are expressed as the fractions of the pore space which they occupy, namely, S_o, S_g, and S_w, with a sum always equal to 1. The variation of the relative permeabilities with the phase saturations is referred to as the permeability-saturation relationship, and is illustrated by Fig. 1 for a mixture of gas and oil flowing through a Nichols Buff sandstone.

In interpreting these curves it is helpful to distinguish between the gas as the nonwetting phase and the oil as the wetting phase, referring to their respective tendencies preferentially to adhere to and wet the internal solid surface of the rock. It then will be observed that the permeability for the wetting phase—oil in the case of Fig. 1—drops rapidly as its saturation decreases from 100%, and falls to zero long before its saturation vanishes. This drop is due to the fact that the initial desaturation of the wetting phase occurs in the larger

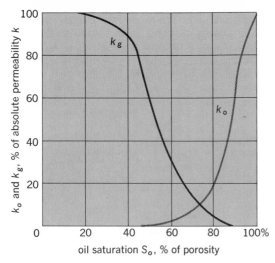

Fig. 1. Gas and oil relative permeability curves for Nichols Buff sandstone ($k = 0.5$). (*After Botset, AIME Trans.*)

pores, which contribute more to the permeability than their proportional volumetric content. Conversely, at higher degrees of desaturation of the wetting phase the latter is left in the finest pores and in disconnected flow channels—the irreducible saturation—permitting negligible flow capacity.

The nonwetting phase—gas—tends to remain in a discontinuously distributed state with zero permeability until sufficient saturation—the equilibrium saturation—is built up for continuity to be established. The larger pore channels so occupied then provide a rapidly rising permeability with increasing gas saturation. Virtually full single-phase permeability is achieved at less than complete liquid desaturation and while the smallest pores are still filled with liquid.

For the more general case where three phases—oil, gas, and water—are flowing, it is found that whereas the permeability to the wetting phase—

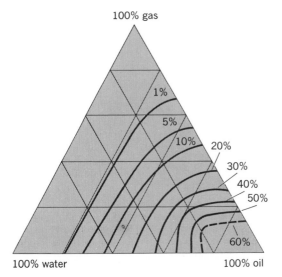

Fig. 2. Curves of constant oil relative permeability in flow of oil, gas, and water through an unconsolidated sand as functions of the fluid saturations. (*After Leverett and Lewis, AIME Trans.*)

water usually—is determined only by its own saturation and qualitatively follows a curve such as that for oil in Fig. 1, the relative permeability to the oil or gas may depend on the distribution as well as amount of the other two phases. These effects are illustrated in Figs. 2 and 3, showing, in triangular plots, the results found for an unconsolidated sand. Curves of this type can indicate many of the gross features of multiphase fluid flow in the porous medium of interest. For example, the nature of composite flow streams which can be maintained in different saturation ranges is illustrated in Fig. 4. It will be seen that simultaneous flow of all three phases in significant amounts will occur only in a very limited range of fluid saturations. By further reference to Figs. 1, 2, and 3 it will be observed that the composite permeability in multiphase flow will generally be but a nominal fraction of that for single-phase flow. The corresponding results for consolidated sands will vary with their individual pore structures and wettability, and even the direction of the saturation or desaturation processes. They may thus differ both qualitatively and quantitatively from the illustrative curves shown here. The latter are to be considered as only indicative of the significant gross characteristics of relative permeability data which ultimately control reservoir fluid flow and oil recovery.

Computing components of flow. The actual fraction of any composite flow stream contributed by a particular phase can be calculated by combining the corresponding Darcy equations. For example, when gas and oil are flowing simultaneously, the fraction f_g of the total volumetric flux q_t represented by the free gas phase is given by Eq. (8), where

$$f_g = \frac{\lambda_g}{\lambda_o + \lambda_g}\left[1 - \frac{\lambda_o}{q_t}\left\{\frac{\partial P_c}{\partial s} - (\rho_o - \rho_g)g\sin\theta\right\}\right] \quad (8)$$

the terms λ_o, λ_g are the oil and gas phase mobilities, that is, the ratio of their permeabilities—the effective values—to their viscosities. P_c is the capillary pressure $p_g - p_o$. The corresponding fraction for the oil phase f_o is simply $1 - f_g$. For oil-water

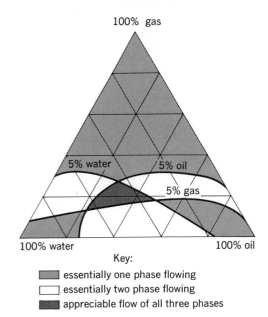

Fig. 4. Composition distribution of three-phase flow streams in unconsolidated sand as functions of fluid saturations. (*After Leverett and Lewis, AIME Trans.*)

flow streams the oil and water fractions are given by the same equation after appropriate changes in the subscripts.

Relative and effective permeabilities are of importance not only in determining the detailed dynamics of the displacement of oil from reservoir rocks but also control the absolute flow rates. In the above equations for rates of production from wells, the permeabilities must be corrected for the connate water even if it is in its irreducible state and immobile, although the steady-state single-phase flow equations will give only approximations of the actual flow magnitudes if both gas and oil are being produced.

Energy and producing mechanisms. Two basic, though elementary, observations underlie the essential principles of reservoir engineering. The first is that movement of viscous fluids such as oil through a reservoir rock involves the consumption of energy. Secondly, the withdrawal of oil from an oil reservoir requires a replacement of its volume in the reservoir space. Considered together these simple facts provide the framework for understanding the various types of oil-producing mechanisms.

Energy required for movement of oil from a reservoir rock into the producing wells may be drawn from four sources: (1) reservoir rock compression, (2) compression of reservoir and surrounding liquids, (3) compression of solution and free gas, and (4) gravity head above levels of withdrawals. Their individual importance depends not only upon the amount of such energy available but also on the effectiveness with which it can be used to displace the oil.

Upon release of the fluid pressure within the pores of a reservoir rock with removal of the oil and gas, the rock matrix will be subjected to increased compressive stress and compaction of the rock mass. However, in consolidated rocks the magnitude of such compaction within the reservoir

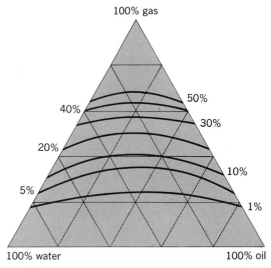

Fig. 3. Curves of constant gas relative permeability in flow of oil, gas, and water through an unconsolidated sand as functions of the fluid saturations. (*After Leverett and Lewis, AIME Trans.*)

itself will usually be too small to play a significant role in the oil expulsion processes. Observable and serious compaction has occasionally occurred in a few unconsolidated sand reservoirs, but in most cases the competence of the overburden apparently makes the compaction effects minor.

The expansion of the connate water within a reservoir, assuming the water is undersaturated, will also generally be a minor factor in oil displacement. A pressure reduction even as high as 3000 psi will lead to an expansion of the water of only about 1%.

Reservoir crudes have compressibilities of the order of 15×10^{-6}/psi, or about five times as great as water. Hence, if the reservoir oil is undersaturated by several thousand psi, its compression energy alone may provide an expansion in its volume and displacement into the producing wells of a few percent of the total volume of the oil in place. Occasionally oil recoveries in this range may be of economic value. Usually, however, the expansion displacement of undersaturated oils is only a supplement to production controlled by other mechanisms.

Water-drive reservoirs. The compression energy of the water in aquifers which adjoin and are in communication with oil reservoirs is often the dominant energy source and mechanism for oil recovery. These are the water-drive reservoirs. Their important characteristics are the volumetric extent and the continuity of the aquifer and its ability to bring water into the oil reservoir fast enough to push the oil out at commercially profitable rates.

To the extent that the aquifers may ultimately outcrop at levels higher than the oil reservoir, the corresponding hydraulic head would, in principle, provide an artesian drive for displacing the oil. Because of the long flow path, however, the rates of flow through the aquifer as a whole will usually be too low to exert an appreciable influence on the reservoir production. On the other hand, it is the very large volumes of water in extended aquifers which make their compression energy and expansion potential important factors in oil displacement. A circular 50-ft-thick aquifer of 50 mi equivalent radius will contain enough water to expand some 1,500,000 bbl for each psi of pressure reduction. Such levels of volumetric expansion could well support and replace the oil withdrawals from a great majority of oil-producing reservoirs. If the adjoining aquifer is of more limited extent or its large-scale continuity is interrupted by faulting or lithologic barriers, its contribution will be correspondingly limited.

A major phase of the study of a water-drive reservoir is the pressure and flow behavior of the aquifer itself. If the water compressibility and such contributory effects from the rock as may occur is taken as a constant c, the water density in the aquifer will be governed by an equation identical with the classical heat conduction equation, Eq. (9). To a very close approximation ρ can be replaced directly by the pressure p.

$$\eta \nabla^2 \rho = \frac{\partial \rho}{\partial t}$$

$$\eta = k/\varphi c \mu$$

(9)

Solutions of the above equations analytically, by

electrical network models, or by digital computers will show how the pressure in the aquifer will react to fluid withdrawals, which, in turn, can be related to the oil and gas production in the adjoining oil reservoir or the rates of water invasion into the latter. Conversely, from observed or assumed pressure histories at the water-oil boundary the rates of flow across the boundary and into the oil reservoir can be calculated. The geometrical and physical properties of the aquifer, about which advance information is often meager, can be determined in an empirical sense by trial and error adjustments so as to make the predicted pressure behavior match that observed in the course of producing the oil reservoir.

Many studies of this type and field observations show that the pressure in water-drive reservoirs is rate sensitive. That is, the average reservoir pressure depends not only on the cumulative oil production but also on the rates at which it has been withdrawn. Sharp increases in production rate will tend to accelerate the pressure decline per unit withdrawal. Cutbacks in withdrawal rate will generally retard the pressure decline and often even lead to buildups in reservoir pressure.

Water-drive reservoirs will permit maintenance of high rates of total fluid withdrawals through most of the economic life, although increasing volumes of water production will continually reduce the net oil rates. The reservoir pressure will tend to stabilize after initial declines which are necessary to induce the water to flow into the oil zone at sufficient rates to replace the oil withdrawals. Gas-oil ratios will rise only moderately during the producing life and in relation to the decline in reservoir pressure.

The water-drive mechanism is of special importance because of the high oil recoveries it often yields. Recovery factors as high as 50% of the initial oil in place are not uncommon, and under very favorable conditions they may be as high as 80%. The main factors controlling the recovery are the uniformity of the oil reservoir body and the viscosity of the reservoir crude. Variations of the permeability in the producing formation may lead to channeling of the invading water through high permeability zones and premature drowning out of the producing wells, so that while the oil displacement efficiency may be high in the invaded strata, the overall average sweep efficiency and recovery will be relatively low. The viscosity of the reservoir crude controls the local displacement efficiency. The latter will be reduced as the oil viscosity increases. Because of this factor water-drive recoveries in reservoirs producing oils of gravity lower than about 20° API may be considerably less than the 50% frequently observed for high-gravity–producing reservoirs [degrees API = (141.5/specific gravity) − 131.5].

The water-drive–producing mechanism controls the production in important reservoirs in all major oil provinces. Many of the large reservoirs in Texas along the Gulf Coast operate under water drives, as does the East Texas Field, the largest in the country. The two main sands in the gigantic Burgan Field in Kuwait are virtually perfect water-drive reservoirs.

Solution-gas drives. The gas dissolved in reservoir crudes is the most common energy source and displacement medium involved in oil production.

When it is the dominant agent for oil recovery, the producing mechanism is termed the solution-gas drive or depletion drive. The decline in reservoir pressure, which necessarily follows any appreciable production of oil and gas, will lead to liberation of solution gas within the pores of the rock and corresponding replacement of the volume of reservoir fluid withdrawn, if the oil is gas-saturated at the initial pressure, as it usually is. If the adjoining aquifer does not then supply an influx of water to provide for continued replacement of the oil withdrawals, the pressure will keep on falling, with continued additional evolution of dissolved gas. Ultimately the pressure and the dissolved gas will be dissipated and the economic life of the reservoir will be terminated.

The reservoir pressures and displacement processes in dissolved gas-drive systems basically are not rate sensitive. They depend only on the reservoir volume of total fluid withdrawals, although the rate and manner of oil production may affect the relative amounts of gas and oil produced and hence the total composite voidage for fixed quantities of oil recovery. As the evolved gas builds up the gas saturation, the permeability to the gas will grow, facilitating its escape to the producing wells without a corresponding increased displacement of the reservoir oil. As a result, after an initial period of rather constant gas-oil ratios, at the level of the initial solution value, the ratio will rise steadily to peaks of the order of 10–20 times as great, and then decline as the contribution of the free gas falls with decreasing pressure. The well- and field-producing capacities will also fall because the driving reservoir pressure drops and the permeability to the oil is reduced with increasing gas saturation.

Except for local effects about the producing wells themselves, the overall history of depletion of a gas-drive reservoir can be predicted by Eq. (10), where r is the fraction, if any, of the produced

$$\frac{dS_o}{dp} = \frac{\alpha S_o + (1 - S_o - S_w)\epsilon + \zeta S_o(\kappa - rR/\gamma)}{1 + \frac{\mu_o}{\mu_g}(\kappa - rR/\gamma)} \quad (10)$$

$$\alpha = \frac{B_g}{B_o}\frac{dR_s}{dp} \quad \epsilon = \frac{-1}{B_g^2}\frac{dB_g}{dp} \quad \zeta = \frac{\mu_o}{\mu_g B_o}\frac{dB_o}{dp}$$

$$\gamma = \frac{\mu_o B_o}{\mu_g B_g} \quad \kappa = \frac{k_g}{k_o}$$

gas which is returned to the reservoir. R, the current gas-oil ratio, can be related to the other variables as in Eq. (11).

$$R = R_s + \gamma\kappa \quad (11)$$

In these equations α, ϵ, ζ, γ, and μ are all functions of the pressure p determined by the properties of the gas and oil. The pertinent rock characteristics enter through κ, the ratio of the gas to oil permeability, as expressed as a function of S_o. The solution of Eq. (10) will show how the current oil saturation S_o in the reservoir declines with falling pressure. Solving Eq. (10) will also give the gas-oil ratio R at the corresponding period. The associated total oil recoveries per acre-foot of productive reservoir at any stage of depletion is given by Eq. (12).

$$N_p = 7758.4\varphi\left(\frac{S_{oi}}{B_{oi}} - \frac{S_o}{B_o}\right) \quad (12)$$

Solutions of Eq. (10) give the typical performance relationship of reservoir pressure and gas-oil ratio versus cumulative production as is observed in actual producing fields. By its construction, in which local well bore effects are ignored, it does not provide for any rate sensitivity of the recoveries.

Except for the mechanism of undersaturated reservoir oil expansion, solution-gas drives are the most inefficient producing systems. This is not because of the lack of sufficient solution-gas energy, but rather because of the internal bypassing of the gas as its saturation is built up so as to escape from the reservoir at high gas-oil ratios and little displacement effectiveness. The increasing oil viscosity, as the pressure declines and the solution gas is evolved, aggravates this effect. Ultimate economic recoveries are 10 to 30+% of the initial oil in place, decreasing generally as the API crude gravity decreases or as the oil viscosity increases.

Solution-gas–drive recovery has been the dominant producing mechanism in many of the older fields developed in the United States in the mid-continent area, West Texas, and California. In recent years appreciation of the low recovery efficiency of this drive has led to the application of fluid injection operations or limitation of the rates of production so as to facilitate potential water drives or gravity segregation assuming greater roles in the recovery mechanism.

Gravity-drainage drives. Gas caps overlying an oil zone contain additional compression energy for oil displacement to supplement that of solution gas. If, as may happen under high production rates and pressure differentials, this gas is permitted to break into the oil zone and join the solution gas flow stream it will be dissipated rapidly and will result in rather limited increased oil recoveries. Its effect will be similar to that of dispersed gas injection directly into the oil zone. If, however, the pressure gradients in the reservoir are so restricted as not to overbalance the gravity differential between the gas and oil, the gas cap will be preserved as a segregated driving piston on the oil zone.

Simple downward drainage by gravity of oil in a vertical column of a porous medium will lead to low residual oil saturations and high recoveries, limited only by the permeability, wettability, and capillary pressure characteristics of the rock. In actual reservoirs, with or without gas caps, it is generally not feasible to simulate pure gravity drainage because the corresponding rates of production will be too low for maximum economic return. The inherent downward flow capacity of the rock will be further restricted by the decreasing permeability to oil as the pressure declines and the solution gas is evolved.

In practice, when the upper part of a reservoir trap contains a gas cap the compression energy of the gas is permitted to supplement that of the gravity head so as to provide rates of withdrawal at economic levels. The gas cap also serves as a surge chamber to retard the pressure decline and hence lessen the rate of gas evolution within the oil zone and the associated effects of reduced oil permeability and increased oil viscosity.

The displacement effectiveness of the expanding gas on the underlying oil zone is decidedly rate sensitive, as may be inferred from Eq. (8). The gravity-drainage mechanism as a whole is likewise affected by the rates of the displacement processes, and becomes less efficient as the latter increase. To achieve the high recovery potential of the gravity-drainage mechanism in a gas cap reservoir, a balance must be made between the beneficial use of the driving pressure in the gas cap to support the desired levels of downflank production and the simultaneous deterioration in the displacement efficiency where the gas has invaded the oil zone. When such a balance is achieved, the gas cap will appear to expand downward as a piston with a relatively sharp gas-oil contact transition zone. Under favorable conditions of gravity-drainage operations, the upward buoyancy force on the gas evolved within the oil zone will overcome the downflank pressure gradients and the gas will migrate upstructure into the gas cap while the oil is flowing downward. Such countercurrent gas migration will aid in maintaining the reservoir pressure as a whole as well as high levels of oil saturation and permeability in the oil zone. Even when there is no initial gas cap, this process can lead to the formation of secondary gas caps with subsequent behavior essentially similar to that of a primary gas cap.

The pressure in gravity-drainage drives, in which effective segregation between the gas and oil is achieved, will decline slowly. Production rates and capacity will hold rather steady except that upstructure wells will be successively shut in as their producing levels are reached by the expanding gas cap. The gas-oil ratios will follow the trend of the solution ratio if downward gas coning is not permitted and the evolved solution gas is allowed to migrate into the gas cap.

To promote the general benefits of maintenance of pressure and production capacity part or all of the gas produced in gravity-drainage reservoirs is often returned to the reservoir through injection wells completed in the gas cap. If enough gas is injected to replace the reservoir withdrawals fully and prevent any pressure decline, the maximum potential of gravity drainage can be achieved, provided it is not nullified by excessive production rates and gas breakthrough. In any case, the higher pressure levels at which the reservoir is depleted will mean that whatever residual reservoir oil does remain undisplaced will have higher shrinkage and will represent less unrecovered stock tank oil than if the pressure had not been maintained.

It is preferable that the gravity-drainage mechanism, where potentially available, be allowed to function throughout a reservoir's producing life. But even when this is not feasible, gravity drainage may still serve to prolong the economic life by resaturating the lower part of the oil zone after its rapid depletion by solution-gas drive. The long-persisting settled production of old fields which have lost their pressure and reservoir gas often reflects the emergence of gravity drainage as a residual source of energy for bringing the oil into the well bores.

The main requirements for the effective development of the gravity-drainage mechanism are high structural relief, long oil column, and good vertical permeability or mobility. When these are present and full advantage is taken of them, recoveries as high as 70–80% of the original oil content can be achieved. Proportionately lower recoveries will be obtained when gravity drainage merely supplements the solution-gas–drive mechanism.

Gravity drainage, often aided by gas injection at the structural crest, has played an especially important role in the production of many of the oil reservoirs in eastern Venezuela. A number of large fields in West Texas and in the Gulf Coast have benefited from gravity drainage. Several of the major fields in Iran also appear to operate with significant gravity segregation.

Reservoir engineering analysis. The primary starting point for the analysis or prediction of the performance of an oil reservoir is its geological structure and environment. This information can only be satisfactorily obtained from wells drilled within the areal confines of the reservoir or its immediate vicinity. Geological, electrical, acoustic, and nuclear logs and the study of cores of the productive rock itself provide the basic data. These, plus determination of the properties of the oil and gas, may suffice to determine the total initial oil and gas contents of the reservoir by applying volumetric Eq. (1).

The real reservoir secrets unfold after the reservoir is placed on production and observations are made on its performance—the history of its production of oil, gas, and water; the history of its pressure; and the distribution of these among the various producing wells. These data combined through material-balance Eq. (2) may give further checks on the initial fluid contents as well as indications of the relative roles being played by the various producing mechanisms.

Quite often at least two or all major types of producing mechanism will contribute appreciably to the composite reservoir behavior, and its analysis will require setting up equations for combination drives. Partial water drives actually occur more frequently than complete water drives. As previously indicated, gravity drainage usually supplements the solution-gas–drive processes, at least to some extent. Even in water drives gravity segregation may be of benefit in minimizing channeling and water coning effects and thus improving the overall sweep efficiency.

The ultimate recoveries are determined by the magnitude of the average residual oil saturation when production is terminated.

Developments in technology. Injection of gas or water to supplement the native energy and oil displacement potential of the reservoir in its original state have become established practices. Such fluid injection operations are now generally undertaken early in the producing life of the reservoir, and as soon as it is determined that otherwise the recoveries will be limited to the inefficient levels of solution-gas drive or reservoir liquid expansion. For the older fields which were substantially depleted before the desirability of pressure maintenance was appreciated, secondary recovery installations have often been made in the form of gas repressuring or water flooding.

The ultimate limitation of oil recovery by presently established methods lies in the fact that the fluids—gas or water—which serve as the displacing phase are basically immiscible with the oil. Their surfaces of contact are therefore well-defined interfaces. Because of the tremendous interfacial area thus distributed throughout the microscopic pores of the reservoir rock, these represent correspondingly large total capillary forces and energies. Except for their beneficial action in inducing imbibition of water in water-wet systems these capillary forces offer resistance to multiphase flow at all saturations, and tend to break up any flowing phase into a discontinuous and immobile distribution as soon as its saturation falls to critical limits. When the latter state is reached, the capillary forces hold the residual oil unrecoverable in spite of continued passage of immiscible displacing fluids such as gas or water. These interactions are empirically expressed by the relative permeability-saturation relationships.

Miscible displacement. If the oil were displaced by a miscible fluid, the interfacial and capillary forces would be eliminated and local displacement efficiencies approaching 100% would result. This principle has long been applied in cycling gas-condensate reservoirs. Here dry liquid-stripped gas is injected into the formation to displace the condensate-containing reservoir gas and at the same time prevent declines in pressure and retrograde condensation and loss of its liquid content. Both gases are mutually miscible and the displacement proceeds without interface formation and capillary forces.

In the case of an oil reservoir, displacement by a miscible liquid such as the liquefied petroleum gases—propane and butane—would achieve similar results. However, to circumvent the economic burden of refilling the whole oil reservoir with these salable liquid products only a relatively small buffer zone or slug of the latter is used—up to 10% of hydrocarbon pore volume—and it in turn is displaced by gas. At pressures of the order of 1500 psi or greater, natural gas will also be miscible with the intermediate hydrocarbons or liquefied petroleum gases, at reservoir temperatures. Thus a continuous phase transition is developed, without the interfaces and capillary forces, from the reservoir oil to the miscible slug and to the final gas displacement phase. Field testing and practical applications of these general principles have shown that under favorable conditions effective sweep and high oil recoveries can be achieved by the miscible displacement process. Quite often, however, it is necessary to contend with a number of disturbing factors, such as the tendency of the miscible fluids to segregate to the upper formation levels because of their lower densities as compared to the oil phase, channeling due to reservoir inhomogeneity, and general miscible phase breakthrough because of the very unfavorable mobility ratio between the displaced oil and the ultimate displacing gas phase. Without corrective measures or variations in the details of the operating technique, little net benefit from the process may result. Very careful planning and close control are therefore required in the application of the miscible displacement principle.

In-situ combustion. A quite different type of technique for improving oil recovery is in-situ combustion. Though suggested many years ago, it has been studied and developed on the basis of modern reservoir engineering principles only rather recently. It has been investigated with special emphasis for application to heavy oil reservoirs where, because of high reservoir oil viscosity and very unfavorable mobility ratios for displacement by gas or water, the latter conventional recovery methods are of low and often noncommercial efficiency.

In essence, in-situ combustion consists of the injection of air into the producing formation to sustain burning of the oil in place, and to provide a flow of heat ahead of the combustion zone to lower the oil viscosity and increase the recovery and producing well productivity. The burning of the oil generates combustion product and vaporized oil gases. These, together with the bank of condensed water vapors, form a composite gas-and-water drive, moving toward the producing wells with the combustion front. In addition, the gases carry heat and raise the temperature of the rocks and fluids ahead of the combustion front, although the rapid attenuation of the temperature wave tends to delay the improvement in well productivity until the fire comes close to the producing wells. The vaporization process immediately ahead of the burning front leaves deposits of heavy oil residue or coke, and these serve as fuel for the final combustion reaction. As a result, the rock through which the fire passes is left essentially clean with all its oil displaced or burned out. About 15% of the original oil may be consumed in this manner, some 85% thus being in principle recoverable in the rock traversed by the fire. Because the heaviest components of the oil are used as fuel, the gravity (in degrees API) of the oil recovered by in-situ combustion may be somewhat raised over that of the original oil. The distribution of temperature and fluids from the air-injection well to the producing well which may occur in a typical in-situ combustion project are indicated in Fig. 5. The tendency of the burning to take place more rapidly in the cen-

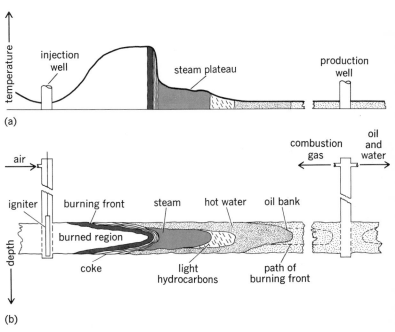

Fig. 5. Schematic diagram of the forward combustion process. (*a*) Temperature distribution 600–1200°F. (*b*) Cross section of formation.

ter of the oil-bearing formation is the result of cooling near its upper and lower boundaries.

The fire may be started by heaters or heating processes developed in injection wells or by spontaneous combustion of the reservoir crude resulting from the exothermic oxidation and absorption of the oxygen in the airstream. The operations are often carried on in pattern distributions of injection and producing wells similar to those used in water flooding.

Commercial success of in-situ combustion requires relatively high porosity and oil saturation so as to hold the ratio of air injection to oil produced down to economically feasible levels. It is also necessary to develop enough gas permeability through the formation to permit sufficient throughflow of air and combustion gases to sustain the burning. Appreciable reservoir bed thicknesses will tend to lessen the effect of heat losses to the top and bottom bounding strata. As in all displacement processes, uniformity of the producing section facilitates achieving high sweep efficiency. However, the rapid advance of the burning front through a limited zone, by gravity segregation of the air or gases or by permeability channeling, will accelerate the transmission of the direct thermal effects to the oil masses near the producing wells, which would otherwise retain their high viscosity and low mobility until late in the recovery life of the reservoir.

A number of field projects have confirmed the basic feasibility of carrying on in-situ combustion in oil reservoirs. For example, at the South Belridge Field in Kern County, Calif., some 50% or more of the 13° API oil in place in the test area was recovered within 18 months after air injection was started.

To expand the scope of applicability of the in-situ combustion principle, consideration has also been given to reversing the "forward" combustion process, as briefly described above, so as to propagate the burning front countercurrent to the flow of injected air and reservoir fluids, the latter being produced behind the front essentially in a vapor phase. While this process can be carried through under laboratory conditions, its practical value appears to be rather limited. Additionally, some recent studies have been focused on the use of in-situ combustion with medium-gravity oil reservoirs, where the burning would serve primarily to give high local recovery efficiency. And some laboratory and field tests indicate that in-situ combustion may be developed to serve as a tertiary recovery process to give a final cleanup of the residual oil which normally is considered to be unrecoverable by established techniques, such as water displacement. However, in these developments, as in all recovery processes, practical applicability will be limited to those conditions of the reservoir itself and its oil content which provide optimum combinations in meeting the requirements for achieving an effective sweep for the displacement mechanism at economically attractive rates and costs.

Steam injection. Another recovery process which is achieving appreciable success and has been expanding rapidly in heavy oil-producing areas is that of steam injection, especially the intermittent type of injection. In the latter, known as the "huff-and-puff" method, superheated steam is forced into the oil formation for a period of days or months, and then the wells are allowed to produce as long as they will do so at a profitable rate. Injection of steam is then resumed, wells are again placed on production, and the cycle is repeated until diminishing returns lead to abandonment.

Thermal methods of recovery, including both combustion and steam injection, have yielded about 200,000 bbl of oil per day in the United States. Moreover, it has been estimated that there have already been located more than 40×10^9 bbl of heavy oil whose recovery will be impractical without resort to thermal techniques.

Formation fracturing. Though not an oil recovery displacement process per se, formation fracturing is an operating practice which has resulted in important recovery increases from many reservoirs and has made exploitation from many others commercially feasible rather than immediate failures.

This involves creating fractures in the productive formation and keeping them propped open with sand or equivalent granular materials. When so prepared, the oil seeps from the tight reservoir matrix into the fractures, and thence into the well. The producing rates are thus raised to more profitable levels, and the total recoveries are increased correspondingly.

The various improvements in oil recovery processes discussed here represent only the major developments which already have some range of economic feasibility or warrant continued study. It is expected that these and the older methods will be materially improved by continued research, and it is hoped that at the same time still more novel and powerful techniques will be brought to light to reduce to a minimum the oil which will have to be abandoned as actually unrecoverable. *See* OIL AND GAS FIELD EXPLOITATION; PETROLEUM SECONDARY RECOVERY.

[MORRIS MUSKAT; HARVEY T. KENNEDY]

Bibliography: J. W. Amyx et al., *Petroleum Reservoir Engineering: Physical Properties*, 1960; F. W. Cole, *Reservoir Engineering Manual*, 1961; B. C. Craft and M. J. Hawkins, Jr., *Applied Petroleum Reservoir Engineering*, 1959; M. Muskat, *The Flow of Homogeneous Fluids Through Porous Media*, 1946; M. Muskat, *Physical Principles of Oil Production*, 1949; S. J. Pirson, *Oil Reservoir Engineering*, 2d ed., 1958.

Petroleum reservoir models

Physical or computational systems designed to simulate the response of petroleum reservoirs to procedures used to produce oil and gas. There are three basic requirements to achieve a useful model: (1) a quantitative description of the relevant properties of the reservoir (often called reservoir formation evaluation), (2) a mathematical description of the mechanics of fluid movement, and (3) a means of solving the resulting equations by mathematical analysis, constructing analogous physical systems, or computational techniques. The purposes of reservoir modeling are basically conservational: The yield of oil from a reservoir depends significantly on the operating procedure, and once depleted an oil reservoir may hold much unrecoverable oil widely dispersed and locked in the pores of the rock by capillary forces. Although theoretically most of such oil is not lost, for practical purposes the cost of recovering it is prohibitive. Modeling enables the selection of procedures to minimize the oil left when for economic reasons the reservoir is abandoned. Obviously this selection also serves the interest of the operator. *See* OIL AND GAS FIELD EXPLOITATION; PETROLEUM RESERVOIR ENGINEERING.

Simple model. The mechanics of fluid flow is complex. Fortunately, for reservoir modeling the flow of a single fluid through the pores of rock can be described simply by Darcy's law, Eq. (1), where

$$q = -\frac{k}{\mu}\left(\frac{\partial p}{\partial x} - \rho g \frac{\partial h}{\partial x}\right) \qquad (1)$$

q is the volume flowing per unit area in the direction of distance variable x, μ the viscosity, p the pressure, ρ the fluid density, g the acceleration of gravity, and h the height above a fixed reference level. The coefficient k is called the permeability and is a property of the rock. For single-phase flow Darcy's law can be combined with the equation of state, $\rho = \rho(p)$, and the statement of mass balance, $\frac{\partial}{\partial x}(\rho v) + \varphi \frac{\partial \rho}{\partial t} = 0$, to yield a partial differential equation which describes the transient behavior completely. If φ (the porosity or fraction of the rock volume containing the flowing fluid), h, k, and μ are constant and $\rho(p)$ follows the usual liquid compressibility relation, $\rho = \text{const } e^{cp}$ where c is the compressibility, then the local description of flow is Eq. (2). This is identical with the transient heat flow equation since

$$\frac{k}{\mu \varphi c}\frac{\partial^2 \rho}{\partial x^2} = \frac{\partial \rho}{\partial t} \qquad (2)$$

ient heat flow equation since $\dfrac{k}{\mu} = p$ plays the role of heat conductivity, φc the role of heat capacity, and ρ the role of heat potential, that is, temperature.

The foregoing process was described in one dimension for simplicity. It can as readily be carried out in two or three dimensions, although in the latter case the effect of $\partial h/\partial x$ must appear. If one found a level, thin reservoir with constant fluid and rock properties, at this point the two-dimensional formulation of Eq. (2) would satisfy the first two requirements of a reservoir model. The rock and fluid properties would be described quantitatively, and there would be a mathematical description of the fluid mechanics locally. Since methods are available to produce solutions to the transient heat equation, there is a complete model for this situation.

This model is simple and often inadequate. However, if a few details of the fluid mechanics are overlooked, it is an approximate model of the water-drive reservoir, a reservoir characterized by having its oil trapped in a very extensive rock containing mostly water. Such rocks are called aquifers. As oil is removed, the water expands to displace it. The displacement pattern can be deduced from the model, and the positions of the oil-water contact can be presented graphically with time. Thus alternative locations for the wells, rates, and effects of possible reinjection of water inadvertently produced with the oil can be ex-

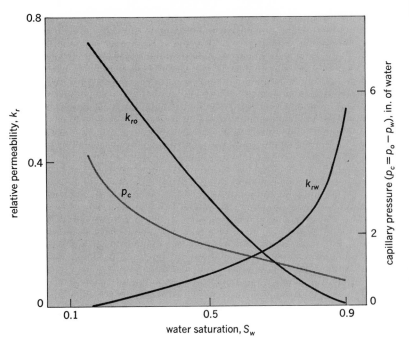

Relative permeabilities and capillary pressure for water and oil in a coarse unconsolidated sand where water saturation is decreasing.

amined. Moreover, if more than one oil reservoir lies in the aquifer, withdrawals in one reservoir produce significant pressure changes in the other. A classic example of modeling to examine such effects is the extensive use of the electric reservoir analyzer (a resistance-capacitor network) to approximate the transient equation describing the Woodbine aquifer, which supplies water for more than a dozen major oil reservoirs in eastern Texas.

A related use of such a model illustrates an interesting interaction between formation evaluation and model application. The rock in the reservoir has been sampled when the wells were drilled. The drilling density in the aquifer is much less, and hence much less is known about the distribution of k and φ in the aquifer. Modifications so the model admits variations in these properties are readily made, and such models have been used in an attempt to deduce the pattern of variation in these properties from pressure-withdrawal history. Unfortunately, data on the nature of rock distant from observation wells are subject to much uncertainty, although in practice the precision needed in such data is not great.

A similar use of modeling in formation evaluation within the reservoir is the well interference test. For data on properties near the wells or near the lines between wells such tests are valuable. Again data deduced from more remote parts of the reservoir are subject to large uncertainties, because the model shows that observed history can be matched all too often by widely differing proposed structures.

Evolution of realistic models. Thus far the major complications of reservoir fluid mechanics have not been considered. When oil is produced from the reservoir, it must be replaced by something. If adequate water exists or if water is injected, the oil may be displaced by water. When oil is removed and is not replaced, the pressure drops, and at some pressure (the bubble point) natural gas components are evolved from solution in the oil, and then oil is replaced by gas. Alternatively, methane or mixtures of methane and propane can be injected either with water or as a bank ahead of water. In any event inevitably the mechanics of the displacement must deal with the simultaneous flow of two or more fluid phases. Furthermore, the phase equilibrium of the multicomponent hydrocarbon system that makes up crude oil is pressure-dependent, so that if a gas phase is present, the phase equilibrium represents a side condition which must be satisfied locally as a function of pressure and composition.

Fortunately, however complicated the structure of porous rock, experimentally it is found that Darcy's law can be generalized as Eq. (3), where

$$q_i = -\frac{kk_{ri}}{\mu_i}\left(\frac{\partial p_i}{\partial x} - p_i g \frac{\partial h}{\partial x}\right) \qquad (3)$$

the subscript i refers to phase i, and k_{ri}, called the relative permeability, is a function of the relative amounts of the phases present. To minimize the complexity, suppose only water w and oil o are present. Then k_{rw} is an increasing function of S_w, the water saturation, which is the fraction of the rock void filled with water. Thus $S_o = 1 - S_w$, and k_{ro} is a decreasing function of S_w. Naturally from Eq. (1), $k_{rw}(1) = k_{ro}(0) = 1$.

In Eq. (3) the subscript appears on pressure. Fluids competing for the channels in the rock are separated by curved interfaces, the curvature is high because the pores are small, and significant pressure differences between phases arise. Experimentally, in a water-oil system $p_o - p_w$ depends on S_w. This function, $p_c(S_w)$, is called the capillary pressure. The illustration shows a plot of k_{ro}, k_{rw}, and p_c for a coarse unconsolidated sand. Unfortunately these functions are not unique functions of S_w, as they depend on the history of S_w at a point. The ones shown are for drainage conditions, that is, decreasing S_w.

To complete the model for this simple water-oil system, the flow equations are combined with the mass balance as before and include the $p_c(S)$ as a side condition. In contrast to the simplicity of Eq. (2), there results a nonlinear partial differential system. No known methods yield explicit solutions for this system, although the mechanics of water-oil flow is by far the simplest encountered in multiphase flow problems. Three approaches can be made to obtain solutions: (1) Simplify the physical description to yield a more tractable system. (2) Build a physical or other analog model whose defining equation is the same (scaling). (3) Obtain approximate solutions by using numerical methods and computing machines.

Meaningful results can be obtained by simplification. If incompressible flow in one direction is assumed and capillary effects are neglected, the system reduces to Eq. (4), a simpler nonlinear equation where $f_w(S)$ is the flowing water fraction,

$$\frac{\partial S_w}{\partial t} = -\frac{q}{\varphi}\frac{\partial f_w}{\partial x} \qquad (4)$$

for example, $f_w = [1 + (k_{ro}\mu_w^{-1}/k_{rw}\mu_o)]$ for x horizontal.

Model building by constructing physical analogs is an often used approach. Here problems are avoided by approximating solutions of messy equations about which too little is known mathematically. Conversely, models are costly, tend not to respond reproducibly, and require physical measurements and long periods to produce a sequence of studies to select optima. Nevertheless, the model theory is well developed, and the models provide good insight and are used extensively as the primary tool of investigation by some researchers. *See* OIL FIELD MODEL.

Computer models. To achieve Eq. (4), capillary effects and multidimensional flow were neglected. The resulting model was shown by two- and three-dimensional physical models to be wholly inadequate to treat stratified reservoirs, that is, ones having communicating layers of differing permeability. During displacement by water complex flow occurs: The water tends to run ahead in the most permeable sand, but the extent is limited by capillary and gravitational forces which cause significant cross-layer interchange. A meaningful model can be constructed by neglecting compressibility. This is expressed in Eqs. (5), where ∇

$$\frac{\nabla \cdot kk_{ro}}{\mu_o}(\nabla p_o - \rho_o g \nabla h) = -\varphi\frac{\partial S_w}{\partial t}$$
$$\frac{\nabla \cdot kk_{rw}}{\mu_w}(\nabla p_w - \rho_w g \nabla h) = \varphi\frac{\partial S_w}{\partial t} \qquad (5)$$

means either $\partial/\partial x$ or $(\partial/\partial x + \partial/\partial y)$.

Finite difference techniques have been used for approximate solutions to Eqs. (5) in one and two dimensions; agreement of result of the latter with measurements was most encouraging. In finite difference techniques the physical region is overlaid by a grid, usually rectangular, and approximations to the desired solutions are defined only at intersections of the grid. Algebraic relations between the variables are defined by approximating derivatives like those in Eqs. (5) by difference quotients. The study of the approximation of desired solutions by finite difference solutions is called numerical analysis.

Since the explosive evolution of computers in the early 1950s, the combined power of numerical analysis, rapid increases in computer capability, and a phenomenal decrease in cost of doing arithmetic has led to increased emphasis on computer models. Much of the necessary complexity in realistic modeling of reservoirs has been attempted. For example, a remarkably bold effort was undertaken to treat simultaneously the flow of three phases and a seven component gas-oil phase equilibrium.

Future models. The significant literature, the power of numerical methods, and increasing computer capabilities might suggest that reservoir modeling on computers is a solved problem. This is far from true since newer and more complex producing procedures continue to appear, and significant problems remain in certain applications to existing procedures. For example, even systems like the one described by Eqs. (5) treated by straightforward finite difference methods for conditions that lead to sharp S_w gradients require an absurd refinement of the grid. This same situation arises in treating miscible displacement, flooding of oil by solvents like propane. A number of not-altogether-successful modifications of finite difference formulations to cope with the movement of steep gradients have been studied.

The serious application of Galerkin methods has also been studied, applied, and found somewhat more effective at treating this problem. Galerkin methods approximate the solution of differential systems, such as Eqs. (5), as linear, time-dependent combinations of a set of space functions called basis function, for example, $S_w = \Sigma \ \alpha_i(t) \cdot v_i(x)$, where $\{v_i\}^N$ is a linearly independent set of functions of the multidimensional space variable x. Putting this approximation and a similar one for p_w into Eqs. (5), performing some minor algebraic manipulation, multiplying by the basis functions one at a time, and integrating over the spatial domain yields a system of ordinary differential equations in $\{\alpha_i(t)\}$. It has been shown that solutions for $\{\alpha_i(t)\}$ yield approximations to systems like Eqs. (5), which in some sense are the best which can be obtained with the rather arbitrary basis set $\{v_i\}$. Since their application appears to be efficient, the Galerkin models promise to play a prominent role in future reservoir modeling.

The final chapter in modeling does not end here. Many involved reservoir processes have been modeled by clever methods too numerous to catalog here. Analysis of methods and the seasoning of application take time. Further, as computers become faster and arithmetic cheaper, more researchers will develop methods. Yearly, an ever wider collection of modeling techniques to meet specific needs becomes available; yet the needs always seem to outnumber the means. It is clear that much work remains before the ultimate goal is achieved, and a model responds exactly like a real reservoir.   [HENRY H. RACHFORD, JR.]

Bibliography: W. J. Bernard and B. H. Caudle, Model studies of pilot waterfloods, *Trans. AIME*, 237:404, 1967; S. E. Buckley and M. C. Leverett, Mechanism of fluid displacement in sands, *Trans. AIME*, 146:107, 1941; J. Douglas, Jr., T. Dupont, and H. H. Rachford, Jr., The application of variational methods to waterflooding problems, *Can. Inst. Mining Metallurgy J.*, in press; J. Douglas, Jr., D. W. Peaceman, and H. H. Rachford, Jr., A method for calculating multidimensional immiscible displacement, *Trans. AIME*, 216:297, 1959; A. O. Garder, Jr., D. W. Peaceman, and A. L. Pozzi, Jr., Numerical calculation of multidimensional displacement by the method of characteristics, *Soc. Petrol. Eng. J.*, 4:26, 1964; H. H. Rachford, Jr., Numerical calculation of immiscible displacement by a moving reference point method, *Soc. Petrol. Eng. J.*, 6:87, 1966; J. F. Roebuck, Jr., et al., The compositional reservoir simulator, Case I: The linear model, *Soc. Petrol. Eng. J.*, 9:115, 1969; A. G. Spillette and R. L. Nielsen, Two-dimensional method for predicting hot waterflood recovery behavior, *Trans. AIME*, 243:627, 1968; H. L. Stone and P. L. T. Brian, Numerical solution of convective transport problems, *Amer. Inst. Chem. Eng. J.*, 9:681, 1963; H. J. Welge, A simplified method of computing oil recovery by gas and water drive, *Trans. AIME*, 195:91, 1952.

Petroleum secondary recovery

The process of removing oil from its native reservoir by the use of supplemental energies after the natural energies causing oil production have been depleted. Petroleum secondary recovery contrasts with primary recovery, which is the oil production resulting from indigenous reservoir energies. Advancing principles of technology and conservation demand that natural energy be supplemented soon after discovery of a reservoir; therefore the best practices combine the primary and secondary recovery periods. However, there are many reservoirs which have been depleted without benefit of supplemental energy, and in the narrowest sense, secondary recovery applies to the further development of these depleted reservoirs.

Secondary recovery was first practiced in the older and shallower petroleum reservoirs of the Appalachian region. It has since spread to all oil-producing regions of the world and has doubled the producing life of some oil fields. When combined with primary recovery, it is applied to deep reservoirs. When practiced on depleted reservoirs, it is generally limited by economic factors to reservoirs shallower than 3000 ft.

Energy supplement and well patterns. Energy is supplemented by introduction of either gas or water under pressure into the reservoir. The use of gas is commonly known as gas-drive or gas-repressuring, the use of water as waterflooding. The injected fluid drives the oil remaining in the reservoir to the vicinity of production wells, from whence it can be lifted to the surface, and also takes up the space within the reservoir previously occupied by the oil.

Two types of wells, injection wells and production wells, are required for secondary recovery operations. Standard patterns have evolved for the arrangement of these wells. Locating wells in patterns permits intensive development of a given land area and ensures the maximum penetration of injected fluid to all parts of the reservoir. In the early history of secondary recovery, a single injection well was surrounded by a large number of production wells. This pattern, known as a circle drive, is still used for gas injection operations. Another pattern, the line drive, is a line of injection wells offset by a line of production wells.

The most common well pattern is the five-spot. Square networks of injection wells and production wells interlock so that each injection well is at the center of a square consisting of four production wells, and each production well is at the center of a square consisting of four injection wells. Another standard pattern is the seven-spot, which consists of injection wells located at the vertices of hexagons with a production well in the center of each hexagon. By exchanging the roles of injection and production wells in the seven-spot pattern, the four-spot pattern is obtained. Nine-spot patterns are five-spot patterns with additional injection wells added at the midpoints of the sides of each injection well square.

The spacing between injection wells and production wells depends upon local physical conditions of the petroleum reservoir and upon economic factors. The resulting well densities range generally from one well per acre to one well per 40 acres. Spacing economics is controlled by the amount of gas or water that can be injected into or produced from a single well. The prediction of the amount of fluid which an injection well will handle is therefore one of the most critical technical points in planning a secondary recovery project.

This amount of fluid depends upon factors such as the permeability of the reservoir rock, the viscosity of the reservoir oil, the fraction of the reservoir pore space that is filled with oil, the thickness of the reservoir formation, the reservoir pressure, and the available surface pressure.

Factors of effectivity. The efficiency of a secondary recovery operation is determined by the effectiveness with which the injected fluid displaces oil from that part of the reservoir which it invades, and the degree to which the injected fluid can be made to invade all parts of the reservoir.

Displacement and retention factors. Even under the most favorable conditions of fluid invasion, it is not possible to replace all the oil in a given segment of reservoir rock. The rock contains a complex and interconnecting assemblage of small channels which are not uniform in shape or in size. Hence it is possible for the invading fluid to bypass and trap some of the oil-containing channels or oil globules. This residual oil is held in place by the strong capillary forces that are operative. On the basis of its oil-retentive properties, a reservoir rock may be classed as either oil-wet or water-wet. In the former, the residual oil may be held as a film or as filling the most minute pore spaces. In the latter, the residual oil may be held as trapped globules or islands within the larger pore spaces.

Fluid segregation problems. Because oil is less dense than water and more dense than gas, there may be a segregation of injected water to the bottom part of the reservoir, or of injected gas to the top part. In either case, the injected fluid will advance toward the production well through only part of the reservoir. Complete entry of the invading fluid to all parts, then, is not possible without production of large amounts of injected fluid from the production well, a procedure which is necessarily costly. The possibility of encountering fluid segregation may lead to the deliberate locating of wells so that injection of gas is to the top of a reservoir structure or injection of water to the bottom, making use of the segregation tendencies. This cannot be done, however, in flat, thin reservoirs.

Reservoir rock properties are seldom uniform, and in particular, the rock may vary in its permeability, that is, in its capacity to conduct fluid. The injected fluid takes the path of least resistance, and by the time it has invaded the higher permeability channels completely, it will have invaded the lower permeability channels only partially. If the permeability channels are sufficiently stratified, special well-completion techniques can be used to promote uniform fluid invasion.

If the resistance to flow is higher for the oil in the reservoir than for the invading fluid, the invading fluid will seek the production well by the most direct flow path. As a result, injected fluid will reach the production well before all of the pattern area has been invaded. The area of a pattern which has been invaded by the time the invading fluid breaks into the production well is termed the areal coverage of the pattern.

Mobility ratio and area extent. The ratio of flow resistances between injected fluid and reservoir oil is known as the mobility ratio. With a mobility ratio of unity, the area coverage for the five-spot pattern is 72.3%; for the seven-spot and four-spot, it is 74%. Each pattern has a characteristic coverage for each mobility ratio. When the flow resistance of

the reservoir oil is extremely low, the coverage will approach unity for any type of pattern.

As the flow resistance of the reservoir oil increases in comparison to that of the invading fluid, the achieved areal coverage will decrease. For this reason, injected gas generally produces a smaller areal coverage than injected water. However, because of the more favorable economic factors for handling gas, the injection of gas can be continued for long periods of time after gas arrives in the production wells. After water arrives at the production well, water injection generally can be continued only until the ratio of water-oil volume reaches 1:100 or less.

Oil recovery and residue. The percentage of oil recovered by secondary methods is a composite result of the effects that have been noted, namely, of the geometrical arrangement of pores and the capillary forces, of the gravity segregation, of the heterogeneous nature of the reservoir, and of the areal coverage that can be achieved. Under the most favorable circumstances, one may expect residual oil following secondary recovery to be as low as 15–20% of the pore space. With some of these factors operating to a disadvantage, residual oil following secondary recovery may be as high as 50% of the pore space. The residual oil is highest where rocks have complex porous structure, where the reservoir oil is quite viscous, or where there are wide variations in reservoir permeability.

Hazards from fluid contamination. In water flooding, considerable attention must be given to the purity of injected water. The reservoir rock acts as a filter to remove suspended material in the well bore and clog the formation. Hence, no suspended material can be permitted. Even bacteria filter on some rock formations and reduce the injection rate. Consequently, treatment for removal of bacteria is often necessary. Highly corrosive waters must also be avoided.

There is always the possibility that chemical interactions will occur between ions contained in the injected water and those present in the native reservoir waters or reservoir minerals. Injected water may also change the structure of reservoir clay material, resulting in reduction of flow capacity.

In the injection of gas, principal attention is paid to the removal of materials that might condense within the reservoir, produce corrosive action on operating equipment, or yield oxidation within the reservoir.

Developments in technology. Advancing technology has continued to seek means to improve secondary recovery. It has long been known that one of the factors which leads to incomplete recovery of oil from a reservoir could be eliminated if the displacing fluid were miscible with the reservoir oil. The miscible displacement process, which consists of replacing the reservoir oil by a fluid with which it will mix completely, was tested in laboratories before being used commercially. Commercial projects consist of injections of liquefied petroleum gas (propane and butane) followed by injections of natural gas into petroleum reservoirs.

Another method for reducing capillary forces is the addition of surface-active materials to a waterflood, removing oil much as one would remove grease with a detergent. The principal disadvantage of using surfactants is that they are adsorbed on the reservoir rock. Their use is not economic unless means can be found for continually desorbing or replacing the surfactant. Other proposed additives in waterflooding are soluble gases such as carbon dioxide.

Thermal and in-place combustion methods have passed the experimental stage and are being utilized in California, Texas, Wyoming, and other states. In these methods a fire, or combustion process, is started in the reservoir at an injection well. By the continued introduction of gas containing oxygen or other material to support combustion, the combustion wave is driven through the reservoir toward the production well. As the combustion wave moves forward, part of the oil is distilled and driven forward, and part is burned to produce the heat necessary for continuing the combustion drive. This process has successfully recovered oil from the Athabaska Tar Sands in an experimental test in northern Alberta, Canada.

A steam-drive test begun in 1957 by Shell Oil Co. in the Mene Grande field in western Venezuela showed results in 1959 for the first effect of steam-soak operations. Steam is utilized in two thermal recovery processes: (1) as a stimulation medium to heat the area of the reservoir around the well bore (called variously steam stimulation, huff-and-puff, cyclic steam injection, and steam soak), and (2) as a displacement medium to drive crude oil to producing wells in a manner similar to that of the waterflood.

Polymer flooding has been utilized in field applications in the United States and other countries. Water-soluble polymers (partially hydrolyzed polyacrylamides, developed as waterflood agents to provide mobility control) have been successful in increasing water viscosity and stability to provide a sealing barrier across the reservoir. One application exceeded by 40% the estimated recovery for an ordinary waterflood.

An oil recovery process called Maraflood, developed by the Marathon Oil Co. in 1962, utilizes a slug of micellar solution that has sufficient viscosity to avoid an unfavorable mobility ratio with the oil and water it displaces. These micellar solutions, composed of hydrocarbons, water, alcohols, and surfactants, miscibly displace crude oil from reservoir rocks. The micellar solution is displaced by a mobility buffer, a bank of reduced-mobility fluid, preferably an aqueous polymer solution. See OIL AND GAS FIELD EXPLOITATION; PETROLEUM RESERVOIR ENGINEERING.

[E. G. DAHLGREN]

Bibliography: American Petroleum Institute, Secondary Recovery of Oil in the United States, 1950; T. C. Frick (ed.), Petroleum Production Handbook, vols. 1 and 2, 1962; M. Muskat, Physical Principles of Oil Production, 1949; National Petroleum Council, Impact of New Technology on the U.S. Petroleum Industry 1946–1965, 1967; C. R. Smith, Mechanics of Secondary Oil Recovery, 1966; P. D. Torrey, Evaluation of United States oil resources as of Jan. 1, 1966, Oil Gas Compact Bull., December, 1966.

Photovoltaic cell

A device that detects or measures electromagnetic radiation by generating a potential at a junction (barrier layer) between two types of material, upon absorption of radiant energy. Typical junctions for photovoltaic cells are silicon–silicon boride, selenium-iron, and copper oxide–copper.

The detection or measurement is performed by connecting the cell directly to a galvanometer, whose reading is a function of the intensity of radiation falling on the cell.

Photovoltaic cells are used as exposure meters in photography and are used in automation to energize sensitive relays. The solar battery, a type of photovoltaic cell, may be used as a source of electricity for portable radios and telephone relays.

A photovoltaic cell has practically no dark current. It is generally not adapted to be used with amplifiers. See PHOTOVOLTAIC EFFECT; SOLAR BATTERY.

[JEAN J. ROBILLARD]

Bibliography: H. Pollack, Photoelectric Control, 1962; Proceedings of the World Symposium on Applied Solar Energy, Stanford Research Institute, 1956.

Photovoltaic effect

A term most commonly used to mean the production of a voltage in a nonhomogenous semiconductor, such as silicon, by the absorption of light or other electromagnetic radiation. In its simplest form, the photovoltaic effect occurs in the common photovoltaic cell, used, for example, in solar batteries and exposure meters. The photovoltaic cell consists of an np junction between two different semiconductors, an n-type material in which conduction is due to electrons, and a p-type material in which conduction is due to positive holes. When light is absorbed near such a junction, new mobile electrons and holes are released, as in photoconduction. An additional feature of a photovoltaic cell, however, is that there is an electric field in the junction region between the two semiconductor types. The released charge moves in this field. This current flows in an external circuit without the need for a battery as required in photoconduction. If the external circuit is broken, an "open-circuit photovoltage" appears at the break.

In certain rather complex electrolytic systems, illumination of the electrodes may give rise to a voltage classed as photovoltaic. See PHOTOVOLTAIC CELL; SOLAR BATTERY.

[L. APKER]

Bibliography: L. Azaroff and J. J. Brophy, Electronic Processes in Materials, 1963; A. Van der Ziel, Solid State Physical Electronics, 1957.

Pinch effect

A name given to manifestations of the magnetic self-attraction of parallel electric currents having the same direction. Since 1952 the pinch effect in a gas discharge has become the subject of intensive study in laboratories throughout the world, since it presents a possible way of achieving the magnetic confinement of a hot plasma (a highly ionized gas) necessary for the successful functioning of a thermonuclear or fusion reactor.

Ampère's law. The law of attraction which describes the interaction between parallel electric currents was discovered by André Marie Ampère in 1820 and can be stated as follows: The force of attraction in dynes per centimeter length between two thin straight wires r cm apart carrying cur-

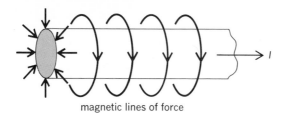

Fig. 1. Pinch pressure on a current-carrying conductor. Arrows at left show direction of pinch pressure.

rents of I_1 and I_2 amperes (amp), respectively, is $I_1 I_2 / 100r$. The law applies equally to the attraction between the individual components of a current in a single wire, in which case, for a cylindrical wire of radius r cm carrying a total current of I amp, it manifests itself as a compression force on the material of the wire (Fig. 1), given by $I^2/200\pi r^2$ dynes/cm². For a uniformly distributed current in the wire, the pressure reaches this value on the axis. *See* AMPERE'S LAW.

For the electric currents of normal experience, this force is small and passes unnoticed, but it is significant that the pressure increases with I^2. At 100,000 amp, the pressure amounts to about 1 atm for a wire of 1-cm radius, but at 10^8 amp, the pressure is about 10^6 atm, which is considerably greater than the pressure produced by the detonation of trinitrotoluene (TNT).

Manifestations. The pinch effect first showed up practically in certain early types of induction electric furnaces in which large low-frequency alternating currents of the order of 100,000 amp were induced at low voltage into a horizontal ring-shaped fused-metal load (Fig. 2). At these currents, the pinch pressure can be larger than the hydrostatic pressure exerted by the fused metal, and as the formula above shows, the pinch pressure increases as the radius of the conductor decreases. Consequently, once the process starts, the pressure at a narrow neck in the ring of fused metal can squeeze out the fluid metal until the neck pinches off completely, cutting off the current. This led to very uneven heating of the charge. The term pinch effect was given to this process by C. Hering in 1907. The technical difficulty was eventually overcome by making the plane of the ring vertical and submerging it deeply below the free surface of the fused metal. The force of the pinch effect has also been known to manifest itself by a crushing of tubular conductors exposed to

large impulsive currents such as occur in lightning strokes or high-power short circuits.

Thermonuclear applications. One of the conditions for the attainment of a profitable balance between energy expended in heating and energy released in fusion from a thermonuclear reaction in a plasma composed of deuterium and tritium (DT, the most favorable case) is that the temperature shall be not less than about 10 kev (1.16 times 10^8 °K). This is an enormous temperature and can be attained and maintained only if the hot plasma is effectively isolated from the material walls of the container by vacuum. The isolation has the double function of preventing cooling by contact with matter at normal temperatures and of preserving the purity of the plasma from foreign atoms which could upset the energy balance. For the plasma to remain confined under these conditions, its outward pressure must be balanced by inward pressure of nonmaterial origin, that is, a magnetic field. A profitable energy balance also depends on the density n of the confined plasma and on τ, the time it is confined: The product of $n\tau$ must exceed a certain minimum, which is 10^{14} ion cm³ sec for DT. *See* FUSION, NUCLEAR; MAGNETOHYDRODYNAMICS; THERMONUCLEAR REACTION.

There are only a limited number of ways in which a magnetic field can be arranged around the plasma to hold it together, and one of these methods is the pinch effect. A fusion reactor using this type of confinement would ideally be a toroidal tube in which the confined plasma would float, the plasma carrying a large electric current induced in it by magnetic induction from a transformer core passing through the axis of the torus. The fundamental equation for the pinch effect in a gas, derived theoretically by W. Bennett in 1934, gives the current I required for the inward pinch pressure to balance the outward gas pressure, as shown in the equation below, where I is the total current in

$$I^2/200 = Nk(T_e + T_i)$$

amperes, N is the number of electrons (also the number of ions) per centimeter length of the pinch (independent of the radius), $k = 1.4 \times 10^{-16}$ erg/°K (Boltzmann's constant), and T_i and T_e are the temperatures in degrees Kelvin of the ions and electrons, respectively.

Experimental studies. In general, two types of apparatus have been used in studies of the pinch effect: (1) straight discharge tubes composed of quartz or porcelain with a metal electrode at each end, intended for short-duration studies, in which the cooling of the plasma by the relatively cold electrodes is slight during the time of the experiment, and (2) toroidal discharge tubes, also composed of quartz or porcelain, in which the pinch is endless and consequently is more effectively confined than in the first type of apparatus, and the current is induced into the discharge by magnetic coupling to a primary winding. In both cases, currents of 50,000–500,000 amp are obtained with gradients of 10–100 volts/cm along the pinch. The primary power source is a charged condenser with a capacity of 4–50 μf, charged to 10–100 kv. The current in the discharge rises rapidly, reaches a peak in a few microseconds, and decays to zero in a damped oscillation.

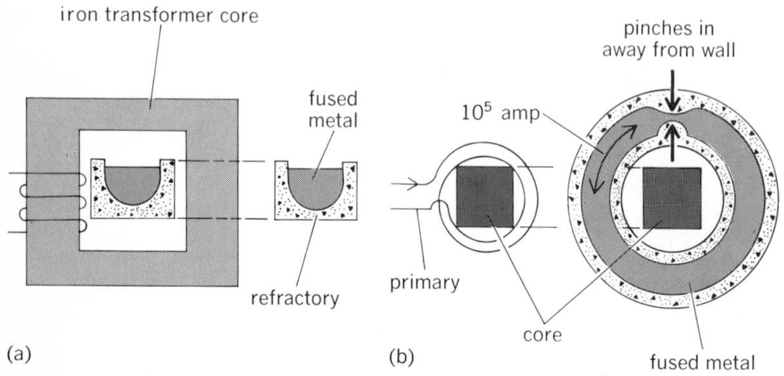

iron transformer core

fused metal

pinches in away from wall

10^5 amp

primary

core

refractory

fused metal

(a)   (b)

Fig. 2. Early type of ring induction electric furnace. (a) Side view. (b) Plan view.

Instability. Characteristically, as can be seen by high-speed photography, the discharge forms at the inner surface of the discharge tube wall and contracts inwardly, forming an intense line on the axis (Fig. 3); the wave usually rebounds slightly; the contracted discharge rapidly develops necks and kinks; and in a few microseconds, all structure is lost in an apparently turbulent glowing gas which fills the tube. Thus, the pinch turns out to be unstable, and the plasma confinement is soon lost by contact with the wall. The cause of the instability is easily seen qualitatively; the pinch confinement can well be described as being caused by the magnetic lines encircling the pinch behaving as slippery rubber bands which are stretched longitudinally but which are in **compression** transversely (Fig. 4). For a uniform cylindrical pinch, the magnetic pinch pressure is everywhere equal to the outward plasma pressure, but at a neck or on the inward side of a kink, the magnetic lines crowd together, creating a higher pressure than the outward gas pressure. Consequently, the neck contracts down further, the kink cuts in on the concave side and bulges out on the convex side, and both perturbations grow.

The instability has a disastrous effect on τ, limiting it to 10^{-6} sec or less in light atom plasmas such as DT. This not only makes it difficult to reach thermonuclear temperature from room temperature in the short time available for heating, but through its effect on $n\tau$ also forces the n required for a positive energy balance to extremely high values.

It must be noted that neutrons have been produced by deuterium pinches in large numbers. For a time (1952–1953), they were believed to be evidence of thermonuclear reaction, but it has since been shown that they are emitted preferentially in certain directions and are associated with the instability of the pinch and the violent accelerations it produces. Such neutrons (phony neutrons) are not a product of thermal collision and are not thermonuclear.

Great efforts have been devoted to overcoming the basic instability of the simple pinch. One such measure was to add an axial magnetic field by means of an external winding round the pinch tube. This might be expected to resist the sausage and kink deformations by stiffening the discharge. Also, the walls of the tube can be made highly conducting; this has the effect of trapping the magnetic field between the pinch and the wall, cushioning and reflecting the moving pinch back to the center.

Levitrons. A much more powerful measure for stabilizing the pinch consists of adding (1) a stiff current-carrying conductor down the axis inside the pinch and (2) a strong longitudinal magnetic field from an exterior winding outside it. The plasma is, in effect, sandwiched and pinched into a tubular region, between magnetic fields having directions differing by 90°. This so-called hard-core pinch shows greatly diminished instability, so much so that, for short time scales ($\sim 10^{-4}$ sec) and straight tubes, it is stable.

Hard-core and levitron pinches have been studied at Lawrence Radiation Laboratory, Livermore, Calif.; Culham, England; and Fontenay-aux-Roses, France, and are associated with S. A. Col-

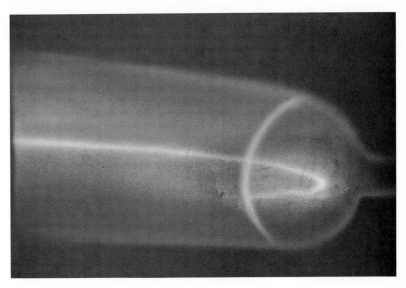

Fig. 3. Xenon pinched discharge in Perhapsatron torus.

gate, H. P. Furth, and P. Rebut, respectively. These geometries are not entirely stable; a more slowly growing tearing mode instability remains. This seems to exhaust the possibility of stabilization of the pinch by static methods, but there remains the possibility of dynamic stabilization by rapidly changing fields. The simple pinch continues to be studied, however. The reason for this is that the reciprocal functional relation between n and τ for power production means that there is always a possibility of achieving a net power output, no matter how much τ is cut down by instabilities, by resorting to very high n. Such plasmas may require heroic extremes of pulse electric power to heat and confine them, but the only real limitation seems to be that the resultant output bursts of thermonuclear power may destroy the machine. For such pulsed systems, the pinch must always rank highly—it is uniquely the most efficient magnetic confinement system expressed in terms of magnetic energy expended per unit of plasma energy confined, and furthermore the pressure of the plasma it confines can exceed the strength of any known material.

The term θ pinch had come into wide usage to denote an important plasma confinement system

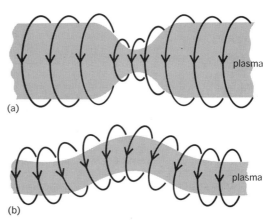

Fig. 4.   Instability. (*a*) Sausage type. (*b*) Kink type.

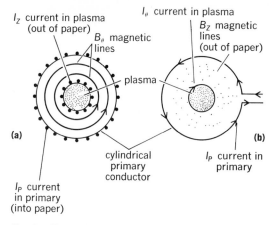

Fig. 5. Two geometries for plasma confinement systems. (*a*) Z pinch, and (*b*) θ pinch.

which relies on the repulsion of oppositely directed currents and which is thus not in accord with the original definition of the pinch effect (self-attraction of currents of the same direction). Plasma confinement systems based on the original pinch effect are known as *z* pinches. Figure 5 defines the two geometries.

Tokamak. The toroidal B_z stabilized pinch became the subject of intense interest when L. Artsimovich, director of the Kurchatov Institute, announced in mid-1969 the achievement of a 20-msec confinement and an ion temperature of 0.5 kev in the T3 Tokamak. Tokamak is essentially a low-density, slow *z* pinch in a torus with a very strong longitudinal field, such that the pinch current I_z is below the Kruskal-Shafranov limit. This means that the helical magnetic lines, resultant from the externally applied B_z, and the internal B_θ of the pinch do not complete one revolution (2π) of the minor axis in going round the major axis of the machine once. This is known theoretically to prevent the growth of certain helical distortions of the plasma. The achievements are listed in the table, together with the target values for achieving a positive power balance and the best achievement of the fast pulsed fusion devices.

It can be seen that Tokamak has surpassed Scylla IV in $n\tau$ by a factor of 3, but is still below it by a factor of 10 in temperature. Tokamak is one of the class of low-β, low-density, quasi-steady-state closed devices (which include Stellarator, Multipole, and Spherator), which has lagged behind the high-β, high-density pulsed devices in reaching high temperatures.

The Tokamak achievement has raised the hopes of those who support the quasi-steady-state closed magnetic bottle approach. The United States Atomic Energy Commission is financing the building of two Tokamaks, one at the Oak Ridge National Laboratory and one at Princeton Plasma

Comparison of Tokamak and Scylla IV

Parameter	Tokamak	Scylla IV	Reactor
T_i (kev)	0.5	5	10
N (ions cm^{-3})	5×10^{13}	4×10^{16}	
τ (sec)	20×10^{-3}	5×10^{-6}	
$N\tau$ (ions cm^{-3}sec)	1×10^{12}	2×10^{11}	10^{14}

Physics Laboratory, by reconstruction of the Model C Stellarator. Meanwhile, the Soviet Union is considering the financing for the construction of a larger Tokamak designed to push $n\tau$ upward by a factor of 10–20. The 0.5-kev temperature in Tokamak is close to the upper limit that can be reached by resistive heating of the plasma by the pinch current, such as has been used as of 1970. Although the restivity observed in T3 is anomalously high, at temperatures > 1 kev, the resistivity would be too low ($R \propto T^{-3/2}$) to be able to deposit sufficient heat in the time available. Some new method of heating to bridge the gap is being sought, and ion cyclotron heating has been suggested.

[JAMES L. TUCK]

Bibliography: C. L. Longmire, J. L. Tuck, and W. B. Thompson (eds.), Plasma physics and thermonuclear research, *Progr. Nucl. Energ.*, Ser. 11, 1963; Theoretical and experimental aspects of controlled nuclear fusion, in *Proc. 2d Int. Conf. Peaceful Uses At. Energ.*, vols. 31–32, 1958; J. L. Tuck, Artsimovich talks about controlled-fusion research, *Phys. Today*, 22(6):54–57, 1969.

Pipeline

A line of piping and the associated pumps, valves, and equipment necessary for the transportation of a fluid. Major uses of pipelines are for the transportation of petroleum, water (including sewage), chemicals, foodstuffs, pulverized coal, and gases such as natural gas, steam, and compressed air. Pipelines must be leakproof and must permit the application of whatever pressure is required to force conveyed substances through the lines. Pipe is made of a variety of materials and in diameters from a fraction of an inch up to 30 ft. Principal materials are steel, wrought and cast iron, concrete, clay products, aluminum, copper, brass, cement and asbestos (called cement-asbestos), plastics, and wood.

Pipe is described as pressure and nonpressure pipe. In many pressure lines, such as long oil and gas lines, pumps force substances through the pipelines at required velocities. Pressure may be developed also by gravity head, as for example in city water mains fed from elevated tanks or reservoirs.

Nonpressure pipe is used for gravity flow where the gradient is nominal and without major irregularities, as in sewer lines, culverts, and certain types of irrigation distribution systems.

Design of pipelines considers such factors as required capacity, internal and external pressures, water- or airtightness, expansion characteristics of the pipe material, chemical activity of the liquid or gas being conveyed, and corrosion.

Most pipe is jointed, although some concrete pipe is monolithically cast in place. The length of the individual sections of pipe and the method of joining them depend upon the pipe material, diameter, weight, and requirements of use. Steel pipe sections are usually joined by welding, couplings, or riveting. Cast-iron pipe may be joined by couplings or, in the case of bell-and-spigot pipe, by filling the space between the bell and the spigot with calked or melted metal such as lead. Flexible-type joints with rubber gaskets are also used for joining cast iron pipe. The rubber gasket is contained in grooves and is ordinarily the sole element making the joint watertight.

Cement-mortar-filled or lead-filled rigid-type bell-and-spigot joints are usually used for joining concrete or vitrified clay sewer pipe. Tongue-and-groove rigid-type mortar-filled joints are often used for concrete pipe in low-pressure installations. The flexible-type joints are most frequently used for asbestos-cement pipe and concrete pipe under higher pressures.

[LESLIE N. MC CLELLAN]

Plutonium

A chemical element, Pu, atomic number 94. Plutonium is a reactive, silvery metal in the transuranium series of elements. The first isotope to be identified was Pu^{238}, produced in cyclotron experiments by G. T. Seaborg, E. M. McMillan, A. C.

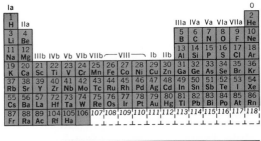

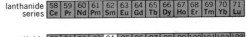

Wahl, and J. Kennedy. The principal isotope of chemical interest is Pu^{239}. It is formed in nuclear reactors by the process shown in Eq. (1). Pu^{239}

$$U^{238} + n \rightarrow U^{239} \xrightarrow[23\ min]{\beta^-} Np^{239} \xrightarrow[2.33\ days]{\beta^-} Pu^{239} \quad (1)$$

decays by α-emission with a half-life of 24,360 years. Its fissionability makes it of importance in nuclear weapons and in nuclear reactors. Minute quantities of Pu^{239} are formed in pitchblende and monazite ores by reaction (1). In pitchblende the uranium-to-plutonium ratio is approximately 10^{11}:1.

Uses. Plutonium is used as a nuclear fuel, to produce radioactive isotopes for research, and as the fissile agent in nuclear weapons. *See* NUCLEAR FUELS; REACTOR, NUCLEAR.

Properties. Like its neighboring elements, uranium and neptunium, plutonium exhibits a variety of valence states in solution and in the solid state. Plutonium metal is highly electropositive. In acid solution the known oxidation states are III, IV, V, and VI. The ions of the IV, V, and VI states are moderately strong oxidizing agents. The ions of the III, IV, and VI states can coexist in 1 M perchloric acid solution. An unstable plutonium(VII) oxidation state, analogous in properties to Np(VII), is reportedly formed by ozone oxidation of Pu(VI) in alkaline solution.

Because the oxidation potentials are so close in value, pure solutions of intermediate oxidation states undergo disproportionation (self-oxidation and reduction reactions). The most important equilibrium is that involving the disproportionation of

Pu(IV), which can be written as Eq. (2), for which

$$3Pu^{4+} + 2H_2O \rightleftharpoons PuO_2^{++} + 2Pu^{3+} + 4H^+ \quad (2)$$

the equilibrium constant is expressed as Eq. (3).

$$K_1 = \frac{[PuO_2^{++}][Pu^{3+}]^2[H^+]^4}{[Pu^{4+}]^3} \quad (3)$$

Here K_1 is calculated from the potentials to be 0.0089 for 1 M acid at 25°C. In 1 M acid the solution resulting from the disproportionation of pure plutonium(IV) would be 72% Pu(IV), 18.6% Pu(III), and 9.3% Pu(VI). Although Pu(V) is unstable in molar perchloric acid, it becomes increasingly stable as the acidity decreases. In 0.1 M acid, appreciable concentration of all four valence states may coexist in solution. The additional equilibrium that must be considered may be written as Eq. (4). The equilibrium constant, K_2, shown in Eq.

$$PuO_2^+ + Pu^{4+} \rightleftharpoons PuO_2^{++} + Pu^{3+} \quad (4)$$

(5), is 13 in 1 M perchlorate solution at 25°C. The

$$K_2 = \frac{[PuO_2^{++}][Pu^{3+}]}{[PuO_2^+][Pu^{4+}]} \quad (5)$$

rate of attainment of the equilibrium of reaction (2) is slow and that of reaction (5) very fast. A constant equilibrium state is not maintained over long periods of time because of the slow reduction in average oxidation number in solution caused by the reaction of the plutonium ions with the α-radiation-induced decomposition products of the water.

In solutions of acids, such as nitric and hydrochloric, whose anions form weak complexes with plutonium ions, the relative stabilities of the different states are little changed. Qualitatively, it is known that univalent anions, with the exception of fluoride, form relatively weak complexes with the ions of all oxidation states. Higher-valent anions form relatively strong complexes. In general, the relative stabilities of complexes with a given anion decreases in the order

$$Pu^{4+} > PuO_2^{++} > Pu^{3+} > PuO_2^+$$

Complex formation will generally stabilize the IV state. Hydrolysis reactions can also markedly affect the relative stabilities of the different states. As in the case of complex formation, the IV state is stabilized by hydrolysis. Polymerization processes are important in the hydrolysis reactions. Plutonium(IV) has been reported to form soluble polymers with molecular weights as high as 10^{10}.

A large amount of information is available on the behavior of plutonium ions when treated with common oxidizing and reducing agents. It is generally found that reactions which involve only changes from the III to IV or V to VI states tend to be rapid. Reactions which involve the formation or destruction of the oxygenated ions of the V or VI states tend to be slow. As examples, oxidation of Pu(III) to Pu(IV) is rapid with bismuthate, bromate, iron (III), dichromate, iodate, permanganate, and cerium(IV) ions. Reduction of Pu(IV) to PU(III) is rapid with iron(II), iodide ion, sulfurous acid, and nitrous acid. Oxidation of Pu(IV) to Pu(VI) is slow with bromate, dichromate, permanganate, and nitrate ions. The rates may be changed markedly by complex-ion formation. In the presence of moderate concentrations of sulfuric acid,

Table 1. Solvents used in the separation of plutonium and uranium from fission products

Solvent (trivial name)	Diluent	Salting agent
Methyl isobutyl ketone (Hexone)	None	$Al(NO_3)_3$
Tri-n-butyl phosphate (TBP)	Kerosine	HNO_3, $Al(NO_3)_3$
Dibutyl ether of ethylene glycol (Carbitol)	None	HNO_3
Dibutyl ether of tetra-ethylene glycol (Pentether)	None or butyl ether	HNO_3
Triglycol dichloride (Trigly)	None	$Al(NO_3)_3$
Thenoyl trifluoroacetone (TTA)	Benzene or toluene	None

Table 2. Distribution coefficients of uranium, plutonium, and fission products from nitrate solutions of 100-day cooled reactor fuel

Solvent	U(VI)	Pu(VI)	Pu(IV)	Pu(III)	Fission products
Hexone	1.5	7.6	1.6×10^{-2}	4.5×10^{-4}	6×10^{-4}
TBP	8.0	0.6	1.5	2×10^{-2}	2×10^{-3}

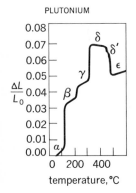

PLUTONIUM

Expansion of high-purity plutonium under conditions of self-heating, $L_0 = 0.5$ in. (*After E. R. Jette*)

for example, oxidation past the IV state is very difficult. Some relatively rapid oxidation reactions are also known. Ceric ion, argentic ion, and bismuthate rapidly oxidize Pu(IV) to Pu(VI).

The ions of the different oxidation states have characteristic colors: Pu^{3+} is blue-violet; Pu^{4+}, yellow-brown; PuO_2^+, reddish; and PuO_2^{++}, pink. Like the rare earths, they also have characteristic absorption spectra with sharp absorption bands. These have been widely used in the analysis of plutonium solutions to determine the amount of each oxidation state present.

Preparation. Methods for the isolation and purification of plutonium make use of the fact that the element can exist in a multiplicity of oxidation states, each differing in chemical properties. Laboratory separation procedures have been devised using carrier, solvent-extraction, and ion-exchange methods. The first plant-scale operations employed the carriers bismuth phosphate and lan-

thanum fluoride. In most recent processes, solvent extraction is employed. It has the advantage that not only the plutonium but also the uranium of the reactor fuel may be readily recovered and decontaminated from fission products. Some of the most important solvents are listed in Table 1. Control of the extraction behavior is obtained by the use of diluents for the solvent, addition of salting agents to the aqueous layer, and the control of solution pH. In Table 1 are listed the diluents and salting agents commonly employed for the different solvents. The behavior of different valence states with these solvents can be illustrated by reference to the relative distribution coefficients, defined as the concentration of the metal in the organic phase divided by the concentration in the aqueous phase (Table 2). The actual values of the distribution coefficients will change with conditions, but the approximate relative values will be maintained.

In the industrial process employing hexone (the redox process), the uranium fuel is dissolved in nitric acid. The solution is oxidized, and the U(VI) and Pu(VI) are coextracted with the fission products. After scrubbing the hexone layer to remove impurities, the solvent is passed over an aluminum nitrate solution containing a reducing agent. The plutonium is removed into the aqueous layer as Pu(III), and the uranium left in the solvent as U(VI). The aqueous layer is then reoxidized and the extraction repeated. By successive cycles of this process the plutonium is purified to the desired degree.

The industrial process employing tri-n-butyl phosphate (TBP) as the solvent (the Purex process) operates in much the same manner. After dissolution of the fuel element, the plutonium is fixed as Pu(IV) and the uranium as U(VI). The nitric acid concentration is adjusted, and the Pu(V) and U(VI) extracted into 30% TBP in kerosine. The solvent is washed with nitric acid to remove impurities. The plutonium is then removed as Pu(III) by scrubbing the solvent with nitric acid containing a reducing agent.

Plutonium metal can be prepared by the reduction of PuF_3 with calcium metal. Plutonium metal

Table 3. Properties of plutonium metal

Phase	Symmetry	Density (20°C), g/cm³	Temperature of phase transition, °C	Linear expansion coefficient* (20°C), $\alpha \times 10^6$	Resistivity (20°C) $\times 10^6$ (ohm-cm)	Temperature coefficient of resistivity†
α	Monoclinic	19.82		50.8	145	−21
			122			
β		17.65		38.0	110.5	−6
			203			
γ	Orthorhombic	17.19 17.14		34.7	110	−5
			319			
δ	Face-centered cubic	15.92		−10.0	103	+7
			453			
δ'	Body-centered tetragonal	16.0		−20	105	+45
			477			
ϵ	Body-centered cubic	16.48		25.7	114	−7
			639.5			
Liquid		16.5		50		

$$*\alpha = \frac{1}{L} \cdot \frac{L}{T} \qquad \dagger \frac{1}{\rho} \cdot \frac{\rho}{t} \times 10^5$$

Table 4. Plutonium halides and oxyhalides

Compound	Color	Melting point, °C	Density at 20°C	Crystal structure
PuF_3	Purple	1425	9.32	Hexagonal
PuF_4	Pale brown	1037	7.0	Monoclinic
PuF_6	Reddish-brown	50.75		Orthorhombic
$PuCl_3$	Green	760	5.70	Hexagonal
$PuBr_3$	Green	681	6.69	Orthorhombic
PuI_3	Green	777	6.92	Orthorhombic
PuOF	Metallic	1635	9.76	Tetragonal
PuOCl	Blue-green		8.81	Tetragonal
PuOBr	Green		9.07	Tetragonal
PuOI	Green		8.46	Tetragonal

Table 5. Some insoluble inorganic compounds of plutonium precipitated from aqueous solution

	Oxidation state		
III	IV	V	VI
$Pu(IO_3)_3$	NH_4PuF_5	$KPuO_2CO_3$	$NaPuO_2(C_2H_3O_2)$
$PuPO_4 \cdot 0.5H_2O$	$Pu(OH)_4 \cdot xH_2O$	$CsF \cdot PuF_5$	
$Pu_2(C_2O_4)_3 \cdot 9H_2O$	$Pu(IO_3)_4$	$2RbF \cdot PuF_5$	
	$PuO_4 \cdot 2H_2O$		
	$Pu(HPO_4)_2 \cdot xH_2O$		
	$Pu(C_2O_4)_2 \cdot 6H_2O$		
	$LiF \cdot PuF_4$		
	$7NaF \cdot 6PuF_4$		

deserves special mention because of its unique properties. It is known to exist in six allotropic forms below the melting point (639°C). Some of its physical properties are given in Table 3. Particularly interesting and puzzling are the contractions which the δ and δ' phases undergo with increasing temperature (see graph). Noteworthy is the fact that for no phase do both the coefficient of thermal expansion and the temperature coefficient of resistivity have the conventional algebraic sign. If the phase expands on heating, the resistance decreases. A number of alloys of plutonium are known — with beryllium, lead, uranium, chromium, manganese, iron, nickel, and osmium.

Principal compounds. A large number of compounds of plutonium have been prepared. Reaction of hydrogen with plutonium metal yields at least two well-defined hydrides, PuH_3 and PuH_2. The common oxide is PuO_2. It is formed by ignition of the hydroxides, oxalates, peroxides, and nitrates of any oxidation state in air at 870–1200°C. It crystallizes in a face-centered-cubic structure (density 11.44 g/cm³). It has been extensively used for gravimetric analyses of plutonium. A number of phases have been reported for the plutonium oxide system in the composition region from $Pu_{1.5}$ to PuO_2, including $\beta = Pu_2O_3$ (hexagonal). One of the most important classes of compounds is made up of the halides. Properties of the known halides and oxyhalides are given in Table 4. The hexafluoride is a low-melting, low-boiling compound of high volatility resembling NpF_6 and PuF_6. It is a strong fluorinating agent. Conditions for the preparation of the fluorides are illustrated by Eqs. (6), (7), and (8). The other halides are prepared by a variety of methods. Treat-

$$PuO_2 + \frac{1}{2}H_2 + 3HF \xrightarrow{600°C} PuF_3 + 2H_2O \quad (6)$$

$$PuO_2 + O_2 + 4HF \xrightarrow{550°C} PuF_4 + 2H_2O + O_2 \quad (7)$$

$$PuF_4 + F_2 \xrightarrow{750°C} PuF_6 \quad (8)$$

ment of PuO_2 with powerful halogenating agents, such as CCl_4, PCl_5, and SCl_2, yields $PuCl_3$. $PuBr_3$ and PuI_3 are conveniently made by the action of the anhydrous gases, HBr and HI, on plutonium metal.

A number of other compounds are known. Among these are the carbides, silicides, sulfides, and nitrides, of interest because of their refractory nature. Among these are PuC, Pu_2C_3, PuN, α-$PuSi_2$, β-$PuSi_2$, PuSi, PuS, and Pu_2S_3-Pu_3S_4.

In addition to those compounds prepared by vacuum line techniques, there are a large number of compounds that have been prepared from solution. The most important of these are given in Table 5.

[JAMES C. HINDMAN]

Bibliography: P. Fields and T. Moeller, (eds.), *Lanthanide/Actinide Chemistry*, 1967; N. N. Krot and A. D. Gel'man, *Dokl. Akad. Nauk SSSR Khimiya*, 177:124, 1967; G. T. Seaborg, *Man-made Transuranium Elements*, 1963.

Power

The time rate of doing work. Like work, power is a scalar quantity, that is, a quantity which has magnitude but no direction. Some units often used for the measurement of power are the watt (1 joule of work per second) and the horsepower (550 ft-lb of work per second). *See* HORSEPOWER; WORK.

Usefulness of the concept. Power is a concept which can be used to describe the operation of any system or device in which a flow of energy occurs. In many problems of apparatus design, the power, rather than the total work to be done, determines the size of the component used. Any device can do a large amount of work by performing for a long time at a low rate of power, that is, by doing work slowly. However, if a large amount of work must be done rapidly, a high-power device is needed. High-power machines are usually larger, more complicated, and more expensive than equipment which need operate only at low power. A motor which must lift a certain weight will have to be larger and more powerful if it lifts the weight rapidly than if it raises it slowly. An electrical resistor must be large in size if it is to convert electrical energy into heat at a high rate without being damaged.

Electrical power. The power P developed in a direct-current electric circuit is $P = VI$, where V is the applied potential difference and I is the current. The power is given in watts if V is in volts and I in amperes. In an alternating-current circuit, $P = VI \cos\phi$, where V and I are the effective values of the voltage and current and ϕ is the phase angle between the current and the voltage. *See* ALTERNATING CURRENT.

Power in mechanics. Consider a force F which does work W on a particle. Let the motion be restricted to one dimension, with the displacement in this dimension given by x. Then by definition the power at time t will be given by Eq. (1). In this

$$P = dW/dt \quad (1)$$

equation W can be considered as a function of either t or x. Treating W as a function of x gives Eq. (2). Now dx/dt represents the velocity v of the

$$P = \frac{dW}{dt} = \frac{dW}{dx}\frac{dx}{dt} \quad (2)$$

particle, and dW/dx is equal to the force F, according to the definition of work. Thus Eq. (3) holds.

$$P = Fv \qquad (3)$$

This often convenient expression for power can be generalized to three-dimensional motion. In this case, if ϕ is the angle between the force $\mathbf{F}$ and the velocity $\mathbf{v}$, which have magnitudes F and v, respectively, Eq. (4) expresses quantitatively the observa-

$$P = \mathbf{F} \cdot \mathbf{v} = Fv \cos \phi \qquad (4)$$

tion that if a machine is to be powerful, it must run fast, exert a large force, or do both.

[PAUL W. SCHMIDT]

Power plant

A means for converting stored energy into work. Stationary power plants such as electric generating stations are located near sources of stored energy, such as coal fields or river dams, or are located near the places where the work is to be performed, as in cities or industrial sites. Mobile power plants for transportation service are located in vehicles, as the gasoline engines in automobiles and diesel locomotives for railroads. Power plants range in capacity from a fraction of a horsepower (hp) to over 1,000,000 kilowatts (kw) in a single unit (Table 1). Large power plants are assembled on location from components made by different manufacturers. Smaller units are mass-produced.

Most power plants convert part of the stored raw energy of fossil fuels into kinetic energy of a spinning shaft. Some power plants harness nuclear energy. Elevated water supply is used in hydro-

electric power plants. For transportation, the plant may produce a propulsive jet instead of the rotary motion of a shaft. Other sources of energy, such as winds, tides, waves, geothermal sources, heat of the sea, nuclear fusion, and solar radiation, are of negligible commercial significance in the generation of power despite their magnitudes.

Table 2 shows the scope of United States power-plant capacity. About a third of the world's electric energy (in kilowatt-hours, kwhr) is generated by the United States public utility systems (Fig. 1), and the installed generating capacity (in kw) of the United States (including Alaska and Hawaii) is about the same as the total of the next four countries (Fig. 2). Figure 3 shows the changing importance of hydroelectric power in the United States, where the kwhr output from hydro has dropped from 30 to 16% of the total electric generation in 25 years' time. These data, coupled with the data of Table 2, reflect the dominant position of thermal power both for stationary service and for the propulsion of land-, water-, and air-borne vehicles. *See* ENERGY SOURCES.

Figures 4 and 5 show heat-balance diagrams for modern steam-electric central stations, conventional and nuclear-fuel-fired, respectively. Many variations are incorporated in power-plant design for the control of efficiency, weight, space, flexibility, reversibility, reliability, life, investment, and operating costs. Rudimentary flow- or heat-balance diagrams for important types of practical power plants are shown in Fig. 6. Figure 7 is a diagram for a by-product-type industrial steam plant which has the double purpose of generating elec-

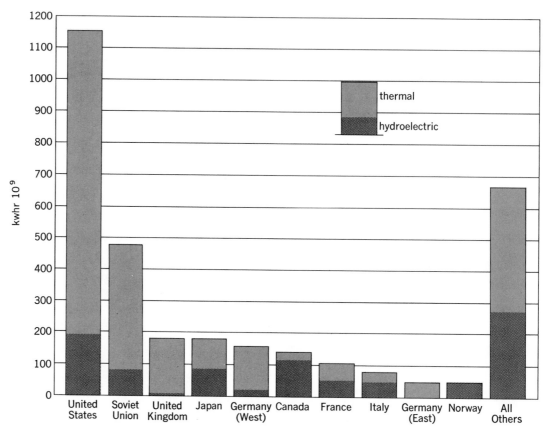

Fig. 1. Electric energy production in 14 highest countries, 1973. Approximately one-third of world electric energy production is from United States plants. (*United Nations*)

Table 1. Representative design and performance data on power plants

Type	Unit size range, kw	Fuel*	Plant weight, lb/kw	Plant volume, ft³/kw	Heat rate, Btu/kwhr
Central station					
Hydro	10,000 – 700,000				
Steam (fossil-fuel – fired)	10,000 – 1,300,000	CO		20 – 50	8,500 – 15,000
Steam (nuclear)	500,000 – 1,200,000	N			10,000 – 12,000
Diesel	1,000 – 5,000	DG			10,000 – 15,000
Combustion turbine	5,000 – 10,000	D'G			11,000 – 15,000
Industrial (by-product) steam	1,000 – 25,000	COGW		50 – 75	4,500 – 6,000
Diesel locomotive	1,000 – 5,000	D	100 – 200	2 – 3	10,000 – 15,000
Automobile	25 – 300	G'	5 – 10	0.1	15,000 – 20,000
Outboard motor	1 – 50	G'	2 – 5	0.1 – 0.5	15,000 – 20,000
Truck	50 – 500	D	10 – 20		12,000 – 18,000
Merchant ship, diesel	5,000 – 20,000	D	300 – 500		10,000 – 12,000
Naval vessel, steam	25,000 – 100,000	DON	25 – 50		12,000 – 18,000
Airplane, reciprocating engine	1,000 – 3,000	G'	1 – 3	0.05 – 0.10	12,000 – 15,000
Airplane, turbojet	3,000 – 10,000	D'	0.2 – 1		13,000 – 18,000

*C, coal; D, diesel fuel; D', distillate; G, gas; G', gasoline; N, nuclear; O, fuel oil (residuum); W, waste.

Table 2. Approximate 1975 installed capacity of United States power plants

Plant type	Capacity, 10⁶ kw
Electric central stations	504
Industrial	40
Agricultural	60
Railroad	70
Marine, civilian	40
Aircraft, civilian	70
Military establishment	2,000 ±
Automotive	10,000 ±
Total	13,000 ±

tric power and simultaneously delivering heating steam by extraction or exhaust from the prime mover. *See* HEAT BALANCE.

Plant load. There is no practical way of storing the mechanical or electrical output of a power plant in the magnitudes encountered in power-plant applications. The output must be generated at the instant of its use. This results in wide variations in the loads imposed upon a plant. The capacity, measured in kw or hp, must be available when the load is imposed. Much of the capacity may be idle during extended periods when there is no demand for output. Hence much of the potential output, measured as kwhr or hp-hr, cannot be generated because there is no demand for output. This greatly complicates the design and confuses the economics of power plants. Kilowatts cannot be traded for kilowatt-hours, and vice versa.

The ratios of average load to rated capacity or to peak load are expressed as the capacity factor and the load factor, Eqs. (1) and (2), respectively. The

$$\text{Capacity factor} = \frac{\text{average load for the period}}{\text{rated or installed capacity}} \quad (1)$$

$$\text{Load factor} = \frac{\text{average load for the period}}{\text{peak load in the period}} \quad (2)$$

range of capacity factors experienced for various types of power plants is given in Table 3.

Variations in loads can be conveniently shown on graphical bases as in Figs. 8 and 9 for public utilities and in Fig. 10 for air and marine propulsion. Rigorous definition of load factor is not possi-

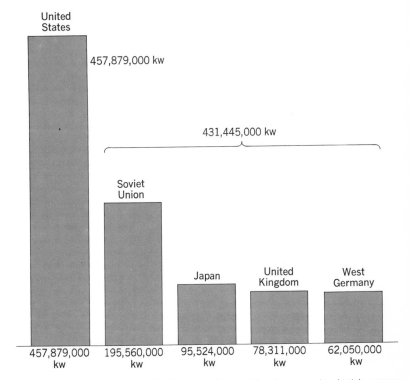

Fig. 2. The five countries with the highest capacities to generate electric power, 1973. (*United Nations*)

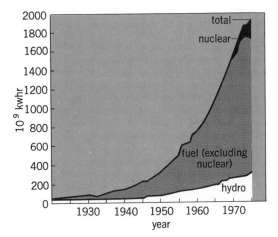

Fig. 3. Electric energy generation in the United States utility industry. (*Edison Electric Institute*)

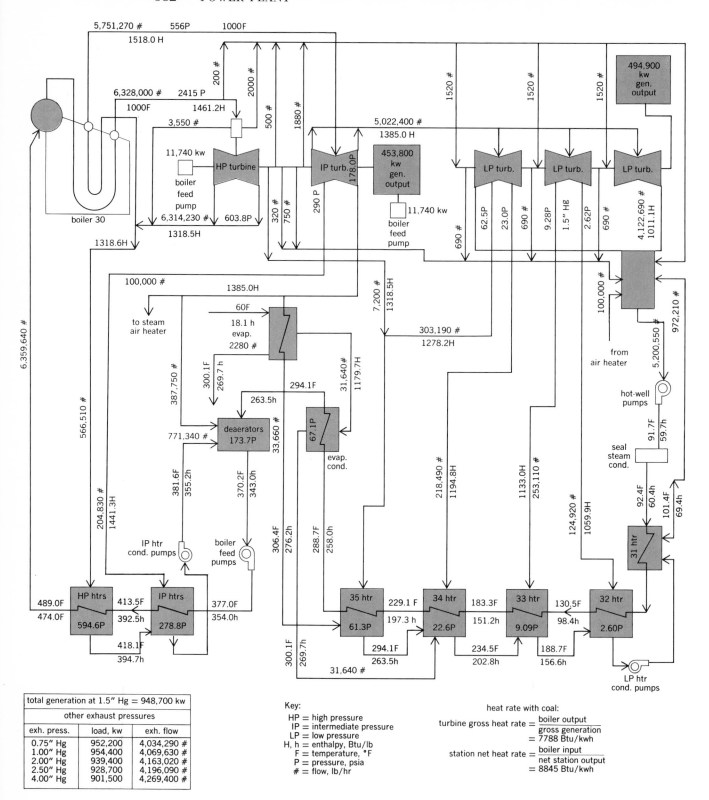

Fig. 4. Heat-balance diagram for a modern steam-electric fossil-fuel–fired power plant (Ravenswood Station, Consolidated Edison Co. of New York). (From D. G. Fink and J. M. Carroll, Standard Handbook for Electrical Engineers, McGraw-Hill, 10th ed., 1968)

Plant efficiency. The efficiency of energy conversion is vital in most power-plant installations. The theoretic power of a hydro plant in kw is $QH/11.8$, where Q is the flow in cubic feet per second and H is the head at the site in feet. Losses in headworks, penstocks, turbines, draft tubes, tailrace, bearings, generators, and auxiliaries will

ble with vehicles like tractors or automobiles because of variations in the character and condition of the running surface.

In propulsion applications, power output may be of secondary import; performance may be based on tractive effort, drawbar pull, thrust, climb, and acceleration.

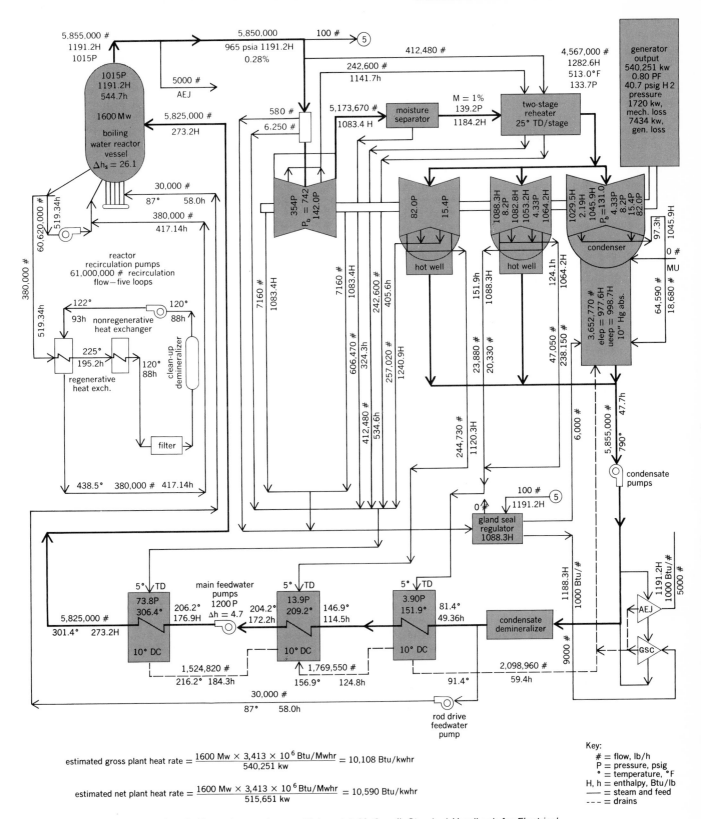

$$\text{estimated gross plant heat rate} = \frac{1600 \text{ Mw} \times 3{,}413 \times 10^6 \text{ Btu/Mwhr}}{540{,}251 \text{ kw}} = 10{,}108 \text{ Btu/kwhr}$$

$$\text{estimated net plant heat rate} = \frac{1600 \text{ Mw} \times 3{,}413 \times 10^6 \text{ Btu/Mwhr}}{515{,}651 \text{ kw}} = 10{,}590 \text{ Btu/kwhr}$$

Fig. 5. Heat-balance diagram for a boiling water reactor (BWR) power plant (Oyster Creek Station, General Electric Co. and General Public Utilities System). (*From D. G.* *Fink and J. M. Carroll, Standard Handbook for Electrical Engineers, McGraw-Hill, 10th ed., 1968*)

reduce the salable output 15–20% below the theoretic in modern installations. The selection of a particular type waterwheel depends on experience with wheels at the planned speed and on the lowest water pressure in the water path. Runners of the reaction type (high specific speed) are suited to low heads (below 500 ft) and the impulse type (low specific speed) to high head service (about 1000 ft). The lowest heads (below 100 ft) are best accommodated by reaction runners of the propeller or the adjustable blade types. Mixed-pressure runners are favored for the intermediate heads (50–500 ft).

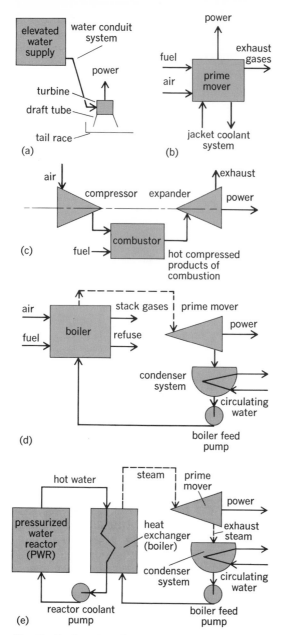

Table 3. Range of capacity factors for selected power plants

	Factor
Public-utility systems, in general	50–70
Chemical or metallurgical plant, three-shift operation	80–90
Seagoing ships, long voyages	70–80
Seagoing ships, short voyages	30–40
Airplanes, commercial	20–30
Private passenger cars	1–3
Main-line locomotives	30–40
Interurban buses and trucks	5–10

which prefers low or net calorific value.

Tables 1 and 3 give performances for selected operations. Figures 11 and 12 reflect the improvement in fuel utilization of the United States electric power industry since 1900. Figure 12 is especially significant, as it shows graphically the impact of technological improvements on the cost of producing electric energy despite the harassing increases in the costs of fuel during the same period. Such technological improvements, however, were unable to compensate for the steep rise in fuel prices that followed the late-1973 Arab oil embargo; this boosted 1974–1975 fuel prices completely out of line with those of earlier years. Figure 13 illustrates the variation in thermal performance as a function of load for an assortment of stationary and marine-propulsion power plants. *See* BRAYTON CYCLE; CARNOT CYCLE; DIESEL CYCLE; OTTO CYCLE; RANKINE CYCLE; THERMODYNAMIC CYCLE.

In scrutinizing data on thermal performance, it should be recalled that the mechanical equivalent of heat (100% thermal efficiency) is 2545 Btu/hp-hr and 3413 Btu/kwhr. Modern steam plants in large sizes (75,000–1,200,000 kw units) and internal combustion plants in modest sizes (1000–5000 kw) have little difficulty in delivering a kwhr for less than 10,000 Btu in fuel (34% thermal efficiency). Lowest fuel consumptions per unit output (8500–9000 Btu/kwhr) are obtained in condensing steam plants with the best vacua, regenerative-reheat cycles using eight stages of extraction feed heating, two stages of resuperheat, primary pressures of 3500 psi (supercritical) and temperatures of 1150°F. An industrial plant generating electric power as a by-product of the process steam load is capable of having a thermal efficiency of 5000 Btu/kwhr.

The atomic power plant substitutes the heat of fission for the heat of combustion, and the consequent plant differs only in the method of preparing the thermodynamic fluid. It is otherwise similar to the usual thermal power plant. Low reactor temperatures lead to the overwhelming preference for steam-turbine rather than gas-turbine cycles. When fluid temperatures can be had above 1200°F, the gas-power cycle will receive more favorable consideration. Otherwise the atomic power plant is essentially a low-pressure, low-temperature steam operation (less than 1000 psi and 600°F).

Power economy. Costs are a significant, and often controlling, factor in any commercial power-plant application. Average costs have little significance because of the many variables, especially load factor. Some plants are short-lived and others long-lived. For example, in most automo-

Fig. 6. Rudimentary flow- or heat-balance diagrams for power plants. (*a*) Hydro. (*b*) Internal combustion. (*c*) Gas-turbine. (*d*) Fossil-fuel–fired. (*e*) Nuclear steam (pressurized water reactor, PWR).

Draft tubes, which permit the unit to be placed safely above flood water and without sacrifice of site head, are essential parts of reaction unit installations. *See* HYDRAULIC TURBINE; WATERPOWER.

With thermal power plants there are the basic limitations of thermodynamics which fix the efficiency of conversion for heat into work. The cyclic standards of Carnot, Rankine, Otto, Diesel, and Brayton are the usual criteria on which heat-power operations are variously judged. Performance of an assembled power plant, from fuel to net salable or usable output, may be expressed as thermal efficiency (%); fuel consumption (lb, pt, or gal per hp-hr or per kwhr); or heat rate (Btu supplied in fuel per hp-hr or per kwhr). American practice uses high or gross calorific value of the fuel to measure heat rate or thermal efficiency and differs in this respect from European practice,

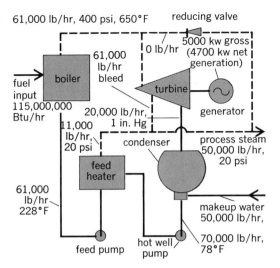

61,000 lb/hr, 400 psi, 650°F

Fig. 7. Heat balance for a by-product industrial power plant delivering both electric energy and process steam.

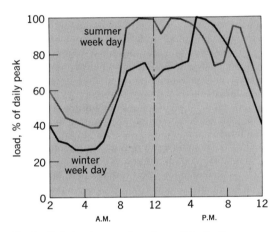

Fig. 8. Daily-load curves for urban utility plant.

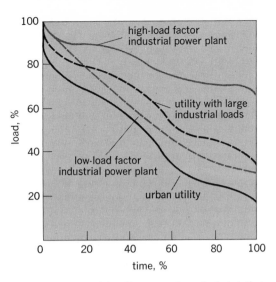

Fig. 9. Annual load-duration curves for selected stationary public utility power plants.

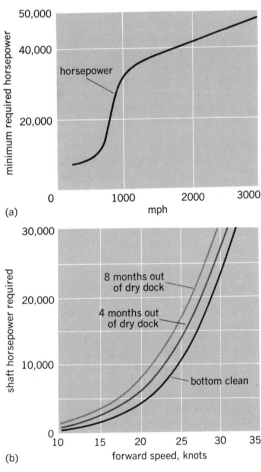

Fig. 10. Air and marine power. (*a*) Minimum power required to drive a 50-ton well-designed airplane in straight level flight at 30,000 ft altitude. (*b*) Power required to drive a ship, showing effect of fouling.

biles, which have short-lived power plants, 100,000 mi and 3000–4000 hr constitute operating life; diesel locomotives, which run 20,000 mi a month with complete overhauls every few years, and large seagoing ships, which register 1,000,000 mi of travel and still give excellent service after 20 years of operation, have long-lived plants; electric central stations of the hydro type remain in service 50 years; and steam plants run round the clock and upward of 8000 hr a year with complete reliability even when 25 years old. Such figures greatly influence costs, Furthermore, costs are open to wide differences of interpretation.

In the effort to minimize cost of electric power to the consumer it is essential to recognize the difference between investment and operating costs, and the difference between average and incremental costs. Plants with high investment (fixed) costs per kw should run at high load factors to spread the burden. Plants with high operating costs (such as fuel) should be run only for the shortest periods to meet peak loads or emergencies. To meet these short operating periods various types of peaking plants have been built. Gas-turbines and pumped-storage plants serve this requirement. In the latter a hydro installation is operated off-peak to pump water from a lower reservoir to an elevated reservoir. On-peak the operation is reversed with water flowing downhill through the prime movers and returning electric energy to the transmission system. High head sites (for example 1000 ft), proximity to transmission lines, and low incremental cost producers (such as nuclear or efficient fossil-fuel–fired plants) are necessary. If 2 kwhr can thus be returned on-peak to the system, for an input of 3 kwhr off-peak, a pumped-storage installation is generally justifiable.

In any consideration of such power-plant installations and operations it is imperative to recognize (1) the requirements of reliability of service and (2) the difference between average and incremental costs. Reliability entails the selection and operation of the proper number and capacity of spares and of their location on the system network. Emergencies, breakdowns, and tripouts are bound to occur on the best systems. The demand for maximum continuity of electric service in modern civilization dictates the clear recognition of the need to provide reserve capacity in all components of the power system. Within that framework the minimum cost to the consumer will be met by the incremental loading of equipment. Incremental loading dictates, typically, that any increase in load should be met by adding that load to the unit then in ser-

vice, which will give the minimum increase in out-of-pocket operating cost. Conversely, for any decrease in load, the unit with the highest incremental production cost should drop that decrease in load. This is a complex technical, economic, and management problem calling for the highest degree of professional competence for its proper solution.

Costs are reflected in the curves of Fig. 14 for a steam-electric, investor-owned, central station in the eastern United States. Fixed costs are based

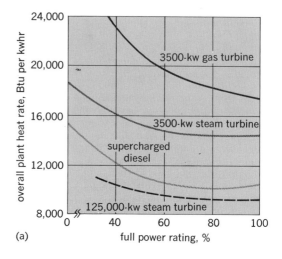

(a)

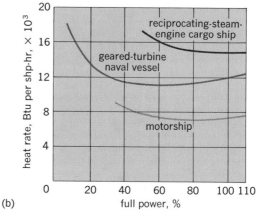

(b)

Fig. 13.   Comparison of heat rates. (a) Stationary power plants. (b) Marine-propulsion plants.

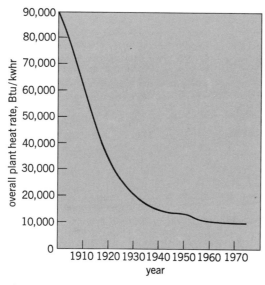

Fig. 11.   Thermal performance of fuel-burning electric-utility power plants in the United States.

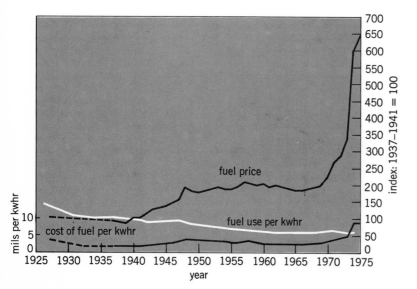

Fig. 12.   The effect of fuel price and efficiency of use upon cost of fuel per kilowatt-hour generated in the United States electric utility industry. (*Edison Electric Institute*)

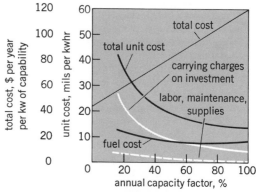

Fig. 14.   Representative costs of power from a large steam-electric generating station as of 1975.

Table 4. Costs of representative power plants in 1975

Cost factors	Steam, central station fossil fuel		Steam, central station nuclear fuel	Hydro		Steam, industrial	
	Large (1,000,000 kw)	Small (50,000 kw)	(1,000,000 kw)	Large (200,000 kw)	Small (20,000 kw)	Large (10,000 kw)	Small (1000 kw)
Investment, $ per kw	250	300	300	300	450	250	325
Fuel cost, cents per 10^6 Btu	80	90	20			100	110
Cost of power, mills per kwhr							
Total cost	15.6	23.7	11.3	10.4	15.4	19.0	24.7
Carrying cost on investment	6.7	9.6	8.0	9.6	14.4	6.7	8.7
Production cost, total	8.9	14.1	3.3	0.8	1.0	12.3	16.0
Fuel	8.0	10.1	2.1			11.5	13.5
Labor, maintenance, supplies, and supervision	0.9	4.0	1.2	0.8	1.0	0.8	2.5

on $250/kw investment with 16% annual carrying charges. If such a plant were government-financed, the annual charges on an investment of $250/kw might be reduced to 8%. Fuel cost is at 80 cents per 1,000,000 Btu. Total cost of power can be conveniently expressed as given below.

$$\text{\$ per kw per year} = K_1 + K_2 + 8760\,K_3 \times \text{capacity factor}$$

where K_1 = capacity charge, $ per year
K_2 = peak prepared-for charge, $ per year
K_3 = energy cost, mills per kwhr

Costs of representative power plants are summarized in Table 4. *See* ELECTRIC POWER GENERATION.

[THEODORE BAUMEISTER; LEONARD M. OLMSTED]

Bibliography: Babcock and Wilcox Co., *Steam: Its Generation and Use,* 1972; T. Baumeister (ed.), *Standard Handbook for Mechanical Engineers,* 7th ed., 1967; W. T. Creager and J. D. Justin, *Hydroelectric Handbook,* 1950; Diesel Engine Manufacturers Association, *Standard Practices,* 1958; D. G. Fink and J. M. Carroll, *Standard Handbook for Electrical Engineers,* 10th ed., 1968; L. C. Lichty, *Combustion Engine Processes,* 1967; H. L. Seward, *Marine Engineering,* vol. 1, 1942, vol. 2, 1944; M. J. Zucrow, *Aircraft and Missile Propulsion,* 1958.

Primary battery

An electric battery designed to deliver only one continuous or intermittent discharge. It cannot be recharged efficiently. Primary batteries are designed to deliver limited amounts of electric energy, determined by the materials used and the size of the cell. When the available energy drops to zero, the battery is usually discarded. Primary batteries may be classified by the type of electrolyte used.

Aqueous-electrolyte batteries. These batteries use solutions of acids, bases, or salts in water as the electrolyte. These solutions have ionic conductivities of the order of 1 mho/cm and practically no electronic conductivity. Practical cells, such as the common Leclanche dry cell and the alkaline-manganese-zinc cell use aqueous electrolytes. Disadvantages of such cells include corrosion of the electrode materials by the electrolyte, a relatively high evaporation rate of water vapor which

can cause cell failure, and the difficulties of preventing leakage. For discussions of examples of cells with aqueous electrolytes *see* BATTERY, ELECTRIC; DRY CELL; MERCURY BATTERY; RESERVE BATTERY.

Solid-electrolyte batteries. These use electrolytes of solid crystalline salts which have predominantly ionic conductivity. The conductivity is small compared with aqueous electrolytes, and the current output is of the order of 10^{-7} amp/in.^{3}

Solid-electrolyte batteries may be classified in two broad categories: (1) cells with solid crystalline salt, such as silver iodide, as the electrolyte; (2) cells with ion-exchange membrane as the electrolyte. In either category, the conductivity must be nearly 100% ionic. Any electronic conductivity causes a continuous discharge of the cell and will limit the stand or shelf life.

A typical cell with solid crystalline salt electrolyte is the lead−lead chloride−silver chloride cell in Fig. 1. Here lead is the anode, lead chloride is the electrolyte, and silver chloride is the cathode. This cell has a potential of 0.49 volt. During discharge, lead is oxidized to lead ion and silver chloride is reduced to silver.

Cells with solid salt electrolyte have been developed into miniature batteries. One type delivers 90−100 volts at 10^{-11} amp, and has a capacity of 1 amp-sec. This is over 10^6 days at 10^{-11} amp. The practical life of the cell is much less but may be as much as 10 years at room temperature. It can be stored at 160°F for at least 30 days and will operate over the range −65 to +165°F. The battery is 3/8 in. in diameter and 1 in. in length. With the increasing use of electronic devices and consequent miniaturization, solid-state batteries delivering low currents are finding newer applications. In addition, the use of electrolytes such as Ag_3SI or MAg_4I_5, (where M is K, Rb, NH_4, or Cs), which have better ionic conductivity than the lead or silver halides previously employed, gives cells with flash currents in the low milliampere range.

An example of a cell with ion-exchange membrane as electrolyte is the zinc−zinc ion exchange membrane, silver ion exchange membrane−silver cell shown in Fig. 2. Physically, the metal electrodes are in contact with the solid membrane which contains two regions. The region adjacent to the zinc is in the zinc ion state. The region adja-

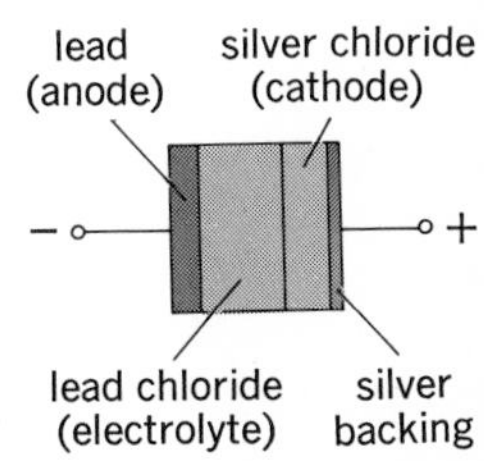

lead (anode)   silver chloride (cathode)

− o   o +

lead chloride (electrolyte)   silver backing

Fig. 1. Typical solid-electrolyte cell with solid crystalline salt electrolyte.

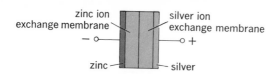

Fig. 2.   Ion-exchange solid-electrolyte bimetallic cell.

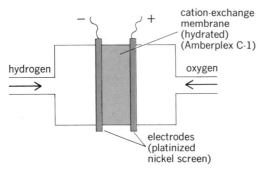

Fig. 3.   Solid-electrolyte cell using an ion-exchange membrane as the electrolyte.

cent to the silver is in the silver ion state. The discharge reaction increases the zinc ion quantity and decreases the silver ion quantity, in proportion to the amount of charge transferred. This cell has a potential of about 1.5 volts.

The zinc-silver cell described has serious shortcomings. The shelf life is poor, indicating internal self-discharge, and the capacity is limited by the available supply of silver ions. In strongly ionized types of ion-exchange material, the volume density of ionizing sites is about 1 equiv/liter, or 0.4 amp-hr/in.³ This is very low compared with metal-oxide cathodes.

A cell with higher capacity can be made by replacing the silver ion exchange material and silver by manganese dioxide plated on an inert metal, such as tantalum. This gives a capacity of about 100 times as much, for equal volume.

Ion-exchange electrolytes are also used with hydrogen and oxygen gas electrodes (Fig. 3). The electrodes consist of platinized metal screens. The electrolyte is a hydrogen ion exchange material. The room temperature emf of this cell is 0.96 volts.

Waxy-electrolyte batteries. These use waxy materials, such as polyethylene glycol, in which a small amount of a salt is dissolved in the molten wax. At room temperatures these materials are solid. The conductivity is small and the current output is limited to about 10^{-6} amp/in.²

Figure 4 shows a battery stack of cells using a waxy electrolyte. The electrodes are sheet zinc

and manganese dioxide. The electrolyte is made of polyethylene glycol in which is dissolved a small amount of zinc chloride. This electrolyte is melted and painted on a paper sheet to form the separator.

A 25-cell stack, built as shown in Fig. 4 and measuring 0.34 in. in length and 0.25 in. in diameter, weighed 1.5 g. A 0.50-in.-diameter stack weighed 6.0 g. The initial open-circuit voltage was 37.5 volts (1.5 volts per cell).

The internal resistance of this cell is high, and it increases as temperature decreases. This high internal resistance limits the usefulness of the cell, but it may be suitable for long-life potential sources of miniature size.

Fused-electrolyte batteries. These use crystalline salts or bases which are solid at room temperature. In use, the cell is heated and maintained at a temperature above the melting point of the electrolyte.

[JACK DAVIS]

Bibliography: C. R. Argue, B. Owens, and I. J. Grace, *Proceedings of the Power Sources Conference*, vol. 22, 1968; W. J. van der Grinten, *J. Electrochem. Soc.*, 103:210C, 1956; R. Jasinski, *High-Energy Batteries*, 1967; K. Lehovec and J. Broder, *J. Electrochem. Soc.*, 101:208, 1954; A. Sator, *Compt. Rend.*, 234:2283, 1952; S. W. Shapiro, *Proceedings of the Power Sources Conference*, vol. 11, 1957.

Propellant

Usually, a combustible substance that produces heat and supplies ejection particles, as in a rocket engine. A propellant is both a source of energy and a working substance; a fuel is chiefly a source of energy, and a working substance is chiefly a means for expending energy. Because the distinction is more decisive in rocket engines, the term propellant is used primarily to describe chemicals carried by rockets for propulsive purposes. *See* AIRCRAFT FUEL; FUEL; THERMODYNAMIC CYCLE.

Propellants are classified as liquid or as solid. Even if a propellant is burned as a gas, it may be carried under pressure as a cryogenic liquid to save space. For example, liquid oxygen and liquid hydrogen are important high-energy liquid bipropellants. Liquid propellants are carried in compartments separate from the combustion chamber; solid propellants are carried in the combustion chamber. The two types of propellants lead to significant differences in engine structure and thrust control. For comparison, the effectiveness of either type of propellant is stated in terms of specific impulse.

LIQUID PROPELLANTS

A liquid propellant releases energy by chemical action to supply motive power for jet propulsion. The three principal types of propellants are monopropellant, bipropellant, and hybrid propellant. Monopropellants are single liquids, either compounds or solutions. Bipropellants consist of fuel and oxidizer carried separately in the vehicle and brought together in the engine. Air-breathing engines carry only fuel and use atmospheric oxygen for combustion. Hybrid propellants use a combination of liquid and solid materials to provide propulsion energy and working substance. Typical liquid propellants are listed in the table. Physical prop-

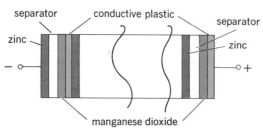

Fig. 4.   Waxy-electrolyte battery stack.

Physical properties of liquid propellants

Propellant	Boiling point, °F	Freezing point, °F	Density, g/ml	Specific impulse,* sec
Monopropellants				
Acetylene	−119	−115	0.62	265
Hydrazine	236	35	1.01	194
Ethylene oxide	52	−168	0.88	192
Hydrogen peroxide	288	13	1.39	170
Bipropellants				
Hydrogen	−423	−436	0.07	
Hydrogen-fluorine	−306	−360	1.54	410
Hydrogen-oxygen	−297	−362	1.14	390
Nitrogen tetroxide	70	12	1.49	
Nitrogen tetroxide-hydrazine	236	35	1.01	290
Red nitric acid	104	−80	1.58	
Red fuming nitric acid – *uns*-dimethyl hydrazine	146	−71	0.78	275

*Maximum theoretical specific impulse at 1000 psi chamber pressure expanded to atmospheric pressure.

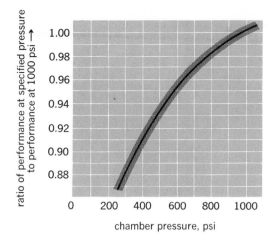

Fig. 1. Graph illustrating approximate effect of chamber pressure on specific impulse.

erties at temperatures from storage to combustion are important. These properties include melting point, boiling point, density, and viscosity. *See* METAL-BASE FUEL.

The availability of large quantities and their high performance led to selection of liquefied gases such as oxygen for early liquid-propellant rocket vehicles. Liquids of higher density with low vapor pressure (see table) are advantageous for the practical requirements of rocket operation under ordinary handling conditions. Such liquids can be retained in rockets for long periods ready for use and are convenient for vehicles that are to be used several times. The high impulse of the cryogenic systems is desirable for rocket flights demanding maximum capabilities, however, such as the exploration of space or the transportation of great weights for long distances.

Performance. Jet propulsion by a reaction engine, using the momentum of the propellant combustion products ejected from the engine, is not limited to atmospheric operation if the fuel reacts with an oxidizer carried with the engine. Performance of the propellant in such an engine depends upon both the heat liberated and the propellant reaction products. Combustion with air of effective fuels for air-breathing engines gives approximately 18,000 Btu/lb, whereas fuels which are more effective in rocket engines may give only 15,000 Btu/lb. A high heat of reaction is most effective with gaseous products which are of low molecular weight.

Performance is rated in terms of specific impulse (occasionally specific thrust), the thrust obtained per pound of propellant used in 1 sec. An alternate measure of performance is the characteristic exhaust velocity. The theoretical characteristic exhaust velocity is determined by the thermodynamic properties of the propellant reaction and its products. Unlike the specific impulse, the characteristic exhaust velocity is independent of pressure, except for second-order effects, such as reactions modifying the heat capacity ratio of the combustion gases.

The relationship between these parameters is given by Eq. (1), in which I_s is specific impulse in seconds, F is thrust, and $\dot{w}$ is flow rate of propellant. The characteristic exhaust velocity c^* is giv-

$$I_s = \frac{F}{\dot{w}} = \frac{c^* C_F}{g} \tag{1}$$

en in feet per second. C_F is the thrust coefficient, and g is the gravitational constant.

The actual exhaust velocity of the combustion gases is given by the product of the characteristic exhaust velocity and the thrust coefficient, $c^* C_F$. The thrust coefficient is a function of the heat capacity ratio of the combustion gases (the ratio of the heat capacity at constant pressure to the heat capacity at constant volume) and of the ratio of the chamber pressure to the exhaust pressure. The heat-capacity ratio of common propulsion gases varies from 1.1 to 1.4.

Increase in the combustion-chamber pressure increases the specific impulse (Fig. 1). Variation in the stoichiometry of the propellant reaction (the oxidizer-fuel ratio) also affects performance. A slightly fuel-rich reaction gives higher performance with common liquid propellants despite the lower heat of reaction because of more favorable working-gas composition. Increase in chamber pressure usually moves the optimum performance point toward the stoichiometric reaction ratio. A nonstoichiometric ratio may be used to give low combustion temperatures if required by the structural materials.

A properly designed engine can give 95–100% of the theoretical performance shown in the table.

Combustion. The energy of liquid propellants is released in combustion reactions which also produce the working fluid for reaction propulsion. The liquids in a bipropellant system may ignite spontaneously on contact, or they may require an ignition device to raise them to ignition temperature. In the first case they are called hypergolic liquids; in the second case, anergolic liquids. Combustion can be initiated with a spark, a hot wire, or an auxiliary hypergolic liquid. Monopropellant combustion, or more properly decomposition, can also be ignited by catalysis with an active surface or by a chemical compound in solution. Ignition of common hypergolic bipropellants occurs in a period of 1–100 msec following initial contact of the liquids. Catalytic quantities of detergents or of certain compounds of metals with several oxidation states,

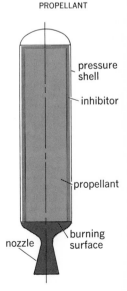

Fig. 2. End-burning grain loaded in rocket.

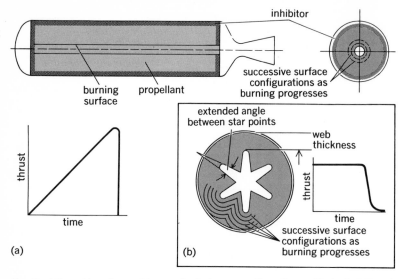

Fig. 3. Internal-burning solid-propellant charge configurations with typical thrust-time characteristics. (a) Cylindrical cavity. (b) Star-shaped cavity for level, or neutral, thrust-time characteristic.

such as vanadium pentoxide, decrease the ignition delay period of specific bipropellants.

The combustion chamber in operation contains a turbulent, heterogeneous, high-temperature-reaction mixture. The liquids burn with droplets of various sizes in close proximity and traveling at high velocity. Larger masses of liquid may be present, particularly at the chamber walls. Very high rates of heat release, of the order of $10^5 - 10^6$ Btu/(min)(ft³)(atm) are encountered.

Oscillations with frequencies of 25 – 10,000 Hz or more may accompany combustion of liquids in jet-propulsion engines. Low-frequency instability (chugging) can result from oscillations coupling the liquid flowing into the combustion chamber with pressure pulses in the chamber. Higher frequencies (screaming) can result from gas oscillations of the acoustic type in the chamber itself.

Engine performance, in contrast to theoretical propellant performance, depends upon effective

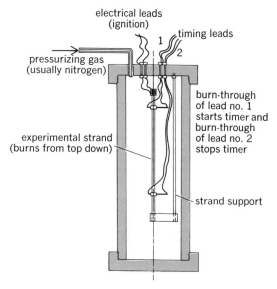

Fig. 4. Strand burner apparatus.

combustor design. Mixing and atomization are essential factors in injection of the propellants into the combustion chamber. Injector and chamber design influence the flow pattern of both liquid and gases in the chamber. The characteristic chamber length L^* is given by Eq. (2), in which V_c is the

$$L^* = V_c/A_T \qquad (2)$$

chamber volume and A_T is the area of the nozzle throat. In general, monopropellants require larger L^* than bipropellants to provide an equal fraction of theoretical performance in a rocket engine, as expected from the slower combustion exhibited by monopropellants.   [STANLEY SINGER]

SOLID PROPELLANTS

A solid propellant is a mixture of oxidizing and reducing materials that can coexist in the solid state at ordinary temperatures. When ignited, a propellant burns and generates hot gas. Although gun powders are sometimes called propellants, the term solid propellant ordinarily refers to materials used to furnish energy for rocket propulsion.

Composition. A solid propellant normally contains three essential components: oxidizer, fuel, and additives. Oxidizers commonly used in solid propellants are ammonium and potassium perchlorates, ammonium and potassium nitrates, and various organic nitrates, such as glyceryl trinitrate (nitroglycerin). Common fuels are hydrocarbons or hydrocarbon derivatives, such as synthetic rubbers, synthetic resins, and cellulose or cellulose derivatives. The additives, usually present in small amounts, are chosen from a wide variety of materials and serve a variety of purposes. Catalysts or suppressors are used to increase or decrease the rate of burning; ballistic modifiers may be used for a variety of reasons, as to provide less change in burning rate with pressure (platinizing agent); stabilizers may be used to slow down undesirable changes that may occur in long-term storage.

Solid propellants are classified as composite or double base. The composite types consist of an oxidizer of inorganic salt in a matrix of organic fuels, such as ammonium perchlorate suspended in a synthetic rubber. The double-base types are usually high-strength, high-modulus gels of cellulose nitrate (guncotton) in glyceryl trinitrate or a similar solvent.

Propellants are processed by extrusion or casting techniques into what are often intricate shapes that are commonly called grains, even though they may weigh many tons. The double-base types and certain high-modulus composites are processed into grains by casting or extrusion, and are then loaded by insertion of the cartridgelike grain or grains into the combustion chamber of the rocket. This technique requires some type of mechanical support to hold the propellant in place in the chamber. Certain types of composite propellants, bonded by elastomeric fuels, can be cast directly into the chamber, where the binder cures to a rubber and the grain is then supported by adhesion to the walls. Most high-performance rockets are made by this case-bonding technique, permitting more efficient use of weight and combustion-chamber volume.

Burning rate. The thrust-time characteristic of a solid-propellant rocket is controlled by the geo-

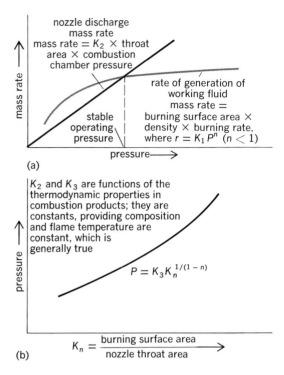

(a)

(b)

$$K_n = \frac{\text{burning surface area}}{\text{nozzle throat area}}$$

Fig. 5. Solid propellant. (a) Stable burning condition. (b) Typical pressure characteristic.

metric shape of the grain. Often it is desired that burning not take place on certain portions of the grain. Such surfaces are then covered with an inert material called an inhibitor or restrictor. Neutral-burning grains maintain a constant surface during burning and produce a constant thrust. Progressive burning grains increase in surface and give an increasing thrust with time. Degressive or regressive grains burn with decreasing surface and give a decreasing thrust.

An end-burning grain is shown in Fig. 2. This type of configuration is neutral, because the surface stays constant while the grain burns forward. For most applications, radial-burning charges which burn outward from the inside perforation are superior because most of the wall area of the chamber can be protected from hot gas generated by combustion. Such protection is a built-in feature of the case-bonded grain; with the cartridge-loaded, inhibited charge, protection is provided by the addition of obturators to prevent gas flow around the outside of the grain.

Figure 3 shows a progressive design called an internal-burning cylinder. Various star-shaped perforations can be used to give neutral or degressive characteristics. By ingenious use of geometry, the thrust-time characteristic can be designed to meet almost any need. Another important neutral-grain design, the uninhibited, internal-external-burning cylinder, is used widely in short-duration applications such as the bazooka rocket weapon, which contains many such grains. Obturation is not necessary because the propellant is burned so rapidly that the walls of the chamber do not rise in temperature sufficiently to cause loss of strength.

Propellant burns at a rate r proportional to a fractional power n of the pressure P as expressed by Eq. (3) in which K_1 is the coefficient of propor-

tionality. This rate may be determined at various pressures by measurements of the burning rate of

$$r = K_1 P^n \tag{3}$$

propellant strands in a strand burner (Fig. 4). If the propellant is to operate properly, the exponent n in the burning-rate equation must be less than 1. As illustrated by Fig. 5a, if $n < 1$, there is a stable operating pressure at which the lines of fluid generation and fluid discharge intersect. Pressure cannot rise above this value because gas would then be discharged at a faster rate than it is generated by burning of propellant. When the propellant meets the requirement $n < 1$, a relationship exists between operating chamber pressure and a design parameter known as K_n, which is the ratio of the propellant burning area to the nozzle throat area. This ratio is illustrated in Fig. 5b.

Specific impulse of solid propellants is normally rated with the rocket operating at chamber pressure of 1000 lb and exhausting through an optimum nozzle into sea-level atmosphere. Under these conditions, solid propellants in use today can give an impulse of about 250, which is near the ceiling imposed by compositions based on ammonium perchlorate and hydrocarbons, and is 5–10% lower than impulses obtainable from liquid oxygen and gasoline.

The lower specific impulse of solid propellants is partly overcome by their densities, which are higher than those of most liquid propellants. In addition, solid-propellant rockets are easy to launch, are instantly ready, and have demonstrated a high degree of reliability. Because they can be produced by a process much like the casting of concrete, there seems to be no practical limit to the size of a solid-propellant rocket.

[H. W. RITCHEY]

Bibliography: J. Humphries, *Rockets and Guided Missiles*, 1956; Princeton University Press, *High Speed Aerodynamics and Jet Propulsion*, vol. 2, 1956, vol. 12, 1959; E. Ring et al., *Rocket Propellant and Pressurization Systems*, 1964; B. Siegel and L. Schieter, *Energetics of Propellant Chemistry*, 1964; G. P. Sutton, *Rocket Propulsion Elements, An Introduction to the Engineering of Rockets*, 2d ed., 1956; F. A. Warren, *Rocket Propellants*, 1959; M. J. Zucrow, *Principles of Jet Propulsion and Gas Turbines*, rev. ed., 1949.

Propulsion

The process of causing a body to move by exerting a force against it. Propulsion is based on the reaction principle, stated qualitatively in Newton's third law, that for every action there is an equal and opposite reaction. A quantitative description of the propulsive force exerted on a body is given by Newton's second law, which states that the force applied to any body is equal to the rate of change of momentum of that body, and the force is exerted in the same direction as the momentum change.

In the case of a vehicle moving in a fluid medium, such as an airplane or a ship, the required change in momentum is generally produced by changing the velocity of the fluid (air or water) passing through the propulsive device or engine. In other cases, such as that of a rocket-propelled

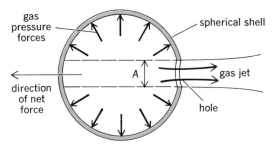

Fig. 1. A shell containing an aperture and saturated with gas at higher pressure than any outside fluid illustrates the principle of propulsion.

vehicle, the propulsion system must be capable of operating without the presence of a fluid medium; that is, it must be able to operate in the vacuum of space. The required momentum change is then produced by using up some of the propulsive device's own mass, called the propellant. *See* PROPELLANT.

Propulsion principle. Any change in momentum, according to Newton's second law, must be exactly equal to the propulsive force exerted on the body. For example, one may examine the simple propulsive means consisting of a hollow sphere with a small hole of area A (Fig. 1). If the sphere is filled with gas whose pressure is higher than that of the surrounding medium, the gas will exert pressure on the inside surface of the hollow sphere, with the exception of the hole, through which gas will rush out into the surroundings. Directly opposite the hole is an area A of the sphere on which the compressed gas does exert a force. In all other radial directions inside the sphere, pressure forces balance each other. Consequently, the net force on the sphere is in the direction shown. To compute the size of the unbalanced force, one can either measure and add or subtract all the pressures at the shell surface (including the gas pressure in the hole), or, more simply, in accordance with Newton's second law, calculate the momentum of the escaping gas jet.

This simple propulsive device shows that a rocket gives more thrust in space than in the atmo-

sphere, because atmospheric pressure tends to increase the pressure in the hole, reducing the net force in the forward direction and, of course, reducing by an equal amount the effluent gas momentum. Also, it shows that although the amount of net thrust can be calculated by using the momentum change of the fluid medium, the thrust itself is produced, as it must be, by something pushing against the body—in this case, the pressure exerted by a gas.

Most propulsion engines are more complicated than the simple spherical rocket discussed here. Consider, for example, an airplane in level flight at constant speed (Fig. 2). Within a rectangular region (the control volume) around the airplane, net forces must be in equilibrium. The sum of all momentum changes occurring across the surfaces of the control volume is balanced by the total of all the forces on the airplane, in accordance with Newton's law.

Some air passing over the wings is given downward momentum, resulting in upward reaction forces, called lift, on the airplane. This air is also given some forward momentum as it is accelerated downward, resulting in a rearward force on the airplane called induced drag. Viscous drag is also produced (in the rearward direction) because the airplane carries some air forward with it through viscous adherence, giving the air forward momentum. Finally, the airplane propulsive means (such as an engine-driven propeller or a turbojet engine) takes in quiescent air, which is thereby accelerated to the same speed as the airplane, providing a rearward force called ram drag, and then the propulsive means discharges the air at a much higher rearward velocity (with high rearward momentum), resulting in a forward force called gross thrust. If all these forces are balanced, the airplane flies at constant speed in a straight line; if not, the airplane is accelerated in whichever direction the forces are unbalanced, thereby providing another momentum change to again bring net forces within the control volume into equilibrium.

Thrust. Propulsion capability is measured in terms of the thrust delivered to the vehicle. In general, the net thrust delivered by any propulsive means, neglecting the effect of ambient pressure, is given by Eq. (1).

$$F = (W_a + W_f) V_e - W_a V \qquad (1)$$

F = net thrust in lb
V_e = velocity of gas leaving the propulsive means relative to the vehicle, in ft/sec
V = flight velocity of the vehicle through still air in ft/sec
W_a = mass flow rate of the ambient medium (air or water) through the propulsive means, in (lb-sec²/sec)/ft
W_f = mass flow rate of fuel (or on-board propellant) through the propulsive means in (lb-sec²/sec)/ft

The first term is the gross thrust; the second is the ram drag. In a rocket engine, no ambient fluid (air or water) is taken into the propulsive means; thus there is no ram drag. The net thrust of a rocket in vacuum is thus given by Eq. (2), where W_f

$$F = W_f V_e \qquad (2)$$

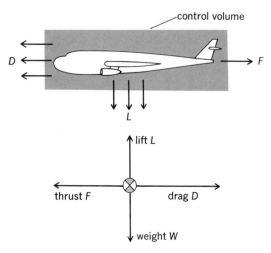

Fig. 2. All forces, including momentum forces, within a control volume about a vehicle are in equilibrium.

represents the total mass flow rate of on-board propellant.

Efficiency. Propulsion efficiency was formerly defined as the useful thrust power divided by the propulsion power developed. Unfortunately, because useful thrust power can only be determined as the thrust multiplied by the flight velocity (FV), and because flight velocity is basically a relative rather than an absolute term, propulsion efficiency has little real meaning as a measure of propulsion effectiveness.

The two terms most generally used to describe propulsion efficiency are thrust specific fuel consumption (SFC) for engines using the ambient fluid (air or water), and specific impulse (I_{sp}) for engines which carry all propulsive media on board. Thrust specific fuel consumption is given by Eq. (3),

$$SFC = \frac{W_f g_0}{F} = g_0 \frac{W_f}{W_a} \frac{1}{\left[\left(1 + \frac{W_f}{W_a}\right) V_e - V\right]}$$

$$\approx g_0 \frac{W_f}{W_a} \frac{1}{V_e - V} \qquad (3)$$

where SFC is expressed in pounds of fuel per second per pound of thrust, g_0 is acceleration due to gravity in ft/sec², and, usually, $W_f/W_a \ll 1$.

Specific impulse is determined similarly (although inversely) from Eq. (4). Obviously,

$$I_{sp} = \frac{F}{W_f g_0} = \frac{V_e}{g_0} \qquad (4)$$

effective propulsion performance is indicated by low values of SFC or high values of I_{sp}.

The energy source for most propulsion devices is the heat generated by the combustion of exothermic chemical mixtures composed of a fuel and an oxidizer. An airbreathing chemical propulsion system generally uses a hydrocarbon such as coal, oil, gasoline, or kerosine as the fuel, and atmospheric air as the oxidizer. A non-air-breathing engine, such as a rocket, almost always utilizes propellants that also provide the energy source by their own combustion. Here the choice is wider: Fuels may be hydrocarbons again, but may also be more efficient low-molecular-weight chemicals such as hydrazine (N_2H_4), ammonia (NH_3), or even liquid hydrogen itself. Oxidizers are usually liquid oxygen, nitrogen tetroxide (N_2O_5), or liquid fluorine.

Where nuclear energy is the source of propulsive power, the heat developed by nuclear fission in a reactor is transferred to a working fluid, which either passes through a turbine to drive the propulsive element such as a propeller, or serves as the propellant itself. Nuclear-powered ships and submarines are accepted forms of transportation. Nuclear-powered airplanes and rockets are still in their early developmental stages. In the case of nuclear-powered vehicles, the concept of specific fuel consumption is no longer valid, because the loss in weight of the nuclear fuel is virtually zero. However, the specific impulse is still a meaningful measure of performance, being based on thrust and propellant flow rate, not on mass consumption of fuel. *See* ELECTROMAGNETIC PROPULSION; TURBINE PROPULSION.

[JERRY GREY]

Bibliography: O. E. Lancaster (ed.), *Jet Propulsion Engines*, 1959; Northrop Institute of Technology, *Powerplants in Aerospace Vehicles*, 3d ed., 1965; M. J. Zucrow, *Aircraft and Missile Propulsion*, vols. 1 and 2, 1958.

Pulse jet

A type of jet engine characterized by periodic surges of thrust. The pulse jet engine was widely known for its use during World War II on the German V-1 missile (Fig. 1). The basic engine cycle was invented in 1908. The inlet end of the engine is provided with a grid to which are attached flap valves. These valves are normally held by spring tension against the grid face and block the flow of air back out of the front of the engine. They can be sucked inward by a negative differential pressure to allow air to flow into the engine. Downstream from the flap valves is the combustion chamber. A fuel injection system is located at the entrance to the combustion chamber. The chamber is also fitted with a spark plug. Following the combustion chamber is a long exhaust duct which provides an inertial gas column.

When the combustion chamber is filled with a mixture of fresh air and fuel, a spark is discharged; it ignites the fuel-air mixture, producing a pressure surge that advances upstream to slam shut the inlet valves and to block off the entrance. Simultaneously, a pressure pulse goes downstream to produce a surge of combustion products out the exhaust duct. Thrust results from the rearward discharge of this gas at high velocity. With the discharge of gas from the combustion chamber, its pressure tends to drop. Inertia causes the column of gas in the exhaust duct to continue to flow rearward even after the explosion pressure in the combustion chamber has been dissipated, and this drops the combustion chamber pressure below atmospheric. As a result, the flap valves open and a fresh charge of air enters the combustion chamber. As this air flows past the fuel nozzles, it receives an injection of fuel and the mixture is then ignited by contact with the hot gas residue from the previous cycle. This causes the mixture to explode and the cycle repeats. Thrust increases with engine speed up to a maximum dependent on design (Fig. 2).

Unlike the ramjet, the pulse jet has an appreciable thrust at zero flight speed. However, as the flight speed is increased, the resistance to the flow of air imposed by the flap valves eventually causes substantial loss in performance and the pulse jet becomes less efficient than the ramjet. *See* RAMJET.

Failure of flap valves and valve seats by fatigue was found to be a problem. Research has been

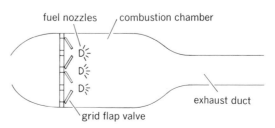

Fig. 1. Diagram of a pulse jet.

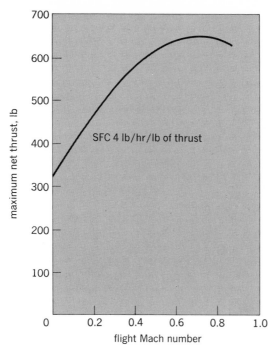

Fig. 2. Effect of flight Mach number on thrust of a German pulse jet; length 137.2 in., diameter 21.6 in.

conducted on valve systems other than that shown in Fig. 1 and on valveless pulse jets.

In addition to their use on the German V-1 buzz-bomb, pulse jets have been used to propel radio-controlled target drones and experimental helicopters. In the latter case, they were mounted on the blade tips for directly driving the rotor. The high fuel consumption, noise, and vibrations generated by the pulse jet limit its scope of applications. *See* PROPULSION.   [BENJAMIN PINKEL]

Pumped storage

A process, also known as hydroelectric storage, for converting large quantities of electrical energy to potential energy by pumping water to a higher elevation, where it can be stored indefinitely and then released to pass through hydraulic turbines and generate electrical energy. An indirect process is necessary because electrical energy cannot be stored effectively in large quantities. Storage is desirable, as the consumption of electricity is highly variable between day and night, between weekday and weekend, as well as among seasons. Consequently, much of the generating equipment needed to meet the greatest daytime load is unused or lightly loaded at night or on weekends. During those times the excess capability can be used to generate energy for pumping, hence the necessity for storage. Normally, pumping energy can be obtained from economical sources, and its value will be upgraded when used for peak loads.

Operation. In a typical operation, night or weekend electrical energy is used to pump water from a lower to a higher elevation, where it is stored as potential energy in the upper reservoir. The water can be retained indefinitely without deterioration or significant loss. During the daylight hours when the loads are greatest, stored water is released to flow from the higher to the lower reservoir through hydraulic-turbine-driven-generators and converted to electrical energy. No water is consumed in either the pumping or generating phase. To provide storage or generation merely requires the transfer of water from one reservoir to the other. Pumped storage installations have attained an overall operating efficiency of about 70%. Projected improvements in equipment design promise an efficiency of 75% or more. Postulating one cycle each of generation and one cycle for pumping per day plus an allowance to change from one mode to the other, the maximum annual generation attainable is 3500 hr.

Description. A typical pumped-storage development is composed of two reservoirs of essentially equal volume situated to maximize the difference in their levels. These reservoirs are connected by a system of waterways along which a pumping-generating station is located (Fig. 1). Under favorable geological conditions, the station will be located underground, otherwise it will be situated on the lower reservoir. The principal equipment of the station is the pumping-generating unit. In United States practice, the machinery is reversible and is used for both pumping and generating; it is designed to function as a motor and pump in one direction of rotation and as a turbine and generator in opposite rotation. Transformers, a substation, switchyard, and transmission line are required to transfer the electrical power to and from the station. *See* ELECTRIC POWER GENERATION.

The lower reservoir may be formed by impounding a stream or by using an existing body of water. Similarly, an upper reservoir may be created by an embankment across a valley or by a circumferential dike. Frequently, these features produce the most significant environmental impact, which is largely land use and visual. The reservoirs are comparatively small, thus affording some latitude in location to minimize unavoidable effects, such as displacement of developed areas, existing roads, and railways. Problems of emission of particulate matter and gases and of water-temperature rise associated with other generating stations do not exist. On the other hand, the reservoirs and surrounding area afford the opportunity to develop recreational facilities such as camp grounds, marinas, boat ramps, picnic areas, and wildlife preserves. In the United States, recreation commands

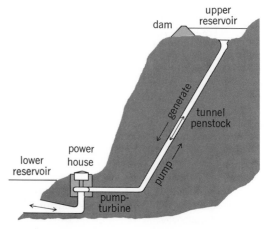

Fig. 1. Schematic of a conventional pumped-storage development.

a high priority, and many existing developments have included recreational facilities.

Economics. The economics of pumped storage are inextricably linked to the system in which it operates. The development must be economically competitive with other types of generation available, namely, nuclear, coal- or oil-fired, gas turbines and hydro. It is a generally accepted practice to evaluate the economy of any plant on the basis of its annual cost to the system. That cost is defined as the sum of fixed costs, operation, maintenance, and fuel. When construction is complete, the fixed costs are established and continue at that level irrespective of the extent of operation. Current variations in fuel prices have a major effect. On the average, coal costs about twice as much as nuclear fuel and oil about 2½ times as much as coal. Analyses based on current construction and fuel cost or on similar costs projected through 1990 show that between 10 and 20% of the generating capacity should be pumped storage. Pumping energy should be supplied either by coal or nuclear fueled plants; oil has not been considered as a fuel because of the forecast incipient shortage and instability of price. In systems having adequate coal and nuclear base-load capacity and gas-turbine peaking capacity, pumped storage may displace the gas turbines on the basis of economy, thereby conserving oil as well. *See the feature article* WORLD ENERGY ECONOMY.

European plants. In European practice, not only system economies receive attention, but also operating requirements and advantages; the latter have materially influenced the design and selection of equipment. Historically, European pumped-storage plants have been used not only to provide peak-load power as in the United States, but also to ensure system stability or frequency control. To accomplish this, the units must have very short response times which will enable them to change mode of operation quickly, to follow changes in load, or to provide emergency generation or load. Such operation is the dominant reason for European preference for three-machine sets, that is, separate pump and separate turbine coupled to a motor-generator, as opposed to the two-machine reversible pump-turbine and motor-generator used in the United States. The extent to which Europeans practice incorporates system stability and frequency regulation is shown by the number of pumping and generating starts per day. In one plant, the daily average starts for pumping was 22 and for generating 32, compared to 1 start for pumping and 1 or possibly 2 per day for generation in typical United States operation.

Prospects. In 1975 there existed in the United States about 11,000 MW of pumped-storage capacity, or about 3% of total capacity, compared to about 10% in the highly developed countries of Western Europe. Given favorable economics, pumped storage should increase as a percentage of overall capacity. One limitation of conventional pumped storage is the need for favorable site conditions accessible electrically to the load centers. Two factors which could further increase interest are conversion of existing hydro developments to pumped storage and the development of the deep underground plant. A number of existing hydro plants could be partially converted to pumped stor-

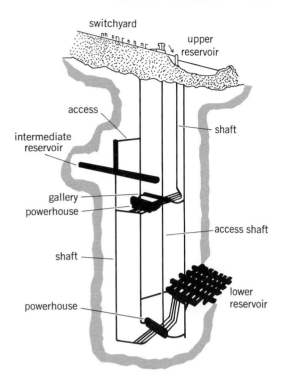

Fig. 2. Schematic of two-stage underground pumped-storage development.

age at a reasonable cost and minimal environmental impact since the reservoirs already exist.

The deep underground concept (Fig. 2) is attracting increased attention because of rising fuel costs as well as construction costs for alternate generation, and the flexibility it affords in selecting a site near load centers. Conceptually, it is similar to conventional pumped storage, having all of the essential features and utilizing similar equipment. There are two notable variations: the upper reservoir is at ground level, while the lower reservoir is a deep underground cavern; and two stages are developed in series to minimize the size and, hence, the construction cost of the lower reservoir. Inherent advantages of the concept are: material reduction in environmental impact since only the upper reservoir is visible, development of heads which utilize the maximum capability of the machines, and elimination of the need for favorable topography. The disadvantages are: the need for large areas to dispose of the excavated material, uncertainties of construction and mining at 3000 ft (900 m) or more below the surface, and substantially increased construction time and costs compared with conventional pumped storage. *See the feature article* EXPLORING ENERGY CHOICES; *see also* HYDROELECTRIC GENERATOR; HYDROELECTRIC POWER; TURBINE.   [DWIGHT L. GLASSCOCK]

Bibliography: E. Comninellis, Ludington Pumped Storage Project, *ASCE J. Power Div.,* 99 (PO1):69–88, May 1973; Engineering Foundation Conference, *Converting Existing Hydroelectric Dams and Reservoirs into Pump Storage Facilities,* American Society of Civil Engineers, 1975; A. Ferreira, *Multipurpose Aspects of the Northfield Mountain Pumped Storage Project,* Paper presented at the International Conference on Pumped Storage Development and Its Environmental Effects, University of Wisconsin, September 1972; J. M. Frohnholzer, Operation of pumped-storage stations in the Federal Republic of Germany, *Water Power*, 25(10):374–384, October 1973; G. M. Karadi, *Pumped Storage Development and Its Environmental Effects: Final Report to the National Science Foundation*, University of Wisconsin, 1974; G. E., Pfafflin, Future trends in hydro pumped-storage equipment, *Proc. Amer. Power Conf.*, 36: 390–402, 1974; J. G. Warnock, and D. C. Willet, Underground reservoirs for high-head pumped-storage stations, *Can. Electr. Ass. Eng. Oper. Div. Trans.* Toronto, vol. 12, pt. 2, Pap. 73-H-107, Mar. 26–30, 1973; (also in *Water Power*, 25(3):81–87, March 1973).

Q (electricity)

Often called the quality factor of a circuit, Q is defined in various ways, depending upon the particular application. In the simple RL and RC series circuits, Q is the ratio of reactance to resistance, as in Eqs. (1), where X_L is the inductive reactance,

$$Q = X_L/R \quad Q = X_C/R \quad \text{(a numerical value)} \quad (1)$$

X_C is the capacitive reactance, and R is the resistance. An important application lies in the dissipation factor or loss angle when the constants of a coil or capacitor are measured by means of the alternating-current bridge.

Q has greater practical significance with respect to the resonant circuit, and a basic definition is given by Eq. (2), where Q_0 means evaluation at res-

$$Q_0 = 2\pi \frac{\text{max stored energy per cycle}}{\text{energy lost per cycle}} \quad (2)$$

onance. For certain circuits, such as cavity resonators, this is the only meaning Q can have.

For the RLC series resonant circuit with resonant frequency f_0, Eq. (3) holds, where R is the total

$$Q_0 = 2\pi f_0 L/R = 1/2\pi f_0 CR \quad (3)$$

circuit resistance, L is the inductance, and C is the capacitance. Q_0 is the Q of the coil if it contains practically the total resistance R. The greater the value of Q_0, the sharper will be the resonance peak.

The practical case of a coil of high Q_0 in parallel with a capacitor also leads to $Q_0 = 2\pi f_0 L/R$. R is the total series resistance of the loop, although the capacitor branch usually has negligible resistance.

In terms of the resonance curve, Eq. (4) holds,

$$Q_0 = f_0/(f_2 - f_1) \quad (4)$$

where f_0 is the frequency at resonance, and f_1 and f_2 are the frequencies at the half-power points.

[BURTIS L. ROBERTSON]

Radiant heating

Any system of space heating in which the heat-producing means is a surface that emits heat to the surroundings by radiation rather than by conduction or convection. The surfaces may be radiators such as baseboard radiators or convectors, or they may be the panel surfaces of the space to be heated. *See* PANEL HEATING AND COOLING.

The heat derived from the Sun is radiant energy. Radiant rays pass through gases without warming them appreciably, but they increase the sensible temperature of liquid or solid objects upon which

they impinge. The same principle applies to all forms of radiant-heating systems, except that convection currents are established in enclosed spaces and a portion of the space heating is produced by convection. The radiation component of convectors can be increased by providing a reflective surface on the wall side of the convector and painting the inside of the enclosure a dead black to absorb heat and transmit it through the enclosure, thus increasing the temperature of that side of the convector exposed to the space to be heated.

Any radiant-heating system using a fluid heat conveyor may be employed as a cooling system by substituting cold water or other cold fluid. This cannot be done with electric radiant-heating systems because, at their present stage of commercial development, they are not reversible; however, experiments on the reversibility of thermocouples may make such a development possible in the future.

[ERWIN L. WEBER/RICHARD KORAL]
Bibliography: American Society of Heating, Refrigerating, and Air Conditioning Engineers, *Guide*, 1962, 1964, 1966, 1967.

Radioactive waste management

The treatment and containment of radioactive wastes. The requirement for radioactive waste management is present to some degree in all nuclear energy operations. Wastes in liquid, solid, or gaseous form are produced in the mining of ore, production of reactor fuel materials, reactor operation, processing of irradiated reactor fuels, and a great variety of related operations. Wastes also result from use of radioactive materials, for example, in research laboratories, industrial operations, and medical research and treatment. The magnitude of waste management operations undoubtedly will increase as the nuclear energy program is further extended and diversified and as a large and widespread nuclear power industry is developed.

In the safe handling and containment of radioactive wastes, the principal objective is the prevention of radiation damage to humans and the environment by controlling the dispersion of radioactive materials. Damage to humans may result from irradiation by external sources or by the intake (by ingestion, by inhalation, or through the skin) of radioactive materials, their passage through the respiratory and gastrointestinal tract, and their partial incorporation into the body. Radioactive waste contaminants in air, water, food, and other elements of the human environment must be kept below specified concentrations for the particular radionuclide or mixture of radionuclides present in the wastes. Liquid or solid waste products containing significant quantities of the more toxic radioactive materials require isolated and permanent containment media from which any potential reentry into human environment would be at tolerable levels. The radioactive materials of major concern are those that may be readily incorporated into the body. Also of concern are those materials that have relatively long half-lives, ranging from a few years to thousands of years.

Waste management operations are focused on those radioisotopes which originate in nuclear reactors. Here the fission products (other chemical

Table 1. Typical materials in high-level liquid waste

Material[b]	Grams per metric ton from various reactor types[a]		
	Light water reactor[c]	High-temperature gas-cooled reactor[d]	Liquid metal fast breeder reactor[e]
Reprocessing chemicals			
Hydrogen	400	3800	1300
Iron	1100	1500	26,200
Nickel	100	400	3300
Chromium	200	300	6900
Silicon	—	200	—
Lithium	—	200	—
Boron	—	1000	—
Molybdenum	—	40	—
Aluminum	—	6400	—
Copper	—	40	—
Borate	—	—	98,000
Nitrate	65,800	435,000	244,000
Phosphate	900	—	—
Sulfate	—	1100	—
Fluoride	—	1900	—
SUBTOTAL	68,500	452,000	380,000
Fuel product losses[f,g]			
Uranium	4800	250	4300
Thorium	—	4200	—
Plutonium	40	1000	500
SUBTOTAL	4840	5450	4800
Transuranic elements[g]			
Neptunium	480	1400	260
Americium	140	30	1250
Curium	40	10	50
SUBTOTAL	660	1440	1560
Other actinides[g]	<0.001	20	<0.001
Total fission products[h]	28,800	79,400	33,000
TOTAL	103,000	538,000	419,000

SOURCE: From K. J. Schneider and A. M. Platt (eds.), *Advanced Waste Management Studies: High-Level Radioactive Waste Disposal Alternatives*, USAEC Rep. BNWL-1900, May 1974.

[a]Water content is not shown; all quantities are rounded.

[b]Most constituents are present in soluble, ionic form.

[c]U-235 enriched pressurized water reactor (PWR), using 378 liters of aqueous waste per metric ton, 33,000 MWd/MT exposure. (Integrated reactor power is expressed in megawatt-days [MWd] per unit of fuel in metric tons [MT].)

[d]Combined waste from separate reprocessing of "fresh" fuel and fertile particles, using 3785 liters of aqueous waste per metric ton, 94,200 MWd/MT exposure.

[e]Mixed core and blanket, with boron as soluble poison, 10% of cladding dissolved, 1249 liters per metric ton, 37,100 MWd/MT average exposure.

[f]0.5% product loss to waste.

[g]At time of reprocessing.

[h]Volatile fission products (tritium, noble gases, iodine, and bromine) excluded.

elements formed by nuclear fragmentation of actinide elements such as uranium or plutonium, and others) accumulate in the nuclear fuel, along with plutonium and other transuranic nuclides. (Transuranic elements are those higher than uranium on the periodic table of chemical elements. They are also called actinide elements.) The concentrations of plutonium are substantially higher than those found in nature, ranging from 10 to 20 kg per metric ton (1 metric ton = 1000 kg) of uranium compared to a high of 17 g per metric ton of uranium in minerals from fumarole areas.

Reprocessing. Fuel discharged from the nuclear reactor is reprocessed to recover uranium and plutonium by chemical dissolution and treatment. During this step, favored treatment processes form high-level waste as an acidic aqueous stream. Other processes are being considered which would produce high-level waste in different forms. This high-level waste contains most of the reactor-produced fission products and actinides, with slight residues of uranium and plutonium (see Table 1). These waste products generate sufficient heat to require substantial cooling and emit large amounts of potentially hazardous ionizing radiation. Because the reprocessing step normally does not dissolve much of the nuclear fuel cladding, high-level waste normally contains only a small amount of the radionuclides formed as activation products within the cladding. This cladding hull waste is managed as a separate solid waste stream, as are several other auxiliary waste streams from the reprocessing plants.

The recovered uranium and plutonium are reused by the nuclear industry by reconstitution into nuclear fuels using plutonium (instead of uranium-235) as the fissile material. The fabrication of such fuels, since they contain both uranium and plutonium mixed oxides (MOX), generates additional wastes that may contain plutonium.

Policy and treatment. The policy of the U.S. Energy Research and Development Administration (ERDA, which includes the former Atomic Energy Commission) is to assume custody of all commercial high-level radioactive wastes and to provide containment and isolation of them in perpetuity. Regulations require that the high-level wastes from nuclear fuels reprocessing plants be solidified within 5 years after reprocessing and

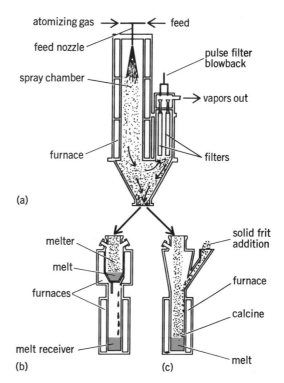

Fig. 1. Spray calcination-vitrification process. (*a*) Spray calciner, producing calcine that drops into either (*b*) continuous silicate glass melter, or (*c*) directly into waste canister vessel for for in-pot melting.

Table 2. Projected accumulation of solidified high-level waste through end of year, 1977–2000

Fiscal year	Volume[a] of waste, m³	Actinide mass, metric tons	Radio-activity,[b,c] MCi[d]	Thermal power,[b,c] MW	Toxicity indices[b,c] Inhalation	Toxicity indices[b,c] Ingestion
1977	190	18	4,600	20	19.62	14.91
1980	550	50	10,200	50	20.26	15.38
1985	1720	190	26,300	140	21.30	15.85
1990	3900	410	50,300	250	21.59	16.18
1995	7650	760	90,500	420	21.71	16.48
2000	13,340	1270	149,000	660	21.86	16.70
Time elapsed after year 2000, years						
10^2			5700	20	21.29	15.55
10^3			30	<1	20.19	12.74
10^4			10	<1	19.70	12.14
10^5			4	<1	18.79	12.38
10^6			1	<1	18.60	11.86

SOURCE: From K. J. Schneider and A. M. Platt (eds.), *Advanced Waste Management Studies: High-Level Radioactive Waste Disposal Alternatives*, USAEC Rep. BNWL-1900, May 1974.

[a]Volume based on 0.057, 0.170, and 0.085 m³ of solidified waste per metric ton of heavy metal for LWR, high-temperature gas-cooled reactor (HTGR), and liquid-metal fast breeder reactor (LMFBR) fuels, respectively.

[b]Waste initially generated 150, 365, and 90 days after spent fuel discharged from LWR, HTGR, and LMFBR units, respectively.

[c]All tritium and noble-gas fission products and 99.9% of iodine and bromine fission products excluded.

[d]1 megacurie (MCi) = 3.7×10^{16} disintegrations per second.

then shipped to a Federal repository within 10 years after reprocessing.

Because of the anticipated increase in the quantities of waste-containing elements or those contaminated with transuranic (TRU) elements, and the long half-life and specific radiotoxicity of these elements, ERDA has also proposed that all transuranic wastes be solidified and transferred to ERDA as soon as practicable, but at most within 5 years after generation.

Both of these policies require that high-level cladding and other transuranic wastes be converted to a solid. A variety of technologies exist for this conversion, including calcination, vitrification, oxidation, and metallurgical smelting, depending on the primary waste.

A typical solidification process, principally for high-level waste, is spray calcination-vitrification. In this process (see Fig. 1) atomized droplets of waste fall through a heated chamber, where flash evaporation results in solid oxide particles. Glassmaking solid frit or phosphoric acid can be added to provide for melting and glass formation in a continuous melter or directly in the vessel that will serve as the waste canister. The molten glass or ceramic is cooled and solidified.

Quantity of waste. The growth of nuclear power in the United States will result in increased quantities of high-level waste. Installed nuclear electrical generating capacity is projected, according to an ERDA Office of Planning and Analysis study, to increase to about 1,200,000 MW by the year 2000. The anticipated volume of solidified high-level waste accumulated by the year 2000 is about 13,000 m³, the result of reprocessing almost 200,000 metric tons of fuel, about 80% of which is associated with light water reactor (LWR) plants. If this amount of solid waste were stacked as a solid cube, the cube would be about 25 m on a side. Approximately 150,000 megacuries (5.5×10^{21} disintegrations per second) of radioactivity and

700 MW of heat will be associated with this projected waste inventory in the year 2000. This heat content is equivalent to about one-third of the waste heat rejected from one LWR generating 1000 MW of electricity.

Table 2 shows the projected accumulation of solidified high-level waste, assuming that 0.5% of the fuel product (uranium and plutonium or thorium) is lost to waste during reprocessing, and that all other actinides are in the waste. The toxicity indices in Table 2 are base 10 logarithms of the quantity in cubic meters of air, for the inhalation index, or in cubic meters of water, for the ingestion hazard index, required to dilute radioactive material to limits stipulated in federal regulations. Beyond the year 2000, fission products (primarily strontium) and transplutonium elements (primarily americium) are the chief potential hazards in drinking water up to about 350 years and 2×10^4 years, respectively. Radioactivity from plutonium losses during reprocessing then becomes the main factor until about 10^6 years. Finally, radioactivity remaining as the result of uranium losses during reprocessing becomes the predominant contribution to the ingestion toxicity index. For comparison with high-level waste projections, the estimated quantities of cladding waste and other TRU waste are shown in Table 3.

Alternative waste management concepts. Scientists are investigating many of the options for separating, treating, and otherwise managing radioactive waste from the time the material is formed in a fission reactor. Key considerations in the route to ultimate storage or disposal or elimination of the material are outlined in Fig. 2.

Constituents of the waste material are a mix of long- and short-lived radioisotopes. Some have radioactive decay half-lives of no more than tens of years, while others must be isolated from the biosphere for many thousands of years. By dividing the high-level waste into actinides and fission prod-

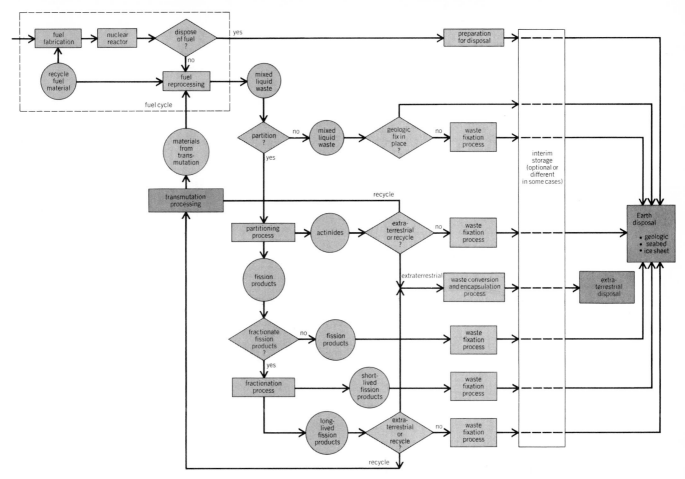

Fig. 2. High-level radioactive waste management options. (K. J. Schneider and A. M. Platt, eds., Advanced Waste Management Studies: High-Level Radioactive Waste Disposal Alternatives, USAEC Rep. BNWL-1900, May 1974)

ucts (a process called partitioning), the materials can be managed as separate classes.

Common to all waste management concepts is the possible need for interim storage in a retrievable surface storage facility. Three concepts have been evaluated by ERDA — a water basin concept, an air-cooled vault concept, and a concept for storage of wastes in sealed casks in the open air. The major differences between these proposals are in radiation shields, containment barriers, heat removal techniques, and relative dependence on utilities and maintenance. The canisters of solidified waste would be retrievable at all times for various waste management options or treatment by future techniques of disposal.

Many short-lived waste components will decay to unimportant radioactivity levels in relatively short periods of time. Their storage in artificial structures can be considered. For the longer-lived and highly toxic actinide fraction of radioactive waste, there appear to be only three basic management options: elimination of waste constituents by transmutation — the conversion to other, less undesirable isotopes by nuclear processes; transport off the Earth; and isolation from the human environment somewhere on Earth for periods of time sufficient to permit natural radioactive decay.

These potential alternative methods for long-term management of high-level radioactive waste provide the framework for a major study by the AEC's Division of Waste Management and Transportation. Included in the comprehensive review is a compilation of information relevant to the technical feasibility; the safety, cost, environmental, and policy considerations; the public response; and research and development needs for various waste management alternatives.

The basic requirement for the suitability of any environment for final storage or disposal of radioactive waste is its capability to safely contain and isolate it until decay has reduced the radioactivity to nonhazardous levels. Geologic formations exist which have been physically and chemically stable for millions of years. Ice sheets appear to offer some potential advantages as a disposal medium remote from the human environment. Both are under study as a potential future alternative. In the

Table 3. Projected accumulation of cladding and transuranic waste through end of year, 1985–2000

Fiscal year	Cladding		Other TRU	
	Volume, m³ × 10⁻³	Actinide mass, metric tons	Volume, m³ × 10⁻³	Actinide mass, metric tons
1985	5	30	161	4
1990	9	55	275	10
1995	19	90	600	28
2000	36	135	1236	63

ice sheet disposal concept, a waste canister would either melt down through the ice sheet to bedrock, or be connected to the surface by cables or chains which would stop its descent through the ice, or be placed in a surface storage facility which would gradually become covered with snow and buried in the ice sheet. Bedrock zones in stable areas of the deep sea floor are the subject of another study concept, as is the process of tectonic plate movement which should carry waste material down into the Earth's mantle from areas known as subduction zones. Also under study is storage in bedrock beneath high-sedimentation-rate areas where major rivers are building deltas into the ocean. Although very high costs per unit of weight would be encountered, an in-depth analysis of extraterrestrial disposal is under way. Also being assessed for possible future applications are nuclear techniques for transmutation of the actinides to isotopes having lower toxicity or shorter half-lives or both. *See* DECONTAMINATION OF RADIOACTIVE MATERIALS; NUCLEAR FUELS REPROCESSING.

[ALLISON M. PLATT]

Bibliography: *Draft Environmental Statement: Management of Commercial High-Level and Transuranium-Contaminated Radioactive Waste*, USAEC Rep. WASH-1539, September 1974; K. J. Schneider and A. M. Platt (eds.), *Advanced Waste Management Studies: High-Level Radioactive Waste Disposal Alternatives*, USAEC Rep. BNWL-1900, May 1974.

Ramjet

The simplest of the air-breathing propulsion engines (Fig. 1). In flight, air enters the front of the diffuser at high velocity. The diffuser is shaped to reduce the airspeed and hence its kinetic energy as it passes through. With an efficient diffuser, the reduction in kinetic energy results in a nearly equal increase in potential energy, in the form of an increase in air pressure. This higher-pressure air enters the combustion chamber, where fuel is continuously injected and burned. The hot gas is then ejected rearwardly through the discharge nozzle at velocity V_J greater than flight speed V_0. To a first approximation, thrust F is given in the equation shown below, where M is defined as the

$$F = M(V_J - V_0)$$

mass of air per second which is flowing through the engine.

Characteristics. The objective of the ramjet cycle is to provide a jet velocity V_J that is considerably greater than the initial velocity V_0. This increase in air velocity represents an increase in kinetic energy. The efficiency of a ramjet in converting the chemical energy in the fuel into kinetic energy of the airstream depends upon the ratio of

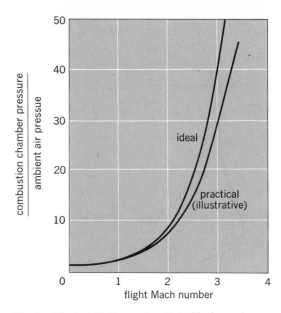

Fig. 2.   Effect of flight speed or flight Mach number on pressure ratio across the inlet diffuser.

the pressure in the combustion chamber to the ambient air pressure. This pressure ratio in turn depends upon the flight speed or, more exactly, upon flight Mach number (Fig. 2).

At zero flight Mach number, there is of course no increase in pressure through the diffuser, and the efficiency of the ramjet is zero. Thus the ramjet has no thrust at takeoff. As the flight speed increases, the pressure ratio and hence efficiency increase, and an increase in thrust and a reduction in specific fuel consumption result (Fig. 3).

At any given flight Mach number, maximum thrust is developed when sufficient fuel is injected into the combustion chamber to consume substantially all of the oxygen in the air passing through. This represents the largest amount of heat that can be introduced into the air. Greater efficiency of utilization of the fuel, however, is obtained when less than the maximum burnable amount is injected. The efficiency of utilization of the fuel is represented by the specific fuel consumption (pounds of fuel consumed per hour per pound of thrust). The higher the efficiency, the lower is the specific fuel consumption. Curves in Fig. 3 are for the two operating conditions of maximum power and maximum efficiency.

Combustion. The fuels used in ramjet engines are hydrocarbons obtained from petroleum. These fuels burn in only a narrow range of fuel-air ratios. They have flame speeds which are considerably lower than the speed at which the air must pass through the combustion chamber to obtain the high thrust per unit of cross-sectional area required for practical applications. Flameholders are therefore located in the combustion chamber in the wake of which the airspeeds are reduced locally to accommodate the low flame speeds. The determination of the configuration and location of these flameholders to provide adequate combustion efficiency without imposing excessive drag on the air is one of the crucial development problems of the ramjet. The location and design of the fuel injection nozzles and the design of the control equipment to provide a suitable fuel-air ratio in

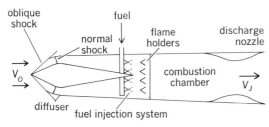

Fig. 1.   Diagram of a ramjet engine.

the vicinity of the flameholder for efficient combustion over the range of flight speeds and altitudes desired in the flight program of a given ramjet vehicle are essential to efficient operation. New high-energy fuels greatly increase the attainable altitude of the ramjet and appreciably reduce the engine length. The absence in this propulsion cycle of moving parts after the burner enables the ramjet to burn metal-based fuels for greater performance.

Takeoff. Because the ramjet has low thrust at low flight speeds, another type of engine is required for takeoff boost. In missiles such as the Bomarc, a rocket is used to take off and accelerate the vehicle to a speed at which the ramjet can take over. In aircraft where successive takeoffs and landings are desired, a turbojet engine can be used for this purpose.

In general, a supersonic ramjet vehicle must be boosted to supersonic flight speeds before the ramjet engines can provide sufficient thrust for propelling the vehicle. By providing diffusers and nozzles in which the configuration and area can be varied, the ramjet can operate efficiently over a wider range of flight speeds and can take over at a lower flight speed. This can result in an appreciable saving in the size of the booster rocket.

The ramjet depends entirely on pressure recovery due to its forward speed. The oblique and normal shock waves at the inlet constitute the first two stages of a compressor deceleration. Control of inlet flow by changing the inlet area through axial translation of the nozzle cone or by varying the back pressure through adjustment of the fuel flow can maintain the optimum positions of the shock waves and thereby minimize aerodynamic drag due to spillage or inadequate capture of air.

Flight speed. Ramjet engines are usually considered for applications in the range of flight Mach numbers between 2.5 and 8, although 8 is not a theoretical upper limit. As the flight Mach number increases above 4, heating of the vehicle by the high air friction becomes progressively a more serious problem, and methods of cooling the structure must be incorporated. At flight Mach numbers of 5 and higher, because of the high gas tempera-

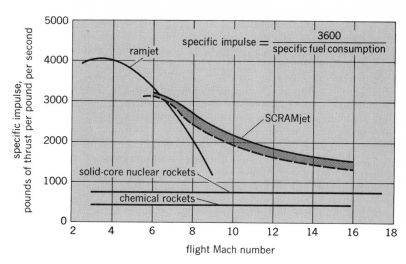

Fig. 4. Comparison of specific impulse of scramjet and ramjet, both burning hydrogen fuel. (*From Air Force Mag., May, 1965*)

tures in the combustion chamber, dissociation of the gases occurs; that is, the combustion does not go to completion and only part of the heat is released. The remainder of the heat can theoretically be released if combustion continues as the gas expands through the discharge nozzle. However, the occurrence of dissociation, the extent of which is influenced by nozzle design, may be a basic practical limitation on the flight speeds attainable by ramjets.

The working fluid (air) can also be heated by a nuclear reactor, thereby freeing the design from limitations which would be imposed by using the working fluid as oxidizer. The air is decelerated, passed directly through a nuclear reactor that occupies the usual combustion chamber space, and then discharged from a conventional convergent-divergent nozzle.

Scramjet. The scramjet (supersonic combustion ramjet) is essentially a ramjet engine intended for flight at hypersonic speeds (that is, above Mach 6), with the gases flowing through the combustion chamber and burning at supersonic speeds. (In the ramjet the velocity through the combustion chamber is subsonic.) By designing the engine inlet diffuser to provide supersonic speeds in the combustion chamber for flight at hypersonic speeds, the problem of extreme air temperature and dissociation in the combustion chamber, which limits the practical flight speed of ramjet engines, is avoided in the scramjet. Thus while ramjets cease to be practical at Mach numbers above 8, scramjets, on the other hand, are believed to be feasible up to Mach 25.

The projected specific impulses of scramjets and ramjets with hydrogen fuel are compared in Fig. 4. The specific impulse, which is the pounds of thrust obtained per pound per second of fuel consumed, is a measure of the efficiency of the engine. The scramjet has a much higher specific impulse than the ramjets at Mach numbers above about 6. The specific impulse of the scramjet is much higher than that of the best chemical rocket and the nuclear rocket. The scramjet is considered the logical engine for aircraft and long-range missiles that cruise in the atmosphere at hypersonic speeds.

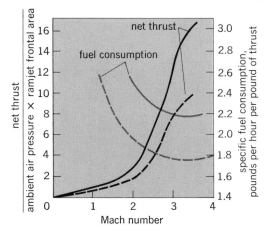

Fig. 3. Effect of flight Mach number on the thrust and the specific fuel consumption of a ramjet engine. Solid lines represent the fuel-air ratio adjusted for maximum thrust; broken lines represent the fuel-air ratio adjusted for maximum efficiency.

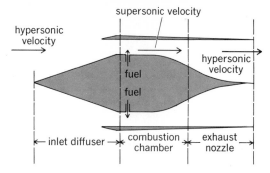

Fig. 5. Diagram of scramjet engine.

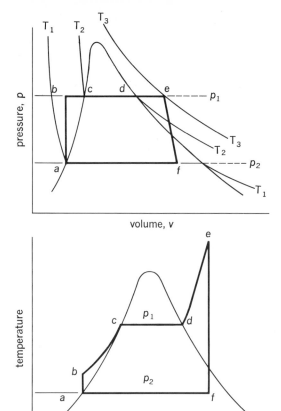

Rankine-cycle diagrams (pressure-volume and temperature-entropy) for a steam power plant using superheated steam. Typically, cycle has four phases.

Various fuels, ranging from hydrocarbons to hydrogen, are being studied for this engine. The specific impulse of the scramjet engine is considerably less with hydrocarbons than with hydrogen, and its flight speeds may be limited to below Mach 15; with hydrogen, flight up to nearly orbital velocity (Mach 26) is believed to be possible.

Because the velocity in the scramjet engine never decreases to subsonic speeds, the diffusers and nozzles have much simpler shapes than their ramjet counterparts; they do not require the converging-diverging contours that characterize ramjet engines. Compare the diagram of a scramjet engine (Fig. 5) with that of the ramjet (Fig. 1). The optimum speed in the combustion chamber is about one-third the vehicle flight speed; that is, for a flight speed of Mach 12 the speed in the combustion chamber is Mach 4.

At hypersonic flight speeds the temperature of the air in the combustion chamber is above the ignition temperature of the fuel. Thus the rate of burning is limited mainly by the rate at which the fuel and air can be mixed. Supersonic combustion of good efficiency has been demonstrated experimentally with hydrocarbons and with hydrogen as fuels. The integration of the diffuser, combustor, and discharge nozzle in an efficient, practical system has yet to be obtained.   [BENJAMIN PINKEL]

Bibliography: *Air Force Mag.*, May, 1965; J. V. Casamassa and R. D. Bent, *Jet Aircraft Power Systems*, 3d ed., 1965; A. Ferri, Review of SCRAMjet propulsion technology, *J. Aircraft* (AIAA), vol. 5, no. 1, 1968.

Rankine cycle

A thermodynamic cycle used as an ideal standard for the comparative performance of heat-engine and heat-pump installations operating with a condensable vapor as the working fluid. Applied typically to a steam power plant, as shown in the illustration, the cycle has four phases: (1) heat addition *bcde* in a boiler at constant pressure p_1 changing water at b to superheated steam at e, (2) isentropic expansion *ef* in a prime mover from initial pressure p_1 to back pressure p_2, (3) heat rejection *fa* in a condenser at constant pressure p_2 with wet steam at f converted to saturated liquid at a, and (4) isentropic compression *ab* of water in a feed pump from pressure p_2 to pressure p_1.

This cycle more closely approximates the operations in a real steam power plant than does the Carnot cycle. Between given temperature limits it offers a lower ideal thermal efficiency for the conversion of heat into work than does the Carnot standard. Losses from irreversibility, in turn, make the conversion efficiency in an actual plant less than the Rankine cycle standard. *See* CARNOT CYCLE; REFRIGERATION CYCLE; THERMODYNAMIC CYCLE.   [THEODORE BAUMEISTER]

Bibliography: T. Baumeister (ed.), *Standard Handbook for Mechanical Engineers*, 7th ed., 1967; J. B. Jones and G. A. Hawkins, *Thermodynamics*, 1960; J. H. Keenan, *Thermodynamics*, 1941.

Reaction turbine

A power-generation prime mover utilizing the steady-flow principle of fluid acceleration, where nozzles are mounted on the moving element. The rotor is turned by the reaction of the issuing fluid jet and is utilized in varying degrees in steam, gas, and hydraulic turbines. All turbines contain nozzles; the distinction between the impulse and reaction principles rests in the fact that impulse turbines use only stationary nozzles, while reaction turbines must incorporate moving nozzles. A nozzle is defined as a fluid dynamic device containing a throat where the pressure of the fluid drops and potential energy is converted to the kinetic form with consequent acceleration of the fluid. For details of the two basic principles of impulse and reaction as applied to turbine design *see* IMPULSE TURBINE.   [THEODORE BAUMEISTER]

Reactor, nuclear

A system utilizing nuclear fission in a controlled and self-sustaining manner. Neutrons are used to fission the nuclear fuel, and the fission reaction produces not only energy and radiation, but also additional neutrons. Thus a neutron chain reaction ensues. A nuclear reactor provides the assembly of materials to sustain and control the neutron chain reaction, to appropriately transport the heat produced from the fission reactions, and to provide the necessary safety features to cope with the radiation and radioactive materials produced by the operation of the nuclear reactor. *See* CHAIN REACTION, NUCLEAR; FISSION, NUCLEAR.

Nuclear reactors are used in a variety of ways as sources for energy, for nuclear radiations, and for special tests and feasibility demonstrations. Since the first demonstration of a nuclear reactor, made beneath the West Stands of Stagg Field at the University of Chicago on Dec. 2, 1942, more than 500 nuclear reactors have been built and operated in the United States. Extreme diversification is possible with the materials available, and reactor dimensions may vary from football size to house size. The rates of energy release for controlled operations may vary from a fraction of a watt to thousands of megawatts. The critical size of a nuclear reactor is governed by the factors affecting the control of the neutron chain reaction, and the thermal output of the reactor is determined by the factors affecting the effectiveness of the coolant in removing the fission energy released.

The generation of electric energy by a nuclear power plant requires the use of heat to produce steam or to heat gases to drive turbogenerators. Direct conversion of the fission energy into useful work is possible, but an efficient process has not yet been realized to accomplish this. Thus, in its operation the nuclear power plant is similar to the conventional coal-fired plant, except that the nuclear reactor is substituted for the conventional boiler.

The rating of a reactor is usually given in kilowatts (kW) or megawatts (MW) thermal, representing the heat generation rate. The net output of electricity of a nuclear plant is about one-third of the thermal output. Significant economical gains have been achieved by building improved nuclear reactors with outputs of about 3000 MW thermal and about 1000 MW electrical. *See* ELECTRIC POWER GENERATION; NUCLEAR POWER.

FUEL AND MODERATOR

The fission neutrons are released at very high energies and are called fast neutrons. The average kinetic energy is 2 MeV, with a corresponding neutron speed of 1/15 the speed of light. Neutrons slow down through collisions with nuclei of the surrounding material. This slowing-down process is made more effective by the introduction of lightweight materials, called moderators, such as heavy water (deuterium oxide), ordinary (light) water, graphite, beryllium, beryllium oxide, hydrides, and organic materials (hydrocarbons). Neutrons that have slowed down to an energy state in equilibrium with the surrounding material are called thermal neutrons. The probability that a neutron will cause the fuel material to fission is greatly enhanced at thermal energies, and thus most reactors utilize a moderator for the conversion of fast neutrons to thermal neutrons. *See* GRAPHITE.

With suitable concentrations of the fuel material, neutron chain reactions also can be sustained at higher neutron energy levels. The energy range between fast and thermal is designated as intermediate. Fast reactors do not have moderators and are relatively small.

Reactors have been built in all three categories. The first fast reactor was the Los Alamos assembly called Clementine, which operated from 1946 to 1953. The fuel core consisted of nickel-coated rods of pure plutonium metal, contained in a 6-in.-diameter (15 cm) mild (low-carbon) steel pot. Coolants for fast reactors may be steam, gas, or liquid metals. Current fast reactors utilize liquid sodium as the coolant and are being developed for breeding and power. An example of an intermediate reactor was the first propulsion reactor for the submarine USS *Seawolf*. The fuel core consisted of enriched uranium with beryllium as a moderator; the original coolant was sodium, and the reactor operated from 1956 to 1959. Examples of thermal reactors are given later.

Fuel composition. Only three isotopes—uranium-235, uranium-233, and plutonium-239—are feasible as fission fuels. However, a very wide selection of materials incorporating these isotopes is available.

Uranium-235. Naturally occurring uranium contains only 0.7% of the fissionable isotope uranium-235, the balance being essentially uranium-238. Uranium with higher concentrations of uranium-235 is called enriched uranium.

Uranium metal is susceptible to irradiation damage, which limits its operating life in a reactor. The life expectancy can be improved somewhat by heat treatment, and considerably more by alloying with elements such as zirconium or molybdenum. Uranium oxide exhibits better irradiation damage resistance and, in addition, is corrosion-resistant in oxidizing media. Ceramics such as uranium oxide have a very low thermal conductivity and lower density than metals, which are disadvantageous in certain applications.

Uranium metal can be fabricated by relatively well-established techniques, provided proper care is taken to prevent oxidation. The metal is melted in vacuum furnaces and can be cast by gravity or injection. Ingots can be rolled or extruded, and relatively complicated shapes can be fabricated. Most commonly, fuel elements are in shape of rods or plates and are fabricated by casting, rolling, or extrusion.

Current light-water-cooled nuclear power reactors utilize uranium oxide as a fuel, with an enrichment of several percent uranium-235. Cylindrical rods are the most common fuel-element configuration. They can be fabricated by compacting and sintering cylindrical pellets which are then assembled into metal tubes which are sealed.

Developmental programs for attaining long-lived solid-fuel elements include studies with uranium oxide, uranium carbide, and other refractory uranium compounds. *See* URANIUM.

Plutonium-239. Plutonium-239 is produced by neutron capture in uranium-238. It is a by-product in power reactors and is becoming increasingly

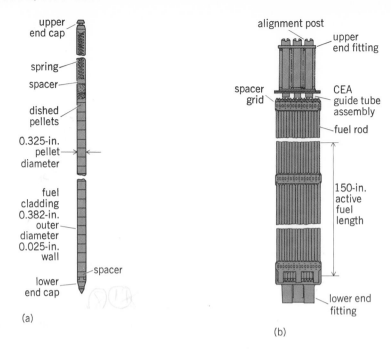

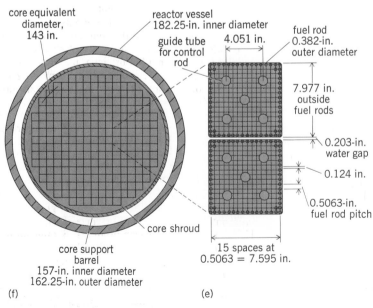

Fig. 1. Arrangement of fuel in the core of a pressurized-water reactor, a typical heterogeneous reactor. (a) Fuel rod; (b) side view (CEA=control element assembly), (c) top view, and (d) bottom view of fuel assembly; (e) cross section of two adjacent fuel assemblies, showing arrangement of fuel rods; (f) cross section of reactor core showing arrangement of fuel assemblies. 1 in.= 25.4 mm. (Combustion Engineering, Inc.)

available as nuclear power production increases. However, plutonium as a fuel is at a relatively early stage of development and in early 1976 the commercial recycle of plutonium from processed spent fuel awaited a Nuclear Regulatory Commission decision.

Plutonium is extremely hazardous to handle because of its biological toxicity and must be fabricated in glove boxes to ensure isolation from operating personnel. It can be alloyed with other metals and fabricated into various ceramic compounds. It is normally used in conjunction with uranium-238; alloys of uranium-plutonium, and mixtures of uranium-plutonium oxides and carbides, are of most interest. Except for the additional requirements imposed by plutonium toxicity, much of the uranium technology is applicable to plutonium. For the light-water nuclear power reactors, the oxide fuel pellets are contained in a zirconium alloy tube. Stainless steel tubes are used for containing the oxide fuel for the fast breeder reactors. *See* PLUTONIUM.

Uranium-233. Uranium-233, like plutonium, does not occur naturally, but is produced by neutron absorption in thorium-232, a process similar to that by which plutonium is produced from uranium-238. Interest in uranium-233 arises from its favorable nuclear properties and the abundance of thorium. However, studies of this fuel cycle are at a relatively early stage.

Uranium-233 also imposes special handling problems because of biological toxicity, but it does not introduce new metallurgical problems. Thorium is metallurgically different, but it has very favorable properties both as a metal and as a ceramic. *See* NUCLEAR FUELS.

Fuel distribution. Fuel-moderator assemblies may be homogeneous or heterogeneous. Homogeneous assemblies include the aqueous-solution-type water boilers and molten-salt-solution dispersions, slurries, and suspensions. The few homogeneous reactors built have been used for limited research and for demonstration of the principles and design features. In the heterogeneous assemblies the fuel and moderator form separate solid or liquid phases, such as solid-fuel elements spaced either in a graphite matrix or in a water phase. Most power reactors utilize an arrangement of closely spaced, solid-fuel rods, about 1/2 in. (13 mm) in diameter and 12 ft (3.7 m) long, in water In the arrangement shown in Fig. 1, fuel rods are arranged in a grid pattern to form a fuel assembly, and over 200 fuel assemblies are in turn arranged in a grid pattern in the reactor core.

The first homogeneous reactor was the Los Alamos Water Boiler, which commenced operations in 1944 at 1/20 watt. Various modifications were carried out to upgrade the thermal output. The aqueous solutions of uranium sulfate and later, uranium nitrate, with enrichments of about 17% were contained in a 1-ft-diameter (0.3 m) sphere. Homogeneous reactors which were used to demonstrate the feasibility of producing electrical power include the Homogeneous Reactor Experiment No. 1 (HRE-1) and HRE-2. HRE-1 operated from 1952 to 1954, generated 140 kW net electrical, contained an aqueous homogeneous solution of UO_2SO_4 with an enrichment in excess of 90%, and

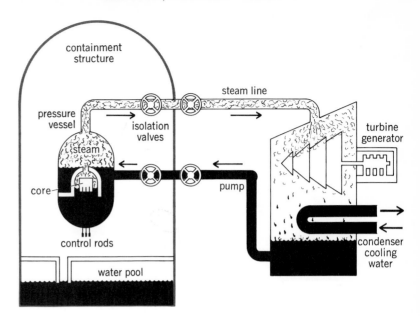

Fig. 2. Boiling-water reactor (BWR). (*Atomic Industrial Forum, Inc.*)

was self-stabilizing because of its large negative coefficient. HRE-2 operated from 1957 to 1961 and generated 300 kW net electrical.

HEAT REMOVAL

The major portion of the energy released by the fissioning of the fuel is in the form of kinetic energy of the fission fragments, which in turn is converted into heat through the slowing down and stopping of the fragments. For the heterogeneous reactors this heating occurs within the fuel elements. Heating also arises through the release and absorption of the radiations from the fission process and from the radioactive materials formed. The heat generated in a reactor is removed by a primary coolant flowing through the reactor.

Heat is not generated uniformly in a reactor. The heat flux decreases axially and radially from a peak at the center of the reactor, or near the center if the reactor is not symmetrical in configuration. In addition, local perturbations in heat generation can occur because of inhomogeneities in the reactor structure. These variations impose special considerations in the design of reactor cooling systems, including the need for establishing variations in coolant flow rate through the reactor to achieve uniform temperature rise in the coolant; avoiding local hot-spot conditions; and avoiding local thermal stresses and distortions in the structural members of the reactor.

Nuclear reactors have the unique thermal characteristic that heat generation continues after shutdown because of fission and radioactive decay of fission products. Significant fission heat generation occurs for only a few seconds after shutdown. Radioactive-decay heating varies with the decay characteristics of the fission products.

Accurate analysis of fission heat generation as a function of time immediately after reactor shutdown requires detailed knowledge of the speed and reactivity worth of the control rods. The longer-term fission-product-decay heating depends upon prior reactor operation. Typical values of the

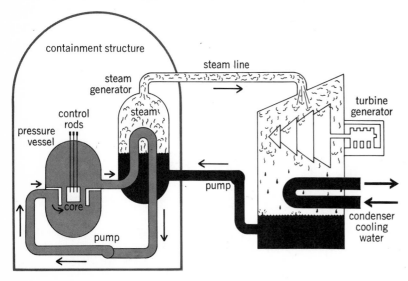

Fig. 3. Pressurized-water reactor (PWR). (*Atomic Industrial Forum, Inc.*)

total heat generation after shutdown (as percent of operating power) are 10–20% after 1 sec, 5–10% after 10 sec, approximately 2% after 10 min, 1.5% after 1 hr, and 0.7% after 1 day.

Reactor coolants. Coolants are selected for specific applications on the basis of their heat-transfer capability, physical properties, and nuclear properties.

Water. Water has many desirable characteristics. It was employed as the coolant in the first production reactors and most power reactors still utilize water as the coolant. In a boiling-water reactor (BWR; Fig. 2) the water is allowed to boil and form steam that is piped to the turbine. In a pressurized-water reactor (PWR; Fig. 3) the coolant water is kept under increased pressure to prevent boiling, and transfers heat to a separate stream of water in a steam generator, changing that water to steam. Figure 4 shows the relation of the core and heat removal systems to the condenser, electrical power system, and waste management system in the Prairie Island Nuclear Plant, which is typical

of plants using pressurized-water reactors. Cool intake water is pumped through hundreds of 1-in.-diameter (25 mm) tubes in the condenser, and the warm water from the condenser is then pumped over cooling towers and returned to the plant. *See* COOLING TOWER; RADIOACTIVE WASTE MANAGEMENT.

For both boiling-water and pressurized-water reactors, the water serves as the moderator as well as the coolant. Both light and heavy water are excellent neutron moderators, although heavy water (deuterium oxide) has a neutron-absorption cross section approximately 1/500 that for light water.

There is no serious neutron-activation problem with pure water; ^{16}N, formed by the (n,p) reaction with ^{16}O (absorption of a neutron followed by emission of a proton), is the major source of activity, but its 7.5-sec half-life minimizes this problem. The most serious limitation of water as a coolant for power reactors is its high vapor pressure. A coolant temperature of 550°F (288°C) requires a system pressure of approximately 1500 psi (10 MPa). This temperature is far below modern power station practice, for which steam temperatures in excess of 1000°F (538°C) have become common. Lower thermal efficiencies result from lower temperatures. Boiling-water reactors operate at about 70 atm (7 MPa), and pressurized-water reactors at 150 atm (15 MPa). The high pressure necessary for water-cooled power reactors imposes severe design problems, which will be discussed later in this article. *See* NUCLEAR REACTION.

Gases. Gases are inherently poor heat-transfer fluids as compared with liquids because of their low density. This situation can be improved by increasing the gas pressure; however, this introduces other problems. Helium is the most attractive gas (it is chemically inert and has good thermodynamic and nuclear properties and has been selected as the coolant for the development of high-temperature gas-cooled reactor (HTGR) systems (Fig. 5), in which the gas transfers heat from the reactor core to a steam generator. Gases are capable of operation at extremely high tempera-

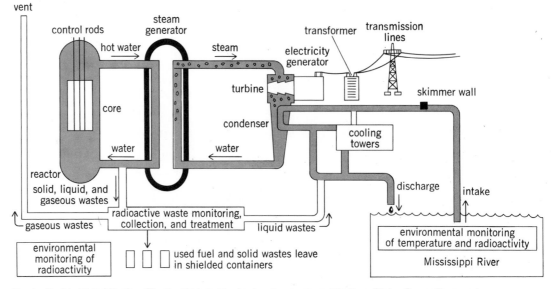

Fig. 4. Prairie Island Nuclear Plant, using pressurized-water reactors. (*Northern States Power Company*)

ture, and they are being considered for special process applications and direct-cycle gas-turbine applications. Hydrogen was used as the coolant for the reactors developed in the Nuclear Engine Rocket Vehicle Application (NERVA) Program, now terminated. Heated gas discharging through the nozzle developed the propulsive thrust.

Organic coolants. Diphenyl and terphenyl possess good neutron-moderating properties and have lower vapor pressures than water. Organic coolants are noncorrosive and relatively inexpensive. Their major disadvantage is dissociation or decomposition under irradiation.

Liquid metals. The alkali metals, in particular, have excellent heat-transfer properties and extremely low vapor pressures at temperatures of interest for power generation. Sodium is the most attractive because of its relatively low melting point (208°F; 98°C) and high heat-transfer coefficient. It is also abundant, commercially available in acceptable purity, and relatively inexpensive. It is not particularly corrosive, provided low oxygen concentration is maintained. Its nuclear properties are excellent for fast reactors. In the liquid metal fast breeder reactor (LMFBR; Fig. 6) sodium in the primary loop collects the heat generated in the core and transfers it to a secondary sodium loop in the heat exchanger, from which it is carried to the steam generator.

Sodium presents an activation problem because ^{24}Na is formed by the absorption of a neutron and is an energetic gamma emitter with a 15-hr half-life. The containing system requires extensive biological shielding, and approximately 2 weeks is required for decay of ^{24}Na activity prior to access to the system for repair or maintenance. Sodium does not decompose, and no makeup is required. Sodium reacts violently with water, imposing severe problems in the design of sodium-to-water steam boilers. The poor lubricating properties of sodium and its reaction with air further complicate the mechanical design of sodium-cooled reactors. The other alkali metals exhibit similar characteristics and appear to be less attractive than sodium. The eutectic alloy of sodium with potassium (NaK), however, has the advantage that it remains liquid at room temperature.

Heavy metals have been considered for use as reactor coolants. Uranium is sufficiently soluble in bismuth at high temperatures to permit a liquid-fuel system. Bismuth also has an extremely small thermal-neutron-absorption cross section. It is a relatively poor heat-transfer fluid and, in addition, the formation of biologically toxic polonium by neutron capture imposes severe leakage restrictions. The high melting point (520°F; 267°C) of bismuth is also a disadvantage. Essentially the same considerations apply to lead-bismuth alloy, except for its more favorable melting point (257°F; 125°C).

Although mercury has seen some application as a heat-transfer fluid, it is not a particularly attractive reactor coolant. As a coolant, mercury has relatively poor heat-transfer and nuclear characteristics and also is toxic and expensive.

Molten salts. Molten salts have been used as reactor coolants because they have favorable high-temperature properties and because mixtures of salts containing fuel permit fluid fuel-coolant sys-

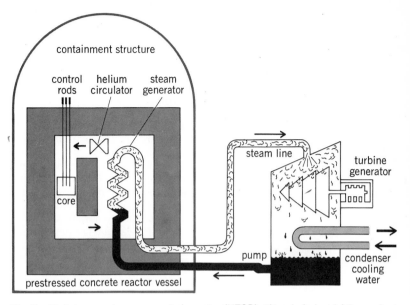

Fig. 5. High-temperature gas-cooled reactor (HTGR). (*Atomic Industrial Forum, Inc.*)

tems. In one small experimental power reactor of this type, a molten mixture of the fluorides of beryllium, lithium, zirconium, and uranium was pumped through channels in a graphite moderator within a reactor vessel in which the fuel salt forms a critical mass and generates energy by fissioning the uranium. Design studies have been made of a larger reactor of this type. The fuel salt would be heated to 1300°F (704°C) and would deliver its heat to sodium fluoroborate in an intermediate loop that would isolate the steam boilers from the radioactive fuel circuit. Steam would be generated in the boilers at a temperature of 1000°F (538°C) and a pressure of 3500 psi (24 MPa) to achieve a thermal efficiency of about 45%. In addition to producing high thermal efficiency, this method has the potential of achieving high neutron efficiency because fission products can be removed continuously from the fluid fuel, thus reducing the fraction of neutrons lost nonproductively by capture in fission

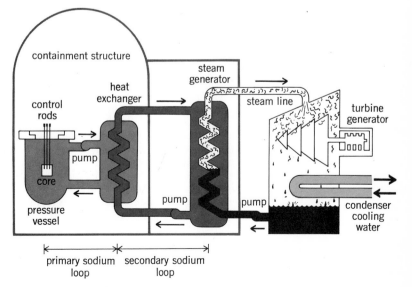

Fig. 6. Liquid metal fast breeder reactor (LMFBR). (*Atomic Industrial Forum, Inc.*)

products. However, the pumping and processing of the intensely radioactive liquid fuel impose special requirements in design, fabrication, and operation.

Fluid flow and hydrodynamics. Because heat removal must be accomplished as efficiently as possible, considerable attention must be given to the fluid-flow and hydrodynamic characteristics of the system.

The heat capacity and thermal conductivity of the fluid at the temperature of operation have a fundamental effect upon the design of the reactor system. The heat capacity determines the mass flow of the coolant required. The fluid properties (thermal conductivity, viscosity, density, and specific heat) are important in determining the surface area required for the fuel to permit transfer of the heat generated at reasonable temperature differences. This, in turn, affects the design of the fuel—in particular, the amount and arrangement of the fuel elements. These factors combine to establish the pumping characteristics of the system because pressure drop and coolant-temperature rise are directly related.

Secondary considerations include other physical properties of the coolant, particularly its vapor pressure. If the vapor pressure is high at the operating temperature, local or bulk boiling of the fluid may occur. This in turn must be considered in establishing the heat-transfer coefficient for the fluid.

Because the coolant absorbs and scatters neutrons, variations in coolant density also affect reactor performance. This is particularly significant in reactors in which the coolant exists in two phases, for example, the liquid and vapor phases in boiling systems. Gases, of course, do not undergo phase change, nor do liquids operating at temperatures well below their boiling point; however, the fluid density does change with temperature and may have an important effect upon the reactor.

Power generation and, therefore, the heat-removal rate are not uniform throughout the reactor. If the mass flow rate of the coolant is uniform throughout the reactor, then unequal temperature rise of the coolant results. This becomes particularly significant in power reactors in which it is desired to achieve the highest possible coolant outlet temperature to attain maximum thermal efficiency of the power cycle. The performance limit of the coolant is set by the temperature in the hottest region or channel of the reactor. Unless the coolant flow rate is adjusted in the other regions of the reactor, the coolant will leave these regions at a lower temperature and thus will reduce the average coolant outlet temperature. In high-performance power reactors, this effect is reduced by orificing the flow in each region of the reactor commensurate with its heat generation. This involves very careful design and analysis of the system. In the boiling-type reactor, this effect upon coolant temperature does not occur because the exit temperature of the coolant is at the saturation temperature for the system. However, the variation in power generation in the reactor is reflected by a difference in the amount of steam generated in the various zones, and orificing is still required to achieve most effective use of coolant flow.

In very-high-performance reactors, the flow rate and consequent pressure drop of the coolant are sufficient to create mechanical problems in the system. It is not uncommon for the pressure drop through the fuel assemblies to exceed the weight of the fuel elements in the reactor, with a resulting hydraulic lifting force on the fuel elements. Often this requires a design arrangement to hold the fuel elements down. Although this problem can be overcome by employing downward flow through the system, it is often undesirable to do so because of shutdown-cooling considerations. It is desirable in most systems to accomplish shutdown cooling by natural-convection circulation of the coolant. If downflow is employed for forced circulation, then shutdown cooling by natural-convection circulation requires a flow reversal, which can introduce new problems.

Thermal stress considerations. The temperature of the reactor coolant increases as it circulates through the reactor. This increase in temperature is constant at steady-state conditions. Fluctuations in power level or in coolant flow rate result in variations in the temperature rise. These are reflected as temperature changes in the coolant exit temperature, which in turn result in temperature changes in the coolant system.

A reactor is capable of very rapid changes in power level, particularly, reduction in power level. Reactors are equipped with mechanisms (reactor scram systems) to ensure rapid shutdown of the system in the event of an operational abnormality.

Therefore, reactor-coolant systems must be designed to accommodate the temperature transients that may occur because of rapid power changes. In addition, they must be designed to accommodate temperature transients that might occur as a result of a coolant-system malfunction, such as pump stoppage. The consequent temperature stresses induced in the various parts of the system are superimposed upon the thermal stresses that exist under normal steady-state operations.

In very-high-performance systems, it is not uncommon for the thermal stresses alone to approach the allowable stresses in the materials of construction. In these cases, careful attention must be given to the transient stresses, and thermal shielding is commonly employed in critical sections of the system. Normally, this consists of a thermal barrier, which, by virtue of its heat capacity and resistance to heat transfer, delays the transfer of heat, thereby reducing the rate of change of temperature and protecting critical system components from thermal stresses.

Thermal stresses are also important in the design of reactor fuel elements. Metals that possess dissimilar thermal-expansion coefficients are frequently required. Heating of such systems gives rise to distortions, which in turn can result in flow restrictions in coolant passages. Careful analysis and experimental verification are often required to avoid such circumstances.

Coolant-system components. The development of reactor systems has necessitated concurrent development of special components for reactor coolant systems. These have been required even for systems employing conventional coolants, such as water or air.

Because of the hazard of radioactivity, leak-tight systems and components are a prerequisite to safe, reliable operation and maintenance. Special problems are introduced by many of the fluids employed as reactor coolants.

More extensive component developments have

been required for the alkali metals (sodium, NaK, and potassium) which are chemically very active and are extremely poor lubricants. Centrifugal pumps employing unique bearings and seals must be specially designed. Liquid metals are excellent electrical conductors, and, in some special cases, electromagnetic-type pumps have been developed. These pumps are completely sealed, contain no moving parts, and derive their pumping action from electromagnetic forces imposed directly on the fluid.

In addition to the variety of special pumps developed for reactor coolant systems, there is a variety of piping-system components and heat-exchange components. As in all flow systems, flow-regulating devices such as valves are required, as well as flow instrumentation to measure and thereby control the systems. Here again, leak-tightness has necessitated the development of special valves with metallic bellows around the valve stem to ensure system integrity. Measurement of flow and pressure has also required the development of sensing instrumentation that is reliable and leak-tight.

Many of these developments have borrowed from other technologies because toxic or flammable fluids are frequently pumped in other applications. In many cases, however, special equipment has been developed specifically to meet the requirements of the reactor systems. An example of this type of development involves the measurement of flow in liquid-metal piping systems. The simple principle of a moving conductor in a magnetic field is employed by placing a magnet around the pipe and measuring the voltage generated by the moving conductor (coolant) in terms of flow rate. Temperature compensation is required, and calibration is critical.

Although the development of nuclear power reactors has introduced many new technologies, no method has yet displaced the conventional steam cycle for converting thermal energy to mechanical energy. Steam is generated either directly in the reactor (direct-cycle boiling reactor) or in auxiliary steam-generation equipment in which steam is generated by transfer of heat to water from the reactor coolant. These steam generators require very special design, particularly when dissimilar fluids are involved. Typical of these problems are the sodium-to-water steam generators in which absolute integrity is essential because of the violent chemical reaction between sodium and water.

CORE DESIGN AND MATERIALS

A typical reactor core for a power reactor consists of the fuel element rods supported by a grid-type structure inside a vessel (Fig. 1).

The primary function of the vessel is to contain the coolant. Its design and materials are determined by such factors as the nature of the coolant (corrosive properties), operating conditions (temperature and pressure), and quantity and configuration of fuel. To complicate vessel design even further, the vessel is pierced by devices which are used for controlling reactor operation, for loading and unloading the fuel, and for coolant entrance and exit.

Design must also take account of thermal stresses caused by temperature differences in the sys-

tem. Another problem is radioactivity induced in core materials because of neutron absorption during reactor operation. This precludes normal maintenance of the equipment and, in some areas, makes repairs virtually impossible. For this reason, an exceptionally high degree of integrity is demanded of this equipment. Reactors have been designed to permit removal of the internals from the vessel; this is difficult, however, and tends to complicate the design of the system.

Structural materials. Structural materials employed in reactor systems must possess suitable nuclear and physical properties and must be compatible with the reactor coolant under the conditions of operation. Some requirements are especially severe because of secondary effects; for example, corrosion limits may be established by the rate of deposition of coolant-entrained corrosion products on critical surfaces rather than by the rate of corrosion of the base material.

The most common structural materials employed in reactor systems are aluminum, stainless steel, and zirconium alloys. Aluminum and zirconium alloys have favorable nuclear properties, whereas stainless steel has favorable physical properties. Aluminum is widely used in low-temperature reactors; zirconium and stainless steel are used in high-temperature reactors. Zirconium is relatively expensive, and its use if therefore confined to applications where neutron absorption is critical.

The 18–8 series stainless steels have been used for structural members in both water-cooled reactors and sodium-cooled reactors because of their corrosion resistance and favorable physical properties at high temperatures. Type 304 and type 347 stainless steel have been used most extensively because of their weldability, machinability, and physical properties. To reduce cost, heavy-walled pressure vessels are normally fabricated from carbon steels and clad on the internal surfaces with a thin layer of stainless steel to provide the necessary corrosion resistance.

Although pressure vessels have been constructed for other industries to meet even more severe service requirements, the complex requirements for reactors have introduced new design and fabrication problems. Of particular importance is the dimensional precision required and the special nozzles and other appurtenances required.

As the size of power reactors has increased, it has become necessary, in some instances, to field-fabricate reactor vessels. This involves field-welding of heavy wall sections and subsequent stress relieving. Prestressed concrete vessels for gas-cooled reaction are also field-fabricated and have the potential capability of being fabricated in much larger sizes than steel vessels.

Research reactors operating at low temperatures and pressures introduce special experimental considerations. The primary objective is to provide the maximum volume of unperturbed neutron flux for experimentation. It is desirable, therefore, to extend the experimental irradiation facilities beyond the vessel wall. This has introduced the need for vessels constructed of materials having a low cross section for neutron capture. Relatively large aluminum reactor vessels with wall sections as thin as practicable have been manufactured for research reactors. Special problems with respect to dimensional stability have necessitated unique supporting structures. The vessel design is complicated further by the variety of openings that must be provided to accommodate experimental apparatus. It is highly desirable to provide access to the reactor proper for experiments and, in many cases, to have apparatus installed in so-called through holes that penetrate the vessel from side to side.

In some instances, stainless steel vessels have been employed for research and test reactors at the sacrifice of some experimental flexibility. The experimental irradiations are performed within the reactor vessel and limited in use if made of the space external to the reactor vessel.

A special problem is introduced by research reactors employing heavy water as a moderator and light water as a coolant. A calandria-type design has been employed, consisting of an all-aluminum multitube container for the heavy water, with additional aluminum tubes connected to separate coolant headers for circulation of the light-water coolant. This arrangement introduces the special problems associated with the multitudinous welds to contain a system within a system, each being tight with respect to leakage to the atmosphere and to the other system.

Fuel cladding. Heterogeneous reactors maintain a separation of fuel and coolant by cladding the fuel. The cladding material must be compatible with both the fuel and the coolant.

The cladding materials must also have favorable nuclear properties. The neutron-capture cross section is most significant because the parasitic absorption of neutrons by these materials reduces the efficiency of the nuclear fission process. Aluminum is a very desirable material in this respect; however, its physical strength and corrosion resistance in water decrease very rapidly above about 300°F (149°C).

Zirconium has very favorable neutron properties, and in addition can be made reasonably corrosion-resistant in high-temperature water. It has found extensive use for water-cooled power reactors. The technology of zirconium and zirconium-base alloys, Zircaloy, has advanced tremendously under the impetus of the various reactor development programs.

Stainless steel is used for the fuel cladding in fast reactors.

CONTROL AND INSTRUMENTATION

The control of reactors requires the measurement and adjustment of the critical condition. A reactor is critical when the rate of production of neutrons equals the rate of consumption in the system. The neutrons are produced by the fission process and are consumed in a variety of ways, including absorption to cause fission, nonfission capture in fissionable materials, capture in fertile materials, capture in structure or coolant, and leakage from the reactor. A reactor is subcritical (power level decreasing) if the number of neutrons produced is less than the number consumed. The reactor is supercritical (power level increasing) if the number of neutrons produced exceeds the number consumed.

Reactors are controlled by adjusting the balance between neutron production and neutron consumption. Normally, neutron consumption is controlled by varying the absorption or leakage of neu-

trons; however, the neutron-generation rate can be controlled by varying the amount of fissionable material in the system.

It is essential to orderly control and management of a reactor that the neutron density be sufficiently high to permit reliable measurement. During reactor startup, a source of neutrons is essential, therefore, to the control and instrumentation of reactor systems. Neutrons are obtained from the photo-neutron effect in materials such as beryllium. Neutron sources consist of a photon (γ-ray) source and beryllium, such as antimony-beryllium. Antimony sources are particularly convenient for use in reactors because the antimony is activated by the reactor neutrons each time the reactor operates.

Control drives and systems. The reactor control system requires the movement of neutron-absorbing rods (control rods) in the reactor under very exacting conditions. They must be arranged to increase reactivity (increase neutron population) slowly and under absolute control. They must be capable of reducing reactivity, both rapidly and slowly.

Normal operation of the control drives can be accomplished manually by the reactor operator or by automatic control systems. Reactor scram (very rapid reactor shutdown) can be initiated automatically by one or more system scram-safety signals, or it can be started manually by depressing a scram button convenient to the operator in the control room.

Control drives are normally electromechanical devices that impart linear or swinging motion to the control rods. They are usually equipped with a relatively slow-speed reversible drive system for normal operational control. Scram is usually effected by a high-speed overriding drive accompanied by unlatching or disconnecting the main drive system. To enhance reliability of the scram system, its operation is usually initiated by deenergizing appropriate electrical circuits. This also automatically produces reactor scram in the event of a system power failure. Hydraulic or pneumatic drive systems, as well as a variety of electromechanical systems, have also been developed.

In addition to the actuating motions required, control-rod-drive systems must also provide accurate indication of the rod positions at all times. Various types of selsyn drive, as well as arrangements of switches and lighting systems, are employed as position indicators. It is possible to provide control-rod-position indication accurate to a few thousandths of an inch.

Reactor instrumentation. Reactor control requires measurement of the reactor condition. Neutron-sensitive ion chambers are used to measure neutron flux. These neutron detectors may be located outside the reactor core, and the flux measurements from the detectors are combined to measure an average flux that is proportional to the average neutron density in the reactor. The chamber current is calibrated against a thermal power measurement and then applied over a wide range of reactor power level. The neutron-sensitive detector system must respond to the lowest neutron flux in the system produced by the neutron source.

Normally, many channels of instrumentation are required to cover the entire operating range. Several channels are required for low-level operation, beginning at the source level, whereas others are required for the intermediate and high-power-level ranges. Ten channels of detectors are not uncommon in reactor systems, and some systems contain a larger number. The total range to be covered is in the range of 7–10 decades of power level.

The chamber current can be employed as a signal, suitably amplified, to operate automatic control-system devices as well as to actuate reactor scram. In addition to absolute power level, rate of change of power level is also an important measurement which is recorded and employed to actuate various alarm and trip circuits. The normal range for the current ion chambers is approximately 10^{-14} to 10^{-4} A. This current is suitably amplified in logarithmic and period amplifiers, and can be measured directly with a galvanometer.

APPLICATIONS

Reactor applications include production of fissionable fuels (plutonium and uranium-233); mobile, stationary, and packaged power plants; research, testing, teaching-demonstration, and experimental facilities; space and process heat; dual-purpose designs; and special applications. The potential use of reactor radiation for sterilization of food and other products, for chemical processes, and for high-temperature applications has been recognized.

Production reactors. Reactor installations at Hanford, WA, and Savannah River, SC, were designed to produce plutonium-239 from uranium-238. Natural uranium is used as the fuel material. The moderator for the reactors at Hanford is graphite, and heavy water is used as the moderator at Savannah River. Water is used as a coolant in the United States production reactors, whereas in the United Kingdom, gas cooling has been the basis for most designs. The thermal, heterogeneous, natural-uranium, graphite-moderated reactors are representative of the largest reactors. The eight graphite-moderated production reactors at Hanford have been shut down, and the remaining operating production reactor, the N Reactor, is a dual-purpose unit producing special nuclear materials as well as steam for a gross power output of 860 MW electrical.

Breeder reactors. The term "converter" is applied to a reactor that converts a fertile material (for example, uranium-238) to a fissionable material (for example, plutonium). A breeder reactor, strictly speaking, produces the same fissionable material that it consumes (for example, it consumes plutonium fuel and at the same time breeds plutonium). The fuel cycle, of course, could be based on fissionable uranium-233 and fertile thorium-232 rather than uranium-238 and plutonium. In popular usage, however, any reactor that has a conversion ratio of over 100% (that is, produces more fuel than it consumes) is called a breeder, even if the fuels that are consumed and produced are different. The Experimental Breeder Reactor I (EBR-I) operated from 1951 to 1964 and was the first reactor to produce electrical power and to demonstrate the feasibility of breeding. The only operating breeder in the United States in 1976 was EBR-II, which demonstrated the use of integral facility for central station power and a closed remote reprocessing and fabrication system for the fuel cycle. EBR-II is used for fast neutron testing

of fuels and materials. Prototype nuclear power breeder reactors are in operation in the United Kingdom, France, and the Soviet Union. Liquid sodium is used for the coolant (Fig. 6). Studies of fast breeder reactors utilizing gas cooling have been undertaken. Breeding with thermal reactors is also possible, and the molten salt breeder reactor (MSBR) concept, for example, has received some consideration. *See* NUCLEAR FUELS REPROCESSING.

Power reactors. Nuclear power reactors are used extensively by the U.S. Navy for propulsion of submarines and surface vessels, and by the nuclear industry for the generation of electrical power. A variety of organizations and public-interest groups have sought to slow or halt the use of the commercial nuclear power reactors.

As of 1975, more than 130 reactors have been operated by the Navy. The total United States military program has involved more than 200 reactors, operable, being built, planned, or shut down. The prototype of the first reactor used for propulsion operated in 1953, and the first reactor-powered submarine, the USS *Nautilus*, was placed in operation in 1955. Water is used as coolant and moderator and is maintained at 2000 psi (14 MPa) to suppress boiling. Two submarines, the USS *Thresher* and the USS *Scorpion*, were lost in the Atlantic in 1963 and 1968, respectively. Pressurized-water reactors are in use and under further development for submarines, cruisers, aircraft carriers, merchant ships, and (in the Soviet Union) icebreakers. The first civilian maritime reactor application (1961) was the nuclear ship *Savannah*, which utilized a pressurized-water reactor rated at 22,000 shaft horsepower (16.4 MW).

The first reactors for central-station power plant prototypes include the pressurized-water reactors—Shippingport Atomic Power Station (Pennsylvania, 231 MW thermal; 60 MW electrical, 1957) and the Atomic Power Station (Obninsk, Soviet Union, 30 MW thermal; 5 MW electrical, 1954); and the gas-cooled reactors—Calder Hall Station (Sellafield, England, originally 180 MW thermal, increased to 210 MW; 35 MW electrical with four reactors, 1956). The Dresden Nuclear Power Station (Morris, IL) is a boiling-water reactor with an output of 700 MW thermal and 208 MW electrical started in 1959. The 175-MW-electrical Yankee plant (Rowe, MA) is a pressurized-water reactor, started in 1960.

As of December 1975, the electrical generating capacity for the commercial nuclear power reactors was 8% of the total United States generating capacity, with nuclear power in some regions furnishing almost 50% of the energy source for the electric power generation. In the United States, 56 nuclear power plants had operating licenses with an electric generating capacity of 38,000 MW; 64 with construction permits, 65,740 MW; 97 plants on order, 110,000 MW; and 17 plants firmly planned, 19,100 MW. High capital costs and slowdowns experienced in use of electrical power, as well as other factors, have contributed to some cancellations and a number of deferrments for 1 to 4 years in nuclear power as well as in fossil power plant constructions. As of June 1975, the status of nuclear power plants outside the United States was 102 plants with operating licenses and with an electrical generating capacity of 29,000 MW; 85

plants with construction permits, 60,000 MW; 70 plants on order, 54,000 MW; and 169 plants firmly planned, 151,000 MW. Nuclear power is used in about 20 countries with planned operations extending to about 30 countries.

Research and test reactors. The research-and-development aspects of a nuclear reactor may be considered from two points of view. One is that the reactor provides experimental irradiation facilities, and the other is that the reactor itself may represent a test of a given design.

Research with reactors covers such activities as measurements of the probabilities of nuclear reactions, shielding measurements, studies of the behavior of materials under neutron and γ-irradiation, and other studies in nuclear physics, solid-state physics, and the life sciences. The irradiation facilities are used extensively for production of isotopes. High-neutron-flux reactors, designed specifically for experimental exposures of materials, are called materials-testing reactors. Reactors built to test design features are called experiments or experimental reactor facilities. Several different types of low-cost reactors, which are called teaching-demonstration reactors, have been promoted to accentuate the teaching aspects. *See* NUCLEAR ENGINEERING.

The four major varieties of research reactors are (1) uranium-fueled, graphite-moderated, air-cooled reactors; (2) uranium-fueled, heavy-water-moderated reactors; (3) enriched-fuel, aqueous-solution-type reactors; and (4) water-moderated, enriched-fuel, pool-type, and tank-type reactors. All the reactors are thermal and, with the exception of the third type, heterogeneous. Both natural and enriched uranium are used in the first two types.

The bulk shielding reactor, or BSR (Oak Ridge, TN, 1950), was the first reactor with the core submerged in an open pool of water—hence the term "swimming-pool reactor." The water is the moderator, coolant, and shield. With forced circulation of water, reactor levels of 1000 kW of heat are possible. Some reactor designs involve the use of a tank instead of a pool. Features of other pool- and tank-type reactors include variability of fuel-element design and configuration, fixed and movable cores, and a lightly pressurized (for tank-type), forced-convection water-cooling system.

The materials-testing reactor, or MTR (1952–1970), was a high-flux irradiation facility designed for studying the behavior of materials for use in power reactors. The maximum neutron fluxes available at 40 MW (thermal) were 5.5×10^{14} thermal neutrons/(cm^2)(s) and 3×10^{14} fast neutrons/(cm^2)(s). Nearly 100 experimental and instrument holes or exposure ports were provided. Other test reactors have been built to accommodate the specialized materials development programs necessary for the continued advancement of the nuclear reactor industry. Included are the engineering test reactor (ETR), 175 MW thermal, in operation since 1957, and the advanced test reactor (ATR), 250 MW thermal, completed in 1967. The ATR provides a flux up to 2.5×10^{15} neutrons/(cm^2)(s).

Test facilities for the fast-breeder-reactor physics program included the zero power reactors (ZPR), zero power plutonium reactors (ZPPR), and the Southwest Experimental Fast Oxide Reactor (SEFOR) reactor. The fast-flux test facility (FFTR)

is designed to provide fast neutron environments for testing fuel and materials for fast reactors.

Among the many thermal research reactors is the high-flux isotope reactor (HFIR) at the Oak Ridge National Laboratory. A principal use of this reactor is the production of transplutonium elements such as berkelium, californium, einsteinium, and fermium.

Experimental reactors. A variety of reactors have been built to test the feasibility of given reactor designs. Reactors already noted include the experimental breeder reactors and the homogeneous reactor experiments. Several types of reactors have been designed and operated under severe power excursions to study reactor stability. Five boiling-water reactor experiments (Borax-1 to -5) have been carried out to study the behavior of such reactors at atmospheric and at elevated pressures and with different kinds of fuel elements, including nonmetallic fuels. Power-excursion experiments have been performed with the homogeneous aqueous-solution-type reactors. For example, kinetic experiment water boiler (KEWB, Canoga Park, CA) has successfully handled a power excursion of 0–530 MW in less than 1 s.

The use of boiling water as a coolant for power-producing reactors was established by the experimental boiling water reactor (EBWR, Argonne National Laboratory, Lemont, IL, 1956) and the Vallecitos boiling water reactor (VBWR, Vallecitos, CA, 1957).

The use of sodium as a high-temperature coolant for power reactors was demonstrated by the sodium-graphite reactor experiment (SRE, 1957–1964).

The feasibility of organics as coolants or coolant-moderators for reactors was studied in the organic moderated reactor experiment (OMRE, 1957–1963. The organic was a polyphenyl compound.

Test reactors for the nuclear engine for the NERVA Program included the Phoebus, NRX, and Kiwi reactors, ranging up to 4000 MW thermal. The adaptation and further testing of the reactors for space vehicles was completed successfully with ground experimental engines (XE).

Among the many other reactor experiments, two additional ones are noted here. The feasibility of the molten-salt-reactor concept has been successfully demonstrated by the molten-salt-reactor experiment (MSRE) (1965–1969). The ultra-high-temperature reactor experiment (UHTREX) (1968–1970) employed helium as a coolant and was designed to operate at 2400°F (1316°C).

Thermoelectric power. In early 1959 the AEC Los Alamos Laboratory announced the first successful production of electricity directly from a reactor core without the use of a heat-transfer medium or conventional generating equipment. The experimental unit operated by means of a thermoelectric process. The thermoelectric medium was cesium vapor, and the heat source was enriched uranium. *See* THERMOELECTRICITY.

Specialized nuclear power units. Nuclear power units are being developed for small electrical outputs, but with special purpose for land, sea, and space applications. A 500-W reactor, SNAP-10A, was orbited in 1965 and operated successfully for 43 days. SNAP reactors are used to supply power for lunar surface experiments left behind by Apollo astronauts. Other systems for nuclear auxiliary power (SNAP) include SNAP-8, a 600-kW, thermal unit, and a series of odd-numbered units employing radioisotopes, such as plutonium-238, curium-242, polonium-210, and promethium-147, for the energy source. Other isotopes being considered are cobalt-60, strontium-90, and thulium-171. *See* NUCLEAR BATTERY.

[HERBERT S. ISBIN]

Bibliography: J. M. Harrer and J. B. Beckerley, *Nuclear Power Reactor Instrumentation Systems Handbook*, National Technical Information Service, TID-25952-P1 and -P2, vol. 1, 1973, vol.2, 1974; *Nuclear Reactors Built, Being Built, or Planned in the United States*, National Technical Information Service, TID-8200, printed twice yearly as of June 30 and December 31; A. Sesonske, *Nuclear Power Plant Design Analysis*, National Technical Information Service, TID-26241, 1973; U.S. Atomic Energy Commission, *The Safety of Nuclear Power Reactors (Light Water-Cooled) and Related Facilities*, WASH-1250, 1973.

Reforming in petroleum refining

A process used for upgrading gasoline by improving its antiknock characteristics (increasing the octane number). It is also widely used for the production of aromatic hydrocarbons for the petrochemical industry.

Catalytic reforming. The process utilizes a supported platinum catalyst and involves a number of different reactions. These reactions convert, or reform, the low-octane-number feed components, such as the paraffins, to components with increased octane number. Eqs. (1)–(5) show typical examples.

Eq. (1) illustrates isomerization (chain branching) of paraffins. Hydrocracking, shown in Eq. (2), is usually held to a minimum because it entails formation of some butane and propane, which boil below gasoline and hence represent a yield loss. The most important reactions in catalytic reforming are those leading to the formation of aromatic hydrocarbons because these are high-octane components as well as valuable petrochemical intermediates. Aromatics are formed by dehydrogenation of six-membered ring cycloparaffins, Eq. (4); rearrangement and dehydrogenation of five-membered ring cycloparaffins, Eq. (3); and cyclization (dehydrocyclization) of paraffins, Eq. (5).

The product from catalytic reforming consists essentially of aromatics and branched paraffins. The higher-boiling fractions are richer in aromatics, whereas the paraffins, particularly in severe reforming, are concentrated in the lower-boiling fractions.

The catalysts commonly employed consist essentially of a platinum dehydrogenation component on an acidic support. The dehydrogenation activity and acid activity of such catalysts are carefully balanced to achieve the highest possible yield at a given product octane number or aromatics concentration. A number of bimetallic catalysts have been introduced in which one or more additional elements, such as rhenium, are used to modify and stabilize the platinum component. This results in greatly improved catalyst stability and, in some cases, improvements in activity and liquid-product yield. Such catalysts have found wide acceptance in the refining industry.

Catalytic reforming units usually have a fractionation section, where the fresh feed is distilled to remove overhead hexane and lower-boiling hydrocarbons and to reject material boiling above the gasoline range (> 400°F). Most modern installations also include a pretreatment section in which sulfur compounds and other impurities are removed by reaction with hydrogen over a hydrotreating catalyst, such as cobalt-molybdena-alumina. This pretreating step also removes from the feed arsenic compounds which otherwise poison the platinum catalyst.

The pretreated C_7 to 400°F fraction is mixed with hydrogen and preheated to the desired temperature, and the reforming is carried out in three or four reactors in series. Intermediate reheating (between stages) is necessary since the overall reaction is endothermic. The reactions are carried out at temperatures of 800–1050°F and pressures of 100–600 psig.

The effluent from the reactors goes through heat exchangers to a separator where the liquid product is separated from the hydrogen and other light gases. It is then sent to a stabilizer to produce a finished gasoline as the bottoms product while removing, as overhead, the propane and butane produced during the reaction. Part of the separator gas (consisting mostly of hydrogen) is withdrawn from the system, and the remainder is recycled.

This recycle of a hydrogen-rich gas is an important feature of catalytic reforming because it acts to suppress those side reactions which tend to form carbonaceous deposits on the catalyst. The usual hydrogen recycle (3–12 moles per mole of hydrocarbon) is sufficient to maintain an active catalyst surface for 6–24 months in normal operation. When a catalyst becomes fouled, the carbonaceous deposits are burned off and the catalyst is then returned to service.

Modern requirements for higher-octane gasoline and higher aromatics yields demand higher-severity reforming, which leads to increased fouling rates despite the use of modern bimetallic catalysts. Many reformers are therefore designed with swing reactors which permit regeneration of the catalyst while the unit remains on-stream. In the most recent version, the catalyst is withdrawn continuously from an operating unit, treated in a separate but integral regenerator, and returned to the reaction section.

The use of platinum-containing catalysts for the reforming of gasoline has grown rapidly since introduction in 1949. In 1974 the catalytic reforming capacity in the Western world was 7,578,000 barrels per day. The United States capacity was in excess of 3,500,000 bpd, representing nearly 40% of total domestic gasoline production.

Aromatics production. Although catalytic reforming is primarily used to upgrade gasoline, the products have also become an exceedingly important source of aromatic hydrocarbons—in fact, the single most important source. These aromatics are used as intermediates in the manufacture of plastics, explosives, detergents, phenols, and other chemicals. The processing schemes are the same, except that somewhat lower pressures are used. The charge stock is a much narrower boiling cut, in the range 200–300°F. The main products include benzene, toluene, and xylenes. The aromatics are concentrated by extraction of the product with a solvent.

Liquid petroleum gas (LPG). LPG consists of propane and butanes and is usually derived from natural gas. In locations where there is no natural gas and gasoline consumption is low, naphtha is converted to LPG by catalytic reforming. The catalyst is modified in the direction of higher acidity, thus promoting the hydrocracking reactions. Under suitable conditions it is possible to convert 40% of the naphtha to LPG, the by-product being high-octane gasoline. *See* CRACKING; HYDROCRACKING; LIQUEFIED PETROLEUM GAS (LPG); PETROLEUM PROCESSING.

[ERNEST L. POLLITZER; VLADIMIR HAENSEL]

Bibliography: L. R. Aalund, *Oil Gas J.* 69(51): 43–60, 1971; Oil and Gas Journal, *Handbook on Catalytic Reforming*, 1966; *Refining Petroleum for Chemicals*, Advances in Chemistry Series 97, pp. 2–37.

Refrigeration cycle

A sequence of thermodynamic processes whereby heat is withdrawn from a cold body and expelled to a hot body. Theoretical thermodynamic cycles consist of nondissipative and frictionless processes. For this reason, a thermodynamic cycle can be operated in the forward direction to produce mechanical power from heat energy, or it can be operated in the reverse direction to produce heat energy from mechanical power. The reversed cycle is used primarily for the cooling effect that it produces during a portion of the cycle and so is called a refrigeration cycle. It may also be used for the heating effect, as in the comfort warming of space during the cold season of the year. *See* HEAT PUMP; THERMODYNAMIC PROCESSES.

In the refrigeration cycle a substance, called the refrigerant, is compressed, cooled, and then expanded. In expanding, the refrigerant absorbs heat from its surroundings to provide refrigeration. After the refrigerant absorbs heat from such a source, the cycle is repeated. Compression raises the temperature of the refrigerant above that of its natural surroundings so that it can give up its heat in a heat exchanger to a heat sink such as air or water. Expansion lowers the refrigerant temperature below the temperature that is to be produced inside the cold compartment or refrigerator. The sequence of processes performed by the refrigerant constitutes the refrigeration cycle. When the refrigerant is compressed mechanically, the refrigerative action is called mechanical refrigeration.

There are many methods by which cooling can be produced. The methods include the noncyclic melting of ice, or the evaporation of volatile liquids, as in local anesthetics; the Joule-Thomson effect, which is used to liquefy gases; the reverse Peltier effect, which produces heat flow from the cold to the hot junction of a bimetallic thermocouple when an external emf is imposed; and the paramagnetic effect, which is used to reach extremely low temperatures. However, large-scale refrigeration or cooling, in general, calls for mechanical refrigeration acting in a closed system.

Reverse Carnot cycle. The purpose of a refrigerator is to extract as much heat from the cold body as possible with the expenditure of as little work as possible. The yardstick in measuring the performance of a refrigeration cycle is the coefficient of performance, defined as the ratio of the heat removed to the work expended. The coefficient of performance of the reverse Carnot cycle is the maximum obtainable for stated temperatures of source and sink. Figure 1 depicts the reverse Carnot cycle on the T-s plane. *See* CARNOT CYCLE.

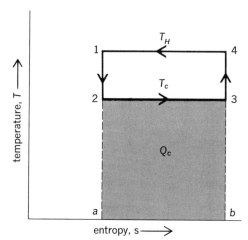

Fig. 1. Reverse Carnot cycle.

The appearance of the cycle in Fig. 1 is the same as that of the power cycle, but the order of the cyclic processes is reversed. Starting from state 1 of the figure, with the fluid at the temperature T_H of the hot body, the order of cyclic events is as follows:

1. Isentropic expansion, 1–2, of the refrigerant fluid to the temperature T_c of the cold body.

2. Isothermal expansion, 2–3, at the temperature T_c of the cold body during which the cold body gives up heat to the refrigerant fluid in the amount Q_c, represented by the area 2–3–b–a.

3. Isentropic compression, 3–4, of the fluid to the temperature T_H of the hot body.

4. Isothermal compression, 4–1, at the temperature T_H of the hot body. During this process, the hot body receives heat from the refrigerant fluid in the amount Q_H represented by the area 1–4–b–a. The difference $Q_H - Q_c$ represented by area 1–2–3–4 is the net work which must be supplied to the cycle by the external system.

Figure 1 indicates that Q_c and the net work rectangles each have areas in proportion to their vertical heights. Thus the coefficient of performance, defined as the ratio of Q_c to net work, is $T_c/(T_H - T_c)$.

The reverse Carnot cycle does not lend itself to practical adaptation because it requires both an expanding engine and a compressor in order to function. Nevertheless, its performance is a limiting ideal to which actual refrigeration equipment can be compared.

Modifications to reverse Carnot cycle. One change from the Carnot cycle which is always made in real vapor-compression plants is the substitution of an expansion valve for the expansion engine. Even if isentropic expansion were possible, the work delivered by the expansion engine would be very small and the irreversibilities present in any real operations would further reduce the work delivered by the expanding engine. The substitution of an expansion valve, or throttling orifice, with constant enthalpy expansion, changes the theoretical performance but little, and greatly simplifies the apparatus. A typical vapor-compression refrigeration cycle is shown in Fig. 2; it is essentially a reverse Rankine cycle. The irreversible adiabatic expansion 1–2 differs only slightly from the vertical isentropic expansion.

Another practical change from the ideal Carnot

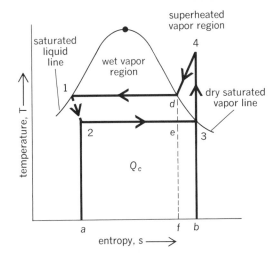

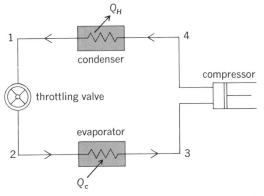

Fig. 2. Vapor-compression refrigeration cycle substitutes valve for expansion engine.

cycle substitutes dry compression 3–4 for wet compression e-d in Fig. 2, placing state 4 in the superheat region above ambient temperature; the process is called dry compression in contrast to the wet compression of the Carnot cycle. Dry compression introduces a second irreversibility by exceeding the ambient temperature, thus reducing the coefficient of performance. Dry compression is usually preferred, however, because it simplifies the operation and control of a real machine. Vapor gives no readily observable signal as it approaches

Ideal refrigeration cycle performance*

Refrigerant	Saturation pressure, psia		Refrigerating effect, Btu/lb	Refrigerant required per ton refrigeration		Compressor horsepower per ton refrigeration	Coefficient of performance
	Evaporator	Condenser		lb/min	cfm		
Carnot— any fluid							6.6
Freon, F12	47	151	49	4.1	3.6	0.89	5.31
Ammonia	66	247	455	0.44	1.9	0.85	5.55
Sulfur dioxide	24	99	135	1.5	4.8	0.83	5.67
Free air	15	75	32	6.2	77	2.7	1.75
Dense air	50	250	32	6.2	23	2.7	1.75
Steam	0.1	1.275	1000	0.2	590	0.91	5.18

*Performance is based on evaporator temperature = 35°F, condenser temperature = 110°F. In vapor-compression cycles, the pressures and temperatures are for saturation conditions; vapor enters the compressor dry and saturated; no subcooling in the condenser.

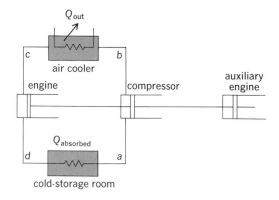

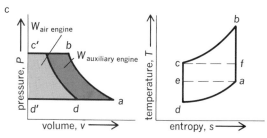

Fig. 3. Schematic arrangement of reverse Brayton, or dense-air, refrigeration cycle.

and passes point *e* in the course of its evaporation, but it would undergo a temperature rise if it accepted heat beyond point 3. This cycle, using dry compression, is the one which has won overwhelming acceptance for refrigeration work.

Reverse Brayton cycle. The reverse Brayton cycle constitutes another possible refrigeration cycle; it was one of the first cycles used for mechanical refrigeration. Before Freon and other condensable fluids were developed for the vapor-compression cycle, refrigerators operated on the Brayton cycle, using air as their working substance. Figure 3 presents the schematic arrangement of this cycle. Air undergoes isentropic compression, followed by reversible constant-pressure cooling. The high-pressure air next expands reversibly in the engine and exhausts at low temperature. The cooled air passes through the cold storage chamber, picks up heat at constant pressure, and finally returns to the suction side of the compressor. *See* BRAYTON CYCLE.

The temperature-entropy diagram, Fig. 3, points up the disadvantage of the dense-air cycle. If the temperature at *c* represents the ambient, then the only way that air can reject a significant quantity of heat along the line *b-c* is for *b* to be considerably higher than *c*. Correspondingly, if the cold body service temperature is *a*, the air must be at a much lower temperature in order to accept heat along path *d-a*. If a reverse Carnot cycle were used with a working substance undergoing changes in state, the fluid would traverse path *a-f-c-e* instead of path *a-b-c-d*. The reverse Carnot cycle would accept more heat along path *e-a* than the reverse Brayton cycle removes from the cold body along path *d-a*. Also, since the work area required by the reverse Carnot cycle is much smaller than the corresponding area for the reverse Brayton cycle, the vapor-compression cycle is preferred in refrigeration practice. *See* THERMODYNAMIC CYCLE.

Comparative performance of refrigerants. The table gives the significant theoretic performance data for a selected group of refrigerants when used in the ideal cycles as outlined above. These data include not only the requisite pressures in the evaporator and condenser for saturation temperatures of 35°F and 110°F, respectively, but also the refrigerating effect, the weight and volume of refrigerant to be circulated, the horsepower required, and the coefficient of performance. The vapor compression cycles presuppose that the refrigerant leaves the evaporator dry and saturated (point 3 in Fig. 2) and leaves the condenser without sub-cooling (point 1 in Fig. 2). The heat absorbed, Q_c, in the evaporator is called the refrigerating effect and is measured in Btu/lb of refrigerant. The removal of heat in the evaporator at the rate of 200 Btu/min is defined as a ton refrigeration capacity. The weight and volume of refrigerant, measured at the compressor suction, and the theoretic horsepower required to drive the compressor per ton refrigeration capacity are also given in the table. These data show a wide diversity of numerical values. Each refrigerant has its advantages and disadvantages. The selection of the most acceptable refrigerant for a specific application is consequently a practical compromise among such divergent data.

[THEODORE BAUMEISTER]

Bibliography: American Society of Heating, Refrigerating and Air-Conditioning Engineers, *Heating, Ventilating, and Air Conditioning Guide,* revised periodically; T. Baumeister (ed.), *Standard Handbook for Mechanical Engineers,* 7th ed., 1967.

Reluctance motor

A synchronous motor which starts as an induction motor and, upon nearing full speed, locks into step with the rotating field and runs at synchronous speed. The stator and rotor windings are similar to those of an induction motor. The rotor is of squirrel-cage construction, to allow induction-motor starting, and has salient-pole projections which provide synchronous operation at full speed. The reluctance motor is built only in small sizes and for situations in which low cost and simplicity are mandatory and efficiency is of little concern. It can be polyphase, but is usually a single-phase motor with a split-phase or capacitor winding for starting. *See* INDUCTION MOTOR; SYNCHRONOUS MOTOR.

[LOYAL V. BEWLEY]

Repulsion motor

An alternating-current (ac) commutator motor designed for single-phase operation. The chief distinction between the repulsion motor and the

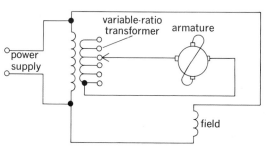

Fig. 2. Doubly excited repulsion motor.

REPULSION MOTOR

power supply

field

armature

Fig. 1. Schematic of a repulsion motor.

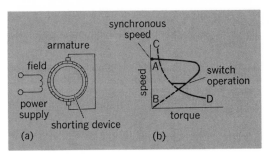

Fig. 3. Repulsion-start, induction-run motor. (a) Schematic diagram. (b) Speed-torque characteristic.

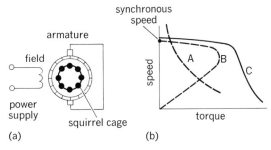

Fig. 4. Repulsion-induction motor. (a) Schematic diagram. (b) Speed-torque characteristic.

single-phase series motors is the way in which the armature receives its power. In the series motor the armature power is supplied by conduction from the line power supply. In the repulsion motor, however, armature power is supplied by induction (transformer action) from the field of the stator winding. See ALTERNATING-CURRENT MOTOR.

The repulsion motor primary or stationary field winding is connected to the power supply. The secondary or armature winding is mounted on the motor shaft and rotates with it. The terminals of the armature winding are short-circuited through a commutator and brushes. There is no electrical contact between the stationary field and rotating armature (Fig. 1).

If the motor is at rest and the field coils are energized from an outside ac source, a current is induced in the armature, just as in a static transformer. If the brushes are in line with the neutral axis of the magnetic field, there is no torque, or tendency to rotate. However, if they are set at a proper angle (generally $15-25°$ from the neutral plane), the motor will rotate.

Repulsion motors may be started with external resistance in series with the motor field, as is done with dc series motors. A more common method is to start the motor with reduced field voltage and increase the voltage as the motor increases speed. This can be done conveniently with a transformer having an adjustable tapped secondary or a variable autotransformer.

It is also possible to doubly excite the motor; that is, the armature may receive its power not only by induction from the stator winding but also by conduction from a transformer with adjustable taps, as shown in Fig. 2.

Repulsion-start, induction-run motor. This motor (Fig. 3a) possesses the characteristics of the repulsion motor at low speeds and those of the induction motor at high speed. It starts as a repulsion motor. At a predetermined speed (generally at about two-thirds of synchronous speed) a centrifugal device lifts the brushes from the commutator and short-circuits the armature coils, producing a squirrel-cage rotor. The motor then runs as an induction motor. In Fig. 3b the curve AB represents the characteristics of an induction motor and curve CD a repulsion motor. The solid curve AD is the combined characteristic of a repulsion-start, induction-run motor. See INDUCTION MOTOR.

Repulsion-induction motor. This motor (Fig. 4a) is very similar to the standard repulsion motor in construction, except for the addition of a second separate high-resistance squirrel cage on the rotor.

Both rotor windings are torque-producing, and the total torque produced is the sum of the individual torques developed in these two windings. In Fig. 4b, curve A is the characteristic of the repulsion-motor torque developed in this motor. Curve B represents the induction-motor torque. Curve C is the combined total torque of the motor. The advantages of this machine are its high starting torque and good speed regulation. Its disadvantages are its poor commutation and high initial cost.

[IRVING L. KOSOW]

Bibliography: I. L. Kosow, *Electric Machinery and Control*, 1964; A. F. Puchstein, T. C. Lloyd, and A. G. Conrad, *Alternating-Current Machines*, 3d ed., 1954.

Reserve battery

A battery which is inert until an operation is performed which brings all the cell components into the proper state and location to become active.

High-energy primary batteries are often limited in their application by poor retention characteristics which are caused by self-discharge between the electrolyte and the active electrode materials. One method of overcoming this problem, particularly when the battery is required to operate at high current levels for relatively short periods of time (minutes or hours), is the use of a reserve battery.

Several types have been developed. In water-activated or electrolyte-activated batteries the water or electrolyte component is not present during storage. It is added just before the cell is put into use. In thermal batteries the electrolyte is a solid at room temperature and has very low conductivity. If the temperature is raised above the melting point, the conductivity of the electrolyte becomes excellent and the cell is capable of delivering significant power.

Water-activated batteries. Practical battery systems have been developed using magnesium anodes against silver chloride or cuprous chloride cathodes (Fig. 1). Cuprous chloride cathodes are less expensive than silver chloride cathodes, but they are also bulkier and less stable, particularly in a humid atmosphere. *Meta*-dinitrobenzene is also finding use as cathode material on account of its high ampere-hour capacity.

The batteries are assembled dry. The active elements may be separated by porous paper or another inert media. Water may be poured into a container holding the elements or may flow continuously through the element. Either fresh or salt water may be used.

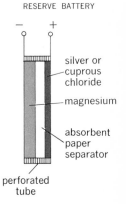

Fig. 1. Schematic of water-activated cell.

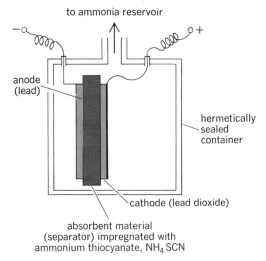

Fig. 2. Schematic of ammonia-vapor-activated reserve-type primary cell.

The most important design factor for all reserve batteries is to ensure that the electrolyte is delivered as quickly as possible at the time of activation, at the same time avoiding chemical short-circuiting of the cells.

Cells with absorbent separators can be activated by immersion. Subsequent operation may either be in air, using only the water retained in the cells, or while immersed. Performance of a two-cell battery immersed in sea water showed an output of 18 watt-hours/pound (whr/lb) when fully discharged in 6 min.

The dry elements stored in a sealed container are capable of indefinite storage life.

Electrolyte-activated batteries. Any cell can be made as a reserve-electrolyte cell. If the electrodes are in place, it is necessary only to add the electrolyte to make a complete cell. In practice, however, the separation of the electrolyte is done only when excessive deterioration would occur during wet storage prior to use. Great ingenuity has been shown in designing complete battery packages in which an aqueous electrolyte is stored in a separate chamber. The package contains a mechanism, which may be operated from a remote location, which drives the electrolyte out of the reservoir and into all the cells of the battery. In general, these packaged batteries have been used only in military applications. The following couples have been used in reserve cells containing electrolytes: Zn/Cu, Pb/PbO_2, Zn/AgO, $Mg/meta$-dinitroben-

zene, Zn/PbO_2, Zn/MnO_2, Cd/PbO_2. However, not all these couples are amenable to heavy rate discharge.

Gas-activated batteries. The liquid-activated batteries previously described have disadvantages which are difficult to overcome; for example, automatic electrolyte-charging equipment may cause intercell shorting. Unless the design is quite complex, it is difficult to avoid flooding or uneven filling. An alternative approach is to introduce a gas which reacts with the spacer material to form a conducting electrolyte.

Boron trifluoride gas reacts with dry, hydrated barium hydroxide to form a highly acid solution containing barium salts, borates, and fluoborates. Ammonia gas reacts with ammonium salts to form a solution having good conductivity (Fig. 2). Suitable electrode couples are Zn/MnO_2 and Pb/PbO_2. These gas-activated batteries are reported to operate well over a wide temperature range.

Thermal batteries. These are also known as heat-activated or fused-electrolyte batteries. Some compounds, such as sodium chloride and potassium hydroxide, show very low conductivity in the solid state at room temperature but very good conductivity in the molten state. For example, a mixture of sodium hydroxide and potassium hydroxide becomes an excellent ionic conductor when heated above 170°C. If a zinc anode and a silver oxide cathode are combined with solid pads of the eutectic mixture, all the elements of a cell are present. The electrolyte has an appreciable amount of entrained moisture, which plays a role in the discharge, but the cell will also work with carefully dried materials. Such cells are capable of high power output for a few minutes, when heated to 200°C or higher. At 1 amp/in.² of positive plate, the cell voltage is 1.16 at 200°C, 1.23 at 250°C, 1.30 at 300°C.

Thermal batteries are capable of operation at very low ambient temperatures, provided that a suitable heat source is available to melt the electrolyte. For ordinary temperatures, they are not advantageous as compared with reserve aqueous-electrolyte types. Because the electrolyte in thermal batteries is inert and nonconductive at normal temperatures, they can be stored indefinitely, acting as primary batteries when the temperature of the electrolyte is raised to that at which they become ionically conductive.

A magnesium and manganese dioxide cell with sodium hydroxide electrolyte can operate for longer discharge times than the zinc–silver oxide cell mentioned previously because of the greater stability of the reactants at high temperatures (Fig. 3).

In addition to the couples already mentioned, successful systems have been employed using Mg, Ca, or Li alloys as anodes and K_2CrO_4, WO_3, MO_3, or $PbCrO_4$ as cathodes.

Thermal batteries are essentially of low energy efficiency because the heat absorbed in melting the electrolyte is not available in the electrical output. Consequently, although their use is restricted to small cell sizes, there is likely to be an increasing demand for thermal cells because of their high energy output, long storage retention over wide temperature ranges, ruggedness, and ability to be used in any position.   [JACK DAVIS]

Bibliography: Burgess Battery Co., *Engineering*

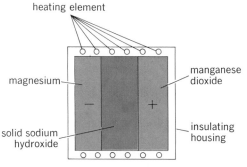

Fig. 3. Schematic of thermal cell.

Manual, 1964; W. J. Hamer and J. P. Schrodt, Investigations of galvanic cells with solid and molten electrolytes, *J. Amer. Chem. Soc.*, 71:2347, 1949; R. Jasinski, *High Energy Batteries*, 1967; J. P. Mullen and P. L. Howard, Characteristics of the silver chloride-magnesium water activated battery, *Trans. Electrochem. Soc.*, 90:529, 1946; C. B. Root and R. A. Sutula, *Proceedings of 22d Power Sources Conference*, May 14–16, 1968; J. P. Schrodt et al., A lead dioxide cell containing various electrolytes, *Trans. Electrochem. Soc.*, 90:405, 1946; R. Schult and W. Stafford, Electrochemical energy sources: Silver oxide/zinc batteries, *Electro-Technol. (NY)*, 67:84, June, 1961; G. W. Vinal, *Primary Batteries*, 1950; U.S. Army Signal Laboratories, *Proceedings of the 11th Annual Battery Research and Development Conference*, 1957; J. C. White, R. T. Pierce, and T. P. Dirkse, Characteristics of the silver oxide-zinc-alkali primary cells, *Trans. Electrochem. Soc.*, 90:467, 1946.

Rotary tool drill

A bit and shaft used for drilling wells. A turntable on the derrick floor rotates a string of hollow steel drill pipe at the bottom of which is a steel bit. The bit grinds the rock. A drilling fluid is pumped down through the drill pipe; the fluid flushes out the rock cuttings and returns up the space between drill string and hole side.

The drilling fluid may be air, water, or, most commonly, mud (a mixture of various clays and chemicals, each having a special function). The mud cools and lubricates the bit, removes cuttings from the hole, and cakes the wall of the hole to prevent caving before steel casing is set. The hydrostatic pressure exerted by the column of mud in the hole prevents blowouts which may result when the bit penetrates a high-pressure oil or gas zone. When the mud reaches the surface, it passes over a vibrating screen to filter out large cuttings. The mud then passes on to a settling tank where smaller particles settle out. The cuttings are examined to determine the type of formation being drilled and for possibilities of oil or gas production. The mud mixture is sucked up from the pit and recirculated by a high-pressure pump. The viscosity, weight, and filtration properties of the mud are altered as drilling proceeds by changing the proportion of its constituents.

Power is transmitted from an engine to a draw works—a winch which drives the rotary table on the derrick floor and also applies power for hoisting or lowering the drill string as shown in the illustration. The string of drill pipe is topped at the surface by a square-sided length of heavy pipe called the kelly. The square shape permits the rotary table to grip and rotate the kelly, and hence the entire drill string, and yet have sufficient freedom so that it can slip vertically through the table as drilling goes deeper. Rotation speeds range from 40 rpm to 500 rpm or more, depending primarily upon the character of the formation being drilled. The drill string usually consists of 30-ft lengths of drill pipe coupled together. On the lower end are heavier-walled lengths of pipe, called drill collars, which help regulate weight on the bit.

The drill string is attached to a swivel suspended from a hook which is connected to a traveling block, or pulley, encased in a frame. The drilling cable runs from the draw works over a crown block at the top of the derrick and down to the traveling block. The mud is pumped through a hose attached to the swivel. An opening in the center of the swivel permits the mud to pass down through the attached drill string.

When the bit has penetrated the distance of a pipe section, drilling is stopped, the string is pulled up to expose the top joint, the kelly is disconnected, a new section added, the kelly attached, the string lowered, and drilling resumed. This process continues until the bit becomes worn out, at which time the entire drill string must be pulled. Pipe is usually disconnected in thribbles, or 90-ft sections of pipe, and stacked in the derrick. The height of the derrick determines whether doubles, thribbles, or fourbles can be stacked. The process continues until the bit reaches the surface. A new bit is attached, and the drilling string reassembled and lowered into the hole. Such round trips may take up to two-thirds of total rig-operating time, depending upon the depth of the hole. In hoisting or lowering the drill string, the swivel is disengaged from the hook. Elevators, or clamps, which grip the pipe securely, are attached. The elevators are also used when the hole is lined with steel casing. In lowering drill pipe or casing, each new section of pipe is lifted from the derrick floor and suspended on the elevator until it is screwed to the preceding joint, just above the hole opening; the entire column is then lowered into the hole. While new sections of pipe or casing are being attached to the elevators, the pipe in the hole is supported in the rotary table by slips, or gripping devices.

Derricks can be skid-, truck-, or trailer-mounted, but larger units used in very deep drilling are assembled on the site. Derricks usually range in height from 66 ft to nearly 200 ft. The derrick floor is set 7–20 ft or more above the ground to provide a basement for control devices, such as blowout preventers, below the rotary table. *See* CABLE-TOOL DRILL; OIL AND GAS WELL DRILLING.

[ADE L. PONIKVAR]

Rural electrification

The generation, distribution, and utilization of electricity in nonurban areas. The almost universal availability and use of electricity throughout rural parts of the United States is the result of a nationwide system of central power-generating stations and cross-country transmission lines and rural-line extensions that distribute current to 99+% of the farms in the United States (Fig. 1). Equally important to the nation's rural population has been the

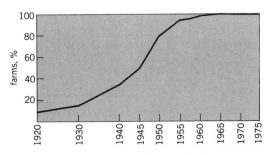

Fig. 1. Percent of farm dwellings in the United States lighted by electricity. (*U.S. Bureau of the Census*)

Fig. 2. Electricity use in the milk house on a dairy farm. Light aids sanitary care of milk utensils, heat is used for water and warming the room, and power from electric motors pumps and cools the milk in a bulk tank. (*Niagara Mohawk Power Corp.*)

development of hundreds of uses for electricity for light, heat, and power.

Development. Use of electric energy in food production started in California in the early 1880s, when pioneering farmers began using current from small water power-driven generating plants for lighting and for pumping water of irrigation. During the next 50 years a number of experimental projects were undertaken by both private and public agencies. Collectively, these investigations demonstrated the economic feasibility of using electricity widely on farms. As a result, national interest developed in the extension of electric lines

to all rural areas. In 1935 the Rural Electrification Administration (REA) was created as one of the approved projects under the Federal Emergency Relief Appropriations Act. The REA was authorized to make loans for the generation, transmission, and distribution of electricity to rural areas. The farmers cooperative was the agency through which most of the loans were made for rural-line construction over the country.

Uses. More than 300 uses of electricity have been developed for the farm and home, and about 150 uses are found in agricultural production. This acceptance of electric energy is largely the result of its economic and laborsaving advantages because it has been established that, for any task a 1-horsepower electric motor can perform, a farmer will require 10 times as much time to accomplish it manually. Electric power is also more versatile and more dependable than other kinds of power.

Dairy farm applications of electricity (Fig. 2) include lighting, pumping water, milking cows, cooling milk, heating water for sanitation, electric fencing, ventilating and cleaning the stable, elevating and drying hay, feeding, ensilage, and handling milk in bulk through pipelines.

Poultry farmers use electricity for lighting the flock to stimulate winter egg production, incubating eggs, brooding chicks, automatic feeding and watering, mechanical egg gathering, cleaning, grading, and cooling eggs, debeaking of birds, mechanical litter removal, and ventilation.

Market gardeners use electricity for irrigating, propagating plants, washing, packing, and cooling vegetables. Florists find electricity useful for heating and pasteurizing soil, lighting to control time of bloom of plants, ventilating, and automatically controlling heating systems which are electrically operated.

The increasing use of electric energy on farms and the resulting decrease in cost to the farmer are shown in Fig. 3. Growth in the use of electricity in the rural United States appears to be almost unlimited. *See* ELECTRIC POWER SYSTEMS.

[CLESSON N. TURNER]

Bibliography: R. H. Brown, *Farm Electrification*, 1956; U. F. Earp, *Rural Electrification Engineering*, 1950; Federal Power Commission, *National Power Survey*, 2 vols., 1964; D. Fink and J. Carroll, *Standard Handbook for Electrical Engineers*, 1968; T. E. Hienton, D. E. Wiant, and O. A. Brown, *Electricity in Agricultural Engineering*, 1958; C. N. Turner (ed.), *Farm Electrical Equipment Handbook*, 1966.

Solar battery

An array of semiconductor *pn* junction devices which converts the radiant energy of sunlight directly and efficiently into electrical energy. The most common form of the solar battery consists of individual crystal silicon photovoltaic cells connected in series and parallel to obtain the required values of current and voltage at the load terminals. Such arrays are expensive, and research is in progress to develop more economical large-area photovoltaic cells.

Solar batteries were the prime power source aboard most artificial space satellites that were designed to remain in space for a considerable time. Terrestrial applications of importance include power sources at remote locations where

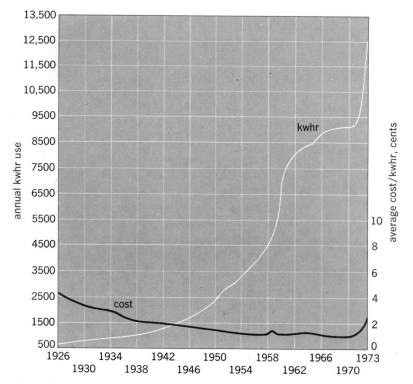

Fig. 3. Trends in annual kilowatt-hour use and cost of electric energy on farms in the United States where little or no irrigation is involved. (*Data for 1926–1959 from Edison Electric Institute, for 1960–1966 from Statistical Reporting Service, USDA*)

conventional sources of electrical energy are either unavailable or too expensive.

Solar radiation. The accepted value of the solar radiation constant (energy falling on 1 cm² at normal incidence outside the Earth's atmosphere) is 1.95 cal/min, or 136 milliwatts (mw). Approximately one-third of this energy is scattered when passing through the Earth's atmosphere so that at noontime on a clear day about 100 mw/cm² (1 kw/m² or 1 kw/yd²) arrive at the surface of the Earth. This is a considerable amount of radiant energy. The cross-sectional area of the Earth is about 50,000,000 mi², and the total radiant energy reaching the atmosphere is 1.8×10^{17} watts. Solar radiation is the source of terrestrial illumination and heat as well as photons for vegetation growth. It can also be used in the generation of electrical power when converted by means of solar batteries. *See* SOLAR ENERGY; SOLAR RADIATION.

Principles of operation. The heart of the silicon solar battery is the *pn* junction formed near the front surface of a plate of silicon, one of the most common elements. The fundamental principles of a *pn* junction are indicated in Fig. 1. Pure silicon to which a trace amount of a fourth-column element such as arsenic has been added is an *n*-type semiconductor so that electric current is carried by free electrons. From each arsenic atom that is added, one electron (unit negative charge) detaches itself, thus becoming free to move, and leaves behind the arsenic atom with unit positive charge bound into the crystal structure. Thus, *n*-type silicon consists of silicon to which are added equal numbers of free electrons and bound positive charge, so that there is no net charge.

By adding minute amounts of an element from the third column of the periodic table, such as boron, aluminum, or gallium, the silicon may be made *p*-type so that electric current is carried by free holes. For example, for each gallium atom that is added one hole (unit positive charge) detaches itself, thus becoming free to move, and leaves behind the gallium atom with unit negative charge bound into the crystal structure. Thus, *p*-type silicon consists of silicon to which are added equal numbers of free positive holes and bound negative charge, so that there is no net charge.

An interface between *p*- and *n*-type silicon is called a *pn* junction. This region contains a permanent dipole charge layer with a high electric field which forces mobile negative charges (electrons) toward the left, and mobile positive charges (holes) toward the right. The free holes and free electrons try to intermix like gases. However, the holes which enter the *n*-type material disappear and leave behind negatively charged gallium atoms, and the electrons which enter the *p*-type material disappear and leave behind positively charged arsenic atoms. These fixed charges constitute an electrical barrier, or field, which prevents the rest of the holes in the *p*-side and the electrons in the *n*-side from mixing.

A cross-sectional view of the silicon *pn* junction solar cell is shown in Fig. 2. The center *n*-type and the *p*-type skin forms a junction about 0.0001 in. below the surface. Photons of light energy from the Sun produce hole-electron pairs near the junction. The built-in electric field forces the holes to the *p*-side and the electrons to the *n*-side: This displacement of free charges results in a voltage

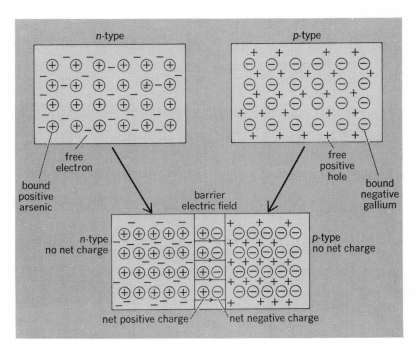

Fig. 1. Operating principles of *pn* junction.

difference between the two regions of the crystal, the *p*-region being plus and the *n*-region minus. When a load is connected at the terminals, an electron current flows in the direction shown by the arrow and useful electrical power is available at the load.

Solar-cell characteristics. The electrical characteristics of a typical silicon *pn* junction solar cell are shown in Fig. 3. Figure 3a shows open-circuit voltage and short-circuit current as a function of light intensity from total darkness to full sunlight (1000 watts/m²). The short-circuit current is directly proportional to light intensity and amounts to 28 milliamps/cm² at full sunlight. The open-circuit voltage rises sharply under weak light and saturates at about 0.6 volt for radiation between 200 and 1000 watts/cm². The variation in power output from the solar cell irradiated by full sun-

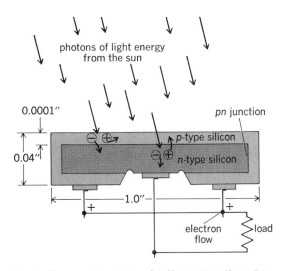

Fig. 2. Cross-sectional view of a silicon *pn* junction solar cell, illustrating the creation of electron pairs by photons of light energy from the Sun.

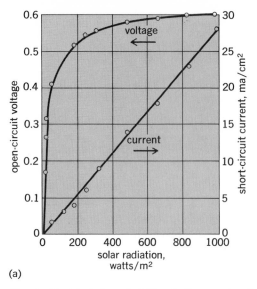

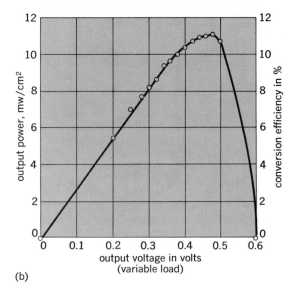

(a)   (b)

Fig. 3. Electrical characteristics of silicon *pn* junction solar cell, at operating temperature of 17°C. (*a*) Variation of open-circuit voltage and short-circuit current with light intensity. (*b*) Variation in power output as load is varied from short to open circuit.

light as its load is varied from short circuit to open circuit is shown in Fig. 3*b*. The maximum power output is about 11 mw/cm² at an output voltage of 0.45 volt.

Under these operating conditions the overall conversion efficiency from solar to electrical energy is 11%. The output power as well as the output current is of course proportional to the irradiated surface area, whereas the output voltage can be increased by connecting cells in series just as in an ordinary chemical storage battery. Experimental samples of silicon solar cells have been produced which operate at efficiencies up to 16%, but commercial cell efficiency is around 10–12% under normal operating conditions.

The spectral relationships in a typical silicon solar cell are shown in Fig. 4. Wavelength in microns is plotted horizontally, and the insert indicates the visible part of the spectrum as well as the ultraviolet and the infrared regions. Curve A represents the distribution of solar energy at the surface of the Earth among the various wavelengths relative to an arbitrary maximum of 1.0. Curve B shows the relative number of photons in solar radiation and is derived from curve A simply by multiplying by the wavelength. Curve C is the

measured short-circuit response curve of a typical silicon solar cell. It shows a maximum at 0.7μ, which is at about the same wavelength as the maximum in the solar photons. Curve C drops to zero at 0.3 and 1.1 μ. The peak can be shifted to the left or right by decreasing the depth of the *pn* junction.

[GERALD L. PEARSON]

Applications. The term solar battery was first applied to the silicon cell invented at Bell Telephone Laboratories to distinguish it from earlier photovoltaic cells producing about an order of magnitude less electrical output. Common usage refers to the earlier photovoltaic cells as sun batteries and reserves the term solar batteries for the more efficient silicon cells. The name also reflects the original intended use of converting solar energy to provide electrical power for remote telephone circuit amplifiers.

Although the silicon solar batteries successfully powered telephone amplifiers in field tests in 1955, their outstanding use at present is for power for satellite communications. The first satellite to use solar cells was *Vanguard 1*, which went into orbit on Mar. 17, 1958. This small satellite carried a 5-mw transmitter powered by six solar batteries. For about 6 years it sent back signals giving valuable information and showing the feasibility of direct use of solar energy for communications in space. Many other satellite uses followed, including *Telstar* and the fixed-position satellites which give worldwide communication channels.

Because of competitive price considerations, terrestrial uses of solar batteries have usually been restricted to low power needs or to inaccessible locations. For example, Japan powered numerous navigation beacons with solar batteries. The U.S. Army uses transistor field radios powered by solar batteries. Transistor radio receivers that run on solar energy can be purchased. However, extended terrestrial applications await a cheaper construction than is presently used.

Construction. A typical *p*-on-*n* solar cell starts as high-grade silicon to which a minute amount of

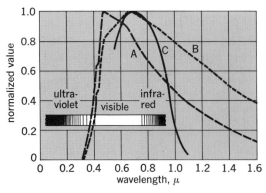

Fig. 4. Spectral relationships in a typical silicon solar cell relative to an arbitrary maximum of 1.

arsenic (about 1 part per million) has been added. Large cylindrical single crystals are grown from a melt by slow withdrawal of a seed crystal held on a quartz rod. The crystals are then sliced into thin slabs using a diamond-bonded wheel flooded with water. After some surface preparation, the slabs are heated to about 1100°C in a boron trichloride and nitrogen atmosphere. This treatment produces a thin surface layer of boron-diffused p-type silicon over an n-type silicon wafer. Cleaning of the surfaces and the attachment of suitable leads to the base material and to the surface layer complete the cell. Because of the greater resistance to radiation damage, most cells currently made for use on satellites are the reverse of the one just described. The base material is a boron-doped p-type slab. Phosphorus is diffused into the surface layer, which becomes n-type.

Energy conversion. Pure silicon (intrinsic) forms a very stable crystal in which nearly all of the electrons are bound in the orbits of silicon atoms and are not free to move about to conduct current. At room temperature, energy of approximately 1.08 electron volts (ev) is needed to free an electron for conduction. If a photon of energy 1.08 ev or more is absorbed in silicon, its energy frees an electron which now possesses 1.08 ev of energy relative to its former bound condition. This is the first and fundamental transformation of solar radiation to electrical energy. But if the electron is subsequently recaptured by a hole, the useful energy is lost. It is the function of the solar cell to direct this useful energy to an external circuit.

The junction between p- and n-type conductors produces a potential difference because the carriers that diffuse from each region into the other constitute a current in the same direction—positive carriers from p- to n-type and negative carriers from n- to p-type. In each case, the carriers are captured by the majority carriers of the new region and are no longer available for conduction in the opposite direction. A voltage difference is developed which opposes the diffusion current and at equilibrium reduces it to equal the small return of the few available minority carriers. It follows that the fewer the minority carriers, the greater will be the potential difference of a pn junction.

Similar diffusion and capture of carriers produce potential differences at contacts made to the separate regions so that the net potential difference around such a circuit is zero in the dark. The absorption of radiant energy at the pn junction supplies additional carriers for the return flow and lowers the potential difference at that junction. Since the contact potentials are not changed, there is now a net potential difference available for external use as long as photons having the proper energy flood the pn region.

Efficiency and temperature dependence. Radiation of wavelength 1.15 μ has just enough energy per photon (1.08 ev) to release electrons in silicon. It follows that any photon of 1.15-μ or shorter wavelength is potentially a source of one electron. The shorter-wave photons carry more energy than is needed and the extra energy is wasted as heat. Analysis of the Sun's spectrum at the Earth shows that 45% of the total solar energy is available for this type of energy conversion. But since a good cell can deliver maximum power at only about 0.5

volt instead of 1.08 volts of a freed electron, a theoretical maximum of about 22% is available. Other losses reduce this to about 11% in good commercial cells. Antireflection surface coatings of SiO help maintain high efficiency. Protective coatings of sapphire decrease the radiation damage of satellite cells.

For terrestrial uses, economy suggests lenses or mirrors to concentrate the radiation. However, internal resistance limits the useful concentration ratio. Also, the increased temperature from concentration lowers the output voltage.

As easily predicted from the above explanation of operation, higher temperature means more thermally produced conduction holes and electrons. All of the junction potentials are lowered by the increase of the carriers available to pass back over the barrier. The open-circuit voltage decreases about 0.002 volt/°C temperature rise. In the other direction, a voltage of 0.99 volt was once measured at liquid nitrogen temperature.

Other materials. Solar cells with efficiencies comparable to those of silicon have been made of several other materials. Two that have been extensively studied are cadmium sulfide and gallium arsenide.   [DARYL M. CHAPIN]

Bibliography: T. S. Moss, *Optical Properties of Semiconductors*, 1961; G. L. Pearson, Conversion of solar to electrical energy, *Amer. J. Phys.*, 25:591, December, 1967; K. D. Smith et al., The solar cells and their mounting, *Bell Syst. Tech. J.*, 42:1765–1816, July, 1963; F. M. Zaren and O. Erway, *Introduction to the Utilization of Solar Energy*, 1963.

Solar collectors

Devices which provide solar heating by collecting the Sun's shortwave radiation (0.3–3.0 μm in wavelength) on a suitably blackened surface which can convert from 90 to 99% of the incident solar radiation into heat. Depending upon the operating temperature of the collector, this heat may be used for producing potable water from brackish or salty sources, for space and service water heating, and, when the temperature is sufficiently high, for absorption refrigeration and power generation. There are two principal types of solar collectors: flat plates (Fig. 1) and concentrators (Fig. 2).

Flat-plate collectors. Flat-plate collectors are usually fixed in a south-facing position (in the Northern Hemisphere) and tilted upward from the horizontal at the angle which is most suitable for the collector's intended function. These collectors are relatively simple in design and easy to build. They function with little or no attention but are limited in the temperatures which they can withstand and the efficiencies which they can attain.

The typical flat-plate collector consists of five components: (1) A transparent cover or covers which can admit from 60 to 90% of the incoming solar radiation, depending upon the quality of the material (usually glass), the number of cover sheets, and the incident angle between the incoming solar rays and the line normal to the collector surface. Glass is highly transparent to the Sun's shortwave radiation but completely opaque to the longwave radiation emitted by the heated collector plate. Glass is also unaffected by the Sun's ultraviolet radiation, which is highly destructive to most plastic films. (2) The collector plate, usually made of copper, aluminum, or steel, and blackened to

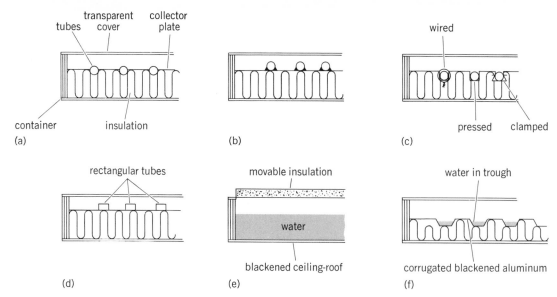

Fig. 1. Flat-plate collector types. (*a*) Tubes in a black plate. (*b*) Tubes bonded to upper surface of black plate. (*c*) Tubes fastened to lower surface of black plate by three methods. (*d*) Rectangular tubes bonded to plate. (*e*) Hay system. (*f*) Thomason system.

convert the incoming shortwave radiation into heat. The surface coating may be either a flat black paint, which absorbs from 95 to 98% of the incoming radiation but also emits a similar percentage of the longwave radiation which would normally come from a blackbody at the collector temperature. Alternatively, a "selective surface" may be used, which absorbs from 90 to 96% of the incoming solar radiation but emits less than 20% of the longwave blackbody radiation. The surface coating must be permanent and able to withstand the high temperatures which will be encountered on hot summer days when there is no circulation of cooling fluid. (3) The tubes or other passages which are integral with or fastened to the collector plate by some process which ensures a good thermal bond between the plate and the fluid which is being heated (Fig. 1 *a–d*). In addition to high thermal conductance, the bonding process must be able to withstand thermal expansion and contraction as well as high temperatures. (4) The insulation, which minimizes loss of heat from the back of the collector plate. Low thermal conductance, resistance to deterioration due to moisture, insects, and so on, and ability to withstand high temperatures without giving off vapors are essential characteristics of a good insulating material. (5) The container, which keeps the entire system watertight and dust-free while permitting the inevitable differential expansion of the collector components. Galvanized sheet steel is often employed for this purpose because of its relatively low cost, although aluminum is also used because of its light weight. Redwood siding and hardboard backs may also be employed where exceptionally long life is not essential.

In the Hay "Sky Therm" system (Fig. 1*e*), the entire horizontal metal roof-ceiling of a structure is covered with a transparent bag containing an 8–10-in. (19.3–25.4 cm) depth of water. No glazing is used, but instead, horizontally movable insulation is employed to expose the water bags for heating during winter days and cooling during summer nights. This system, adaptable primarily to flat-roofed structures, can provide year-round comfort without auxiliary energy in hot, dry desert climates.

The Thomason (Solaris) system, shown in Fig. 1*f*, uses blackened corrugated aluminum under a single sheet of glass to trap the solar rays, while a shallow layer of water, flowing downward along the corrugations, carries the collected heat to a storage tank.

Air and water heating. Solar air heaters are usually simpler in design and lower in cost than solar water heaters. In a number of systems the shortwave solar radiation is absorbed by finned or corrugated metal plates, by matrices of fabric or perforated metal, or by overlapping glass plates, alternatively clear and blackened. Air is blown over these surfaces or allowed to flow by natural circulation, and the collected heat can then be stored in rock beds. The performance of flat-plate collectors generally depends upon the angle of incidence between the Sun's rays and the collector and upon the difference in temperature between the collector plate and the outdoor air. Collection efficiency for a south-facing collector usually reaches its maximum value at noon or shortly thereafter. For low-temperature collectors, with a 50°F (9.5°C) temperature difference between the collector plate and the air, efficiency will range from 50 to 60% from 9:30 A.M. to 2:30 P.M. on a clear July day, depending upon the number of cover glasses and the degree of selectivity of the collector surface.

If the temperature difference is raised to 100°F (36°C) under the same conditions, the efficiency will fall to the 20–40% range.

For unglazed swimming-pool heaters designed for summer use, the efficiency can be very high, since the air temperature may well be warmer than the collector surface, and the water will gain heat from both the Sun and the air. Heating of uncovered pools in winter, however, is difficult to accomplish. Plastic collectors may be useful in this application because of the low pressures and temperatures which are encountered when they are in service. *See* SOLAR HEATING.

Concentrators. Concentrators focus a large area of radiation from the solar disk upon a small absorbing surface by means of flat or curved reflecting or refracting surfaces. The absorbed heat is removed by a flowing or evaporating fluid. Concentrators can reach extremely high temperature, but since they generally can use only the Sun's direct rays they must follow the Sun's apparent motion across the sky (15° per hour).

Concentrating collectors were employed long before the flat-plate types came into use. A. Mouchot used a sun-following reflective cone (Fig. 2*a*) in Paris to generate steam and to produce ice in September 1878, while J. Ericsson used a paraboloidal concentrator (Fig. 2*b*) to run a hot-air engine in New York at about the same time. C. G. Abbot used highly polished paraboloidal troughs (Fig. 2*c*) to concentrate the Sun's rays on glass-enclosed metal pipes as early as 1908. Shuman and Boys used very long horizontal reflective parabolic troughs (Fig. 2*d*) to concentrate the Sun's rays on a glass-enclosed steam generator at Meadi, near Cairo, in 1913. The Somor pump (Fig. 2*e*) demonstrated in Phoenix, AZ, in 1955, used flat-plate collectors with reflecting aluminum wings to concentrate enough solar radiation to boil sulfur diox-

ide and thus drive a small pump.

The most promising system for large-scale power generation is that proposed by the Martin-Marietta Company (Fig. 2*f*), in which a very large number of slightly curved mirrors focus the Sun's rays into the opening of a large steam generator. Similar in principle to the device used by Archimedes 2000 years ago to set fire to the sails of a Roman fleet, the Martin-Marietta system can be expanded to produce 100,000 kW from steam power plants under desert sunshine.

F. Trombe has led in the development of concentrating collectors which use fixed paraboloids, ranging in diameter from 5 to 95 ft (1.5 to 29 m). Solar radiation is reflected into these concentrators by means of flat heliostats, mounted to the north of the concentrators, so that they can turn and tilt to follow the Sun and reflect its rays into the paraboloids (Fig. 2*g*). *See* SOLAR ENERGY.

[JOHN I. YELLOTT]

Bibliography: E. A. Christie, Spectrally selective surfaces for solar energy collectors, *ISES Conference, Melbourne*, Pap. no. 7/81, 1970; *Energy Primer, Solar, Water, Wind and Bio Fuels*, Portola Institute, Menlo Park, CA, pp. 6–8, 1975; H. R. Hay and J. I. Yellott, Natural air conditioning with roof ponds and movable insulation, *ASHRAE Trans.*, vol. 75, pt. 1, p. 165, 1969; J. I. Yellott, Solar energy utilization for heating and cooling, ch. 59, *ASHRAE Handbook of Applications*, 1974.

Solar cooling

The use of solar radiation to produce cooling by means of thermal processes. The oldest and most widely used of these processes is absorption refrigeration, first demonstrated with the Sun as the heat source by A. Mouchot in 1878. Another solar-powered cooling process which has been tested with some success uses vapor jets to produce evaporation of fluids such as water or fluorinated hydrocarbons and thus to cool them. This process, although shown to be feasible, has such a low coefficient of performance (COP) that it has not been competitive with the absorption system. The COP is the ratio of the cooling effect in Btu/hr (1 Btu = 1055 J) or kW to the energy input, expressed in the same units.

A third system, which awaits the development of high-performance solar collectors capable of producing steam or other vapors at temperatures in the 500°F (246°C) range, proposes to use small turbines or other expansion engines operating on the Rankine cycle. The power output would then be used to drive a conventional compression-refrigeration machine. The use of solar batteries to drive heat-pump refrigerators is also under study. *See* HEAT PUMP; SOLAR BATTERY.

Nocturnal processes. A very important source of cooling is available at night when the Sun is not shining, in the form of nocturnal radiation to the sky. Evaporation of moisture from horizontal surfaces at night has also been shown to be a very effective cooling process. These two processes, which were used to produce ice in Persia and other arid regions in ancient times, utilize the absence rather than the presence of the Sun, but, when combined with the "Sky Therm" solar heating system, they can produce comfortable indoor conditions in arid areas during both summer and winter.

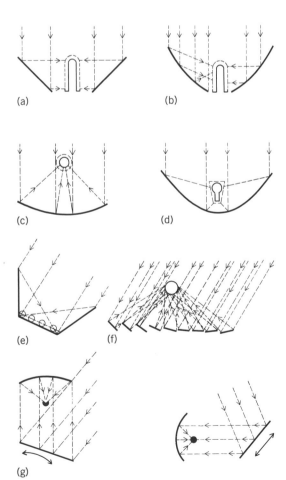

Fig. 2. Concentrating collector types. (*a*) Mouchot. (*b*) Ericsson. (*c*) Abbot. (*d*) Shuman and Boys. (*e*) Somor. (*f*) Martin-Marietta. (*g*) Trombe.

Flat-plate collectors. The conventional solar cooling system uses high-performance flat-plate collectors to produce as much hot water as possible during the daylight hours so as to operate the refrigeration plant during the periods of peak demand (usually in the early afternoon), and any excess heat, still at temperatures near 200°F (88°C), is stored for use at night; an auxiliary source of heat must be available for use when the supply of solar radiation is inadequate.

Absorption cooling. The two combinations of absorbant and refrigerant which are currently in use are (1) lithium bromide and water and (2) water and ammonia. Both of these systems are widely used for commercial air conditioning in very large units (hundreds of tons of refrigeration, where a ton of refrigeration represents heat removal at the rate of 12,000 Btu/hr). These systems employ steam at 250°–350°F (115°–167°C) as the heat supply; such temperatures are readily available as waste heat from steam plants, diesel engines, or gas turbines. However, these high temperatures cannot yet be obtained from fixed flatplate collectors with acceptable efficiency, and the concentrating collectors which can produce steam at suitable temperatures were still in the development stage in 1975.

Water-ammonia system. Water-ammonia absorption systems for solar cooling have been studied for many years. A tank full of cold water has the ability to absorb large quantities of ammonia vapor, and the evaporation of some of the ammonia produces temperatures low enough to chill brine well below the freezing point of water (32°F or 0°C). The water, having absorbed its quota of ammonia, is then pumped into a second tank, called the generator, where it is heated, preferably by energy derived from solar collectors, thus driving the ammonia out of solution. Cool water, usually provided by an evaporative cooling tower, must be used to condense the gaseous ammonia back to liquid form, and also to cool the water so that it can again serve as an effective absorbant. The temperature in the heating portion of the system can be as low as 160°F (67°C), as demonstrated by tests run over a long period of time. Well water, returned to the Earth after it has been used, is employed for cooling the condenser in these tests, but cooling towers are more widely used as heat sinks to dispose of both the heat absorbed by the cooling system and that put into the cycle from the solar collectors.

The COP of this system is usually considerably below 1.0, but the advantage of the water-ammonia system over the more efficient compression system is the relatively small amount of pumping power which is required. Most city building codes require that the ammonia portion of the system be located outside of the building because of the toxicity of NH_3.

Lithium bromide–water system. The most widely used solar cooling systems have lithium bromide as the absorbant and the water as the refrigerant. A cool (85–90°F or 28–30.5°C) concentrated solution of lithium bromide can absorb large quantities of water, and the evaporation of some of the water from the chiller tank cools the remainder down to 45–50°F (68–94°C), which is adequate for air-conditioning purposes. As the lithium bromide solution becomes saturated with the absorbed water vapor, it is pumped to the generator where it is heated to 190–210°F (83–94°C) to drive off the water vapor. This passes into an air-cooled or water-cooled condenser, and the liquid water is returned through an expansion valve to the chiller tank. The rich lithium bromide solution must also be cooled, and so relatively cool water or air must also be available for this phase of the process.

The COP of lithium bromide–water systems operating between a generator temperature of 195°F (86°C) and a condenser temperature of 85°F (28°C) will generally not exceed 0.6, which means that, to produce 5 tons of refrigeration (heat removal of 60,000 Btu hr), at least 100,000 Btu of solar heat must be collected at temperatures approaching 200°F (88°C). This requires a large area of high-performance collector surface, as well as a means either of storing excess heat collected during the day, or of producing more cold water than is needed when the solar input is at its maximum, thus providing cold-side storage for use at night.

The advantage of solar-activated absorption cooling lies in the fact that the process has been shown to be technically feasible. The disadvantages are the high cost of the large area of high-performance collector which is required and the large amount of auxiliary heat which must be provided on hot, cloudy days. *See* SOLAR COLLECTORS; SOLAR ENERGY.

[JOHN I. YELLOTT]

Bibliography: H. R. Hay and J. I. Yellott, Natural air conditioning with roof ponds and movable insulation, *ASHRAE Trans.* vol. 75, pt. 1, p. 165, 1969; J. I. Yellott, Solar energy utilization for heating and cooling, *ASHRAE Handbook of Applications*, ch. 59, 1974.

Solar energy

The energy transmitted from the Sun. This energy is in the form of electromagnetic radiation. The Earth receives about one-half of one-billionth of the total solar energy output. In 1971, based on radiation measurements in space, the National Aeronautics and Space Administration proposed a new space solar constant of 1353 watts per square meter (W/m²), and a standard spectral irradiance in W/m² over a small range of wavelengths (bandwidth) centered at the wavelength (in millionths of meters, or μm) shown in Fig. 1. Accordingly, the solar radiation energy in the ultraviolet is 105.8 W/m² (7.82% of the solar constant), in the

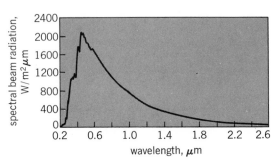

Fig. 1.   NASA standard spectral irradiance at 1 astronomical unit (AU) and a solar constant of 1353 W/m². (*From Solar Electromagnetic Radiation, NASA, SP-8005, May 1971*)

visible 640.4 W/m² (47.33%), and in the infrared 606.8 W/m² (44.85%). The solar radiation energy output is essentially constant. However, because of the ellipticity of the Earth's orbit, the solar constant varies between 1398 W/m² at the winter solstice and 1308 W/m² at the summer solstice, or 3% about the mean value. Based on its cross-sectional area, the rotating Earth receives therefore $751 \cdot 10^{15}$ kWhr annually.

The following are the most frequently used metric units for the radiative input area: langley (1 l = 1 cal/cm² = 0.001163 Whr/cm² = 4.186 J/cm²); calorie (1 cal/cm² min = 0.0697 W/cm² = 1 l/min = 0.00418 J/cm² min); kilowatt-hour (1 kWhr/m² = 860,000 cal/m² = 86.2 l = 3.6 · 10⁶ J/m²); and joule (1 J = 0.239 cal = 2.78 · 10⁻⁴ Whr = 0.239 l-cm²).

Actually, passage through the atmosphere splits the radiation reaching the surface into a direct and a diffuse component, and reduces the total energy through selective absorption by dry air molecules, dust, water molecules, and thin cloud layers, while heavy cloud coverage eliminates all but the diffuse radiation. Figure 2 specifies the conditions for a surface perpendicular under a clear sky at mid- and low latitudes within 4 hr on either side of high noon. If these conditions were to prevail for 12 hr each day of the year (4383 hr), the energy received would lie between some 4200 and 5200 kWhr/m² yr. Actually, the number of sunshine hours even in high-insolation areas ranges from 78 to 89% of the possible, resulting in a reduction in radiative energy ("solar crude") received to the values shown in Fig. 2. Since atmospheric absorption and scattering increase strongly at low solar elevation, the average solar crude received in the most favorable areas by a horizontal surface is about 2550 kWhr/m² yr or 2.55 terawatt-hours per square kilo-

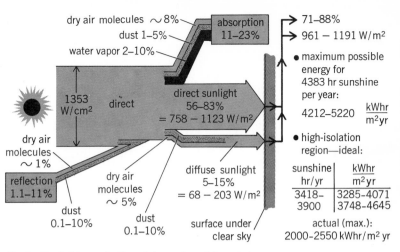

Fig. 2. Sunlight penetration of atmosphere (clear sky).

meter per year (TWhr/km² yr). Figure 3 shows the global distribution of the average solar radiation energy incident on a horizontal surface. By following the Sun's diurnal and seasonal motion, thus facing it from near-sunrise to sunset, an instrument such as a heliostat can attain values between 3 and 3.5 TWhr/km² yr.

On a global basis, about 50% of the total incident radiation of $751 \cdot 10^{15}$ kWhr/yr is reflected back into space by clouds, 15% by the surface, and about 5.3% is absorbed by bare soil. Of the remaining 29.7%, only about 1.7% ($3.79 \cdot 10^{15}$ kWhr/yr) is absorbed by marine vegetation and 0.2% ($4.46 \cdot 10^{14}$ kWhr/yr) by land vegetation. By far the largest portion is used to evaporate water and lift it into the atmosphere. The evaporation

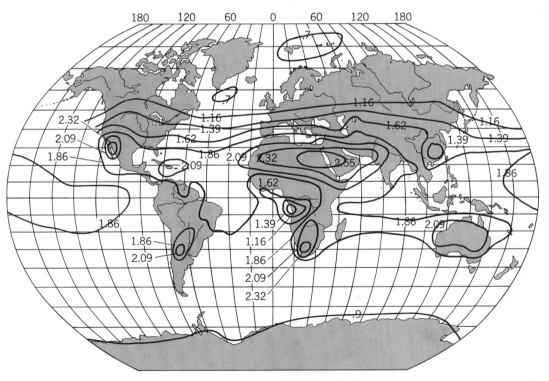

Fig. 3. Global distribution of the average annual solar radiation energy incident on a horizontal surface at the ground. Measurement units are terawatt-hours per square kilometer per year.

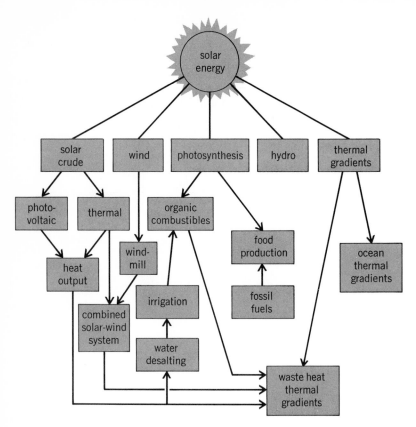

Fig. 4. Basic energy systems in solar power generation.

In 1974, the U.S. Congress established the Energy Research and Development Agency (ERDA) and charged it with the development of energy conservation techniques and new technologies for extracting less readily accessible fossil energy reserves (such as shale oil) and for broadening the use of coal (coal gasification and liquefaction); with the advancement of nuclear (fission) technology; and with spearheading new energy technologies, geothermal, fusion, and all forms of solar energy. The basic energy systems associated with solar power generation are surveyed in Fig. 4.

Waterpower. Of the world power consumption, only about 2% is derived from waterpower, while contributions by wind power are negligible. In Europe, about 23% of available waterpower is utilized, in North America about 22%. This contrasts sharply with the lower utilization level of waterpower in other parts of the globe.

Waterpower is continuously resupplied by the Sun. Its use does not diminish a given reservoir as in the case of fossil fuels (oil, coal, gas). Waterpower is concentrated solar power and is more regularly available. It can be readily regulated and stored in the form of reservoirs. Today hydroelectric conversion efficiency from waterpower to electric power approaches 80%, compared to about 33% conversion efficiency from coal or oil. Utilization of waterpower avoids air pollution.

These are compelling reasons for increasing utilization of the world's waterpower. It is entirely within man's grasp to raise the hydroelectric power supply by a factor of 10, to some 6,400,000,000 kWhr, by the year 2000, benefiting primarily the developing areas. The main obstacle is the availability of investment capital to build the large installations required and, in some cases, the transmission facilities to distant load centers.

The development of waterpower is being advanced particularly in the Soviet Union and the People's Republic of China. Large unused waterpower resources exist still in New Guinea, Africa, South America, and Greenland. *See* WATERPOWER.

Another form of waterpower generation is through utilization of the thermal gradient in oceans, which serve as the storage system for a vast amount of solar energy. The temperature difference between surface and bottom waters is a potentially very large source of electric power. Tropical regions, particularly between 10°N and 10°S latitude, are especially suitable, because relatively high surface temperatures provide a larger temperature gradient. Temperature differences of at least 20–23°C are available between surface and depth throughout the year, wind speeds are moderate (25 knots or less; no hurricanes), and currents are below 1 knot at all depths. One method (Fig. 5) to extract energy is to heat a suitable working fluid (for example, propane or ammonia) which is evaporated in the warm surface water and ducted to an underwater turbine, where it is allowed to expand, driving a turbogenerator system, and to condense in cool depth-waters. From there it is returned to the surface, and the process is repeated (closed Rankine cycle).

Wind power. Wind is the next largest solar-derivative power, after solar crude itself and the energy contained in the oceans. Its utilization is environmentally even more benign than fresh

energy is radiated into space by vapor condensation to clouds. The solar energy spent to lift the water can be partly recovered in the form of waterpower (hydraulic energy). Solar energy can be utilized in the form of heat, organic chemical energy through photosynthesis, and wind power, and also in the form of photovoltaic power (generating electricity by means of solar cells). The two greatest problems in utilizing solar energy are its low concentration and its irregular availability due to the diurnal cycle and to seasonal and climatic variations. Improved technology and investment capital are also needed.

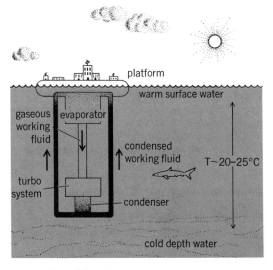

Fig. 5. Closed Rankine cycle, ocean thermal energy conversion system.

waterpower, because no dams and land floodings are involved.

The bulk of wind power, which lies in the upper troposphere and the lower stratosphere, is not accessible to present-day technological potential. However, there are large areas with moderate to strong surface winds in the United States, particularly along the Aleutian chain, through the Great Plains, and along portions of the East and West Coasts. This is shown in Fig. 6 for the contiguous 48 states. A study conducted at Oklahoma State University showed that the average wind energy in the Oklahoma City area is about 0.2 kW/m² (18.5 W/ft²) of area perpendicular to the wind direction. This is roughly equivalent to the solar energy received by the same area, averaging the sunlight over 24 hr per day, all seasons, and all weather conditions. However, in contrast to solar energy, the wind energy could be converted at an efficiency of some 40%. *See* WIND POWER.

Agricultural utilization. The basis of the biological utilization of solar energy is the process called photosynthesis, in which solar energy provides the power within plants to convert carbon dioxide (CO_2) and water (H_2O) into sugars (carbohydrates) and oxygen. The prime conversion mechanism is the chlorophyll molecule. Organic-chemical solar energy conversion operates at very low efficiency of 0.1–0.2%; that is, for every light quantum used, 1000–500 quanta are reflected by the vegetation. However, research on algae, especially the alga *Chlorella*, has shown that higher efficiencies can be achieved. On the basis of extensive experimentation, the practically achievable yield of "chlorella farms," using sunlight as the energy source, has been estimated to be at least of the order of 35 tons of dry algae per acre. This corresponds to about 0.6% efficiency and compares very favorably even with the highest agricultural yields (10–15 tons per acre), let alone the much

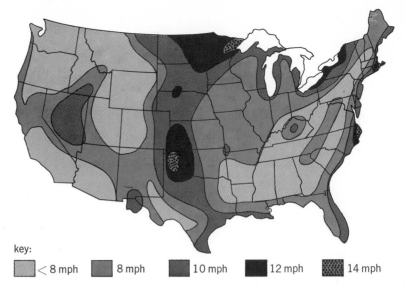

key:
< 8 mph   8 mph   10 mph   12 mph   14 mph

Fig. 6. Average surface wind velocities in high-wind regions of the 48 contiguous states.

lower yields in less developed countries (2–2.5 tons per acre). By building large algae farms on nonarable ground, a growing portion of the solar energy presently absorbed by bare soil, that is, about 5.3% or 400×10^{14} kWhr per annum, could be utilized for the production of organic matter for food and for conversion into synthetic liquid fuels as a complement to the world's oil supply. Again, capital and local or regional requirements determine the feasibility of such endeavors.

Industrial utilization. In Japan, the United States, the southern Soviet Union, and other countries, solar energy is utilized for drying of fruits and vegetables by means of solar-heated air. Water evaporation in solar stills is an attractive

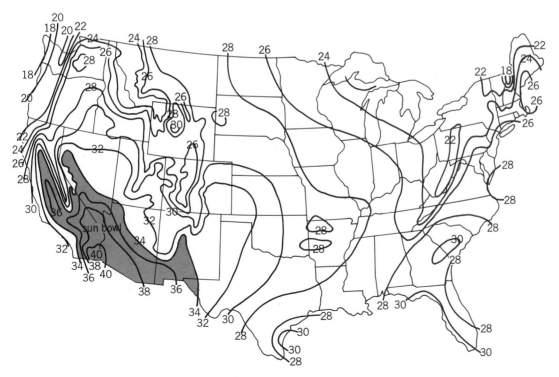

Fig. 7. National climatic center annual sunshine hours with "sun bowl" accented; values in hundreds.

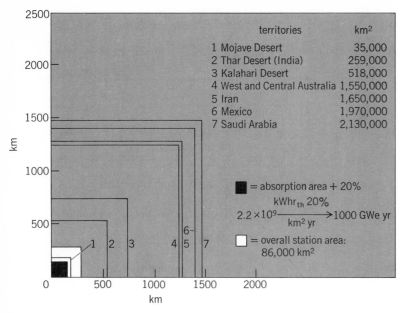

Fig. 8. Comparison of a 1000 GWe-yr medium-temperature solar power complex with high-insolation territories (1 GWe-yr = 8766 × 10⁶ kWhr).

territories	km²
1 Mojave Desert	35,000
2 Thar Desert (India)	259,000
3 Kalahari Desert	518,000
4 West and Central Australia	1,550,000
5 Iran	1,650,000
6 Mexico	1,970,000
7 Saudi Arabia	2,130,000

For solar-electric power plants, the annual sunshine hours should be as large as possible, and the humidity, which causes absorption and scattering, should be low. With its "sun bowl" (Fig. 7), the United States is the only major industrial country with high-insolation (2–2.5 TWhr/km² yr) territory within its borders; the highest values lie around 2.2 TWhr/km² yr (Fig. 3). The table shows that if in such high-insolation area, the incident solar radiation over 1000 km² is converted to electric energy at only 20% efficiency, the output is equivalent to the annual consumption of 734×10^6 bbl of crude oil or 161×10^6 metric tons of coal.

At a received solar energy level of 2.2×10^9 kWhr/km² yr and 20% conversion efficiency to electricity, a 100-GWe-yr plant requires 20,000 km² net collector area. With an arrangement of the linear parabolic reflectors such that they can follow the Sun from 30° elevation when rising to 30° elevation when setting, the total occupied area is larger. Adding 20% reserve, the collector area is 24,000 km², or about 28% of the overall occupied area of 86,000 km². Such a system would be highly modularized and would require an area that is small (Fig. 8). At 90% thermal efficiency and 40% electric efficiency, that is, a total efficiency of 36%, the required collector area is reduced from 24,000 to 13,300 km² and the overall occupied area is about 48,000 km². The desert area west of Phoenix, AZ, alone covers some 70,000 km² and is only a fraction of the overall available land in the Southwest.

The energy budget of a solar station is compared with that of average desert ground in Fig. 9. Of the solar irradiation, the desert surface reflects about 40% and retains 60%, whereas the absorber area retains about 90%. Thus, 30% more of the incident solar energy is trapped. Of this, about two-thirds appears as electric energy and one-third as true, that is, extrinsic, heat production at 20% overall conversion. At 30% overall conversion from solar crude to electricity, no extrinsic heat would be generated. At still higher conversion efficiency, the area occupied by the power station would be cooled, compared to the natural environment, rather than slightly heated as with the 20% system.

The overall unused thermal energy is concentrated in the power station's thermal storage system and ultimately in its electric conversion system. If the electric conversion system uses cooling water rather than air cooling, advantage can be taken of the large amount of thermal energy concentrated in the water. For this, two alternatives are available: desalination, and power generation by temperature-gradient utilization.

Three methods can be used to generate solar-electric power: (1) the solar-thermal distributed receiver system (STDS); (2) the solar-thermal central receiver system (STCRS); and (3) the photovoltaic system (PVS). In each case, the overall system may serve as backup, that is, operating only when the sun shines, equipped only with a minor energy storage capacity (for example, for 1 hr full output) to bridge temporary cloud coverage; alternatively, the overall system may include a conventional fuel system to replace solar energy at night or during cloudy days; and finally, the independent solar-electric system includes a storage system to ensure continuous power-generating capacity based on solar energy only.

In the STDS, solar radiation is absorbed over a

method only where fresh water is extremely expensive, as in isolated arid regions with ready access to brackish or sea water, or under special conditions such as emergency provisions for downed flyers or astronauts, or for shipwrecked sailors. Chile, Greece, Australia, and Israel are operating and developing solar stills. Outputs may run as high as 0.15 gal of fresh water per square foot per day. Generally, it is of the order of 0.1 to 0.12 gal per square foot per day.

The most important direct use of solar radiation can be divided into two basic categories: solar heating and cooling for residential and commercial buildings, and electric power generation. As of 1975–1976, the four principal systems for buildings—water heating, space heating, space cooling, and combined system—were at widely different stages of development. Only water-heating systems had reached commercial readiness. *See* SOLAR COOLING; SOLAR HEATING.

Comparison of energy sources in terms of electric energy output at the bus-bar

Energy source	at electric conversion efficiency of	yields the following electric energy in kilowatt-years*
1000 metric tons of oil	35%	490 = 1.0
1000 tons of coal†	35%	312 = 0.637
1000 tons of enriched uranium in light-water reactor	31.7%	28,500,000 = 58,160
1000 tons of plutonium in liquid-metal fast-breeder reactor	40.4%	36,000,000 = 73,470
1000 km² solar absorber area‡ (solar crude: 2.2 × 10⁹ kWhr/km² yr)	20%	50,300,000§ = 102,700

*The numbers at the right are indices, based on the yield of 1000 tons of oil as unity (1.0). For example, 1000 km² solar absorber area under specified conditions yields 102,700 times as much electric energy as 1000 tons of oil, and 1000 tons of coal yields 63.7% of the energy from 1000 tons of oil.

†Based on mean heating value of 26×10^6 Btu/ton.

‡Since 1000 km² solar crude yields the equivalent of 102.7×10^6 tons of oil and since a ton of oil corresponds to 7.14 British barrels, the yield of the 1000 km² corresponds to $102.7 \times 10^6 \times 7.14 = 734 \times 10^6$ bbl. Coal yield equivalent per 1000 km² solar follows from 102,700/0.637 = 161,000 per 1000 tons of coal, or the yield of 1000 km² corresponds to the yield of 161×10^6 tons of coal.

§United States production in 1969 = 177.1×10^6 kW-yr.

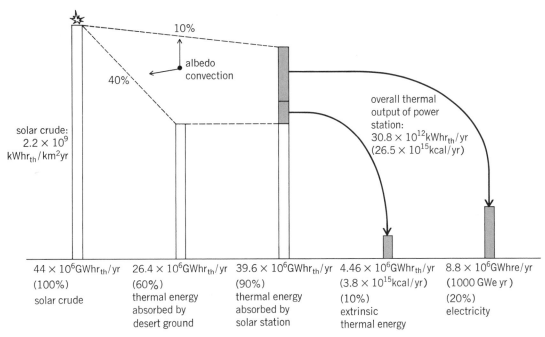

solar crude:
2.2 × 10⁹
$kWhr_{th}/km^2yr$

10%

albedo
convection

40%

overall thermal
output of power
station:
30.8 × 10¹² $kWhr_{th}$/yr
(26.5 × 10¹⁵ kcal/yr)

44 × 10⁶ $GWhr_{th}$/yr	26.4 × 10⁶ $GWhr_{th}$/yr	39.6 × 10⁶ $GWhr_{th}$/yr	4.46 × 10⁶ $GWhr_{th}$/yr	8.8 × 10⁶ GWhre/yr
(100%)	(60%)	(90%)	(3.8 × 10¹⁵ kcal/yr)	(1000 GWe yr)
solar crude	thermal energy absorbed by desert ground	thermal energy absorbed by solar station	(10%) extrinsic thermal energy	(20%) electricity

Fig. 9.   Typical energy budget of a 1000 GWe-yr solar power station.

large area covered with flat-plate (nonconcentrating) collectors or parabolic-trough concentrators, which focus the sunlight on a heat pipe carrying the working fluid (Fig. 10). The heat pipes are covered with selective coating characterized by high absorptivity to solar radiation and low emissivity at the temperature of the heat pipe. The flat-plate collector operates at turbine inlet temperatures of 250–500°F (121–260°C). With the parabolic trough, temperatures between 550–1000°F (288–538°C) at turbine inlet are attainable. The higher the temperature, the higher the efficiency and the smaller the land area needed for a given power level, but the more expensive is the system, especially the piping and the coating.

In the STCRS, sunlight is concentrated on a receiver by a large number of mirrors designed to follow the Sun (heliostats). The receiver is a heater located atop a tower (Fig. 11) which is served by a certain collector area. The power output of such a module depends on the collector array area which, in turn, determines the tower height. The horizontal distance of the farthest mirror from the foot of the tower is about twice the tower height, which may range from 250 to 500 m. A given plant may have an arbitrary number of these modules.

Both the STDS and STCRS require large areas, due to the nature of the energy source. It is possible, of course, to subdivide the STDS into small modules, each with a small standardized electric power station, and to collect the electric current in an overall power-conditioning station. In the STCRS, the working fluid can be heated to higher temperatures, since the collector area acts as a giant parabolic mirror consisting of individual facets (heliostats). The higher temperatures attainable with the STCRS yield higher overall efficiencies (25–35%), or 150,000–200,000 kWe per square kilometer of *collector* area (not total area covered), compared to about 1.5 km² for the same power output in an STDS. The working fluid can be water (steam), sodium, or helium (in the order of

increasing system temperature). In the simplest case, superheated steam is generated in a heater atop the tower. The steam is ducted to the ground and used in a high-pressure and low-pressure turbogenerator system.

To provide power during off-radiation hours, part of the energy generated during sunshine hours must be stored. This means that part of a solar power module's (SPM) power output, or the output of entire modules, is not available during sunshine hours. Several options are available for storage: the energy can be stored as heat, as mechanical energy (pumping water to elevated storage basins), or as chemical energy (for example, electrolytic decomposition of water to hydrogen and oxygen). It may be possible to design the central receiver towers so that wind generators can be attached, causing the towers to generate power in the absence of sunshine as well. Figure 12 compares the cost effects of several storage modes in terms of magnitude of storage capacity (hours of full power output) and type of storage.

In the photovoltaic system (PVS), solar cells are used to produce the necessary electric energy. They are spread over a large area, as in a distributed system, to collect sunlight for the desired electric output which is in direct current. This is advantageous if long-distance transmission is

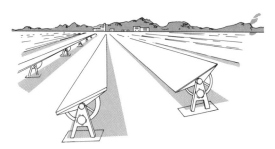

Fig. 10.   Solar-thermal distributed power station system.

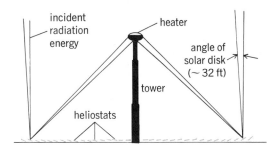

Fig. 11. Solar-thermal central receiver system.

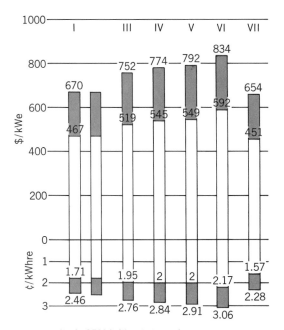

basic SPM (without storage):
area 2.6 km²
collector area 1.3 km²
number of towers 2
sunshine hours per year/day = 3800/10.4
concentration and cycle efficiency = 0.3
oil equivalent (0.35 cycle efficiency) annually:
    182,000 tons
    1.3 × 10⁶ bbl

I	basic SPM	
II	SPM with thermal storage capacity	– 1 hr
III		– 5.2 hr
IV		– 10.4 hr
V	SPM with flywheel storage capacity	– 5.2 hr
VI		– 10.4 hr
VII	SPM with wind power storage capacity	– 1 hr

▨ uncertainty range

Fig. 12. Solar power module (SPM), performance data, and cost summary for different energy storage types and capacities. (*K. A. Ehricke*).

involved, since losses are lower than in transmission of alternating current. For most uses, however, the dc must be converted into ac, and if the output is fed into existing distribution grids, dc-to-ac conversion equipment must be available at the plant.

The principal advantage of the PVS over the thermal systems is the absence of moving parts

and fluids at high temperatures. No cooling is needed and probably no Sun-tracking. Its major disadvantages are its comparatively low efficiency (probably below 20% even after extensive development) and the fact that thermal storage cannot be used. Considerable development is required before the system can be economically competitive with the thermal systems.

Solar energy and space. Extensive use of solar energy can be anticipated in spacecraft and by extraterrestrial communities in orbit or on the Moon. Solar cells are the primary power supply for unmanned spacecraft. They are the most attractive source of electric power for manned space stations. In the 1970s and 1980s, solar cells will power tiny electric thrust units, propelling unmanned probes into the asteroid belt and on other far-flung missions in the solar system. On the Moon, undiminished solar energy can be soaked up by solar cells and solar concentrators during 14 days for immediate use, for storage in large banks of rechargeable batteries, and in fuel cells to be used during the long lunar night.

Another important relation between space and solar energy is provided by weather satellites. With improving long-term weather prediction, the practical aspects of utilizing solar energy on Earth will improve also. Advancements in the industrial utilization of space beyond the application satellites will make it possible to intercept large amounts of solar radiation in space for use on Earth or in extraterrestrial production facilities. For terrestrial applications, sunlight can be transmitted optically by reflectors (Space Light), or it can be converted to electric energy in space and transmitted to Earth via microwave beam (power generation satellites, or PGS). Moreover, electric energy which is generated at one location on Earth can be transmitted by means of microwave beam to a distant load center via power relay satellite (PRS).

The general objective of Space Light is to transmit sunlight to selected areas of the Earth's night side. The key objectives can be divided into three categories: low light level for night illumination (Lunetta satellite); strong light (up to half the Sun's brightness) for stimulating food growth by enhancing photosynthetic production (photosynthetic production enhancement, or PSPE, Soletta); and light up to about 70% of the Sun's brightness as a "night sun" to provide around-the-clock solar energy for industrial purposes in a selected area of about 90,000 km².

Lunetta is designed to illuminate with the brightness of 10 to 100 full moons on a clear night (about ⅙ of that on a cloudy night), where one full moon nominally corresponds to 1/400,000 of the Sun's brightness on a clear day. In agriculture Lunetta provides the necessary brightness for sowing and harvesting at night. Lunetta's light can also accelerate large construction projects in remote areas or during long polar nights. The Lunetta can be economically placed in a 4-hr inclined orbit (about 1 earth-radius altitude) and still illuminate a fixed area (minimum of 2800 km²) if the reflector is rotated appropriately. To ensure an 8-hr illuminative period per night for a given area, six Lunettas, totaling 4.2 km² in area, are required.

The PSPE Soletta provides daylight extension (dusk and dawn illumination), especially in northern regions, and brief nighttime illumination. In a 4-hr orbit, a PSPE Soletta of a nominal 40% of full solar brightness would consist of a "swarm" of reflectors of a total area of 1040 km², all trained on the same focal area.

The night sun Soletta is in a fixed position with respect to the focal area, which preferably is located within 30° latitude on either side of the Equator. This means that the Soletta stays near the zenith of the irradiated area and does not rise or set like the Sun. Consequently, even at about 60% solar brightness the overall energy delivered per night by the Soletta is about equal to the energy delivered by the Sun per day (at equal sky conditions). The overall reflecting area of the Soletta swarm, therefore, is about 54,000 km². A solar power station in this area could operate around the clock, as if in space.

PRS and PGS are part of an overall system in which the primary electric-power station is located either on the ground or in space. The electricity is converted to microwave energy which is shaped to a controlled beam by a transmitter array antenna, which in turn is trained on a large receiver array where the microwave energy is reconverted to dc electricity by rectifiers. In the absence of optical line-of-sight connection between the transmitter and the receiver, a relay is required to redirect the beam.

Many costly technological problems must be solved before either optical- or microwave-beam systems can be realized. In addition, research on possible environmental effects of large numbers of microwave beams, each carrying millions of kilowatts, must be completed and evaluated for added technological requirements.

[KRAFFT A. EHRICKE]

Bibliography: F. Daniels, *Direct Use of the Sun's Energy*, 1965; J. A. Duffie and W. A. Beckman, *Solar Energy Processes*, 1974; K. A. Ehricke, *Space Industrial Productivity: New Options for the Future*, Future Space Programs, Hearings of House Subcommittee on Space Science and Applications, 1975; K. A. Ehricke, *The Power Relay Satellite*, Rockwell International Rep. E74-3-1, 1974; J. A. Eibling, R. E. Thomas, and B. A. Landry, *An Investigation of Multiple-Effect Evaporation of Saline Waters by Steam for Solar Radiation*, Report from Battelle Memorial Institute to U.S. Department of Interior, December 1953; F. C. Fuglister, *Atlantic Ocean Atlas of Temperature and Salinity Profiles and Data from the International Geophysical Year of 1957–58*, Woods Hole Oceanographic Institute, 1960; P. E. Glaser, The satellite solar power station, *Proc. IEEE: International Microwave Symposium*, 1973; D. S. Halacy, Jr., *The Coming Age of Solar Energy*, 1964; National Aeronautics and Space Administration, *Feasibility Study of a Satellite Solar Power Station System*, NASA CR-2357, 1974; National Aeronautics and Space Administration, *Solar Electromagnetic Radiation*, NASA SP-8005, May 1971; N. Robinson, *Solar Radiation*, 1966; U.S. Department of Health, Education and Welfare, *Proceedings of a Symposium on Biological Effects and Health Implications of Microwave Radiation*, Pub. Health Serv. Rep. BRH/DBE 70, 1969; C. Zener, Solar sea power, *Phys. Today*, 1973.

Solar heating

Conversion of solar radiation into heat for technological purposes. Energy from the Sun reaches the Earth as electromagnetic radiation in the wavelength band 0.3–3.0 μm. In theory, all this energy can be absorbed by suitable surfaces and converted into heat ranging from 70 to 6000°F (21 to 3871°C). In practice, the efficiency of conversion depends largely upon the difference in temperature between the absorbing surface and the surrounding environment. *See* SOLAR ENERGY.

Solar distillation. The oldest technological application of solar heating is the distillation of sea water to produce salt or potable water. In the production of salt, sea water is confined in shallow ponds where the heat produced by absorption of solar radiation evaporates all of the water, leaving a residue of sodium chloride and other salts.

To produce potable water from contaminated or salt-laden sources, a shallow blackened box is partially filled with the available water and provided with a roof-shaped transparent cover made of glass or certain plastic films (Fig. 1). About 90%

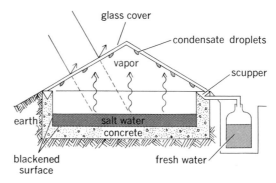

Fig. 1. Roof-type solar still.

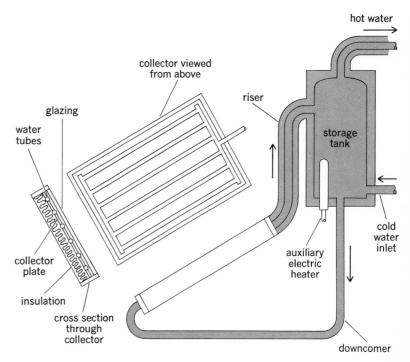

Fig. 2. Thermosyphon solar water heater.

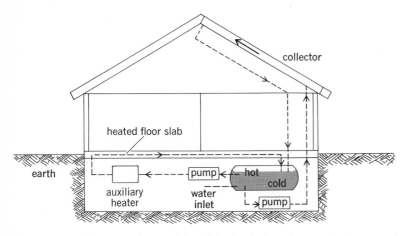

Fig. 3. Conventional solar house with roof heat collectors, basement storage, and radiant heating.

of the incident solar radiation is transmitted into the still where the Sun's rays are absorbed and the water is heated. The transparent cover is essentially opaque to the long-wave radiation emitted from the warm water, and so most of the solar heat is trapped within the still. Some of the water is evaporated and the resulting vapor condenses on the undersurface of the cover. The resulting droplets of condensate trickle down into scuppers which lead to a fresh-water storage vessel. In a clear, sunny climate, the average production of such stills can reach 1 gal of fresh water per day for each 10 ft² of irradiated area. Many stills of this type are now in use in the arid Greek islands in the eastern Mediterranean and in the desert regions of central Australia.

Domestic hot-water heaters. Hot water for domestic purposes can be provided at low cost by the thermosyphon solar heater (Fig. 2). The heat collector is a flat plate of blackened metal (steel, aluminum, or copper), to which tubes are attached in any manner which provides good thermal contact. The tubes are connected to headers at the top and bottom of the collector plate, which is insulated to minimize loss of heat from the rear surface. The collector is glazed with a glass or plastic cover to trap the Sun's heat and to reduce heat loss to the atmosphere.

An elevated storage tank is connected to the

bottom header by a downcomer pipe, while a riser runs from the upper header to the top of the tank. A cold-water line is attached to the bottom of the tank, and the hot-water outlet comes from the tank top. When the Sun shines on the collector, much of the radiation is absorbed and water in the tubes is heated. The hot water rises into the tank, to be replaced by cold water, and this process continues as long as the collector plate is irradiated. Water temperatures up to 180°F (81°C) can be attained in summer, and on clear winter days in sunny areas, a tankful of water can be had at 120°F (49°C). The average production rate under favorable circumstances can exceed 1 gal of hot water per square foot of collector surface. Thousands of thermosyphon water heaters are now in use in Israel, North Africa, Japan, and Australia.

Solar house heating. Under favorable conditions, the south-facing roof area (in the Northern Hemisphere) of a house receives enough solar radiation during the sunlit hours of a typical winter day to supply the heat requirements for the entire 24-hr period. To make use of this heat, the roof must be covered with a suitable solar radiation absorber, and a flow medium, usually either air or water, must be used to convey the absorbed heat to a storage device from which it can be withdrawn at night.

In the system shown in Fig. 3, solar water heaters similar in principle to that shown in Fig. 2 are mounted on the south-facing roof of a dwelling, and the heated water is stored in an insulated tank in the basement. When heat is required at night, this hot water is used in a conventional heating system, to which a fuel-burning or electric heater is connected for use during prolonged periods of cold, cloudy weather. *See* HEATING, COMFORT; HOT-WATER HEATING SYSTEM.

A development in solar house heating which also incorporates natural cooling is shown in Fig. 4. The roof of this house is flat, and the metal ceilings of the rooms support shallow ponds of water. Movable horizontal panels of foamed insulation are provided above these ponds. In the winter, the panels are moved away from the ponds during the daylight hours to allow the water to be warmed by the Sun, and then the panels are slid back over the ponds at night to retain the absorbed heat. The rooms are warmed by heat radiated from their warm ceilings.

In summer, the timing of the panel-moving operation is reversed, since the panels are moved away from the ponds at night to allow the water to be cooled by evaporation and by radiation to the sky. At daybreak, the panels are slid back over the cool water to keep off the rays of the Sun, and comfort is maintained within the house by the effect of the cool ceilings and by circulating some of the water from the roof ponds through small fan-coil units.

The feasibility of this system of natural air conditioning in hot dry areas has been demonstrated by a year-long test conducted in Phoenix, AZ, during 1967–1968 and at Atascadero, CA, during 1974–1975.

Solar cooking. Cooking can be accomplished by the use of several types of concentrating solar collectors. The solar oven (Fig. 5) uses metallic wings to reflect sunshine through a double glass cover into an insulated and blackened box. If the oven is

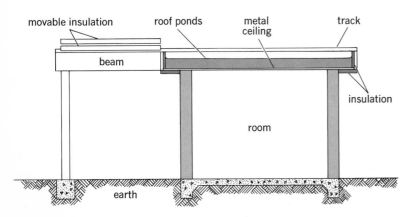

Fig. 4. Naturally air-conditioned house with roof ponds for solar heating in the winter and for cooling in the summer.

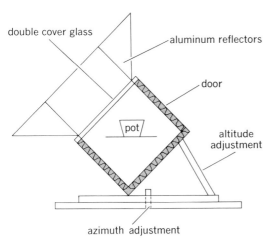

double cover glass

aluminum reflectors

door

altitude adjustment

pot

azimuth adjustment

Fig. 5. Oven-type solar cooker.

kept pointed toward the Sun during the midday hours, the interior temperature can readily attain 350–400°F (176–204°C), which is adequate for most baking operations. Higher temperatures can be reached by the reflector cooker shown in Fig. 6, in which the paraboloidal reflector concentrates the reflected solar radiation upon the bottom and sides of the cooking pot. Both types of solar cooker are feasible for cooking noon meals in sunny locations, but cost and sociological considerations have kept them from making any real impact in those parts of the world where lack of fuel and abundance of sunshine indicate that they should be most useful.

Solar power. Both steam and hot-air engines can be operated by solar energy through the use of suitable concentrators and heat absorbers. John Ericsson, designer of the *Monitor*, operated such engines in New York shortly after the Civil War, and very large solar steam plants were built in California, Arizona, and Egypt between 1900 and 1914. None of these engines survived, because the advent of small, inexpensive, and reliable gasoline engines and electric motors rendered them obsolete.

Interest in heliothermal power plants has been revived by the space program and its needs for increasingly large solar power-generating systems. In space, the supply of solar radiation is markedly greater and more reliable than on the surface of the Earth, and there is no inexpensive alternative source of power.

The power systems which are currently in the design stage for use on Earth employ steam generators mounted on towers 200–400 ft high. Large numbers of slightly curved heliostats concentrate the Sun's rays into the boiler cavity to produce steam at 1000 psi superheated to 1000°F (538°C). Air-cooled condensers are contemplated to eliminate the need for cooling water.

[JOHN I. YELLOTT]

Bibliography: AIA Research Corporation, *Solar-Oriented Architecture*, 1975; ASHRAE, *Handbook of Fundamentals*, 1972; ASHRAE, *Utilization*, chap. 59, 1974; Portola Institute, *Energy Primer*, chap. 1, 1975; J. I. Yellott, *Solar Energy Utilization for Heating and Cooling*, National Science Foundation, 1975.

Solar radiation

The electromagnetic radiation and particles (electrons, protons, and rarer heavy atomic nuclei) emitted by the Sun. Electromagnetic energy has been observed over the whole spectrum with wavelengths varying from 0.1 A (10^{-9}cm) to 30 km. The bulk of the energy is in the spectrum of visible light (4000–8000 A). The solar spectrum doubtless extends far beyond the observed limits in both directions. The total power is 3.86×10^{33} ergs/sec.

The Sun also emits a continuous stream of electrons with shorter bursts of electron and proton showers sufficiently intense to affect the ionization of the upper terrestrial atmosphere. These sporadic particles have energies from a few thousand to a few billion electron volts. The lower-energy particles are much more abundant, but those of high energy are sufficient to occasionally damage the solid-state circuitry of spacecraft. The physical mechanism of high-energy particle emission is not understood, but is closely associated with the more energetic forms of solar activity. *See* SOLAR ENERGY.

[JOHN W. EVANS]

Space charge

The net electric charge within a given volume. If both positive and negative charges are present, the space charge represents the excess of the total positive charge diffused through the volume in question over the total negative charge. Since electric field lines end on electric charge, the space-charge density ρ may also be defined in terms of the divergence of the electric field $\mathbf{E}$ or the Laplacian of the electric potential V by Eq. (1)

$$-\frac{4\pi\rho}{\epsilon} = -\text{div } \mathbf{E} = \nabla^2 V = \frac{\partial^2 V}{\partial x^2} + \frac{\partial^2 V}{\partial y^2} + \frac{\partial^2 V}{\partial z^2} \quad (1)$$

(Poisson's equation). Here ϵ is the dielectric constant of the medium and x, y, and z are rectangular coordinates defining the position of a point in space. If, under the influence of an applied field, the charge carriers acquire a drift velocity v, the space charge becomes j/v, where j is the current density. For current carried by both positive and negative carriers, such as positive ions and electrons, the space charge density is given by Eq. (2).

$$\rho = j_+/v_+ - j_-/v_- \quad (2)$$

Here the subscripts $+$ and $-$ indicate the current density and drift velocity for the positive and negative carriers, respectively. Thus a relatively small current of slow-moving positive ions can neutralize the space charge of a much larger current of high-velocity electrons.

[EDWARD G. RAMBERG]

Steam engine

A machine for converting the heat energy in steam to mechanical energy of a moving mechanism, for example, a shaft. The steam engine dominated the industrial revolution and made available a practical source of power for application to stationary or transportation services. The steam power plant could be placed almost anywhere, whereas other means of power generation were more restricted, experiencing such site limitations as an elevated water supply, wind, animal labor, and so on. The steam engine can utilize any source of heat in the

SOLAR HEATING

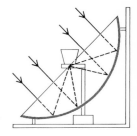

Fig. 6. Reflector-type solar cooker.

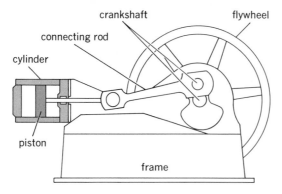

Fig. 1. Principal parts of horizontal steam engine.

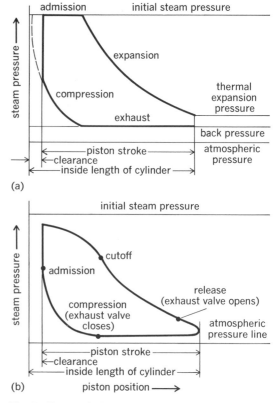

(a)

(b)

Fig. 3. Events during one cycle of a piston operation. (a) In ideal engine. (b) As depicted on the indicator card of a noncondensing steam engine.

form of steam from a boiler. It was developed in sizes which ranged from that of children's toys to 25,000 hp, and it was adaptable to steam pressures up to 200 psi. It reached its zenith in the 19th century in stationary services such as drives for pumping plants; drives for air compressor and refrigeration units; power supply for factory operations with shafting for machine shops, rolling mills, and sawmills; and drives for electric generators as electrical supply systems were perfected. Its adaptability to portable and transportation services rested largely on its development of full-rated torque at any speed from rest to full throttle; its speed variability at the will of the operator; and its reversibility, flexibility, and dependability under the realities of stringent service requirements. These same features favored its use for many stationary services such as rolling mills and mine hoists, but the steam engine's great contribution was in the propulsion of small and large ships, both naval and merchant. Also, in the form of the steam locomotive, the engine made the railroad the practical way of land transport. Most machine elements known today had their origin in the steam engine: cylinders, pistons, piston rings, valves and valve gear crossheads, wrist pins, connecting rods, crankshafts, governors, and reversing gears.

The 20th century saw the practical end of the steam engine. The steam turbine with (1) its high speed (for example, 3600± rpm); (2) its utilization of maximum steam pressures (2000–5000 psi), maximum steam temperatures (1100°±F), and highest vacuum (29± in. Hg); and (3) its large size (1,000,000±kw) led to such favorable weight, bulk, efficiency, and cost features that it replaced the steam engine as the major prime mover for electric generating stations. The internal combustion engine, especially the high-speed automotive types which burn volatile (gasoline) or nonvolatile (diesel) liquid fuel, offers a self-contained, flexible, low-weight, low-bulk power plant with high thermal efficiency that has completely displaced the steam locomotive with the diesel locomotive and marine steam engines with the motorship and motorboat. It is the heart of the automotive industry, which produces in a year 10,000,000± vehicles that are powered by engines smaller than 1000 hp. Because of the steam engine's weight and speed limitations, it was excluded from the aviation field, which has become the exclusive preserve of the internal combustion piston engine or the gas turbine. *See* DIESEL ENGINE; GAS TURBINE; INTERNAL

COMBUSTION ENGINE; STEAM TURBINE; TURBINE.

Cylinder action. A typical steam reciprocating engine consists of a cylinder fitted with a piston (Fig. 1). A connecting rod and crankshaft convert the piston's to-and-fro motion into rotary motion. A flywheel tends to maintain a constant-output angular velocity in the presence of the cyclically changing steam pressure on the piston face. A D slide valve admits high-pressure steam to the cylinder and allows the spent steam to escape (Fig. 2). The power developed by the engine depends upon the pressure and quantity of steam admitted per unit time to the cylinder.

Indicator card. The combined action of valves and piston is most conveniently studied by means of a pressure-volume diagram or indicator card (Fig. 3). The pressure-volume diagram is a thermodynamic analytical method which traces the sequence of phases in the cycle. It may be an idealized operation (Fig. 3a), or it may be an actual picture of the phenomena within the cylinder (Fig. 3b) as obtained with an instrument commonly known as a steam engine indicator. This instrument, in effect, gives a graphic picture of the pressure and volume for all phases of steam admission, cutoff, expansion, release, exhaust, and compression. It is obtained as the engine is running and shows the conditions which prevail at any instant within the cylinder. The indicator is a useful instrument not only for studying thermodynamic performance, but for the equally important operating knowledge of inlet and exhaust valve leakage and losses, piston ring tightness, and timing correctness. *See* THERMODYNAMIC PROCESSES.

STEAM ENGINE

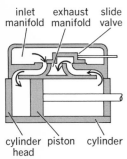

Fig. 2. Single-ported slide valve on counterflow double-acting cylinder.

The net area of the indicator card shows, thermodynamically, the work done in the engine cylinder. By introducing the proper dimensional quantities, power output can be measured. Thus if the net area within the card is divided by the length, the consequent equivalent mean height is the average pressure difference on the piston during the cycle and is generally called the mean effective pressure p, usually expressed in lb/in.² With a cylinder dimension of piston area a (in in.²), length of piston stroke l (in ft), and n equal to the number of cycles completed per minute, the equation given below holds. It has often been referred to as

$$\text{Indicated horsepower} = \frac{plan}{33,000}$$

the most important equation in mechanical engineering.

Engine types. Engines are classified as single- or double-acting, and as horizontal (Fig. 1) or vertical depending on the direction of piston motion. If the steam does not fully expand in one cylinder, it can be exhausted into a second, larger cylinder to expand further and give up a greater part of its initial energy. Thus, an engine can be compounded for double or triple expansion. In counterflow engines, steam enters and leaves at the same end of the cylinder; in uniflow engines, steam enters at the end of the cylinder and exhausts at the middle.

Steam engines can also be classed by functions, and are built to optimize the characteristics most desired in each application. Stationary engines drive electric generators, in which constant speed is important, or pumps and compressors, in which constant torque is important. Governors acting through the valves hold the desired characteristic constant. Marine engines require a high order of safety and dependability.

Valves. The extent to which an actual steam piston engine approaches the performance of an ideal engine depends largely on the effectiveness of its valves. The valves alternately admit steam to the cylinder, seal the cylinder while the steam expands against the piston, and exhaust steam from the cylinder. The many forms of valves can be grouped as sliding valves and lifting valves (Fig. 4). *See* CARNOT CYCLE; THERMODYNAMIC CYCLE.

D valves (Fig. 2) are typical sliding valves where admission and exhaust are combined. A common sliding valve is the rocking Corliss valve; it is driven from an eccentric on the main shaft like other valves but has separate rods for each valve on the engine. After a Corliss valve is opened, a latch automatically disengages the rod and a separate dashpot abruptly closes the valve. Exhaust valves are closed by rods, as are other sliding valves.

Lifting valves are more suitable for use with high-temperature steam. They, too, are of numerous forms. The poppet valve is representative.

Valves are driven through a crank or eccentric on the main crankshaft. The crank angle is set to open the steam port near dead center, when the piston is at its extreme position in the cylinder. The angle between valve crank and connecting-rod crank is slightly greater than 90°, the excess being the angle of advance. So that the valves will open and close quickly, they are driven at high velocity with consequently greater travel than is necessary to open and close the ports. The additional travel

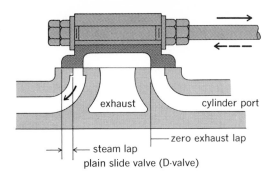

plain slide valve (D-valve)

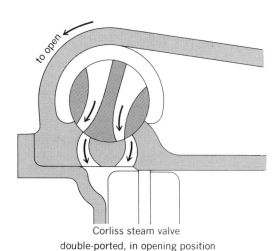

Corliss steam valve
double-ported, in opening position

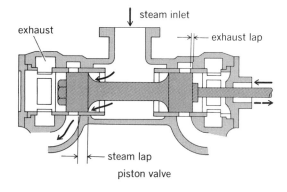

piston valve

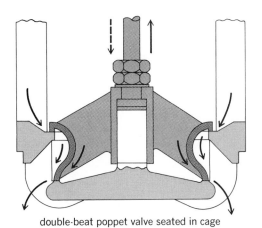

double-beat poppet valve seated in cage

Fig. 4. Steam engine valves in closed positions. The arrows indicate the path steam will travel when the valves are open.

of a sliding valve is the steam lap and the exhaust lap. The greater the lap, the greater the angle of advance to obtain the proper timing of the valve action.

Engine power is usually controlled by varying the period during which steam is admitted. A shifting eccentric accomplishes this function or, in releasing Corliss and in poppet valves, the eccentric is fixed and cutoff is controlled through a governor to the kickoff cams or to the latch that allows the valves to be closed by their dashpots.

For high engine efficiency, the ratio of cylinder volume after expansion to volume before expansion should be high. The volume before expansion into which the steam is admitted is the volumetric clearance. It may be determined by valve design and other structural features. For this reason, valves and ports are located so as not to necessitate excessive volumetric clearance.

[THEODORE BAUMEISTER]

Bibliography: T. Baumeister (ed.), *Standard Handbook for Mechanical Engineers*, 7th ed., 1967.

Steam heating

A heating system that uses steam generated from a boiler. The steam heating system conveys steam through pipes to heat exchangers, such as radiators, convectors, baseboard units, radiant panels, or fan-driven heaters, and returns the resulting condensed water to the boiler. Such systems normally operate at pressure not exceeding 15 pounds per square inch gage (psig), and in many designs the condensed steam returns to the boiler by gravity because of the static head of water in the return piping. With utilization of available operating and safety control devices, these systems can be designed to operate automatically and safely with minimum maintenance and attention.

One-pipe system. In a one-pipe steam heating system, a single main serves the dual purpose of supplying steam to the heat exchanger and conveying condensate from it. Ordinarily, there is but one connection to the radiator or heat exchanger, and this connection serves as both the supply and return; separate supply and return connections are sometimes used. Because steam cannot flow through the piping or into the heat exchanger until all the air is expelled, it is important to provide automatic air-venting valves on all exchangers and at the ends of all mains. These valves may be of a type which closes whenever steam or water comes in contact with the operating element but which also permits air to flow back into the system as the pressure drops. A vacuum valve closes against subatmospheric pressure to prevent return of air.

Two-pipe system. A two-pipe system is provided with two connections from each heat exchanger, and in this system steam and condensate flow in separate mains and branches (Fig. 1). A vapor two-pipe system operates at a few ounces above atmospheric pressure, and in this system a thermostatic trap is located at the discharge connection from the heat exchanger which prevents steam passage, but permits air and condensation to flow into the return piping.

When the steam condensate cannot be returned by gravity to the boiler in a two-pipe system, an alternating return lifting trap, condensate return pump, or vacuum return pump must be used to force the condensate back into the boiler. In a

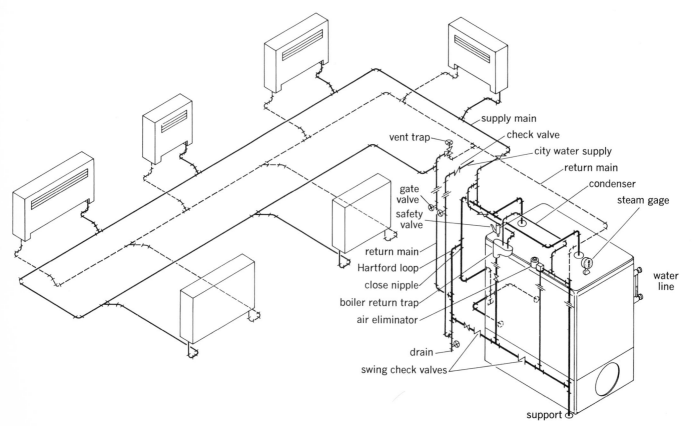

Fig. 1. Two-pipe up-feed system with automatic return trap.

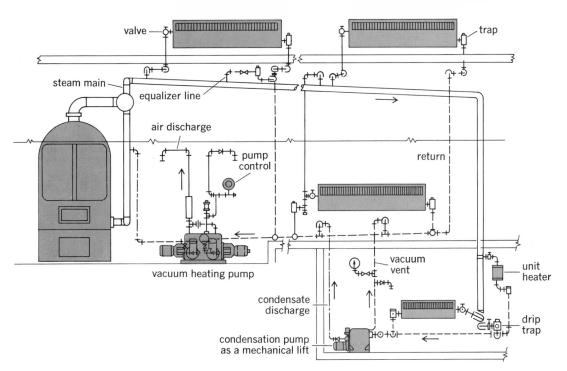

Fig. 2. Layout of a vacuum heating system with condensation and vacuum pumps.

condensate return-pump arrangement, the return piping is arranged for the water to flow by gravity into a collecting receiver or tank, which may be located below the steam-boiler waterline. A motor-driven pump controlled from the boiler water level then forces the condensate back to the boiler.

In large buildings extending over a considerable area, it is difficult to locate all heat exchangers above the boiler water level or return piping. For these systems a vacuum pump is used that maintains a suction below atmosphere up to 25± in. (max) of mercury in the return piping, thus creating a positive return flow of air and condensate back to the pumping unit. Subatmospheric systems are similar to vacuum systems, but in contrast provide a means of partial vacuum control on both the supply and return piping so that the steam temperature can be regulated to vary the heat emission from the heat exchanger in direct proportion to the heat loss from the structure.

Figure 2 depicts a two-pipe vacuum heating system which uses a condensation pump as a mechanical lift for systems where a part of the heating system is below the boiler room. Note that the low section of the system is maintained under the same vacuum conditions as the remainder of the system. This is accomplished by connecting the vent from the receiver and pump discharge to the return pipe located above the vacuum heating pump.

With the wide acceptance of all-year air conditioning, low-pressure steam boilers have been used to produce cooling from absorption refrigeration equipment. With this system the boiler may be used for primary or supplementary steam heating as diagrammed in Fig. 3.

Exhaust from gas- or oil-driven turbines or engines may be used in waste heat boilers or separators, along with a standby boiler to produce steam for a heating system as indicated in Fig. 4.

Another source of steam for heating is from a

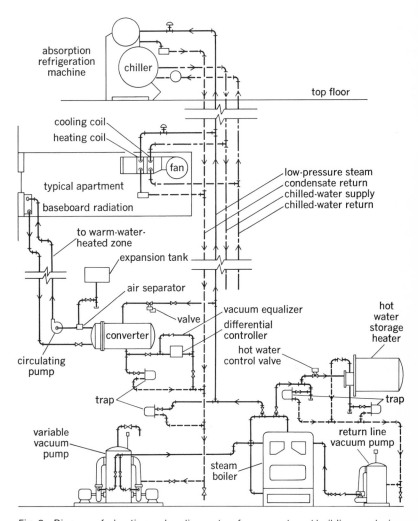

Fig. 3. Diagram of a heating and cooling system for an apartment building employing a low-pressure steam boiler.

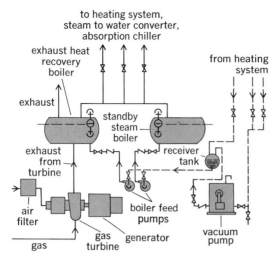

Fig. 4.   Steam system using waste heat.

high-temperature water source (350–450°F) using a high-pressure water to low-pressure steam heat exchanger. *See* HEATING, COMFORT; OIL BURNER.

[JOHN W. JAMES]

Bibliography: J. R. Allen, J. H. Walker, and J. W. James, *Heating and Air Conditioning*, 6th ed., 1946; R. H. Emerick, *Heating Handbook*, 1964; E. B. Woodruff and H. B. Lammers, *Steam-Plant Operation*, 3d ed., 1967.

Steam turbine

A machine for generating mechanical power in rotary motion from the energy of steam at temperature and pressure above that of an available sink. By far the most widely used and most powerful turbines are those driven by steam. In the United States well over 85% of the electrical energy consumed is produced by steam-turbine-driven generators. By the mid-1970s, over 25,000

Mw (1 Mw = 1341 hp) of steam turbine capacity for electrical power generation was shipped in the United States in a single typical year. In the United States, electrical power consumption doubles about every 10 to 15 years. Accordingly, steam turbine shipments of about 40,000 Mw per year are forecast by the mid 1980s. Individual turbine ratings historically have tended to follow the increasing capacity trend but are now reaching limits imposed by material and machine design considerations. The largest unit shipped during the 1950s was rated 500 Mw. Units rated about 1100 Mw were in service by the close of the 1960s, and ratings up to 1300 Mw were seeing frequent application in the 1970s. Units of all sizes, from a few horsepower to the largest, have their applications. Manufacturers of steam turbines are located in every industrial country.

Until the 1960s essentially all steam used in turbine cycles was raised in boilers burning fossil fuels (coal, oil, and gas) or, in minor quantities, certain waste products. The 1960s marked the beginning of the introduction of commercial nuclear power. About 50% of the steam turbine capacity ordered from 1965 to 1975 was designed for steam from nuclear reactor steam supplies. Approximately 10% of the power generated in 1975 was from nuclear steam plants, about 75% from fossil fuel-fired steam plants, and the balance from other sources.

Turbine parts. Figure 1 shows a small, simple mechanical-drive turbine of a few horsepower. It illustrates the essential parts for all steam turbines regardless of rating or complexity: (1) a casing, or shell, usually divided at the horizontal center line, with the halves bolted together for ease of assembly and disassembly; it contains the stationary blade system; (2) a rotor carrying the moving buckets (blades or vanes) either on wheels or drums, with bearing journals on the ends of the rotor; (3) a set of bearings attached to the casing to support the shaft; (4) a governor and valve system for regulating the speed and power of the turbine by controlling the steam flow, and an oil system for lubrication of the bearings and, on all but the smallest machines, for operating the control valves by a relay system connected with the governor; (5) a coupling to connect with the driven machine; and (6) pipe connections to the steam supply at the inlet and to an exhaust system at the outlet of the casing or shell.

Applications. Steam turbines are ideal prime movers for driving machines requiring rotational mechanical input power. They can deliver constant or variable speed and are capable of close

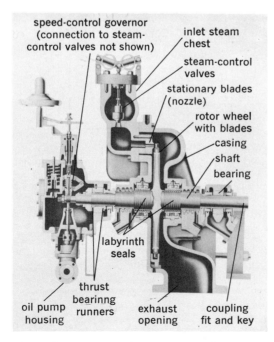

Fig. 1.   Cutaway of small, single-stage steam turbine. (*General Electric Co.*)

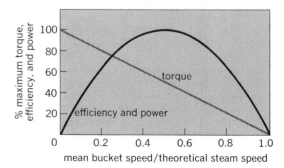

Fig. 2.   Illustrative stage performance versus speed.

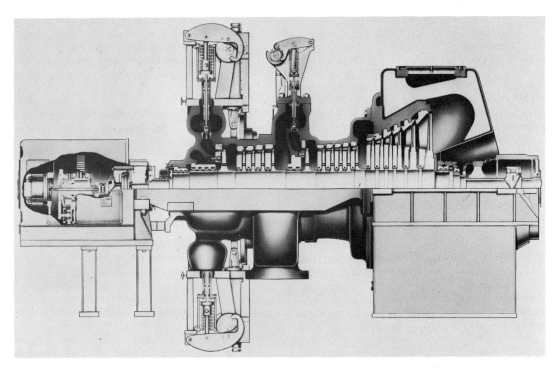

Fig. 3. Cross-section view of single-automatic-extraction condensing steam turbine. (*General Electric Co.*)

speed control. Drive applications include centrifugal pumps, compressors, ship propellers, and, most important, electric generators.

The turbine shown in Fig. 1 is a small mechanical-drive unit. Units of this general type provide 10–1000 hp with steam at 100–600 pounds per square inch gage (psig) inlet pressure and temperatures to 800°F. [See keys to Figs. 6 and 7 for conversion factors from United States customary to metric (SI) units.] These and larger multistage machines drive small electric generators, pumps,

blowers, air and gas compressors, and paper machines. A useful feature is that the turbine can be equipped with an adjustable-speed governor and thus be made capable of producing power over a wide range of rotational speeds. In such applications efficiency varies with speed (Fig. 2), being 0 when the rotor stalls at maximum torque and also 0 at the runaway speed at which the output torque is 0. Maximum efficiency and power occur where the product of speed and torque is the maximum.

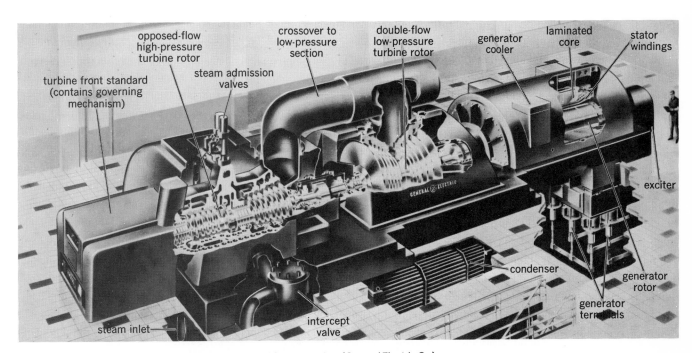

Fig. 4. Partial cutaway view of 3600-rpm fossil-fuel turbine generator. (*General Electric Co.*)

Many industries need steam at one or more pressures (and consequently temperatures) for heating and process work. Frequently it is more economical to raise steam at high pressure, expand it partially through a turbine, and then extract it for process, than it would be to use a separate boiler at the process steam pressure. Figure 3 is a cross section through an industrial automatic extraction turbine. The left set of valves admits steam from the boiler at the flow rate to provide the desired electrical load. The steam flows through five stages to the controlled extraction point. The second set of valves acts to maintain the desired extraction pressure by varying the flow through the remaining 12 stages. Opening these internal valves increases the flow to the condenser and lowers the controlled extraction pressure.

Industrial turbines are custom-built in a wide variety of ratings for steam pressures to 2000-psig, for temperatures to 1000°F, and in various combinations of nonextracting, single and double automatic extraction, noncondensing and condensing. Turbines exhausting at or above atmospheric pressure are classed as noncondensing regardless of what is done with the steam after it leaves the turbine. If the pressure at the exhaust flange is less than atmospheric, the turbine is classed as condensing.

Turbines in sizes to about 75,000 hp are used for ship propulsion. The drive is always through reduction gearing (either mechanical or electrical) because the turbine speed is in the range of 4000–10,000 rpm, while 50–200 rpm is desirable for ship propellers. Modern propulsion plants are designed for steam conditions to 1450 psig and 950°F with resuperheating to 950°F. Fuel consumption rates as low as 0.4 lb of oil per shaft-horsepower-hour are achieved.

Central station generation of electric power provides the largest and most important single application of steam turbines. Ratings smaller than 50 Mw are seldom employed today; newer units are rated as large as 1300 Mw. Large turbines for electric power production are designed for the efficient use of steam in a heat cycle that involves extraction of steam for feedwater heating, resuperheating of the main steam flow (in fossil-fuel cycles), and exhausting at the lowest possible pressure economically consistent with the temperature of the available condenser cooling water.

In fossil-fuel-fired cycles, steam pressures are usually in the range of 1800–3500 psig and tend to increase with rating. Temperatures of 950–1050°F are used, with 1000°F the most common. Single resuperheat of the steam to 950–1050°F is almost universal. A second resuperheating is occasionally employed. Figure 4 shows a typical unit designed for fossil-fuel steam conditions. Tandem-compound double-flow machines of this general arrangement are applied over the rating range of 100–400 Mw. Initial steam flows through the steam admission valves and passes to the left through the high-pressure portion of the opposed flow rotor. After resuperheating in the boiler it is readmitted through the intercept valves and flows to the right through the reheat stages, then crosses over to the double-flow low-pressure rotor, and exhausts downward to the condenser.

The water-cooled nuclear reactor systems common in the United States provide steam at pressures of about 1000 psig, with little or no initial superheat. Temperatures higher than about 600°F are not available. Further, reactor containment and heat-exchanger considerations preclude the practical use of resuperheating at the reactor. The boiling-water reactor, for example, provides steam to the turbine cycle at 950-psig pressure and the saturation temperature of 540°F. Such low steam conditions mean that each unit of steam flow through the turbine produces less power than in a fossil-fuel cycle. Fewer stages in series are needed but more total flow must be accommodated for a given output. Nuclear steam conditions often produce a turbine expansion with water of condensation present throughout the entire steam path. Provisions must be made to control the adverse effects of water: erosion, corrosion, and efficiency loss. In consequence of the differences in steam conditions, the design for a nuclear turbine differs considerably from that of a turbine for fossil fuel application. The former tend to be larger, heavier, and more costly. For example, at 800 Mw, a typical fossil-fuel turbine can be built to run at 3600 rpm, is about 90 ft long, and weighs about 1000 tons. A comparable unit for a water-cooled nuclear reactor requires 1800 rpm, is about 125 ft long, and weighs about 2500 tons. With their electric generators, and at 1975 price levels, the unit for fossil fuel was priced at about $25,000,000 while the nuclear turbine cost about $35,000,000. Savings in fuel cost offset the greater capital costs of the nuclear plants, and the result is full economic competitiveness. See REACTOR, NUCLEAR.

Figure 5 represents a large nuclear turbine generator suitable for ratings of 1000 to 1300 Mw. Steam from the reactor is admitted to a double-flow high-pressure section, at the left, through four parallel pairs of stop and control valves, not

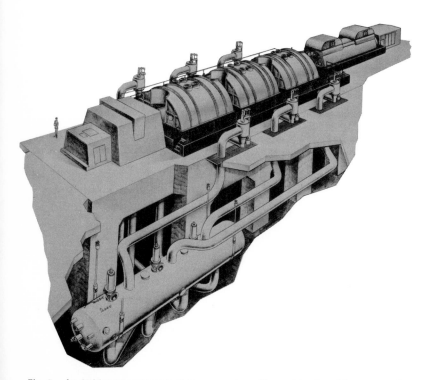

Fig. 5.   An 1800-rpm nuclear turbine generator with combined moisture separator and two-stage steam reheater. (*General Electric Co.*)

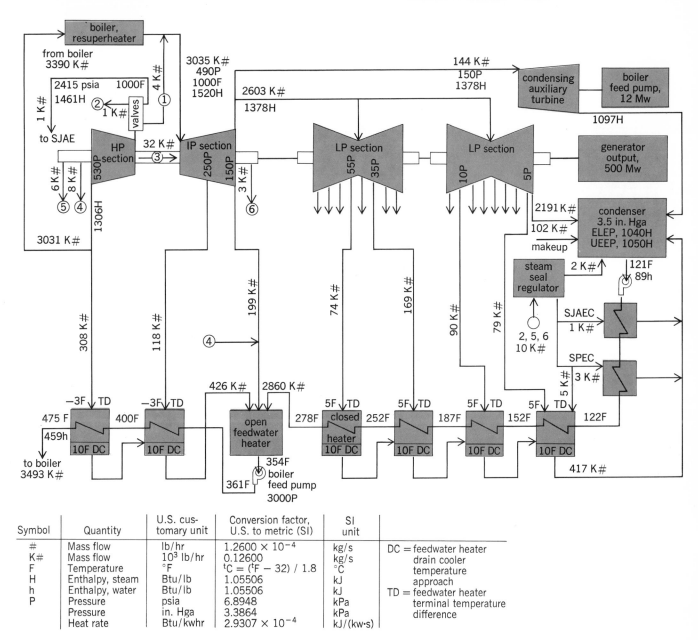

Fig. 6. Typical fossil-fuel steam turbine cycle.

Symbol	Quantity	U.S. customary unit	Conversion factor, U.S. to metric (SI)	SI unit	
#	Mass flow	lb/hr	1.2600×10^{-4}	kg/s	DC = feedwater heater
K#	Mass flow	10^3 lb/hr	0.12600	kg/s	drain cooler
F	Temperature	°F	$t°C = (t°F - 32) / 1.8$	°C	temperature
H	Enthalpy, steam	Btu/lb	1.05506	kJ	approach
h	Enthalpy, water	Btu/lb	1.05506	kJ	TD = feedwater heater
P	Pressure	psia	6.8948	kPa	terminal temperature
	Pressure	in. Hga	3.3864	kPa	difference
	Heat rate	Btu/kwhr	2.9307×10^{-4}	kJ/(kw·s)	

shown. The stop valves are normally fully open and are tripped shut to prevent dangerous overspeed of the unit in the event of loss of electrical load combined with control malfunction. The control valves regulate output by varying the steam flow rate. The steam exhausting from the high-pressure section is at 150 to 200 psia and contains about 13% liquid water by weight. The horizontal cylinder alongside the foundation is one of a pair of symmetrically disposed vessels performing three functions. A moisture separator removes most of the water in the entering steam. Two steam-to-steam reheaters follow. Each is a U-tube bundle which condenses heating steam within the tubes and superheats the main steam flow on the shell side. The first stage uses heating steam extracted from the high-pressure turbine. The final stage employs reactor steam which permits reheating to near initial temperature. Alternate cycles employ reheat with initial steam only or moisture separation alone with

no reheat. Reheat enhances cycle efficiency at the expense of increased investment and complexity. The final choice is economic, with two-stage steam reheat, as shown, selected most frequently.

Reheated steam is admitted to the three double-flow low-pressure turbine sections through six combined stop-and-intercept intermediate valves. The intermediate valves are normally wide open but provide two lines of overspeed defense in the event of load loss. Exhaust steam from the low-pressure sections passes downward to the condenser, not shown in Fig. 5.

Turbine cycles and performance. Figures 6 and 7 are representative fossil-fuel and nuclear turbine thermodynamic cycle diagrams, frequently called heat balances. Heat balance calculations establish turbine performance guarantees, provide data for sizing the steam supply and other cycle components, and are the basis for designing the turbine generator.

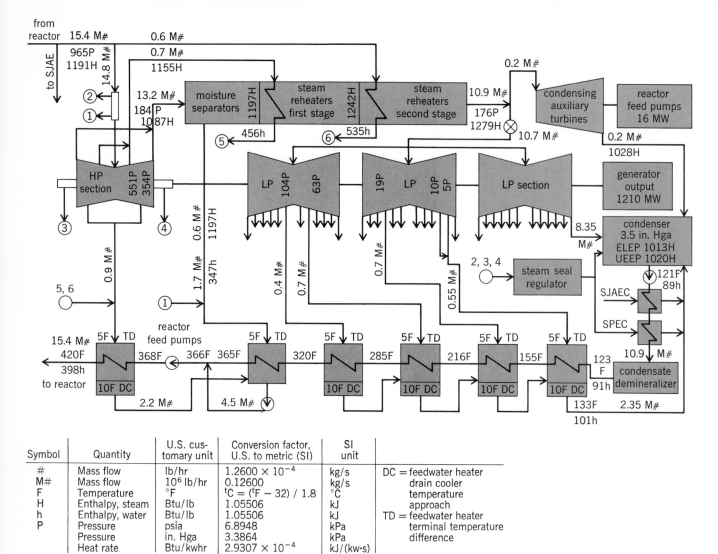

Fig. 7. Typical nuclear steam turbine cycle.

Symbol	Quantity	U.S. customary unit	Conversion factor, U.S. to metric (SI)	SI unit	
#	Mass flow	lb/hr	1.2600×10^{-4}	kg/s	DC = feedwater heater
M#	Mass flow	10^6 lb/hr	0.12600	kg/s	drain cooler
F	Temperature	°F	$^tC = (^tF - 32) / 1.8$	°C	temperature
H	Enthalpy, steam	Btu/lb	1.05506	kJ	approach
h	Enthalpy, water	Btu/lb	1.05506	kJ	TD = feedwater heater
P	Pressure	psia	6.8948	kPa	terminal temperature
	Pressure	in. Hga	3.3864	kPa	difference
	Heat rate	Btu/kwhr	2.9307×10^{-4}	kJ/(kw·s)	

The fossil-fuel cycle (Fig. 6) assumes a unit rated 500 Mw, employing the standard steam conditions of 2400 psig (2415 psia) and 1000°F, with resuperheat to 1000°F. As can be seen in the upper left corner, the inlet conditions correspond to a total heat content, or enthalpy, of 1461 Btu/lb of steam flow. A flow rate of 3,390,000 lb/hr is needed for the desired output of 500 Mw (500,000 kw). For efficiency considerations the regenerative feedwater heating cycle is used. Eight heaters in series are employed so that water is returned to the boiler at 475°F and 459 Btu/lb enthalpy, rather than at the condenser temperature of 121°F. Because of the higher feedwater temperature, the boiler adds heat to the cycle at a higher average temperature, more closely approaching the ideal Carnot cycle, in which all heat is added at the highest cycle temperature. The high-pressure turbine section exhausts to the resuperheater at 530 psia pressure and 1306 Btu/lb enthalpy. The reheat flow of 3,031,-000 lb/hr returns to the reheat or intermediate turbine section at 490 psia pressure and 1520 Btu/lb enthalpy. These data are sufficient to calculate the turbine heat rate, or unit heat charged against the turbine cycle. The units are Btu of heat added in

the boiler per hour per kilowatt of generator output. Considering both the initial and reheat steam, the heat rate is given by Eq. (1).

Turbine heat rate
$$= \frac{3,390,000\,(1461 - 459) + 3,031,000\,(1520 - 1306)}{500,000}$$
$$= 8090 \text{ Btu/kwhr} \qquad (1)$$

The typical power plant net heat rate is poorer than the turbine heat rate because of auxiliary power required throughout the plant and because of boiler losses. Assuming 3% auxiliary power (beyond the boiler-feed pump power given in the turbine cycle in Fig. 6) and 90% boiler efficiency, the net plant heat rate is given by Eq. (2).

Net plant heat rate
$$= \frac{3,390,000\,(1461 - 459) + 3,031,000\,(1520 - 1306)}{500,000\,\big((100 - 3)/100\big)(90/100)}$$
$$= 9270 \text{ Btu/kwhr} \qquad (2)$$

The heat rates of modern fossil-fuel plants fall in

the range of 8600–10,000 Btu/kwhr. Considering that the heat-energy equivalent of 1 kwhr is 3412 Btu, the thermal efficiency of the example is given by Eq. (3).

$$\eta_t = (3412/9270)\,100 = 37\% \qquad (3)$$

Figure 6 shows 2,191,000 lb/hr of steam exhausting from the main unit to the condenser at an exhaust pressure of 3.5 in. of mercury absolute. The theoretical exhaust enthalpy (ELEP), without considerations of velocity energy loss and friction loss between the last turbine stage and the condenser, is 1040 Btu/lb. The actual used energy end point (UEEP) is 1050 Btu/lb. The exhaust heat at the condenser pressure is thermodynamically unavailable and is rejected as waste heat to the plant's surroundings. The exhaust steam is condensed at a constant 121°F and leaves the condenser as water at 89 Btu/lb enthalpy.

On a heat-rate basis, this cycle rejects heat to the condenser at the approximate rate given by Eq. (4).

Net station condenser heat rejection rate

$$= \frac{\begin{array}{l} 2{,}191{,}000\,(1050-89) \\ +\,144{,}000\,(1097-89) \\ +\,417{,}000\,(100-89) \end{array}}{500{,}000\,(0.97)}$$

$$= 4650 \text{ Btu/kwhr} \qquad (4)$$

If evaporating cooling towers are used, each pound of water provides about 1040 Btu cooling capacity, which is equivalent to a required minimum cooling-water flow rate of 4.5 lb/kwhr. The cooling-water needs of a large thermal plant are a most important consideration in plant site selection.

The nuclear cycle (Fig. 7) assumes a unit rated 1210 Mw and the steam conditions of the boiling-water reactor. Many similarities can be seen to Fig. 6. The major differences include moisture separation and steam reheating and the lack of need for an intermediate pressure element. The low steam conditions are apparent. The consequent turbine heat rate is given in Eq. (5).

Turbine heat rate

$$= \frac{15{,}400{,}000\,(1191-398)}{1{,}210{,}000}$$

$$= 10{,}090 \text{ Btu/kwhr} \qquad (5)$$

A typical nuclear plant also requires about 3% auxiliary power beyond the reactor feed-pump power already included in Fig. 7. The equivalent boiler efficiency approaches 100% however, and leads to the equivalent net plant heat rate given by Eq. (6).

Net plant heat rate

$$= \frac{15{,}400{,}000\,(1191-398)}{1{,}210{,}000\,[\,(100-3)/100\,]}$$

$$= 10{,}400 \text{ Btu/kwhr} \qquad (6)$$

The corresponding thermal efficiency is given by Eq. (7).

$$\eta_t = (3412/10{,}400)\,100 = 33\% \qquad (7)$$

Heat is rejected at the condenser at a rate given approximately by Eq. (8).

Net station condenser heat rejection rate

$$= \frac{\begin{array}{l} 8{,}350{,}000\,(1020-89) \\ +\,200{,}000\,(1028-89) \\ +\,2{,}350{,}000\,(101-89) \end{array}}{1{,}210{,}000\,(0.97)}$$

$$= 6810 \text{ Btu/kwhr} \qquad (8)$$

Comparison of the heat rates shows that the nuclear cycle requires about 12% more input heat than does the fossil-fuel cycle and rejects about 46% more heat to the condenser, thus requiring a correspondingly larger supply of cooling water. In its favor, the nuclear cycle consumes heat priced at about half that from coal or one quarter that from oil, and is essentially free of rejection to the atmosphere of heat and combustion products from the steam supply.

Turbine classification. Steam turbines are classified (1) by mechanical arrangement, as single-casing, cross-compound (more than one shaft side by side), or tandem-compound (more than one casing with a single shaft); (2) by steam flow direction (axial for most, but radial for a few); (3) by steam cycle, whether condensing, noncondensing, automatic extraction, reheat, fossil fuel, or nuclear; and (4) by number of exhaust flows of a condensing unit, as single, double, triple flow, and so on. Units with as many as eight exhaust flows are in use.

Often a machine will be described by a combination of several of these terms.

The least demanding applications are satisfied by the simple single-stage turbine of Fig. 1. For large power output and for the high inlet pressures and temperatures and low exhaust pressures which are required for high thermal efficiency, a single stage is not adequate. Steam under such conditions has high available energy, and for its efficient utilization the turbine must have many stages in series, where each takes its share of the total energy and contributes its share of the total output. Also, under these conditions the exhaust volume flow becomes large, and it is necessary to have more than one exhaust stage to avoid a high velocity upon leaving and consequent high kinetic energy loss. Figure 5 is an example of a large nuclear turbine generator which has six exhaust stages in parallel.

Machine considerations. Steam turbines are high-speed machines whose rotating parts must be designed for high centrifugal stress. Difficult stress problems are found in long last-stage blading, hot inlet blading, wheels, and rotor bodies.

Casing or shell stresses. The casings or shells at the high-pressure inlet end must be high-strength pressure vessels to contain the internal steam pressure. The design is made more difficult by the need for a casing split at the horizontal center line for assembly. The horizontal flange and bolt design must be leakproof. Shell design problems lead to the use of small-diameter, high-speed turbines at high pressure, and the use of double shell construction (Fig. 4).

Rotor buckets or blades. Turbine buckets must be strong enough to withstand high centrifugal, steam bending, and vibration forces. Buckets must be designed so that their resonant natural frequencies avoid the vibration stimulus frequencies of the steam forces, or are strong enough to withstand the vibrations.

Sealing against leakage. It is necessary to minimize to the greatest possible extent the wasteful leakage of steam along the shaft both at the ends and between stages. The high peripheral velocities between the shaft and stationary members preclude the use of direct-contact seals. Seals in the form of labyrinths with thin, sharp teeth on at least one of the members are utilized. In normal operation these seals do not touch one another, but run at close clearance. In the case of accidental contact, the sharp teeth can rub away without distorting the shaft.

Vibration and alignment. Shaft and bearings should be free of critical speeds in the turbine operating range. The shaft must be stable and remain in balance.

Governing. Turbines usually have two governors, one to control speed and a second, emergency governor to limit possible destructive overspeed. The speed signal is usually mechanical or electrical. A power relay control, usually hydraulic, converts speed signals to steam valve position. Great reliability is required.

Lubrication. The turbine shaft runs at high surface speed; consequently its bearings must be continuously supplied with oil. At least two oil pumps, a main pump and a standby driven by a separate power source, are usually provided on all but the smallest machines. A common supply of oil is often shared between the governing hydraulic system and the lubrication system.

Aerodynamic design. The vane design for highest efficiency, especially for the larger sizes of turbines, draws upon modern aerodynamic theory. Classic forms of impulse and reaction buckets merge in the three-dimensional design required by modern concepts of loss-free fluid flow. To meet the theoretical steam flow requirements, vane sections change in shape along the bucket. To minimize centrifugal forces on the vanes and their attachments, long turbine buckets are tapered toward their tips. See CARNOT CYCLE; HEAT BALANCE; TURBINE.

[FREDERICK G. BAILY]

Bibliography: F. G. Baily, K. C. Cotton, and R. C. Spencer, Predicting the performance of large steam turbine-generators operating with saturated and low superheat steam conditions, *Combustion*, 3(3):8–13, 1967; R. L. Bartlett, *Steam Turbine Performance and Economics*, 1958; P. H. Knowlton, Jr., Steam turbines, in T. Baumeister (ed.), *Standard Handbook for Mechanical Engineers*, sec. 9, pp. 59–82, 7th ed., 1967; J. K. Salisbury, *Steam Turbines and Their Cycles*, 1950; B. G. A. Strotzki, Steam turbines, *Power*, 106(6):S1–S40, 1962.

Storage battery

An assembly of identical voltaic cells in which the electrochemical action is reversible so that the battery may be recharged by passing a current through the cells in the opposite direction to that of discharge. While many nonstorage batteries have a reversible process, only those that are economically rechargeable are classified as storage batteries. *See* BATTERY, ELECTRIC; PRIMARY BATTERY.

Storage batteries, sometimes known as electric accumulators or secondary batteries, have two general classifications: lead-acid and alkaline. Active materials and electrolytes for both classes of batteries will be explained later. The table gives an approximate comparison of the several principal types of storage battery couples in terms of output per unit weight and unit volume.

Some of the important uses of storage batteries are to start gasoline and diesel engines; to operate communications circuits; switch tripping and closing in power-generating and -handling systems; emergency lighting; emergency power both with and without conversion to alternating current; railway car lighting and air conditioning; rapid transit car controls; marine power systems; power for underwater exploratory vehicles and submarines; to activate photographic and portable sound systems as well as portable TV and radio; and various military applications.

LEAD-ACID STORAGE BATTERY

The lead-acid type of storage battery is so classified because the electrolyte is an acid and the plates are largely lead. The positive active material is lead peroxide and the negative active material is lead sponge. The active materials are supported by grids made of lead alloys.

The lead-acid battery maintains a preeminent place among all commercial types of storage batteries in volume of manufacture.

Principles of operation. A great many types of lead-acid cells are produced, but all have certain features in common. One is the open-circuit cell electromotive force (emf), which exists between a positive lead peroxide (PbO_2) electrode and a negative sponge lead (Pb) electrode when the two are immersed in sulfuric acid electrolyte ($H_2SO_4 + H_2O$). This value is independent of the quantities of lead peroxide, lead, or electrolyte present but does vary with temperature and sulfuric acid (H_2SO_4) concentration. At 25°C the emf varies from 2.050 volts with acid at 1.200 sp gr to 2.148 volts with acid at 1.300 sp gr. The relatively small variation with temperature is given in millivolts/°C over a range 0–40°C, as 0.30 for 1.200 sp gr electrolyte, 0.22 for 1.250 sp gr, 0.19 for 1.280 sp gr, and 0.18 for 1.300 sp gr.

Equation (1) represents the cell reactions insofar

$$PbO_2 + Pb + 2H_2SO_4 \underset{\text{Charge}}{\overset{\text{Discharge}}{\rightleftharpoons}}$$

$$2PbSO_4 + 2H_2O \quad (1)$$

as beginning and end materials are concerned. It is

Comparison of the principal types of storage battery

Battery type	Volts per cell	Energy, whr/lb	Density, whr/in.³
Lead-acid	2.0	10–15	0.6–1.3
Nickel-iron	1.2	10–14	0.6–1.0
Nickel-cadmium	1.2	8–11	0.4–0.8
Nickel-cadmium sintered	1.2	10–13	1.0
Silver-zinc	1.5	20–100	3
Silver-cadmium	1.1	15–50	2.5

known as the double-sulfate theory, since lead sulfate ($PbSO_4$) is formed at both electrodes.

Equation (1) can be split into Eqs. (2) and (3),

$$PbO_2 + 2H^+ + H_2SO_4 + 2e^- \underset{\text{Charge}}{\overset{\text{Discharge}}{\rightleftharpoons}}$$

$$PbSO_4 + 2H_2O \quad \text{(at positive)} \quad (2)$$

$$Pb + SO_4^- \underset{\text{Charge}}{\overset{\text{Discharge}}{\rightleftharpoons}}$$

$$PbSO_4 + 2e^- \quad \text{(at negative)} \quad (3)$$

indicating the reactions at the two electrodes.

On discharge the overall effect is a reduction of PbO_2 at the positive electrode and an oxidation of Pb at the negative electrode, accompanied by sulfation in both cases. In charging, a counter voltage is imposed on the cell terminals, and current is forced through the cell in a direction opposite to that in which the cell discharges. This reverses the ionic movements in relation to the electrodes and, in effect, reverses the cell reactions. On discharge the electrolyte specific gravity decreases, and on charge it increases. Specific gravity serves as a measure of the sulfuric acid concentration and thus as an index of state of charge.

Reactants. For a given quantity of electricity, such as ampere-hours, the three reactants PbO_2, Pb, and H_2SO_4 take part in the reaction in amounts governed by Faraday's law. Thus, for a 1 amp-hr discharge, 3.866 g of sponge lead are converted to $PbSO_4$, 4.463 g PbO_2 are converted to $PbSO_4$, and 3.660 g of H_2SO_4 are consumed.

A cell constructed to contain exactly the amounts of reactants given above, however, would not yield 1 amp-hr of capacity even under optimum practical conditions. Action at each electrode is slowed drastically when the concentration of H_2SO_4 in the electrolyte approaches a low figure because it is required in the electrode reaction. But even if ample H_2SO_4 were present, 1 amp-hr of capacity still would not be attained, since there will always remain an appreciable amount of PbO_2 or Pb, or both, in the solid electrodes, which cannot be reached by the electrolyte. The capacity attained in practice divided by what should be obtained in principle from the amount of reactants present is known as the utilization coefficient. This coefficient varies with types of cells, rate of discharge, and temperature. Unfortunately, it is a low value even under the best of conditions.

This problem is aggravated by the coating of nonconducting sulfate that forms on the active materials. Another is the diminishing conductivity of the electrolyte as the H_2SO_4 content decreases.

The utilization coefficient is decreased by the use of high-current rates. At higher current densities, the electrode reaction is concentrated at the surface of the plates. As a result, the pore openings at the plate surfaces become blocked with sulfate, restricting conduction and diffusion to the interiors of plates. Plates destined for high-discharge current densities are therefore made relatively thin. By substituting many thin plates for a few thick plates containing the same amounts of active materials, the utilization coefficient at high rates, and hence the capacity attainable, will be increased.

Figure 1 shows a typical set of volt-time curves for for different discharge rates of a lead-acid cell, illustrating the variations from a 1-hr rate at high discharge current to a 10-hr rate at low discharge current. Figure 2 illustrates the decrease in ampere-hour capacity with increase in discharge current. If a short time is allowed for diffusion after a high-rate discharge, more of the unused possible capacity of the cell becomes available.

Cell temperature has an appreciable effect on capacity, largely because the viscosity of the electrolyte changes. Thus the diffusion of H_2SO_4 is retarded at low temperatures and the capacity is lowered.

Also, the capacity is decreased if the acid concentration becomes too low. On the other hand experience has shown that negative plates do not function well if the full-charge specific gravity is over 1.300, although positive plates operate more efficiently in high specific gravity. The usual range of full-charge specific gravity is 1.200–1.280, the choice depending on the application of the cell, the ambient operating temperature, and susceptibility of the cell to self-discharge. Specific gravity is usually determined by a hydrometer. It can also be measured by chemical methods.

Cell construction. Aside from cost, first consideration must be given to the kind of service for which a cell is destined, and second consideration

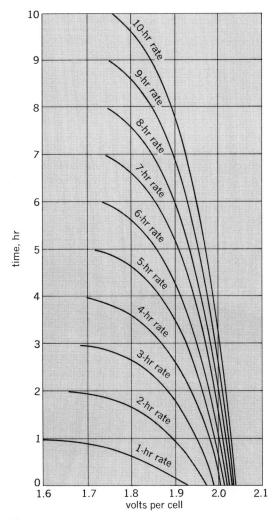

Fig. 1. Typical volt-time curves for a lead-acid cell for various discharge rates. (*ESB Inc.*)

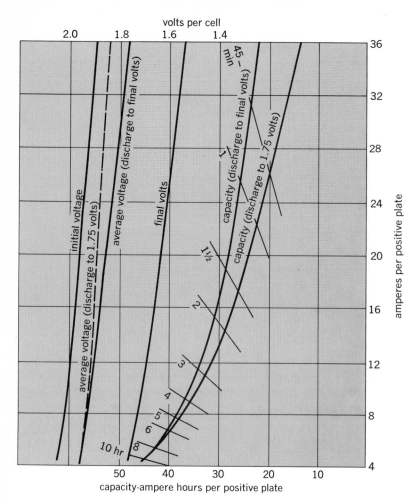

Fig. 2. Rated-discharge curves for a lead-acid cell. (*ESB Inc.*)

cium. The grid is pasted with a slurry of lead oxide, sulfuric acid, and water. This is followed by a processing which finally converts the active material to PbO_2 for positive plates (chocolate-brown color) and to sponge lead for negative plates (gray color).

The negative plates are always sponge lead whether used with pasted positive plates or with positive plates of other types.

The life of a lead-acid battery plate is closely related to the thickness of the metal bars used in the positive plate. In applications such as engine cranking where light weight and high-rate performance are of more importance than life, the battery plates are usually made as thin as possible. Thicker plates are used where long life and reliability are more important than first cost, space, and weight. The thinnest plates in use are about 0.05 in. thick; the thickest plates range up to 0.75 in.

Manchester plate. These plates consist of a heavy alloy grid with circular openings into which pure lead "buttons" are pressed. These buttons are made from lead tape by crimping and rolling to develop a large surface area (Fig. 4). A forming agent in dilute sulfuric acid electrochemically forms a layer of PbO_2 on the surface of the button. Manchester plates are usually mounted in a cell with pasted negatives and in a relatively large quantity of low-gravity acid.

The cells are heavy and bulky. They are used in stationary installations, as for telephones, switch operation, and large emergency lighting, where they are "floated" on a line of constant voltage or trickle-charged with a constant current and are only occasionally discharged. Under such conditions of service, Manchester plates give exceptionally long life.

Gould spun plates. This type of positive plate, shown in Fig. 5, is manufactured from heavy sheet lead by passing the plate between disks which cause the lead to flow in between them to form leaves and spaces. After PbO_2 is formed on this developed surface, the plates are assembled with pasted negative plates and used in substantially the same types of service as Manchester plates. An advantage of this plate is elimination of antimony, hence local action, from the cell construction. This advantage is usually gained at some sacrifice of life.

Tubular-type plate. In this positive plate the active material is held in a porous-walled tube with a central alloy spine as conductor. The tube is made of felted or woven chemically inert fibers. This plate has many applications but is particularly successful where the service calls for repeated or routine deep-discharge cycles, such as in industrial trucks and mine locomotives.

Freezing of electrolyte. The freezing points of the usual range of sulfuric acid electrolytes at full-charge specific gravities (from −52°C for 1.250 at 15°C to −70°C for 1.300 at 15°C) are well below most arctic temperatures, but the end-of-discharge specific gravities can result in freezing points above arctic temperatures unless precautions are taken. With the proper choice of separator, a high-gravity acid can be used in severe arctic conditions without detriment to the negative electrodes and yet can have a relatively high end-of-discharge gravity.

to design features that reduce operational troubles. A compromise, for example, between life and weight or between life and cost, is usually required.

Pasted plates. In the most familiar type of plate construction, the basic structural number is a die-cast grid, such as the ladder type shown in Fig. 3. This grid is made of a lead alloy containing, for example, (3−11%) antimony or (.01−0.1%) cal-

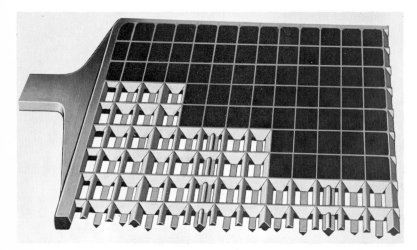

Fig. 3. Typical ladder-type grid showing a portion of it pasted. (*ESB Inc.*)

Charging. For fast, yet efficient and noninjurious, charging, the modified constant-potential method is recommended. A high current rate is used until a voltage, such as 2.38, is obtained. This voltage is then maintained with a decreasing current until the finishing rate recommended by the manufacturer is reached. The finishing rate is continued to the end of the charge. Several methods are used for setting the end of the charge, the best known being arrival at a constant potential, arrival at a constant specific gravity, or charging a certain percent of ampere-hours in excess of the ampere-hours that have been taken out. A lengthy but efficient charge can be made using the finishing rate from the start. A two-step charge can be made using first a high current and then the finishing rate, the change being made automatically by a voltage relay or ampere hour meter.

Cell containers are currently made of hard rubber or plastic. For automobile, railway, and motive power batteries the cell containers are made of highly shock-resistant materials such as semi-hard rubber, modified polystyrene, or polyporpylene. Stationary batteries use jars made of clear modified polystyrene. For extreme shock resistance such as submarine service, rubber-lined polyester fiber-glass jars are used.

Battery troubles and remedies. Some of the important battery troubles are corrosion of the grid, shedding of active materials, and self-discharge.

Corrosion. Gradual wearing away of the electrode grid containing the active PbO_2 results in subsequent disintegration of the plate. Cells subjected to repeated deep discharges and overcharges are particularly prone to this trouble. The antimony used in the grid alloy aids in resisting corrosion under certain conditions. The addition of small percentages of arsenic and silver increases this resistance to a notable extent.

Shedding of active materials. This usually pertains to cells that are subjected to overcharging, whereby gas formation strains and loosens particles of material near the surface of the plate. Unless retainers, such as slotted rubber or plastic or mats of glass fiber, are used against the positive plates, active material may drop to the bottom as sediment and cause short circuits between plates. The material may also be carried by gas streams to the top of the cell to pile up and short-circuit there.

Self-discharge or local action. Self-discharge of negative plates is caused by deposition of certain metals on the plate to form a voltaic couple with the sponge lead. Since these couples are actually local short circuits, which have an emf in excess of the hydrogen overvoltage on the metal deposit, an evolution of hydrogen ensues, and the adjacent lead is sulfated. Metals most frequently producing local action include antimony, copper, silver, and, less frequently, tin, arsenic, bismuth, platinum, and nickel.

Other metals, such as iron and manganese, whose salts readily exist in solution in two stages of oxidation, can reduce the positive plate by diffusion or convection and oxidize the negative plate with $PbSO_4$ formation at both electrodes.

If self-discharge of the negative electrodes becomes rapid or is allowed to act over a long period, or if a cell stands for some time in a discharged

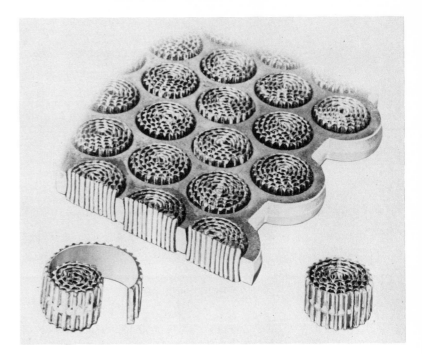

Fig. 4. Section of Manchester plate with detail of lead button. (*ESB Inc.*)

Fig. 5. Gould spun Plante positive plate. (*Gould-National Batteries, Inc.*)

condition, the sulfate crystals become large, hard, and difficult to reduce to lead.

Undercharging. This causes buckling, or warping, of plates and sulfation. This condition is usually due to unequal work on the two sides of the positive plates, resulting from unequal electrolytic attack or unequal expansion of active materials.

Overcharging. This causes corrosion, buckling, washing, and overheating. It is caused by high

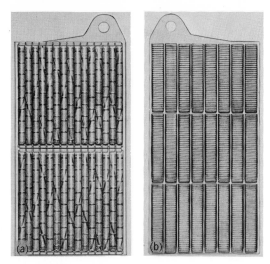

Fig. 6. Plate construction of nickel-iron cell. (a) Positive plate. (b) Negative plate. (ESB Inc.)

charging rates, which should be tapered off as the battery becomes charged.

Densification of negative material. A deficiency of certain organic and inorganic compounds that are used as additives to the active material permits coalescence of lead particles with a consequent loss of porosity. Compounds added to the negative plate material to prevent this are commonly called expanders.

Separator shorts. The separators between positive and negative plates may be oxidized through contact with PbO_2, permitting lead bridges to form between positive and negative plates and thus short-circuiting the cell.

Sealed lead-acid cells. Ways have been found to operate smaller lead-acid cells in a completely sealed container. These batteries are built with capacities up to about 10 amp-hr. They are used for portable TV, electric hand tools, and so on. In order to keep these batteries operable, it is necessary to use a very carefully controlled charge.

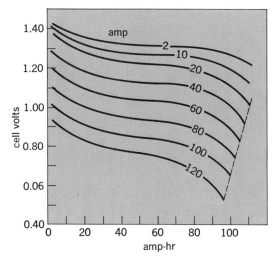

Fig. 7. Typical volt-time curves of nickel-iron alkaline cells for various discharge rates. (*From D. G. Fink and J. M. Carroll, eds., Standard Handbook for Electrical Engineers, 10th ed., McGraw-Hill, 1968*)

ALKALINE-TYPE STORAGE BATTERY

The alkaline-type storage battery is so classified because the electric energy is obtained from chemical action of an alkaline solution. One type of battery has positive plates of some nickel compound and negative plates of iron. Another type uses a nickel compound and cadmium. A third uses silver oxide and zinc.

Nickel-iron alkaline cell. This battery is composed of cells having a hydrated nickel oxide and iron in an alkaline solution. It was invented by Thomas Edison early in the 20th century. The positive active material in this cell is a higher oxide or hydroxide of nickel. The negative material is fine iron powder. The electrolyte is 1.200 sp gr (at 15°C) potassium hydroxide, to which a little lithium hydroxide is sometimes added.

The chemical behavior of the nickel-iron cell is shown in Eq. (4).

$$2NiOOH \cdot H_2O + Fe \xrightleftharpoons[\text{Charge}]{\text{Discharge}}$$

$$2Ni(OH)_2 + Fe(OH)_2 \quad (4)$$

The nickel hydrate formed by charging the battery is not an exact chemical compound. Directly after charging, it contains some excess dissolved oxygen. The dissolved oxygen is not tightly held and is released in the 10–24-hr period following the charge. It is electrically active, and a battery discharged immediately after charge will have a greater output than if it stands until the oxygen is lost.

The KOH electrolyte supplies ions for conductivity, but unlike the lead-acid battery, the concentration of electrolyte in alkaline cells (nickel-iron, nickel-cadmium, silver-zinc, and silver-cadmium) does not undergo any net change in the chemical action of the cell. As a consequence, the specific gravity of the electrolyte does not change and cannot be used to indicate the state of charge of an alkaline battery, as in the case of the lead acid battery. However, it also means that the gravity stays up at all times, and the battery is much less susceptible to accidental damage from freezing than lead acid.

In the tubular nickel-iron cell a perforated steel tube is tightly packed with alternate layers of nickel hydrate and thin nickel flake to provide electrical conductivity. The layers are thin and there are about 300 layers in a 4-in. tube.

The negative material is packed into long pockets of perforated sheet steel. The pockets are laced together and pressed to form a single structural member. The top rail and bottom rail are welded in place to complete the plate. Positive and negative plates are shown in Fig. 6. Figure 7 shows typical discharge curves of a 100-amp-hr nickel-iron cell.

It has been found desirable to charge the tubular nickel-iron cell at comparatively high rates (10–20 amp for a 100-amp-hr battery) in order to reduce the time required for a full charge. However, great numbers of tubular-iron cells are floated at rates of 0.002–0.004 milliampere per ampere-hour of capacity with excellent results.

The open-circuit potential of the negative iron electrode is very close to the hydrogen potential. This makes the electrode susceptible to rather high local action or self-discharge. The battery when on open circuit or on charge continually gives off hydrogen. Therefore nickel-iron cells must be well ventilated. They require higher float currents and more frequent watering than most other types of cell.

Manufacturers warn against operating nickel-iron cells above 46°C. Also, depending somewhat on discharge rate, capacities drop rapidly below a critical temperature of about 2.2°C.

Nickel-cadmium alkaline cells. Equation (5), the chemical equation for the nickel-cadmium cell, is exactly the same as the one for the previous cell

$$2NiOOH \cdot H_2O + Cd \xrightleftharpoons[\text{Charge}]{\text{Discharge}}$$

$$2Ni(OH)_2 + Cd(OH)_2 \quad (5)$$

type except for the use of cadmium instead of iron. The earlier remarks about positive electrode and electrolyte apply equally to this cell. The cadmium negative electrode differs from the iron negative in that its potential is below the hydrogen potential. Therefore, the cadmium electrode is completely inert to the electrolyte. It requires almost no float current to keep charged, and consequently the water consumption and float charge currents are extremely low.

The original nickel-cadmium cell, now known as the "pocket" type, was invented by Waldmar Jungner at about the same time that Edison invented the tubular cell. In this cell the positive and negative plates are of the same construction as that described for the iron electrode used in the tubular-iron cell. The positive electrode uses graphite as conductor instead of the nickel flake used in the tubular cell.

The pocket nickel-cadmium battery is widely used for emergency power use. The cadmium electrode has extremely low stand loss, and it can be kept at a state of full charge with very little maintenance. In most forms it is not well suited for cycle service. It complements the tubular-iron cell. Nickel-cadmium cells may be floated at voltages of 1.40–1.45 per cell.

After an emergency discharge it is desirable to fully charge the battery before it is shifted to the float circuit.

Sintered plate cells. During World War II the Germans developed a sintered-plate type of nickel-cadmium cell. Extremely fine nickel powder, obtained from decomposition of nickel carbonyl, is sintered in a mold around a nickel or nickel-plated screen. For positive plates these plaques are impregnated with a nickel salt (usually nitrate) and processed to produce nickel hydrate in the pores. Plaques for the negative electrodes are impregnated with a cadmium salt (nitrate or chloride) and processed in a manner like that for the positive.

The electrolyte is a solution of KOH made with specific gravities ranging 1.240–1.300.

Sintered-plate cells are displacing the original types, being superior in several respects. They have less internal resistance and a higher utilization coefficient, and they perform better at both higher and lower temperatures.

They are especially suited to extremely-high-rate discharges, low-temperature operation, and other severe applications. They are used for air-craft and diesel starting and for many military services.

Sealed cells. It has been found that the smaller sizes of nickel-cadmium cells can be operated in the fully sealed state. Sealed cells are made from very small hearing-aid sizes up to the larger flashlight sizes. In order to work in the sealed state, the cell must have a very limited amount of electrolyte, and the ratio and relative states of charge of positive and negative plates must be carefully controlled. Containers are made of nickel-plated steel or plastic. In the flashlight types the plates and separators are often rolled up in a spiral coil. They have many of the features of sintered-plate cells.

Charging is not critical. It can be done rapidly and efficiently by constant-current, constant-potential, and modified constant-potential methods; gassing begins around 1.47 volts, and when using normal charge rate (5-hr), the end voltage will be 1.75.

Typical discharge and charge curves are shown in Figs. 8 and 9.

Silver oxide–zinc alkaline cell. Silver oxide positive plates and sponge-zinc negative plates came into use during the late 1940s. They have high ampere-hour and watt-hour capacities per unit of volume or weight. A high-specific-gravity KOH solution, up to 1.450, has been found advantageous in minimizing local action. The cell reaction can be expressed as Eq. (6).

$$AgO + Zn + H_2O \xrightleftharpoons[\text{Charge}]{\text{Discharge}} Ag + Zn(OH)_2 \quad (6)$$

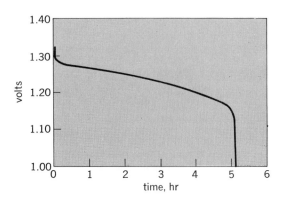

Fig. 8. Typical discharge curve for sintered-plate nickel-cadmium cell. (*ESB Inc.*)

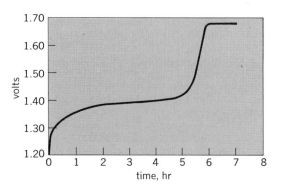

Fig. 9. Typical charge curve for sintered-plate nickel-cadmium cell. (*ESB Inc.*)

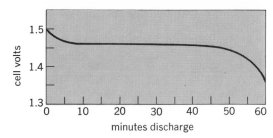

Fig. 10. Typical discharge curve exhibited by a silver oxide–zinc cell. (*ESB Inc.*)

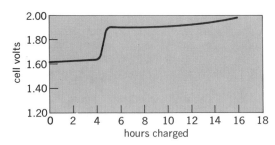

Fig. 11. Typical charge curve exhibited by a silver oxide–zinc cell. (*ESB Inc.*)

Charging can be accomplished by a constant-current or modified constant-potential charge, as long as the cell voltage does not exceed 2.1 volts at any time. Typical discharge and charge curves are shown in Figs. 10 and 11.

Silver oxide–zinc cells are used both as primary and secondary cells for military use and for nonmilitary applications where battery power with minimum weight is an essential consideration.

Freezing of alkaline electrolyte. The use of high-gravity KOH electrolyte for nickel-cadmium and silver oxide–zinc cells eliminates freezing under severe arctic conditions. High-specific-gravity electrolyte cannot be used with nickel-iron cells.

Venting of storage cells. Venting must be provided for all storage cells to permit escape of local-action gas or gas generated in the charging process. The only exceptions are the special sealed cells, in which gassing is held to a minimum and any hydrogen or oxygen generated is recombined through catalysis.

The provision for escape of gas has led to numerous devices to prevent spillage of electrolyte from cells in aircraft and other applications.

[W. W. SMITH]

Bibliography: J. T. Crennell and F. M. Lea, *Alkaline Accumulators*, 1928; S. U. Falk and A. J. Salkind, *Alkaline Storage Batteries*, 1969; D. G. Fink and J. M. Carroll (eds.), *Standard Handbook for Electrical Engineers*, 10th ed., 1968; G. W. Vinal, *Storage Batteries*, 4th ed., 1955.

Substitute natural gas (SNG)

Any synthesized high-Btu-value gas, usually manufactured by the chemical conversion of a hydrocarbon fossil fuel, that is chemically and physically interchangeable with the natural gas sold today by utilities. *See* NATURAL GAS.

The principal feedstocks used or anticipated for producing substitute natural gas (SNG) are coal, oil shale, petroleum, and petroleum-related products; however, SNG might also be produced by using vegetation or biological waste as feedstocks.

Specifications. SNG is distinguished from other manufactured gases by its high methane content—usually above 85%, and by its high heating value—above 900 Btu/standard cubic foot (SCF), or 33.5 megajoule/m³. Minimum specifications for SNG were set by the Office of Coal Research (OCR), formerly in the Department of the Interior, now incorporated into the U. S. Energy Research and Development Administration (ERDA). The specifications include a pipeline delivery pressure of 1000 psig, minimum heating value of 900 Btu/SCF, 0.1% maximum CO content, 10 grains/100 SCF (2.289 grams/m³) total sulfur, 7 lb/10⁶ ft³ (3.18 kg/28,317 m³) water, specific gravity (air standard) of 0.59–0.62, 5% maximum total inert constituents, and hydrocarbon dew point of −40°F (−40°C) at 1000 psig. *See* HEATING VALUE.

In addition, an interchangeability index, to determine the compatibility of SNG with appliances that burn natural gas, has been compiled by the American Gas Association (AGA) on the basis of SNG calorific value, specific gravity, flame velocity, and flame stability. The index is used to predict such undesirable flame features as lifting, flashback, and yellow tipping.

Commercial production. As a result of early research supported by the AGA and OCR, a number of advanced American processes are emerging that are expected to provide the basis for commercial production of SNG from both coal and oil shale beginning before 1985. Actually, however, gas manufactured from coal preceded the worldwide use of natural gas, and was the basis for the United States gas industry until pipeline technology developments in the 1940s permitted nationwide conversion to less expensive natural gas. Historically, gas, manufactured from coal in unaerated retorts, was used for lighting as early as the 17th century. After the gas industry conversion to processed wellhead gas in the 1940s, manufactured gases in the United States served for the most part to supply utility peakload needs. In this instance, an SNG was produced by pyrolysis of oil on hot checker-brick. *See* COAL GASIFICATION.

The resurgence of interest in synthetic gas by the United States gas industry in the mid-1970s resulted from predictions and experience of diminishing domestic natural gas producibility and gas reserves, and rising fuel consumption rates, plus the reexamination of the commodity and political values of energy after the 1973 Organization of Petroleum Exporting Countries (OPEC) oil embargo. As part of a United States goal termed Project Independence, large-scale commercial SNG production is now advocated along with other interim fossil and fissionable fuel energy sources to make the nation independent of foreign fuel sources until fission, fusion, solar energy, and other resources can be advanced to large-scale commercial application. *See* ENERGY SOURCES.

Feedstocks. Alternative liquid and solid feedstocks have been proposed for SNG production. Preference is based on the technological status of a conversion process, feedstock availability and cost, and the economics of conversion. Preferably, the feedstock should have a low sulfur content, and a low carbon-hydrogen weight ratio approaching that of the methane end product. A feedstock of this type is desirable because: (1) high-carbon liquid feedstocks tend to result in coke deposition that can plug the gasifier apparatus and can interfere with catalyst performance (such feedstocks require greater quantities of expensive hydrogen to produce methane); and (2) sulfur deactivates most gasification process catalysts and is a potential pollutant. On these bases, the preferred feedstocks are light (low-molecular-weight) petroleum fractions, which include propane, butane, ethane, naphtha, and natural gasolines. These feedstocks, however, being required in increasing amounts which must be largely imported, represent a solution that is contrary to the stated goal of Project Independence. In terms of domestic availability, abundant coal and oil shale are the most practical feedstocks for long-term domestic SNG production.

Coal is found in 30 of the 50 states in beds that often traverse natural-gas pipelines. The estimated recoverable domestic coal reserves total more than 1400×10^9 short tons (1270×10^{12} kg), and estimated recoverable oil shale reserves in the Green River Basin alone total more than 1000×10^9 bbl of syncrude. Understandably, then, the first United States commercial SNG plants have been designed to reform naphtha and liquid petroleum gas

feedstocks while emerging coal and oil shale gasification processes are perfected. *See* COAL; LIQUEFIED PETROLEUM GAS (LPG); OIL SHALE.

Reforming processes. The three principal light-hydrocarbon reforming processes of current (1975) interest are the Catalytic Rich Gas (CRG, developed by the British Gas Council), Gasynthan (Lurgi GmbH), and Methane Rich Gas (MRG, Japan Gasoline and Universal Oil Products) processes. Each makes methane by reacting light hydrocarbons with low-temperature steam (800–1000°F, or 425–540°C), over a fixed bed of nickel-based catalyst to produce methane, and carbon dioxide (which is later removed).

Despite the simplicity and economy of light-hydrocarbon reforming, the shortage of such feedstocks and their cost make coal by far the greatest potential domestic source of SNG through new processes. Whether coal is converted to a gas or a liquid in a process depends on the nature of process conditions.

Coal conversion. Four principal conversion routes for coal are possible to produce a gaseous or liquid fuel convertible to the principal SNG constituent—methane:

1. Thermal cracking (pyrolysis) thermally breaks chemical bonds of the molecules in coal, generating a broad spectrum of lower-molecular-weight products. *See* CRACKING.

2. Solvation dissolves the coal in a solvent that transfers hydrogen to the molecules in coal, which

Summary of characteristics for conceptual and emerging coal-to-SNG processes

Process	Type of process*	Feedstock and technique	Gasifier stages	Reactor parameters Temp., °F	Reactor parameters Press., psig	Gas-solids contacting scheme	Nitrogen barrier	Fraction of total methane produced directly	Process status
Koppers-Totzek	G	Pulverized coal in entrained flow	1	2700–3300	Atm.+	Dilute phase co-current slagging	Oxygen plant	Minor	Commercial gasifier (for synthesis gas)
Wellman-Galusha	G	3/16–5/16-in. coal lock hopper	1	1000–1200	Atm.+	Moving-bed countercurrent	Oxygen plant	Minor	Commercial gasifier (air-blown, for low-Btu gas)
Winkler	G	Crushed 0–3/8-in. coal screw feeder	1	1500–1850	Atm.+	Fluidized-bed	Oxygen plant	Minor	Commercial gasifier (for synthesis gas)
Texaco	G	Pulverized coal in entrained slurry flow	1	2700–3300	1000	Dilute phase co-current	Oxygen plant	Minor	Gasifier commercial on oil; pilot plant on coal
Union Carbide/Battelle	G	−35 mesh coal with ≅ 6 mesh hot agglomerates in lock hoppers	1 + regenerator	1600–1800	200	Fluidized-bed	Heat carrier	20%†	Pilot plant under construction
COGAS	G	Direct-fed char from pyrolysis stage	1 + regenerator	1600–1700	Atm.+	Fluidized-bed	Heat carrier	Minor	Pilot plant
Lurgi	G/HD	1/4–1 1/4-in. coal lock hopper	1	1150–1600	350–450	Fixed-bed countercurrent	Oxygen plant	40%+†	Commercial gasifier (for synthesis gas)
IGT HYGAS–EG	H	−8 mesh coal in slurry	2 + hydrogen Source	1200–1400 (1st stage)	1100	Dilute phase co-current (1st stage)	Electrical process	64%	Pilot plant
IGT HYGAS–SO	G/H						Oxygen plant	58%	Pilot plant
IGT HYGAS–SI	H			1600–1800 (2d stage)		Fluidized-bed (2d stage)	Iron oxide solids	78%	Pilot plant
CO₂ acceptor	G/HD	<1/8-in. crushed lignite and subbituminous only, through lock hopper	1 + regenerator	1575 (gasifier) 1900 (regenerator)	150 150	Fluidized-bed	Dolomite	55%†	Pilot plant
BI-GAS	G/HD	70%, −200 mesh pulverized coal; lock hopper slurry	2	3000 (1st stage) 1700 (2d stage)	1000 1000	Dilute phase co-current (1st stage) Cyclone slagging gasifier (2d stage)	Oxygen plant	50%†	Pilot plant nearing completion
Synthane	G/H	−60 mesh in lock hopper slurry	2	1800 (1st stage) 1800 (2d stage)	1000 1000	Dilute phase counter-current (1st stage) Fluidized-bed (2d stage)	Oxygen plant	65%†	Pilot plant nearing completion

*G = gasification, H = hydrogasification, HD = hydrodevolatilization.
†Estimated or calculated from limited available data.

break up, resulting in lower-molecular-weight stable hydrocarbons. Based on the process, this reaction is carried out with or without a catalyst, and with or without free hydrogen being present. Methane is produced during this processing, and product oils can subsequently be reformed to SNG. *See* COAL LIQUEFACTION.

3. Gasification reacts the coal with steam, producing a synthesis gas that, after suitable processing, can be converted to methane over a nickel catalyst. Oxygen can be added with the steam to react with a portion of the coal to supply the necessary heat of reaction, or heat can be introduced by other indirect means.

4. Hydrogasification reacts the coal with hydrogen or hydrogen-steam mixtures, usually at high pressures (30–70 atm; 1 atm = 101,325 N/m²), and produces methane directly.

The amount of methane produced directly in these various routes depends on the process conditions, mainly pressure (high, giving increased methane content) and temperature (low, giving high methane content); on the method of contacting the coal with the reacting gases; and on the feed composition.

On leaving the reactor, the hot product gas, in addition to methane, will have steam, hydrogen, carbon monoxide, and carbon dioxide as principal constituents, combined with impurities, which may include ammonia, nitrogen, hydrogen cyanide, hydrogen sulfide, oils, and tars. After the impurities are removed, the gas is upgraded to SNG by reacting the carbon oxides and hydrogen over a catalyst to produce additional methane.

Of the four routes outlines for fuel conversion to SNG, principal interest focuses on the conversion of coal to SNG by direct hydrogasification, or by gasification followed by methanation, or by both.

New processes. Emerging processes now being developed in the United States vary principally on the following points: (1) the technique of introducing heat while excluding nitrogen; (2) the technique of gas-solids contacting, by which gases are placed in contact with solids; and (3) the technique of high-carbon utilization. These differences account for different reactor concepts. The principal candidate processes for commercial coal-to-SNG conversion are compared in the table.

Each of these processes employs a series of basic steps, which are: (1) coal receiving and storage; (2) coal processing into the size required for gasification, and often into a form, such as coal oil or coal-water slurry, for convenient introduction into the gasification reactor; (3) conversion to a gas; (4) gas cleaning and cooling; (5) shift conversion of carbon monoxide in the gas to carbon dioxide, with the generation of more hydrogen (usually a 3:1 hydrogen-to-carbon monoxide ratio is sought prior to catalytic methanation); (6) acid-gas removal and purification, taking out carbon dioxide and gaseous sulfur compounds, for the most part as hydrogen sulfide; (7) methanation, to maximize the heating value of the gas by reacting hydrogen and carbon monoxide to form additional methane; and (8) drying and compressing the gas (if required) to meet pipeline requirements mentioned above.

Domestic coal. The various emerging American processes that are listed along with existing technology in the table are all designed specifically to produce SNG and to utilize domestic coal as a

feedstock. American coal generally differs from European coal, for which the older gasification processes were designed. Eastern United States coals have somewhat higher sulfur contents and may tend to agglomerate; therefore, these new processes must take advantage of the most modern mechanical and chemical engineering concepts to overcome these objectionable features.

Hygas process. The only United States process to have reached an advanced operational pilot plant stage by May 1975 is the Institute of Gas Technology's HYGAS proeess, sponsored jointly by the AGA and ERDA. In 2 years of pilot plant operation, the process has achieved many milestones, including the first production of SNG on a large scale in continuous plant operation using coal as a feedstock. The HYGAS process is basically a two-stage high-pressure (1100 psi) process that feeds any rank of coal, utilizing dilute-phase gas-solids contacting in the first stage at 1200–1400°F (650–760°C), and fluidized-bed gas-solids contacting in the second stage at 1600–1800°F (870–980°C), as well as in the steam-oxygen char gasifier in which the necessary hydrogen is produced.

Underground processes. In-place processes are being considered again for the production of gas from coal. By this technique the coal is fractured in place, and the gasification reactions are conducted underground. Because of difficulties in controlling these underground reactions and associated gas flow, this concept—although used for low-Btu gas production in Russia with varying degrees of success—does not at this time offer as promising an approach as do aboveground techniques. Work on in-place processes is continuing, however, in an attempt to achieve a practical solution.

Oil shale conversion. Oil shale is the second most likely feedstock for commercial SNG production; however, the concept of producing SNG from oil shale is not as advanced in development as the coal-based processes. Two approaches can be taken. In one concept, the shale is first retorted to convert the kerogen in the shale to oil, which is subsequently gasified by techniques already discussed. The second approach, and the one that holds the most promise, is the direct hydrogasification of the kerogen in the shale to directly form a high-methane content gas. Because of the differences between kerogen and coal, considerable oil is coproduced when producing gas by this route. *See the feature article* OUTLOOK FOR FUEL RESERVES.   [FRANK C. SCHORA]

Bibliography: Institute of Gas Technology data on oil shale and coal reserves, developed by H. R. Linden and J. D. Parent, in F. C. Schora et al., Progress in coal gasification, *IEEE Power Engineering Society Conference*, Anaheim, CA, July 14–19, 1974; G. Long, *Amer. Gas Ass. Mon.*, 54(6): 31–33, 1972; H. H. Lowry, Pyrolytic reactions of coal, *Chemistry of Coal Utilization, Supplementary Volume*, pp. 340–395, 1963; Office of Coal Research, formerly of the U.S. Department of the Interior, now in the U.S. Energy Research and Development Administration, *Standard for Acceptable Quality Pipeline Gas from Coal*, Apr. 23, 1965; F. C. Schora et al., Conversion of oil shale by controlled hydrogasification, *3d Conference on Natural Gas Research and Technology*, Dallas, Mar. 6–8, 1974; F. C. Schora, B. S. Lee, and J.

Huebler, The HYGAS process, *12th World Gas Conference*, Nice, France, 1973; Synthetic Gas–Coal Task Force, *Final Report: The Supply-Technical Advisory Task Force*, Supply-Technical Advisory Committee, National Gas Survey, Federal Power Commission, April 1973; R. Vener and F. C. Schora, Survey of the current gasification research and development in progress in the U.S., *Economic Commission for Europe*, November 1975.

Superport

An oil export terminal where many oil tankers can be loaded simultaneously. A striking example is located in the Persian Gulf off the coast of Kharg Island, Iran. Designed to store and ship oil produced in fields on the Iranian mainland, it consists of large-diameter storage tanks, pipelines, and tanker mooring and loading facilities. At the Kharg Island Superport it is possible to load up to 12 tankers simultaneously, ranging in size up to 500,000 deadweight tons (1 DWT = 1017 kg).

Design and construction. From its initial design in 1958, the terminal has developed and been constructed in stages (Fig. 1). The original installation, commissioned in 1960, consisted of a 2,700,000-barrel (10^3 bbl = 2.36 m^3) tank farm, a 1500-ft (1 ft = 0.3 m) causeway and small-craft harbor, and a 2500-ft pile-supported steel trestle, connected to a deep-water pier 2016 ft long for mooring and loading tankers. An expansion program completed late in 1966 raised the tank farm capacity to approximately 8,000,000 bbl, widened the trestle, and lengthened the pier to 6000 ft overall. Further expansion added more tank farm storage and a sea island structure, providing two berths, one for 500,000-DWT tankers.

In the mid-1950s, the Iranian government reached an agreement with a consortium consisting of the world's major oil companies to explore, produce, and develop the oil fields in an agreed area of Iran. During 1957, preliminary planning for a suitable terminal progressed. At first, locations along the coast of the mainland were explored, principally in the northeastern corner of the Persian Gulf, because of the proximity to the fields. Eventually, the natural attributes of a terminal site off Kharg Island became apparent. It was relatively close to the mainland (27 mi; 1 mi = 1.6 km) and could be supplied by pipeline from the fields. It had deep water close to the island for pier facilities, and there was adequate land elevation on the island for storage to allow the tankers to be loaded by gravity flow. The east side of the island was chosen for the tanker marine terminal facilities because protection is provided by the island against the prevailing northwesterly winds and waves.

The facilities engineered for the original terminal included a continuous pier for berthing four tankers, two on each side, located in 65 ft of water. A trestle containing a roadway and oil pipelines connects the pier to shore through a causeway constructed where water depth is shallower. A breakwater arm at the deep end of the causeway provides a harbor for smaller vessels, such as tugs and launches used in mooring tankers. Onshore, 12 tanks were erected for crude oil storage. Pipelines, 36 in. (1 in. = 2.5 cm) in diameter, connect the tank farm with the loading positions on the pier (Fig. 1).

Considerable study resulted in selecting the 100,000-DWT tanker as the design ship for the

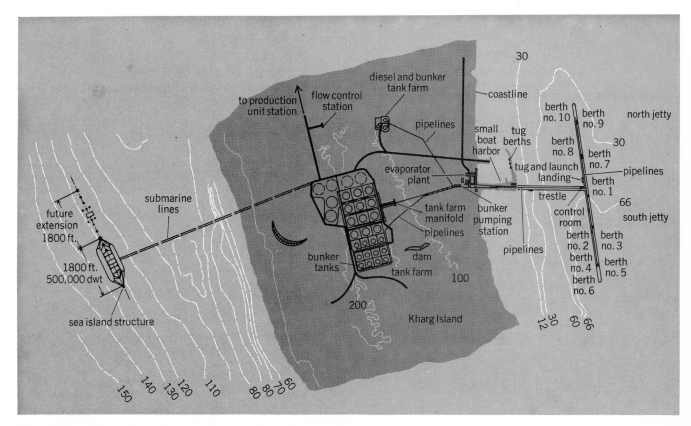

Fig. 1. Plot plan of Kharg Export Terminal. Water depths and land elevation are in feet.

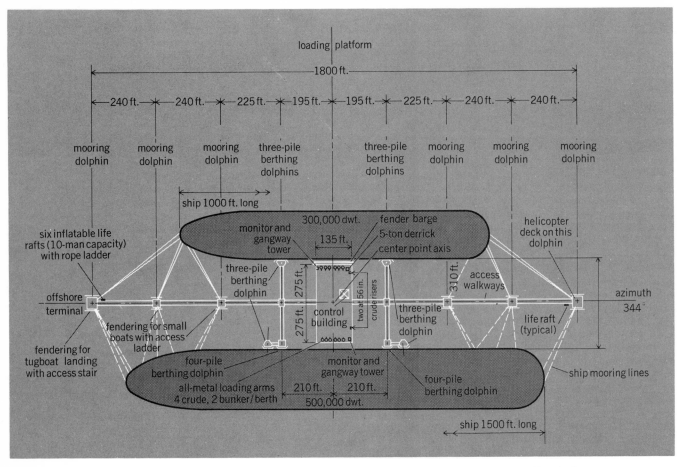

Fig. 2. Plan of the arrangement of structures for the sea island portion of Kharg Export Terminal.

terminal. Sufficient capacity in the design allowed future expansion to handle 150,000-DWT tankers. This was a major decision, as the largest vessel afloat at the time was approximately 80,000 DWT and drew 45 ft of water fully loaded. The parameters selected added 1500 ft to the distance from shore for the location of the pier and placed it in a 65-ft water depth to accommodate the loaded draft of 150,000-DWT tankers. The original design provided a continuous berthing face for tankers to approach the pier, as berthing was conceived to be without the use of tugs. The energy of impact on berthing could, therefore, be applied anywhere along the length of the pier for the contact length between tanker and fender system. The resulting design was a continuous steel superstructure of transverse trusses at regular intervals connected in the horizontal plane with bracing. The trusses sit on pairs of steel batter piles which provide the reaction for the forces of impact. A fender system of rubber blocks acting in shear and placed along the pier face absorbs the energy of impact. Horizontal and vertical steel members with timber cladding and supported by vertical piles are the contact surface with the tanker, and distribute the berthing impacts to fenders and piles. The structure is all steel with timber decking. No concrete is used except for foundation structures onshore.

Terminal expansion. The original terminal was expanded as additional capacity was required to meet growing world oil demand. The existing pier was lengthened at both ends to allow 10 ships to berth simultaneously. Eleven 500,000-bbl storage tanks were added to the tank farm. Additional pipelines, 48 in. in diameter, were added to the delivery system. The expansion provided the terminal with a capacity to deliver 337,500 barrels per hour (bph) to the ships berthed at the pier, and a maximum of 75,000 bph to a tanker. A 100,000-DWT tanker could be loaded in 10 hr, and with about 4 hr for berthing, deballasting, documentation, and deberthing, it could be turned around in a total of 14 hr. Tankers of up to 200,000 DWT could be accommodated at the outside berths under controlled conditions of berthing and with suitable weather and tide levels. The export capacity of the terminal with this expansion was rated at 3,250,000 barrels per day (bpd).

To keep pace with increasing oil demands, additional upgrading was required and resulted in improvements to the energy-absorbing capacity of the fender system and localized dredging. These improvements allowed 250,000-DWT tankers to be loaded at selected berths and raised the capacity of the terminal to approximately 3,750,000 bpd.

Crude oil supply areas, especially in the Middle East, were under continuing pressure to increase export throughout the late 1960s and into the 1970s. The rapid advancement in the size of tankers was making obsolete many existing facilities. The original pier facility of Kharg Island for loading tankers had reached the limit of expansion possibilities. No extensions to the pier were practical, and a new generation of tankers had to be accommodated—the 500,000-DWT class. Also, the loading rate had to be substantially increased to keep the turn-around time in port of these tankers within reasonable limits to avoid demurrage costs.

New facilities. Three sites were considered for new facilities. All had 100 ft of water depth available so that they could berth vessels with drafts of 90 ft, the best estimate of the draft for the 500,000-DWT tanker. One site was east of the existing 10-berth pier on the east side of Kharg Island. Another was east of the nearby island of Khargu, which afforded some protection from the prevailing northwesterly swells. A third site was west of Kharg Island but exposed to the prevailing winds. However, it was much closer to shore and the existing tank farm than the other sites, and was selected.

Single-point mooring (SPM) systems were considered when studying alternate tanker berths. However, two berths were necessary, and as SPM needs more sea area per berth to allow for a 360° mooring swing from critical water depths and obstructions, the submarine lines became long and the system uneconomic. A fixed-structure island design was chosen. Two berths were immediately available in a small area closest to shore. Also, it best suited the very high loading rates (30,000 tons per hour; 200,000 bph) for which the new facilities were to be designed. Although the site was exposed to the prevailing northwesterly winds and it was estimated there could be as many as 60 days a year in which the berths were unavailable due to unfavorable weather, it proved economic to adopt this site and the sea island concept.

Sea island. The sea island structure is located in 105 ft of water and has an overall length of 1800 ft. The arrangement of the structure as shown in Fig. 2 reflects the latest principles with regard to design of such facilities, that is, placing structures only as necessary for function, either berthing, mooring, loading, or communication. It is influenced by the fact that tankers are brought alongside by tugs pushing a stopped tanker sideways. The central platform is 275 ft by 135 ft with two pairs of berthing dolphins, three on either side of the central platform. The overall spacing of the outer mooring dolphins is based on a maximum vessel length of 1500 ft. The four inner dolphins are conveniently arranged to accept breasting lines from a range of ship sizes. It is expected that the outer berth will be used for ships from 500,000 to about 250,000 DWT while the inner berth will be used for ships from 300,000 to about 150,000 DWT. Vessels smaller than that are generally berthed at the 10-berth pier on the eastern side of the island. In spacing the breasting dolphins, the distance was made as large as possible consistent with the length of the parallel sides of the tankers. This increases the distance from the center of mass of the tanker to the point of contact with the structure, thus decreasing the impact.

The spacing between the berthing faces of outer and inner berth was calculated to limit the maximum vertical angle of a breasting line from the tanker to no more than 30° through all stages of tide.

The design of the berthing dolphins was controlled by the energy of impact of the tankers. The design speed perpendicular to the berth was taken as 6 in. per second, increased by the tidal current

set onto the outside berthing face, to a total of 8.5 in. per second. This energy results in extreme berthings of the 500,000-DWT vessel, and the berthing dolphin is designed up to the level of yield stresses in the material. Normal berthings are expected to be no greater than 50% of this energy, resulting in lower working stresses.

A scale model of the tanker was tested for random sea conditions, winds, and current to determine loads from mooring lines. The tanker was ballasted for deadweight tonnages varying from 20 to 100%. Tests showed that a tanker at the inner berth produced the maximum mooring dolphin loads of 300 tons (1 ton force = 8896 N) when remaining alongside in seas having a 6-ft significant wave height and westerly winds of up to 40 knots (20 m/s).

Six storage tanks, each having a capacity of 1,000,000 bbl, were added to the tank farm. Oil is delivered to the sea island through two parallel 78-in.-diameter lines running aboveground to the west shoreline. Here they reduce to 56-in. diameter and continue to the central platform of the sea island as buried submarine pipelines.

Piping on the central platform is manifolded to deliver crude to each berth from the incoming submarine lines through turbine meters. Four 24-in. all-metal loading arms load oil to tankers at rates to 30,000 tons per hour (200,000 bph) at the outer berth and four 16-in. arms allow loading at a rate of 15,000 tons (100,000 bph) at the inner berth. The Kharg Island terminal, with all its facilities, can now export in excess of 5,000,000 bpd.

The terminal at Kharg Island provides a fine example of a high-capacity export facility engineered to the highest state of the art in each development phase.

With continuing dependence on crude oil as a source of world energy, and the ability of Iran to supply this oil, Kharg will remain a major terminal with future expansions as necessary. *See* PIPE-LINE.

[JOSEPH A. FERENZ]

Bibliography: C. L. Bretschneider, A theory of waves of finite height, *7th Conference of Coastal Engineering*, 1961; P. Bruun, *Port Engineering*, 1973; J. R. Morrison et al., The force exerted by surface waves on piles, *Petrol. Trans.*, 1950; A. D. Quinn, *Design and Construction of Ports and Marine Structures*, 2d ed., 1972.

Synchronous motor

An alternating-current (ac) motor which operates at a fixed synchronous speed proportional to the frequency of the applied ac power. A synchronous machine may operate as a generator, motor, or condenser depending only on its applied shaft torque (whether positive, negative, or zero) and its excitation. There is no fundamental difference in the theory, design, or construction of a machine intended for any of these roles, although certain design features are stressed for each of them. In use, the machine may change its role from instant to instant. For these reasons it is preferable not to set up separate theories for synchronous generators, motors, and condensers. It is better to establish a general theory which is applicable to all three and in which the distinction between them is merely a difference in the direction of the currents and the sign of the torque angles. *See* ALTERNAT-

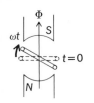

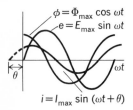

Fig. 1. Single-phase, two-pole synchronous machine.

ING-CURRENT GENERATOR; ALTERNATING-CURRENT MOTOR.

Basic theory. A single-phase, two-pole synchronous machine is shown in Fig. 1. The coil is on the pole axis at time $t=0$, and the sinusoidally distributed flux ϕ linked with the coil at any instant is given by Eq. (1), where ωt is the angular displacement of the coil and Φ_{max} is the maximum value of

$$\phi = \Phi_{max} \cos \omega t \qquad (1)$$

the flux. This flux will induce in a coil of N turns an instantaneous voltage e, given by Eq. (2). The

$$e = -N\frac{d\phi}{dt} = \omega N\Phi_{max} \sin \omega t = E_{max} \sin \omega t \qquad (2)$$

effective (rms) value E of this voltage is given by Eq. (3).

$$E = \frac{E_{max}}{\sqrt{2}} = \sqrt{2}\,\pi f N\Phi_{max} = 4.44\, f N\Phi_{max} \qquad (3)$$

If the impedance of the coil and its external circuit of resistance R and reactance X is given by Eq. (4), there will flow a current, with a value given

$$\mathbf{Z} = R \pm jX = Z\underline{/\pm\theta} \qquad (4)$$

by Eq. (5), in which the phase angle θ is taken positive for a leading current.

$$\mathbf{I} = \frac{\mathbf{E}}{\mathbf{Z}} = \frac{E}{Z}\underline{/\mp\theta} \qquad (5)$$

itive for a leading current. This current will develop a sinusoidal space distribution of armature reaction as in Eq. (6). If this single-phase mmf is

$$A = 0.8NI_{max} \sin(\omega t + \theta) \qquad (6)$$

expressed as a space vector and resolved into direct (in line with the pole axis) A_d and quadrature A_q components, it is given by Eq. (7).

$$\begin{aligned}\mathbf{A} &= A_d + jA_q \\ &= 0.4\,NI_{max}\{[\sin\theta + \sin(2\omega t + \theta)] \\ &\quad + j[\cos\theta - \cos(2\omega t + \theta)]\}\end{aligned} \qquad (7)$$

In a three-phase machine with balanced currents, the phase currents are given by Eqs. (8).

$$\begin{aligned}i_a &= I_{max} \sin(\omega t + \theta) \\ i_b &= I_{max} \sin(\omega t + \theta - 120°) \\ i_c &= I_{max} \sin(\omega t + \theta - 240°)\end{aligned} \qquad (8)$$

Upon writing Eq. (7) for ωt, $\omega t - 120°$, and $\omega t - 240°$, respectively, and adding, Eq. (9) results for the polyphase armature reaction.

$$\mathbf{A} = A_d + jA_q = 1.2NI_{max}(\sin\theta + j\cos\theta) \qquad (9)$$

Equation (10) gives the three-phase power of the

$$P = 3EI \cos\theta \qquad (10)$$

machine, and Eq. (11) gives the developed torque.

$$T = \frac{P}{\omega} = \frac{3}{\omega} EI \cos\theta \qquad (11)$$

The above equations constitute the essential description of the synchronous generator. The same equations apply for a motor if the currents are reversed, that is, by changing the sign of the current I. They may also be interpreted in the form of vector diagrams, and show the two cases of a smooth-rotor and a salient-pole machine.

Smooth-rotor synchronous machine. In the smooth-rotor machine, the reluctance of the magnetic path is essentially the same in either the direct or quadrature axes. In Fig. 2a let the flux Φ be selected as reference vector and drawn vertically. Then comparing Eqs. (1) and (2) it is seen that the induced voltage E_f lags the flux by 90°. By Eq. (5) the current I lags the voltage by an angle θ for an inductive circuit, and by Eq. (9) causes a constant mmf of armature reaction A in phase with the current. This armature reaction causes a flux ϕ_a, stationary in space with respect to the field poles, which in turn induces a voltage E_a lagging it by 90°. The two induced voltages E_f (due to the field flux Φ) and E_a (due to the armature reaction flux ϕ_a) combine vectorially to give the resultant voltage E'. But the terminal voltage V is less than E' by the resistance and reactance drops, $R\mathbf{I}$ and $jx_l\mathbf{I}$ in the winding, and Eq. (12) applies.

$$\mathbf{V} = \mathbf{E}' - (R + jx_l)\mathbf{I} \qquad (12)$$

The leakage reactance drop $jx_l\mathbf{I}$ lags the current by 90° as does the armature reaction voltage E_a. If a fictitious reactance of armature reaction x_a is introduced to account for E_a, it is obvious that Eq. (12) may be rewritten to give Eqs. (13), in which

$$\begin{aligned}\mathbf{V} &= \mathbf{E}_f - jx_a\mathbf{I} - (R + jx_l)\mathbf{I} \\ &= \mathbf{E}_f - R\mathbf{I} - j(x_a + x_l)\mathbf{I} \\ &= \mathbf{E}_f - (R + jX_s)\mathbf{I}\end{aligned} \qquad (13)$$

$X_s = x_a + x_l$ is called the synchronous reactance of the machine.

Salient-pole synchronous machine. In a similar fashion the vector diagram for a salient-pole machine, Fig. 2b, may be set up. Here the effects of saliency result in proportionately different armature reaction fluxes in the direct and quadrature axes, thereby necessitating corresponding direct, X_d, and quadrature, X_q, components of the synchronous reactance. The angle δ in Fig. 2 is called the torque angle. It is the angle between the field-induced voltage E_f and the terminal voltage V and is positive when E_f is ahead of V.

The foregoing equations and vector diagrams were established for a generator. A motor may be regarded as a generator in which the power component of the current is reversed 180°, that is, becomes an input instead of an output current. The motor vector diagram is shown in Fig. 3. Here the torque angle δ is reversed, since V is ahead of E_f in a motor (it was behind in the generator). Therefore a motor differs from a generator in two essential respects: (1) The currents are reversed, and (2) the torque angle has changed sign. As a result the power input, Eq. (10), for a motor is negative, or has become a power output, and the torque is reversed in sign.

When the current I is 90° out of phase with the terminal voltage V the torque angle δ is nearly zero, being just sufficient to account for the power lost in the resistance.

Therefore, a synchronous machine is a generator, motor, or condenser depending on whether its torque angle δ is positive, negative, or zero. For these conditions the output current is respectively at an angle less than ±90°, greater than ±90°, or essentially ±90° with respect to the terminal voltage. Furthermore, depending on this power-factor angle, the field-induced voltage E_f may be greater

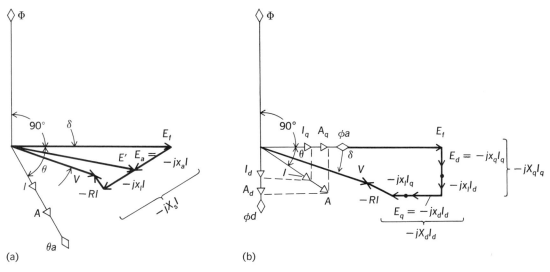

Fig. 2. Vector diagrams of synchronous generators. (a) Smooth-rotor machine. (b) Salient-pole machine.

(overexcited) or less (underexcited) than the terminal voltage V, and the machine may be made to take either leading or lagging currents.

Synchronous condenser. A synchronous condenser can be made to draw a leading current and to behave like a capacitance by overexciting its field. Or, it will draw a lagging current on underexcitation. This characteristic thus presents the possibility of power-factor correction of a power system by adjusting the field excitation. A machine so employed at the end of a transmission line permits a wide range of voltage regulation for the line. One used in a factory permits the power factor of the load to be corrected. Of course a synchronous motor can also be used for power-factor correction, but since it must also carry the load current, its power-factor correction capabilities are more limited than for the synchronous condenser.

Power equations. The power P_g and reactance power Q_g of a round-rotor synchronous generator is given by Eq. (14), in which $\tan \alpha = R/X_s$ and $Z_s = R + jX_s$.

$$P_g + jQ_g = \left(\frac{VE_f}{Z_s} \sin (\delta - \alpha) + \frac{RE_f^2}{Z_s^2}\right)$$
$$+ j\left(\frac{VE_f}{Z_s} \cos (\delta - \alpha) - \frac{X_s E_f^2}{Z_s^2}\right) \quad (14)$$

For a round-rotor motor the torque angle δ is negative, and the gross mechanical power output (including windage and friction) is given by Eq. (15).

$$P_m = \frac{VE_f}{Z_s} \sin (\delta + \alpha) - \frac{RE_f^2}{Z_s^2} \cong \frac{VE_f}{Z_s} \sin \delta \quad (15)$$

For a salient-pole machine, neglecting resistance, Z_s is equal to the direct-axis reactance X_d, α is zero, and P_m is given by Eq. (16). Thus the

$$P_m = \frac{VE_f}{X_d} \sin \delta + V^2 \frac{X_d - X_q}{2X_d X_q} \sin 2\delta \quad (16)$$

power or torque depends essentially on the product of the terminal and induced voltages and sine of the torque angle δ; but in the case of the salient-pole machine there is also a second harmonic term

which is independent of the excitation voltage E_f. This term, the so-called reluctance power, vanishes for nonsaliency when $X_d = X_q$. The small synchronous motors used in some electric clocks and other low-torque applications depend solely on this reluctance torque. *See* RELUCTANCE MOTOR.

Excitation characteristics. The so-called V curves of a synchronous motor are curves of armature current plotted against field current with power output as parameter. Usually a second set of curves with input power factor (pf) as parameter is superimposed on the same plot. Such curves (Fig. 4), where armature current is plotted against generated voltage, can be determined from design calculations or from test; they yield a considerable amount of data on the performance of the motor. Thus, given any two of the four variables E_f, I, pf, P, the remaining two may be easily determined, as well as the conditions of maximum power, constant pf, minimum excitation, stability limit, and so forth.

Circle diagrams. Voltage equation (13) and current equation (5) can be combined in such a fashion as to yield Eq. (17), which is the equation of

$$I^2 = \frac{V^2}{Z_s^2} + \frac{E_f^2}{Z_s^2} - 2\frac{V}{Z_s}\frac{E_f}{Z_s} \cos \delta \quad (17)$$

a set of circles with offset center and with different radii (E_f/Z_s). The locus of these circles is the current.

A companion set of circles can be developed giv-

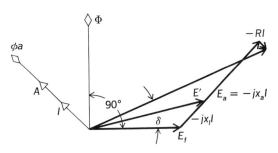

Fig. 3. Vector diagram of synchronous motor.

ing the locus of I as a function of its pf angle for different values of constant developed power.

These two sets of circles are shown in Fig. 5. Such circle diagrams relate the power, pf angle, armature current, torque angle, and excitation.

Losses and efficiency. The losses in a synchronous motor comprise the copper losses in the field, armature, and amortisseur windings; the exciter and rheostat losses of the excitation system; the core loss due to hysteresis and eddy currents in the armature core and teeth and in the pole face; the stray loss due to skin effect in conductors; and the mechanical losses due to windage and friction. The efficiency of the motor is then given by Eq. (18).

$$\text{Eff} = \frac{\text{output}}{\text{input}} = \frac{\text{output}}{\text{output} + \text{losses}} \quad (18)$$

Mechanical oscillations. A synchronous motor subjected to sudden changes of load, or when driving a load having a variable torque (for example, a reciprocating compressor), may oscillate about its mean synchronous speed. Under these conditions the torque angle δ does not remain fixed, but varies. As a result the four separate torques expressed in Eq. (19) act on the machine rotor. The

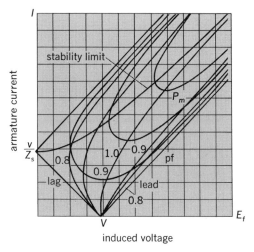

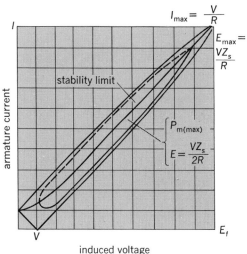

Fig. 4. V curves (armature current versus induced voltage) of synchronous motor.

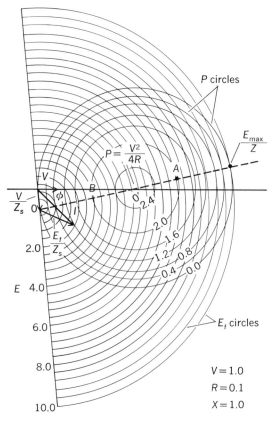

Fig. 5. Circle diagram of synchronous motor.

$$\begin{pmatrix} \text{Synchronous} \\ \text{motor torque} \\ \text{Eq. (16)} \end{pmatrix} + \begin{pmatrix} \text{induction motor} \\ \text{torque of} \\ \text{amortisseur} \end{pmatrix}$$

$$= \begin{pmatrix} \text{torque to} \\ \text{overcome} \\ \text{inertia} \end{pmatrix} + \begin{pmatrix} \text{torque} \\ \text{required} \\ \text{by the load} \end{pmatrix} \quad (19)$$

possibility exists that cumulative oscillations will build up and cause the motor to fall out of step.

Starting of synchronous motors. Synchronous motors are provided with an amortisseur (squirrel-cage) winding embedded in the face of the field poles. This winding serves the double purpose of starting the motor and limiting the oscillations or hunting. During starting, the field winding is either closed through a resistance, short-circuited, or opened at several points to avoid dangerous induced voltages. The amortisseur winding acts exactly as the squirrel-cage winding in an induction motor and accelerates the motor to nearly synchronous speed. When near synchronous speed, the field is excited, and the synchronous torque pulls the motor into synchronism. During starting, Eq. (19) applies, since all four types of torque may be present. Of course, up to the instant when the field is excited the portion of the synchronous motor torque depending on E_f does not exist, although the reluctance torque will be active.

Other methods of starting have been used. If the exciter is direct-connected and a dc source of power is available, it may be used to start the synchronous motor. In the so-called supersynchronous motor the stator is able to rotate in bearings of its own, and is provided with a brake band. For stator allowed to come up to nearly synchronous speed by virtue of the amortisseur windings; the field is then excited and the stator brought to synchronous speed, the rotor remaining stationary. Then as the brake band is tightened, the torque on the rotor causes it to accelerate while the speed of the stator correspondingly slackens; finally the stator comes to rest and is locked by the brake band. In this way maximum synchronous motor torque is made available for acceleration of the load. For other types of synchronous motors *see* HYSTERESIS MOTOR; RELUCTANCE MOTOR.

[LOYAL V. BEWLEY]

Bibliography: L. V. Bewley, *Alternating-current Machinery*, 1949; D. G. Fink and J. M. Carroll (eds.), *Standard Handbook for Electrical Engineers*, 10th ed., 1968; A. S. Langsdorf, *Theory of Alternating-current Machinery*, 2d ed., 1955; M. Liwschitz-Garik and C. C. Whipple, *Electric Machinery: A-C Machines*, vol. 2, 1946; A. F. Puchstein, T. C. Lloyd, and A. G. Conrad, *Alternating-current Machines*, 3d ed., 1954.

Temperature

A concept related to the flow of heat from one object or region of space to another. The term refers not only to the senses of hot and cold but to numerical scales and thermometers as well. Fundamental to the concept of temperature are the absolute scale and absolute zero and the relation of absolute temperatures to atomic and molecular motions.

The sensations of hot and cold are as old as man, but numbers for temperatures, such as 100°C and −15°F, have been used for only about 300 years. By the 17th century, science had developed to the point that, to fully describe the properties of matter, a numerical, quantitative scale of temperature differences was needed. For example, in 1756 Joseph Black in Scotland discovered that ice does not change temperature when it melts. Almost all substances behave this way; also, the melting temperature depends on the purity of the substance. Thus one reason for devising a thermometer (literally, a meter for temperature) was that with it the composition of matter could be studied.

Thermometers do not measure a special physical quantity. They measure length (as of a mercury column) or pressure or volume (with the gas thermometer at the National Bureau of Standards) or electrical voltage (with a thermocouple). The basic fact is that, if a mercury column has the same length when touching two different, separated objects, when the objects are placed in contact no heat will flow from one to the other.

Empirical scales. The numbers on thermometer scales are merely historical choices; they are not scientifically fundamental. The most widely used scales are the Fahrenheit (°F) and the Celsius (°C). The Centigrade scale with 0° assigned to ice water (ice point) and 100° assigned to water boiling under one atmosphere pressure (steam point) was formerly used, but it has been succeeded by the Celsius scale, defined in a different way than the Centigrade scale. However, on the Celsius scale the temperatures of the ice and steam points differ by only a few hundredths of a degree from 0° and 100°, respectively. Fig. 1 shows how the Celsius and Fahrenheit scales compare and how they fit onto the absolute scales. To convert between the

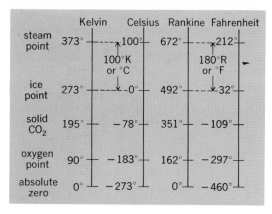

Fig. 1. Comparisons of Kelvin, Celsius, Rankine, and Fahrenheit temperature scales. Temperatures are rounded off to nearest degree. (*M. W. Zemansky, Temperatures Very Low and Very High, Van Nostrand, 1964*)

Celsius and Fahrenheit scales, use $T(°F) = 32 + 9T(°C)/5$.

In 1848 William Thompson, of Scotland, following ideas of Sadi Carnot, of France, stated the concept of an absolute scale of temperature in terms of measuring amounts of heat flowing between bodies, rather than measuring lengths or pressures. Most important, Thompson conceived of a body which would not give up any heat and which was at an absolute zero of temperature. Experiments have shown that, in terms of Celsius and Fahrenheit degrees, absolute zero corresponds to −273.15°C and −459.7°F. Two absolute scales, shown in Fig. 1, are used, and are called the Kelvin, °K (now designated K), after Thompson's honorary title, and the Rankine, °R, to honor a Scottish engineer.

Importance in engineering. The importance of the absolute scales in engineering is their relation to the efficiency of ideal, frictionless Carnot engines. The efficiency is the ratio of the work performed to the heat taken in. The Carnot engine operates by taking in heat from a source at temperature T_1 and exhausting heat at temperature T_2

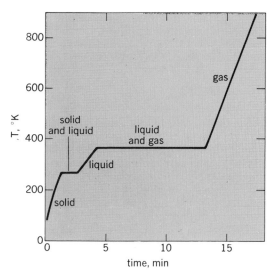

Fig. 2. Temperature of 1 g of H_2O, starting at 0°K, with a constant heat input of 1 cal/sec.

less than T_1. The efficiency is $1 - T_2/T_1$. The efficiency of the Carnot engine operated between the boiling point ($T_1 = 373°K$) and the ice point ($T_2 = 273°K$) of water is 0.268. A real engine with friction would have an efficiency less than this, but the concept is nevertheless of great importance in engineering involving engines and heat transfer.

Importance in basic science. In basic science the importance of the absolute scale is in its relation to the motions of atoms and molecules, whether vibrations, as in solids and liquids, or straight path flights with collisions, as in gases. There are two important facts here: (1) There is a definite distribution of motions. For example, in a gas, even though the motions are chaotic and a particular molecule changes velocity after each collision with another molecule, at any instant a definite number of molecules have a particular velocity. The interesting thing is that it cannot be said that the gas is at a definite temperature unless the molecules have this definite distribution of velocities, although different small portions of the gas may have definite, though different, temperatures. The same idea holds for the distribution of vibration frequencies in solids and liquids. (2) A body has a minimum amount of motion energy. It was supposed in the 19th century that this minimum was zero energy, but modern theories and experiments show that the minimum is greater than zero. A body in its lowest energy state cannot give out heat and is at absolute zero.

Temperature measurements are useful for indicating the energy state of a body. Figure 2 shows how the temperature of 1g of H_2O changes in time as heat is added at the constant rate of 1 cal/sec, assuming that no heat is lost by the H_2O to its surroundings. The different rates of increase of temperature indicate that the heat is going into different kinds of molecular motion.

It is a fundamental principle that no object can actually be at $0°K$. The lowest temperature that has been reached is about $0.000001°K$, in a small bundle of copper wires. The highest temperature reached on Earth has been about $100,000,000°K$, in nuclear explosions. *See* CARNOT CYCLE; THERMODYNAMIC PRINCIPLES.

[ROLAND A. HULTSCH]

Bibliography: C. M. Herzfeld (ed.), *Temperature: Its Measurement and Control in Science and Industry*, vol. 3, 1962; K. Mendelssohn, *The Quest for Absolute Zero*, 1966; M. W. Zemansky, *Temperatures Very Low and Very High*, 1964.

Thermal spring

A spring with water temperature substantially above the average temperature of springs in the region in which it occurs. The average temperature of springs is ordinarily within a few degrees of the mean annual temperature of the atmosphere. Thus waters of thermal springs range in temperature from as low as 60°F, in an area where normal groundwater has a temperature of 40–50°F, to well above the boiling point.

The two main considerations in the origin of thermal springs are the source of the water and the source of the heat. The water may be ordinary groundwater that percolates slowly downward, is heated by the Earth's normal thermal gradient (the temperature of the Earth normally increases about 1°F for each 50–100 ft of depth), and then returns

to the surface without losing all the added heat. The water of thermal springs may be in part juvenile, a product of the crystallization or recrystallization of rock at depth. Since juvenile water is virtually certain to become mixed with connate or meteoric water on its way to the surface, there are no thermal springs whose water can be demonstrated to be wholly juvenile.

Investigations of Warm Springs, Ga., and of other thermal waters in the eastern United States indicate that the water entered the aquifer by normal recharge from precipitation, percolated deep into the Earth by reason of the geologic structure, and there received its heat before returning to the surface. On the other hand, the springs in Yellowstone Park, Wyo., Steamboat Springs, Nev., and many other localities in the western United States may derive part of their water and much of their heat from bodies of superheated rocks, perhaps in the last stages of cooling from the molten state. Many of the springs in the western United States discharge water that is near the boiling point.

Where spring water is above the boiling point it has been tapped to provide steam for power production. Such power installations are found in New Zealand, Italy, and California. Hot springs have also been used to heat homes and swimming pools. *See* GEOTHERMAL POWER.

[ALBERT N. SAYRE/RAY K. LINSLEY]

Thermionic power generator

A device in which heat energy is directly converted to electric energy, frequently called a thermionic converter. The free electrons of good electric conductors flow around suitably arranged conducting paths to create the infinity of useful applications of electricity. At normal temperatures the escape of these electrons from the conducting material can hardly be detected, but at higher temperatures (from 1000 to 2500°K) large numbers of electrons do escape from a heated conductor. This is called thermionic emission of electrons.

Two metallic elements, an emitter and a collector, are the minimum needed for a thermionic converter. The thermionic electron emitter must be capable of yielding electrons to the space that separates the emitter from the electron collector. The collector must be operated at a significantly lower temperature than the emitter so that the collector does not also emit electrons. The general term that describes such a thermionic device is thermionic diode.

Classification. If the space between the two elements of the diode is evacuated sufficiently so that the residual gas has no significant influence on the flow of electrons from the emitter to the collector, the device is known as a vacuum thermionic converter. Electrons are negatively charged particles and thus repel each other. The presence of electrons in transit between the emitter and the collector can, therefore, interfere seriously with the free flow of additional charges and thus set up a space-charge limitation on the current density and the efficiency attainable. *See* SPACE CHARGE.

Two methods are used to minimize the space-charge limitation. One depends on a diode construction with fantastically close spacing—of the order of 5 μ or approximately 0.0002 in. The second method is to introduce an ionized gas. The number density of the ions of the gas must be equal to or locally greater than the electron density in order to neutralize the negative space charge otherwise present. Since the ions used are positively charged, the net charge can be zero even though a high density of electrons is present to provide the means of conduction from the emitter to the collector. The term plasma is applied to a medium in which the net electric charge is zero. A thermionic converter that depends on the presence of an ionized gas to give good conduction in the space between the emitter and the collector is known as a plasma thermionic converter.

Emitter and collector properties. The maximum possible current density in any diode depends on the temperature of the emitter and on the ease of electron removal. The work function of a substance is a direct measure of the energy per electron required for its removal. The emitter work function is defined as the energy difference between the Fermi level within the conductor and the potential energy of an electron at rest just outside the conductor. If current flows through any conductor, the value of the current is generally directly proportional to a measured voltage difference at the ends of the conductor. This voltage difference is equal to the voltage displacement of the Fermi levels at each end of the conductor. In a thermionic converter under actual operating conditions, the Fermi level of the collector must therefore be negative with respect to the Fermi level of the emitter for electric power to be delivered to an external circuit. The work function of the collector must be as small as possible in order to make the voltage available as large as possible. *See* WORK FUNCTION (THERMODYNAMICS).

These points are illustrated in the figure, known as a motive diagram, by which energy relations may be shown. The difference in potential between the Fermi level of the emitter and its surface potential is represented by the vertical arrow designated ϕ_1 and is equal to the emitter work function. For illustrative purposes, the surface potential of the collector is set at the same energy value as that of the emitter, and the Fermi level is positive with respect to this point by the amount of ϕ_2, which is the work function of the collector. Thus, if ϕ_2 is smaller than ϕ_1 and the surface potentials under operating conditions are practically equal, an output voltage designated by V_o will appear at the terminals of the converter. This voltage can be used to drive current through the external load and, under the circumstances illustrated, this voltage is equal to the difference between the emitter work function and the collector work function.

Two conditions of operation are illustrated, that of the vacuum-type converter and that of the preferred plasma converter. In the plasma converter the current transported across the space between the emitter and the collector could be nearly equal to the maximum possible current density J available at the emitter, which is given by Eq. (1), where

$$J = 120 T_1^2 \, e^{-(\phi_1 q/kT_1)} \quad \text{amp/cm}^2 \qquad (1)$$

T_1 is the temperature of the emitter in degrees Kelvin, ϕ_1 is the work function, q is the charge on the electron, and k is Boltzmann's constant. If space charge is present, as in the vacuum diode, then the energy difference represented by ϕ_m must be used in place of ϕ_1 in Eq. (1) to determine the

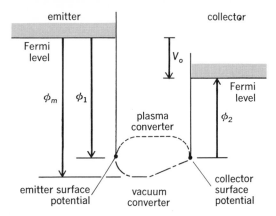

Motive diagram showing energy relations in vacuum and plasma thermionic converters.

maximum current that will be available as the output current of the converter. In the plasma converter there is no inhibiting action, and a current may approach the full emitter current available. This is important because the product of the current and output voltage is the power delivered to the external circuit by the converter.

Inspection of Eq. (1) shows that the ratio (ϕ_1/T_1) must be as small as is practical in order to achieve a high current density. In order not to sacrifice output voltage, this desirable result can best be obtained by having as high a temperature as is possible. Refractory materials such as tungsten, molybdenum, and tantalum can be operated at high temperatures. All these metals have relatively high work-function values unless the emitter surface is partially covered by an electropositive metal, such as cesium. The cesium plasma converter has great promise of being an efficient device.

In the temperature range of 550–650°K, the vapor pressure of cesium changes from 1 to 10 torr. Associated with a thermionic converter, a cesium reservoir maintained within this range of temperature can supply enough cesium to an emitter surface so that even the refractory materials will have work functions as low as 3 ev. Cesium ions are produced at the heated emitter surface in sufficient quantity to neutralize the electron space charge and give a motive function qualitatively represented by that for the plasma converter of the illustration. The adsorption of cesium on the colder collector surface serves to lower its work function to a value of about 1.8 ev, or even less under well-controlled conditions.

Since a low-work-function collector is necessary for an efficient thermionic converter, a correspondingly low temperature must be maintained at the collector to stop the back emission of electrons, which would produce a reduction in current. A satisfactory estimate of the collector temperature T_2 needed to limit back current to less than 2% of the forward current is given by Eq. (2), where ϕ_2 is the work function of the collector.

$$T_2 = T_1 \frac{\phi_2}{\phi_1 + 2.6 \times 10^{-4} T_1} \qquad (2)$$

State of development. Although many engineering details remain to be worked out, it is anticipated that a high-temperature plasma converter will be capable of delivering to an external circuit power corresponding to a density at the emitter of not less than 10 watts/cm² and probably not greater than 40 watts/cm². This operation will be done at an efficiency of approximately 20%, measured in terms of the heat actually delivered to the emitter structure. This heat may be obtained in space vehicles from solar radiation received on a reflector and a suitably designed concentrator. Since applications of this type are so important, research and development related to the direct conversion of heat to electricity by thermionic converters is well warranted.

[W. B. NOTTINGHAM/T. F. STRATTON]
Bibliography: R. O. Jenkins and W. C. Trodden, *Electron and Ion Emission from Solids*, 1965; J. Kaye and J. A. Welsh (eds.), *Direct Energy Conversion*, 1968; N. W. Snyder, Energy conversion for space power, in M. Summerfield (ed.), *Progress in Astronautics and Rocketry*, vol. 3, 1961.

Thermocouple

A device that uses the voltage developed by the junction of two dissimilar metals to measure temperature difference. Two wires of dissimilar metals welded together at the ends make up the basic thermocouple (Fig. 1). One junction, called the sensing or measuring junction, is placed at the point where temperature is to be measured. The other junction, called the reference or cold junction, is maintained at a known reference temperature. The voltage developed between the two junctions is approximately proportional to the difference between the temperatures of the two junctions, and may be measured by including a suitable voltmeter in the circuit, as in Fig. 2. *See* THERMOELECTRICITY.

Although all dissimilar metals exhibit the thermoelectric effect, only a few are in wide use. The major characteristics which make certain metals

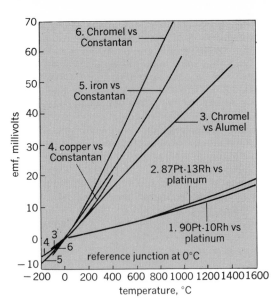

Fig. 3. Temperature-emf curves for thermocouples.

or combinations of metals outstanding for this purpose are (1) stability or reproducibility, the emf does not change rapidly with time; (2) constant or controllable composition, small impurities or changes in composition of wire from end to end or lot to lot can result in varying or nonidentical emf curves; (3) corrosion resistance, wires do not deteriorate or change properties in oxidizing or reducing atmospheres; (4) sensitivity, the emf generated per degree temperature change is large; (5) range, the couple can be used through a broad range of temperatures; (6) ruggedness, tough but easily worked metals give good service; and (7) cost.

The temperature-emf relation of a homogeneous thermocouple is a definite physical property and does not depend upon the details of the apparatus. In a homogeneous thermocouple each element is homogeneous in both chemical composition and physical condition throughout its length. The temperature-emf curves for the six most commonly used pairs of wire materials are shown in Fig. 3,

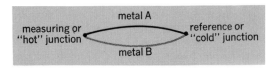

Fig. 1. Elements of a thermocouple.

THERMOCOUPLE

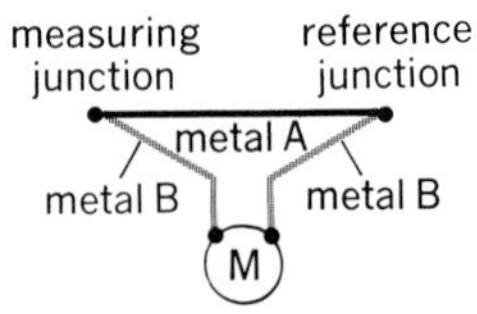

Fig. 2. Thermocouple circuit for measuring metal temperatures.

Temperature limits and corrosion characteristics of thermocouples*

Positive element	Negative element	Temperature range, °C	Influence of temperature and gas atmospheres
90% Pt, 10% Rh; 87% Pt, 13% Rh	Platinum	0 to 1450	Resistance to oxidizing atmosphere very good; resistance to reducing atmosphere poor; platinum corrodes easily above 1000°C and should be used in gastight ceramic protecting tube
Chromel-P	Alumel	−200 to 1100	Resistance to oxidizing atmosphere good to very good; resistance to reducing atmosphere poor; affected by sulfur, reducing or sulfurous gas, SO₂, and H₂S
Iron	Constantan	−200 to 750	Oxidizing and reducing atmospheres have little effect on accuracy, best used in dry atmospheres; resistance to oxidation good up to 400°C, poor above 700°C; resistance to reducing atmosphere good up to 400°C; protection from oxygen, moisture, and sulfur required
Copper	Constantan	−200 to 350	Subject to oxidation and alteration above 400°C due to copper, above 600°C due to Constantan wire; contamination of copper affects calibration greatly; resistance to oxidizing atmosphere good; resistance to reducing atmosphere good; requires protection from acid fumes
Chromel-P	Constantan	−100 to 1000	Chromel attacked by sulfurous atmosphere; resistance to oxidation good; resistance to reducing atmosphere poor

*From D. M. Considine (ed.), *Process Instruments and Controls Handbook*, McGraw-Hill, 1957.

and the normal temperature ranges and corrosion characteristics are listed in the table.

As implied in the preceding paragraph, the emf generated by a thermocouple is independent of the size of wire, and thus small fine wires may be used so that the mass and thermal lag of the sensing junction can be very low. However, small fine wires have limited mechanical strength, and generally such junctions are encased in protective tubing. Thus the fast-response feature of the fine-wire junction can be masked by the protective tubing. In most cases a balance has to be found between a fast response, a long service life, and the protection needed at the place of use.

Usually, extension leads are used to connect the measuring junction with the reference junction and voltmeter when the lengths required are over about 10 ft. The extension wire may be alloyed to match the thermoelectric characteristics of the thermocouple wires or copper wire used (Fig. 4a). When the reference junction can be located close to the measuring junction, copper-wire leads to the meter can be used as required (Fig. 4b and c).

A common practice is to maintain the reference junction at 0°C, or 32°F, by an ice and water bath. Alternatively, the reference junction may be maintained at a carefully controlled temperature somewhat above atmospheric temperature, and the instrument may be zeroed by adding the corresponding voltage. Other instruments allow the reference-junction temperature to vary with ambient conditions within the voltage-measuring instrument and to provide a calibrated electrical or mechanical manual adjustment so that with proper adjustment the instrument reading is correct. Most of the newer instruments that are used with only one type of thermocouple have automatic reference-junction compensation, and their scale or chart is calibrated directly in temperature units.

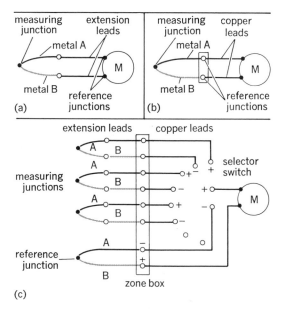

Fig. 4. Thermocouple circuits. (a) Single thermocouple with reference junction(s) at a distance from measuring junction. (b) Single thermocouple circuit with reference junction close to measuring junction. (c) Multiple thermocouple system with reference junction at a distance from both measuring junction and meter.

Any method of measuring small voltages accurately may be used for thermocouple voltages. The millivoltmeter and the potentiometer in various forms are in wide use. Precision measurements with thermocouples (say to 0.1°F) are possible in certain ranges with specially calibrated units. On industrial applications, errors of 5°F and larger occur, but the actual magnitude depends upon the thermocouple materials, the temperature level, the lead wires, the compensation, the voltage-measuring system, and the installation. Although thermocouple systems have many limitations, they are particularly advantageous for applications requiring remote indication or recording, for those on which the measuring junction must be replaced relatively frequently, and for those requiring measurements in the temperature range between 800 and 2400°F. In the laboratory and in experimental work, the thermocouple is frequently a convenient substitute for the more accurate, but less rugged, liquid-in-glass thermometer.

The thermopile is a number of thermocouples connected in series or parallel. The series circuit provides a higher sensitivity (greater emf per degree) than one thermocouple alone and is often selected for this reason. Either the series or parallel circuit may be used for obtaining an indication which approaches the average of the several temperatures.

[HOWARD S. BEAN]

Bibliography: D. M. Considine (ed.), *Process Instruments and Controls Handbook*, 1957; W. F. Coxon, *Temperature Measurement and Control*, 1960; W. F. Roeser and H. T. Wensel, Methods of testing thermocouples and thermocouple materials, *NBS J. Res.*, 14:768, 1935.

Thermodynamic cycle

A procedure or arrangement in which one form of energy, such as heat at an elevated temperature from combustion of a fuel, is in part converted to another form, such as mechanical energy on a shaft, and the remainder is rejected to a lower temperature sink as low-grade heat.

Common features of cycles. A thermodynamic cycle requires, in addition to the supply of incoming energy, (1) a working substance, usually a gas or vapor; (2) a mechanism in which the processes or phases can be carried through sequentially; and (3) a thermodynamic sink to which the residual heat can be rejected. The cycle itself is a repetitive series of operations.

There is a basic pattern of processes common to power-producing cycles. There is a compression process wherein the working substance undergoes an increase in pressure and therefore density. There is an addition of thermal energy from a source such as a fossil fuel, a fissile fuel, or solar radiation. There is an expansion process during which work is done by the system on the surroundings. There is a rejection process where thermal energy is transferred to the surroundings. The algebraic sum of the energy additions and abstractions is such that some thermal energy is converted into mechanical work. *See* HEAT.

A steam cycle that embraces a boiler, a prime mover, a condenser, and a feed pump is typical of the cyclic arrangement in which the thermodynamic fluid, steam, is used over and over again. An

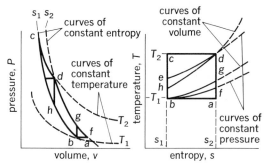

cycles are, in the order of decreasing efficiency:
Carnot cycle (a-b-c-d-a)
Brayton cycle (b-e-d-f-b)
Diesel cycle (b-e-d-g-b)
Otto cycle (b-h-d-g-b)

Comparison of principal thermodynamic cycles.

alternative procedure, after the net work flows from the system, is to employ a change of mass within the system boundaries, the spent working substance being replaced by a fresh charge ready to repeat the cyclic events. The automotive engine and the gas turbine illustrate this arrangement of the cyclic processes, called an open cycle because new mass enters the system boundaries and the spent exhaust leaves it.

The basic processes of the cycle, either open or closed, are heat addition, heat rejection, expansion, and compression. These processes are always present in a cycle even though there may be differences in working substance, the individual processes, pressure ranges, temperature ranges, mechanisms, and heat transfer arrangements.

Air-standard cycle. It is convenient to study the various power cycles by using an ideal system such as the air-standard cycle as illustrated. This is an ideal, frictionless mechanism enveloping the system, with a permanent unit charge of air behaving in accordance with the perfect gas relationships.

The unit air charge is assumed to have an initial state at the start of the cycle to be analyzed. Each process is assumed to be perfectly reversible, and all effects between the system and the surroundings are described as either a heat transfer or a mechanical work term. At the end of a series of processes, the state of the system is the same as it was initially. Because no chemical changes take place within the system, the same unit air charge is conceivably capable of going through the cyclic processes repeatedly.

Whereas this air-standard cycle is an idealization of an actual cycle, it provides an amenable method for the introductory evaluation of any power cycle. Its analysis defines the upper limits of performance toward which the actual cycle performance may approach. It defines trends, if not absolute values, for both ideal and actual cycles. The air-standard cycle can be used to examine such cycles as the Carnot and those applicable to the automobile engine, the diesel engine, the gas turbine, and the jet engine.

Cyclic standards. Many cyclic arrangements, using various combinations of phases but all seeking to convert heat into work, have been proposed by many investigators whose names are attached

to their proposals, for example, the Diesel, Otto, Rankine, Brayton, Stirling, Ericsson, and Atkinson cycles. All proposals are not equally efficient in the conversion of heat into work. However, they may offer other advantages which have led to their practical development for various applications. Nevertheless, there is one overriding limitation on efficiency. It is set by the dictates of the Carnot cycle, which states that no thermodynamic cycle can be projected whose thermal efficiency exceeds that of the Carnot cycle between specified temperature levels for the heat source and the heat sink. Many cycles may approach and even equal this limit, but none can exceed it. This is the uniqueness of the Carnot principle and is basic to the second law of thermodynamics on the conversion of heat into work. *See* BRAYTON CYCLE; CARNOT CYCLE; DIESEL CYCLE; OTTO CYCLE; THERMODYNAMIC PROCESSES.

[THEODORE BAUMEISTER]

Bibliography: T. Baumeister (ed.), *Standard Handbook for Mechanical Engineers*, 7th ed., 1967; J. B. Jones and G. A. Hawkins, *Engineering Thermodynamics*, 1960; J. H. Keenan, *Thermodynamics*, 1941; M. W. Zemansky, *Heat and Thermodynamics*, 1957.

Thermodynamic principles

Laws governing the conversion of energy from one form to another. Among the many consequences of these laws are relationships between the properties of matter and the effects of changes in pressure, temperature, electric field, magnetic field, and composition. The great practicality of the science arises from the foundations of the subject. Thermodynamics is based upon observations of common experience that have been formulated into the thermodynamic laws. From these few laws all of the remaining laws of the science are deducible by purely logical reasoning. There is a choice as to which few are considered independent laws, from which the remainder may be derived. A modern tendency is to choose basic laws or postulates that are different from those first discovered. Some of these choices are most useful in that the derivation of the remainder may be accomplished very efficiently. However, those laws that arose from the historical development will be discussed here since they are less abstract and lend themselves to a clearer physical interpretation.

One may say that the whole development of thermodynamic principles was completed when three state functions, the absolute temperature T, the internal energy U, and the entropy S, were defined. The zeroth law formalizes the concept of temperature, the first law defines the internal energy, and the second law brings in the concept of entropy as well as the absolute scale of temperature. Finally, the third law describes the behavior of entropy and internal energy as the absolute temperature approaches zero.

For exposition, it is necessary to define a few terms. A *system* is that part of the physical world under consideration. The rest of the world is the *surroundings*. An *open system* may exchange mass, heat, and work with the surroundings. A *closed system* may exchange heat and work but not mass with the surroundings. An *isolated system* has no exchange with the surroundings. A closed or isolated system is sometimes referred to as a body.

Those parts of a system spatially uniform and homogeneous are called *phases*. For example, a liquid together with its vapor may be considered a two-phase system. Systems may be made quite elaborate when required, but since focus is on the thermal properties, single-phase isotropic systems not acted upon by electric or magnetic fields are considered so that the only force allowed is that generated by a uniform normal pressure. Such a restriction is not a basic limitation on the generality of thermodynamics but is simply a pedagogical device.

Specification of equilibrium state. The material properties of concern to thermodynamics are the macroscopic properties such as temperature, pressure, volume, concentration, surface tension, and viscosity. Molecular properties such as interatomic distances are not used. The state of a system is specified by all of the macroscopic properties together with their spatial variation. It is a fact of experience, however, that an isolated system approaches a particularly simple terminal state such that the properties are constant and spatially uniform. This simple state is called an equilibrium state. If one confines attention to a given quantity of a single-phase system, the equilibrum state is completely specified by $r+1$ of its properties, where r is the number of components. For a single-component, single-phase system not subject to magnetic or electric fields, one may fix two properties such as pressure and volume; all the remaining properties such as viscosity, surface tension, and so forth then assume fixed values. In other words, any macroscopic property of the system may be expressed as a function of the pressure and volume.

Temperature. It is within the scope of thermodynamics to refine the primitive notion of hotness and coldness into an operational and precise concept of temperature. The equilibrium states of a single-component, single-phase fluid provide a starting point. For such a fluid the equilibrium state is defined by fixing two of its properties. For example, one could construct a mercury-in-glass thermometer that has its pressure held constant; then only one other property could be varied independently. If the volume (height of mercury) is observed at any equilibrium state, there is a 1:1 correspondence between it and any other property excepting the pressure. The degree of hotness is one of these properties.

Also note that thermal equilibrium between different systems exists. For example, if the mercury thermometer is placed in contact with a body of quiet water, the mercury will either expand or contract. The volume change of the mercury will eventually stop, and the properties of the mercury will be constant, indicating an equilibrium state; moreover, the water will also have the constant properties of an equilibrium state. Bodies in thermal equilibrium are said to have the same temperature. Thus, one arrives at a method of measuring the temperature of a body of water.

Now suppose there is a large body of water in thermal contact through a wall with a large body of another fluid such as alcohol, both at equilibrium. It is an experimental fact that the mercury-in-glass thermometer will register the same volume when placed in either of the fluids. This fact is most important if one is to attach meaningful numbers

to temperature. This fact of experience is designated as the zeroth law of thermodynamics: If two bodies A and B are separately in thermal equilibrium with C, then A and B are in equilibrium with each other. Thus, a useful empirical temperature measurement based upon the volume of mercury under constant pressure is established.

But the empirical temperature scale is unfortunately unique to the choice of the fluid. The mercury-in-glass thermometer is calibrated by bringing it into equilibrium with an ice-water mixture and then with boiling water, all under a pressure of 1 atm. The two mercury levels are marked 0° and 100°, respectively, and linear interpolation is used to assign numbers between the two fixed points. If one constructs and calibrates another thermometer but uses another fluid instead of mercury, one would find that the numerical values between the two fixed points do not agree. Indeed, water cannot be used as a working fluid at all since it has a minimum value of volume at 4°C and yields anomalous readings in this region. It is necessary to choose some fluid as a standard and calibrate all other thermometers by comparison with the standard. The second law of thermodynamics removes this dependence upon a particular material by defining an absolute temperature scale that is independent of the working fluid. Meanwhile the empirical temperature serves a useful operational purpose.

One could have used a low-pressure gas as the thermometric fluid. The volume could have conveniently been held constant, and the pressure of the fluid would have had a 1:1 correspondence with temperature. Here one would have found that all gases at low pressures yield the same temperature scale. This ideal gas temperature scale proves to be identical with the thermodynamic scale of the second law.

In summary, the relationship between the properties of the equilibrium state, the notion of thermal equilibrium, and the zeroth law have been used to establish the property of temperature. It may be noted in passing that temperature is a state property, and for a given mass of a single-phase, single-component fluid the temperature is a function of pressure and volume as in Eq. (1). This can be inverted as in Eqs. (2).

$$t = t(p,V) \tag{1}$$

$$p = p(t,V)$$
$$V = V(p,t) \tag{2}$$

Internal energy. Thermodynamics does not define the concepts of energy or work but adopts them from the other macroscopic sciences of mechanics and electromagnetism. Also, the conservation of energy is taken as axiomatic. Therefore, if an isolated system is formed from any part of the world, a definite amount of energy will be trapped in the system. The energy resides in the kinetic and potential energy of the trapped molecules. The trapped energy is of a definite quantity because the isolated system cannot gain or lose energy to the surroundings and remains constant because of the conservation principle. This trapped energy is called the internal energy U.

Because of the conservation of energy, the internal energy of a closed system can be altered only by an exchange of energy with the surroundings.

There are only three modes by which the exchange can occur: by mass transfer, heat transfer, or work exchange. So for a closed (no mass transfer), adiabatic (no heat transfer) system the change in internal energy ΔU is equal to the work done by the surroundings on the system, as defined by Eq. (3).

$$\Delta U = W_{AD} \tag{3}$$

Here a convention has been adopted that work done on the system is positive. There is a great mass of experimental information where work has been done on a closed system enclosed within adiabatic walls. Among these are experiments performed by J. Joule more than a century ago. He caused work to be done on an adiabatically enclosed mass of water in several different ways. A measured amount of work was used to drive an agitator in the water, to create an electric current which was then passed through a coil in the water, to compress a gas in a cylinder immersed in the water, and to rub metal blocks together in the water. In Joule's experiments the same temperature increase was always obtained with the same expenditure of work. It may be concluded from Joule's experiments that the expenditure of a given quantity of work always causes the same change of state regardless of how the work is carried out. Both W_{AD} and ΔU are independent of the path. It is concluded that U is a state function. So for a single-phase, single-component fluid Eq. (4) is

$$W_{AD} = U_2 - U_1 = \Delta U \tag{4}$$

written, where U_2 and U_1 depend only on the final and initial state, respectively. Also one may write Eqs. (5).

$$U = U(p,V) \qquad U = (t,p) \tag{5}$$

It is known from experience that the same change in state of a system can be effected by either supplying work to the system in an adiabatic enclosure or by contacting the system through a conducting wall with a higher temperature system. The latter method is a different means of transferring energy than work and is termed heat and given the symbol Q. Measuring the amount of work required to cause the same change in state as an amount of heat enables one to express heat quantities in terms of work quantities. For example, 1 calorie is taken to be the amount of heat necessary to raise 1 g of water 1°C at 15°C and 1 atm. The same change in state can be effected by 4.186×10^7 ergs of work. Therefore, Eq. (6) holds.

$$1 \text{ calorie} = 4.186 \times 10^7 \text{ ergs} \tag{6}$$

The first law of thermodynamics may now be derived by the useful device of a composite system. Imagine a very large system of water that transfers neither heat nor work to the surroundings. Within this large system there is a small system of a cylinder of gas in thermal contact with the water and having a piston connected to the outside. Work can be done on the small system, and it can in turn interchange heat with the large system called a reservoir. Using subscripts s for the small system and r for the reservoir, the process of doing work can be described as in Eq. (7). But $\Delta U_r =$

$$W_s = \Delta U_s + \Delta U_r \tag{7}$$

$Q_r = -Q_s$. Therefore, for the small system that is exchanging both heat and work with its surroundings, one writes Eq. (8), omitting the subscript.

$$\Delta U = Q + W \tag{8}$$

This is the first law of thermodynamics and states that the algebraic sum of heat and work during a process is equal to the change in the state function U. The term $(Q + W)$ is therefore independent of the path taken between the two states. One could, for example, cause 1 g of water to undergo the change in state as in notation (9) by

$$15°C, 1 \text{ atm} \rightarrow 16°C, 1 \text{ atm} \tag{9}$$

supplying 1 calorie of heat and no work or by doing 4.186×10^7 ergs of work alone, or one could do a great deal of work and abstract all of this energy in the form of heat excepting 1 calorie. Thus, although $(Q + W)$ is independent of the path, neither Q nor W by itself is independent of the path.

It is important to realize that U is a state function and a property of the system whereas W and Q are not. The work, as well as the heat, simply represents energy in transit. Once the energy is in the system, it is not possible to determine whether it came from heat transfer or work transfer; it is simply internal energy.

The differential form of the first law is given by Eq. (10), where q and w represent small quantities.

$$dU = q + w \tag{10}$$

In general, one may not treat q or w as well-behaved differential coefficients dQ and dW. However, if the change is such that either q or w depends only on the initial and final states and not on the path, Eq. (11) may be correctly written. If two

$$dU = dQ + dW \tag{11}$$

terms are independent of the path, the third must be also. Obviously, if either q or w is zero, one may properly write Eqs. (12). There is a third case

$$\begin{aligned} dU &= dW \\ dU &= dQ \end{aligned} \tag{12}$$

where neither q nor $w = 0$, but nevertheless $dU = dQ + dW$ is still proper. Before treating this interesting case one needs to develop the notion of a reversible process.

Reversible and irreversible processes. Any process that occurs in nature is in agreement with the first law, but many processes permissible by the first law never occur. It has already been noted that systems approach an equilibrium state if left to themselves. There is an overwhelming preference for processes to proceed in one direction. Consider Joule's experiment. A falling weight caused a paddle to do work on an adiabatically enclosed body of water. The total effect of the experiment was to increase the internal energy of the water and to lower the weight. The surroundings remained unchanged. The water temperature increased and the volume increased slightly. There is no way one can reverse this process, that is, restore the water to its original state and raise the weight to its original height without also making some additional change in the surroundings. The process is irreversible.

Consider some other processes occurring within an adiabatic enclosure by examining only the initial and final states. Two blocks of copper are initially at different temperatures and finally at the same temperature which is intermediate between the two initial temperatures. A gas is initially filling just half of a container and finally the whole of the container. Again these processes are irreversible; that is, they cannot be reversed without causing some permanent change in the surroundings. The reverses of these processes do not violate the first law; therefore, there must be some condition other than the conservation of energy which is obeyed by those processes which actually take place.

If one were presented with a description of only the initial and final states of these irreversible processes, as was done in the last two examples, one could unerringly decide from the description which state was initial and which was final. The direction of the process is entirely determined by the nature of the states. It may be expected, therefore, that there is some state function that shows which state precedes the other. The function which tells whether a process is possible or not is the entropy S and will be derived from the information that some adiabatic processes are impossible.

Thermodynamics makes use of an idealization, called a reversible process, that is a limiting case of the natural or irreversible process. The reversible process may be defined as one which can be completely reversed without leaving more than a vanishingly small change in the surroundings. It is a consequence of the definition that a reversible process proceeds through a succession of equilibrium states and may be reversed by an infinitesimal change in the external conditions. Imagine having a cylinder of gas fitted with a frictionless piston. If the piston is moved so slowly that pressure gradients are absent, the gas will be in an equilibrium state at all times. The difference between the gas pressure and the external pressure needs only to be infinitesimal in order to move the frictionless piston. Under the rather restrictive conditions of a reversible process, Eq. (13) may be

$$dW = -p\,dV \qquad (13)$$

quite properly written, where p is the gas pressure and V is the gas volume. The first law now may be written as Eq. (14).

$$dQ = dU + p\,dV \qquad (14)$$

Entropy. The discussion on irreversible processes has led to the second law of thermodynamics, which is just a general statement of the idea that there is a preferred direction for a given process. There are many physical statements of the second law, all being equivalent and leading to the same mathematical statement. The statement of R. Clausius is: "It is *not* possible that, at the end of a cycle of changes, heat has been transferred from a colder to a hotter body without producing some other effect." Lord Kelvin's statement is: "It is *not* possible that, at the end of a cycle of changes, heat has been extracted from a reservoir and an equal amount of work has been produced without producing some other effect."

A specific example of Kelvin's statement may be useful. Work can be converted continuously and completely into heat. For example, work could be expended on rubbing blocks in a large mass of water. The blocks would become infinitesimally

hotter than the water and transfer energy to the water by heat flow. The process could be continued indefinitely with the only effect being a complete conversion of work into heat. If, however, heat is converted from the large water reservoir completely into work, some other effect occurs. For example, a gas within a cylinder can be expanded reversibly causing a transfer of heat from the bath to the gas. All of the heat extracted from the bath is converted into work. However, the gas, in this process, has changed its state since its volume is larger. The gas cannot be returned to its original state without undoing the conversion of heat into work already accomplished.

The most efficient way of developing the mathematical consequences of the second law is to proceed from Caratheodory's principle, which can be either taken as another physical expression of the second law or derived from the Clausius or Kelvin statement. Caratheodory's principle is: "In the neighborhood of any equilibrium state of a system there are states which are not accessible by an adiabatic process."

Caratheodory used this principle together with a mathematical theorem that he developed to infer the existence of a state function S and an integrating factor $1/T$, where T is the thermodynamic temperature such that Eq. (15) holds for a reversible

$$dQ_{\text{REV}} = T\,dS \qquad (15)$$

change. The state function S is called the entropy. It can also be shown that the entropy in an adiabatic system increases for an irreversible change and remains constant for a reversible change as in Eq. (16). The implication is that entropy increases for a

$$\Delta S_{\text{AD}} \geq 0 \qquad (16)$$

natural change until equilibrium is reached, and then it remains constant at its maximum value.

The first part of the mathematical statement of the second law allows one to write one of the most important thermodynamic equations, Eq. (17). Al-

$$dU = T\,dS - p\,dV \qquad (17)$$

though this equation was derived for reversible changes, it is valid for all changes. All the quantities are functions of state. Therefore, for a change between two states the integral of the equation will be valid even if the path is not reversible. In other words, for a change from a state characterized by (p_1, T_1) to a state characterized by (p_2, T_2), the values of ΔU, ΔV, and ΔS will all have definite values dependent only upon the two states and independent of how the change came about. From this equation are obtained some of the most fruitful applications of thermodynamics to physical problems.

The second part of the mathematical statement is a concise summary of physical statements on the direction of processes. As a simple example, consider pure heat transfer to a body. The heat transfer causes a definite change of state such that $dQ = dU$. The definite change of entropy is then given by Eq. (18). If heat dQ is transferred from a

$$dS = \frac{dQ}{T} \qquad (18)$$

body at temperature T_2 to a body at temperature T_1, the change in entropy is given by Eq. (19).

$$dS = dS_1 + dS_2$$

$$= \frac{dQ}{T_1} - \frac{dQ}{T_2} \qquad (19)$$

$$= \frac{dQ(T_2 - T_1)}{T_1 T_2}$$

Since dS must be positive or zero, $T_2 > T_1$. Therefore heat flows from the hotter body to the colder body.

Also contained in the second part of the entropy statement is the key idea of equilibrium. The equilibrium state of an adiabatic or isolated system is characterized by entropy being at its maximum value consistent with the physical constraints. Therefore, equilibrium states can be determined by setting $dS = 0$. Also, for a maximum in entropy, $d^2S < 0$. This latter condition leads to the notion of stability that is important in the study of phase equilibrium.

When a system is in thermal contact with its surroundings, the entropy of the system may decrease. For example, a gas being compressed isothermally decreases its entropy, but a greater increase of entropy occurs in the surroundings. The total entropy change is always positive. Clausius stated the first and second laws of thermodynamics as: "The energy of the world is constant. The entropy of the world tends toward a maximum."

By way of completeness the third law of thermodynamics needs comments. In the main body of thermodynamics one is mostly interested in changes of entropy and internal energy between states. However, the third law defines an absolute scale for entropy: The entropy of all perfect crystalline solids is zero at absolute zero temperature. The third law is used primarily in classical thermodynamics for the calculation of absolute entropies which combined with thermochemical data permits the calculation of chemical equilibrium. The foundations of the third law of thermodynamics, however, are to be found in molecular theory and therefore require a statistical mechanical treatment.

Summary. By way of summary, for a closed system all of the fundamentals of thermodynamics are contained in notation (20).

$$dU = q + w$$

$$dS = \frac{dQ}{T} \qquad \text{reversible change}$$

$$dS \geq 0 \qquad \text{for an isolated system} \qquad (20)$$

$$dU = TdS - pdV$$

The equations are applicable when work is restricted to volume changes only. But the generalization to include changes of polarization, magnetization, surface area, and so forth is quite straightforward. Also, the equations are not applicable to systems that involve irreversible chemical changes, but here too the extension to include these situations presents no difficulty. In actual application it is convenient to define other state functions in terms of those already introduced, but no additional basic principles are needed. *See* ENTHALPY.

[WILLIAM F. JAEP]

Bibliography: H. B. Callen, *Thermodynamics*, 1960; K. Denbigh, *The Principles of Chemical Equilibrium*, 1964; E. A. Guggenheim, *Thermodynamics*, 1967; J. G. Kirkwood and I. Oppenheim, *Chemical Thermodynamics*, 1961; A. B. Pippard, *Classical Thermodynamics*, 1964; W. C. Reynolds, *Thermodynamics*, 1965; M. W. Zemansky, *Heat and Thermodynamics*, 1957.

Thermodynamic processes

Changes of any property of an aggregation of matter and energy, accompanied by thermal effects. The participants in a process are first identified as a system to be studied; the boundaries of the system are established; the initial state of the system is determined; the path of the changing states is laid out; and, finally, supplementary data are stated to establish the thermodynamic process. These steps will be explained in the following paragraphs. At all times it must be remembered that the only processes which are allowed are those compatible with the first and second laws of thermodynamics: Energy is neither created nor destroyed, and entropy always increases.

A system and its boundaries. To evaluate the results of a process, it is necessary to know the participants that undergo the process, and their mass and energy. A region, or a system, is selected for study, and its contents determined. This region may have both mass and energy entering or leaving during a particular change of conditions, and these mass and energy transfers may result in changes both within the system and within the surroundings which envelop the system.

As the system undergoes a particular change of condition, such as a balloon collapsing due to the escape of gas or a liquid solution brought to a boil in a nuclear reactor, transfers occurring in mass and energy can be evaluated at the boundaries of the arbitrarily defined system under analysis.

A question that immediately arises is whether a system such as a tank of compressed air should have boundaries which include or exclude the metal walls of the tank. The answer depends upon the aim of the analysis. If its aim is to establish a relationship among the physical properties of the gas, such as to determine how the pressure of the gas varies with the gas temperature at constant volume, then only the behavior of the gas is involved; the metal walls do not belong within the system. However, if the problem is to determine how much externally applied heat would be required to raise the temperature of the enclosed gas a given amount, then the specific heat of the metal walls, as well as that of the gas, must be considered, and the system boundaries should include the walls through which the heat flows to reach the gaseous contents. In the laboratory, regardless of where the system boundaries are taken, the walls will always play a role and must be reckoned with.

State of a system. To establish the exact path of a process, the initial state of the system must be determined. This initial thermodynamic state or condition of the system is characterized by a definite pressure, temperature, volume, and other such dimensions. If a dimension is reproducible, it is called a property of the system. If enough properties of a system are fixed the state of the system and the remaining physical properties are also fixed. The number of properties required to specify the state of a system depends upon the complexity of the system. Whenever a system changes from one state to another, then a process occurs.

After an initial state of a system is established, several questions may be asked. What makes a system change from one state to another? What driving potentials cause a process to occur, or cause a change of state of the system? How far forward will a process go? These questions will be approached by first discussing the fundamental aspects of the equilibrium of a system.

Whenever an unbalance occurs in an intensive property such as temperature, pressure, or density, either within the system or between the system and its surroundings, the force of the unbalance initiates a process that causes a change of state. Examples of an unbalanced potential property that can initiate a change of state are the unequal molecular concentration of different gases within a single rigid enclosure, a difference of temperature across the system boundary, a difference of pressure normal to a nonrigid system boundary, or a difference of electrical potential across the system boundaries. The direction of the change of state caused by the unbalanced force is such as to reduce the unbalanced driving potential. All rates of changes of state tend to decelerate as this driving potential is decreased.

Equilibrium. The decelerating rate of change implies that all states move toward new conditions of equilibrium. When there are no longer any balanced forces acting within the boundaries of a system or between the system and its surroundings, then no mechanical changes can take place, and the system is said to be in mechanical equilibrium. A system in mechanical equilibrium, such as a mixture of hydrogen and oxygen, under certain conditions might undergo a chemical change. However, if there is no net change in the chemical constituents, then the mixture is said to be in chemical as well as in mechanical equilibrium.

If all parts of a system in chemical and mechanical equilibrium attain a uniform temperature and if, in addition, the system and its surroundings are at the same temperature, then the system has also reached a condition of thermal equilibrium.

Whenever a system is in mechanical, chemical, and thermal equilibrium, so that no mechanical, chemical, or thermal changes can occur, the system is in thermodynamic equilibrium. The state of equilibrium is the state of maximum entropy; it is the state of maximum probability. The particular state of thermodynamic equilibrium reached by a system may be described by its properties or dimensions. In fact, only when the system is in equilibrium may its state be defined; then it can be described by specifying a limited number of observables. Equilibrium is a state of dynamical equilibrium with rates of reaction equal to each other. By the third law of thermodynamics, at the absolute zero of temperature the entropy of a system in equilibrium is zero. In general, those systems considered in thermodynamics can include not only mixtures of material substances but also mixtures of matter and all forms of energy. For example, one could consider the equilibrium between a gas of charged particles and electromagnetic radiation contained in an oven.

Process path. If under the influence of an unbalanced intensive factor the state of a system is altered, then the change of state of the system is described in terms of the end states, or of the difference between the initial and final physical properties.

The path of a change of state is the locus of the whole series of states through which the system passes when going from an initial to a final state. It is possible to duplicate the path that a system follows by several different methods. For example, liquid in a container can undergo a particular temperature rise, describing a particular path from initial to final state. The change of state along this definite path may be produced either by heat transfer alone, by vigorous stirring of the liquid, or by combination of the two. Thus, the exact process causing a change of state must be particularized by more than merely presenting the path traversed and must be described by the method which induces the change. To be adequate in defining the process, the description of the method must include at least the heat or the work entering the system during the process.

There are several corollaries from the above descriptions of systems, boundaries, states, and processes. First, all properties are identical for identical states. Second, the change in a property between initial and final states is independent of path or processes. The third corollary is that a quantity whose change is fixed by the end states and is independent of the path is a point function or a property. However, it must be remembered that by the second law of thermodynamics not all states are available (possible final states) from a given initial state and not all conceivable paths are possible in going toward an available state.

Pressure-volume-temperature diagram. Whereas the state of a system is a point function, the change of state of a system, or a process, is a path function. Various processes or methods of change of a system from one state to another may be depicted graphically as a path on a plot using thermo-

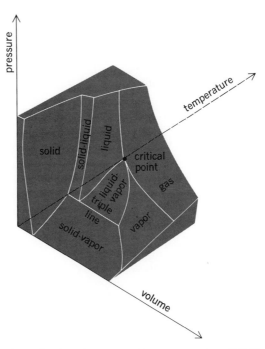

Fig. 1. Portion of pressure-volume-temperature (*P-V-T*) surface for a typical substance.

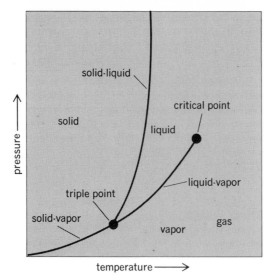

Fig. 2.   Portion of equilibrium surface projected on pressure-temperature (*P-T*) plane.

dynamic properties as coordinates.

Several variable properties are frequently and conveniently measured: pressure, volume, and temperature, any two being the selected independent variables and the third being the dependent variable. To depict the relationship among these physical properties of the particular working substance, these three variables may be used as the coordinates of a three-dimensional space. The resulting surface is a graphic presentation of the equation of state for this working substance, and all possible equilibrium states of the substance lie on this *P-V-T* surface. The *P-V-T* surface may be extensive enough to include all three phases of the working substance: solid, liquid, and vapor.

Because a *P-V-T* surface represents all equilibrium conditions of the working substance, any line on the surface represents a possible reversible process, or a succession of equilibrium states.

The portion of the *P-V-T* surface shown in Fig. 1 typifies most real substances; it is characterized

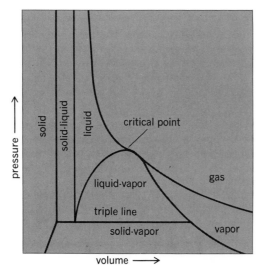

Fig. 3.   Portion of equilibrium surface projected on pressure-volume (*P-V*) plane.

by contraction of the substance on freezing. Going from the liquid surface to the liquid-solid surface onto the solid surface involves a decrease in both temperature and volume. Water is one of the few exceptions to this condition; it expands upon freezing, and its resultant *P-V-T* surface is somewhat modified where the solid and liquid phases abut.

Gibbs' phase rule is defined in Eq. (1). Here *f* is

$$f = c - p + 2 \qquad (1)$$

the degree of freedom; this integer states the number of intensive properties (such as temperature, pressure, and mole fractions or chemical potentials of the components) which can be varied independently of each other and thereby fix the particular equilibrium state of the system (see discussion under Temperature-entropy diagram, below). Also, *p* indicates the number of phases (gas, liquid, or solid) and *c* the number of component substances in the system. Consider a one-component system (a pure substance) which is either in the liquid, gaseous, or solid phase. In equilibrium the system has two degrees of freedom; that is, two independent thermodynamic properties must be chosen to specify the state. Among the properties of a thermodynamic substance which can be quantitatively evaluated are the pressure, temperature, specific volume, internal energy, enthalpy, entropy, viscosity, and electrical resistivity. From among these properties, any two may be selected. If these two prove to be independent of each other, when the values of these two properties are fixed, the state is determined and the values of all the other properties are also fixed. A one-component system with two phases in equilibrium (such as liquid in equilibrium with its vapor in a closed vessel) has $f = 1$; that is, only one intensive property can be independently specified. Also, a one-component system with three phases in equilibrium has no degree of freedom. Examination of Fig. 1 shows that the three surfaces (solid-liquid, solid-vapor, and liquid-vapor) are generated by lines parallel to the volume axis. Moving the system along such lines (constant pressure and temperature) involves a heat exchange and a change in the relative proportion of the two phases. Note that there is an entropy increment associated with this change.

One can project the three-dimensional surface onto the *P-T* plane as in Fig. 2. The triple point is the point where the three phases are in equilibrium. When the temperature exceeds the critical temperature (at the critical point), only the gaseous phase is possible. The gas is called a vapor when it can coexist with another phase (at temperatures below the critical point). The *P-T* diagram for water would have the solid-liquid curve going upward from the triple point to the left (contrary to the ordinary substance pictured in Fig. 2). Then the property so well known to ice skaters would be evident. As the solid-liquid line is crossed from the low-pressure side to the high-pressure side, the water changes from solid to liquid: Ice melts upon application of pressure.

Work of a process. The three-dimensional surface can also be projected onto the *P-V* plane to get Fig. 3. This plot has a special significance: The area under any reversible path on the pure fluid part of this plane represents the work done during the process. The fact that this P-V area represents useful work can be demonstrated by the following example.

Let a gas undergo an infinitesimal expansion in a cylinder equipped with a frictionless piston, and let this expansion perform useful work on the surroundings. The work done during this infinitesimal expansion is the force multiplied by the distance through which it acts, as in Eq. (2), wherein dW is

$$dW = F \, dl \qquad (2)$$

an infinitesimally small work quantity, F is the force, and dl is the infinitesimal distance through which F acts.

But force F is equal to the pressure P of the fluid times the area A of the piston, or PA. However, the product of the area of the piston times the infinitesimal displacement is really the infinitesimal volume swept by the piston, or $A \, dl = dV$, with dV equal to an infinitesimal volume. Thus Eq. (3) is valid. The work term is found by integration, as in Eq. (4).

$$dW = P A \, dl = P \, dV \qquad (3)$$

$$_1W_2 = \int_1^2 P \, dV \qquad (4)$$

Figure 4 shows that the integral represents the area under the path described by the expansion from state 1 to state 2 on the P-V plane. Thus, the area on the P-V plane represents work done during this expansion process.

Temperature-entropy diagram. Energy quantities may be depicted as the product of two factors: an intensive property and an extensive one. Examples of intensive properties are pressure, temperature, and voltage; extensive ones are volume, current, and mass. Thus, in differential form, work has been presented as the product of a pressure exerted against an area which sweeps through an infinitesimal volume, as in Eq. (5). Note that as a

$$dW = P \, dV \qquad (5)$$

gas expands, it is doing work on its environment. However, the energy change experienced by a wire which is stretched an infinitesimal amount is given by Eq. (6), where F is the tensile force and l is elon-

$$dW = -F \, dl \qquad (6)$$

gation. The minus sign indicates a behavior opposite to that of the gas. The increase in the length of the wire is caused by external work done on the wire.

By extending this approach, one can depict transferred heat as the product of an intensive property, temperature, and a distributed or extensive property defined as entropy, for which the symbol is S. *See* ENTROPY.

If an infinitesimal quantity of heat dQ is transferred during a reversible process, this process may be expressed mathematically as in Eq. (7),

$$dQ = T \, dS \qquad (7)$$

with T being the absolute temperature and dS the infinitesimal entropy quantity.

Furthermore, a plot of the change of state of the system undergoing this reversible heat transfer can be drawn on a plane in which the coordinates are absolute temperature and entropy (Fig. 5). The total heat transferred during this process equals

the area between this plotted line and the horizontal axis.

Reversible processes. Not all energy contained in or associated with a mass can be converted into useful work. Under ideal conditions only a fraction of the total energy present can be converted into work. The ideal conversions which retain the maximum available useful energy are reversible processes.

Characteristics of a reversible process are that the working substance is always in thermodynamic equilibrium and the process involves no dissipative effects such as viscosity, friction, inelasticity, electrical resistance, or magnetic hysteresis. Thus, reversible processes proceed quasistatically so that the system passes through a series of states of thermodynamic equilibrium, both internally and with its surroundings. This series of states may be traversed just as well in one direction as in the other.

If there are no dissipative effects, all useful work done by the system during a process in one direction can be returned to the system during the reverse process. When such a process is reversed so that the system returns to its starting state, it must leave an effect on the surroundings since, by the second law of thermodynamics, in energy conversion processes the form of energy is always degraded. Part of the energy of the system (including heat source) is transferred heat from a higher temperature to a lower temperature. The heat energy in a lower-temperature heat sink cannot be recovered. To return the system (including heat source and sink) to its original state, then, requires more energy than the useful work done by the system during a process in one direction. Of course, if the process were purely a mechanical one with no thermal effects, then both the surroundings and system could be returned to their initial states. *See* CARNOT CYCLE; THERMODYNAMIC CYCLE.

It is impossible to satisfy the conditions of a quasistatic process with no dissipative effects; a reversible process is an ideal abstraction which is not realizable in practice but is useful for theoretical calculations. An ideal reversible engine operating between hotter and cooler bodies at the temperatures T_1 and T_2, respectively, can put out $(T_1 - T_2)/T_1$ of the transferred heat energy as useful work.

There are four reversible processes wherein one of the common thermodynamic parameters is kept constant. The general reversible process for a closed or nonflow system consisting of an ideal gas is described as a polytropic process.

Irreversible processes. Actual processes of a gas deviate from the idealized situation of a quasistatic process devoid of dissipative effects. The extent of the deviation from ideality is correspondingly the extent of the irreversibility of the process.

Real expansions take place in finite time, not infinitely slowly, and these expansions occur with friction of rubbing parts, turbulence of the fluid, pressure waves sweeping across and rebounding through the cylinder, and finite temperature gradients driving the transferred heat. These dissipative effects, the kind of effects that make a pendulum or yo-yo slow down and stop, also make the work output of actual irreversible expansions less than the maximum ideal work of a corresponding

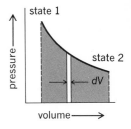

Fig. 4. Area under path in P-V plane is work done by expanding gas against piston.

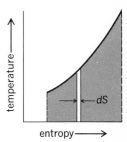

Fig. 5. Heat transferred during a process is area under path in temperature-entropy (T-S) plane.

reversible process. For a reversible process, as stated earlier, the entropy change is given by $dS = dQ/T$. For an irreversible process even more entropy is produced (turbulence and loss of information) and there is the inequality $dS > dQ/T$.

Vapor process. Figure 1 presents a three-dimensional surface in pressure-volume-temperature coordinates. This surface contains all equilibrium states that the working substance can occupy. In the portion of the surface outside the liquid-vapor dome, where temperature is high and pressure is low, the working substance behaves as though it were a perfect gas. Attractive forces appear to be negligible between any two of the molecules in a low-pressure, rarefied gas.

In the region just outside the dome, the gas is referred to as a vapor. When a gaseous system follows a process path which approaches the liquid-vapor dome on the thermodynamic surface (that is, when higher pressures or lower temperatures are attained), attractive forces appear among the more crowded, slower-moving molecules, and the behavior differs from that of a rarefied gas (ideal gas). If the path of the process cuts through the liquid-vapor dome, there is a change of phase and some of the vapor condenses to form liquid droplets in the gaseous vapor. However, if the pressure is high enough (above the critical point pressure), the cooling process can convert a gas continuously into the liquid state without going through the process of condensation.

If heat is added at constant pressure to a mixture of liquid droplets and steam vapor, the resulting constant pressure process includes an expansion of the vapor, doing work, as the droplets of liquid vaporize at a constant temperature and pressure. When no liquid droplets are left, additional heat input expands the vapor, but this expansion is now accompanied by a temperature rise, which produces a heated vapor.

If a rigid tank is filled with such a heated vapor and if heat is removed from the tank, the system undergoes a constant volume process. No work is performed in such a process, but the transferred heat is exactly equal to the reduction in internal energy stored in the system. This constant-volume cooling causes the vapor to lose both pressure and temperature until the state path crosses the saturation dome. After that, further cooling produces not only a reduction of pressure and temperature, but also some condensation and "rain" within the rigid tank.   [PHILIP E. BLOOMFIELD]

Bibliography: S. W. Angrist and L. G. Hepler, *Order and Chaos: Laws of Energy and Entropy,* 1967; H. A. Bent, *The Second Law,* 1965; M. Mott-Smith, *The Concept of Energy Simply Explained,* 1964; F. W. Sears, *An Introduction to Thermodynamics,* 1953; J. T. Vanderslice, H. W. Schamp, Jr., and E. A. Mason, *Thermodynamics,* 1966.

Thermoelectric power generator

A device that converts heat energy directly into electric energy by using the Seebeck effect. A thermoelectric generator is composed of at least two dissimilar materials, one junction of which is in contact with a heat source and the other junction of which is in contact with a heat sink.

The power converted from heat to electricity is dependent upon the materials used, the tempera-

tures of the heat source and sink, the electrical and thermal design of the thermocouple, and the load of the thermocouple.

Theory and operation. All thermoelectric effects are related to the transport properties of electrons in materials. The most important transport parameter for analysis of thermoelectric phenomena is called the Seebeck coefficient, which is the open-circuit voltage per unit temperature difference between the hot and cold junctions. The Seebeck coefficient is also called the thermoelectric power. For definition and typical values *see* THERMOELECTRICITY.

A single two-element generator is shown in Fig. 1. One leg is n-type semiconductor material; the other is p-type material. The effective composite parameters of this simple generator are the total Seebeck coefficient of the junction S, the total internal resistance r, and the total thermal conductivity K. If the Seebeck coefficients of each leg S_n and S_p, the electrical resistivities of each leg ρ_n and ρ_p, and the thermal conductivities of each leg k_n and k_p are all assumed to be independent of temperature, the composite parameters can be defined as

$$S = S_p - S_n = |S_p| + |S_n|$$
$$r = \frac{\rho_n l_n}{A_n} + \frac{\rho_p l_p}{A_p}$$
$$K = \frac{k_n A_n}{l_n} + \frac{k_p A_p}{l_p}$$

where l_n, l_p and A_n, A_p refer to the length and area of the n-type material.

The thermocouple of Fig. 1 operating as a generator has a heat source of constant temperature T_h, a heat sink at temperature T_c, and an electrical load of resistance R. The efficiency η of the generator is the power out I^2R divided by the heat in Q_{in}, which consists of the Peltier heat $S T_h I$ plus the conduction heat $K(T_h - T_c)$ less one-half of the

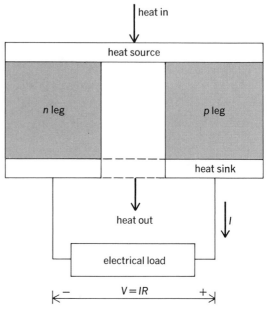

Fig. 1. Simple thermoelectric generator.

Joule heat I^2r liberated in the thermocouple legs:

$$\eta = \frac{P_{out}}{Q_{in}} = \frac{I^2R}{ST_hI + K(T_h - T_c) - \frac{1}{2}I^2r}$$

This efficiency does not consider losses in maintaining the temperature T_h and is therefore not a total efficiency including heat-source losses.

The ratio of load resistance R to internal resistance r is defined as $m = R/r$; the open-circuit voltage is $V_o = S(T_h - T_c)$; and the current is $I = V_o/(R + r)$. Using these quantities the efficiency expression becomes

$$\eta = \frac{T_h - T_c}{T_h}\left[\frac{\dfrac{m}{m+1}}{1 + \left(\dfrac{m+1}{ZT_h}\right) - \dfrac{1}{2}\dfrac{T_h - T_c}{T_h}\left(\dfrac{1}{m+1}\right)}\right]$$

where Z, the figure of merit, is

$$Z = \frac{S^2}{Kr}$$

If efficiency is optimized by selecting m to give optimum loading, the efficiency expression becomes

$$\eta = \frac{T_h - T_c}{T_h}\left[\frac{M - 1}{M + \dfrac{T_c}{T_h}}\right]$$

where $M = m\bigg|_{\frac{d\eta}{dm} = 0} = \sqrt{\dfrac{Z(T_h - T_c)}{2}}$

This efficiency for an optimum load consists of a Carnot efficiency

$$\eta_C = \frac{T_h - T_c}{T_h}$$

and a device efficiency

$$\eta_d = \left[\frac{M - 1}{M + \dfrac{T_c}{T_h}}\right]$$

The device efficiency η_d will be a maximum value for the largest value of M. For a fixed T_h and T_c this requires a maximum value of Z. Z depends upon the material parameters S_p, k_p, ρ_p, S_n, k_n, ρ_n and the dimensions of the two legs A_p, l_p, A_n, l_n. Maximizing Z with respect to X (the area-to-length ratio of the legs) gives

$$Z\bigg|_{\frac{dZ}{dX} = 0} = \frac{(S_p - S_n)^2}{[(k_p\rho_p)^{1/2} + (k_n\rho_n)^{1/2}]^2}$$

when $\left[\dfrac{k_n\rho_n}{k_p\rho_p}\right]^{1/2} = \dfrac{A_pl_n}{A_nl_p} = X$

For the optimum area-to-length ratio of the legs, the figure of merit Z depends only upon the specific properties of the thermoelectric materials. The effect of Z in determining the efficiency is shown in Fig. 2, where η is plotted versus Z and T_h for a fixed cold temperature $T_c = 300°K$.

In general, the parameters S, k, and ρ are not independent of temperature, and in fact the tem-

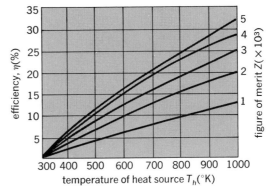

Fig. 2. Relation of efficiency η to the temperature of the heat source T_h for a constant temperature T_c of the heat sink of 300°K. Different curves are for different values of the figure of merit Z.

perature dependence of the n and p legs may differ radically. The simple figure of merit shown above does not apply to the temperature-dependent case. The general solution of the thermocouple with temperature-dependent parameters is difficult even if parameters and temperature dependence are known, and requires numerical methods.

Several approximate methods of treating temperature-dependent parameters have been proposed. A. H. Boerdijk has shown that there is no gain in performance by varying the cross section of the elements along their length. A. F. Ioffe has proposed that replacing $(S_p - S_n)$, $k_p\rho_p$, and $k_n\rho_n$ by their average over the temperature range in the figure of merit equation of the temperature-independent thermocouple is a reasonable approximation. C. Zener has proposed that the thermocouple with temperature-dependent parameters be treated by considering an infinite series of differential thermocouples with each material matched to its optimum load.

To check the Ioffe and Zener approximations against several exact calculations, B. Sherman, R. R. Heikes, and R. W. Ure, Jr., assumed three hypothetical thermocouples and calculated the efficiency by all three methods, using numerical calculations on a digital computer. These calculations show that both approximate methods—average parameters (Ioffe) and infinite staging (Zener)—lead to reasonably correct results, yielding accuracy of the order of ±10%.

State of the art. The present state of the art in materials development indicates that existing thermoelectric p and n materials operate from 300 to 1300°K and yield an overall theoretical thermal efficiency of 18%. The most widely used generator material is lead telluride, which has a maximum figure of merit Z of approximately 1.5×10^{-3} reciprocal degrees Kelvin. It can be doped to produce both p- and n-type material and has a useful temperature range of about 300–700°K. In segmented couples at the low-temperature end, bismuth telluride and its alloys are sometimes used. Couples of bismuth telluride and its alloys can be obtained both as raw materials and as finished couples. Lead telluride is similarly available commercially.

Thermoelectric generators have been built in sizes up to 5 kw. Primary energy sources are hydrocarbon fuels, radioisotopes, and solar energy.

The maximum theoretical thermal efficiency for materials over a temperature range of 300–1300°K is approximately 18%. The best actual thermal efficiency of a physically constructed device is between 6 and 10% and is obtained by operation between 300 and 950°K. The overall conversion efficiency of hydrocarbon-fueled generators is 2–3%; and overall efficiency of radioisotope generators is approximately 5%. Specific powers of 12 watts/lb have been obtained, but advances in electrical contacts indicate that higher values are obtainable.

Major problems still exist in the development of materials with higher figures of merit that are capable of operation at higher temperature. Even with present materials, there are severe engineering problems involving such items as low electrical and thermal contact resistances, mechanical strength to withstand thermal and mechanical shocks, minimum heat losses due to packaging of thermoelements, efficient fuel combustion, heat transfer from heat source to thermoelements to heat sink, packaging to minimize weight, and long-term contamination of thermoelements by diffusion.

The state of the art can probably best be summarized by stating that thermoelectric generators have been built and are under construction for many special applications.

[DAVID C. WHITE]

Bibliography: A. H. Boerdijk, Contribution to a general theory of thermocouples, *J. Appl. Phys.*, 30(7):1080–1083, 1959; I. B. Cadoff and E. Miller (eds.), *Thermoelectric Materials and Devices*, 1960; S. F. DeGroot, *Thermodynamics of Irreversible Processes*, 1952; C. A. Domenicali, Irreversible thermodynamics of thermoelectricity, *Rev. Mod. Phys.*, 26(2):237–275, 1954; P. H. Egli (ed.), *Thermoelectricity*, 1960; A. F. Ioffe, *Semiconductor Thermoelements and Thermoelectric Cooling*, 1957; B. Sherman, R. R. Heikes, and R. W. Ure, Jr., Calculation of efficiency of thermoelectric devices, *J. Appl. Phys.*, 31(1):1–16, 1960.

Thermoelectricity

The direct conversion of heat into electrical energy, or the reverse, in solid or liquid conductors by means of three interrelated phenomena—the Seebeck effect, the Peltier effect, and the Thomson effect—including the influence of magnetic fields upon each. The Seebeck effect concerns the electromotive force (emf) generated in a circuit composed of two different conductors whose junctions are maintained at different temperatures. The Peltier effect refers to the reversible heat generated at the junction between two different conductors when a current passes through the junction. The Thomson effect involves the reversible generation of heat in a single current-carrying conductor along which a temperature gradient is maintained. Specifically excluded from the definition of thermoelectricity are the phenomena of Joule heating and thermionic emission.

The three thermoelectric effects are described in terms of three coefficients: the absolute thermoelectric power (or thermopower) S, the Peltier coefficient Π, and the Thomson coefficient μ, each of which is defined for a homogeneous conductor at constant temperature. These coefficients are connected by the Kelvin relations, which convert complete information about one into complete information about all three. It is therefore necessary to measure only one of the three coefficients; usually the thermopower S is chosen. The combination of electrical resistivity, thermal conductivity, and thermopower is sufficient to provide a complete description of the electronic transport properties of conductors for which the electric current and heat current are linear functions of both the applied electric field and the temperature gradient.

Thermoelectric effects have significant applications in both science and technology and show promise of more importance in the future. Studies of thermoelectricity in metals and semiconductors yield information about electronic structure and about the interactions between electrons and both lattice vibrations and impurities. Practical applications include the measurement of temperature, generation of power, cooling, and heating. Thermocouples are widely used for temperature measurement, providing both accuracy and sensitivity. Research has been undertaken concerning the direct thermoelectric generation of electricity using the heat produced by nuclear reactors. Cooling units using the Peltier effect have been constructed in sizes up to those of home refrigerators. Development of thermoelectric heating has also been undertaken.

SEEBECK EFFECT

In 1821 T. J. Seebeck discovered that when two different conductors are joined into a loop, and a temperature difference is maintained between the two junctions, an emf will be generated. Such a loop is called a thermocouple, and the emf generated is called a thermoelectric (or Seebeck) emf.

Measurements. The magnitude of the emf generated by a thermocouple is standardly measured by using the system shown in Fig. 1. Here the contact points between conductors A and B are called junctions. Each junction is maintained at a well-controlled temperature (either T_0 or T_1) by immersion in a bath or connection to a heat reservoir. This bath or reservoir is indicated by the dashed rectangles. From each junction, conductor A is brought to a measuring device, usually a potentiometer. When the potentimeter is balanced, no current flows, thereby allowing direct measurement of the open-circuit emf, undiminished by resistive losses and unperturbed by spurious effects arising from Joule heating or from Peltier heating and cooling at the junctions. This open-circuit emf is the thermoelectric emf.

Equations. According to the experimentally established law of Magnus, for homogeneous conductors A and B the thermoelectric emf depends only upon the temperatures of the two junctions and not upon either the shapes of the samples or the detailed forms of the temperature distributions along them. This emf can thus be described by the symbol $E_{AB}(T_0,T_1)$. According to both theory and experiment, if one of the conductors, say B, is a superconductor in its superconducting state, it makes no contribution to E_{AB}. That is, when B is superconducting, $E_{AB}(T_0,T_1)$ is determined solely by conductor A and can be written as $E_A(T_0,T_1)$. It is convenient to express this emf in terms of a property which depends upon only a single temperature. Such a property is the absolute thermoelectric

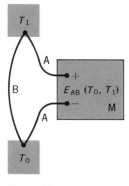

Fig. 1. Diagram of apparatus usually used for measuring thermoelectric (Seebeck) emf $E_{AB}(T_0,T_1)$. M is an instrument for measuring potential.

power (or, simply, thermopower) $S_A(T)$, defined so that Eq. (1) is valid. If $E_A(T,T+\Delta T)$ is known—for example, from measurements involving a superconductor—then $S_A(T)$ can be determined from Eq. (2). If Eq. (1) is valid for any homogeneous

$$E_A(T_0,T_1) = \int_{T_0}^{T_1} S_A(T)\,dT \qquad (1)$$

$$S_A(T) = \lim_{\Delta T \to 0} \frac{E_A(T,T+\Delta T)}{\Delta T} \qquad (2)$$

conductor, then it ought to apply to both sides of the thermocouple shown in Fig. 1. Indeed, it has been verified experimentally that the emf $E_{AB}(T_0,T_1)$ produced by a thermocouple is just the difference between the emfs, calculated using Eq. (1), produced by its two arms. This result can be derived as follows. Employing the usual sign convention, to calculate $E_{AB}(T_0,T_1)$, begin at the cooler bath, labeled T_0, integrate $S_A(T)\,dT$ along conductor A up to the warmer bath labeled T_1; and then return to T_0 along conductor B by integrating $S_B(T)\,dT$. This circular excursion produces $E_{AB}(T_0,T_1)$, given by Eq. (3). Inverting the last integral in Eq. (3) gives Eq. (4), which, from Eq. (1), can be rewritten as Eq. (5). Alternatively, combining the two integrals in Eq. (4) gives Eq. (6). Defining S_{AB} according to Eq. (7) then yields Eq. (8).

$$E_{AB}(T_0,T_1) = \int_{T_0}^{T_1} S_A(T)\,dT + \int_{T_1}^{T_0} S_B(T)\,dT \qquad (3)$$

$$E_{AB}(T_0,T_1) = \int_{T_0}^{T_1} S_A(T)\,dT - \int_{T_0}^{T_1} S_B(T)\,dT \qquad (4)$$

$$E_{AB}(T_0,T_1) = E_A(T_0,T_1) - E_B(T_0,T_1) \qquad (5)$$

$$E_{AB}(T_0,T_1) = \int_{T_0}^{T_1} [S_A(T) - S_B(T)]\,dT \qquad (6)$$

$$S_{AB}(T) = S_A(T) - S_B(T) \qquad (7)$$

$$E_{AB}(T_0,T_1) = \int_{T_0}^{T_1} S_{AB}(T)\,dT \qquad (8)$$

Equation (6) shows that $E_{AB}(T_0,T_1)$ can be calculated for a given thermocouple whenever the thermopowers $S_A(T)$ and $S_B(T)$ are known for its two constituents over temperature range T_0 to T_1. By convention, the signs of $S_A(T)$ and $S_B(T)$ are chosen so that, if the temperature difference $T_1 - T_0$ is taken small enough for $S_A(T)$ and $S_B(T)$ to be presumed constant, $S_A(T) > S_B(T)$ when the emf $E_{AB}(T_0,T_1)$ has the polarity shown in Fig. 1.

Results of equations. These equations lead directly to the following experimentally and theoretically verified results.

Uniform temperature. In a circuit kept at a uniform temperature throughout, $E=0$ even though the circuit may consist of a number of different conductors. This follows directly from Eq. (8), since $dT=0$ everywhere throughout the circuit. It follows also from thermodynamic reasoning. If E did not equal 0, the circuit could drive an electric motor and make it do work. But the only source of energy would be heat from the surroundings, which, by assumption, are at the same uniform temperature as the circuit. Thus, a contradiction with the second law of thermodynamics would result.

Homogeneous conductor. A circuit composed of a single, homogeneous conductor cannot produce a thermoelectric emf. This follows from Eq. (6) when $S_B(T)$ is set equal to $S_A(T)$. It is important to

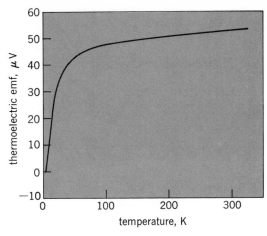

Fig. 2. The thermoelectric emf of a thermocouple formed from pure annealed and pure cold-worked copper. The cold junction reference temperature is 4.2 K. (*From R. H. Kropschot and F. J. Blatt, Thermoelectric power of cold-rolled pure copper, Phys. Rev., 116:617–620, 1959*)

emphasize that, in this context, homogeneous means perfectly uniform throughout. A sample made of an isotropic material can be inhomogeneous either because of small variations in chemical composition or because of strain. Figure 2 shows the thermoelectric emf generated by a thermocouple in which one arm is a cold-rolled copper (Cu) sample, and the other arm is the same material after annealing to remove the effects of the strain introduced by the cold-rolling. Figure 3 shows how the addition of impurities can change the thermopower of a pure metal. An additional effect can occur in a noncubic material. As illustrated in Fig. 4, two samples cut in different directions from a noncubic single crystal may be thermoelectrically different even if each sample is highly homogeneous. A thermocouple formed from these two samples will generate a thermoelectric emf.

If material B is superconducting, so that $S_B = 0$,

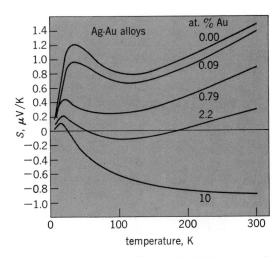

Fig. 3. The thermopower S from 0 to 300 K for pure silver (Ag) and a series of dilute silver-gold (Au) alloys. (*From R. S. Crisp and J. Rungis, Thermoelectric power and thermal conductivity in the silver-gold alloy system from 3–300°K, Phil. Mag., 22:217–236, 1970*)

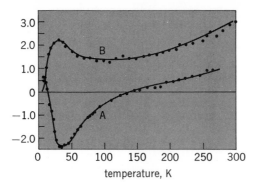

Fig. 4. The thermopower S of zinc (Zn) parallel (A) and perpendicular (B) to the hexagonal axis. *(From V. A. Rowe and P. A. Schroeder, Thermopower of Mg, Cd and Zn between 1.2° and 300°K, J. Phys. Chem. Sol., 31:1–8, 1970)*

Eq. (5) reduces to $E_{AB}(T_0,T_1) = E_A(T_0,T_1)$, as assumed above.

Source of emf. Finally, Eq. (6) makes it clear that the source of the thermoelectric emf in a thermocouple lies in the bodies of the two materials of which it is composed, rather than at the junctions. This serves to emphasize that thermoelectric emfs are not related to the contact potential or Volta effect, which is a potential difference across the junction between two different metals arising from the difference between their Fermi energies. The contact potential is present even in the absence of temperature gradients or electric currents.

PELTIER EFFECT

In 1834 J. C. A. Peltier discovered that, when an electric current passes through two different conductors connected in a loop, one of the two junctions between the conductors cools, and the other warms. If the direction of the current is reversed, the effect also reverses: the first junction warms, and the second cools. In 1853 Quintus Icilius showed that the rate of heat output or intake at each junction is directly proportional to the current i. The Peltier coefficient Π_{AB} is defined as the heat generated per second per unit current flow through the junction between materials A and B. By convention, Π_{AB} is taken to be positive when cooling occurs at the junction through which current flows from conductor A to conductor B. Quintus Icilius's result guarantees that the Peltier coefficient is independent of the magnitude of the current i. Additional experiments have shown that it is also independent of the shapes of the conductors. It therefore depends only upon the two materials and the temperature of the junction, and can be written as $\Pi_{AB}(T)$ or, alternatively, $\Pi_A(T) - \Pi_B(T)$, where Π_A and Π_B are the Peltier coefficients for materials A and B respectively. The second form emphasizes that the Peltier coefficient is a bulk property which can be defined for a single conductor.

Because of the small amount of heat transfer associated with the Peltier effect, as well as complications resulting from the simultaneous presence of Joule heating and the Thomson effect, $\Pi_{AB}(T)$ is usually difficult to measure accurately, and has therefore rarely been carefully studied. Rather, its value is usually determined from the Kelvin relations, using experimental values for S_{AB}.

THOMSON EFFECT AND KELVIN RELATIONS

When an electric current passes through a conductor which is maintained at a constant temperature, heat is generated at a rate proportional to the square of the current. This is called Joule heat, and its magnitude for any given material is determined by the electrical resistivity of the material. In 1854 William Thomson (Lord Kelvin), in an attempt to explain discrepancies between experimental results and a relationship between Π_{AB} and S_{AB} which he had derived from thermodynamic analysis of a thermocouple, postulated the existence of an additional reversible generation of heat when a temperature gradient is applied to a current-carrying conductor. This heat, called Thomson heat, is proportional to the product of the current and the temperature gradient. It is reversible, in the sense that the conductor changes from a generator of Thomson heat when the direction of either the current or the temperature gradient (but not both at once) is reversed. By contrast, Joule heating is irreversible, in that heat is generated for both directions of current flow.

The magnitude of Thomson heat generated (or absorbed) is determined by the Thomson coefficient μ. Using reasoning based upon equilibrium thermodynamics, Thomson derived results equivalent to Eqs. (9) and (10), called the Kelvin (or Kelvin-Onsager) relations.

$$\frac{\Pi_A}{T} = S_A \tag{9}$$

$$\frac{\mu_A}{T} = \frac{dS_A}{dT} \tag{10}$$

Here, μ_A is the Thomson coefficient, defined as the heat generated per second per unit current flow per unit temperature gradient when current flows through conductor A in the presence of a temperature gradient. Equation (10) can be integrated to give Eq. (11), in which the third law of

$$S_A(T) = \int_0^T \frac{\mu_A(T')}{T'} dT' \tag{11}$$

thermodynamics has been invoked to set $S_A(0) = 0$. By using Eq. (11), $S_A(T)$ can be determined from measurements on a single conductor. In practice, however, accurate measurements of μ_A are very difficult to make; therefore, they have been carried out for only a few metals—most notably lead (Pb)—which then serve as standards for determining $S_B(T)$ by using measurements of $S_{AB}(T)$ in conjunction with Eq. (7).

Long after the Thomson heat was observed and the Kelvin relations were verified experimentally, debate raged over the validity of the derivation employed by Thomson. But the theory of irreversible processes, developed by L. Onsager in 1931, and by others, yields the same equations and thus provides them with a relatively firm foundation.

THERMOPOWERS OF METALS AND SEMICONDUCTORS

Since the Kelvin relations provide recipes for calculating any two of the thermoelectric coeffi-

Table 1. The absolute thermoelectric power S of pure lead between 0 and 300 K*

T (K)	$S(\mu V/K)$†	T (K)	$S(\mu V/K)$†
0	0	60	-0.77_9
5	0	70	-0.78_4
7.5	-0.22_1	80	-0.79_4
8	-0.25_7	90	-0.82_4
8.5	-0.29_7	100	-0.86_5
9	-0.34_3	113.2	-0.91
10	-0.43_4	133.2	-0.96
11	-0.51_6	153.2	-1.02
12	-0.59_3	173.2	-1.06
14	-0.70_6	193.2	-1.10_5
16	-0.77_1	213.2	-1.15
18	-0.78_{45}	233.2	-1.18
20	-0.78_4	253.2	-1.21
30	-0.77_4	273.2	-1.25
40	-0.76_4	293.2	-1.27_5
50	-0.77_4		

*From J. W. Christian et al., Thermoelectricity at low temperatures: VI. A redetermination of the absolute scale of thermo-electric power of lead, *Proc. Roy. Soc.*, A245:213–221, 1958.

†Subscripts indicate figures which are uncertain.

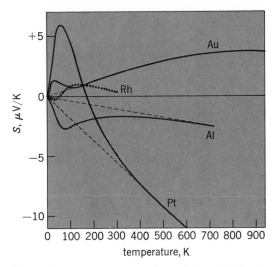

Fig. 5. The thermopower S of the metals gold (Au), aluminum (Al), platinum (Pt), and rhodium (Rh) as a function of temperature. The differences between the solid curves for Pt, Al, and Au and the broken lines indicate the magnitude of the phonon-drag component S_g.

cients, S, Π, and μ, from the third, only one of the three coefficients need be measured to determine the thermoelectric properties of any given material. Although there are some circumstances under which one of the other two coefficients may be preferred, because of ease and accuracy of measurement it is almost always the thermopower S which is measured.

Reference materials. Because S must be measured by using a thermocouple, the quantity determined experimentally is $S_A - S_B$, the difference between the thermopowers of the two conductors constituting the couple. Only when one of the arms of the thermocouple is superconducting, and therefore has zero thermopower, can the absolute thermopower of the other arm be directly measured. At temperatures up to about 18 K it is possible to use the superconducting alloy Nb_3Sn for conductor B and thereby determine S_A. For higher temperatures no superconducting wires are generally available. It is thus necessary to have a standard thermoelectric material to use above 18 K. For historical reasons, the reference material for temperatures up to 293 K has been chosen to be Pb. Based upon Thomson coefficient measurements made in the early 1930s, the thermopower of Pb has been calculated from Eq. (11) to have the values given in Table 1. All accepted values for S in this temperature range are ultimately traceable to this table. A redetermination of the Thomson coefficient of Pb has been undertaken. For temperatures above 293 K, no standard exists for which S is as well known as for Pb, but platinum (Pt) is often used because of its high melting temperature, resistance to chemical attack, and availability in high purity.

Temperature variation. Figure 5 shows the variation with temperature of the thermopowers of four different pure metals. The data for the three metals gold (Au), aluminum (Al), and platinum (Pt) are typical of those for most simple metals and for some transition metals as well. The thermopower S consists of a slowly varying portion which in-

creases approximately linearly with absolute temperature, upon which a "hump" is superimposed at lower temperatures. In analyzing these results, S is written as the sum of two terms, as in Eq. (12), where S_d, called the electron-diffusion

$$S = S_d + S_g \qquad (12)$$

component, is the slowly varying portion, and S_g, called the phonon-drag component, is the hump. For some transition metals, on the other hand, the behavior of S is more complicated, as illustrated by

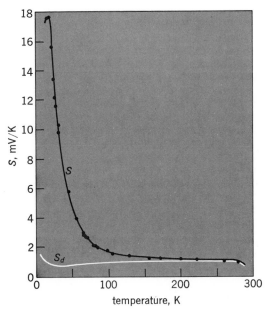

Fig. 6. The thermopower S of p-type germanium (Ge) (1.5×10^{14} acceptors per cubic centimeter) and calculated value for the electron-diffusion thermopower S_d. (*From C. Herring, The role of low-frequency phonons in thermoelectricity and thermal conductivity, Proc. Int. Coll. 1956, Garmisch-Partenkirchen, Vieweg. Braunschweig, p. 184, 1958*)

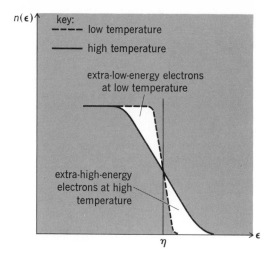

Fig. 7. The variation with energy ϵ of the number of conduction electrons $n(\epsilon)$ in a metal in the vicinity of the Fermi energy η for two different temperatures. A small variation of η with temperature has been neglected.

the data for rhodium (Rh) in Fig. 5. Figure 6 shows comparable data for a simple p-type semiconductor, illustrating that the separation of S into S_d and S_g is still valid.

Theory. When a small temperature difference ΔT is established across a conductor, heat is carried from its hot end to its cold end by the flow of both electrons and phonons (quantized lattice vibrations). If the electron current is constrained to be zero, for example, by the insertion of a high resistance measuring device in series with the conductor, the electrons will redistribute themselves in space so as to produce an emf along the conductor. This is the thermoelectric emf. If the phonon flow could somehow be turned off, this emf would be just $S_d\Delta T$. However, the phonon flow cannot be turned off, and as the phonons move down the sample, they interact with the electrons and "drag" them along. This produces an additional contribution to the emf, $S_g\Delta T$.

Source of S_d. The conduction electrons in a metal are those having energies near the Fermi energy η. Only these electrons are important for thermoelectricity. As illustrated in Fig. 7, the energy distribution of these electrons varies with the temperature of the metal. When it is at high temperatures, a metal has more high-energy electrons, and less low-energy electrons, than when it is at low temperatures. This means that if a temperature gradient is established along a metal sample,

the total number of electrons will remain constant, but the hot end will have more high-energy electrons than the cold end, and the cold end will have more low-energy electrons. The high-energy electrons will then diffuse toward the cold end, and the low-energy electrons will diffuse toward the hot end. However, in general, the diffusion rate is a function of electron energy, and thus a net electron current will result. This current will cause electrons to pile up at one end of the metal (usually the cold end) and thereby produce an emf which opposes the further flow of electrons. When the emf becomes large enough, the current will be reduced to zero. This is the thermoelectric emf arising from electron diffusion. Essentially the same argument is applicable also to semiconductors, except that in this case the conduction electrons (or holes) are those electrons just above (or below) the band gap.

S_d for a metal. For a completely free-electron metal, S_d would be given by Eq. (13), where k is

$$S_d = \frac{\pi^2}{2}\frac{k}{e}\left(\frac{kT}{\eta}\right) \qquad (13)$$

Boltzmann's constant, e is the charge on an electron, T is the absolute (Kelvin) temperature, and η is the Fermi energy of the metal. According to Eq. (13), S_d should be negative—since e is a negative quantity—and should increase linearly with T. In Table 2 the predictions of Eq. (13) are compared with experiment for a number of the most free-electron-like metals. Equation (13) is not very satisfactory, in several cases even predicting the wrong sign. To understand the thermopowers of real metals, it is necessary to use a more sophisticated model which takes into account interactions between the electrons in the metal and the crystal lattice, as well as scattering of the electrons by impurities and phonons. The proper generalization of Eq. (13) is Eq. (14), where $\sigma(\epsilon)$ is a generalized

$$S_d = \frac{\pi^2 k^2 T}{3e}\left[\frac{\partial \ln \sigma(\epsilon)}{\partial \epsilon}\right]_\eta \qquad (14)$$

conductivity defined so that $\sigma(\eta)$ is the experimental electrical conductivity of the metal, and the logarithmic derivative with respect to the energy ϵ is to be evaluated at $\epsilon = \eta$. Equation (14) is able to account, at least in principle, for all the deviations of experiment from Eq. (13). If the logarithmic derivative is negative, S_d will be positive; S_d will differ in magnitude from Eq. (13) if the logarithmic derivative does not have the value $(3/2)\eta^{-1}$; and $S_d(T)$ will deviate from a linear dependence on T if the logarithmic derivative is temperature-dependent. Research interest in S_d in metals centers upon understanding changes in S_d resulting from alloying with both magnetic and nonmagnetic impurities, strain, application of pressure, and application of magnetic fields. In some cases the changes can be dramatic. Figure 8 shows that the addition of very small amounts of the magnetic impurity iron (Fe) can produce enormous changes in S_d for copper (Cu). Sample 1 (in which the deviation of the thermopower from zero is too small to be seen with the chosen scales) is most representative of pure copper because the iron is present as an oxide and is thus not in "magnetic form." Figure 9 shows that, at low temperatures, application

Table 2. Comparison between theoretical values for S and experimental data

Metal	Thermopower S (μV/K)	
	Theoretical values at 0°C according to Eq. (13)	Experimental data at approximately 0°C
Lithium (Li)	−2	+11
Sodium (Na)	−3	− 6
Potassium (K)	−5	−12
Copper (Cu)	−1.5	+ 1.4
Gold (Au)	−2	+ 1.7
Aluminum (Al)	−0.7	− 1.7

of a magnetic field H to Al can cause S_d to change sign. (To obtain a temperature-independent quantity, S_d has been divided by the absolute temperature T. In order to remove the effects of varying impurity concentrations, H has been divided by $\rho(4.2)nec$, where $\rho(4.2)$ is the sample resistivity at 4.2 K, n is the number of electrons per unit volume in the sample, and c is the speed of light.) Figure 10 illustrates the significant changes which occur in S when a metal melts. Substantial effort has been devoted to the study of thermoelectricity in liquid metals and liquid metal alloys.

S_d for a semiconductor. Equation (13) is appropriate for a free-electron gas which obeys Fermi-Dirac statistics. The conduction electrons in a metal obey these statistics. However, there are so few conduction electrons in a semiconductor that, to a good approximation, they can be treated as though they obey a different statistics — Maxwell-Boltzmann statistics. For electrons obeying these statistics, S_d is given by Eq. (15), which predicts

$$S_d = \frac{3}{2}\frac{k}{e} \qquad (15)$$

that S_d will be temperature-independent and will have the value $S_d = -130 \times 10^{-6}$ V/K. For a p-type extrinsic semiconductor, in which the carriers are approximated as free holes, S_d would be just the negative of this value. An examination of the data of Fig. 6 reveals that S_d is very nearly independent of temperature, but is considerably larger than predicted by Eq. (15). Again, a complete understanding of the thermopowers of semiconductors requires the generalization of Eq. (15). The appropriate generalizations are different for single-band and multiband semiconductors, the latter being considerably more complicated. For a single-band (extrinsic) semiconductor, the generalization is relatively straightforward and yields predictions for S_d which, in agreement with experiment, vary slowly with temperature and are several times larger than the prediction of Eq. (15). (The curve for S_d in Fig. 6 is calculated from this generalization.). Experimental interest in the thermopower of semiconductors concerns topics similar to those for metals. In addition, the large magnitudes of the thermopowers of semiconductors continue to spur efforts to develop materials better suited for electrical power generation and thermoelectric cooling.

Source of S_g. Unlike the behavior of S_d, which is determined in both metals and semiconductors primarily by the properties of the charge carriers, the behavior of S_g is determined in both cases primarily by the properties of the phonons. At low temperatures, phonons scatter mainly from electrons or impurities rather than from other phonons. The increase in S_g with increasing temperature shown in Figs. 5 and 6 results from an increasing number of phonons being available to drag the electrons along. However, at higher temperatures the phonons begin to scatter more frequently from each other. At sufficiently high temperatures, phonon-phonon scattering becomes dominant, the electrons are no longer dragged along, and S_g falls off in magnitude. Interest in phonon drag is based on such questions as whether it is the sole source of the humps shown in Figs. 5 and 6, how it changes upon the addition of

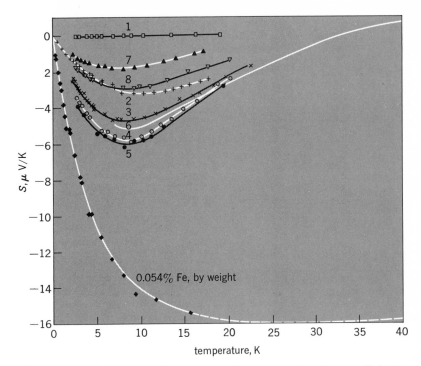

Fig. 8. The low-temperature thermopowers of various samples of copper (Cu) containing very small concentrations of iron (Fe). Specific compositions of samples 1–8 are unknown. (*From A. V. Gold et al., The thermoelectric power of pure copper, Phil. Mag., 5:765–786, 1960*)

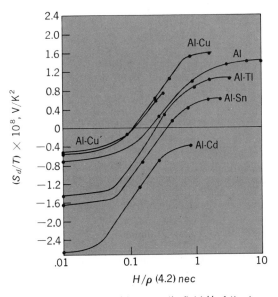

Fig. 9. The variation with magnetic field H of the low-temperature electron-diffusion thermopower S_d of aluminum (Al) and various dilute aluminum-based alloys. Sample labeled Al-Cu' is a second sample of Al-Cu. (*From R. S. Averback C. H. Stephan, and J. Bass, Magnetic field dependence of the thermopower of dilute aluminum alloys, J. Low Temp. Phys., 12:319–346, 1973*)

impurities, and how it varies in the presence of a magnetic field.

APPLICATIONS

The most important practical application of thermoelectric phenomena is in the accurate measurement of temperature. The phenomenon in-

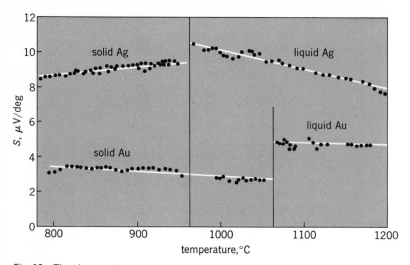

Fig. 10. The changes in the thermopowers of gold (Au) and silver (Ag) upon melting. (*From R. A. Howe and J. E. Enderby, The thermoelectric power of liquid Ag-Au, Phil. Mag., 16:467–476, 1967*)

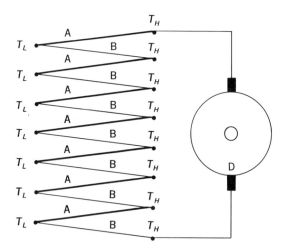

Fig. 11. Thermopile, a battery of thermocouples connected in series; D is a device appropriate to the particular application.

volved is the Seebeck effect. Of lesser importance are the direct generation of electric power by application of heat (also involving the Seebeck effect) and thermoelectric cooling and heating (involving the Peltier effect).

A basic system suitable for all four applications is illustrated schematically in Fig. 11. Several thermocouples are connected in series to form a thermopile, a device with increased output (for power generation or cooling and heating) or sensitivity (for temperature measurement) relative to a single thermocouple. The junctions forming one end of the thermopile are all at the same low temperature T_L, and the junctions forming the other end are at the high temperature T_H. The thermopile is connected to a device D which is different for each application. For temperature measurement, the temperature T_L is fixed, for example, by means of a bath; the temperature T_H becomes the running temperature T, which is to be measured; and the device is a potentiometer for measuring the thermoelectric emf generated by the thermopile. For power generation, the temperature T_L is fixed by

connection to a heat sink; the temperature T_H is fixed at a value determined by the output of the heat source and the thermal conductivity of the thermopile; and the device is whatever is to be run by the electricity which is generated. For heating or cooling, the device is a current generator which passes current through the thermopile. If the current flows in the proper direction, the junctions at T_H will heat up, and those at T_L will cool down. If T_H is fixed by connection to a heat sink, thermoelectric cooling will be provided at T_L. Alternatively, if T_L is fixed, thermoelectric heating will be provided at T_H. Such a system has the advantage that at any given location it can be converted from a cooler to a heater merely by reversing the direction of the current.

Temperature measurement. In principle, any material property which varies with temperature can serve as the basis for a thermometer. In practice, the two properties most often used for precision thermometry are electrical resistance and thermoelectric emf. Thermocouples are widely employed to measure temperature in both scientific research and industrial processes. In the United States alone, several hundred tons of thermocouple materials are produced annually.

Construction of instruments. In spite of their smaller thermopowers, metals are usually preferred to semiconductors for precision temperature measurement because they are cheaper, are easier to fabricate into convenient forms such as thin wires, and have more reproducible thermoelectric properties. With modern potentiometric systems, standard metallic thermocouples provide temperature sensitivities adequate for most needs; small fractions of a degree Celsius are routinely obtained. If greater sensitivity is required, several thermocouples can be connected in series to form a thermopile (Fig. 11). A 10-element thermopile provides a temperature sensitivity 10 times as great as that of each of its constituent thermocouples. However, the effects of any inhomogeneities are also enhanced 10 times.

The thermocouple system standardly used to measure temperature is shown in Fig. 12. It consists of wires of three metals A, B, and C, where C is usually the metal copper. The junction between the wires of metals A and B is located at the temperature to be measured T. Each of these two wires is joined to a wire of metal C at the reference temperature T_0. The other ends of the two wires of metal C are connected to the potentiometer at room temperature T_r. Integrating the appropriate thermopowers around the circuit of Fig. 12 yields the total thermoelectric emf E in terms of the separate emfs generated by each of the four pieces of wire, as given in Eq. (16).

$$E = E_A (T_0, T_1) - E_B (T_0, T_1)$$
$$+ E_{C1} (T_0, T_r) - E_{C2} (T_0, T_r) \quad (16)$$

If the two wires C_1 and C_2 have identical thermoelectric characteristics, the last two terms in this expression cancel, and, with the use of Eq. (5), Eq.

$$E = E_{AB}(T_0, T_1) \quad (17)$$

(17) results. That is, two matched pieces of metal C produce no contribution to the thermoelectric emf

of the circuit shown in Fig. 12, provided their ends are maintained at exactly the same two temperatures. This means that it is not necessary to use either of the sometimes expensive metals making up the thermocouple to go from the reference-temperature bath to the potentiometer. This portion of the circuit can be constructed of any uniform, homogeneous metal. Copper is often used because it is inexpensive, is available in adequate purity to ensure uniform, homogeneous samples when handled with care, can be obtained in a wide variety of wire diameters, and can be either spot-welded or soldered to the ends of the thermocouple wires. Special low-thermal emf alloys are available for making solder connections in thermocouple circuits.

Choice of materials. Characteristics which make a thermocouple suitable as a general-purpose thermometer include adequate sensitivity over a wide temperature range, stability against physical and chemical change under various conditions of use and over extended periods of time, availability in a selection of wire diameters, and moderate cost. No single pair of thermocouple materials satisfies all needs. Platinum versus platinum–10% rhodium can be used up to 1700°C. A combination of the two alloys chromel versus alumel gives greater sensitivity and an emf which is very closely linear with temperature, but this thermocouple cannot be used to so high a temperature. A combination of copper versus the alloy constantan also has high sensitivity above room temperature, and maintains adequate sensitivity down to as low as 15 K. For temperatures of 4 K or lower, special gold-cobalt alloys versus copper or gold-iron alloys versus chromel are used.

Thermocouple tables. To use a thermocouple composed of metals A and B as a thermometer, it is necessary to know how $E_{AB}(T_0,T)$ varies with temperature T for some reference temperature T_0. According to Eq. (6), $E_{AB}(T_0,T_1)$ can be determined for any two temperatures T_0 and T_1 if both $S_A(T)$ and $S_B(T)$ are known for all temperatures between T_0 and T_1. Once $S_A(T)$ and $S_B(T)$ are known, it is possible to construct a table of values for $E_{AB}(T_0,T)$ using any arbitrary reference temperature T_0. Such tables are available for the thermocouples mentioned above, and for some others as well, usually with a reference temperature of 0°C. A table of $E_{AB}(T_0,T)$ for one reference temperature T_0 can be converted into a table for any other reference temperature T_1 merely by subtracting a constant value $E_{AB}(T_0,T_1)$ from each entry in the table to give Eq. (18). Here $E_{AB}(T_0,T_1)$ is a positive

$$E_{AB}(T_1,T) = E_{AB}(T_0,T) = E_{AB}(T_0,T_1) \quad (18)$$

quantity when T_1 is greater than T_0 and when $S_{AB}(T)$ is positive between T_0 and T_1.

Other uses. Thermoelectric systems made from even the best available materials have the disadvantages of relatively low efficiencies and concomitant high cost per unit of output. Their use in power generation, heating, and cooling has therefore been largely restricted to situations in which these disadvantages are outweighed by such advantages as small size, low maintenance due to lack of moving parts, quiet performance, light weight, and long life.

Figure of merit. A measure of the utility of a given thermoelectric material for power generation, cooling, or heating at a given temperature T is provided by a dimensionless parameter zT, where z is called the figure of merit. The dimensionless parameter zT is given by Eq. (19), where S is the

$$zT = \frac{S^2 \sigma T}{\kappa} \quad (19)$$

thermopower of the material, σ is its electrical conductivity, and κ is its thermal conductivity. The largest values for zT are attained in semimetals and highly doped semiconductors, which are therefore the materials normally used in practical thermoelectric devices. As illustrated in Fig. 13, for most materials z varies substantially with temperature. The best available thermoelectric materials, such as lead-telluride (Pb-Te) and bismuth-telluride (Bi-Te), have values of z as large as 2 to $4 \times 10^{-3} K^{-1}$ at their best temperatures. Unfortunately, however, z falls off substantially at both higher and lower temperatures. Combining these materials into thermocouples therefore results in values of z which average less than $1 \times 10^{-3} K^{-1}$ over a temperature range sufficiently wide to be useful. Such values of z yield conversion efficiencies of only a few percent. A commercially competitive efficiency of about 30% would require a constant value of $z = 5 \times 10^{-3} K^{-1}$ over the temperature range 300–1000 K. See THERMOELECTRIC POWER GENERATOR.

Just as the figures of merit for single materials vary with temperature, so do the figures of merit for thermocouples formed from two such materials. This means that one thermocouple can be better than another at one temperature but less effective at a second temperature. To take maximum advantage of the different properties of different couples, thermocouples are often cascaded as shown in Fig. 14. Cascading produces power gen-

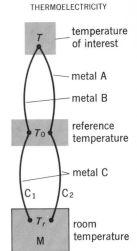

THERMOELECTRICITY

Fig. 12. The thermocouple system standardly used to measure temperature; M is a measuring device, usually a potentiometer, which is at room temperature.

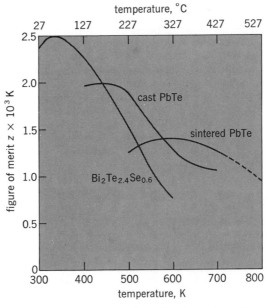

Fig. 13. Temperature variation of the figure of merit for some *n*-type semiconductors. (*From R. R. Heikes and R. W. Ure, Thermoelectricity: Science and Engineering, p. 538, Interscience Publishers, 1961*)

THERMOELECTRICITY

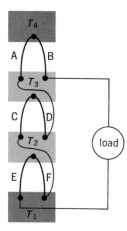

Fig. 14. Three-level cascade consisting of three different thermocouples (A versus B, C versus D, and E versus F) at four temperatures (T).

eration in stages, the higher temperature of each stage being determined by the heat rejected from the stage above. Thus, in Fig. 14 the highest and lowest temperatures T_4 and T_1 are fixed by connection to external reservoirs, whereas the middle temperatures T_3 and T_2 are determined by the properties of the materials. By cascading, a series of thermocouples can be used simultaneously in the temperature ranges where their figures of merit are highest. Cascaded thermocouple systems have achieved conversion efficiencies as high as 10–15%.

The quantity of importance in power generation is the figure of merit of a thermocouple rather than the separate figures of merit of its constituents. Although at least one constituent should have a high figure of merit, two constituents with high figures of merit do not necessarily guarantee that the figure of merit of the thermocouple will be high. For example, if the thermopowers of the two constituents are the same, the figure of merit of the couple will be zero.

Thermoelectric generators. A thermoelectric generator requires a heat source and a thermocouple. Kerosine lamps and firewood have been used as heat sources in producing a few watts of electricity in locations where electricity was otherwise unavailable. In the future, sunlight may also be used. Radioactive sources, especially strontium-90, have provided the heat to activate small, rugged thermoelectric batteries for use in lighthouses, in navigation buoys, at isolated weather stations or oil platforms, in spaceships, and in heart pacemakers. Small nuclear batteries have been operating pacemakers implanted in humans since 1970. One such battery, powered by Pu^{238} and using a bismuth telluride thermopile module, supplies a few tenths of a volt over a design lifetime of more than 10 years. Nuclear-powered batteries for medical use must be designed to remain intact following the maximum credible accident. Capabilities such as retention of integrity after crushing by 1 ton, or impact at 50 m/s, or saltwater corrosion for centuries, or cremation at temperatures up to 1300°C for half an hour are required. Investigation of the feasibility of thermoelectric generation using the copious heat generated by nuclear reactors has also been undertaken. Here, one major problem lies in the development of efficient thermoelectric materials capable of operating for a long time at the high temperatures which are encountered. *See* NUCLEAR BATTERY; NUCLEAR POWER.

Peltier cooling. With available materials, thermoelectric refrigerators suitable for use in homes are more expensive and less efficient than standard vapor-compression-cycle refrigerators. Their use is thus largely restricted to situations in which lower maintenance, increased life, or quiet performance are essential, or in situations (such as in space vehicles or artificial satellites) in which the compressor type of refrigerator is impractical. A number are in use in hotels and other large facilities. A typical unit having about 50 liters' capacity requires a dc power input of 40 W, has a refrigerative capacity of 20 kcal/hr (23 W), and a cooling time of 4–5 hr.

For lower temperatures, the proper choice of thermoelectric materials and the use of cascading can result in a reduction in temperature at the coldest junctions of as much as 150°C. Tempera-

ture drops of 100°C have been obtained in single crystals of the semimetal bismuth through use of the thermomagnetic Ettingshausen effect.

Small cooling units with capacities of 10 W or less have been developed for miscellaneous applications such as cold traps for vacuum systems, cooling controls for thermocouple reference junctions, cooling devices for scientific equipment such as infrared detectors, and cold stages on microscopes or on microtomes used for sectioning cooled tissues. However, the real commercial success of thermoelectric refrigeration appears to await development of thermocouple materials with higher figures of merit.

Thermoelectric heating. As noted earlier, a thermoelectric heater is nothing more than a thermoelectric refrigerator with the current reversed. No large heaters have been marketed. However, various small household convenience devices have been developed, such as a baby-bottle cooler-warmer which cools the bottle until just before feeding time and then automatically switches to a heating cycle to warm it, and a thermoelectric hostess cart. *See* ELECTRICITY.

[JACK BASS]

Bibliography: American Institute of Physics, *Temperature, Its Measurement and Control in Science and Industry*, vol. 1, 1941, and vol. 2, 1955; R. D. Barnard, *Thermoelectricity in Metals and Alloys*, 1972; F. J. Blatt et al., *Thermoelectric Power of Metals*, 1976; T. C. Harman and H. M. Honig, *Thermoelectric and Thermomagnetic Effects and Applications*, 1967; R. R. Heikes and R. W. Ure, Jr., *Thermoelectricity: Science and Engineering*, 1961; A. F. Joffe, The revival of thermoelectricity, *Sci. Amer.*, 199(5):31–37, 1958; D. K. C. MacDonald, *Thermoelectricity: An Introduction to the Principles*, 1962; A. C. Smith, J. F. Janak, and R. B. Adler, *Electronic Conduction in Solids*, 1967.

Thermonuclear reaction

A nuclear fusion reaction which occurs between various nuclei of the light elements when they are constituents of a gas at very high temperatures. Thermonuclear reactions, the source of energy generation in the Sun and the stable stars, are utilized in the fusion bomb. *See* FUSION, NUCLEAR.

Thermonuclear reactions occur most readily between isotopes of hydrogen (deuterium and tritium) and less readily among a few other nuclei of higher atomic number. At the temperatures and densities required to produce an appreciable rate of thermonuclear reactions, all matter is completely ionized; that is, it exists only in the plasma state. Thermonuclear fusion reactions may then occur within such an ionized gas when the agitation energy of the stripped nuclei is sufficient to overcome their mutual electrostatic repulsions, allowing the colliding nuclei to approach each other closely enough to react. For this reason, reactions tend to occur much more readily between energy-rich nuclei of low atomic number (small charge) and particularly between those nuclei of the hot gas which have the greatest relative kinetic energy. This latter fact leads to the result that, at the lower fringe of temperatures where thermonuclear reactions may take place, the rate of reactions varies exceedingly rapidly with temperature.

The reaction rate may be calculated as follows: Consider a hot gas composed of a mixture of two energy-rich nuclei, for example, tritons and deuterons. The rate of reactions will be proportional to the rate of mutual collisions between the nuclei. This will in turn be proportional to the product of their individual particle densities. It will also be proportional to their mutual reaction cross section σ and relative velocity v. Thus Eq. (1) gives the

$$R_{12} = n_1 n_2 \langle \sigma v \rangle_{12} \text{ reactions/(cm}^3)(\text{sec}) \quad (1)$$

rate of reaction. The quantity $\langle \sigma v \rangle_{12}$ indicates an average value of σ and v obtained by integration of these quantities over the velocity distribution of the nuclei (usually assumed to be maxwellian). Since the total density $n = n_1 + n_2$, then if the relative proportions of n_1 and n_2 are maintained, R_{12} varies as the square of the total nuclear particle density.

The thermonuclear energy release per unit volume is proportional to the reaction rate and the energy release per reaction, as in Eq. (2).

$$P_{12} = R_{12} W_{12} \text{ergs/ (cm}^3)(\text{sec}) \quad (2)$$

If this energy release, on the average, exceeds the energy losses from the system, the reaction can become self-perpetuating. *See* MAGNETOHYDRODYNAMICS; NUCLEAR REACTION; PINCH EFFECT.

[RICHARD F. POST]

Bibliography: S. Glasstone and R. H. Lovberg, *Controlled Thermonuclear Reactions*, 1960.

Thermostat

An instrument which directly or indirectly controls one or more sources of heating and cooling to maintain a desired temperature. To perform this function a thermostat must have a sensing element and a transducer. The sensing element measures changes in the temperature and produces a desired effect on the transducer. The transducer converts the effect produced by the sensing element into a suitable control of the device or devices which affect the temperature.

The most commonly used principles for sensing changes in temperature are (1) unequal rate of expansion of two dissimilar metals bonded together (bimetals), (2) unequal expansion of two dissimilar metals (rod and tube), (3) liquid expansion (sealed diaphragm and remote bulb or sealed bellows with or without a remote bulb), (4) saturation pressure of a liquid-vapor system (bellows), and (5) temperature-sensitive resistance element.

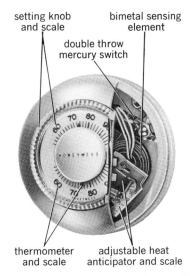

Fig. 1. Typical heat-cool thermostat. (*Honeywell Inc.*)

The most commonly used transducers are (1) switches that make or break an electric circuit, (2) potentiometer with a wiper that is moved by the sensing element, (3) electronic amplifier, and (4) pneumatic actuator.

The most common thermostat application is for room temperature control. Figure 1 shows a typical on-off heating-cooling room thermostat. In a typical application the thermostat controls a gas valve, oil burner control, electric heat control, cooling compressor control, or damper actuator.

To reduce room temperature swings, high-performance on-off thermostats commonly include a means for heat anticipation. The temperature swing becomes excessive if thermostats without heat anticipation are used because of the switch differential (the temperature change required to go from the break to the make of the switch), the time lag of the sensing element (due to the mass of the thermostat) in sensing a change in room temperature, and the inability of the heating system to respond immediately to a signal from the thermostat.

To reduce this swing, a heater element (heat anticipator) is energized during the on period. This causes the thermostat to break prematurely. Figure 2 shows a comparison of the room temperature variations when a thermostat with and without heat anticipation is used.

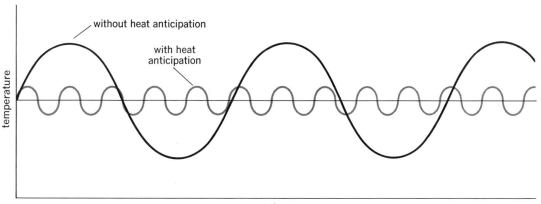

Fig. 2. Comparison of temperature variations using a timed on-off thermostat with and without heat anticipation.

The same anticipation action can be obtained on cooling thermostats by energizing a heater (cool anticipator) during the off period of the thermostat. Room thermostats may be used to provide a variety of control functions, such as heat only; heat-cool; day-night, in which the night temperature is controlled at a lower level; and multistage, in which there may be one or more stages of heating, or one or more stages of cooling, or a combination of heating and cooling stages.

Thermostats are also used extensively in safety and limit application. Thermostats are generally of the following types: insertion types that are mounted on ducts with the sensing element extending into a duct, immersion types that control a liquid in a pipe or tank with the sensing element extending into the liquid, and surface types in which the sensing element is mounted on a pipe or similar surface. *See* HEATING, COMFORT; OIL BURNER.

[NATHANIAL ROBBINS, JR.]

Bibliography: American Society of Heating, Refrigeration, and Air-Conditioning Engineers, *Guide and Data Book, Systems and Equipment*, 1967; J. E. Haines, *Automatic Control of Heating and Air Conditioning*, 1953; V. C. Miles, *Thermostatic Control*, 1965.

Tidal power

Tidal-electric power is obtained by utilizing the recurring rise and fall of coastal waters in response to the gravitational forces of the Sun and the Moon. Marginal marine basins are enclosed with dams, making it possible to create differences in the water level between the ocean and the basins. The oscillatory flow of water filling or emptying the basins is used to drive hydraulic turbines which propel electric generators.

Large amounts of electric power could be developed in the world's coastal regions having tides of sufficient range, although even if fully developed this would amount to only a small percentage of the world's potential water (hydroelectric) power. The estimated annual energy production at known tidal sites in the world amounts to about 1.2×10^{12} kWhr. This amount is equivalent to approximately two-thirds of the annual production of electrical energy in the United States. Nevertheless, tidal-electric power may become locally important, particularly because it produces no air or thermal pollution, consumes no exhaustible resource, and produces relatively minor impacts on the environment.

The use of ocean tides for power purposes dates back to the tidal mills in Europe during the Middle Ages and to those in America during colonial times. Two tidal developments producing electric power are now in operation. The Rance development in northwestern France was completed in 1967. It has an installed capacity of 240,000 kW in 24 units and is capable of producing about 500×10^6 kWhr annually. The experimental 400-kW Kislayaguba development, located north of Murmansk in the Soviet Union was completed in 1969.

Tidal range requirements. Tidal range is measured as the difference in level between the successive high and low waters. Although there are variations at certain locations in the intervals between successive high tides, at most places the tides reach the highest levels at intervals of about 12 hr 25 min. The tidal ranges vary from day to day. The highest tides, known as spring tides, occur twice monthly near the time of the new moon and the full moon when the Sun and Moon are in line with the Earth. The lowest tides, known as neap tides, occur midway between the spring tides when the Sun and Moon are at right angles with the Earth. The highest spring tides occur near the time of the equinoxes in the spring and fall of the year. Except for variations caused by meteorological changes, the tides are predictable and follow similar patterns from year to year.

Large tidal ranges occur when the oscillation of the ocean tides is amplified by relatively shallow bays, inlets, or estuaries. There is a limited number of locations where the tidal ranges are sufficiently large to be considered favorable for power development. The largest tidal ranges in the world, reaching a maximum of over 50 ft (1 ft = 0.3 m), are said to occur in the Bay of Fundy in Canada. Other locations with large maximum tidal ranges are the Severn Estuary in Great Britain, 45 ft; the Rance Estuary in France, 40 ft; Cook Inlet in Alaska, 33 ft; and the Gulf of California in Mexico, 30 ft. Large tidal ranges also occur at locations in Argentina, India, Korea, Australia, and on the northern coast of the Soviet Union.

Choosing the site. In order to be favorable for power development, a site not only must have a large tidal range, but it must also be capable of storing large amounts of water for energy production with minimum dam and dike construction. The site should also be in reasonably close proximity to population centers for the purpose of minimizing transmission requirements. The arrangement of facilities depends on the site conditions, and a number of types of development have been proposed. *See* DAM.

Single-pool scheme. The simplest type of development would be a single-pool scheme, consisting of dams and dikes to separate the pool from the ocean, sluiceways to fill the pool, and a power plant containing hydroelectric generating units. Water would be stored in the pool during high tide and subsequently released through the power plant during low tide. Additional energy could be produced if generation took place during both the filling and emptying of the pool. Since the power would be produced in accordance with the lunar cycle of the tides, much of the output would occur during periods of low power demands. Also, there would be no generation during portions of the operating cycle. Thus, there would be no firm power production.

A variation of the single-pool scheme, adopted for the existing Rance development, involves the installation of units capable of generating or pumping in either direction. In addition to generating during both the filling and emptying phases, the plant could use energy generated elsewhere to increase the available head by pumping into or out of the pool. Although power could not be produced continuously, a certain amount of dependable power could be provided by pumping into the pool and holding the water for release during peak-load periods. However, such operation would reduce the total amount of energy generated.

Two-pool scheme. A two-pool development

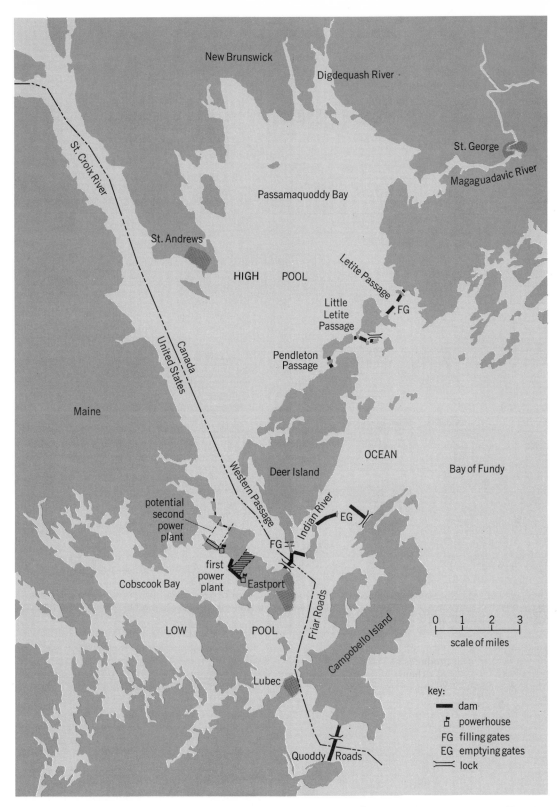

Fig. 1. Tidal power project, selected plan, general arrangement. (*From S. L. Udall, The International Passamaquoddy Tidal Power Project and Upper Saint John River* *Hydroelectric Power Development: A report to President John F. Kennedy, July 1963*)

would include a high pool which would be filled through sluices or filling gates during high tides, a low pool which would discharge through other sluices or gates during low tides, and a power plant utilizing the head between the two pools to gener-

ate power. Some power could be produced continuously. This is the type of development proposed for the Passamaquoddy site on the United States – Canadian border (Fig. 1). Additional power could be provided for peak-load use, at the ex-

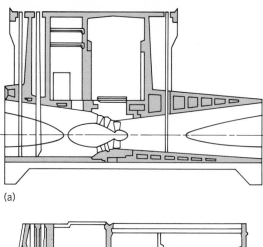

(a)

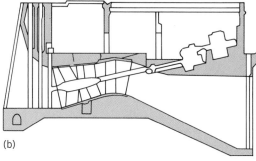

(b)

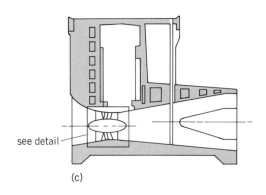

see detail

(c)

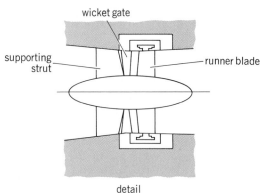

wicket gate

supporting
strut

runner blade

detail

Fig. 2. General arrangements for generating units. (a) Bulb turbine. (b) Tube turbine. (c) Straight-flow (rim-type) turbine. (*From R. H. Clark, Fundy Tidal Power, Energy Int., 9(11):21–26, November 1972*)

pense of some loss in energy production, by holding the high pool near the high-tide level and the low pool near the low-tide level until the power is required. The installation of reversible pumping-generating units would permit pumping from the low pool to the high pool during off-peak hours to provide greater head and power production during peak-load periods.

Generating units. Three types of units are considered suitable for tidal power installations. The Rance development contains bulb-type units (Fig. 2a), each of which consists of a horizontal-shaft turbine connected to a generator, with both housed in a metal bulb-shaped casing. Each unit is supported on struts in the horizontal water passage. These units can be operated for pumping and generation in both directions. The tube-type unit (Fig. 2b), considered for installation at the proposed Passamaquoddy development, consists of an axial-flow turbine connected by an inclined shaft to a generator located outside of the water passage. By means of a gear arrangement, the speed of the generator is increased and its size is reduced. The units can be used for both generating and pumping. Designs also have been prepared for a straight-flow turbine unit (Fig. 2c) in which the rotor of the generator would be mounted as a rim on the tips of the turbine blades and would turn in a sealed recess in the water conduit. The stator would be located in the dry area surrounding the rotor recess.

Major elements of cost in developing tidal power include the construction of dams and powerhouses. Difficult and costly construction may be involved in building dams in deep water with swift currents produced by the tides. An innovative scheme was used in building the powerhouse for the Kislayaguba development. The plant was constructed in the dry, floated to the site, and sunk onto a prepared foundation.

Future developments. Although only two tidal power plants are now in operation, studies and plans have been made for developing other sites. In the Soviet Union, plans have been announced for building a 6×10^6 kW plant at Mezenskaya on the White Sea. A number of plans have been prepared for developing power in the Severn Estuary. Detailed studies have been made for several tidal power developments near the head of the Bay of Fundy. The most favorable site would have an installed capacity of over 2×10^6 kW.

The Passamaquoddy site on the Bay of Fundy, where the maximum tidal range is 26 ft, has been studied at various times over half a century. Detailed plans completed in 1961 by the International Joint Commission, United States and Canada, as subsequently modified by the U.S. Department of the Interior, provide for an installation of 500,000 kW, capable of an annual generation of approximately 1.9×10^9 kWhr. The studies to date have not found the development of this site to be economically justified. *See* ELECTRIC POWER GENERATION; HYDROELECTRIC POWER.

[GEORGE G. ADKINS]

Bibliography: R. H. Clark, *Energy Int.*, 9(11): 21–26, November 1972; J. Cotillon *Water Power*, 26(10):314–322, October 1974; E. Jeffs, *Energy Int.*, 11(12):19–21, December 1974; U.S. Department of Commerce, National Oceanic and Atmospheric Administration, National Ocean Survey, *Tide Tables, East Coast, North and South America (including Greenland)*, and *Tide Tables, West Coast of North and South America (including the Hawaiian Islands)*; E. M. Wilson, *Underwater J.*, 5(4):175–186, August 1973.

Transformer

An electrical component used to transfer electric energy from one alternating-current (ac) circuit to another by magnetic coupling. Essentially, it consists of two or more multiturn coils of wire placed in close proximity to cause the magnetic field of one to link the other. In general, the transformer accomplishes one or more of the following between two circuits: (1) a difference in voltage magnitude, (2) a difference in current magnitude, (3) a difference in phase angle, (4) a difference in impedance level, and (5) a difference in voltage insulation level, either between the two circuits or to ground.

Transformers are used to meet a wide range of requirements. Pole-type distribution transformers supply relatively small amounts of power to residences. Power transformers are used at generating stations to step up the generated voltage to high levels for transmission. The transmission voltages are then stepped down by transformers at the substations for local distribution. Instrument transformers are used to measure voltage and currents accurately. Audio- and video-frequency transformers must function over a broad band of frequencies. Radio-frequency transformers transfer energy in narrow frequency bands from one circuit to another.

A power transformer consists of two or more multiturn coils wound on a laminated iron core. At least one of these coils serves as the primary winding.

Principle of operation. When the primary of a power transformer is connected to an alternating voltage, it produces an alternating flux in the core. The flux generates a primary electromotive force, which is essentially equal and opposite to the voltage supplied to it. It also generates a voltage in the other coil or coils, one of which is called a secondary. This voltage generated in the secondary will supply alternating current to a circuit connected to the terminals of the secondary winding. A current in the secondary winding requires an additional current in the primary. The primary current is essentially self-regulated to meet the power (or volt-ampere) demand of the load connected to the secondary terminals. Thus in normal operation, energy (or volt-amperes) can be transferred from the primary to the secondary electromagnetically.

Figure 1 shows a transformer with a primary of N_1 turns and a secondary of N_2 turns. A primary voltage V_1 causes a current I_1 to flow through the coil. Since all quantities shown are alternating, the arrows indicate only instantaneous polarities.

The magnetic flux ϕ set up by the primary consists of two components. One part passes completely around the magnetic circuit defined by the iron core, thus linking the secondary coil. This is the mutual flux ϕ_m. The second part is a smaller component of flux that links only the primary coil. This is the primary leakage flux ϕ_{l1}. If the secondary circuit is completed through a load, a secondary current I_2 flows and in turn creates a secondary leakage flux ϕ_{l2}. These leakage fluxes contribute to the impedance of the transformer. If the leakage flux is small, the coupling between primary and secondary is said to be close. The use of an iron core decreases the leakage flux by providing a low-reluctance path for the flux. *See* MAGNETIC CIRCUITS.

In a power transformer the voltage drops due to winding resistance and leakage are small; therefore V_1 and V_2 are essentially in phase (or 180° out of phase, depending on the choice of polarity). Since the no-load current is small, I_1 and I_2 are essentially in phase (or 180° out of phase). Therefore, Eq. (1) applies, and the voltage ratio is expressed

$$V_1 I_1 \simeq V_2 I_2 \qquad (1)$$

by Eq. (2), in which a is the transformation ratio.

$$\frac{V_1}{V_2} \simeq a \qquad (2)$$

Substituting Eq. (2) into Eq. (1) demonstrates that the current ratio is inversely proportional to the transformation ratio, as in Eq. (3). A transform-

$$\frac{I_1}{I_2} \simeq \frac{1}{a} \qquad (3)$$

er therefore may be used to step up or down a voltage from a level V_1 to a level V_2 according to the transformation ratio a. Simultaneously the current will be transformed inversely proportional to a.

Equation (1) may be rewritten in the form of Eq. (4).

$$I_1^2 \frac{V_1}{I_1} \simeq I_2^2 \frac{V_2}{I_2} \qquad (4)$$

Since V_2/I_2 is the impedance Z_2 of the load on the secondary and V_1/I_1 is the impedance Z_1 of the load as measured on the primary, Eq. (5) applies.

$$I_1^2 Z_1 \simeq I_2^2 Z_2 \qquad (5)$$

Equation (5) may be rewritten in the form of Eq. (6).

$$\frac{Z_1}{Z_2} \simeq \left(\frac{I_2}{I_1}\right)^2 \simeq a^2 \qquad (6)$$

The transformer is thus capable of transforming circuit impedance levels according to the square of the transformation ratio; this property is used in telephone, radio, television, and audio systems.

The transmission of power from primary coil to secondary coil is via the magnetic flux. The flux is proportional to the ampere turns in either coil. Since the power in each coil is nearly the same, Eqs. (7) and (8) are obtained. The transformation ratio is therefore approximately equal to the turns ratio.

$$N_1 I_1 \simeq N_2 I_2 \qquad (7)$$

$$\frac{N_1}{N_2} \simeq \frac{I_2}{I_1} \simeq a \qquad (8)$$

Construction. Transformer cores are made of special alloy steels rolled to approximately 0.014

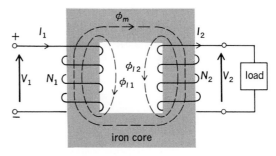

Fig. 1. Basic transformer.

TRANSFORMER

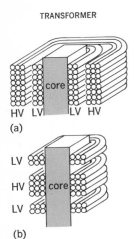

Fig. 2. Winding arrangements.
(a) Concentric.
(b) Interleaved.

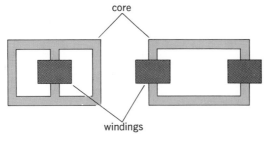

Fig. 3. Location of windings in single-phase cores.

in. thick. These thin sheets, or laminations, are stacked to form the transformer core, each sheet being insulated from the others to reduce unwanted eddy-current loss. The steel is heat-treated to obtain low hysteresis loss, low exciting current, and low sound level.

Copper conductors are used almost universally. Conductor wires are round in smaller transformers, and rectangular in larger ones.

The conductors are insulated with special paper or cotton covering, with enamel, or with a combination of both. Large outdoor transformers are immersed in oils to obtain good electrical insulation within small spacings and to provide a cooling medium. When lightweight or nonflammable materials are important, transformers may be made with compressed gases as the insulating and cooling medium. Increasing the pressure raises the dielectric strength. The gas is pumped through the

transformer and through a gas-to-air heat exchanger for cooling.

The low-voltage (LV) winding is usually in the form of a cylinder next to the core. The high-voltage (HV) winding, also cylindrical, surrounds the LV windings as in Fig. 2a. These windings are often described as concentric windings. The number of turns N may be obtained from Eq. (9), where

$$E = \frac{fBAN}{22,500} \tag{9}$$

E is rms voltage, f frequency in cps, B maximum flux density in kilolines/in.^{2}, and A cross-sectional area of the iron core in square inches.

Some manufacturers use a winding arrangement having coils adjacent to each other along the core leg as in Fig. 2b. The coils are wound in the form of a disk, with a group of disks for the LV winding stacked alternately with a group of disks for the HV windings. This construction is referred to as interleaved windings.

The core sheets are stacked sheet by sheet to form the desired cross-sectional area. The closed magnetic circuit typically has joints between adjacent sheets, but cores of moderate cross section may be made with a long continuous sheet which has been coiled up to give the required cross section. Passages may be provided between groups of sheets for circulation of the cooling oil.

For single-phase transformers (Fig. 3), the HV and LV coils may be on one leg of a core, with the return path in one, two, or more other legs. The total area of the return legs is equal to that of the main leg. An alternative construction has two legs, each with half of the primary windings and half of the secondary windings.

Figure 4 shows a typical three-phase transformer core with coils. A typical three-phase core has three legs, with the HV and LV windings for one phase on each leg. The yokes of the core connect between the two outer legs and the middle leg on top and bottom. This core-type construction is shown in Fig. 5a. The iron in another construction that is sometimes used (shell type) is as shown in Fig. 5b. Either concentric windings or interleaved windings may be used with either core.

The core and coils are placed in a steel tank with openings for the electrical connections to the windings, and for the cooling equipment.

Cooling. Small transformers are self-cooled. Radiation, conduction, and convection from the tank or from radiating surfaces remove the heat generated by the power losses of the transformer. On larger units, fans are sometimes added to the radiating surfaces. A transformer may have one rating with a basic method of cooling and a higher rating with supplemental cooling. Pumps may be added to give further cooling. An oil-to-air heat exchanger with finned tubes is used on the very large units. This equipment has a pump for circulating oil and fans for forcing the air against the heat exchanger. Water cooling may be used with cooling coils or with an oil-to-water heat exchanger having an oil pump.

Characteristics. The service conditions for a particular transformer are considered by the designer in choosing materials and the arrangement of parts.

Fig. 4.   Three-phase core and coils, rated at 50,000 kva, 115,000 volts.

The final design then may be measured by test with respect to a number of characteristics.

No-load loss. The sum of the hysteresis and eddy loss in the iron core is the no-load loss.

Exciting current. The exciting current is that supplied to the transformer at no load when operating at rated voltage. This current energizes the core and supplies the no-load loss. Owing to the characteristic shape of the *B-H* curve of iron, the current is not a true sine wave, but has higher frequency harmonics. In a typical power transformer the exciting current is so small (usually less than 1%) that I_2 is approximately $(N_1/N_2)I_1$. In this sense the ampere-turns in the two windings are said to balance.

Load loss. This is the sum of the copper loss, due to the resistance of the windings (I^2R loss), plus the eddy-current loss in the winding, plus the stray loss (loss due to flux in metallic parts of the transformer adjacent to the windings, the flux resulting from current in the windings).

Total loss and efficiency. The total loss in a transformer is the sum of the no-load and full-load losses. Representative values for a 20,000-kva, three-phase, 115-kv power transformer are no-load loss, 42 kw; load loss, 85 kw; and total loss, 127 kw. Equation (10) expresses the efficiency of a trans-

$$\text{Efficiency} = \frac{\text{output in kw}}{\text{input in kw}} = \frac{\text{output}}{\text{output} + \text{losses}} \quad (10)$$

former. For this transformer the efficiency is 20,000/20,127, or 99.37%.

Voltage ratio. This is the ratio of voltage on one winding to the voltage on another winding at no load. It is the same as the turns ratio.

Impedance. Consider a transformer having equal turns in the primary and secondary windings. If one side is connected to a generator and the other side to a typical power system load, the voltage measured on the load side will be less than that on the generator side, by the amount of the impedance drop through the transformer.

Impedance is measured by connecting the secondary terminals together (short-circuited) and applying sufficient voltage to the primary terminals to cause rated current to flow in the primary winding. The transformer impedance in ohms equals the primary voltage divided by the primary current. Impedance is usually referred to the transformer kva and kv base and given as percent impedance, as in Eq. (11). Percent reactance is usually close in value to percent impedance, since the percent resistance, given by Eq. (12), is small.

$$\% \text{ impedance} = \frac{1}{10} \frac{\text{kva}}{(\text{kv})^2} \times \text{ohms} \quad (11)$$

$$\% \text{ resistance} = \frac{\text{load loss in kva}}{\text{kva rating}} \times 100 \quad (12)$$

Typical values for a 20,000-kva, three-phase, 115-kv self-cooled power transformer are resistance, 0.4% and impedance, 7.5%.

Regulation. Regulation is the change in output (secondary) voltage that occurs when the load is reduced from rated value to zero, with the primary impressed terminal voltage maintained constant. This is usually expressed as a percent of rated output voltage at full load (E_{FL}), as in Eq. (13),

$$\% \text{ regulation} = \frac{E_{NL} - E_{FL}}{E_{FL}} \times 100 \quad (13)$$

where E_{NL} is the output voltage at no load. When a transformer supplies a capacitive load, the power factor may cause a higher full-load voltage than no-load voltage.

Cooling. Temperature tests (heat run tests) are made by operating the transformer with total losses until the temperatures are constant. In the United States the standard winding rise is 55°C over a 30°C air ambient.

Insulation. Sufficient insulation strength must be built into a transformer so that it can withstand normal operation at its rated voltage and system voltage transients due to lightning and switching surges.

Audio sound. The iron core lengthens and shortens because of magnetostriction during each voltage cycle, giving rise to a hum having a frequency twice that of the voltage. This and other frequencies may cause mechanical vibrations in different parts of the transformer due to resonance.

Taps. The application of a transformer to a power system involves a correct choice of turns ratio for average operating conditions, and the selection of proper taps to obtain improved voltage levels when average conditions do not prevail.

Tap changers are frequently used in the HV winding to give plus or minus two $2\frac{1}{2}$% taps (5% above and 5% below rated voltage). These taps may be changed only when the transformer is de-energized, that is, when the service is interrupted.

Tap changing under load. A special motor-driven tap changer is used to permit tap changing when the transformer is energized and carrying full load. One of its simpler forms is shown in Fig. 6. The transformer taps are brought to a tap changer having two sets of fingers A and B. Initially these are on the same tap. When a change is required, a contactor C opens, and A moves to the next lower tap. C now closes. Next D opens and B moves down to the same tap as A. The current, which initially divided half-and-half through A and B, has changed, first to be all in B, then partly in A and partly in B, then all in A, and finally half-and-half in A and B. E is a center-tapped reactor, which limits the current when A and B are not on a common tap.

This equipment is essential where a constant voltage is required under changing loads. It is fre-

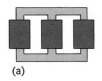

(a)

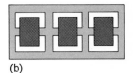

(b)

Fig. 5. Typical three-phase cores showing location of windings. (*a*) Core type. (*b*) Shell type.

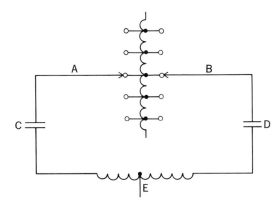

Fig. 6. Typical circuit for tap changing under load.

quently applied with a tap range of plus or minus 10% of rated voltage. It may be made to operate automatically, maintaining a specified voltage at a predetermined point remote from the transformer.

Tap changing under load equipment is used on power transformers supplying residential loads, where variations in voltage would adversely affect the use of lights and appliances. It is also used for chemical and industrial processes, such as on pot lines for the manufacture of aluminum.

Parallel operation. Two transformers may be operated in parallel (primaries connected to the same source and secondaries connected to the same load) if their turns ratios and per unit impedances are essentially equal. A slight difference in turns ratio would cause a relatively large out-of-phase circulating current between the two units and result in power losses and possible overheating.

Phase transformation. Polyphase power may be changed from 3-phase to 6-phase, 3-phase to 12-phase, and so forth, by means of transformers. This is of value in the power supply to rectifiers, where the greater number of phases results in a smoother dc voltage wave.

Overloads. Transformers have a capacity for loading above their rating. Such factors as low ambient temperature and type of load carried may be used to increase the continuous load possible on a given transformer. In emergencies it is possible to increase the load further for short times with a calculable loss of transformer life. Such a load would permit, for instance, a 50% overload for 2 hr following full load.   [J. R. SUTHERLAND]

Transmission lines

A system of conductors suitable for conducting electric power or signals between two or more termini. For example, commercial-frequency electric power transmission lines connect electric generating plants, substations, and their loads. Telephone transmission lines interconnect telephone subscribers and telephone exchanges. Radio-frequency transmission lines transmit high-frequency electric signals between antennas and transmitters or receivers. In this article the theory of transmission lines is considered first, followed by its application to power transmission lines.

Although only a short cord is needed to connect an electric lamp to a wall outlet, the cord is, properly speaking, a transmission line. However, in the electrical industry the term transmission line is applied only when both voltage and current at one line terminus may differ appreciably from those at another terminus as a result of the electrical properties of the line. Transmission lines are described either as electrically short if the difference between terminal conditions is attributable simply to the effects of conductor series resistance and inductance, or to the effects of a shunt leakage resistance and capacitance, or to both; or as electrically long when the properties of the line result from traveling-wave phenomena.

TRANSMISSION-LINE THEORY

Depending on the configuration and number of conductors and the electric and magnetic fields about the conductors, transmission lines are described as open-wire transmission lines, coaxial

transmission lines, cables, or waveguide transmission lines.

Open-wire transmission lines. Open-wire lines may comprise a single wire with an earth (ground) return or two or more conductors. The conductors are supported at more or less evenly spaced points along the line by insulators, with the spacing between conductors maintained as nearly uniform as feasible, except in special-purpose tapered transmission lines, discussed later in this section.

Open-wire construction is used for communication or power transmission whenever practical and permitted, as in open country and where not prohibited by ordinances.

Open-wire lines are economical to construct and maintain and have relatively low losses at low and medium frequencies. Difficulties arise from electromagnetic radiation losses at very high frequencies and from inductive interference, or crosstalk, resulting from the electric and magnetic field coupling between adjacent lines accompanying the characteristic field configuration (Fig. 1).

Coaxial transmission lines. A coaxial transmission line comprises a conducting cylindrical shell, solid tape, or braided conductor surrounding an isolated, concentric, inner conductor which is solid, stranded, or (in certain video cables and delay cables) helically wound on a plastic or ferrite core. The inner conductor is supported by ceramic or

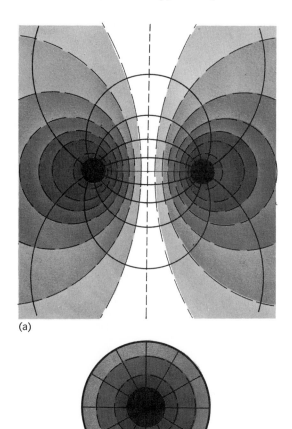

(a)

(b)

Fig. 1. Electric (solid lines) and magnetic (dashed lines) fields about two-conductor (a) open-wire and (b) coaxial transmission lines in a plane normal to the conductors, for continuous and low-frequency currents.

plastic beads or washers in air- or gas-dielectric lines, or by a solid polyethylene or polystyrene dielectric.

The purpose of this construction is to have the shell prevent radiation losses and interference from external sources. The electric and magnetic fields shown in Fig. 1b are nominally confined to the space inside the outer conductor. Some external fields exist, but may be reduced by a second outer sheath.

Coaxial lines are widely used in radio, radar, television, and similar applications.

Sheathed cables. Also termed shielded cables, these comprise two or more conductors surrounded by a conducting cylindrical sheath, commonly supported by a continuous solid dielectric. The sheath provides both shielding and mechanical protection.

Coaxial lines, sheathed cables, or shielded cables are often termed simply cables.

Traveling waves. When electric power is applied at a terminus of a transmission line, electromagnetic waves are launched and guided along the line. The steady-state and transient electrical properties of transmission lines result from the superposition of such waves, termed direct waves, and the reflected waves which may appear at line discontinuities or at load terminals.

Principal mode. When the electric and magnetic field vectors are perpendicular to one another and transverse to the direction of the transmission line, this condition is called the principal mode or the transverse electromagnetic (TEM) mode. The principal-mode electric- and magnetic-field configurations about the conductors are essentially those of Fig. 1. Modes other than the principal mode may exist at any frequency for which conductor spacing exceeds one-half of the wavelength of an electromagnetic wave in the medium separating the conductors. Such high-frequency modes are called waveguide transmission modes.

In a uniform (nontapered) transmission line, the voltage or current applied at a sending terminal determines the shape of the initial voltage or current wave. In a line with negligible losses the transmitted shape remains unchanged. When losses are present, the shape, unless sinusoidal, is altered, because the phase velocity and attenuation vary with frequency.

If a wave shape is sinusoidal, the voltage and current decay exponentially as a wave progresses. The voltage or current, at a distance x from the sending end, is decreased in magnitude by a factor of $\epsilon^{-\alpha x}$, where ϵ is the Napierian base (2.718), and α is called the attenuation constant. The voltage or current at that point lags behind the voltage or current at the sending end by the phase angle βx, where β is called the phase constant.

The attenuation constant α and the phase constant β depend on the distributed parameters of the transmission line, which are (1) resistance per unit length r, the series resistance of a unit length of both going and returning conductors; (2) conductance per unit length g, the leakage conductance of the insulators, conductance due to dielectric losses, or both; (3) inductance per unit length l, determined as flux linkages per unit length of a line of infinite extent carrying a constant direct current; and (4) capacitance per unit length c, de-

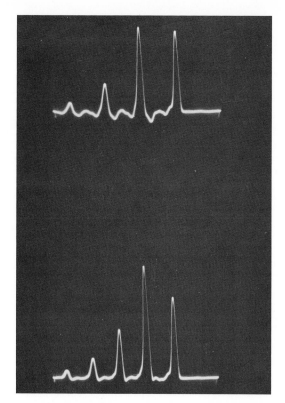

Fig. 2. Typical transient phenomena in a transmission line. These are oscillographic recordings of voltage as a function of time at the sending end of a 300-m transmission line with the receiving end open-circuited. Time increases from right to left; the first (right-hand) pulse is delivered by a generator, equivalent to an open circuit, so that a new forward wave results from each reflected wave arriving at the sending end. At the end of each 2-μsec interval, an echo arrives from the receiving end. In the upper trace, minor discontinuities in the line at intermediate points result in intermediate echos. Intermediate discontinuities are minimized in lower trace.

termined from charge per unit length of a line of infinite extent with constant voltage applied.

The values of α and β may be found from complex equation (1), where j is the notation for the

$$\alpha + j\beta = \sqrt{(r + j2\pi fl)(g + j2\pi fc)} \qquad (1)$$

imaginary number $\sqrt{-1}$, and f is the frequency of the alternating voltage and current. The complex quantity $\alpha + j\beta$ is often called the propagation constant γ. Since $r + 2\pi fl$ is the impedance z per unit length of line, and $g + 2\pi fc$ is the admittance y per unit length of line, the equation for the propagation constant is often written in the form of Eq. (2). The velocity at which a point of constant phase

$$\gamma = \sqrt{zy} \qquad (2)$$

is propagated is called the phase velocity v, and is equal to $2\pi f/\beta$. For negligible losses in the line (when r and g are approximately zero) the phase velocity is $1/\sqrt{lc}$, which is also the velocity of electromagnetic waves in the medium surrounding the transmission-line conductors.

The distributed inductance and resistance of the lines may be modified from their dc values because of skin effect in the conductors. This effect,

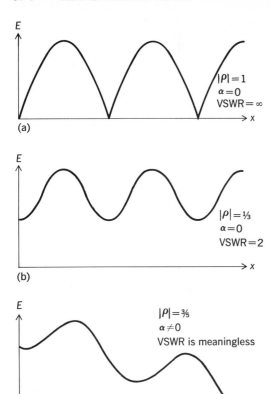

Fig. 3. Voltage distribution under sinusoidal steady-state conditions on a section of transmission line, illustrating three standing-wave conditions: (a) line with negligible losses, reflection coefficient of unity, (b) line with negligible losses, reflection coefficient of one-third, and (c) line with finite losses, reflection coefficient of three-fifths. Position of the voltage wave in each case is dependent on angle of phasor value of reflection coefficient. In each case a current maximum (not shown) appears at a voltage minimum in the wavelength.

which increases with frequency and conductor size, is usually, but not always, negligible at power frequencies.

Characteristic impedance. The ratio of the voltage to the current in either the forward or the reflected wave is the complex quantity Z_0, called the characteristic impedance.

When line losses are relatively low, that is, when relationships (3) apply, the characteristic impedance is given by Eq. (4), and is a quantity nearly

$$r \ll 2\pi fl$$
$$g \ll 2\pi fc \qquad (3)$$

ance is given by Eq. (4), and is a quantity nearly

$$Z_0 = \sqrt{l/c} \qquad (4)$$

independent of frequency (but not exactly so since both l and c may be somewhat frequency-dependent). The magnitude of Z_0 is used widely, at high frequencies, to identify a type of transmission line such as 50-ohm line, 200-ohm line, and the popular 300-ohm antenna lead-in line used with television antennas.

Distortionless line. Transmission lines used for communications purposes should be as free as possible of signal waveshape distortion. Two types

of distortion occur. One is a form of amplitude distortion due to line attenuation, which varies with the signal frequency. The other, delay distortion, occurs when the component frequencies of a signal arrive at the receiving end at different instants of time. This occurs because the velocity of propagation along the line is a function of the frequency.

Theoretically, a distortionless line can be devised if the line parameters are adjusted so that $r/g = l/c$. In practice this is approached by employing loading circuits. Under these conditions the propagation constant is given by Eq. (5).

$$\gamma = \alpha + j\beta = \sqrt{r/g}(g + j2\pi fc) \qquad (5)$$

The attenuation constant α is $\sqrt{rg}$, which is independent of frequency f. Therefore, there will be no frequency distortion.

The phase constant β is $2\pi f\sqrt{lc}$ which depends upon frequency. The velocity of propagation along any transmission line is $2\pi f/\beta$, and for the distortionless line this becomes $1/\sqrt{lc}$. Thus the velocity of propagation is independent of frequency, and there will be no delay distortion.

Transmission-line equations. The principal-mode properties of the transmission-line equations are described by Eqs. (6) and (7), in which e and i

$$\frac{\partial e}{\partial x} = -\left(ri + l\frac{\partial i}{\partial t}\right) \qquad (6)$$

$$\frac{\partial i}{\partial x} = -\left(ge + c\frac{\partial e}{\partial t}\right) \qquad (7)$$

are instantaneous values of voltage and current, respectively, x is distance from the sending terminals, and t is time.

For steady-state sinusoidal conditions, the solutions of these equations are given by Eqs. (8) and (9) for voltage E and current I at a distance x from

$$E = E_s \cosh \gamma x - I_s Z_0 \sinh \gamma x \qquad (8)$$

$$I = I_s \cosh \gamma x - \frac{E_s}{Z_0} \sinh \gamma x \qquad (9)$$

the sending end in terms of voltage E_s and current I_s at the sending end. In Eqs. (8) and (9) $Z_0 = \sqrt{(r + j2\pi fc)/(g + j2\pi fc)}$. All values of current and voltage in these and the following equations are complex.

In terms of receiving-end voltage E_r and current I_r, these solutions are given by Eqs. (10) and (11), where x is now the distance from the receiving end.

$$E = E_r \cosh \gamma x + I_r Z_0 \sinh \gamma x \qquad (10)$$

$$I = I_r \cosh \gamma x + \frac{E_r}{Z_0} \sinh \gamma x \qquad (11)$$

Reflection coefficient. If the load at the receiving end has an impedance Z_r, the ratio of reflected voltage to direct voltage, known as the reflection coefficient ρ, is given by Eq. (12).

$$\rho = \frac{Z_r - Z_0}{Z_r + Z_0} \qquad (12)$$

When the load impedance is equal to Z_0, the reflection coefficient is zero. Under this condition the line is said to be matched.

Pulse transients. The transient solutions of Eqs. (6) and (7) are dependent on the particular problem involved. Typical physical phenomena with pulse transients are shown in Fig. 2. The characteristic time delay in transmission is often advantageously employed in radar systems and other pulse-signal systems.

Standing waves. The superposition of direct and reflected waves under sinusoidal conditions in an unmatched line results in standing waves (Fig. 3).

Voltage standing-wave ratio. When losses are negligible, successive maxima are approximately equal; under this condition a quantity, the voltage standing-wave ratio, abbreviated VSWR, is defined by Eq. (13).

$$\text{VSWR} = \frac{V_{max}}{V_{min}} \qquad (13)$$

Power standing-wave ratio. This quantity, abbreviated PSWR, is equal to $(\text{VSWR})^2$. Measurements of voltage magnitude and distribution on a line of known characteristic impedance Z_0 can be used to determine the magnitude and phase angle of an unknown impedance connected at its receiving end. Lines adapted for such impedance measurements, known as standing-wave lines, are widely used.

Transmission-line circuit elements. The impedance Z_s at the sending end of a loss-free section of transmission line that has a length d, in terms of its receiving-end impedance Z_r, is given by Eq. (14).

$$Z_s = \frac{Z_r \cos \beta d + j Z_0 \sin \beta d}{\cos \beta d + j(Z_r/Z_0) \sin \beta d} \qquad (14)$$

This equation describes the property of a length of line which transforms an impedance Z_r to a new impedance Z_s. In the simple cases, in which Z_r is a short circuit or open circuit, Z_s is a reactance. Various lengths of line may be used to replace more conventional capacitors or inductors. These properties are widely applied at high frequencies, where suitable values of βx require only physically short lengths of line.

Tapered transmission lines. Transmission lines with progressively increasing or decreasing spacing are used as impedance transformers at very high frequencies and as pulse transformers for pulses of millimicrosecond duration. Although tapers designed to produce exponential-varying parameters, as in the exponential line, are most common, a number of other tapers are useful.

[EVERARD M. WILLIAMS]

POWER TRANSMISSION LINES

In an electric power system the facility used to transfer large amounts of power from one location to a distant location is termed a power transmission line. Techniques of power transmission are presented in this section.

Power transmission lines are distinguished from subtransmission and distribution lines by their higher voltages, greater power capabilities, and greater lengths. With the exception of a few high-voltage dc lines for satisfying special requirements, power transmission lines employ three-phase alternating currents. Such lines require three conductors. The standard frequency in the United States is 60 hertz (Hz). In Europe it is 50 Hz, while in the rest of the world both of these frequencies are used. For transmitting large amounts of

power over long distances, high voltages are necessary. Standard transmission voltages in the United States are 69, 115, 138, 161, 230, 345, 500, and 765 kilovolts (kV). These figures refer to the nominal effective voltages between any two of the three conductors. The line conductors are usually placed overhead, supported by poles or towers; however, they may form part of an underground or underwater cable. *See* ELECTRIC DISTRIBUTION SYSTEMS.

Requirements of transmission. Power transmission systems must be reliable, have good voltage regulation and adequate power capability, and be capable of economical operation.

Reliability. This requirement is met by sturdy construction, by protection against overvoltages, by rapid automatic disconnection of accidentally short-circuited lines, by suitable transmission layouts, and by automatic rapid reconnection of lines experiencing only transitory faults.

Good voltage regulation. When the load voltage does not vary appreciably as the load increases from no load to full load, the regulation is said to be good. The inherent voltage regulation depends mainly on the inductive reactance of the line and the power factor of the load. If the inherent regulation is unsatisfactory, the voltage can be controlled by switched shunt capacitors or synchronous condensers connected at the load.

Power capability. The maximum power that can be transmitted, with due regard to limitations imposed by losses, temperature of the conductors, voltage regulation, and system stability, is the power capability of the line. It varies approximately as the square of the voltage.

Economy. Fulfillment of this requirement depends on a balance between low first cost and low operating cost, including cost of power loss. The principal loss is the I^2R loss in the conductors.

Constants. From a knowledge of the size and type of conductors and the spacing between them, one can obtain the values of series resistance r and inductive reactance x per phase per unit length of line and of shunt capacitive susceptance b and leakage conductance g per phase per unit length of line. All of these values are multiplied by the length of the line, giving constants R, X, B, and G, respectively. These are then combined to give the complex impedance, admittance, hyperbolic angle, and characteristic impedance, Eqs. (15)–(18), respectively.

$$Z = R + jX \qquad (15)$$

$$Y = G + jB \qquad (16)$$

$$\theta = \sqrt{ZY} \qquad (17)$$

$$Z_0 = \sqrt{\frac{Z}{Y}} \qquad (18)$$

The approximate value of characteristic impedance for a line with low losses, given by $Z_0 = \sqrt{l/c}$, is often called the surge impedance and is real. The power carried by a transmission line is often expressed in terms of its natural power or surge-impedance loading (SIL), which is defined by Eq. (19), where E_n is the nominal voltage, and Z_0 is the

$$P_n = \left| \frac{E_n^2}{Z_0} \right| \qquad (19)$$

surge impedance. Units of kilovolts for E_n, ohms

TRANSMISSION LINES

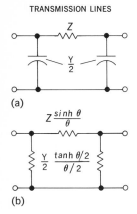

Fig. 4. Lumped-constant representations of a single-phase transmission line or of one phase of a three-phase line. (a) Nominal π. (b) Equivalent π.

for Z_0, and megawatts for P_n are convenient for power lines. If E_n is the voltage between conductors, P_n is the three-phase power.

If a line is operating at its surge-impedance loading, there are no standing waves (Fig. 3), but the graph of voltage magnitude versus distance along the line is flat. In addition, the reactive power E^2B produced by the shunt capacitance is balanced by the reactive power I^2X consumed by the series inductance, and at every point of the line the current is in phase with the voltage.

Equivalent circuit. This circuit indicates lumped values which represent values distributed along the line. A short line can be represented adequately by its nominal π circuit shown in Fig. 4a. Here the shunt admittance Y, actually distributed uniformly along the line, is assumed to be lumped and divided into two equal parts, one at each end of the line. For a long line the theoretically exact equivalent π circuit should be used. Each of its branch impedances is calculated by multiplying the corresponding branch impedance of the nominal π by a correction factor which is given in Fig. 4b.

By use of equivalent circuits, the steady-state electrical performance of a line can be calculated by the ordinary theory of ac circuits with lumped constants. These circuits can be combined with circuits representing series capacitors, shunt reactors, transformers, and loads. If a complicated network is to be studied, it can be represented by a low-power model in a network analyzer wherein each line is represented by its π circuit or can be solved on a digital computer by use of a suitable power-flow program.

Alternating-current overhead lines. An overhead transmission line consists of a set of conductors, usually bare, which are supported at a specified distance apart and with specified clearances from the ground and from the supporting structures.

Routes. Lower-voltage transmission lines are usually built along highways, whereas higher-voltage lines are put on a special right of way, cleared of trees and brush. Such routes are often chosen from results of aerial surveys.

Supporting structures. Lower-voltage overhead lines are usually supported by wooden poles and higher-voltage lines by wooden H frames or steel towers. Rigid steel towers give the greatest strength and reliability. The higher the voltage, the greater must be the spacing between conductors and the clearance from conductor to ground. The farther apart the towers are placed, the greater is the sag of the conductors and the taller and stronger the towers must be. Figure 5 shows some typical structures.

The towers shown with vertical strings of suspension insulators are tangent towers or suspension towers. Dead-end towers, used at the ends of a line, and angle towers, used at large angles in the line, have almost horizontal strings of insulators. The center conductor of Fig. 5f is supported by V strings, which prevent the conductor from swinging sideways in a cross wind, keeping it from the grounded tower.

Insulators. Conductor supports, or insulators (Fig. 6), are generally made of glazed porcelain or of glass. On lower-voltage lines, they are usually of the pin or post type. On higher-voltage lines, they are of the suspension type, consisting of several units connected by swivel joints. The number of units per string depends on the desired impulse flashover voltage, but is not proportional to it, because the voltage does not divide equally between the several units.

Insulators exposed to industrial dust deposits or to salt spray will, when moist from fog, carry leakage currents which may lead to flashovers at normal operating voltage. Remedies are the use of special fog-type insulators with deeper corrugations on their lower sides for increasing the length of leakage paths, occasional washing of insulators by a stream of water drops from a nozzle, and coating of insulators with silicone grease.

Conductors. For overhead lines, conductors are usually bare, multilayered, concentrically stranded aluminum cables. Adjacent layers of strands are spiraled in opposite directions. Additional tensile strength is provided usually by a core of steel strands (Fig. 7a) or sometimes by inclusion of strands of a strong aluminum alloy. These two types of conductor are known by the abbreviations ACSR (aluminum cable, steel reinforced) and ACAR (aluminum cable, alloy reinforced), respectively. If the conductor diameter required for low corona loss is greater than that of an ordinary stranded conductor having the cross-sectional area required for the desired resistance, special "expanded" conductors are used. Some of these have a paper filler between the steel core and the outer layers of aluminum strands; others have two filler layers each of four large aluminum strands (Fig. 7b).

Many very high voltage lines (400 kV and above) use multiple, or bundle, conductors, each consisting of two, three, or four stranded subconductors connected one to another through metal spacers and hung from the same insulators (Fig. 5f). Bundle conductors have the advantages of lower inductive reactance; higher natural power; lower corona loss, radio interference, and audible noise; and better cooling than single conductors of the same total cross-sectional area.

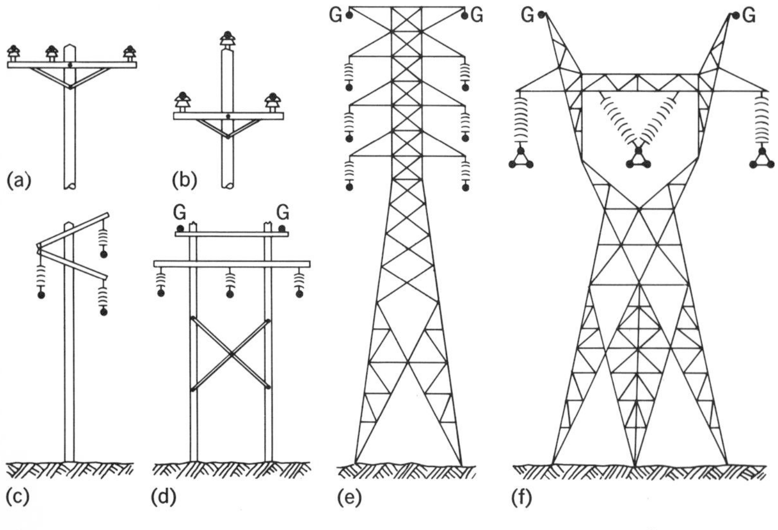

Fig. 5. Supporting structures for electric power transmission lines: (a-d) wood poles; (e, f) steel towers. Structures in a–d and f are for single-circuit lines; e is for double-circuit lines. In a and b, pin-type insulators are used; in c–f, suspension insulators are used. Structures in d–f have ground wires (G) above the line conductors.

Splices in large conductors are usually made with metal sleeves squeezed over the butted ends of the conductor by hydraulic jacks.

Sag and tension. Conductors between adjacent supports hang in a curve called a catenary (Fig. 8). For a given length of span, the greater the tension in the conductor, the smaller is the sag. High mechanical tension is desirable to reduce sag and thus to permit use of longer spans or shorter towers, while maintaining adequate ground clearance. However, the tension must not exceed the tensile yield strength of the conductor under the worst condition, which occurs under a combination of low temperature (causing shortening) simultaneously with the thickest coating of ice on the conductor and the strongest wind.

Vibration. At times the wind causes the conductors to vibrate with low amplitude and audible frequency. This vibration bends the conductor where it is clamped to the insulators and eventually may produce fatigue breakage. A device with duplex weights is used on some lines to reduce conductor vibration. One or more of these dampers are fastened to the conductors several feet from the insulator clamp (Fig. 9).

Sleet. An ice-covered conductor acts as an air foil and is lifted by the wind, so that "dancing" occurs, of such amplitude that one conductor may strike another, producing a short circuit. Conductors should be located so that contact will not be made. Formation of ice is prevented if the current heats the conductor sufficiently. Some power companies make a practice of periodically taking endangered lines out of service and sending high currents through them.

Corona. When the voltage gradient, or electric field strength, at the surface of the conductor exceeds the breakdown gradient of air, the air near the conductor surface becomes ionized. This condition, called corona, is evidenced by a visible glow at night and by a buzzing noise.

Corona results in a loss of power, interference with radio reception, and audible noise, all of which increase rapidly with voltage. Transmission lines are normally operated at a voltage near that at which corona becomes appreciable. The larger the conductor diameter and the greater the number of subconductors, the higher the operating voltage may be.

Inductive coordination. If a telephone line runs near and parallel to a power line for some distance, the high currents and voltages in the power line may induce currents and voltages in the telephone line. These signals may be comparable in strength to the telephone signals and thus produce objectionable noise in telephone receivers. The worst noise is produced by magnetic coupling from harmonic currents having a ground-return path. The coupling between the two lines can be reduced by greater physical separation, by transposition of the telephone wires, and by shielding, such as that provided by grounded cable sheaths.

Inspection and fault location. Transmission lines should have both periodic general inspections and special immediate inspections of points where short circuits have occurred, to detect damage, such as broken insulators, which might impair the reliability of the line. Faults can be located approximately by electrical measurements made from the ends of the line and then exactly by visual

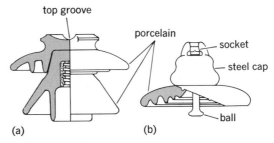

Fig. 6. Insulators used on transmission lines. (*a*) Large two-piece pin-type insulator. (*b*) One unit of a suspension insulator. (*From H. Pender and W. A. Del Mar, Electrical Engineers' Handbook, vol. 1, 4th ed., Wiley, 1949*)

patrol of the vicinity. *See* CIRCUIT (ELECTRICITY).

Lightning protection. Lightning is the most detrimental factor affecting the reliability of electric power service, but its damaging effects have been greatly reduced by proper design. Lightning striking a transmission line momentarily impresses a very high voltage on the line, causing spark-over to ground, usually at an insulator. Power current then follows the spark path, producing an arc, which constitutes a short circuit and which can be extinguished only by disconnecting the faulted line from the rest of the power network. Lines built where severe thunderstorms are prevalent are equipped with overhead ground wires (Fig. 5*d–f*) for intercepting the lightning stroke and leading it to ground at the nearest tower.

Switching surges. Another source of overvoltage, which has become important enough on extra-high-voltage lines (500 kV and above) to determine their insulation levels, is switching. The transient overvoltages caused by reenergization of a line which still has trapped charges left from a recent deenergization may be as high as 3.5 times normal line-to-ground crest voltage. By use of circuit breakers which energize the line through one step of series resistance before making direct connection from the power source to the line, the overvoltage can be reduced to about twice normal. By use of two or three steps of decreasing resistance, the overvoltage can be limited to a still lower value, say 1.5. The values cited are representative, but actual overvoltages vary with the length of line section, the time during which the resistors are inserted, the time span from closure of the first pole of the breaker to the last pole, and so on.

Another source of transient overvoltage similar to a switching surge is a short circuit from one line

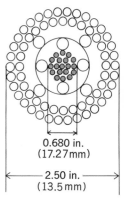

(a)

(b)

0.680 in. (17.27 mm)

2.50 in. (13.5 mm)

Fig. 7. Cross sections of typical overhead conductors. (*a*) Steel-reinforced aluminum cable (ACSR) with 19 steel and 42 aluminum strands. (*b*) Expanded ACSR. (*Aluminum Company of America*)

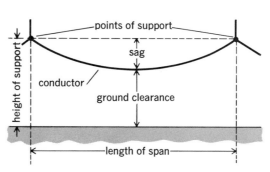

Fig. 8. Catenary curve assumed by one span of a flexible conductor with supports at equal elevations.

Fig. 9. Stockbridge dampers on transmission-line conductor. (*From H. Pender and W. A. Del Mar, Electrical Engineers' Handbook, vol. 1, 4th ed., Wiley, 1949*)

conductor to ground. The overvoltage, having a crest value up to 2.0 times normal crest voltage, appears on the conductor whose voltage phase leads that of the faulted conductor.

Overhead power line constants. Overhead 60-Hz power transmission lines with single conductors per phase have series inductive reactance of about 0.8 ohm/mi, shunt capacitive susceptance of about 5 micromhos/mi, and surge impedance of about 400 ohms. For lines with two-conductor bundles per phase these become, respectively, 0.6 ohm/mi, 6.7 micromhos/mi, and 300 ohms. For any 60-Hz open-wire line, the phase constant β is about 0.0020 radian/mi.

Power capability. The economic loading of an overhead power line is usually in the range of 1.0 to 1.5 times its natural power (SIL loading), depending on conductor size. At much higher loadings, the losses become excessive. The table gives typical values of natural power and economic loading of single-circuit, three-phase, 60-Hz lines.

Cost of transmission. The cost of building a transmission line is very nearly proportional to the voltage and to the length of the line. Its power capability is almost proportional to the square of the voltage. Consequently, the cost per unit of power varies directly as the distance and inversely as the square root of the power. If the amount of power to be transmitted is quadrupled, it can be transmitted twice as far for the same unit cost. This explains why it is not economical to transmit for a long distance unless a large quantity of power is involved.

Reactive compensation. There are two kinds of reactive compensation: shunt and series. In shunt compensation the distributed shunt capacitive susceptance B_C of the transmission line is partially or entirely compensated by the addition of lumped inductive susceptance B_L in the form of shunt reactors. The net shunt susceptance is $B = B_C - B_L$. The degree of shunt compensation is the ratio B_L/B_C. The net charging current is reduced in proportion of 1 minus this ratio. Shunt compensation is needed principally on extra-high-voltage overhead lines (500 kV and above) when they are lightly loaded and on long cables.

In series compensation the distributed series inductive reactance X_L of the transmission line itself is partially compensated by the addition of lumped series capacitive reactance X_C in the form of series capacitors. The net series reactance is $X = X_L - X_C$. The degree of series compensation is the ratio X_C/X_L.

Series compensation is used for several different purposes. The first is to reduce the voltage fluctuation (flicker) due to rapidly changing loads, as in electric furnaces and the starting of large motors. The second purpose is to obtain proper division of current between transmission lines connected in parallel. If two such lines have conductors of different cross section, the total losses can be reduced by connecting a capacitor in series with the line having the larger conductors. The capacitor should have such capacitive reactance that the ratio of net reactance to resistance is the same for both lines.

The third purpose of series compensation is to increase the distance that a given amount of power can be transmitted, to increase the power that can be transmitted over a given distance, or, more generally, to increase the possible product of power and distance. Analysis of the equivalent circuits of Fig. 4 for a line with negligible losses shows that the power P transmitted for terminal voltages of fixed magnitudes E_s and E_r but of variable phase difference δ varies according to Eq. (20), where X

$$P = \frac{E_s E_r}{X} \sin \delta \qquad (20)$$

is the series inductive reactance of the line (Z in Fig. 4a) or, more exactly, for a long line, of the horizontal branch of the equivalent π of Fig. 4b. This equation shows that maximum power is obtained at $\delta = 90°$, for which $\sin \delta = 1$. Considerations of stability require that the line be operated with δ less than 90°, and experience indicates that it can be operated prudently at $\delta = 30°$, $\sin \delta = 0.5$. With any fixed values of the terminal voltages and fixed maximum value of $\sin \delta$, there is a maximum value of PX, the product of power and series reactance. Since the frequency is fixed in practice, the series reactance of an overhead line of given frequency is directly proportional to the length of line d, and hence there is a limit to the product Pd (in megawatt-miles) of an uncompensated line. Assuming the economic power to be 1.25 P_n, $f = 60$ Hz, and $\sin \delta = 0.5$, the limiting distance for that power is about 200 mi.

In the past, this limitation was not seriously felt because transmission for distances much greater than 200 mi between points of maintained voltage was generally not economical in comparison with supplying power from generating stations nearer to the load centers. However, the use of electric power has grown to a point where transmission of large amounts of power over greater distances (perhaps up to 1000 mi) is economical in some circumstances. Transmission of power over such distances becomes technically feasible through series compensation.

Underground ac lines. Insulated cables are used in congested areas where the cost of right of way for overhead lines would be excessive, in city streets where overhead lines would be too unsightly or hazardous, in and around power stations, and for crossing wide bodies of water. About 1% of the

Natural and economic loading of single-circuit, three-phase, 60-Hz lines

Voltage, kV	Surge impedance, ohms	Natural loading, MW	Economic loading, MW
69	400	12	15
138	400	47	60
230	400	130	170
345	350	340	430
500	300	830	1000
700	300	1600	2000
1000	250	4000	5000

total transmission mileage of the United States is underground, located mostly in congested urban areas. Because the cost of underground transmission is so much higher than that of overhead transmission, it is used only where necessary. The cost of constructing an underground line ranges from 8 times (at 69 kV) to 20 times (at 500 kV) that of an overhead line of the same length and power capability.

Cables are now made for alternating voltages up to 500 kV, and cables for 700 kV are being developed.

The conductors of most cables are of stranded copper, insulated by wrapping with layers of paper tape saturated with mineral oil. The thickness of the insulation depends on the voltage, and varies from 0.285 in. in 69-kV cables to 1.34 in. for 500-kV cables. The dielectric constant of this insulation is from 3.5 to 3.7.

Solid and oil-filled cables. Low-voltage cables are of the solid type, in which the only oil is that put in the paper during manufacture. The disadvantage of this type of cable for high voltages is that small voids may form between layers of paper and that corona in such voids causes the insulation to deteriorate, leading eventually to puncture. Therefore, high-voltage cables are provided with oil under pressure.

In the low-pressure oil-filled cable, there are oil channels in the center of the conductor of single-conductor cables or between the insulated conductors of three-conductor cables. Oil reservoirs, connected to the cable at intervals, keep the cable full of oil at a pressure of from 3 to 20 pounds per square inch (psi) in spite of contraction and expansion caused by changes in cable temperature.

The solid and low-pressure oil-filled types of cables have a lead sheath surrounding the insulation to keep moisture out of the cable and the oil in. Such cables are pulled into concrete or fiber ducts which give mechanical protection. At intervals of 600 ft or less, these ducts terminate in underground chambers called manholes, where sections of cables are spliced together.

Pipe-type cables. In these, three single-conductor, paper-insulated cables are pulled into a buried stell pipe, which is later filled with oil at high pressure (200 psi). Manholes may be spaced as far apart as 1/2 mi. Oil reservoirs and pumps are required. Power capability ranges from about 225 MVA at 138 kV to 650 MVA at 550 kV.

Compressed-gas-insulated cables. Each of the three isolated phases consists of two coaxial aluminum tubes separated mechanically by epoxy insulating spacers and electrically by these spacers and the compressed gas. The inner tube is the main conductor, while the outer tube, which is grounded, normally carries little current but serves as a sheath, contains the gas, and gives some mechanical protection. Expansion joints incorporated in the central tube allow differential thermal expansion. The gas cable is shipped in factory-assembled sections about 40 ft long. These are then welded together in the field. The outer surface of the sheath is provided with a protective coating to prevent corrosion. Additional protection against corrosion may be given by a cathodic protection such as that used on long pipes for oil or gas. The compressed gas currently used is sulfur hexafluoride (SF_6) at pressures of 45 to 60 psi. It combines excellent chemical stability with high electric breakdown strength, good heat-transfer characteristics, and low dielectric constant compared to paper and oil. Cables employing SF_6 can be designed to withstand rated voltage even with the gas pressure reduced to normal atmospheric pressure. Research is under way to find other gases (or mixtures) as good as SF_6 but less expensive. Power ratings range from 600 MVA at 230 kV to 11,000 MVA at 765 kV.

Low-temperature (cryogenic) cables. Cryogenic cables are being investigated. The resistance of aluminum or copper decreases as the temperature decreases. If a cable is cooled by circulating liquid nitrogen at temperature 77 K through a hollow conductor, for example, the power loss (I^2R) in the conductor is greatly reduced. However, the power required for refrigeration of the coolant partly offsets the saving in I^2R loss. Good thermal insulation of such a cable is obviously necessary.

Certain special alloys become superconductive at very low temperatures (below 18.2 K for niobium-tin), and their resistance to direct current becomes zero, leaving only the power for refrigeration as the loss of the cable. With alternating current, there is also some loss in the conductors but much less than that at or above room temperature. If the current or the temperature should exceed its critical value, the conduction would become resistive. For this reason, the superconductive alloy is either plated onto the surface of an ordinary metal or is made in fine filaments embedded in the ordinary metal. Thus, if the alloy should lose its superconductivity, the ordinary metal could carry the current for a long enough time to permit reduction of the current before melting of the conductors could occur.

Terminations. Connections of cables to overhead lines or to substations are made through potheads which provide oil seals and longitudinal as well as radial insulation. High-voltage potheads are encased in porcelain with a corrugated outer surface and are similar in appearance to the insulating bushings used on transformers.

Splices in high-voltage cables must be made with great care and require many hours of skilled labor. The distance between splices depends largely upon the length of cable that can be wound on one shipping reel.

Losses and power capability. The power capability of a cable is limited by the rise of temperature in the conductor and adjacent insulation, because too high a temperature will char the insulation and cause its breakdown. The temperature rise depends on the power losses in the cable and on the rate of conduction of heat from the cable into the surrounding soil. Whereas the loss in the conductor is the only important loss in an overhead line, the cable has also a considerable dielectric loss, which increases with voltage because of the increasing thickness of the insulation. Also, whereas the overhead conductor is bare and directly exposed to the cooling air, the heat produced by losses in a cable must pass through the insulation, the duct walls (if ducts are used) and a considerable thickness of earth before reaching cool earth or air. Thus the permissible current in a cable is less than that in an overhead conductor of equal resistance, and the permissible power for a given voltage is correspondingly less. This is the apparent power, $S = \sqrt{P^2 + Q^2}$, where P is the active power and Q is the reactive power.

Improved cooling. Since the power capability of cables is limited by their temperature rise, capability can be increased by improved heat removal. Among the means that have been used—but so far only to limited extent—are (1) backfilling the trench in which cable or cable ducts are buried with material of better heat conductivity than that of the original soil, (2) burying cooling pipes in the earth near the cables and circulating water or other fluid through these pipes, and (3) circulating the oil of pipe-type cables through heat exchangers.

Charging current and critical length. Because of the closer spacing between conductors and the higher dielectric constant of the oil-paper insulation, cables have a much higher shunt capacitance per unit length than do overhead lines. Hence for a given voltage the charging current and the charging reactive power of cables are correspondingly higher. Compressed-gas-insulated cables have an advantage in this respect.

The critical length of a cable is that length for which the charging current equals the rated current. A cable longer than the critical length would be overloaded at the sending end even if nothing were connected to the receiving end.

Shunt compensation. The limitation of length or active-power capability due to charging current can be raised by connecting inductive reactors in shunt with the cable at its terminals and at intermediate points. The total shunt current taken by the reactors should be approximately equal in magnitude to the charging current of the whole length of the cable but in phase opposition to it. The economic spacing of shunt reactors on a 60-Hz cable would be between 5 and 10 mi.

Shunt compensation of power cables is seldom, if ever, used. Most underground cables are too short to require it.

Submarine ac cables. These are used in crossing rivers, bays, or straits too wide for overhead spans and to transmit power to offshore islands. They are mostly of the solid type. Water pressure prevents formation of voids and the natural cooling is good. Length of uncompensated cables is limited by charging current to about 25 mi, and shunt compensation is deemed impractical because of the additional complications in laying the cables and in retrieving them when repairs are needed. The lead sheath is protected by an armor of steel wires, sometimes covered with jute. Submarine cables are liable to damage by trawling and by dragging of ship's anchors. Shore sections, used in shallow water and across beaches, usually differ in diameter and amount of armor from the deep-water sections.

Direct-current lines. Although most electric power transmission is by alternating current, there is an increasing number of direct-current transmission lines. These require converter stations at both ends to connect the line to an ac system.

Overhead lines. Bipolar overhead lines are similar in construction to overhead three-phase lines except that they have only two conductors instead of three. For the same conductor size and insulation level, a dc line can carry the same power on two conductors that a three-phase line can carry on three conductors; and the cost of the dc line is about two-thirds the cost of the corresponding ac line. In some places, two monopolar lines spaced well apart (about 1½ mi) are used instead of one bipolar line for the sake of improved reliability.

Since internal overvoltages are somewhat less on dc lines than on ac lines, lower insulation levels are used for the same crest voltage to ground. Under these conditions leakage currents at normal operating voltage become more important, especially when insulators are dirty from industrial wastes and moist from fog. For this reason, insulators for dc lines are usually of special design, having a higher ratio of length of leakage path to flashover distance.

Cable lines. Most of the dc lines built before 1969 where wholly or partly submarine cables. Since that date the trend has been toward overhead lines, although several cases have been considered for underground cables to bring power to metropolitan areas. Direct-current cables have no charging current and therefore are not subject to the limitation on their length that applies to ac cables. In addition, a dc cable has no dielectric loss and can safely withstand a higher direct voltage than root-mean-square alternating voltage. As a result, a dc cable can carry about six times as much power as the rated apparent power when the same cable is used for alternating current. Single-conductor solid cables are used.

Ground return. The resistance of the ground to direct current is very much lower than to alternating current, being essentially only that in the vicinity of the ground electrodes. The use of ground or sea return for a monopolar line saves most of the cost of one conductor and of its power loss. A bipolar dc line can operate with one pole and ground return while there is a fault on the other pole of the line or of the terminal equipment.

[EDWARD W. KIMBARK]

Bibliography: American Radio Relay League, *ARRL Antenna Book*, revised periodically; C. C. Barnes, *Electrical Cables*, 1974; *Electrical Transmission and Distribution Reference Book*, 4th ed., 1950; Federal Power Commission, *National Power Survey*, 1970; Federal Power Commission, *Underground Power Transmission*, April, 1966; D. G. Fink and J. M. Carroll, *Standard Handbook for Electrical Engineers*, 10th ed., 1968; E. W. Kimbark, *Direct Current Transmission*, vol. 1, 1971; E. W. Kimbark, *Electrical Transmission of Power and Signals*, 1949; R. W. P. King, *Transmission-Line Theory*, 1955; H. Pender and W. A. Del Mar (eds.), *Electrical Engineers' Handbook*, vol. 1, 4th ed., 1949; E. R. Schatz and E. M. Williams, Pulse transients in exponential transmission lines, *Proc. IRE*, vol. 38, pt. 2, 1950; W. D. Stevenson, Jr., *Elements of Power System Analysis*, 2d ed., 1962; B. M. Weedy, *Electric Power Systems*, 2d ed., 1972; E. M. Williams and J. B. Woodford, Jr.,

Turbine

A machine for generating rotary mechanical power from the energy in a stream of fluid. The energy, originally in the form of head or pressure energy, is converted to velocity energy by passing through a system of stationary and moving blades in the turbine. Changes in the magnitude and direction of the fluid velocity are made to cause tangential forces on the rotating blades, producing mechanical power via the turning rotor.

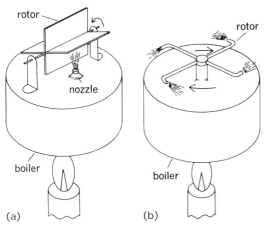

Turbine principles. (a) Impulse. (b) Reaction.

The fluids most commonly used in turbines are steam, hot air or combustion products, and water. Steam raised in fossil fuel-fired boilers or nuclear reactor systems is widely used in turbines for electrical power generation, ship propulsion, and mechanical drives. The combustion gas turbine has these applications in addition to important uses in aircraft propulsion. Water turbines are used for electrical power generation. Collectively, turbines drive over 95% of the generating capacity in the world. *See* GAS TURBINE; HYDRAULIC TURBINE; STEAM TURBINE; TURBOJET.

Turbines effect the conversion of fluid to mechanical energy through the principles of impulse, reaction, or a mixture of the two. Illustration *a* shows the impulse principle. High-pressure fluid at low velocity in the boiler is expanded through the stationary nozzle to low pressure and high velocity. The blades of the turning rotor reduce the velocity of the fluid jet at constant pressure, converting kinetic energy (velocity) to mechanical energy. *See* IMPULSE TURBINE.

The reaction principle is shown in illustration *b*. The nozzles are attached to the moving rotor. The acceleration of the fluid with respect to the nozzle causes a reaction force of opposite direction to be applied to the rotor. The combination of force and velocity in the rotor produces mechanical power. *See* REACTION TURBINE.

[FREDERICK G. BAILY]

Turbojet

A propulsion engine used in many high-speed military and commercial aircraft.

Operating principle. In a turbojet, as illustrated, air approaches the inlet diffuser at a relative velocity equal to the flight speed. In passing through the diffuser the velocity of the air is decreased and its pressure is increased. The air pressure is increased further as it passes through the compressor. In the combustion chamber a steady stream of fuel is injected into the air and combustion takes place continuously. The high-pressure hot gas passes through the turbine nozzles, which direct it at high velocity against the buckets on the turbine wheel, thereby causing the wheel to rotate. The turbine wheel drives the compressor to which it is connected through a shaft. This is the sole function of the turbine.

After the hot gas leaves the turbine, it is still at a

high temperature and at a pressure considerably above atmospheric. The hot gas is discharged rearward through the exhaust nozzle of the engine at a high velocity. *See* BRAYTON CYCLE.

The thrust obtained is equal to the overall increase in momentum of the gas as a result of its passage through the engine. This thrust is given by the equation below, where M is the mass flow of gas per second through the engine, V_j is the exhaust jet velocity, and V_o is the airplane velocity.

$$F = M(V_j - V_o)$$

Compressors. Two types of compressors are used on turbojet engines—centrifugal-flow and axial-flow compressors. The centrifugal compressor is the simpler of the two and was used on the early versions of the engine. For example, it was used on the first British engine designed by Frank Whittle, which was the forerunner of a series of early American engines, namely, the I-A, I-16, and I-40 (which became the J-33).

The trend has been consistently toward the application of the axial-flow compressor because of its greater efficiency and greater air-handling capacity per unit frontal area, in spite of its complexity and fragility. Engines such as the J-57, the J-75, and the J-79 are of this type.

Compressor stall. One of the problems associated with the axial-flow compressor is that of designing the stator vanes to direct the air into the rotor vanes at the proper angle at all rotational speeds. If the air is directed at a vane at too sharp an angle, it will not follow the vane but will break away. Loss in pressure and efficiency occurs and even possibly vibration of the blade.

Axial-flow compressors are designed for the full-power condition. At engine rotative speeds below 70% of maximum, the compressors on some engines develop a rotating stall. One or more stall cells or areas of stall form around the compressor; in these cells the flow strikes the blades at an improper angle of attack. The stalled areas move around the compressor, and each time one passes a given blade, the blade is subject to an impact. If these impacts come in resonance with a natural frequency of the blade, severe and sometimes destructive vibration may result.

The problems of low efficiency and stall at low rotative speeds are particularly severe in compressors with many stages. One solution is to separate the compressor into two sections, each mounted on a separate shaft coaxial with the other and driv-

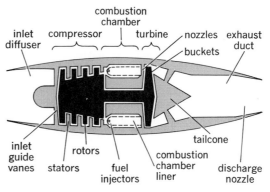

Diagram of an axial-flow subsonic turbojet engine.

Maximum continuous rating of Pratt and Whitney J-57 engine

Flight speed, knots	Sea level		25,000 ft		45,000 ft	
	Thrust, lb	Specific fuel consumption, lb/(hr)(lb)	Thrust, lb	Specific fuel consumption, lb/(hr)(lb)	Thrust, lb	Specific fuel consumption, lb/(hr)(lb)
0	9500	0.76				
200	8000	0.87	4700	0.82	2005	0.85
400	7650	0.99	4750	0.92	2250	0.93
600	7600	1.07	5000	0.98	2500	0.98

en by separate shaft coaxial with the other and driven by separate turbines. An engine of this type, called a two-spool engine, is the J-57. Another solution to these problems is to provide a mechanism for adjusting the angular setting of the stator vanes with change in rotative speed. The J-79 engine uses this method. Interstage bleedoff of air at low rotative speeds is also used to assist in combating stall on some engines, as in the J-57. Other schemes have also been investigated.

Foreign object damage. Another serious problem has been the ingestion of foreign objects by the engine. A foreign object passing through an axial compressor may cause blade failure or may produce a nick in a blade, which can become the nucleus of a fatigue failure. The failure of one blade can set off an avalanche of failures as it passes through the compressor. Compressor failures have been caused by objects picked up from the external environment, by objects shaken loose from the engine, or by objects left by mechanics. These later two sources can be eliminated by careful design and maintenance.

Foreign objects have been sucked up from runways during takeoff and landing with the assistance of vortexes generated by the engine. Forward-directed air jets, which operate during takeoff to break up these vortexes, have been suggested. Screens are also used to stop the larger particles. However, particles smaller than the screen mesh can get through. Screens have the disadvantages that their resistance to the airflow imposes a loss in performance; furthermore, they provide a surface to which ice can adhere in an icing atmosphere. For this reason, they are designed to be retracted once the airplane is aloft. By keeping runways clean and by application of these various aids, the possibility of foreign object damage can be greatly reduced.

Control. The following are some conditions encountered in turbojet operations which are associated with engine control:

1. Overtemperature of the combustion gas during start-up can damage turbine buckets.

2. Overspeed of the turbine and overtemperature of the combustion gas at maximum thrust can overstress turbine wheels and buckets.

3. Compressor surge during acceleration of the engine may cause damaging vibration of compressor blades.

4. Flameout (cessation of combustion) can occur when engine speed is reduced at high altitude.

5. Flameout and also compressor surge can occur when an engine is jockeyed in the course of landing the airplane.

Automatic engine controls are designed to relieve the pilot of the task of avoiding these undesir-

able operating conditions. The earliest automatic controls incorporated only a maximum rotative speed governor and an ambient-pressure-sensing element that adjusted fuel flow for change in altitude.

The trend has been toward more sophisticated automatic controls that handle a greater number of the control requirements at the expense of greater complexity.

Typical performance. For the purpose of illustrating the performance of turbojet engines, the maximum continuous performance of an engine is displayed in the table. *See* PROPULSION.

[BENJAMIN PINKEL]

Turboprop

A gas turbine power plant producing shaft power for aircraft using a propeller. The turboprop engine has basic components similar to those of a turbojet: compressor, combustor, and turbine. In addition, it has a power turbine. This power turbine extracts usable shaft power from the engine mass flow and drives a conventional propeller through a reduction gear. Figure 1 illustrates the simplest possible turboprop configuration. Here the power turbine is rigidly connected to the main engine rotor; hence this type is commonly called a single-shaft engine. This has certain disadvantages in regard to starting, partial-load fuel consumption, and windmilling drag. *See* GAS TURBINE; TURBINE.

Figure 2 shows a more advanced turboprop type. Here the power turbine has no mechanical connection with the main engine rotor and consequently operates at a different rotational speed, this type being called a free-turbine engine. This arrangement improves the operating characteristics and, especially, the cruise efficiency under partial-load conditions. This form of engine is also used to

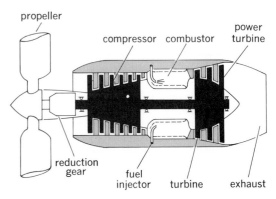

Fig. 1. Turboprop engine with connected power turbine.

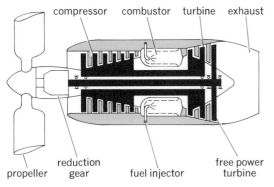

Fig. 2. Turboprop engine with free power turbine.

drive helicopter rotors and is then called a gas turbine.

Performance characteristics. Compared to the turbojet and, to a lesser degree, to the turbofan, the turboprop offers lower fuel consumption and a higher takeoff thrust. It has low engine noise level. Its propellers can be reversed to shorten the landing run. For these reasons the turboprop is an excellent power plant for aircraft in which these qualities are important. Its disadvantages are heavier weight and increased complexity and maintenance cost. Because of propeller characteristics, the turboprop usually reaches peak operating efficiency at lower cruise speeds than the turbofan and is, therefore, better suited for transports in the speed range below 450 mph, although it is basically possible to reach high subsonic and even supersonic flight speeds with a turboprop. At high altitudes the turboprop achieves lower cruise fuel consumption levels than those of the best reciprocating engines (approximately 0.34 lb of fuel per equivalent shaft horsepower-hour).

Design considerations. For maximum efficiency the turboprop engine should have a high pressure ratio (between 8 and 14) and high turbine-inlet temperatures (2000–2400°F). Its specific fuel consumption decreases with rising turbine-inlet temperatures, in contrast to the turbojet cycle.

The design limitations of the turboprop are similar to those of the turbojet, and it is subject to similar problems. *See* TURBOJET.

Its high pressure ratio leads to compressor stall under adverse operating conditions and requires appropriate measures for control; interstage bleed, variable stators, or a twin spool rotor are examples.

Control problems. The turboprop has the same control variables as the turbojet, as well as a number of additional ones. These include fuel flow, affecting rotor speed, and variable stator position or bleed valve position, depending upon the system selected. In addition, there is the propeller pitch control which, with fuel flow, affects the speed of the power turbine. It is generally desirable to maintain gas generator speed at a high level and to select the desired horsepower level through proper adjustment of fuel flow and propeller pitch. The free power turbine configuration allows propeller rotational speed to be an additional variable.

Compensation for altitude and ram air density is required as in the turbojet engine.

Regeneration. The efficiency of a turboprop engine can be increased through use of a regenera-

tor, especially on engines with only a moderate pressure ratio. Early designs attempted to incorporate this feature. Advances in compressor technology made it possible to obtain almost equivalent fuel consumption levels through high pressure ratios and high turbine-inlet temperatures at lower overall complexity and weight, so that regenerators are not widely used in modern turboprop engines. *See* BRAYTON CYCLE.

[PETER G. KAPPUS]

Bibliography: J. V. Casamassa and R. D. Bent, *Jet Aircraft Power Systems,* 3d ed., 1965; V. C. Finch, *Jet Propulsion-Turboprops,* 1955; J. G. Keenan, *Elementary Theory of Gas Turbines and Jet Propulsion,* 1946; A. W. Morley, *Aircraft Propulsion,* 1953.

Uranium

A chemical element, U, atomic number 92, atomic weight 238.03, one of the actinide series, in which the $5f$ electron shell is being filled. The valence electron configuration is $5f^3 6d 7s^2$. Uranium was isolated in 1789 by M. H. Klaproth in a sample of pitchblende from Saxony. In 1841 E. M. Peligot showed that the "semimetallic" element obtained by Klaproth was actually the dioxide. Peligot succeeded in preparing the metal by reduction of uranium tetrachloride with potassium. In 1896 A. H. Becquerel discovered that uranium undergoes radioactive decay. Recognition of the nuclear-fission phenomenon by Otto Hahn and F. Strassmann in 1939 vaulted uranium from a position of relative obscurity to a role of major importance.

Uranium in nature is a mixture of three isotopes: ^{234}U (0.0054%), ^{235}U (0.7204%), and ^{238}U (99.2742%). Half-lives of the three isotopes are 2.45×10^5 yr, 7.04×10^6 yr, and 4.47×10^9 yr, respectively. U^{235} undergoes fission with slow neutrons to release large amounts of energy; U^{238} absorbs slow neutrons to form U^{239}, which in turn decays to fissile Pu^{239} by the emission of two β-particles. Other isotopes of uranium ranging in mass from U^{227} through U^{240} have been prepared by radioactive processes.

Natural occurrence. Uranium is believed to be concentrated largely in the Earth's crust, where the average concentration is 4 parts per million (ppm). For comparison, the Earth's crust contains 0.1 ppm silver and 0.5 ppm mercury. Basic rocks (basalts) contain less than 1 ppm uranium, whereas acidic rocks (granites) may have 8 ppm or more. Estimates for sedimentary rocks are 2 ppm, and for ocean water, 0.001 ppm. The total uranium content of the Earth's crust to a depth of 25 km is

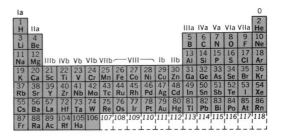

Table 1. Uranium minerals

Mineral	Chemical composition	Color	Specific gravity, 20°C	Typical occurrence
Uraninite	UO_2 (contains Th, rare earths)	Black	8–10.6	Arendal, Norway
Pitchblende (var.)	UO_{2+x}	Black	6–8	Shinkolobwe, Zaire
Euxenite-polycrase	$(Y,Ca,Ce,U,Th)(Nb,Ta,Ti)_2O_6$	Dark brown	4–6	Nipissing, Ontario
Samarskite	$(Y,Ca,Fe,U,Th)(Nb,Ta)_2O_6$	Black	5–6	Mitchell Co., N.C.
Brannerite	$(Y,Ca,Fe,U,Th)_3(Ti,Si)_5O_{16}$	Black	4–5	Blind River, Ontario
Davidite	$(Fe,Ce,U)(Ti,Fe,V,Cr)_3(O,OH)_7$	Black	4–5	Rum Jungle, Australia
Coffinite	$USiO_4$	Black	5–6	Colorado Plateau
Carnotite	$K_2(UO_2)_2(VO_4)_2 \cdot xH_2O$	Yellow	3–5	Colorado Plateau
Tyuyamunite	$Ca(UO_2)_2(VO_4)_2 \cdot xH_2O$	Yellow	3–4	Ferghana, Turkestan
Autunite	$Ca(UO_2)_2(PO_4)_2 \cdot xH_2O$	Greenish-yellow	3–4	Autun, France
Torbernite	$Cu(UO_2)_2(PO_4)_2 \cdot xH_2O$	Green	3–4	Erzgebirge, Saxony
Uranophane	$Ca(UO_2)_2Si_2O_7 \cdot xH_2O$	Greenish-yellow	3–4	Congo Republic

calculated to be 10^{17} kg; the oceans may contain 10^{13} kg of uranium.

Several hundred uranium-containing minerals have been identified, but only a few are of commercial importance. Table 1 summarizes data on some of the more important minerals. All uranium minerals contain lead, resulting from radioactive decay of uranium. Uraninite, as found in pegmatites, usually occurs in rather small amounts which are of little economic significance. The euxenite-polycrase series, brannerite, and davidite are complex primary pegmatite minerals. Pitchblende, a variety of uraninite found in hydrothermal veins, is the most important mineral of uranium. It is usually poorly crystalline, contains very little thorium or rare earths, and is frequently found associated with sulfide minerals. Coffinite, first identified in 1951, has become recognized as an important uranium mineral on the Colorado Plateau. The remaining compounds are secondary minerals which have been formed by means of oxidation and weathering of primary uranium minerals. Near-surface uranium mineralization invariably consists of oxidized ore.

Prior to 1942, uranium was obtained principally as a by-product of radium mining operations. With the discovery of nuclear fission and the potential of atomic power, the possession of uranium reserves became vitally important. Uranium reserves at less than \$22/kg of U_3O_8 for that part of the world for which statistics are available are estimated at about 2×10^9 kg U_3O_8, and those for the United States at about 10^9 kg U_3O_8. Deposits containing as little as 0.1% uranium are being mined. Some of the largest occurrences are the sandstone-impregnated Colorado Plateau deposits, the Blind River conglomerates in Ontario, and the reefs of the Witwatersrand (South Africa) from which uranium is produced as a by-product of the gold industry. The vein deposits at Great Bear Lake and Lake Athabaska are also important sources of uranium, but Shinkolobwe, Zaire, deposits are now virtually exhausted. In addition to these occurrences, extensive reserves of low-grade ore (0.005–0.02% uranium) exist in phosphate deposits (Florida, Soviet Union, and North Africa), bituminous shales (Soviet Union, Sweden, and Tennessee), and lignites (the Dakotas).

For a discussion of extraction methods *see* URANIUM METALLURGY.

Uranium metal. Uranium is a very dense, strongly electropositive, reactive metal, ductile and malleable, but a poor conductor of electricity. It is most conveniently prepared by reduction of a halide (UF_4) with calcium or magnesium in a sealed vessel at 1200–1400°C. The steps involved in preparation of the metal from uranyl nitrate are summarized by Eqs. (1)–(4).

$$UO_2(NO_3)_2 \cdot 6H_2O \xrightarrow{500°C}$$
$$UO_3 + 2NO_2 + \tfrac{1}{2}O_2 + 6H_2O \quad (1)$$

$$UO_3 + H_2 \xrightarrow{700°C} UO_2 + H_2O \quad (2)$$

$$UO_2 + 4HF \xrightarrow{550°C} UF_4 + 2H_2O \quad (3)$$

$$UF_4 + 2Mg \xrightarrow{1200°C} U + 2MgF_2 \quad (4)$$

Some of the physical and thermal properties of uranium are listed in Table 2. The metal exists in three crystalline modifications. α-Uranium (25–668°C) is orthorhombic ($a = 2.854$, $b = 5.869$, $c = 4.956$ A) with four atoms per unit cell and a density of 19.04. Its structure is interpreted as a distorted hexagonal lattice containing corrugated sheets of uranium atoms. The β phase (668–775°C) is a complex tetragonal structure ($a = 10.754$ A, $c = 5.623$ A) with 30 atoms per cell, and a density of 18.13 at 720°C. γ-Uranium (775–1132°C) is body-centered cubic ($a = 3.525$ A) with two atoms per cell. Its density is 18.06 at 805°C. The β form can be stabilized at room temperature by the addition of small amounts of chromium, the γ form with molybdenum.

The unique nature of the room-temperature, α structure curtails solid solution of uranium with

Table 2. Physical and thermal properties of uranium

Property	Value
Melting point	1132 ± 1°C
Boiling point	3818°C
Vapor pressure, between 1630 and 1970 K	$\log p_{mm} = -\dfrac{23{,}330}{T} + 8.583$
Heat of fusion	10.6 kJ/mole
Heat of transition $\alpha \to \beta$	2.9 kJ/mole
Heat of transition $\beta \to \gamma$	4.8 kJ/mole
Enthalpy, at 25°C	6.364 kJ/mole
Heat capacity, at 25°C	27.658 J/(°C) (mole)
Entropy, at 25°C	50.15 J/(°C) (mole)
Thermal conductivity, at 25°C	0.23 J/(cm-s) (°C)
Electrical resistivity, at 25°C	30 μohm/cm

many other metals. Extensive solid solution without compound formation has been found only with molybdenum and niobium. Aluminum, beryllium, bismuth, cadmium, cobalt, gallium, germanium, gold, indium, iron, lead, manganese, mercury, nickel, tin, titanium, zinc, and zirconium all form one or more intermetallic compounds with uranium. Chromium, magnesium, silver, tantalum, thorium, tungsten, vanadium, calcium, sodium, and some of the rare-earth metals form neither compounds nor extensive solid solutions. The uranium alloys are of great interest in nuclear technology because the pure metal is chemically active and anisotropic, and has poor mechanical properties. Alloys can also be useful in diluting enriched uranium for reactors, and in providing liquid fuels.

Uranium reacts with nearly all nonmetallic elements and their binary compounds. Table 3 lists a number of its reactions. Uranium dissolves in hydrocloric acid to leave a black residue of uranium hydroxy-hydride. Addition of fluosilicate prevents formation of the residue. Nitric acid dissolves the metal, but nonoxidizing acids such as sulfuric, phosphoric, or hydrofluoric react very slowly. Uranium metal is inert toward alkalies, but addition of peroxide causes formation of water-soluble peruranates.

Uranium ions in solution. Four oxidations states of uranium exist in solution, but only two, U^{IV} and U^{VI}, are stable. Aqueous solutions of U^{3+} are red, and may be prepared by dissolving a trivalent halide or by electrolytic reduction of tetravalent uranium. Trivalent uranium reduces water to form U^{4+} and hydrogen.

Green solutions of tetravalent uranium are readily oxidized by air. Although the ion, U^{4+}, is extensively hydrolyzed, its existence in acidic solutions has been proved. The first hydrolysis step is believed to give the monomeric ion $U(OH)^{3+}$.

Pentavalent uranium exists in aqueous solutions as the ion UO_2^+. It is unstable in solution because of disproportionation to U^{4+} and UO_2^{++}, but decomposes at the lowest rate in the pH range of 2–4.

Hexavalent uranium, as the uranyl ion (UO_2^{++}), is the most stable oxidation state. The yellow uranyl solution can be reduced with moderately strong reducing agents such as Zn, Sn^{++}, Ti^{3+}, or sodium hydrosulfite. Hydrolysis of UO_2^{++} leads first to the formation of UO_2OH^+. Further hydrolysis gives polymeric ions of the type $UO_2[(OH)_2UO_2]_n^{++}$, involving sheetlike complexes with double hydroxyl bridges.

Extensive studies have been made of complex formation of UO_2^{++}, and to a lesser extent U^{4+}, with many anions (for example, fluoride, chloride, bromide, thiocyanate, nitrate, bisulfate, sulfate, acetate, phosphate, oxalate, citrate, carbonate, and acetyl acetonate). In general, the strength of the complex is inversely proportional to the strength of the acid from which the complexing anion is derived.

The oxidation potentials of uranium in one molar acid solution are as in notation (5) and in one molar basic solution as in notation (6).

$$U \xrightarrow{+1.80 \text{ volts}} U^{3+} \xrightarrow{+0.63 \text{ volt}}$$
$$U^{4+} \xrightarrow{-0.58 \text{ volt}} UO_2^+ \xrightarrow{-0.06 \text{ volt}} UO_2^{++}$$
$$\xrightarrow{-0.32 \text{ volt}} \tag{5}$$

$$U \xrightarrow{2.17 \text{ volts}} U(OH)_3 \xrightarrow{2.14 \text{ volts}}$$
$$U(OH)_4 \xrightarrow{0.62 \text{ volt}} UO_2(OH)_2 \tag{6}$$

Hydride. Uranium reacts reversibly with hydrogen to form the compound UH_3 at 200–300°C. The dissociation pressure of hydrogen over the hydride is 1 atm at 430°C. UH_3 is often pyrophoric, and is used to make highly reactive powdered metal.

Oxides. The uranium-oxygen system is complicated by extensive areas of solid solution, especially at elevated temperatures. The four most important oxides of uranium are UO_2, U_4O_9, U_3O_8, and UO_3. Uranium monoxide exists as a gaseous species at high temperatures, but is not stable as a solid phase below 1800°C. Several tetragonal oxides in the composition range $UO_{2.25}$–$UO_{2.40}$ have been prepared by low-temperature (150–300°C) oxidation of UO_2. At atmospheric pressure the tetragonal phases disproportionate to U_4O_9 and U_3O_8 at 500°C, but they are stabilized by high pressure. Uranium dioxide is brown (melting point ~ 2800°C); it crystallizes in the fluorite structure ($a = 5.469$ A), and is readily oxidized by air above 100°C. A metastable hydrate, $U(OH)_4$, is obtained by precipitation of U^{4+} from aqueous solutions. Ignition of any uranium oxide to 750°C in air leads to the formation of U_3O_8. This oxide is olive-green to black in color and has an orthorhombic structure. The trioxide exists in an amorphous form at least six crystalline modifications, one of which has been prepared only at pressure above 15,000 atm. Decomposition temperatures in air vary between 400 and 650°C. The yellow (γ) trioxide, prepared by thermal decomposition of uranyl nitrate, is the most stable. Reduction to the dioxide may be accomplished with hydrogen or carbon monoxide at 400–600°C. Uranium trioxide reacts with water to form $UO_2(OH)_2 \cdot H_2O$, three crystal forms of $UO_2(OH)_2$, and $U_3O_8(OH)_2$. Because of the amphoteric nature of the trioxide, the compounds may also be classed as acids, that is, H_2UO_4 and $H_2U_3O_{10}$. An insoluble peroxide ($UO_4 \cdot 2H_2O$) is obtained by the addition of hydrogen peroxide to uranyl solutions. The peroxide cannot be dehydrated without decomposition.

Table 3. Chemical reactions of uranium metal

Reactant (reaction temperature, °C)†	Products	Heat of formation of underlined product, kJ/mole at 25°C
H_2(250)	$\alpha, \beta \underline{UH_3}$	−127
O_2(100–350)	$\underline{UO_2}, U_3O_8$	−3572
F_2(250)	$\underline{UF_6}$	−2186
Cl_2(500)	$\underline{UCl_4}, UCl_5, UCl_6$	−1051
Br_2(650)	$\underline{UBr_4}$	−826
I_2(350)	$\underline{UI_3}, UI_4$	−529
B(1650)	$\underline{UB_2}, UB_4, UB_{12}$	−148
C(1500)	$\underline{UC}, U_2C_3, \underline{UC_2}$	−82
Si(1700)	$U_3Si, U_3Si_2, \underline{USi}, USi_2, USi_3$	−80
N_2(500)	$UN, U_2N_3, \underline{UN_2}$	−709
P(400–1100)*	$UP, \underline{U_3P_4}, UP_2$	−316
S(400)	$\underline{US}, US_2$	−306
H_2O(100)	$\underline{UO_2}$	−1084
H_2S(500)*	$\underline{US}, US_2$	−502
HF(350)*	$\underline{UF_4}$	−1882
HCl(300)*	$\underline{UCl_3}$	−893
NH_3(700)	$\underline{UN}, UN_2$	−303
CH_4(650)*	$\underline{UC}$	−97
CO(750)	UO_2, UC	
CO_2(600)	UO_2, UC	
NO(400)	U_3O_8	
N_2O_4(25)	$UO_2(NO_3) \cdot 2NO_2$	

*Powdered metal.   †Values are given for massive metal.

Table 4. Preparation of uranium halides

Compound	Reaction
UF_3	$UF_4 + Al \rightarrow UF_3 + AlF \uparrow$ (900°C)
UF_4	$UO_2 + HF \rightarrow UF_4$ (550°C)
U_4F_{17}, U_2F_9, UF_5	$xUF_4 + yUF_6 \rightarrow U_4F_{17}, U_2F_9, UF_5$ (250°C)
UF_6	$UF_4 + F_2 \rightarrow UF_6$ (350°C)
UCl_3, UBr_3, UI_3	$UH_3 + HX \rightarrow UX_3$ (350°C)
UCl_4	$UO_3 + Cl_2C{=}CClCCl_3 \rightarrow UCl_4$ (210°C)
UCl_5	$UCl_4 + Cl_2 \rightarrow UCl_5$ (500°C)
UCl_6	$UCl_5 \rightarrow UCl_6 + UCl_4$ (125°C in vac.)
UBr_4, UI_4	$U + X_2 \rightarrow UX_4$ (550°C)

Ternary oxide systems of tetra- and pentavalent uranium with alkaline-earth, rare-earth, and some group IV metal oxides have been investigated. Extensive regions of solid solution based on the fluorite UO_2 structure are generally encountered. Hexavalent uranium forms ternary oxides (uranates) exhibiting a wide range of composition with other metals. These compounds are insoluble in water, but readily soluble in acids. The monouranate, $M_2^IUO_4$, and diuranate, $M_2^IU_2O_7$, are most frequently encountered, but more complex polyuranates have been identified. The uranates may best be prepared by air ignition of the oxides or salts in the proper proportions, as in Eqs. (7) and (8). Precipitation from uranyl solutions is un-

$$6Na_2CO_3 + 2U_3O_8 + O_2 \rightarrow 6Na_2UO_4 + 6CO_2 \quad (7)$$

$$CaO + UO_3 \rightarrow CaUO_4 \quad (8)$$

satisfactory, since hydrolysis generally precludes the isolation of stoichiometric compounds. Addition of hydrogen peroxide and alkali to uranyl solutions leads to the formation of soluble peroxyuranates, for example, $Na_4UO_8 \cdot xH_2O$.

Halides. The uranium halides constitute an important group of compounds. Uranium tetrafluoride is an intermediate in the preparation of the metal and of the hexafluoride. Uranium hexafluoride, the most volatile compound of uranium, is used in the isotope separation of U^{235} and U^{238}. The halide volatilities increase in the order $UX_3 < UX_4 < UX_5 < UX_6$. Reactions for the preparation of uranium halides are summarized in Table 4. Uranium hexafluoride boils at 56.54°C, and melts at 64°C (1140 mm pressure). It is a reactive substance and a strong fluorinating agent. Equipment for containing the compound may be constructed of copper, nickel, aluminum, or fluorine-containing polymers (Teflon, Kel-F). The hexafluoride reacts with water to form UO_2F_2. The chlorides, bromides, and iodides are hygroscopic and soluble in water. The penta- and hexachlorides are also soluble in some nonpolar solvents (CCl_4, CS_2). The halides will react with oxygen at elevated temperatures to form uranyl compounds and ultimately U_3O_8.

Uranyl salts. Uranyl nitrate is obtained as the hexahydrate (UNH) from dilute nitric acid solutions, as the trihydrate from concentrated nitric acid, and as the dihydrate from fuming nitric acid. The lower hydrates may also be obtained by careful dehydration of the hexahydrate. Anhydrous uranyl nitrate and U^{IV} nitrate do not exist as

identifiable compounds. Uranyl nitrate is very soluble in water and is probably the most frequently encountered compound in uranium chemistry.

Uranyl sufate, also very soluble in water, is precipitated from solution as the trihydrate. A monohydrate is formed by careful dehydration or by equilibration in water at 180°C. The anhydrous sulfate is formed by heating the hydrate to 300°C. Phase diagrams of uranyl nitrate, uranyl sulfate, uranyl fluoride, uranyl carbonate, and uranyl phosphate have been constructed. Some of the important uranyl salts of organic acids are the formate, $UO_2(HCOO)_2 \cdot H_2O$; the acetate, $UO_2(CH_3COO)_2 \cdot 2H_2O$; and the oxalate; $UO_2C_2O_4 \cdot 3H_2O$.

Uranium in nonaqueous solvents. The solubility of uranium salts, especially uranyl nitrate, in certain organic solvents can be used to separate U from other metal ions by solvent extraction. Numerous types of liquid (aqueous)–liquid (organic) separations techniques are employed. Distribution of uranium between the two phases may be controlled by (1) "salting" the aqueous phase with mineral acids or their salts, (2) pH changes in the aqueous phase, or (3) using different types of organic extractants. The three types of organic phases are: (1) a neutral extractant such as diethyl ether, methyl isobutyl ketone (hexone), or tributyl phosphate (TBP); (2) an acidic extractant such as octanoic acid or thenoyl trifluoroacetone; and (3) a basic extractant such as tri-n-octylamine. In some instances the organic phase may be diluted with a "carrier" diluent such as kerosine. The most frequently used extraction procedure is the purification of uranium with TBP dissolved in kerosine or hexane. The extraction is based on the formation of a neutral unionized complex according to Eq. (9).

$$UO_2^{++}(aq) + 2NO_3^{-}(aq) + 2TBP(org) \rightleftharpoons$$

$$UO_2(NO_3)_2 \cdot 2TBP(org) \quad (9)$$

Analysis for uranium. Numerous procedures are available for the quantitative determination of uranium. Macro quantities of uranium are analyzed by gravimetric or volumetric methods. Gravimetric procedures utilize U_3O_8 ignited in air to 750°C or 8-hydroxyquinolate. Volumetric methods are based upon reduction of uranium to U^{IV} with lead, followed by titration with an oxidizing agent such as potassium dichromate, ceric sulfate, or potassium bromate. Small amounts of uranium may be determined by coulometric, polarographic, colorimetric, or fluorescence methods. *See* FISSION, NUCLEAR; NUCLEAR FUELS; NUCLEAR POWER; NUCLEAR REACTION. [HENRY R. HOEKSTRA]

Bibliography: K. W. Bagnall, *The Actinide Elements*, 1972; D. Brown, *Halides of the Lanthanides and Actinides*, 1968; E. H. P. Cordfunke, *The Chemistry of Uranium*, 1969; *MTP International Review of Science: Inorganic Chemistry*, vol. 7: *Lanthanides and Actinides*, 1972; D. F. Peppard, Liquid-liquid extraction of the actinides, *Annu. Rev. Nucl. Sci.*, 21:365, 1971.

Uranium metallurgy

The processing treatments for the production of uranium concentrates and the recovery of pure uranium compounds. These operations are numerous because of the great variety in the nature of uranium minerals and associated materials, and the wide limits of concentration that exist in the naturally occurring ores. Recovery of the uranium requires chemical processing; however, preliminary treatment of the ore may involve a roasting operation, a physical or chemical concentration, or a combination of these treatments.

Although roasting can bring about significant chemical change, it can also improve the filtering and settling characteristics of an ore for subsequent processing. Only a limited number of deposits are sufficiently rich in uranium-bearing minerals and have physical characteristics that make a preliminary physical concentration of the uranium a feasible operation by present techniques. Chemical concentration is most common as a preliminary treatment.

In general, one of two leaching treatments is used as the initial step in chemical concentration. One of these is acid leaching and the other carbonate leaching of the ore. The choice depends on the nature of the ore, which determines largely the efficiency and cost of the process employed.

Acid leaching is carried out universally with sulfuric acid. The finely ground ore is treated with the dilute acid, and usually an oxidant is added to make sure that the uranium will go into solution as the uranyl ion. When an ore is carbonate leached, the action of sodium carbonate on the uranium forms a water-soluble uranyl tricarbonate ion. This solution is stabilized by adding bicarbonate and by keeping the uranium in the oxidized state. From either the acid or carbonate leaching process, the uranium can be recovered in concentrated form with only minor amounts of impurities.

The concentrate, whether obtained by chemical or physical means, is treated chemically to give a uranyl nitrate solution that can be further purified by solvent extraction. The impurities remain in the aqueous phase while the uranium is extracted by the organic phase. The pure uranium is then stripped from the organic phase by means of water, and this high-purity uranium can be recovered as nitrate crystals or precipitated from solution. Pure oxides and other compounds can then be prepared. In large-scale processing, the nitrate is decomposed thermally to give uranium trioxide, UO_3, which is subsequently reduced with hydrogen to form uranium dioxide, UO_2. Uranium tetrafluoride, UF_4, is prepared by treating UO_2 with hydrogen fluoride, HF, gas.

Uranium metal can be obtained from its halides by fused-salt electrolysis or by reduction with more reactive metals. The reaction of UO_2 with calcium yields metal of fair quality.

The electrolysis of uranium tetrafluoride in a fused-salt bath of sodium and calcium chlorides has been used to prepare tons of high-purity metal. The metal deposits as small granules on a molybdenum cathode. In the calcium reduction of UO_2, the ingredients of the charge are mixed and placed in an inert atmosphere, and are heated to a temperature near 1000°C to form finely divided uranium and calcium oxide. Calcium hydride can be employed instead of calcium in this process. The metal from either electrolysis or the calcium-UO_2 process is leached, pressed, and sintered or melted to give solid uranium.

The largest tonnages of uranium metal of good quality have been prepared from uranium tetrafluoride by reduction with calcium or magnesium. The charge, consisting of a mixture of the finely divided UF_4 (green salt) and the reducing metal in granular form, is placed in a refractory-lined vessel, and the reaction is initiated. The conditions are such that the temperature reached by the reaction in either case (with calcium or with magnesium) is sufficient to melt the products — uranium and the calcium or magnesium fluoride. The dense molten uranium collects by gravity in a pool under the molten slag where it remains during solidification. The recovered piece of metal (referred to as a biscuit) is quite pure. In the reduction employing calcium, the reaction heat is such that the charge can be ignited at room temperature, and the products will be sufficiently molten to separate in the reaction vessel. If magnesium is employed, a closed vessel (bomb) is used, and the whole charge is preheated in a heat-soaking pit until the charge ignites. The addition of heat before ignition is necessary because of the smaller heat of the magnesium reaction so that fusion and good separation of the uranium metal and slag will result.

Uranium metal ingots can be readily prepared by vacuum melting the uranium biscuits in graphite crucibles and casting in graphite molds. The metal can be fabricated by conventional methods with due consideration for its chemical reactivity and allotropic transformations. The low-ductility, β-crystalline form of uranium is usually avoided in fabrication processes. *See* NUCLEAR FUELS; NUCLEAR FUELS REPROCESSING; URANIUM.

[HARLEY A. WILHELM]

Bibliography: C. A. Hampel, *Rare Metals Handbook*, 1961; H. A. Wilhelm, The preparation of uranium metal by the reduction of uranium tetrafluoride with magnesium, *Proceedings of the 1st Geneva Atoms for Peace Conference*, 8:162–174, 1955; W. D. Wilkinson, *Uranium Metallurgy*, vols. 1 and 2, 1962.

Warm-air heating system

In a general sense, a heating system which circulates warm air. Under this definition both a parlor stove and a steam blast coil circulate warm air. Strictly speaking, however, a warm-air system is one containing a direct-fired furnace surrounded by a bonnet through which air circulates to be heated (see illustration).

When air circulation is obtained by natural gravity action, the system is referred to as a gravity warm-air system. If positive air circulation is provided by means of a centrifugal fan (referred to in the industry as a blower), the system is referred to as a forced-air heating system.

Direct-fired furnaces are available for burning of solid, liquid, or gaseous fuels, although in recent years oil and gas fuels have been most commonly used. Furnaces have also been designed which have air circulating over electrical resistance heaters. A completely equipped furnace-blower package consists of furnace, burner, bonnet, blower, filter, and accessories. The furnace shell is usually of welded steel. The burner supplies a positively metered rate of fuel and a proportionate amount of air for combustion. A casing, or jacket, encloses the furnace and provides a passage for the air to be circulated over the heated furnace shell. The cas-

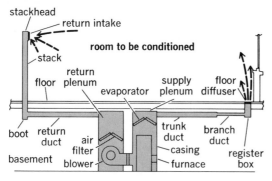

Air passage in a warm-air duct system. (*From S. Konzo, J. R. Carroll, and H. D. Bareither, Winter Air Conditioning, Industrial Press, 1958*)

ing is insulated and contains openings to which return-air and warm-air ducts can be attached. The blower circulates air against static pressures, usually less than 1.0-in. water gage. The air filter removes dust particles from the circulating air. The most common type is composed of 1- to 2-in. thick fibrous matting, although electrostatic precipitators are sometimes used. *See* OIL BURNER.

Accessories to assure effective operation include automatic electrical controls for operation of burner and blower and safety control devices for protection against (1) faulty ignition of burner and (2) excessive air temperatures.

Ratings of warm-air furnaces are established from tests made in laboratories under industry-specified conditions. The tests commonly include heat-input rate, bonnet capacity, and register delivery. Heat-input rate is the heat released inside the furnace by the combustion of fuel, in Btu/hr. Bonnet capacity refers to the heat transferred to the circulating air, in Btu/hr. Register delivery is the estimated heat available at the registers in the room after allowance for heat loss from the ducts has been made, in Btu/hr.

The recommended method for selection of a furnace is to estimate the total heat loss from the structure under design weather conditions, including the losses through the floor and from the basement, and to choose a furnace whose bonnet capacity rating is equal to, or greater than, the total design heat loss.

The complete forced-air heating system consists of the furnace-blower package unit; the return-air intake, or grille, together with return-air ducts leading from the grille to the return-air plenum chamber at the furnace; and the supply trunk duct and branch ducts leading to the registers located in the different spaces to be heated.

The forced-air system is no longer confined to residential installations. The extreme flexibility of the system, as well as the diversity of furnace types, has resulted in widespread use of the forced-air furnace installations in the following types of installations, both domestic and commercial: residences with basement, crawl space, or with concrete floor slab; apartment buildings with individual furnaces for each apartment; churches with several furnaces for different zones of the building; commercial buildings with summer-winter arrangements; and industrial buildings with individual furnace-duct systems in each zone. *See* HEATING, COMFORT.   [SEICHI KONZO]

Bibliography: American Society of Heating, Refrigerating, and Air-Conditioning Engineers, *ASHRAE Guide and Data Book*, 1969; B. H. Jennings and S. R. Lewis, *Air Conditioning and Refrigeration*, 4th ed., 1958; S. Konzo, J. R. Carroll, and H. D. Bareither, *Winter Air Conditioning*, 1958; National Warm Air Heating and Air Conditioning Association, *Heat Gain Calculation and System Design for Summer Air Conditioning*, manual no. 11.

Waste heat management

The systematic process of minimizing the release of energy to the environment by placing priority emphasis on technical sophistication in the use of energy-related machines and products over short-term economic return.

For more than a century, physical growth and financial expansion have been generally regarded as synonymous with progress. Many of the notions of modern resources management continue to reflect this idea. Certainly, there is no single segment of Western culture that has contributed more to the advancement of man on Earth than energy technology, yet all the achievements will be of little value if we fail in our management of energy.

In order to evaluate the potential effects of the present course of action in energy management, it is important to understand that energy demand and energy release to the environment are synonymous. All of the energy converted from nuclear or fossil sources appears in the environment as heat to be added to the Earth's thermal budget. Hence, in a realistic discussion of total energy demand, it is necessary to consider the subject in terms of the thermal energy released to satisfy the demand. In this context, because of prime mover cycle limits, the projected growth of electrical power assumes a positon of paramount importance.

ENERGY DEMAND FORECAST

Energy demand projections for the contiguous United States vary widely, depending on the assumptions of the investigator, ranging from 130 to 190×10^{15} Btu per year. Persons who place a high degree of faith in economic elasticity tend to forecast on the low side, with the implicit hope that increasing technological sophistication will in some undefined way reduce the energy demand. Persons who represent energy interests tend to forecast higher figures. Table 1 represents the author's best view of the many forecasts available. *See the feature article* ENERGY CONSUMPTION.

Current study. Using these forecasts and other prominent forecasts, this author, with the assistance of other Battelle Northwest staff members, conducted a study of the breakdown of energy requirements by major expected use. This study is the basis for the interpretive comments which follow. Figure 1 summarizes the results of the study. The projections of energy allocations on a per capita basis or by gross national product are derived from this summary.

Figure 2, which is based on a total United States population of 260×10^6 by the year 2000, places the expected overall energy usage per capita in perspective. Energy usage per capita is expected to increase by some 210% over 1960 estimates, and electrical energy (kilowatts at the bus-bar) is expected to increase by 1050% in the same period.

Table 1. United States demand for energy resource by major sources in 1973 and estimated demand in 1985 and 2000

Resource	1973	1985	2000
Petroleum (includes natural gas liquids)			
Million barrels	5,982	6,500	7,190
Million barrels per day	16.4	17.8	19.7
Trillion Btu	34,700	37,700	41,700
Percent of gross energy inputs	45.9	39.2	30.8
Natural gas (includes gaseous fuels)			
Billion cubic feet	22,868	21,705	16,472
Trillion Btu	23,600	22,400*	17,000*
Percent of gross energy inputs	31.2	23.3	12.6
Coal (bituminous, anthracite, lignite)			
Thousand short tons	515,500	736,973	901,170
Trillion Btu	13,500	19,300	23,600
Percent of gross energy inputs	17.9	20.1	17.4
Hydropower, utility			
Billion kilowatt-hours	286	385	570
Trillion Btu	2,900	3,600	5,300
Percent of gross energy inputs	3.8	3.7	3.9
Nuclear power			
Billion kilowatt-hours	83.5	1,224	4,435
Trillion Btu	900	13,200	47,800
Percent of gross energy inputs	1.2	13.7	35.3
Total gross energy inputs, trillion Btu	75,600	96,200	135,300

*Not including synthetic gas estimated at 20% of coal usage in 2000.

The growth shown in Fig. 1 will occur primarily in the use of electricity to replace traditional heat sources in the conversion industries and for space heating. Because of the high cost of air conditioning and lighting, better design is expected to reduce the per capita consumption in large buildings. This will be accomplished largely by the elimination of incandescent lighting and low-efficiency fluorescent lighting, and by the employment of regenerative systems to reduce heating and cooling loads.

People-related energy requirements present some interesting contrasts, as shown by Table 2. The largest energy by far is released by the private motor car. Even total conversion to electric automobiles would cut the total energy release to the environment by only a factor of two because a 50% thermal efficiency appears to be barely achievable under current energy policies by the year 2000.

Of particular interest to the chemical and conversion industry is the projection of excellent progress in reducing of unit heat (energy) requirements for the production of steel by the basic oxygen process and other related new reduction processes. However, total energy use by industry is expected to increase about twice as fast as people-related use (Fig. 3). Also of interest is our projection that the use of exothermic energy in the production of petroleum will show a decrease of about 7% to about 590,000 Btu/barrel by the year 2000.

Legislation. The electrical energy requirements are expected to be the predominant growth segment in the next 30 years, rising from 21.5% of all energy requirements in 1965 to 52% of all requirements by 2000. It is also in this segment that legislation has delayed research and development in the multipurpose use of related materials in the production of electrical energy to the greatest extent. The Holding Company Act of 1935 limited the effectiveness of mergers in the utility field as a means of integrating the transmission system of the industry. The act has also served to concen-

trate the electrical power industry's research and development on conversion processes which have a single purpose and are limited to the direct interest of the industry as constituted under the act of 1935. This is in contrast to other industries, where aggressive development of related business opportunities has permitted a wide cross-fertilization of technology. *See* ELECTRICAL UTILITY INDUSTRY.

The modern pulverized coal furnace-boiler, for example, has followed a uniquely single-purpose path. This type of furnace is commonly used to convert the chemical energy in coal to heat energy in steam for electrical generation. Ever since its

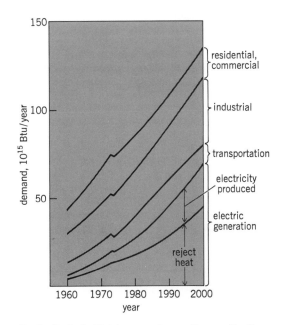

Fig. 1. Projected total energy demand by use. Top three curves do not include electrical energy consumption. (*From Nuclear Power Growth 1974–2000, USAEC, WASH-1139*)

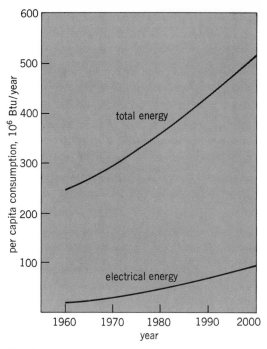

Fig. 2.   Relative growth of energy use by decades through 2000.

adoption in the early 1920s as a means of creating an ash so fine as to be readily dispersed to the atmosphere, to avoid the sooty particles from incomplete carbon combustion prevalent from stoker systems of that era, designers of this combustion system have thought of it as an incinerator instead of as a chemical reactor which it really should more properly emulate.

The chemical industry is not hampered in this way. Combustion processes are common in the oil refinery business and a majority of the fuels and major petroleum products are involved in high-temperature transformations using flame technology. As a result, the chemical industry is in a unique position to retain valuable side streams of otherwise useless materials and recycle them for other in-plant processes or sell them.

However, even chemical processes are cheaper to operate with relatively pure input materials, and the availability of relatively cheap sulfur from

Table 2. People-related heat release rates

Parameter	Rate, Btu/hr
One human being at rest	450
Electric lighting for a typical office	
Incandescent	2.500
Fluorescent	900
Theoretical (100% efficient)	82
Air conditioning for typical office	600
Heating to maintain 25°F ΔT for	
Typical office	
Electrical resistance*	1,000
Heat pump*	350
Furnace, oil or gas (clean)	1,600
4000-lb automobile at 70 mph	
Internal combustion (piston)	750,000
Electric*	150,000
Electric (including losses)	450,000

*Does not include generation or transmission losses. Total energy and resulting heat are about three times these values at efficiency of conversion predicted through the year 2000.

underground deposits has not produced a high incentive for aggressive recycling of the various sulfur by-products. The low cost of sulfur is attributable not only to large supplies of pure materials, but to national policy on depletion allowances and related incentives to the mining and conversion industry. In general, such policies do not encourage the use of scrap or reclaimed materials. During the 1930s and continuing through World War II, the simple pragmatism of more and more production at lower and lower prices held full sway. In that period, annual sulfur production (estimated at 10×10^6 tons) was closely matched by the discharge of wasted sulfur as SO_2 from the utility industry of about 7×10^6 tons. Other more valuable chemicals, such as vanadium, follow similar trends. Arthur M. Squires of New York University has suggested that the single-purpose development of the pulverized coal furnace would have taken an entirely different turn if the patents that led to the fluid-bed reactor, which became standard in the petroleum industry, had been adopted by the utility industry.

While these two industries—utilities, on the one hand, and the chemical industry, on the other—have been perfecting their individual processes, the availability of cheap, relatively pure input fuels and ores has declined. Nevertheless, because of the availability of foreign oil and more recently natural gas in liquid form, research in the use of native resources such as coal and oil shale for the production of liquid and gaseous fuels has not made great progress. See COAL; LIQUEFIED NATURAL GAS (LNG); OIL SHALE.

Similarly, policies which depress fuel prices in relation to other indexes of productivity have given little incentive for advanced research and development of energy conversion. However, in the recent past, an impending gas shortage has created new impetus to the early resolution of the price-supply question for that fuel. See ENERGY CONVERSION.

WASTE HEAT AND THE ATMOSPHERE

Many persons have argued that the energy release per square foot in the United States is so low that there is no cause for any long-range concern over the potential effects of thermal release to the environment. On the basis of a total United States land area of 84×10^{12} ft^2, it can be computed that the total release of electricity as heat to the environment in 1970 amounted to about 0.19 W/ft^2, or only about 0.018% of the Sun's averaged incident energy on the surface of the land and water of the country. However, this number fails to take into account the approximately 2 W of heat that is released for each watt of electrical power generated. Moreover, no matter what the efficiency, eventually all 3 W of heat is released as waste heat to the environment. In addition, the argument fails to include the large contribution of thermal releases from transportation, industry, and residential heating. In 1970 such contributions were about 10 times the amount of energy used in electrical systems.

Forecasted to the year 2000, the figures for energy release per square foot become quite significant. For example, in the energy forecast of Fig. 1, the total energy release per square foot increases

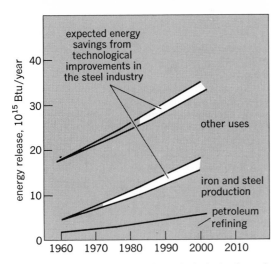

Fig. 3. Nonelectrical energy release by industry through 2000.

from 0.22 W/ft² in 1970 to 0.66 W/ft² in 2000. If the energy release in key metropolitan areas is related to local areas and climatic conditions, enough energy would be concentrated to cause a mean air temperature increase of 13°F in Los Angeles and 5°F in the New York–Washington corridor (Fig. 4). These rates of release approach 50% of the solar incident energy.

The effects are more pronounced in the city than in the country, largely because of the increased roughness of the surface resulting from the erection of structures and the changes in albedo resulting from the modifications of the surface by man (Table 3). A number of investigators, such as Helmut E. Landsberg and Steven R. Hanna, have developed empirical models of the effects of increased roughness of the city landscape. However, much more research on this subject is needed to arrive at a clear and unequivocal concept which can be used to generate constructive action.

City and country. As poorly conducting and evaporating vegetable surfaces (rural areas) are replaced by well conducting, essentially dry surfaces of high heat capacity and low albedo (urban areas), a higher thermal balance is favored. As thermal releases increase from energy decay, and as human sources (including metabolic sources), industry, and transportation activity become more concentrated, and as the surface roughness of the urban landscape increases, a relatively stable low-pressure area develops which prevents normal nocturnal circulation. The lack of circulation concentrates the effluents of the urban aggregation. The energy budget of the metropolitan area is thus greatly affected by the prodigious production of pollutants. As a result, ultraviolet radiation is much reduced, and net incoming radiation is concentrated in upper air masses instead of being distributed. These redistributed fractions of the energy spectrum then react with the trapped chemical ejecta of our man-made volcano to produce a fantastic array of photochemical products known to be injurious to all higher plants and animals. *See* ATMOSPHERIC POLLUTION.

Development of policy. Despite a lack of clear agreement on the numerical modeling of such "heat islands" and their microclimates, the fol-

lowing observations appear to be a sound basis for the development of a policy for waste heat management:

1. When examined on an entirely global perspective, the foreseeable effects of the release of energy and the by-products of energy conversion do not provoke the need for crisis thinking in the institutional management of resources. The single exception is particle emission in the upper atmosphere and stratosphere, trophosphere, and ionosphere.

2. Particle emission and changes in the opacity of the atmosphere resulting from artificial cloud seeding and from heat islands are not likely to cause serious dislocations in the weather patterns, rainfall, and general surface energy budget of the Earth.

3. On a regional scale of 360,000 mi² (10⁶ km²), distinctly observable patterns are related to the release of products of combustion, moisture, and thermal energy (the end state of all useful energy of interest to this discussion). These patterns are increasing in size and intensity. One can predict with certainty that unless accelerating urbanization is controlled, climatic changes will increase from the present purely local scale to a regional scale and will appreciably affect the existing ecological balance. Known climatic changes include increased cloudiness, reduction of winds, concentration of pollutants (smog), drastic changes in runoff of rainfall, and profound, but as yet quantitatively unmodeled, effects of the immense quantities of gaseous and suspended materials ejected to the metropolitan atmosphere.

The role of the urban heat island in stabilizing and concentrating chemical pollutants and particulates released by various combustion processes is becoming increasingly apparent. Ulrich H. Czapski has suggested that the climatic effects in and near cities may in fact be more a result of the

Table 3. Climatic changes produced by cities

Parameter	Comparison with rural environs
Radiation	
Total on horizontal	15–20% less
Ultraviolet, winter	30% less
Ultraviolet, summer	5% less
Cloudiness	
General cloud cover	5–15% more
Fog, winter	100% more
Fog, summer	30% more
Precipitation	
Total amounts	5–10% more
Heavy rains over 0.2 in./day	10% more
Temperature	
Annual mean	1–1.5°F more
Winter minimum	2–3°F more
Relative humidity	
Annual mean	4–6% less
Winter	2% less
Summer	8% less
Wind speed	
Annual mean	20–30% less
Extreme gusts	12–20% less
Calms	5–20% less
Contaminants	
Dust and particulate	10 times more
Sulfur dioxide	5 or more times
Carbon dioxide	10 times more
Carbon monoxide	25 times more

SOURCE: H. E. Landsberg, *City Air: Better or Worse?* 1961.

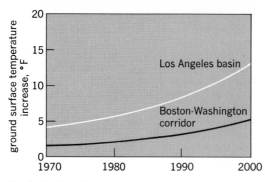

Fig. 4. Regional temperature elevations computed from energy forecasts, based on all energy and waste heat sources.

added heat than of the condensation and freezing nuclei to which they are usually attributed. Czapski predicts severe consequences of large latent and sensible heat emissions: Cumulus clouds prevail most of the time downwind from a large power plant or aggregation of power plants; rainfall is increased downwind for a considerable distance; severe thunderstorms, and even tornadoes, can be caused in very unstable weather by dry and clear heat.

A BASIS FOR CONCERN

The microclimatic effects described here make it clear that waste heat management may be one of the seriously underestimated problems of our century. There is not only a lack of appreciation for the size of the problem but also a lack of understanding on the part of many people, including those with technical training, of the pervasive nature of heat itself.

It is time to begin to stress the fact that all energy produced and consumed modifies the environment and pollutes it. This occurs regardless of how efficient the conversion is or how many uses of "waste heat" are cascaded. Norman H. Brooks has put it this way: "Heat as a pollutant has a unique characteristic. Because of the basic laws of thermodynamics there is no treatment as such; any efforts to concentrate it (by heat pumps) simply require more mechanical energy, which means more waste heat is generated at the power plants. Other types of environmental pollution can be alleviated by various processes, which usually consume power and ultimately produce waste heat. Thus, heat is an ultimate residual of society's activities." *See* HEAT PUMP.

Currently, water-quality criteria for streams are stated in terms which do not relate to thermal modification of the atmosphere. But as the thermal islands related to metropolitan areas expand in size, streams will be affected. Therefore any policy regarding the increase of thermal releases to the metropolitan atmosphere is in fact involved in the modification of regional stream systems. An outstanding example of the influence of metropolitan air temperature on streams occurs in the Chicago area, where a 30-year record of Illinois waterway temperatures, even when corrected for thermal releases of the Commonwealth Edison Co. and other industry, still shows a 3°F increase. This increase is a result of a general increase in the mean air temperature of the Chicago metropolitan area.

WASTE HEAT AND ENERGY IN WATER

If waters are warmed, life activity in the waters is increased. The rate at which organic materials decompose and the rate at which microorganisms die and are replaced by succeeding generations are therefore generally increased. The rate at which oxygen is consumed and the rate at which carbon dioxide is evolved go up with the increasing temperature. Conversely, the solubility of gases declines. In addition, the overall process of self-purification of streams tends to be augmented, although this is an extremely general statement. *See* WATER POLLUTION.

Microorganisms. Observations of plankton and microorganisms passing through industrial processes involving sudden temperature increases usually show a rather dramatic and usually fatal effect. The death of plankton in turn may produce a regional rise in biochemical oxygen demand as a result of the accelerated formation of decaying materials. It is also true that dissolved oxygen concentration caused by natural reaeration of streams increases geometrically at about 1½%/°C. However, even as reaeration occurs, the dissolved oxygen concentration of the water at saturation declines, and the rate at which organic matter consumes oxygen also rises. The net result is that, during colder seasons, oxygen deficits appear earlier in the downstream flow of a warmer stream than in its thermally polluted equivalent.

Although small increases in water temperature tend to provide a more congenial condition for the multiplication of microorganisms in a high-nutrient media, natural water usually provides a rather lean diet. Many tests have indicated that increased stream temperatures improve the rate at which disease-causing microorganisms are eliminated in lakes and streams. This relation between microorganism count and temperature is currently receiving considerable attention, which should lead to additional sophistication in separation of the variables affecting reactions in water. It may also lead to rejuvenation of microorganism counting as a measure of stream pollution.

Drinking water. Along these lines, it is interesting to examine the effects of changing temperature on the effectiveness of chlorine as a disinfectant for potable water. These effects have been studied for many years by many industrial and public authorities. It is generally true that greatly increased quantities of chlorine are required as temperature decreases in the range of 25 to 4°C.

However, both the pH of the water and the presence of contaminants play a part in determining the overall activity of disinfectants. Furthermore, there are few cases on record, if any, that give much insight into the effects, other than seasonal, on a standardized pathogen with only temperature as the variable parameter. Since the formation of free chlorine residual is also temperature dependent, the question of chemistry versus biology remains unresolved.

Of considerable concern to the public consumer of water is the matter of taste. Generally, increased water temperature increases the taste problems that result from the contaminants remaining after treatment. Algae are the culprits, but not all species are equally offensive. The grass-green algae are most abundant in water supplies

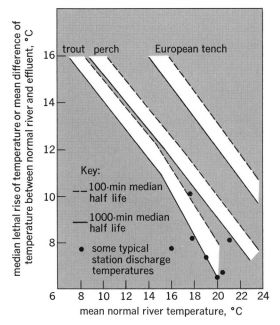

Fig. 5. Effect of temperature rise on lifetime of three fish species.

more will continue to be done as thermal problems are better defined. Figure 5 shows the results of a statistical study conducted in the United Kingdom. The shaded area between the broken lines for each species represents the range in which a temperature rise (vertical axis) above the normal water temperature (horizontal axis) will affect the lifetime of the fish. Figure 5 shows conclusively the better adaptability of the perch-type fish forms over the salmonoid forms, such as trout, and the high adaptability of the carplike species, of which the European tench is an example. The points at lower right indicate the discharge temperatures of some major power-generating stations and their relationship to the life expectancy of these three statistical groups.

Cooling water passed through power station condensers is frequently warmed as much as 10°C (18°F). Poorly mixed effluents of this type disturb normal river conditions, particularly in streams with restricted flows. In addition, these heated effluents, by virtue of high temperatures—and temperature fluctuations alone—are lethal to trapped trout and even to coarse fish acclimatized to normal river temperature. However, small free-living fish are affected to a lesser extent, and larger fish seem to be able to swim away from warmer regions without difficulty. Sudden temperature changes, nevertheless, are problems to all species.

The effect of temperature on fish increases dramatically when dissolved oxygen is low and CO_2 is high. This is another case where proper definition of temperature as a parameter is necessary to draw meaningful and objective conclusions.

HYDRO POWER AND THERMAL EFFECTS

The impoundment of streams for hydroelectric purposes has a number of significant effects on the water quality. Reservoirs which are relatively deep tend to produce cooling of summer extremes and warming of winter lows with relatively small changes to the average temperature. However, the timing of thermal maximums and minimums is

during the summer but are seldom found in winter. The seasonal distribution of the blue-greens is similar but their maximum growth occurs later in the summer, and they often show a great increase after a period of sustained warm weather. Since the blue-greens seem to be the biggest problem in regard to both taste and oxygen depletion in high-nutrient waters, improved understanding of the role of temperature in the acclimatization of these species is essential.

Fish. Perhaps the most highly publicized effect of temperature on life forms is that on fish. A substantial amount of research has gone into the relationship of temperature to the propagation of desirable and undesirable species. Undoubtedly, much

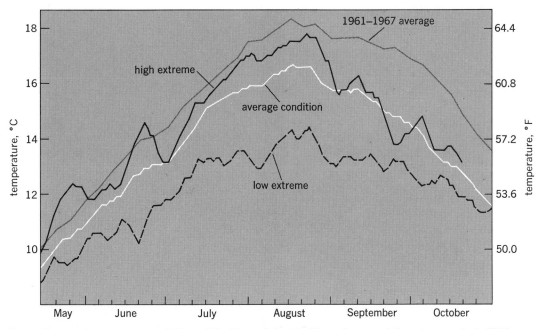

Fig. 6. Expected temperature variations of the Upper Columbia River after completion of reservoirs in 1975.

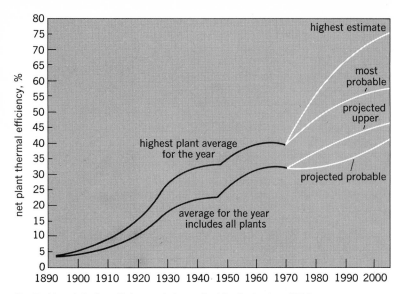

Fig. 7. Trends in the efficiency of steam electric plants since 1890.

changed from the normal seasonal cycle. The larger the system, the more severe the modification of timing. Associated with this change in timing are variations in the total annual available water flow.

Reactions which tend to favor reducing conditions and oxygen depletion are enhanced in the lower depths of larger reservoirs, even those with relatively low pollution burdens, because of the high degree of organic loading which occurs. On the other hand, in relatively shallow reservoirs on river main stems such as the Columbia and the St. Lawrence, such effects are minimal. The reservoir regime is relatively homogeneous, although other problems, such as gas supersaturation, can result from the entraining effect of spillway discharges.

As an example of the extent of thermal modifica-

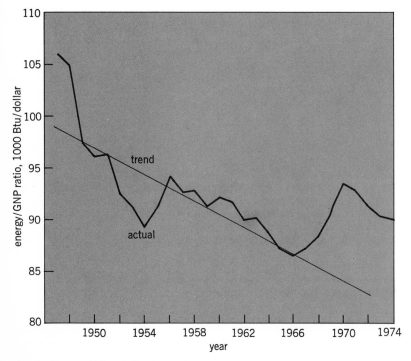

Fig. 8. Energy/GNP ratio for 1947–1974.

tion in a hydroelectric system, Fig. 6 shows the changes in Upper Columbia River temperatures expected from the Columbia Treaty projects. The projects will impound some 20×10^6 acre-ft of water and increase the monthly mean flow of the normally low-flow months by some 20,000 cfs. Temperature reductions forecast for the system in the summer months are the equivalent of some 25,000 MW of thermal energy. Figure 6 shows that August temperatures will average 2°C (3.6°F) lower under normal water conditions, and range downward to 4°C (7.2°F) lower in years with excess water. In critical years, temperatures will approach historical maximums and the timing of the peaks will move forward to the middle of June, in contrast to present conditions where peaks are expected in the middle of September. *See* HYDROELECTRIC POWER.

The data from this and a large number of other carefully studied cases show that the management of streams for the purpose of relating thermal history to power needs is feasible and, in fact, is of vital necessity.

EQUIPMENT DEVELOPMENT TRENDS

In the face of the staggering increase in total energy requirements forecast for the next 30 years, it seems logical that equipment, principally electrical generation prime movers, would reflect increasing emphasis on extending available fuel resources. The facts show the contrary to be the case. *See the feature article* OUTLOOK FOR FUEL RESERVES.

Historical trends. Figure 7 shows the historical trends of efficiency in the utilization of energy in the power industry. Commencing in the late 1800s, the efficiency of thermal plants rose steadily, paced by fuel cost/supply balances that favored the growth of technical efficiency. This progress was seriously affected by the lack of research and development during the 1929–1939 depression and came to a complete halt during World War II. After the war, paced by inflating fuel prices, technical efficiency showed a sharp recovery until about 1960, when a number of factors combined to slow and eventually reverse the rising trend. Others concerned with the economic effects of deterioration in technical efficiency have compiled relationships that show very unfavorable trends in the use of energy per dollar of gross national product (GNP) (Fig. 8), contrary to the historical trend since World War II.

Causes. There are several reasons for the lack of technical progress. First of all, reliance in the recent past on the economy of scale has tended to distort traditional measurements of technical efficiency. This is because insufficient experience is available to fully weigh the economic consequences of unit size over a statistically significant operating period. Economic policies which have led to relatively cheap fuel prices with respect to plant investment and interest costs have had a large impact on these trends.

Misguided policies. Policies which tend to arbitrarily require off-stream cooling cause further deterioration of technical efficiency. Recent practices which favor the production of high-speed single-shaft prime movers, instead of cross-compound systems, produce machines which are optimized for purposes other than total energy conservation. Similarly, the relatively low-efficiency light-water reactors place additional emphasis on capital optimization at the cost of large increases in energy consumption and release. Such practices have culminated in "high back-end loading" of single-shaft machines to reduce capital cost, meanwhile incurring higher energy consumption. (High back-end loaded machines are generally used on peaking plants or plants served by cooling towers in order to reduce the sensitivity of the system to water temperature changes. However, once such a machine is built and the operating parameters set, the net losses cannot be reduced by the later use of cooler water because of the inherent losses in the terminal blading of such machines.)

Price policies. Figure 8 reveals that policies which preserve low energy costs in contradiction to the overall inflationary trend cause energy waste and exacerbate the potential energy shortage and the environmental impacts of energy production and waste heat. The data of Table 4 suggest strongly that raw fuel prices are seriously undervalued in terms of gross national product and that increased costs to the consumer, though they may be burdensome, are a correct direction for energy prices to move in order to correct the situation. Thus, as a first principle in the management of waste heat, prices must be allowed to increase enough to make it worthwhile to conserve energy and promote incentives for research and development on improved efficiency in prime movers and energy-consuming equipment. *See the feature article* WORLD ENERGY ECONOMY.

The cost push effect of the economy of scale has yet to be fully evaluated on an industry-wide basis.

The opening of turbine blading clearances to improve reliability, high back-end loading, and other practices are appealing for short-term financial optimization—but all have serious long-term effects on fuel reserves and the environment. Cooling towers, too, can be included in the list, despite current policies advocating them.

CAUSE FOR OPTIMISM

In the face of all these adverse trends, it is important to avoid defeatist attitudes. Quite to the contrary, there is reason for optimism about the future. However, the amount of wishful thinking and false notions about energy and energy policy are so pervasive that it is imperative that we conduct a realistic examination of open alternatives. In the following paragraphs, I will suggest the most promising alternatives, all based on the philosophy that economic elasticity will balance usage and provide a test of reality.

Whenever possible, environmental enhancement on a broad scale should be a principal consideration in all future utility planning. Interim measures adopted as expedients without a long-range goal should be rejected as soon as such measures are revealed to be incorrect or misleading.

Simulation modeling should be used for the advance planning of alternatives to improve coordination of water resources and energy technology. This will provide more systematic use of water for low flow augmentation, recreational development, enhancement of fisheries, irrigation, and new development for cities, including interbasin diversions.

Both short- and long-run solutions to the waste heat problem should fully recognize the capacity of

Table 4. Energy consumption, gross national product, and the energy/GNP ratio for 1947–1974

Year	Aggregate energy consumption		GNP		Energy/GNP ratio	
	Amount, Btu × 10¹²	Change from preceding year, %	Amount 1958 dollars, × 10⁹	Change from preceding year, %	Amount, 1000 Btu/dollar	Change from preceding year, %
1947	32,870	—	309.9	—	106.1	—
1948	33,994	+2.5	323.7	+4.5	105.0	−1.0
1949	31,604	−7.0	324.1	+0.1	97.5	−7.1
1950	34,153	+8.1	355.3	+9.6	96.1	−1.4
1951	36,913	+8.1	383.4	+7.9	96.3	+0.2
1952	36,576	−0.9	395.1	+3.1	92.6	−3.8
1953	37,697	+3.1	412.8	+4.5	91.3	−1.4
1954	36,360	−3.5	407.0	−1.4	89.3	−2.2
1955	39,956	+9.9	438.0	+7.6	91.2	+2.1
1956	42,007	+5.1	446.1	+1.8	94.2	+3.3
1957	41,920	−0.2	525.5	+1.4	92.6	−1.7
1958	41,493	−1.0	447.3	−1.1	92.8	+0.2
1959	43,411	+4.6	475.9	+6.4	91.2	−1.7
1960	44,960	+3.6	487.7	+2.5	92.2	+1.0
1961	45,573	+1.4	497.2	+1.9	91.7	−0.5
1962	47,620	+4.5	529.8	+6.6	89.9	−2.0
1963	49,649	+4.3	551.0	+4.0	90.1	+0.2
1964	51,515	+3.8	581.1	+5.5	88.7	−1.6
1965	53,785	+4.4	617.8	+6.3	87.1	−1.8
1966	56,948	+5.9	658.1	+6.5	86.5	−0.7
1967	58,868	+3.4	675.2	+2.6	87.2	+0.8
1968	62,448	+6.1	707.2	+4.7	88.3	+1.3
1969	65,832	+5.4	727.1	+2.8	90.5	+2.5
1970	67,099	+2.0	722.7	−0.7	92.8	+2.5
1971	68,700	+2.4	745.4	+3.1	92.2	−0.6
1972	72,100	+4.9	790.7	+6.1	91.2	−1.1
1973	75,561	+4.8	837.6	+5.9	90.2	−1.1
1974	72,260	−4.4	803.7	−4.0	89.9	−0.3

water in dispersing heat whenever these uses can be shown to have long-range public benefits.

Fuel cell concept. New systems deserve more attention than they are currently receiving. An example is the electrically interconnected neighborhood fuel cell concept with full recovery of thermal energy for water heating and other domestic uses. In conjunction with better insulation, improved lighting efficiency, and more efficient air-conditioning equipment employing air, such systems could provide improved sophistication in home living with total energy use reduced by as much as 75% over present uncoordinated systems. The fuel gas supplies needed for the fuel cell concept are much easier to provide, and gas service corridors require an order of magnitude less space than electrical transmission lines for equivalent energy-transmission rates. It is entirely possible that the average residential load could be held to as little as 3500 kWhr per year using a fully coordinated optimized residential energy system. This is only one-quarter of the present load in the Pacific Northwest as reported by the Bonneville Power Administration. Electrical systems of the present high-voltage, alternating-current type are ideal for high industrial and commercial loads and would be retained for service to the transportation and conversion industries as the total energy source. In this way, transmission lines would avoid residential areas. The development of improved utilization technology could make surprising reductions in predicted requirements for energy. *See* FUEL CELL; FUEL GAS.

New equipment. The efficiency and flexibility of equipment available to the utilities need order-of-magnitude improvements. Presently, such equipment is far too limited by convention, standards, and lack of fundamental research. Equipment offered by manufacturers should reflect less attention to quick response, near-term considerations, and regulatory requirements and should focus more attention on long-term, national objectives. Equipment designs that minimize energy consumption and that decrease environmental impact should be encouraged by suitably appropriate economic incentives.

Basic research need. Also needed is a closer identification of the specific changes associated with the impact of energy on the environment. Almost all of the existing study programs place so little emphasis on the specific identification of temperature as an independent variable that it is obscured in search for an oversimplified solution. Separation of the variables, with emphasis on temperature, would properly recognize the interdisciplinary nature of heat and provide background data of real usefulness.

Finally, the question of energy in the environment must be faced from an integrated, sociological viewpoint. The impact of increased industrialization, along with the predicted need for recreational facilities, provides the technical basis for a rational synthesis. What are needed are appropriate guiding factors, a constructive attitude, and a sincere appreciation of the total broad needs of the community. *See the feature article* PROTECTING THE ENVIRONMENT.

[ROBERT T. JASKE]

Water pollution

Any change in natural waters which may impair their further use, caused by the introduction of organic or inorganic substances, or a change in temperature of the water. The growth of population and the concomitant expansion in industrial and agricultural activities have rapidly increased the importance of the field of water-pollution control. In the attack on environmental pollution, higher standards for water cleanliness are being adopted by state and Federal governments, as well as by interstate organizations.

WASTE MATERIALS

Ancient man joined into groups for protection. Later, he formed communities on watercourses or the seashore. The waterway provided a convenient means of transportation, and fresh waters provided a water supply. The watercourses then became receivers of his waste water along with contaminants. As industries developed, they added their discharges to those of the community. When the concentration of added substances became dangerous to man or so degraded the water that it was unfit for further use, water-pollution control began. With development of wide areas, pollution of surface water became more critical because wastewater of an upstream community became part of the water supply of a downstream one.

Serious epidemics of waterborne diseases such as cholera, dysentery, and typhoid fever were caused by underground seepage from privy vaults into town wells. Such direct bacterial infections through water systems can be traced back to the late 18th century, even though the germ or bacterium as the cause of disease was not proved for nearly another century. The well-documented epidemic of the Broad Street Pump in London during 1854 resulted from direct leakage from privies into the hand-pumped well which provided the neighborhood water supply. There were 616 deaths from cholera among the users of the well within 40 days.

Eventually, abandoning wells in such populated locations and providing piped water to buildings improved public health. Further, sewers for drainage of wastewater were constructed, but then infections between communities rather than between the residents of a single community became apparent. Modern public health protection is provided by highly refined and well-controlled plants both for the purification of the community water supply and treatment of the wastewater.

Relation to water supply. Water-pollution control is closely allied with the water supplies of communities and industries because both generally share the same water resources. There is great similarity in the pipe systems that bring water to each home or business property, and the systems of sewers or drains that subsequently collect the wastewater and conduct it to a treatment facility. Treatment should prepare the flow for return to the environment so that the receiving watercourse will be suitable for beneficial uses such as general recreation, and safe for subsequent use by downstream communities or industries.

The volume of wastewater, the used water that must be disposed of or treated, is a factor to be considered. Depending on the amount of water used for irrigation, the amount lost in pipe leakage, and the extent of water metering, the volume of wastewater may be 70–130% of the water drawn from the supply. In United States cities, wastewater quantities are usually 75–200 gal per capita daily. The higher figure applies to large cities with old systems, limited metering, and comparatively cheap water; the lower figure to smaller communities with little leakage and good metering. Probably the average in the United States for areas served by sewers is 125–150 gal of wastewater per person per day. Of course, industrial consumption in larger cities increases per capita quantities.

Related scientific disciplines. The field of water-pollution control encompasses a part of the broader field of sanitary or environmental engineering. It includes some aspects of chemistry, hydrology, biology, and bacteriology, in addition to public administration and management. These scientific disciplines evaluate problems and give the civil and sanitary engineer basic data for the designing of structures to solve the problems. The solutions usually require the collection of domestic and industrial wastewaters and treatment before discharge into receiving waters.

Self-purification of natural waters. Any natural watercourse contains dissolved gases normally found in air in equilibrium with the atmosphere. In this way fish and other aquatic life obtain oxygen for their respiration. The amount of oxygen which the water holds at saturation depends on temperature and follows the law of decreased solubility of gases with a temperature increase. Because water temperature is high in summer, oxygen dissolved in the water is then at a low point for the year.

Degradable or oxidizable substances in wastewaters deplete oxygen through the action of bacteria and related organisms which feed on organic waste materials, using available dissolved oxygen for their respiration. If this activity proceeds at a rate fast enough to depress seriously the oxygen level, the natural fauna of a stream is affected; if the oxygen is entirely used up, a condition of oxygen exhaustion occurs which suffocates aerobic organisms in the stream. Under such conditions the stream is said to be septic and is likely to become offensive to the sight and smell.

Domestic wastewaters. Domestic wastewaters result from the use of water in dwellings of all types, and include both water after use and the various waste materials added: body wastes, kitchen wastes, household cleaning agents, and laundry soaps and detergents. The solid content of such waste water is numerically low and amounts to less than 1 lb per 1000 lb of domestic wastewater. Still, the character of these waste materials is such that they cause significant degradation of receiving waters, and they may be a major factor in spreading waterborne diseases, notably typhoid and dysentery.

Characteristics of domestic wastewater vary from one community to another and in the same community at different times. Physically, community wastewater usually has the grayish colloidal appearance of dishwater, with floating trash apparent. Chemically, it contains the numerous and complex nitrogen compounds in body wastes, as

Table 1. General nature of industrial wastewaters

Industry	Processes or waste	Effect
Brewery and distillery	Malt and fermented liquors	Organic load
Chemical	General	Stable organics, phenols, inks
Dairy	Milk processing, bottling, butter and cheese making	Acid
Dyeing	Spent dye, sizings, bleach	Color, acid or alkaline
Food processing	Canning and freezing	Organic load
Laundry	Washing	Alkaline
Leather tanning	Leather cleaning and tanning	Organic load, acid and alkaline
Meat packing	Slaughter, preparation	Organic load
Paper	Pulp and paper manufacturing	Organic load, waste wood fibers
Steel	Pickling, plating, and so on	Acid
Textile manufacture	Wool scouring, dyeing	Organic load, alkaline

well as soaps and detergents and the chemicals normally present in the water supply. Biologically, bacteria and other microscopic life abound. Wastewaters from industrial activities may affect all of these characteristics materially.

Industrial wastewaters. In contrast to the general uniformity of substances found in domestic wastewaters, industrial wastewaters show increasing variation as the complexity of industrial processes rises. Table 1 lists major industrial categories along with the undesirable characteristics of their wastewaters.

Because biological treatment processes are ordinarily employed in water-pollution control plants, large quantities of industrial wastewaters can interfere with the processes as well as the total load of a treatment plant. The organic matter present in many industrial effluents often equals or exceeds the amount from a community. Accommodations for such an increase in the load of a plant should be provided for in its design.

Discharge directly to watercourses. The industrial revolution in England and Germany and the subsequent similar development in the United States increased problems of water-pollution control enormously. The establishment of industries caused great migrations to the cities, the immediate result being a great increase in wastes from both population and industrial activity. For some years discharges were made directly to watercourses, the natural assimilative power of the receiving water being used to a level consistent with the required cleanliness of the watercourse. Early dilution ratios required for this method are shown in Table 2. Because of the more rapid absorption of oxygen from the air by a turbulent stream, it has a

Table 2. Dilution ratios for waterways

Type	Stream flow, ft³/sec/1000 population
Sluggish streams	7–10
Average streams	4–7
Swift turbulent streams	2–4

high rate of reaeration and a low dilution ratio; the converse is true of slow-flowing streams.

Development of treatment methods. With the passage of time, the waste loads imposed on streams exceeded the ability of the receiving water to assimilate them. The first attempts at wastewater treatment were made by artificially providing means for the purification of wastewaters as observed in nature. These forces included sedimentation and exposure to sunlight and atmospheric oxygen, either by agitated contact or by filling the interstices of large stone beds intermittently as a means of oxidation. However, practice soon outstripped theory because bacteriology was only then being born and there were many unknowns about the processes.

In later years testing stations were set up by municipalities and states for experimental work. Notable among these were the Chicago testing station and one established at Lawrence by the state of Massachusetts, a pioneer in the public health movement. From the results of these direct investigations, practices evolved which were gradually explained through the mechanisms of chemistry and biology in the 20th century.

The treatment methods are classed into (1) primary treatment, essentially physical removal of coarse materials by bar screens and plain sedimentation, and (2) secondary treatment, which consists of oxidation processes usually carried out by biological processes. In effect, these processes have performed, within a treatment plant, the improvement in water quality wrought by natural waters, and so have prevented the serious degradation of the receiving water. However, as the burdens of population and industry have increased, still more refined treatments have become necessary. In forcing these processes to greater removals of organic matter, tertiary treatment processes have been evolved to the point that plant fertilizing constituents of nitrogen and phosphorus have been found to cause adverse effects on the receiving water. Accordingly, methods for their reduction have been perfected and are applied in locations where justified. In water-short areas, such effluents are useful in many industrial processes or in irrigation, conserving the naturally clean sources for human use.

Thermal pollution. An increasing amount of attention has been given to thermal pollution, the raising of the temperature of a waterway by heat discharged from the cooling system or effluent wastes of an industrial installation. This rise in temperature may sufficiently upset the ecological balance of the waterway to pose a threat to the native life-forms. This problem has been especially noted in the vicinity of nuclear power plants. Thermal pollution may be combated by allowing the wastewater to cool before it empties into the waterway. This is often accomplished in large cooling towers. *See* WASTE HEAT MANAGEMENT.

Current status. The modern water-pollution engineer or chemist has a wealth of published information, both theoretical and practical, to assist him. While research necessarily will continue, he can draw on established practices for the solution to almost any problem. A challenging problem has been the handling of radioactive wastes. Reduction in volume, containment, and storage consti-

tute the principal attack on this problem. Because of the fundamental characteristics of radioactive wastes, the development of other methods seems unlikely.

Federal aid. Because of public demands and the actions of state legislatures and the Congress of the United States, there has been a surge of interest in, and a demand for, firm solutions to water-pollution problems. Although the Federal government granted aid for construction of municipal treatment plants as an employment relief measure in the 1930s, no comprehensive Federal legislation was enacted until 1948. This was supplemented by a major change in 1956, when the United States government again offered grants to municipalities to assist in the construction of water-pollution control facilities. These grants were further extended to small communities for the construction of both water and sewer systems.

Since 1965, Federal activity in water-pollution control has advanced from a minor activity in the Public Health Service, through the Water Pollution Control Administration in the Department of the Interior, to a major activity in the Environmental Protection Agency. In the 1972 act (P. L. 92-500) Congress authorized a massive attack on municipal pollution problems by a grant-in-aid program eclipsing any previous effort. Federal funds for wastewater treatment plant construction during 1970–1975 are given in Table 3.

State and Federal regulations are increasing constantly in severity. This tendency is expected to continue until the problem of water pollution is brought under complete control. Even then, water quality will be monitored to make certain that actual control is achieved on a day-to-day or even an hour-to-hour basis. The increase in activity in water-pollution control is apparent, and the 86,000 United States workers in the municipal field in 1974 are expected to increase 50% by 1980 and to double by 1985.   [RALPH E. FUHRMAN]

OIL SPILL

The problem of oil spillage came to the public's attention following the grounding of the tanker *Torrey Canyon* in March 1967 at the southwest coast of England near the entrance to the English Channel. Subsequent major oil spills such as the Santa Barbara channel California oil spill in January 1969 have further raised the level of concern until today the terminology "oil spill" has become a household word. Few problems have had greater impact on the petroleum industry than those associated with oil spills. This industry in the United States is faced with the problem of supplying the

Table 3. Federal funds for wastewater treatment plant construction

Fiscal year	Authorized, 10^9	Actual allocation, 10^9	Obligation, nearest 10^9	Outlays, 10^9
1970	1	0.8	0.438	0.176
1971	1.25	1	1.168	0.478
1972	2	2	0.872	0.413
1973	5	2	2.995	0.684
1974	6	3	2.661	1.564
1975	7	4	4.2 (est.)	2.3 (est.)

Table 4. Oil refining capacity and production by region*

Region	Refining capacity, 10^3 bpd		1973 oil production, 10^3 bpd
	1-1-73	1-1-74	
Middle East	2,758	2,882	21,158
Western Europe	16,826	18,110	370
United States and Canada	14,812	15,171	10,987

*Data from *International Petroleum Encylopedia*, Petroleum Publishing Co., 1974.

ever-increasing demand for oil and petroleum products to customers who are demanding that the oil be supplied without a risk of oil spills. The magnitude of this problem can be appreciated by reference to Table 4, which shows the refining capacity and oil production by region. In the Middle East at the end of 1973, the refining capacity was 2,882,000 barrels per day (bpd; 1 bbl = 0.159 m³), compared with a production rate in 1973 of 21,158,000 bpd. Contrast this with Western Europe, where only 370,000 bpd were produced in 1973, and the refining capacity at the end of 1973 was 18,110,000 bpd. This large disparity between the production and use of oil throughout the world results in the requirement that enormous quantities of oil be transported large distances, primarily by tanker.

To counter the threat of environmental damage as a result of oil spills, extensive research is being performed in the United States by private industry as well as by the Environmental Protection Agency and the Coast Guard. This research is primarily directed at developing methods to combat oil spills which minimize the damage to the environment. Treating the spilled oil with dispersants was the primary method used to fight oil spills at the time of the *Torrey Canyon*. Dispersants cause oil to spread farther and disperse in a manner similar to the way soap removes oil from one's hands, allowing the oil to be emulsified and washed away with the water. The dispersants used during the *Torrey Canyon* cleanup effort were not developed specifically for use in waters containing marine life and contained aromatic solvents which are toxic. Since that time specific dispersants less toxic to marine life and biota have been developed. Today it is generally accepted that the most extensive damage to marine life resulting from the *Torrey Canyon* incident was caused by the excessive use of dispersants in the coastal zone. In fact, the areas of the shore where dispersants were not used, but which were heavily polluted with oil alone, showed very minimum damage according to J. E. Smith, director of the Plymouth Laboratory, who has studied the biological effects of the *Torrey Canyon* oil spill. At present in the United States regulations severely limit the use of dispersants, and research efforts place emphasis on containment and recovery of oil by mechanical means.

Effects of oil pollution. When oil is spilled on water, it spreads rapidly over the surface. The forces which cause the oil to spread include the force of gravity, which results in the lighter oil seeking constant level by spreading horizontally on the heavier water. A second force is the surface-

tension force, which acts at the edge of the oil slick as shown in Fig. 1. It is the surface-tension force which can result in the oil spreading to a thickness approaching a monomolecular layer. This limiting thickness is almost never achieved in large oil spills, however, because the oil interfacial surface tensions change, and the net surface tension becomes negative.

The interfacial surface-tension forces are changed as a result of the natural processes which affect the oil. One of the most important natural processes is the evaporation of the oil. Evaporation occurs rapidly, the rate depending upon the nature of the oil, the rate of thinning of the slick, wave intensity, strength of the wind, temperature, and so forth. Crude oil is a mixture of a very large number of components, each with its own properties. The most volatile components evaporate first, but with all crude oils there will undoubtedly be a residue left which is virtually involatile. In addition to evaporation, some of the oil goes into solution with the water, some is oxidized, and some is utilized by microorganisms. The most important of these processes, and the one receiving the most extensive research, is the process of microbial degradation. Many microorganisms present in seas, fresh-water lakes, and rivers have a great capacity to utilize hydrocarbons. The hydrocarbons are used as an energy source and are incorporated into new cell mass. Seeding oil slicks with special bacterial cultures has been suggested to accelerate the rate of microbial decay. However, the rate of microbial degradation of oil is limited not only by the quantity of organisms but by the availability of the oxygen and nutrients needed to support the metabolic process. Acceleration of the natural process by adding nutrients such as phosphorus or nitrogen compounds, particularly in open seas where the nutrients are not naturally available, is presently being considered. Unfortunately the rate of bacterial degradation (accelerated or natural) of floating oil is slow and is therefore not effective if the oil is threatening a coastline.

As with most types of problems, the short-term deleterious effects associated with an oil spill are better understood than the long-term effects. Marine birds, especially diving birds, appear to be the most vulnerable of the living resources to the effects of oil spillage. Harm to birds from contact with oil is reported to be a result of breakdown of

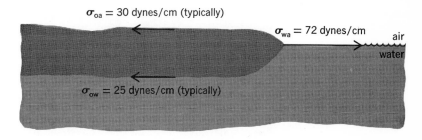

Fig. 1. Cross section of oil-water-air interface of a spreading oil slick showing the relevant surface-tension forces; 1 dyne is equivalent to 10^{-5} N.

the natural insulating oils and waxes which shield the birds from water, as well as due to plumage damage and ingestion of oil. Efforts to cleanse or rehabilitate birds have been generally unsuccessful because of the excessive stress that the bird experiences. If treatment is prolonged for any reason, most if not all of the birds will die. Shellfish are another segment of marine life directly affected by oil spillage in the coastal zone. Many shellfish have a relatively high tolerance to oil, but their flesh can become tainted for a period subsequent to heavy pollution. Shellfish are particularly vulnerable to most chemical dispersants. Fish are not generally affected by an oil spill because of their mobility which allows them to avoid heavily contaminated areas. The effects of oil on the marine food chain which consists of plants, bacteria, and small organisms is not well understood because of its complexity and because of the wide fluctuations that occur naturally and are independent of the effects of oil. In contrast to the ecological damage, the damage caused by an oil spill which is associated with recreational beach areas, coastline areas used for water sports, and areas where personal property such as docks or boats are located is well understood. There is a large expense associated with the loss of the use of these areas for even a few days.

Cleanup procedures. Experience in attempting to clean up an oil spill has shown that no perfect method exists for all situations. Cleanup methods must be evaluated and chosen on a case-by-case basis. Spills can be more easily dealt with if they are confined to a small area on the water surface. At present, however, confinement devices have not been developed which are successful in all sit-

uations. The methods being studied and which are being used to dispose of oil floating on the surface of the sea include mechanical removal of the floating oil, the use of absorbents to facilitate the removal of the oil, sinking the oil, dispersion of the oil, and burning the oil.

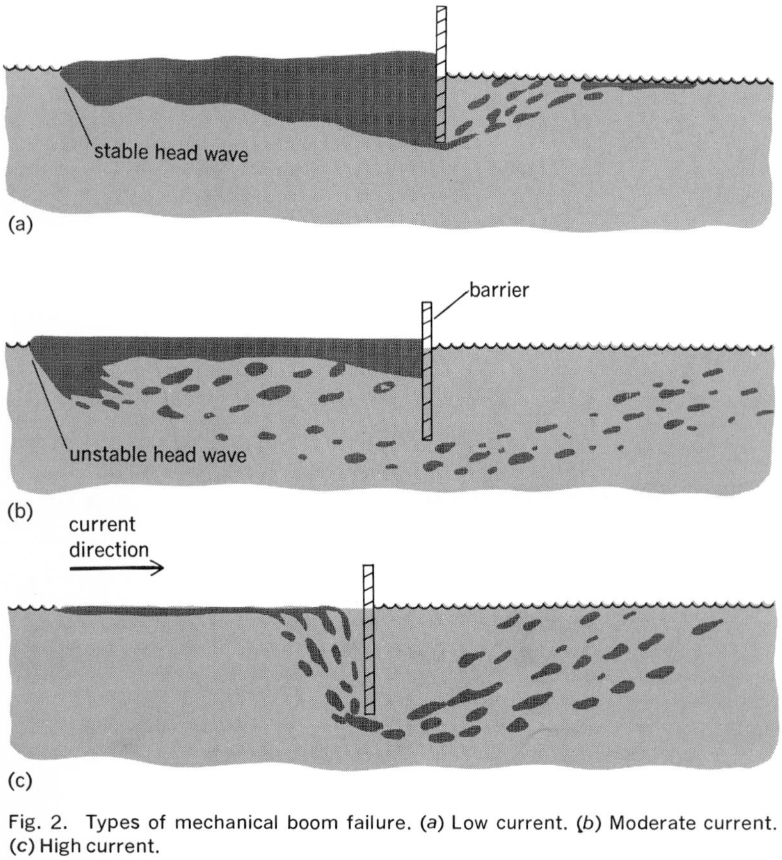

Mechanical removal of oil. The primary ingredient of many mechanical oil spill cleanup systems is the use of mechanical booms or barriers. Containment of the oil spill at its source is the most important single action which can be taken when the oil spill is first detected. The use of booms has only had limited success to date because at moderate currents as low as 1 knot many booms have failed by allowing oil to pass below the boom. Figure 2 shows the types of failure which booms can experience in the presence of currents. At low currents, approximately 1.5 knots or less, oil will pile down against the barrier until the boom reaches its capacity. Additional oil will cause the boom to fail as shown in the figure. At higher currents the situation is less manageable since the boom will fail before a large quantity of oil has been collected within the barrier. For the low-current situation booms will perform satisfactorily if the oil is continually skimmed from the region in front of the barrier. Booms also are satisfactory for directing or sweeping oil, provided the angle between the boom and the current or drift direction of the oil is small so that oil does not accumulate along the boom. In addition, the relative velocity of the water at right angles to the boom must be less than the critical velocity for boom failure.

The performance of skimmers depends upon the thickness of the oil. When the film thickness is below ¼ in., many techniques require pumping large amounts of water and very small amounts of oil. However, once the oil-and-water mixture is removed from the water surface, the separation of the oil from the water is easily accomplished by gravity when the oil and water are allowed to settle in a tank. Skimmers now available generally fall into one of two categories. The first, mechanical surface skimming, removes the top layers of the water and oil from the surface. These devices suffer particularly in wave action, where they gulp large amounts of water unless some provision is provided to allow the weir or suction port to follow the water surface. A second type of skimmer operates on the principle of selective wetting of a surface by oil rather than by water. Rotating metal disks or conveyor belts dip into the water surface through the oil slick. When the moving surface is drawn from the water, a surface layer of oil is removed.

Absorbents. Absorbents are used to facilitate the cleaning up of oil spills. When they are applied to the slick, they absorb the oil and prevent it from spreading, and when the absorbent material is removed from the water, the oil is removed. A class of absorbents which is commonly used consists of natural materials such as peat moss, straw, sawdust, pine bark, talc, and perlite. A second class of absorbents is derived from synthetics or plastics such as high-molecular-weight polyethylene and polystyrene, polypropylene, and polyurethane. Of all synthetic absorbents, polyurethane is generally accepted to be the most promising. The natural absorbents are generally less expensive and are attractive whenever there is a chance

Fig. 2. Types of mechanical boom failure. (a) Low current. (b) Moderate current. (c) High current.

of losing the absorbent material. The natural absorbents are either inert or biodegrade more quickly than the synthetic materials. Alternatively the synthetic material has a greater buoyancy and a higher affinity for oil. One of the problems with absorbents is distributing them in large enough quantities on the slick. Unless the absorbents can be applied and recovered from the shore, special equipment must be available. One of the most recent concepts including absorbents involves recycling the absorbent material by wringing the oil from the absorbent and returning the absorbent material to the slick. Synthetic absorbents are most suitable for this application, and a feasibility study of such a system for operation in offshore conditions is presently being conducted by the Environmental Protection Agency.

Sinking the oil. A sinking agent which consisted of 3000 tons of calcium carbonate with about 1% of sodium stearate was applied to an oil slick which originated from the *Torrey Canyon* and reportedly resulted in the sinking of about 20,000 tons of oil. The oil was sunk in the Bay of Biscay off the coast of France in 60 to 70 fathoms of water. The sinking of the oil prevented the French coast from being contaminated, and after a period of 14 months no sign of the oil was found. Other materials such as specially treated sand, fly ash, and similar synthetic material have also been used to sink oil. Opinion is still divided as to the possible environmental effects of treating the oil with a dense material and sinking it. Opposition to sinking centers around the fact that sinking the oil reduces the contact surfaces between the oil and air and between the oil and water by preventing natural diffusion of the oil. Hardening of the oil subsequent to sinking would lead to a more persistent and concentrated pollution of the sea bed compared with a lower level of more dispersed pollution on the surface. In any case, utilization of this technique would be most advantageous in deeper waters outside the heavy fishing zones and where there will be a minimum of adverse effects to productive biological life in the coastal zones.

Dispersants. A dispersant is a substance which, when applied to an oil slick, causes the oil to spread farther and disperse. A dispersant contains a surfactant, a solvent, and a stabilizer. The solvent usually comprises the bulk of the dispersant and enables the surface-active agent or surfactant to mix with, and penetrate into, the oil slick and thus form an emulsion. The stabilizer fixes the emulsion and prevents it from coalescing once it is formed. The process of dispersion of the slick is shown in Fig. 3. Dispersion is similar to sinking in that it simply displaces the oil from the water surface rather than removing it from the water altogether. Dispersion has one advantage over sinking in that it increases the slick surface area and allows a rapid increase in the rate of microbial decomposition. Dispersants are not useful in coastal regions because the process of dispersion leads to an increased extent of contamination. In addition, the most effective dispersants use solvents which are toxic to marine life. To reduce this toxic effect, reduced effectiveness must be accepted. Primarily due to the question of toxicity, the use of dispersants in the open sea appears doubtful, although their use there has potential pending additional study and field data.

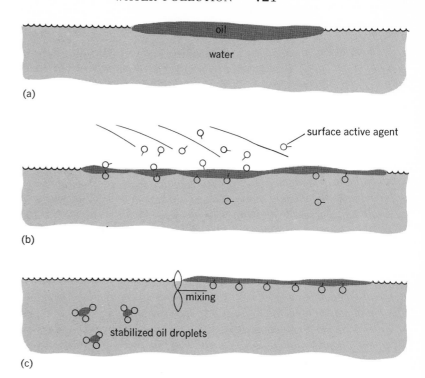

Fig. 3.   Mechanism of oil slick dispersion. (*a*) Initial slick. (*b*) Application of chemical dispersant. (*c*) Mixing forms droplets which become stable emusion.

Burning. Burning oil slicks on the open sea has generally met with little success because the more volatile light ends evaporate quickly from the oil slick. Also, the water generally can remove heat faster than it can be created to support the combustion. Various attempts have been made to treat oil slicks to facilitate burning. The most promising of these techniques involves spreading a wicking material over the oil slick which acts to physically separate the flame from the water. Wicking agents also aid in confining the fire to a particular location, but air pollution must be expected when oil is burned in this fashion.

Prevention of oil spills. Although oil spills can probably not be eliminated entirely, steps are being taken to reduce the probability that they will occur. The United States Environmental Protection Agency and others have sponsored studies that apply reliability engineering principles to the problem and that will result in recommending procedures to be adopted to reduce the oil-spill threat. Steps are also being taken to quickly discover oil spills and to develop methods to continually monitor the water where spills are likely so that, in the event of a spill, it can be discovered quickly and proper action can be taken. The United States Coast Guard is sponsoring considerable research toward the development of remote sensing techniques. These studies have shown that, using radar and passive microwave techniques, large areas can be surveyed on a 24-hr basis even in adverse weather conditions. For the purpose of cleaning up oil spills, 67 private cooperatives are now in operation throughout the United States. These cooperatives will become more numerous in the future. Cooperatives operate in coordination with the Coast Guard, the Army Corps of Engineers, and the Environmental Protection Agency. Assessing the ability of these cooperatives, or of any group for that matter, to clean up a large spill is difficult, but it appears that near the shore mechanical techniques can be used effectively to remove oil from the water. In the offshore situation, oil recovery is more complicated and will depend upon the given situation and other environmental conditions encountered. Most current research efforts are directed toward developing systems which will operate effectively in offshore conditions.

[ROBERT A. COCHRAN]

Bibliography: G. M. Fair et al., *Elements of Water Supply and Wastewater Disposal*, 1971; Federal Water Quality Administration, *Santa Barbara Oil Pollution, 1969: A Study of the Biological Effects of the Oil Spill Which Occurred at Santa Barbara, California, in 1969*, University of California, Santa Barbara, October 1970; R. E. McKinney, *Microbiology for Sanitary Engineers*, 1962; N. L. Nemerow, *Liquid Waste of Industry*, 1971; *Proceedings of the 1975 Conference on Prevention and Control of Oil Pollution*, San Francisco, Mar. 25–27, 1975, American Petroleum Institute, 1975; C. N. Sawyer and P. L. McCarty, *Chemistry for Sanitary Engineers*, 1962; J. E. Smith (ed.) *"Torrey Canyon" Pollution and Marine Life: A Report by the Plymouth Laboratory of the Marine Biological Association of the United Kingdom*, 1968; J. Snow, *Mode of Communication of Cholera*, 1855; Water Pollution Control Federation, *Careers in Water Pollution Control*.

Waterpower

Power developed from movement of masses of water. Such movement is of two kinds: (1) the falling of streams through the force of gravity, and (2) the rising and falling of tides through lunar (and solar) gravitation. While that part of solar energy expended to lift water vapor against Earth gravity is a minute fraction of the total, the absolute amount of energy theoretically recoverable from resulting streams is an enormous but unknown quantity. Of this, but a tiny portion is actually suitable for harnessing.

WATER RESOURCES

The capacity of world waterpower plants in use at the end of 1966 was about 230×10^6 kw, which produced in that year about 956×10^9 kwhr. This is about 27% of the total world electric power generated. Of this, the United States accounted for about one-fifth of the total. As of May 1974, the total conventional hydropower developed in the contiguous United States was 54,885 Mw, provided by about 1400 installations. Almost one-half of this hydropower was installed in Washington, Oregon, and California. Under construction were installations of some 6878 Mw capacity, of which 90% was in the western United States.

In 1965 the waterpower plants of Europe totaled 105×10^6 kw; North America, 69×10^6; Asia, 29×10^6, leaving but 17×10^6 for the remainder of the world. Thus the principal waterpower development of the future can be expected in Africa, Asia, South America, and island countries, whose potential has been little investigated.

The contribution of waterpower installations to the nation's electric power supply at the beginning of World War II was about 30%. While the output from hydroelectric plants has grown (to 260×10^9 kwhr in 1973), their contribution to the total electric power has dropped to about 14%, because steam-electric plants have grown at a much more rapid rate.

As of May 1974, the Federal Power Commission estimated that by the end of 1993 the total developed capacity of waterpower installations could be 78,587 Mw. However, that figure does not include a number of potential sites that could at some future time be considered. In theory, the Federal Power Commission estimates an eventual maximum potential of 180,000 Mw, most of the undeveloped sites being in the Pacific Northwest and Alaska. Only a fraction of that will be developed for a variety of reasons. The most attractive sites have already been utilized. Hydro plants, with their initial high cost and generally long distances from major load centers, must compete with the large, efficient, fuel-fired stations, and the burgeoning, economical, large nuclear plants. Large dam sites usually must be justified not alone on the value of the power developed, but also on the benefits from flood control, irrigation, and recreation. Problems of migrating fish, conservation, and preservation of esthetic values are also factors. On the other hand, waterpower developments add greatly to power-system flexibility in meeting peak and emergency loads. Modern excavation and tunneling techniques are lowering construction costs. The economies of lowhead sites are improved by the new, efficient, axial-flow turbines of the tubular type. *See* ENERGY SOURCES.

Silting. The capacity of hydro plants cannot be counted on for perpetuity because of gradual filling of reservoirs with sediment. This effect is serious for irrigation, flood control, and navigation. Even when a lake behind a power dam becomes filled completely with silt, electric power can be generated on the run-of-the-river flow, although output would vary with stream flow.

The rate of silting varies widely with drainage basins. Because the Columbia River carries comparatively little silt, the reservoirs at Grand Coulee and Bonneville dams should have lives of many hundreds of years. The Colorado River, on the other hand, is muddy. In the first 13.7 years after Hoover Dam went into operation in 1935, 1,424,000 acre-feet of silt was dumped into Lake Mead. That is equivalent to a layer 1 ft deep over 223 mi². This inflow of silt has been diminished by about 22% by the construction of other dams upstream, for example, the Glen Canyon Dam. It is now expected that Lake Mead will have a useful life of more than 500 years.

Pumped storage. The most significant waterpower development of the 1960s was the rapid growth of pumped-storage hydroelectric systems. In these schemes, water is pumped from a stream or lake to a reservoir at a higher elevation. Pumping up to a storage reservoir is most commonly done by reversing the hydraulic turbine and generator. The generator becomes a motor driving the turbine as a pump. Power is drawn from the power system at night or on weekends when demand is low. It is not practical to shut down large, high-temperature steam stations or nuclear units for a few hours at night or even over a weekend. Because they must run anyway, the cost of pumping power is low, whereas the power generated from pumped storage at peak periods is valuable. Also, the pumped-storage system provides a means of supplying power quickly in an emergency situation, for example, during the failure of a large steam or nuclear unit. A pumped-storage system can be changed over from pumping to generation in 2 to 5 min.

Pumped-storage systems are not new. They date from 1928, but by 1972 only 13 pumped-storage plants of a capacity greater than 50 Mw had been installed in the United States. These had a combined capacity of 3718.5 Mw. As of May 1974, 12 more were under construction, which would add 8522 Mw of installed capacity. The Federal Power Commission lists 16 more pumped-storage sites, with 15,111.5 Mw potential, as projected. *See* PUMPED STORAGE.

Tidal power. A portion of the kinetic energy of the rotation of the Earth appears as ocean tides. The mean tide of all the oceans has been calculated as 2.1 ft, and the mean power as $54,000 \times 10^6$ hp or, on a yearly basis, the equivalent of 36×10^{12} kwhr. Unfortunately, only a minute amount of this is likely to be harnessed for use. For tidal sites to be of sufficient engineering interest, the fall would have to be at least 15 ft. There are few such falls, and some of these are in remote areas. The only tide-power sites that have received serious attention are on the Severn River in England, the Rance River and Mont St. Michel in northern France, the San José and Deseado rivers of Argentina, the Petitcodiac and Memramcook estuaries in the Bay of Fundy, Canada, the Passamaquoddy River where Maine joins New Brunswick, Canada, and, lately, the Cambridge Gulf of Western Australia.

The Passamaquoddy site, with a potential of 1800 Mw (peak), is the only important tidal-power prospect in the United States. However, as late as 1974 engineers did not consider its electrical output to be economically competitive with power produced by other means.

A second major handicap to tidal power is that, with a simple, single-basin installation, power is available only when there is a several-feet difference between levels in the sea and the basin. Thus, firm power is not available. Also, periods of generation occur in consonance with the tides—not necessarily when power is needed.

The only major tidal power plant in operation is the one near the mouth of the Rance River in Normandy, France. This plant operates on 40-ft tides. It began operation in November, 1966. It will consist, when complete, of twenty-four 10-Mw turbine-generator units of novel design. The system embodies a reservoir into which sea water is pumped during off-peak hours. Turbines are then run as pumps, power being drawn from the French electrical grid. The plant is expected to produce 540×10^6 kwhr annually, including a significant amount of firm power.

Tidal power is an appealing and dramatic technique, and some other large plants will be constructed. However, the total contribution of the tides to the world's energy supply will be miniscule. *See* TIDAL POWER.

[CHARLES A. SCARLOTT]

APPLICATION

The basic relation for power P in kilowatts from a hydrosite is $P = QH/11.1$, where Q is water flow in ft³/sec, and H is head in feet. Actual power will be less as occasioned by inefficiencies such as (1) hydraulic losses in conduit and turbines; (2) mechanical losses in bearings; and (3) electrical losses in generators, station use, and transmission. Overall efficiency is always high, usually in excess of 80% to the station bus bars.

Choice of site. The competitive position of a hydro project must be judged by the cost and reliability of the output at the point of use or market. In most hydro developments, the bulk of the investment is in structures for the collection, control, regulation, and disposal of the water. Electrical transmission frequently adds a substantial financial burden because of remoteness of the hydrosite from the market. The incremental cost for waterwheels, generators, switches, yard, transformers, and water conduit is often a smaller fraction of the total investment than is the cost for the basic structures, real estate, and transmission facilities. Long life is characteristic of hydroelectric installations, and the annual carrying charges of 6–12% on the investment are a minimum for the power field. Operating and maintenance costs are lower than for other types of generating stations.

The fundamental elements of potential power, as given in the equation above, are runoff Q and head H. Despite the apparent basic simplicities of the relation, the technical and economic development of a hydrosite is a complex problem. No two sites are alike, so that the opportunity for standardization of structures and equipment is nearly nonexistent. The head would appear to be a simple surveying problem based largely on topography. However, geologic conditions, as revealed by core drillings, can eliminate an otherwise economically desirable site. Runoff is complicated, especially when records of flow are inadequate. Hydrology is basic to an understanding of water flow and its variations. Runoff must be related to precipitation and to the disposal of precipitation. It is vitally influenced by climatic conditions, seasonal changes, temperature and humidity of the atmosphere, meteorological phenomena, character of the watershed, infiltration, seepage, evaporation, percolation, and transpiration. Hydrographic data are essential to show the variations of runoff over a period of many years. Reservoirs, by providing storage, reduce the extremes of flow variation, which are often as high as 100 to 1 or occasionally 1000 to 1.

Economic factors. The economic factors affecting the capacity to be installed, which must be evaluated on any project, include load requirements, runoff, head, development cost, operating cost, value of output, alternative methods of generation, flood control, navigation, rights of other industries on the stream (such as fishing and lumbering), and national defense. Some of these factors are components of multipurpose developments with their attendant problems in the proper allocation of costs to the several purposes. The prevalence of government construction, ownership, and operation, with its subsidized financial formulas which are so different from those for investor-owned projects, further complicates economic evaluation. Many people and groups are parties of interest in the harnessing of hydrosites, and stringent government regulations prevail, including those of the U.S. Corps of Engineers, Federal Power Commission, Bureau of Reclamation, Geological Survey, and Securities and Exchange Commission.

Capacity. Prime capacity is that which is continuously available. Firm capacity is much larger and is dependent upon interconnection with other power plants and the extent to which load curves permit variable-capacity operation. The incremental cost for additional turbine-generator capacity is small, so that many alternatives for economic development of a site must be considered. The alternatives include a wide variety of base load, peak load, run-of-river, and pumped-storage plants. All are concerned with fitting installed capacity, runoff, and storage to the load curve of the power system and to give minimum cost over the life of the installation. In this evaluation it is essential clearly to distinguish capacity (kw) from energy (kwhr) as they are not interchangeable. In any practical evaluation of water power in this electrical era it should be recognized that the most favorable economics will be found with an interconnected electric system where the different methods of generating power are complementary as well as competitive.

As noted above, there is an increasing tendency in many areas to allocate hydro capacity to peaking service and to foster pumped-water storage for the same objective. Pumped storage, to be practical, requires the use of two reservoirs for the storage of water—one at considerably higher elevation, say, 500 to 1000 ft. A reversible pump-turbine operates alternatively (1) to raise water from the lower to the upper reservoir during off-peak periods, and (2) to generate power during peak-load periods by letting the water flow in the opposite direction through the turbine. Proximity of favorable sites on an interconnected electrical transmission system reduces the investment burden. Under such circumstances the return of 2 kwhr on-peak for 3 kwhr pumping off-peak has been demonstrated to be an attractive method of economically utilizing interconnected fossil-fuel, nuclear-fuel, and hydro power plants. *See* ELECTRIC POWER GENERATION; HYDRAULIC TURBINE; POWER PLANT; REACTOR, NUCLEAR.

[THEODORE BAUMEISTER]

Bibliography: Annual Statistical Report for 1969, pt. 2, *Elec. World*, 171:77, Feb. 24, 1969; H. K. Barrows, *Water Power Engineering*, 3d ed., 1943; T. Baumeister (ed.), *Standard Handbook for Mechanical Engineers*, 7th ed., 1967; W. P. Creager and J. D. Justin, *Hydroelectric Handbook*, 2d ed., 1950; Department of Agriculture, *Summary of Reservoir Sediment Deposition Surveys Made in the U.S. through 1960*, USDA Misc. Publ. no. 964, 1964; Department of the Interior, *Rate of Sediment Accumulation Drops at Lake Mead*, 1967; Edison Electric Institute, *Statistical Year Book of the Electric Utility Industry for 1967*, no. 35, 1968; Federal Power Commission, *Development of Pumped Storage Facilities in the United States*, 1972, *Hydroelectric Power Resources of the United States*, Jan. 1, 1972, *The Role of Hydroelectric Developments in the Nation's Power Supply*, May 1974, *World Power Data*, 1966; D. G. Fink and J. M. Carroll, *Standard Handbook for Electrical Engineers*, 10th ed., 1968; French stem the tides, *Elec. World*, 166:17, Nov. 7, 1966; Pumped storage power, at last, comes into its own, *Eng. News Rec.*, 182:22–25, Jan. 2, 1969; W. H. Hunt, Pumped storage: A major hydro power resource, *Civil Eng.*, 38:48–53, March, 1968; Rance tidal power station, *Smokeless Air*, no. 145, pp. 174–176, spring, 1968: W. A. Schoales, Prospects of tidal power in western Australia, *Elec. World*, 167:61–63, Feb. 20, 1967: Worldwide pumped storage projects, *Power Eng.*, p. 58, Oct., 1968.

Watt-hour meter

An electricity meter which measures and registers the integral, with respect to time, of the power in the circuit in which it is connected. In effect, it is an electric motor, the torque of which is proportional to the electric power in the circuit. The speed of the rotor is proportional to the torque, making each revolution of the rotor a measurement in watt-hours. Summation of the watt-hours is accomplished by gearing a counter, or register, to the rotor. *See* MOTOR, ELECTRIC.

The basic elements of a watt-hour meter are the stator, rotor, retarding magnet or magnets, register, and meter housing.

Principle of operation. Watt-hour meters may be classified into three types, according to fundamental differences in principle of operation.

Mercury type. This type is used for measuring energy on a dc circuit. It differs from other types in that the driven portion of the rotor consists of a radially slotted copper disk immersed in mercury. The load current flows diametrically through the disk, interacting with a flux produced by the line-voltage electromagnet to cause the disk to rotate.

Mercury-type meters are readily applied to high-current loads, since they are used with shunts and since only a portion of the load current passes through the current circuit of the meter.

Commutator type. This is also used for measuring energy on a dc circuit. It may also be used on ac circuits if all windings are of air-core construction. This meter is a shunt-type motor. The field coils, part of the stator, produce a field proportional to the load current. The armature is mounted on the rotor and is energized by the line voltage through a commutator and brushes, producing a rotor torque proportional to the power in the circuit.

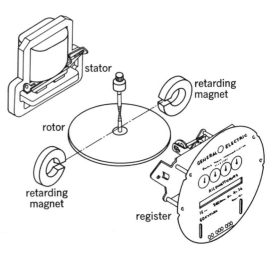

Fig. 1. The basic elements of the induction-type watt-hour meter. (*General Electric Co.*)

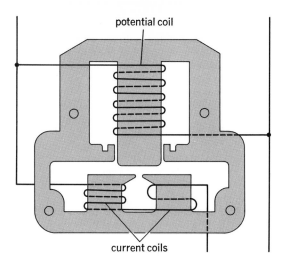

Fig. 2. Stator of induction-type watt-hour meter, showing the coils. (*General Electric Co.*)

Induction type. This is the common meter found in homes. It is used for measuring energy on an ac circuit. Figure 1 is a schematic sketch showing the basic elements of an induction meter. Figure 2 shows the stator in more detail.

The potential-circuit winding of the stator is made highly inductive to obtain a quadrature-time relationship between the potential-circuit and the current-circuit working fluxes in the disk air gap. These fluxes, displaced in time and space, produce a rotor-disk torque proportional to the circuit power. Retarding magnets control the rotor speed, making it proportional to the power. The register, which is geared to the rotor, records the watt-hours.

This type of meter has been developed to a high degree of accuracy under extreme environmental conditions and over great ranges of load and voltage. Although it has the appearance of simplicity and low cost, its magnetic circuitry is extremely complex. Its calibration is stable, and there are almost no maintenance requirements.

Multistator, or polyphase, watt-hour meters employ the same principle of operation as do single-stator meters. Magnetic interference between stators and the need for an adjustment to balance the torques of all stators are added complications of multistator meters.

Special watt-hour meters. Many types of watt-hour meter are available for special needs and applications.

The switchboard-type meter is used for industrial or central-station applications. It differs from the basic types only in housing construction.

The totalizing meter records in one meter the energy used in two or more circuits. This meter may have four or more stators acting on a single rotor.

The portable watt-hour meter standard is a specially developed, high-accuracy watt-hour meter having a multiplicity of current and voltage circuits. It is generally used in the meter shops of utilities and in the field for testing watt-hour meters.

A combination watt-hour meter and time switch, consisting of a standard single-phase meter and a time switch combined in one housing, is used on water heaters. The time switch opens the main heater circuit during predetermined peak-load conditions.

A combination watt-hour and demand meter is used to indicate the maximum demand in addition to recording watt-hours. This device is made in two different constructions. The thermal type combines in a single housing a thermal-demand indicator with a single or multistator watt-hour meter. The mechanical, or integrated-demand, type consists of a single or multistator watt-hour meter equipped with a demand register instead of a conventional watt-hour register. The demand register has, in addition to the watt-hour register parts, a timing means and a mechanism for integrating the energy consumed over the demand interval.

A watt-hour meter with a contact device is used for measurement of demand, particularly when large blocks of energy are involved. A demand meter, located externally to the watt-hour meter, is actuated through a contact device contained in the watt-hour meter and geared to the rotor shaft. *See* ELECTRIC POWER MEASUREMENT.

[GEORGE R. STURTEVANT; J. ANDERSON]

Bibliography: American National Standard, *Code for Electricity Metering*, C-12, 1969; Edison Electric Institute, *AEIC-EEI-NEMA Standards for Watthour Meters*, Publ. no. MSJ-10; Edison Electric Institute, *Electrical Metermen's Handbook*, 7th ed., 1965; F. K. Harris, *Electrical Measurements*, 1952; I. F. Kinnard, *Applied Electrical Measurements*, 1956; F. W. Kirk and N. R. Rimboi, *Instrumentation*, 1962; A. E. Knowlton, *Electric Power Metering*, 1934.

Wattmeter

An instrument that measures electric power. For a complete discussion of the power in various types of electric circuits *see* ELECTRIC POWER MEASUREMENT.

A variety of wattmeters is available to measure the power in ac circuits. They are generally classified by names descriptive of their operating principles. Determination of power in dc circuits is almost always done by separate measurements of voltage and current. However, some of the instruments described will also function in dc circuits.

Electrodynamic wattmeter. Probably the most useful instrument in the measurement of ac power at commercial frequencies is the indicating (deflecting) electrodynamic wattmeter. It is similar in principle to the double-coil dc ammeter or voltmeter in that it depends on the interaction of the fields of two sets of coils, one fixed and the other movable. The moving coil is suspended, or pivoted, so that it is free to rotate through a limited angle about an axis perpendicular to that of the fixed coils. As a single-phase wattmeter, the moving (potential) coil, usually constructed of fine wire, carries a current proportional to the voltage applied to the measured circuit, and the fixed (current) coil carries the load current. This arrangement of coils is due to the practical necessity of designing current coils of relatively heavy conductors to carry large values of current. The potential coils can be lighter because the operating current is limited to low values.

If i_1 is the instantaneous current in the potential coil and i_2 is the instantaneous current in the current coil, and there is no iron or disturbing magnet-

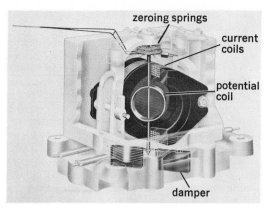

Fig. 1. Single-element electrodynamic wattmeter. (*Weston Instruments, Inc.*)

ic field due to current in neighboring conductors, then Eq. (1) holds. Since i_1 is proportional to the

$$\text{Instantaneous torque} = k(i_1 i_2) \qquad (1)$$

instantaneous voltage e across the circuit, Eq. (2) is valid.

$$\text{Instantaneous torque} = k(e i_2) \qquad (2)$$

The moving system, however, is designed with sufficient inertia that it is unable to follow the rapid alternations of the alternating current, but it will rotate, opposed by light zeroing springs, to a position corresponding to the average torque, which is proportional to the average power being supplied. A modern wattmeter of the electrodynamic type is shown in Fig. 1.

To avoid a loss of power in the instrument due to the current flowing in the potential coil, a fine-wire coil is wound with the current coil and connected in series with the potential coil. Its effect is to can-

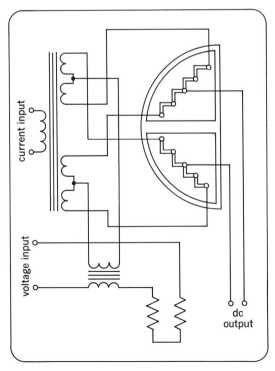

Fig. 2. Thermal wattmeter circuit.

cel the magnetic effect of the potential coil current from the field of the current coils. If this compensation is correct, with the load circuit open, the instrument will read zero.

The presence of inductance in the potential circuit would normally introduce phase displacement between voltage and current, which is theoretically inadmissible. Simple tuning with capacity would introduce frequency error, but capacity in series with the moving coil and shunted by noninductive resistance is found to reduce the net reactance to a tolerably low value.

Other disturbing effects are those of ambient temperature, eddy currents in metal structures close to the moving coils, transformer effect due to mutual induction between current and potential coils, and skin effect in which a change in distribution of current in the current coil causes frequency error. In precision instruments these effects are taken care of as much as possible by proper choice of materials and by shielding.

This type of wattmeter is capable of great accuracy within its range and is commonly used as a laboratory standard. It has the great advantage of being usable as a transfer instrument; that is, it can be calibrated with direct current and used for ac measurements. It is useful chiefly at power frequencies. In modern practice, the ranges of voltage are limited to about 150 volts, though external multipliers to extend the voltage range may be used. Maximum current is usually about 20 amp. Shunting of current coils is not recommended. The preferred method of extending the range of both current and voltage is through current and potential transformers, by which the voltages and current in the load circuit are reduced to nominal values by definite transformation ratios without introducing appreciable phase errors, or, if such errors are introduced, their values are known and may be accounted for.

Thermal wattmeter. The thermal wattmeter is a versatile means of power measurement, since its operation is based on the heating effect of current, the I^2R relationship. It is applicable to both direct current and alternating current and is usable at frequencies up into the radio range without regard to waveform. Various electronic wattmeters for audio and radio power measurements are largely based on this principle. Thermocouples are commonly used as transducers, and the derived voltage is available for the common types of dc indicators and for potentiometers of either the manual or automatic self-balancing type. A limitation is that the thermal wattmeter must be calibrated against another wattmeter. Also, it is relatively slow in response and may be affected by temperature.

A typical system used in commercial power measurements is shown in Fig. 2. The transducer elements are placed in the semicylindrical structure shown, which is divided into two compartments designed to limit cooling by convection. Each compartment contains a resistance-thermocouple assembly, which is made up of a series of thermocouples having the cold junctions a mounted on short metal posts and the hot junctions b suspended in air. The mounting posts are electrically insulated from, but in close thermal association with, the base of the structure. The materials of the thermocouples are so selected and proportioned to obtain not only a suitable value of thermoelectric effect but also an ohmic resistance, which may be heated by the load current. Thus, in each of these identical elements, a thermoelectric potential is developed with a polarity dependent on the arrangement of the thermoelectric pairs and a magnitude proportional to the square of the current that is flowing. The load current is applied to the primary of a current transformer with two identical secondary windings I_1 and I_2, which are differentially connected to the transducer elements. The line voltage is connected in series with a resistance to a potential transformer E whose secondary voltage is tapped into the midpoints of secondaries of the current transformer. If the circuit elements are symmetrical, the temperature rise of the free ends of the thermocouples is proportional to the power dissipated in each element or to the squares of the corresponding currents. These values are different, since they depend on sum and difference terms, and thus the value and polarities of the thermoelectric elements are such that the dc potential existing between the midpoints of the two elements is proportional to the respective temperatures and, therefore, to the squares of the respective currents and, in turn, to the power in the ac circuit.

As with other methods, the range of voltage and current is extended through the use of additional current and potential transformers in the circuit external to the equipment shown in Fig. 2.

Electrostatic wattmeter. The elementary quadrant electrometer has been adapted for power measurements. A schematic diagram of this instrument is shown in Fig. 3. The mechanism consists of two quadrants a and b charged by the voltage drop across a noninductive shunt resistance R_1 through which the load current passes. The line voltage is applied between the moving vane and one of the quadrants. In the circuit shown, the voltage to the moving vane is taken from the voltage divider R_3. This, in combination with the series resistance R_2, is a common means for providing compensation for power losses in the shunt. All resistances are noninductive. The deflection θ of the indicator is proportional to the average power, as defined by Eq. (3), where T is the period of the alternating wave.

$$K\theta = \frac{1}{T}\int_0^T ei\,dt \qquad (3)$$

The method is unique among others in use in that it is a voltage method rather than one of the more usual current methods. It has the advantages of (1) the possibility of wide ranges of measured current and voltage, (2) readings essentially unaffected by ambient conditions, and (3) measurements generally free of errors due to frequency or waveform.

Its disadvantages are that it is relatively insensitive, being limited by the voltage drop permitted across the shunt, and that it requires careful screening to eliminate all possible effects of charged bodies in the vicinity. It also must be calibrated at the voltages at which it is used. Moreover, the movement is generally low in torque and weight and hence relatively delicate mechanically.

This instrument has found its chief use in the laboratory for standardizing purposes and in capacity testing where small values of power, low power factors, and high voltages are involved.

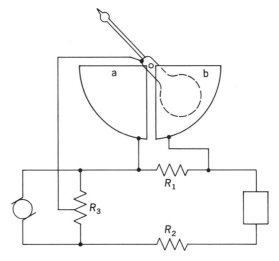

Fig. 3.   Electrostatic wattmeter circuit.

Polyphase wattmeter. The instruments thus far considered are designed for single-phase power measurement. In polyphase circuits, the total power is the algebraic sum of the power in each phase. This summation is assisted by simple modifications of single-phase instruments.

For example, an electrodynamic wattmeter may contain a second coil system similar to the coils of a single-phase meter, with the second potential coil on the same shaft as the potential coil in the first system. The two systems are mounted in the same case and are designed to have matched characteristics, but care is taken that there is no magnetic interaction between them (Fig. 4). The deflection is proportional to the sum of the torques of the two elements; thus, total power is read from the instrument scale. Electrical connections are the same as for two single-phase wattmeters.

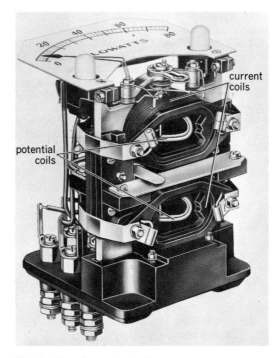

Fig. 4.   Two-element, polyphase electrodynamic wattmeter. (*Weston Instruments, Inc.*)

The thermal wattmeter is also commonly made in a two-element system, the second element being identical with the first and the two transducer networks being connected so that the combined dc output is the algebraic sum of the ac power in the two systems.

[DONALD B. SINCLAIR]

Bibliography: C. V. Drysdale and A. C. Jolley, *Electrical Measuring Instruments*, pt. 1, 2d ed., 1952; E. Frank, *Electrical Measurement Analysis*, 1959; F. K. Harris, *Electrical Measurements*, 1952; F. A. Laws, *Electrical Measurements*, 2d ed., 1938; F. E. Terman and J. M. Pettit, *Electronic Measurements*, 2d ed., 1952.

Well logging (oil and gas)

The technique of lowering specialized electronic tools into drill holes and obtaining continuous recordings of physical characteristics of subsurface formations. Analysis of these data gives geological information and permits detection and evaluation of oil and gas deposits.

From the earliest well logs to the present logs, one of the main uses has been well-to-well correlation. Variations in shaliness, liminess, or water content are reflected by distinctive curve patterns on many types of well logs. These patterns can be correlated to geological formations or horizons and are recognizable in well logs from holes drilled over wide areas. Geologists use the logs to work out subsurface geology, thus planning where subsequent wells should be drilled. *See* OIL AND GAS WELL DRILLING.

The second main use of logs is for hydrocarbon detection. For this, a proper suite or combination of modern logs is generally required. The choice of the logs to be used depends on several factors: the type of fluid in the borehole (that is fresh-water mud, brine, or oil); the lithology of the formations to be studied; and the nature of the hydrocarbon (that is, gas, oil, tar, "oil shale"). Figure 1 illustrates a typical logging suite for fresh-water mud, where the aim is to detect and evaluate the oil and gas zones.

Well logging devices. Improvements in logging devices have permitted accurate determination of the various factors required to evaluate oil or gas reservoirs. These devices are designed to eliminate the influence of the borehole itself on the measurement. Likewise, better accuracy in thin beds is obtainable. These logs can be classified as resistivity types, porosity-sensitive types, and auxiliary logs.

Resistivity systems. In soft sand–shale formations, true resistivity (R_t) is obtained by an induction device. An insulated coil generates an alternating electromagnetic field, the shape of which is controlled by auxiliary coils to provide vertical focusing and to minimize the effect from the borehole and the formation close by. Eddy currents are set up in the formation. Their strengths are proportional to the conductivity of the formation. These currents create a secondary alternating electromagnetic field, which is detected by a receiver coil. The tool thus senses changes in conductivity, but usually this is converted to its reciprocal, resistivity, for presentation and interpretation. Often a normal resistivity curve is recorded along with the induction curve.

The induction device is designed for deep penetration, and thus mainly responds to the formation that is not contaminated by invasion of drilling fluids. When logging in formations which tend to invade, that is, to be penetrated by the liquid from the drilling fluid, such as well-consolidated rocks, another induction curve is added. This one is designed for moderate rather than deep penetration. In addition, a focused resistivity device (the Laterolog-8) which has even smaller penetration, is simultaneously recorded with the two induction measurements. Comparison of the three permits estimation of the depth of invasion and thereby correction for its effect on the deep induction measurement. Figure 1a and b show typical curve responses for this logging system.

Focused resistivity systems generally employ seven or more electrodes placed symmetrically on an insulated mandrel. These are the Laterolog types. The current emitted from the central electrode is forced to flow into the formation by sending a regulated bucking current through the outer electrodes. The spacings can be varied to regulate the depth of penetration. For example, the Dual Laterolog system gives a deep and a medium penetration curve, plus a very shallow penetration curve of the same type, but from a small pad which is applied to the wall of the hole. Again, the three measurements permit invasion depth estimation and correction for its effect. The Dual Laterolog system is primarily used when brine muds are used to drill the hole. For fresh-mud logging, the Laterolog-8 device is placed on the Dual Induction mandrel, being designed to have only limited penetration.

Induction devices are better adapted to low or moderate resistivity logging. They also provide the only way to measure resistivity when oil or air fills the borehole. Laterolog devices are recommended when resistivities are high. They require a water-base mud to conduct the current to and from the formation.

Porosity systems. The space in subsurface rocks which contain fluids (water, oil, or gas) is called porosity. It is denoted by either the decimal fraction or the percentage of a unit volume of formation. While resistivity can be converted to porosity by empirical methods, the porosity-sensitive devices give a more direct way of obtaining it. Two radioactivity methods (density logging and neutron logging) and an acoustic method are commonly used.

The density logging method uses a source (cesium-137) of gamma rays which interact with the formation by Compton scattering. The scattered lower-energy gamma rays are sensed by a pair of detectors spaced above the source. The logarithm of the counting rate is proportional to the bulk density, so that the curve can be scaled in bulk density in grams per cubic centimeter. The hole influence is largely eliminated by use of dual detectors and by placing the source and shielded detectors on a sidewall skid.

In neutron logging, when high-energy neutrons bombard a formation, the neutrons are slowed down mainly by hydrogen. The resulting flux of low-energy (thermal or epithermal) neutrons is inversely proportional to the concentration of hydrogen. Since the concentration of hydrogen in water and in oil is essentially equal, the detected counting rate is often rescaled in porosity units.

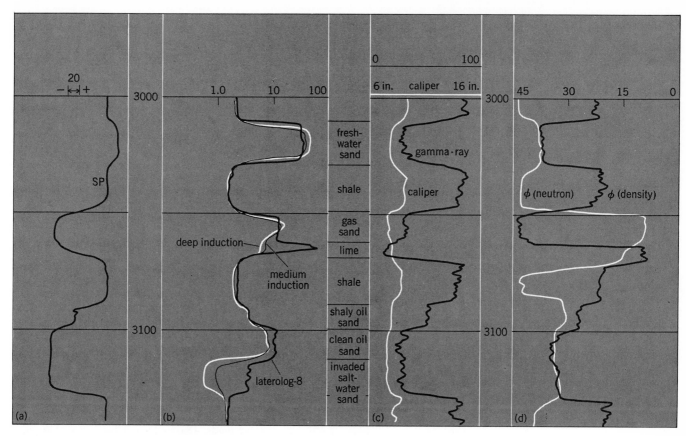

Fig. 1. Typical log responses in fresh-water drilling mud. (a) Spontaneous potential (SP), millivolts; (b) resistivity, ohms·m²/m; (c) Gamma-ray, API units; and (d) porosity (assuming sandstone).

This is the one device effective in cased boreholes, as steel is nearly transparent to neutrons.

For acoustic logging, ultrasonic pulses are emitted from a transducer. These create pressure waves which are transmitted by the hole fluid into the formation. These waves then travel at a rate dependent on the rock minerals and the included pore fluids. A pair of receivers, either 1 or 2 ft (30 or 60 cm) apart, sense the time difference as the wave passes by. Thus the log is scaled in interval transit time (Δt) in microseconds per foot. Sonic velocity in feet per second in the formation equals $10^6/\Delta t$.

Important auxiliary curves. In this group are curves which are generally recorded with resistivity or porosity logs. None of them requires energizing the formation; the information is there for the recording. Spontaneous potential (SP) occurs in holes filled with water or water-base muds. The SP gives a way of distinguishing sands or porous carbonates from shales (Fig. 1a). Also, its magnitude is related to the shale content and, in shale-free formations, to the relative ionic concentrations of the borehole fluid compared to the formation water. Since the borehole fluid can be directly measured, the SP provides a way of determining the salinity and thus the resistivity (R_w) of the formation water (Fig. 1b). To record an SP, only an electrode in the hole and a ground are required, plus scaling controls in the recorder. It is recorded along with resistivity logs.

The natural gamma-ray intensity of the rocks penetrated (Fig. 1c) is also recorded in most boreholes, generally being presented with a porosity log (Fig. 1d). Gamma rays also are used to separate sands or carbonate rocks from shales, the latter having higher radioactivity. The gamma-ray curve may be used to estimate the percent of shale. Gamma rays are practically unaffected by the material in the borehole and are only moderately reduced by steel casing. They provide excellent correlation curves for long-distance correlation and tying in open-hole to cased-hole depths.

A caliper of the drill hole diameter changes is also normally run with porosity logs. It aids in correcting logging data when the hole is enlarged. It also is used to identify a buildup of mud filter cake, which is a sign that the formation is permeable.

Applications for oil and gas location. Detection and evaluation of oil and gas deposits in subsurface formations require borehole measurements of several factors. Besides determining the top and bottom of the pay zone, data are needed on the intergranular pore space (porosity, ϕ) and the hydrocarbon saturation (fraction of the pore space which contains oil or gas), and a way of verifying permeability of the formation (to establish that oil or gas will be producible). All of these can be obtained by using suitable borehole logs. The choice of the logging suite depends on borehole conditions and on the characteristics of the reservoir rock. *See* PETROLEUM GEOLOGY.

Resistivity in hydrocarbon evaluation. The level of the resistivity of a subsurface bed is largely dependent on the quantity and salinity of the interstitial water present. The resistivity of dry mineral grains and of oil and gas is nearly infinite. The resistivity of the formation water (R_w) is usually very low, decreasing as its salinity increases. For a given salinity, a water-saturated rock with 10% of pore space will have the same resistivity as a rock with 20% of pore space which contains half water and half oil or gas. Therefore, resistivity measurements locate hydrocarbons only when some knowledge of porosity is available. In addition, the resistivity of the interstitial water is required for computation of the degree of hydrocarbon saturation. Since it is the water fraction which controls the resistivity of the rock (R_t), the occurrence of oil or gas is implied when the water saturation (S_w) is computed to be less than unity. An empirical relation often used is shown in Eq. (1). The hydrocar-

$$S_w = \frac{1}{\phi}\sqrt{\frac{R_w}{R_t}} \qquad (1)$$

bon saturation (S_h), which at this point may be either oil or gas, is equal to $1 - S_w$.

Determination of interstitial water resistivity. To determine the value of R_w, several avenues are available. The direct way is to obtain a produced water sample, measure its resistivity at a given temperature, then adjust this value for the temperature of the formation. The value of R_w can also be accurately predicted from a chemical analysis of produced water. There are two logging methods for obtaining an R_w value: (1) Spontaneous potential analysis. The SP magnitude is related to the resistivity contrast of the borehole fluid with the formation water. Since the borehole fluid can be directly measured at the surface, the formation water resistivity is easily computed. (2) Use of combined resistivity and porosity data in a water-bearing formation. When $S_w = 1.0$, Eq. (1) may be written in the form of Eq. (2). Here, R_0 is the forma-

$$R_w = \phi^2 R_0 \qquad (2)$$

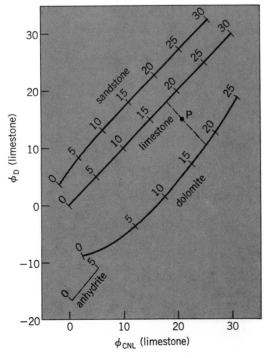

Fig. 2. Porosity and lithology determination for FDC density and CNL neutron logs in water-filled holes. *(From Schlumberger Well Services, Log Interpretation: vol. I, Principles, Schlumberger Limited, 1972)*

tion resistivity when it is at 100% water saturation (R_0 is the formation resistivity when hydrocarbons are present).

Porosity evaluation. Formation porosity provides the potential storage capacity for oil and gas; thus, it is of vital concern. Porosity is equally required for oil and gas detection through the use of Eq. (1), and for R_w evaluation by Eq. (2). Three types of measurements are used for porosity derivation: bulk density, acoustic transit time, and neutron hydrogen index. For best utility, each parameter measured is converted to a porosity index, assuming that the mineral making up the rock is either quartz or calcite. The equations are Eq. (3), giving the density-derived porosity (ϕ_D)

$$\phi_D = \frac{\rho_{ma} - \rho_b}{\rho_{ma} - \rho_f} \qquad (3)$$

where ρ_{ma} = grain density of the mineral (2.65 g/cm³ for quartz, 2.71 g/cm³ for calcite), ρ_b = measured bulk density, g/cm³, and ρ_f = water density (1.0 in fresh-water mud; a higher value, such as 1.1, in salty muds); and Eq. (4), giving the

$$\phi_S = \frac{\Delta t - \Delta t_{ma}}{\Delta t_f - \Delta t_{ma}} \qquad (4)$$

acoustic-derived (sonic) porosity (ϕ_S) where Δt_{ma} = mineral transit time, μsec/ft (47.5 for calcite, 55.5 for quartz), Δt = formation transit time, μsec/ft, and Δt_f = water transit time, μsec/ft (189 for fresh-water muds, 185 for brine-filled holes). The third index, neutron porosity (ϕ_N), is calibrated from count-rate data obtained by placing the device in "neutron test pits" constructed of water-filled blocks of sandstone or limestone with various porosities. The neutron porosity is also somewhat sensitive to the minerals making up the rocks.

When two or three of these porosity index logs are available, they supplement each other and enhance the evaluation of a given formation. The main applications are:

1. Gas detection. From resistivity and porosity alone, Eq. (1) allows determination of the hydrocarbon saturation, but does not indicate whether it is oil or gas. When recorded with the correct lithology assumption, the comparison of ϕ_D versus ϕ_N identifies gas because its presence drives ϕ_D to an abnormally high value and ϕ_N to an abnormally low value (see Fig. 1b). Also, ϕ_S can be used to compare with ϕ_N and is, in fact, a preferred method if the formation is shaly.

2. Lithology identification. When the lithology is other than that assumed, the computed values of ϕ_D, ϕ_S, or ϕ_N are altered and different than true porosity (ϕ). Since the matrix parameters (log values when porosity is zero) are well known for each, a cross-plot method or computer analysis will yield correct porosity as well as lithology information. For example, Fig. 2 shows a cross-plot chart for modern density and neutron devices. It is assumed that log data have been scaled for limestone rocks. If the density reads 15% and the neutron 21% (point P), the true porosity is 18% and the rock is composed of about 40% dolomite and 60% limestone.

3. In sand-shale regions, porosity cross-plots are used to evaluate the amount of clay present and give better effective porosity values. When clay is present, the saturation equation, Eq. (1), is not

applicable. A more complex relation which includes correction for the volume of clay is required.

[ROBERT P. ALGER]

Bibliography: F. A. Alekseev (ed.), *Soviet Advances in Nuclear Geophysics*, 1965; E. J. Lynch, *Formation Evaluation*, 1962; S. J. Pirson, *Geologic Well Log Analysis*, 1970; S. J. Pirson, *Handbook of Well Log Analysis: For Oil and Gas Formation Evaluation,* 1963; Schlumberger Well Services, *Log Interpretation*: vol. I, *Principles*, 1972, vol. II, *Applications*, 1974; M. R. Wyllie, *Fundamentals of Well Log Interpretation*, 3d ed., 1963.

Wet cell

A primary cell in which there is a substantial amount of free electrolyte in liquid form. Important examples are the Lalande or caustic soda cell, the air-depolarized alkaline cell, the Weston standard cell, and the organic electrolyte cell.

Lalande cell. This cell (Fig. 1) uses a zinc anode, a cupric oxide cathode, and an electrolyte of sodium hydroxide in aqueous solution (caustic soda).

The amalgamated zinc electrodes are cast as flat plates or as hollow cylinders with thin sections, which corrode through as the copper oxide electrode approaches exhaustion. The bright copper color of the cathode can be seen through these openings to warn the user of approaching exhaustion of the cell.

The cathode is made by molding cupric oxide into flat plates or hollow cylinders. The oxide is mixed with a binder, pressed, and roasted. As it is used, the cupric oxide is partially reduced to metallic copper, which greatly increases the conductivity of the cathode.

The electrolyte is a solution of sodium hydroxide in water of good purity. The specific gravity is about 1.21. Normally, the surface of the electrolyte is covered with a layer of oil which retards evaporation of water and absorption of carbon dioxide from the atmosphere.

The anode reaction in the Lalande cell is the oxidation of zinc to form zinc oxide, which dissolves in the electrolyte to form sodium zincate. With sufficient electrolyte, no solid phase forms until the cell is nearly exhausted. Then precipitation occurs. Because of this reaction, it is necessary to provide about 8 ml of 1.21 sp gr solution per rated ampere-hour. The zinc must be mounted near the top of the electrolyte to prevent premature cutting off of the discharge by anodic polarization.

The cathodic reaction is the reduction of cupric oxide to metallic copper. For this reason, the plate always has the appearance of metallic copper.

The cell potential is 0.95–1.0 volt, but the closed-circuit voltage on normal discharge starts at about 0.65 and decreases slowly to a cutoff voltage of about 0.50. Commercial cells are intended for relatively long continuous service. A 500-amp-hr light-duty cell has an output rating of 1.75 amp at 70°F. A heavy-duty cell of the same ampere-hour rating is capable of continuous output at 6.5–12.0 amp. Unit energy output is about 1.1 kwhr/ft³ of cell.

At temperatures below 70°F, full output (ampere-hours) can be obtained at a reduced current. Current is reduced by 40% at 40°F, by 67% at 20°F, and by 83% at 0°F.

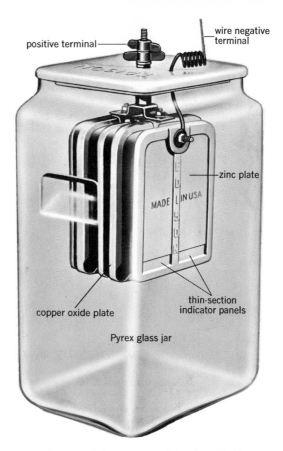

Fig. 1. Lalande-Edison copper oxide primary battery.

Air-depolarized alkaline cell. This cell (Fig. 2) uses a zinc anode, a porous carbon cathode exposed to air on one face, and an alkaline electrolyte. The carbon cathode utilizes atmospheric oxygen. Its zinc anode and alkaline electrolyte are like those of the Lalande cell, but it has twice the operating voltage and twice the watt-hour output of an equal-size Lalande cell.

The cathodic reactions have been shown to be Eqs. (1)–(3). The oxygen reacts to form hydrogen

$$2e + O_2 + H_2O \rightarrow O_2H^- + OH^- \qquad (1)$$

$$O_2H^- + OH^- + 2H^+ \rightarrow H_2O_2 + H_2O \qquad (2)$$

$$H_2O_2 \rightarrow H_2O + \tfrac{1}{2}O_2 \qquad (3)$$

peroxide, which decomposes readily to water and oxygen. The net amount of oxygen, then, is 0.3 g/amp-hr. At standard temperature and pressure, 210 ml of oxygen is consumed per ampere-hour.

The porous carbon has been reported to have an apparent density of only 0.65 with a porosity of 60%. To function properly, the inner surfaces should be dry. To resist penetration of electrolyte, the pore size must be small and the surfaces partially waterproofed by impregnation with paraffin. The cathode acts as a pump, drawing oxygen from the air. If too great a current is drawn, the pressure in the pore may drop sufficiently to allow electrolyte penetration. This reduces the activity of the carbon and may cause cell failure. Hence it is important that the cell should not be overloaded. For a railway cell, rated at 500 amp-hr, the recommended continuous drain is 2.0 amp at temperatures above 45°F. This is a current density of 3.1 amp/ft². Much higher current densities can be ob-

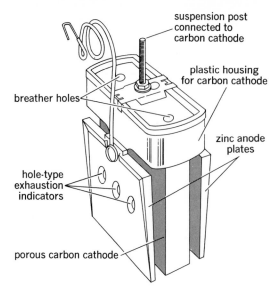

Fig. 2.   Construction of air-depolarized alkaline cell.

tained with porous carbon electrodes by special design.

The cell open-circuit voltage is 1.46. The railway cell at room temperature (75°F), rated at 500 amp-hr, will deliver 2 amp at an average of 1.13 volts to a cutoff of 1.05 volts. On 3-amp intermittent signal test, the cell delivers rated capacity of 1.09 volts, average at 75°F, 0.98 volt average at 32°F.

Air-depolarized cells are now available in sizes up to 2500 amp-hr. The 2500-amp-hr cell delivers 5.64 kwhr/ft³ and 0.074 kwhr/lb when discharged at low rates.

Weston standard cell. The Weston cell of 1893 has become the accepted standard of electromotive force (emf). The Weston normal or saturated cadmium cell has an emf of 1.01864 absolute volts at 20°C. When purified materials are used, cells having the same emf to within a few microvolts may be made. These cells maintain their emf very well. Reference standards, in daily use for many years, are remarkably constant. These standard cells are made with materials of spectroscopic purity and are maintained under diffuse light in a thermostatically controlled oil bath in a room maintained at 25°C and 60% relative humidity.

The cell uses a two-phase amalgam of cadmium as the anode. For a 10% amalgam, one part of cadmium by weight and nine parts of mercury are required. These materials may be combined either by heating them together or by electrolytic deposition of cadmium into mercury. At ordinary temperatures a liquid phase is in equilibrium with a solid phase. This gives a very stable potential which depends only on the temperature.

The cathode is mercurous sulfate, Hg_2SO_4, in contact with mercury.

The electrolyte is a saturated solution of cadmium sulfate in equilibrium with the solid phase $CdSO_4 \cdot \frac{8}{3}H_2O$. In some cells, sulfuric acid is added in order to prevent the hydrolysis of mercurous sulfate.

The cell is usually made of glass in the form of an H, as in Fig. 3. Platinum-wire leads are sealed in the base of each arm. Mercury, carefully purified, is placed at the bottom of one arm and a

10% cadmium amalgam is placed, while warm and in a single phase, in the other arm. When the amalgam has cooled and separated into two phases, crystals of mercurous sulfate are placed above the mercury and crystals of cadmium sulfate are placed above the amalgam. A saturated solution of cadmium sulfate is then added to about 2–3 mm above the crossbar, and the cell is hermetically sealed.

The best-saturated cells of this type may be measured to the ten-millionth part of a volt at specified temperatures which must be known to within 0.01°C.

The saturated Weston cell has a relatively large temperature coefficient of emf. For portable use, it is general practice to use a cell with an unsaturated electrolyte. This has a temperature coefficient which is only one-fourth as great as that of the saturated cell.

Standard cells are not intended as power sources. They should be used only for comparison of voltages. Ordinary voltmeters put too heavy a drain on the cell for any reliable voltage measurements.

Organic electrolyte cell. A different class of cells is that based on the use of particularly reactive metals (Li, Ca, Mg) in conjunction with organic electrolytes. The best-known type in this class is the lithium-cupric fluoride cell, theoretically capable of delivering over 700 whr/lb, more than three times the capacity of the highly ranked Zn-AgO cell.

The lithium anode is usually made in sheet form, but variants using lithium powder trapped in appropriate grids are also known.

The cathode is a mixture of CuF_2 and various conductive materials, most often graphite and carbon black either together or separately, in order to obtain the electronic conductivity that CuF_2 does not possess.

The electrolyte considered most compatible with these electrodes is lithium perchlorate (1 M) in propylene carbonate or butyrolactone.

The cell works with 80% cathodic current efficiency, delivering 25% of its energy to the 2-volt end point, the initial voltage being 3.2 volts. This represents 80 whr/lb, that is, 320 whr/lb available energy. On the other hand the current density is only 0.5–3 milliamperes/cm², too little except for special applications. Nonetheless, owing to their

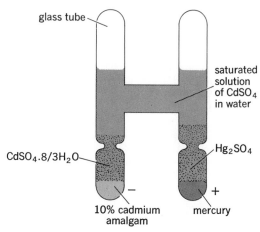

Fig. 3.   Schematic of Weston saturated standard cell.

considerable thermodynamic energy density, the organic electrolyte cells remain the great hope for the future.

For the present, the most favored anode material is lithium which, besides providing fully reversible anodic reaction, possesses the exclusive properties of a low specific gravity (0.534) and a high electrode potential (3.045 volts). In fact after beryllium, lithium has the highest energy availability per atom and is by far the lightest metal known.

The cathode problem is still under investigation, and at least two other salts, CoF_3 and $NiCl_2$, are claimed to be superior, working with cathodic current efficiencies close to 100%.

The critical problem remains the electrolyte because of the difficulty of combining high specific conductivity, low viscosity, hydrophobicity, chemical inertness toward electrodes, and adequate electrolytic properties. The conditions of high specific conductivity (2.85×10^{-2} ohm^{-1} cm^{-1}) and low viscosity (0.35 centipoise) are fulfilled by acetonitrile. Unfortunately acetonitrile reacts powerfully with lithium. The compromise solution at this stage is propylene carbonate, 8 times more viscous and 30 times less conductive than acetonitrile but chemically less aggressive.

Another problem in which progress is expected is the construction of the electrodes. The techniques presently used, sintering in a grid or dispersing the cathode powder in a porous structure under the protection of filter paper or some ion-exchange membrane, seem to be only a step in an evolutionary process. Finally, the short shelf-life, due to uncontrolled ionic circulation and dendrite formation, is a problem that must be improved before organic electrolyte cells enter the industrial production stage.   [L. ROZEANU; JACK DAVIS]

Bibliography: D. P. Boden, Electrolytes for non-aqueous batteries, *Proceedings of the 20th Annual Power Sources Conferences*, p. 63, 1966; K. H. M. Braeuer, Organic electrolyte, high energy density batteries, *Proceedings of the 20th Annual Power Sources Conferences*, p. 57, 1966; W. E. Elliot et al., Active metal anode-electrolyte systems, *Proceedings of the 20th Annual Power Sources Conferences*, p. 67, 1966; G. W. Heise, E. A. Schumacher, and C. R. Fisher, The air-depolarized primary cell with caustic alkali electrolyte II, *Trans. Electrochem. Soc.*, 92:173, 1947; G. W. Vinal, *Primary Batteries*, 1950.

Wind power

Kinetic energy in the Earth's atmosphere used to perform useful work. Total atmospheric wind power is of the order of 10^{14} kW. Annual kinetic energy is of the order of 10^{17} kWhr (1 kWhr = 3.60×10^6 J). Practical land-based wind generators could extract as much as 10^{14} kWhr of energy per year worldwide. Energy, and thus productivity of winds, varies markedly with geographic location. Annual energy available to a conversion machine at a site is very reproducible ($\pm 15\%$ variability). Annual average power per square meter of vertical area at a 10-m height across and over water near the United States is estimated in Fig. 1. Southern Wyoming shows the greatest chance for productive wind power systems on land (>400 W/m²), and the edge of the shelf off New England shows >800 W/m².

Wind power has lifted water for centuries, but there is new interest in wind-powered irrigation in the United States. Electricity was generated by wind power in 1880. Thousands of small wind generators and water pumps worked in the United States as late as 1940. Subsidized rural electrification, low fossil-fuel prices, and desire for more powerful farm machines caused near extinction of wind power systems in this country. In other nations, simple wind power systems are still the key to material sufficiency. The exponential energy appetite of the industrialized world consuming finite resources and the exponential growth in pollution associated with energy production has renewed interest in wind power.

Water lifting. Water spilled on agricultural soil by a random wind is still the best example of a storage subsystem buffering between the wind and the energy consumer. Most large-scale irrigation

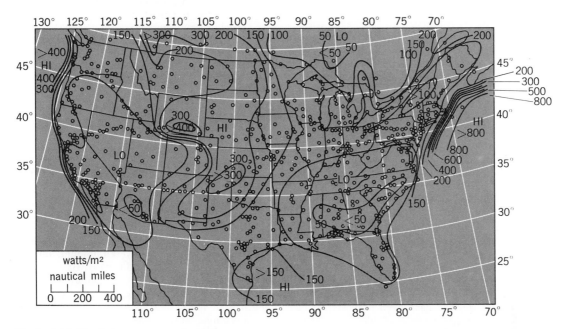

Fig. 1. Available wind power—annual average.

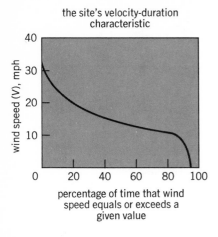

the site's velocity-duration characteristic

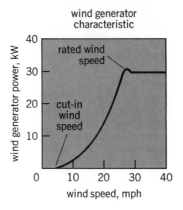

wind generator characteristic

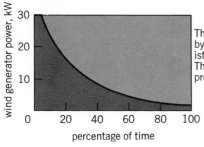

The wind generator's characteristic weighted by the site's velocity-duration characteristic yields a power duration curve. The area under that curve measures productivity of that machine at that site.

Fig. 2. Productivity of a wind generator is a function of wind regime, height, the machine's aero-mechanical-electrical characteristics, and the delivered product.

pumping in the United States today consumes natural gas, but wind power pumping could come back. Wind-powered pump-back could expand the capacity of hundreds of smaller hydroelectric installations.

Heating. At least 20% of United States energy is consumed in heating buildings. In colder climates, heating demand frequently matches high winds. Wind-generated electricity feeding thermal storage units, in some instances combined with solar thermal collectors feeding adjacent storage units, offers excellent potential for reducing fuel oil consumption. Large-scale wind-generator systems located off the Atlantic coast could take over a large fraction of building heating load in the largest urban areas.

Wind-generated electricity. Complete sets of hardware combining wind generators and electric storage batteries large enough to supply 500–1000 kWhr/month electricity on demand can be purchased. Where the generator can be placed at moderate height into a strong and persistent wind regime, delivered electricity is economically competitive. In milder wind regimes, electricity is still expensive compared with most utility prices. Larger centralized wind electricity systems sharing larger storage subsystems have a greater chance of being economic. When winds produce more electricity than the market demands, the energy must be stored, then recalled and used when demand exceeds windpower. Wind generators in large numbers located in productive winds and equipped with hydrogen generation, storage, and reconversion devices have been proposed. It has been estimated that winds available to the United States could generate as much as 2×10^{12} kWhr of firm power on demand, equivalent to the total 1975 United States electricity consumption. *See* ELECTRIC POWER GENERATION; ELECTRICAL UTILITY INDUSTRY; STORAGE BATTERY.

Mechanics. A windmill is a rotating machine capable of interchanging (extracting) momentum with particles of air mass that flow through its swept area. Power available in the wind in a swept area varies with the size of that area, the density of the air, and the square of the velocity. Energy extractable from an oncoming wind stream over a period of time varies as the size of swept area, density, and cube of the velocity, as described in Eq. (1), where K.E. is the kinetic energy available,

$$K.E. = kDV^3 \qquad (1)$$

k is the density of air, D is the sweeping blades' diameter, and V is the average wind speed. At sea level, K.E. = 0.000935 D^2V^3 lb-ft/s, corresponding to 1.7×10^{-6} D^2V^3 horsepower, for D in feet, and V in feet per second. Adolph Betz showed that no more than 59% of the energy in an oncoming stream tube of wind could be extracted without bypassing the momentum exchanger. The ability of any windmill to approach that 59% maximum extraction capability is thus an excellent indicator of its performance, and has been named the coefficient of performance, C_p. Thus, the realizable power P_R from a windmill is as shown in Eq. (2).

$$P_R = C_p \times 1.7 \times 10^{-6}D^2V^3 \text{ horsepower} \qquad (2)$$

A C_p as high as 0.48 has been observed for a modern high-speed propeller-type wind machine, whereas the sheet-metal-bladed American Fan Mill for water pumping seldom achieves a C_p in excess of 0.30. Coefficient of performance is related to the aerodynamic features of a machine, particularly to the tip speed ratio, defined by Eq. (3).

$$\text{Tip speed ratio} = \frac{\text{tangential speed of blade tip}}{\text{speed of oncoming wind}} \qquad (3)$$

Different types of wind machines work best at their own optimum tip speed ratio. The American Fan Mill wants a tip speed ratio near 1; the very-high-speed twisted and tapered two- or three-bladed propeller type wants a ratio between 7 and 12. Vertical-axis machines of the S-rotor (Savonius) type, cross-flow vertical-bladed machines, and the Darrieus twirling rope (troposkien) rotor operate best at a ratio between 1 and 2. It has been suggested that any configuration of horizontal-axis machine can be given characteristics that will permit it to have a high C_p, but economics and other practical considerations seem to favor the three-bladed twisted and tapered propeller type operating at a tip speed ratio of about 8. Advocates of the vertical-axis Darrieus type hope to prove that type most economic, however.

System and economic considerations. The utility and competitiveness of a wind power system depend upon the wind regime and height at which windmills are placed, size and characteristics of the machine, nature of delivered product, and productivity of that product. Figure 2 shows how a wind regime can be characterized by a velocity duration curve (for any specific time period) and the overall power out versus wind speed of the wind machine (in this case a generator of electricity). If the desired product is simply raw electrical energy, the product of delivered voltage times delivered amperage suffices. Figure 2 shows how a power duration curve measures that productivity. Wind systems delivering other products would be assessed differently, but with the same philosophy. Wind systems can thus be compared with one another and with other systems capable of delivering the same product. *See* ENERGY SOURCES.

[WILLIAM HERONEMUS]

Bibliography: A. Betz, *Windmills in the Light of Modern Research*, Nat. Advis. Comm. Aeronaut. Nat. Mem. no. 474 (available from NTIS), 1928; F. R. Eldridge (ed.), *Proceedings of the 2d Workshop on Wind Energy Conversion Systems*, Washington, DC, 1975; H. Glauert, Windmills and fans, in W. F. Durand (ed.), *Aerodynamic Theory*, vol. 4, 1935; E. W. Golding, *Generation of Electricity by Windpower*, 1955; W. E. Heronemus, Pollution free energy from offshore winds, in *Preprints of the 8th Annual Conference and Exposition, Marine Technology Society*, Washington, DC, September, 1972; P. C. Putnam, *Power from the Wind*, 1948.

Work

In physics, the term work refers to the transference of energy that occurs when a force is applied to a body that is moving in such a way that the force has a component in the direction of the body's motion. Thus work is done on a weight that is being lifted, or on a spring that is being stretched or compressed, or on a gas that is undergoing compression in a cylinder.

When the force acting on a moving body is constant in magnitude and direction, the amount of work done is defined as the product of just two factors: the component of the force in the direction of motion, and the distance moved by the point of application of the force. Thus the defining equation for work W is Eq. (1), where f and s are the

$$W = f \cos \phi \cdot s \qquad (1)$$

magnitudes of the force and displacement, respectively, and ϕ is the angle between these two vector quantities (Fig. 1). Because $f \cos \phi \cdot s = f \cdot s \cos \phi$, work may be defined alternatively as the product of the force and the component of the displacement in the direction of the force. In Fig. 2 the work of the constant force f when the application point moves along the curved path from P to P', and therefore undergoes the displacement $\overline{PP'}$, is $f \cdot \overline{PP'} \cos \phi$, or $f' \overline{PE}$.

Work is a scalar quantity. Consequently, to find the total work done on a moving body by several different forces, the work of each may be computed separately and the ordinary algebraic sum taken.

Examples and sign conventions. Suppose that a car slowly rolls forward a distance of 30 ft along a straight driveway while a man pushes on it with a constant magnitude of 50 pounds of force (50 lbf) and let Eq. (1) be used to compute the work W done under each of the following circumstances: (1) If the man pushes straight forward, in the direction of the car's displacement, then $\phi = 0°$, $\cos \phi = 1$, and $W = 50$ lbf $\times 1 \times 30$ ft $= 1500$ foot-pounds of force (ft-lbf); (2) if he pushes in a sideways direction making an angle ϕ of 60° with the displacement, then $\cos 60° = 0.50$ and $W = 750$ ft-lbf; (3) if he pushes against the side of the car and therefore

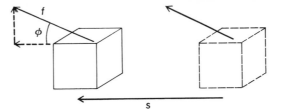

Fig. 1. Work of constant force f is $fs \cos \phi$.

at right angles to the displacement, $\phi = 90°$, $\cos \phi = 0$, and $W = 0$; (4) if he pushes or pulls backward, in the direction opposite to the car's displacement, $\phi = 180°$, $\cos \phi = -1$, and $W = -1500$ ft-lbf.

Notice that the work done is positive in sign whenever the force or any component of it is in the same direction as the displacement; one then says that work is being done *by* the agent exerting the force (in the example, the man) and *on* the moving body (the car). The work is said to be negative whenever the direction of the force or force component is opposite to that of the displacement; then work is said to be done *on* the agent (the man) and *by* the moving body (the car). From the point of view of energy, an agent doing positive work is losing energy to the body on which the work is done, and one doing negative work is gaining energy from that body.

Units of work and energy. These consist of the product of any force unit and any distance unit. Units in common use are the foot-pound, the foot-poundal, the erg, and the joule. The product of any power unit and any time unit is also a unit of work or energy. Thus the horsepower-hour (hp-hr) is equivalent, in view of the definition of the horsepower, to 550 ft-lbf/sec $\times$ 3600 sec, or 1,980,000 ft-lbf. Similarly, the watt-hour is 1 joule/sec $\times$ 3600 sec, or 3600 joule; and the kilowatt-hour is 3,600,000 joule. *See* HORSEPOWER.

Work of a torque. When a body which is mounted on a fixed axis is acted upon by a constant torque of magnitude τ and turns through an angle θ (radians), the work done by the torque is $\tau\theta$.

Work principle. This principle, which is a generalization from experiments on many types of machines, asserts that, during any given time, the work of the forces applied to the machine is equal to the work of the forces resisting the motion of the

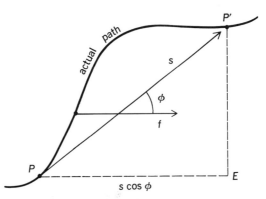

Fig. 2. The work done in traversing any path connecting points P and P' is $f \cdot \overline{PE}$, assuming the force f to be constant in magnitude and direction.

WORK

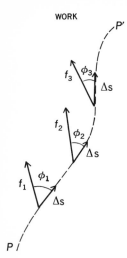

Fig. 3. Work done by a variable force.

machine, whether these resisting forces arise from gravity, friction, molecular interactions, or inertia. When the resisting force is gravity, the work of this force is mgh, where mg is the weight of the body and h is the vertical distance through which the body's center of gravity is raised. Note that if a body is moving in a horizontal direction, h is zero and no work is done by or against the gravitational force of the Earth. If a person holds an object or carries it across level ground, he does no net work against gravity; yet he becomes fatigued because his tensed muscles continually contract and relax in minute motions, and in walking he alternately raises and lowers the object and himself.

The resisting force may be due to molecular forces, as when a coiled elastic spring is being compressed or stretched. From Hooke's law, the average resisting force in the spring is $-\frac{1}{2}ks$, where k is the force constant of the spring and s is the displacement of the end of the spring from its normal position; hence the work of this elastic force is $-\frac{1}{2}ks^2$.

If a machine has any part of mass m that is undergoing an acceleration of magnitude a, the resisting force $-ma$ which the part offers because of its inertia involves work that must be taken into account; the same principle applies to the resisting torque $-I\alpha$ if any rotating part of moment of inertia I undergoes an angular acceleration α.

When the resisting force arises from friction between solid surfaces, the work of the frictional force is $-\mu f_n s$, where μ is the coefficient of friction for the pair of surfaces, f_n is the normal force pressing the two surfaces together, and s is the displacement of the one surface relative to the other during the time under consideration. The frictional force μf_n and the displacement s giving rise to it are always opposite in direction ($\phi = 180°$).

The work done by any conservative force, such as a gravitational, elastic, or electrostatic force, during a displacement of a body from one point to another has the important property of being path-independent: Its value depends only on the initial and final positions of the body, not upon the path traversed between these two positions. On the other hand, the work done by any nonconservative force, such as friction due to air, depends on the path followed and not alone on the initial and final positions, for the direction of such a force varies with the path, being at every point of the path tangential to it. *See* FORCE.

Since work is a measure of energy transfer, it

can be calculated from gains and losses of energy. It is useful, however, to define work in terms of forces and distances or torques and angles because these quantities are often easier to measure than energy changes, especially if energy changes are produced by nonconservative forces.

Work of a variable force. If the force varies in magnitude and direction along the path $\overline{PP'}$ of its point of application, one must first divide the whole path into parts of length Δs, each so short that the force component $f \cos \phi$ may be regarded as constant while the point of application traverses it (Fig. 3). Equation (1) can then be applied to each small part and the resulting increments of work added to find the total work done. Various devices are available for measuring the force component as a function of position along the path. Then a work diagram can be plotted (Fig. 4). The total work done between positions s_1 and s_2 is represented by the area under the resulting curve between s_1 and s_2 and can be computed by measuring this area, due allowance being made for the scale in which the diagram is drawn.

For an infinitely small displacement ds of the point of application of the force, the increment of work dW is given by Eq. (2), a differential expres-

$$dW = f \cos \phi \, ds \qquad (2)$$

sion that provides the most general definition of the concept of work. In the language of vector analysis, dW is the scalar product of the vector quantities $\mathbf{f}$ and $d\mathbf{s}$; Eq. (2) then takes the form $dW = \mathbf{f} \cdot d\mathbf{s}$. If the force is a known continuous function of the displacement, the total work done in a finite displacement from point P to point P' of the path is obtained by evaluating the line integral in Eq. (3).

$$W = \int_P^{P'} f \cos \phi \, ds = \int_P^{P'} \mathbf{f} \cdot d\mathbf{s} \qquad (3)$$

When a variable torque of magnitude τ acts on a body mounted on a fixed axis, the work done is given by $W = \int_{\theta_1}^{\theta_2} \tau \, d\theta$, where $\theta_2 - \theta_1$ is the total angular displacement expressed in radians. *See* ENERGY.   [LEO NEDELSKY]

Bibliography: R. Benumof, *Concepts in Physics*, 1965; G. P. Harnwell and G. J. F. Legge, *Physics: Matter, Energy and the Universe*, 1967; E. M. Rogers, *Physics for the Inquiring Mind*, 1960.

Work function (thermodynamics)

A thermodynamic function, also called the work content, Helmholtz free energy, or by the European school, simply the free energy. It is defined as the internal energy E of a system minus the temperature-entropy product, TS, and has a characteristic value for each state of a system. In an isothermal process, the maximum work which can be done by a system is equal to the decrease in its work function. When only work due to expansion against a fluid pressure is possible, as in ordinary chemical reactions, a spontaneous process at constant temperature and volume is characterized by a decrease in the work function, whereas the corresponding criterion for equilibrium is that the work function for the system should be at a minimum. *See* FREE ENERGY.

[PAUL BENDER]

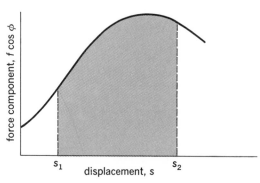

Fig. 4. A work diagram.

McGRAW-HILL ENCYCLOPEDIA OF ENERGY

List of Contributors

List of Contributors

A

Adkins, George G. *Chief, Department of River Basins, Bureau of Power, Federal Power Commission.* TIDAL POWER.

Alger, Robert P. *New Services, Schlumberger Well Services, Houston.* WELL LOGGING (OIL AND GAS).

Altman, Dr. David. *Division Vice President and Technical Director, United Technology Center, Sunnyvale, CA.* METAL-BASE FUEL.

Anderson, J. *Manager, Meter and Instrument Business Department, General Electric Company, Somersworth, NH.* ELECTRIC ENERGY MEASUREMENT; WATT-HOUR METER—in part.

Antonides, Lloyd E. *Development Engineer, Fenix and Scisson, Inc., Tulsa.* MINING MACHINERY.

Apker, Dr. L. *General Electric Research Laboratory, Schenectady.* PHOTOVOLTAIC EFFECT.

Archie, G. E. *Manager (retired), Petroleum Engineering Research, Shell Development Company, Houston.* PETROLEUM ENGINEERING.

Aron, Dr. Walter. *Science Application, Inc., Palo Alto.* ELECTRICITY.

B

Baily, Frederick G. *Large Steam Turbine–Generator Department, General Electric Company, Schenectady.* STEAM TURBINE; TURBINE.

Bass, Prof. Jack. *Department of Physics, Michigan State University.* THERMOELECTRICITY.

Batchelder, Howard R. *Consulting Chemical Engineer, Battelle Memorial Institute, Columbus, Ohio.* FUEL, FOSSIL.

Baumeister, Theodore. *Consulting Editor, and Stevens Professor Emeritus of Mechanical Engineering, Columbia University.* BRAYTON CYCLE; CARNOT CYCLE; DIESEL CYCLE; DIESEL ENGINE; ENERGY CONVERSION; HEAT PUMP; HYDRAULIC TURBINE; IMPULSE TURBINE; OTTO CYCLE; POWER PLANT—in part; RANKINE CYCLE; REACTION TURBINE; REFRIGERATION CYCLE; STEAM ENGINE; THERMODYNAMIC CYCLE; WATER POWER—in part.

Bean, Howard S. *Consultant on Fluid Metering, Liquids and Gases, Sedona, AZ.* THERMOCOUPLE.

Beaty, H. Wayne. *"Electrical World," McGraw-Hill Publishing Company, New York.* CONDUCTOR, ELECTRIC.

Bellum, Donald P. *Pit Superintendent, Nevada Mines Division, Kennescott Copper Corporation, McGill.* MINING, OPEN-PIT.

Bender, Prof. Paul J. *Professor of Physical Chemistry, University of Wisconsin.* FREE ENERGY; INTERNAL ENERGY; WORK FUNCTION (THERMODYNAMICS).

Bernhardt, Carl P. *Executive Assistant (retired), Indusrial Electronics Department, Westinghouse Electric Corporation, Baltimore.* INDUCTION HEATING.

Bewley, Dr. Loyal V. *Dean (retired), College of Engineering, Lehigh University.* HYSTERESIS MOTOR; RELUCTANCE MOTOR; SYNCHRONOUS MOTOR.

Binning, Dr. Robert C. *Associate Director, Dayton Laboratory, Monsanto Research Corporation.* HEATING VALUE—in part.

Blair, Lewis N. *Anaconda Company, Butte.* ENVIRONMENTAL ENGINEERING—in part.

Bloomfield, Prof. Philip E. *Department of Physics, University of Pennsylvania, and City College, City University of New York.* THERMODYNAMIC PROCESSES.

Bolt, Prof. J. A. *Department of Mechanical Engineering, University of Michigan.* GAS TURBINE—in part.

Boyle, Dr. R. W. *Geological Survey of Canada, Department of Energy, Mines and Resources, Ottawa.* GEOCHEMICAL PROSPECTING.

Brewer, Dr. Charles P. *Shell Development Corporation, Emeryville, CA.* HYDROCRACKING.

Brigham, Harry S. *Executive Vice President, Dixilyn Corporation, New Orleans.* OIL AND GAS WELL COMPLETION.

Bromley, Dr. D. Allan. *Professor and Chairman, Department of Physics, and Director, Wright Nuclear Structure Laboratory, Yale University.* NUCLEAR PHYSICS.

Brzustowski, Dr. J. A. *Department of Mechanical Engineering, University of Waterloo, Ontario.* COMBUSTION OF LIGHT METALS.

Buechner, Prof. William W. *Department of Physics, Massachusetts Institute of Technology.* NUCLEAR RADIATION; NUCLEAR REACTION.

Burnett, Peter G. *Petroleum Engineer, Chicago.* OIL AND GAS STORAGE.

C

Cady, Dr. Gilbert H. *Consulting Coal Geologist, Urbana, IL.* COAL; LIGNITE; PEAT.

Cain, Dr. Stanley A. *Director, Institute for Environmental Quality, and Charles Lathrop Pack Professor, Department of Resource Planning and Conservation, University of Michigan.* CONSERVATION OF RESOURCES.

Cambel, Dr. Ali B. *Executive Vice President for Academic Affairs, Wayne State University.* GAS DYNAMICS—in part.

Campbell, Harold E. *Senior Engineer, Power Distribution Systems Engineering, General Electric Company, Schenectady.* ELECTRIC DISTRIBUTION SYSTEMS.

Chapin, Dr. Daryl M. *Bell Telephone Laboratories, Murray Hill, NJ.* SOLAR BATTERY—in part.

Cheatham, Dr. J. B., Jr. *Professor of Mechanical Engineering and Chairman of Department of Mechanical and Aerospace Engineering and Materials Science, Rice University.* OIL AND GAS WELL DRILLING.

Chenault, Roy L. *Chief Research Engineer (retired), Oilwell Division, U.S. Steel Corporation.* OIL AND GAS FIELD EXPLOITATION.

Childs, Dr. Orlo E. *Texas Technical University.* FUEL, LIQUID.

Chinitz, Dr. Wallace. *Cooper Union for the Advancement of Science and Art, School of Engineering, New York.* CHEMICAL FUEL.

Cleveland, Prof. Laurence Fuller. *Department of Electrical Engineering, Northeastern University.* DIRECT-CURRENT MOTOR.

Cochran, Robert A. *Shell Development Company, Bellaire Research Center, Houston.* WATER POLLUTION—in part.

Conrad, Prof. Albert G. *Dean Emeritus and Professor of Electrical Engineering, College of Engineering, University of California, Santa Barbara.* INDUCTION MOTOR.

Cooper, Dr. Benjamin S. *Committee on Interior and Insular Affairs, U.S. Senate.* U.S. POLICIES AND POLITICS—feature.

Cooper, Franklin D. *Bureau of Mines, U.S. Department of the Interior.* COAL LIQUEFACTION.

Corlett, A. V. *Mining Engineer, Kingston, Ontario.* MINING, UNDERGROUND; MINING SAFETY.

Coroniti, S. C. *Climatic Impact Assessment Program, Office of the Secretary of Transportation, Washington, DC.* ATMOSPHERIC POLLUTION—in part.

Crittenden, Dr. Charles V. *Geographer, Economic Development Administration, U.S. Department of Commerce.* GAS FIELD AND GAS WELL.

Crowson, Brig. Gen. Delmar L. *Formerly, Director, Office of Safeguards and Materials Management, U.S. Atomic Energy Commission.* NUCLEAR MATERIALS SAFEGUARDS.

D

Dahlgren, E. G. Ty. *Oil and Gas Consultant, Oklahoma City.* PETROLEUM SECONDARY RECOVERY.

Davis, Dr. J. *Vice President, Research and Development, Bright Star Industries, Inc., Clifton, NJ.* DRY CELL—in part; FUEL CELL—in part; MERCURY BATTERY; NUCLEAR BATTERY—in part; PRIMARY BATTERY; RESERVE BATTERY; WET CELL.

Dean, H. Clark. *Harza Engineering Company, Consulting Engineers, Chicago.* HYDROELECTRIC POWER—in part.

Dowd, James J. *Mining and Preparation Section, Bureau of Mines, U.S. Department of the Interior.* MINING, STRIP.

Duckworth, Dr. Henry E. *Department of Physics, University of Manitoba.* NUCLEUS, ATOMIC.

Dukek, W. G. *Exxon Research and Engineering Company, Linden, NJ.* JET FUEL; PETROLEUM PRODUCTS.

E

Eckels, Dr. Arthur R. *Department of Electrical Engineering, North Carolina State University.* DYNAMIC BRAKING; DYNAMO; DYNAMOMETER; MOTOR-GENERATOR SET.

Ehricke, Dr. Krafft A. *North American Space Operations, Rockwell International, El Segundo, CA.* ELECTROMAGNETIC PROPULSION; MAGNETOGAS DYNAMICS; SOLAR ENERGY.

Evans, Dr. John W. *Director, Sacramento Peak Observatory, Air Force Cambridge Research Laboratories, Sunspot, NM.* SOLAR RADIATION.

Everetts, Dr. John, Jr. *Professor of Architectural Engineering, Pennsylvania State University.* DEHUMIDIFIER.

Ewing, John I. *Lamont-Doherty Geological Observatory, Columbia University, Palisades, NY.* MARINE GEOLOGY—in part.

Ewing, Dr. Maurice. *Lamont-Doherty Geological Observatory, Palisades, NY.* MARINE GEOLOGY—in part.

F

Ferenz, Joseph A. *Vice President, Frederic R. Harris, Inc., Consulting Engineers, Great Neck, NY.* SUPERPORT.

Field, Joseph H. *Benfield Corporation, Pittsburgh.* BERGIUS PROCESS; FISCHER-TROPSCH PROCESS.

Fisher, Dr. John C. *Electric Power Research Institute, Palo Alto.* ENERGY CONSUMPTION—feature; ENERGY FLOW.

Forsyth, R. B. *Market Development Manager, Union Carbide Corporation, New York.* GRAPHITE—in part.

Foster, Prof. Mark G. *Department of Electrical Engineering, University of Virginia.* MAGNETIC CIRCUITS.

Francis, Arthur W. *Linde Division, Union Carbide Corporation, New York.* LIQUEFIED NATURAL GAS (LNG); OXYGEN.

Fremed, Raymond F. *Burson-Marsteller Associates, New York.* HEAT EXCHANGER.

Fuhrman, Dr. Ralph E. *Assistant Director, National Water Commission, Arlington.* WATER POLLUTION—in part.

G

Gambs, Gerard C. *Vice President, Ford, Bacon, and Davis, Inc., New York.* ENERGY SOURCES.

Gaylord, W. M., Jr. *Union Carbide Corporation, New York.* GRAPHITE—in part.

Gibson, Dr. John E. *Dean of Engineering, Oakland University.* ELECTROMAGNET.

Glasscock, Dwight L. *Charles T. Main, Inc., Southeast Tower, Prudential Center, Boston.* PUMPED STORAGE.

Goodheart, Prof. Clarence Francis. *Chairman, Department of Electrical Engineering, Union College, Schenectady.* CIRCUIT (ELECTRICITY).

Gordon, Dr. William E. *Associate Professor of Physical Chemistry, Pennsylvania State University, and Consultant.* EXPLOSION AND EXPLOSIVE; FLAME.

Gray, G. Ronald. *Director, Syncrude Canada Ltd., Edmonton, Alberta.* OIL MINING; OIL SAND.

Greensfelder, Dr. Bernard D. *Deceased; formerly, Director of Oil Research, Shell Development Company, Emerville, CA.* CRACKING—in part.

Greenspan, Martin. *National Bureau of Standards.* DAMPING.

Gregory, Dr. G. R. *Department of Forestry, School of Natural Resources, University of Michigan.* FOREST RESOURCES.

Grey, Jerry. *President, Greyad Corporation, Princeton.* PROPULSION.

Grobecker, Dr. Alan J. *Project Manager, Climatic Impact Assessment Program, Office of the Secretary of Transportation, Washington, DC.* ATMOSPHERIC POLLUTION—in part.

H

Haensel, Dr. Vladimir. *Vice President and Director of Research, Universal Oil Products Company, Des Plaines, IL.* REFORMING IN PETROLEUM REFINING—in part.

Halbouty, Dr. Michel T. *Consulting Geologist and Petroleum Engineer, Houston.* NATURAL GAS—in part.

Halkias, Dr. Christos C. *Associate Professor of Electrical Engineering, Columbia University.* IMPEDANCE MATCHING.

Harbaugh, Dr. John W. *Department of Geology, Stanford University.* PETROLEUM GEOLOGY.

Harza, Richard D. *Vice President, Harza Engineering Company, Consulting Engineers, Chicago.* HYDROELECTRIC POWER—in part.

Hazen, Ronald McKean. *Consultant, Indianapolis.* AIRCRAFT ENGINE; AIRCRAFT ENGINE, RECIPROCATING.

Heronemus, Prof. William. *Department of Engineering, University of Massachusetts.* WIND POWER.

Hewson, Dr. E. W. *Chairman, Department of Atmospheric Sciences, Oregon State University.* ATMOSPHERIC POLLUTION—in part.

Hibbard, Robert R. *Technical Consultant for Propulsion Chemistry Division, Lewis Research Center, National Aeronautics and Space Administration, Cleveland.* AIRCRAFT FUEL.

Hill, James E. *Assistant Director (retired), Mining Research, U.S. Bureau of Mines.* MINING EXCAVATION.

Hindman, Dr. James C. *Argonne National Laboratory.* PLUTONIUM.

Hockett, Richard S. *Research Physicist, Dayton Labora-*

tory, Monsanto Research Corporation. HEATING VALUE—in part.

Hodges, Dr. Laurent. *Physics Department, Iowa State University.* ENERGY STORAGE; ENERGY TRANSMISSION.

Hoekstra, Dr. Henry R. *Chemistry Division, Argonne National Laboratory.* URANIUM.

Hoffman, Harold L. *Refining Editor, "Hydrocarbon Processing Magazine."* PETROLEUM PROCESSING—in part.

Holm, Prof. Jens T. *Webb Institute of Naval Architecture, Glen Cove, NY.* MARINE ENGINE.

Hubbert, Dr. M. King. *Research Geophysicist, U.S. Department of the Interior Geological Survey, Reston, VA.* OUTLOOK FOR FUEL RESERVES—feature.

Huebler, Dr. Jack. *Senior Vice President, Institute of Gas Technology, IIT Center, Chicago.* FUEL GAS.

Huizenga, Dr. John R. *Nuclear Structure Research Laboratory, University of Rochester.* FISSION, NUCLEAR.

Hultsch, Dr. Roland A. *Department of Physics, University of Missouri.* TEMPERATURE.

Hunt, R. A. *Plant and Substation Engineering Department, Cleveland Electric Illuminating Company.* ELECTRIC POWER SUBSTATION.

Hynes, Lee Powers. *Consulting Engineer, Electrical and Mechanical Engineering, Haddonfield, NJ.* HEATING, ELECTRIC.

I

Ingram, William T. *Consulting Engineer, Whitestone, NY.* AIR-POLLUTION CONTROL.

Isbin, Prof. Herbert S. *Department of Chemical Engineering, University of Minnesota.* NUCLEAR POWER; REACTOR, NUCLEAR.

J

Jackson, Dr. William D. *U.S. Energy Research and Development Administration.* MAGNETOHYDRODYNAMIC POWER GENERATOR.

Jaep, William F. *Central Research Department, Experimental Station, E. I. du Pont de Nemours and Company, Wilmington.* ENTROPY—in part; HEAT BALANCE; THERMODYNAMIC PRINCIPLES.

James, John W. *Vice President, Research, McDonnell and Miller, Inc., Chicago.* STEAM HEATING.

Jaske, Dr. Robert T. *Battelle Pacific Northwest Laboratories, Richland, WA.* WASTE HEAT MANAGEMENT.

Jensen, Dr. Howard B. *Research Supervisor, Laramie Energy Research Center, ERDA.* OIL SHALE—in part.

Johnson, Prof. Walter C. *Department of Electrical Engineering, Princeton University.* ELECTROMAGNETIC WAVES, TRANSMISSION OF.

Johnston, Dr. Francis J. *Department of Chemistry, University of Georgia.* FIRE.

Jones, Prof. Lawrence W. *Harrison M. Randall Laboratory of Physics, University of Michigan.* HYDROGEN-FUELED TECHNOLOGY.

Judd, Prof. William B. *Professor of Rock Mechanics, School of Civil Engineering, Purdue University, and formerly, Chairman, NAS-NAE U.S. National Committee on Rock Mechanics.* ENGINEERING GEOLOGY.

Just, Prof. Evan. *Department of Mining and Geology, Stanford University.* MINING.

K

Kaplan, Dr. Louis. *Senior Chemist, Argonne National Laboratory.* HYDROGEN.

Kaplan, Robert. *Vice President in Charge of Engineering, Automatic Burner Corporation, Chicago.* OIL BURNER.

Kappus, Peter G. *Flight Propulsion Laboratory, General Electric Company, Cincinnati.* TURBOPROP.

Kerr, Dr. William. *Chairman, Department of Nuclear Engineering, University of Michigan.* NUCLEAR ENGINEERING.

Kessler, George W. *Vice President, Engineering and Technology, Power Generation Division, Babcock and Wilcox Company, Barberton, OH.* ECONOMIZER, BOILER.

Kimbark, Dr. Edward W. *Head, Systems Analysis Group, Bonneville Power Administration.* TRANSMISSION LINES—in part.

Konzo, Prof. Seichi. *Department of Mechanical Engineering, University of Illinois.* WARM-AIR HEATING SYSTEM.

Koral, Richard L. *"Building Systems Design," New York.* AIR CONDITIONING; AIR COOLING; PANEL HEATING AND COOLING—in part; RADIANT HEATING—in part.

Kosow, Dr. Irving L. *Staten Island Community College.* ALTERNATING-CURRENT MOTOR—in part; REPULSION MOTOR.

Kuo, C. H. *Department of Chemical Engineering, Mississippi State University.* NATURAL GAS AND SULFUR PRODUCTION.

L

Landshoff, Dr. Rolf K. *Consulting Scientist, Lockheed Missiles and Space Company, Palo Alto.* MAGNETOHYDRODYNAMICS.

Lane, James A. *Oak Ridge National Laboratory.* ATOMIC ENERGY; NUCLEAR FUELS.

Lane, Dr. John C. *Ethly Corporation, Ferndale, MI.* OCTANE NUMBER.

Lewis, Dr. Bernard. *Combustion and Explosives Research, Inc., Pittsburgh.* COMBUSTION.

Linsley, Prof. Ray K. *Professor of Civil Engineering, Stanford University.* THERMAL SPRING—in part.

Lord, Prof. Richard C. *Spectroscopy Laboratory, Massachusetts Institute of Technology.* INFRARED RADIATION.

Luckenbach, Edward C. *Exxon Research and Engineering Company, Exxon Engineering-Petroleum Department, Florham Park, NJ.* CRACKING—in part.

Luebbers, Dr. Ralph H. *Professor of Chemical Engineering, University of Missouri.* HEAT TRANSFER.

M

McClellan, Leslie N. *Consulting Editor, Engineering Consultants, Inc., Denver.* PIPELINE.

McCormack, Congressman Michael M. *U.S. House of Representatives.* PROTECTING THE ENVIRONMENT—feature.

MacCoull, Neil. *Lecturer in Mechanical Engineering, Columbia University.* INTERNAL COMBUSTION ENGINE.

McGrain, Preston. *Assistant State Geologist, Kentucky Geological Survey, University of Kentucky.* OIL FIELD WATERS.

McKetta, Dr. John J. *Professor of Chemical Engineering, University of Texas.* PETROLEUM PROCESSING—in part.

McNish, Alvin G. *Chief, Meteorology Division, National Bureau of Standards, Chevy Chase, MD.* ELECTRICAL UNITS AND STANDARDS.

Manning, Dr. Kenneth V. *Professor Emeritus, Pennsylvania State University.* ELECTROMAGNETISM; INDUCTION, ELECTROMAGNETIC; INDUCTION, MAGNETIC; MAGNETIC FIELD; MAGNETOMOTIVE FORCE.

Markus, John. *Consultant, Sunnyvale, CA.* EDISON BATTERY.

Morgan, Dr. Karl Z. *Director, Health Physics Division, Oak Ridge National Laboratory.* DECONTAMINATION OF RADIOACTIVE MATERIALS.

Moser, J. F., Jr. *Formerly, Esso Research Laboratories, Humble Oil Refining Company.* COKING IN PETROLEUM REFINING.

Muskat, Dr. Morris. *Gulf Oil Corporation, Coral Gables.* PETROLEUM RESERVOIR ENGINEERING—in part.

N

Nedelsky, Prof. Leo. *Professor of Physical Science, University of Chicago.* CONSERVATION OF ENERGY—in part; ENERGY—in part; WORK.

Nottingham, Prof. W. B. *Deceased; formerly, Professor Emeritus of Physics, Massachusetts Institute of Technology.* THERMIONIC POWER GENERATOR—in part.

O

Olmsted, Leonard M. *Technical Consultant, South Orange, NJ.* ELECTRIC POWER SYSTEMS; ELECTRICAL UTILITY INDUSTRY; POWER PLANT—in part.

Ott, Paul A., Jr. *Automotive Engineering Consultant, Mount Clemens, MI.* AUTOMOTIVE ENGINE.

P

Pake, Dr. George E. *Vice President, Xerox Corporation, and General Manager, Xerox Palo Alto Research Center.* FORCE.

Park, Prof. Charles F., Jr. *Department of Earth and Planetary Sciences, Massachusetts Institute of Technology.* MINERAL RESOURCES CONSERVATION.

Pearson, Dr. Gerald L. *Stanford Electronics Laboratories.* SOLAR BATTERY—in part.

Perry, Stephen F. *Exxon Research, Florham Park, NJ.* DEWAXING OF PETROLEUM.

Phelan, Prof. Richard M. *Department of Mechanical Systems and Design, Cornell University.* EFFICIENCY; MACHINE.

Phinney, John A. *Consolidation Coal Company, Library, PA.* FUEL, SYNTHETIC.

Pinkel, Benjamin. *Engineering Sciences Department, Rand Corporation, Santa Monica.* PULSE JET; RAMJET; TURBOJET.

Platt, Allison M. *Manager, Nuclear Waste Technology Department, Battelle Pacific Northwest Laboratories, Richland, WA.* RADIOACTIVE WASTE DISPOSAL.

Pollitzer, Dr. Ernest L. *Associate Director of Research, Corporate Research Center, Universal Oil Products Company, Des Plaines, IL.* REFORMING IN PETROLEUM REFINING—in part.

Ponikvar, Ade L. *Formerly, "Modern Plastics," McGraw-Hill Publications Company, New York.* CABLE-TOOL DRILL; ROTARY TOOL DRILL.

Post, Dr. Richard F. *Lawrence Radiation Laboratory, University of California, Livermore.* FUSION, NUCLEAR; LAWSON CRITERION; THERMONUCLEAR REACTION.

Priester, G. B. *Principal Engineer, Electric Engineering Department, Baltimore Gas and Electric Company.* HEATING, COMFORT.

Pryke, John K. M. *Slocum and Fuller, New York.* CENTRAL HEATING.

Puchstein, Albert Frederick. *Consulting Engineer, Columbus, OH.* ALTERNATING-CURRENT MOTOR—in part; MOTOR, ELECTRIC.

Putz, Dr. T. J. *Westinghouse Electric Corporation, Philadelphia.* GAS TURBINE—in part.

R

Rachford, Prof. Henry H. *Department of Mathematics, Rice University.* PETROLEUM RESERVOIR MODELS.

Ramberg, Dr. Edward G. *Radio Corporation of America Laboratories, Princeton.* SPACE CHARGE.

Ramey, Dr. Robert L. *Professor of Electrical Engineering, University of Virginia.* ELECTRICAL MEASUREMENTS—in part.

Rees, Jack. *Corporate Services Staff, Exxon Research and Engineering Company, Florham Park, NJ.* OIL ANALYSIS.

Reilly, Dr. James D. *Vice President, Consolidation Coal Company, Pittsburgh.* COAL MINING.

Remde, Dr. Harry. *Chief, Basic Physics Research Section, Johns-Manville Research and Engineering Center, Manville, NJ.* INSULATION, HEAT.

Ries, Dr. Harold C. *Stanford Research Institute.* ALKYLATION, PETROLEUM; KEROSINE.

Riggs, Harold C. *Manager (retired), Marketing New Product Development, Electric Storage Battery Company, Philadelphia.* BATTERY, ELECTRIC.

Ritchey, Dr. H. W. *President, Thiokol Chemical Corporation, Ogden, UT.* PROPELLANT.

Robbins, Nathaniel, Jr. *Director of Engineering, Residential Division, Honeywell Inc., Minneapolis.* THERMOSTAT.

Robertson, Prof. Burtis Lowell. *Professor of Electrical Engineering (retired), University of California, Berkeley.* Q (ELECTRICITY).

Robillard, Dr. Jean J. *Vice President, Director of Research, General Transistor Corporation.* PHOTOVOLTAIC CELL.

Rockett, Frank H. *Engineering Consultant, Charlottesville, VA.* ENGINE; ENTROPY—in part; OIL FURNACE.

Roller, Dr. Duane E. *Deceased; formerly, Harvey Mudd College.* CONSERVATION OF ENERGY—in part; ENERGY—in part.

Rosenberg, Leon T. *Senior Consultant, Generator Design, Power Generation, Installation and Service Department, Allis-Chalmers Manufacturing Company, Milwaukee.* ALTERNATING-CURRENT GENERATOR; ELECTRIC ROTATING MACHINERY; GENERATOR, ELECTRIC; HYDROELECTRIC GENERATOR.

Rozeanu, Prof. L. *Department of Material Science, Technion, Israel Institute of Technology.* FUEL CELL; NUCLEAR BATTERY; WET CELL—all in part.

S

Sayre, Dr. Albert N. *Deceased; formerly, Consulting Groundwater Geologist, Behre Dolbear and Company.* THERMAL SPRING—in part.

Scarlott, Charles A. *Manager of Publications Department (retired), Stanford Research Institute.* WATER-POWER—in part.

Schaefer, Dr. M. B. *Institute of Marine Resources, University of California, La Jolla.* MARINE RESOURCES.

Schmerling, Dr. Louis. *Research Associate, Universal Oil Products Company, Des Plaines, IL.* BUTANE; METHANE; PETROLEUM.

Schmidt, Dr. Paul W. *Department of Physics, University of Missouri.* HORSEPOWER; POWER.

Schoonmaker, G. R. *Vice President, Production-Exploration, Marathon Oil Company, Findlay, OH.* OIL AND GAS, OFFSHORE.

Schora, Frank C. *Vice President, Process Research Institute of Gas Technology, ITT Center, Chicago.* SUBSTITUTE NATURAL GAS (SNG).

Schwieger, Robert G. *Associate Editor, "Power," McGraw-Hill Publishing Company, New York.* COMBUSTION TURBINE.

Sebald, Joseph F. *Consulting Engineer, and President, Heat Power Products Corporation, Bloomfield, NJ.* COOLING TOWER.

Sell, Dr. Heinz G. *Metals Development Section, Westinghouse Lamp Divisions, Bloomfield, NJ.* BLACKBODY; EMISSIVITY; GRAYBODY; HEAT RADIATION—all in part.

Shannon, Dr. Hugh F. *Products Research Division, Exxon Research and Engineering Company, Linden, NJ.* GASOLINE.

Sheriff, Robert E. *Seiscom Delta, Inc., Houston.* GEOPHYSICAL EXPLORATION.

Siegmund, C. W. *Exxon Research and Engineering Company, Linden, NJ*. FUEL OIL.

Silfvast, Dr. William T. *Holmdel Labs, Holmdel, NJ*. LASER-INDUCED FUSION.

Sinclair, Dr. Donald B. *President (retired), General Radio Company, Concord, MA*. ELECTRIC POWER MEASUREMENT; ELECTRICAL MEASUREMENTS—in part; WATTMETER.

Skilling, Prof. Hugh Hildreth. *Department of Electrical Engineering, Stanford University*. ALTERNATING CURRENT.

Smith, Dr. John Ward. *Research Supervisor, Laramie Energy Research Center, ERDA*. OIL SHALE—in part.

Smith, W. W. *Technical Adviser, Legal Department, ESB Inc., Philadelphia*. STORAGE BATTERY.

Smythe, Dr. William R. *Department of Physics, California Institute of Technology*. ELECTROMAGNETIC RADIATION; ELECTROMAGNETIC WAVE.

Souders, Dr. Mott. *Formerly, Director, Oil Development, Shell Oil Company, Emeryville, CA*. CRACKING—in part; DISTILLATE FUEL; LIQUEFIED PETROLEUM GAS (LPG).

Spindler, John C. *Anaconda Company, Butte*. ENVIRONMENTAL ENGINEERING—in part.

Spinrad, Dr. Bernard I. *Senior Physicist, Applied Physics Division, Argonne National Laboratory*. NUCLEONICS.

Starr, Dr. Eugene C. *U.S. Department of the Interior, Bonneville Power Administration, Portland*. ELECTRIC POWER GENERATION.

Steindler, Martin J. *Chemical Engineering Division, Argonne National Laboratory*. NUCLEAR FUELS REPROCESSING.

Stewart, Dr. John W. *Department of Physics, University of Virginia*. CONDUCTION (ELECTRICITY); CURRENT (ELECTRICITY); DIRECT CURRENT; ELECTRODYNAMICS.

Storch, Dr. Henry H. *Deceased; formerly, Assistant Professor of Chemistry, New York University*. DESTRUCTIVE DISTILLATION—in part.

Stratton, Dr. Thomas F. *Los Alamos Scientific Laboratory*. THERMIONIC POWER GENERATOR—in part.

Sturges, Frank C. *President, Pennsylvania Drilling Company, Pittsburgh*. BORING AND DRILLING (MINERAL); CORE DRILLING.

Sturtevant, George R. *Manager (retired), Engineering Meter Department, General Electric Company*. WATT-HOUR METER—in part.

Sutherland, J. R. *Power Transformer Department, General Electric Company, Pittsfield, MA*. TRANSFORMER.

T

Teasley, Robert E., Jr. *Cummins Engine Company, Columbus, IN*. DIESEL FUEL.

Thompson, Jack R. *U.S. Army Corps of Engineers, Office of the Secretary of the Army, Department of Army*. DAM.

Thomson, Dr. Robb M. *Office of Programs, U.S. Department of Commerce, National Bureau of Standards*. EXPLORING ENERGY CHOICES—feature.

Tuck, Dr. James L. *Associate Division Leader, Los Alamos Scientific Laboratory*. PINCH EFFECT.

Turner, Prof. Clesson N. *Professor of Agricultural Engineering, Project Leader of N.Y. Farm Electrification Council, Cornell University*. RURAL ELECTRIFICATION.

U

Unger, Walter H. *Assistant Director, Anaconda Company, Butte*. ENVIRONMENTAL ENGINEERING—in part.

W

Waddington, Prof. Thomas C. *Department of Chemistry, University of Durham*. CHEMICAL ENERGY.

Wainwright, Howard W. *Coal Research Center, U.S. Bureau of Mines, Morgantown, WV*. DESTRUCTIVE DISTILLATION—in part.

Walsh, Dr. Peter J. *Department of Physics, Fairleigh Dickinson University*. BLACKBODY; EMISSIVITY; GRAYBODY; HEAT RADIATION—all in part.

Watson, Prof. William W. *Professor Emeritus of Physics, Yale University*. MASS DEFECT.

Weast, Dr. Robert C. *Vice President, Research, Consolidated Natural Gas Service Company, Cleveland*. COAL GASIFICATION.

Weaver, Dr. Paul. *(Retired) Texas A&M College*. OIL FIELD MODEL.

Weber, Erwin L. *Deceased; formerly, Trust Department, National Bank of Commerce, Seattle*. HOT-WATER HEATING SYSTEM; PANEL HEATING AND COOLING—in part; RADIANT HEATING—in part.

Weber, Prof. Harold C. *Department of Chemical Engineering, Massachusetts Institute of Technology*. BRITISH THERMAL UNIT (BTU); ENTHALPY; HEAT; HEAT CAPACITY.

Weil, Robert T., Jr. *Dean, School of Engineering, Manhattan College*. DIRECT-CURRENT GENERATOR.

West, Dr. William. *Eastman Kodak Company, Rochester*. ABSORPTION OF ELECTROMAGNETIC RADIATION.

Wheeler, Prof. John A. *Department of Physics, Joseph Henry Laboratories, Princeton University*. CHAIN REACTION, NUCLEAR; CRITICAL MASS.

White, Prof. David C. *Department of Electrical Engineering, Massachusetts Institute of Technology*. THERMOELECTRIC POWER GENERATOR.

Whitelaw, Dr. Robert L. *Mechanical Engineering Department, Virginia Polytechnic Institute*. GEOTHERMAL POWER.

Wilhelm, Dr. Harley A. *Associate Director, Institute for Atomic Research and Ames Laboratory*. URANIUM METALLURGY.

Wilhelm, John K. *Assistant Administrator, U.S. Department of State Agency for International Development*. WORLD ENERGY ECONOMY—feature.

Wilkinson, Prof. Denys H. *Department of Nuclear Physics, Oxford University*. BINDING ENERGY, NUCLEAR.

Williams, Prof. Everard M. *Department of Electrical Engineering, Carnegie-Mellon University*. TRANSMISSION LINES—in part.

Winch, Prof. Ralph P. *Department of Physics, Williams College*. CHARGE, ELECTRIC.

Wylie, Prof. E. Benjamin. *Department of Civil Engineering, University of Michigan*. NATURAL GAS—in part.

Y

Yellott, Dr. John I. *Director, Yellott Solar Energy Laboratory, Phoenix*. SOLAR COLLECTORS; SOLAR COOLING; SOLAR HEATING.

McGRAW-HILL ENCYCLOPEDIA OF ENERGY

Appendix

Appendix

The Appendix discusses three measurement systems — U.S. Customary, metric, and International systems — and provides conversion tables. It also describes usage of the Fahrenheit, Celsius, and Kelvin temperature scales, and lists the chemical elements with their symbols and atomic numbers.

U.S. Customary System and the metric system

Scientists and engineers have been using two major systems of units in measurement. These are commonly called the U.S. Customary System (inherited from the British Imperial System) and the metric system.

In the U.S. Customary System the units yard and pound with their divisions, such as the inch, and multiples, such as the ton, are basic. The metric system was evolved during the 18th century and has been adopted for general use by most countries. Nearly everywhere it is used for precise measurements in science. The meter and kilogram with their multiples, such as the kilometer, and fractions, such as the gram, are basic to the metric system.

In the U.S. Customary System, units of the same kind are related almost at random. For example, there are the units of length, the inch, yard, and mile. In the metric system the relationships between units of the same kind are strictly decimal (millimeter, meter, and kilometer).

However, to complicate matters in scientific writing, there is no uniformity within each of these two systems as to the choice of units for the same quantities. For example, the hour or the second, the foot or the inch, and the centimeter or the millimeter could be chosen by a scientist as the unit of measurement for the quantities time and length.

Introduction of the International System, or SI

To simplify matters and to make communication more understandable, an internationally accepted system of units is coming into use. This is termed the International System of Units, which is abbreviated SI in all languages.

Fundamentally the system is metric with the base units derived from scientific formulas or natural constants. For example, the meter in the SI is defined as the length equal to 1 650 763.73 wavelengths in vacuum of the radiation corresponding to the transition between the electronic energy levels $2p_{10}$ and $5d_5$ of the krypton-86 atom. Previously, in the metric system, the meter was

Introduction of the International System, or SI (cont.)

defined as the distance between two marks on a specific metal bar.

In a similar way the second in the SI is defined as the duration of 9 192 631 770 periods of the radiation corresponding to the transition between two hyperfine levels of the ground state of the cesium-133 atom.

Interestingly, the kilogram, the SI unit of mass, is still the mass of the kilogram kept at Sèvres, France. However, it is possible that eventually the unit will be redefined in terms of atomic mass.

Although the SI is increasing in usage by scientists and engineers, there are some units in everyday use which will probably remain, for example, minute, hour, day, degree (angle), and liter. The point should be made, however, that these terms will not be employed in a scientific context if the SI is fully adopted.

Because of their extremely common use among scientists, several units are still permitted in conjunction with SI units, for example, the electron volt, gauss, barn, and curie. In time their usage might be phased out.

One further point is that in October, 1967, the Thirteenth General Conference of Weights and Measures decided to name the SI unit of thermodynamic temperature "kelvin" (symbol K) instead of "degree Kelvin" (symbol °K). For example, the notation is 273 K and not 273°K.

The base units and derived units of the SI are shown in **Table 1** and **Table 2.**

In the SI the prefixes differ from a unit in steps of 10^3. A list of prefix terms, symbols, and their factors is given in **Table 3**. Some examples of the use of these prefixes follow:

$$1000 \text{ m} = 1 \text{ kilometer} \quad = 1 \text{ km}$$

$$1000 \text{ V} = 1 \text{ kilovolt} \quad = 1 \text{ kV}$$

$$1\,000\,000 \ \Omega = 1 \text{ megohm} \quad = 1 \text{ M}\Omega$$

$$0.000\,000\,001 \text{ s} = 1 \text{ nanosecond} = 1 \text{ ns}$$

Only one prefix is to be employed for a unit. For example:

$$1000 \text{ kg} = 1 \text{ Mg} \quad \text{not } 1 \text{ kkg}$$

$$10^{-9} \text{ s} = 1 \text{ ns} \quad \text{not } 1 \text{ m}\mu\text{s}$$

$$1\,000\,000 \text{ m} = 1 \text{ Mm} \quad \text{not } 1 \text{ kkm}$$

Also, when a unit is raised to a power, the power applies to the whole unit including the prefix. For example:

$$\text{km}^2 = (\text{km})^2 = (1000 \text{ m})^2 = 10^6 \text{ m}^2 \quad \text{not } 1000 \text{ m}^2$$

Table 1. Base units of the International System

Quantity	Name of unit	Unit symbol
length	meter	m
mass	kilogram	kg
time	second	s
electric current	ampere	A
temperature	kelvin	K
luminous intensity	candela	cd
amount of substance	mole	mol

Table 2. Derived units of the International System

Quantity	Name of unit	Unit symbol or abbreviation, where differing from basic form	Unit expressed in terms of basic or supplementary units*
area	square meter		m^2
volume	cubic meter		m^3
frequency	hertz	Hz	s^{-1}
density	kilogram per cubic meter		kg/m^3
velocity	meter per second		m/s
angular velocity	radian per second		rad/s
acceleration	meter per second squared		m/s^2
angular acceleration	radian per second squared		rad/s^2
volumetric flow rate	cubic meter per second		m^3/s
force	newton	N	$kg \cdot m/s^2$
surface tension	newton per meter, joule per square meter	$N/m, J/m^2$	kg/s^2
pressure	newton per square meter, pascal	$N/m^2, Pa$	$kg/m \cdot s^2$
viscosity, dynamic	newton-second per square meter, pascal-second	$N \cdot s/m^2, Pa \cdot s$	$kg/m \cdot s$
viscosity, kinematic	meter squared per second		m^2/s
work, torque, energy, quantity of heat	joule, newton-meter, watt-second	$J, N \cdot m, W \cdot s$	$kg \cdot m^2/s^2$
power, heat flux	watt, joule per second	$W, J/s$	$kg \cdot m^2/s^3$
heat flux density	watt per square meter	W/m^2	kg/s^3
volumetric heat release rate	watt per cubic meter	W/m^3	$kg/m \cdot s^3$
heat transfer coefficient	watt per square meter kelvin	$W/m^2 \cdot K$	$kg/s^3 \cdot K$
heat capacity (specific)	joule per kilogram kelvin	$J/kg \cdot K$	$m^2/s^2 \cdot K$
capacity rate	watt per kelvin	W/K	$kg \cdot m^2/s^3 \cdot K$
thermal conductivity	watt per meter kelvin	$W/m \cdot K, \dfrac{Jm}{s \cdot m^2 \cdot K}$	$kg \cdot m/s^3 \cdot K$
quantity of electricity	coulomb	C	$A \cdot s$
electromotive force	volt	$V, W/A$	$kg \cdot m^2/A \cdot s^3$
electric field strength	volt per meter		V/m
electric resistance	ohm	$\Omega, V/A$	$kg \cdot m^2/A^2 \cdot s^3$
electric conductivity	ampere per volt meter	$A/V \cdot m$	$A^2 \cdot s^3/kg \cdot m^3$
electric capacitance	farad	$F, A \cdot s/V$	$A^3 \cdot s^4/kg \cdot m^2$
magnetic flux	weber	$Wb, V \cdot s$	$kg \cdot m^2/A \cdot s^2$
inductance	henry	$H, V \cdot s/A$	$kg \cdot m^2/A^2 \cdot s^2$
magnetic permeability	henry per meter	H/m	$kg \cdot m/A^2 \cdot s^2$
magnetic flux density	tesla, weber per square meter	$T, Wb/m^2$	$kg/A \cdot s^2$
magnetic field strength	ampere per meter		A/m
magnetomotive force	ampere		A
luminous flux	lumen	lm	$cd \cdot sr$
luminance	candela per square meter		cd/m^2
illumination	lux, lumen per square meter	$lx, lm/m^2$	$cd \cdot sr/m^2$

*Supplementary units are: plane angle, radian (rad); solid angle, steradian (sr).

Table 3. Prefixes for units in the International System

Prefix	Symbol	Power	Example
tera	T	10^{12}	
giga	G	10^9	
mega	M	10^6	megahertz (MHz)
kilo	k	10^3	kilometer (km)
hecto	h	10^2	
deca	da	10^1	
deci	d	10^{-1}	
centi	c	10^{-2}	
milli	m	10^{-3}	milligram (mg)
micro	μ	10^{-6}	microgram (μg)
nano	n	10^{-9}	nanosecond (ns)
pico	p	10^{-12}	picofarad (pf)
femto	f	10^{-15}	
atto	a	10^{-18}	

Introduction of the International System, or SI (cont.)

Some common units defined in terms of SI units are given in **Table 4** (the definitions in the fourth column are exact).

Table 4. Some common units defined in terms of SI units

Quantity	Name of unit	Unit symbol	Definition of unit
length	inch	in.	2.54×10^{-2} m
mass	pound (avoirdupois)	lb	0.45359237 kg
force	kilogram-force	kgf	9.80665 N
pressure	atmosphere	atm	101325 N $\cdot$ m^{-2}
pressure	torr	Torr	$(101325/760)$ N $\cdot$ m^{-2}
pressure	conventional millimeter of mercury*	mmHg	$13.5951 \times 980.665 \times 10^{-2}$ N $\cdot$ m^{-2}
energy	kilowatt-hour	kWh	3.6×10^6 J
energy	thermochemical calorie	cal	4.184 J
energy	international steam table calorie	cal$_{IT}$	4.1868 J
thermodynamic temperature (T)	degree Rankine	°R	$(5/9)$ K
customary temperature (t)	degree Celsius	°C	$t(°C) = T(K) - 273.16$
customary temperature (t)	degree Fahrenheit	°F	$t(°F) = T(°R) - 459.68$
radioactivity	curie	Ci	3.7×10^{10} s^{-1}
energy†	electron volt	eV	eV $\approx 1.60219 \times 10^{-19}$ J
mass†	unified atomic mass unit	u	u $\approx 1.66057 \times 10^{-27}$ kg

*The conventional millimeter of mercury, symbol mmHg (not mm Hg), is the pressure exerted by a column exactly 1 mm high of a fluid of density exactly 13.5951 g · cm^{-3} in a place where the gravitational acceleration is exactly 980.665 cm · s^{-2}. The mmHg differs from the Torr by less than 2×10^{-7} Torr.
†These units defined in terms of the best available experimental values of certain physical constants may be converted to SI units. The factors for conversion of these units are subject to change in the light of new experimental measurements of the constants involved.

Conversion factors for the measurement systems

Because it will take some years for all scientists and engineers to convert to the SI, the Encyclopedia has retained the U.S. Customary and metric systems, but has incorporated SI units when preparation of the text permitted. Conversion factors between the three measurement systems are given in **Table 5** for some prevalent units; in each of the subtables the user proceeds as follows:

To convert a quantity expressed in a unit in the left-hand column to the equivalent in a unit in the top row of a subtable, multiply the quantity by the factor common to both units.

The factors have been carried out to seven significant figures, as derived from the fundamental constants and the definitions of the units. However, this does not mean that the factors are always known to that accuracy. Numbers followed by ellipses are to be continued indefinitely with repetition of the same pattern of digits. Factors written with fewer than seven significant digits are exact values. Numbers followed by an asterisk are definitions of the relation between the two units.

Table 5. Conversion factors for the U.S. Customary System, metric system, and International System

A. UNITS OF LENGTH

Units	cm	m	in.	ft	yd	mile
1 cm	= 1	0.01*	0.3937008	0.03280840	0.01093613	6.213712×10^{-6}
1 m	= 100.	1	39.37008	3.280840	1.093613	6.213712×10^{-4}
1 in.	= 2.54*	0.0254	1	0.08333333...	0.02777777...	1.578283×10^{-5}
1 ft	= 30.48	0.3048	12.*	1	0.3333333...	$1.893939... \times 10^{-4}$
1 yd	= 91.44	0.9144	36.	3.*	1	$5.681818... \times 10^{-4}$
1 mile	$= 1.609344 \times 10^{5}$	1.609344×10^{3}	6.336×10^{4}	5280.*	1760.	1

B. UNITS OF AREA

Units	cm²	m²	in.²	ft²	yd²	mile²
1 cm²	= 1	10^{-4}*	0.1550003	1.076391×10^{-3}	1.195990×10^{-4}	3.861022×10^{-11}
1 m²	$= 10^{4}$	1	1550.003	10.76391	1.195990	3.861022×10^{-7}
1 in.²	= 6.4516*	6.4516×10^{-4}	1	6.944444×10^{-3}...	7.716049×10^{-4}	2.490977×10^{-10}
1 ft²	= 929.0304	0.09290304	144.*	1	0.1111111...	3.587007×10^{-8}
1 yd²	= 8361.273	0.8361273	1296.	9.*	1	3.228306×10^{-7}
1 mile²	$= 2.589988 \times 10^{10}$	2.589988×10^{6}	4.014490×10^{9}	2.78784×10^{7}*	3.0976×10^{6}	1

continued

Conversion factors for the measurement systems (cont.)

Table 5. Conversion factors for the U.S. Customary System, metric system, and International System (cont.)

C. UNITS OF VOLUME

Units	m³	cm³	liter	in.³	ft³	qt	gal
1 m³	= 1	10^6	10^3	6.102374×10^4	35.31467	1.056688×10^3	264.1721
1 cm³	= 10^{-6}	1	10^{-3}	0.06102374	3.531467×10^{-5}	1.056688×10^{-3}	2.641721×10^{-4}
1 liter	= 10^{-3}	1000.*	1	61.02374	0.03531467	1.056688	0.2641721
1 in.³	= 1.638706×10^{-5}	16.38706*	0.01638706	1	5.787037×10^{-4}	0.01731602	4.329004×10^{-3}
1 ft³	= 2.831685×10^{-2}	28316.85	28.31685	1728.*	1	2.992208	7.480520
1 qt	= 9.46353×10^{-4}	946.353	0.946353	57.75	0.0342014	1	0.25
1 gal (U.S.)	= 3.785412×10^{-3}	3785.412	3.785412	231.*	0.1336806	4.*	1

D. UNITS OF MASS

Units	g	kg	oz	lb	metric ton	ton
1 g	= 1	10^{-3}	0.03527396	2.204623×10^{-3}	10^{-6}	1.102311×10^{-6}
1 kg	= 1000.	1	35.27396	2.204623	10^{-3}	1.102311×10^{-3}
1 oz (avdp)	= 28.34952	0.02834952	1	0.0625	2.834952×10^{-5}	$5. \times 10^{-4}$
1 lb (avdp)	= 453.5924	0.4535924	16.*	1	4.535924×10^{-4}	0.0005
1 metric ton	= 10^6	1000.*	35273.96	2204.623	1	1.102311
1 ton	= 907184.7	907.1847	32000.	2000.*	0.9071847	1

E. UNITS OF DENSITY

Units	g · cm⁻³	g · l.⁻¹, kg · m⁻³	oz · in.⁻³	lb · in.⁻³	lb · ft⁻³	lb · gal⁻¹
1 g · cm⁻³	= 1	1000.	0.5780365	0.03612728	62.42795	8.345403
1 g · l.⁻¹, kg · m⁻³	= 10^{-3}	1	5.780365×10^{-4}	3.612728×10^{-5}	0.06242795	8.345403×10^{-3}
1 oz · in.⁻³	= 1.729994	1729.994	1	0.0625	108.	14.4375
1 lb · in.⁻³	= 27.67991	27679.91	16.	1	1728.	231.
1 lb · ft⁻³	= 0.01601847	16.01847	9.259259×10^{-3}	5.7870370×10^{-4}	1	0.1336806
1 lb · gal⁻¹	= 0.1198264	119.8264	4.749536×10^{-3}	4.3290043×10^{-3}	7.480519	1

Table 5. Conversion factors for the U.S. Customary System, metric system, and International System (cont.)

F. UNITS OF PRESSURE

Units	Pa, $N \cdot m^{-2}$	$dyn \cdot cm^{-2}$	bar	atm	$kg(wt) \cdot cm^{-2}$	mmHg (Torr)	in. Hg	lb (wt) $\cdot$ in.$^{-2}$
1 Pa, 1 N $\cdot$ m^{-2}	= 1	10	10^{-5}	9.869233×10^{-6}	1.019716×10^{-5}	7.500617×10^{-3}	2.952999×10^{-4}	1.450377×10^{-4}
1 dyn $\cdot$ cm^{-2}	= 0.1	1	10^{-6}	9.869233×10^{-7}	1.019716×10^{-6}	7.500617×10^{-4}	2.952999×10^{-5}	1.450377×10^{-5}
1 bar	= 10^5*	10^6	1	0.9869233	1.019716	750.0617	29.52999	14.50377
1 atm	= 101325.0*	1013250.	1.013250	1	1.033227	760.	29.92126	14.69595
1 kg (wt) $\cdot$ cm^{-2}	= 98066.5	980665.	0.980665	0.9678411	1	735.5592	28.95903	14.22334
1 mmHg (Torr)	= 133.3224	1333.224	1.333224×10^{-3}	1.3157895×10^{-3}	1.3595099×10^{-3}	1	0.03937008	0.01933678
1 in. Hg	= 3386.388	33863.88	0.03386388	0.03342105	0.03453155	25.4	1	0.4911541
1 lb (wt) $\cdot$ in.$^{-2}$	= 6894.757	68947.57	0.06894757	0.06804596	0.07030696	51.71493	2.036021	1

G. UNITS OF ENERGY†

Units	g mass (energy equiv)	J	int J	cal	cal$_{IT}$	Btu$_{IT}$	kW hr	hp hr	ft-lb (wt)	cu ft-lb (wt) in.$^{-2}$	l.-atm
1 g mass (energy equiv)	= 1	8.987552×10^{13}	8.986069×10^{13}	2.148076×10^{13}	2.146640×10^{13}	8.518555×10^{10}	2.496542×10^{7}	3.347918×10^{7}	6.628878×10^{13}	4.603388×10^{11}	8.870024×10^{11}
1 J	= 1.112650×10^{-14}	1	0.999835	0.2390057	0.2388459	9.478172×10^{-4}	$2.777777... \times 10^{-7}$	3.725062	0.7375622	5.121960×10^{-3}	9.869233×10^{-3}
1 int J	= 1.112834×10^{-14}	1.000165	1	0.2390452	0.2388853	9.479735×10^{-4}	2.778236×10^{-7}	3.725676×10^{-7}	0.7376839	5.122805×10^{-3}	9.870862×10^{-3}
1 cal	= 4.655328×10^{-14}	4.184*	4.183310	1	0.9993312	3.965667×10^{-3}	$1.1622222... \times 10^{-6}$	1.558562×10^{-6}	3.085960	2.143028×10^{-2}	0.04129287
1 cal$_{IT}$	= 4.658443×10^{-14}	4.1868*	4.186109	1.000669	1	3.968321×10^{-3}	1.163000×10^{-6}	1.559609×10^{-6}	3.088025	2.144462×10^{-2}	0.04132050
1 Btu$_{IT}$	= 1.173908×10^{-11}	1055.056	1054.882	252.1644	251.9958*	1	2.930711×10^{-4}	3.930148×10^{-4}	778.1693	5.403953	10.41259
1 kW hr	= 4.005540×10^{-8}	3600000.*	3599406.	860420.7	859845.2	3412.142	1	1.341022	2655224.	18439.06	35529.24
1 hp hr	= 2.986931×10^{-8}	2684519.	2684077.	641615.6	641186.5	2544.33	0.7456998	1	1980000.*	13750.	26494.15
1 ft-lb (wt)	= 1.508551×10^{-14}	1.355818	1.355594	0.3240483	0.3238315	1.285067×10^{-3}	3.766161×10^{-7}	$5.050505... \times 10^{-7}$	1	$6.944444... \times 10^{-3}$	0.01338088
1 cu ft-lb (wt) in.$^{-2}$	= 2.172313×10^{-12}	195.2378	195.2056	46.66295	46.63174	0.1850497	5.423272×10^{-5}	$7.272727... \times 10^{-5}$	144.*	1	1.926847
1 l.-atm	= 1.127393×10^{-12}	101.3250	101.3083	24.21726	24.20106	0.09603757	2.814583×10^{-5}	3.774419×10^{-5}	74.73349	0.5189825	1

†The electrical units are those in terms of which certification of standard cells, standard resistances, and so forth, is made by the National Bureau of Standards. Unless otherwise indicated, all electrical units are absolute.

Units of temperature in measurement systems

Temperature is a basic physical quantity. It is a measure of the thermal energy of random motion of particles in a system. As such it has been chosen as one of the basic quantities in the SI. It is to be treated as are the units length, mass, time, electric current, and luminous intensity. In the SI the unit of length is the meter, the unit of time the second, and so on. The question arises as to the choice of the unit of temperature in the SI.

In the past it was customary to refer to scales of temperature, for example, the Celsius and Fahrenheit scales. On the Celsius scale, 0 designates the freezing point (ice point) and 100 the boiling point (steam point) of water. Corresponding numbers on the Fahrenheit scale are 32 and 212. There are 100 units between the ice point and steam point on the Celsius scale, and 180 units between these points in the Fahrenheit system.

By measuring the volume changes of a gas within the 100-unit interval of the ice point and steam point of water on the Celsius scale, it was found that a numerical value could be assigned for a basic unit of temperature. Careful measurement of this ice-steam interval in a gas thermometer determined that the ice point of water should be assigned the value of 273.15 kelvins. The unit of temperature was thus called the kelvin with the symbol K. Further experiments led to the decision to define the kelvin in the SI along the same lines but in terms of the triple point of water. This is the temperature and pressure at which ice, liquid water, and water vapor coexist at equilibrium. The triple point was chosen because it was a more reproducible value than the ice point.

This change led to the SI definition of temperature in terms of the triple point of water, which contains exactly 273.16 kelvins.

It follows that the Celsius temperature (°C) is an intermediate scale. It is useful in defining Kelvin temperature in the SI. Celsius temperature (t) is related to Kelvin temperature (K) as follows:

$$t_{\text{ice point}} = 0°C$$

$$t_{\text{steam point}} = 100°C$$

$$0\,K = -273.16°C$$

A summary of the conventions in the SI as proposed in the Thirteenth General Conference of Weights and Measures pertaining to temperature units is given below.

1. The unit of SI temperature is the kelvin, symbol K.

2. The word "scale" is not to be used except in terms of measurement of temperature between certain fixed points on the Celsius scale.

3. The terms "thermodynamic scale" or "absolute scale" are not to be used to describe temperature. The degree sign is to be eliminated with the symbol K.

4. When Celsius temperatures are used (°C), it is understood that the temperature unit is the kelvin.

Not all scientists and engineers have adopted the SI of temperature terminology. For this reason the contributors to the Encyclopedia have retained the term "scale" in relation to thermodynamic temperature. Furthermore, many engineers in the United States still use the Fahrenheit system in discussing practical engineering systems.

In converting Fahrenheit (°F) to Celsius (°C) the following formula applies.

$$°C = \frac{°F - 32°}{1.8}$$

Units of temperature in measurement systems (cont.)

In converting Celsius to Fahrenheit the following formula can be used.

$$°F = (°C \times 1.8) + 32°$$

In changing from Celsius terminology (t) to kelvin units (K) the following formula can be used.

$$K = t + 273.16$$

McGRAW-HILL ENCYCLOPEDIA OF ENERGY

Index

Index

*Asterisks indicate page references
to article titles.*

A

Abbot, C. G. 625
Abrasive drilling 115
Absolute entropy 267–268
Absorption coefficient 68,
  70–71
Absorption of electromagnetic
  radiation 67–74*
  absorption and emission
    coefficients 70–71
  absorption measurement 68
  dispersion 71–74
  fluorescence 70
  laws of absorption 67–68
  luminescence 70
  phosphorescence 70
  physical nature 69–70
  scattering 68–69
  selective reflection 74
Absorption refrigeration 78,
  625, 626
Abyssal floor 435
Acceleration (force) 287–288
Accumulator batteries 111
Acidizing (oil and gas well
  completion) 523
ACRS see Advisory Committee
  on Reactor Safeguards

Advisory Committee on Reactor
  Safeguards (ACRS) 491
AEC see Atomic Energy
  Commission
Aerosols (air pollutants) 79
Agricola, Georgius 325
Air: Carnot cycle 119
  radioactive decontamination
    178
Air-cleaning devices 80–81
Air conditioning 74–76*
  calculation of loads 74–75
  comfort 74
  heat pump 353
  process 74
  solar house heating 634
  spaces 76
  steam boiler 639
Air conditioning systems 75–76
  air washer 77
  built-up 75
  central 75
  field-erected 75
  incremental 75
  split 75
  unitary 75
Air-cooled aircraft engine 85,
  86, 87
  fixed-radial 85

Air-cooled aircraft engine
  —cont.
  opposed 87
  two-row radial 86
Air-cooled condensers 159
Air-cooled heat exchanger 159,
  351
Air cooling 76–78*
  air washer 77
  coils 77–78
  evaporative 76–77
  ice as heat sink 78
  refrigeration heat sink 78
  use of well water 78
Air-depolarized alkaline cell
  731–732
Air-monitoring instruments
  82–83
Air percussion drilling 115
Air pollution see Atmospheric
  pollution
Air-pollution control 51, 78–84*
  air-quality control 83
  collection of contaminants
    109
  containment of contaminants
    109
  dispersion of contaminants
    109

Air-pollution control—cont.
  Federal legislation 54
  fossil-fueled station
    regulations 229–230
  methods 109–110
  prevention of pollution 51,
    109
  sources of pollution 78–79
Air Quality Act (1967) 109–110,
  268
Air-standard cycle
  (thermodynamics) 666
Air-standard engine 119
Air temperature: air
  conditioning 74–76*
  air cooling 76–78*
  heating, comfort 360–362*
Air-to-air heat pump 353
Air washer 77
Air-water central air
  conditioning system 76
Aircraft engine 84*
  air-cooled 85, 86, 87
  gas turbine 318, 319
  jet fuel 408*
  liquid-cooled 85–86
  pulse jet 593–594*
  ramjet 600–602*
  turbojet 701–702*

N